NEW DEVELOPMENTS IN THE THEORY OF KNOTS

ADVANCED SERIES IN MATHEMATICAL PHYSICS

Also in the Series

Volume 1: Mathematical Aspects of String Theory
edited by S-T Yau (*Harvard*)
Volume 2: Bombay Lectures on Highest Weight Representations of Infinite Dimensional Lie Algebras
by V G Kac (*MIT*) & A K Raina (*Tata*)
Volume 3: Kac-Moody and Virasoro Algebras: A Reprint Volume for Physicists
edited by P Goddard (*Cambridge*) & D Olive (*Imperial College*)
Volume 4: Harmonic Mappings, Twistors and σ-Models
edited by P Gauduchon (*CNRS, Paris*)
Volume 5: Geometric Phases in Physics
edited by A Shapere & F Wilczek (*IAS, Princeton*)
Volume 7: Infinite Dimensional Lie Algebras and Groups
edited by V Kac (*MIT*)
Volume 8: Introduction to String Field Theory
by W Siegel (*SUNY*)
Volume 9: Braid Group, Knot Theory and Statistical Mechanics
edited by C N Yang (*SUNY*) & M L Ge (*Nankai Inst*)
Volume 10: Yang-Baxter Equation in Integrable Systems: A Reprint Volume
edited by M Jimbo (*Kyoto*)

Advanced Series in Mathematical Physics Vol. 11

NEW DEVELOPMENTS IN THE THEORY OF KNOTS

Editor

Toshitake Kohno
Nagoya University

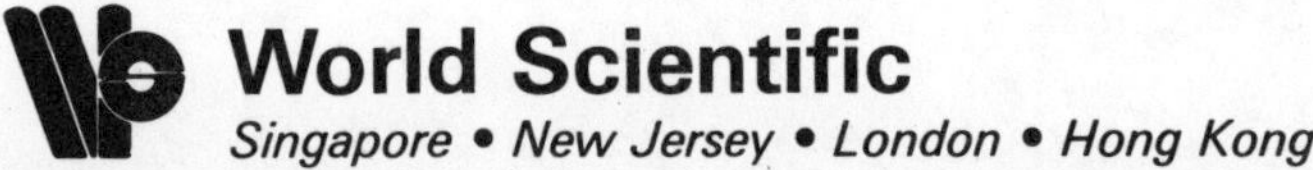

Published by

World Scientific Publishing Co. Pte. Ltd.

P O Box 128, Farrer Road, Singapore 9128

USA office: 687 Hartwell Street, Teaneck, NJ 07666

UK office: 73 Lynton Mead, Totteridge, London N20 8DH

The editor and publisher are grateful to the authors and the following publishers for their assistance and permission to reproduce the reprinted papers found in these volumes:

American Mathematical Society (*Bull. Amer. Maths. Soc., Contemporary Maths. Soc.*)
Ann. Inst. Fourier
Cambridge University Press (*Maths. Proc. Camb. Phil. Soc.*)
C R Academy Science Paris
Elsevier Science Publishers BV (*Adv. Stud. Pure Math., Nuclear Physics*)
Kobe University (*Kobe J. Math.*)

l'Enseignement Mathematique
Osaka University (*Osaka J. Math.*)
Pacific Journal of Mathematics (*Pacific J. Math.*)
Pergamon Press (*Topology*)
Physical Society of Japan (*J. Phys. Soc. Japan*)
Plenum Publishing (*Plenum Press*)
Princeton University Press (*Ann. of Math.*)

Springer-Verlag (*Comm. Math. Phys., Invent. Math., Mathematische Annalen*)

While every effort has been made to contact the publishers of reprinted papers prior to publication, we have not been successful in a few cases. Where we could not contact the publishers, we have acknowledged the source of the material. Proper credit will be given to these publishers in future editions of this work after permission is granted.

Library of Congress Cataloging-in-Publication data is available.

Preface

The present volume is a collection of reprints related to recent developments in the theory of knots arising from the discovery of the Jones polynomial. The papers are grouped into six chapters. A relation between new link invariants and statistical mechanics was already implicit in the original work of V. Jones. A systematic study of constructing link invariants from solutions to the Yang-Baxter equation has been pursued afterwards, and this new progress also revealed a striking relation among link invariants, quantum groups and monodromy of conformal field theory. On the other hand, the Jones polynomial and its relatives have also presented a powerful tool for classical problems in knot theory which were not accessible before this discovery.

Since articles concerning these subjects are scattered in journals of many different domains including both mathematics and physics, we hope that this volume is helpful for the readers to get a perspective on these new developments. A reprint volume on the Yang-Baxter equation is also being prepared by M. Jimbo. One finds a rather extensive bibliography at the end of the present volume, in which we tried to cover the following subjects:

(1) general aspects on braid groups and mapping class groups;
(2) new link polynomials and their applications;
(3) relations among link polynomials, the Yang-Baxter equation and quantum groups;
(4) monodromy representations of conformal field theory.

Concerning classical literatures on knot theory and braid groups until about 1985, the readers may refer to the references in the books [31] and [62]. As for the Yang-Baxter equation, solvable lattice models and quantum groups, a more extensive bibliography can be found at the end of the reprint volume edited by M. Jimbo. Although we hope that the collection of references to the works on new link polynomials is at least dense, we still feel that we could not give a proper credit to many important contributions to the above subjects. We would like to apologize for that to the authors.

December 1989

T. Kohno

CONTENTS

MONODROMY OF BRAID GROUPS, A RELATION WITH CONFORMAL FIELD THEORY

PIONEERING WORKS ON NEW KNOT POLYNOMIALS

BULLETIN (New Series) OF THE
AMERICAN MATHEMATICAL SOCIETY
Volume 12, Number 1, January 1985

A POLYNOMIAL INVARIANT FOR KNOTS VIA VON NEUMANN ALGEBRAS[1]

BY VAUGHAN F. R. JONES[2]

A theorem of J. Alexander [**1**] asserts that any tame oriented link in 3-space may be represented by a pair (b, n), where b is an element of the n-string braid group B_n. The link L is obtained by closing b, i.e., tying the top end of each string to the same position on the bottom of the braid as shown in Figure 1. The closed braid will be denoted $b^\wedge$.

Thus, the trivial link with n components is represented by the pair $(1, n)$, and the unknot is represented by $(s_1 s_2 \cdots s_{n-1}, n)$ for any n, where $s_1, s_2, \ldots, s_{n-1}$ are the usual generators for B_n.

The second example shows that the correspondence of (b, n) with $b^\wedge$ is many-to-one, and a theorem of A. Markov [**15**] answers, in theory, the question of when two braids represent the same link. A Markov move of type 1 is the replacement of (b, n) by (gbg^{-1}, n) for any element g in B_n, and a Markov move of type 2 is the replacement of (b, n) by $(bs_n^{\pm 1}, n+1)$. Markov's theorem asserts that (b, n) and (c, m) represent the same closed braid (up to link isotopy) if and only if they are equivalent for the equivalence relation generated by Markov moves of types 1 and 2 on the *disjoint* union of the braid groups. Unforunately, although the conjugacy problem has been solved by F. Garside [**8**] within each braid group, there is no known algorithm to decide when (b, n) and (c, m) are equivalent. For a proof of Markov's theorem see J. Birman's book [**4**].

The difficulty of applying Markov's theorem has made it difficult to use braids to study links. The main evidence that they might be useful was the existence of a representation of dimension $n - 1$ of B_n discovered by W. Burau in [**5**]. The representation has a parameter t, and it turns out that the determinant of $1-$(Burau matrix) gives the Alexander polynomial of the closed braid. Even so, the Alexander polynomial occurs with a normalization which seemed difficult to predict.

In this note we introduce a polynomial invariant for tame oriented links via certain representations of the braid group. That the invariant depends only on the closed braid is a direct consequence of Markov's theorem and a certain trace formula, which was discovered because of the uniqueness of the trace on certain von Neumann algebras called type II_1 factors.

Notation. In this paper the Alexander polynomial Δ will always be normalized so that it is symmetric in t and t^{-1} and satisfies $\Delta(1) = 1$ as in Conway's tables in [**6**].

Received by the editors August 15, 1984.
1980 *Mathematics Subject Classification.* Primary 57M25; Secondary 46L10.
[1]Research partially supported by NSF grant no. MCS-8311687.
[2]The author is a Sloan foundation fellow.

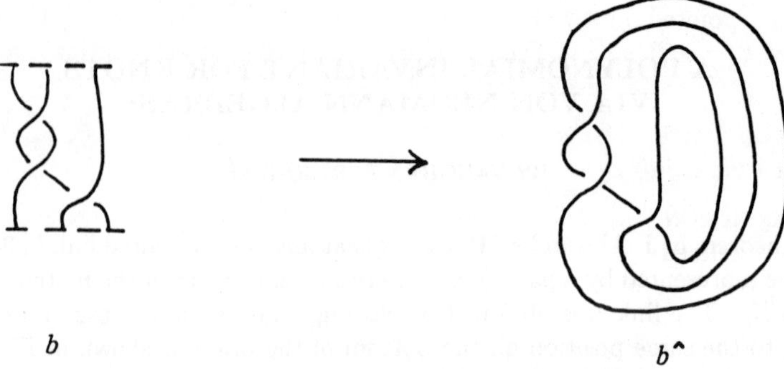

FIGURE 1

While investigating the index of a subfactor of a type II_1 factor, the author was led to analyze certain *finite-dimensional* von Neumann algebras A_n generated by an identity 1 and n projections, which we shall call $e_1, e_2, \ldots, e_n$. They satisfy the relations

(I) $e_i^2 = e_i, e_i^* = e_i$,

(II) $e_i e_{i\pm 1} e_i = t/(1+t)^2 e_i$,

(III) $e_i e_j = e_j e_i$ if $|i-j| \geq 2$.

Here t is a complex number. It has been shown by H. Wenzl [**24**] that an arbitrarily large family of such projections can only exist if t is either real and positive or $e^{\pm 2\pi i/k}$ for some $k = 3, 4, 5, \ldots$. When t is one of these numbers, there exists such an algebra for all n possessing a trace $\mathrm{tr}\colon A_n \to \mathbf{C}$ completely determined by the normalization $\mathrm{tr}(1) = 1$ and

(IV) $\mathrm{tr}(ab) = \mathrm{tr}(ba)$,

(V) $\mathrm{tr}(we_{n+1}) = t/(1+t)^2 \, \mathrm{tr}(w)$ if w is in A_n,

(VI) $\mathrm{tr}(a^*a) > 0$ if $a \neq 0$

(note $A_0 = \mathbf{C}$).

Conditions (I)–(VI) determine the structure of A up to $*$-isomorphism. This fact was proved in [**9**], and a more detailed description appears in [**10**]. Remember that a finite-dimensional von Neumann algebra is just a product of matrix algebras, the $*$ operation being conjugate-transpose.

For real t, D. Evans pointed out that an explicit representation of A_n on $\mathbf{C}^{2n+2}$ was discovered by H. Temperley and E. Lieb [**23**], who used it to show the equivalence of the Potts and ice-type models of statistical mechanics. A readable account of this can be found in R. Baxter's book [**2**]. This representation was rediscovered in the von Neumann algebra context by M. Pimsner and S. Popa [**18**], who also found that the trace tr is given by the restriction of the Powers state with $t = \lambda$ (see [**18**]).

For the roots of unity the algebras A_n are intimately connected with Coxeter groups in a way that is far from understood.

The similarity between relations (II) and (III) and Artin's presentation of the n-string braid group,

$$\{s_1, s_2, \ldots, s_n \colon s_i s_{i+1} s_i = s_{i+1} s_i s_{i+1},\ s_i s_j = s_j s_i \text{ if } |i-j| \geq 2\},$$

was first pointed out by D. Hatt and P. de la Harpe. It transpires that if one defines $g_i = \sqrt{t}(te_i - (1 - e_i))$, the g_i satisfy the correct relations, and one obtains representations r_t of B_n by sending s_i to g_i.

THEOREM 1. *The number* $(-(t+1)/\sqrt{t})^{n-1}\,\mathrm{tr}(r_t(b))$ *for* b *in* B_n *depends only on the isotopy class of the closed braid* $b^\wedge$.

DEFINITION. If L is a tame oriented classical link, the *trace invariant* $V_L(t)$ is defined by

$$V_L(t) = (-(t+1)/\sqrt{t})^{n-1}\,\mathrm{tr}(r_t(b))$$

for any (b, n) such that $b^\wedge = L$.

The Hecke algebra approach shows the following.

THEOREM 2. *If the link* L *has an odd number of components,* $V_L(t)$ *is a Laurent polynomial over the integers. If the number of components is even,* $V_L(t)$ *is* $\sqrt{t}$ *times a Laurent polynomial.*

The reader may have observed that the von Neumann algebra structure (i.e., the $*$ operation) and condition (VI) are redundant for the definition of $V_L(t)$. This explains why V_L can be extended to all values of t except 0. However, it must be pointed out that for positive t and the relevant roots of unity, the presence of *positivity* gives a powerful method of proof.

The trace invariant depends on the oriented link but not on the chosen orientation. Let $L^\sim$ denote the mirror image link of L.

THEOREM 3. $V_{L^\sim}(t) = V_L(1/t)$.

Thus, the trace invariant can be used to detect a lack of amphicheirality. It seems to be very good at this. A glance at Table 1 shows that it distinguishes the trefoil knot from its mirror image and hence, via Theorem 6, it distinguishes the two granny knots and the square knot.

CONJECTURE 4. If L is not amphicheiral, $V_{L^\sim} \neq V_L$.

There is some evidence for this conjecture, but only \$10 hangs on it. In this direction we have the following result, where b is in B_n, b_+ is the sum of the positive exponents of b, and b_- is the (unsigned) sum of the negative ones in some expression for b as a word on the usual generators.

THEOREM 5. *If* $b_+ - 3b_- - n + 1$ *is positive, then* $b^\wedge$ *is not amphicheiral.*

For $b_- = 0$, i.e., positive braids, this follows from a recent result of L. Rudolf [**21**]. Also, if the condition of the theorem holds, we conclude that $b^\wedge$ is not the unknot. This is similar in kind to a recent result of D. Bennequin [**3**].

The connected sum of two links can be handled in the braid group provided one pays proper attention to the components being joined. Let us ignore the subtleties and state the following (where # denotes the connected sum).

THEOREM 6. $V_{L_1 \# L_2} = V_{L_1} V_{L_2}$.

As evidence for the power of the trace invariant, let us answer two questions posed in [**4**]. Both proofs are motivated by the fact, shown in [**10**], that $r_t(B_n)$ is sometimes finite.

THEOREM 7. *For every n there are infinitely many words in B_{n+1} which give close braids inequivalent to closed braids coming from elements of the form $Us_n^{-1}Vs_n$, where U and V are in B_n.*

Explicit examples are easy to find; e.g., all but a finite number of powers of $s_1^{-1}s_2s_3$ will do.

THEOREM 8 (SEE [4 P. 217, Q. 8]). *If b is in B_n and there is an integer k greater than 3 for which $b \in \ker r_t$, $t = e^{2\pi i/k}$, then $b^\wedge$ has braid index n.*

Here the braid index of a link L is the smallest n for which there is a pair (b, n) with $b^\wedge = L$. The kernel of r_t is not hard to get into for these values of t.

COROLLARY 9. *If the greatest common divisor of the exponents of $b \in B_n$ is more than 1, then the braid index of $b^\wedge$ is n.*

More interesting examples can be obtained by using generators and relations for certain finite groups; e.g., the finite simple group of order 25,920 (see [**10**, **7**]). In general, the trace invariant can probably be used to determine the braid index in a great many cases.

Note also that the trace invariant detects the kernel of r_t.

THEOREM 10. *For $t = e^{2\pi i/k}, k = 3, 4, 5, \ldots$, $V_{b^\wedge}(t) = (-2\cos\pi/k)^{n-1}$ if and only if $b \in \ker r_t$ (for $b \in B_n$).*

COROLLARY 11. *For transcendental t, $b \in \ker r_t$ if and only if $V_{b^\wedge}(t) = (-(t+1)/\sqrt{t})^{n-1}$.*

For transcendental t, r_t is very likely to be faithful.

There is an alternate way to calculate V_L without first converting L into a closed braid. In [**6**] Conway describes a method for rapidly computing the Alexander polynomials of links inductively. In fact, his first identity suffices in principle—see [**11**]. This identity is as follows.

Let L^+, L^-, and L be links related as in Figure 2, the rest of the links being identical. Then $\Delta_{L^+} - \Delta_{L^-} = (\sqrt{t} - 1/\sqrt{t})\Delta_L$.

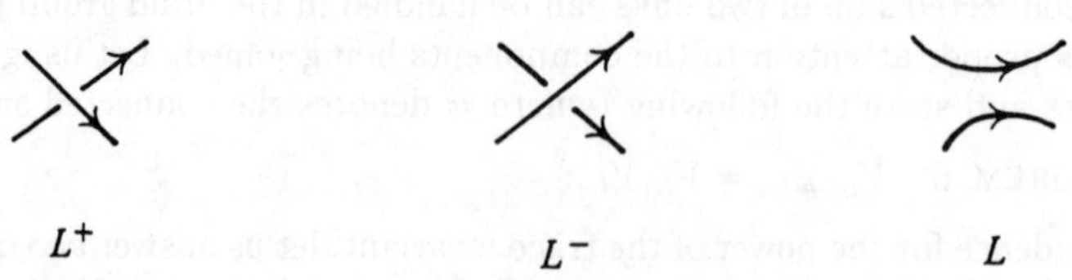

L^+ L^- L

FIGURE 2

For the trace invariant we have

THEOREM 12. $1/tV_{L^-} - tV_{L^+} = (\sqrt{t} - 1/\sqrt{t})V_L$.

COROLLARY 13. *For any link* L, $V_L(-1) = \Delta_L(-1)$.

That the trace invariant may always be calculated by using Theorem 12 follows from the proof of the same thing for the Alexander polynomial. We urge the reader to try this method on, say, the trefoil knot.

The special nature of the algebras A_n when t is a relevant root of unity can be exploited to give information about V_L at these values.

THEOREM 14. *If K is a knot then* $V_K(e^{2\pi i/3}) = 1$.

THEOREM 15. $V_L(1) = (-2)^{p-1}$, *where p is the number of components of L.*

A more subtle analysis at $t = 1$ via the Temperley-Lieb-Pimsner-Popa representation gives the next result.

THEOREM 16. *If K is a knot then* $d/dtV_K(1) = 0$.

It is thus sensible to simplify the trace invariant for knots as follows.

DEFINITION 17. If K is a knot, define W_K to be the Laurent polynomial

$$W_K(t) = (1 - V_K(t))/(1 - t^3)(1 - t).$$

Amphicheirality is less obvious for W. In fact, $W_{K^\sim}(t) = 1/t^4 W_K(1/t)$. It is amusing that for W the unknot is 0 and the trefoil is 1. The connected sum is also less easy to see in the W picture. For the record the formula is

$$W_{K_1 \# K_2} = W_{K_1} + W_{K_2} - (1-t)(1-t^3)W_{K_1}W_{K_2}.$$

COROLLARY 18. $\Delta_K(-1) \equiv 1$ *or* $5 \pmod 8$.

When $t = i$ the algebras A_n are the complex Clifford algebras. This together with a recent result of J. Lannes [**13**] allows one to show the following.

THEOREM 19. *If K is a knot the Arf invariant is of K is $W_K(i)$.*

COROLLARY 20. $\Delta_K(-1) = 1$ *or* $5 \pmod 8$ *when the Arf invariant is* 0 *or* 1, *respectively.*

This is an alternate proof of a result in Levine [**14**]; also see [**11**, p. 155]. Note also that Corollary 20 allows one to define an Arf invariant for links as $V(i)$. It may be zero and is always plus or minus a power of two otherwise.

The values of V at $e^{\pi i/3}$ are also of considerable interest, as the algebra A_n is then related to a kind of cubic Clifford algebra. Also, in this case, $r_t(B_n)$ is always a finite group, so one can obtain a rapid method for calculating $V(t)$ without knowing V completely. We have included this value of V in the tables. Note that it is always in $1 + 2\mathbf{Z}(e^{i\pi/3})$.

There is yet a third way to calculate the trace invariant. The decomposition of A_n as a direct sum of matrix algebras is known [**10**], and H. Wenzl has explicit formulae for the (irreducible) representations of the braid group in each direct summand. So in principle this method could always be used. This brings in the Burau representation as a direct summand of r_t. For 3 and 4 braids this allows one to deduce some powerful relations with the

Alexander polynomial. An application of Theorem 16 allows one to determine the normalization of the Alexander polynomial in the Burau matrix for proper knots, and one has the following formulae.

THEOREM 21. *If b in B_3 has exponent sum e, and $b^\wedge$ is a knot, then*

$$V_{b^\wedge}(t) = t^{e/2}(1 + t^e + t + 1/t - t^{e/2-1}(1 + t + t^2)\Delta_{b^\wedge}(t)).$$

THEOREM 22. *If b in B_4 has exponent sum e, and $b^\wedge$ is a knot, then*

$$t^{-e}V(t) + t^eV(1/t) = (t^{-3/2} + t^{-1/2} + t^{1/2} + t^{3/2})(t^{e/2} + t^{-e/2}) - (t^{-2} + t^{-1} + 2 + t + t^2)\Delta(t)$$

(*where $V = V_{b^\wedge}$ and $\Delta = \Delta_{b^\wedge}$*).

These formulas have many interesting consequences. They show that, except in special cases, e is a knot invariant. They also give many obstructions to being closed 3 and 4 braids.

COROLLARY 23. *If K is a knot and $|\Delta_K(i)| > 3$, then K cannot be represented as a closed* 3 *braid.*

Of the 59 knots with 9 crossings or less which are known not to be closed 3 braids, this simple criterion establishes the result for 43 of them, at a glance.

COROLLARY 24. *If K is a knot and $\Delta_K(e^{2\pi i/5}) > 6.5$, then K cannot be represented as a closed* 4 *braid.*

For $n > 4$ there should be no simple relation with the Alexander polynomial, since the other direct summands of r_t look less and less like Burau representations.

In conclusion, we would like to point out that the q-state Potts model could be solved if one understood enough about the trace invariant for braids resembling certain braids discovered by sailors and known variously as the "French sinnet" (sennit) or the "tresse anglaise", depending on the nationality of the sailor. See [**21**, p. 90].

The author would like to thank Joan Birman. It was because of a long discussion with her that the relation between condition (V) and Markov's theorem became clear.

Tables. A single example should serve to explain how to read the tables. The knot 8_8 has trace invariant

$$t^{-3}(-1 + 2t - 3t^2 + 5t^3 - 4t^4 + 4t^5 - 3t^6 + 2t^7 - t^8).$$

Its W invariant is

$$t^{-3}(1 - t + 2t^2 - t^3 + t^4).$$

A braid representation for it is

$$s_1^{-1}s_2s_1^2s_3^{-1}s_2^2s_3^{-2} \quad \text{in } B_4.$$

Also note that $w = e^{\pi i/3}$.

ADDED IN PROOF. The similarity between the relation of Theorem 12 and Conway's relation has led several authors to a two-variable generalization of V_L. This has been done (independently) by Lickorish and Millett, Ocneanu, Freyd and Yetter, and Hoste.

TABLE 1. The trace invariant for prime knots to 8 crossings.

knot	braid rep.	p_0	pol(V)	V(w)	p_0	pol(W)
0_1	1	0	1	1	0	0
3_1	1^3	1	101-1	$i\sqrt{3}$	0	1
4_1	$12^{-1}12^{-1}$	-2	1-11-11	-1	-2	-1
5_1	1^5	2	101-11-1	-1	0	1101
5_2	$2^21^{-1}21^2$	1	1-12-11-1	-1	0	101
6_1	$12^{-1}13^{-1}23^{-1}2^{-1}$	-4	1-11-22-11	$i\sqrt{3}$	-4	-10-1
6_2	$1^{-1}21^{-1}2^3$	-1	1-12-22-21	1	-1	-11-1
6_3	$1^{-1}2^21^{-2}2$	-3	-12-23-22-1	1	-3	1-11
7_1	1^7	3	101-11-11-1	-1	0	1111101
7_2	$1^{-1}3^321^23^{-1}2$	1	1-12-22-11-1	1	0	10101
7_3	$2^512^21^{-2}$	2	1-12-23-21-1	1	0	110201
7_4	$3^21^{-1}23^{-1}21^22$	1	1-23-23-21-1	-1	0	10201
7_5	$2^312^41^{-2}$	2	1-13-33-32-1	-1	0	1102-11
7_6	$3^{-1}1^{-1}2113^{-1}2^{-3}$	-6	-12-34-33-21	-1	-6	1-12-1
7_7	$13^{-1}23^{-1}21^{-1}23^{-1}2$	-3	-13-34-43-21	$-i\sqrt{3}$	-3	1-21-1
8_1	$12^{-1}3^{-1}214^{-2}\cdot 3^{-1}2^{-1}4$	-6	1-11-22-22-11	1	-6	-10-10-1
8_2	$2^51^{-1}21^{-1}$	0	1-12-23-32-21	-1	1	1-11-1
8_3	$123^{-1}4^{-1}3^{-1}2.1^{-1}3^2243^{-1}2^{-2}$	-4	1-12-33-32-11	-3	-4	-10-20-1
8_4	$11132^{-1}3^{-2}12^{-1}$	-3	1-12-33-33-21	-1	-3	-10-21-1
8_5	$1112^{-1}1112^{-1}$	0	1-13-33-43-21	$i\sqrt{3}$	1	1-21-1
8_6	$3^{-2}12^{-1}132^{-3}$	-7	1-23-44-43-11	1	-7	-11-21-1
8_7	$2^{-2}12^{-1}1^4$	-2	-12-24-44-32-1	1	-2	1-12-11
8_8	$1^{-1}2113^{-1}223^{-2}$	-3	-12-35-44-32-1	1	-3	1-12-11
8_9	$2^31^{-1}21^{-3}$	-4	1-23-45-43-21	1	-4	-11-21-1
8_{10}	$221^{-2}2^31^{-1}$	-2	-12-35-45-42-1	$i\sqrt{3}$	-2	1-13-11
8_{11}	$12^{-2}32^{-1}3^{-2}12^{-1}$	-7	1-23-55-44-21	$-i\sqrt{3}$	-7	-11-22-1
8_{12}	$12^{-1}34^{-1}34^{-1}\cdot 213^{-1}2^{-1}$	-4	1-24-55-54-21	-1	-4	-11-31-1
8_{13}	$1123^{-1}21^{-1}3^{-2}2$	-3	-13-45-55-32-1	-1	-3	1-22-11
8_{14}	$11221^{-1}3^{-1}23^{-1}2$	-1	1-24-56-54-31	-1	-1	-12-22-1
8_{15}	$111231^{-1}2^332^{-1}$	2	1-25-56-64-31	$-i\sqrt{3}$	0	1103-22-1
8_{16}	$112^{-1}112^{-1}12^{-1}$	-2	-13-46-66-53-1	1	-2	1-23-21
8_{17}	$21^{-1}21^{-1}221^{-2}$	-4	1-35-67-65-31	1	-4	-12-32-1
8_{18}	$(12^{-1})^3$	-4	1-46-79-76-41	3	-4	-13-33-1
8_{19}	121212^21	3	10100-1	$-i\sqrt{3}$	0	11111
8_{20}	21^321^{-3}	-1	-12-12-11-1	$i\sqrt{3}$	-1	101
8_{21}	$2^312^21^{-2}21^{-1}$	1	2-23-32-21	$i\sqrt{3}$	0	1-11-1

TABLE 2. The trace invariant for some divers knots and links.

link	p_0	pol(V)	V(w)	p_0	pol(W)	braid rep.
10_{141} (1)	-2	1-23-34-32-21	$i\sqrt{3}$	-2	-11-11-1	$2^{-4}112221$
KT (2)	-4	-12-22001-22-21	1	-4	1-110-11-1	$1113323^{-1}1^{-2}2.1^{-1}3^{-1}2^{-1}$
C (3)	-4	-12-22001-22-21	1	-4	1-110-11-1	$22213^{-1}2^{-2}1.2^{-1}13^{-1}$
2^2_1	1/2	-10-1	$-i$			1^2
4^2_1 (4)	3/2	-10-11-1	$-i$			1^4
4^2_1 (4)	1/2	-11-10-1	i			$12^{-1}122$
5^2_1	-7/2	1-21-21-1	i			$12^{-1}12^{-2}$
6^2_1	5/2	-10-11-11-1	$\sqrt{3}$			1^6
6^2_2	3/2	-11-22-21-1	$-i$			2221121^{-1}
6^2_3	-3/2	-12-22-31-1	$\sqrt{3}$			$21^{-1}23^{-1}2123^{-1}$
H_1 (5)	1/2	-11-10-1	i			$12^{-1}122$
H_2 (5)	-3/2	1-10-10-1	$-i$			$12^{-3}122$
W (6)	-3/2	-11-21-21	$-i$			$11221^{-1}2^{-2}$
6^3_1	-1	1-13-13-21	$i\sqrt{3}$			$221^{-1}221^{-1}$
6^3_2	-3	-13-24-23-1	1			$12^{-1}12^{-1}12^{-1}$
6^3_3	2	10102	1			122122
A (7)	5/2	-10-32-34-22-1	$3i$			11222333
B (7)	5/2	-10-32-34-22-1	$3i$			11122333

Table Notes
(1) Compare 8_5 which has the same Alexander polynomial.
(2) The Kinoshita-Terasaka knot with 11 crossings. See [**12**].
(3) This is Conway's knot with trivia Alexander polynomial. See [**20**].
(4) Same link, different orientation.
(5) These links have homeomorphic complements.
(6) The Whitehead link.
(7) Two composite links with the same trace invariant.

REFERENCES

1. J. W. Alexander, *A lemma on systems of knotted curves,* Proc. Nat. Acad. **9** (1923), 93–95.
2. R. J. Baxter, *Exactly solved models in statistical mechanics,* Academic Press, London, 1982.
3. D. Bennequin, *Entrelacements et structures de contact,* These, Paris, 1982.
4. J. Birman, *Braids, links and mapping class groups,* Ann. Math. Stud. **82** (1974).
5. W. Burau, *Uber Zopfgruppen und gleichsinning verdrillte Verkettunger,* Abh. Math. Sem. Hanischen Univ. **11** (1936), 171–178

6. J. H. Conway, *An enumeration of knots and links.* Computational Problems in Abstract Algebra, Pergamon, New York, 1970, pp. 329–358.

7. H. Coxeter, *Regular complex polytopes,* Cambridge Univ. Press, 1974.

8. F. Garside, *The braid group and other groups.* Quart J. Math. Oxford Ser. **20** (1969), 235–254.

9. V. F. R. Jones, *Index for subfactors,* Invent. Math. **72** (1983), 1–25.

10. ——, *Braid groups, Hecke algebras and type* II_1 factors, Japan-U.S. Conf. Proc. 1983.

11. L. H. Kaufman, *Formal knot theory,* Math. Notes, Princeton Univ. Press, 1983.

12. S. Kinoshita and H. Terasaka, *On unions of knots,* Osaka Math. J. **9** (1959), 131–153.

13. J. Lannes, *Sur l'invariant de Kervaire pour les noeuds classiques,* École Polytechnique, Palaiseau, 1984 (preprint).

14. J. Levine, *Polynomial invariants of knots of codimension two,* Ann. of Math. (2) **84** (1966), 534–554.

15. A. A. Markov, *Uber die freie Aquivalenz geschlossener Zopfe,* Mat. Sb. **1** (1935), 73–78.

16. K. Murasugi, *On closed 3-braids.* Mem. Amer. Math. Soc.

17. K. A. Perko, *On the classification of knots,* Proc. Amer. Math. Soc. **45** (1974), 262–266.

18. M. Pimsner and S. Popa, *Entropy and index for subfactors,* INCREST, Bucharest, 1983 (preprint).

19. R. T. Powers, *Representations of uniformly hyperfinite algebras and the associated von Neumann algebras,* Ann. of Math. (2) **86** (1967), 138–171.

20. D. Rolfsen, *Knots and links,* Publish or Perish Math. Lecture Ser., 1976.

21. L. Rudolph, *Nontrivial positive braids have positive signature,* Topology **21** (1982), 325–327.

22. S. Svensson, *Handbook of Seaman's ropework,* Dodd, Mead, New York, 1971.

23. H. N. V. Temperley and E. H. Lieb, *Relations between the percolation and colouring problem and other graph-theoretical problems associated with regular planar lattices: some exact results for the percolation problem,* Proc. Roy. Soc. (London) (1971), 251–280.

24. H. Wenzl, *On sequences of projections,* Univ. of Pennsylvania, 1984 (preprint).

DEPARTMENT OF MATHEMATICS, UNIVERSITY OF PENNSYLVANIA, PHILADELPHIA, PENNSYLVANIA 19104

MATHEMATICAL SCIENCES RESEARCH INSTITUTE, 2223 FULTON STREET, ROOM 3603, BERKELEY, CALIFORNIA 94720

BULLETIN (New Series) OF THE
AMERICAN MATHEMATICAL SOCIETY
Volume 12, Number 2, April 1985

A NEW POLYNOMIAL INVARIANT OF KNOTS AND LINKS[1]

BY P. FREYD, D. YETTER; J. HOSTE;
W. B. R. LICKORISH, K. MILLETT; AND A. OCNEANU

The purpose of this note is to announce a new isotopy invariant of oriented links of tamely embedded circles in 3-space.

We represent links by plane projections, using the customary conventions that the image of the link is a union of transversely intersecting immersed curves, each provided with an orientation, and undercrossings are indicated by broken lines. Following Conway [**6**], we use the symbols L_+, L_0, L_- to denote links having plane projections which agree except in a small disk, and inside that disk are represented by the pictures of Figure 1.

Conway showed that the one-variable Alexander polynomials of L_+, L_0, L_- (when suitably normalized) satisfy the relation

$$\Delta_{L_+}(t) - \Delta_{L_-}(t) + (t^{1/2} - t^{-1/2})\Delta_{L_0}(t) = 0.$$

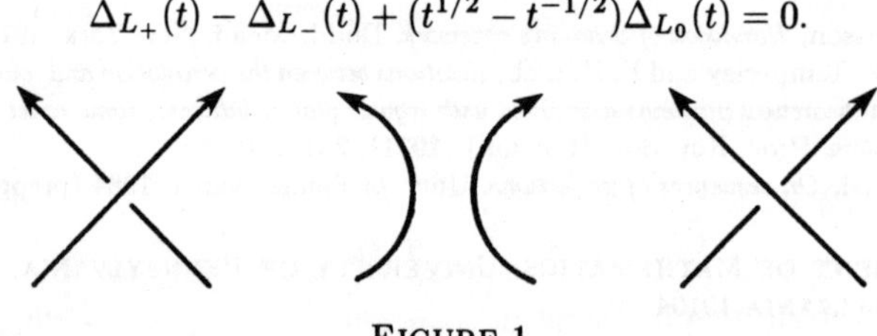

FIGURE 1

Received by the editors January 14, 1985.
1980 *Mathematics Subject Classification*. Primary 57M25.

[1] *Editor's Note.* The editors received, virtually within a period of a few days in late September and early October 1984, four research announcements, each describing the same result—the existence and properties of a new polynomial invariant for knots and links. There was variation in the approaches taken by the four groups and variation in corollaries and elaboration. These were: *A new invariant for knots and links* by Peter Freyd and David Yetter; *A polynomial invariant of knots and links* by Jim Hoste; *Topological invariants of knots and links*, by W. B. R. Lickorish and Kenneth C. Millett, and *A polynomial invariant for knots: A combinatorial and an algebraic approach*, by A. Ocneanu.

It was evident from the circumstances that the four groups arrived at their results completely independently of each other, although all were inspired by the work of Jones (cf. [**10**], and also [**8, 9**]). The degree of simultaneity was such that, by common consent, it was unproductive to try to assess priority. Indeed it would seem that there is enough credit for all to share in.

Each of these papers was refereed, and we would have happily published any one of them, had it been the only one under consideration. Because the alternatives of publication of all four or of none were both unsatisfying, all have agreed to the compromise embodied here of a paper carrying all six names as coauthors, consisting of an introductory section describing the basics written by a disinterested party, and followed by four sections, one written by each of the four groups, briefly describing the highlights of their own approach and elaboration.

Recently, Vaughan Jones [**10**] constructed a new polynomial invariant satisfying the relation

$$tV_{L_+}(t) - t^{-1}V_{L_-}(t) + (t^{1/2} - t^{-1/2})V_{L_0}(t) = 0.$$

Our invariant, which can be regarded either as a nonhomogeneous polynomial in two variables or a homogeneous polynomial in three variables, generalizes both the Alexander-Conway [**2, 6**] and the Jones polynomials.

MAIN THEOREM. *There is a unique function P from the set of isotopy classes of tame oriented links to the set of homogeneous Laurent polynomials of degree 0 in x, y, z such that*

(1) $xP_{L_+}(x,y,z) + yP_{L_-}(x,y,z) + zP_{L_0}(x,y,z) = 0$,

(2) $P_L(x,y,z) = 1$ *if L consists of a single unknotted component.*

REMARKS. (1) If L consists of n unlinked and unknotted components, then

$$P_L(x,y,z) = (-(x+y)/z)^{n-1}.$$

(2) The Alexander-Conway polynomial of L is

$$\Delta_L(t) = P_L(1, -1, t^{1/2} - t^{-1/2}),$$

and the Jones polynomial is

$$V_L(t) = P_L(t, -t^{-1}, t^{1/2} - t^{-1/2}).$$

(3) Since $P_L(x,y,z)$ is homogeneous, it can be viewed (in many ways) as a polynomial in two nonhomogeneous variables. A convenient way to do this is to set $\mathcal{P}_L(l,m) = P_L(l, l^{-1}, m)$. In this notation the basic relation is $l\mathcal{P}_{L_+}(l,m) + l^{-1}\mathcal{P}_{L_-}(l,m) + m\mathcal{P}_{L_0}(l,m) = 0$.

(4) Reversing the orientation of $\mathbf{R}^3$ has the effect

$$P_L(x,y,z) \leftrightarrow P_L(y,x,z), \qquad \rho_L(l,m) \leftrightarrow \rho_L(l^{-1},m).$$

However, neither polynomial is changed by simultaneous reversal of orientations on all components of L.

(5) If L is a connected sum of links L_1 and L_2, then $P_L = P_{L_1}P_{L_2}$.

(6) The following examples are easily computed by using the recurrence relation and the values of P for unlinks.

$L =$ 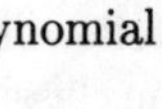 $: P_L = yz^{-1} + x^{-1}y^2z^{-1} - x^{-1}z,$

$L =$ 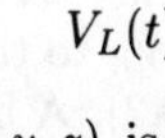 $: P_L = x^{-2}z^2 - 2x^{-1}y - x^{-2}y^2,$

$L =$ 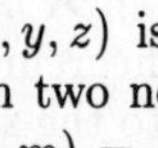 $: P_L = y^{-2}z^2 - 2xy^{-1} - x^2y^{-2},$

$L =$ 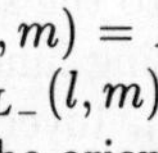 $: P_L = x^{-1}y^{-1}z^2 - xy^{-1} - x^{-1}y - 1.$

We wish to express our appreciation to Vaughan Jones for communicating his results to us. The particular viewpoints of the various authors are outlined below.

THE FREYD-YETTER APPROACH

In order to reduce the unique existence of the homfly invariant to a word problem, let R be the ring of Laurent polynomials on x, y, z; W_n the set of words on $n-1, \dots, 2, 1, -1, \dots, -n+1$ (no zero); W the disjoint union of the W_n's; F the free R-module generated by W. An element of W is denoted $[u]_n$, where $u \in W_n$. $R \to F$ denotes the map that sends 1 to $[\ ]_1$. $F \to M$ denotes the quotient module obtained by imposing the following relations, in which $u, v \in W_n$ and $0 < b < a < n$:

(A0) $[ua(-a)v]_n \equiv [uv]_n \equiv [u(-a)av]_n$,
(A1) $[uabv]_n \equiv [ubav]_n$ if $b < a-1$,
(A2) $[ua(a-1)av]_n \equiv [u(a-1)a(a-1)v]_n$,
(M1) $[uv]_n \equiv [vu]_n$,
(M2) $[unv]_{n+1} \equiv [uv]_n \equiv [u(-n)v]_{n+1}$,
(P1) $x[uav]_n + y[u(-a)v]_n + z[uv]_n \equiv 0$.

LEMMA. *$R \to F \to M$ is an isomorphism.*

This lemma implies the Main Theorem by the following argument. Each element of W_n describes a braid on n strands. By the classic theorem of Artin [**1**], the map $W \to F \to M$ sends two elements of W_n to the same element of M if they describe equivalent braids (A0), (A1), (A2). A braid gives rise to a link by joining its top and bottom. By the classic theorem stated by Markov [**12**] and proved by Birman [**3**], two braids are sent to the same element in M if they induce the same link (M1), (M2). The resulting function from links to M satisfies condition one of the Main Theorem (P1). Invert the isomorphism to obtain an R-valued function that satisfies both conditions.

To prove the lemma consider the following six *directed* substitution rules (still with $u, v \in W_n$, $0 < b < a < n$):

(S1) $[uabv]_n \to [ubav]_n$ for $b < a-1$,
(S2) $[ua(a-1)\cdots(b+1)bav]_n \to [u(a-1)a(a-1)\cdots(b+1)bv]_n$,
(S3) $[u(-a)v]_n \to -xy^{-1}[uav]_n - y^{-1}z[uv]_n$,
(S4) $[uaav]_n \to -x^{-1}y[uv]_n - x^{-1}z[uav]_n$,
(S5) $[unv]_{n+1} \to [uv]_n$,
(S6) $[uv]_{n+1} \to -(x+y)z^{-1}[uv]_n$.

LEMMA. *These rules obey the DCC: that is, there is no infinite sequence of correct applications starting with a single form.*

LEMMA. *Suppose that f is a form and that it is possible to apply a rule to obtain the form g and to make another application (of possibly another rule) to obtain the form h. Then there is a form j, a sequence of applications from g to j, and a sequence of applications from h to j.*

An inductive proof then shows that starting with any form f and arbitrarily applying the rules, one inevitably arrives at a *unique* "terminal form" $T(f)$ on which no rule applies.

LEMMA. *A form is terminal iff it is an R-multiple of* $[\ ]_1$.

The Main Theorem is thus established with

LEMMA. $T(f) = T(g)$ *iff* $f \equiv g$.

For the most part these four lemmas have straightforward (but tedious) proofs. The DCC lemma needs a "complexity" measurement on forms guaranteed to decrease under each of the substitution rules. The next lemma reduces to 21 separate cases, one for each pair of rules. Each is proven by the simple expedient of applying in sequence the unique rule that applies. The characterization of terminal forms is achieved by finding at least one rule that applies to $[u]_n$ for $n > 1$. The last lemma is easy if (M1) is deleted from the relations. ((M1) is in fact a consequence of the other relations.) The proof that $T[uv]_k = T[vu]_k$ quickly reduces to the case $T[nuv]_{n+1} = T[uvn]_{n+1}$, where u and v are descending words of positive elements, u starting with $n-1$, v with n. If v is empty or a singleton or if u is empty, a mindless verification works. Separate arguments are needed for the two remaining cases (u of length one, u of length greater than one). If long computational proofs are acceptable, a mindless approach works for both (as it must if these lemmas are true): Apply any sequence of applicable substitution rules.

THE HOSTE APPROACH

We define P as follows. Given any oriented link L first choose some projection S of L and then "resolve" S by first "changing" and "smoothing" some crossing in S, and then again changing and smoothing a crossing in each of the two projections that result from S, and so on until only unlinks remain. Next use property (1) of the Main Theorem and the values of P for the unlinks given in Remark (1) to obtain a value of P for S. We prove that the polynomial so produced is well defined and depends only on the isotopy class of L.

To avoid difficulties that arise from the fact that S can be resolved in infinitely many ways, we begin with a slightly different definition of P. Given S, first order the components of S and also distinguish a point on each component. We may then gain more control on the resolution of S by first giving a rule whereby an ordering and pointing is induced on projections obtained from S, and then demanding that the resolution ends, not just in unlinks, but in "descending" projections. (A pointed ordered projection is descending if one never crosses over one's path while traversing the components in the given order and direction starting on each at the distinguished point.) Now use this "distinguished" resolution to compute P of S, denoted P_S.

We next show P_S is well defined—i.e., independent of the choice of ordering, pointing, and distinguished resolution, by induction on the number of crossings in S. To prove the inductive step we first show the choice of resolution is immaterial. This is made possible by the pointing and ordering scheme, which essentially implies that any two resolutions must eventually change all the same crossings, but perhaps in different orders. Employing the inductive hypothesis, it remains only to prove that (1) changing a crossing and immediately changing it back has no effect, and (2) changing the ith and then the jth crossings has the same effect as first changing the jth and then the ith crossings. Next we show that P_S is unchanged if S is repointed. This can be reduced to the case where S is descending and one distinguished point is moved forward along its component past one crossing. Finally, we prove P_S is independent of the ordering of S. This is the hardest step, although we can first reduce to the case where S has two components and is descending, but not with respect to its own ordering. Now we show that S can be transformed into a split descending projection (one where the components do not cross each other and, hence, the ordering does not matter) by a finite sequence of Reidemeister moves that never increase the number of crossings (so we can make use of the inductive hypothesis) and, furthermore, preserve P_S.

Once P_S is known to be well defined, it is not hard to show that it is preserved by Reidemeister moves. Thus it depends only on the isotopy class of L. At this point we finally know that any distinguished resolution of the unlink produces the unlink polynomial, and from this we can prove that any resolution can be used to compute P, not just ones that are distinguished with respect to some choice of pointing and ordering of S.

The Lickorish and Millett Approach

Once the announced Main Theorem is believed, the only difficulty is proving that $\mathcal{P}_L$ can be well defined; uniqueness, calculations, and Remark (4) are easy. Our combinatorial proof has simplicity: Let $\mathcal{L}_n$ be all planar projections of oriented links of at most n crossings with components *ordered* and *base-pointed.* If $L \in \mathcal{L}_n$ let $\alpha L \in \mathcal{L}_n$ be L-with-changed-crossings so that αL *ascends* from the base point of the first component onwards (thus, following along the components of αL in the given order, always beginning at base points, a crossing is always first encountered as an underpass). Initially $\mathcal{P}$ is defined as a function $\mathcal{P}: \bigcup_n \mathcal{L}_n \to \mathbf{Z}[l^{\pm 1}, m^{\pm 1}]$. For $L \in \bigcup_n \mathcal{L}_n$ define $\mathcal{P}_{\alpha L} = \mu^{cL-1}$, where $\mu = -(l + l^{-1})m^{-1}$, and L has cL components. Inductively, assume that on $\mathcal{L}_{n-1}$, $\mathcal{P}$ has been defined, is independent of base points and orders, and

$$(*) \qquad l\mathcal{P}_{L_+} + l^{-1}\mathcal{P}_{L_-} + m\mathcal{P}_{L_0} = 0.$$

For L in $\mathcal{L}_n$ define $\mathcal{P}_L$ to be the polynomial obtained from $\mathcal{P}_{\alpha L}$ using $(*)$ (and the inductive definition) on a sequence of crossing changes that creates αL from L. The choice of sequence is irrelevant, $(*)$ holds in $\mathcal{L}_n$, and $\mathcal{P}_L$ is unchanged by those Reidemeister moves that remain *within* $\mathcal{L}_n$. If now βL denotes the new ascending element of $\mathcal{L}_n$ formed from L by reference to a new set of base points and component order, the induction step is completed by showing that $\mathcal{P}_{\beta L}$, as calculated from $\mathcal{P}_{\alpha L}$, is μ^{cL-1}. A straightforward argument on sliding base points along components shows (using the prescribed value of μ) that this is so if only base points are changed; a change of component order requires a very careful use of the *permitted* Reidemeister moves on

βL to change βL to another ascending projection with fewer crossings, thus determining $\mathcal{P}_{\beta L}$ by induction. This done, $\mathcal{P}_L$ is defined independent of base point and ordering; it is invariant under all Reidemeister moves, and the Main Theorem is proved.

The entire dependence of $\mathcal{P}_L$ on $(*)$ demands a redevelopment of Conway's skein theory **[7]** in the two-variable context. At once this new version of linear skein theory produces the elementary results that $\mathcal{P}_{L_1 \# L_2} = \mathcal{P}_{L_1}\mathcal{P}_{L_2} = \mu^{-1}\mathcal{P}_{L_1 \sqcup L_2}$. Applied to two-string tangles **[11]**, it shows that $\mathcal{P}_L$ is unchanged by mutation (rotation through π of such a tangle) in L. Thus the two eleven-crossing knots (of distinct genera) for which $\Delta(t) = 1$ have the same $\mathcal{P}(l, m)$, which is, by calculation, not trivial (so $\mathcal{P}_L$ depends on more than the Alexander module of L). Also, mutation shows that for a pretzel knot $K(a_1, \ldots, a_r)$, $\mathcal{P}_K$ is unchanged by permutation of the a_i. Further, if S and T are two-string tangles, if $S + T$ is

and S^N and S^D are the polynomials for

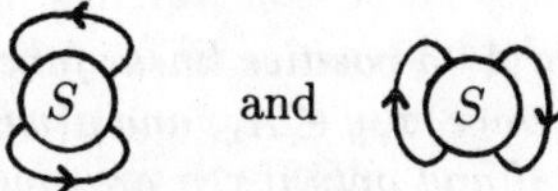

then

$$(1-\mu^2)(S+T)^N = (S^N T^D + S^D T^N) - \mu(S^N T^N + S^D T^D).$$

This can be much applied in producing formulae for the polynomials of rational and other arborescent links. Two knots of Birman **[4]** with distinct signatures have the same polynomial. Thus if signature is a skein invariant, then $\mathcal{P}$ is not injective on the set of skein-equivalence classes of oriented links. Two-variable analogues exist for most of the other skein-theoretic results given in **[7]**.

Tables will be produced (with computer aid) of the values of $\mathcal{P}$ for oriented links of low crossing number. At present the literature lacks an agreed notation and even a set of diagrams for *oriented* links. Manual calculations of the polynomials proceed via various neat recursion formulae for particular types (e.g., rational and pretzel) of links. Calculations of M. B. Thistlethwaite show that if K is the knot 11_{388} (in the notation of K. Perko) and $\overline{K}$ is its mirror image, then $\mathcal{P}_K \neq \mathcal{P}_{\overline{K}}$, but both $V_K = V_{\overline{K}}$ and $\Delta_K = \Delta_{\overline{K}}$.

The Ocneanu Approach

We define for each permutation in S_n a layered braid in the n-string braid group B_n. We show combinatorially that under the equivalence relation in the Main Theorem, any braid decomposes uniquely as a linear combination of layered braids. With the coefficients of the decomposition we construct invariants to conjugation in B_n, and obtain the invariant P_L of the closed braid as an invariant to the Markov moves **[3]**.

We then study the algebraic structure of the invariant P_L. Linear combinations of braids in B_n have an algebra structure in which layered braids form a basis. The algebra turns out to be the Hecke algebra H_n with Dynkin diagram A_n [5]. In this way we explicitly construct a family of finite-dimensional irreducible representations of B_n, indexed by n-Young diagrams. The invariant appears as a weighted trace, explicitly determined via the S-functions associated to the discrete Euler β-function [**13**].

We then study the values of the variables for which the representation has additional orthogonality and positivity structure.

THEOREM. *Let $\tau \in \mathbf{C}$. The free C^*-algebra A_τ over $\mathbf{C}$ generated by $\{e_k | k = 1, 2, \ldots\}$, satisfying $e_k^2 = e_k^* = e_k$ and*

$$e_k e_{k+1} e_k - \tau e_k = e_{k+1} e_k e_{k+1} - \tau e_{k+1}, \qquad k \in N,$$

$$e_k e_l = e_l e_k \quad \text{if } |k-l| \geq 2,$$

is nonzero if and only if $\tau \in [0, \frac{1}{4}] \cup \{(4\cos(\pi/n))^{-1},\ n = 3, 4, 5, \ldots\}$. □

THEOREM. *Let $\alpha, \theta \in \mathbf{C}$, and let*

$$\tau = (4\cosh^2\theta)^{-1}, \qquad \eta = \tfrac{1}{2}(1 - \tanh\alpha \tan\theta).$$

Consider on the C^*-algebra A_τ a positive linear functional $\mu = \mu_{\tau,\eta}$ satisfying $\mu(xy) = \mu(yx)$, i.e., μ is a trace, $x, y \in A_\tau$, and $\mu(we_{n+1}) = \eta\mu(w)$, $w \in A_{\tau,n}$. Such a functional μ exists if and only if τ, η are given by one of the following values of α, θ:

(i) $\theta \in \mathbf{R}^+$, $\alpha \in \mathbf{R}$ *(the normal range)*;
(ii) $\theta \in \mathbf{R}^+$, $\alpha = k\theta + \pi i/2$, $k \in \mathbf{Z}\backslash\{0\}$ *(the exceptional lines)*;
(iii) $\theta = \pi i/m$, $m = 3, 4, 5, \ldots$; $\alpha = (k/m + 1/2)\pi i$, $k = 1, 2, \ldots, [(m-1)/2]$ *(the exceptional points)*,

or, if $0 \leq \eta \leq 1$ and $\tau = \eta(1-\eta)$ (the limit line).

The trace μ is then uniquely determined and the corresponding weak closure $R_{\tau,\eta}$ of A_τ is the hyperfinite II_1 factor for $\eta \neq 0, 1$, and $\mathbf{C}$ for $\eta = 0, 1$. □

For α, θ as above consider the representation $\gamma = \gamma_{\alpha,\theta,\eta}$ of the braid group B_n into $R_{\tau,\eta}$ given by

$$\gamma(\sigma_k) = i(e^{-\alpha-\theta} e_k - e^{-\alpha+\theta}(1 - e_k)), \qquad k = 1, \ldots, n-1.$$

Consider on $R_{\tau,\eta}$ the trace

$$\phi = \phi_{\alpha,\theta} = (i \cosh\alpha \operatorname{csch}\theta)^{n-1} \mu_{\tau,\eta}.$$

THEOREM. *For $b \in B_n$ the invariant polynomial $P_{\hat{b}}$ of the closure $\hat{b}$ is given by $P_{\hat{b}}(e^\alpha, -e^{-\alpha}, 2i\sinh\theta) = \phi(\gamma(b))$.*

P_L is given by a positive trace of a unitary representation precisely at the exceptional points. The Jones invariant corresponds to the line $k = 2$ and the Alexander invariant to $k = 0$; the Arf invariant to $m = 4$, $k = 2$. For $k = 1$, $P_L \equiv 1$, and for $k = -1$, $P_L = \pm 1$. This shows that for any knot K, $P_K - 1$ is divided by $P_{\text{trefoil}} - 1$.

We have implemented a program which computes the coefficients of P_L.

REMARK. H. Wenzl has independently obtained the structure of the simple modules of the Hecke algebras by studying subfactors of the hyperfinite II_1 factor and has computed the index and entropy of the subfactor generated by $e_2, e_3, \ldots$ in $R_{\alpha,\theta}$.

REFERENCES

1. E. Artin, *Theorie der Zopfe*, Hamburg Abh 9, pp. 47–72.

2. J. W. Alexander, *Topological invariants of knots and links*, Trans. Amer. Math. Soc. **30** (1928), 275–306.

3. J. Birman, *Braids, links and mapping class groups*, Ann. Math. Studies **82** (1974).

4. ——, *On the Jones polynomial of closed 3-braids* (to appear).

5. N. Bourbaki, *Groupes et algèbres de Lie*, Chaps. 4, 5, 6, no. 1337, Hermann, Paris, 1960–1972.

6. J. H. Conway, *An enumeration of knots and links*, Computational Problems in Abstract Algebra, Pergamon, N.Y., 1970, pp. 329–358.

7. C. Giller, *A family of links and the Conway calculus*, Trans. Amer. Math. Soc. **270** (1982), 75–109.

8. V. F. R. Jones, *Index for subfactors*, Invent. Math. **72** (1983), 1–25.

9. ——, *Braid groups, Hecke algebras and type* II_1 *factors*, Proc. Japan-U.S. Conf., 1983 (to appear).

10. ——, *A polynomial invariant for knots via von Neumann algebras*, Bull. Amer. Math. Soc. **12** (1985), 103–112.

11. W. B. R. Lickorish, *Prime knots and tangles*, Trans. Amer. Math. Soc. **267** (1981), 321–332.

12. A. A. Markov, *Über die freie Aequivalenz geschlossener Zopfe*, Mat. Sb. **1** (1935), 73–78.

13. A. Wasserman, Thesis, Univ. of Pennsylvania, 1981.

DEPARTMENT OF MATHEMATICS, UNIVERSITY OF PENNSYLVANIA, PHILADELPHIA, PENNSYLVANIA 19146 (Current address of P. Freyd)

DEPARTMENT OF MATHEMATICS, CLARK UNIVERSITY, WORCESTER, MASSACHUSETTS 01610 (Current address of D. Yetter)

DEPARTMENT OF MATHEMATICS, RUTGERS UNIVERSITY, NEW BRUNSWICK, NEW JERSEY 08903 (Current address of J. Hoste)

DEPARTMENT OF MATHEMATICS, UNIVERSITY OF CAMBRIDGE, CAMBRIDGE, ENGLAND (Current address of W. B. R. Lickorish)

DEPARTMENT OF MATHEMATICS, UNIVERSITY OF CALIFORNIA AT SANTA BARBARA, SANTA BARBARA, CALIFORNIA 93106 (Current address of K. Millett) AND MATHEMATICAL SCIENCES RESEARCH INSTITUTE, BERKELEY, CALIFORNIA 94720

MATHEMATICAL SCIENCES RESEARCH INSTITUTE, BERKELEY, CALIFORNIA 94720 (Current address of A. Ocneanu)

Annals of Mathematics, 126 (1987), 335–388

Hecke algebra representations of braid groups and link polynomials

By V. F. R. Jones

Abstract

By studying representations of the braid group satisfying a certain quadratic relation we obtain a polynomial invariant in two variables for oriented links. It is expressed using a trace, discovered by Ocneanu, on the Hecke algebras of type A. A certain specialization of the polynomial, whose discovery predated and inspired the two-variable one, is seen to come in two inequivalent ways, from a Hecke algebra quotient and a linear functional on it which has already been used in statistical mechanics. The two-variable polynomial was first discovered by Freyd-Yetter, Lickorish-Millet, Ocneanu, Hoste, and Przytycki-Traczyk.

0. Introduction

This paper initiates the detailed study of representations of Artin's braid groups B_n which arise from the Hecke algebras of type A_{n-1}. There appears to be no direct understanding of these representations so our approach will be via generators and relations. The braid group B_n has a presentation $\langle \sigma_1, \ldots, \sigma_{n-1} | \sigma_i \sigma_{i+1} \sigma_i = \sigma_{i+1} \sigma_i \sigma_{i+1}, i = 1, 2, \ldots, n-2, \ \sigma_i \sigma_j = \sigma_j \sigma_i, |i-j| \geq 2 \rangle$ and the Hecke algebra $H(q, n)$ of type A_{n-1} has a presentation $\langle g_1, \ldots, g_{n-1} | g_0^2 = (q-1) g_i + q, \ i = 1, \ldots, n-1, \ g_i g_{i+1} g_i = g_{i+1} g_i g_{i+1}, i = 1, 2, \ldots, n-2, \ g_i g_j = g_j g_i, \ |i-j| \geq 2 \rangle$, where q is a parameter. Our attitude will be that q is a complex number which may take any value. Thus for each $q \neq 0$, B_n has a representation inside $H(q, n)$ obtained by sending σ_i to g_i.

There seems to be no a priori reason why these representations should be of interest but we shall show that in fact they are. For generic q it is possible that they are faithful.

The geometric picture of braids gives relations with links in 3-space. We shall see how two such relations are connected with linear functionals on the Hecke algebras. The first is a trace defined, by Ocneanu, inductively from the relations $\mathrm{tr}(ab) = \mathrm{tr}(ba)$, $\mathrm{tr}(1) = 1$, and $\mathrm{tr}(x g_n) = z \, \mathrm{tr}(x)$ where $x \in H(q, n)$

and $H(q, n)$ is embedded in $H(q, n+1)$ by identifying the g_i's. The parameter z is another complex number independent of q. The trace will be a two-variable polynomial invariant of oriented links discovered independently by Lickorish and Millet, Freyd and Yetter, Ocneanu, and Hoste ([14]). One of its specializations is the classical Alexander polynomial of [2]. We refer to [22] for a treatment of it from a different point of view.

The other linear functional is more difficult to define and lives on a quotient of the Hecke algebra in which the relation

$$g_i g_{i+1} g_i + g_i g_{i+1} + g_{i+1} g_i + g_i + g_{i+1} + 1 = 0$$

is satisfied. It corresponds to unoriented links via the theory of plats ([7]). Remarkably, the invariant so defined is a specialization of the one coming from the trace and was discovered first as an invariant of oriented links in [16]. This second linear functional occurs in statistical mechanics as the partition function in the Potts and "ice-type" models (see [4], [42]).

A topological interpretation of these invariants is lacking at present. In this direction it would seem very important to understand the representations of the braid groups in $H(q, n)$ in a more intrinsic manner, in particular the meaning of the parameters q and z. This might also show how to use the other Hecke algebras (not of type A_{n-1}), and their rich representation theory, in some field related to knots.

While very few of our results will use the theory of von Neumann algebras, it should be pointed out that they were the starting point of this work and continue to motivate many of the results. A deeper understanding of subfactors of finite index (see [17], [18]) will almost certainly clarify many of the topological questions.

The author would like to single out Joan Birman among the many recipients of his thanks. Her contribution to this new topic has been of inestimable importance.

I. Braids and links

Braids are formed when n points on a horizontal plane are connected by n strings to n points on another horizontal plane directly below the first n points. The strings are not allowed to go back upwards at any point in their travel. The braid group B_n on n strings is the group formed by appropriate isotopy classes of braids with the obvious concatenation operation. The n points may be supposed to lie on a single straight line which gives rise to an obvious preferred embedding of B_n in B_{n+1} and to a preferred set of generators called $\sigma_1, \sigma_2, \ldots, \sigma_{n-1}$ given by the following picture:

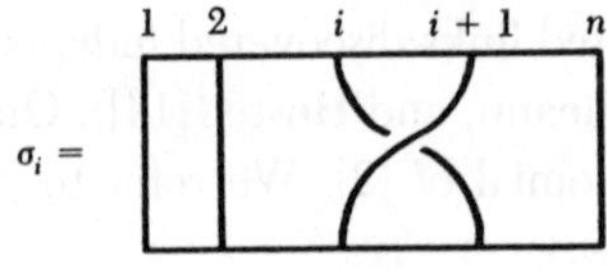

FIGURE 1.1

One may easily convince oneself that the σ_i's satisfy

$$\sigma_i\sigma_{i+1}\sigma_i = \sigma_{i+1}\sigma_i\sigma_{i+1}, \tag{1.2}$$

$$\sigma_i\sigma_j = \sigma_j\sigma_i \qquad \text{if } |i-j| \geq 2. \tag{1.3}$$

That (1.2) and (1.3) give a presentation of B_n was proved by E. Artin. The importance of Artin's result for us is that, to construct a representation of B_n it suffices to find matrices satisfying (1.2) and (1.3). As a general reference on braids, see [6].

Given a braid $\alpha \in B_n$ one may form the oriented link $\hat{\alpha}$, called the closure of α, in a manner adequately described by the following diagram:

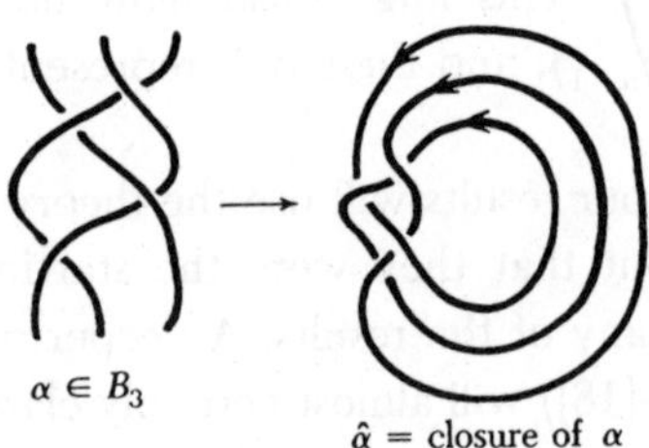

FIGURE 1.4

A result of J. Alexander asserts that any (tame) oriented link is isotopic to the closure of some braid, but attempts to exploit braids to study knots run into the following serious problem: The representation of a link L as a closed braid is highly non-unique. Fortunately, A. A. Markov found purely algebraic necessary and sufficient conditions for braids $\alpha \in B_n$ and $\beta \in B_m$ to have isotopic closures. This is stated in terms of "Markov moves" as follows. A Markov move of *type* I is changing $\alpha \in B_n$ to $\beta\alpha\beta^{-1} \in B_n$ for any $\beta \in B_n$, and a Markov move of *type* II is changing $\alpha \in B_n$ to $\alpha\sigma_n^{\pm 1} \in B_{n+1}$, or the inverse of this operation. Markov's theorem, whose first published proof appears in [6], is that if $\alpha \in B_n$ and $\beta \in B_m$ have isotopic closures then there is a finite sequence of Markov moves of types I and II which takes α to β. Unfortunately, Markov's theorem is not easy to apply directly as the sequence of moves may be long and go through several different braid groups.

There are more ways to obtain links from braids than by closing them as above. The plat method is discussed in detail in Section 14 and recently more general closures are being considered (see [11]).

2. The Burau representation

If t is a non-zero complex number let β_i, for $1 \leq i \leq n-1$, be the $n \times n$ matrix

$$\begin{pmatrix} \ddots & & & & & 0 \\ & 1 & & & & \\ & & \ddots & & & \\ & & & 1-t & t & \\ & & & 1 & 0 & \\ & & & & & 1 \\ 0 & & & & & & \ddots \end{pmatrix}$$

where $1-t$ is the $i-i$ entry. One may easily check that $\beta_i\beta_{i+1}\beta_i = \beta_{i+1}\beta_i\beta_{i+1}$ and $\beta_i\beta_j = \beta_j\beta_i$ if $|i-j| \geq 2$. Sending σ_i to β_i defines the (non-reduced) Burau representation of B_n. It clearly leaves invariant the subspace of C^n of all vectors whose entries add up to 0 and on the quotient by this subspace a convenient $(n-1) \times (n-1)$ expression is

$$b_1 = \begin{pmatrix} -t & 0 & & & \\ & & & & 0 \\ -1 & 1 & & & \\ & & 1 & & \\ & & & \ddots & \\ & 0 & & & 1 \end{pmatrix}, \quad b_{n-1} = \begin{pmatrix} 1 & & & & 0 \\ & 1 & & & \\ & & \ddots & & \\ & & & 1 & -t \\ 0 & & & 0 & -t \end{pmatrix},$$

$$b_i = \begin{pmatrix} 1 & & & & & & \\ & 1 & & & & & \\ & & 1 & -t & 0 & & \\ & & 0 & -t & 0 & & \\ & & 0 & -1 & 1 & & \\ & & & & & 1 & \\ & & & & & & 1 \end{pmatrix}$$

$2 \leq i \leq n-2$ where the diagonal $-t$ is in the $i-i$ position.

These are the most simple non-trivial representations of the braid groups and are of fundamental importance. Except when $n = 2$ or 3, it is unknown at the time of writing if these representations are faithful, even for $t = 2$.

Burau recognized that his represention was related to closed braids. He knew that if $\alpha \in B_n$ and ψ is the reduced Burau representation then

$\det(1 - \psi(\alpha))$ is $(1 + t + \cdots + t^{n-1})$ times the Alexander polynomial of the link $\hat{\alpha}$. He showed this by connecting the matrix $\psi(\alpha)$ with a known way of calculating the Alexander polynomial from a presentation of the fundamental group of the complement of the link. In Section 7 we give a new proof of this result from an entirely different definition of the Alexander polynomial.

For positive braids there is also a mechanical interpretation of the Burau matrix: Lay the braid out flat and make it into a bowling alley with n lanes, the lanes going over each other according to the braid. If a ball travelling along a lane has probability t of falling off the top lane (and continuing in the lane below) at every crossing then the (i, j) entry of the (non-reduced) Burau matrix is the probability that a ball bowled in the ith lane will end up in the jth. Thus in particular every polynomial entry of the Burau matrix of a positive braid will be nonnegative for $0 \leq t \leq 1$.

One should think of the (reduced) Burau representation of B_n as deforming the fundamental representation of the Coxeter group of type A_{n-1} to a pseudo-reflection, since the Burau matrices of the generators σ_i have eigenvalues $-t$ with multiplicity 1 and 1 with multiplicity $n - 2$.

3. Representations of symmetric groups

A braid α defines in an obvious way a permutation on its end points. This homomorphism from B_n to S_n simply corresponds to adding the relation $\sigma_i^2 = 1$ to (1.2) and (1.3). Thus it is to be expected that some family of representations of B_n will be related to representations of symmetric groups. (Note in particular that putting $t = 1$ in the (non-reduced) Burau representation gives the representation of S_n as permutations of basis vectors of an n-dimensional space.) For this reason we devote this section to a discussion of some features of the representation theory of the symmetric groups.

As always for a finite group, irreducible representations (called irreps) are in one-to-one correspondence with conjugacy classes of the group, though not in a natural way. Conjugacy classes of S_n are indexed by *partitions* of n corresponding to the periods of the disjoint cycles of the permutation. These can be represented diagrammatically as follows.

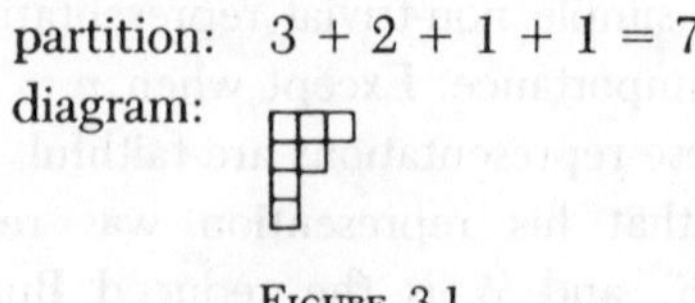

FIGURE 3.1

The length of the rows in the diagram is made to be non-decreasing. We will call such a diagram a *Young diagram*.

Thus irreps of S_n are indexed by Young diagrams. An explicit expression of the representation coming from a diagram is possible but will not be useful in this paper. What we shall describe is how a given irrep of S_n decomposes when it is restricted to S_{n-1}. In fact with some thought this rule suffices to construct the representation inductively. It is very simple: Given an irrep π of S_n with Young diagram Y, its restriction to S_{n-1} is the direct sum of (one copy of) each representation of S_{n-1} obtained from Y by removing a node so as to obtain a *Young diagram*.

Example 3.2. The representation of S_8 restricts to the direct sum of and and of S_7.

We can now build up by induction all the irreps of all the symmetric groups. As this will be extremely important let us record the picture up to S_5.

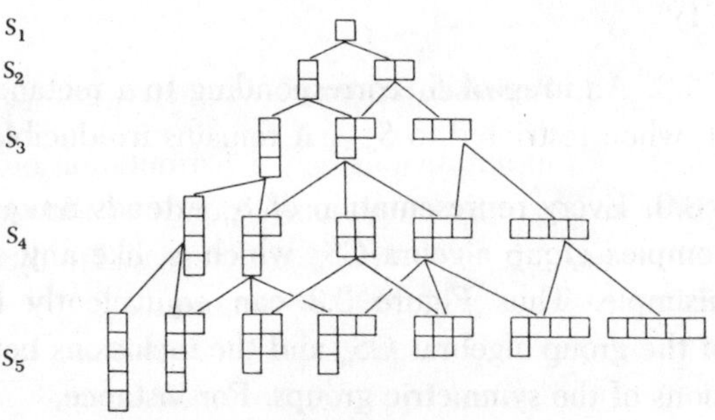

FIGURE 3.3

One is to imagine this diagram continuing indefinitely downwards. The lines connecting different rows represent the restrictions of representations. Several remarks are in order:

Remark 3.4. There is an ambiguity in assigning diagrams to representations created by the obvious row-column symmetry of Figure 3.3. To fix this problem it suffices to do it for S_2. We use □□ to represent the trivial representation of S_2 and the parity. On the representation level the row-column symmetry corresponds to tensoring by the one-dimensional parity irrep of S_n.

Remark 3.5. The dimension of the representation associated with a given diagram is, by induction, given by the number of descending paths from the initial □ to the diagram in question. But there is a closed formula for the dimension known as the "hook length" formula. We will need this procedure, so here is how it works: record in each box its hook length, i.e. the number of boxes horizontally to the right and vertically below the box in question, e.g. for 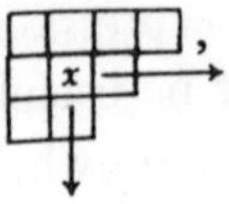

, the box marked x has hook length 3. The dimension of the representation is then $n!$ divided by the product of the hook lengths, e.g.

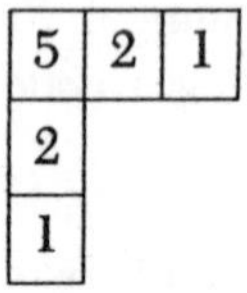

$$\text{dimension} = \frac{5!}{2 \cdot 2 \cdot 5} = 6$$

Remark 3.7. The characters of non-trivial conjugacy classes may also be calculated from the Young diagram. We record here the following fact which was known to Frobenius: in an irrep of S_n the trace of an n-cycle is non-zero if and only if the diagram has at most one corner, i.e. is of the form 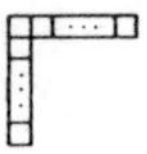

and then it is $(-1)^{\alpha}$.

Remark 3.8. An irrep of S_n corresponding to a rectangular tableau has the property that, when restricted to S_{n-1} it remains irreducible.

Remark 3.9. Every representation of S_n extends uniquely to a representation of the complex group algebra $\mathbf{C}S_n$ which is, like any complex finite group algebra, semisimple. Thus Figure 3.3 can equivalently be considered as a description of the group algebras $\mathbf{C}S_n$ and the inclusions between them induced by the inclusions of the symmetric groups. For instance,

$$\mathbf{C}S_3 \cong \mathbf{C} \times M_2(\mathbf{C}) \times \mathbf{C}\,.$$

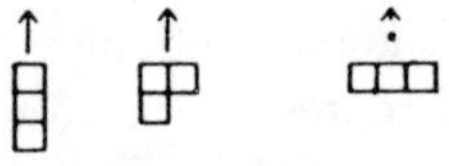

The irreps of S_n define the simple components of $\mathbf{C}S_n$ and the lines of Figure 3.3 show how the simple components of $\mathbf{C}S_n$ fit inside those of $\mathbf{C}S_{n+1}$.

4. Hecke algebras

The various generators σ_i of the braid groups are all conjugate, so that all one-dimensional representations of a braid group are classified by non-zero scalars. One could obtain all two-dimensional representations but we will see that a much richer structure emerges if we try to describe all representations of B_n in which the σ_i's have *at most two eigenvalues*. Writing g_i for the image of σ_i, under such a representation we must have an equation of the form $g_i^2 + ag_i + b = 0$ where a and b are scalars. It seems that two constants are involved but by modifying g_i by a fixed constant we may eliminate one. It is convenient to express the relation as $g_i^2 = (q-1)g_i + q$ (q a scalar). Let us call such a representation *quadratic*.

Thus knowledge of representations of B_n in which the g_i's have $\leq$ two eigenvalues is the same as knowledge of the algebra $H(q, n)$ with presentation on generators $g_1, g_2, \ldots, g_{n-1}$, and relations

$$g_i^2 = (q-1)g_i + q, \tag{4.1}$$

$$g_i g_{i+1} g_i = g_{i+1} g_i g_{i+1}, \tag{4.2}$$

$$g_i g_j = g_j g_i \quad \text{if } |i-j| \geq 2. \tag{4.3}$$

Now recall that the symmetric group S_n has a presentation on $s_1, \ldots, s_{n-1}$, $s_i^2 = 1$, $s_i s_{i+1} s_i = s_{i+1} s_i s_{i+1}$, $s_i s_j = s_j s_i$ if $|i-j| \geq 2$. Imagine trying to reduce words on the g_i's and s_i's to words of minimal length. The point is that relation (4.1) is as good as $s_i^2 = 1$ for this purpose. Thus a system of reduced words on the s_i's for S_n will furnish a basis for $H(q, n)$, by simply writing g_i for s_i. A convenient such basis is

$$\left\{ (g_{i_1} g_{i_1 - 1} \cdots g_{i_1 - k_1})(g_{i_2} \cdots g_{i_2 - k_2}) \cdots (g_{i_p} g_{i_p - 1} \cdots g_{i_p - k_p}) \,\middle|\, 1 \leq i_1 < i_2 < \cdots < i_p \leq n-1 \right\}. \tag{4.4}$$

We see that the dimension of $H(q, n)$ is $n!$. (In fact one must also show that the algebra does not collapse, but this is well established (see [12]). It would suffice to find an expression for the multiplication law in this basis and prove associativity.)

But we can do much better than this. For if we put $q = 1$ in (4.1) we see that $H(1, n)$ is the group algebra $\mathbf{C}S_n$ which, as we noted before, is semisimple. Now semisimplicity is an open condition; so if we change q slightly from 1, $H(q,n)$ will remain semisimple. Not only this, but the whole structure of $H(q, n)$ and the inclusions $H(q, n) \subseteq H(q, n+1)$ (defined from (4.4) in the obvious way) must remain the same under this deformation, at least for n not too large, depending on how far q is from 1. We conclude:

THEOREM 4.5. *For q close to* 1, *the simple $H(q, n)$ modules (or quadratic irreps of B_n) are in one-to-one correspondence with Young diagrams. Their decomposition rule and hence their dimensions are the same as for S_n.*

The above arguments are somewhat imprecise but can be made good without too much trouble. See [12], pages 54, 55 and 56.

In the same way that Figure 3.3 determines the representations of S_n, so it determines the representations of $H(q, n)$. Wenzl in [46] has written down explicit and demonstrably irreducible representations of $H(q, n)$ for each Young diagram. Once equipped with these formulae one may dispense with all the subtleties of the proof of Theorem 4.5, as there are sufficient representations to show that $\dim_{\mathbf{C}}(H(q,n)) = n!$. One is also able to make precise "sufficiently close to 1", for the formulae of [46] only require one to be able to invert q and $1 - q^p$ for the appropriate p. Thus the conclusion of Theorem 4.5 is true provided q is not a root of unity or zero. We do not wish to imply that the roots of unity are uninteresting. In fact they are probably of more importance than other values and will occur often later.

The algebra defined by (4.1), (4.2) and (4.3) is called the Hecke algebra of type A_{n-1} since the defining relations fit into the Coxeter-Dynkin picture. They arise in the study of representations of $GL(n,q)$ where they define the centralizer of the natural representation on the set of flags. Other Hecke algebras exist for other Coxeter-Dynkin diagrams and it would be nice to know if any of the ideas of this paper can be suitably modified for them.

We will always consider $H(q,n)$ as embedded in $H(q,n+1)$ via (4.4) and the representation of B_n inside $H(q,n)$ will be denoted π, so that $\pi(\sigma_i) = g_i$. We will often abbreviate $H(q,n)$ to H_n.

Note 4.6. The symmetry of Figure 3.3 corresponds to the automorphism $g_i \mapsto -qg_i^{-1}$ of the Hecke algebra in the obvious way.

Note 4.7. We decided in Remark 3.4 to let a Young diagram with one row correspond to the trivial irrep of S_n and a diagram with one column to the parity irrep. When this convention is extended to $q \neq 1$, we find that the "trivial" irrep of $H(q,n)$ must be defined by $g_i \mapsto q$ and the "parity" irrep by $g_i \mapsto -1$. This agrees with note 4.6.

5. Ocneanu's trace on $H(q,n)$

The following theorem is inspired directly from von Neumann algebras where normalized traces are building blocks of the theory.

THEOREM 5.1 (Ocneanu [14]). *For every $z \in \mathbf{C}$ there is a linear trace* tr *on $\bigcup_{n=1}^{\infty} H(q,n)$ uniquely defined by*

1) $\operatorname{tr}(ab) = \operatorname{tr}(ba)$.
2) $\operatorname{tr}(1) = 1$.
3) $\operatorname{tr}(xg) = z\operatorname{tr}(x)$ for $x \in H(q,n)$.

Proof. The first thing to observe is that the map $C: H_n \oplus H_n \otimes_{H_{n-1}} H_n \Rightarrow H_{n+1}$ given by $C(x \oplus y_1 \otimes y_2) = x \oplus y_1 g_n y_2$ is an isomorphism of $H_n - H_n$ bimodules. This follows by considering the set (4.4) and noting that any word for $H(q,n+1)$ contains g_n at most once. Surjectivity is immediate and injectivity follows from a dimension count.

Now we are free to define a linear functional inductively from the formulae $\mathrm{tr}(1) = 1$ and $\mathrm{tr}(xg_n y) = z\,\mathrm{tr}(xy)$ for $x, y \in H(q, n)$. The problem is then to show property 1). By induction we may suppose it for $a, b \in H_n$. Now if S is a subset of an algebra A which generates it as an algebra, then to show that a linear functional is a trace it suffices to show $f(xs) = f(sx)$ for all $s \in S$ and $x \in A$. Applying this where $S = H_n \cup \{g_n\}$, we see that the only case which does not follow trivially from the definition of tr is $\mathrm{tr}(g_n x g_n y) = \mathrm{tr}(x g_n y g_n)$ for $x, y \in H_n$. But by the remark at the beginning of the proof it suffices to consider the four cases:

a) $x \in H_{n-1}$, $y \in H_{n-1}$,
b) $x = ag_{n-1}b$ where $a, b \in H_{n-1}$ and $y \in H_{n-1}$,
c) same as b) with the roles of x and y reversed,
d) $x = ag_{n-1}b$, $y = cg_{n-1}d$ where $a, b, c, d \in H_{n-1}$.

Case a) is trivial since g_n commutes with H_{n-1} so that we need only consider cases b) and d).

b)
$$\begin{aligned}\mathrm{tr}(g_n a g_{n-1} b g_n y) &= \mathrm{tr}(a g_n g_{n-1} g_n b y)\\ &= \mathrm{tr}(a g_{n-1} g_n g_{n-1} b y)\\ &= z\,\mathrm{tr}\big(a g_{n-1}^2 b y\big) \qquad \text{(definition of tr)}\\ &= (q-1) z\,\mathrm{tr}(a g_{n-1} b y) + qz\,\mathrm{tr}(aby)\end{aligned}$$

and

$$\begin{aligned}\mathrm{tr}(a g_{n-1} b g_n y g_n) &= \mathrm{tr}(a g_{n-1} b g_n^2 y)\\ &= (q-1)\mathrm{tr}(a g_{n-1} b g_n y) + q\,\mathrm{tr}(a g_{n-1} b y)\\ &= z(q-1)\mathrm{tr}(a g_{n-1} b y) + qz\,\mathrm{tr}(aby).\end{aligned}$$

d)
$$\begin{aligned}\mathrm{tr}(g_n a g_{n-1} b g_n c g_{n-1} d) &= z\,\mathrm{tr}\big(a g_{n-1}^2 b c g_{n-1} d\big) \qquad \text{(as above)}\\ &= z(q-1)\mathrm{tr}(a g_{n-1} b c g_{n-1} d) + z^2 q\,\mathrm{tr}(abcd)\end{aligned}$$

and

$$\begin{aligned}\mathrm{tr}(a g_{n-1} b g_n c g_{n-1} d g_n) &= z\,\mathrm{tr}\big(a g_{n-1} b g_{n-1}^2 d\big)\\ &= z(q-1)\mathrm{tr}(a g_{n-1} b c g_{n-1} d) + z^2\mathrm{tr}(abcd).\end{aligned}$$

Q.E.D.

Remark 5.2. This proof differs from Ocneanu's original proof. He used a different basis for $H(q, n)$ corresponding to a different section for the natural homomorphism from B_n to S_n, which he called "layered" braids. The trace of a layered braid is a very simple expression but it is more difficult to prove property 1). On the other hand, we have used the basis (4.4) and the expression for the trace of one of these basis elements is quite complicated. It would be interesting to develop the theory using the basis $\{C_w\}$ of [20].

It should be apparent from the proof that properties 1), 2), and 3) suffice to calculate the trace of any element of $H(q, n)$. To emphasize this let us do a sample calculation, that of the trace of $g_2g_1g_3g_2$ which is of minimal length in the braid group:

$$\begin{aligned}\operatorname{tr}(g_2g_1g_3g_2) &= \operatorname{tr}(g_2^2g_1g_3) && \text{(property 1)}\\ &= z\operatorname{tr}(g_2^2g_1) && \text{(property 3)}\\ &= z(q-1)\operatorname{tr}(g_2g_1) + zq\operatorname{tr}(g_1)\\ &= (z^2(q-1)+zq)\operatorname{tr}(g_1)\\ &= z^3(q-1)+z^2q && \text{(properties 3 and 2).}\end{aligned}$$

But another method of calculating tr is available. We know that the Hecke algebra is a direct product of matrix algebras, indexed by Young diagrams. The trace of Theorem 5.1 will be determined by its restrictions to these matrix algebras and a trace on a matrix algebra is necessarily a multiple of the usual trace. Thus if we know the scaling factors, called "weights", associated with each diagram we may calculate the trace of any element of the Hecke algebra by decomposing it according to the matrix algebras and taking the weighted sum of the traces.

The weights have been calculated by Ocneanu. They are Schur functions as we shall now describe (see [27]). Note that, given the solution, one only has to verify two things: first that the weights do indeed give a well-defined trace, i.e. the restriction to $H(q, n)$ of the trace defined on $H(q, n+1)$ by the weights is correct, and second, that it satisfies the Markov property $\operatorname{tr}(wg_n) = z\operatorname{tr}(w)$ for $w \in H(q, n)$. For this see [46]. We shall content ourselves simply to give the answer.

If Y is a Young diagram let π_Y denote the ensuing irreducible braid group representation. We must start with a Young diagram Y and specify a function W_Y of q and z which gives the trace of a minimal idempotent in the simple component of $H(q, n)$ specified by that Young diagram. First of all, define $S(q, z)$ as follows (where $w = 1 - q + z$): superimpose the Young diagram on the following diagram

$w-z$	$w-qz$	$w-q^2z$	$\cdots$	$\cdots$
$qw-z$	$qw-qz$	$qw-q^2z$	$\cdots$	$\cdots$
q^2w-z	q^2w-qz	$\cdots$	$\cdots$	
$\vdots$	$\vdots$	$\vdots$		
$\vdots$	$\vdots$	$\vdots$		

FIGURE 5.3

and let $S(q, z)$ be the product of the terms covered by Y. For instance, if $Y =$ 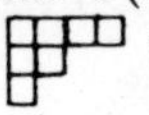 then

$$S(q, z) = (w - z)(w - qz)(w - q^2z)(w - q^3z) \times (qw - z)(qw - qz)(q^2w - z).$$

Now define $Q(q)$ as follows: First fill in the hook lengths as if calculating the dimension of the irrep of S_n corresponding to Y. Now replace each integer appearing, say m, by $1 - q^m$. Then $Q(q)$ is the product of these terms. For instance if Y is as before we form

6	4	2	1
3	1		
1			

and $Q(q) = (1 - q)(1 - q^2)(1 - q^4)(1 - q^6)(1 - q)(1 - q^3)(1 - q)$. The promised formula is then

(5.4) $$W_Y(q, z) = S(q, z)/Q(q).$$

If Y is a Young diagram let tr_Y be the trace on the Hecke algebra obtained by evaluating the usual trace (sum of the diagonal entries) on the image of a Hecke algebra element in the representation π_Y. Then the significance of (5.4) is the "Fourier transform" formula:

(5.5) $$\mathrm{tr}(x) = \sum_Y W_Y(q, z)\mathrm{tr}_Y(x).$$

Thus to calculate Ocneanu's trace of a word on the g_i's in this fashion we need only explicit matrices representing the g_i's for each π_Y. The explicit formulae of [46] are not well adapted for calculations as they involve square roots of certain polynomials. Much better from this point of view is the paper of Kazhdan and Lusztig ([20]) which gives a method of calculating, in principle, an extremely simple form of the g_i's in terms of what they call W-graphs. But a simple way to go from a Young diagram to a W-graph seems to be lacking.

There should be analogues of Ocneanu's trace for Hecke algebras other than those of type A_n. Some work has been done in this direction by B. Seifert in [39].

For knot theory it will be more convenient to change variables. We will use $\lambda = (1 - q + z)/qz = w/qz$, so that $z = -(1 - q)/(1 - \lambda q)$, $w = -\lambda q(1 - q)/(1 - \lambda q)$. The only change to the formula is that $S(q, z)$ becomes

$S(q, -(1-q)/(1-\lambda q))$ and is obtained by superimposing the Young diagram on the following version of Figure 5.3:

$$\begin{array}{cccc} 1-\lambda q & q-\lambda q & q^2-\lambda q & \cdots \\ 1-\lambda q^2 & q-\lambda q^2 & q^2-\lambda q^2 & \cdots \\ 1-\lambda q^3 & q-\lambda q^3 & q^2-\lambda q^3 & \cdots \\ 1-\lambda q^4 & \vdots & \vdots & \\ \vdots & & & \end{array}$$

FIGURE 5.6

and then multiplying by $((1-q)/(1-\lambda q))^n$, there being n nodes in the Young diagram. Let $R_Y(q, \lambda)$ be the product of the relevant terms in Figure 5.6, before multiplying by $((1-q)/(1-\lambda q))^n$.

Note 5.7. A final note is in order. Let us relate this section to Section 2. If $\psi(\sigma_i)$ are the reduced Burau matrices it is easy to check that $(-\psi(\sigma_i))^2 = (t-1)(-\psi(\sigma_i)) + t$ so that sending g_i to $-\psi(\sigma_i)$ gives a representation of the Hecke algebra with $q = t$. Since the reduced Burau representation is irreducible for generic t, this Hecke algebra representation must correspond to one of those already defined. Either by looking at $q = 1$ or by counting eigenvalues (see Lemma 9.1) we deduce that, for the n-string braid group, the representation has the following Young diagram: (a two-column diagram, n nodes)

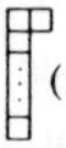

6. A two-variable knot polynomial

The key observation here is the similarity between condition 3) of Theorem 5.1 and the Markov moves of type II of Section 2. We call such traces Markov traces. The most natural way to obtain the invariant is to normalize the g_i's so that both type II Markov moves affect the trace in the same way; so let θ satisfy $\mathrm{tr}(\theta g_i) = \mathrm{tr}((\theta g_i)^{-1})$. Then

$$\theta^2 = \mathrm{tr}(g_i^{-1})/\mathrm{tr}(g_i) = \mathrm{tr}(g_i/q - (1 - 1/q))/z = (1 - q + z)/qz.$$

This is the quantity we have called λ. Thus $\mathrm{tr}(\sqrt{\lambda}\, g_i) = \mathrm{tr}((\sqrt{\lambda}\, g_i)^{-1})$ and $\mathrm{tr}(\sqrt{\lambda}\, g_i) = z\sqrt{\lambda} = -\sqrt{\lambda}(1-q)/(1-\lambda q)$.

It is now immediate that if we represent B_n by π_λ, $\pi_\lambda(\sigma_i) = \sqrt{\lambda}\, g_i \in H(q, n)$, then the function of q and λ given by

$$\left(-(1-\lambda q)/\sqrt{\lambda}(1-q)\right)^{n-1}\mathrm{tr}(\pi_\lambda(\alpha)), \text{ for } \alpha \in B_n,$$

depends only on the link $\hat{\alpha}$. The representation π ($\pi(\sigma_i) = g_i$) has the advantage of only involving the variable q; so we define:

Definition 6.1. The two-variable invariant $X_L(q, \lambda)$ of the oriented link L is the function

$$X_L(q, \lambda) = \left(-\frac{1 - \lambda q}{\sqrt{\lambda}(1 - q)}\right)^{n-1} (\sqrt{\lambda})^e \operatorname{tr}(\pi(\alpha))$$

where $\alpha \in B_n$ is any braid with $\hat{\alpha} = L$, e being the exponent sum of α as a word on the σ_i's and π the representation of B_n in $H(q, n)$, $\sigma_i \mapsto g_i$.

PROPOSITION 6.2. *To each oriented link L (up to isotopy) there is a Laurent polynomial $P_L(t, x)$ in the two variables t and x such that, if λ and q satisfy $t = \sqrt{\lambda}\sqrt{q}$, $x = (\sqrt{q} - 1/\sqrt{q})$ then $P_L(t, x) = X_L(q, \lambda)$. Moreover, $P_L(t, x)$ is uniquely defined by the "Skein rule": If L_+, L_- and L_0 are links that have projections identical, except in one crossing where they are as in Figure* 6.3:

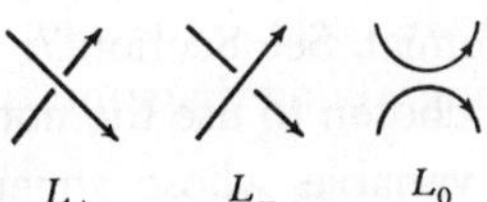

FIGURE 6.3

then $t^{-1}P_{L_+} - tP_{L_-} = xP_{L_0}$.

(Note: This two-variable knot polynomial was discovered by Lickorish and Millett, Freyd and Yetter, Ocneanu, and Hoste in [14]. Another common choice of variables is $l = it^{-1}$, $m = ix$ where $i^2 = -1$.)

Proof. We begin by investigating the behavior of X under skein moves (see Figure 6.3). Note that if one chooses a single crossing of an oriented link projection one may turn it into braid form so that the crossing becomes a term σ_i (or σ_i^{-1} depending on whether it is a positive or negative crossing) in an expression of the braid as a word on the σ_i's. Thus after a Markov move of type I we may assume $L_+ = \widehat{\alpha\sigma_i^2}$, $L_- = \hat{\alpha}$ and $L_0 = \widehat{\alpha\sigma_i}$ for some $\alpha \in B_n$. By (4.1) we see $\operatorname{tr}(\pi(\alpha\sigma_i^2)) - q \operatorname{tr}(\pi(\alpha)) = (q - 1)\operatorname{tr}(\pi(\alpha\sigma_i))$. Let the exponent sum of α be

e and multiply this equation by $T(\sqrt{\lambda})^{e+1}/\sqrt{q}$ with

$$T=\left(-(1-\lambda q)/\sqrt{\lambda}(1-q)\right)^{n-1}.$$

Then

$$\frac{T}{\sqrt{q}\sqrt{\lambda}}(\sqrt{\lambda})^{e+2}\mathrm{tr}\left(\pi\left(\alpha\sigma_i^2\right)\right)-\sqrt{q}\sqrt{\lambda}\,T(\sqrt{\lambda})^{e}\mathrm{tr}(\pi(\alpha))$$
$$=\left(\sqrt{q}-1/\sqrt{q}\right)T(\sqrt{\lambda})^{e+1}\mathrm{tr}\left(\pi\left(\alpha\sigma_i\right)\right);$$

so by the definition of X, $t^{-1}X_{L_+}-tX_{L_-}=xX_{L_0}$.

That the skein relation suffices to calculate $P_L(t,x)$ uniquely follows from an induction implicit in [13] (see also [22]). This induction associates a Laurent polynomial in t and x with any skein decomposition of L. But provided we can find $\sqrt{\lambda}$ and $\sqrt{q}$ with $t=\sqrt{\lambda}\sqrt{q}$ and $x=\sqrt{q}-1/\sqrt{q}$, the value of $P_L(t,x)$ can only depend on L since, as we have seen, $X_L(q,\lambda)$ is a link invariant and by induction they are equal. There is an open set of values of x and t for which $\sqrt{\lambda}$ and $\sqrt{q}$ exist; so since $P_L(x,t)$ is a Laurent polynomial, it can only depend on L and not on the skein decomposition.

Notes. 1) It is interesting that the values of t and x which do not admit corresponding q and λ values define precisely the specialization of P_L which is the Alexander-Conway polynomial. See Section 7.

2) We have deliberately chosen to use the notation $\sqrt{\lambda}$ rather than defining things in terms of another variable whose square would replace λ. This is because X_L is more than a Laurent polynomial in $\sqrt{\lambda}$—it is a power of $\sqrt{\lambda}$ times a Laurent polynomial in λ. So if L has an odd number of components the square root disappears.

Example 6.4. The (right-handed) trefoil is given by the closure of the braid $\sigma_1^3\in B_2$. Thus

$$X_L(q,\lambda)=\left(-\frac{1-\lambda q}{\sqrt{\lambda}(1-q)}\right)(\sqrt{\lambda})^3\mathrm{tr}\left(g_1^3\right).$$

But by (4.1), $g_1^3=(q^2-q+1)g_1+q(q-1)$ so that

$$\begin{aligned}X_L(q,\lambda)&=\left(\frac{\lambda(1-\lambda q)}{1-q}\right)\left((q^2-q+1)\frac{(1-q)}{1-\lambda q}q(q-1)\right)\\&=\lambda\left(1+q^2-\lambda q^2\right)\\&=(\lambda q)\left(\left(\sqrt{q}-1/\sqrt{q}\right)^2+2-\lambda q\right)\\&=(2t^2-t^4)+t^2x^2.\end{aligned}$$

Example 6.5. The figure-8 knot is given by the closure of the braid $\sigma_1\sigma_2^{-1}\sigma_1\sigma_2^{-1} \in B_3$. So

$$X_L(q,\lambda) = \frac{(1-\lambda q)}{\lambda(1-q)^2}\operatorname{tr}\left(g_1g_2^{-1}g_1g_2^{-1}\right)$$

$$= \frac{(1-\lambda q)^2}{\lambda(1-q)^2}\operatorname{tr}\left(\frac{1}{q^2}g_1g_2g_1g_2 + \frac{1}{q}\left(\frac{1}{q}-1\right)g_1^2g_2\right.$$

$$\left. + \frac{1}{q}\left(\frac{1}{q}-1\right)g_1g_2g_1 + \left(\frac{1}{q}-1\right)^2 g_1^2\right)$$

$$= \frac{(1-\lambda q)^2}{q^2\lambda(1-q)^2}\operatorname{tr}\left(g_1^3g_2 - 2\frac{(1-q)^2}{(1-\lambda q)}\operatorname{tr}(g_1^2) + (1-q)^2\operatorname{tr}(g_1^2)\right)$$

$$= \frac{1}{q^2\lambda}\left(-\operatorname{tr}(g_1^3)\frac{1-\lambda q}{1-q} + (1-\lambda q)(1-\lambda q-2)\operatorname{tr}(g_1^2)\right)$$

$$= \frac{1}{q^2\lambda}\left(\frac{1-q}{1-\lambda q}\,\frac{1-\lambda q}{1-q}(1+q^2-\lambda q^2)\right.$$

$$\left. -(1-\lambda q)(1+\lambda q)\left(\frac{(1-q)^2}{1-\lambda q}+q\right)\right) \quad \text{(by (6.4))}$$

$$= \frac{1}{q^2\lambda}(1+q^2-\lambda q^2-(1+\lambda q)(1-q+q^2-\lambda q^2))$$

$$= \frac{1}{q\lambda}(1-\lambda(1-q+q^2)+\lambda^2q^2)$$

$$= \frac{1}{t^2}(1-t^2(x^2+1)+t^2) = -1-x^2+\frac{1}{t^2}+t^2.$$

Example 6.6. The n-component unlink is given by the closure of the braid $1 \in B_n$ so that

$$X_L(q,\lambda) = \left(-\frac{1-\lambda q}{\sqrt{\lambda}(1-q)}\right)^{n-1}$$

and

$$P_L(t,x) = \left(\frac{t-t^{-1}}{x}\right)^{n-1}.$$

The above calculations are probably no simpler than those of a skein induction but they illustrate an important point. If one is interested in closed braids for a fixed B_n, the invariants can be computed quite rapidly no matter how many crossings there are. Just write the braid as a word on the σ_i's and expand the product in terms of the basis (4.4). The number of algebraic operations will be proportional to the length of the word. In practice the space required for the basis elements is quite large and 7 or 8 strings seems to be as far as one can go on a present-day average-sized computer. But this already allows one to compute the invariant for many knots and links whose computation is impossible by the direct skein theoretic method for which calculation time grows exponentially with the number of crossings. A program such as the one outlined above has been implemented by Morton and Short who have thus been able to answer in the negative the important question: Is the invariant of $\hat{\alpha}$ determined by the characteristic polynomial of the Burau matrix of α?

Example 6.7. *Connected sums*. We claim that if L_1 and L_2 are oriented links then $X_{L_1 \# L_2}(q, \lambda) = X_{L_1}(q, \lambda) X_{L_2}(q, \lambda)$. This is regardless of which components of L_1 and L_2 one chooses to perform the connected sum.

The claim follows trivially from two facts: (i) If $L_1 = \hat{\alpha}_1$, $\alpha_1 \in B_n$, and $L_2 = \hat{\alpha}_2$, $\alpha_2 \in B_m$, then $L_1 \# L_2 = \alpha_1 \Sigma^{n-1}(\alpha_2) \in B_{n+m}$ where Σ is the shift map on the inductive limit of the B_n's, $\Sigma(\sigma_i) = \sigma_{i+1}$. For instance, the connected sum of the figure-8 and the trefoil is $\hat{\alpha}$ where $\alpha = \sigma_1 \sigma_2^{-1} \sigma_1 \sigma_2^{-1} \sigma_3^3 \in B_4$ (see [6]). (ii) If w_1 is a word on $1, g_1, \ldots, g_n$ and w_2 is a word on $1, g_{n+1}, \ldots, g_{n+m}$ then $\operatorname{tr}(w_1 w_2) = \operatorname{tr}(w_1)\operatorname{tr}(w_2)$. The easiest way to prove this is to note that $x \mapsto \operatorname{tr}(x w_2)$ defines a non-normalized Markov trace on $H_{n+1}(q)$. Normalizing it one obtains the result by the uniqueness of a Markov trace. See [22] for the skein theoretical proof.

Example 6.8. *Reversing the orientation*. Reversing all the arrows in Figure 6.3 preserves the diagram; so by (6.2) we may conclude that $P_L = P_{L'}$ if L' is the oriented link obtained from L by reversing all the orientations. Note though that if the orientations of individual components are reversed, P_L may change drastically. Compare this with Corollary 13.16.

The result $P_L = P_{L'}$ can be seen on the Hecke algebra level as well. The symmetry of the presentation of the braid group (1.2) and (1.3) implies the existence of an antiautomorphism θ of B_n which sends σ_i to σ_i. It is geometrically trivial that $\widehat{\theta(\alpha)} = (\hat{\alpha})'$. By formulae (4.1), (4.2) and (4.3), θ also defines an antiautomorphism of $H(q, n)$ which preserves the trace (uniqueness of a Markov trace). Thus $X_L = X_{L'}$.

This result may be construed as a negative result about the polynomial: It cannot detect (at least in any simple way) the difference between two orientations of a knot. But it is to be expected that a more refined analysis of the Markov relation in the Hecke algebra will be more successful. One might try, for instance, to let q and z be elements of a finite field so that $\pi(B_n)$ is finite for all n.

Example 6.9. *Mirror images.* One of the useful features of P_L is that it is very sensitive to taking the mirror image of a link.

If L is an oriented link let $\tilde{L}$ be the oriented link obtained by viewing L in a mirror or equivalently, reversing all the crossings in some projection of L. It is obvious from Figure 6.3 that $P_{\tilde{L}}(t, x) = P_L(t^{-1}, -x)$. We shall show this using the Hecke algebra. Although this method is less easy, it is quite revealing. If $\alpha \in B_n$ then the mirror image of $\hat{\alpha}$ is $\theta(\alpha^{-1}) \in B_n$ but as we have seen we may take $\alpha^{-1} \in B_n$. Thus if $L = \hat{\alpha}$ and e is the exponent sum of α then

$$X_{\tilde{L}}(q, \lambda) = \left(-\frac{(1-\lambda q)}{\sqrt{\lambda}\,(1-q)} \right)^{n-1} (\sqrt{\lambda})^{-e} \operatorname{tr}(\pi(\alpha^{-1})).$$

PROPOSITION 6.10. $X_{\tilde{L}}(q, \lambda) = X_L(1/q, 1/\lambda)$.

Proof. Let $f(q, \lambda)$ be $\operatorname{tr}(\pi(\alpha))$. Write $\pi(\alpha)$ as a product with each g_i written $(q+1)e_i - 1$ where $e_i = (1+g_i)/(1+q)$. The e_i satisfy

$$(6.11) \qquad e_1^2 = e_i$$

$$(6.12) \qquad e_i e_{i+1} e_i - q/(1+q)^2 e_i = e_{i+1} e_i e_{i+1} - q/(1+q)^2 e_{i+1}$$

$$(6.13) \qquad e_i e_j = e_j e_i \qquad \text{if } |i-j| \geq 2$$

$$(6.14) \qquad \operatorname{tr}(e_i) = \frac{q(1-\lambda)}{(1+q)(1-\lambda q)}.$$

Then $\pi(\alpha^{-1})$ will be the same expression on the e_i's simply with q replaced by q^{-1}. Both the expressions $q/(1+q)^2$ and $q(1-\lambda)/(1+q)(1-\lambda q)$ are invariant under the change of variables $q \mapsto 1/q$, $\lambda \mapsto 1/\lambda$. Moreover, relations (6.11), (6.12) and (6.13) are equivalent, for $q \neq -1$, to (4.1), (4.2) and (4.3), so they suffice to calculate the trace of any word on the e_i's, which will be a sum of powers of q times powers of $q(1-\lambda)/(1+q)(1-\lambda q)$. Thus $\operatorname{tr}(\pi(\alpha)^{-1}) = f(1/q, \lambda) = f(1/q, 1/\lambda)$. Finally, $(1-\lambda q)/\sqrt{\lambda}\,(1-q)$ is invariant under the change of variables and $(\sqrt{\lambda})^e$ becomes $(\sqrt{\lambda})^{-e}$, which completes the proof. Q.E.D.

Scholium 6.15 (Morton, Franks-Williams [28], [47]). *If* $|e| > n-1$, $\hat{\alpha}$ *is not amphicheral* ($\alpha \in B_n$).

Proof. We saw in the proof of Proposition 6.10 that $\operatorname{tr}(\pi(\alpha))$, as a function of q and λ, has a finite limit, for fixed $q \neq 0$, when $\lambda \to 0$ (relation (6.14)).

Suppose $e > 0$ and $e > n - 1$. Then as $\lambda \to 0$ we see that $X_L(q, \lambda) \to 0$. But since X_L is a Laurent polynomial in $\sqrt{\lambda}$, this means that only strictly positive powers of λ can occur; hence $X_L \neq X_{\bar{L}}$. If $e < 0$, use the same formula for $X_{\bar{L}}$. Q.E.D.

Note 6.16. L. Rudolph has used this result to show that the figure-8 knot cannot be represented by a quasi-positive braid, i.e., one which is a product of conjugates of σ_i's ([38]).

Note 6.17. Bennequin had shown in [5] that $\hat{\alpha}$ is non-trivial whenever $|e| > n - 1$. Scholium 6.15 gives a rather different proof of this.

Scholium 6.18. *Let $\alpha \in B_n$ and let q_+ and q_- be the orders of the poles at infinity and zero respectively in $P_{\hat{\alpha}}(q, \lambda)$. Let e_+ be the sum of the positive exponents of the σ_i's in α and e_- be the sum of the negative ones. Then $e_+ > 0$ implies $q_+ \leq e_+ - 1$, and $e_- < 0$ implies $q_- \leq e_- + 1$.*

Proof. Fix λ and let $q \to \infty$. The expression $(1 - \lambda q)/\sqrt{\lambda}(1 - q)$ of Definition 6.1 has a finite limit as $q \to \infty$. Each of the terms in $\operatorname{tr}(\pi(\alpha))$ is either 1 or of the form $q^a \operatorname{tr}(w)$ where w is a product of the e_i's and $a \leq e_+$. Calculating the trace of w using the usual inductive procedure will yield a sum of non-empty products of traces of e_i's multiplied by various powers of $q/(1 + q)^2$ using (6.10)–(6.13). All such terms will go to zero as $q \to \infty$ at least as rapidly as $1/q$. Thus the most rapid growth possible at infinity in q is $q^{e_+ - 1}$ $(e_+ \geq 0)$ provided $e_+ > 0$. The same argument applied to α^{-1} yields the full result. Q.E.D.

7. The Alexander polynomial

The Alexander polynomial of a link L will, for the purposes of this paper, be defined as the specialization $\Delta_L(t) = X_L(t, 1/t) = P_L(1, \sqrt{t} - 1/\sqrt{t})$. That such a polynomial exists follows from Proposition 6.2. From Definition 6.1 of $X_L(q, \lambda)$ it would appear that one cannot calculate the Alexander polynomial using the trace of Hecke algebra representations. Indeed for these values of the parameters $\operatorname{tr}(g_i)$ does not make sense. But we will show that if one uses the weighted sum of the traces as in Section 4, the singularity at $\lambda = 1/q$ disappears and one may evaluate Δ_L. A bonus of this approach is an alternative proof of the fact that Δ is given by the Burau matrix.

If Y is a Young diagram, the contribution of Y to $X_L(q, \lambda)$ (for $|q| \neq 1$, $\lambda \neq 1/q$) is

$$(\sqrt{\lambda})^e \left(-\frac{1}{\sqrt{\lambda}}\right)^{n-1} \left(\frac{1 - q}{1 - \lambda q}\right) \frac{R_Y(q, \lambda)}{T_Y(q)} \operatorname{trace}(\pi_Y(\alpha))$$

where $R_Y(q, \lambda)$ is as calculated from Figure 5.6, and "trace" is the usual trace, sum of the diagonal elements. Since the term in the top left corner of Figure 5.6 is $1 - \lambda q$, the limit exists as $\lambda \to 1/q$; so we may evaluate ΔL. (Here $\alpha \in B_n$, $\hat{\alpha} = L$, $e =$ exponent sum of α.) But we also see from Figure 5.6 that if Y is not of the form of Figure 7.1

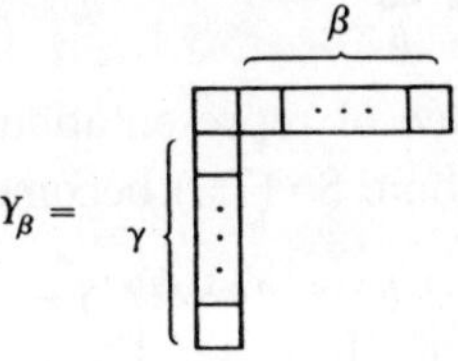

FIGURE 7.1

with $\beta + \gamma + 1 = n$, the contribution is zero since the (2, 2) position of Figure 5.6 is $q - q^2\lambda$. We find that for Y as in Figure 7.1, the term

$$(1 - q/1 - \lambda q)R_Y(q, \lambda)/Q_Y(\lambda)$$

is

$$(-1)^\beta \frac{(1-q)(1-q^2)\ldots(1-q^\alpha)(1-q)\ldots(1-q^\beta)}{(1-q)\ldots(1-q^\alpha)(1-q^2)(1-q^3)\ldots(1-q^\beta)(1-q^n)}$$

$$= (-1)^\beta \frac{(1-q)}{(1-q^n)}.$$

Thus we obtain the formula

$$(7.2)\quad \Delta_L(t) = (-1)^{n-1}\left(\frac{1}{t}\right)^{(e-n+1)/2} \frac{1-t}{1-t^n} \sum_{\beta=0}^{n-1} (-1)^\beta \operatorname{trace}\left(\pi_{Y_\beta}(\alpha)\right).$$

We shall now identify the representations π_{Y_β}, up to sign, with the exterior powers of π_{Y_1} which we shall call π_1. Consider $f_i = (-1)^{\beta-1}\Lambda^\beta \pi_1(\sigma_i)$. Since $\pi_1(\sigma_i)$ has eigenvalues q and -1, f_i does also; so $f_i^2 = (q-1)f_i + q$. Thus the assignment $\rho(g_i) = f_i$ defines a representation of $H(q, n)$. To see how many irreducible representations it contains it suffices to look at the case $q = 1$, i.e. representations of the symmetric groups. But then we know that π_1 is the tensor product of the signature representation of S_n with the $(n-1)$-dimensional irrep of S_n contained in the n-dimensional permutation representation. But it is well known that the βth exterior powers of this representation are irreducible and have Young diagram $Y_{\beta'}$ for some β'. So ρ is irreducible and corresponds to a Young diagram $Y_{\beta'}$. To find out the value of β' it suffices to count the multiplicities of the eigenvalues q and -1. One deduces that ρ is equivalent to π_{Y_β} (see Lemma 9.1).

Thus we can rewrite (7.2) as

$$(7.3)\qquad \Delta_L(t) = (-1)^{n-1}\left(\frac{1}{t}\right)^{(e-n+1)/2} \times \frac{1-t}{1-t^n}\sum_\beta (-1)^\beta (-1)^{e(\beta-1)}\mathrm{trace}\left(\Lambda^\beta \pi_1(\alpha)\right).$$

But by (5.7), up to equivalence of representations, $\pi_1(\sigma_i) = -\psi(\sigma_i)$, ψ being the reduced Burau representation. So (7.3) becomes

$$(7.4)\qquad \Delta_L(t) = (-1)^{e-n+1}\left(\frac{1}{t}\right)^{(e-n+1)/2}\left(\frac{1-t}{1-t^n}\right)\sum_\beta (-1)^k \mathrm{trace}\left(\Lambda^k\psi(\alpha)\right)$$

or

$$\Delta_L(t) = (-1/\sqrt{t})^{e-n+1}\left(\frac{1-t}{1-t^n}\right)\det(1-\psi(\alpha)).$$

This also makes clear the normalization of the Alexander polynomial as calculated from the Burau representation. For this see also [8].

8. Special formulae for closed 3 and 4 braids

For 3 and 4 braids, Figure 3.3 and Notes 4.6 and 5.6 show that almost all the Hecke algebra representations are essentially Burau representations. (Indeed, even the two-dimensional irrep of B_4 corresponding to the Young diagram $\yng(2,2)$ is the composition of the irrep $\yng(2,1)$ of B_3 with the map $\sigma_1 \to \sigma_1$, $\sigma_2 \to \sigma_2$, $\sigma_3 \to \sigma_1$ of B_4 onto B_3). Thus by Section 7 we expect close relationships for closed 3 and 4 braids between the polynomials of Section 6 and the Alexander polynomial. We shall derive these relations for knots only, leaving the case of links up to the reader.

8.1. *Closed* 3-*braids*. The Hecke algebra $H(q,3)$ is, for generic q, the direct sum of three components corresponding to Young diagrams $\yng(3)$, $\yng(2,1)$ and $\yng(1,1,1)$. So (5.5) becomes, for $x \in H(q,3)$,

$$(8.2)\qquad \mathrm{tr}(x) = \left(\frac{1}{1-\lambda q}\right)^3\left[\frac{(1-\lambda q)(q-\lambda q)(q^2-\lambda q)}{(1-q)(1-q^2)(1-q^3)}\mathrm{tr}_{\yng(3)}(x) + \frac{(1-\lambda q)(1-\lambda q^2)(q-\lambda q)}{(1-q)(1-q)(1-q^3)}\mathrm{tr}_{\yng(2,1)}(x) + \frac{(1-\lambda q)(1-\lambda q^2)(1-\lambda q^3)}{(1-q)(1-q^2)(1-q^3)}\mathrm{tr}_{\yng(1,1,1)}(x)\right].$$

Now if $\alpha \in B_3$ has exponent sum e, and $\hat{\alpha}$ is a knot, then e is even. Also the reduced Burau representation ψ is two-dimensional, so that we have

$$\text{(8.3)} \qquad \det(1 - \psi(\alpha)) = 1 - \operatorname{trace}(\psi(\alpha)) + \det(\psi(\alpha)).$$

Moreover, $\det \psi(\alpha) = (-q)^e$ so that by (7.4), $\operatorname{trace}(\psi(\alpha)) = 1 + q^e - q^{e/2}(1 + q + q^{-1})\Delta_{\hat{\alpha}}(q)$. Combining (8.3), (8.2), (5.7), (4.7) and (6.1) we obtain

$$\text{(8.4)} \quad X_{\hat{\alpha}}(q, \lambda) = (\sqrt{\lambda})^{e-2}\left[q^{e+2}\frac{(1-\lambda)(q-\lambda)}{(1-q^2)(1-q^3)} + q\frac{(1-\lambda)(1-\lambda q^2)}{(1-q)(1-q^3)} \right.$$
$$\times\left(1 + q^e - q^{e/2}(1 + q + q^{-1})\Delta_{\hat{\alpha}}(q)\right)$$
$$\left. + \frac{(1-\lambda q^2)(1-\lambda q^3)}{(1-q^2)(1-q^3)} \right].$$

In particular $\Delta_{\hat{\alpha}}$ and e determine $X_{\hat{\alpha}}(q, \lambda)$ and hence $P_{\hat{\alpha}}(t, x)$. Birman shows in [8] that $\Delta_{\hat{\alpha}}$ and e do not suffice to determine the type of the knot $\hat{\alpha}$. We invite the reader to try out (8.4) on a few test cases.

8.5. *Closed 4-braids*. For closed 4-braids, $P_{\hat{\alpha}}$ is no longer determined by e and the Alexander polynomial but there is a relation for knots which is sometimes useful. Note now that for $\alpha \in B_4$, e will be odd and there will be a sign difference between $\psi(\alpha)$ and $\pi_{(2,1,1)}(\alpha)$.

Let $w_1, \ldots, w_5$ be the weights for $Y = (4), (3,1), (2,2), (2,1,1)$ and $(1,1,1,1)$, respectively, using (4.6), (4.7), (5.7) and (5.5), and (6.1) becomes, after multiplication by $\lambda^{(1-e)/2}$,

$$\text{(8.6)} \quad \left(\frac{1}{\lambda}\right)^{(e-1)/2} X_{\hat{\alpha}}(q, \lambda) = -\frac{1}{\lambda}\left(\frac{1-\lambda q}{1-q}\right)^3 \left(q^e w_1 + q^e \operatorname{Tr}(\psi(\alpha^{-1}))\, w_2\right.$$
$$\left. + \operatorname{Tr}(\pi_{(2,2)}(\alpha))\, w_3 - \operatorname{Tr}(\psi(\alpha))\, w_4 - w_5 \right).$$

Doing the same for α^{-1} and using Proposition 6.9 we obtain

$$\text{(8.7)} \quad (\lambda)^{(e+1)/2} X_{\hat{\alpha}}\left(\frac{1}{q}, \frac{1}{\lambda}\right) = -\frac{1}{\lambda}\left(\frac{1-\lambda q}{1-q}\right)^3$$
$$\times\left(q^{-e} w_1 + q^{-e}\operatorname{Tr}(\psi(\alpha))\, w_2 + \operatorname{Tr}(\pi_{(2,2)}(\alpha))\, w_3\right.$$
$$\left. - \operatorname{Tr}(\psi(\alpha^{-1}))\, w_4 - w_5 \right).$$

Multiplying (8.7) by q^e and adding we get

$$(8.8)\quad \left(\frac{1}{\lambda}\right)^{(e-1)/2} X_{\hat{\alpha}}(q,\lambda)+(\lambda)^{(e+1)/2}X_{\hat{\alpha}}\left(\frac{1}{q},\frac{1}{\lambda}\right)$$

$$=-\frac{1}{\lambda}\left(\frac{1-\lambda q}{1-q}\right)^3\big[(1+q^e)w_1+(w_2-w_4)\mathrm{Tr}(\psi(\alpha)+q^e\psi(\alpha_{-1}))$$

$$+w_3\mathrm{Tr}(\pi_{\boxplus}(\alpha)+q^e\pi_{\boxplus}(\alpha^{-1}))-w_5(1+q^e)\big].$$

But the coefficient of w_3 vanishes since $\pi_{\boxplus}$ is two-dimensional and one may easily check that $\det(\pi_{\boxplus}(\sigma_i))=-q$ so that $\det(\pi_{\boxplus}(\alpha))=(-q)^e$. And for any 2×2 invertible matrix A, $\mathrm{Tr}(A)=\det(A)\mathrm{Tr}(A^{-1})$.

Also for any invertible 3×3 matrix A we have $\det(1-A)=1-\mathrm{Tr}(A)+\det(A)\mathrm{Tr}(A^{-1})-\det A$, so that we may rewrite the right-hand side of (8.8) using (7.4) as (remember e is odd and $\det\psi(\sigma_i)=-q$),

$$(8.9)\quad -\frac{1}{\lambda}\left(\frac{1-\lambda q}{1-q}\right)^3\big[(w_1-w_5)(1+q^e)+(w_2-w_4)$$

$$\times\left(1+q^e-\left(1+q+q^2+q^3\right)q^{(e-3)/2}\Delta_{\hat{\alpha}}(q)\right)\big]$$

Finally, using Figure 5.6 we obtain, for a 4-braid knot $\hat{\alpha}$, $\alpha\in B_4$,

$$(8.10)\quad \left(\frac{1}{\lambda}\right)^{(e-1)/2} X(q,\lambda)+\lambda^{(e+1)/2}q^e X\left(\frac{1}{q},\frac{1}{\lambda}\right)$$

$$=\frac{-1}{\lambda(1-q^2)(1-q^3)(1-q^4)}$$

$$\times\big[\{q^3(1-\lambda)(q-\lambda)(q^2-\lambda)-(1-\lambda q^2)(1-\lambda q^3)(1-\lambda q^4)\}(1+q^e)$$

$$+q(1+q+q^2)(1-\lambda q^2)(1-\lambda)(q^2-1)(1+\lambda q)$$

$$\times\{1+q^e-(1+q+q^2+q^3)q^{(e-3)/2}\Delta_{\hat{\alpha}}(q)\}\big].$$

This rather complicated formula has a more manageable form in certain specializations (see §12).

Formulae (8.4) and (8.10) lend some weight to the possibility that the exponent sum in a minimal braid representation is a knot invariant.

9. Torus knots

We shall calculate $X_L(q,\lambda)$ for torus knots. It will be clear from the calculation how one can also write down a formula for torus links but the answer is already rather complicated for knots, so we shall content ourselves with that.

A *torus link* of type (m, n) is the closure of the braid $(\sigma_1\sigma_2 \dots \sigma_{n-1})^m \in B_n$ (by various symmetries we may suppose $m, n \in \mathbf{N}$). The link is a knot if and only if m and n are relatively prime. A torus link of type (n, n) is represented by $\Delta^2 = (\sigma_1\sigma_2 \dots \sigma_{n-1})^n \in B_n$. The calculation of the polynomial for torus knots relies heavily on the easy fact that Δ^2 is in the center of B_n.

Let us assume that $q \in \mathbf{R}^+$. This allows us to use the structure of $H(q, n)$ outlined in Section 4. If Y is a Young diagram with n nodes, let π_Y be as usual, let $h_i = \pi_Y(\sigma_i)$ and $e_i = (1 + h_i)/(1 + q)$ and let $d = \dim \pi_Y$.

LEMMA 9.1. *We have $e_i^2 = e_i$ and the rank of the idempotent e_i is the number of descending paths on Figure 3.3 from the diagram $\square\!\square$ to Y. In particular, for Y_β of Figure 7.1,*

$$\operatorname{rank}(e_i) = \binom{\beta + \gamma - 1}{\gamma}.$$

Proof. All the generators $\sigma_i \in B_n$ are conjugate in B_n which means that all the e_i's have the same rank (= trace) in π_Y. So it suffices to calculate the rank of e_1. But the lines in Figure 3.3 denote the restriction of the representation B_n to B_{n-1}. Thus the assertion about rank(e_i) as the number of descending paths follows by induction and our convention that $\square\!\square$ corresponds to the representation $\pi_{\square\!\square}(\sigma_1) = q$ so that $\pi_{\square\!\square}(e_1) = 1$. The explicit formula is obtained by counting. Q.E.D.

LEMMA 9.2.

$$\dim(\pi_{Y_\beta}) = \binom{\gamma + \beta}{\gamma}.$$

Proof. Use the hook length formula. Q.E.D.

LEMMA 9.3. *If Y is a Young diagram with* $\dim \pi_Y = d$ *and* $\operatorname{rank} e_i = r$, *then*

$$\pi_Y\big((\sigma_1\sigma_2 \dots \sigma_{n-1})^n\big) = q^{rn(n-1)/d}\mathrm{id}_{\pi_Y}.$$

Proof. The representation π_Y is irreducible and $(\sigma_1\sigma_2 \dots \sigma_{n-1})^n = \Delta^2$ is in the center of B_n; so its image is a scalar. Moreover, since $h_i = qe_i - (1 - e_i)$ it follows that $\det(h_i) = \pm q^r$, hence $\det(\pi(\Delta^2)) = q^{rn(n-1)}$. Thus $\pi_Y(\Delta^2) = \omega q^{rn(n-1)/d}$ for some dth root of unity ω. But by [46] or [20] we know that one may write down explicit formulae for $\pi_Y(\sigma_i)$ which depend continuously on q. Thus the root of unity ω varies continuously with q; so we may evaluate it by putting $q = 1$. But then π_Y is a representation of S_n and $\pi_Y(\sigma_1 \dots \sigma_{n-1})$ represents an n-cycle, so that $\omega = 1$. Q.E.D.

Remark. It is also clear from the proof of Lemma 9.3 that $rn(n-1)/d$ is an integer.

LEMMA 9.4. *The matrix $q^{-r(n-1)/d}\pi_Y(\sigma_1 \ldots \sigma_{n-1})$ is diagonalizable and has the same eigenvalues, with the same multiplicities, as an n-cycle in the representation of S_n corresponding to Y.*

Proof. By Lemma 9.3 we know that $q^{-r(n-1)/d}\pi_Y(\sigma_1 \ldots \sigma_{n-1})$ is an nth root of unity so it is diagonalizable. The multiplicities of the various nth roots of unity, ω, as eigenvalues, depend continuously on q since they can be expressed in the form Trace$((1/n)\sum_{i=0}^{n-1}\omega^i \pi_Y(\sigma_1 \ldots \sigma_{n-1})^i)$. As before they can be evaluated by putting $q = 1$. Q.E.D.

COROLLARY 9.5. *If m and n are relatively prime and Y gives an irrep of S_n for which the trace of an n-cycle is zero, then* trace$(\pi_Y(\sigma_1 \ldots \sigma_{n-1})^m) = 0$.

Proof. All n-cycles in S_n are conjugate; so if A is a matrix representing an n-cycle in the irrep of S_n, trace$(A^m) = 0$. But by Lemma 9.4, $\pi_Y(\sigma_1 \ldots \sigma_{n-1})$ is conjugate to a scalar multiple of A. Q.E.D.

The following formula now follows immediately from Remark 3.7, (5.5) and Figure 5.6 (remember $(m, n) = 1$):

$$\text{(9.6)} \quad \operatorname{tr}\left((g_1 \cdots g_{n-1})^m\right) = \sum_{\substack{\gamma+\beta+1=n \\ \alpha, \beta \geq 0}} (-1)^{\gamma} q^{\beta m} \frac{q^{\gamma(\gamma+1)/2}}{[\gamma]![\beta]!(1 - q^{\gamma+\beta+1})} \times \left(\frac{1-q}{1-\lambda q}\right)^n \prod_{i=-\gamma}^{\beta} (q^i - \lambda q)$$

(we have used the fact that the trace of $\pi_Y(\sigma_1 \ldots \sigma_{n-1})^m$ is $(-1)^{\gamma} q^{r(n-1)m/d}$).

From (9.6) and Definition 6.1 we get the following:

THEOREM 9.7. *Let K be a torus knot of type m, n. Then*

$$X_K(q, \lambda) = \left(\frac{1-q}{1-q^n}\right) \frac{\lambda^{(n-1)(m-1)/2}}{1-\lambda q} \sum_{\substack{\gamma+\beta+1=n \\ \gamma, \beta \geq 0}} (-1)^{\beta} q^{\beta m + \gamma(\gamma+1)/2} \times \frac{\prod_{i=-\gamma}^{\beta} (q^i - \lambda q)}{[\gamma]![\beta]!}.$$

Since the formula is so involved, let us do a few explicit checks.

Check 9.8. The unknot: $n = 2$, $m = 1$.

$$X_K(q, \lambda) = \frac{1}{(1+q)(1-\lambda q)} \left(\frac{q(q^{-1} - \lambda q)(1 - \lambda q)}{1-q} - q \frac{(1-\lambda q)(q - \lambda q)}{1-q} \right) = \frac{1}{1-q^2}(1 - \lambda q^2 - q^2 + \lambda q^2) = 1.$$

Check 9.9. The trefoil: $n = 2$, $m = 3$.

$$X_K(q, \lambda)$$
$$= \frac{\lambda}{(1+q)(1-\lambda q)}\left(\frac{q(q^{-1}-\lambda)(1-\lambda q)}{1-q} - \frac{q^3(1-\lambda q)(q-\lambda q)}{1-q}\right)$$
$$= \frac{\lambda}{1-q^2}(1 - \lambda q^2 - q^4 + \lambda q^4) = \lambda(1 + q^2 - \lambda q^2)$$

which checks with Example 6.4.

Thus we can be confident of the following formula for 3, m torus knots:

(9.10)

$$X_K(q, \lambda) = \frac{\lambda^{m-1}}{(1-q^3)(1-q^2)}((1-\lambda q^3)(1-\lambda q^2)$$
$$- q^{m+1}(1+q)(1-\lambda q^2)(1-\lambda) + q^{2m+2}(1-\lambda)(q-\lambda)).$$

Note. The formula of (9.7) is not obviously symmetric in m and n. However, professor G. Andrews has kindly supplied me with a combinatorial proof of this symmetry using the Heine transform.

10. Mapping class groups

The problem of classification of closed 3-manifolds can be reduced via Heegard decompositions to the study of the mapping class groups (= diffeomorphism groups modulo the connected component of the identity) of closed orientable surfaces of arbitrary genus. It would be significant if one could find representations of these groups and an invariant via the Reidemeister–Singer theorem ([36]) as we have done for links via Markov's theorem. We have not yet succeeded but we would like to describe some progress towards that goal.

Presentations for the mapping class groups are known (see [6] and [45]). Let us at least explain a known set of generators which owe a lot to Lickorish.

Figure 10.1 depicts a surface of arbitrary genus n. There are $2n + 1$ broken curves $c_1, \dots, c_{2n+1}$ and one exceptional curve, called d. It is known that Dehn twists (see [6]) about these curves generate the mapping class groups. Let θ_i denote the (isotopy class of a) Dehn twist about c_i. It is not hard to see that $\theta_i\theta_{i+1}\theta_i = \theta_{i+1}\theta_i\theta_{i+1}$ and $\theta_i\theta_j = \theta_j\theta_i$ if $|i - j| \geq 2$. Thus the mapping class group is generated by a homomorphic image of the $2n + 2$-string braid group and one further element. An obvious question is whether any of the Hecke algebra representations give representations of this homomorphic image. The answer is yes as we shall see.

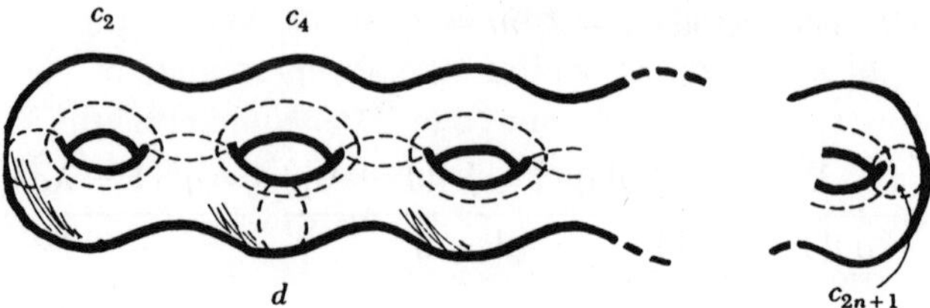

FIGURE 10.1

One may consider the surface of genus n as a branched cover of the 2-sphere, branched at $2n$ points, via an obvious involution of the surface. It is a result of Birman and Hilden in [9] that this branched cover actually gives rise to a homomorphism from the mapping class group generated by the above θ_i's to the mapping class group $M(0, 2n)$ of the sphere minus $2n$ points. A natural presentation of $M(0, m)$ is known. It is (see [6], p. 164) $M(0, m) = \langle \omega_1, \ldots, \omega_{m-1} | \omega_i\omega_j = \omega_j\omega_i$ if $|i - j| \geq 2$, $\omega_i\omega_{i+1}\omega_i = \omega_{i+1}\omega_i\omega_{i+1}$, $\omega_1\omega_2 \ldots \omega_{m-2}\omega_{m-1}^2\omega_{m-2} \ldots \omega_1 = 1$, $(\omega_1\omega_2 \ldots \omega_{m-1})^m = 1\rangle$. The homomorphism from the subgroup described above is obtained by sending θ_i to ω_i. The kernel of this map is of order 2, so a powerful representation of $M(0, m)$ will be useful. Now our question becomes: For which Young diagrams on m nodes does the corresponding Hecke algebra representation of B_m pass to $M(0, m)$ (viewed as a quotient, $\sigma_i \to \omega_i$)?

The clue is that in the braid group B_m one has $\sigma_1\sigma_2 \ldots \sigma_{m-2}\sigma_{m-1}^2 \ldots \sigma_1 = (\sigma_1\sigma_2 \ldots \sigma_{m-1})^m(\sigma_2 \ldots \sigma_{m-1})^{-(m-1)}$ so that the relation $\omega_1\omega_2 \ldots \omega_{m-1}^2\omega_{m-2} \ldots \omega_1 = 1$ can be replaced by $(\omega_2\omega_3 \ldots \omega_{m-1})^{m-1} = 1$. Now we know that for a fixed Young diagram Y, π_Y can be adjusted to π_Y' by an appropriate power of q so that $\pi_Y'(\sigma_1\sigma_2 \ldots \sigma_{m-1})^m = 1$. This was done explicitly in Section 9. The essential observation now is that if Y is rectangular the restriction of π_Y to B_{m-1} is still irreducible. This follows immediately from the interpretation of the lines on Figure 3.3. But the element $(\sigma_2\sigma_3 \ldots \sigma_{m-1})^{m-1}$ is in the center of B_{m-1} (thought of as the subgroup of B_m generated by $\sigma_2, \sigma_3, \ldots, \sigma_{m-1}$). Thus $\pi_Y(\sigma_2 \ldots \sigma_{m-1})^{m-1}$ will be a multiple of the identity. Let us check that in fact $\pi_Y'(\sigma_2 \ldots \sigma_{m-1})^{m-1} = 1$. By Lemma 9.3 we see that $\pi_Y'(\sigma_i)$ is $q^{-r/d}\pi_Y(\sigma_i)$. But r and d are the same for π_Y and its restriction to B_{m-1} since Y is rectangular, so that $\pi_Y'(\sigma_2 \ldots \sigma_{m-1})^{m-1} = 1$. Now we have the following:

THEOREM 10.2. *Let Y be a Young diagram and let π_Y' be the corresponding representation of B_m, adjusted as above so that $\pi_Y'(\sigma_1 \ldots \sigma_{m-1})^m = 1$. Then π_Y' defines a representation of $M(0, m)$ via $\omega_i \mapsto \pi_Y'(\sigma_i)$ if and only if Y is rectangular.*

Proof. We have already proved the "if" part. Now if Y is not rectangular, $\pi_Y|_{B_{m-1}}$ will split as the direct sum of several representations corresponding to Young diagrams $\tilde{Y}$ of the form Y minus one node. Numbering the $\tilde{Y}$s from 1 to k, let r_i and d_i be the corresponding dimensions. Then $d = \Sigma_{i=1}^k d_i$, $r = \Sigma_{i=1}^k r_i$. But the only way to have $\pi'_Y(\sigma_2 \ldots \sigma_{m-1})^m = 1$ is for r_i/d_i to be r/d for all i. A combinatorial argument shows that this is impossible if $k > 1$. Q.E.D.

Thus for every integer m which is not a prime we have constructed non-trivial representations of $M(0, m)$ with a parameter q. The primes seem to be more difficult to deal with. Certain special values of q give representations of $M(0, m)$ in the context of the (k, l) tableaux of [46], but their interest is, at this stage, unclear. It is entirely possible that the representations with parameter q are faithful. This is closely linked to the question of faithfulness of the Burau representation.

As described above we also get representations of subgroups of mapping class groups of closed orientable surfaces, but this begs the important question of whether they can be extended in some way to the whole mapping class group.

However, in genus two, the group generated by the θ_i's *is* the whole mapping class group so that we do obtain representations of this group $M(2, 0)$. Up to symmetry there is only one rectangular tableau on 6 nodes, so in fact there is really only one representation. Here is a choice of matrices corresponding to ⊞ which, when multiplied by $q^{-2/5}$, give a representation

$$\theta_1: \begin{pmatrix} -1 & 0 & 0 & 0 & q \\ 0 & -1 & 1 & 0 & 0 \\ 0 & 0 & q & 0 & 0 \\ 0 & 0 & 1 & -1 & 0 \\ 0 & 0 & 0 & 0 & q \end{pmatrix} \qquad \theta_2: \begin{pmatrix} q & 0 & 0 & 0 & 0 \\ 0 & q & 0 & 0 & 0 \\ 0 & q & -1 & 0 & 0 \\ 1 & 0 & 0 & -1 & 0 \\ 1 & 0 & 0 & 0 & -1 \end{pmatrix}$$

$$\theta_3: \begin{pmatrix} -1 & 0 & 0 & q & 0 \\ 0 & -1 & 1 & 0 & 0 \\ 0 & 0 & q & 0 & 0 \\ 0 & 0 & 0 & q & 0 \\ 0 & 0 & 1 & 0 & -1 \end{pmatrix} \qquad \theta_4: \begin{pmatrix} q & 0 & 0 & 0 & 0 \\ 1 & -1 & 0 & 0 & 0 \\ 0 & 0 & -1 & 0 & q \\ 1 & 0 & 0 & -1 & 0 \\ 0 & 0 & 0 & 0 & q \end{pmatrix}$$

$$\theta_5: \begin{pmatrix} -1 & q & 0 & 0 & 0 \\ 0 & q & 0 & 0 & 0 \\ 0 & 0 & q & 0 & 0 \\ 0 & 0 & 1 & -1 & 0 \\ 0 & 0 & 1 & 0 & -1 \end{pmatrix}.$$

One could also have written down matrices using [46] but the ones given above have the advantage of being defined over $\mathbf{Z}[q, q^{-1}]$. They were obtained using the Kazhdan-Lusztig formalism of W-graphs.

An element of interest in this mapping class group is the Dehn twist about the curve in Figure 10.3.

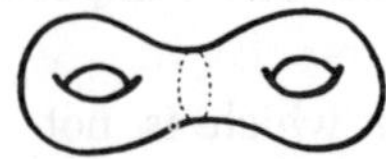

FIGURE 10.3

It is represented by the matrix

$$\begin{pmatrix} q^6 & 9 & 0 & 0 & 0 \\ 0 & q^6 & 0 & 0 & 0 \\ 0 & 0 & q^6 & 0 & 0 \\ -1+q^2-q^3+q^5 & q-q^2+q^4-q^5 & -1+q^2-q^3+q^5 & 1 & q-q^2+q^4-q^5 \\ 0 & 0 & 0 & 0 & q^6 \end{pmatrix}.$$

This shows that the representation is highly non-trivial on the Torelli group—the normal closure of the above element. Note that when $q = -1$ this matrix is the identity.

Presumably one may use the above representation to settle most questions about genus 2 Heegaard splittings rather mechanically by specializing q to be in a finite field.

11. The V polynomial

The discovery of $P_L(l, m)$ was preceded by the discovery of one of its specializations $V_L(t)$, different from the Alexander polynomial; see [16]. It satisfies the skein relation

$$(11.1) \qquad 1/tV_{L_+} - tV_{L_-} = (\sqrt{t} - 1/\sqrt{t})V_{L_0}$$

where L_+, L_-, and L_0 are as in Figure 6.3. Thus,

$$V_L(t) = P_L(it^{-1}, -i(\sqrt{t} - 1/\sqrt{t})) = X_L(t, t).$$

However, V_L has retained its interest in spite of its more powerful generalization. Ironically, one reason is simply because it only has one variable which makes it easier to work with. Different models for it have been recently given by Kauffman [19], which have led to solutions of some old problems in knot theory; see [19], [32], [43]. It is also shorter to calculate, for reasons that will become clear.

We will also see in the next section that V has the surprising property of being essentially an invariant of unoriented links. There is every indication that a topological understanding of P will only be found once V has been understood in its own right. For these reasons we would like to devote the next three sections to the study of V_L from the Hecke algebra point of view. In particular we will give full proofs of most of the results of [16].

It is unclear from the skein picture why V_L (or even Δ_L for that matter) is an interesting specialization. For this we look back to the origins of this work. In fact the Hecke algebra relations (4.1), (4.2) and (4.3) were not the author's original motivation. In work on type II_1 factors we discovered $*$-algebras A_n with generators $1, e_1, e_2, \ldots, e_n$ and relations

(11.2) $$e_i^* = e_i, \quad e_i^2 = e_i$$

(11.3) $$e_i e_{i\pm 1} e_i = \tau e_i$$

(11.4) $$e_i e_j = e_j e_i \quad \text{if } |i-j| \geq 2$$

(where τ is a real number related to the II_1 factors). These relations are not, for all τ, a presentation of A_n, whose actual structure is decided by the existence on it of a trace (which we will call tr by abuse of notation), for which the associated sesquilinear form $\langle a, b\rangle = \text{tr}(ab^*)$ is positive definite. This trace is uniquely determined by the following Markov property:

(11.5) $$\text{tr}(xe_{n+1}) = \tau\,\text{tr}(x) \quad \text{if } x \in A_n.$$

Now let t be such that $\tau^{-1} = 2 + t + t^{-1}$ and set $g_i = te_i - (1 - e_i)$. Then $e_i^2 = e_i$ is equivalent to $g_i^2 = (t-1)e_i + t$ and $e_i e_{i+1} e_i = \tau e_i$ is equivalent to

(11.6) $$g_i g_{i+1} g_i + g_i g_{i+1} + g_{i+1} g_i + g_i + g_{i+1} + 1 = 0.$$

Thus in particular the g_i's satisfy (4.1), (4.2) and (4.3) with $t = q$. But (11.6) does not follow from these relations. Thus the algebra A_n is a quotient of $H(t, n+1)$. To understand which quotient, it suffices, for generic t, to look at the case $t = 1$. It is well known that the representations of the symmetric group for which $s_i s_{i+1} s_i + s_i s_{i+1} + s_{i+1} s_i + s_i + s_{i+1} + 1 = 0$ (the s_i's are transpositions as in Section 4) are those whose Young diagrams have at most two columns. One obtains the following diagram for A_n, whose meaning is the same as Figure 3.3 was for the Hecke algebras, where the Young diagrams are replaced by the dimensions of the corresponding representations.

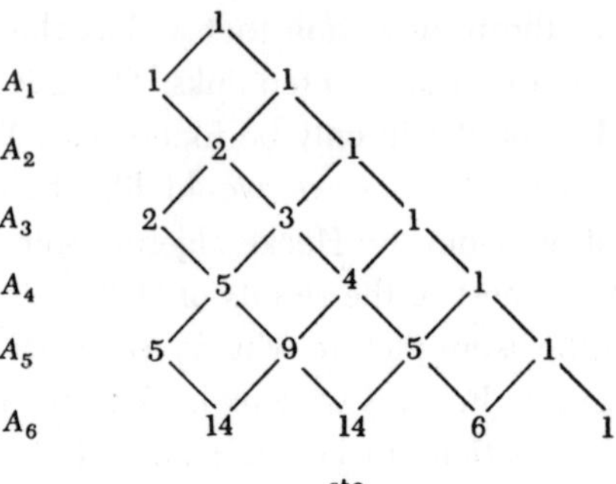

FIGURE 11.7

We see that dim A_n is the Catalan number

$$\frac{1}{n+2}\binom{2n+2}{n+1}.$$

When $t = e^{\pm 2\pi i/n}$ the Hecke algebra is not semisimple, but a C^*-algebra must be, so that some collapsing occurs. The 1's on the extreme right of each level of Figure 11.7 die out, together with their progeny, as explained in [17], [18]. This creates representations of mapping class groups as described in Section 10.

It is clear that the trace of (11.5) is related to the Markov trace of Section 5. In fact for only one value of z does the trace of Section 5 pass to the quotient A_n. This value may be evaluated in two different ways. The first is to take the trace of (11.6). One finds $z\,\mathrm{tr}(g_1^2) + 2z\,\mathrm{tr}(g_i) + 2z + 1 = 0$ or $(z(1+t)+1)(z+1) = 0$. The value $z = -1$ is of little interest so we have $z = -1/(1+t)$. This gives $\mathrm{tr}(e) = t/(1+t)^2 = \tau$. The other way to derive this is by noting that the trace must be zero on simple Hecke algebra quotients whose Young diagrams have more than two columns. By Figure 5.6 we see that this will be the case if $q^2 - \lambda q = 0$, or $\lambda = q$, which is the same as $z = -1/(1+t)$.

Thus we are led to consider $V_L(t) = X_L(t,t)$ which satisfies (11.1). Of course V_L can be defined entirely within the A_n's. We will use π_0 to be the representation of B_n into A_{n-1} given by $\pi_0(\sigma_i) = te_i - (1 - e_i)$. Then if $\alpha \in B_n$ has exponent sum e one has

$$V_L(t) = \left(-\frac{t+1}{\sqrt{t}}\right)^{n-1} (\sqrt{t})^e \mathrm{tr}(\pi_0(\alpha)). \tag{11.8}$$

Relations (11.2)–(11.5) suffice to calculate $\mathrm{tr}(\pi_0(\alpha))$ but since

$$\frac{1}{(n+2)}\binom{2n+2}{n+1}$$

grows more slowly than $(n+1)!$, computer calculation of V_L, as outlined in Section 6, is feasible for closed braids on more strings than for P_L.

Let us record some specializations to V_L of results obtained earlier for P_L.

Proposition 11.9. *Let n and m be relatively prime and let K be an (n, m) torus knot. Then*

$$V_K(t) = \frac{t^{(n-1)(m-1)/2}}{(1-t^2)}\left(1 - t^{m+1} - t^{n+1} + t^{n+m}\right).$$

Proof. Putting $q = t$ in the formula of Theorem (9.7) causes the sum to collapse to only two terms, those with $\beta = 0$ and 1. The answer follows by calculation. Q.E.D.

Note. It would be possible to avoid the algebraic complexity of the proof of (9.7) by rerunning the argument on A_n alone. The two terms in the sum correspond to the two right-hand terms on each line of Figure 11.7. The weights are easily obtained from Figure 5.7 and also appear in [17], [18] and [44]. Moreover, if (9.7) is considered a reasonable formula one could write down a formula for V_L where L is a general torus link.

Proposition 11.10 ([16], Theorem 21). *If $\alpha \in B_3$ is such that $\hat{\alpha}$ is a knot and the exponent sum of α is e then*

$$V_{\hat{\alpha}}(t) = t^{e/2}\left(1 + t^e + t + 1/t - t^{e/2-1}(1 + t + t^2)\Delta_{\hat{\alpha}}(t)\right).$$

Proof. Putting $q = \lambda = t$ in (8.4) gives the result. Q.E.D.

Proposition 11.11 ([16], Theorem 22). *With notation as in Proposition 11.10 except that $\alpha \in B_4$,*

$$t^{-e}V_{\hat{\alpha}}(t) + t^e V_{\hat{\alpha}}(1/t) = (t^{-3/2} + t^{-1/2} + t^{1/2} + t^{3/2})(t^{e/2} + t^{-e/2}) - (t^{-2} + t^{-1} + 2 + t + t^2)\Delta_{\hat{\alpha}}(t).$$

Proof. Put $t = \lambda = q$ in (8.10), multiply by $(1/t)^{(e+1)/2}$ and simplify. Q.E.D.

We have discussed various explicit formulae for the irreducible representations of the Hecke algebras but there are many other natural representations which are not necessarily irreducible. One which plays a prominent role in the algebraic theory of the V polynomial was first discovered by Temperley and Lieb in [42] in connection with the Potts and ice-type models of statistical mechanics. It was rediscovered by Pimsner and Popa in [34]. We use the Pimsner–Popa formalism. The infinite tensor product of 2×2 matrices, $\otimes_{i=1}^{\infty} M_2(\mathbf{C})$, has an obvious shift endomorphism σ defined by $\sigma(x \otimes 1 \otimes 1 \otimes \dots) = 1 \otimes x \otimes 1 \otimes 1 \dots$. Let

$$e = \left\{ \frac{t}{1+t} e_{11} \otimes e_{22} + \frac{\sqrt{t}}{1+t}(e_{21} \otimes e_{12} + e_{12} \otimes e_{21}) + \frac{1}{1+t} e_{22} \otimes e_{11} \right\} \otimes 1 \otimes 1 \otimes \dots$$

where e_{ij} are matrix units for $M_2(\mathbf{C})$. One may easily check that if $e_i = \sigma^i(e)$ then relations (11.2)–(11.4) hold for $t \in \mathbf{R}^+$ where $*$ is the usual conjugate transpose. The ensuing representation, call it θ, of A_n for all n is faithful for generic t. For positive real t this follows easily from [17].

It is interesting that the trace tr on A_n is *not* given by the restriction of the (normalized) trace to $\theta(A_n)$. In fact, the relevant linear functional on $\otimes_{i=1}^{\infty} M_2(\mathbf{C})$ is the Powers state ϕ_t (see [35]) defined by

$$\phi_t(x_1 \otimes x_2 \otimes \cdots \otimes x_n \otimes 1 \otimes 1 \dots) = \mathrm{TR}((h \otimes h \otimes \cdots \otimes h)(x_1 \otimes \cdots \otimes x_n))$$

where TR is the non-normalized trace (sum of diagonal elements) on $M_{2^n}(\mathbf{C})$ and h is the 2×2 matrix

$$\begin{pmatrix} 1/(1+t) & 0 \\ 0 & t/(1+t) \end{pmatrix}.$$

One may check in fact that $\phi_t(xy) = \phi_t(yx)$ for $x \in \bigcup_n \theta(A_n)$ and $y \in T$. The Markov property $\phi_t(\theta(xe_{n+1})) = t/(1+t)^2 \phi_t(\theta(x))$ is straightforward. Let us call this representation (of A_{n+1}, $H(q,n)$ or B_n) the PPTL representation.

Another representation for $\tau = 1/n$, $n = 2, 3, 4, \dots$, can be obtained by iterated crossed products. If $\mathbf{CZ}_n$ is the group algebra, then $\hat{\mathbf{Z}}_n$ ($\cong \mathbf{Z}_n$) acts on it in the obvious way. So one may form $\mathbf{CZ}_n \rtimes \hat{\mathbf{Z}}_n$. Iterating this procedure one obtains an algebra generated by u_k's, $k = 1, 2, \dots$ with $u_k^n = 1$, $u_k u_{k+1} = e^{2\pi i/n} u_{k+1} u_k$ and $u_k u_j = u_j u_k$ if $|k - j| \geq 2$. By Takesaki duality [41] one knows that, for even k, the algebra generated by $u_1, \dots, u_k$ is the $p^{k/2} \times p^{k/2}$ matrices. Setting

$$e_k = \frac{1}{n} \sum_{i=0}^{n-1} u_k^i$$

one obtains a faithful representation of A_k. An explicit matrix form may be found in [4]. Here the Markov trace is the usual normalized trace.

12. The values $V_L(e^{2\pi i/n})$, $n = 1, 2, 3, 4, 6, 10$

In [17] it was shown that the only values of τ for which the trace on A_n is positive definite ($\mathrm{tr}(a^*a) > 0$ for $a \neq 0$) are the values $\tau = (4\cos^2\pi/n)^{-1}$, $n = 3, 4, 5, \ldots$. These values correspond to $t = e^{\pm 2\pi i/n}$ so it was expected that these values of V_L should be somehow special, as A_n at these values of τ provided entirely new examples of subalgebras of von Neumann algebras. We remind the reader that these are also values of q (except for $n = 1$) for which the Hecke algebra is not always semisimple.

Although many of the following results can be deduced from the skein relation, we shall indicate the algebraic proofs as well since the skein relation is not very suggestive.

(12.1) $V_L(1)$: When $t = 1$ the PPTL representation of the braid group factors through the symmetric group and is given, up to sign, by permutations of the tensor product components of $\mathbf{C}^2 \otimes \mathbf{C}^2 \otimes \ldots$. The Powers state is in this case only the normalized trace. Also $\sqrt{t} + 1/\sqrt{t} = 2$ and $\mathrm{tr}(g_i) = 2$. It follows immediately that $V_L(1) = (-2)^{c-1}$ where c is the number of components of the link L (Theorem 15 of [16]). Note that the value $t = 1$ is not allowed in the two variable polynomial, $V_L(1)$ being the limit of $X_L(q, \lambda)$ as $(q, \lambda) \to (1, 1)$ along the curve $q = \lambda$.

(12.2) $d/dt\, V_L(1)$: Here the skein theoretic proof is much simpler and the algebraic proof does not lead to any more understanding. Suffice it to say that the following result may be obtained non-inductively (though not easily) by differentiating the formula for V_L coming from the PPTL representation and the Powers state, putting $t = 1$ and using the permutation method of (12.1). The result is

$$V_L'(1) = -3(-2)^{c-2} \quad (\text{total linking number of } L).$$

(This result was also noticed by Murakami in [30].)

Proof. Differentiating the skein relation and putting $t = 1$ gives $V_{L_+}'(1) - V_{L_-}'(1) = V_{L_0}(1) + V_{L_+}(1) + V_{L_-}(1)$. If the crossing involves only one component of the link $V_{L_+}'(1) = V_{L_-}(1) = -(1/2)V_{L_0}(1)$ by (12.1) so that in this case $V_{L_+}'(1) = V_{L_-}'(1)$. If the crossing involves two components, $V_{L_+}(1) = V_{L_-}(1) = -2V_{L_0}(1) = (-2)^{c-1}$ so that in this case $V_{L_+}'(1) - V_{L_-}'(1) = -3(-2)^{c-2}$. The result follows by induction Q.E.D.

The higher derivatives of V_L at 1 seem interesting as the difference between the nth derivatives of V_{L_+} and V_{L_-} will not involve the nth derivative of V_{L_0}. We record here the simple fact that changing a positive crossing of a knot to a negative one will change $V_K''(1)$ by six times the linking number of the two-component link formed by eliminating the crossing. A similar argument with Δ shows $V_K''(1) = -3\Delta_K''(1)$.

(12.3) $V_L(-1)$: The formula $V_L(-1) = \Delta_L(-1)$ follows immediately from their skein relations.

(12.4) $V_L(e^{2\pi i/3})$: Define the representation of A_n when $t = e^{2\pi i/3}$ $(\tau = 1)$ by sending e_i to 1. The trace on A_n that this defines satisfies the Markov property trivially; so since $\sqrt{t}\,\sigma_i$ is sent to -1 we have, by (11.8), $V_L(e^{2\pi i/3}) = (-1)^{c-1}$ where L has C components (see 14 of [16]).

Proposition 12.5. *If K is a knot then* $1 - V_K(t) = W_K(t)(1-t)(1-t^3)$ *for some Laurent polynomial* $W_K(t)$.

Proof. By (12.1) and (12.4), $1 - V_K$ is divisible by $1 - t^3$ and by (12.2) it is further divisible by $1 - t$. Q.E.D.

Thus extraneous information is recorded in $V_K(t)$. We have chosen to record the simplified $W_K(t)$ in Table 15.9.

(12.6) $V_K(i)$: It has been shown by Murakami [31] and Lickorish-Millet [23] that

$$V_L(i) = \begin{cases} 0 & \text{unless } \operatorname{arf}(L) \text{ exists} \\ (-2\sqrt{2})^{c-1}(-1)^{\operatorname{arf}(L)} & \text{otherwise.} \end{cases}$$

The author first proved the result for knots by observing that the calculus of Lannes in [21] when applied to a braid projection of a knot follows the same pattern as calculating the trace of the image of the braid in the iterated crossed product representation at the end of Section 11 when $n = 2$.

(12.7) $V_L(e^{i\pi/3})$: This value suggested itself as interesting because of the special nature of the algebra A_n when $\tau = 1/3$. It is shown in [18] that the image of B_n is finite, and the groups are identified for all n in [10]. Birman conjectured that $V_L(e^{i\pi/3}) = i^k(\sqrt{3})^{k_2}$ for integers k_1 and k_2 depending on L. The situation is completely clarified in papers by Lickorish-Millet [23] and Lipson [26] where $V_L(e^{i\pi/3})$ is calculated in terms of a Seifert form for L. A braid-representation version is also possible. This has been begun by the author and D. Goldschmidt in [15].

(12.8) $V_L(e^{i\pi/5})$: The values $q = e^{\pm i\pi/5}$ have a peculiar feature in the Hecke algebra picture: $\pi(B_n)$ is finite when $n \leq 3$ but infinite otherwise. In fact, it was shown in [18] that $\pi(\sigma_1\sigma_2\sigma_3^{-1})$ has infinite order. Thus $V_L(e^{i\pi/5})$ can be used as a test for being a 3-braid—the set of all possible values for closed 3-braids is finite and could easily be written down.

A question posed in [6] is whether every $\alpha \in B_{n+1}$ is conjugate to an element of the form $\alpha_1\sigma_n\alpha_2\sigma_n^{-1}$ with α_1 and $\alpha_2 \in B_n$. For α of this form we have

$\pi_0(\alpha) = \pi_0(\alpha_1)((t+1)e_n - 1)\pi_0(\alpha_2)((t^{-1}+1)e_n - 1)$ and expanding and simplifying we find that $\mathrm{tr}(\pi_0(\alpha)) = 1/\tau\, \mathrm{tr}(e_n\pi_0(\alpha_1)e_n\pi_0(\alpha_2))$. Letting $n = 3$ and $t = e^{i\pi/5}$ we see that the set of V_L values of elements of the above form is finite, so to answer the question in the negative (even with conjugacy replaced by Markov equivalence), it suffices to exhibit infinitely many values of $V_L(e^{i\pi/5})$ with $L = \hat{\alpha}$, $\alpha \in B_4$. But in [18] it is established that the Burau matrix of $\sigma_1\sigma_2\sigma_3^{-1}$ has an eigenvalue which is not a root of unity. On the other hand, in the representation on the extreme left of the A_3 level of Figure 11.7, $e_1 = e_3$ so that $\sigma_1\sigma_2\sigma_3^{-1}$ is the same as $\sigma_1\sigma_2\sigma_1^{-1}$ which has finite order. Thus in the weighted sum version of the trace, the contributions of the first and third representations on the A_3 line are periodic whereas that of the second is aperiodic. So infinitely many powers of $\sigma_1\sigma_2\sigma_3^{-1}$ are not Markov equivalent to 4-braids of the special form.

Finally, note that for $t = e^{\pm 2\pi i/n}$ except those above, one cannot expect any especially simple behavior. In fact, for each such t, $\{|V_L(t)|\ |L$ is a closed 3-braid$\}$ is dense in the interval $[0, 4\cos^2\pi/n]$, which we shall prove in Section 14. For other roots of unity the situation is unclear at this point.

13. The plat approach

A plat is a closed, *unoriented* link $\tilde{\alpha}$ that is formed when a braid $\alpha \in B_{2m}$ is closed according to the prescription of Figure 13.1.

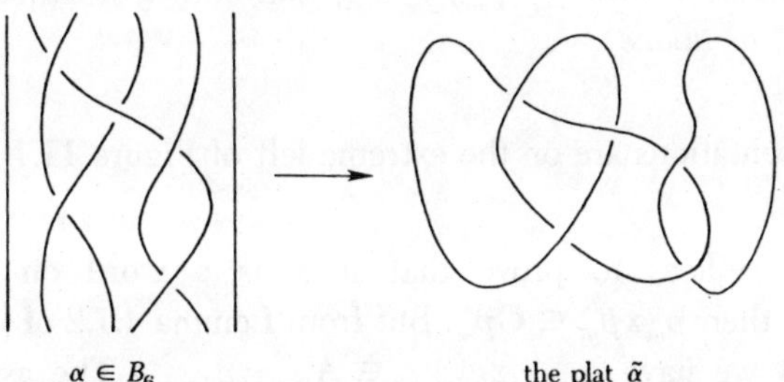

FIGURE 13.1

Any unoriented link is of the form $\tilde{\alpha}$ for some (α, m), $\alpha \in B_{2m}$ (see [6]). Birman has proved the following analogue of Markov's theorem which is essentially contained in [7].

THEOREM 13.2 (Birman). *For each m let C_m be the subgroup of B_{2m} generated by the set $X \cup Y \cup Z$ where*

$$X = \{\sigma_{2i-1}; \quad i = 1, 2, \ldots, m\},$$
$$Y = \{\sigma_{2i}\sigma_{2i-1}\sigma_{2i+1}\sigma_{2i}; \quad i = 1, 2, \ldots, m-1\},$$
$$Z = \{\sigma_{2i}\sigma_{2i-1}\sigma_{2i+1}^{-1}\sigma_{2i}^{-1}; \quad i = 1, 2, \ldots, m-1\}.$$

Also let the map S_k: $B_{2k} \to B_{2k+2}$ *be defined by* $S_k(\alpha) = \alpha\sigma_{2k}$. *Then*

(a) *If* $\alpha \in B_{2m}$, $\widetilde{x\alpha y} = \tilde{\alpha}$ *for* $x, y \in C_m$ *and* $\widetilde{S_m(\alpha)} = \tilde{\alpha}$ (= *means isotopic*).

(b) *The equivalence relation on* $\coprod_m B_{2m}$ *generated by the two "moves"* $\alpha \mapsto x\alpha y$ ($\alpha \in B_{2m}$, $x, y \in C_m$) *and* $\alpha \leftrightarrow S_m(\alpha)$, $\alpha \in B_{2m}$ *is the same as the equivalence relation given by isotopy of the plats.*

It is not obvious that the plat closure should be related to the Hecke algebra in any way similar to the closure of Section 1. In fact it seems impossible to use all the information of the Hecke algebra in the plat picture. But pp. 192–194 of [6] and Section 10 give a clue as to how to use the Hecke algebra: the plat closure of $\alpha \in B_{2m}$ really only depends on the image of α in the mapping class group of the 2-sphere minus $2m$ points. Thus one should look first at quadratic representations whose Young diagrams are of the form 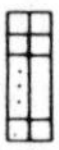 , which we know occur as summands of the representation π_0 of Section 11. The following lemma is thus suggestive (A_n are as in Section 11).

LEMMA 13.3. (i) *In* A_{2m-1}, *the idempotent* $p_m = e_1 e_3 \cdots e_{2m-1}$ *is minimal, i.e.* $p_m A_{2m-1} p_m \subseteq \mathbf{C} p_m$.

(ii) *If* ρ_Y *is the irrep of* A_{2m-1} *corresponding to the Young diagram (with at most two columns)* Y *then* $\rho_Y(p_m) \neq 0$ *if and only if* Y *is rectangular with two columns (i.e. as above).*

These representations are on the extreme left of Figure 11.7.

Proof. (i) It suffices to prove that if x is a word on the e_i's, $i = 1, 2, \ldots, 2m-1$, then $p_m x p_m \in \mathbf{C} p_m$. But from Lemma 4.1.2 of [17] and (11.2), (11.3) and (11.4) we have $e_{2m-1} x e_{2m-1} \in A_{2m-3} e_{2m-1}$. The assertion follows by induction.

(ii) Since p_m is minimal there is at most one Y for which $\rho_Y(p_m) \neq 0$. By induction and Figure 11.7 we see that Y must be either $Y_1 =$ 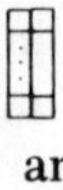 or $Y_2 =$  But the Markov trace of p_m is τ^m and the weights for Y_1 and Y_2 are different (use Figure 5.4 with $q = \lambda$), that of Y_1 being τ^m. Q.E.D.

COROLLARY 13.4. *Let* Y *be the Young diagram* . *Then* $\rho_Y(e_1) = \rho_Y(e_3)$.

Proof. Both $\rho_Y(e_1)$ and $\rho_Y(e_3)$ are minimal idempotent in the two-dimensional representation ρ_Y. They both dominate the minimal idempotent $\rho_Y(e_1 e_3) = \rho_Y(p_2)$. Q.E.D.

Remark 13.5. In general the number of nodes in the second (smaller) column of Y is the largest m for which the product of a commuting subset of size m of the e_i's is non-zero in the corresponding representation of A_n. For instance, $e_ie_j = 0$ if $|i-j| \geq 2$ in the Burau representation (tensored with parity) which characterizes it among Hecke algebra representations.

Definition 13.6. The linear functional $\phi\colon \bigcup_m A_{2m-1} \to \mathbf{C}$ is defined by $p_m x p_m = \phi(x) p_m$ for $x \in A_{2m-1}$.

Note that in order for ϕ to be well-defined we need to show that if $p_{m+1} x p_{m+1} = \lambda p_{m+1}$ and $p_m x p_m = \mu p_m$ then $\lambda = \mu$. This is clear from the definition of p_n. Thus ϕ is also defined by $\lim_{m\to\infty}(p_m x p_m - \phi(x) p_m) = 0$ (see [1]).

Note further that the proof of Lemma 13.3(i) was constructive. It gives an algorithm for calculating $\phi(x)$. The following formula will also be useful.

PROPOSITION 13.7. *For* $x \in A_{2m-1}$, $\phi(x) = 1/\tau^m \operatorname{tr}(x p_m)$.

Proof. Just take the trace of the defining relation for ϕ and use $\operatorname{tr}(p_m) = \tau^m$. Q.E.D.

The next result gives the connection of ϕ with plats.

LEMMA 13.8. *Let* $g_i = te_i - (1 - e_i) \in A_n$. *Then*

(i) $p_m g_{2i-1}^{\pm 1} = g_{2i-1}^{\pm 1} p_m = t^{\pm 1} p_m$ for $i = 1, 2, \ldots, m$.

(ii) $p_m (g_{2i} g_{2i-1} g_{2i+1} g_{2i}) = (g_{2i} g_{2i-1} g_{2i+1} g_{2i}) p_m = t p_m$ for $i = 1, 2 \ldots m-1$

(iii) $p_m (g_{2i} g_{2i-1} g_{2i+1}^{-1} g_{2i}^{-1}) = (g_{2i} g_{2i-1} g_{2i+1}^{-1} g_{2i}^{-1}) p_m = p_m$ for $i = 1, 2 \ldots m-1$.

Proof. (i) is trivial. For (ii) and (iii) note first that it suffices to consider the case $m = 2$ since all the e_i's except two are irrelevant to the calculation.

Let f be the minimal central idempotent of A_3 corresponding to $\boxplus$. Then $p_3 = p_3 f$ by Lemma 13.3 so that $p_3 g_2 g_1 g_3^{-1} g_2^{-1} = p_3 f g_2 g_1 g_3^{-1} g_2^{-2} = p_3$ by Corollary 13.4, which proves (iii).

For (ii) note that $(g_2 g_1 g_3 g_2) g_1 (g_2^{-1} g_1^{-1} g_3^{-1} g_2^{-1}) = g_3$ follows from the braid group relations as does the formula with 1 and 3 interchanged. Thus it is immediate that $g_2 g_1 g_3 g_2$ commutes with $e_1 e_3 = p_3$. Hence $p_3 g_2 g_1 g_3 g_2 = p_3 g_2 f g_1 g_3 g_2 p_3 = p_3 g_2 g_1^2 g_2 p_3 = \phi(g_2 g_1^2 g_2) p_3$. Now it remains to calculate $\phi(g_2 g_1^2 g_2) = 1/\tau^2 \operatorname{tr}(e_1 e_3 g_2 g_1^2 g_2) = 1/\tau \operatorname{tr}(e_1 g_2 g_1^2 g_2)$. But $g_1^2 = (t-1) g_1 + t$

so that

$$\begin{aligned}\phi(g_2g_1^2g_2) &= \frac{t-1}{\tau}\operatorname{tr}(e_1g_2g_1g_2) = \frac{t}{\tau}\operatorname{tr}(e_1g_2^2)\\ &= (t-1)t^2\operatorname{tr}(g_2) + t\operatorname{tr}((t^2-1)e_2+1)\\ &= \frac{-t^3+t^2+t^3+t}{1+t} = t. \qquad \text{Q.E.D.}\end{aligned}$$

COROLLARY 13.9. *Suppose $\alpha \in B_{2m}$ and $x, y \in C_m$ (the group of Theorem 13.2); then $\phi(x\alpha y) = t^k\phi(\alpha)$ for some $k \in \mathbf{Z}$.*

We now take care of the stabilizing move S_m.

LEMMA 13.10. *Let $x \in A_{2m-1}$. Then*

$$\phi(xg_{2m}) = -\frac{1}{(1+t)}\phi(x).$$

Proof. By Proposition 13.7,

$$\phi(xg_{2m}) = \frac{1}{\tau^{m+1}}\operatorname{tr}(xg_{2m}p_me_{2m+1}) = \frac{1}{\tau^m}\operatorname{tr}(g_{2m})\operatorname{tr}(xp_m) = -\frac{1}{(1+t)}\phi(x).$$

Q.E.D.

Combining Corollary 13.9, Lemma 13.10, and Theorem 13.2 we obtain the following:

THEOREM 13.11. *Let $\alpha \in B_{2m}$ and $\beta \in B_{2n}$ be such that $\tilde{\alpha}$ and $\tilde{\beta}$ are isotopic. Then there exists $k \in \mathbf{Z}$ such that*

$$(-(t+1))^{m-1}\phi(\pi_0(\alpha)) = t^k(-(t+1))^{n-1}\phi(\pi_0(\beta)).$$

Proof. By Corollary 13.9 and Theorem 13.2 it suffices to check that $(-(t+1))^{m-1}\phi(\pi_0(\alpha)) = (-(t+1))^m\phi(\pi_0(S_m(\alpha)))$. But this is Lemma 13.10. Q.E.D.

Thus we have a (potentially) new invariant for unoriented links. We shall show that it is in fact only an unnormalized version of $V_{\hat{\alpha}}$. The idea of the proof is that a closed braid is really a special kind of plat, where the closing scheme is changed as in Figure 13.12.

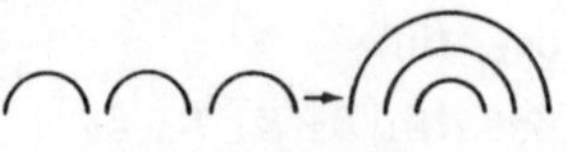

FIGURE 13.12

The two theories corresponding to the different types of closure are equivalent under conjugation by an element Ω which we shall make explicit below. We begin with a simple lemma.

LEMMA 13.13. *Let* $\psi\colon \bigcup_n A_n \to \mathbf{C}$ *be a linear functional with* $\psi(1) = 1$, $\psi(xe_{n+1}y) = \tau\psi(xy)$ *if* $x, y \in A_n$. *Then* $\psi = \mathrm{tr}$.

Proof. By induction assume $\psi = \mathrm{tr}$ on A_n. Then $\psi = \mathrm{tr}$ on the subspace of A_{n+1} spanned by A_n and all elements of the form $xe_{n+1}y$, which is all of A_{n+1} by [17]. Q.E.D.

THEOREM 13.14. *Let L be an oriented link and let* $\alpha \in B_{2m}$ *be such that* $\tilde{\alpha} = L$ *as unoriented links. Then there is a* $k \in \mathbf{R}$, $2k \in \mathbf{Z}$ *with* $V_L(t) = t^k(-(t+1))^{m-1}\phi(\pi_0(\alpha))$.

Proof. We know there is an n and a $\beta \in B_n$ for which $\hat{\beta} = \tilde{\alpha}$. Let

$$\Omega_n = (\sigma_2\sigma_3 \dots \sigma_{2n-1})(\sigma_3\sigma_4 \dots \sigma_{2n-2}) \dots (\sigma_n\sigma_{n+1}) \in B_{2n}.$$

As shown in [6], $\widetilde{\Omega_n\beta\Omega_n^{-1}} = \hat{\beta} = \tilde{\alpha}$ where β is considered as an element of B_{2n} in the usual way (equality meaning isotopy as unoriented links). By Theorem 13.11 and the definition of $V_{\hat{\beta}}$ it suffices to show that $\phi(\pi_0(\Omega_n\beta\Omega_n^{-1})) = \mathrm{tr}(\pi_0(\beta))$. In fact we claim that if ψ on $\bigcup_n A_n$ is defined by

$$\psi(x) = \phi\big(\pi_0(\Omega_n)x\pi_0(\Omega_n)^{-1}\big) \quad \text{for } x \in A_{n-1}, \text{ then } \psi = \mathrm{tr}.$$

The first thing to check is that ψ is well defined; i.e.

$$\phi\big(\pi_0(\Omega_{n+1})x\pi_0(\Omega_{n+1})^{-1}\big) = \phi\big(\pi_0(\Omega_n)x\pi_0(\Omega_n)^{-1}\big) \quad \text{for } x \in A_{n-1}.$$

But the formulae $\Omega_p\sigma_i\Omega_p^{-1} = \sigma_{2i-1}^{-1}\sigma_{2i}\sigma_{2i-1}$ for $1 \le i \le p-1$ are easily checked in the braid groups. They imply

$$(13.15) \qquad \pi_0(\Omega_p)e_i\pi_0(\Omega_p^{-1}) = g_{2i-1}^{-1}e_{2i}g_{2i-1} \quad \text{for } 1 \le i \le p-1.$$

So conjugation by $\pi_0(\Omega_{n+1})$ has the same effect on A_{n-1} as conjugation by $\pi_0(\Omega_n)$, and ψ is well-defined.

Now $\psi(1) = 1$ is obvious so by Lemma 13.13 we only have to show that $\psi(xe_ny) = \tau\psi(xy)$ if $x, y \in A_{n-1}$. We proceed by induction on n. Using Proposition 13.7 we have

$$\begin{aligned}
\psi(xe_ny) &= \phi\big(\pi_0(\Omega_{n+1})xe_ny\pi_0(\Omega_{n+1})^{-1}\big)\\
&= \phi\big(\pi_0(\Omega_n)x\pi_0(\Omega_n)^{-1}g_{2n-1}^{-1}e_{2n}g_{2n-1}\pi_0(\Omega_n)y\pi_0(\Omega_n)^{-1}\big) \quad \text{(by (13.15))}\\
&= \tau^{-n-1}\mathrm{tr}\big(p_{n+1}\pi_0(\Omega_n)x\pi_0(\Omega_n)^{-1}g_{2n-1}^{-1}e_{2n}g_{2n-1}\pi_0(\Omega_n)y\pi_0(\Omega_n)^{-1}\big)\\
&= \tau^{-n+1}\mathrm{tr}\big(p_n\pi_0(\Omega_n)xy\pi_0(\Omega_n)^{-1}\big) \quad \text{(by (11.5) used twice)}\\
&= \tau\big(\tau^{-n}\phi\big(\pi_0(\Omega_n)xy\pi_0(\Omega_n)^{-1}\big)\big)\\
&= \tau\psi(xy).
\end{aligned}$$

Q.E.D.

FIGURE 13.18

COROLLARY 13.16 (see [24]). *If L and L' are two oriented links which are isotopic as unoriented links, there is a $k \in \mathbf{Z}$ such that*

$$V_L(t) = t^k V_{L'}(t).$$

Note that the value of k is easily determined in terms of linking numbers by (12.2).

COROLLARY 13.17 (see [11], [25]). *Let L_+, L_- and L_∞ be unoriented links identical except at one crossing where they are as in Figure 13.18. Then there are numbers k_1 and k_2, $2k_i \in \mathbf{Z}$, such that*

$$V_{L_+} - t^{k_1} V_{L_-} = t^{k_2}(1-t)V_{L_\infty}.$$

Proof. Changing a link into a plat is rather easy, being achieved simply by threading local maxima and minima through the link. This can be done so as to leave alone any portion of the link that already looks like a plat, in particular a single crossing. Thus there are braids $\alpha_1, \alpha_2 \in B_{2m}$ such that $L_- = \alpha_1\sigma_k\alpha_2$, $L_+ = \alpha_1\sigma_k^{-1}\alpha_2$, $L_0 = \alpha_1\alpha_2$ (look at Figure 13.18 sideways). But by Eq. (4.1), $\pi_0(\alpha_1\sigma_k\alpha_2) - t\pi_0(\alpha_1\sigma_k^{-1}\alpha_2) = (t-1)\pi_0(\alpha_1\alpha_2)$. Taking ϕ on both sides of the equation gives the answer by Theorem 13.14. Q.E.D.

The values of k_1 and k_2 can be determined by calculus. For instance, if L_+ and L_- are knots and L_∞ is also a knot, we have $V'_{L_+}(1) - k_1V_{L_-}(1) - V'_{L_-}(1) = -V_{L_\infty}(1)$ so that by (12.1) and (12.2), $k_1 = 1$ and $V_{L_+} - tV_{L_-} = t^{k_2}(1-t)V_{L_\infty}$. Differentiating again gives $V''(1) - V''(1) = -2k_2$ or $k_2 = -(V''_{L_+}(1) - V''_{L_-}(1))/2$. But by the comments at the end of (12.2), $V''_{L_+}(1) - V''_{L_-}(1)$ is -6 times the linking number of the oriented link L_0 formed by eliminating the crossing according to the orientation. Thus we obtain in the above situation $V_{L_+} - tV_{L_-} = t^{3lk(L_0)}(1-t)V_{L_\infty}$. This is easily checked on the right-handed trefoil, where L_- and L_∞ are both unknots.

We see by induction that V_L is determined by certain linking numbers of two-component links obtained by successively eliminating crossings of L so as to obtain knots. A similar statement is true for links.

Corollary 13.16 is somewhat surprising (Lickorish has shown that Corollaries 13.16 and 13.17 are essentially the same) since the polynomial P_L, and even the Alexander polynomial, usually change wildly if the orientation of a link is altered.

Note 13.19. A plat $\tilde{\alpha}$ with $\alpha \in B_{2m}$ is called an m-bridge link. By (ii) of Lemma 13.3 and Figure 11.7, to calculate V_L for an m-bridge link it suffices to know a

$$\frac{1}{m+1}\binom{2m}{m}$$

matrix representing α and an expression for the matrix p_m. For 2-bridge links, one is dealing with 2×2 matrices and the representation of B_4 is given by

$$\sigma_1, \sigma_3 \to \begin{pmatrix} t & 0 \\ 1 & -1 \end{pmatrix}, \qquad \sigma_2 \mapsto \begin{pmatrix} -1 & t \\ 0 & t \end{pmatrix} \quad \text{and} \quad p_1 = \begin{pmatrix} 1 & 0 \\ 1/1+t & 0 \end{pmatrix}$$

so that $V_{\hat{\alpha}}$, for $\alpha \in B_4$, is obtained by calculating the 2×2 matrix of $\tilde{\alpha}$ and adding up the terms in the first row with weights $1 + t$ and 1 (the answer is correct up to a power of t which must be determined).

What is more interesting is that we can give the formula for 3-bridge links, since the representation in question is for $Y = \begin{array}{|c|c|}\hline & \\ \hline & \\ \hline & \\ \hline\end{array}$, which appears explicitly (with $q \leftrightarrow t$ and $\theta_i \leftrightarrow \sigma_i$) in Section 10. One may easily calculate $p_2 = e_1 e_3 e_5 = (1+t)^{-3}(1+g_1)(1+g_3)(1+g_5)$ and one finds that $V_{\hat{\alpha}}$ is, up to a power of t, obtained by calculating the 5×5 matrix representing α and adding up the third row, with weights t, $1+t$, $(1+t)^2$, $1+t$, $1+t$. On a computer this calculation is very rapid for indefinitely complicated 3-bridge links.

Note 13.20. There is a remarkable connection here with statistical mechanics. In fact, the algebras A_n had been used by Temperley and Lieb in [42] to partially solve a statistical mechanical model known as the Potts model. Indeed the linear functional ϕ essentially defines the partition function. A solution of the Potts model would be an explicit expression for $f(x, y)$ defined as

$$\lim_{n\to\infty} \frac{1}{n^2} \log\Big(\phi\Big(\big((1+xe_1)(1+xe_3)\dots(1+xe_{n-1})(y+e_2) \times (y+e_4)\dots(y+e_n)\big)^n\Big)\Big).$$

The Potts model is defined for a system of "atoms" arrayed on the vertices of a regular lattice in $\mathbf{R}^2$. This is how the relation with closed braids (especially regular knot projections) comes about. The algebras A_n are a calculational device known as transfer matrices. In [19], Kauffman eliminates the need for braids by defining a Potts model on an arbitrary link projection (see also Chapter 12 of [4] and [42]).

14. Positivity considerations

In the context encountered by the author, the algebras A_n possessed more structure. They themselves were C^*-algebras and the trace tr was the data necessary to complete $\bigcup_n A_n$ to become a von Neumann algebra. More precisely, for each positive real t and $t \in \{e^{2\pi i/n} | n = 3, 4, 5, \dots\}$ there is a unique C^*-algebra A (infinite dimensional if $t \neq e^{\pm 2\pi i/3}$) generated by non-zero projections e_i $(i = 1, 2, \dots)$ satisfying (11.2), (11.3) and (11.4) together with a faithful trace tr uniquely defined by (11.2). The values $t = e^{\pm 2\pi i/n}$ and $t \in \mathbf{R}^+$ are distinguished on the braid group level by the obvious fact that for $t = 1$ or $e^{\pm 2\pi i/n}$, $\pi_0(\sigma_i)$ is unitary, but is not unitary otherwise.

The G.N.S. construction ([3]) is to make A into a pre-Hilbert space with scalar product $\langle a, b \rangle = \mathrm{tr}(ab^*)$. Then A itself acts faithfully by left multiplication on the Hilbert space completion $\mathscr{H}_{\mathrm{tr}}$ and the e_i's are realized as projections onto closed subspaces. The situation is thus very geometric and relations (11.2)–(11.4) can be thought of as defining special configurations of subspaces. Note that they are *not* (for $t \neq 1$) the same configurations as those arising in Coxeter-Dynkin theory, though there are relations; see [12].

The von Neumann algebra in question is the closure of A (on $\mathscr{H}_{\mathrm{tr}}$) in the topology of pointwise convergence. Similar considerations apply to the linear functional ϕ.

PROPOSITION 14.1. *For positive real t and $t \in \{e^{2\pi i/n} | n = 3, 4, 5, \dots\}$, the linear functional ϕ of Section* 13 *is a* state, *i.e.* $\phi(x^*x) \geq 0$, $\phi(1) = 1$.

Proof. Elements of the form a^*a are positive in a C^*-algebra and since $p_m = p_m^*$, $p_m x^* x p_m$ is positive. Q.E.D.

In fact, it is easy to see that ϕ is a pure state; i.e., in the G.N.S. representation (as above), A acts irreducibly.

While many of the early uses of the C^*-structure have been improved upon (e.g. Theorem 5 of [16] is much weaker than Scholium 6.15), there are still some results that have no other proof and the C^*-techniques are often useful in quick growth estimates. To illustrate the technique we prove an inequality which, while not terribly sharp, is certainly in the right direction.

PROPOSITION 14.2. *Let $\alpha \in B_n$ have an exponent sum e and let e_+ be the sum of all positive exponents of σ_i's in α. Let V_+ be the largest power of t in $V_{\hat{\alpha}}$. Then* $V_+ \leq (n - 1 + e)/2 + e_+$.

Proof. For $t \in \mathbf{R}$, $t \geq 1$, $\|g_i\| = t$ $(g_i = te_i - (1 - e_i))$ and $\|ab\| \leq \|a\|\,\|b\|$ so that $\|\pi_0(\alpha)\| \leq t^{e+}$ since $\|g_i^{-1}\| = 1$. When $t \to \infty$ the result follows that from formula (11.8) and $|\mathrm{tr}(x)| \leq \|x\|$. Q.E.D.

Note 14.3. One could deduce something about the degree of the polynomial part of V from Proposition 14.2 but Murasugi [32] and Thistlethwaite [43] have shown that this degree is in fact a lower bound for the crossing number of the link L. The C^*-von Neumann picture is more complicated for the whole Hecke algebra but the theory is adequately explained in [14], [46].

Positivity might be useful in questions of faithfulness of representations by the following results.

PROPOSITION 14.4. *If* $\alpha \in B_n$ *then* $\alpha \in \ker \pi_0$ *for* $t = \pm e^{2\pi i/k}$, $k = 3, 4, 5, \ldots$ *if and only if* $V_{\hat{\alpha}}(e^{2\pi i/k}) = (-2\cos \pi/k)^{n-1}$ (*see Theorem* 10 *of* [16]).

Proof. Saying $V_{\hat{\alpha}}(e^{2\pi i/k}) = (-2\cos \pi/k)^{n-1}$ is the same as saying $\mathrm{tr}(\pi_0(\alpha)) = 1$. But π_0 is unitary, so that by the "equality" part of Cauchy–Schwarz, $\pi_0(\alpha) = 1$. Q.E.D.

COROLLARY 14.5. *If* $\alpha \in B_n$ *then* $\alpha \in \ker \pi_0$ (*for generic* t) *if and only if* $V_{\hat{\alpha}}(t) = (-(t+1)/\sqrt{t})^{n-1}$.

Proof. All entries in the PPTL representation are rational functions of $\sqrt{t}$ and so are determined by their values at $e^{2\pi i/k}$, $k = 3, 4, 5, \ldots$. Q.E.D.

Finally we use positivity to prove an assertion of Section 12.

PROPOSITION 14.6. *For each* t *let* $A_{t,n} = \{|V_L(t)|\,|L = \hat{\alpha},\ \alpha \in B_n\}$. *Then for* $t = e^{\pm 2\pi i/k}$, $k \in \mathbf{Z}^+$, $k \notin \{1, 2, 3, 4, 6, 10\}$ *and* $n \geq 3$, $\overline{A_{t,n}} = [0, (2\cos \pi/k)^{n-1}]$ (*where the bar denotes closure*).

Proof. That $A_{t,n} \subseteq [0, (2\cos \pi/k)^{n-1}]$ follows from the Cauchy–Schwarz inequality and formula (11.8). Now it is shown in [18] (essentially because of the paucity of finite subgroups of SO(3)) that for t of the above form, $\psi(B_3)$ is infinite, ψ being the Burau representation. Thus also $\psi(\Gamma)$ is infinite for $\Gamma < B_3$, $[B_3: \Gamma] < \infty$. Now by [40], ψ may be normalized by changing $\psi(\sigma_1)$, $\psi(\sigma_2)$ by a root of unity so that $\psi(B_3) \subseteq \mathrm{SU}(2)$, but then if Γ is the kernel of any map from B_3 to a finite cyclic group, $\psi(\Gamma)$ will be an infinite subgroup of SU(2). Its closure must be a Lie group with non-trivial connected sets. In particular the connected component of the identity of $\overline{\psi(\Gamma)}$ must contain -1. But if Γ is chosen correctly, $\mathrm{tr}(\psi(\gamma))$ will equal $\mathrm{tr}(\pi_{\boxplus}(\gamma))$ for $\gamma \in \Gamma$ and $\mathrm{tr}(\pi_{\boxminus}(\gamma)) = 1$. Thus using the weights for the third row of Figure 11.7, which are $(4\cos^2\pi/k)^{-1}$ and $1 - (2\cos^2\pi/k)^{-1}$, we see that the connected component G of the identity of $\overline{\pi_0(\Gamma)}$ must contain an element of (Markov) trace zero. But then $x \mapsto |\mathrm{tr}(x)|$

is a continuous function from G to $[0, 1]$ which contains 1 and 0. This argument proves the assertion for $n = 3$, but for $n \geq 4$ the only change is in the normalization since tr is compatible with the inclusions of the A_n's. Q.E.D.

COROLLARY 14.7. *For t as above,* $\overline{\{|V(t)| \, |L \text{ a link}\}}$ *is* $[0, \infty)$.

Note. The corollary is also true for $t = e^{2\pi i/10}$ by a special argument; indeed, Proposition 14.6 holds for $n \geq 4$ for $t = e^{2\pi i/10}$.

15. Braid index; bridge number; tables

THEOREM 15.1 (Morton, Franks-Williams [28], [47]). *Let L be an oriented link with polynomial $X_L(q, \lambda)$. Let d_+ be the degree of the largest power of λ in X_L and d_- be the smallest. If $\alpha \in B_n$ has $\hat{\alpha} = L$ then*

(MFW) $$n \geq d_+ - d_- + 1.$$

Proof. Let α have exponent sum e. Then

$$X_L = \left(-\frac{1 - \lambda q}{\sqrt{\lambda}(1 - q)} \right)^{n-1} (\sqrt{\lambda})^e \operatorname{tr}(\pi(\alpha)).$$

By induction on the Markov property $\operatorname{tr}(\pi(\alpha))$ is an honest polynomial in $z = -(1 - q)/(1 - \lambda q)$. In particular for fixed q it has a finite limit as $\lambda \to \infty$. Thus $|X_L| \leq (\text{const})\lambda^{(n+e-1)/2}$ as $\lambda \to \infty$. So $d \leq (n + e - 1)/2$. Considering α^{-1} and by Proposition 6.10, $-d_- \leq (n - e - 1)/2$. Thus $d_+ - d_- \leq n - 1$. Q.E.D.

The *braid index* of L is by definition the smallest n for which there is an $\alpha \in B_n$ with $\hat{\alpha} = L$.

The MFW inequality is a wonderful thing. A careful look at the proof of Theorem 16.1 shows that the inequality, viewed as a lower bound for the braid index, should be fairly good. But the author was totally unprepared for what he found in compiling Table 15.9. Of the more than 270 knots on ten crossings or less up to symmetry, the MFW inequality is sharp on all but five (9_{42}, 9_{49}, 10_{150}, 10_{132} and 10_{156}) of them! In the Lickorish-Millet variables d_+ and d_- are the largest and smallest powers of l^2. The MFW lower bound can be read immediately off the tables, for instance, in the notation of [22],

$$P_{9_{22}} = (-1 - 4[-4] - 2)(2\,6\,[6]\,1)(-1 - 4[-2])(1\,[0]),$$

so the braid index of 9_{22} is ≥ 4 since there are four non-zero terms in the first $(\ldots)$. In an interesting improvement of the MFW inequality, Morton shows in [29] that the number $d_+ - d_- + 1$ is in fact a lower bound for the number of Seifert circles in any regular projection of L.

It is useful to have other tools available in case the MFW inequality fails. A convenient one comes from Section 8.

PROPOSITION 15.2. *If K is a knot and $|\Delta_K(i)| > 3$, then K is not a closed 3-braid (see 23 of [16]).*

Proof. By (12.6), $V_K(i) = \pm 1$. So by (8.4) and (11.2), $|1 + i^e - i^{e/2}\Delta(i)| = 1$ if K is a closed 3-braid; so we would have $|\Delta_K(i)| \leq 3$. Q.E.D.

The basic inequality coming from positivity is the following.

PROPOSITION 15.3. *If L is a closed n-braid then*

$$\left|V_L(e^{2\pi i/k})\right| \leq (2\cos\pi/k)^{n-1}, \qquad k = 3,4,5,\ldots.$$

Proof. See 14.6.

COROLLARY 15.4. *If $\alpha \in \ker \pi_0$, $\alpha \in B_n$, for $t = e^{2\pi i/k}$, $k = 3,4,5,\ldots$, then the braid index of α is n.*

Proof. $|V_{\hat{\alpha}}(e^{2\pi i/k})| = (2\cos\pi/k)^{n-1}$; so by Proposition 15.3, $\hat{\alpha}$ cannot be a closed p-braid for $p < n$. Q.E.D.

COROLLARY 15.5. *Suppose $\alpha \in B_n$ is a product of conjugates of terms of the form $\sigma_i^{k_i}$ with $\mathrm{GCD}(k_i) \geq 2$. Then the braid index of $\hat{\alpha}$ is n.*

Proof. If $\mathrm{GCD}(k_i)$ is even then $\hat{\alpha}$ is an n-component link, so obviously has braid index $\leq n$; otherwise $\pi_0(\alpha) = \pm 1$. So by 16.3 we are through, as in Corollary 15.4. Q.E.D.

By [6], the minimal bridge number of a link L is the smallest n for which L is of the form $\tilde{\alpha}$ for $\alpha \in B_{2n}$. We have been unable to find sharp general results on the bridge number using V or P but Proposition 14.1 does give results formally the same as Proposition 15.3 and Corollaries 15.4, 15.5.

PROPOSITION 15.6. *Suppose L is a plat on $2n$ strings; then*

$$\left|V_L(e^{2\pi i/k})\right| \leq (2\cos\pi/k)^{n-1}.$$

Proof. Imitate 15.3 using ϕ instead of tr. Q.E.D.

COROLLARY 15.7. *If $a \in B_{2n}$ is in $\ker \pi_0$ for $t = e^{2\pi i/k}$, $k = 3,4,5,\ldots$, then the bridge number of $\tilde{\alpha}$ is n.*

Proof. See Corollary 15.4. Q.E.D.

COROLLARY 15.8. *Suppose $\alpha \in B_{2n}$ is a product of conjugates of terms of the form $\sigma_i^{k_i}$ with $\mathrm{GCD}(k_i) \geq 2$. Then the bridge number of $\tilde{\alpha}$ is n.*

Proof. See Corollary 15.5. Q.E.D.

TABLE 15.9. The following table gives the braid index, a braid expression, amphicheirality information, and the W polynomial of Proposition 12.5 for all unoriented prime knots up to 10 crossings. The last column may also contain a reference number for any special comments, which appear after the table. For amphicheirality, the last column is left blank if the knot is not amphicheiral and this is detected by V. An entry "A" means the knot is amphicheiral and "N" means that it is not, but V fails to detect it (so also does P for knots on ≤ 10 crossings).

Some care was taken to ensure that the braid words are as simple as possible but no adequate technique is yet available to give proofs except in special cases. Less care was taken with the 10-crossing knots.

The knot enumeration is as in [13], [37] with one or two corrections. Either the knot drawn in [37] or its mirror image is recorded, chosen according to whose V has the least power of t^{-1}. No confusion need arise as the braid word completely specifies the knot. Braid indices come from the MFW inequality unless otherwise indicated.

All coincidences of V are recorded.

The following example explains how to read the table: The knot 8_8 has W polynomial $t^{-3}(1 - t + 2t^2 - t^3 + t^4)$.

Knot		Braid Word	$P_0(W)$	W	A/N
3_1	2	1^3	0	1	
4_1	3	$12^{-1}12^{-1}$	-2	-1	
5_1	2	1^5	0	1101	4
5_2	3	$1^2 2^2 1^{-1} 2$	0	101	
6_1	4	$1^{-1}21^{-1}32^{-1}32$	-2	$-10-1$	
6_2	3	$1^{-1}21^{-1}2^3$	-1	$-11-1$	
6_3	3	$1^{-1}2^2 1^{-2} 2$	-3	$1-11$	A
7_1	2	1^7	0	1111101	
7_2	4	$1^{-1}3^3 21^2 3^{-1} 2$	0	10101	
7_3	3	$1^2 21^{-1} 2^4$	0	110201	
7_4	4	$1^2 23^2 1^{-1} 23^{-1} 2$	0	10201	
7_5	3	$1^4 21^{-1} 2^2$	0	$1102-11$	
7_6	4	$12^{-1}1^{-2}32^3 3$	-1	$-12-11$	
7_7	4	$13^{-1}23^{-1}21^{-1}23^{-1}2$	-3	$1-21-1$	
8_1	5	$1^{-1}232^{-1}1^{-1}4^2 324^{-1}$	-2	$-10-10-1$	
8_2	3	$1^{-1}2^5 1^{-1}2$	1	$1-11-1$	
8_3	5	$1^{-2}2^{-1}14^2 34^{-1}2^{-1}3$	-4	$-10-20-1$	A
8_4	4	$1^3 32^{-1}3^{-2}12^{-1}$	-3	$-10-21-1$	
8_5	3	$1^3 2^{-1}1^3 2^{-1}$	1	$1-21-1$	
8_6	4	$1^{-1}21^{-1}3^{-1}2^3 3^2$	-1	$-11-21-1$	
8_7	3	$1^4 2^{-2}12^{-1}$	-2	$1-12-11$	

Knot		Braid Word	$P_0(W)$	W	A/N
8_8	4	$1^{-1}21^23^{-1}2^23^{-2}$	−3	1 − 12 − 11	5
8_9	3	$1^{-1}21^{-3}2^3$	−4	−11 − 21 − 1	A
8_{10}	3	$1^{-1}2^21^{-2}2^3$	−2	1 − 13 − 11	
8_{11}	4	$1^{-1}2^23^{-1}23^21^{-1}2$	−1	−12 − 21 − 1	
8_{12}	5	$12^{-1}34^{-1}34^{-1}213^{-1}2^{-1}$	−4	−11 31 − 1	A
8_{13}	4	$1^223^{-1}21^{-1}3^{-2}2$	−3	1 − 22 − 11	
8_{14}	4	$1^22^21^{-1}3^{-1}23^{-1}2$	−1	−12 − 22 − 1	
8_{15}	4	$1^22^{-1}13^22^23$	0	1103 − 22 − 1	
8_{16}	3	$1^22^{-1}1^22^{-1}12^{-1}$	−2	1 − 23 − 21	
8_{17}	3	$1^{-1}21^{-1}2^21^{-2}2$	−4	−12 − 32 1	A
8_{18}	3	$12^{-1}12^{-1}12^{-1}12^{-1}$	−4	$-13-33^{-1}$	A
8_{19}	3	121212^21	0	11111	
8_{20}	3	$1^321^{-3}2$	−1	101	
8_{21}	3	$12^{-2}1^22^3$	0	1 − 11 − 1	
9_1	2	1^9	0	1112111101	
9_2	5	$1234^334^{-1}213^{-1}2^{-1}$	0	1010101	
9_3	3	$12^{-1}1^62^2$	0	111120201	
9_4	4	$1^{-1}321^223^42^{-1}$	0	11020201	
9_5	5	$1^221^{-1}32^{-1}34^234^{-1}2$	0	1020201	
9_6	3	$1^22^21^52^{-1}$	0	11112 − 12 − 11	
9_7	4	$1^323^21^{-1}2^33^{-1}$	0	1102 − 12 − 11	
9_8	5	$12^{-1}31^{-2}4^{-1}32^{-1}3^243$	−3	−11 − 22 − 11	
9_9	3	$1^32^{-1}1^42^2$	0	11112 − 13 − 11	
9_{10}	4	$1^{-1}21^22^33^{-1}23^2$	0	1103 − 1301	
9_{11}	4	$1^{-1}23^{-1}21^{-1}2^432$	1	2 − 13 − 11	
9_{12}	5	$12^{-1}1^{-2}32^34^234^{-1}$	−1	−12 − 22 − 11	
9_{13}	4	$1^223^{-1}21^{-1}3^22^3$	0	1103 − 13 11	
9_{14}	5	$14^23^{-1}23^{-1}23^{-1}1^{-1}4^{-1}2^232^{-1}$	−3	1 − 22 − 21 − 1	
9_{15}	5	$12^{-1}132^{-1}43^{-1}4^23$	−1	−12 − 23 − 11	
9_{16}	3	$1^22^{-1}1^32^4$	0	11112 − 23 − 21	
9_{17}	4	$1^332^{-1}12^{-1}3^{-1}2^{-1}12^{-1}$	−3	−11 − 32 − 21	
9_{18}	4	$1^23^{-1}21^{-1}2^23^22^2$	0	1103 − 23 − 11	
9_{19}	5	$1^221^{-1}3^{-1}4^{-1}3^{-1}23^{-1}243^{-1}$	−4	−11 − 32 − 21	
9_{20}	4	$12^23^{-1}21^{-1}23^{-1}2^3$	1	2 − 23 − 21	
9_{21}	5	$1^{-2}34^{-1}32^{-1}134^22^2$	−1	−13 − 23 − 11	
9_{22}	4	$12^{-1}3^32^{-1}31^{-1}2^{-1}32^{-1}$	−3	−11 − 33 − 21	
9_{23}	4	$12^{-1}1^223^323^{-1}2$	0	1103 − 23 − 21	
9_{24}	4	$13^22^{-1}132^{-3}$	−4	−12 − 33 − 11	
9_{25}	5	$1^{-1}21^{-1}4^{-1}3^{-1}24^234^22^23^{-1}$	−1	−12 − 33 − 21	
9_{26}	4	$1^{-1}2^3321^{-1}21^{-1}3^{-1}2$	−2	1 − 23 − 32 − 1	
9_{27}	4	$1^{-1}21^{-2}3^{-1}21^{-1}2^232$	−4	−12 − 33 − 21	6
9_{28}	4	$1^23^22^{-2}132^{-1}$	−2	1 − 24 − 32 − 1	
9_{29}	4	$1^{-1}23^{-1}21^{-1}23^{-1}2^2$	−3	−12 − 34 − 21	
9_{30}	4	$13^{-2}23^{-1}1^{-1}21^23^{-1}2$	−4	−12 − 43 − 21	
9_{31}	4	$1^{-1}2^23^{-1}2^21^{-1}231^{-1}2$	−2	1 − 24 − 33 − 1	
9_{32}	4	$1^{-1}231^{-1}21^{-1}3^223^{-1}2$	−2	1 − 34 − 42 − 1	
9_{33}	4	$1^{-3}21^{-1}2^2312^{-1}3$	−4	−13 − 44 − 21	
9_{34}	4	$12^{-1}32^{-1}12^{-1}312^{-1}$	−4	−13 − 54 − 31	
9_{35}	5	$13^24^{-1}2^{-1}123^223^{-1}43^{-1}2$	0	1020301	

Knot		Braid Word	$P_0(W)$	W	A/N
9_{36}	4	$1^{-1}2^{3}3^{-1}23^{2}1^{-1}23$	1	2 − 23 − 11	
9_{37}	5	$12^{-1}32^{-1}31^{-1}43^{-1}2^{-1}432^{-1}$	− 4	− 11 − 42 − 21	
9_{38}	4	$1^{2}23^{2}2^{2}1^{-1}23^{-1}2$	0	1104 − 34 − 21	
9_{39}	5	$1^{-1}3^{-1}243^{2}1^{-1}2^{2}342^{-1}$	− 1	− 13 − 34 − 21	
9_{40}	4	$132^{-1}312^{-1}132^{-1}$	− 2	1 − 45 − 53 − 1	
9_{41}	5	$1^{-1}2^{2}4^{-1}34^{-1}2^{-1}13^{-1}234^{-1}32$	− 3	1 − 23 − 32 − 1	
9_{42}	4	$1^{3}32^{-1}31^{-2}2^{-1}$	− 3	− 10 − 1	N, 1
9_{43}	4	$121^{2}2^{2}32^{-1}12^{-1}3^{-1}$	1	1 − 11	
9_{44}	4	$1^{-1}21^{-1}32^{-1}32^{2}3^{-1}$	− 2	− 11 − 11	
9_{45}	4	$12^{-1}132^{3}32^{-1}$	0	1 − 12 − 11	
9_{46}	4	$1321^{-1}3^{-1}2132^{-1}$	0	− 10 − 1	
9_{47}	4	$1^{-1}231^{-1}21^{-1}232$	− 2	1 − 22 − 2	
9_{48}	4	$1^{2}2^{-1}32^{2}1^{-1}3^{-1}23^{-1}2$	− 1	− 13 − 12	
9_{49}	4	$1^{2}2^{2}321^{-1}2^{2}32^{-1}$	0	1103 − 12	1
10_{1}	6	$1^{2}4^{-1}5^{-1}3^{-1}21^{-1}345^{-1}432$	− 2	− 10 − 10 − 10 − 1	
10_{2}	3	$12^{-1}1^{7}2^{-1}$	0	10101 − 11 − 1	
10_{3}	6	$13^{-1}2^{-1}15^{-2}234^{-1}5324^{-1}$	− 4	− 10 − 20 − 20 − 1	
10_{4}	5	$1^{2}4^{-3}232^{-1}1^{-1}4^{-1}32$	− 5	− 11 − 20 − 20 − 1	
10_{5}	3	$1^{-2}21^{-1}2^{6}$	− 1	102 − 12 − 11	
10_{6}	4	$12^{-1}31^{-1}2^{-1}1^{-1}32^{6}$	1	1 − 22 − 21 − 1	
10_{7}	5	$13^{-1}2^{-1}1234^{-1}34^{-1}3^{2}2$	− 1	− 12 − 22 − 21 − 1	
10_{8}	4	$1^{5}32^{-1}13^{-2}2^{-1}$	− 2	− 10 − 11 − 21 − 1	
10_{9}	3	$1^{-3}21^{-1}2^{5}$	− 3	− 11 − 22 − 21 − 1	
10_{10}	5	$1^{-1}3^{-1}43^{2}21^{2}3^{-1}24^{-1}3^{-1}23^{-1}$	− 3	1 − 22 − 22 − 11	
10_{11}	5	$1^{-2}3^{-1}2^{-1}143^{3}2^{-1}34$	− 3	− 10 − 32 − 31 − 1	
10_{12}	4	$1^{-2}21^{-1}3^{2}2^{4}3^{-1}$	− 2	1 − 13 − 23 − 11	
10_{13}	6	$1^{2}3^{-1}21^{-1}3^{-1}45^{2}435^{-1}2^{-2}4^{-1}$	− 4	− 11 − 42 − 31 − 1	
10_{14}	4	$1^{-1}21^{-1}3^{2}2^{4}3^{-1}2$	1	2 − 24 − 32 − 1	
10_{15}	4	$1^{-1}32^{-1}3^{4}2^{-1}1^{-2}2$	− 4	1 − 12 − 23 − 11	
10_{16}	5	$12^{-1}1^{-2}32^{-1}4^{-1}34^{2}3^{2}$	− 3	− 11 − 32 − 31 − 1	
10_{17}	3	$1^{4}2^{-1}12^{-4}$	− 5	1 − 12 − 22 − 11	A
10_{18}	5	$13^{2}4^{2}3^{-1}2^{-1}41^{-2}32^{-1}$	− 3	− 11 − 33 − 32 − 1	
10_{19}	4	$1^{-2}2^{-1}13^{4}2^{-1}32^{-1}$	− 4	1 − 12 − 33 − 21	
10_{20}	5	$1^{-1}3^{-1}2^{3}34^{2}34^{-1}1^{-1}2$	− 1	− 11 − 21 − 21 − 1	
10_{21}	4	$1^{-1}21^{-1}3^{2}23^{-1}2^{4}$	1	2 − 23 − 21 − 1	
10_{22}	4	$1^{-3}23^{2}1^{-1}2^{3}3^{-1}$	− 4	− 11 − 32 − 31 − 1	8
10_{23}	4	$1^{-1}21^{-1}2^{3}3^{-1}21^{-1}3^{2}$	− 2	1 − 24 − 33 − 11	
10_{24}	5	$1^{-1}23^{-1}2^{2}1^{2}234^{-1}34^{-1}$	− 1	− 12 − 33 − 31 − 1	
10_{25}	4	$1^{-1}3^{2}2^{3}3^{-1}2^{2}1^{-1}2$	1	2 − 34 − 42 − 1	7
10_{26}	4	$12^{2}3^{-1}2^{-3}12^{-1}1^{2}32^{-1}$	− 4	− 12 − 43 − 31 − 1	
10_{27}	4	$1^{-2}23^{-1}23^{4}1^{-1}2$	− 2	1 − 24 − 44 − 21	
10_{28}	5	$1^{2}23^{-1}21^{-1}4^{-1}34^{-2}32$	− 3	1 − 23 − 23 − 11	
10_{29}	5	$12^{-1}34^{-1}32^{-1}1^{-1}2^{-1}32^{-1}43^{3}$	− 3	− 11 − 43 − 42 − 1	
10_{30}	5	$1^{2}23^{-1}21^{-1}3^{2}43^{-1}23^{-1}4^{-1}2$	− 1	− 13 − 34 − 32 − 1	
10_{31}	5	$1^{2}2^{2}1^{-1}43^{-1}4^{-2}3^{-1}23^{-1}$	− 5	1 − 13 − 33 − 21	
10_{32}	4	$1^{-1}23^{-1}21^{2}3^{-3}2^{2}$	− 4	− 12 − 44 − 32 − 1	
10_{33}	5	$12^{-1}34^{-1}32^{-1}1^{-2}2^{-1}34^{2}$	− 5	1 − 23 − 43 − 21	A
10_{34}	5	$1^{-1}4^{-2}3^{2}21^{2}4^{-1}3^{-1}23$	− 3	1 − 12 − 12 − 11	
10_{35}	6	$1^{-1}23^{-1}23^{-1}45^{-1}45^{-1}12^{-1}34^{-1}2^{2}$	− 4	− 11 − 32 − 31 − 1	8

Knot		Braid Word	$P_0(W)$	W	A/N
10_{36}	5	$1^23^{-1}2^234^{-1}34^{-1}1^{-1}32$	-1	$-12-23-22-1$	
10_{37}	5	$143^{-1}4^{-2}23^{-2}21^22^{-1}$	-5	$1-13-33-11$	A
10_{38}	5	$1^243^{-1}2^{-2}3^34^23^{-1}1^{-1}2$	-1	$-12-33-32-1$	
10_{39}	4	$1^{-1}21^{-1}2^33^{-1}23^3$	1	$2-34-32-1$	
10_{40}	4	$1^221^{-1}2^23^{-1}213^{-2}$	-2	$1-25-44-21$	
10_{41}	5	$12^{-1}32^{-1}1^{-1}4^{-1}32^{-1}3432^{-1}3^2$	-3	$-12-44-42-1$	9
10_{42}	5	$12^{-1}1^{-2}34^{-1}2^234^{-2}3$	-5	$1-34-54-21$	
10_{43}	5	$1^{-1}23^{-1}4^{-1}23^{-2}123^{-1}4^{-1}32^2$	-5	$1-24-44-21$	A, 10
10_{44}	5	$1^{-1}2^{-1}32^{-1}34^2123^{-1}2^{-2}43$	-3	$-12-45-43-1$	
10_{45}	5	$13^{-1}23^{-1}243^{-1}1^{-1}23^{-1}24^{-1}3^{-1}2$	-5	$1-34-64-31$	A
10_{46}	3	$1^{-1}2^51^{-1}2^3$	0	$101-11-21-1$	
10_{47}	3	$1^52^{-1}1^22^{-2}$	-1	$103-13-11$	
10_{48}	3	$1^{-2}2^41^{-3}2$	-5	$1-13-23-11$	N
10_{49}	4	$1^432^3312^{-1}$	0	$11113-24-32-1$	
10_{50}	4	$1^{-1}2^23^{-1}23^21^{-1}2^3$	1	$2-33-31-1$	
10_{51}	4	$1^223^{-2}1^{-1}2^23^{-1}2^2$	-2	$1-25-34-11$	
10_{52}	4	$1^{-2}2^{-1}3^22^{-1}3^312^{-1}$	-4	$1-13-34-21$	
10_{53}	5	$12^{-1}3^2421^{-1}3^22^{-1}31^{-1}4^22^2$	0	$1104-35-32-1$	
10_{54}	4	$1^{-2}2^{-1}3^22^{-1}3^31^{-1}2$	-4	$1-13-23-11$	
10_{55}	5	$1^{-1}43421^323^24^{-1}32^{-1}$	0	$1103-24-32-1$	
10_{56}	4	$1^{-1}2^23^22^{-1}31^{-1}2^3$	1	$2-34-42-1$	7
10_{57}	4	$1^323^{-2}1^{-1}2^23^{-1}2$	-2	$1-25-45-21$	
10_{58}	6	$125^{-1}43^{-1}23^{-1}4^{-1}5^{-1}12^{-1}34$	-4	$-11-43-42-1$	
10_{59}	5	$12^241^{-2}3^{-1}43^{-1}2^{-1}34^23^{-1}$	-3	$-12-45-42-1$	11
10_{60}	5	$1^{-2}2^212^{-1}34^{-1}32^{-1}32^{-1}43$	-4	$-13-55-42-1$	12
10_{61}	4	$12^{-1}1^{-2}3^32^{-1}3^3$	-2	$-10-21-21-1$	
10_{62}	3	$1^{-2}2^31^{-1}2^4$	-1	$103-23-11$	
10_{63}	5	$1^223^21^{-1}43^{-1}2^{-1}34^232$	0	$1103-24-22-1$	
10_{64}	3	$1^{-3}2^31^{-1}2^3$	-3	$-11-33-31-1$	
10_{65}	4	$132^231^{-1}2^{-1}3^31^{-1}2^{-2}$	-2	$1-24-34-11$	
10_{66}	4	$1^223^{-1}2^212^{-1}3^42$	0	$11113-35-43-1$	
10_{67}	5	$1^{-1}24^{-1}3^{-1}24^{-1}1^23^22^243^{-1}$	-1	$-12-34-32-1$	
10_{68}	5	$1^{-1}23^{-1}4^21^22^{-1}32^24^{-1}3^{-2}$	-3	$1-23-33-11$	
10_{69}	5	$13423^{-1}42^{-2}1^{-2}32^3$	-2	$1-35-55-21$	
10_{70}	5	$12^{-1}3^31^{-2}2^24^{-1}34^{-1}$	-3	$-11-44-42-1$	
10_{71}	5	$1^24^{-2}2^{-1}3^{-1}2^23^{-1}4^{-1}231^{-1}2$	-5	$1-24-54-21$	13, N
10_{72}	4	$1^32^{-1}1^223^{-1}23^{-1}2$	1	$2-35-43-1$	
10_{73}	5	$12^{-1}3^214^{-1}32^{-1}3432^{-1}$	-2	$1-35-54-21$	14
10_{74}	5	$1^{-1}423^{-1}24^{-1}3^{-1}21^24^{-1}3^22$	-1	$-13-34-31-1$	
10_{75}	5	$12^{-1}132^{-1}3432^{-1}34^{-1}2^{-1}$	-4	$-13-45-42-1$	
10_{76}	4	$1^22^{-1}1^{-1}3^32^{-1}12^{-1}3^3$	1	$1-33-42-1$	
10_{77}	4	$1^{-1}2^33^{-2}1^22^23^{-1}$	-2	$1-14-34-21$	
10_{78}	5	$1^{-2}432^{-1}13^24^{-1}32^33$	1	$3-35-32-1$	
10_{79}	3	$1^32^{-2}1^22^{-3}$	-5	$1-14-34-11$	A
10_{80}	4	$1^223^{-1}2^21^{-1}2^332^2$	0	$11113-35-42-1$	
10_{81}	5	$1^{-2}2^{-2}1^{-1}3^{-1}24^23^24$	-5	$1-25-55-21$	15, A
10_{82}	3	$1^42^{-2}12^{-1}12^{-1}$	-3	$-12-34-32-1$	
10_{83}	4	$1^{-1}23^{-1}23^{-2}21^223^{-1}$	-4	$-13-55-42-1$	12, 2
10_{84}	4	$1^32^23^{-1}21^{-1}3^{-1}23^{-1}$	-2	$1-25-55-31$	

Knot		Braid Word	$P_0(W)$	W	A/N
10_{85}	3	$1^{-1}2^{2}1^{-1}21^{-1}2^{4}$	−1	1 − 13 − 33 − 21	
10_{86}	4	$1^{2}23^{-1}2^{2}3^{-1}21^{-1}3^{-1}2$	−2	1 − 35 − 54 − 21	2, 14
10_{87}	4	$1^{3}3^{-1}23^{-1}1^{-1}23^{-1}23^{-1}$	−4	− 12 − 45 − 43 − 1	
10_{88}	5	$1^{2}2^{-1}4^{-1}312^{-1}32^{-1}43^{-1}1^{-1}23^{-1}$	−5	1 − 35 − 75 − 31	A
10_{89}	5	$123^{-1}21^{-1}423^{-1}243^{-1}4$	−2	1 − 46 − 65 − 21	
10_{90}	4	$1^{2}32^{-2}1^{2}2^{-1}12^{-1}31^{-1}2^{-1}$	−4	− 12 − 54 − 42 − 1	
10_{91}	3	$1^{3}2^{-2}12^{-2}12^{-1}$	−5	1 − 24 − 44 − 21	N, 10
10_{92}	4	$1^{3}2^{2}3^{-1}21^{-1}23^{-1}2$	1	3 − 46 − 53 − 1	
10_{93}	4	$1^{-2}3^{2}2^{-1}312^{-1}3^{2}2^{-1}$	−4	1 − 23 − 44 − 21	
10_{94}	3	$1^{3}2^{-1}1^{2}2^{-2}12^{-1}$	−3	− 12 − 44 − 42 − 1	9
10_{95}	4	$1^{-2}23^{-1}21^{-1}2^{2}3^{2}2$	−2	1 − 36 − 55 − 21	
10_{96}	5	$123^{-1}42132^{-1}3^{-2}43^{-1}$	−4	− 13 − 65 − 52 − 1	
10_{97}	5	$1^{2}2^{-1}34^{-1}2^{2}1^{-1}3^{-1}23^{2}4^{-1}2$	−1	− 13 − 46 − 43 − 1	
10_{98}	4	$1^{-1}2^{2}3^{2}21^{-1}2^{2}3^{-1}2$	1	3 − 45 − 52 − 1	
10_{99}	3	$1^{-2}21^{-2}2^{2}1^{-1}2^{2}$	−5	1 − 25 − 45 − 21	A
10_{100}	3	$1^{3}2^{-1}1^{2}2^{-1}1^{2}2^{-1}$	−1	1 − 14 − 34 − 21	
10_{101}	5	$1^{-1}32^{2}1432^{3}3^{2}43^{-1}1^{-1}2^{-1}$	0	1104 − 36 − 43 − 1	
10_{102}	4	$12^{2}3^{-1}21^{-1}3^{-1}213^{-2}$	−4	− 12 − 44 − 42 − 1	
10_{103}	4	$1^{2}2^{2}3^{-1}2^{2}3^{-1}1^{-1}23^{-1}$	−2	1 − 25 − 44 − 21	
10_{104}	3	$1^{2}2^{-3}1^{2}2^{-1}12^{-1}$	−5	1 − 24 − 54 − 21	13, N
10_{105}	5	$1^{-1}2^{-1}32^{-1}4132^{-1}3^{2}4^{2}3^{-1}2^{-1}$	−3	− 12 − 56 − 53 − 1	
10_{106}	3	$1^{3}2^{-2}1^{2}2^{-1}12^{-1}$	−3	− 12 − 45 − 42 − 1	11
10_{107}	5	$1432^{-1}3^{2}43^{-1}2^{-2}1^{-1}2^{-1}32^{-1}$	−5	1 − 35 − 65 − 21	
10_{108}	4	$1^{-2}3^{2}2^{-1}13^{2}2^{-1}32^{-1}$	−4	1 − 23 − 43 − 21	
10_{109}	3	$1^{-2}2^{2}1^{-2}2^{2}1^{-1}2$	−5	1 − 25 − 55 − 21	15, A
10_{110}	5	$13^{-1}23^{-1}4^{-1}32^{3}12^{-1}3^{-1}4^{-1}2$	−3	− 12 − 55 − 52 − 1	
10_{111}	4	$1^{2}2^{2}3^{-1}2^{2}1^{-1}23^{-1}2$	1	3 − 45 − 42 − 1	
10_{112}	3	$1^{3}2^{-1}12^{-1}12^{-1}12^{-1}$	−3	− 13 − 46 − 43 − 1	
10_{113}	4	$1^{3}23^{-1}21^{-1}23^{-1}23^{-1}$	−2	1 − 36 − 76 − 41	
10_{114}	4	$1^{-1}2^{-1}32^{-1}32^{-2}1^{2}3^{-1}23^{2}$	−4	− 13 − 56 − 43 − 1	
10_{115}	5	$1^{-1}234^{-1}32^{-1}1^{-1}4342^{-1}32^{-2}$	−5	1 − 36 − 76 − 31	A
10_{116}	3	$1^{2}2^{-1}12^{-1}12^{-1}1^{2}2^{-1}$	−3	− 13 − 56 − 53 − 1	
10_{117}	4	$1^{-1}23^{-1}21^{2}23^{-1}2^{2}3^{-1}$	−2	1 − 36 − 66 − 31	
10_{118}	3	$1^{2}2^{-1}12^{-2}12^{-1}12^{-1}$	−5	1 − 35 − 65 − 31	A
10_{119}	4	$1^{2}23^{-2}21^{-1}3^{-1}23^{-1}2$	−4	− 13 − 66 − 53 − 1	
10_{120}	5	$12^{2}34^{-1}132^{-1}31^{-1}4^{2}3^{2}23^{-1}$	0	1105 − 57 − 53 − 1	
10_{121}	4	$1^{-1}23^{-1}23^{-1}21^{2}23^{-1}2$	−2	1 − 47 − 76 − 31	
10_{122}	4	$12^{-1}32^{-1}32^{-1}1^{-1}2^{-1}321^{-1}2^{2}$	−4	− 13 − 57 − 54 − 1	
10_{123}	3	$12^{-1}12^{-1}12^{-1}12^{-1}12^{-1}$	−5	1 − 46 − 86 − 41	A
10_{124}	3	$12^{5}12^{3}$	0	1112111	
10_{125}	3	$1^{-1}2^{-3}1^{-1}2^{5}$	−4	10101	N
10_{126}	3	$12^{-3}12^{5}$	0	20201	
10_{127}	3	$1^{5}21^{-2}2^{2}$	0	11 − 12 − 21 − 1	
10_{128}	4	$1^{2}32^{2}32^{2}1^{-1}32$	0	1111201	
10_{129}	4	$1^{-2}32^{-2}32^{2}132^{-1}$	−3	1 − 12 − 11	5
10_{130}	4	$1^{-1}23^{-3}21^{2}23^{2}$	−1	10201	
10_{131}	4	$1^{-1}23^{3}21^{2}23^{-2}$	0	1 − 12 − 21 − 1	
10_{132}	4	$1^{-1}231^{3}2^{-3}32$	0	1101	1, 4, 18
10_{133}	4	$1^{-1}323^{2}21^{2}3^{-1}2^{-1}13^{-1}2$	0	101 − 11 − 1	

Knot		Braid Word	$P_0(W)$	W	A/N
10_{134}	4	$12^212^3321^{-1}2^23^{-1}$	0	11112 − 12 − 1	
10_{135}	4	$12^232^{-1}1^{-3}2^{-1}3^2$	− 3	2 − 23 − 11	
10_{136}	4	$1^223^{-1}21^{-1}2^23^{-1}2^{-2}$	− 3	− 11 − 11	
10_{137}	5	$1^232^{-1}12^{-1}31^{-1}4^{-1}32^{-1}32^{-1}4$	− 2	− 11 − 21 − 1	16
10_{138}	5	$1^232^{-1}12^{-1}341^{-1}3^{-1}23^{-1}24^{-1}$	− 3	− 11 − 32 − 2	
10_{139}	3	$1^22^31^2212$	0	111211101	
10_{140}	4	$12^{-1}13^{-3}23^32$	1	101	
10_{141}	3	$1^{-3}21^22^{-1}1^22$	− 2	− 11 − 11 − 1	
10_{142}	4	$1^{-1}3^323^31^22$	0	1111202	
10_{143}	3	$1^{-1}2^21^{-2}21^22^2$	0	2 − 12 − 11	
10_{144}	4	$1^232^21^{-1}23^{-2}12^{-1}$	− 1	− 22 − 32 − 1	17
10_{145}	4	$1^221^{-1}3^{-1}2132^23$	0	1101101	
10_{146}	4	$1^22^{-1}3^{-2}2^212^{-1}32^{-1}$	− 3	1 − 22 − 21	
10_{147}	4	$1^{-1}2^{-2}3^22^21^22^{-1}1^{-1}3^{-1}2$	− 3	− 11 − 22 − 1	
10_{148}	3	$1^22^31^{-2}21^{-1}2$	0	2 − 13 − 11	
10_{149}	3	$1^{-2}2^31^32^2$	0	11 − 13 − 32 − 1	
10_{150}	4	$1^22^2321^{-1}2^{-1}3^22^{-1}$	1	2 − 22 − 1	18
10_{151}	4	$1^23^{-1}2^2132^{-2}3^{-1}2$	− 2	1 − 24 − 22	
10_{152}	3	$1^22^31^32^2$	0	1112111 − 11 − 1	
10_{153}	4	$1^223^{-2}2^{-2}13^{-1}2^2$	− 4	101101	
10_{154}	4	$12^{-1}3^22^21^232^2$	0	111111 − 11 − 1	
10_{155}	3	$1^321^{-2}21^{-2}2$	− 2	− 11 − 21 − 1	16
10_{156}	4	$1^{-1}232^231^{-2}23^{-1}2$	− 2	1 − 23 − 21	18
10_{157}	3	$1^{-1}2^21^22^{-1}1^22^2$	0	11 − 14 − 33 − 1	
10_{158}	4	$1^2231^{-1}2^{-2}32^{-2}3$	− 4	− 12 − 42 − 2	
10_{159}	3	$12^{-1}12^{-1}12^312^{-1}$	0	2 − 23 − 21	
10_{160}	4	$123^21^22^{-1}12^{-1}13^{-1}$	1	2 − 12	
10_{161}	3	$1^{-1}21^22^31^22$	0	11111101	3
10_{163}	4	$1^{-2}32^{-1}3^212^232^{-1}$	− 1	− 22 − 31 − 1	
10_{164}	4	$12^{-1}31^{-1}21^22^{-1}12^{-1}3$	− 2	1 − 34 − 32	
10_{165}	4	$1^{-1}23^{-2}12^{-1}12^{-1}132$	− 3	2 − 33 − 21	
10_{166}	4	$1^223^{-1}21^{-1}23^21^{-1}2$	0	1 − 13 − 22 − 1	

Notes:

1. MFW braid index inequality not sharp. Answer, if known, from [33].
2. These two knots are interchanged from Rolfsen's table, to make the Alexander polynomials correct.
3. This is the same as 10_{162} of [37].
4. Compare $5_1 - 10_{132}$ same V, same P.
5. Compare $8_8 - 10_{129}$ same V, same P.
6. Compare $8_{16} - 10_{156}$ same V, same P.
7. Compare $10_{25} - 10_{56}$ same V, same P.
8. Compare $10_{22} - 10_{35}$ same V, different Δ.
9. Compare $10_{41} - 10_{94}$ same V, different Δ.
10. Compare $10_{43} - 10_{91}$ same V, different Δ.
11. Compare $10_{59} - 10_{106}$ same V, different Δ.
12. Compare $10_{60} - 10_{83}$ same V, different Δ.
13. Compare $10_{71} - 10_{104}$ same V, different Δ.
14. Compare $10_{73} - 10_{86}$ same V, different Δ.
15. Compare $10_{81} - 10_{109}$ same V, different Δ.
16. Compare $10_{137} - 10_{155}$ same V, different Δ.
17. The uppermost crossing of [37] has been changed.
18. Braid index obtained by Morton by showing braid index of a 2 cable = 8.

University of California, Berkeley

References

[1] C. Akemann, J. Anderson, and G. Pedersen, Excising states of C^*-algebras, MSRI preprint (1985).

[2] J. Alexander, Topological invariants of knots and links, Trans. AMS **30** (1928), 275–306.

[3] W. Arveson, *An Invitation to C^*-Algebra*, Graduate Texts in Mathematics. (Springer-Verlag, 1976).

[4] R. Baxter, *Exactly Solved Models in Statistical Mechanics*, Academic Press, London, 1982.

[5] D. Bennequin, Entrelacements et structures de contact, Thèse, Paris (1982).

[6] J. Birman, *Braids, Links and Mapping Class Groups*, Ann. Math. Stud. **82** (1974).

[7] ______, On the stable equivalence of plat representations of knots and links, Can. J. Math. **28** (1976), 264–290.

[8] ______, On the Jones polynomial of closed 3-braids, Invent. Math. **81** (1985), 287–294.

[9] J. Birman and H. Hilden, Isotopies of homeomorphisms of Riemann surfaces and a theorem about Artin's braid group, Ann. of Math. **97** (1973), 424–439.

[10] J. Birman and B. Wajnryb, Markov classes in certain finite quotients of Artin's braid group, Columbia University preprint (1986).

[11] J. Birman and T. Kanenobu, Jones' braid-plat formula and a new surgery triple, Columbia University preprint.

[12] N. Bourbaki, "Groupes et Algèbres de Lie, IV, V, VI," Masson, Paris (1982).

[13] J. Conway, An enumeration of knots and links, in *Computational Problems in Abstract Algebra*, Pergamon Press, New York (1970), 329–358.

[14] P. Freyd, D. Yetter, J. Hoste, W. Lickorish, K. Millet, and A. Ocneanu, A new polynomial invariant of knots and links, Bull. AMS **12** (1985), 183–312.

[15] D. Goldschmidt and V. Jones, The metaplectic invariant of a link, U.C. Berkeley preprint (1986).

[16] V. Jones, A polynomial invariant for knots via von Neumann algebras, Bull. AMS **12** (1985), 103–111.

[17] ______, Index for subfactors, Invent. Math. **72** (1983), 1–25.

[18] ______, Braid groups, Hecke algebras and type II_1 factors, in *Geometric Methods in Operator Algebras*, Proc. US-Japan Seminar (1986), 242–273.

[19] L. Kauffman, State models and the Jones polynomial, to appear in Topology.

[20] D. Kazdan and G. Lusztig, Representations of Coxeter groups and Hecke algebras, Invent. Math. **53** (1979), 165–184.

[21] J. Lannes, Sur l'invariant de Kervaire pour les noeuds classiques, Ecole Polytechnique, Palaiseu, preprint (1984).

[22] W. Lickorish and K. Millet, A polynomial invariant of knots and links, Topology **26** (1987), 107–141.

[23] ______, Some evaluations of link polynomials, Comment. Math. Helv. **61** (1986), 349–359.

[24] ______, The reversing result for the Jones polynomial, Pacific J. Math. **124** (1986), 173–176.

[25] W. Lickorish, A relationship between link polynomials, Math. Proc. Camb. Phil. Soc. **100** (1986), 109–112.

[26] A. Lipson, An evaluation of a link polynomial, Proc. Cambridge Philos. Soc. **100** (1986), 361–364.

[27] D. Littlewood, *The Theory of Group Characters*, Oxford University Press (1940).

[28] H. Morton, Closed braid representations for a link and its 2-variable polynomial, preprint, Liverpool (1985).

[29] H. Morton, Seifert circles and knot polynomials, Math. Proc. Cambridge Philos. Soc. **99** (1986), 107–110.
[30] H. Murakami, A note on the first derivative of the Jones polynomial, preprint (1985).
[31] ______, A recursive calculation of the Arf invariant of a link, preprint (1985).
[32] K. Murasugi, Jones polynomials and classical conjectures in knot theory, to appear in Topology.
[33] ______, "On Closed 3-braids," Memoirs 151, AMS (1974).
[34] M. Pimsner and S. Popa, Entropy and index for subfactors, Ann. Sci. Ec. Norm. Sup. **19** (1986), 57–106.
[35] R. Powers, Representations of uniformly hyperfinite algebras and the associated von Neumann algebras, Ann. of Math. **86** (1967), 138–171.
[36] K. Reidemeister, Zür dreidimensionalen topologie, Abh. Math. Sem. Univ. Hamburg **9** (1933), 189–194.
[37] D. Rolfsen, *Knots and Links*, Publish or Perish Press, Berkeley (1976).
[38] L. Rudolph, Algebraic functions and closed braids, Topology **22** (1983), 191–202.
[39] B. Seifert, The spherical traces for Hecke algebras associated to type II_1 factors, preprint, IHES.
[40] C. Squier, The Burau representation is unitary, preprint (1982).
[41] M. Takesaki, Duality for crossed products and the structure of von Neumann algebras of type III, Acta Math. **131** (1973), 249–310.
[42] H. Temperley and E. Lieb, Relations between the percolation and colouring problem..., Proc. Roy. Soc. (London) (1971), 251–280.
[43] M. Thistlethwaite, A spanning tree expansion of the Jones polynomial, to appear in Topology.
[44] A. Wassermann, Automorphic actions of compact groups on operator algebras, Thesis. Univ. of Pennsylvania (1981).
[45] B. Wajnryb, A simple presentation for the mapping class group of an orientable surface, preprint (1984).
[46] H. Wenzl, Representations of Hecke algebras and subfactors, Thesis, Univ. of Pennsylvania (1985).
[47] J. Franks and R. Williams, Braids and the Jones-Conway polynomial, preprint (1985).

(Received April 4, 1986)

FURTHER WORKS, A RELATION WITH STATISTICAL MECHANICS AND THE YANG-BAXTER EQUATION

Journal of the Physical Society of Japan
Vol. 56, No. 3, March, 1987, pp. 839–842

LETTERS

Knot Invariants and the Critical Statistical Systems

Yasuhiro AKUTSU and Miki WADATI*

Institute of Physics, Kanagawa University, Rokkakubashi, Kanagawa-ku, Yokohama 221
**Institute of Physics, College of Arts and Sciences, University of Tokyo, Meguro-ku, Tokyo 153*

(Received December 4, 1986)

A new invariant polynomial for knots and links is constructed from a solvable vertex model describing a critical statistical system. Various implications and the possible generalizations are discussed in connection with the recent development in the study of critical phenomena in two dimensions.

Recent development[1-7] in the study of two-dimensional statistical systems is quite exciting. The conformal bootstrap program[2] revealed rich physical and mathematical structures of the statistical systems at criticality. An even more exciting fact[3] is that a hierarchy of models[1] which are exactly solvable, both in the critical and the off-critical regions, gives a realization of the conformal field theory. Based on the discovery[4-7] of numerous (infinite!) hierarchies of solvable models, we have reached an intriguing conjecture[6] that all the universality classes in two dimensions are exhausted by such an infinite number of solvable models.

Meanwhile, we met a totally unexpected link between mathematics and physics, namely, knot theory and the critical statistical system. In the construction of a new polynomial invariant[8] for knots and links which is more powerful than the well-known Alexander polynomial,[9] Jones utilized a C^*-algebra $A_{q,n}$ generated by $\{1, e_1, e_2, \cdots, e_n\}$. The generators $\{e_i\}$ are essentially identical with the Temperley-Lieb operators[10] and satisfy the following relations:

$$e_i^* = e_i, \quad e_i^2 = e_i, \tag{1a,b}$$

$$e_i e_j = e_j e_i \quad (\text{for } |i-j| \geqq 2), \tag{1c}$$

$$e_i e_{i\pm 1} e_i = q^{-1} e_i. \tag{1d}$$

The Temperley-Lieb algebra appears in the construction of the transfer matrices of the two-dimensional critical statistical systems such as the ferroelectric models (6-vertex models),[11] and the self-dual Potts models,[10,12] which are exactly solvable. An important fact is that for the existence of the normalized trace associated with the algebra $A_{q,n}$ in the $n \to \infty$ limit ("thermodynamic limit"), the value of the parameter q(the state number of the Potts model) is restricted to the following two cases:

$$\text{(a) } q \geqq 4, \text{ (b) } q = 4\cos^2(\pi/k) \quad (k=3, 4, \cdots). \tag{2a,b}$$

We found[13] that these two cases correspond to the possible cases of the central charge c in the Virasoro algebra[14] characterizing the conformal field theory.

The case (2a) has another importance for knot theory. Let us introduce a parameter t by

$$q = 2 + t^{-1} + t. \tag{3}$$

For the case (2a), the parameter t is a free (positive) parameter. Jones constructed a polynomial in t, a new topologically invariant quantity for knots and links, by using a representation of the *braid group*[15] in terms of the *Hecke algebra*[16] $H(t, n)$ constructed from $A_{q,n-1}$, whose generators $\{g_i\}$ are defined by

$$g_i = (t+1)e_i - 1. \tag{4}$$

Thus, various important problems in physics (i.e. critical phenomena in two dimensions, completely integrable systems) and mathematics (i.e. infinite dimensional Lie algebra, von Neumann algebra, knot theory) are closely interrelated. In most cases where such an 'interaction' between the two fields occurs mathematics is usually applied to physics. In this paper, we shall reverse the order of application. Namely, we shall construct a further new polynomial invariant for knots and links from the Boltzmann weights of a solvable vertex model describing a critical statistical system.

Let us give a brief summary of knot theory. A *knot* is a closed non-self-intersecting line embedded in three dimensions. This is a special case of a more generalized object, called the *link*, which is an assembly of knots with mutual entanglements. The most fascinating but also the most difficult problem in knot theory is the complete classification of knots and links. An important mathematical object related to the link is the *braid* and the *braid group*. Consider n strings with each end tied to the 'bars'. We have a trivial configuration (trivial n-braid) where no intersection between the strings is present (Fig. 1). A general n-braid is constructed from this trivial n-braid by successive applications of the operation b_i ($i=1, \cdots, n-1$), which produces an intersection between the i-th string and the $(i+1)$-th string, such that the $(i+1)$-th string crosses *above* the i-th string (Fig. 2). The inverse operation b_i^{-1} makes the $(i+1)$-th string cross *below* the i-th string. The generators $\{b_i\}$ define a group, the braid group, which is denoted by B_n. Topological equivalence between seemingly different expressions of a braid is guaranteed by the following relations (Fig. 3):

$$b_i b_j = b_j b_i \quad (\text{for } |i-j| \geqq 2), \tag{5a}$$

$$b_i b_{i+1} b_i = b_{i+1} b_i b_{i+1}. \tag{5b}$$

Under the relation (5), each topologically equivalent class of braids is identified with an element in B_n. Hence, any n-braid is expressed

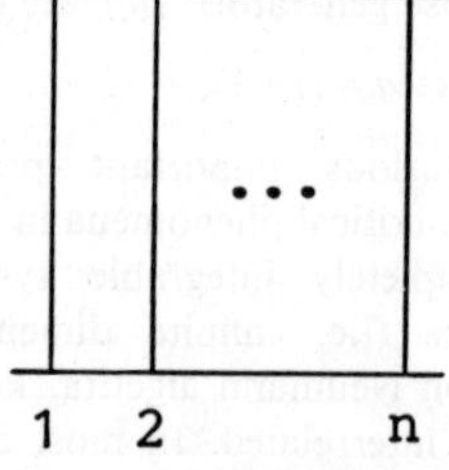

Fig. 1. A trivial n-braid.

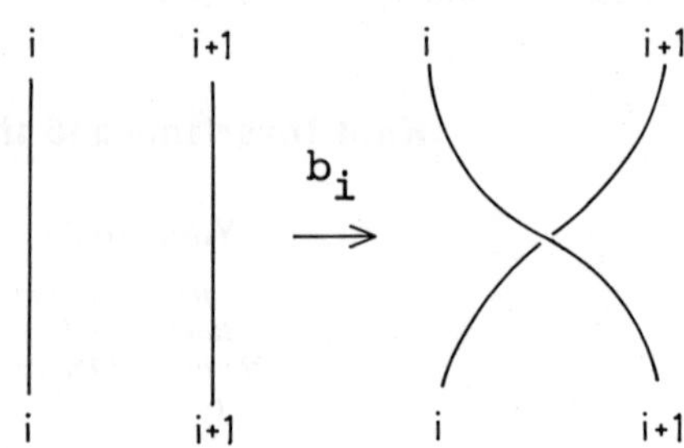

Fig. 2. An operation by a braid group element b_i.

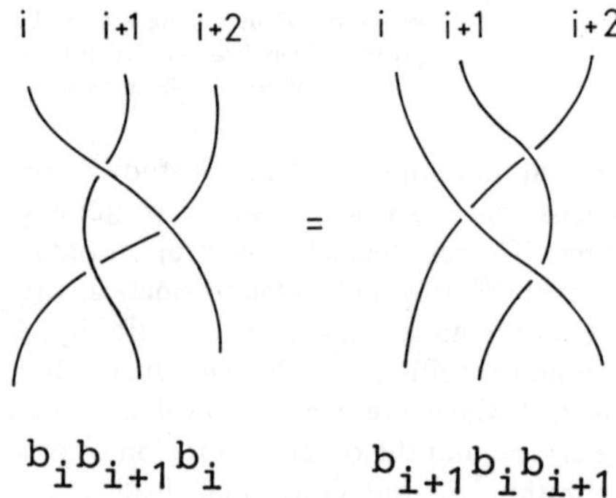

Fig. 3. Two topologically equivalent braids.

as a word on B_n (*e.g.*, $b_1 b_2 b_3 b_2^{-1} b_1$). In physics, the braid group has been applied to explain the appearance of particles with fractional statistics in two dimensions.[17] For knot theory, the braid group plays an essential role, since any link is represented as a closed braid.[18] Although we have many different representations of a link by a braid, there is an important theorem[19] that equivalent braids expressing the same link are related by successive applications of *moves*, called the *Markov moves* of types I and II (Fig. 4):

$$\text{type I: } AB \longrightarrow BA \quad (\text{for } A, B \in B_n).$$

$$\text{type II: } A \longrightarrow Ab_n,\ A \longrightarrow Ab_n^{-1},$$

$$(\text{for } A \in B_n \text{ and } b_n, Ab_n, Ab_n^{-1} \in B_{n+1}).$$

Hence, the *link polynomial* α, a topological invariant for links, must satisfy the following conditions:

$$\alpha(AB) = \alpha(BA) \quad (A, B \in B_n), \tag{6a}$$

$$\alpha(Ab_n) = \alpha(Ab_n^{-1}) = \alpha(A) \quad (A \in B_n). \tag{6b}$$

To construct α, it is convenient to find a linear

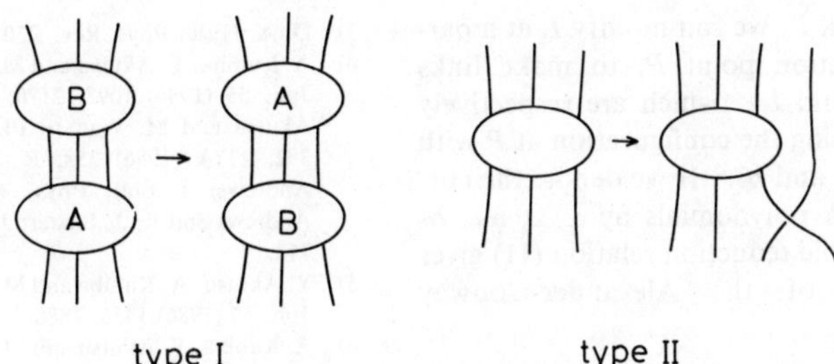

Fig. 4. Two types (I and II) of Markov moves.

functional, called the *Markov trace*, denoted by ϕ, having the following properties (Markov properties):

$$\phi(AB)=\phi(BA) \quad (A, B\in B_n), \tag{7a}$$

$$\phi(Ab_n)=\tau\cdot\phi(A), \quad \phi(Ab_n^{-1})=\bar{\tau}\cdot\phi(A) \quad (A\in B_n), \tag{7b}$$

where the parameters τ and $\bar{\tau}$ are given by

$$\tau=\phi(b_i),\ \bar{\tau}=\phi(b_i^{-1}) \quad \text{(for all } i\text{)}. \tag{8}$$

Then, the invariant polynomial α is given by

$$\alpha(A)=(\tau\cdot\bar{\tau})^{-(n-1)/2}\left(\frac{\bar{\tau}}{\tau}\right)^{e/2}\cdot\phi(A), \tag{9}$$

where e is the exponent sum of b_i's appearing in the braid A(*e.g.*, for $A=b_1b_2b_3b_2^{-1}b_1$, $e=1+1+1-1+1=3$).

In the construction of the Jones polynomial, the representation of B_n by the Hecke algebra $H(t, n)$ is essential. The generators $\{g_i\}$ satisfy both the defining relation (5) of the braid group (b_i should be interpreted as g_i) and the following *quadratic reduction relation*:

$$g_i^2=(t-1)g_i+t, \tag{10}$$

which gives the combinatorial definition of the Jones polynomial via the Alexander-Conway relation[9,20] (or the skein relation).

Our strategy to find a new link polynomial is to construct a different representation B_n' of the braid group using the Boltzmann weights of a solvable vertex model.[21,22] The difference from the Hecke algebra representation is that the reduction relation (10) is replaced by the following cubic relation:

$$g_i^3=(t^3-t^2+1)g_i^2+(t^5-t^3+t^2)g_i-t^5. \tag{11}$$

We obtained an explicit expression of the generator g_i ($\in B_n'$) in a sum of tensor products of 3×3 matrices as

$$g_i=\sum_{klpq} w_{kl,pq}I^{(1)}\otimes I^{(2)}\otimes\cdots\otimes I^{(i-1)}\otimes E_{kl}^{(i)}\otimes E_{pq}^{(i+1)}\otimes I^{(i+2)}\otimes\cdots, \tag{12}$$

where $I^{(i)}$ is the identity acting on the i-th position, and E_{kl} is a matrix such that $(E_{kl})_{ij}=\delta_{ki}\delta_{lj}$. The non-zero 'weights' $w_{kl,pq}$ are the following:

$$\begin{aligned}
&w_{11,11}=w_{33,33}=1,\\
&w_{12,21}=w_{21,12}=w_{23,32}=w_{32,23}=-t,\\
&w_{22,11}=w_{33,22}=1-t^2,\\
&w_{13,31}=w_{31,13}=t^2,\\
&w_{22,22}=t,\\
&w_{23,21}=w_{32,12}=t^{5/2}-t^{1/2},\\
&w_{33,11}=t^3-t^2-t+1.
\end{aligned} \tag{13}$$

It is straightforward to verify eqs. (5) and (11).

For the Markov trace ϕ, we found that it is given by a direct generalization of the Powers state:[23,24]

$$\begin{aligned}
&\phi(A)=Tr(HA) \quad (\text{for } A\in B_n'),\\
&H=h^{(1)}\otimes h^{(2)}\otimes\cdots\otimes h^{(i)}\otimes\cdots,\\
&h^{(i)}=\mathrm{diag}(1, t, t^2)/(1+t+t^2) \quad (\text{for all } i).
\end{aligned} \tag{14}$$

It is also straightforward to check the Markov properties (eq. (7)) of ϕ, with

$$\tau=1/(1+t+t^2),\ \bar{\tau}=t^2/(1+t+t^2). \tag{15}$$

Putting eq. (15) into eq. (9), we have a new link polynomial α as

$$\alpha(A)=(1+t+t^{-1})^{n-1}t^e\cdot\phi(A). \tag{16}$$

For a given link L, we can modify L at a particular intersection point P, to make links L_{++}, L_{+}, L_0 and L_{-}, which are respectively given by replacing the configuration at P with that of b^2, b, 1 and b^{-1}. If we denote the corresponding link polynomials by α_{++}, α_{+}, α_0 and α_{-}, the cubic reduction relation (11) gives an analogue of the Alexander-Conway relation:

$$\alpha_{++}=t(t^3-t^2+1)\alpha_{+}+t^2(t^5-t^3+t^2)\alpha_0-t^8\alpha_{-}. \tag{17}$$

We have applied our new link polynomial to several examples to confirm that ours is indeed a new link polynomial different from the Jones'. This is, of course, the consequence of the different initial braid group representation.

The construction procedure of braid group representations from the Boltzmann weights of solvable models (both vertex and IRF) at criticality can be generalized. Since we have infinite hierarchies of such solvable models,[4-7,22] we shall have an infinite sequence of new link polynomials. Then, with some multivariable extensions,[25] more detailed classification of knots and links may be possible.

From our braid group representation, we can define an analogue of the Temperley-Lieb operators. Hence, it is an interesting problem how the Temperley-Lieb equivalence is generalized. Relation to the conformal (or super-conformal) field theory and interpretation in terms of the von Neumann algebra theory are also quite interesting subjects.

The authors thank Mr. Atsuo Kuniba for his helpful discussions.

References

1) G. E. Andrews, R. J. Baxter and P. J. Forrester: J. Stat. Phys. **35** (1984) 193.
2) A. A. Belavin, A. M. Polyakov and A. B. Zamolodchikov: Nucl. Phys. **B241 [FS]** (1984) 333; D. Friedan, Z. Qiu and S. Shenker: Phys. Rev. Lett. **52** (1984) 1575.
3) D. A. Huse: Phys. Rev. **B30** (1984) 3908.
4) A. Kuniba, Y. Akutsu and M. Wadati: J. Phys. Soc. Jpn. **55** (1986) 1092, 2170, 3338; A. Kuniba, Y. Akutsu and M. Wadati: Phys. Lett. **116A** (1986) 382; **117A** (1986) 358; R. J. Baxter and G. E. Andrews: J. Stat. Phys. **44** (1986) 249; G. E. Andrews and R. J. Baxter: J. Stat. Phys. **44** (1986) 713.
5) Y. Akutsu, A. Kuniba and M. Wadati: J. Phys. Soc. Jpn. **55** (1986) 1466, 1880.
6) A. Kuniba, Y. Akutsu and M. Wadati: J. Phys. Soc. Jpn. **55** (1986) 2605.
7) Y. Akutsu, A. Kuniba and M. Wadati: J. Phys. Soc. Jpn. **55** (1986) 2907; E. Date, M. Jimbo, T. Miwa and M. Okado: Lett. Math. Phys. **12** (1986) 209.
8) V. F. R. Jones: Bull. Amer. Math. Soc. **12** (1985) 103.
9) J. W. Alexander: Trans. Amer. Math. Soc. **30** (1928) 275.
10) H. N. V. Temperley and E. H. Lieb: Proc. Soc. London A **322** (1971) 251.
11) E. H. Lieb and F. Y. Wu: in *Phase Transitions and Critical Phenomena*, ed. C. Domb and M. S. Green (Academic Press, London, 1972) Vol. 1, p. 331.
12) R. J. Baxter, S. B. Kelland and F. Y. Wu: J. Phys. **A9** (1976) 397.
13) A. Kuniba, Y. Akutsu and M. Wadati: J. Phys. Soc. Jpn. **55** (1986) 3285.
14) M. Virasoro: Phys. Rev. **D1** (1970) 2933.
15) E. Artin: Ann. Math. **48** (1947) 101.
16) N. Bourbaki: *Groupes et Algèbres de Lie* (Hermann, Paris, 1968) Chap. 4.
17) Y. S. Wu: Phys. Rev. Lett. **52** (1984) 2103.
18) J. W. Alexander: Proc. Nat. Acad. **9** (1923) 93.
19) A. A. Markov: Recueil Math. Moscov (1935) 73.
20) J. H. Conway: in *Computational Problems in Abstract Algebra*, ed. J. Leech (Pergamon press, London, 1969) p. 329.
21) A. B. Zamolodchikov and V. A. Fateev: Sov. J. Nucl. Phys. **32** (1980) 298.
22) K. Sogo and Y. Akutsu and T. Abe: Prog. Theor. Phys. **70** (1983) 730, 739.
23) M. Pimsner and S. Popa: unpublished.
24) R. T. Powers: Ann. Math. **86** (1967) 138.
25) P. Freyd, D. Yetter, J. Hoste, W. B. R. Lickorish, K. Millett and A. Ocneanu: Bull. Amer. Math. Soc. **12** (1985) 239.

Note added in proof—After the submission of the paper, we have found a general method to construct link polynomials using the exactly solvable q-state models in ref. 22. The result in this paper corresponds to the $q=3$ case.

Journal of the Physical Society of Japan
Vol. 56, No. 9, September, 1987, pp. 3039–3051

Exactly Solvable Models and New Link Polynomials. I. N-State Vertex Models

Yasuhiro AKUTSU and Miki WADATI†

Institute of Physics, Kanagawa University, Rokkakubashi, Kanagawa-ku, Yokohama 221
†Institute of Physics, College of Arts and Sciences, University of Tokyo, Komaba, Meguro-ku, Tokyo 153

(Received April 15, 1987)

Presented is a general method to construct representations of the braid group, a basic object in the knot theory, from the Boltzmann weights of the exactly solvable models in statistical mechanics at criticality. The method is applied to a class of N-state vertex models (N=2, 3 and 4) to have explicit braid group representations. Furthermore, the Markov traces are introduced and a sequence of new link polynomials, topological invariants for knots and links, is constructed.

§1. Introduction

Recent development in the theory of quantum completely integrable systems provides us a unified treatment of various exactly solvable models in $1+1$ dimensional field theory and in 2-dimensional classical statistical mechanics.[1] The key point is that to each model we can associate a commuting family of transfer matrices which are the generators of an infinite number of conserved quantities. The commutability condition is called the Yang-Baxter relation.

The Yang-Baxter relation takes several forms depending on the types of models under consideration. For the $1+1$ dimensional field theory, the Yang-Baxter relation is the factorization condition[2] for the many-body S-matrices and is called the factorization equation. Denoting the scattering amplitude for the process $(i, j)\to(k, l)$ by $S^{ik}_{jl}(u)$, the factorization equation reads as

$$\sum_{\alpha,\beta,\gamma} S^{\beta q}_{\gamma r}(v)S^{\alpha p}_{k\gamma}(u+v)S^{i\alpha}_{j\beta}(u) = \sum_{\alpha,\beta,\gamma} S^{\alpha p}_{\beta q}(u)S^{i\alpha}_{\gamma r}(u+v)S^{j\beta}_{k\gamma}(v), \tag{1.1}$$

where u and v are spectral parameters (or the rapidities) (Fig. 1). For 2-dimensional statistical mechanics, we have two types of models, the vertex model and the IRF (Interaction Round a Face) model.[3] Some of the IRF models are equivalent to the vertex models through the Wu-Kadanoff-Wegner transformation.[4] It is now well-known[5] that any factorized S-matrix can be interpreted as the Boltzmann weight of a solvable vertex model, and the factorization equation is the commutability condition of the transfer matrices of the vertex model. For the IRF models, the commutability condition is called the star-triangle relation[3] which reads as

$$\sum_c w(b, d, c, a; u)w(a, c, f, g; u+v)w(c, d, e, f; v) = \sum_c w(a, b, c, g; v)w(b, d, e, c; u+v)w(c, e, f, g; u). \tag{1.2}$$

Here $w(a, b, c, d; u)$ denotes the Boltzmann weight for the spin configuration (a, b, c, d) round a face.

The Boltzmann weight satisfying (1.1) or (1.2) defines a sequence of operators[3] $X_1(u)$, $X_2(u), \cdots, X_i(u), \cdots$, which satisfy the relations:

$$X_i(u)X_j(v)=X_j(v)X_i(u), \quad \text{for } |i-j|\geqq 2, \tag{1.3a}$$

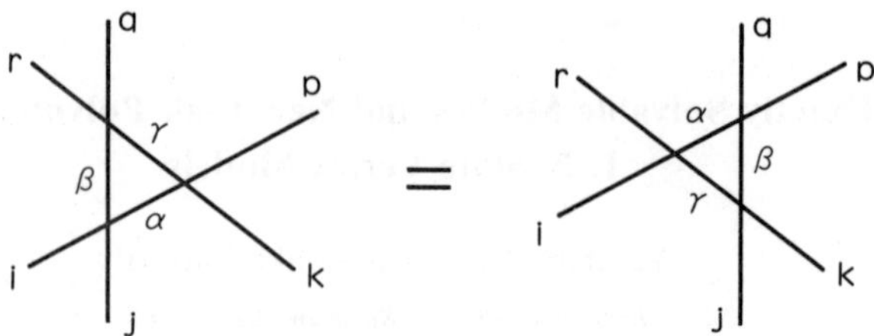

Fig. 1. Schematic explanation of the factorization equation. Time direction is upward.

$$X_i(u)X_{i+1}(u+v)X_i(v)$$
$$=X_{i+1}(v)X_i(u+v)X_{i+1}(u). \quad (1.3b)$$

In particular, relation (1.3b) corresponds to the Yang-Baxter relation (1.1) or (1.2). In the vertex model, for example, we have the following expression of the operator $X_i(u)$:

$$X_i(u)=\sum_{klmn} S_{ln}^{km}(u)\cdot I^{(1)}\otimes I^{(2)}\otimes\cdots$$
$$\otimes E_{nk}^{(i)}\otimes E_{ml}^{(i+1)}\otimes I^{(i+2)}\otimes\cdots, \quad (1.4)$$

where $I^{(j)}$ is the identity acting at the j-th position and E_{nk} is a matrix such that $(E_{nk})_{pq}=\delta_{np}\delta_{kq}$. In statistical mechanics, the operator $X_i(u)$ represents an element of the diagonal to diagonal transfer matrix.

By regarding the Yang-Baxter relation as a functional equation, we have a powerful method to find a new solvable model. As a remarkable result of the method, we should note the recent discovery[6-11] of an infinite number of solvable hierarchies which are expected to exhaust the whole universality classes in two dimensions.

An important object in mathematics always has an important physical significance and vice versa. We have observed[12] close connections among the exactly solvable models, the conformal field theory,[13,14] partition theory of natural numbers,[7-9,15-17] infinite dimensional Lie algebra[18,19] and von Neumann algebra.[20,21] In the recent discovery of a new polynomial invariant[21] (Jones polynomial) for knots and links, we can also find a close connection between physics and mathematics. An algebra $A_{q,n}$ appearing in the construction of the Jones polynomial is equivalent to the Temperley-Lieb algebra,[22] a well-known object in statistical mechanics, which describes the transfer matrices of the critical Potts model and the 6-vertex model.[23,24] The algebra $A_{q,n}$ is generated by $\{1, e_1, e_2, \cdots, e_n\}$ satisfying the following relations:

$$e_i^*=e_i, \quad (1.5a)$$
$$e_i^2=e_i, \quad (1.5b)$$
$$e_ie_j=e_je_i \quad (\text{for} \quad |i-j|\geqq 2), \quad (1.5c)$$
$$e_ie_{i\pm 1}e_i=q^{-1}e_i. \quad (1.5d)$$

The algebra $A_{q,n}$ is utilized to have a representation (Hecke algebra[25] representation) of the braid group[26] which is a fundamental object in the knot theory. Let us introduce a parameter t by

$$q=2+t^{-1}+t, \quad (1.6)$$

and a generator g_i by

$$g_i=(t+1)e_i-1 \quad (e_i\in A_{q,n-1}). \quad (1.7)$$

Then the Hecke algebra $H(t, n)$ is generated by $\{g_i\}$. The generators satisfy both the defining relation of the braid group (see §2) and the following quadratic relation:

$$g_i^2=(t-1)g_i+t. \quad (1.8)$$

Physically, the braid group gives a natural explanation for the appearance of the particles with fractional statistics in two dimensions.[27]

In most cases, the Jones polynomial is more powerful than the classic Alexander polynomial.[28] It has been proved that the Jones polynomial and the Alexander polynomial correspond to special cases of the two-variable extension[29] of the Jones polynomial. An important fact is that these polynomials admit combinatorial (or recursive) definitions through the Alexander-Conway relation[28,30] (or skein relation), which is a direct consequence of the Hecke algebra representation of the braid group.

The Jones polynomial and its two-variable extension are not 'complete'; there exist infinitely many different links which have the same polynomial.[31,32] It is clear that this incompleteness partly stems from the Alexander-Conway relation, or equivalently, the Hecke algebra representation of the braid group. It is, then, quite natural to expect that if we utilize a totally different braid group representation which does not admit the Alexander-Conway relation, we should have a different and a more powerful link polynomial. Indeed, the present authors have found a general method to construct such representation and a new link polynomial.[33]

The aim of this series of papers is to extend the relation between the exactly solvable models and the link polynomials. In this paper, the first of the series, we present a sequence of new link polynomials, based on the general method to obtain a braid group representation from a solvable model.

This paper is organized as follows. In §2, a brief summary of the knot theory is presented. The role of the braid group is explained. In §3, a general prescription to make a braid group representation from the Boltzmann weights satisfying the Yang-Baxter relation is presented. The reason why the starting solvable model should be 'critical' is explained. In §4, the prescription presented in §3 is applied to the N-state vertex model to have a sequence of braid group representations. The Hecke algebra representation is straightforwardly obtained as the $N=2$ case. In §5, the Markov traces associated with new braid group representations are given, from which new link polynomials, more powerful than the Jones polynomial, are constructed. The last section is devoted to the summary of this paper.

§2. Knots and Links as Closed Braids

A *knot* is a closed self-avoiding string which is embedded in three dimensions (Fig. 2). As a more generalized object we have a *link* which is an assembly of knots with mutual entanglements (Fig. 3). An important mathematical object related to the link is the *braid* and the *braid group*.[26] Consider n strings with each end tied to the upper and the

Fig. 2. A knot. Fig. 3. A link.

lower bars. There is a configuration called a *trivial n-braid*, where no intersection between the strings is present (Fig. 4). We have an operation b_i ($i=1,\cdots,n-1$) which produces an intersection between the $(i+1)$-th string and the i-th string such that the i-th string crosses *above* the $(i+1)$-th string (Fig. 5). To be precice, we should note that the number i does not refer to an individual string but to the *position* of the string. Hence, the term 'i-th string' should be interpreted as the string at the i-th position of the upper bar. The inverse operation b_i^{-1} makes the i-th string cross *below* the $(i+1)$-th string (Fig. 5). The generators $\{b_i\}$ define the braid group, which is denoted by B_n. A general n-braid is constructed by applying an element in B_n to the trivial n-braid. Then, by regarding the trivial n-braid as the identity operation in B_n, we can identify any

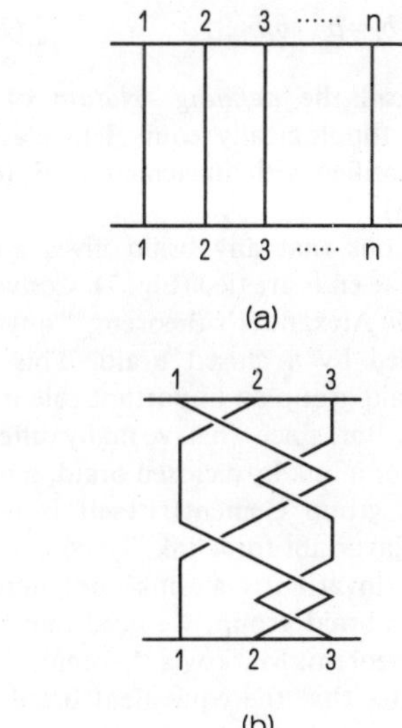

Fig. 4. Examples of braids: (a) trivial n-braid, (b) a nontrivial 3-braid ($b_1 b_2^2 b_1 b_2^2$).

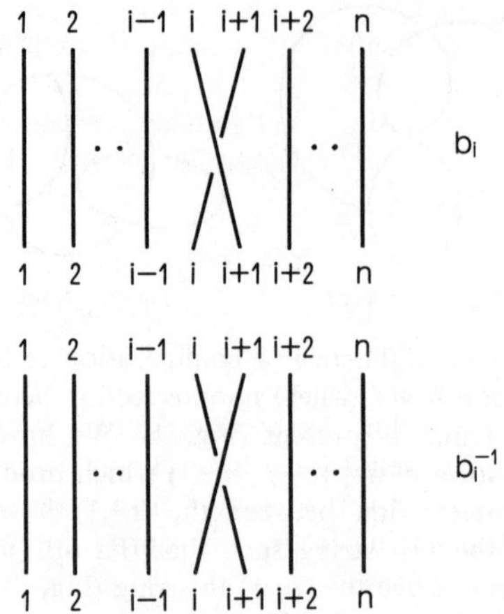

Fig. 5. Operations b_i and b_i^{-1}.

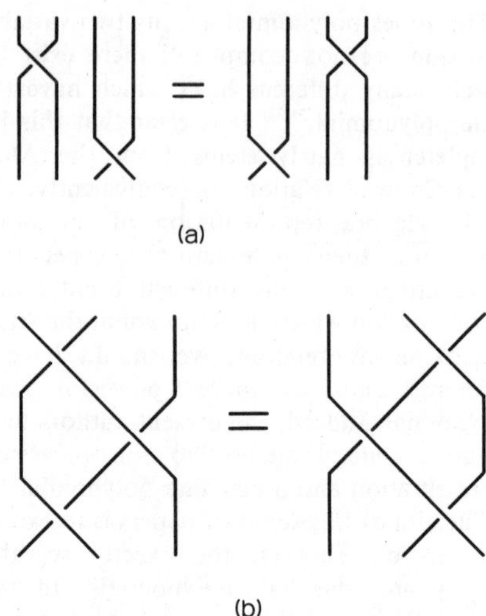

Fig. 6. Topologically equivalent braids.

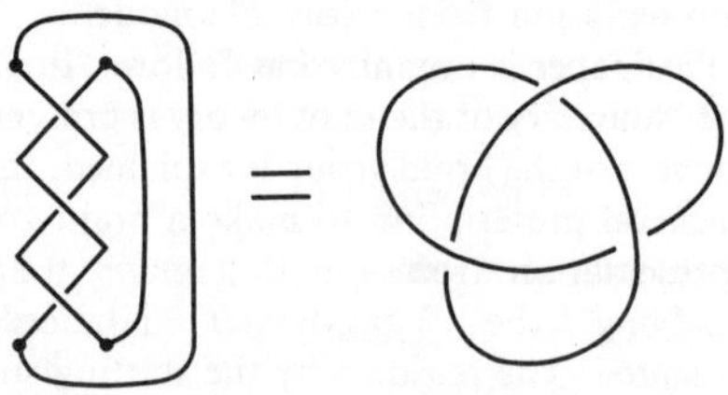

Fig. 7. A closed braid which corresponds to b_1^3.

element in B_n as an n-braid.

We are interested in the topologically equivalent classes of n-braids. To guarantee the topological equivalence between seemingly different expressions of a braid in terms of a braid group element, the following conditions are known to be necessary and sufficient (Fig. 6):

$$b_i b_j = b_j b_i \quad (\text{for } |i-j| \geqq 2), \tag{2.1a}$$

$$b_i b_{i+1} b_i = b_{i+1} b_i b_{i+1}, \tag{2.1b}$$

which we call the *defining relation* of B_n. Then, each topologically equivalent class of braids is identified with an element in B_n (*e.g.*, $b_1 b_2 b_3 b_2^{-1} b_1$).

It is obvious that any braid gives a link when opposite ends are tied (Fig. 7). Conversely, due to the Alexander's theorem,[34] any link is represented by a closed braid. This fact gives the braid group an important role in the knot theory. But, since we have many different expressions of a link by a closed braid, a braid (or a braid group element) iteself is not a topological invariant for a link. To construct a topological invariant, a link polynomial, based on the braid group, we need one more important theorem, Markov's theorem.[35] The theorem states that the equivalent braids expressing the same link are mutually related by successive applications of moves, called the *Markov moves* of type I and II (Fig. 8):

$$\text{type I:} \quad AB \to BA \quad (\text{for } A, B \in B_n), \tag{2.2a}$$

$$\text{type II:} \quad A \to Ab_n, \quad A \to Ab_n^{-1}, \tag{2.2b}$$

$$(\text{for } A \in B_n \text{ and } b_n, Ab_n, Ab_n^{-1} \in B_{n+1}).$$

Hence, to obtain a link polynomial we should first construct a suitable representation B_n' of the braid group B_n and then find a quantity defined on B_n' which is invariant under the Markov moves.

§3. Construction of Braid Group Representations

We should remark the close similarity between (1.3) and (2.1) (see also the similarity between Fig. 1 and Fig. 6). The only difference is that in (1.3) we have spectral parameters as the arguments of the operators. Then, to have a braid group representation, we must get rid of the spectral parameters. The simplest way is to

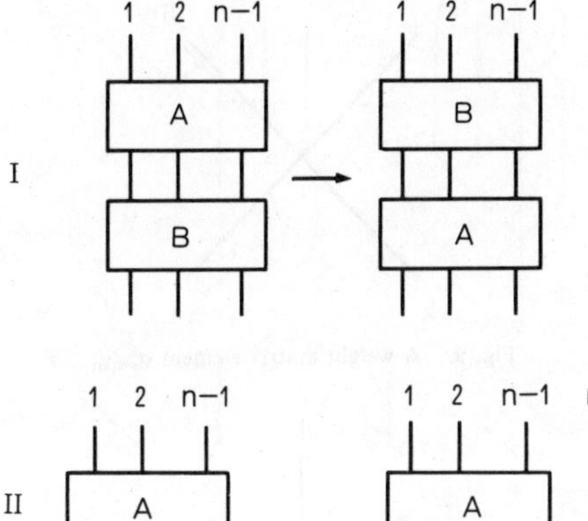

Fig. 8. Two types (I and II) of Markov moves.

set all the spectral parameters to a certain constant value. Namely, we should require

$$u=u+v=v. \tag{3.1}$$

An obvious solution is

$$u=v=0. \tag{3.2}$$

Then, we have a 'trivial' braid group representation $B_n^{(0)}$ by identifying the braid group generator b_i as

$$b_i \to X_i(0). \tag{3.3}$$

We call this representation 'trivial' because almost all the solutions to the Yang-Baxter relation are obtained under the 'standard initial condition'

$$X_i(0)=c\cdot I, \tag{3.4}$$

where c is a constant and I is the identity operator. Hence, the resulting representation is one-dimensional and is not of much interest. We should note that eq. (3.1) has more 'possible' solutions. We may have a solution

$$u=u+v=v=\infty, \tag{3.5}$$

if, after suitable normalization, the following limit exists:

$$\lim_{u\to\infty} X_i(u)=g_i. \tag{3.6}$$

It is simple to see that these operators $\{g_1, g_2, \cdots, g_{n-1}\}$ satisfy the defining relations (2.1) for the braid group B_n, by taking the $u\to\infty$ limit of both sides in (1.3). Then we may have a non-trivial representation of B_n through the identification

$$b_i \to g_i. \tag{3.7}$$

There actually exist solvable models which have the limit (3.6). A typical example is the 6-vertex model. Unfortunately, not all of the known solutions for the Yang-Baxter relation admit the well-defined limit (3.6). For example, for the 8-vertex model[36] parametrized by the elliptic functions, we cannot choose a definite $u\to\infty$ direction since the elliptic functions have double periodicity in the complex u plane. Then we do not have a well-defined operator g_i in (3.6). We should remark here that the 8-vertex model is more 'general' than the 6-vertex model in the sense that the latter corresponds to a special limit of the former. The 6-vertex model, parametrized by the trigonometric or hyperbolic functions, corresponds to the special limit of the 8-vertex model where one of the periods for the elliptic functions tends to infinity (i.e., degenerate case). Then taking the limit $u\to\infty$ along this infinite period, we may have a well-defined operator g_i in (3.6) for the case of the 6-vertex model. In physics, the 6-vertex model corresponds to the 8-vertex model *at criticality*.[3]

All the known solutions to the Yang-Baxter relation are classified into three cases depending on the function of parametrization: (a) elliptic (b) trigonometric or hyperbolic and (c) rational. It is a general fact that any model in case (a) at criticality corresponds to a model in case (b) (this correspondence holds also for cases (b) and (c)). Then, to obtain a braid group representation from a solvable model as general as possible, we should use a model which describes a certain critical statistical system and is parametrized by the trigonometric or the hyperbolic functions.[3,37] This gives a natural explanation for the following: Temperley-Lieb algebra for the transfer matrices of the 6-vertex model and the critical (self-dual) Potts model appears in the construction of the braid group representation by Jones.

For definiteness, in what follows, we consider only the case of solvable vertex models whose Boltzmann weights are parametrized by the hyperbolic functions. We assume the unitarity condition

$$\sum_{p,q} S_{pk}^{ql}(-u)S_{jq}^{ip}(u)=\delta_{ik}\delta_{jl}. \tag{3.8}$$

Let us define a quantity $\sigma_{nm,kl}$ as the $u\to+\infty$ limit of the Boltzmann weight $S_{ln}^{km}(u)$:

$$\lim_{u\to+\infty} S_{ln}^{km}(u)=\sigma_{nm,kl}. \tag{3.9}$$

The matrix $(\sigma_{nm,kl})$ is referred to as the *weight matrix*. It acts on the space of neighboring edge states; the quantity $\sigma_{nm,kl}$ stands for the matrix element between the edge states (k, l) and (n, m) (Fig. 9). Then, from (1.4), we have a representation $B_n^{(+)}$ of the braid group B_n, whose generators $\{g_1, g_2,\cdots, g_n\}$ are given by

$$\begin{aligned} g_i&=\lim_{u\to+\infty} X_i(u)\\ &=\sum_{klmn}\sigma_{nm,kl}\cdot I^{(1)}\otimes I^{(2)}\otimes\cdots\\ &\quad\otimes E_{nk}^{(i)}\otimes E_{ml}^{(i+1)}\otimes I^{(i+2)}\otimes\cdots\otimes I^{(n)}. \end{aligned}\tag{3.10}$$

From the same vertex model, we have another representation $B_n^{(-)}$ generated by $\{g_1', g_2',\cdots, g_{n-1}'\}$, by taking the $u\to-\infty$ limit of the Boltzmann weights:

$$g_i'=\lim_{u\to-\infty} X_i(u). \tag{3.11}$$

But, due to the unitarity (3.8) which amounts to

$$X_i(-u)=X_i(u)^{-1}, \tag{3.12a}$$

$$g_i'=g_i^{-1}, \tag{3.12b}$$

the representation $B_n^{(-)}$ is just the one generated by $\{g_1^{-1}, g_2^{-1},\cdots, g_{n-1}^{-1}\}$. Hence, the essential information is contained in the representation $B_n^{(+)}$.

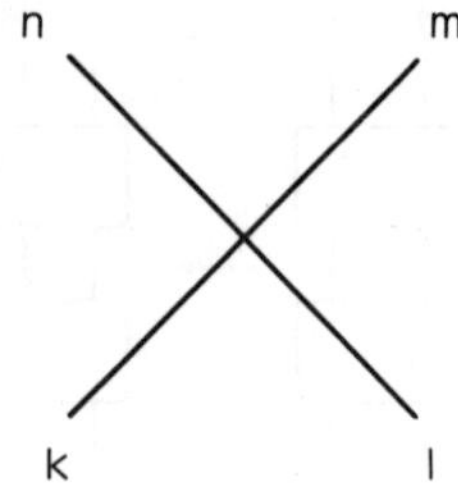

Fig. 9. A weight matrix element $\sigma_{nm,kl}$.

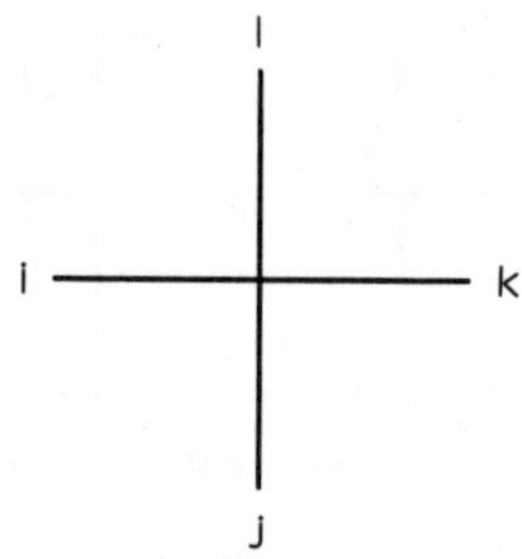

Fig. 10. A Boltzmann weight of the N-state vertex model.

§4. Application to the N-State Vertex Model

We know many models with hyperbolic (or trigonometric) parametrization. Among them, we take a series of vertex models proposed by Sogo, Akutsu and Abe.[38] This series includes the 6-vertex model by Lieb and Wu and the 19-vertex model by Zamolodchikov and Fateev[39] as the first two in the series.

We denote the Boltzmann weights of the N-state vertex model satisfying the Yang-Baxter relation (or the factorization equation) by $\{S_{jl}^{ik}(u)\}$ (Fig. 10). The edge variables i, j, k and l take the following values:

$$i, j, k, l=-s, -s+1,\cdots, s-1, s, \tag{4.1}$$

where s is related to the state number N by

$$N=2s+1. \tag{4.2}$$

The most important constraint imposed on the model is the charge (or spin momentum) conservation condition

$$S_{jl}^{ik}(u)=0 \text{ unless } i+j=k+l. \tag{4.3}$$

Also the model has the CPT invariances

$$\begin{aligned} S_{jl}^{ik}(u)&=S_{\bar{j}\bar{l}}^{\bar{i}\bar{k}}(u) \quad (C \text{ invariance}, \bar{i}=-i, \text{etc.})\\ &=S_{ik}^{jl}(u) \quad (P \text{ invariance})\\ &=S_{lj}^{ki}(u) \quad (T \text{ invariance}), \end{aligned}\tag{4.4}$$

and the 'crossing symmetry'

$$S_{jl}^{ik}(u)=S_{\bar{j}\bar{l}}^{k\bar{l}}(\lambda-u), \tag{4.5}$$

with some parameter λ which is often called the crossing point of the spectral parameter. Since the unitarity factor $\rho(u)$ (see Appendix) is given by

$$\rho(u)=S_{ss}^{ss}(u)\,S_{ss}^{ss}(-u), \tag{4.6}$$

we can make the model satisfy the unitarity condition (3.8) by renormalizing all the Boltzmann weights as

$$S_{jl}^{ik}(u)/S_{ss}^{ss}(u)\to S_{jl}^{ik}(u). \tag{4.7}$$

Explicit parametrizations of the weights are given in the Appendix, for $N=2$, 3 and 4 cases.

For the purpose of constructing a braid group representation from the N-state vertex model, there is one more important feature possessed by the model. From the symmetric model satisfying (4.4), we can make an asymetric model through the 'symmetry-breaking transformation'[38)]

$$S_{jl}^{ik}(u)\to\tilde{S}_{jl}^{ik}(u)=\alpha_{ij,kl}(u)\cdot\beta_{ij,kl}\cdot\gamma_{ij,kl}\cdot S_{jl}^{ik}(u), \tag{4.8}$$

where each constituent transformation is given by

$$\alpha_{ij,kl}(u)=\exp\,[\mu(k-i-l+j)u], \tag{4.9a}$$

$$\beta_{ij,kl}=\exp\,[\nu(l-i-k+j)], \tag{4.9b}$$

$$\gamma_{ij,kl}=\exp\,[\omega(kl-ij)]. \tag{4.9c}$$

In the above, μ, ν and ω are free parameters. Due to the charge conservation condition (4.3), the resulting asymmetrized Boltzmann weights $\{\tilde{S}_{jl}^{ik}(u)\}$ also satisfy the factorization equation. We note that the transformation does not affect the unitarity condition.

Now we are in a position to construct the braid group representations. We apply the formula (3.10) to the N-state vertex model which is asymmetrized by the transformation (4.8)–(4.9). Since only the factor $\alpha_{ij,kl}(u)$ affects the construction in an essential manner, we set $\beta_{ij,kl}=\gamma_{ij,kl}=1$ (or equivalently, $\nu=\omega=0$) for simplicity.

As an illustration, we will take the $N=2$ case (6-vertex model). In the limit $u\to+\infty$, the asymmetrized Boltzmann weights have the following asymptotic forms:

$$\tilde{S}_{1/2\,1/2}^{1/2\,1/2}(u)=1, \tag{4.10a}$$

$$\tilde{S}_{-1/2\,-1/2}^{-1/2\,-1/2}(u)=1, \tag{4.10b}$$

$$\tilde{S}_{-1/2\,-1/2}^{\;\;1/2\;\;\;1/2}(u)=-\mathrm{e}^{\lambda}, \tag{4.10c}$$

$$\tilde{S}_{\;\;1/2\;\;\;1/2}^{-1/2\,-1/2}(u)=-\mathrm{e}^{\lambda}, \tag{4.10d}$$

$$\tilde{S}_{-1/2\;\;\;1/2}^{\;\;1/2\,-1/2}(u)=-2\sinh\lambda\cdot\mathrm{e}^{\lambda}\,\mathrm{e}^{(-2\mu-1)u}, \tag{4.10e}$$

$$\tilde{S}_{\;\;1/2\,-1/2}^{-1/2\;\;\;1/2}(u)=-2\sinh\lambda\cdot\mathrm{e}^{\lambda}\,\mathrm{e}^{(2\mu-1)u}. \tag{4.10f}$$

Then, for the existence of the well-defined limit $u\to\infty$, the value of μ is restricted as

$$-1\leqq 2\mu\leqq 1. \tag{4.11}$$

According to the limiting behavior, we have three cases to consider:

case (1): $\mu=1/2$ and

$\tilde{S}_{-1/2\;\;\;1/2}^{\;\;1/2\,-1/2}$ tends to be zero, others non-zero,

case (2): $\mu=-1/2$ and

$\tilde{S}_{\;\;1/2\,-1/2}^{-1/2\;\;\;1/2}$ tends to be zero, others non-zero,

case (3): $-1<2\mu<1$ and

both $\tilde{S}_{-1/2\;\;\;1/2}^{\;\;1/2\,-1/2}$ and $\tilde{S}_{\;\;1/2\,-1/2}^{-1/2\;\;\;1/2}$ tend to be zero.

As will be seen below, cases (1) and (2) give the Hecke algebra representation of the braid group which leads to the Jones polynomial. Also, judging from the number of remaining non-zero elements, these cases seem to be more interesting than case (3). In fact, we have checked that the braid group representation corresponding to case (3) does not lead to an interesting link polynomial. Hence, we consider only cases (1) and (2). Moreover, since case (2) is recovered by $u\to-\infty$ limit with suitable redefinition of parameters (*e.g.*, $\lambda\to-\lambda$), we shall consider only case (1) in the following.

We set $\mu=1/2$ and introduce a parameter t as

$$t=e^{2\lambda}. \tag{4.12}$$

Due to the charge conservation condition, the weight matrix $(\sigma_{nm,kl})$ defined in (3.9) is decomposed into blocks according to the 'total charge' $k+l=n+m=c$. By $\sigma^{(c)}$, we denote the submatrix acting in the sector of total charge c. Then the weight matrix is given as follows:

$$\sigma^{(1)}=(\sigma_{1/2\,1/2,\,1/2\,1/2})=(1), \tag{4.13a}$$

$$\sigma^{(0)}=\begin{pmatrix}\sigma_{\;1/2\,-1/2,\,1/2\,-1/2} & \sigma_{\;1/2\,-1/2,\,-1/2\,1/2}\\ \sigma_{-1/2\;\;1/2,\,1/2\,-1/2} & \sigma_{-1/2\;\;1/2,\,-1/2\,1/2}\end{pmatrix}=\begin{pmatrix}0 & -t^{1/2}\\ -t^{1/2} & 1-t\end{pmatrix}, \tag{4.13b}$$

$$\sigma^{(-1)}=(\sigma_{-1/2\,-1/2,\,-1/2\,-1/2})=(1). \tag{4.13c}$$

As can be verified by direct calculations, the resulting expression of the braid group generator g_i given by (3.10) satisfies the following 'reduction relation':

$$g_i^2=(1-t)g_i+t. \tag{4.14}$$

This relation can be also derived from the eigenvalue equation for $\sigma^{(0)}$ which leads to

$$(g_i-1)(g_i+t)=0. \tag{4.15}$$

Hence, the braid group representation we have obtained is, after the substitution $g_i\to -g_i$, the Hecke algebra representation (1.8). Our construction procedure for the braid group representation is direct since we do not need an intermediate algebra $A_{q,n}$.

By similar considerations, we can apply the formula (3.10) to the model with $N\geqq 3$. We start with the symmetric Boltzmann weights presented in the Appendix. The asymmetrization factor is chosen to be the same as in the $N=2$ case; we set

$$\mu=1/2,$$
$$\nu=0,$$
$$\omega=0, \tag{4.16}$$

in (4.7). In what follows, we present the explicit forms of weight matrices for $N=3$ and $N=4$ cases. The results are given in terms of the total charge submatrices with the parameter t defined by (4.12).

$N=3$ case

$$\sigma^{(2)}=(\sigma_{11,11})=(1), \tag{4.17a}$$

$$\sigma^{(1)}=\begin{pmatrix}\sigma_{10,10} & \sigma_{10,01}\\ \sigma_{01,10} & \sigma_{01,01}\end{pmatrix}=\begin{pmatrix}0 & -t\\ -t & 1-t^2\end{pmatrix}, \tag{4.17b}$$

$$\sigma^{(0)}=\begin{pmatrix}\sigma_{1-1,1-1} & \sigma_{1-1,00} & \sigma_{1-1,-11}\\ \sigma_{00,1-1} & \sigma_{00,00} & \sigma_{00,-11}\\ \sigma_{-11,1-1} & \sigma_{-11,00} & \sigma_{-11,-11}\end{pmatrix}=\begin{pmatrix}0 & 0 & t^2\\ 0 & t & t^{5/2}-t^{1/2}\\ t^2 & t^{5/2}-t^{1/2} & t^3-t^2-t+1\end{pmatrix}, \tag{4.17c}$$

$$\sigma^{(-1)}=\begin{pmatrix}\sigma_{0-1,0-1} & \sigma_{0-1,-10}\\ \sigma_{-10,0-1} & \sigma_{-10,-10}\end{pmatrix}=\sigma^{(1)}, \tag{4.17d}$$

$$\sigma^{(-2)}=(\sigma_{-1-1,-1-1})=\sigma^{(2)}. \tag{4.17e}$$

These expressions have already been given in our earlier paper[33] with indices (1, 2, 3) instead of (1, 0, −1).

$N=4$ case

$$\sigma^{(3)}=(\sigma_{\frac32\frac32,\frac32\frac32})=(1), \tag{4.18a}$$

$$\sigma^{(2)}=\begin{pmatrix}\sigma_{\frac32\frac12,\frac32\frac12} & \sigma_{\frac32\frac12,\frac12\frac32}\\ \sigma_{\frac12\frac32,\frac32\frac12} & \sigma_{\frac12\frac32,\frac12\frac32}\end{pmatrix}=\begin{pmatrix}0 & -t^{3/2}\\ -t^{3/2} & 1-t^3\end{pmatrix}, \tag{4.18b}$$

$$\sigma^{(1)}=\begin{pmatrix}\sigma_{\frac32-\frac12,\frac32-\frac12} & \sigma_{\frac32-\frac12,\frac12\frac12} & \sigma_{\frac32-\frac12,-\frac12\frac32}\\ \sigma_{\frac12\frac12,\frac32-\frac12} & \sigma_{\frac12\frac12,\frac12\frac12} & \sigma_{\frac12\frac12,-\frac12\frac32}\\ \sigma_{-\frac12\frac32,\frac32-\frac12} & \sigma_{-\frac12\frac32,\frac12\frac12} & \sigma_{-\frac12\frac32,-\frac12\frac32}\end{pmatrix}=\begin{pmatrix}0 & 0 & t^3\\ 0 & t^2 & \sqrt{(1-t)(1-t^3)(1+t)t}\\ t^3 & \sqrt{(1-t)(1-t^3)(1+t)t} & (1-t^2)(1-t^3)\end{pmatrix}, \tag{4.18c}$$

$$\sigma^{(0)}=\begin{pmatrix}\sigma_{\frac32-\frac32,\frac32-\frac32} & \sigma_{\frac32-\frac32,\frac12-\frac12} & \sigma_{\frac32-\frac32,-\frac12\frac12} & \sigma_{\frac32-\frac32,-\frac32\frac32}\\ \sigma_{\frac12-\frac12,\frac32-\frac32} & \sigma_{\frac12-\frac12,\frac12-\frac12} & \sigma_{\frac12-\frac12,-\frac12\frac12} & \sigma_{\frac12-\frac12,-\frac32\frac32}\\ \sigma_{-\frac12\frac12,\frac32-\frac32} & \sigma_{-\frac12\frac12,\frac12-\frac12} & \sigma_{-\frac12\frac12,-\frac12\frac12} & \sigma_{-\frac12\frac12,-\frac32\frac32}\\ \sigma_{-\frac32\frac32,\frac32-\frac32} & \sigma_{-\frac32\frac32,\frac12-\frac12} & \sigma_{-\frac32\frac32,-\frac12\frac12} & \sigma_{-\frac32\frac32,-\frac32\frac32}\end{pmatrix}$$

$$=\begin{pmatrix}0 & 0 & 0 & -t^{9/2}\\ 0 & 0 & -t^{5/2} & t^2(1-t^3)\\ 0 & -t^{5/2} & (1-t^2)(t+t^2) & -t^{1/2}(1-t^2)(1-t^3)\\ -t^{9/2} & t^2(1-t^3) & -t^{1/2}(1-t^2)(1-t^3) & (1-t)(1-t^2)(1-t^3)\end{pmatrix}, \tag{4.18d}$$

$$\sigma^{(-1)}=\begin{pmatrix} \sigma_{\frac{1}{2}-\frac{3}{2},\frac{1}{2}-\frac{3}{2}} & \sigma_{\frac{1}{2}-\frac{3}{2},-\frac{1}{2}-\frac{1}{2}} & \sigma_{\frac{1}{2}-\frac{3}{2},-\frac{3}{2}\frac{1}{2}} \\ \sigma_{-\frac{1}{2}-\frac{1}{2},\frac{1}{2}-\frac{3}{2}} & \sigma_{-\frac{1}{2}-\frac{1}{2},-\frac{1}{2}-\frac{1}{2}} & \sigma_{-\frac{1}{2}-\frac{1}{2},-\frac{3}{2}\frac{1}{2}} \\ \sigma_{-\frac{3}{2}\frac{1}{2},\frac{1}{2}-\frac{3}{2}} & \sigma_{-\frac{3}{2}\frac{1}{2},-\frac{1}{2}-\frac{1}{2}} & \sigma_{-\frac{3}{2}\frac{1}{2},-\frac{3}{2}\frac{1}{2}} \end{pmatrix}=\sigma^{(1)}, \tag{4.18e}$$

$$\sigma^{(-2)}=\begin{pmatrix} \sigma_{-\frac{1}{2}-\frac{3}{2},-\frac{1}{2}-\frac{3}{2}} & \sigma_{-\frac{1}{2}-\frac{3}{2},-\frac{3}{2}-\frac{1}{2}} \\ \sigma_{-\frac{3}{2}-\frac{1}{2},-\frac{1}{2}-\frac{3}{2}} & \sigma_{-\frac{3}{2}-\frac{1}{2},-\frac{3}{2}-\frac{3}{2}} \end{pmatrix}=\sigma^{(2)}, \tag{4.18f}$$

$$\sigma^{(-3)}=(\sigma_{-\frac{3}{2}-\frac{3}{2},-\frac{3}{2}-\frac{3}{2}})=\sigma^{(3)}. \tag{4.18g}$$

Substituting these expressions into (3.10) we have explicit representation of the braid group generator g_i. From the eigenvalue equations for the total charge submatrices with the largest size, we can write down the reduction relations for the submatrices. It is very interesting to see that remaining smaller-size submatrices satisfy the same reduction relation. We next summarize the reduction relations for the generator g_i.

N=3 case

$$(g_i-1)(g_i+t^2)(g_i-t^3)=0, \tag{4.19}$$

or equivalently,

$$g_i^3=(1-t^2+t^3)g_i^2+(t^2-t^3+t^5)g_i-t^5. \tag{4.20}$$

N=4 case

$$(g_i-1)(g_i+t^3)(g_i-t^5)(g_i+t^6)=0, \tag{4.21}$$

or equivalently,

$$\begin{aligned} g_i^4=&(1-t^3+t^5-t^6)g_i^3\\ &+t^3(1-t^2+t^3+t^5-t^6+t^8)g_i^2\\ &+t^8(-1+t-t^3+t^6)g_i-t^{14}. \end{aligned} \tag{4.22}$$

The extension to general N can be done in a similar manner.

§5. Markov Traces and New Link Polynomials

We have succeeded in constructing a braid group representation B_n' generated by $\{g_1, g_2, \cdots, g_{n-1}\}$. The remaining task is to find a topological invariant of links defined on B_n'. As was mentioned in §2, the desired quantity must be invariant under the Markov moves (type I and II). That is, the *link polynomial* α, a topological invariant for links, must satisfy the conditions:

$$\alpha(AB)=\alpha(BA) \quad (A, B\in B_n'), \tag{5.1a}$$

$$\alpha(Ag_n)=\alpha(Ag_n^{-1})=\alpha(A) \quad (A\in B_n'). \tag{5.1b}$$

This quantity is easily constructed if we can find a linear functional defined on B_n', called the *Markov trace*[20,21] denoted by ϕ which has the following properties (Markov properties):

$$\phi(AB)=\phi(BA) \quad (A, B\in B_n'), \tag{5.2a}$$

$$\phi(Ag_n)=\tau\cdot\phi(A),\ \phi(Ag_n^{-1})=\bar{\tau}\cdot\phi(A)\ (A\in B_n'), \tag{5.2b}$$

with the parameters τ and $\bar{\tau}$ being given by

$$\tau=\phi(g_i),\ \bar{\tau}=\phi(g_i^{-1}) \quad (\text{for all } i). \tag{5.3}$$

Then the invariant polynomial $\alpha(A)$, $A\in B_n'$, is given by

$$\alpha(A)=(\tau\cdot\bar{\tau})^{-(n-1)/2}\left(\frac{\bar{\tau}}{\tau}\right)^{e(A)/2}\cdot\phi(A), \tag{5.4}$$

where $e(A)$ is the exponent sum of g_i's appearing in the braid A (e.g., for $A=g_1g_2g_3g_2^{-1}g_1$, $e(A)=1+1+1-1+1=3$). The property (5.1a) directly follows from (5.2a). The property (5.2b) is easily verified as follows:

$$\begin{aligned} \alpha(Ag_n)&=(\tau\cdot\bar{\tau})^{-(n+1-1)/2}\left(\frac{\bar{\tau}}{\tau}\right)^{e(Ag_n)/2}\cdot\phi(Ag_n)\\ &=(\tau\cdot\bar{\tau})^{-(n+1-1)/2}\left(\frac{\bar{\tau}}{\tau}\right)^{e(Ag_n)/2}\cdot\tau\cdot\phi(A)\\ &=\left[(\tau\cdot\bar{\tau})^{-1/2}\left(\frac{\bar{\tau}}{\tau}\right)^{1/2}\cdot\tau\right](\tau\cdot\bar{\tau})^{-(n-1)/2}\\ &\quad\times\left(\frac{\bar{\tau}}{\tau}\right)^{e(A)/2}\cdot\phi(A)\\ &=(\tau\cdot\bar{\tau})^{-(n-1)/2}\left(\frac{\bar{\tau}}{\tau}\right)^{e(A)/2}\\ &\quad\times\phi(A)=\alpha(A). \end{aligned} \tag{5.5}$$

Note that (5.2b) and $e(Ag_n)=e(A)+1$ have been used in the second and the third line, respectively.

We found[33] that the Markov trace $\phi(\cdot)$ associated with our braid group representation is given in the form

$$\phi(A)=\mathrm{Tr}(\mathrm{HA}), \tag{5.6}$$

where Tr stands for the ordinary trace (sum of diagonal elements) and the matrix H is given by a tensor product of a diagonal matrix h:

$$H=h\otimes h\otimes\cdots\otimes h\otimes\cdots, \tag{5.7}$$

with

$$(h)_{pq}=t^{-p}\delta_{pq}\bigg/\left(\sum_{k=-s}^{s}t^{-k}\right). \tag{5.8}$$

The denominator in (5.8) gives the proper normalization

$$\phi(I_n)=1\quad (I_n\text{: identity in } B_n'). \tag{5.9}$$

The Markov trace defined by (5.6)–(5.8) is a generalization of the Powers state.[40,41] To show that $\phi(\cdot)$ indeed has the Markov property, we first note that the type I Markov property (5.2a) reads

$$\mathrm{Tr}(\mathrm{HAB})=\mathrm{Tr}(\mathrm{HBA}). \tag{5.10}$$

To prove this, it is sufficient to show that

$$[H, g_i]=0. \tag{5.11}$$

In terms of the elements of the weight matrix, eq. (5.11) is written as

$$t^{-p}t^{-q}\sigma_{pq,kl}=\sigma_{pq,kl}t^{-k}t^{-l}, \tag{5.12}$$

which is easily verified by using the charge conservation condition ($p+q=k+l$). The type II Markov property (5.2b) can be satisfied if the following condition holds:

$$\sum_{l}\sigma_{kl,kl}t^{-l}\text{ is independent of } k. \tag{5.13}$$

This condition can be verified by direct calculations. Further, the parameters τ and $\bar{\tau}$ are found to be

$$\tau(t)=1/(1+t+t^2+\cdots+t^{N-1}), \tag{5.14a}$$

$$\bar{\tau}(t)=t^{N-1}/(1+t+t^2+\cdots+t^{N-1})=\tau(1/t). \tag{5.14b}$$

Then the link polynomial α, constructed from the N-state vertex model, is explicitly given by

$$\alpha(A)=[t^{-(N-1)/2}(1+t+t^2+\cdots+t^{N-1})]^{n-1}\times[t^{(N-1)/2}]^{e(A)}\phi(A).\quad (A\in B_n') \tag{5.15}$$

From the reduction relation (4.14), the Alexander-Conway relation for the Jones polynomial (the $N=2$ case) is derived in the following manner. Consider a braid group element A expressing a link and which has an intersection represented by a generator g_i. The element A has the form:

$$A=A'g_iA''\quad (A, A''\in B_n'). \tag{5.16}$$

We rewrite the relation (4.14) as

$$g_i=(1-t)+t\cdot g_i^{-1}, \tag{5.17}$$

by operating g_i^{-1} on both sides of (4.14). Substituting the expression (5.17) into (5.16), we have

$$A=(1-t)A'A''+t\cdot A'g_i^{-1}A''. \tag{5.18}$$

Due to the linear property of the functional $\phi(\cdot)$, we obtain

$$\phi(A)=(1-t)\phi(A'A'')+t\cdot\phi(A'g_i^{-1}A''). \tag{5.19}$$

To translate the relation (5.19) into the relation between the link polynomials, we note the relation

$$\begin{aligned}e(A)&=e(A'A'')+1\\&=e(A'g_i^{-1}A'')+2.\end{aligned} \tag{5.20}$$

Then from (5.5), we arrive at

$$\begin{aligned}\alpha(A)&=(1-t)\left(\frac{\bar{\tau}}{\tau}\right)^{1/2}\alpha(A'A'')+t\left(\frac{\bar{\tau}}{\tau}\right)\alpha(A'g_i^{-1}A'')\\&=(1-t)t^{1/2}\alpha(A'A'')+t^2\alpha(A'g_i^{-1}A''),\end{aligned} \tag{5.21}$$

or, equivalently, the Alexander-Conway relation

$$\alpha(L_+)=(1-t)t^{1/2}\alpha(L_0)+t^2\alpha(L_-). \tag{5.22}$$

In (5.22), by L_+, L_0 and L_-, we have denoted links which have the configuration of g_i, g_i^0 ($=I$: identity) and g_i^{-1}, respectively, at an intersection. For $N\geqq 3$ cases, we do not have the Alexander-Conway relation (5.22) since the different reduction relations hold. Instead, we have a new generalized relation for each $N\geqq 3$ as shown below which we call the N-th order Alexander-Conway relation.

$N=3$

$$\alpha(L_{2+})=t(1-t^2+t^3)\alpha(L_+)$$
$$+t^2(t^2-t^3+t^5)\alpha(L_0)-t^8\alpha(L_-). \tag{5.23}$$

$N=4$

$$\alpha(L_{3+})=t^{3/2}(1-t^3+t^5-t^6)\alpha(L_{2+})$$
$$+t^6(1-t^2+t^3+t^5-t^6+t^8)\alpha(L_+)$$
$$+t^{9/2}t^8(-1+t-t^3+t^6)\alpha(L_0)$$
$$-t^{20}\alpha(L_-). \tag{5.24}$$

In the above, by L_{3+}, L_{2+}, L_+, L_0 and L_- we have denoted links which have, at a particular intersection, the configuration represented by g_i^3, g_i^2, g_i^1, g_i^0 and g_i^{-1}, respectively.

§6. Summary

In this paper, the first of a series, we have shown that general braid group representation can be obtained from solvable models in statistical mechanics, both vertex and IRF models, by making the spectral parameter u tend to infinity. We have also shown that the existence of the well-defined $u\to\infty$ limit requires the starting model to be critical and to be parametrized by trigonometric or hyperbolic functions. To have explicit examples of the braid group representations, the method has been applied to a sequence of the N-state vertex models. With the Markov traces constructed by generalizing the Powers state, we have obtained a sequence of new link polynomials.

Our new link polynomials presented in this paper are different from the Jones polynomial since ours do not obey the Alexander-Conway relation. Ours are more powerful than the Jones polynomial and its two-variable extension. It should be remembered that Jones polynomial is rederived as the $N=2$ case. In general, we believe that a link polynomial corresponding to a larger N is more powerful. To be more concrete, even in the simplest $N=3$ case, our link polynomial can discriminate the Birman's examples[31] which the Jones polynomial and its two-variable extension cannot. Details will be the subject of the next paper of the series.

As for the application of the general construction procedure of the braid group, we have restricted ourselves to the cases of vertex models. For IRF models, we have also an infinite sequence of solvable models[6-11] to which our method can be directly applied. Construction of new link polynomials from these IRF models will be presented elsewhere.

Acknowledgments

The authors wish to express their sincere thanks to Professor Kazuhiko Aomoto, Nagoya University. In 1981, he suggested to the authors the similarity between the Yang-Baxter relation and the braid group. The authors also thank Atsuo Kuniba and Tetsuo Deguchi for helpful discussions.

Appendix: Boltzmann Weights for the N-State Vertex Models

In this Appendix, for the reader's convenience, we list the explicit form of Boltzmann weights for the first three models in the series of the N-state vertex models. The parametrizations given below are slightly modified from the original ones[38,39] in order to satisfy the crossing symmetry (4.5).

$N=2$ ($s=1/2$) case

$$S^{1/2\;1/2}_{1/2\;1/2}(u)=S^{-1/2\;-1/2}_{-1/2\;-1/2}(u)=\sinh(\lambda-u), \tag{A·1}$$
$$S^{\;\;1/2\;\;\;1/2}_{-1/2\;-1/2}(u)=S^{-1/2\;-1/2}_{\;\;1/2\;\;\;1/2}(u)=\sinh u, \tag{A·2}$$
$$S^{\;\;1/2\;-1/2}_{-1/2\;\;\;1/2}(u)=S^{-1/2\;\;\;1/2}_{\;\;1/2\;-1/2}(u)=\sinh\lambda. \tag{A·3}$$

$N=3$ ($s=1$) case

$$S^{1\,1}_{1\,1}(u)=S^{-1\,-1}_{-1\,-1}(u)=\sinh(\lambda-u)\sinh(2\lambda-u), \tag{A·4}$$
$$S^{\;\;1\;\;\;1}_{-1\,-1}(u)=S^{-1\,-1}_{\;\;1\;\;\;1}(u)=\sinh u\sinh(\lambda+u), \tag{A·5}$$
$$S^{\;\;1\,-1}_{-1\;\;\;1}(u)=S^{-1\;\;\;1}_{\;\;1\,-1}(u)=\sinh\lambda\sinh 2\lambda, \tag{A·6}$$
$$S^{1\,1}_{0\,0}(u)=S^{0\,0}_{1\,1}(u)=S^{-1\,-1}_{\;\;0\;\;\;0}(u)=S^{\;\;0\;\;\;0}_{-1\,-1}(u)=\sinh u\sinh(\lambda-u), \tag{A·7}$$
$$S^{1\,0}_{0\,1}(u)=S^{0\,1}_{1\,0}(u)=S^{-1\;\;0}_{\;\;0\,-1}(u)=S^{\;\;0\,-1}_{-1\;\;\;0}(u)=\sinh 2\lambda\sinh(\lambda-u), \tag{A·8}$$
$$S^{0\;\;\;1}_{0\,-1}(u)=S^{0\,-1}_{0\;\;\;1}(u)=S^{\;\;1\,0}_{-1\,0}(u)=S^{-1\,0}_{\;\;1\,0}(u)=\sinh 2\lambda\sinh u, \tag{A·9}$$
$$S^{0\,0}_{0\,0}(u)=\sinh\lambda\sinh 2\lambda-\sinh u\sinh(\lambda-u). \tag{A·10}$$

N=4 (s=3/2) case

$$S^{3/2\,3/2}_{3/2\,3/2}(u)=S^{-3/2\,-3/2}_{-3/2\,-3/2}(u)=\sinh(\lambda-u)\sinh(2\lambda-u)\sinh(3\lambda-u), \tag{A·11}$$

$$S^{3/2\,3/2}_{-3/2\,-3/2}(u)=S^{-3/2\,-3/2}_{3/2\,3/2}(u)=\sinh u\sinh(\lambda+u)\sinh(2\lambda+u), \tag{A·12}$$

$$S^{3/2\,-3/2}_{-3/2\,3/2}(u)=S^{-3/2\,3/2}_{3/2\,-3/2}(u)=\sinh\lambda\sinh 2\lambda\sinh 3\lambda, \tag{A·13}$$

$$S^{3/2\,3/2}_{1/2\,1/2}(u)=S^{1/2\,1/2}_{3/2\,3/2}(u)=S^{-3/2\,-3/2}_{-1/2\,-1/2}(u)=S^{-1/2\,-1/2}_{-3/2\,-3/2}(u)=\sinh u\sinh(\lambda-u)\sinh(2\lambda-u), \tag{A·14}$$

$$S^{3/2\,3/2}_{-1/2\,-1/2}(u)=S^{-1/2\,-1/2}_{3/2\,3/2}(u)=S^{-3/2\,-3/2}_{1/2\,1/2}(u)=S^{1/2\,1/2}_{-3/2\,-3/2}(u)=\sinh u\sinh(\lambda-u)\sinh(\lambda+u), \tag{A·15}$$

$$S^{3/2\,1/2}_{1/2\,3/2}(u)=S^{1/2\,3/2}_{3/2\,1/2}(u)=S^{-3/2\,-1/2}_{-1/2\,-3/2}(u)=S^{-1/2\,-3/2}_{-3/2\,-1/2}(u)=\sinh 3\lambda\sinh(\lambda-u)\sinh(2\lambda-u), \tag{A·16}$$

$$\begin{aligned}S^{3/2\,1/2}_{-1/2\,1/2}(u)&=S^{-1/2\,1/2}_{3/2\,1/2}(u)=S^{-3/2\,-1/2}_{1/2\,-1/2}(u)=S^{1/2\,-1/2}_{-3/2\,-1/2}(u)=S^{1/2\,3/2}_{1/2\,-1/2}(u)\\&=S^{1/2\,-1/2}_{1/2\,3/2}(u)=S^{-1/2\,-3/2}_{-1/2\,1/2}(u)=S^{-1/2\,1/2}_{-1/2\,-3/2}(u)\\&=2\sqrt{\sinh\lambda\sinh 3\lambda}\cosh\lambda\sinh u\sinh(\lambda-u),\end{aligned} \tag{A·17}$$

$$S^{3/2\,1/2}_{-3/2\,-1/2}(u)=S^{-3/2\,-1/2}_{3/2\,1/2}(u)=S^{1/2\,3/2}_{-1/2\,-3/2}(u)=S^{-1/2\,-3/2}_{1/2\,3/2}(u)=\sinh 3\lambda\sinh u\sinh(\lambda+u), \tag{A·18}$$

$$S^{3/2\,-1/2}_{-1/2\,3/2}(u)=S^{-1/2\,3/2}_{3/2\,-1/2}(u)=S^{-3/2\,1/2}_{1/2\,-3/2}(u)=S^{1/2\,-3/2}_{-3/2\,1/2}(u)=\sinh 2\lambda\sinh 3\lambda\sinh(\lambda-u), \tag{A·19}$$

$$S^{3/2\,-1/2}_{-3/2\,1/2}(u)=S^{-3/2\,1/2}_{3/2\,-1/2}(u)=S^{-1/2\,3/2}_{1/2\,-3/2}(u)=S^{1/2\,-3/2}_{-1/2\,3/2}(u)=\sinh 2\lambda\sinh 3\lambda\sinh u, \tag{A·20}$$

$$S^{1/2\,1/2}_{1/2\,1/2}(u)=S^{-1/2\,-1/2}_{-1/2\,-1/2}(u)=\sinh(\lambda-u)[\sinh 2\lambda\sinh 3\lambda-\sinh u\sinh(\lambda-u)], \tag{A·21}$$

$$S^{1/2\,1/2}_{-1/2\,-1/2}(u)=S^{-1/2\,-1/2}_{1/2\,1/2}(u)=\sinh u[\sinh 2\lambda\sinh 3\lambda-\sinh u\sinh(\lambda-u)], \tag{A·22}$$

$$S^{1/2\,-1/2}_{-1/2\,1/2}(u)=S^{-1/2\,1/2}_{1/2\,-1/2}(u)=\sinh 2\lambda[\sinh\lambda\sinh 3\lambda-2\cosh\lambda\sinh u\sinh(\lambda-u)]. \tag{A·23}$$

The above Boltzmann weights satisfy the equation

$$\sum_{p,q}S^{ql}_{pk}(-u)S^{ip}_{jq}(u)=\rho(u)\cdot\delta_{ik}\delta_{jl}, \tag{A·24}$$

where the unitarity factor $\rho(u)$ is given by

$$\begin{aligned}\rho(u)&=\prod_{k=1}^{N-1}(\sinh^2 k\lambda-\sinh^2 u)\\&=S^{ss}_{ss}(-u)S^{ss}_{ss}(u).\end{aligned} \tag{A·25}$$

References

1) L. D. Faddeev: Sov. Sci. Rev. Math. Phys. **C1** (1981) 107; H. B. Thacker: Rev. Mod. Phys. **53** (1981) 253; P. P. Kulish and E. K. Sklyanin: Lecture Notes in Physics (Springer-Verlag, New York, 1982) Vol. 151, p. 61; M. Wadati: in *Dynamical Problems in Soliton Systems*, ed. S. Takeno (Springer-Verlag, 1985) p. 68.
2) M. Karowski, H. J. Thun, T. T. Truong and P. H. Weisz: Phys. Lett. **67B** (1977) 321; A. B. Zamolodchikov and A. B. Zamolodchikov: Ann. Phys. (N.Y.) **120** (1979) 253; K. Sogo, M. Uchinami, A. Nakamura and M. Wadati: Prog. Theor. Phys. **66** (1981) 1284; K. Sogo, M. Uchinami, Y. Akutsu and M. Wadati: Prog. Theor. Phys. **68** (1982) 508.
3) R. J. Baxter: *Exactly Solved Models in Statistical Mechanics* (Academic Press, London, 1982).
4) F. Y. Wu: Phys. Rev. **B4** (1971) 2312; L. P. Kadanoff and F. J. Wegner: Phys. Rev. **B4** (1971) 3989.
5) A. B. Zamolodchikov: Commun. Math. Phys. **69** (1979) 165.
6) G. E. Andrews, R. J. Baxter and P. J. Forrester: J. Stat. Phys. **35** (1984) 193.
7) A. Kuniba, Y. Akutsu and M. Wadati: J. Phys. Soc. Jpn. **55** (1986) 1092, 2170 and 3338; A. Kuniba, Y. Akutsu and M. Wadati: Phys. Lett. **116A** (1986) 382; **117A** (1986) 358. R. J. Baxter and G. E. Andrews: J. Stat. Phys. **44** (1986) 249; G. E. Andrews and R. J. Baxter: J. Stat. Phys. **44** (1986) 713.
8) Y. Akutsu, A. Kuniba and M. Wadati: J. Phys. Soc. Jpn. **55** (1986) 1466; 1880.
9) A. Kuniba, Y. Akutsu and M. Wadati: J. Phys. Soc. Jpn. **55** (1986) 2605.
10) Y. Akutsu, A. Kuniba and M. Wadati: J. Phys. Soc. Jpn. **55** (1986) 2907; E. Date, M. Jimbo, T. Miwa and M. Okado: Lett. Math. Phys. **12** (1986) 209; E. Date, M. Jimbo, T. Miwa and M. Okado: Phys. Rev. **B35** (1987) 2105; E. Date, M. Jimbo, A. Kuniba, T. Miwa and M. Okado: preprint.
11) M. Wadati and Y. Akutsu: *Exactly Solvable Models in Statistical Mechanics*, to appear in Lecture Notes in Physics, ed. M. Lakshmanan (Springer-Verlag).
12) A. Kuniba, Y. Akutsu and M. Wadati: J. Phys. Soc. Jpn. **55** (1986) 3285.
13) A. A. Belavin, A. M. Polyakov and A. B. Zamolodchikov: Nucl. Phys. **B241** [FS] (1984) 333; D. Friedan, Z. Qiu and S. Shenker: Phys. Rev. Lett. **52** (1984) 1575.
14) D. A. Huse: Phys. Rev. **B30** (1984) 3908.
15) G. E. Andrews: *The Theory of Partitions* (Addison-Wesley, 1976).
16) R. J. Baxter: J. Phys. **A13** (1980) L61.

17) C. A. Tracy: preprint, 1986.
18) J. Lepowsky and R. L. Wilson: Proc. Nat. Acad. Sci. USA **78** (1981) 7254.
19) M. Virasoro: Phys. Rev. **D1** (1970) 2933.
20) V. F. R. Jones: Invent. Math. **72** (1983) 1.
21) V. F. R. Jones: Bull. Amer. Math. Soc. **12** (1985) 103.
22) H. N. V. Temperley and E. H. Lieb: Proc. Soc. London A **322** (1971) 251.
23) E. H. Lieb and F. Y. Wu: in *Phase Transitions and Critical Phenomena*, ed. C. Domb and M. S. Green, (Academic Press, London, 1972) Vol. 1, p. 331.
24) R. J. Baxter, S. B. Kelland and F. Y. Wu: J. Phys. **A9** (1976) 397.
25) N. Bourbaki: *Groupes et algebres de Lie* (Hermann, Paris, 1968) Chap. 4.
26) E. Artin: Ann. Math. **48** (1947) 101.
27) Y. S. Wu: Phys. Rev. Lett. **52** (1984) 2103.
28) J. W. Alexander: Trans. Amer. Math. Soc. **30** (1928) 275.
29) O. Freyd, D. Yetter, J. Hoste, W. B. R. Lickorish, K. Millett and A. Ocneanu: Bull. Amer. Math. Soc. **12** (1985) 239.
30) J. H. Conway: in *Computational Problems in Abstract Algebra*, ed. J. Leech (Pergamon Press, London, 1969) p. 329.
31) J. S. Birman: Invent. Math. **81** (1985) 287.
32) T. Kanenobu: Math. Ann. **275** (1986) 555.
33) Y. Akutsu and M. Wadati: J. Phys. Soc. Jpn. **56** (1987) 839.
34) J. W. Alexander: Proc. Nat. Acad. **9** (1923) 93.
35) A. A. Markov: Recueil Math. Moscov (1935) 73.
36) R. J. Baxter: Ann. Phys. (N.Y.) **70** (1972) 193.
37) R. J. Baxter: Physica **18D** (1986) 321.
38) K. Sogo, Y. Akutsu and T. Abe: Prog. Theor. Phys. **70** (1983) 730; 739.
39) A. B. Zamolodchikov and V. A. Fateev: Sov. J. Nucl. Phys. **32** (1980) 298.
40) R. T. Powers: Ann. Math. **86** (1967) 138.
41) M. Pimsner and S. Popa: unpublished.

Journal of the Physical Society of Japan
Vol. 56, No. 10, October, 1987, pp. 3464–3479

Exactly Solvable Models and New Link Polynomials. II. Link Polynomials for Closed 3-Braids

Yasuhiro AKUTSU, Tetsuo DEGUCHI† and Miki WADATI†

Institute of Physics, Kanagawa University, Rokkakubashi, Kanagawa-ku, Yokohama 221

†Institute of Physics, College of Arts and Sciences, University of Tokyo, Meguro-ku, Tokyo 153

(Received May 14, 1987)

Using a generalized Alexander-Conway relation derived from a three-state exactly solvable model in statistical mechanics, new invariant polynomials for knots and links are explicitly evaluated. It is shown that the invariant polynomials for closed 3-braids are obtained recursively. It is also shown that the invariant polynomials are more powerful than the Jones polynomials.

§1. Introduction

Recently, the study of exactly solvable models in statistical mechanics has made remarkable progress.[1,2] By solving the Yang-Baxter relation (the star-triangle equation), many solvable models have been found.[3-7] In fact, there exists an infinite number (at least $\infty\times\infty$ number) of exactly solvable models in two-dimensional statistical mechanics,[8,9] which we call the Grand Hierarchy.[5,8]

This development should be of great significance both in physics and in mathematics. In previous papers,[10,11] we have found a novel connection between exactly solvable models in statistical mechanics and the knot theory in mathematics.[12,13] A general method has been obtained for constructing invariant polynomials for knots and links (link polynomials) from exactly solvable models. The present paper is devoted to a study of the $N=3$ case, where N is the number of states of the model. We explicitly evaluate link polynomials up to the closed 3-braids.

The paper is organized as follows. To make the paper self-contained and unify the notation, we briefly summarize the salient points of the theory[10,11] in §2. In §3, we present several formulae which are useful for later discussions. We show, in §4, that the link polynomials for closed 3-braids can be calculated by the recursion formulae. In §5, two special closed 3-braids by Birman[14] are examined. We find that the new link polynomials for the two closed braids are different. This implies that our new link polynomials are more powerful than the Jones polynomials[15] even in the $N=3$ case. A list of link polynomials is given in the last section.

§2. Link Polynomials and Solvable Models

We will briefly summarize the theory of new link polynomial[10,11] based on the exactly solvable models in statistical mechanics.

We want to classify the configurations of one-dimensional objects (strings) in three-dimensional space. A knot is a closed string which does not intersect with itself. An assembly of knots with mutual entanglements is called a link. Knots and links appear in many branches of physics (polymers, dislocation lines, magnetic fluxes, etc.) and even in daily life (shoe strings, plaited threads).

Consider n strings whose ends are fixed to base points on upper and lower bars (Fig. 1). The configuration of the n strings is called an n-braid. A general n-braid is constructed by successive applications of operator b_i (Fig. 2). The $(n-1)$ generators, $b_1, b_2, \cdots, b_{n-1}$, define the braid group B_n satisfying

$$b_i b_{i+1} b_i = b_{i+1} b_i b_{i+1},$$
$$b_i b_j = b_j b_i, \quad |i-j| \geqq 2. \qquad (2.1)$$

Any link can be represented by a closed braid. Since there are many different representations of a link by a closed braid, the following

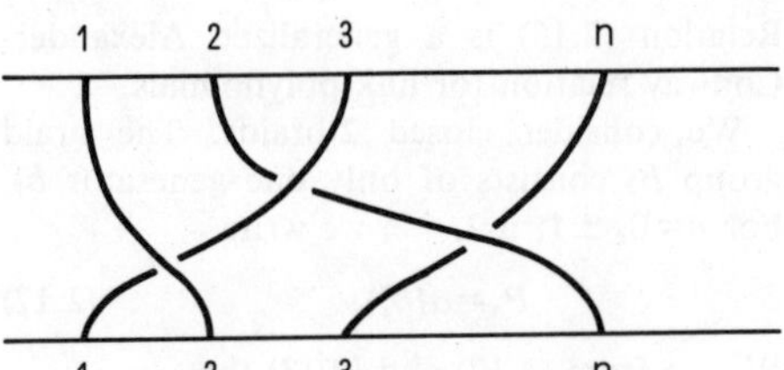

Fig. 1. An example of n braids.

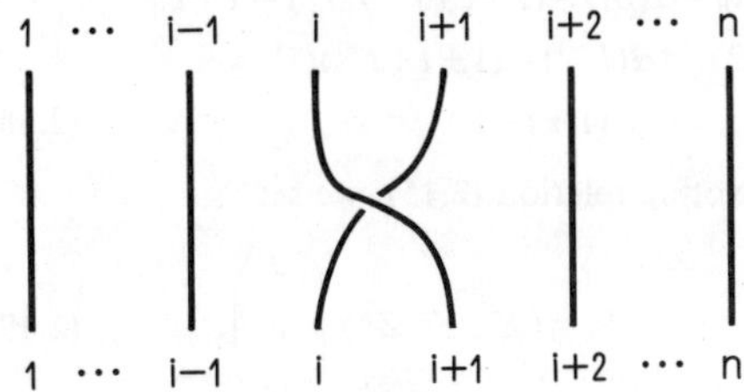

Fig. 2. Generator b_i.

theorem by A. A. Markov is important. The equivalent braids expressing the same link are mutually transformed by successive applications of two types of operations (type I and type II Markov moves):

$$\text{I.}\quad AB \longrightarrow BA \ (A, B\in B_n),$$
$$\text{II.}\quad A \longrightarrow Ab_n, \quad A \longrightarrow Ab_n^{-1}, \quad (A\in B_n,\ b_n\in B_{n+1}). \tag{2.2}$$

Therefore, the link polynomial α which is a topological invariant for links should satisfy the relations

$$\text{I.}\quad \alpha(AB)=\alpha(BA),$$
$$\text{II.}\quad \alpha(A)=\alpha(Ab_n)=\alpha(Ab_n^{-1}). \tag{2.3}$$

The link polynomial α is easily found if we can construct a functional ϕ (the Markov trace) defined on B_n which has the following properties:

$$\text{I.}\quad \phi(AB)=\phi(BA),$$
$$\text{II.}\quad \phi(Ab_n)=\tau\cdot\phi(A), \quad \phi(Ab_n^{-1})=\bar{\tau}\cdot\phi(A), \tag{2.4}$$

where τ and $\bar{\tau}$ are defined by

$$\tau=\phi(b_i), \quad \bar{\tau}=\phi(b_i^{-1}), \quad \text{for all } i. \tag{2.5}$$

With the Markov trace ϕ, α is given by

$$\alpha(A)=(\tau\cdot\bar{\tau})^{-(n-1)/2}\left(\frac{\bar{\tau}}{\tau}\right)^{e/2}\phi(A), \tag{2.6}$$

where constant e is the exponent sum of b_i's in $A\in B_n$.

We have reported two findings in previous papers[10,11] which lead to successful constructions of new link polynomials. First, a representation of the braid group B_n is derived from the Boltzmann weights (the S matrices) of exactly solvable models. Second, the Markov trace is explicitly introduced. Our method is general enough to start with the N-state solvable model, but for the following, we restrict ourselves to the $N=3$ case.

We construct the weight function $w_{kl,pq}$ from the Boltzmann weight S_{kl}^{ij} of the 3-state model;[16]

$$w_{kl,pq}=\lim_{u\to\infty} 4\, e^{3\lambda-2u}\, e^{1/2(k+l-p-q)u} S_{qk}^{lp}(u). \tag{2.7}$$

The explicit forms of $w_{kl,pq}$ are given in refs. 10 and 11. Note that in this paper b_i's are used for both generators and their representations. In terms of the weight function $w_{kl,pq}$, a representation of B_n is given by

$$b_i=\sum_{klpq} w_{kl,pq} I^{(1)}\otimes\cdots\otimes I^{(i-1)}\otimes E_{kl}^{(i)}\otimes E_{pq}^{(i+1)} \otimes I^{(i+2)}\otimes\cdots. \tag{2.8}$$

Here $\otimes$ represents the direct product of 3×3 matrices, $I^{(i)}$ the identity matrix acting on the i-th position, and E_{kl} the matrix whose elements are $(E_{kl})_{ij}=\delta_{ki}\delta_{lj}$. The Markov trace $\phi(A)$ associated with our braid group representation is given by

$$\phi(A)=\mathrm{Tr}\,(HA), \tag{2.9}$$

where Tr stands for the trace of matrix, and the matrix H has the form

$$H=h^{(1)}\otimes h^{(2)}\otimes\cdots\otimes h^{(i)}\otimes\cdots, \tag{2.10}$$

with

$$h^{(i)}=\frac{1}{1+t+t^2}\begin{pmatrix}1&0&0\\0&t&0\\0&0&t^2\end{pmatrix}, \quad t=e^{2\lambda}. \tag{2.11}$$

Using the Markov trace (2.9), we have

$$\tau=\phi(b_i)=\frac{1}{1+t+t^2},$$
$$\bar{\tau}=\phi(b_i^{-1})=\frac{t^2}{1+t+t^2}. \quad (2.12)$$

Then we find that the link polynomial $\alpha(A)$ is expressed as

$$\alpha(A)=(1+t+t^{-1})^{n-1}t^e\phi(A),\ A\in B_n, \quad (2.13)$$

where e is the exponent sum of b_i's in A.

With the explicit forms of $w_{kl,pq}$, we obtain a cubic relation for b_i:

$$b_i^3=(t^3-t^2+1)b_i^2+(t^5-t^3+t^2)b_i-t^5. \quad (2.14)$$

This relation helps to reduce the powers of the generators. We introduce braids L_{3+}, L_{2+}, L_+ and L_0 (Fig. 3). Noting the exponent sum e and the cubic relation (2.14), we have

$$\begin{aligned}\alpha(L_{3+})=&(X+Y+Z)\alpha(L_{2+})\\ &-(XY+YZ+ZX)\alpha(L_+)\\ &+XYZ\alpha(L_0),\end{aligned} \quad (2.15)$$

where

$$X=t,\quad Y=-t^3,\quad Z=t^4. \quad (2.16)$$

Relation (2.15) is a generalized Alexander-Conway relation for link polynomials.

We consider closed 2-braids. The braid group B_2 consists of only one generator b_1. For $n=0, \pm1, \pm2, \cdots$, we write

$$P_n=\alpha(b_1^n). \quad (2.17)$$

We see from (2.12) and (2.13) that

$$\begin{aligned}&P_1=\alpha(b_1)=(1+t+t^{-1})t(1+t+t^2)^{-1}=1,\\ &P_0=\alpha(b_1^0)=(1+t+t^{-1})\cdot1\cdot1=1+t+t^{-1},\\ &P_{-1}=\alpha(b_1^{-1})=(1+t+t^{-1})t^{-1}\\ &\quad\cdot t^2(1+t+t^2)^{-1}=1.\end{aligned} \quad (2.18)$$

Noting relation (2.15), we set

$$P_n=(X^n\ Y^n\ Z^n)\begin{pmatrix}u\\ v\\ w\end{pmatrix}. \quad (2.19)$$

Then, we have

$$\begin{pmatrix}P_{-1}\\ P_0\\ P_1\end{pmatrix}=D\begin{pmatrix}u\\ v\\ w\end{pmatrix}, \quad (2.20)$$

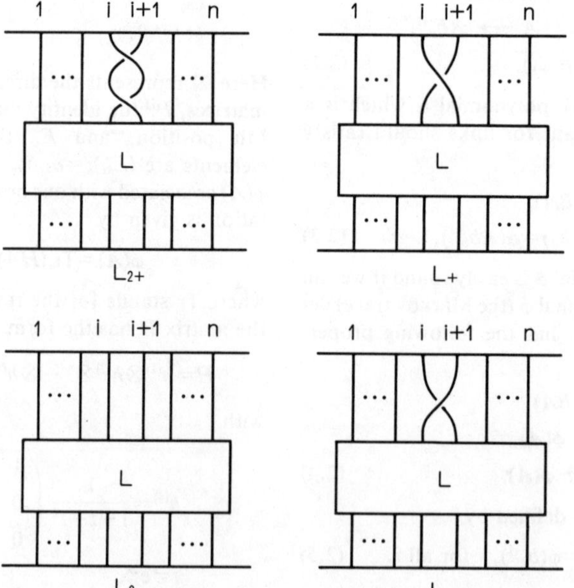

Fig. 3. Braids L_{3+}, L_{2+}, L_+ and L_0.

where the matrix D is given by

$$D=\begin{pmatrix} X^{-1} & Y^{-1} & Z^{-1} \\ 1 & 1 & 1 \\ X & Y & Z \end{pmatrix}. \tag{2.21}$$

$$\begin{aligned} P_n &= \alpha(b_1^n) \\ &= (X^n\ Y^n\ Z^n)D^{-1}\begin{pmatrix} P_{-1} \\ P_0 \\ P_1 \end{pmatrix} \\ &= (-1)^n t^{3n} + \frac{(t^{n-2}+t^{n+2})+(t^{n-1}+t^{n+1})+(t^n+t^{4n})}{1+t+t^{-1}}. \end{aligned} \tag{2.22}$$

This result for closed 2-braids will be used in later sections.

§3. Useful Formulae

To evaluate link polynomials for closed 3-braids, we prepare several formulae which turn out to be very useful. The 3-braid group B_3 consists of two generators b_1 and b_2. In the following formulae, l, m, n, l_i, and m_i are 0, $\pm 1, \pm 2, \cdots$.

1) From the property of the braid group

$$b_1 b_2 b_1 = b_2 b_1 b_2, \tag{3.1}$$

we have

$$b_1 b_2 b_1^n = b_2^n b_1 b_2, \tag{3.2a}$$

$$b_2 b_1 b_2^n = b_1^n b_2 b_1, \tag{3.2b}$$

$$b_1^n (b_1 b_2)^{-1} = (b_1 b_2)^{-1} b_2^n, \tag{3.2c}$$

$$b_2^n (b_2 b_1)^{-1} = (b_2 b_1)^{-1} b_1^n, \tag{3.2d}$$

and

$$\Delta^{2n} b_1^m = b_1^m \Delta^{2n}, \tag{3.3a}$$

$$\Delta^{2n} b_2^m = b_2^m \Delta^{2n}, \tag{3.3b}$$

$$\Delta^{2n+1} b_1^m = b_2^m \Delta^{2n+1}, \tag{3.3c}$$

$$\Delta^{2n+1} b_2^m = b_1^m \Delta^{2n+1}, \tag{3.3d}$$

where

$$\Delta = b_1 b_2 b_1 = b_2 b_1 b_2. \tag{3.4}$$

2) From (3.1) and the property of the Markov move I

$$\alpha(AB) = \alpha(BA), \tag{3.5}$$

we have

Therefore, we obtain link polynomials for closed 2-braids;

$$\alpha(b_1^{l_1} b_2^{m_1} b_1^{l_2} b_2^{m_2} \cdots) = \alpha(b_2^{l_1} b_1^{m_1} b_2^{l_2} b_1^{m_2} \cdots), \tag{3.6a}$$

$$\alpha(\Delta^{2n} b_1^{l_1} b_2^{m_1} \cdots b_1^{l_p} b_2^{m_p}) = \alpha(\Delta^{2n} b_2^{m_1} b_1^{l_2} \cdots b_2^{m_p} b_1^{l_1}), \tag{3.6b}$$

$$\alpha(\Delta^2 b_1^l b_2^m) = \alpha(b_1 b_2^{l+2} b_1 b_2^{m+2}), \tag{3.7a}$$

$$\alpha(\Delta^4 b_1^l b_2^m) = \alpha(b_1 b_2^3 b_1 b_2^{l+3} b_1 b_2^{m+3}), \tag{3.7b}$$

$$\alpha(\Delta^6 b_1^l b_2^m) = \alpha(b_1 b_2^3 b_1 b_2^{l+4} b_1 b_2^3 b_1 b_2^{m+4}), \tag{3.7c}$$

$$\alpha(\Delta^8 b_1^l b_2^m) = \alpha(b_1 b_2^3 b_1 b_2^5 b_1 b_2^3 b_1 b_2^{l+4} b_1 b_2^{m+4}), \tag{3.7d}$$

$$\alpha(\Delta^{4n}) = \alpha((b_1 b_2^3)^{3n}), \tag{3.8a}$$

$$\alpha((b_1^2 b_2^2)^n) = \alpha(\Delta^{2n} b_1^{-2n}), \tag{3.8b}$$

and in particular,

$$\alpha\left(\Delta^{2n} \cdot \prod_{i=1}^{p} (b_1^{l_i} b_2^{m_i})\right) = \alpha\left(\Delta^{2(n-p)} \cdot \prod_{i=1}^{p} (b_1 b_2^3 b_1 b_2^{l_i+3} b_1 b_2^{m_i+3})\right). \tag{3.9}$$

3) When we use the generalized Alexander-Conway relation (2.15), we may obtain an important recursion formula. Let us define the following quantities

$$\begin{aligned} Q_{l_1 \cdots l_p m_1 \cdots m_p} &= Q^{l_1 \cdots l_p}_{m_1 \cdots m_p} \\ &= \alpha(b_1^{l_1} b_2^{l_2} b_1^{l_3} \cdots b_2^{l_p} b_1^{m_1} \cdots b_2^{m_p}), \quad p;\ \text{even} \\ &= \alpha(b_1^{l_1} \cdots b_2^{l_{p-1}} b_1^{l_p} b_2^{m_m} \cdots b_2^{m_p}), \quad p;\ \text{odd} \end{aligned} \tag{3.10}$$

$$Q = Q^{a_1 \cdots a_p}_{b_1 \cdots b_p}, \quad a_i, b_i = -1, 0, 1, \tag{3.11}$$

$$V_n = v_n^\alpha = \begin{pmatrix} X^n \\ Y^n \\ Z^n \end{pmatrix}, \quad \alpha = -1, 0, 1, \tag{3.12}$$

$$V_{l_1\cdots l_p}=V_{l_1}\otimes V_{l_2}\otimes\cdots\otimes V_{l_p}, \tag{3.13}$$

$$D(p)=D_1\otimes D_2\otimes\cdots\otimes D_p, \tag{3.14}$$

$$D_i=\begin{pmatrix} X^{-1} & Y^{-1} & Z^{-1} \\ 1 & 1 & 1 \\ X & Y & Z \end{pmatrix},\quad i=1,2,\cdots,p. \tag{3.15}$$

Noting relation (2.15), we set

$$\begin{aligned} Q_{l_1\cdots l_p m_1\cdots m_p} &= {}^tV_{l_1\cdots l_p}\hat{Q}V_{m_1\cdots m_p} \\ &= \sum_{\alpha_i=-1,0,1}\cdots\sum_{\beta_i=-1,0,1}\hat{Q}^{\alpha_1\cdots\alpha_p}_{\beta_1\cdots\beta_p}v^{\alpha_1}_{l_1}\cdots v^{\alpha_p}_{l_p}v^{\beta_1}_{m_1}\cdots v^{\beta_p}_{m_p}, \end{aligned} \tag{3.16}$$

where the superscript t indicates the transpose of the matrix. Since matrices Q and $\hat{Q}$ are related as

$$Q=D(p)\hat{Q}^tD(p), \tag{3.17}$$

we arrive at

$$Q_{l_1\cdots l_p m_1\cdots m_p}={}^tV_{l_1\cdots l_p}D(p)^{-1}Q^tD(p)^{-1}V_{m_1\cdots m_p}. \tag{3.18}$$

This relation is a recursion formula which gives the general expression for $Q_{l_1\cdots l_p m_1\cdots m_p}$ in terms of Q with low powers of b_1 and b_2.

4) From the above formulae and (2.15), we can show that

$$\begin{aligned} &\alpha(b_1b_2^{-1}b_1b_2^{-1}b_1^lb_2^m) \\ &=\frac{1}{c}(Q_{1,2,l+1,m-1}-aP_{l+1}P_{m+1}-bQ_{2,-1,l,m}), \end{aligned} \tag{3.19a}$$

$$\begin{aligned} &\alpha(b_1^{-1}b_2b_1^{-1}b_2b_1^lb_2^m) \\ &=\frac{1}{c}(Q_{2,1,l-1,m+1}-aP_{l+1}P_{m+1}-bQ_{-1,2,l,m}), \end{aligned} \tag{3.19b}$$

where a, b and c are defined by

$$\begin{aligned} a&=X+Y+Z=t-t^3+t^4, \\ b&=-(XY+YZ+ZX)=t^4-t^5+t^7, \\ c&=XYZ=-t^8, \end{aligned} \tag{3.20}$$

and P_n has been defined in (2.22).

5) From the braid group representation b_i and the properties of the Markov trace, we find for the general braid group B_n that

$$\alpha(\{b_i^{-1}\})(t)=\alpha(\{b_i\})(t^{-1}), \tag{3.21}$$

$$\alpha(\{b_i\})(t)=\alpha(\{b_{n-i}\})(t), \tag{3.22}$$

$$\alpha(\{b_i\}')(t)=\alpha(\{b_i\})(t). \tag{3.23}$$

Here $\{b_i\}'$ represents the braid which is obtained from $\{b_i\}$ by reversing the order of the generators; for instance, $\alpha(b_1b_2^2b_3)(t)=\alpha(b_3b_2^2b_1)(t)$. We remark that (3.21) relates the braid with its mirror image.

§4. Recursion Relations for Link Polynomials

In this section, we shall show that the link polynomial for any closed 3-braid can be obtained recursively. More precisely, a knowledge of three quantities, τ, $\bar{\tau}$ and Q_{1212}, and the recursion relations among link polynomials is sufficient to evaluate all the link polynomials for closed 3-braids. The following is a proof of the above statement.

We first note that any element of B_3 can be written in the form[14)]

$$T^{np}_{l_1m_1\cdots l_pm_p}=\alpha\left((b_1b_2b_1)^{2n}\cdot\prod_{i=1}^{p}(b_1^{l_i}b_2^{m_i})\right), \tag{4.1}$$

where $p=0,1,2,\cdots$, and l_i, m_i, $n=0,\pm1,\pm2,\cdots$. It is easy to see that

$$T^{01}_{lm}=\alpha(b_1^lb_2^m)=P_l\cdot P_m. \tag{4.2}$$

We have the following relations for $n\geqq 0$;

$$\begin{pmatrix} T^{n1}_{-2-2} & T^{n1}_{-2-1} & T^{n1}_{-20} \\ T^{n1}_{-1-2} & T^{n1}_{-1-1} & T^{n1}_{-10} \\ T^{n1}_{0-2} & T^{n1}_{0-1} & T^{n1}_{00} \end{pmatrix}=\begin{pmatrix} T^{n-11}_{20} & T^{n-11}_{21} & T^{n-11}_{22} \\ T^{n-11}_{21} & T^{n-11}_{31} & T^{n-11}_{41} \\ T^{n-11}_{22} & T^{n-11}_{41} & T^{n-12}_{1212} \end{pmatrix}, \tag{4.3a}$$

$$\begin{pmatrix} T^{-n1}_{00} & T^{-n1}_{01} & T^{-n1}_{02} \\ T^{-n1}_{10} & T^{-n1}_{11} & T^{-n1}_{12} \\ T^{-n1}_{20} & T^{-n1}_{21} & T^{-n1}_{22} \end{pmatrix}=\begin{pmatrix} T^{-n+12}_{-1-2-1-2} & T^{-n+11}_{-4-1} & T^{-n+11}_{-2-2} \\ T^{-n+11}_{-4-1} & T^{-n+11}_{-3-1} & T^{-n+11}_{-2-1} \\ T^{-n+11}_{-2-2} & T^{-n+11}_{-2-1} & T^{-n+11}_{-20} \end{pmatrix}, \tag{4.3b}$$

$$\begin{pmatrix} T^{n\,2}_{l\,m\,-2\,-2} & T^{n\,2}_{l\,m\,-2\,-1} & T^{n\,2}_{l\,m\,-2\,0} \\ T^{n\,2}_{l\,m\,-1\,-2} & T^{n\,2}_{l\,m\,-1\,-1} & T^{n\,2}_{l\,m\,-1\,0} \\ T^{n\,2}_{l\,m\,0\,-2} & T^{n\,2}_{l\,m\,0\,-1} & T^{n\,2}_{l\,m\,0\,0} \end{pmatrix} = \begin{pmatrix} T^{n-1\,2}_{l\,m+1\,2\,-1} & T^{n-1\,1}_{1+2\,m+1} & T^{n\,1}_{1-2\,m} \\ T^{n-1\,1}_{l+1\,m+2} & T^{n-1\,1}_{l+m+3\,1} & T^{n\,1}_{l-1\,m} \\ T^{n\,1}_{l\,m-2} & T^{n\,1}_{l\,m-1} & T^{n\,1}_{l\,m} \end{pmatrix}, \tag{4.4a}$$

$$\begin{pmatrix} T^{-n\,2}_{l\,m\,0\,0} & T^{-n\,2}_{l\,m\,0\,1} & T^{-n\,2}_{l\,m\,0\,2} \\ T^{-n\,2}_{l\,m\,1\,0} & T^{-n\,2}_{l\,m\,1\,1} & T^{-n\,2}_{l\,m\,1\,2} \\ T^{-n\,2}_{l\,m\,2\,0} & T^{-n\,2}_{l\,m\,2\,1} & T^{-n\,2}_{l\,m\,2\,2} \end{pmatrix} = \begin{pmatrix} T^{-n\,1}_{l\,m} & T^{-n\,1}_{l\,m+1} & T^{-n\,1}_{l\,m+2} \\ T^{-n\,1}_{l+1\,m} & T^{-n+1\,1}_{l+m-3\,-1} & T^{-n+1\,1}_{l+1\,m-2} \\ T^{-n\,1}_{1+2\,m} & T^{-n+1\,1}_{1+2\,m-1} & T^{-n+1\,2}_{l\,m-1\,-2\,1} \end{pmatrix}, \tag{4.4b}$$

$$\begin{aligned} &\begin{pmatrix} T^{n\,p+1}_{l_1m_1\cdots l_pm_p\,-2\,-2} & T^{n\,p+1}_{l_1m_1\cdots l_pm_p\,-2\,-1} & T^{n\,p+1}_{l_1m_1\cdots l_pm_p\,-2\,0} \\ T^{n\,p+1}_{l_1m_1\cdots l_pm_p\,-1\,-2} & T^{n\,p+1}_{l_1m_1\cdots l_pm_p\,-1\,-1} & T^{n\,p+1}_{l_1m_1\cdots l_pm_p\,-1\,0} \\ T^{n\,p+1}_{l_1m_1\cdots l_pm_p\,0\,-2} & T^{n\,p+1}_{l_1m_1\cdots l_pm_p\,0\,-1} & T^{n\,p+1}_{l_1m_1\cdots l_pm_p\,0\,0} \end{pmatrix} \\ &\qquad = \begin{pmatrix} T^{n-1\,p+1}_{l_1m_1\cdots l_pm_p+1\,2\,-1} & T^{n-1\,p}_{l_1+2\,m_1\cdots l_pm_p+1} & T^{n\,p}_{l_1-2\,m_1\cdots l_pm_p} \\ T^{n-1\,p}_{l_1+1\,m_1\cdots l_pm_p+2} & T^{n-1\,p}_{l_1+m_p+2\,m_1\cdots l_p+1\,1} & T^{n\,p}_{l_1-1\,m_1\cdots l_pm_p} \\ T^{n\,p}_{l_1m_1\cdots l_pm_p-2} & T^{n\,p}_{l_1m_1\cdots l_pm_p-1} & T^{n\,p}_{l_1m_1\cdots l_pm_p} \end{pmatrix}, \end{aligned} \tag{4.5a}$$

$$\begin{aligned} &\begin{pmatrix} T^{-n\,p+1}_{l_1m_1\cdots l_pm_p\,0\,0} & T^{-n\,p+1}_{l_1m_1\cdots l_pm_p\,0\,1} & T^{-n\,p+1}_{l_1m_1\cdots l_pm_p\,0\,2} \\ T^{-n\,p+1}_{l_1m_1\cdots l_pm_p\,1\,0} & T^{-n\,p+1}_{l_1m_1\cdots l_pm_p\,1\,1} & T^{-n\,p+1}_{l_1m_1\cdots l_pm_p\,1\,2} \\ T^{-n\,p+1}_{l_1m_1\cdots l_pm_p\,2\,0} & T^{-n\,p+1}_{l_1m_1\cdots l_pm_p\,2\,1} & T^{-n\,p+1}_{l_1m_1\cdots l_pm_p\,2\,2} \end{pmatrix} \\ &\qquad = \begin{pmatrix} T^{-n\,p}_{l_1m_1\cdots l_pm_p} & T^{-n\,p}_{l_1m_1\cdots l_pm_p+1} & T^{-n\,p}_{l_1m_1\cdots l_pm_p+2} \\ T^{-n\,p}_{l_1+1\,m_1\cdots l_pm_p} & T^{-n+1\,p}_{l_1+m_p-2\,m_1\cdots l_p-1\,-1} & T^{-n+1\,p}_{l_1-1\,m_1\cdots l_pm_p-2} \\ T^{-n\,p}_{l_1+2\,m_1\cdots l_pm_p} & T^{-n+1\,p}_{l_1-2\,m_1\cdots l_pm_p-1} & T^{-n+1\,p+1}_{l_1m_1\cdots l_pm_p-1\,-2\,1} \end{pmatrix}. \end{aligned} \tag{4.5b}$$

With respect to (4.4) and (4.5), we have

$$T^{n-1\,2}_{l\,m+1\,2\,-1} = (a^2+b)T^{n-1\,1}_{l\,m} + (ab+c)T^{n-2\,1}_{l+m+4\,1} + acT^{n-2\,1}_{l+2\,m+2}, \tag{4.6a}$$

$$T^{-n+1\,2}_{l\,m-1\,-2\,1} = \frac{b^2-ac}{c^2}\,T^{-n+1\,1}_{l\,m} + \frac{ab+c}{c^2}\,T^{-n+2\,1}_{l+m-4\,-1} - \frac{b}{c^2}\,T^{-n+2\,1}_{l-2\,m-2}, \tag{4.6b}$$

$$T^{n-1\,p+1}_{l_1m_1\cdots l_pm_p+1\,2\,-1} = (a^2+b)T^{n-1\,p}_{l_1m_1\cdots l_pm_p} + (ab+c)T^{n-2\,p}_{l_1m_1+3\,m_1\cdots l_p+1\,1} + acT^{n-2\,p}_{l_1+2\,m_1\cdots l_pm_p+2}, \tag{4.7a}$$

$$T^{-n+1\,p+1}_{l_1m_1\cdots l_pm_p-1\,-2\,1} = \frac{b^2-ac}{c^2}\,T^{-n+1\,p}_{l_1m_1\cdots l_pm_p} + \frac{ab+c}{c^2}\,T^{-n+2\,p}_{l_1+m_p-3\,m_1\cdots l_p-1} - \frac{b}{c^2}\,T^{-n+2\,p}_{l_1-2\,m_1\cdots l_p\,m_p-2}, \tag{4.7b}$$

where a, b and c have been defined in (3.20).

Using relations (4.3), (4.4) and (4.6), we can express $T^{n\,p}_{l_1m_1\cdots l_pm_p}$ for $p=0, 1, 2$ and $n=0, \pm 1, \pm 2, \cdots$ in terms of τ, $\bar{\tau}$, Q_{1212} and $Q_{-1-2-1-2}$. Then, by applying relations (4.5) and (4.7), we obtain the expressions of $T^{n\,p}_{l_1m_1\cdots l_pm_p}$ for $p=3, 4, \cdots$ and $n=\pm 1, \pm 2, \cdots$ in terms of τ, $\bar{\tau}$, Q_{1212}, and $Q_{-1-2-1-2}$. Since $Q_{-1-2-1-2}$ is evaluated from Q_{1212} by the use of relation (2.15), we arrive at the conclusion that all $T^{n\,p}_{l_1m_1\cdots l_pm_p}$ defined in (4.1), for $n=0, \pm 1, \pm 2 \cdots$ and $p=0, 1, 2, \cdots$, are expressed in terms of the three quantities, τ, $\bar{\tau}$ and Q_{1212}, τ and $\bar{\tau}$ of which have been defined in (2.12) and Q_{1212} in (3.10).

§5. Birman's Example

We shall consider two braids (Fig. 4),

$$\begin{aligned} A &= (b_1b_2b_1)^4 b_1^{-12} b_2^6, \\ B &= b_1^{-6} b_2^{12}. \end{aligned} \tag{5.1}$$

Birman has proved[14] that two closed-braids in (5.1) cannot be distinguished by the Jones polynomial. In fact, if we use the $N=2$ case corresponding to Jones theory where

$$g_i^2 = (1-t)g_i + t, \quad t = e^{2\lambda}, \tag{5.2}$$

we obtain

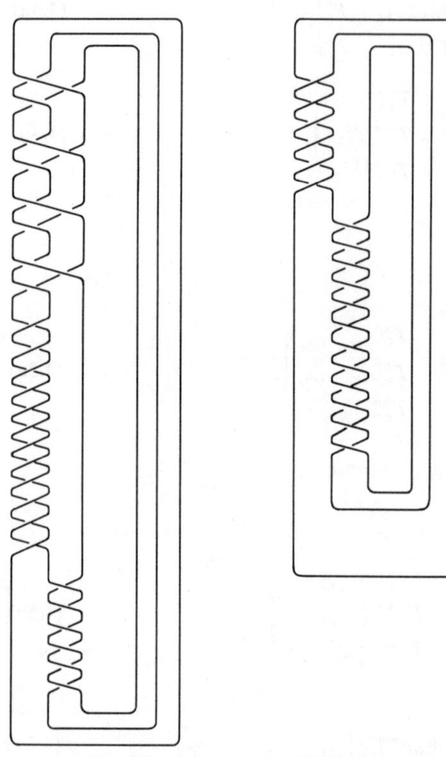

Fig. 4. Closed braids, $A=(b_1b_2b_1)^4b_1^{-12}b_2^6$ and $B=b_1^{-6}b_2^{12}$.

$$
\begin{aligned}
\alpha(A)=\alpha(B)&\\
=(&t^{18}-t^{17}+2t^{16}-3t^{15}+4t^{14}-5t^{13}+6t^{12}\\
&-6t^{11}+6t^{10}-6t^9+6t^8-5t^7+6t^6-4t^5\\
&+4t^4-3t^3+2t^2-t+1)/t^3.
\end{aligned}
\tag{5.3}
$$

On the other hand, $\alpha(A)$ and $\alpha(B)$ are different in the $N=3$ case. We have

$$
\begin{aligned}
\alpha(B)=P_{-6}P_{12}=(&t^{54}-t^{53}+2t^{51}-2t^{50}-t^{49}\\
&+4t^{48}-3t^{47}-2t^{46}+6t^{45}-4t^{44}-2t^{43}\\
&+8t^{42}-4t^{41}-4t^{40}+10t^{39}-6t^{38}-5t^{37}\\
&+12t^{36}-7t^{35}-5t^{34}+12t^{33}-7t^{32}-5t^{31}\\
&+13t^{30}-7t^{29}-5t^{28}+12t^{27}-7t^{26}-5t^{25}\\
&+12t^{24}-6t^{23}-6t^{22}+12t^{21}-5t^{20}-6t^{19}\\
&+12t^{18}-4t^{17}-6t^{16}+10t^{15}-3t^{14}-6t^{13}\\
&+8t^{12}-2t^{11}-5t^{10}+6t^9-t^8-3t^7+4t^6\\
&-2t^4+2t^3-t+1)/t^{12}.
\end{aligned}
\tag{5.4}
$$

We may evaluate $\alpha(A)$ as follows. Using formula (3.7b) and the generalized Alexander-Conway relation (2.15), we have

$$
\begin{aligned}
\alpha(A)&=\alpha(\Delta^4b_1^{-12}b_2^9)=\alpha(b_1b_2^3b_1b_2^{-9}b_1b_2^9)\\
&=a\alpha(b_1b_2^2b_1b_2^{-9}b_1b_2^{-9})\\
&\quad+b\alpha(b_1b_2b_1b_2^{-9}b_1b_2^9)\\
&\quad+c\alpha(b_1^2b_2^{-9}b_1b_2^9).
\end{aligned}
\tag{5.5}
$$

Each term in (5.5) may be calculated as

$$
\begin{aligned}
\alpha(b_1b_2^2b_1b_2^{-9}b_1b_2^9)=P_2P_3&\\
=t^3(&t^{15}-t^{14}-t^{13}+2t^{12}-2t^{11}-t^{10}+3t^9-2t^8-t^7+3t^6-t^5+2t^3+1),
\end{aligned}
$$

$$
\begin{aligned}
&\alpha(b_1b_2b_1b_2^{-9}b_1b_2^9)\\
&=Q_{1\,10\,1\,-8}\\
&=(t^{54}-t^{53}+2t^{51}-2t^{50}-t^{49}+4t^{48}-3t^{47}-2t^{46}+7t^{45}-4t^{44}-4t^{43}+9t^{42}-3t^{41}-6t^{40}+9t^{39}-3t^{38}\\
&\quad-5t^{37}+9t^{36}-6t^{35}-4t^{34}+14t^{33}-10t^{32}-7t^{31}+18t^{30}-9t^{29}-10t^{28}+19t^{27}-7t^{26}-12t^{25}+19t^{24}\\
&\quad-6t^{23}-10t^{22}+15t^{21}-4t^{20}-7t^{19}+11t^{18}-5t^{17}-4t^{16}+11t^{15}-7t^{14}-5t^{13}+11t^{12}-4t^{11}-6t^{10}\\
&\quad+8t^9-t^8-4t^7+4t^6-2t^4+2t^3-t+1)/t^{16},
\end{aligned}
$$

$$
\begin{aligned}
&\alpha(b_1^2b_2^{-9}b_1b_2^9)\\
&=Q_{1\,9\,2\,-9}\\
&=(t^{57}-2t^{56}+4t^{54}-5t^{53}-t^{52}+9t^{51}-8t^{50}-5t^{49}+16t^{48}-9t^{47}-10t^{46}+20t^{45}-8t^{44}-12t^{43}+20t^{42}\\
&\quad-11t^{41}-9t^{40}+25t^{39}-20t^{38}-10t^{37}+38t^{36}-26t^{35}-20t^{34}+48t^{33}-23t^{32}-28t^{31}+48t^{30}-18t^{29}\\
&\quad-30t^{28}+45t^{27}-14t^{26}-27t^{25}+40t^{24}-14t^{23}-20t^{22}+36t^{21}-18t^{20}-16t^{19}+36t^{18}-18t^{17}-20t^{16}
\end{aligned}
$$

$$+34t^{15}-10t^{14}-21t^{13}+24t^{12}-3t^{11}-14t^{10}+14t^{9}-2t^{8}-8t^{7}+8t^{6}-t^{5}-5t^{4}+4t^{3}-2t+1)/t^{20}. \tag{5.6}$$

Therefore, we obtain

$$\begin{aligned}\alpha(A)&=aP_2P_3+bQ_{1101-8}+cQ_{192-9}\\&=(t^{54}-t^{53}+2t^{51}-2t^{50}-t^{49}+4t^{48}-2t^{47}-2t^{46}+4t^{45}-3t^{44}+5t^{42}-7t^{41}+t^{40}+11t^{39}-11t^{38}\\&\quad-4t^{37}+16t^{36}-8t^{35}-9t^{34}+14t^{33}-4t^{32}-8t^{31}+11t^{30}-3t^{29}-7t^{28}+12t^{27}-5t^{26}-7t^{25}+13t^{24}\\&\quad-7t^{23}-6t^{22}+14t^{21}-7t^{20}-9t^{19}+16t^{18}-3t^{17}-10t^{16}+10t^{15}+t^{14}-5t^{13}+4t^{12}-t^{11}-t^{10}+4t^{9}\\&\quad-4t^{8}-t^{7}+5t^{6}-2t^{5}-2t^{4}+3t^{3}-t+1)/t^{10}.\end{aligned} \tag{5.7}$$

We see that $\alpha(B)$ in (5.4) and $\alpha(A)$ in (5.7) are different.

This shows that the $N=3$ theory is more powerful than the $N=2$ theory. We expect that a set of link polynomials for $N=2, 3, 4, \cdots$ will provide us with a systematic method

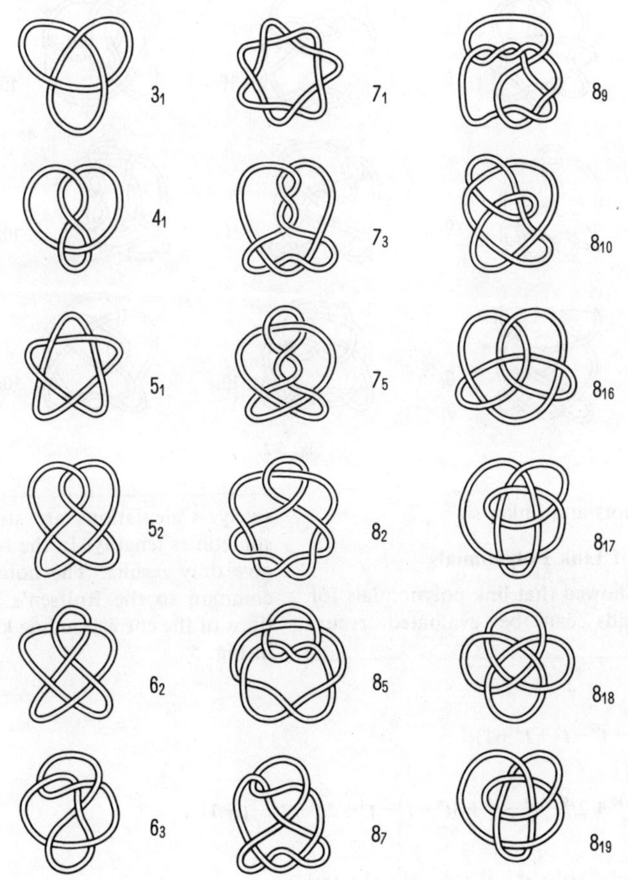

Fig. 5-1.

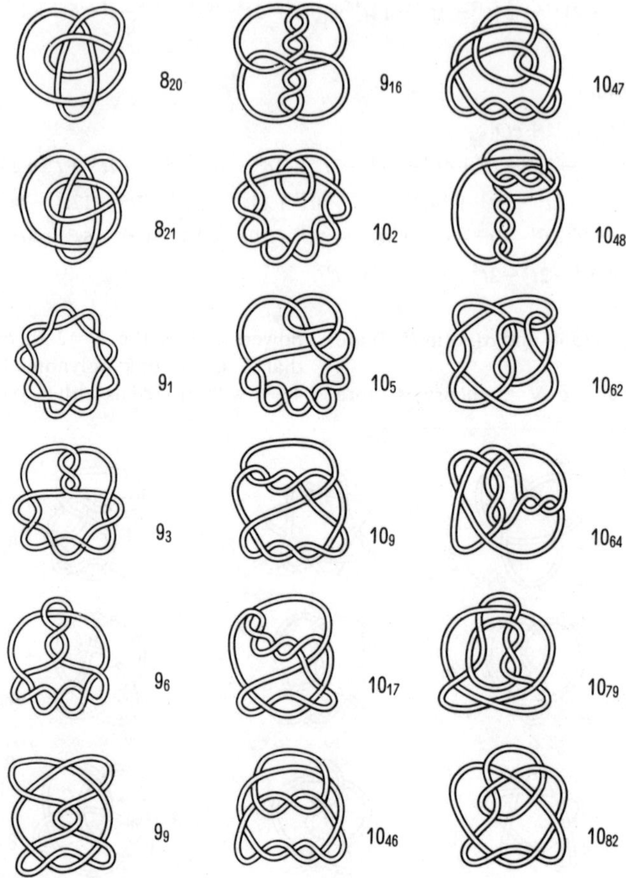

Fig. 5-2.

to classify knots and links.

§6. A List of Link Polynomials

In §4, we showed that link polynomials for closed 3-braids can be evaluated recursively. Calculations are straightforward but sometimes lengthy. In the following, we shall give only results. The notation for knots is common to the Rolfsen's.[13] The configurations of the corresponding knots are described in Fig. 5.

3_1; b_1^3

$$(t^9-t^8-t^7+t^6-t^5+t^3+1)t^2$$

4_1; $b_1b_2^{-1}b_1b_2^{-1}$

$$(t^{12}-t^{11}-t^{10}+2t^9-t^8-t^7+3t^6-t^5-t^4+2t^3-t^2-t+1)/t^6$$

5_1; b_1^5

$$(t^{15}-t^{14}+t^{12}-2t^{11}+t^9-t^8+t^6-t^5+t^3+1)t^4$$

5_2; $b_1b_2^2b_1^2b_2^{-1}$

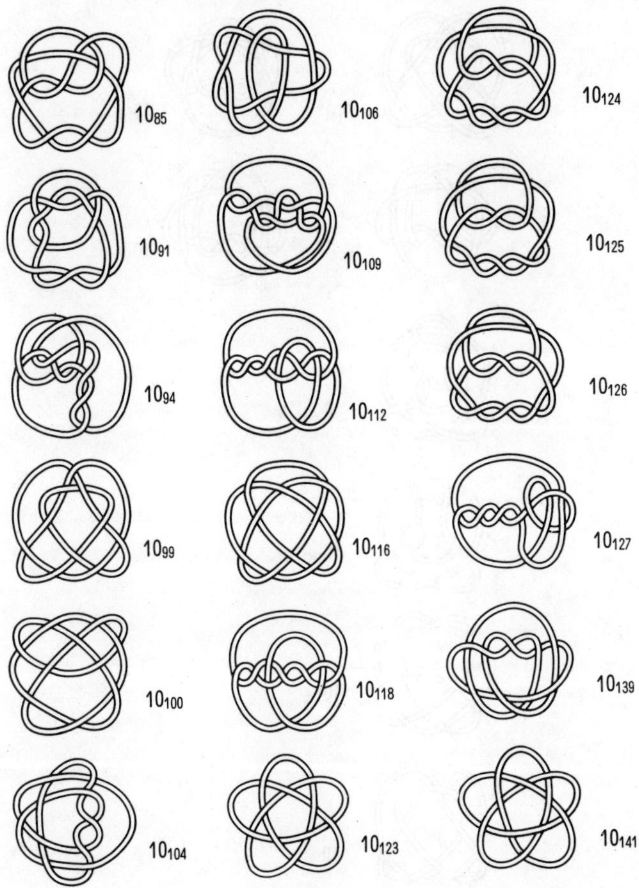

Fig. 5-3.

$(t^{15}-t^{14}-t^{13}+2t^{12}-t^{11}-2t^{10}+3t^9-t^8-3t^7+4t^6-t^5-2t^4+3t^3-t+1)t^2$

6_2; $b_1b_2^{-1}b_1^3b_2^{-1}$

$(t^{18}-2t^{17}+4t^{15}-5t^{14}+6t^{12}-6t^{11}+6t^9-5t^8-t^7+5t^6-3t^5-t^4+3t^3-t^2-t+1)/t^4$

6_3; $b_1b_2^{-1}b_1^2b_2^{-2}$

$(t^{18}-2t^{17}-t^{16}+5t^{15}-4t^{14}-3t^{13}+9t^{12}-5t^{11}-5t^{10}$

$+11t^9-5t^8-5t^7+9t^6-3t^5-4t^4+5t^3-t^2-2t+1)/t^9$

7_1; b_1^7

$(t^{21}-t^{20}+t^{18}-t^{17}-t^{14}+t^{12}-t^{11}+t^9-t^8+t^6-t^5+t^3+1)t^6$

7_3; $b_1b_2^2b_1^{-1}b_2^5$

$(t^{21}-t^{20}+2t^{18}-3t^{17}-t^{16}+5t^{15}-5t^{14}-2t^{13}+8t^{12}-6t^{11}-2t^{10}+7t^9$

$-4t^8-2t^7+5t^6-2t^5-2t^4+3t^3-t+1)t^4$

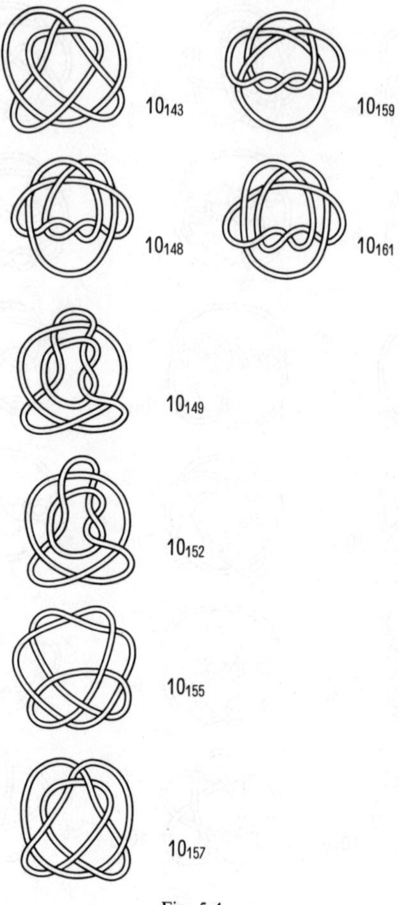

Fig. 5-4.

Fig. 5. Closed 3-braids, $3_1 \sim 10_{161}$.

7_5; $b_1 b_2^4 b_1^{-2} b_2^3$

$$(t^{21}-2t^{20}+5t^{18}-6t^{17}-2t^{16}+11t^{15}-9t^{14}-4t^{13}+14t^{12}-10t^{11}-5t^{10}$$
$$+13t^9-7t^8-5t^7+9t^6-3t^5-3t^4+4t^3-t+1)t^4$$

8_2; $b_1 b_2^{-1} b_1^5 b_2^{-1}$

$$(t^{24}-2t^{23}+3t^{21}-4t^{20}+2t^{19}+3t^{18}-6t^{17}+3t^{16}+4t^{15}-7t^{14}+2t^{13}+5t^{12}$$
$$-7t^{11}+t^{10}+5t^9-5t^8+5t^6-3t^5-t^4+3t^3-t^2-t+1)/t^2$$

8_5; $b_1 b_2^{-3} b_1 b_2^{-3}$

$$(t^{24}-t^{23}-t^{22}+4t^{21}-t^{20}-5t^{19}+7t^{18}-9t^{16}+8t^{15}+3t^{14}-12t^{13}+7t^{12}+6t^{11}$$
$$-12t^{10}+5t^9+7t^8-10t^7+3t^6+5t^5-6t^4+2t^3+t^2-2t+1)/t^{22}$$

8_7; $b_1 b_2^{-1} b_1^2 b_2^{-4}$

$$(t^{24}-2t^{23}-t^{22}+5t^{21}-4t^{20}-4t^{19}+10t^{18}-3t^{17}-9t^{16}+14t^{15}-2t^{14}-14t^{13}$$

$+16t^{12}-16t^{10}+14t^9+t^8-12t^7+9t^6+t^5-6t^4+4t^3-2t+1)/t^{17}$

8_9; $b_1b_2^{-3}b_1^3b_2^{-1}$

$(t^{24}-2t^{23}+5t^{21}-6t^{20}-2t^{19}+12t^{18}-10t^{17}-7t^{16}+20t^{15}-12t^{14}-11t^{13}+25t^{12}$
$-11t^{11}-12t^{10}+20t^9-7t^8-10t^7+12t^6-2t^5-6t^4+5t^3-2t+1)/t^{12}$

8_{10}; $b_1b_2^{-2}b_1^2b_2^{-3}$

$(t^{24}-2t^{23}-t^{22}+6t^{21}-5t^{20}-6t^{19}+14t^{18}-5t^{17}-14t^{16}+21t^{15}-2t^{14}-21t^{13}+23t^{12}$
$+2t^{11}-24t^{10}+20t^9+3t^8-19t^7+12t^6+3t^5-9t^4+4t^3+t^2-2t+1)/t^{17}$

8_{16}; $b_1^3b_2^3b_1^{-2}b_2^{-2}$

$(t^{24}-3t^{23}+2t^{22}+6t^{21}-15t^{20}+7t^{19}+19t^{18}-32t^{17}+8t^{16}+32t^{15}-41t^{14}+4t^{13}$
$+38t^{12}-37t^{11}-3t^{10}+35t^9-25t^8-8t^7+24t^6-10t^5-8t^4+10t^3-t^2-3t+1)/t^7$

8_{17}; $b_1^2b_2^3b_1^{-3}b_2^{-2}$

$(t^{24}-3t^{23}+t^{22}+9t^{21}-14t^{20}-3t^{19}+28t^{18}-25t^{17}-14t^{16}+47t^{15}-29t^{14}-25t^{13}$
$+55t^{12}-25t^{11}-29t^{10}+47t^9-14t^8-25t^7+28t^6-3t^5-14t^4+9t^3+t^2-3t+1)/t^{12}$

8_{18}; $b_1b_2^{-1}b_1b_2^{-1}b_1b_2^{-1}b_1b_2^{-1}$

$(t^{24}-4t^{23}+2t^{22}+13t^{21}-21t^{20}-4t^{19}+41t^{18}-38t^{17}-20t^{16}+69t^{15}-43t^{14}-36t^{13}$
$+81t^{12}-36t^{11}-43t^{10}+69t^9-20t^8-38t^7+41t^6-4t^5-21t^4+13t^3+2t^2-4t+1)/t^{12}$

8_{19}; $b_1^{-1}b_2^{-3}b_1^{-1}b_2^{-3}$

$(t^{17}+t^{14}+t^{11}-t^{10}-t^7-t^4+t^3-t+1)/t^{23}$

8_{20}; $b_1b_2^3b_1b_2^{-3}$

$(t^{17}-t^{16}-t^{15}+2t^{14}-t^{13}-2t^{12}+2t^{11}-2t^9+2t^8+t^7-2t^6+t^5+2t^4-2t^3+t^2+t-1)/t^2$

8_{21}; $b_1b_2^{-2}b_1^2b_2^3$

$(t^{19}-2t^{18}-t^{17}+5t^{16}-3t^{15}-4t^{14}+8t^{13}-2t^{12}-8t^{11}$
$+10t^{10}-t^9-10t^8+10t^7-8t^5+6t^4+t^3-4t^2+2t+1)t$

9_1; b_1^9

$(t^{27}-t^{26}+t^{24}-t^{23}+t^{21}-t^{20}-t^{19}+t^{18}-t^{17}+t^{15}-t^{14}+t^{12}-t^{11}+t^9-t^8+t^6-t^5+t^3+1)t^8$

9_3; $b_1b_2^{-6}b_1^{-2}b_2^{-1}$

$(t^{27}-t^{26}+3t^{24}-2t^{23}-2t^{22}+5t^{21}-2t^{20}-4t^{19}+6t^{18}-t^{17}-5t^{16}+6t^{15}$
$-6t^{13}+5t^{12}+t^{11}-6t^{10}+5t^9-4t^7+3t^6-t^5-2t^4+2t^3-t+1)/t^{33}$

9_6; $b_1^2b_2^2b_1^{-1}b_2^5$

$(t^{27}-2t^{26}+4t^{24}-5t^{23}+t^{22}+7t^{21}-11t^{20}+3t^{19}+11t^{18}-16t^{17}+3t^{16}+14t^{15}-16t^{14}$
$+t^{13}+14t^{12}-13t^{11}-2t^{10}+12t^9-8t^8-4t^7+9t^6-3t^5-3t^4+4t^3-t+1)t^6$

9_9; $b_1b_2^{-1}b_1^4b_2^4$

$(t^{27}-2t^{26}+t^{25}+4t^{24}-8t^{23}+3t^{22}+10t^{21}-18t^{20}+6t^{19}+17t^{18}-25t^{17}+6t^{16}+20t^{15}$
$-25t^{14}+2t^{13}+21t^{12}-19t^{11}-2t^{10}+17t^9-11t^8-4t^7+10t^6-4t^5-3t^4+4t^3-t+1)t^6$

9_{16}; $b_1b_2^{-3}b_1^{-4}b_2^{-2}$

$(t^{27}-t^{26}+5t^{24}-4t^{23}-5t^{22}+15t^{21}-7t^{20}-17t^{19}+27t^{18}-4t^{17}-32t^{16}+35t^{15}+3t^{14}$
$-42t^{13}+35t^{12}+10t^{11}-42t^{10}+28t^9+12t^8-31t^7+16t^6+7t^5-14t^4+6t^3+2t^2-3t+1)/t^{33}$

10_2; $b_1b_2^{-1}b_1^7b_2^{-1}$

$t^{30}-2t^{29}+3t^{27}-4t^{26}+t^{25}+4t^{24}-4t^{23}+4t^{21}-4t^{20}+4t^{18}-5t^{17}+t^{16}+4t^{15}-6t^{14}$
$+t^{13}+5t^{12}-6t^{11}+t^{10}+5t^{9}-5t^{8}+5t^{6}-3t^{5}-t^{4}+3t^{3}-t^{2}-t+1$

10_5; $b_1b_2^{-6}b_1^2b_2^{-1}$
$(t^{30}-2t^{29}-t^{28}+5t^{27}-4t^{26}-4t^{25}+10t^{24}-3t^{23}-9t^{22}+13t^{21}-t^{20}-12t^{19}+13t^{18}+t^{17}$
$-14t^{16}+11t^{15}+3t^{14}-13t^{13}+9t^{12}+3t^{11}-11t^{10}+8t^{9}+t^{8}-8t^{7}+7t^{6}-5t^{4}+4t^{3}-2t+1)/t^{25}$

10_9; $b_1^{-1}b_2^{-6}b_1^2b_2^3$
$(t^{30}-2t^{29}+5t^{27}-6t^{26}-2t^{25}+12t^{24}-9t^{23}-8t^{22}+19t^{21}-8t^{20}-15t^{19}+24t^{18}-5t^{17}$
$-21t^{16}+25t^{15}-t^{14}-23t^{13}+22t^{12}-18t^{10}+15t^{9}-10t^{7}+8t^{6}-5t^{4}+4t^{3}-2t+1)/t^{20}$

10_{17}; $b_1b_2^{-4}b_1^4b_2^{-1}$
$(t^{30}-2t^{29}+4t^{27}-6t^{26}+10t^{24}-11t^{23}-3t^{22}+20t^{21}-16t^{20}-10t^{19}+30t^{18}-18t^{17}-16t^{16}+35t^{15}$
$-16t^{14}-18t^{13}+30t^{12}-10t^{11}-16t^{10}+20t^{9}-3t^{8}-11t^{7}+10t^{6}-6t^{4}+4t^{3}-2t+1)/t^{15}$

10_{46}; $b_1b_2^{-5}b_1b_2^{-3}$
$(t^{30}-t^{29}-t^{28}+4t^{27}-t^{26}-5t^{25}+7t^{24}+t^{23}-9t^{22}+7t^{21}+4t^{20}-12t^{19}+6t^{18}+6t^{17}-12t^{16}+4t^{15}$
$+7t^{14}-9t^{13}+2t^{12}+6t^{11}-6t^{10}+t^{9}+4t^{8}-5t^{7}+2t^{6}+3t^{5}-5t^{4}+2t^{3}+t^{2}-2t+1)/t^{30}$

10_{47}; $b_1b_2^{-5}b_1^2b_2^{-2}$
$(t^{30}-2t^{29}-t^{28}+6t^{27}-5t^{26}-6t^{25}+14t^{24}-4t^{23}-15t^{22}+19t^{21}+t^{20}-22t^{19}+20t^{18}+7t^{17}-26t^{16}$
$+17t^{15}+11t^{14}-25t^{13}+11t^{12}+12t^{11}-20t^{10}+8t^{9}+7t^{8}-13t^{7}+7t^{6}$
$+2t^{5}-7t^{4}+4t^{3}+t^{2}-2t+1)/t^{25}$

10_{48}; $b_1b_2^{-3}b_1^4b_2^{-2}$
$(t^{30}-2t^{29}+5t^{27}-7t^{26}-t^{25}+14t^{24}-16t^{23}-5t^{22}+30t^{21}-25t^{20}-15t^{19}+47t^{18}-28t^{17}-26t^{16}+56t^{15}$
$-24t^{14}-30t^{13}+48t^{12}-14t^{11}-27t^{10}+31t^{9}-3t^{8}-19t^{7}+14t^{6}+2t^{5}-9t^{4}+4t^{3}+t^{2}-2t+1)/t^{15}$

10_{62}; $b_1b_2^{-3}b_1^2b_2^{-4}$
$(t^{30}-2t^{29}-t^{28}+6t^{27}-5t^{26}-7t^{25}+15t^{24}-3t^{23}-18t^{22}+23t^{21}+3t^{20}-29t^{19}+24t^{18}+11t^{17}-36t^{16}$
$+19t^{15}+19t^{14}-34t^{13}+12t^{12}+20t^{11}-27t^{10}+7t^{9}+13t^{8}-16t^{7}+5t^{6}+5t^{5}$
$-7t^{4}+3t^{3}+t^{2}-2t+1)/t^{25}$

10_{64}; $b_1b_2^{-3}b_1^3b_2^{-3}$
$(t^{30}-2t^{29}+6t^{27}-7t^{26}-4t^{25}+18t^{24}-13t^{23}-16t^{22}+35t^{21}-13t^{20}-33t^{19}+48t^{18}-5t^{17}-48t^{16}$
$+50t^{15}+5t^{14}-53t^{13}+42t^{12}+9t^{11}-42t^{10}+26t^{9}+9t^{8}-22t^{7}+10t^{6}+5t^{5}-8t^{4}+3t^{3}+t^{2}-2t+1)/t^{20}$

10_{79}; $b_1^2b_2^{-2}b_1^3b_2^{-3}$
$(t^{30}-2t^{29}+t^{28}+5t^{27}-11t^{26}+2t^{25}+21t^{24}-30t^{23}-5t^{22}+53t^{21}-48t^{20}-24t^{19}+85t^{18}-53t^{17}$
$-44t^{16}+99t^{15}-44t^{14}-53t^{13}+85t^{12}-24t^{11}-48t^{10}+53t^{9}-5t^{8}$
$-30t^{7}+21t^{6}+2t^{5}-11t^{4}+5t^{3}+t^{2}-2t+1)/t^{15}$

10_{82}; $b_1^5b_2^{-3}b_1^{-2}b_2^2$
$(t^{30}-3t^{29}+t^{28}+7t^{27}-12t^{26}+4t^{25}+17t^{24}-31t^{23}+11t^{22}+36t^{21}-58t^{20}+13t^{19}+60t^{18}-75t^{17}$
$+5t^{16}+74t^{15}-70t^{14}-9t^{13}+71t^{12}-49t^{11}-20t^{10}+54t^{9}-23t^{8}$
$-23t^{7}+30t^{6}-4t^{5}-14t^{4}+9t^{3}+t^{2}-3t+1)/t^{10}$

10_{85}; $b_1^3b_2^{-2}b_1^{-2}b_2^5$
$(t^{30}-3t^{29}+2t^{28}+5t^{27}-12t^{26}+8t^{25}+8t^{24}-24t^{23}+21t^{22}+9t^{21}-40t^{20}+32t^{19}+17t^{18}-54t^{17}$

$+31t^{16}+29t^{15}-57t^{14}+20t^{13}+36t^{12}-47t^{11}+6t^{10}+35t^{9}-30t^{8}$
$-5t^{7}+25t^{6}-11t^{5}-8t^{4}+10t^{3}-t^{2}-3t+1)/t^{5}$

10_{91}; $b_1^4b_2^{-3}b_1^{-3}b_2^2$
$(t^{30}-3t^{29}+t^{28}+8t^{27}-15t^{26}+2t^{25}+30t^{24}-40t^{23}-7t^{22}+71t^{21}-64t^{20}-32t^{19}+113t^{18}-71t^{17}$
$-60t^{16}+132t^{15}-58t^{14}-73t^{13}+114t^{12}-31t^{11}-66t^{10}+72t^{9}$
$-5t^{8}-43t^{7}+30t^{6}+5t^{5}-17t^{4}+7t^{3}+2t^{2}-3t+1)/t^{15}$

10_{94}; $b_1^{-3}b_2^3b_1^2b_2^{-4}$
$(t^{30}-3t^{29}+t^{28}+10t^{27}-15t^{26}-6t^{25}+36t^{24}-27t^{23}-31t^{22}+70t^{21}-24t^{20}-67t^{19}+94t^{18}-9t^{17}$
$-97t^{16}+98t^{15}+11t^{14}-104t^{13}+79t^{12}+23t^{11}-81t^{10}+46t^{9}+20t^{8}-43t^{7}+19t^{6}$
$+9t^{5}-15t^{4}+6t^{3}+2t^{2}-3t+1)/t^{20}$

10_{99}; $b_1^3b_2^3b_1^{-3}b_2^{-3}$
$(t^{30}-3t^{29}+2t^{28}+8t^{27}-19t^{26}+5t^{25}+37t^{24}-53t^{23}-7t^{22}+91t^{21}-85t^{20}-40t^{19}+146t^{18}-93t^{17}$
$-75t^{16}+171t^{15}-75t^{14}-93t^{13}+146t^{12}-40t^{11}-85t^{10}+91t^{9}-7t^{8}-53t^{7}+37t^{6}+5t^{5}$
$-19t^{4}+8t^{3}+2t^{2}-3t+1)/t^{15}$

10_{100}; $b_1^3b_2^{-1}b_1^2b_2^{-1}b_1^2b_2^{-1}$
$(t^{30}-4t^{29}+5t^{28}+4t^{27}-21t^{26}+25t^{25}+3t^{24}-56t^{23}+73t^{22}-2t^{21}-111t^{20}+127t^{19}+5t^{18}-158t^{17}$
$+149t^{16}+24t^{15}-172t^{14}+131t^{13}+42t^{12}-146t^{11}+83t^{10}+49t^{9}-93t^{8}+31t^{7}$
$+37t^{6}-37t^{5}+t^{4}+14t^{3}-5t^{2}-2t+1)/t^{5}$

10_{104}; $b_1^{-2}b_2^3b_1^{-2}b_2b_1^{-1}b_2$
$(t^{30}-3t^{29}+2t^{28}+7t^{27}-18t^{26}+6t^{25}+33t^{24}-49t^{23}-5t^{22}+84t^{21}-78t^{20}-36t^{19}+134t^{18}-86t^{17}$
$-68t^{16}+154t^{15}-70t^{14}-84t^{13}+133t^{12}-37t^{11}-76t^{10}+83t^{9}-7t^{8}-46t^{7}+33t^{6}+3t^{5}$
$-16t^{4}+8t^{3}+t^{2}-3t+1)/t^{15}$

10_{106}; $b_1^{-3}b_2^2b_1^{-2}b_2b_1^{-1}b_2$
$(t^{30}-3t^{29}+t^{28}+10t^{27}-16t^{26}-6t^{25}+40t^{24}-30t^{23}-36t^{22}+81t^{21}-28t^{20}-80t^{19}+111t^{18}-9t^{17}$
$-115t^{16}+115t^{15}+15t^{14}-124t^{13}+92t^{12}+29t^{11}-97t^{10}+53t^{9}$
$+26t^{8}-50t^{7}+20t^{6}+11t^{5}-16t^{4}+6t^{3}+2t^{2}-3t+1)/t^{20}$

10_{109}; $b_1^{-2}b_2^2b_1^{-2}b_2^2b_1^{-1}b_2$
$(t^{30}-3t^{29}+2t^{28}+8t^{27}-20t^{26}+6t^{25}+40t^{24}-60t^{23}-7t^{22}+105t^{21}-98t^{20}-45t^{19}+169t^{18}-108t^{17}$
$-87t^{16}+195t^{15}-87t^{14}-108t^{13}+169t^{12}-45t^{11}-98t^{10}+105t^{9}$
$-7t^{8}-60t^{7}+40t^{6}+6t^{5}-20t^{4}+8t^{3}+2t^{2}-3t+1)/t^{15}$

10_{112}; $b_1^3b_2^{-1}b_1b_2^{-1}b_1b_2^{-1}b_1b_2^{-1}$
$(t^{30}-4t^{29}+3t^{28}+9t^{27}-22t^{26}+13t^{25}+30t^{24}-69t^{23}+34t^{22}+73t^{21}-137t^{20}+47t^{19}+127t^{18}-182t^{17}$
$+33t^{16}+161t^{15}-176t^{14}+t^{13}+158t^{12}-128t^{11}-29t^{10}+122t^{9}$
$-65t^{8}-39t^{7}+66t^{6}-17t^{5}-24t^{4}+20t^{3}-5t+2)/t^{10}$

10_{116}; $b_1^2b_2^{-1}b_1^2b_2^{-1}b_1b_2^{-1}b_1b_2^{-1}$
$(t^{30}-4t^{29}+4t^{28}+8t^{27}-26t^{26}+20t^{25}+30t^{24}-81t^{23}+46t^{22}+80t^{21}-155t^{20}+52t^{19}+143t^{18}-199t^{17}$
$+27t^{16}+182t^{15}-186t^{14}-12t^{13}+176t^{12}-129t^{11}-42t^{10}+129t^{9}$
$-58t^{8}-47t^{7}+63t^{6}-10t^{5}-25t^{4}+15t^{3}+2t^{2}-4t+1)/t^{10}$

10_{118}; $b_1^{-1}b_2b_1^{-3}b_2^{-2}b_1^3b_2^2$

$(t^{30}-4t^{29}+3t^{28}+11t^{27}-27t^{26}+8t^{25}+52t^{24}-76t^{23}-8t^{22}+130t^{21}-122t^{20}-55t^{19}+208t^{18}-134t^{17}$
$-107t^{16}+241t^{15}-107t^{14}-134t^{13}+208t^{12}-55t^{11}-122t^{10}+130t^9$
$-8t^8-76t^7+52t^6+8t^5-27t^4+11t^3+3t^2-4t+1)/t^{15}$

10_{123}; $b_1b_2^{-1}b_1b_2^{-1}b_1b_2^{-1}b_1b_2^{-1}b_1b_2^{-1}$

$(t^{30}-5t^{29}+5t^{28}+15t^{27}-41t^{26}+14t^{25}+80t^{24}-121t^{23}-10t^{22}+206t^{21}-197t^{20}-85t^{19}+331t^{18}$
$-215t^{17}-169t^{16}+383t^{15}-169t^{14}-215t^{13}+331t^{12}-85t^{11}-197t^{10}+206t^9-10t^8-121t^7$
$+80t^6+14t^5-41t^4+15t^3+5t^2-5t+1)/t^{15}$

10_{124}; $b_1^{-1}b_2^{-5}b_1^{-1}b_2^{-3}$

$(t^{21}+t^{18}+t^{15}-t^{14}+t^{12}-t^{11}-t^{10}+t^9-t^8-t^7+t^6-t^4+t^3-t+1)/t^{29}$

10_{125}; $b_1b_2^3b_1b_2^{-5}$

$(t^{24}-t^{23}-t^{22}+2t^{21}-t^{20}-2t^{19}+2t^{18}-2t^{16}+2t^{15}-t^{13}+2t^{12}+t^9-t^8+t^7-t^5+t^3-t^2-t+1)/t^{11}$

10_{126}; $b_1b_2^{-3}b_1b_2^5$

$t^{23}-t^{22}+2t^{20}-3t^{19}-t^{18}+5t^{17}-4t^{16}-4t^{15}+8t^{14}-3t^{13}-7t^{12}+10t^{11}-t^{10}-9t^9+10t^8$
$-7t^6+6t^5+t^4-4t^3+2t^2+t-1$

10_{127}; $b_1b_2^{-2}b_1^2b_2^5$

$(t^{25}-2t^{24}+4t^{22}-6t^{21}+11t^{19}-12t^{18}-4t^{17}+22t^{16}-17t^{15}-9t^{14}+28t^{13}-17t^{12}$
$-12t^{11}+25t^{10}-11t^9-12t^8+17t^7-4t^6-9t^5+8t^4+t^3-4t^2+2t+1)t^3$

10_{139}; $b_1^{-1}b_2^{-3}b_1^{-3}b_2^{-3}$

$(t^{25}+t^{22}+t^{19}-2t^{18}+2t^{16}-2t^{15}-t^{14}+3t^{13}-3t^{11}+2t^{10}+t^9-4t^8$
$+2t^7+t^6-3t^5+t^4+2t^3-t^2-t+1)/t^{33}$

10_{141}; $b_1b_2^3b_1^2b_2^{-4}$

$(t^{24}-2t^{23}-t^{22}+5t^{21}-3t^{20}-4t^{19}+8t^{18}-t^{17}-8t^{16}+7t^{15}+t^{14}-9t^{13}+6t^{12}$
$+4t^{11}-8t^{10}+3t^9+6t^8-6t^7+t^6+3t^5-3t^4+t^3-t+1)/t^6$

10_{143}; $b_1b_2^{-3}b_1^3b_2^3$

$t^{23}-2t^{22}+5t^{20}-6t^{19}-3t^{18}+12t^{17}-7t^{16}-10t^{15}+18t^{14}-4t^{13}-17t^{12}+21t^{11}-t^{10}-21t^9+18t^8$
$+2t^7-16t^6+11t^5+4t^4-8t^3+4t^2+2t-2$

10_{148}; $b_1^2b_2^3b_1^{-2}b_2b_1^{-1}b_2$

$t^{23}-2t^{22}+6t^{20}-7t^{19}-5t^{18}+17t^{17}-9t^{16}-16t^{15}+26t^{14}-5t^{13}-26t^{12}+29t^{11}+t^{10}-30t^9+26t^8$
$+5t^7-24t^6+15t^5+6t^4-12t^3+4t^2+3t-2$

10_{149}; $b_1^3b_2^3b_1^{-2}b_2^2$

$(t^{25}-3t^{24}+t^{23}+8t^{22}-14t^{21}+27t^{19}-29t^{18}-10t^{17}+50t^{16}-37t^{15}-22t^{14}+61t^{13}-34t^{12}-29t^{11}$
$+55t^{10}-21t^9-28t^8+36t^7-6t^6-18t^5+14t^4+2t^3-6t^2+2t+1)t^3$

10_{152}; $b_1^3b_2^3b_1^2b_2^2$

$(t^{28}-2t^{27}-t^{26}+5t^{25}-3t^{24}-5t^{23}+8t^{22}+t^{21}-10t^{20}+8t^{19}+4t^{18}-11t^{17}+6t^{16}+5t^{15}-9t^{14}+t^{13}$
$+5t^{12}-4t^{11}-3t^{10}+3t^9+t^8-3t^7+t^6+t^5+t^3+1)t^8$

10_{155}; $b_1^{-3}b_2^{-1}b_1^2b_2^{-1}b_1^2b_2^{-1}$

$(t^{24}-t^{23}+t^{22}-5t^{20}+7t^{19}+t^{18}-11t^{17}+11t^{16}+3t^{15}-15t^{14}+9t^{13}+8t^{12}-15t^{11}+5t^{10}+11t^9$

$-13t^8+10t^6-7t^5-3t^4+6t^3-t^2-2t+1)/t^{18}$

10_{157}; $b_1^{-2}b_2^{-3}b_1^{-2}b_2b_1^{-1}b_2$

$(t^{25}+2t^{24}-8t^{23}+3t^{22}+19t^{21}-26t^{20}-7t^{19}+51t^{18}-41t^{17}-28t^{16}+79t^{15}-43t^{14}-47t^{13}+88t^{12}$

$-33t^{11}-53t^{10}+72t^9-15t^8-43t^7+40t^6-21t^4+12t^3+2t^2-4t+1)/t^{28}$

10_{159}; $b_1b_2^3b_1b_2^{-1}b_1b_2^{-1}b_1b_2^{-1}$

$t^{23}-3t^{22}+10t^{20}-11t^{19}-9t^{18}+28t^{17}-13t^{16}-27t^{15}+42t^{14}-6t^{13}-43t^{12}+46t^{11}+3t^{10}-49t^9$

$+39t^8+10t^7-39t^6+22t^5+11t^4-19t^3+6t^2+5t-3$

10_{161}; $b_1^2b_2^3b_1^2b_2b_1^{-1}b_2$

$(t^{25}-t^{24}-t^{23}+2t^{22}-2t^{20}+t^{19}+t^{18}-t^{17}-t^{16}+t^{15}-2t^{13}+2t^{12}-2t^{10}+t^9+t^8-t^7+t^5+1)t^6$

In summary, we again remark that the above link polynomials are evaluated using the recursion formulae in §3 and 4 with the knowledge of three quantities, τ, $\bar{\tau}$ and Q_{1212}, defined in (2.12) and (3.10).

References

1) R. J. Baxter: *Exactly Solved Models in Statistical Mechanics* (Academic Press, London, 1982).
2) M. Wadati and Y. Akutsu: *Exactly Solvable Models in Statistical Mechanics*, to appear in Lecture Notes in Physics, ed. M. Lakshmanan (Springer-Verlag).
3) A. Kuniba, Y. Akutsu and M. Wadati: J. Phys. Soc. Jpn. **55** (1986) 1092, 2166 and 3285. Phys. Lett. **116A** (1986) 382; **117A** (1986) 358.
4) Y. Akutsu, A. Kuniba and M. Wadati: J. Phys. Soc. Jpn. **55** (1986) 1466 and 1880.
5) A. Kuniba, Y. Akutsu and M. Wadati: J. Phys. Soc. Jpn. **55** (1986) 2605.
6) G. E. Andrews, R. J. Baxter and P. J. Forrester: J. Stat. Phys. **35** (1984) 193.
7) R. J. Baxter and G. E. Andrews: J. Stat. Phys. **44** (1986) 249 and 713.
8) Y. Akutsu, A. Kuniba and M. Wadati: J. Phys. Soc. Jpn. **55** (1986) 2907.
9) E. Date, M. Jimbo, T. Miwa and M. Okado: Lett. Math. Phys. **12** (1986) 209.
10) Y. Akutsu and M. Wadati: J. Phys. Soc. Jpn. **56** (1987) 839.
11) Y. Akutsu and M. Wadati: J. Phys. Soc. Jpn. **56** (1987) 3039. This paper is part I of the series.
12) J. S. Birman: *Braids, Links, and Mapping Class Groups* (Princeton University Press, 1974).
13) D. Rolfsen: *Knots and Links* (Publish or Perish Inc., Berkeley, 1976).
14) J. S. Birman: Invent. Math. **81** (1985) 287.
15) V. F. R. Jones: Bull. Amer. Math. Soc. **12** (1985) 103.
16) K. Sogo, Y. Akutsu and T. Abe: Prog. Theor. Phys. **70** (1983) 730 and 739.

Note added in proof—The authors would like to express their sincere thanks to Professor Kunio Murasugi for valuable comments which were extremely useful in correcting the polynomials in §5.

Journal of the Physical Society of Japan
Vol. 57, No. 3, March, 1988, pp. 757–776

Exactly Solvable Models and New Link Polynomials. III. Two-Variable Topological Invariants

Tetsuo DEGUCHI, Yasuhiro AKUTSU† and Miki WADATI

Institute of Physics, College of Arts and Science, University of Tokyo, Komaba, Meguro-ku, Tokyo 153
†Institute of Physics, Kanagawa University, Rokkakubashi, Kanagawa-ku, Yokohama 221

(Received October 9, 1987)

New link polynomials, reported in I and II of the series, are extended into those with two variables. A concept of composite string is introduced. It is shown that the composite string representation and the generalized Ocneanu's trace lead to a sequence of two-variable link polynomials. In addition, algebraic aspects of the composite string representation are studied in some detail.

§1. Introduction

Recently we developed a general method to construct link polynomials, topological invariants for knots and links, from exactly solvable models in statistical mechanics.[1-4] The method was applied to the N-state vertex model[5] and a sequence of new link polynomials was derived. In this paper, as a continuation of previous work, we extend the new link polynomials into those with two variables.

Let us start from the braid and the braid group. Braids are formed when n points on a horizontal line are connected by n strings to n points on another horizontal line directly below the first n points. Trivial n-braid is a configuration where no intersection between the strings is present. A nontrivial n-braid is constructed from the trivial n-braid by successive application of operator b_i, $i=1, 2, \cdots, n-1$ (Fig. 1). A set of generators, $b_1, b_2, \cdots, b_{n-1}$, defines the braid group B_n. By regarding the trivial n-braid as the identity operation

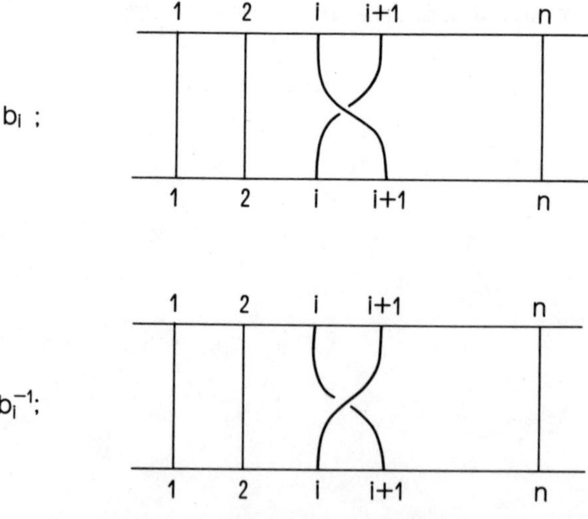

Fig. 1. Operation b_i and its inverse b_i^{-1}.

in B_n, we can identify any element in B_n as an n-braid. To guarantee the topological equivalence between different expressions of a braid in terms of braid group elements, Artin[6] proved that the following conditions are necessary and sufficient;

$$b_i b_j = b_j b_i, \quad |i-j| \geq 2, \tag{1.1a}$$

$$b_i b_{i+1} b_i = b_{i+1} b_i b_{i+1}. \tag{1.1b}$$

We call them the defining relation of B_n. Relation (1.1) is shown schematically in Fig. 2.

In previous papers,[1-4] we reported a prescription to have a braid group representation from the Boltzmann weights which satisfy the Yang-Baxter relation. We considered the N-state vertex model whose Boltzmann weights are denoted by $S_{lp}^{km}(u)$. The formula for a representation g_i of b_i is

$$g_i = \lim_{u\to\infty} \sum_{klmp} S_{lp}^{km}(u) I^{(1)} \otimes I^{(2)} \otimes \cdots \otimes E_{pk}^{(i)} \otimes E_{ml}^{(i+1)} \otimes I^{(i+2)} \otimes \cdots \otimes I^{(n)}. \tag{1.2}$$

Here, $I^{(j)}$ is the identity acting at the j-th position and E_{pk} is a matrix such that $(E_{pk})_{ab} = \delta_{pa}\delta_{kb}$. Note that a suitable renormalization for $S_{lp}^{km}(u)$ is necessary[2] before taking the limit $u\to\infty$. These operators satisfy the defining relation of the braid group B_n;

$$g_i g_j = g_j g_i, \quad |i-j| \geq 2, \tag{1.3a}$$

$$g_i g_{i+1} g_i = g_{i+1} g_i g_{i+1}. \tag{1.3b}$$

In addition, for the $N=2$ case where the model is the 6-vertex model, we have the reduction relation

$$g_i^2 = (1-t) g_i + t. \tag{1.4}$$

The group algebra generated by the operators $g_1, g_2, \cdots, g_{n-1}$, which satisfy (1.3) and (1.4) is equivalent to the Hecke algebra $H(t, n)$.[7]

We now explain the outline of the present paper. In §2, by introducing a "composite string" with projectors (symmetrizers), we give a systematic method to construct the braid group representation. In §3, we use the composite string representation $B_n^{[s]}$, $N=2s+1$, of the braid group to obtain the two variable extension of the new link polynomials. The algebra of $B_2^{[s]}$ is studied in §4. We show that I_i, $S_i^{(1)}, \cdots, S_i^{(N-1)}$, and G_i yield a closed algebra (definitions of G_i and $S_i^{(l)}$ are in §2 and 4, respectively). In §5, for the $N=3$ and $N=4$ cases, we calculate two-variable invariants for closed 2-braids. The last section is devoted to summary and discussion. Throughout the paper we often use diagrams to clarify discussion since analytic expressions are sometimes lengthy and not transparent.

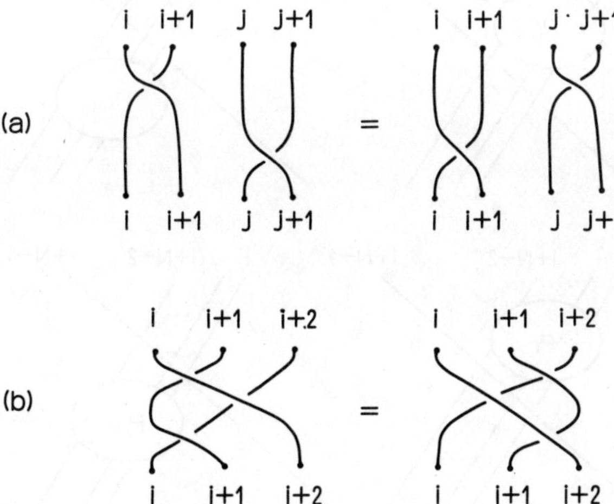

Fig. 2. Braid group. (a) $b_i b_j = b_j b_i$, $|i-j| \geq 2$, (b) $b_i b_{i+1} b_i = b_{i+1} b_i b_{i+1}$.

§2. Composite String Representation

Formula (1.2) establishes a direct relation between the braid group representation and the Boltzmann weight of the N-state vertex model. In this section, we propose an alternative approach. By introducing a "composite string," we present a systematic method to construct the braid group representations.

Suppose that a set of generators, $\{g_i\}$, satisfying (1.3) and (1.4) is known. We form a "composite string" by combining $(N-1)$ strings and attaching a projector P at each end (Fig. 3). This process corresponds to a procedure of making higher spin S-matrices from lower spin S-matrices.[8] For the moment (the explicit form of the projector will be given shortly), we assume that the projector P_i has the properties;

(i) $P_i^2=P_i$, (2.1)

(ii) P_i consists of generators $g_i, g_{i+1}, \cdots, g_{i+N-3}$. (2.2)

The projector P_i plays a role to select a spin s of the representation, where s is related to the state number N by

$$s=\frac{1}{2}(N-1). \tag{2.3}$$

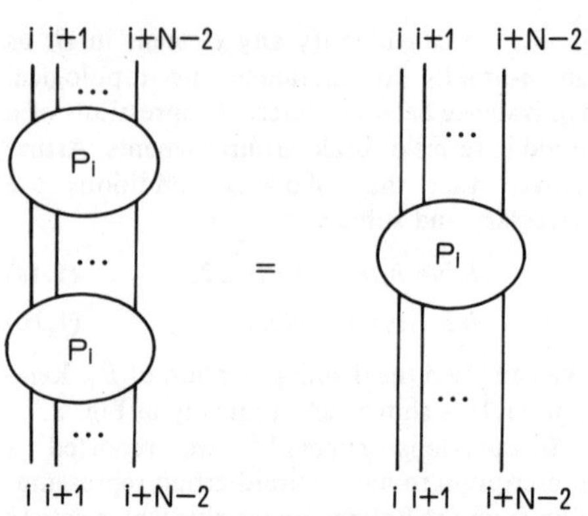

Fig. 3. A composite string. Since $P_i^2=P_i$, the two diagrams are equivalent.

Due to relation (1.3b) and property (ii), the projector passes through the string by the rules (Fig. 4);

(iii)
$$P_i(g_{i+N-2}g_{i+N-3}\cdots g_i)=(g_{i+N-2}g_{i+N-3}\cdots g_i)P_{i+1}, \tag{2.4a}$$
$$P_i(g_{i+N-2}^{-1}g_{i+N-3}^{-1}\cdots g_i^{-1})=(g_{i+N-2}^{-1}g_{i+N-3}^{-1}\cdots g_i^{-1})P_{i+1}. \tag{2.4b}$$

Let us denote the spin s representation of B_n

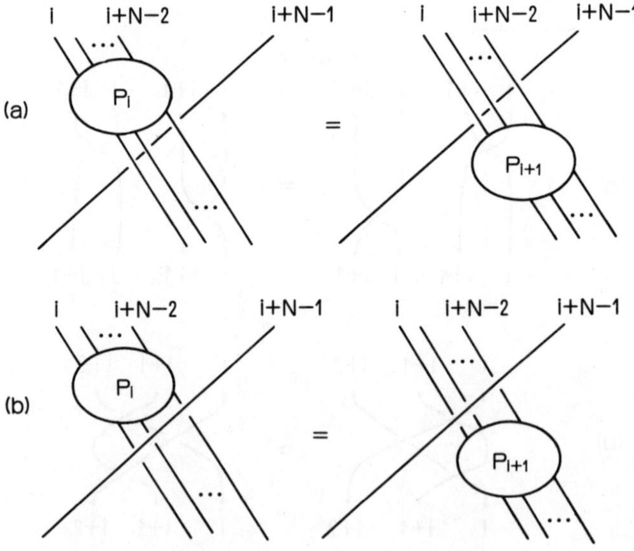

Fig. 4. Property (iii) of the projector.

by $B_n^{[s]}$. The composite string representation $B_n^{[s]}$ is given as follows. For notational simplicity, we sometimes use

$$k \equiv N-1. \tag{2.5}$$

We prepare n sets of k strings (in total $k \times n$ strings) and introduce an operator $g_i^{(l)}$ by

$$g_i^{(l)} = g_{ik+1-l} g_{ik+2-l} \cdots g_{(i+1)k-l}, \quad l = 1, 2, \cdots, N-1. \tag{2.6}$$

The generator G_i of $B_n^{[s]}$ is expressed as (Fig. 5);

$$G_i = P_{(i-1)k+1}^{(N)} P_{ik+1}^{(N)} \bar{G}_i^{(N)} P_{(i-1)k+1}^{(N)} P_{ik+1}^{(N)}, \tag{2.7a}$$

where

$$\bar{G}_i^{(N)} = g_i^{(1)} g_i^{(2)} \cdots g_i^{(N-1)}. \tag{2.7b}$$

We now construct the projector $P_i^{(N)}$. It is easy to see that

$$P_i^{(3)} = \frac{1}{1+t}(t+g_i) \tag{2.8}$$

satisfies properties (i), (ii) and (iii). We find that the projector $P_i^{(N)}$ for $N \geqq 3$ is derived through a recursion formula;

$$P_i^{(N)} = P_i^{(N-1)} h_{i+N-3}^{(N)} P_i^{(N-1)}, \quad P_i^{(2)} \equiv 1, \tag{2.9}$$

where

$$h_j^{(N)} = \frac{\tau_{N-2}}{\tau_{N-1}} \left(\frac{t^{N-2}}{\tau_{N-2}} + g_j \right), \tag{2.10}$$

with

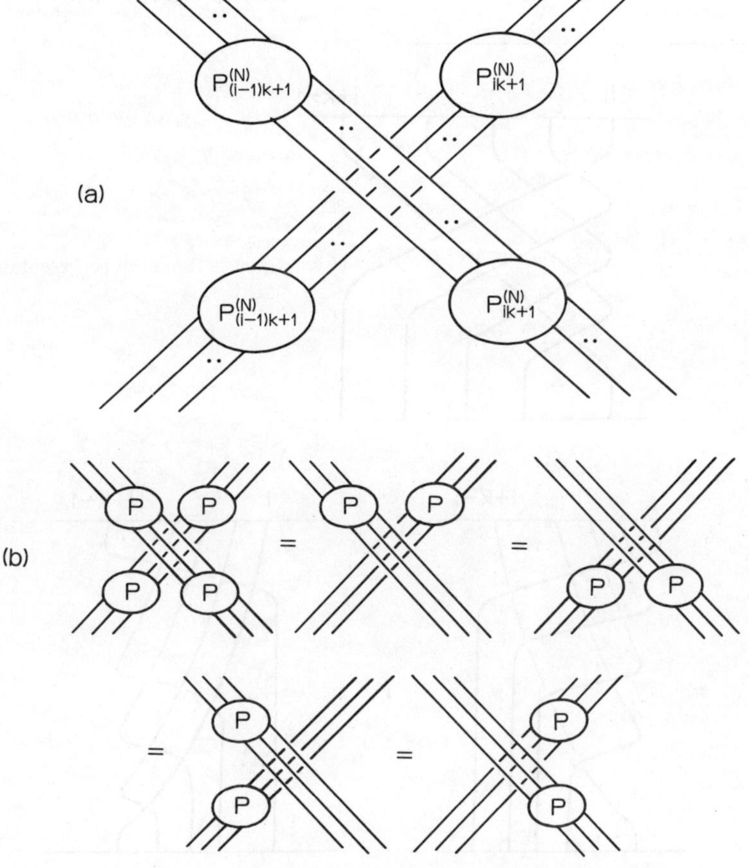

Fig. 5. (a) Generator G_i of $B_n^{[s]}$. (b) Equivalent diagrams.

$$\tau_m=1+t+t^2+\cdots+t^{m-1}. \tag{2.11}$$

The recursion relation (2.9) gives, for instance,

$$P_i^{(4)}=\frac{1}{1+t}(t+g_i)\cdot\frac{1+t}{1+t+t^2}\left(\frac{t^2}{1+t}+g_{i+1}\right)\cdot\frac{1}{1+t}(t+g_i)=\frac{1}{(1+t)(1+t+t^2)}\times\{t^3+t^2(g_i+g_{i+1})+t(g_ig_{i+1}+g_{i+1}g_i)+g_ig_{i+1}g_i\}. \tag{2.12}$$

The projector P_i defined by (2.9) satisfies not only properties (i), (ii), (iii) but also

(iv) $P_i g_i=P_i$, (2.13)

(v) $P_i\Delta_i^2=P_i$, (2.14)

where

$$\Delta_i=(g_ig_{i+1}\cdots g_{i+k-2})\times(g_ig_{i+1}\cdots g_{i+k-3})\cdots(g_i). \tag{2.15}$$

The operator Δ_i is sometimes called a half-twist Δ_i[9] (see Fig. 6). As will be shown in the next section, property (v) of the projector P_i assures the Markov property II (cf. (3.3)).

The proofs for the recursion formula (2.9) and the properties of the projectors are given in the Appendix.

We summarize the composite string representation $B_n^{[s]}$. The generators $G_1, G_2,\cdots, G_{n-1}$ are defined by (2.7) with the projector P_i given by (2.9). Figure 7 illustrates that the generators $\{G_i;\ i=1, 2,\cdots, n-1\}$ satisfy the defining relation of the braid group;

$$G_iG_j=G_jG_i,\quad |i-j|\geq 2, \tag{2.16a}$$

$$G_iG_{i+1}G_i=G_{i+1}G_iG_{i+1}. \tag{2.16b}$$

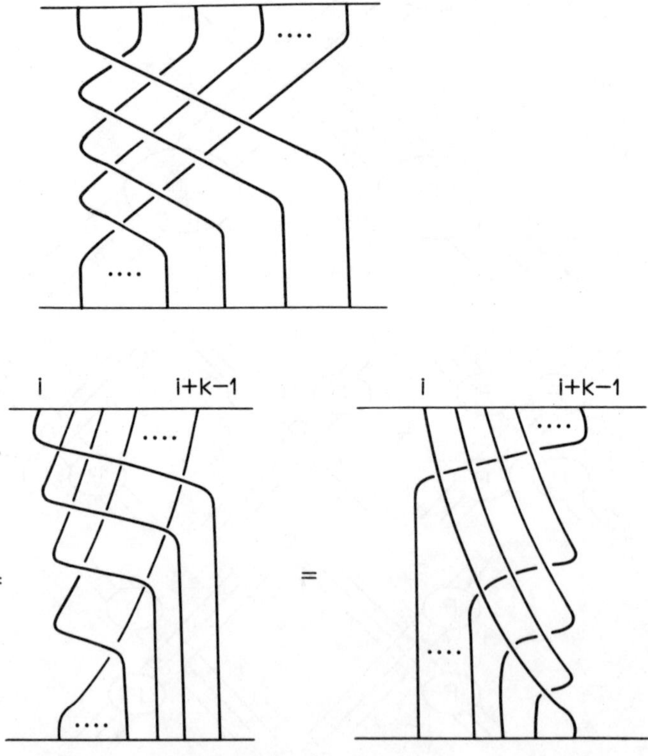

Fig. 6. Half-twist Δ_i.

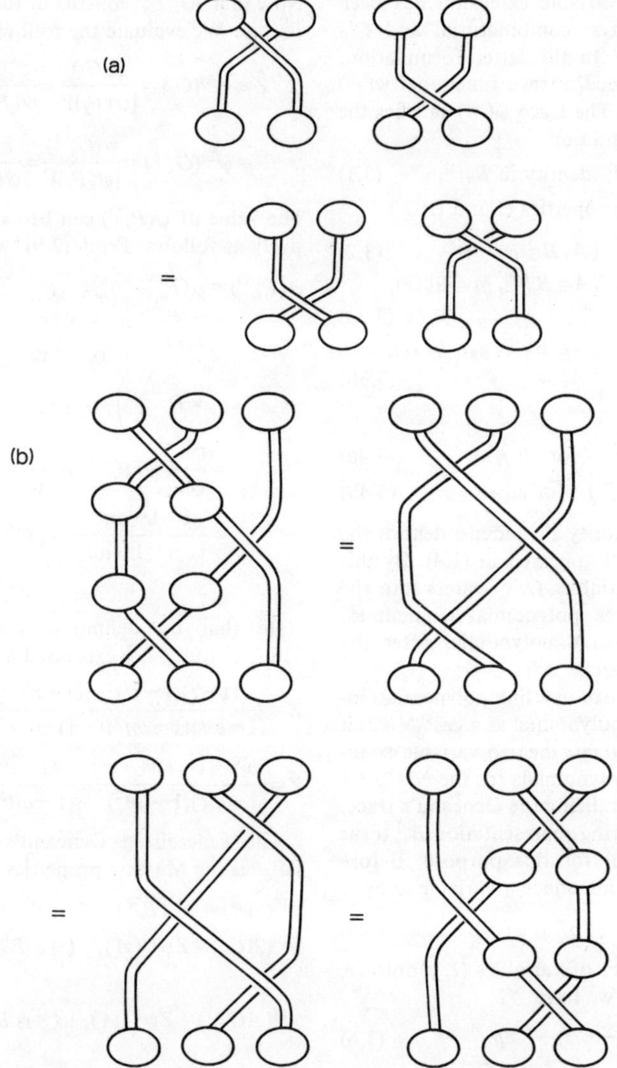

Fig. 7. The generators G_i, $i=1, 2, \cdots, n-1$, of $B_n^{[s]}$ satisfy (a) $G_iG_j=G_jG_i$, $|i-j|\geq 2$ and (b) $G_iG_{i+1}G_i=G_{i+1}G_iG_{i+1}$. Diagrams are for $N=3$. Circles represent the projectors.

Note that when the relations or the equations are valid for any N (any $s=(N-1)/2$), we do not write subscript N. For instance, P_i in (2.1), (2.2), (2.4), (2.13) and (2.14) means $P_i^{(N)}$ for any integer $N\geq 3$.

§3. Two-Variable Link Polynomials

We shall now use the composite string representation $B_n^{[s]}$ which was introduced in §2 to extend the new link polynomials[1-4] into those with two variables. For the Jones polynomial,[10] which is $N=2s+1=2$ in our theory, the two-variable extension has been made in two ways; combinatorial and C^*-algebraic ways.[11)] In the latter formulation, Ocneanu introduced a trace functional $\psi(\bullet)$ defined on $B_n^{[1/2]}$. The trace $\psi(\bullet)$ satisfies the normalization condition

$$\psi(I)=1, \quad I;\text{ identity in } B_n^{[1/2]}, \tag{3.1}$$

and the Markov properties (I and II)

$$\psi(AB)=\psi(BA) \quad (A, B\in B_n^{[1/2]}), \tag{3.2}$$

$$\psi(Ag_n)=z\psi(A) \quad (A\in B_n^{[1/2]}, g_n\in B_{n+1}^{[1/2]}), \tag{3.3a}$$

$$\psi(Ag_n^{-1})=\bar{z}\psi(A) \quad (A\in B_n^{[1/2]}, g_n\in B_{n+1}^{[1/2]}), \tag{3.3b}$$

where

$$z=\psi(g_j) \quad \text{for all } j, \tag{3.4a}$$

$$\bar{z}=\psi(g_j^{-1}) \quad \text{for all } j. \tag{3.4b}$$

Note that the quantity z is independent of the parameter t which appeared in (1.4). By this way a pair of variables (t, z) enters into the two-variable Jones polynomial (sometimes, called the HOMFLY polynomial after the names of its researchers[11)]).

Since a sequence of new link polynomials includes the Jones polynomial as a case $N=2$, it is natural to investigate the two-variable extension of the link polynomials for the $N=2s+1 \geq 3$ cases by generalizing the Ocneanu's trace. The composite string representation $B_n^{[s]}$ turns out to be helpful for this purpose. Before showing this, we introduce a variable ω by

$$\omega=\bar{z}/z=\psi(g_j^{-1})/\psi(g_j), \tag{3.5}$$

and change the set of variables (t, z) into (t, ω).[7)] From (1.4), we have

$$g_j^{-1}=(1-t^{-1})+t^{-1}g_j, \tag{3.6}$$

and then

$$\bar{z}=(1-t^{-1})+t^{-1}z. \tag{3.7}$$

Solving (3.5) and (3.7), we obtain

$$z=\frac{1-t}{1-\omega t}, \quad \bar{z}=\frac{\omega(1-t)}{1-\omega t}. \tag{3.8}$$

Let us define the generalized Ocneanu's trace $\psi^{[s]}(\bullet)$ by

$$\psi^{[s]}(A)=\frac{\psi(A)}{[\psi(P_j)]^n}, \quad A\in B_n^{[s]}. \tag{3.9}$$

Note that $A\in B_n^{[s]}$ consists of the generators in $B_{(N-1)n}^{[1/2]}$. We evaluate the following quantities;

$$Z\equiv\psi^{[s]}(G_j)=\frac{\psi(G_j)}{[\psi(P_j)]^2}=\frac{z^{N-1}}{\psi(P_j)}, \tag{3.10a}$$

$$\bar{Z}\equiv\psi^{[s]}(G_j^{-1})=\frac{\psi(G_j^{-1})}{[\psi(P_j)]^2}=\frac{\bar{z}^{N-1}}{\psi(P_j)}. \tag{3.10b}$$

The value of $\psi(P_j^{(N)})$ can be calculated recursively as follows. From (2.9), we have

$$\begin{aligned}\psi(P_i^{(N)})&=\psi(P_i^{(N-1)}h_{i+N-3}^{(N)}P_i^{(N-1)})\\&=\psi\left(P_i^{(N-1)}\frac{\tau_{N-2}}{\tau_{N-1}}\left(\frac{t^{N-2}}{\tau_{N-2}}+g_{i+N-3}\right)\times P_i^{(N-1)}\right)\\&=\frac{t^{N-2}}{\tau_{N-1}}\psi(P_i^{(N-1)})+\frac{\tau_{N-2}}{\tau_{N-1}}z\psi(P_i^{(N-1)})\\&=\frac{1}{\tau_{N-1}}\frac{1-\omega t^{N-1}}{1-\omega t}\psi(P_i^{(N-1)}),\ N\geq 3.\end{aligned} \tag{3.11}$$

Note that by definition $\psi(P_i^{(2)})=\psi(I)=1$. Thus, Z and $\bar{Z}$ are expressed as

$$Z=\frac{(1-t)(1-t^2)\cdots(1-t^{N-1})}{(1-\omega t)(1-\omega t^2)\cdots(1-\omega t^{N-1})}, \tag{3.12a}$$

$$\bar{Z}=\frac{\omega^{N-1}(1-t)(1-t^2)\cdots(1-t^{N-1})}{(1-\omega t)(1-\omega t^2)\cdots(1-\omega t^{N-1})}. \tag{3.12b}$$

The generalized Ocneanu's trace $\psi^{[s]}(\bullet)$ satisfies the Markov properties (I and II);

$$\psi^{[s]}(AB)=\psi^{[s]}(BA), \quad (A, B\in B_n^{[s]}), \tag{3.13}$$

$$\psi^{[s]}(AG_n)=Z\psi^{[s]}(A), \quad (A\in B_n^{[s]}, G_n\in B_{n+1}^{[s]}), \tag{3.14a}$$

$$\psi^{[s]}(AG_n^{-1})=\bar{Z}\psi^{[s]}(A), \quad (A\in B_n^{[s]}, G_n\in B_{n+1}^{[s]}). \tag{3.14b}$$

From the properties of the Ocneanu's trace, (3.13) is rather trivial. Markov property II (3.14a) is depicted in Fig. 8. We tie two right-ends of G_n shown in box S of Fig. 8 (this operation corresponds to taking the trace). We deform a tied braid as shown in Fig. 9. Note that for each string the trace is the Ocneanu's trace (not the generalized one). Therefore, we have a z factor for each Markov move II of strings and obtain

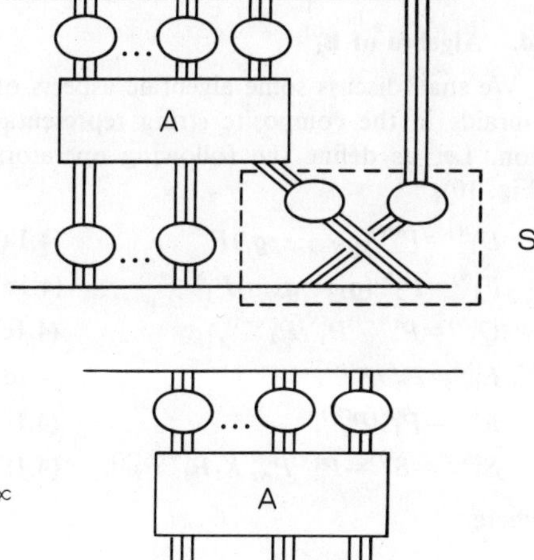

Fig. 8. Markov property II for the composite spin representation $B_n^{[s]}$ and the generalized Ocneanu's trace $\psi^{[s]}(\,\bullet\,)$.

$$\begin{aligned}\psi(AG_n)&=z^{N-1}\psi(AP_{k(n-1)+1}\Delta^2_{k(n-1)+1}P_{k(n-1)+1})\\&=z^{N-1}\psi(AP_{k(n-1)+1})\\&=z^{N-1}\psi(A),\ (A\in B_n^{[s]}).\end{aligned}\tag{3.15}$$

We have used (2.14) and Δ_i has been defined in (2.15). Combining (3.15) and (3.9), we confirm that

$$\begin{aligned}\psi^{[s]}(AG_n)&=\frac{\psi(AG_n)}{[\psi(P)]^{n+1}}\\&=\frac{z^{N-1}}{\psi(P_j)}\frac{\psi(A)}{[\psi(P_j)]^n}=Z\psi^{[s]}(A),\\&\qquad (A\in B_n^{[s]}).\end{aligned}\tag{3.16}$$

With the generalized Ocneanu's trace we find that the two-variable link polynomial $\alpha_\omega^{[s]}(\,\bullet\,)$ is expressed as

$$\alpha_\omega^{[s]}(A)=(\bar{Z}Z)^{-\frac{n-1}{2}}\left(\frac{\bar{Z}}{Z}\right)^{\frac{1}{2}e(A)}\psi^{[s]}(A),\ (A\in B_n^{[s]}),\tag{3.17}$$

where $e(A)$ is the exponent sum of the generators in $B_n^{[s]}$. In fact, we can prove that $\alpha_\omega^{[s]}(\,\bullet\,)$ is invariant under the Markov moves (I

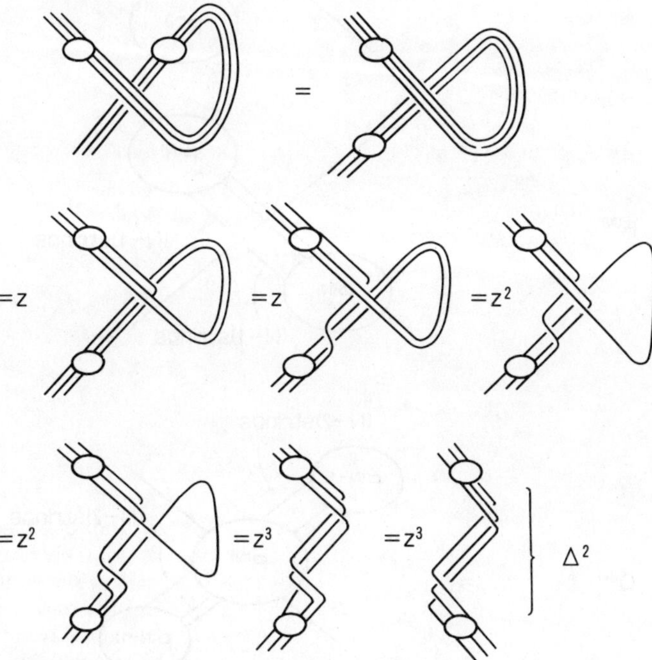

Fig. 9. Deformation of a tied braid to prove the Markov move II (3.14a). When the deformation is substituted into box S of Fig. 8, we obtain (3.15). To simplify the diagram, $N=4$ case is shown.

and II);

$$\alpha_\omega^{[s]}(AB)=\alpha_\omega^{[s]}(BA),\ (A, B\in B_n^{[s]}), \quad (3.18a)$$

$$\alpha_\omega^{[s]}(AG_n)=\alpha_\omega^{[s]}(AG_n^{-1})=\alpha(A),$$
$$(A\in B_n^{[s]},\ G_n\in B_{n+1}^{[s]}). \quad (3.18b)$$

To conclude this section we give two comments. First, the two-variable link polynomial $\alpha_\omega^{[s]}(\cdot)$ reduces to the new link polynomial[1-4] when we set $\omega=t$. Second, a crucial point of two-variable extension is that there exists a trace functional having the Z-factor (cf. (3.14)) which is independent of the parameter t.

§4. Algebra of $\mathbf{B}_2^{[s]}$

We shall discuss some algebraic aspects of 2-braids in the composite string representation. Let us define the following operators (Fig. 10);

$$D_1^{(N)}=P_1^{(N)}(g_{N-1}\cdots g_1)P_2^{(N)}, \quad (4.1a)$$

$$F_1^{(N)}=P_2^{(N)}(g_1\cdots g_{N-1})P_1^{(N)}, \quad (4.1b)$$

$$Q_1^{(N)}=P_1^{(N-1)}P_1^{(N)}P_2^{(N-1)}, \quad (4.1c)$$

$$L_1^{(N)}=P_2^{(N)}P_1^{(N)}, \quad (4.1d)$$

$$R_1^{(N)}=P_1^{(N)}P_2^{(N)}, \quad (4.1e)$$

$$S_1^{(N,r)}=S_1^{(r)}=P_1^{(N)}P_N^{(N)}K_1P_1^{(N)}P_N^{(N)}, \quad (4.1f)$$

where

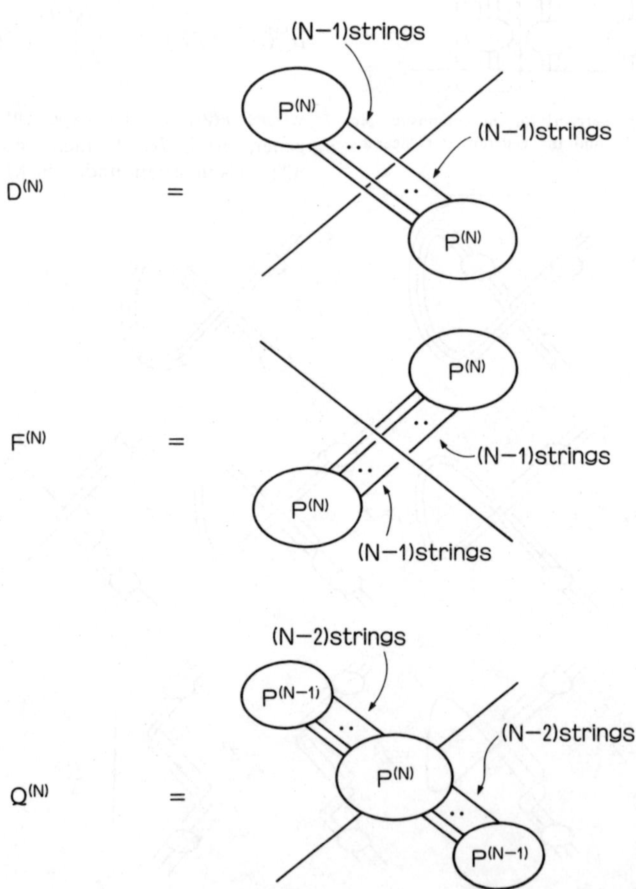

Fig. 10

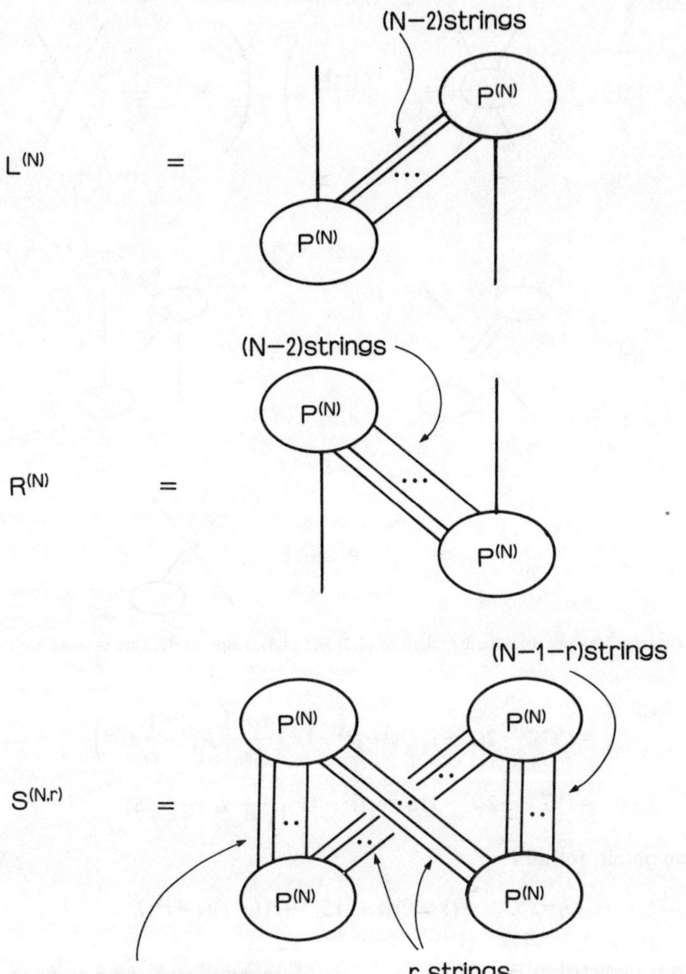

Fig. 10. Diagrams for operators $D_1^{(N)}$, $F_1^{(N)}$, $L_1^{(N)}$, $Q_1^{(N)}$, $R_1^{(N)}$ and $S_1^{(N,r)} \equiv S_1^{(r)}$. Note that $D^{(2)} = S^{(2,1)} = S^{(1)}$.

$$K_1 = (g_{N-1}g_N \cdots g_{N+r-2})(g_{N-2}g_{N-1} \cdots g_{N+r-3}) \cdots (g_{N-r}g_{N-r+1} \cdots g_{N-1}). \tag{4.1g}$$

Since the algebraic calculation is lengthy, we simplify the presentation by using graphs and by considering the $N=3$ case. From the expression (4.1), we have (Fig. 11)

$$Q_1^{(3)} = \frac{\tau_2 - 1}{\tau_2} R_1^{(2)} + \frac{1}{\tau_2} D_1^{(2)}, \tag{4.2}$$

$$D_1^{(3)} = (1-t^2) R_1^{(3)} + t^2 F_1^{(3)-1}. \tag{4.3}$$

The following relations are proved graphically in Fig. 12;

$$\begin{aligned} G_1^{-1} &= (F_1^{(3)-1})^2 = (1-t^{-2})^2 R_1^{(3)2} + 2t^{-2}(1-t^{-2}) R_1^{(3)} D_1^{(3)} + t^{-4} D_1^{(3)2} \\ &= (1-t^{-2})^2 Q_1^{(3)} + 2t^{-2}(1-t^{-2}) D_1^{(2)} + t^{-4} G_1, \end{aligned} \tag{4.4}$$

$$G_1 = t^4 G_1^{-1} - 2(t^2-1) D_1^{(2)} - (t^2-1)^2 Q_1^{(3)}$$

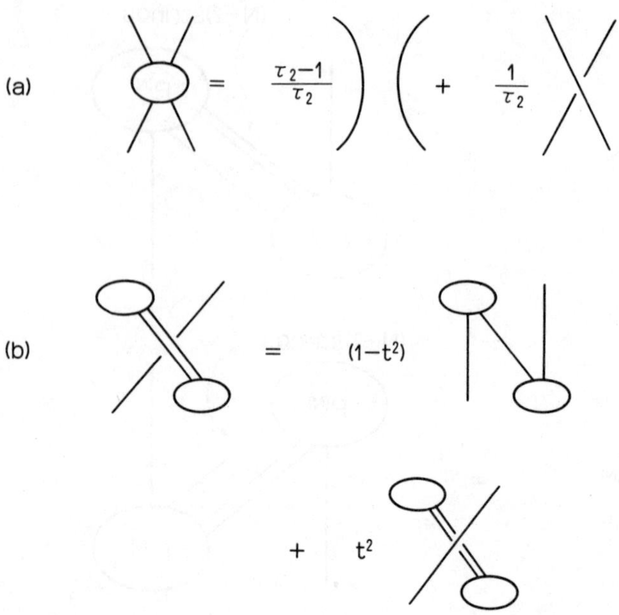

Fig. 11. Diagrams (a) and (b) describe relations (4.2) and (4.3), respectively. Circles stand for the projectors.

$$=t^4G_1^{-1}-2(t^2-1)D_1^{(2)}-(t^2-1)^2\left(\frac{\tau_2-1}{\tau_2}R_1^{(2)}+\frac{1}{\tau_2}D_1^{(2)}\right)$$

$$=t^4G_1^{-1}-2(t^2-1)S_1^{(1)}-(t^2-1)^2\left(\frac{t}{1+t}I_1+\frac{1}{1+t}S_1^{(1)}\right). \tag{4.5}$$

Therefore we obtain for all i

$$G_i=t^4G_i^{-1}+(1-t^2)(1+t)S_i^{(1)}-t(1-t)(1-t^2)I_i. \tag{4.6}$$

This relation is depicted in Fig. 13.

Moreover, we find that a set of operators I_i, $S_i^{(1)}$ and G_i makes a closed algebra;

$$G_i^2=t^4I_i+t(1+t)(1-t^2)S_i^{(1)}+(1-t)(1-t^2)G_i,$$

$$S_i^{(1)2}=\frac{t^3}{(1+t)^2}I_i+\frac{2t+t^2-t^3}{(1+t)^2}S_i^{(1)}+\frac{1}{(1+t)^2}G_i,$$

$$S_i^{(1)}G_i=G_iS_i^{(1)}=tS_i^{(1)}+(1-t)G_i. \tag{4.7}$$

It is interesting to observe that a set of operators I_i, $\hat{G}_i=G_i/t\sqrt{-1}$ and E_i, where

$$E_i\equiv(G_i-1)(G_i+t^2)/t^3(1-t^2)$$

$$=-t^{-2}(tI_i-(1+t)S_i+G_i), \tag{4.8}$$

gives an algebra of Birman and Wenzl[12] and Murakami[13] for the Kauffman polynomial[14] when we set $l=\sqrt{-1}/t^2$ and $m=(1-t^2)/(\sqrt{-1}t)$ in their notation and when we use the matrix representation.[2]

The above analysis is generalized for all $N\geq 3$. By using the relation

$$Q_i^{(N)}=\frac{\tau_{N-1}-1}{\tau_{N-1}}R_i^{(N-1)}+\frac{1}{\tau_{N-1}}D_i^{(N-1)}, \tag{4.9}$$

$$D_i^{(N)}=(1-t^{N-1})R_i^{(N)}+t^{N-1}(F_i^{(N)})^{-1}, \tag{4.10}$$

$$(R_i^{(N)})^l(D_i^{(N)})^{N-1-l}=(Q_i^{(N)})^{l-1}(D_i^{(N-1)})^{N-1-l}, \tag{4.11}$$

we show that

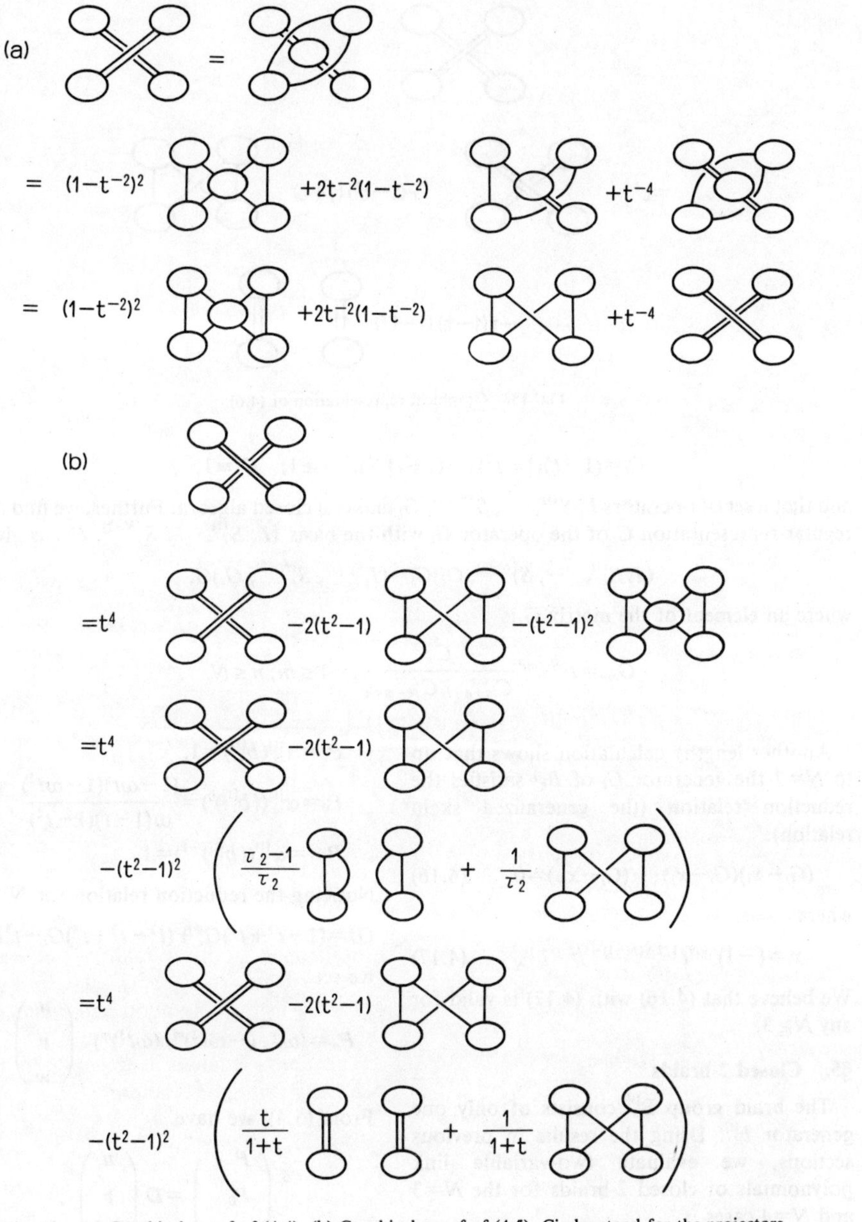

Fig. 12. (a) Graphical proof of (4.4). (b) Graphical proof of (4.5). Circles stand for the projectors.

$$G_i = t^{(N-1)^2} G_i^{-1} + \sum_{r=0}^{N-2} (-1)^r t^{\frac{r}{2}(r+1)} \frac{C_N^2}{C_{r+2} C_{N-1-r}^2} S_i^{(N-2-r)}, \tag{4.12}$$

where the constant C_r is given by

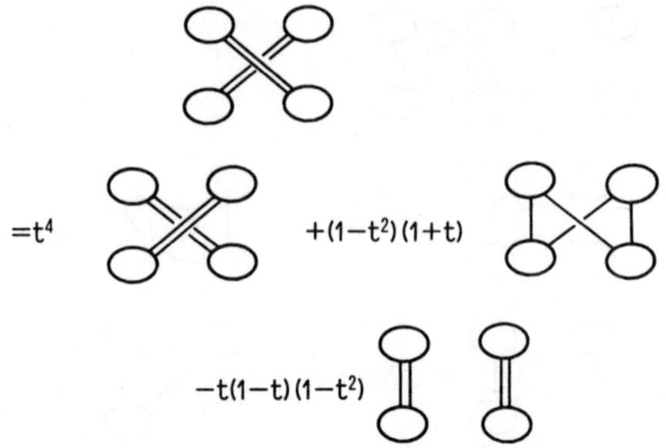

$=t^4$ $+(1-t^2)(1+t)$ $-t(1-t)(1-t^2)$

Fig. 13. Graphical representation of (4.6).

$$C_r=(1-t)(1-t^2)\cdots(1-t^{r-1}),\quad r\geq 1,\quad C_1\equiv 1, \tag{4.13}$$

and that a set of operators I_i, $S_i^{(1)},\cdots,S_i^{(N-2)}$, G_i makes a closed algebra. Further, we find that the regular representation $\hat{G}$ of the operator G_i with the basis $\{I_i, S_i^{(1)},\cdots, S_i^{(N-2)}, G_1\}$ is given by

$$(I, S_i^{(1)},\cdots,S_i^{(N-2)}, G_i)G_i=(I,\cdots,S_i^{(N-2)},G_i)\hat{G}, \tag{4.14}$$

where an element of the matrix $\hat{G}$ is

$$\hat{G}_{mn}=t^{(N-m)^2}\frac{C_n^2}{C_{m+n-N}C_{N-m+1}^2},\quad 1\leq m,n\leq N. \tag{4.15}$$

Another lengthy calculation shows that up to $N=7$ the generator G_i of $B_n^{[s]}$ satisfies the reduction relation (the generalized skein relation):

$$(G_i-x_1)(G_i-x_2)\cdots(G_i-x_N)=0, \tag{4.16}$$

where

$$x_r=(-1)^{r+N}t^{1/2N(N-1)-1/2r(r-1)}. \tag{4.17}$$

We believe that (4.16) with (4.17) is valid for any $N\geqq 3$.

§5. Closed 2-braids

The braid group $B_2^{[s]}$ consists of only one generator $b_1^{[s]}$. Using the results in previous sections, we evaluate two-variable link polynomials of closed 2-braids for the $N=3$ and $N=4$ cases.

1) $N=3$ $(s=1)$

We write

$$P_m^{(1)}=\alpha_\omega^{[1]}((b_1^{[1]})^m),\ m=0,\pm1,\pm2,\cdots. \tag{5.1}$$

We see from (3.17) that

$$P_1=\alpha_\omega^{[1]}(b_1^{[1]})=1,$$
$$P_0=\alpha_\omega^{[1]}((b_1^{[1]})^0)=\frac{(1-\omega t)(1-\omega t^2)}{\omega(1-t)(1-t^2)},$$
$$P_{-1}=\alpha_\omega^{[1]}((b_1^{[1]})^{-1})=1. \tag{5.2}$$

Noticing the reduction relation for $N=3$,

$$G_i^3=(1-t^2+t^3)G_i^2+(t^2-t^3+t^5)G_i-t^5I_i, \tag{5.3}$$

we set

$$P_m=(\omega^m,(-\omega t^2)^m,(\omega t^3)^m)\begin{pmatrix}u\\v\\w\end{pmatrix}. \tag{5.4}$$

From (5.4), we have

$$\begin{pmatrix}P_{-1}\\P_0\\P_1\end{pmatrix}=D\begin{pmatrix}u\\v\\w\end{pmatrix}, \tag{5.5}$$

where the matrix D is given by

$$D=\begin{pmatrix}\omega^{-1} & (-\omega t^2)^{-1} & (\omega t^3)^{-1}\\ 1 & 1 & 1\\ \omega & -\omega t^2 & \omega t^3\end{pmatrix}. \tag{5.6}$$

Then, we obtain two-variable link polynomial for closed 2-braids as follows:

$$P_m^{(1)}(\omega,t)=\alpha_\omega^{[1]}((b_i^{[1]})^m)=(\omega^m(-\omega t^2)^m(\omega t^3)^m)D^{-1}\begin{pmatrix}P_{-1}\\ P_0\\ P_1\end{pmatrix}=\omega^m\frac{(1-\omega t^3)(1-\omega t^4)}{\omega(1-t^3)(1-t^4)}$$

$$+(-\omega t^2)^m\frac{t(1-\omega)(1-\omega t^3)}{\omega(1-t)(1-t^4)}+(\omega t^3)^m\frac{t^2(1-\omega)(1-\omega t)}{\omega(1-t^2)(1-t^3)}\,. \tag{5.7}$$

When $\omega=t$, expression (5.7) reduces to a previous result;[3])

$$P_m(t)=(-1)^m t^{3m}+\frac{t^{m-2}+t^{m+2}+t^{m-1}+t^{m+1}+t^m+t^{4m}}{1+t+t^{-1}}\,. \tag{5.8}$$

2) $N=4$ $(s=3/2)$

The calculation becomes rather lengthy in this case. We start from the relation

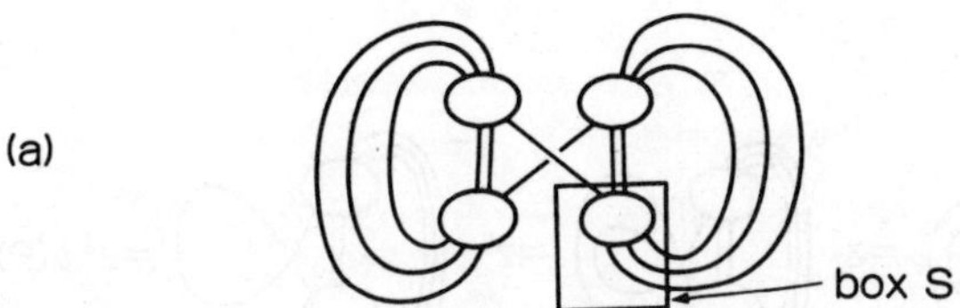

Fig. 14(a)~(d).

$$G_i^2=(1-t)(1-t^2)(1-t^3)G_i+t(1+t)(1-t^3)^2S_i^{(2)}+t^4(1+t+t^2)(1-t^3)S_i^{(1)}+t^9I_i, \tag{5.9}$$

which is derived from (4.14) with (4.15). Taking the generalized Ocneanu's trace $\psi^{[3/2]}(\,\bullet\,)$ on both sides of (5.9), we have

$$\psi^{[3/2]}((b_i^{[3/2]})^2)=(1-t)(1-t^2)(1-t^3)Z+t(1+t)(1-t^3)^2\psi^{[3/2]}(S_i^{(2)})$$
$$+t^4(1+t+t^2)(1-t^3)\psi^{[3/2]}(S_i^{(1)})+t^9. \tag{5.10}$$

Since

$$\psi^{[3/2]}(S_i^{(l)})=\psi(S_i^{(l)})/[\psi(P_i)]^2, \tag{5.11}$$

we calculate Ocneanu's trace, $\psi(S_i^{(l)})$, $l=1, 2$. The evaluation of $\psi(S_i^{(1)})$ is schematically shown in Fig. 14. The result is

$$\psi(S_i^{(1)})=\frac{1}{\tau_2\tau_3}(t^3z+2t^2z^2+2tz^3+z^3(1-t)$$
$$+z^2t)\psi(P_i)$$

$$=\frac{z((1+t)z+t^2)(z+t)}{(1+t)(1+t+t^2)}\psi(P_i). \tag{5.12}$$

Similarly, we have

$$\psi(S_i^{(2)})=\frac{z^2((1+t)z+t^2)}{1+t+t^2}\psi(P_i). \tag{5.13}$$

Therefore, from (5.10) ~ (5.13), we obtain

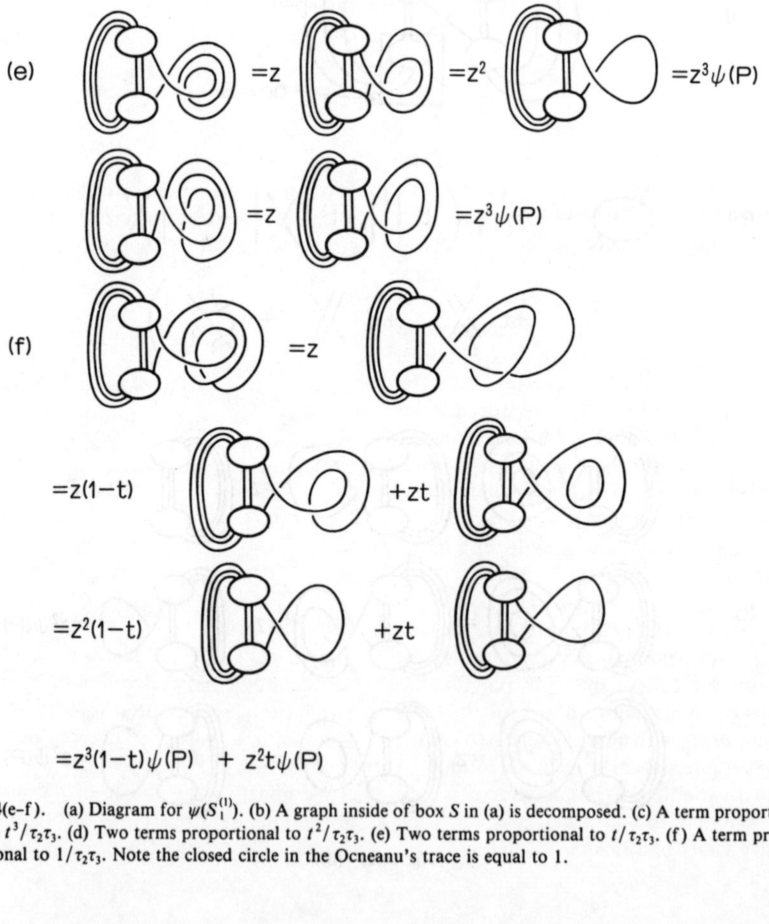

(e) =z =z^2 =$z^3\psi$(P)

=z =$z^3\psi$(P)

(f) =z

=z(1−t) +zt

=z^2(1−t) +zt

=z^3(1−t)ψ(P) + $z^2t\psi$(P)

Fig. 14(e-f). (a) Diagram for $\psi(S_1^{(1)})$. (b) A graph inside of box S in (a) is decomposed. (c) A term proportional to $t^3/\tau_2\tau_3$. (d) Two terms proportional to $t^2/\tau_2\tau_3$. (e) Two terms proportional to $t/\tau_2\tau_3$. (f) A term proportional to $1/\tau_2\tau_3$. Note the closed circle in the Ocneanu's trace is equal to 1.

$$\psi^{[3/2]}((b_i^{[3/2]})^2)=\frac{z(1-t^3)}{\psi(P_i)}\left[(1-t)(1-t^2)z^2+\frac{t}{1+t}(t^4+(1+t-t^2)z)(t^2+(1+t)z)\right]+t^9. \tag{5.14}$$

In general, we can show that

$$\begin{aligned}\psi^{[3/2]}((b_i^{[3/2]})^m)=&\frac{(1-\omega t^4)(1-\omega t^5)(1-\omega t^6)}{(1+t^2)(1+t^3)(1+t+t^2+t^3+t^4)(1-\omega t)(1-\omega t^2)(1-\omega t^3)}\\&+(-t^3)^m\frac{t(1-\omega)(1-\omega t^4)(1-\omega t^5)}{(1+t^2)(1+t^3)(1-\omega t)(1-\omega t^2)(1-\omega t^3)}\\&+(t^5)^m\frac{t^2(1+t+t^2)^2(1-\omega)(1-\omega t^4)}{(1+t)(1+t^2)(1+t+t^2+t^3+t^4)(1-\omega t^2)(1-\omega t^3)}\\&+(-t^6)^m\frac{t^3(1-\omega)}{(1+t)(1+t^2)(1-\omega t^3)}\,.\end{aligned} \tag{5.15}$$

Thus, we obtain the following two-variable link polynomial for closed 2-braids:

$$\begin{aligned}P_m^{(3/2)}(\omega,t)\equiv\alpha_\omega^{[3/2]}((b_i^{[3/2]})^m)=&\,\omega^{3m/2}\frac{(1-\omega t^4)(1-\omega t^5)(1-\omega t^6)}{\omega^{3/2}(1-t^4)(1-t^5)(1-t^6)}\\&+(-\omega^{3/2}t^3)^m\frac{t(1-\omega)(1-\omega t^4)(1-\omega t^5)}{\omega^{3/2}(1-t)(1-t^4)(1-t^6)}\\&+(\omega^{3/2}t^5)^m\frac{t^2(1+t+t^2)^2(1-\omega)(1-\omega t)(1-\omega t^4)}{\omega^{3/2}(1+t)(1-t^3)(1-t^4)(1-t^5)}\\&+(-\omega^{3/2}t^6)^m\frac{t^3(1-\omega)(1-\omega t)(1-\omega t^2)}{\omega^{3/2}(1-t^2)(1-t^3)(1-t^4)}\,.\end{aligned} \tag{5.16}$$

§6. Summary and Discussion

In this paper we have presented a two-variable extension of the new link polynomials. By combining $(N-1)$ strings into a composite string and projecting it to spin $s=(N-1)/2$ subspace, we have obtained the composite string representation $B_n^{[s]}$ from the braid group $B_{n(N-1)}^{[1/2]}$. This procedure is straightforward and does not utilize a specific representation of $B_n^{[1/2]}$. With the composite string representation and the generalized Ocneanu's trace, we have been led to an infinite sequence of two-variable link polynomials corresponding to $s=1/2$, 1, 3/2, 2, $\cdots$.

The two-variable Jones polynomial (HOMFLY polynomial) corresponds to the $N=2$ case. We believe that link polynomials with a larger N are more powerful since the composite string with a large spin $s=(N-1)/2$ has many "internal degrees of freedom." A set of two-variable polynomials for $N=2$, 3, 4, 5, $\cdots$ provides us with a systematic method to classify knots and links.

We may ask about other possibilities for forming composite strings. To examine these possibilities, let us define a "decorated" generator G_i (Fig. 15) by

$$G_i=\alpha_{(i-1)(N-1)+1}\beta_{i(N-1)+1}\bar{G}_i\gamma_{i(N-1)+1}\delta_{(i-1)(N-1)+1}, \tag{6.1}$$

where $\bar{G}_i$ has been defined in (2.7b). In order to satisfy the Markov move II (Fig. 8), the "decorations" α_i, β_i, γ_i and δ_i can not be arbitrary. We find from Fig. 9 that there are two possibilities;

$$\alpha_i\beta_i\gamma_i\delta_i\Delta_i^2=P_i, \tag{6.2}$$

or

$$\alpha_i\beta_i\gamma_i\delta_i\Delta_i^2=I_i. \tag{6.3}$$

The former case corresponds to our theory. In particular, we have chosen $\alpha_i=\beta_i=\gamma_i=\delta_i=P_i$. The latter case has been recently studied by Murakami[15] with a choice $\alpha_i=\Delta_i^{-2}$, $\beta_i=\gamma_i=\delta_i=I_i$.

We find that the choice of the projector is not unique. Various types of projectors with different symmetries are possible as far as the

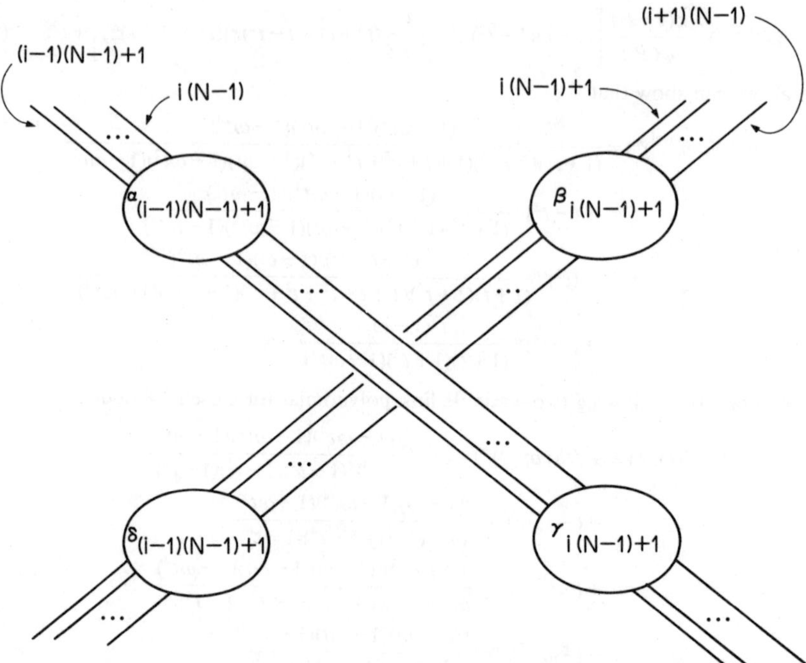

Fig. 15. A "decorated" generator (a general composite string) G_i.

following relation holds;

$$P_i\Delta_i^2=\alpha P_i, \quad \alpha;\ \text{constant}. \tag{6.4}$$

In this paper, we have chosen the symmetrizer in the sense that at $t=1$ it satisfies $P_i^{(+)} g_i=+P_i^{(+)}$. We may also choose the anti-symmetrizer satisfying $P_i^{(-)}g_i=-P_i^{(-)}$ at $t=1$. We have checked that $P_i^{(+)}$ and $P_i^{(-)}$ give equivalent results since they are transformed into each other by $t\to t^{-1}$ and $g_i\to -t^{-1}g_i$. More generally, we can construct the projectors with mixed symmetries. For example, for $N=4$, we find that

$$P_i=\{t^{\frac{3}{2}}+t^{\frac{1}{2}}(1-t)g_i-tg_{i+1}+(-1+t-t^{\frac{1}{2}})g_ig_{i+1}+g_ig_{i+1}g_i\}/t^{\frac{1}{2}}(1+t+t^2), \tag{6.5}$$

is also the projector satisfying $P_i\Delta_i^2=t^3P_i$. This kind of operator is known as the Young operator[16] in the theory of symmetric groups.

In conclusion, we would like to emphasize that we have a sequence of two-variable link polynomials corresponding to a sequence of exactly solvable models in statistical mechanics. The underlying concept is the Yang-Baxter relation. While the Yang-Baxter relation is a sufficient condition for exactly solvable models (integrable models), it is an extension of the braid group. These two different significances of the celebrated Yang-Baxter relation offer a key to why two different subjects, theory of integrable systems and knot theory, are closely related.

Appendix

We shall show that the projector $P_i^{(N)}$ defined by (2.9) satisfies the properties (i) ~ (v) (see eqs. (2.1), (2.2), (2.4), (2.13) and (2.14)). To do this, we prepare four propositions.

Let the generators of symmetric group S_n be 1 (unit element), $s_1, s_2, \cdots, s_{n-1}$. We introduce a product $\sigma_m^{(l)}$ of "length" l by

$$\begin{aligned}\sigma_m^{(l)}&=s_{m-l+1}s_{m-l+2}\cdots s_m \quad \text{for} \quad l=1,2,\cdots,m,\\ &=1 \quad \text{for} \quad l=0.\end{aligned} \tag{A·1}$$

Proposition 1 Any element of symmetric

group S_n is uniquely expressed in terms of products of $\sigma_m^{(l)}$;

$$S_n=\{\sigma_{n-1}^{(l_{n-1})}\sigma_{n-2}^{(l_{n-2})}\cdots\sigma_1^{(l_1)};\ 0\le l_i\le i,\ i=1,2,\cdots,n-1\}. \tag{A·2}$$

[Proof] We use a method of induction. It is true for $n=2$ since S_2 consists of 1 and s_1. Suppose that (A·2) holds for n. Since $\sigma_n^{(l)}$ can be written as

$$\sigma_n^{(l)}=\begin{pmatrix}1\cdots n-l & n+1-l & n+2-l\cdots n & n+1\\ 1\cdots n-l & n+2-l & n+3-l\cdots n+1 & n+1-l\end{pmatrix},$$

we see that for an element α in S_n, $\sigma_n^{(l)}\alpha$ changes $n+1$ into $n+1-l$. Therefore, if $l\neq l'$, then $\sigma_n^{(l)}\alpha \neq\sigma_n^{(l')}\beta$ for any elements α and β in S_n. This indicates that we have a decomposition; $S_{n+1}=\bigcup_{l=0}^{n}\sigma_n^{(l)}S_n$, from which we arrive at

$$S_{n+1}=\{\sigma_n^{(l_n)}\sigma_{n-1}^{(l_{n-1})}\cdots\sigma_1^{(l_1)};\ 0\le l_i\le i,\ i=1,2,\cdots,n\}.$$

Q.E.D.

We define some notation. We set

$$p\equiv\sigma_{n-1}^{(l_{n-1})}\sigma_{n-2}^{(l_{n-2})}\cdots\sigma_1^{(l_1)}. \tag{A·3}$$

The exponent sum of $e(p)$ is given by

$$e(p)=\sum_{i=1}^{n-1}l_i. \tag{A·4}$$

So far, the discussion is restricted to S_n. We now extend it to the Hecke algebra $H(t,n)$. Let $\theta_n^{(l)}$ be a product of generators 1, g_1, $g_2,\cdots,g_{n-1}$ of $H(t,n)$ whose "length" is l;

$$\theta_m^{(l)}=g_{m-l+1}g_{m-l+2}\cdots g_m \quad \text{for } l=1,2,\cdots,m,$$
$$=1 \quad \text{for } l=0. \tag{A·5}$$

Corresponding to the product p in (A·3), we write

$$g(p)=\theta_{n-1}^{(l_{n-1})}\theta_{n-2}^{(l_{n-2})}\cdots\theta_1^{(l_1)}. \tag{A·6}$$

We prepare one more definition. An "inversion" $I_V^{(n)}$ is the operation which changes the generators $g_1,g_2,\cdots,g_{n-1}$ in such a way that

$$I_V^{(n)};\ g_1\to g_{n-1},\ g_2\to g_{n-2},\ \cdots,\ g_{n-1}\to g_1. \tag{A·7}$$

The inversion $I_V^{(n)}$ has the properties;

$$I_V^{(n)}(f(g_i)h(g_j))=I_V^{(n)}f(g_i)\cdot I_V^{(n)}h(g_j),$$
$$f(g_i)\in H(t,n),\quad h(g_j)\in H(t,n), \tag{A·8}$$

and

$$I_V^{(n)}g(p)=g(S_V^{(n)}pS_V^{(n)}), \tag{A·9}$$

where $S_V^{(n)}$ is a permutation given by

$$S_V^{(n)}=\begin{pmatrix}1 & 2\cdots n\\ n & n-1\cdots 1\end{pmatrix}. \tag{A·10}$$

Proposition 2 Let U_n be an element of $H(t, n+1)$ defined by the recursion relation;

$$U_n=t^n+U_{n-1}g_n \quad \text{with} \quad U_0\equiv 1,$$
$$=t^n+[t^{n-1}+\cdots+\{t^2+(t+g_1)g_2\}\cdots g_{n-1}]g_n. \tag{A·11}$$

Then, U_n is expanded as

$$U_n=\sum_{l=0}^{n}t^{n-l}\theta_n^{(l)}=t^n+t^{n-1}g_n+t^{n-2}g_{n-1}g_n+\cdots+g_1g_2\cdots g_n. \tag{A·12}$$

[Proof] We use a method of induction. It is true for $n=1$ since $U_1=t+g_1=\sum_{l=0}^{n}t^{1-l}\theta_n^{(l)}$. We assume (A·12) for n. Then we have

$$U_{n+1}=t^{n+1}+U_ng_{n+1}=t^{n+1}+\sum_{l=0}^{n}t^{n-l}\theta_n^{(l)}g_{n+1}$$
$$=t^{n+1}+\sum_{l=1}^{n+1}t^{n+1-l}\theta_{n+1}^{(l)}=\sum_{l=0}^{n+1}t^{n+1-l}\theta_{n+1}^{(l)}.$$

Q.E.D.

Proposition 3 The operator U_n given by (A·11) satisfies

$$U_{n-1}U_{n-2}\cdots U_1=\sum_{p\in S_n}t^{1/2\,n(n-1)-e(p)}g(p), \tag{A·13}$$

where p and $e(p)$ have been defined in (A·3) and (A·4), respectively.

[Proof] We use a method of induction. It is true for $n=2$ since $U_1=t+g_1=\sum_{p\in S_2}t^{1-e(p)}g(p)$. We assume (A·13) for n. From (A·12) and (A·13), we have

$$U_nU_{n-1}\cdots U_1=\sum_{l=0}^{n} t^{n-l}\theta_n^{(l)}$$

$$\cdot\sum_{p\in S_n} t^{1/2\,n(n-1)-e(p)}g(p)$$

$$=\sum_{l=0}^{n}\sum_{p\in S_n} t^{1/2\,n(n+1)-(e(p)+l)}\theta_n^{(l)}g(p). \qquad (A\cdot 14)$$

Substituting the identities:

$$\theta_n^{(l)}g(p)=\theta_n^{(l)}\theta_{n-1}^{(l_{n-1})}\cdots\theta_1^{(l_1)}$$

$$=g(\sigma_n^{(l)}\sigma_{n-1}^{(l_{n-1})}\cdots\sigma_1^{(l_1)})$$

$$=g(\sigma_n^{(l)}p),$$

$$e(\sigma_n^{(l)}p)=l+e(p),$$

into (A·14), we obtain

$$U_nU_{n-1}\cdots U_1=\sum_{l=0}^{n}\sum_{p\in S_n} t^{\frac{1}{2}n(n+1)-e(\sigma_n^{(l)}p)}g(\sigma_n^{(l)}p).$$

By proposition 1, we have $S_{n+1}=\{\sigma_n^{(l)}p;\ l=0, 1,\cdots, n,\ p\in S_n\}$ and then rewriting $\sigma_n^{(l)}p$ by p in S_{n+1}, we arrive at

$$U_nU_{n-1}\cdots U_1=\sum_{p\in S_{n+1}} t^{\frac{1}{2}n(n+1)-e(p)}g(p).$$

Q.E.D.

Proposition 4 A product of the operators, $U_{n-1}U_{n-2}\cdots U_1$, is invariant under the inversion $I_V^{(n)}$ defined in (A·7);

$$I_V^{(n)}(U_{n-1}U_{n-2}\cdots U_1)=U_{n-1}U_{n-2}\cdots U_1. \qquad (A\cdot 15)$$

[Proof] By using proposition 3 and collecting the terms with the same length, we have

$$U_{n-1}U_{n-2}\cdots U_1=\sum_{p\in S_n} t^{\frac{1}{2}n(n-1)-e(p)}g(p)=\sum_{l=0}^{\frac{1}{2}n(n-1)} t^{\frac{1}{2}n(n-1)-l}\sum_{\substack{p\in S_n\\ e(p)=l}} g(p).$$

Since we have

$$I_V(\sum_{\substack{p\in S_n\\ e(p)=l}} g(p))=\sum_{\substack{p\in S_n\\ e(p)=l}} I_V(g(p))=\sum_{\substack{p\in S_n\\ e(p)=l}} g(S_VpS_V)=\sum_{\substack{S_VpS_V\in S_n\\ e(S_VpS_V)=l}} g(p)=\sum_{\substack{p\in S_n\\ e(p)=l}} g(p),$$

we obtain (A·15).
Q.E.D.

Now we shall prove the properties of the projector P_i.

Theorem 1 For the projector $P_i^{(N)}$ as defined by (2.9), the following relations hold;

$$P_1^{(N)}=\frac{1}{\tau_2\tau_3\cdots\tau_{N-1}}U_{N-2}U_{N-3}\cdots U_1, \qquad (A\cdot 16)$$

$$P_1^{(N)}=I_V(P_1^{(N)}), \qquad (A\cdot 17)$$

$$P_1^{(N)}P_1^{(N)}=P_1^{(N)}. \qquad (A\cdot 18)$$

Here and hereafter, I_V is used for $I_V^{(N-1)}$, and $\tau_m=1+t+\cdots t^{m-1}$.

[Proof] We prove (A·16)~(A·18) by a method of induction. For $N=3$, since $P_1^{(3)}=(t+g_1)/(1+t)$, we can show (A·16)~(A·18) by direct calculations. We assume that (A·16)~(A·18) hold for $n\leq N$. Since

$$P_1^{(N+1)}=P_1^{(N)}h_{N-1}^{(N+1)}P_1^{(N)}=P_1^{(N)}\frac{\tau_{N-1}}{\tau_N}\left(\frac{t^{N-1}}{\tau_{N-1}}+g_{N-1}\right)P_1^{(N)}$$

$$=\frac{1}{\tau_N}(t^{N-1}P_1^{(N)}+\tau_{N-1}P_1^{(N)}g_{N-1})P_1^{(N)}$$

$$=\frac{1}{\tau_N}(t^{N-1}+U_{N-2}P_1^{(N-1)}g_{N-1})P_1^{(N)}$$

$$=\frac{1}{\tau_N}(t^{N-1}+U_{N-2}g_{N-1})P_1^{(N)}$$

$$=\frac{1}{\tau_N}U_{N-1}P_1^{(N)}=\frac{1}{\tau_2\tau_3\cdots\tau_N}U_{N-1}U_{N-2}\cdots U_1,$$

(A·16) is proved. Next, by using (A·16) and proposition 4, we have

$$I_V(P_1^{(N+1)})=\frac{1}{\tau_N\tau_{N-1}\cdots\tau_2}I_V(U_{N-1}\cdots U_1)$$
$$=\frac{1}{\tau_N\tau_{N-1}\cdots\tau_2}U_{N-1}\cdots U_1$$
$$=P_1^{(N+1)}.$$

Then, (A·17) is proved. Lastly, by using the identities;

$$h_1^{(N+1)}P_1^{(N+1)}=P_1^{(N+1)},$$
$$h_{N-1}^{(N+1)}=I_V(h_1^{(N+1)}),\quad P_{N-1}^{(N+1)}=I_V(P_1^{(N+1)}),$$

we obtain

$$P_1^{(N+1)}P_1^{(N+1)}=P_1^{(N)}h_{N-1}^{(N+1)}P_1^{(N)}h_{N-1}^{(N+1)}P_1^{(N)}$$
$$=P_1^{(N)}h_{N-1}^{(N+1)}P^{(N+1)}$$
$$=P_1^{(N)}I_V(h_1^{(N+1)})I_V(P_1^{(N+1)})$$
$$=P_1^{(N)}I_V(h_1^{(N+1)}P_1^{(N+1)})$$
$$=P_1^{(N)}I_V(P_1^{(N+1)})=P_1^{(N)}P_1^{(N+1)}$$
$$=P_1^{(N+1)}.$$

The last equality follows from the form of the projector P_i. Thus, (A·18) is proved.
Q.E.D.

Theorem 2 For $1\leq i\leq N-2$, the following holds;

$$P_1^{(N)}g_i=P_1^{(N)}. \qquad (A\cdot 19)$$

[Proof] We use a method of induction. For $N=3$, we have $P_1^{(3)}g_1=P_1^{(3)}$ since $P_1^{(3)}=(t+g_1)/(1+t)$. We assume (A·19) for N. For $1\leq i\leq N-2$, we have

$$P_1^{(N+1)}g_i=P_1^{(N)}h_{N-1}^{(N+1)}P_1^{(N)}g_i=P_1^{(N)}h_{N-1}^{(N+1)}P_1^{(N)}$$
$$=P_1^{(N+1)}.$$

For $i=N-1$, we have

$$P_1^{(N+1)}g_{N-1}=P_1^{(N+1)}I_V(g_1)=I_V(P_1^{(N+1)})I_V(g_1)$$
$$=I_V(P_1^{(N+1)}g_1)=I_V(P_1^{(N+1)})=P_1^{(N+1)}.$$

Thus, (A·19) is proved.
Q.E.D.

Theorem 3 The projector P_i constructed by (2.9) and (2.10) satisfies the properties that

(1) P_i consists of generators $g_i, g_{i+1}, \cdots, g_{i+N-3}$.

(2) $P_i(g_{i+N-2}g_{i+N-3}\cdots g_i)=(g_{i+N-2}g_{i+N-3}\cdots g_i)P_{i+1}$,

$P_i(g_{i+N-2}^{-1}g_{i+N-3}^{-1}\cdots g_i^{-1})=(g_{i+N-2}^{-1}g_{i+N-3}^{-1}\cdots g_i^{-1})P_{i+1}$.

(3) $P_i\Delta_i^2=P_i$,

where

$$\Delta_i=(g_ig_{i+1}\cdots g_{i+k-2})\cdot(g_ig_{i+1}\cdots g_{i+k-3})\cdots g_i.$$

[Proof] From the construction, (1) is obvious and from (1) we also see (2) is obvious. Due to theorem 2, we obtain (3). Q.E.D.

Thus we have proved that the projector P_i defined by (2.9) with (2.10) and (2.11) satisfies properties (i) ~ (v).

References

1) Y. Akutsu and M. Wadati: J. Phys. Soc. Jpn. **56** (1987) 839.
2) Y. Akutsu and M. Wadati: J. Phys. Soc. Jpn. **56** (1987) 3039. This paper is part I of the series and contains references for earlier developments.
3) Y. Akutsu, T. Deguchi and M. Wadati: J. Phys. Soc. Jpn. **56** (1987) 3464. This paper is part II of the series.
4) Y. Akutsu and M. Wadati: to appear in Commun. Math. Phys. A part of the present result was reported.
5) K. Sogo, Y. Akutsu and T. Abe: Prog. Theor. Phys. **70** (1983) 730 and 739.
6) E. Artin: Ann. Math. **48** (1947) 101.
7) V. F. R. Jones: Hecke algebra representations of braid groups and link polynomials, preprint, 1986.
8) P. P. Kulish and E. K. Sklyanin: Lecture Notes in Physics (Springer-Verlag, New York, 1982) Vol. 151, p. 61.
9) J. S. Birman: *Braids, Links, and Mapping Class Groups*, (Princeton University Press, Princeton, New Jersey, 1974) p. 70.
10) V. F. R. Jones: Bull. Amer. Math. Soc. **12** (1985) 103.
11) P. Freyd, D. Yetter, J. Hoste, W. B. R. Lickorish, K. Millett and A. Ocneanu: Bull. Amer. Math. Soc. **12** (1985) 239.
12) J. S. Birman and H. Wenzl: Braids, link polynomials and a new algebra, preprint, 1987.
13) J. Murakami: The Kauffman polynomial of links and representation theory, preprint, 1987.
14) L. Kauffman: An invariant of regular isotopy, preprint, 1986.
15) J. Murakami: The parallel version of link invariants, preprint, 1987.
16) M. Hamermesh: *Group Theory and its Application to Physical Problems*, (Addison-Wesley, Reading, Massachusets, 1962) p. 243.

Journal of the Physical Society of Japan
Vol. 57, No. 4, April, 1988, pp. 1173–1185

Exactly Solvable Models and New Link Polynomials. IV. IRF Models

Yasuhiro AKUTSU, Tetsuo DEGUCHI† and Miki WADATI†

Institute of Physics, Kanagawa University, Rokkakubashi, Kanagawa-ku, Yokohama 221
†Institute of Physics, College of Arts and Sciences, University of Tokyo, Komaba, Meguro-ku, Tokyo 153

(Received December 15, 1987)

We present a general method to construct link polynomials, invariants for knots and links, from the exactly solvable IRF (Interaction Round a Face) models in statistical mechanics which satisfy the Yang-Baxter relation.

§1. Introduction

The theory of quantum completely integrable systems has set a unified framework for studying various exactly solvable models in (1+1)-dimensional field theory and in 2-dimensional classical statistical mechanics.[1-5] The central idea is that to each sôlvable model we can associate a family of commuting transfer matrices. The commutability condition is generally called the Yang-Baxter relation.[6-7]

It is instructive to begin with the original usage of the Yang-Baxter relation in 1+1 dimensional quantum many-body problem.[6] We denote the scattering amplitude for the collision process $(i, j)\rightarrow(k, l)$ by $S_{jl}^{ik}(u)$ where u is the rapidity difference. The Yang-Baxter relation in this context is the self-consistency condition for the factorized many-body S-matrices and is referred to as the factorization equation.[8-11] It reads as

$$\sum_{\alpha,\beta,\gamma} S_{\gamma r}^{\beta q}(v) S_{k\gamma}^{\alpha p}(u+v) S_{j\beta}^{i\alpha}(u) = \sum_{\alpha,\beta,\gamma} S_{\beta q}^{\alpha p}(u) S_{\gamma r}^{i\alpha}(u+v) S_{k\gamma}^{j\beta}(v). \tag{1.1}$$

In 2-dimensional statistical mechanics, we have two types of models:[12]

1) vertex model,
2) IRF (Interaction Round a Face) model.

It is known that factorized S-matrices can be interpreted as the Boltzmann weights of a solvable vertex model.[13] Then, the factorization equation (1.1) becomes the commutability condition of the transfer matrices for the vertex model.

For the IRF model, the Yang-Baxter relation is again the commutability condition of the transfer matrices and is often called the star-triangle relation. We denote the Boltzmann weight for a "spin" configuration (a, b, c, d) round a face by $w(a, b, c, d; u)$ where u is the spectral parameter (Fig. 1). The spin variables can be interpreted as the heights of surface and as the occupation numbers of adsorbed molecules on surface. The IRF model is referred to as SOS (solid-on-solid) model in the former case and as generalized lattice gas model in the latter case. The star-triangle relation reads as (Fig. 2)

$$\sum_c w(b, d, c, a; u) w(a, c, f, g; u+v) w(c, d, e, f; v) = \sum_c w(a, b, c, g; v) w(b, d, e, c; u+v) w(c, e, f, g; u). \tag{1.2}$$

By solving (1.2), many solvable IRF models have been found.[14-22] A collection of these IRF models was named the (Super) Grand Hierarchy.[18]

While the Yang-Baxter relation is a sufficient condition for exactly solvable

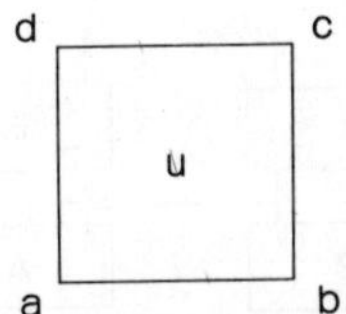

Fig. 1. Boltzmann weight $w(a, b, c, d; u)$.

models (integrable models), it defines an algebra which is an extension of the braid group. These two different significances of the celebrated Yang-Baxter relation offer a key why two different subjects, theory of integrable systems and knot theory, are closely related.[23-27] To make the paper self-contained and unify the notations, we briefly summarize the relevant mathematics.[28-30]

A knot is a closed string which does not cross with itself. An assembly of knots with mutual entanglements is called a link. Knots and links are defined by using the braid and the braid group as follows. Braids are formed when n points on a upper horizontal line are connected by n strings to n points directly below the first n points on a lower horizontal line. Trivial n-braid is a configuration where no intersection is present (Fig. 3). By successive applications of operation b_i and its inverse b_i^{-1} (Fig. 4), a general n-braid is constructed from the trivial n-braid. A set of generators $b_1, b_2, \cdots, b_{n-1}$, define the braid group B_n.[31] The topological equivalence between seemingly different expressions of a braid in terms of braid group elements is guaranteed by conditions:

$$b_i b_j = b_j b_i, \quad |i-j| \geqq 2, \tag{1.3a}$$

$$b_i b_{i+1} b_i = b_{i+1} b_i b_{i+1}. \tag{1.3b}$$

This is the defining relation of B_n. Then, each topologically equivalent class of the braids is identified with an element in the braid group.

A closed braid is formed by tying opposite ends of a braid. Alexander proved that any link is represented by a closed braid.[32] This fact gives the braid group a fundamental role in the knot theory. However, the representation of a link as a closed braid is not unique. Therefore, the following theorem due to A. A. Markov[33] is important. The equivalent braids expressing the same link are transformed into each other by successive applications of two types of operations, type I and type II Markov moves (Fig. 5);

I. $AB \to BA \ (A, B \in B_n)$

II. $A \to Ab_n,\ A \to Ab_n^{-1},\ (A \in B_n,\ b_n \in B_{n+1})$. (1.4)

Hence, a link polynomial, a topological invariant for knots and links, can be found in two steps. We first construct a suitable representation of the braid group B_n and then find a Markov move invariant quantity defined on the representation.

Recently we developed a general method to construct link polynomials from exactly solvable models in statistical mechanics. The method was applied to the N-state vertex models ($N=2, 3, 4, 5, \cdots$) and an infinite sequence of new link polynomials was derived.[23-25] Two-variable extensions of the new link polynomials were also reported.[26,27] The $N=2$ case of the new link polynomials and their two-variable extensions corresponds to the famous Jones polynomial[34] and its 2-variable extension.[35] As we announced in

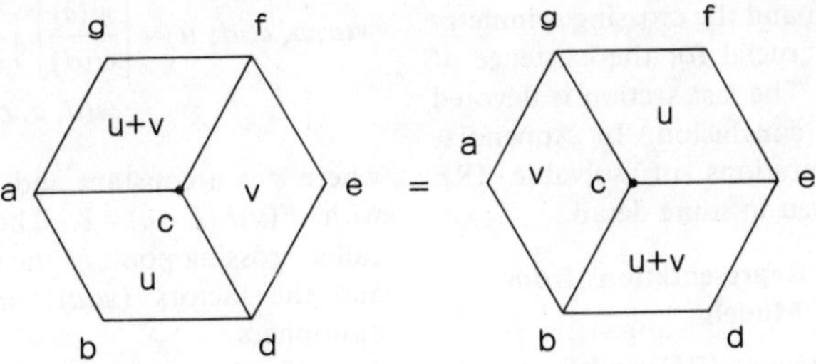

Fig. 2. The star-triangle relation (the Yang-Baxter relation). The dot denotes the summation.

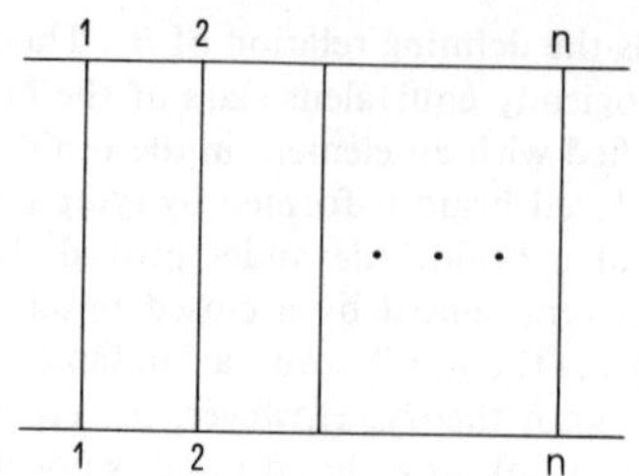

Fig. 3. Trivial *n*-braid.

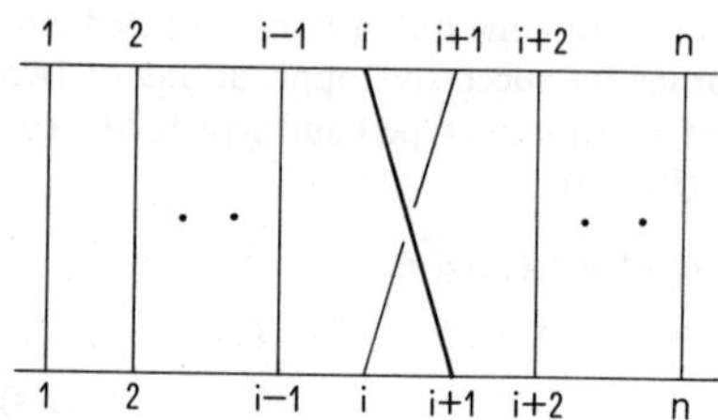

b_i

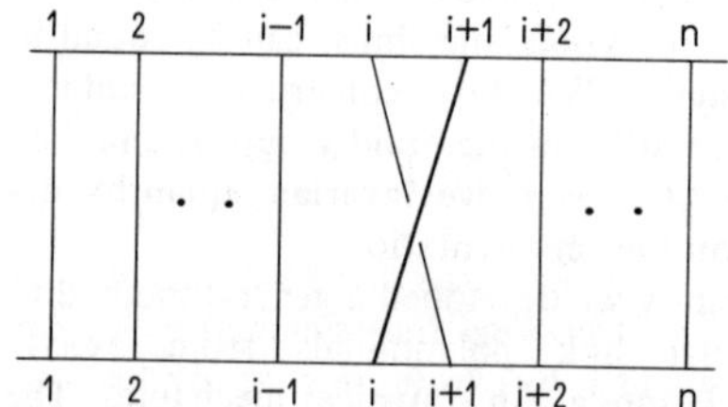

b_i^{-1}

Fig. 4. The operation b_i and its inverse b_i^{-1}.

I

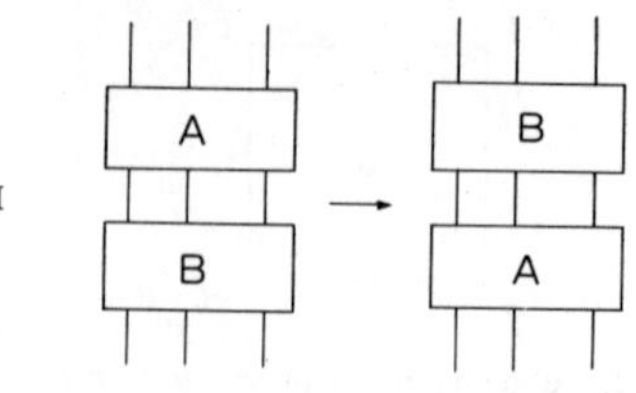

II

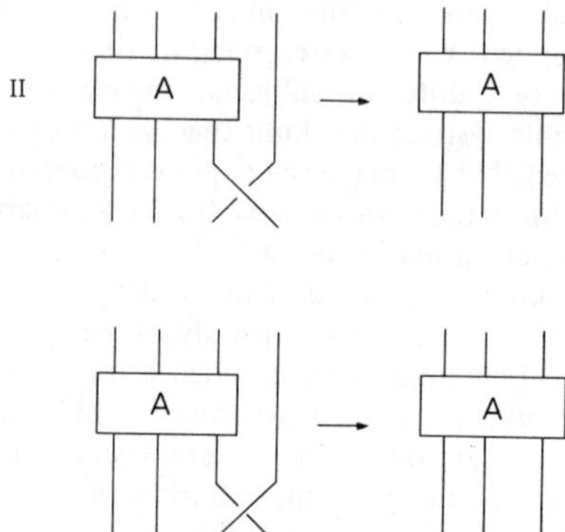

Fig. 5. Markov moves, Type I and Type II.

earlier works,[23-27] application of our method to the IRF models is straightforward, as we shall describe in this paper.

The outline of the paper is the following. In §2, we present a procedure to construct the braid group representations from the Boltzmann weights of solvable IRF models. In §3, we introduce the Markov trace which leads to the link polynomial. We show that the unitarity condition and the crossing symmetry of the model are crucial for the existence of the Markov trace. The last section is devoted to summary and conclusion. In Appendix, graphical representations of solvable IRF models are discussed in some detail.

§2. Braid Group Representations from Solvable IRF Models

2.1 *Basic properties of IRF models*

We assume that the Boltzmann weights $\{w(a, b, c, d; u)\}$ satisfying the star-triangle relation (1.2) have additional relations.

1) *standard initial condition*

$$w(a, b, c, d; u=0)=\delta_{ac}, \tag{2.1}$$

where δ_{ac} is the Kronecker's delta.

2) *unitarity condition* (Fig. 6)

$$\sum_e w(e, c, d, a; -u)w(b, c, e, a; u)=\delta_{bd}. \tag{2.2}$$

3) *crossing symmetry* (Fig. 7)

$$w(a, b, c, d; u)=\left[\frac{\psi(a)}{\psi(d)}\right]^{\eta}\left[\frac{\psi(c)}{\psi(b)}\right]^{1-\eta} F(u) \cdot w(d, a, b, c; \lambda-u), \tag{2.3}$$

where η is a constant and $F(u)$ is a function with $F(u)F(\lambda-u)=1$. The parameter λ is called crossing point of the spectral parameter and the factors $\{\psi(a)\}$ are called crossing multipliers.

4) *reflection symmetry* (Fig. 8)

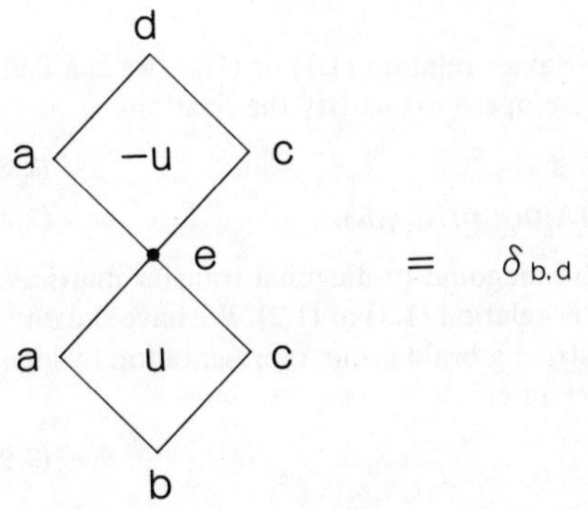

Fig. 6. Unitarity condition. The dot denotes the summation.

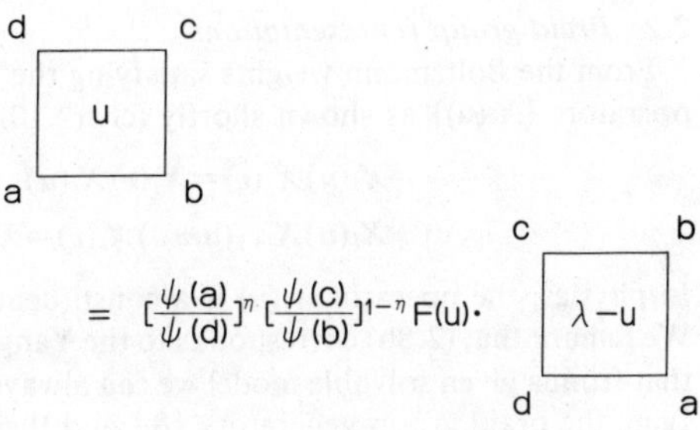

Fig. 7. Crossing relation.

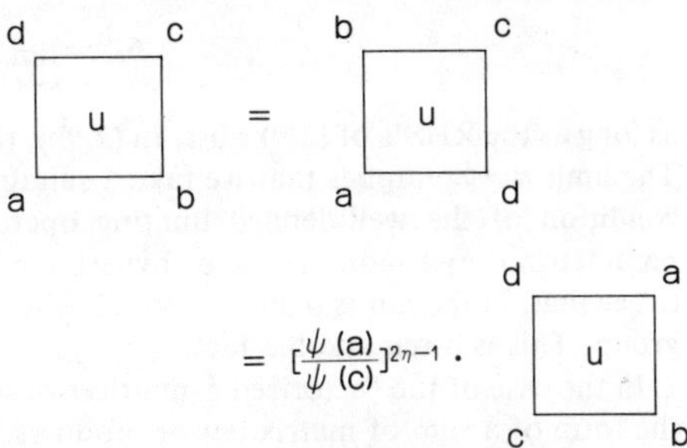

Fig. 8. Reflection symmetry.

$$w(a, b, c, d; u) = w(a, d, c, b; u)$$
$$= \left[\frac{\psi(a)}{\psi(c)}\right]^{2\eta-1} w(c, b, a, d; u). \tag{2.4}$$

These relations are IRF analogues of the corresponding relations for the factorized S-matrices (the vertex models). In particular, the reflection symmetry corresponds to the CPT invariances for the S-matrices.[24] Although most of the known solutions to the star-triangle relation satisfy (2.3) and (2.4) with $\eta=1/2$, we can always 'asymmetrize' the weights so as to satisfy (2.3) and (2.4) for arbitrary η. Denoting the 'symmetric' weights with $\eta=1/2$ by $\{w_s(a, b, c, d; u)\}$, we get the asymmetrized weights $\{w(a, b, c, d; u)\}$ through the transformation

$$w_s(a, b, c, d; u) \longrightarrow w(a, b, c, d; u) = \left[\frac{\psi(a)}{\psi(c)}\right]^{\eta-1/2} w_s(a, b, c, d; u). \tag{2.5}$$

It is easy but important to see that the transformation (2.5) is compatible with the star-triangle relation (1.2); if the weights $\{w_s(a, b, c, d; u)\}$ satisfy (1.2), so do the transformed weights $\{w(a, b, c, d; u)\}$. Further, we have similar symmetry-breaking transformations[36] as those for the vertex models which are also compatible with the star-triangle relation:

$$w(a, b, c, d; u) \longrightarrow \tilde{w}(a, b, c, d; u) = \alpha_{abcd}(u) \cdot \beta_{abcd} \cdot \gamma_{abcd} \cdot w(a, b, c, d; u), \tag{2.6}$$

where

$$\alpha_{abcd}(u) = \exp\{[-p(a)+p(b)-p(c)+p(d)]u\}, \tag{2.7a}$$

$$\beta_{abcd} = \frac{q(a)}{q(c)}, \tag{2.7b}$$

$$\gamma_{abcd} = \exp\{\omega[(b-c)(c-d)-(a-d)(b-a)]\}, \tag{2.7c}$$

with arbitrary functions $p(\cdot)$ and $q(\cdot)$, and arbitrary parameter ω. It should be noted that the transformation (2.5) is a special case of (2.7b). These transformations are sometimes useful in constructing an interesting braid group representation from a given IRF model.

2.2 Braid group representations

From the Boltzmann weights satisfying the Yang-Baxter relation (1.1) or (1.2), we can define operators $\{X_i(u)\}$ as shown shortly (cf. (2.10)). These operators satisfy the relations

$$X_i(u)X_j(v)=X_j(v)X_i(u),\ |i-j|\geqq 2, \tag{2.8a}$$

$$X_i(u)X_{i+1}(u+v)X_i(v)=X_{i+1}(v)X_i(u+v)X_{i+1}(u). \tag{2.8b}$$

In physics, the operator $X_i(u)$ is a constituent of the diagonal-to-diagonal transfer matrices.[12] We remark that (2.8b) corresponds to the Yang-Baxter relation (1.1) or (1.2). We have shown[23-27] that from a given solvable model we can always construct a braid group representation by identifying the braid group generators $\{b_i\}$ and their inverses as

$$b_i=\lim_{u\to\infty} X_i(u), \tag{2.9a}$$

$$b_i^{-1}=\lim_{u\to\infty} X_i(-u), \tag{2.9b}$$

as long as the RHS's of (2.9) exist. In (2.9b), the unitarity condition has been taken into account. The limit $u\to\infty$ implies that we take a suitable direction in the complex u-plane. The existence condition of the well-defined limiting operator is that the starting model is critical and is parametrized by trigonometric or hyperbolic functions. With the identification (2.9) it is simple to see that, in the limits u and $v\to\infty$, (2.8) is nothing but the defining relations (1.3) of the braid group. This is a remarkable fact.

In the case of the factorized S-matrices or vertex models, the operators $\{X_i(u)\}$ are defined in the form of a sum of matrix tensor products.[23,24] In the case of IRF models, $X_i(u)$ $(i=1, 2, \cdots, n-1)$ is defined as[12]

$$[X_i(u)]_{l_0,l_1,l_2,\cdots,l_n}^{l_0',l_1',l_2',\cdots,l_n'}=\delta_{l_0'l_0}\delta_{l_1'l_1}\cdots\delta_{l_{i-1}'l_{i-1}}w(l_i, l_{i+1}, l_i', l_{i-1}; u)\cdot\delta_{l_{i+1}'l_{i+1}}\cdots\delta_{l_n'l_n}, \tag{2.10}$$

where n corresponds to the 'horizontal' size of the system (see Fig. 9). We assume that the Boltzmann weights $\{w(a, b, c, d; u)\}$ have limits denoted by $\{\sigma(a, b, c, d)\}$:

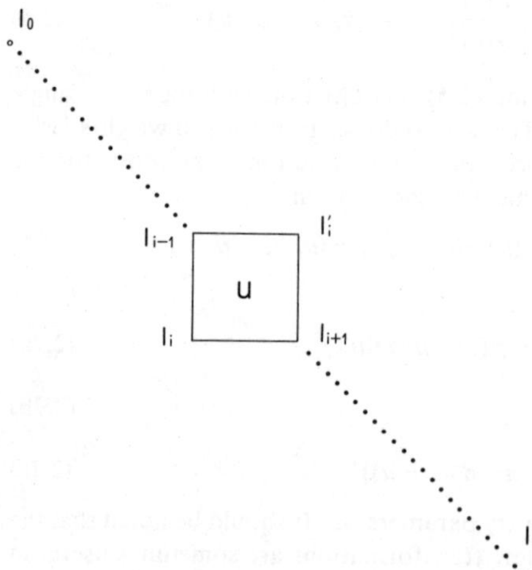

Fig. 9. Operator $X_i(u)$.

$$\sigma(a, b, c, d)=\lim_{u\to\infty} w(a, b, c, d; u), \tag{2.11}$$

with a suitable choice of the direction in the complex u-plane. Then the braid group generator $b_i(i=1, 2, \cdots, n-1)$ is expressed in terms of weight matrix $\{\sigma(a, b, c, d)\}$

$$[b_i]_{l_0,l_1,l_2,\cdots,l_n}^{l_0',l_1',l_2',\cdots,l_n'}=\delta_{l_0'l_0}\delta_{l_1'l_1}\cdots \times\delta_{l_{i-1}'l_{i-1}}\sigma(l_i, l_{i+1}, l_i', l_{i-1}) \times\delta_{l_{i+1}'l_{i+1}}\cdots\delta_{l_n'l_n}. \tag{2.12}$$

The above expressions (2.11) and (2.12) might require further explanation. Recall that, in most of the solvable IRF models, there exist constraints imposed on the spin configurations round a face. Hence $\{X_i(u)\}$ or $\{b_i\}$ should be regarded as operators which act on a 'constrained Hilbert space'. For example, the expression of the identity operator $I^{(n)}(\in B_n)$

$$[I^{(n)}]_{l_0,l_1,l_2,\cdots,l_n}^{l_0',l_1',l_2',\cdots,l_n'}=\delta_{l_0'l_0}\delta_{l_1'l_1}\cdots\delta_{l_{n-1}'l_{n-1}}\delta_{l_n'l_n}, \tag{2.13}$$

is subject to the constraint imposed on the neighboring indices. The constraint depends on the model: In the case of the eight-vertex

SOS model (see below), only the configurations with $|l_i-l_{i-1}|=1$ $(i=1, 2, \cdots, n)$ are allowed. We should note that the constraint can be automatically taken into account by the identification

$$I^{(n)}=X_1(0)\cdot X_2(0)\cdots X_{n-1}(0), \tag{2.14}$$

where the standard initial condition (2.1) has been considered. Accordingly, precise identification of the braid group generator is

$$b_i=I^{(n)}\cdot[\lim_{u\to\infty} X_i(u)] \\ =[\lim_{u\to\infty} X_i(u)]\cdot I^{(n)}. \tag{2.15}$$

Here we have used a fact that $X_i(u)$ and $X_j(0)$ commute;

$$[X_i(u), X_j(0)]=0, \quad \text{for any } i \text{ and } j. \tag{2.16}$$

We further note that the identifications (2.14) and (2.15) hold also for the case of factorized S-matrices or vertex models, and fit well with the composite braid interpretation of the higher-spin representation of B_n.[26,27]

2.3 *Examples*

We apply the general formula (2.9) or (2.15) to interesting IRF models and show a procedure to obtain the weight matrices $\{\sigma(a, b, c, d)\}$.

a) *Unrestricted Eight-Vertex SOS Model*

In the SOS model, the spin variables are usually called *heights*. The unrestricted eight-vertex SOS (8VSOS) model[14] means that each height variable l_i takes any integer value

$$l_i=0, \pm 1, \pm 2, \cdots, \pm\infty, \tag{2.17}$$

under the constraint

$$|l_i-l_j|=1, \text{ for nearest-neighbor sites } i \text{ and } j. \tag{2.18}$$

Originally, the Boltzmann weights of the unrestricted 8VSOS model are parametrized in terms of the elliptic theta functions. In the 'critical limit' that the nome of the theta functions is set to be zero, we have

$$w(a, b, c, d; u)=\delta_{ac}+\delta_{bd}\cdot\frac{\sin u}{\sin(\lambda-u)} \cdot\frac{\psi(a)^\eta\psi(c)^{1-\eta}}{\psi(b)}, \tag{2.19}$$

where the crossing point λ is arbitrary and the crossing multiplier is given by

$$\psi(a)=\sin(a\lambda+\omega_0), \tag{2.20}$$

for arbitrary value of ω_0. By taking the limit $u\to i\infty$ in (2.19), we obtain

$$\sigma(a, b, c, d)=\delta_{ac}-\delta_{bd}\,e^{-i\lambda}\cdot\frac{\psi(a)^\eta\psi(c)^{1-\eta}}{\psi(b)}. \tag{2.21}$$

b) *General Graph-State IRF Models*

The constraint imposed on the IRF spin configurations can be expressed as a graph.[21,37] For instance, the unrestricted 8VSOS model corresponds to a one-dimensional lattice with an infinite length. For any given graph we can always obtain a solvable IRF model with a constraint expressed by the graph, by constructing the Temperley-Lieb operators[38] (see Appendix). In this class of models, the Boltzmann weights (resp. the weight matrices) have the same form as (2.19) (resp.(2.21)) with suitable substitution of the crossing point λ and the crossing multipliers $\{\psi(a)\}$.

c) *Models in the Gordon-Generalization Hierarchy*

As a generalization of the hard hexagon model,[12] there exists a hierarchy of solvable IRF models called 'Gordon-generalization hierarchy'.[15] The k-state model in the hierarchy, the $GG(k)$ model, is characterized by the following constraints on spins $\{\alpha_i\}$:

$$\alpha_i=0, 1, 2, \cdots, k-1, \\ 0\leqq\alpha_i+\alpha_j\leqq k-1, \text{ for nearest-neighbor sites } i \text{ and } j. \tag{2.22}$$

The parametrization of the Boltzmann weights has been given in terms of the elliptic theta functions. To have a braid group representation, we set nome to be zero. For general k, the crossing point λ and the crossing multipliers $\{\psi(a)\}$ are given by

$$\psi(a)=\sin[(k-a)\lambda], \tag{2.23}$$

$$\lambda=\frac{\pi}{2k+1}. \tag{2.24}$$

Basic properties (2.1)–(2.4) are satisfied if we normalize each Boltzmann weight by $w(0, k-1, 0, k-2; u)$.

§3. Markov Traces and Link Polynomials

3.1 Markov traces for IRF models

Once we have a braid group representation, the remaining task left for us is to construct a functional $\phi(\ \cdot\ )$ called the Markov trace. It must have the proper normalization $\phi(I^{(n)})=1$ for the identity operator $I^{(n)}\in B_n$, and satisfy the following properties; type I and II Markov properties:

type I

$$\phi(AB)=\phi(BA),\ \text{for}\ A,\ B\in B_n, \tag{3.1}$$

type II

$$\phi(Ab_n)=\tau\cdot\phi(A),$$
$$\phi(Ab_n^{-1})=\bar{\tau}\cdot\phi(A),$$
$$\text{for}\ A\in B_n,\ b_n\in B_{n+1}, \tag{3.2}$$

where τ and $\bar{\tau}$ are some constants. With the Markov trace $\phi(\ \cdot\ )$, link polynomial $\alpha(\ \cdot\ )$, a quantity invariant under the Markov moves (1.4), is given by

$$\alpha(A)=(\tau\cdot\bar{\tau})^{-(n-1)/2}\cdot\left(\frac{\bar{\tau}}{\tau}\right)^{e(A)/2}\cdot\phi(A),\quad (A\in B_n), \tag{3.3}$$

where $e(A)$ is the exponent sum of the generators $\{b_i\}$ appearing in the braid A.

We shall show that $\phi(\ \cdot\ )$ is expressed as a weighted sum of the diagonal elements within the 'constrained space' discussed in §2.2. This definition of $\phi(\ \cdot\ )$ generalizes the Powers' state for the vertex model case. For definiteness, let us take the unrestricted 8VSOS model and the braid group representation derived from it. Recall that there is imposed the following constraint on the height variables:

$$l_i=l_j\pm 1,\quad \text{for nearest-neighbor sites}\ i\ \text{and}\ j. \tag{3.4}$$

In the language of the diagonal-to-diagonal transfer-matrix approach, this constraint amounts to the restriction of the 'Hilbert space' on which the transfer matrices operate. That is, we are considering a state space $\tilde{H}$ which is given by

$$\tilde{H}=\text{space spanned by}\ \{|l_0,\ l_1,\ \cdots,\ l_k,\ \cdots\rangle\}, \tag{3.5a}$$

with

$$l_i=l_{i-1}\pm 1\quad (i=1,\ 2,\ \cdots), \tag{3.5b}$$
$$l_i=0,\ \pm 1,\ \pm 2,\ \cdots,\ \pm\infty\ (i=0,\ 1,\ 2,\ \cdots). \tag{3.5c}$$

For any element A in the braid group B_n which has a matrix representation

$$A=(A_{\{l'\},\{l\}})=(A^{l'_0,\,l'_1,\,l'_2,\,\cdots,\,l'_n}_{l_0,\,l_1,\,l_2,\,\cdots,\,l_n}), \tag{3.6}$$

we define the 'constrained trace' $\tilde{\mathrm{T}}\mathrm{r}(A)$ by

$$\tilde{\mathrm{T}}\mathrm{r}(A)=\tilde{\sum_{\substack{l_1,\,l_2,\,\cdots\,l_n\\ l_0:\ \text{fixed}}}} A^{l_0,\,l_1,\,l_2,\,\cdots,\,l_n}_{l_0,\,l_1,\,l_2,\,\cdots,\,l_n}\cdot\frac{\psi(l_n)}{\psi(l_0)}. \tag{3.7}$$

In the above and hereafter, the symbol $\tilde{\Sigma}$ represents the summation under the constraint (3.5b), and the quantity $\psi(l)$ is the crossing multiplier. Then, the Markov trace $\phi(\ \cdot\ )$ is given by

$$\phi(A)=\tilde{\mathrm{T}}\mathrm{r}\,(A)/\tilde{\mathrm{T}}\mathrm{r}\,(I^{(n)}),\ (I^{(n)}\text{: identity}). \tag{3.8}$$

The type I Markov property (3.1) of $\phi(\ \cdot\ )$ is equivalent to the condition

$$\tilde{\mathrm{T}}\mathrm{r}\,(AB)=\tilde{\mathrm{T}}\mathrm{r}\,(BA),\quad (A,\ B\in B_n), \tag{3.9}$$

which immediately follows from the definition (3.7). It is easy to see that the type II Markov property (3.2) is satisfied if the following relations hold:

$$\sum_{b\sim a}\sigma(a,\ b,\ a,\ c)\psi(b)=\chi(\lambda)\psi(a)\quad (\text{independent of}\ c), \tag{3.10a}$$

$$\sum_{b\sim a}\psi(b)=\xi(\lambda)\psi(a), \tag{3.10b}$$

Here $\chi(\lambda)$ and $\xi(\lambda)$ are constants, and the summation is over all heights b admissible to the nearest-neighbor height a. In fact, from these relations, we have for $A\in B_n$

$$\tilde{\mathrm{T}}\mathrm{r}(Ab_n)=\tilde{\sum_{\substack{l_1,\,l_2,\,\cdots\,l_n,\,l_{n+1}\\ l_0:\ \text{fixed}}}} A^{l_0,\,l_1,\,l_2,\,\cdots,\,l_n}_{l_0,\,l_1,\,l_2,\,\cdots,\,l_n}\cdot\frac{\psi(l_n)}{\psi(l_0)}\sigma(l_n,\ l_{n+1},\ l_n,\ l_{n-1})\cdot\frac{\psi(l_{n+1})}{\psi(l_n)}=\chi(\lambda)\,\tilde{\mathrm{T}}\mathrm{r}\,(A), \tag{3.11a}$$

$$\tilde{\mathrm{Tr}}\,(I^{(n+1)})=\sum_{\substack{l_1,l_2,\cdots l_n,l_{n+1}\\ l_0:\ \text{fixed}}}^{\sim}\frac{\psi(l_1)}{\psi(l_0)}\cdot\frac{\psi(l_2)}{\psi(l_1)}\cdots\frac{\psi(l_n)}{\psi(l_{n-1})}\cdot\frac{\psi(l_{n+1})}{\psi(l_n)}$$

$$=\xi(\lambda)^{n+1}$$

$$=\xi(\lambda)\cdot\tilde{\mathrm{Tr}}\,(I^{(n)}), \qquad (3.11b)$$

giving rise to

$$\phi(Ag_n)=\frac{\chi(\lambda)}{\xi(\lambda)}\cdot\phi(A). \qquad (3.12)$$

By using the explicit form (2.21) of the weight matrices $\{\sigma(a,b,c,d)\}$, we can verify (3.10) yielding

$$\chi(\lambda)=\mathrm{e}^{i\lambda}, \qquad (3.13a)$$

$$\xi(\lambda)=2\cos\lambda. \qquad (3.13b)$$

The τ-factor appearing in the type II Markov property (3.2) is given by

$$\tau=\frac{\chi(\lambda)}{\xi(\lambda)}$$

$$=\frac{1}{1+t}, \qquad (3.14)$$

with

$$t=\mathrm{e}^{-2i\lambda}. \qquad (3.15)$$

Similary, we obtain

$$\bar{\tau}=\frac{t}{1+t}. \qquad (3.16)$$

Before proceeding further, we shall make a few remarks on the Markov trace we have constructed. First, the parameter ω_0 entering in the crossing multiplier $\psi(l)=\sin(l\lambda+\omega_0)$ does not appear in (3.10)~(3.11), implying ω_0-independence of $\tilde{\mathrm{Tr}}\,(\cdot)$. This also leads to l_0-independence of $\tilde{\mathrm{Tr}}\,(\cdot)$ since a change $\omega_0\to\omega_0-k\lambda$ (k: integer) is equivalent to the parallel shift of all heights $l_i\to l_i+k$ $(i=0,\cdots n)$. Second, in discussing the type II Markov property, we are considering the natural inclusion $B_1\subset B_2\subset\cdots\subset B_n\subset B_{n+1}\cdots$ and a sequence of trace functional $\phi^{(1)}(\cdot),\phi^{(2)}(\cdot),\cdots,\phi^{(n)}(\cdot),\phi^{(n+1)}(\cdot),\cdots$, where $\phi^{(k)}(\cdot)$ represents the trace for B_k defined in (3.8). Then the Markov trace $\phi(\cdot)$ should be interpreted as a set $\{\phi^{(k)}(\cdot);\ k=1,2,\cdots\}$. It should be noted that the following property which is a consequence of (3.10b) assures such interpretation: $\phi^{(n)}(A)=\phi^{(n')}(A)$, for any n, n' with $n'\geqq n$ and for any element A in B_n.

The above construction procedure is straightforwardly repeated for any IRF models. All we have to do is to check the properties (3.10a) and (3.10b). In fact, these properties and l_0-independence hold for all the models presented in §2.3. For general graph-state models which we have described in Appendix, the τ-factors are given by (3.14) and (3.16) with suitable choice of λ and t. For the $GG(k)$ model, we have

$$\tau=\frac{\sin\lambda}{\sin k\lambda}\,\mathrm{e}^{i(k-1)\lambda}=\frac{1-t}{1-t^k}=\frac{1}{1+t+\cdots+t^{k-1}}, \qquad (3.17a)$$

$$\bar{\tau}=\frac{\sin\lambda}{\sin k\lambda}\,\mathrm{e}^{-i(k-1)\lambda}=\frac{1-t^{-1}}{1-t^{-k}}=\frac{t^{k-1}}{1+t+\cdots+t^{k-1}}, \qquad (3.17b)$$

$$\lambda=\frac{\pi}{2k+1}. \qquad (3.17c)$$

3.2 *Unitarity condition and the extended Markov property*

The relations (3.10a, b) are deduced from a relation

$$\sum_{b\sim a} w(a,b,a,c;u)\psi(b)=H(u)\psi(a)$$

$$\text{(independent of } c\text{)}, \qquad (3.18)$$

where $H(u)$ is a model-dependent function. Let us call the property (3.18) the extended Markov property, because the limits $u\to 0$ and $u\to\infty$ reproduce (3.2) with the τ-factors given by

$$\tau=\frac{H(i\infty)}{H(0)},\quad \bar{\tau}=\frac{H(-i\infty)}{H(0)}. \qquad (3.19)$$

For various IRF models, we have verified the extended Markov property (3.18). For example, we have for the general graph-state IRF models including the 8VSOS model

$$H(u)=\frac{\sin(2\lambda-u)}{\sin(\lambda-u)}, \tag{3.20}$$

and for the $GG(k)$ model

$$H(u)=\frac{\sin(k\lambda-u)}{\sin(\lambda-u)}. \tag{3.21}$$

Let us now show that the extended Markov property is a direct consequence of the unitarity condition (2.2). For this purpose, we make the crossing transformation on the unitarity condition:

$$\sum_e w(e,c,d,a;\,-u)w(b,c,e,a;\,u)=\sum_e\left[\frac{\psi(e)}{\psi(a)}\right]^\eta\left[\frac{\psi(d)}{\psi(c)}\right]^{1-\eta}F(-u)\cdot w(a,e,c,d;\,\lambda+u)$$
$$\times\left[\frac{\psi(b)}{\psi(a)}\right]^\eta\left[\frac{\psi(e)}{\psi(c)}\right]^{1-\eta}F(u)\cdot w(a,b,c,e;\,\lambda-u)$$
$$=\delta_{bd}. \tag{3.22}$$

From (3.22), we have

$$\sum_e w(a,e,c,d;\,\lambda+u)\psi(e)\cdot w(a,b,c,e;\,\lambda-u)=\frac{1}{F(u)F(-u)}\cdot[\psi(c)^{1-\eta}\psi(a)^\eta]^2\cdot\frac{1}{\psi(b)}\cdot\delta_{bd}. \tag{3.23}$$

Using (2.19), we have

LHS of (3.23)

$$=\sum_e w(a,e,c,d;\,\lambda+u)\psi(e)\cdot\delta_{ac}$$
$$+\left[\delta_{ac}-\delta_{bd}\frac{\sin(\lambda+u)}{\sin(u)}\cdot\frac{\psi(a)^\eta\psi(c)^{1-\eta}}{\psi(b)}\right]\frac{\sin(\lambda-u)}{\sin(u)}\cdot\psi(a)^\eta\psi(c)^{1-\eta}, \tag{3.24a}$$

and

$$F(u)=\frac{\sin u}{\sin(\lambda-u)}. \tag{3.24b}$$

Substituting (3.24) into (3.23) and putting $a=c$ and $b=d$, we obtain

$$\sum_e w(a,e,a,b;\,\lambda+u)\psi(e)=-\frac{\sin(\lambda-u)}{\sin u}\cdot\psi(a), \tag{3.25}$$

which gives the relation (3.18) after a change $u\to u-\lambda$.

3.3 Relation to the generalized Powers' state

We shall show that the limit $\omega_0\to i\infty$ for the unrestricted 8VSOS model reproduces the Markov trace, known also as the Powers' state, for the case of vertex model. In this limit, the weights $\{w(a,b,c,d;\,u)\}$ depend only *on relative heights*; we have 'translational invariance' $w(a+l,b+l,c+l,d+l;\,u)=w(a,b,c,d;\,u)$ for any integer l. Then, application of the Wu-Kadanoff-Wegner transformation (Fig. 10) leads to the equivalence between the unrestricted 8VSOS model (with $\omega_0\to i\infty$) and the six-vertex model:

$$S_{jl}^{ik}(u)=w(0,j,j-k,j-k-l=-i;\,u), \tag{3.26}$$

or,

$$w(a,b,c,d;\,u)=S_{b-ac-d}^{a-db-c}(u). \tag{3.27}$$

With this correspondence, the star-triangle relation becomes the factorization equation (1.1). Hence the corresponding braid group representations are equivalent. To see that our trace functional $\phi(\cdot)$ with $\omega_0\to i\infty$ reduces to the Powers' state, we substitute (3.27) into (3.18) to have

$$\sum_{b-a}S_{b-aa-c}^{a-cb-a}(u)\cdot e^{-i\lambda(b-a)}=\text{const}, \tag{3.28}$$

where we have taken into account

$$\psi(a)\sim e^{-i(\lambda a+\omega_0)},\quad(\omega_0\to i\infty). \tag{3.29}$$

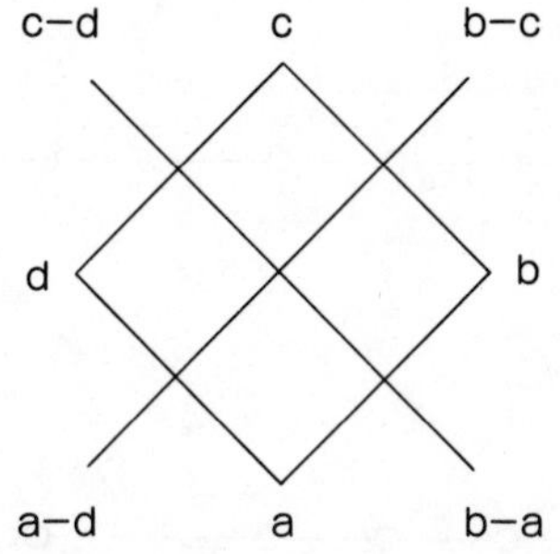

Fig. 10. Wu-Kadanoff-Wegner transformation.

Taking the $u \to i\infty$ limit, we find that the relation (3.28) becomes the required relation for the type II Markov property for the vertex model (see (5.13) in paper I of the series[24]). We should note here that the above correspondence is not confined to the unrestricted 8VSOS model. It also holds for other 'unrestricted' IRF models and vertex models with charge conservation symmetry.[36] An example is the fusion[19] or the Z-invariant generalization[20] of the unrestricted 8VSOS model.

§4. Conclusion

In this paper, the fourth of the series, we have shown that link polynomials, topological invariants for knots and links, can be constructed from the exactly solvable IRF (Interaction Round a Face) models in statistical mechanics. By exactly solvable model we mean that the Boltzmann weights of the model are solutions to the Yang-Baxter relation. This result has been announced in our previous works.[23-27] The unrestricted 8VSOS model has been discussed in detail, but our method does not rely on its special features. The only model-dependent quantities are $\chi(\lambda)$ and $\xi(\lambda)$ in (3.11), or $H(u)$ in the extended Markov property (3.18). In Appendix, we have shown that corresponding to a graph we can make a solvable IRF model (a graph-state model). Then, it is clear that we obtain a new link polynomial from each solvable IRF model. A final remark is in order. The existence of the Markov trace, more precisely, the property of the Markov trace under the type II Markov move, is guaranteed by the unitarity condition of the model. Since the unitarity condition follows from the Yang-Baxter relation, the discussion given in §3.2 is applicable to most of solvable vertex models and all solvable IRF models.

In conclusion, we would like to emphasize that for both vertex models (equivalently, S-matrices) and IRF models which satisfy the Yang-Baxter relation, we have found a general method to construct link polynomials. Extensions into two-variable polynomials are straightforward as shown in refs. 26 and 27. It is interesting to note that vertex and IRF models have their own advantages. For pictorial and physical presentations, vertex models in particular S-matrices are suitable. On the other hand, IRF models seem to be appropriate for the generalization of the theory. Firstly, charge conservation property[24] which is necessary for the type I Markov move invariance is automatically set in IRF models. Second, the exactly solvable IRF models are abundant as we called them the (Super) Grand hierarchy.[18] We believe that research using advantages of both types of models will give a deeper understanding to the classification problem of knots and links.

Acknowledgements

We would like to express our sincere thanks to Professor Kunio Murasugi, University of Toronto, for his valuable comments and continous encouragement on our series of works. This work is partially supported by Grant-in-Aids for Scientific Research Fund from the Ministry of Education, Science and Culture (62740212 and 61540271).

Appendix: Exactly Solvable Graph-State IRF Models

It is known that using the Temperley-Lieb algebra[38] $\{U_i\}$ defined by the relations

$$U_i^2 = q^{1/2} U_i, \qquad (\text{A}\cdot 1\text{a})$$

$$U_i U_{i\pm 1} U_i = U_i, \qquad (\text{A}\cdot 1\text{b})$$

we can always obtain a solvable model satisfying (2.8) by constructing the operator $X_i(u)$ as

$$X_i(u) = I + \frac{\sin u}{\sin(\lambda - u)} U_i, \qquad (\text{A}\cdot 2\text{a})$$

with

$$q=4\cos^2\lambda. \quad (A\cdot 2b)$$

We call a class of solvable models with the Temperley-Lieb algebra structure *TL class*. It is known that the partition function of each model in the TL class is equal to that of the 6-vertex model with corresponding value of λ (Temperley-Lieb equivalence). This equivalence however may not hold for critical behaviors of the models. For the purpose of clarifying the universality class structure of the TL class, it is, then, important to make explicit representations of Temperley-Lieb algebra and to construct as many solvable models as possible.

We shall consider IRF models in the TL class including, for instance, the critical hard hexagon model[39] and the critical unrestricted/restricted 8VSOS models.[40] We look for representation of the TL operator U_i in a similar form as the diagonal-to-diagonal transfer matrix $X_i(u)$. That is, we introduce u-independent weights $\{Y(a, b, c, d)\}$ and make a substitution

$$w(a, b, c, d; u)\rightarrow Y(a, b, c, d), \quad (A\cdot 3)$$

in the expression (2.10) of $X_i(u)$ to obtain U_i.

Let us express the constraint on the IRF spin configurations using a graph. For example the unrestricted and restricted SOS models are represented by straight lines shown in Fig. 11 Each point of the graph corresponds to a possible spin value of the model and any pair of the points is connected if it is an admissible pair of spin values. To each point or imaginary 'site' a we associate a 'wave amplitude' $\psi(a)$ and consider an eigenvalue problem for the wave function $\psi(\,\cdot\,)$ on this imaginary one-dimensional lattice. The 'wave equation' can be written as

$$-\bar{\Delta}\psi(a)=E\psi(a), \quad (A\cdot 4)$$

where $\bar{\Delta}$ is a discrete Laplacian operator defined by

$$\bar{\Delta}\psi(a)=\psi(a+1)+\psi(a-1)-2\psi(a), \quad (A\cdot 5)$$

or equivalently

$$\sum_{b\sim a}\psi(b)=\Lambda\psi(a), \quad (A\cdot 6)$$

where the summation is over all admissible values of b. We regard the wave amplitudes

(a)

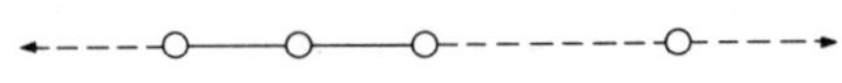

(b)

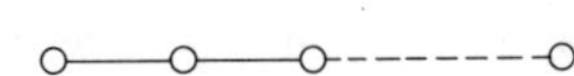

Fig. 11. Graphical representations of (a) unrestricted 8VSOS model and (b) restricted 8VSOS model.

$\{\psi(a)\}$ as crossing multipliers and put

$$Y(a, b, c, d)=\delta_{bd}\frac{\psi(a)^{1-\eta}\psi(c)^{\eta}}{\psi(b)}, \quad (A\cdot 7)$$

with an arbitrary parameter η. Then, it is easy to see that the resulting operator U_i satisfies (A·1) with

$$\Lambda=q^{1/2}$$
$$=2\cos\lambda. \quad (A\cdot 8)$$

Following the procedure described above, Pasquier[21] considered models corresponding to Dynkin diagrams. Note however that we need not restrict the graphs to the Dynkin diagrams. For any graph, we can always construct $\{U_i\}$ by solving the eigenvalue problem (A·6) and using the eigenfunction as the crossing multiplier. Thus the Boltzmann weights of the resulting solvable model are given by

$$w(a, b, c, d; u)=\delta_{ac}+\delta_{db}\cdot\frac{\sin u}{\sin(\lambda-u)}\cdot\frac{\psi(a)^{1-\eta}\psi(c)^{\eta}}{\psi(b)}, \quad (A\cdot 9)$$

where the 'spins' a, b, c, d correspond to sites of the graph. For example, we have solvable IRF models expressed by the graphs shown in Fig. 12.

Some models in the TL class can be extended to 'non-critical' ones where the Boltzmann weights are parametrized in terms of the elliptic theta functions. All the models shown in Figs. 11 and 12 (except Fig. 12(e)) are such examples. In particular, the half-infinite 8VSOS model and the periodic 8VSOS model are obtained from the non-critical unrestricted

(a)

(b)

=

=

(c)

=

=

(d)

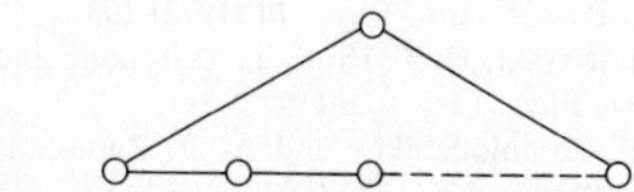

(e)

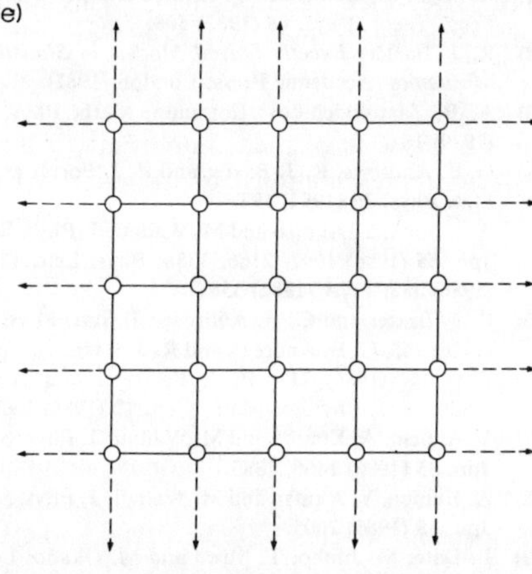

Fig. 12. Graph-state IRF models: (a) half-infinite 8VSOS model, (b) hard hexagon model,[12] (c) special S_2-generalization[17] ($D^{(1)}$ type), (d) periodic 8VSOS model, (e) a two-dimensional graph.

8VSOS model by suitably choosing the crossing point and the parameter ω_0. It is interesting to observe that the unrestricted 8VSOS model, even the non-critical one, plays the role of 'plane wave solution' to the star-triangle relation. In this view, the crossing point λ corresponds to the wave number. Then the discretization of λ occuring in the case of the restricted 8VSOS model (also in the case of the periodic 8VSOS model) is naturally explained as that of an ordinary one-dimensional wave confined within a box.

As a final remark, we point out that if we choose $\eta=0$ and perform the symmetry-breaking transformation (2.7b) with $q(a) = \exp[i(\pi/2)a]$, the resulting asymmetrized unrestricted 8VSOS model has the property of 'edge-height independence'[20] in applying the procedure called fusion[19] or the Z-invariant generalization[20] to make new models. In terms of the braid group, this procedure corresponds to the composite string representation of the braid group presented in our previous papers.[26,27] As noted there, the projectors needed in the construction of composite string representation can be chosen in many ways. So, various generalized models can be constructed with these projectors.

References

1) L. D. Faddeev: Sov. Sci. Rev. Math. Phys. C1 (1981) 107.
2) H. B. Thacker: Rev. Mod. Phys. **53** (1981) 253.
3) P. P. Kulish and E. K. Sklyanin: *Lecture Notes in Physics* (Springer-Verlag, New York, 1982) Vol. 151, p. 61.
4) M. Wadati: in *Dynamical Problems in Soliton Systems*, ed. S. Takeno (Springer-Verlag, 1985) p. 68.
5) M. Wadati and Y. Akutsu: *Exactly Solvable Models in Statistical Mechanics, Lecture Notes in Physics*, ed. M. Lakshmanan (Springer-Verlag, 1988).
6) C. N. Yang and C. P. Yang: Phys. Rev. **150** (1966) 321, 327.

7) R. J. Baxter: Ann. Phys. **70** (1972) 193.
8) M. Karowski, H. J. Thun, T. T. Truong and P. H. Weisz: Phys. Lett. **67B** (1977) 321.
9) A. B. Zamolodchikov and A. B. Zamolodchikov: Ann. Phys. (N.Y.) **120** (1979) 253.
10) K. Sogo, M. Uchinami, A. Nakamura and M. Wadati: Prog. Theor. Phys. **66** (1981) 1284.
11) K. Sogo, M. Uchinami, Y. Akutsu and M. Wadati: Prog. Theor. Phys. **68** (1982) 508.
12) R. J. Baxter: *Exactly Solved Models in Statistical Mechanics* (Academic Press, London, 1982).
13) A. B. Zamolodchikov: Commun. Math. Phys. **69** (1979) 165.
14) G. E. Andrews, R. J. Baxter and P. J. Forrester: J. Stat. Phys. **35** (1984) 193.
15) A. Kuniba, Y. Akutsu and M. Wadati: J. Phys. Soc. Jpn. **55** (1986) 1092, 2166, 3338; Phys. Lett. **116A** (1986) 382; **117A** (1986) 358.
16) R. J. Baxter and G. E. Andrews: J. Stat. Phys. **44** (1986) 245; G. E. Andrews and R. J. Baxter: J. Stat. Phys. **44** (1986) 713; P. J. Forrester and G. E. Andrews: J. Phys. A. Math. Gen. **19** (1986) L923.
17) Y. Akutsu, A. Kuniba and M. Wadati: J. Phys. Soc. Jpn. **55** (1986) 1466, 1880.
18) A. Kuniba, Y. Akutsu and M. Wadati: J. Phys. Soc. Jpn. **55** (1986) 2605.
19) E. Date, M. Jimbo, T. Miwa and M. Okado: Lett. Math. Phys. **12** (1986) 209; Phys. Rev. **B35** (1987) 2105.
20) Y. Akutsu, A. Kuniba and M. Wadati: J. Phys. Soc. Jpn. **55** (1986) 2907.
21) V. Pasquier: J. Phys. A. Math. Gen. **20** (1986) L217, L221.
22) A. Kuniba and T. Yajima: to be published in J. Phys. A. (Math. Gen.) **21** (1988) 519.
23) Y. Akutsu and M. Wadati: J. Phys. Soc. Jpn. **56** (1987) 839.
24) Y. Akutsu and M. Wadati: J. Phys. Soc. Jpn. **56** (1987) 3039.
25) Y. Akutsu, T. Deguchi and M. Wadati: J. Phys. Soc. Jpn. **56** (1987) 3464.
26) T. Deguchi, Y. Akutsu and M. Wadati: to be published in J. Phys. Soc. Jpn. **57** (1988) 753.
27) Y. Akutsu and M. Wadati: Commun. Math. Phys. (in press).
28) J. S. Birman: *Braids, Links and Mapping Class Group* (Princeton University Press, 1974).
29) D. Rolfsen: *Knots and Links* (Publish or Perish Inc., Berkeley, 1976).
30) L. H. Kauffman: *On Knots* (Princeton University Press, 1987).
31) E. Artin: Ann. Math. **48** (1947) 101.
32) J. W. Alexander: Proc. Nat. Acad. **9** (1923) 93.
33) A. A. Markov: Recuell Math. Moscov (1935) 73.
34) V. F. R. Jones: Bull. Amer. Math. Soc. **12** (1985) 103.
35) P. Freyd, D. Yetter, J. Hoste, W. B. R. Lickorish, K. Millett and A. Ocneanu: Bull. Amer. Math. Soc. **12** (1985) 239.
36) K. Sogo, Y. Akutsu and T. Abe: Prog. Theor. Phys. **70** (1983) 730, 739.
37) Y. Akutsu and M. Wadati: Read at the Sectional Meeting of Phys. Soc. Jpn., Sendai September, 1987.
38) H. N. V. Temperley and E. H. Lieb: Proc. R. Soc. London **A322** (1971) 251.
39) R. J. Baxter: J. Stat. Phys. **28** (1982) 1.
40) A. Kuniba, Y. Akutsu and M. Wadati: J. Phys. Soc. Jpn. **55** (1986) 3285.

Journal of the Physical Society of Japan
Vol. 57, No. 6, June, 1988, pp. 1905–1923

Exactly Solvable Models and New Link Polynomials. V. Yang-Baxter Operator and Braid-Monoid Algebra

Tetsuo DEGUCHI, Miki WADATI and Yasuhiro AKUTSU†

Institute of Physics, College of Arts and Sciences, University of Tokyo, Komaba, Meguro-ku, Tokyo 153
†Institute of Physics, Kanagawa University, Rokkakubashi, Kanagawa-ku, Yokohama 221

(Received February 1, 1988)

Yang-Baxter operator is shown to be a fundamental object which relates theory of solvable models to theory of knots and links. First, general properties of Yang-Baxter operators are investigated. Second, a method to construct composite Yang-Baxter operators is explicitly shown. Lastly, from Yang-Baxter operators with crossing symmetry, braid-monoid algebras are derived. It is emphasized that the factorized S-matrices and their graphical illustrations link two approaches, algebraic and combinatorial, in the knot theory.

§1. Introduction

In a series of papers,[1-6] we have reported a general method to construct link polynomials from the exactly solvable models. It is known that the S-matrices of the exactly solvable models satisfy the Yang-Baxter relation[7,8] (Fig. 1):

$$\sum_{\alpha,\beta,\gamma} S_{\gamma r}^{\beta q}(v) S_{k\gamma}^{\alpha p}(u+v) S_{j\beta}^{i\alpha}(u) = \sum_{\alpha,\beta,\gamma} S_{\beta q}^{\alpha p}(u) S_{\gamma r}^{i\alpha}(u+v) S_{k\gamma}^{j\beta}(v). \quad (1.1)$$

Here the S-matrix element $S_{jl}^{ik}(u)$ denotes the scattering amplitude for the process $(i, j) \to (k, l)$ (Fig. 2) and u, v are the spectral parameters (or rapidities). The S-matrices which satisfy the Yang-Baxter relation are referred to as factorized S-matrices.[9)]

Let us introduce an operator $X_i(u)$ which is constructed from the factorized S-matrices:

$$X_i(u) = \sum_{k,l,m,n} S_{ln}^{km}(u) \cdot I^{(1)} \otimes I^{(2)} \otimes \cdots \otimes E_{nk}^{(i)} \otimes E_{ml}^{(i+1)} \otimes I^{(i+2)} \otimes \cdots, \quad i=1, 2, 3, \cdots, \quad (1.2)$$

where E_{nk} is a matrix whose matrix elements are $(E_{nk})_{ab} = \delta_{na}\delta_{kb}$. We call the operator $X_i(u)$ Yang-Baxter operator, or in short, YB operator. The YB operators $\{X_i(u)\}$ satisfy the following relations:

$$X_i(u) X_{i+1}(u+v) X_i(v) = X_{i+1}(v) X_i(u+v) \times X_{i+1}(u), \quad (1.3a)$$

$$X_i(u) X_j(v) = X_j(v) X_i(u), \quad \text{for} \quad |i-j| \geq 2. \quad (1.3b)$$

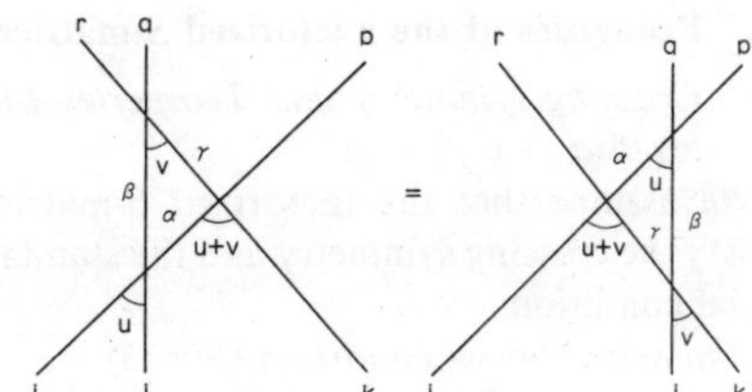

Fig. 1. Schematic explanation of the Yang-Baxter relation. Time direction is upward.

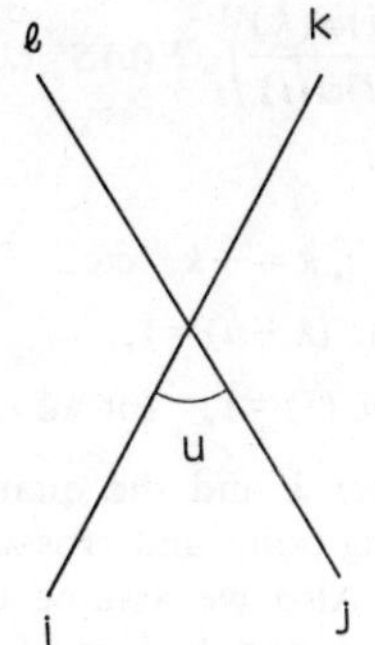

Fig. 2. S-matrix element $S_{jl}^{ik}(u)$.

The relation (1.3a) is nothing but the Yang-Baxter relation in operator form. In the context of statistical mechanics, the YB operator $X_i(u)$ is a constituent of the diagonal-to-diagonal transfer matrix. The YB operator $X_i(u)$ which satisfies (1.3) is the most fundamental object common to quantum completely integrable systems and solvable models in statistical mechanics. In this paper, we discuss general properties of the YB operators and construct the braid-monoid algebras from them.

The outline of the paper is the following. In §2, we discuss general properties of the factorized S-matrices. We show that the Temperley-Lieb algebra is derived from the factorized S-matrices with the crossing symmetry and the standard initial condition. In §3, we construct the Yang-Baxter operators $\{X_i(u)\}$ from the representation of the Temperley-Lieb algebra. Using them, we obtain composite Yang-Baxter operators $\{Y_i(u)\}$. In §4, we relate the composite YB operators $\{Y_i(u)\}$ with the braid-monoid algebra. The last section is devoted to a summary of the paper.

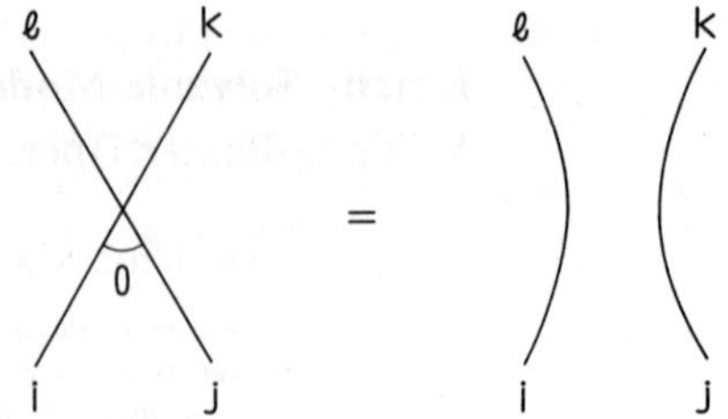

Fig. 3. Standard initial condition.

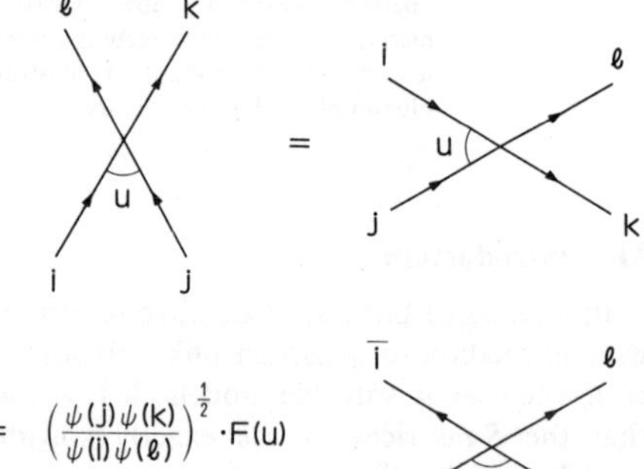

Fig. 4. Crossing symmetry. The s-channel scattering is transformed into the crossing channel scattering. A particle going backward in time is considered as the antiparticle going forward in time.

§2. Properties of the Factorized S-matrices

2.1 Crossing symmetry and Temperley-Lieb algebra

We assume that the factorized S-matrices satisfy the crossing symmetry and the standard initial condition.

1) *standard initial condition* (Fig. 3)

$$S_{jl}^{ik}(0)=\delta_{il}\delta_{jk}. \tag{2.1}$$

2) *crossing symmetry* (Fig. 4)

$$S_{jl}^{ik}(u)=\left(\frac{\psi(j)\psi(k)}{\psi(i)\psi(l)}\right)^{1/2}F(u)S_{k\bar{i}}^{j\bar{l}}(\lambda-u), \tag{2.2a}$$

where

$$\bar{i}=-i,\ \bar{k}=-k \quad \text{etc.}, \tag{2.2b}$$

$$F(u)F(\lambda-u)=1, \tag{2.2c}$$

$$\psi(i)\psi(\bar{i})=1, \quad \text{for all } i. \tag{2.2d}$$

The parameter λ and the quantity $\psi(i)$ are called crossing point and crossing multiplier, respectively. Also we assume that the state variables i, j, k and l of the S-matrix $S_{jl}^{ik}(u)$ take the following values:

$$i, j, k, l=-s, -s+1, \cdots, s-1, s, \tag{2.3}$$

where the "spin" (or "charge") s is related to the state number M by

$$M=2s+1. \tag{2.4}$$

It is interesting to interpret the S-matrices with the crossing symmetry as the scattering amplitudes of the relativistic particles.[10] When we regard the variable i as the charge of a particle, $\bar{i}$ is considered as the charge of the antiparticle. The crossing symmetry implies that the s-channel scattering is transformed into the crossing-channel scattering. The crossing symmetry also implies that the S-matrices are invariant under the CPT transformation:

$$S_{jl}^{ik}(u)=S_{\bar{l}\bar{j}}^{\bar{k}\bar{i}}(u). \tag{2.5}$$

The crossing symmetry leads to an important consequence: The YB operators which have the crossing symmetry and the standard initial condition satisfy the Temperley-Lieb algebra at the crossing point λ. Set the spectral parameter $u=\lambda$, then, from (2.1) and (2.2a) we have

$$S_{jl}^{ik}(\lambda)=\left(\frac{\psi(j)\psi(k)}{\psi(i)\psi(l)}\right)^{1/2}F(\lambda)S_{\bar{k}\bar{i}}^{jl}(0)$$
$$=\left(\frac{\psi(j)\psi(k)}{\psi(i)\psi(l)}\right)^{1/2}F(\lambda)\delta_{l\bar{k}}\delta_{\bar{i}\bar{j}}. \tag{2.6}$$

We define an operator U_i by

$$U_i=\frac{1}{F(\lambda)}X_i(\lambda). \tag{2.7}$$

From (2.6) and (1.2) we have

$$U_i=\sum_{k,l,m,n}\left(\frac{\psi(l)\psi(m)}{\psi(k)\psi(n)}\right)^{1/2}\delta_{l\bar{k}}\delta_{m\bar{n}}I^{(1)}\otimes\cdots$$
$$\otimes E_{nk}^{(i)}\otimes E_{ml}^{(i+1)}\otimes\cdots. \tag{2.8}$$

It is easy to show that the operator U_i satisfies the Temperley-Lieb algebra:[11]

$$U_i^2=q^{\frac{1}{2}}U_i, \tag{2.9a}$$
$$U_iU_{i\pm1}U_i=U_i, \tag{2.9b}$$
$$U_iU_j=U_jU_i,\quad \text{for}\quad |i-j|\geq 2, \tag{2.9c}$$

where

$$q^{1/2}=\sum_a \psi^2(a). \tag{2.10}$$

It is noted that in the above derivation we have used only the crossing symmetry and the standard initial condition. There are many solutions of the Yang-Baxter equation which satisfy the crossing symmetry and the standard initial condition. We shall give some examples in Appendix A.

2.2 *Unitarity condition*

The factorized S-matrices which have the standard initial condition satisfy the unitarity condition (or the inversion relation in statistical mechanics[12]). The following is proof. From (1.3) and (2.1) we have[13]

$$X_i(u)X_i(-u)=X_i(u)X_{i+1}(0)X_i(-u)$$
$$=X_{i+1}(-u)X_i(0)X_{i+1}(u)$$
$$=X_{i+1}(-u)X_{i+1}(u), \tag{2.11a}$$
$$=X_{i+1}(-u)X_{i+2}(0)X_{i+1}(u)$$
$$=X_{i+2}(u)X_{i+1}(0)X_{i+2}(-u)$$
$$=X_{i+2}(u)X_{i+2}(-u). \tag{2.11b}$$

Since the YB operator $X_i(u)$ in (1.2) is expressed in terms of the sum of direct products of matrices, (2.11b) implies that

$$X_i(u)X_i(-u)=C(u)\cdot I, \tag{2.12}$$

where $C(u)$ is a function of the spectral parameter u. From (2.11a) and (2.12) we get

$$C(u)=C(-u). \tag{2.13}$$

The relation (2.12) is the unitarity condition (or the inversion relation) in operator form.

From (2.12) and (1.2) we have

$$C(u)=\sum_{p,q}S_{ql}^{pl}(u)S_{lp}^{lq}(-u). \tag{2.14}$$

Note that $C(u)$ is independent of the value of l. We take $u=\lambda$ and without loss of generality, set $l=s$. Then, we have

$$C(\lambda)=\sum_{p,q}S_{qs}^{ps}(\lambda)S_{sp}^{sq}(-\lambda)$$
$$=\sum_{p,q}S_{\bar{s}\bar{p}}^{qs}(0)S_{sp}^{sq}(-\lambda)\cdot F(\lambda)\left(\frac{\psi(q)\psi(s)}{\psi(p)\psi(s)}\right)^{1/2}$$
$$=\sum_{p,q}\delta_{s\bar{s}}\delta_{q\bar{p}}S_{sp}^{sq}(-\lambda)\cdot F(\lambda)\psi(q)$$
$$=0, \tag{2.15}$$

where the crossing symmetry has been used. From (2.12), (2.13) and (2.15) we obtain an important relation:

$$X_i(\lambda)X_i(-\lambda)=X_i(-\lambda)X_i(\lambda)=0. \tag{2.16}$$

Thus, the YB operators $X_i(-\lambda)$ and $X_i(\lambda)$ are orthogonal.

§3. Composite Yang-Baxter Operators

3.1 *Yang-Baxter operator in terms of Temperley-Lieb operator*

In the previous section we derived the Temperley-Lieb algebra from the factorized S-matrices with the crossing symmetry and the standard initial condition. Reversely from the Temperley-Lieb algebra we shall construct the factorized S-matrices with the crossing symmetry and the standard initial condition. We assume the following relation between the crossing multipliers $\{\psi(i)\}$ and the crossing point μ:

$$2\,\mathrm{ch}\,\mu=\sum_a \psi^2(a). \tag{3.1}$$

With some functions $\rho(u)$ and $f(u)$, we define an operator $\tilde{X}_i(u)$[14] as

$$\tilde{X}_i(u)=\rho(u)(I+f(u)U_i). \tag{3.2}$$

Here U_i is the Temperley-Lieb operator defined in (2.8) which satisfies (2.9) with a relation:

$$q^{1/2}=2\,\mathrm{ch}\,\mu. \tag{3.3}$$

In order that the operator $\tilde{X}_i(u)$ be the YB operator, function $f(u)$ should satisfy

$$\begin{aligned}f(u+v)=&f(u)+f(v)+2\,\mathrm{ch}\,\mu\cdot f(u)f(v)\\&+f(u)f(v)f(u+v).\end{aligned} \tag{3.4}$$

With the initial condition:

$$f(0)=0, \qquad \rho(0)=1, \tag{3.5}$$

we find a solution as follows,

$$f(u)=\frac{\mathrm{sh}\,u}{\mathrm{sh}\,(\mu-u)}, \quad \rho(u)=\frac{\mathrm{sh}\,(\mu-u)}{\mathrm{sh}\,\mu}. \tag{3.6}$$

Remark that $\rho(u)$ is introduced in order that the operator $\tilde{X}_i(u)$ has the crossing symmetry with $F(u)=1$. Hereafter we use λ in place of μ. There should be no confusion about this change of notation.

It is sometimes convenient to use an operator $X_i(u)$ defined by

$$X_i(u)=I+f(u)U_i. \tag{3.7}$$

With the solution $f(u)$ in (3.6), the operator $X_i(u)$ also satisfies the Yang-Baxter relation with the standard initial condition. It has the crossing symmetry with the same crossing point λ and has the same crossing multiplier $\psi(i)$ as the operator $\tilde{X}_i(u)$ with $F(u)$ being given by $f(u)$.

3.2 Construction of composite Yang-Baxter operators

Let us explain a method to construct the composite YB operator. "Composite factorized *S*-matrices" describe the scattering of composite particles with higher spin. From (3.2), (3.6) and (2.16), we have

$$\tilde{X}_i(-\lambda)+\tilde{X}_i(\lambda)=\rho(-\lambda)I, \tag{3.8a}$$
$$\tilde{X}_i(-\lambda)\tilde{X}_i(\lambda)=\tilde{X}_i(\lambda)\tilde{X}_i(-\lambda)=0. \tag{3.8b}$$

Noticing (3.8) we define projectors by

$$P_i^+=\frac{1}{\rho(-\lambda)}\tilde{X}_i(-\lambda), \tag{3.9a}$$
$$P_i^-=\frac{1}{\rho(-\lambda)}\tilde{X}_i(\lambda), \tag{3.9b}$$

where

$$\rho(-\lambda)=2\,\mathrm{ch}\,\lambda=q^{1/2}. \tag{3.10}$$

From the relations (3.8) we have the properties of the projectors:

$$\begin{aligned}&(P_i^+)^2=P_i^+, \quad (P_i^-)^2=P_i^-,\\&P_i^+P_i^-=P_i^-P_i^+=0,\\&P_i^++P_i^-=I.\end{aligned} \tag{3.11}$$

The projectors P_i^+ and P_i^- are symmetrizer and anti-symmetrizer, respectively.[5] From (3.9), (3.2) and (3.7) we have the relations between the Temperley-Lieb operators and the projectors:

$$P_i^+=X_i(-\lambda)=I-q^{-1/2}U_i, \tag{3.12a}$$
$$P_i^-=q^{-1/2}U_i. \tag{3.12b}$$

The Yang-Baxter relation (1.3) gives the following "pass-through" condition[15] (Fig. 5):

$$P_i^+X_{i+1}(u-\lambda)X_i(u)=X_{i+1}(u)X_i(u-\lambda)P_{i+1}^+, \tag{3.13a}$$

$$P_i^-X_{i+1}(u+\lambda)X_i(u)=X_{i+1}(u)X_i(u+\lambda)P_{i+1}^-. \tag{3.13b}$$

We first construct 2-particle composite YB operator $Y_i(u)$. In the 2-particle composition, identity operator I^S for the symmetrizer is given by (Fig. 6)

$$I^S=P_1^+P_3^+\cdots P_{2i-1}^+P_{2i+1}^+\cdots. \tag{3.14}$$

Here and hereafter, superscript *S* denotes that the operator is related to the symmetrizer. We have 2-particle composite YB operator $Y_i^S(u)$

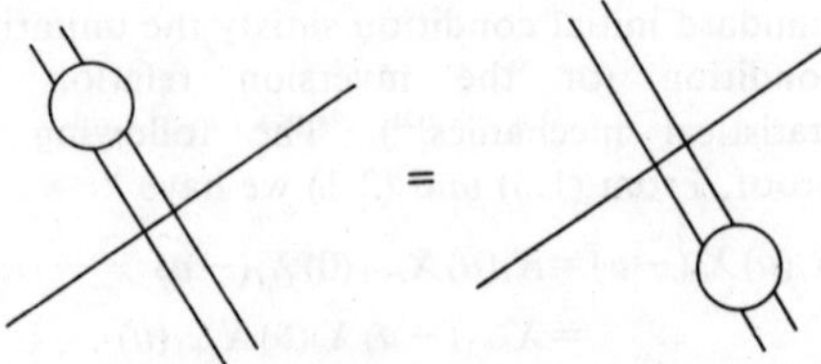

Fig. 5. Pass-through condition. Circles denote projectors.

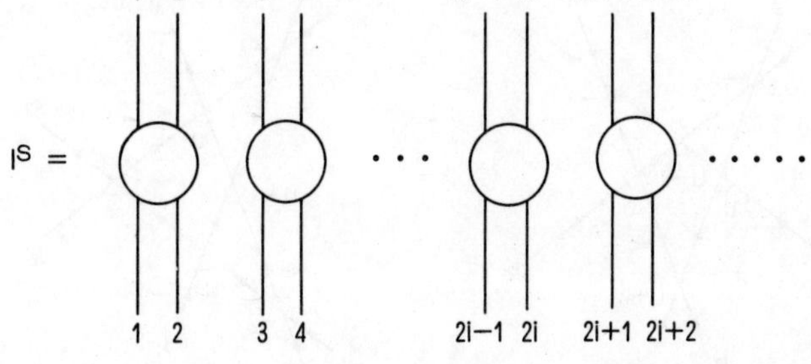

Fig. 6. Identity operator I^S for symmetrizer in the 2-particle composition. Circles denote symmetrizers.

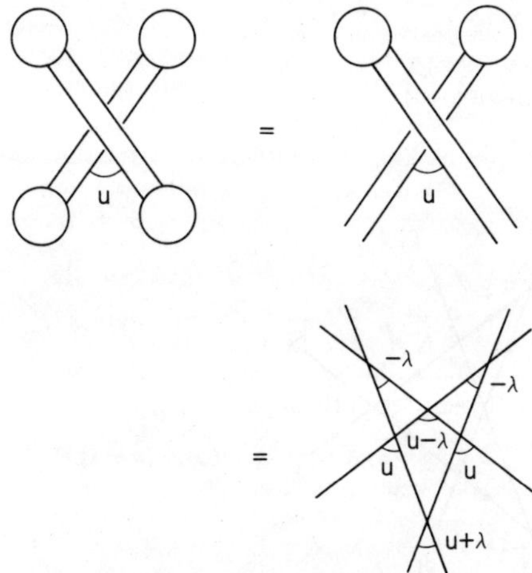

Fig. 7. Composite YB operator $Y_i^S(u)$ constructed from the YB operators $X_i(u)$.

from the YB operators $X_i(u)$ (Fig. 7):

$$\begin{aligned} Y_i^S(u)&=P_{2i-1}^+P_{2i+1}^+X_{2i}(u-\lambda)X_{2i-1}(u)X_{2i+1}(u)\\ &\quad\times X_{2i}(u+\lambda)P_{2i-1}^+P_{2i+1}^+I^S\\ &=X_{2i-1}(-\lambda)X_{2i+1}(-\lambda)X_{2i}(u-\lambda)\\ &\quad\times X_{2i-1}(u)X_{2i+1}(u)X_{2i}(u+\lambda). \end{aligned} \tag{3.15}$$

For notational simplicity, we sometimes omit I^S in the expression of $Y_i^S(u)$. The operators $\{Y_i(u)\}$ satisfy the Yang-Baxter relation since the YB operators $\{X_i(u)\}$ do. Physically, the composite YB operator $Y_i(u)$ describes the scattering of composite particles. The composite factorized S-matrices $\{\hat{S}_{jl}^{ik}(u)\}$ have the crossing symmetry with the crossing point λ. It is clear from the fact that each matrix element has the crossing symmetry. In fact, we have (Fig. 8)

$$\begin{aligned} \hat{S}_{jl}^{ik}(u)&=\left(\frac{\hat{\psi}(j)\hat{\psi}(k)}{\hat{\psi}(i)\hat{\psi}(l)}\right)^{1/2}F^2(u)F(u-\lambda)F(u+\lambda)\\ &\quad\times\hat{S}_{\bar{k}\bar{i}}^{jl}(\lambda-u), \end{aligned} \tag{3.16a}$$

where

$$i=(i_1, i_2),\quad \bar{i}=(-i_1, -i_2), \tag{3.16b}$$

$$\hat{\psi}(i)=\psi(i_1)\psi(i_2). \tag{3.16c}$$

Note that the composite factorized S-matrices also have the standard initial condition. We show it by operator calculations. From (2.9), (3.7), (3.4) and (3.12b) we have

$$\begin{aligned} Y_i^S(u)&=P_{2i-1}^+P_{2i+1}^+(I+f(u-\lambda)U_{2i})(I+f(u)U_{2i-1})(I+f(u)U_{2i+1})(I+f(u+\lambda)U_{2i})P_{2i-1}^+P_{2i+1}^+\\ &=P_{2i-1}^+P_{2i+1}^+(I+\{f(u-\lambda)+f(u+\lambda)+2\,\mathrm{ch}\,\lambda\cdot f(u-\lambda)f(u+\lambda)+2f(u)f(u-\lambda)f(u+\lambda)\}\,U_{2i}\\ &\quad+f(u)^2f(u-\lambda)f(u+\lambda)U_{2i}U_{2i-1}U_{2i+1}U_{2i})P_{2i-1}^+P_{2i+1}^+\\ &=P_{2i-1}^+P_{2i+1}^+(I+\{f(2u)+(2f(u)-f(2u))f(u-\lambda)f(u+\lambda)\}\,U_{2i}\\ &\quad+f(u-\lambda)f(u)^2f(u+\lambda)U_{2i}U_{2i-1}U_{2i+1}U_{2i})P_{2i-1}^+P_{2i+1}^+. \end{aligned} \tag{3.17}$$

Thus we have,

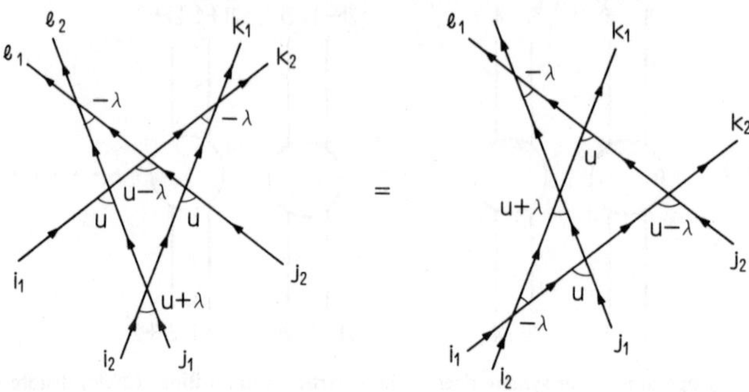

$$= \left(\frac{\psi(j_1)\psi(j_2)\psi(k_1)\psi(k_2)}{\psi(i_1)\psi(i_2)\psi(\ell_1)\psi(\ell_2)} \right)^{\frac{1}{2}} \cdot F(u)^2 F(u-\lambda) F(u+\lambda)$$

$\times$

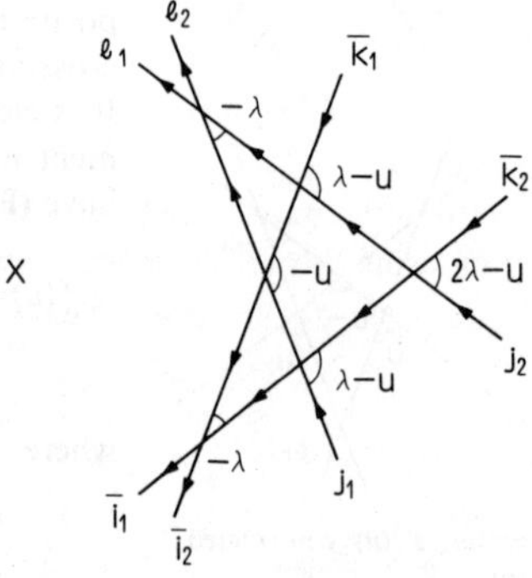

Fig. 8. Crossing symmetry of the composite factorized S-matrices.

$$Y_i^S(0)=P_{2i-1}^+ P_{2i+1}^+, \tag{3.18}$$

and then find that $Y_i^S(0)$ is the identity operator in the 2-particle composition. The same discussion also holds for a 2-particle composite YB operator $Y_i^A(u)$ defined by

$$\begin{aligned} Y_i^A(u) &= P_{2i-1}^- P_{2i+1}^- X_{2i}(u+\lambda) X_{2i-1}(u) X_{2i+1}(u) X_{2i}(u-\lambda) P_{2i-1}^- P_{2i+1}^- \\ &= \frac{1}{\rho(-\lambda)^2} \tilde{X}_{2i-1}(\lambda) \tilde{X}_{2i+1}(\lambda) X_{2i}(u+\lambda) X_{2i-1}(u) X_{2i+1}(u) X_{2i}(u-\lambda). \end{aligned} \tag{3.19}$$

The superscript A denotes that the operator is related to the anti-symmetrizer.

We generalize the above discussion to $(N-1)$-particle composite YB operator. From (3.7), (3.12b) and (3.10) we have

$$X_i(u) = I + 2\,\mathrm{ch}\,\lambda \cdot f(u) P_i^- . \tag{3.20}$$

Then we have the following property:

$$X_i(u)X_i(-\lambda)=(I+2\operatorname{ch}\lambda\cdot f(u)P_i^-)P_i^+$$
$$=P_i^+$$
$$=X_i(-\lambda),\quad \text{for}\quad u\neq\lambda. \tag{3.21}$$

Using the property (3.21) and the Yang-Baxter relation we can construct the symmetrizer $P_i^{S(N)}$ from the YB operators $\{X_i(u)\}$. The symmetrizer $P_i^{S(N)}$ means the symmetric projector of interacting $(N-1)$ particles and is defined recursively as

$$P^{S(N)}_{i\cdots i+N-3}=P^{S(N-1)}_{i\cdots i+N-4}\cdot X_{i+N-3}(-(N-2)\lambda)\cdot P^{S(N-1)}_{i\cdots i+N-4}, \tag{3.22a}$$

with

$$P_i^{S(3)}=P_i^+. \tag{3.22b}$$

We shall use $P_i^{S(N)}$ in place of $P^{S(N)}_{i\cdots i+N-3}$ to simplify the notation. The symmetrizers satisfy the pass-through condition since they are composed of the YB operators $\{X_i(u)\}$:

$$P_{i+1}^{S(N)}X_i(u)X_{i+1}(u-\lambda)\cdots X_{i+N-2}(u-(N-2)\lambda)$$
$$=X_i(u-(N-2)\lambda)\cdots X_{i+N-3}(u-\lambda)X_{i+N-2}(u)P_i^{S(N)}. \tag{3.23}$$

For example we have in the $N=4$ case

$$P_{i+1}^{S(4)}X_i(u)X_{i+1}(u-\lambda)X_{i+2}(u-2\lambda)=X_i(u-2\lambda)X_{i+1}(u-\lambda)X_{i+2}(u)P_i^{S(4)}. \tag{3.24}$$

The symmetrizer $P_i^{S(N)}$ is idempotent:

$$(P_i^{S(N)})^2=P_i^{S(N)}. \tag{3.25}$$

The proof is given in Appendix B.

Since the symmetrizers satisfy the pass-through condition (3.23) we can construct the composite YB operators $Y_i(u)$ by using them. We define "composite YB operator" $Y_i(u)$ as

$$Y_i(u)=P^{S(N)}_{(i-1)k+1}P^{S(N)}_{ik+1}\bar{Y}_i^{(N)}(u)P^{S(N)}_{(i-1)k+1}P^{S(N)}_{ik+1}, \tag{3.26a}$$

where

$$\bar{Y}_i^{(N)}(u)=\hat{X}_i^{(1)}(u)\hat{X}_i^{(2)}(u)\cdots\hat{X}_i^{(N-1)}(u), \tag{3.26b}$$
$$\hat{X}_i^{(l)}(u)=X_{ik+1-l}(u-(N-1-l)\lambda)X_{ik+2-l}(u-(N-2-l)\lambda)\cdots X_{(i+1)k-l}(u-(1-l)\lambda),$$
$$\text{for}\quad l=1,2,\cdots N-1, \tag{3.26c}$$

and

$$k=N-1. \tag{3.26d}$$

The operator $Y_i(u)$ has the crossing symmetry with the crossing point λ and the standard initial condition. For the crossing symmetry it is clear from the fact that each matrix element has the crossing symmetry (see (3.16)). As for the standard initial condition we shall give the proof in Appendix C.

A set of generators $\{g_i\}$ defined by

$$g_i=\lim_{u\to\infty}X_i(u)=\lim_{u\to\infty}(I+f(u)U_i)=1-e^{\lambda}\cdot U_i, \tag{3.27}$$

satisfies the defining relations for the Hecke algebra:[16,17]

$$g_i^2=(1-t)g_i+t, \tag{3.28a}$$
$$g_ig_{i+1}g_i=g_{i+1}g_ig_{i+1}, \tag{3.28b}$$
$$g_ig_j=g_jg_i,\quad \text{for}\quad |i-j|\geq 2, \tag{3.28c}$$

where

$$t=e^{2\lambda}. \tag{3.28}$$

The symmetrizer and the anti-symmetrizer can be expressed in terms of the representation of the generator g_i. For instance, in the case of $N=3$ (2-particle composition) they are

$$P_i^+=P_i^{S(3)}=\frac{1}{1+t}(t+g_i), \tag{3.30a}$$

$$P_i^- = P_i^{A(3)} = \frac{1}{1+t}(1-g_i). \tag{3.30b}$$

The symmetrizer $P_i^{S(N)}$ constructed from the YB operators $X_i(u)$ is equivalent to the symmetrizer $P_i^{(N)}$ defined by[5)]

$$P_i^{(N)} = P_i^{(N-1)} \cdot h_{i+N-3}^{(N)} \cdot P_i^{(N-1)}, \tag{3.31a}$$

with

$$P_i^{(3)} = \frac{1}{1+t}(t+g). \tag{3.31b}$$

Because we have

$$\begin{aligned} h_i^{(N)} &= \frac{\tau_{N-2}}{\tau_{N-1}}\left(\frac{t^{N-2}}{\tau_{N-2}} + g_i\right) \\ &= 1 - t^{1/2} \cdot \frac{\tau_{N-2}}{\tau_{N-1}} U_i \\ &= 1 + f(-(N-2)\lambda) U_i \\ &= X_i(-(N-2)\lambda), \end{aligned} \tag{3.32}$$

where

$$\tau_N = 1 + t + \cdots + t^{N-1}. \tag{3.33}$$

Hereafter we write $h_i^{S(N)}$ in place of $h_i^{(N)}$.

3.3 *Projectors in Hecke algebra*

As shown in (3.27) and (3.28) the Temperley-Lieb algebra gives a representation of the Hecke algebra. But the reverse is not true; if we put

$$g_i = 1 - t^{1/2} A_i, \tag{3.34}$$

a necessary and sufficient condition for g_i to satisfy (3.28) is

$$A_i^2 = (t^{1/2} + t^{-1/2}) A_i, \tag{3.35a}$$

$$A_i A_{i+1} A_i - A_i = A_{i+1} A_i A_{i+1} - A_{i+1}, \tag{3.35b}$$

$$A_i A_j = A_j A_i, \quad \text{for} \quad |i-j| \geq 2. \tag{3.35c}$$

It is remarked that operator $X_i(u)$ defined by

$$X_i(u) = I + f(u) A_i, \tag{3.36}$$

is also the YB operator. In the following discussion, only the properties (3.28) of the Hecke algebra will be used.

We construct anti-symmetrizers. This can be done in the same way as the symmetrizers. From (3.20) we have

$$\begin{aligned} X_i(u) P_i^- &= (I + 2\,\mathrm{ch}\,\lambda \cdot f(u) P_i^-) P_i^- \\ &= (I + 2\,\mathrm{ch}\,\lambda \cdot f(u)) P_i^-. \end{aligned} \tag{3.37}$$

Then, an operator defined by (prime on the operator is nothing to do with differentiation)

$$X_i'(u) = \frac{X_i(u)}{1 + 2\,\mathrm{ch}\,\lambda \cdot f(u)}, \tag{3.38}$$

satisfies

$$X_i'(u) P_i^- = P_i^-. \tag{3.39}$$

We define the anti-symmetrizers $P_i^{A(N)}$ recursively as

$$\begin{aligned} P_{i\cdots i+N-3}^{A(N)} = P_{i\cdots i+N-4}^{A(N-1)} \cdot X'_{i+N-3}((N-2)\lambda) \\ \cdot P_{i\cdots i+N-4}^{A(N-1)}, \end{aligned} \tag{3.40a}$$

with

$$P_i^{A(3)} = P_i^- = X_i'(\lambda). \tag{3.40b}$$

To simplify the notation we write $P_i^{A(N)}$ in place of $P_{i\cdots i+N-3}^{A(N)}$. The anti-symmetrizer satisfies the pass-through condition:

$$\begin{aligned} P_{i+1}^{A(N)} X_i(u) X_{i+1}(u+\lambda) \cdots X_{i+N-2}(u+(N-2)\lambda) \\ = X_i(u+(N-2)\lambda) \cdots X_{i+N-3}(u+\lambda) \\ \times X_{i+N-2}(u) P_i^{A(N)}. \end{aligned} \tag{3.41}$$

We can prove in the same way as the symmetrizer that the anti-symmetrizer is idempotent:

$$(P_i^{A(N)})^2 = P_i^{A(N)}. \tag{3.42}$$

Note that in the Hecke algebra the symmetrizer and the anti-symmetrizer are transformed into each other by the following transformation[5)]

$$t \longrightarrow t^{-1}, \qquad g_i \longrightarrow -t^{-1} g_i. \tag{3.43}$$

Let $h_i^{A(N)}$ denote the quantity obtained from $h_i^{S(N)}$ by the transformation, then, we have from (3.31)

$$P_i^{A(N)} = P_i^{A(N-1)} \cdot h_{i+N-3}^{A(N)} \cdot P_i^{A(N-1)}, \tag{3.44}$$

where

$$h_i^{A(N)} = \frac{\tau_{N-2}}{\tau_{N-1}}\left(\frac{1}{\tau_{N-2}} - g_i\right) = X_i'((N-2)\lambda). \tag{3.45}$$

It is interesting to observe that A_i satisfies the relations:

$$A_i A_{i+1} A_i = A_i + (t^{1/2} + t^{-1/2})(t + 1 + t^{-1}) P_{i,i+1}^{A(4)}, \tag{3.46a}$$

$$\begin{aligned} A_{i+1} A_i A_{i+1} = A_{i+1} + (t^{1/2} + t^{-1/2})(t + 1 + t^{-1}) \\ \times P_{i,i+1}^{A(4)}. \end{aligned} \tag{3.46b}$$

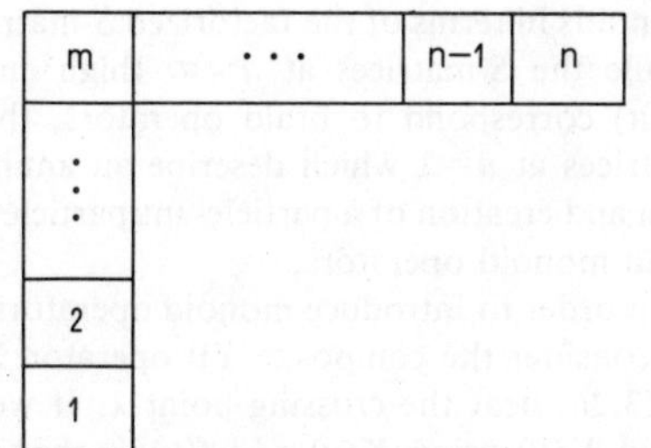

Fig. 9. Young diagram corresponding to the projector P in (3.47).

The fact that anti-symmetrizer $P_{i,i+1}^{A(4)}$ does not vanish in the case of the Hecke algebra assures the existence of higher order anti-symmetrizers $P_i^{A(N)}$ for $N \geq 4$.

We have other projectors constructed from the generators of the Hecke algebra. For example corresponding to the Young diagram in Fig. 9 we have

$$p = \frac{r_2 r_{n-m+1}}{r_3 r_{n-m} - r_2 r_{n-m-1}} P_{m\cdots n-1}^{S(n-m+2)} P_{1\cdots m-1}^{A(m+1)}, \tag{3.47a}$$

where

$$r_l = t^{-(l-1)/2}(1+t+\cdots+t^{l-1}), \tag{3.47b}$$

In the limit $t \to 1$, P becomes the Young operator defined in the symmetric group.[18] We can construct the composite YB operator which has the symmetry (Fig. 9), because the projector P satisfies the pass-through condition. Note that, because of the mixed-symmetry of the Young diagram, differences of rapidities of the internal particles are not equal. We shall prove in Appendix D that P is idempotent.

We have simple examples[19] of the matrix representation of the operator A_i:

$$A_i = \sum_{\substack{k=1\\k\neq l}}^{M} \sum_{l=1}^{M} I^{(1)} \otimes \cdots \otimes (t^{\varepsilon(k,l)/2} E_{kk}^{(i)} \otimes E_{ll}^{(i+1)} + E_{kl}^{(i)} \otimes E_{lk}^{(i+1)}) \otimes I^{(i+2)} \otimes \cdots, \tag{3.48}$$

where

$$\varepsilon(k,l) = \begin{cases} -1 & \text{if } k<l \\ 1 & \text{if } k>l. \end{cases} \tag{3.49}$$

Note that the state number M is arbitrary. The factorized S-matrices corresponding to A_i describe the scattering amplitudes of the particles which have the $SU(M)$ quantum number. The $su(M)$ algebra has $su(2)$ subalgebras in itself and the representation (3.48) has a kind of 'crossing symmetry' with respect to these $su(2)$ subalgebras. For example, let us consider the flavour $SU(3)$ of u, d and s quarks. Restricted to u and d quarks the corresponding $SU(2)$ symmetry is called as isospin symmetry or I-spin.[20] Similarly, we have U-spin and V-spin. When we think I-spin, the 'antiparticle' of u quark is d quark. The factorized S-matrices constructed from the matrix representation of A_i have these $SU(2)$ spin crossing symmetries. As a matter of fact, we have obtained the representation (3.48) from Babelon-de Vega-Viallet's solution[19] applying symmetry breaking transformation[2] and using formula (2.7) with $t=e^{2\lambda}$ determined by $SU(2)$ crossing symmetries.

3.4 Rational solutions and symmetric group

We have used hyperbolic solutions of the Yang-Baxter relation but the similar results also hold in the case of rational solutions. The YB operator of a rational solution has the following operator form (see Appendix A):

$$X_i(u) = I - \frac{u}{\lambda}\sigma_i, \tag{3.50}$$

where σ_i are the permutation operators. From (3.50) and (2.16) we have

$$X_i(\lambda) + X_i(-\lambda) = 2I, \tag{3.51a}$$

$$X_i(\lambda)X_i(-\lambda) = X_i(-\lambda)X_i(\lambda) = 0. \tag{3.51b}$$

Then we obtain as in (3.9)

$$P_i^+ = \frac{1}{2}X_i(-\lambda) = \frac{1}{2}(I+\sigma_i), \tag{3.52a}$$

$$P_i^- = \frac{1}{2}X_i(\lambda) = \frac{1}{2}(I-\sigma_i). \tag{3.52b}$$

The operators P_i^+ and P_i^- are the symmetrizer and the anti-symmetrizer of the symmetric group, respectively. Higher order symmetrizers and anti-symmetrizers can be constructed from the YB operator $X_i(u)$ in the same way as the case of hyperbolic solutions. We give an example of the $N=4$ case:

$$P^{S(4)}_{i,i+1}=\frac{1}{2}X_i(-\lambda)\cdot\frac{1}{3}X_{i+1}(-2\lambda)\cdot\frac{1}{2}X_i(-\lambda)$$
$$=\frac{1}{2}X_{i+1}(-\lambda)\cdot\frac{1}{3}X_i(-2\lambda)\cdot\frac{1}{2}X_{i+1}(-\lambda)$$
$$=\frac{1}{6}(I+\sigma_i+\sigma_{i+1}+\sigma_i\sigma_{i+1}+\sigma_{i+1}\sigma_i$$
$$+\sigma_i\sigma_{i+1}\sigma_i). \tag{3.53}$$

§4. Braids and Monoids

4.1 S-matrices and monoid diagram

As discussed in §2, the factorized S-matrices with the crossing symmetry satisfy the Temperley-Lieb algebra at the crossing point. This fact may be explained as follows. The crossing symmetry transformation changes the scattering channel to the crossing channel (Fig. 4). This corresponds to the observation of the scattering diagram from a 90°-rotated direction. Then, the scattering at $u=\lambda$ in the scattering channel can be considered as the scattering at $u=0$ in the crossing channel (Fig. 10). On the other hand, the Temperley-Lieb operator can be considered as the monoid diagram.[21,22] The coincidence of two pictures suggests that the S-matrices with the crossing symmetry satisfy the Temperley-Lieb algebra at the crossing point. We now have extremely interesting interpretations of braids and

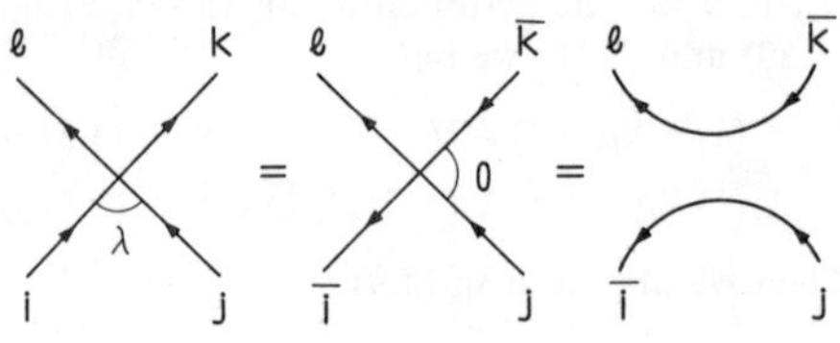

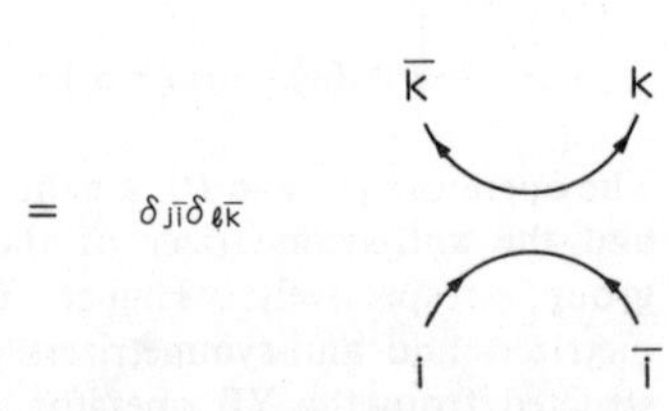

Fig. 10. Scattering at $u=\lambda$ in the scattering channel can be considered as the scattering $u=0$ at in the crossing channel. We omit the crossing multipliers and the function $F(u)$.

monoids in terms of the factorized S-matrices. While the S-matrices at $u=\infty$ (high energy limit) correspond to braid operators, the S-matrices at $u=\lambda$ which describe an annihilation and creation of a particle-antiparticle pair yield monoid operators.

In order to introduce monoid operators, let us consider the composite YB operator $Y_i(u)$ in (3.26) near the crossing point λ. If we expand $Y_i(u)$ using $X_i(u)=I+f(u)U_i$, the terms which are not annihilated by the symmetrizers can survive in the expansion. In the limit $u\to\lambda$, $f(u)$ diverges and $f(u-\lambda)$ tends to be zero. We find that only the coefficient of $\bar{E}^{(N)}_i$ (defined shortly) is divergent in this limit. Hence, we obtain

$$\lim_{u\to\lambda}\rho(u)Y_i(u)=a_N P^{S(N)}_{(i-1)k+1}P^{S(N)}_{ik+1}\bar{E}^{(N)}_i$$
$$\times P^{S(N)}_{(i-1)k+1}P^{S(N)}_{ik+1}, \tag{4.1a}$$

where

$$\bar{E}^{(N)}_i=\hat{E}^{(1)}_i\hat{E}^{(2)}_i\cdots\hat{E}^{(N-1)}_i, \tag{4.1b}$$
$$\hat{E}^{(l)}_i=U_{ik+1-l}U_{ik+2-l}\cdots U_{(i+1)k-l}, \tag{4.1c}$$
$$k=N-1, \tag{4.1d}$$

and

$$a_N=\lim_{u\to\lambda}\rho(u)\prod_{n=-(N-2)}^{(N-2)}(f(u+n\lambda))^{N-1-|n|}, \tag{4.2a}$$
$$=\frac{\mathrm{sh}\,(N-1)\lambda}{\mathrm{sh}\,\lambda}. \tag{4.2b}$$

4.2 Braid-Monoid algebra

We discuss braid-monoid algebras constructed from the YB operators $X_i(u)$ and $Y_i(u)$. We set

$$t=e^{2\lambda},\qquad q^{1/2}=2\,\mathrm{ch}\,\lambda=t^{1/2}+t^{-1/2}. \tag{4.3}$$

We first consider the $N=2$ case. The braid and monoid operators are defined by (Fig. 11)

$$g_i=X_i(\infty),\quad g_i^{-1}=X_i(-\infty),$$
$$I=X_i(0),\quad E_i=\lim_{u\to\lambda}\rho(u)X_i(u), \tag{4.4}$$

where $X_i(u)$ is defined in (3.7). Then, we have the following relations:

$$(g_i-1)(g_i+t)=0, \tag{4.5a}$$
$$g_i=1-t^{1/2}E_i,\quad g_i^{-1}=1-t^{-1/2}E_i, \tag{4.5b}$$
$$E_i^2=(t^{1/2}+t^{-1/2})E_i, \tag{4.5c}$$

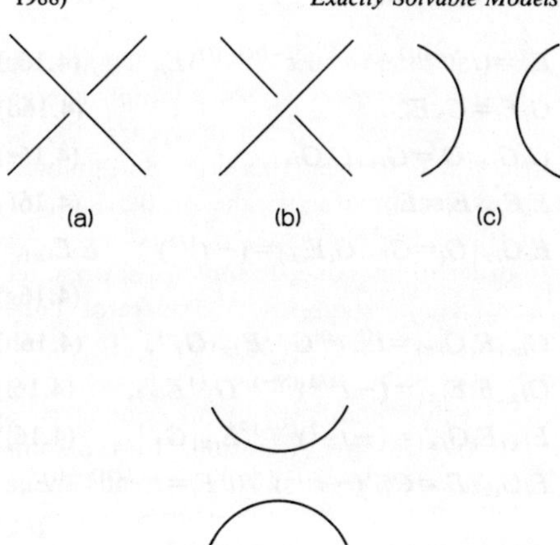

Fig. 11. Braid and monoid diagrams: (a) braid operator, (b) inverse of braid operator, (c) identity, (d) monoid operator.

$$g_iE_i=E_ig_i=-tE_i, \quad (4.5d)$$
$$g_ig_{i+1}g_i=g_{i+1}g_ig_{i+1}, \quad (4.5e)$$
$$E_iE_{i\pm1}E_i=E_i, \quad (4.5f)$$
$$E_ig_{i\pm1}g_i=g_{i\pm1}g_iE_{i\pm1}=-t^{1/2}E_iE_{i\pm1}, \quad (4.5g)$$
$$g_{i\pm1}E_ig_{i\pm1}=tg_i^{-1}E_{i\pm1}g_i^{-1}, \quad (4.5h)$$
$$g_{i\pm1}E_iE_{i\pm1}=-t^{1/2}g_i^{-1}E_{i\pm1}, \quad (4.5i)$$
$$E_{i\pm1}E_ig_{i\pm1}=-t^{1/2}E_{i\pm1}g_i^{-1}, \quad (4.5j)$$
$$E_ig_{i\pm1}E_i=t^{-1/2}E_i, \quad (4.5k)$$
$$g_ig_j=g_jg_i, \quad \text{for} \quad |i-j|\geq 2, \quad (4.5l)$$
$$E_iE_j=E_jE_i, \quad \text{for} \quad |i-j|\geq 2. \quad (4.5m)$$

Among (4.5a) ~ (4.5d), only two relations are independent. The relations (4.5h) ~ (4.5k) are derived from the relations (4.5d) ~ (4.5g). If we set

$$\hat{g}=t^{-1/4}g, \quad \hat{E}=-E, \quad A=t^{-1/4}, \quad (4.6)$$

we find that

$$\hat{g}_i=A+A^{-1}\hat{E}, \quad \hat{g}_i^{-1}=A^{-1}+A\hat{E},$$
$$\hat{E}_i^2=-(A^2+A^{-2})\hat{E}. \quad (4.7)$$

These relations have been used in the calculation of the Kauffman's Bracket polynomial[21] which is equivalent to the Jones polynomial.[16] We list up all the relations corresponding to (4.5a) ~ (4.5m):

$$(\hat{g}_i-A)(\hat{g}_i+A^{-3})=0, \quad (4.8a)$$
$$\hat{g}_i=A+A^{-1}\hat{E}_i, \quad \hat{g}_i^{-1}=A^{-1}+A\hat{E}_i, \quad (4.8b)$$
$$\hat{E}_i^2=-(A^2+A^{-2})\hat{E}_i, \quad (4.8c)$$
$$\hat{g}_i\hat{E}_i=\hat{E}_i\hat{g}_i=-A^{-3}\hat{E}_i, \quad (4.8d)$$
$$\hat{g}_i\hat{g}_{i+1}\hat{g}_i=\hat{g}_{i+1}\hat{g}_i\hat{g}_{i+1}, \quad (4.8e)$$
$$\hat{E}_i\hat{E}_{i\pm1}\hat{E}_i=\hat{E}_i, \quad (4.8f)$$
$$\hat{E}_i\hat{g}_{i\pm1}\hat{g}_i=\hat{g}_{i\pm1}\hat{g}_i\hat{E}_{i\pm1}=\hat{E}_i\hat{E}_{i\pm1}, \quad (4.8g)$$
$$\hat{g}_{i\pm1}\hat{E}_i\hat{g}_{i\pm1}=\hat{g}_i^{-1}\hat{E}_{i\pm1}\hat{g}_i^{-1}, \quad (4.8h)$$
$$\hat{g}_{i\pm1}\hat{E}_i\hat{E}_{i\pm1}=\hat{g}_i^{-1}\hat{E}_{i\pm1}, \quad (4.8i)$$
$$\hat{E}_{i\pm1}\hat{E}_i\hat{g}_{i\pm1}=\hat{E}_{i\pm1}\hat{g}_i^{-1}, \quad (4.8j)$$
$$\hat{E}_i\hat{g}_{i\pm1}\hat{E}_i=-A^3\hat{E}_i, \quad (4.8k)$$
$$\hat{g}_i\hat{g}_j=\hat{g}_j\hat{g}_i, \quad \text{for} \quad |i-j|\geq 2, \quad (4.8l)$$
$$\hat{E}_i\hat{E}_j=\hat{E}_j\hat{E}_i, \quad \text{for} \quad |i-j|\geq 2. \quad (4.8m)$$

Next, we consider the $N=3$ case (2-particle composition). We define the braid and monoid operators by

$$G_i=Y_i(\infty), \quad G_i^{-1}=Y_i(-\infty),$$
$$I=Y_i(0), \quad E_i=\lim_{u\to\lambda}\rho(u)Y_i(u)/a_3, \quad (4.9)$$

where $Y_i(u)$ is the composite YB operator defined in (3.26) and a_3 in (4.2). We have the following relations:

$$(G_i-1)(G_i+t^2)(G_i-t^3)=0, \quad (4.10a)$$
$$E_i=\frac{(t+1+t^{-1})}{(t^3-1)(t^3+t^2)}(G_i-1)(G_i+t^2), \quad (4.10b)$$
$$E_i^2=(t+1+t^{-1})E_i, \quad (4.10c)$$
$$G_iE_i=E_iG_i=t^3E_i, \quad (4.10d)$$
$$G_iG_{i+1}G_i=G_{i+1}G_iG_{i+1}, \quad (4.10e)$$
$$E_iE_{i\pm1}E_i=E_i, \quad (4.10f)$$
$$E_iG_{i\pm1}G_i=G_{i\pm1}G_iE_{i\pm1}=t^2E_iE_{i\pm1}, \quad (4.10g)$$
$$G_{i\pm1}E_iG_{i\pm1}=t^4G_i^{-1}E_{i\pm1}G_i^{-1}, \quad (4.10h)$$
$$G_{i\pm1}E_iE_{i\pm1}=t^2G_i^{-1}E_{i\pm1}, \quad (4.10i)$$
$$E_{i\pm1}E_iG_{i\pm1}=t^2E_{i\pm1}G_i^{-1}, \quad (4.10j)$$
$$E_iG_{i\pm1}E_i=t^{-1}E_i, \quad (4.10k)$$
$$G_iG_j=G_jG_i, \quad \text{for} \quad |i-j|\geq 2, \quad (4.10l)$$
$$E_iE_j=E_jE_i, \quad \text{for} \quad |i-j|\geq 2. \quad (4.10m)$$

If we define

$$\hat{G}=\frac{G}{it}, \quad \hat{E}=-E, \quad (4.11)$$

then $\hat{G}$ and $\hat{E}$ satisfy the relations of the Birman-Wenzl-Murakami[23,24] algebra for the Kauffman polynomial[22] with

$$l=it^{-2}, \qquad m=\frac{1-t^2}{it}, \tag{4.12}$$

in their notations. We have already reported this result.[4,5)]

Similarly, we define operators G_i and E_i for the $N=4$ case (3-particle composition):

$$G_i=Y_i(\infty), \quad G_i^{-1}=Y_i(-\infty),$$
$$I=Y_i(0), \quad E_i=\lim_{u\to\lambda}\rho(u)Y_i(u)/a_4, \tag{4.13}$$

and we obtain the following relations:

$$(G_i-1)(G_i+t^3)(G_i-t^5)(G_i+t^6)=0, \tag{4.14a}$$
$$E_i=\frac{(t^{3/2}+t^{1/2}+t^{-1/2}+t^{-3/2})}{(-t^6-1)(-t^6+t^3)(-t^6-t^5)} \times(G_i-1)(G_i+t^3)(G_i-t^5), \tag{4.14b}$$
$$E_i^2=(t^{3/2}+t^{1/2}+t^{-1/2}+t^{-3/2})E_i, \tag{4.14c}$$
$$G_iE_i=E_iG_i=-t^6E_i, \tag{4.14d}$$
$$G_iG_{i+1}G_i=G_{i+1}G_iG_{i+1}, \tag{4.14e}$$
$$E_iE_{i\pm1}E_i=E_i, \tag{4.14f}$$
$$E_iG_{i\pm1}G_i=G_{i\pm1}G_iE_{i\pm1}=-t^{9/2}E_iE_{i\pm1}, \tag{4.14g}$$
$$G_{i\pm1}E_iG_{i\pm1}=t^9G_i^{-1}E_{i\pm1}G_i^{-1}, \tag{4.14h}$$
$$G_{i\pm1}E_iE_{i\pm1}=-t^{9/2}G_i^{-1}E_{i\pm1}, \tag{4.14i}$$
$$E_{i\pm1}E_iG_{i\pm1}=-t^{9/2}E_{i\pm1}G_i^{-1}, \tag{4.14j}$$
$$E_iG_{i\pm1}E_i=t^{-3/2}E_i, \tag{4.14k}$$
$$G_iG_j=G_jG_i, \quad \text{for} \quad |i-j|\geq 2, \tag{4.14l}$$
$$E_iE_j=E_jE_i, \quad \text{for} \quad |i-j|\geq 2. \tag{4.14m}$$

For general $N\geq 2$ we define the operators G_i and E_i as follows:

$$G_i=Y_i(\infty), \tag{4.15a}$$
$$G_i^{-1}=Y_i(-\infty), \tag{4.15b}$$
$$I=Y_i(0), \tag{4.15c}$$
$$E_i=\lim_{u\to\lambda}\rho(u)Y_i(u)/a_N, \tag{4.15d}$$

where $Y_i(u)$ is defined in (3.26) and a_N in (4.2). From (4.1) and (4.15d) we obtain

$$E_i=P^{S(N)}_{(i-1)k+1}P^{S(N)}_{ik+1}\bar{E}_i^{(N)}P^{S(N)}_{(i-1)k+1}P^{S(N)}_{ik+1}. \tag{4.15e}$$

The operators G_i and E_i can be considered to satisfy the following relations:

$$(G_i-C_1)(G_i-C_2)\cdots(G_i-C_N)=0, \tag{4.16a}$$
$$E_i=\frac{(t^{(N-1)/2}+\cdots+t^{-(N-1)/2})}{(C_N-C_1)\cdots(C_N-C_{N-1})}(G_i-C_1)\cdots \times(G_i-C_{N-1}), \tag{4.16b}$$
$$E_i^2=(t^{(N-1)/2}+\cdots+t^{-(N-1)/2})E_i, \tag{4.16c}$$
$$G_iE_i=C_NE_i, \tag{4.16d}$$
$$G_iG_{i+1}G_i=G_{i+1}G_iG_{i+1}, \tag{4.16e}$$
$$E_iE_{i\pm1}E_i=E_i, \tag{4.16f}$$
$$E_iG_{i\pm1}G_i=G_{i\pm1}G_iE_{i\pm1}=(-t^{1/2})^{(N-1)^2}E_iE_{i\pm1}, \tag{4.16g}$$
$$G_{i\pm1}E_iG_{i\pm1}=t^{(N-1)^2}G_i^{-1}E_{i\pm1}G_i^{-1}, \tag{4.16h}$$
$$G_{i\pm1}E_iE_{i\pm1}=(-t^{1/2})^{(N-1)^2}G_i^{-1}E_{i\pm1}, \tag{4.16i}$$
$$E_{i\pm1}E_iG_{i\pm1}=(-t^{1/2})^{(N-1)^2}E_{i\pm1}G_i^{-1}, \tag{4.16j}$$
$$E_iG_{i\pm1}E_i=C_N^{-1}(-t^{1/2})^{(N-1)^2}E_i=t^{-1/2(N-1)}E_i, \tag{4.16k}$$
$$G_iG_j=G_jG_i, \quad \text{for} \quad |i-j|\geq 2, \tag{4.16l}$$
$$E_iE_j=E_jE_i, \quad \text{for} \quad |i-j|\geq 2, \tag{4.16m}$$

where for $r=1, 2, \cdots, N$,

$$C_r=(-1)^{r+1}t^{N(N-1)/2-(N-r)(N-r+1)/2}. \tag{4.16n}$$

It is remarked that the relations (4.16h) ~(4.16k) are derived from the relations (4.16d)~(4.16g). We have confirmed the relations (4.16) up to $N=7$ explicitly using the operators $\{U_i\}$. These calculations are rather complicated. Here, we shall give another proof for the relations (4.16a)~(4.16m).

We can show that if the generalized skein relation (4.16a) holds for arbitrary N (note that the relation (4.16a) has been proved up to $N=7$[5)]), then the other relations (4.16b) ~(4.16k) are to be proved by using the algebra in the $N=2$ case and the properties[5)] of the symmetrizers. The point of the proof is that the braid-monoid operators G_i, E_i and the symmetrizers are all constructed from the operators of the algebra in the $N=2$ case. So we can decompose the operators G_i and E_i and the symmetrizer P_i for higher N into those for $N=2$ to prove the relations (4.16).

The outline of the proof is the followings. First we show that the symmetrizers P in the operator E_i moves as

$$\begin{aligned} E_i&=P_iP_{i+1}\bar{E}_i^{(N)}P_iP_{i+1}=P_iP_{i+1}\bar{E}_i^{(N)}P_i \\ &=P_iP_{i+1}\bar{E}_i^{(N)}P_{i+1}=P_{i+1}\bar{E}_i^{(N)}P_{i+1} \\ &=P_i\bar{E}_i^{(N)}P_{i+1}, \end{aligned} \tag{4.17}$$

where

$$P_i=P^{S(N)}_{(i-1)k+1}. \tag{4.18}$$

This property is verified recursively with N using the following relations:

$$X_i(-(N-2)\lambda)=1-\frac{r_{N-2}}{r_{N-1}}U_i, \qquad (4.19a)$$

where

$$r_N=t^{-(N-1)/2}\tau_N$$
$$=t^{-(N-1)/2}+t^{-(N-3)/2}+\cdots+t^{(N-1)/2}, \qquad (4.19b)$$

and

$$U_i=q^{1/2}P_i^-. \qquad (4.19c)$$

We have depicted a proof of (4.17) in Fig. 12. Next, using the relations (4.17), (4.19) and the relations (4.5) we show the relations (4.16c), (4.16d), (4.16f) and (4.16g) recursively with respect to N. Proof is shown in Figs. 13–16 respectively. To save the space, pictures for the $N=3$ case have been drawn. Finally, we show (4.16b). From (4.16d), the operator E_i is an eigenvector of G_i with eigenvalue C_N. From (4.16a), we find that the dimension of the vector space of G_i is N and the eigenvalues are non-degenerate. Then, E_i must have the form:

$$E_i=\text{const.}\times(G_i-C_1)(G_i-C_2)\cdots(G_i-C_{N-1}). \qquad (4.20)$$

The constant factor in (4.20) is determined from (4.16c). The relations (4.16h) ~ (4.16k) are derived from (4.16d) ~ (4.16g) and we have finished the proof.

4.3 *Connection to Reidemeister moves*

As in the $N=2$ case, we normalize the

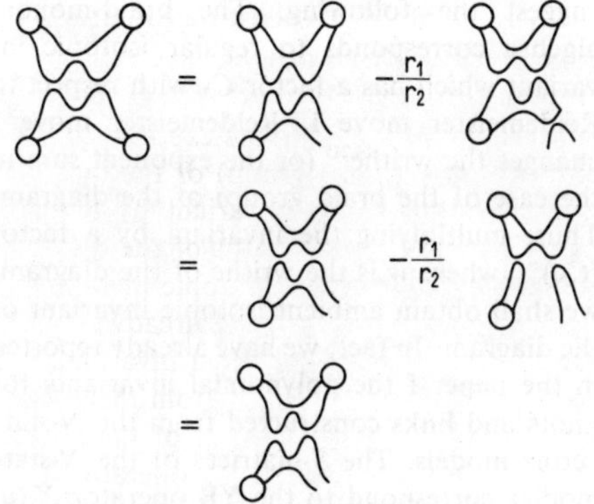

Fig. 12. Graphical proof of first equality of (4.17) for the $N=3$ case.

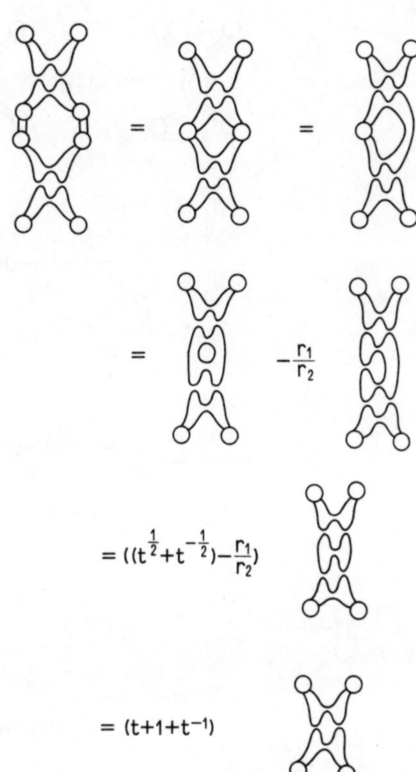

Fig. 13. Graphical proof of (4.16c) for the $N=3$ case.

operators G_i and E_i as

$$\hat{G}_i=(t^{-1/4})^{(N-1)^2}G_i, \qquad (4.21a)$$
$$\hat{E}_i=(-1)^{N-1}E_i. \qquad (4.21b)$$

Then we have the following relations:

$$(\hat{G}_i-\hat{C}_1)(\hat{G}_i-\hat{C}_2)\cdots(\hat{G}_i-\hat{C}_N)=0, \qquad (4.22a)$$

$$\hat{E}_i=(-1)^{N-1}\frac{(t^{(N-1)/2}+\cdots+t^{-(N-1)/2})}{(\hat{C}_N-\hat{C}_1)\cdots(\hat{C}_N-\hat{C}_{N-1})}\times(\hat{G}_i-\hat{C}_1)\cdots(\hat{G}_i-\hat{C}_{N-1}), \qquad (4.22b)$$

$$\hat{E}_i^2=(-1)^{N-1}(t^{(N-1)/2}+\cdots+t^{-(N-1)/2})\hat{E}_i, \qquad (4.22c)$$

$$\hat{G}_i\hat{E}_i=\hat{E}_i\hat{G}_i=\hat{C}_N\hat{E}_i, \qquad (4.22d)$$

$$\hat{G}_i\hat{G}_{i+1}\hat{G}_i=\hat{G}_{i+1}\hat{G}_i\hat{G}_{i+1}, \qquad (4.22e)$$

$$\hat{E}_i\hat{E}_{i\pm1}\hat{E}_i=\hat{E}_i, \qquad (4.22f)$$

$$\hat{E}_i\hat{G}_{i\pm1}\hat{G}_i=\hat{G}_{i\pm1}\hat{G}_i\hat{E}_{i\pm1}=\hat{E}_i\hat{E}_{i\pm1}, \qquad (4.22g)$$

$$\hat{G}_{i\pm1}\hat{E}_i\hat{G}_{i\pm1}=\hat{G}_i^{-1}\hat{E}_{i\pm1}\hat{G}_i^{-1}, \qquad (4.22h)$$

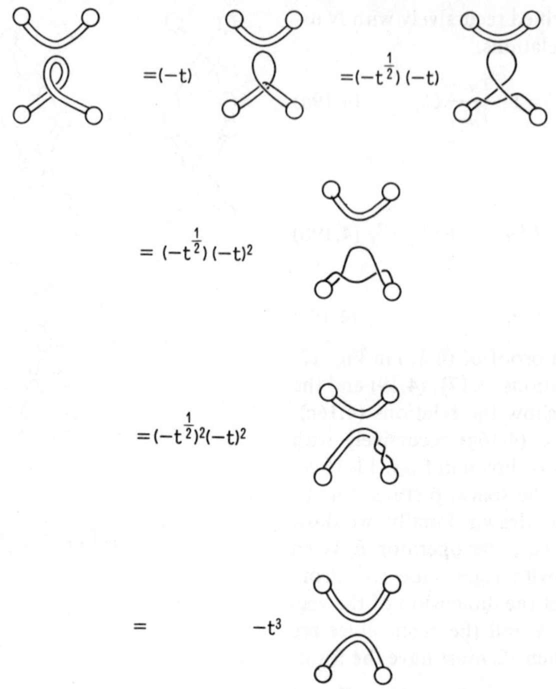

$=(-t)$ $=(-t^{\frac{1}{2}})(-t)$

$=(-t^{\frac{1}{2}})(-t)^2$

$=(-t^{\frac{1}{2}})^2(-t)^2$

$=$ $-t^3$

Fig. 14. Graphical proof of (4.16d) for the $N=3$ case.

$$\hat{G}_{i\pm1}\hat{E}_i\hat{E}_{i\pm1}=\hat{G}_i^{-1}\hat{E}_{i\pm1}, \tag{4.22i}$$

$$\hat{E}_{i\pm1}\hat{E}_i\hat{G}_{i\pm1}=\hat{E}_{i\pm1}\hat{G}_i^{-1}, \tag{4.22j}$$

$$\hat{E}_i\hat{G}_{i\pm1}\hat{E}_i=\hat{C}_N^{-1}\hat{E}_i, \tag{4.22k}$$

$$\hat{G}_i\hat{G}_j=\hat{G}_j\hat{G}_i, \quad \text{for} \quad |i-j|\geq 2, \tag{4.22l}$$

$$\hat{E}_i\hat{E}_j=\hat{E}_j\hat{E}_i, \quad \text{for} \quad |i-j|\geq 2, \tag{4.22m}$$

where for $r=1, 2, \cdots, N$,

$$\begin{aligned}\hat{C}_r&=C_r\cdot(t^{-1/4})^{(N-1)^2}\\&=(-1)^{r-1}t^{(N^2-1)/4-(N-r)(N-r+1)/2}.\end{aligned} \tag{4.22n}$$

Let us discuss the braid-monoid algebras from the viewpoint of the Reidemeister moves. There are three basic types of Reidemeister moves I, II and III[21] (Fig. 17). It was proved by Reidemeister that the equivalent link diagrams expressing the ambient isotopic links are mutually related by successive applications of the Reidemeister moves. Furtheremore, the Reidemeister moves II and III are generators of regular isotopy.[21] We apply the concept of the Reidemeister moves for the diagram of the braid-monoid algebra. Absence of constant factors in the relations (4.22e)~(4.22j) implies the invariance of the braid-monoid diagram under the regular isotopic moves. Only the relations (4.22d) and (4.22k) which correspond to Reidemeister move I have a factor $\hat{C}_N$. Thus normalized relations (4.22a)~(4.22n) suggest the following: The braid-monoid algebra corresponds to regular isotopic invariant which has a factor $\hat{C}_N$ with respect to Reidemeister move I. Reidemeister move I changes the writhe[21] (or the exponent sum in the case of the braid group) of the diagram. Thus, multiplying the invariant by a factor $(\hat{C}_N)^{-w}$ where w is the writhe of the diagram, we shall obtain ambient isotopic invariant of the diagram. In fact, we have already reported in the paper I the polynomial invariants for knots and links constructed from the N-state vertex models. The S-matrices of the N-state models correspond to the YB operator $X_i(u)$ ($N=2$) and the composite YB operator ($N\geq 3$). Therefore, the braid group representations ob-

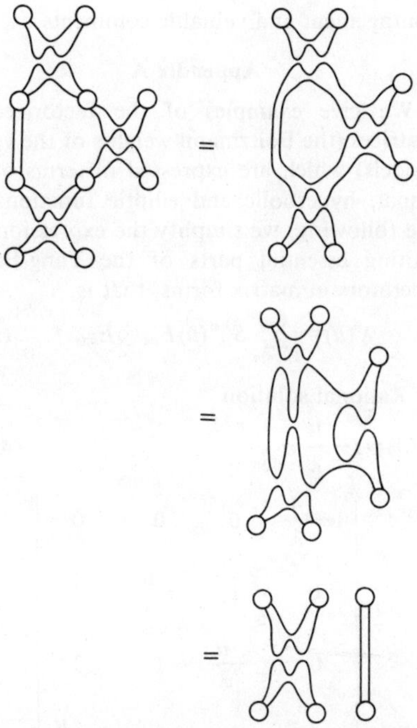

Fig. 15. Graphical proof of (4.16f) for the $N=3$ case.

tained from the N-state vertex models satisfy the braid-monoid algebras (4.16) if we introduce the monoid operators by (4.16b). It is an interesting problem whether there exist multi-variable extensions of the braid-monoid algebras (4.16).

§5. Summary

We summarize the results of the paper.

(1) The Yang-Baxter operators $\{X_i(u)\}$ constructed from the factorized S-matrices satisfy the following relations:

$$X_i(u)X_{i+1}(u+v)X_i(v)=X_{i+1}(v)X_i(u+v) \times X_{i+1}(u), \tag{5.1a}$$

$$X_i(u)X_j(v)=X_j(v)X_i(u), \quad \text{for} \quad |i-j|\geq 2. \tag{5.1b}$$

This fact is fundamental in our theory[1-6] and repeated here due to its importance. The Yang-Baxter operators (YB operators in short) generalize the generators of the braid group. The braid operators are expressed in terms of the YB operators:

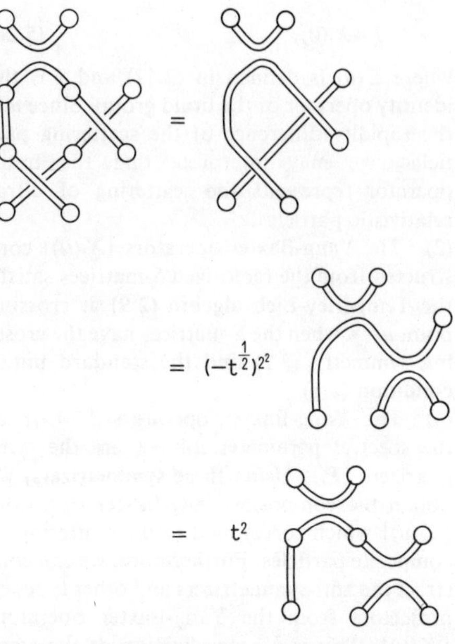

Fig. 16. Graphical proof of (4.16g) for the $N=3$ case.

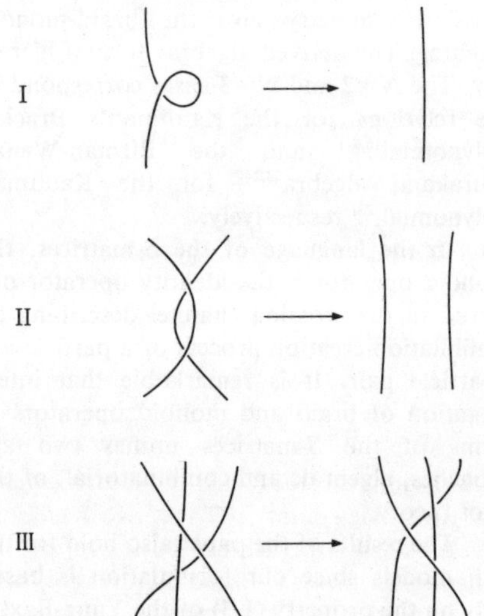

Fig. 17. Reidemeister moves I, II and III.

$$b_i=\lim_{u\to\infty} X_i(u)/\sqrt{C(u)}, \tag{5.2a}$$

$$b_i^{-1}=\lim_{u\to\infty} X_i(-u)/\sqrt{C(u)}, \tag{5.2b}$$

$$I=X_i(0), \tag{5.2c}$$

where $C(u)$ is defined in (2.12) and I is the identity operator of the braid group. Since u is the rapidity difference of the scattering particles, we may interprete that the braid operator represents the scattering of ultra-relativistic particles.

(2) The Yang-Baxter operators $\{X_i(u)\}$ constructed from the factorized S-matrices satisfy the Temperley-Lieb algebra (2.9) at crossing point $u=\lambda$ when the S-matrices have the crossing symmetry (2.2) and the standard initial condition (2.1).

(3) The Yang-Baxter operators $\{X_i(u)\}$ at the spectral parameter $u=-\lambda$ are the symmetrizers $\{P_i\}$. Using these symmetrizers, we obtain the composite Yang-Baxter operators $\{Y_i(u)\}$ which correspond to the scattering of composite particles. Furthermore, we can construct the anti-symmetrizers and other types of projectors from the Yang-Baxter operators $\{X_i(u)\}$. This is a generalization of the composite string representation.[5]

(4) The Yang-Baxter operators with the crossing symmetry give the braid-monoid algebras. The derived algebras form a hierarchy. The $N=2$ and $N=3$ cases correspond to the relations for the Kauffman's Bracket polynomial[16,21] and the Birman-Wenzl-Murakami algebra[23,24] for the Kauffman polynomial,[22] respectively.

(5) In the language of the S-matrices, the monoid operator is the identity operator observed in the crossing channel describing an annihilation-creation process of a particle-antiparticle pair. It is remarkable that interpretation of braid and monoid operators in terms of the S-matrices unifies two approaches, algebraic and combinatorial, of the knot theory.

(6) The results of the paper also hold for the IRF models since our formulation is based only on the property (1.3) of the Yang-Baxter operator $X_i(u)$.

Acknowledgements

The authors would like to express their sincere thanks to Professors C. N. Yang, K. Murasugi, L. H. Kauffman, C. A. Tracy, A. Jaffe and K. Aomoto for continuous encouragement and valuable comments.

Appendix A

We give examples of the factorized S-matrices (the Boltzmann weights of the vertex models) which are expressed in terms of rational, hyperbolic and elliptic functions. In the following, we simplify the expressions by writing essential parts of the Yang-Baxter operators in matrix forms, that is,

$$X(u)=\sum_{k,l,m,n} S_{ln}^{km}(u)E_{nk}\otimes E_{ml}. \tag{A·1}$$

(i) Rational solution

$$X(u)=I-\frac{u}{\lambda}\sigma \tag{A·2a}$$

$$=\begin{bmatrix} 1-\frac{u}{\lambda} & 0 & 0 & 0 \\ 0 & 1 & -\frac{u}{\lambda} & 0 \\ 0 & -\frac{u}{\lambda} & 1 & 0 \\ 0 & 0 & 0 & 1-\frac{u}{\lambda} \end{bmatrix}. \tag{A·2b}$$

This solution satisfies the standard initial condition at $u=0$ and the crossing symmetry (crossing point λ) with

$$\psi(k)=e^{i\pi k}, \qquad k=\pm\frac{1}{2}, \tag{A·3a}$$

and

$$F(u)=-1. \tag{A·3b}$$

(ii) Hyperbolic solution (3-state case)

$$X(u)=I+f(u)U \tag{A·4a}$$

$$=\begin{bmatrix} 1&0&0&0&0&0&0&0&0 \\ 0&1&0&0&0&0&0&0&0 \\ 0&0&a&0&b&0&c&0&0 \\ 0&0&0&1&0&0&0&0&0 \\ 0&0&b&0&d&0&e&0&0 \\ 0&0&0&0&0&1&0&0&0 \\ 0&0&c&0&e&0&g&0&0 \\ 0&0&0&0&0&0&0&1&0 \\ 0&0&0&0&0&0&0&0&1 \end{bmatrix}, \tag{A·4b}$$

where

$$a=1+f(u)\gamma^{-1},\quad b=+f(u)\gamma^{-1/2},\quad c=+f(u),$$
$$d=1+f(u),\quad e=+f(u)\gamma^{1/2},\quad g=1+f(u)\gamma, \tag{A·4c}$$

and

$$f(u)=\frac{\operatorname{sh}(u)}{\operatorname{sh}(\lambda-u)}. \tag{A.4d}$$

This solution satisfies the standard initial condition at $u=0$ and the crossing symmetry (crossing point λ) with

$$\psi(k)=\gamma^k,\quad k=-1,0,1, \tag{A·5a}$$
$$2\operatorname{ch}\lambda=\gamma+1+\gamma^{-1}, \tag{A·5b}$$

and

$$F(u)=f(u). \tag{A·5c}$$

(iii) Eight vertex model

$$X(u)=\begin{bmatrix} a & 0 & 0 & d \\ 0 & b & c & 0 \\ 0 & c & b & 0 \\ d & 0 & 0 & a \end{bmatrix}, \tag{A·6a}$$

where

$$a=\frac{\operatorname{sn}(\lambda-u)}{\operatorname{sn}\lambda},\quad b=1,$$
$$c=\frac{\operatorname{sn}u}{\operatorname{sn}\lambda},\quad d=k\operatorname{sn}u\operatorname{sn}(\lambda-u),$$

k: modulus. (A·6b)

This solution satisfies the standard initial condition at $u=0$ and the crossing symmetry (crossing point λ) with

$$\psi(j)=1,\quad j=\pm\frac{1}{2}, \tag{A·7a}$$

and

$$F(u)=1. \tag{A·7b}$$

Appendix B

We shall prove (3.25). In the case of $N=4$, a calculation shows that

$$\begin{aligned}(P_1^{S(4)})^2&=\{X_1(-\lambda)X_2(-2\lambda)X_1(-\lambda)\}^2\\
&=\{X_1(-\lambda)X_2(-2\lambda)X_1(-\lambda)\}X_1(-\lambda)X_2(-2\lambda)X_1(-\lambda)\\
&=\{X_1(-\lambda)X_2(-2\lambda)X_1(-\lambda)\}X_2(-2\lambda)X_1(-\lambda)\\
&=\{X_2(-\lambda)X_1(-2\lambda)X_2(-\lambda)\}X_2(-2\lambda)X_1(-\lambda)\\
&=\{X_2(-\lambda)X_1(-2\lambda)X_2(-\lambda)\}X_1(-\lambda)\\
&=\{X_1(-\lambda)X_2(-2\lambda)X_1(-\lambda)\}X_1(-\lambda)\\
&=X_1(-\lambda)X_2(-2\lambda)X_1(-\lambda)=P_1^{S(4)}.\end{aligned} \tag{B·1}$$

In general, we prove (3.25) by a method of induction. It is true for $N=2$. Suppose the relation $(P_i^{S(N-1)})^2=P_i^{S(N-1)}$. Then making use of (3.21), (3.22) and the Yang-Baxter relation we have

$$\begin{aligned}(P_i^{S(N)})^2&=P_i^{S(N-1)}X_{i+N-3}(-(N-2)\lambda)P_i^{S(N-1)}\cdot P_i^{S(N-1)}X_{i+N-3}(-(N-2)\lambda)P_i^{S(N-1)}\\
&=P_{i\cdots i+N-4}^{S(N-1)}X_{i+N-3}(-(N-2)\lambda)P_{i\cdots i+N-4}^{S(N-1)}\cdot X_{i+N-3}(-(N-2)\lambda)P_i^{S(N-1)}\\
&=P_{i+1\cdots i+N-3}^{S(N-1)}X_i(-(N-2)\lambda)P_{i+1\cdots i+N-3}^{S(N-1)}\cdot X_{i+N-3}(-(N-2)\lambda)P_i^{S(N-1)}\\
&=P_{i+1\cdots i+N-3}^{S(N-1)}X_i(-(N-2)\lambda)P_{i+1\cdots i+N-3}^{S(N-1)}\cdot P_i^{S(N-1)}\\
&=P_{i\cdots i+N-4}^{S(N-1)}X_{i+N-3}(-(N-2)\lambda)P_{i\cdots i+N-4}^{S(N-1)}\cdot P_{i\cdots i+N-4}^{S(N-1)}\\
&=P_{i\cdots i+N-4}^{S(N-1)}X_{i+N-3}(-(N-2)\lambda)P_{i\cdots i+N-4}^{S(N-1)}\\
&=P_{i\cdots i+N-3}^{S(N)},\end{aligned} \tag{B·2}$$

which completes the proof.

Appendix C

We show that the composite YB operator $Y_i(u)$ also has the standard initial condition. In the limit $u\to\lambda$, $Y_i(u)$ has the form shown in (4.1a). Note that $X_i(u)$ defined in (3.7) has the crossing symmetry (2.2) with $F(u)=f(u)=\text{sh}\,u/\text{sh}(\lambda-u)$. For the composite operator, a factor

$$\prod_{n=-(N-2)}^{(N-2)} (f(u+n\lambda))^{N-1-|n|}, \tag{C·1}$$

appears in place of $F(u)=f(u)$ in the crossing symmetry. The operator E_i in (4.15e) is nothing but the "composite operator" of U_i so it has matrix representation expressed as (2.8) where the spin variables are considered as those of composite particles. From the matrix representation (2.8) of the composite YB operator at crossing point λ and the crossing symmetry of the composite S-matrix, we conclude that the composite operator $Y_i(u)$ has the standard initial condition.

Appendix D

We shall prove that the operator P defined in (3.47) is idempotent; $P^2=P$. We define operator $\tilde{P}$ by

$$\tilde{P}=P^{S(n-m+2)}_{m\cdots n-1}\cdot P^{A(m+1)}_{1\cdots m-1}. \tag{D·1}$$

We sometimes denote $n-m+1$ by q. Using (3.21), (3.22), (3.25) and the Yang-Baxter relation, we find that

$$\begin{aligned}
(\tilde{P})^2&=P^{S(n-m+2)}_{m\cdots n-1}P^{A(m+1)}_{1\cdots m-1}\cdot P^{S(n-m+2)}_{m\cdots n-1}P^{A(m+1)}_{1\cdots m-1}\\
&=P^{S(q+1)}_{m\cdots n-1}\cdot P^{A(m)}_{1\cdots m-2}h^{A(m+1)}_{m-1}P^{A(m)}_{1\cdots m-2}\cdot P^{S(q)}_{m\cdots n-2}h^{S(q+1)}_{n-1}P^{S(q)}_{m\cdots n-2}\cdot P^{A(m)}_{1\cdots m-2}h^{A(m+1)}_{m-1}P^{A(m)}_{1\cdots m-2}\\
&=P^{S(q+1)}_{m\cdots n-1}\cdot P^{A(m)}_{2\cdots m-1}h^{A(m+1)}_{1}P^{A(m)}_{2\cdots m-1}\cdot P^{S(q)}_{m+1\cdots n-1}h^{S(q+1)}_{m}P^{S(q)}_{m+1\cdots n-1}\cdot P^{A(m)}_{2\cdots m-1}h^{A(m+1)}_{1}P^{A(m)}_{2\cdots m-1}\\
&=P^{S(q+1)}_{m\cdots n-1}\{P^{A(m)}_{2\cdots m-1}h^{A(m+1)}_{1}P^{A(m)}_{2\cdots m-1}\}P^{-}_{m-1}\cdot\{P^{S(q)}_{m+1\cdots n-1}h^{S(q+1)}_{m}P^{S(q)}_{m+1\cdots n-1}\}\\
&\quad\cdot P^{-}_{m-1}\{P^{A(m)}_{2\cdots m-1}h^{A(m+1)}_{1}P^{A(m)}_{2\cdots m-1}\}\\
&=P^{S(q+1)}_{m\cdots n-1}\{P^{A(m)}_{2\cdots m-1}h^{A(m+1)}_{1}P^{A(m)}_{2\cdots m-1}\}\cdot P^{S(q)}_{m+1\cdots n-1}\{P^{-}_{m-1}h^{S(q+1)}_{m}P^{-}_{m-1}\}P^{S(q)}_{m+1\cdots n-1}\\
&\quad\cdot\{P^{A(m)}_{2\cdots m-1}h^{A(m+1)}_{1}P^{A(m)}_{2\cdots m-1}\}\\
&=P^{S(q+1)}_{m\cdots n-1}\{P^{A(m)}_{2\cdots m-1}h^{A(m+1)}_{1}P^{A(m)}_{2\cdots m-1}\}\cdot\{P^{-}_{m-1}h^{S(q+1)}_{m}P^{-}_{m-1}\}P^{S(q)}_{m+1\cdots n-1}\cdot\{P^{A(m)}_{2\cdots m-1}h^{A(m+1)}_{1}P^{A(m)}_{2\cdots m-1}\}\\
&=P^{S(q+1)}_{m\cdots n-1}\{P^{A(m)}_{2\cdots m-1}h^{A(m+1)}_{1}P^{A(m)}_{2\cdots m-1}\}\left\{\frac{r_3r_{q-1}-r_2r_{q-2}}{r_2r_q}P^{-}_{m-1}-\frac{r_3r_{q-1}}{r_2r_q}P^{A(4)}_{m-1,m}\right\}\\
&\quad\cdot P^{S(q)}_{m+1\cdots n-1}\{P^{A(m)}_{2\cdots m-1}h^{A(m+1)}_{1}P^{A(m)}_{2\cdots m-1}\}\\
&=P^{S(q+1)}_{m\cdots n-1}P^{S(q)}_{m+1\cdots n-1}\{P^{A(m)}_{2\cdots m-1}h^{A(m+1)}_{1}P^{A(m)}_{2\cdots m-1}\}\cdot\left(\frac{r_3r_{q-1}-r_2r_{q-2}}{r_2r_q}\right)\cdot\{P^{A(m)}_{2\cdots m-1}h^{A(m+1)}_{1}P^{A(m)}_{2\cdots m-1}\}\\
&=\frac{r_3r_{q-1}-r_2r_{q-2}}{r_2r_q}P^{S(q+1)}_{m\cdots n-1}\{P^{A(m)}_{2\cdots m-1}h^{A(m+1)}_{1}P^{A(m)}_{2\cdots m-1}\}\cdot\{P^{A(m)}_{2\cdots m-1}h^{A(m+1)}_{1}P^{A(m)}_{2\cdots m-1}\}\\
&=\frac{r_3r_{q-1}-r_2r_{q-2}}{r_2r_q}P^{S(q+1)}_{m\cdots n-1}P^{A(m+1)}_{1\cdots m-1}P^{A(m+1)}_{1\cdots m-1}\\
&=\frac{r_3r_{q-1}-r_2r_{q-2}}{r_2r_q}\tilde{P}.
\end{aligned} \tag{D·2}$$

Thus, $P=\{r_2r_q/(r_3r_{q-1}-r_2r_{q-2})\}\,\tilde{P}$ satisfies $P^2=P$.

References

1) Y. Akutsu and M. Wadati: J. Phys. Soc. Jpn. **56** (1987) 839.

2) Y. Akutsu and M. Wadati: J. Phys. Soc. Jpn. **56** (1987) 3039. This paper is part I of the series and contains references for earlier developments.

3) Y. Akutsu, T. Deguchi and M. Wadati: J. Phys. Soc. Jpn. **56** (1987) 3464. This paper is part II of the series.
4) Y. Akutsu and M. Wadati: (to appear in) Commun. Math. Phys.
5) T. Deguchi, Y. Akutsu and M. Wadati: J. Phys. Soc. Jpn. **57** (1988) 757. This paper is part III of the series.
6) Y. Akutsu, T. Deguchi and M. Wadati: J. Phys. Soc. Jpn. **57** (1987) 1173. This paper is part IV of the series.
7) C. N. Yang: Phys. Rev. Lett. **19** (1967) 1312.
8) R. J. Baxter: Ann. Phys. **70** (1972) 193.
9) M. Karowski, H. J. Thun, T. T. Truong and P. H. Weisz: Phys. Lett. **67B** (1977) 321; A. B. Zamolodchikov and A. B. Zamolodchikov: Ann. Phys. (USA) **120** (1979) 253; K. Sogo, M. Uchinami, A. Nakamura and M. Wadati: Prog. Theor. Phys. **66** (1982) 1284; K. Sogo, M. Uchinami, Y. Akutsu and M. Wadati: Prog. Theor. Phys. **68** (1982) 508.
10) For example, see K. Nishijima: *Fundamental Particles* (W. A. Benjamin, 1964) p. 137.
11) H. N. V. Temperley and E. H. Lieb: Proc. R. Soc. London **A322** (1971) 251.
12) R. J. Baxter: *Exactly Solved Models in Statistical Mechanics* (Academic Press, London, 1982).
13) J. H. H. Perk and F. Y. Wu: Physica **138A** (1986) 100.
14) A. Kuniba, Y. Akutsu and M. Wadati: J. Phys. Soc. Jpn. **55** (1986) 3285.
15) P. P. Kulish and E. K. Sklyanin: Lecture Notes in Physics (Springer-Verlag, New York, 1082) Vol. 151, p. 61. See also, V. G. Drinfeld: Soviet Math. Dokl. **32** (1985) 254.
16) V. F. R. Jones: Bull. Amer. Math. Soc. **12** (1985) 103.
17) N. Bourbaki: *Groupes et algebras de Lie* (Hermann, Paris, 1968) Chap. 4.
18) M. Hamermesh: *Group Theory and its Application to Physical Problems* (Addison-Wesley., 1962) p. 243.
19) O. Babelon, H. J. de Vega and C. M. Viallet: Nucl. Phys. **B190** (1981) 542.
20) C. Itzykson and J. Zuber: *Quantum Field Theory* (McGraw-Hill, 1980) p. 513.
21) L. H. Kauffman: *On Knots* (Princeton University Press, 1987).
22) L. H. Kauffman: Topology **26** (1987) 395. L. H. Kauffman and P. Vogel: Link Polynomials and a Graphical Calculus, preprint.
23) J. S. Birman and H. Wenzl: Braids, Link Polynomials and a new Algebra, preprint.
24) J. Murakami: The Kauffman Polynomial of Links and Representation Theory, preprint.

Topology Vol. 26, No. 3, pp. 395–407, 1987

STATE MODELS AND THE JONES POLYNOMIAL

LOUIS H. KAUFFMAN

(*Received in revised form* 1 *September* 1986)

§1. INTRODUCTION

IN THIS PAPER I construct a state model for the (original) Jones polynomial [5]. (In [6] a state model was constructed for the Conway polynomial.)

As we shall see, this model for the Jones polynomial arises as a normalization of a regular isotopy invariant of unoriented knots and links, called here the *bracket polynomial*, and denoted $\langle K \rangle$ for a link projection K. The concept of regular isotopy will be explained below. The bracket polynomial has a very simple state model.

In §2 (Theorem 2.10) I use the bracket polynomial to prove (via Proposition 2.9 and an observation of Kunio Murasugi) that the number of crossings in a connected, reduced alternating projection of a link L is a topological invariant of L. (A projection is reduced if it has no isthmus in the sense of Fig. 5.) In other words, *any two connected, reduced alternating projections of the link* L *have the same number of crossings.* This is a remarkable application of our technique. It solves affirmatively a conjecture going back to the knot tabulations of Tait, Kirkman and Little over a century ago (see [16], [9], [10]).

Along with this application to alternating links, we also use the bracket polynomial to obtain a necessary condition for an alternating reduced link diagram to be ambient isotopic to its mirror image (Theorem 3.1). One consequence of this theorem is that a reduced alternating diagram with twist number greater than or equal to one-third the number of crossings is necessarily chiral.

The paper is organized as follows. In §2 the bracket polynomial is developed, and its relationship with the Jones polynomial is explained. This provides a self-contained introduction to the Jones polynomial and to our techniques. The last part of §2 contains the applications to alternating knots, and to bounds on the minimal and maximal degrees of the polynomial. §3 contains the results about chirality of alternating knots. §4 discusses the structure of our state model in the case of braids. Here the states have an algebraic structure related to Jones's representation of the braid group into a Von Neumann Algebra.

§2. BRACKET INVARIANT

We first describe a general scheme of calculation from unoriented knot and link diagrams. This scheme associates a polynomial in three variables A, B and d to each diagram. It is well-defined on equivalence classes of diagrams. Two diagrams are *equivalent* if their underlying planar graphs are equivalent under orientation preserving homeomorphisms of the plane. Note that the Reidemeister moves change the graphical structure.

We distinguish three relations on diagrams: equivalence (as above), ambient isotopy and regular isotopy. Two diagrams are *ambient isotopic* if one can be obtained from the other by a sequence of Reidemeister moves of type I, type II and type III (see Fig. 1) plus equivalence as

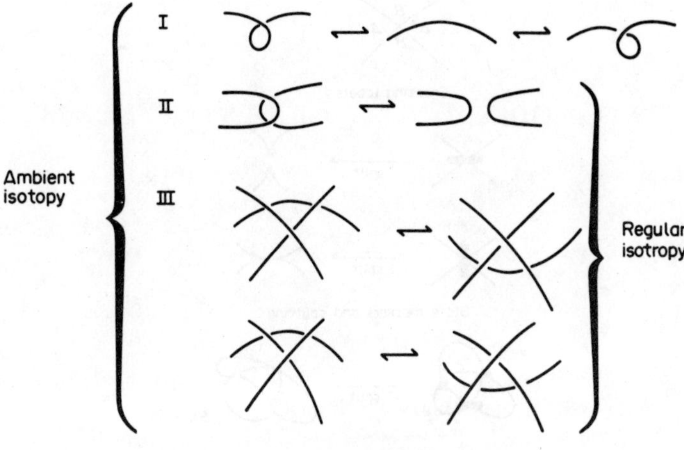

Fig. 1.

defined above. Two diagrams are *regularly isotopic* if they are *ambient isotopic without the use of the type I move*. Regular isotopy turns out to be a convenient concept for us.

DEFINITION 2.1. *Let K be an unoriented knot or link diagram. Let* $\langle K \rangle$ *be the element of the ring* $Z[A, B, d]$ *defined by means of the rules*:

(i) $\langle 0 \rangle = 1$
(ii) $\langle 0 \cup K \rangle = d\langle K \rangle$, K not empty
(iii) $\langle \times \rangle = A\langle \asymp \rangle + B\langle)(\rangle$.

Remark. A formula may involve the bracket and a few small diagrams. *These small figures represent larger diagrams that differ only as indicated in the small diagrams.*

The bracket, $\langle K \rangle$, is well defined on diagrams, but it is not invariant under any of the Reidemeister moves. It is the purpose of this section to determine relations among A, B, d so that $\langle K \rangle$ becomes invariant under Reidemeister moves. We will obtain invariance under types II and III, hence the use of regular isotopy.

Some comments are in order about the rules: Rule (i) says that $\langle K \rangle$ takes the value 1 on a single unknotted circle diagram. Rule (ii) says that $\langle K \rangle$ is multiplied by d in the presence of a disjoint circular component. This component can surround other parts of the diagram. Rule (iii) applies to diagrams that differ locally at the site of a single crossing. We can use rule (iii) to keep expanding the formulas until we reach diagrams consisting of disjoint unions of circles (Jordan curves in the plane). Rules (ii) and (iii) then imply that the value of $\langle K \rangle$ on a disjoint collection of circles is d raised to one less than the cardinality of the collection.

Note that rule (iii) entails the formula

$$\langle \times \rangle = B\langle \asymp \rangle + A\langle)(\rangle.$$

In fact, we can create a mnemonic for this expansion by labelling the crossings as shown in Fig. 2. *This label* A *marks the two local regions swept out by turning the overcrossing line counterclockwise until it coincides with the undercrossing line.*

For the expansion formula (iii) we can indicate which way a crossing is to be split by scoring a *marker* on it that connects the two regions that will be joined by the splitting. See Fig. 2.

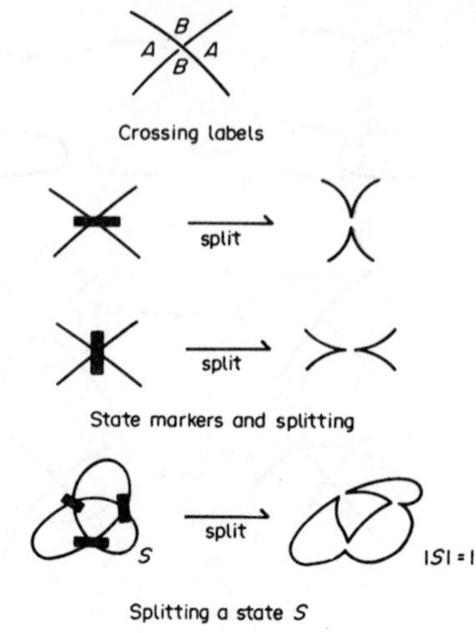

Fig. 2.

If U is the underlying planar graph for K, then a *state of* U is a choice of splitting marker for every vertex of U. Again see Fig. 2. I choose to call the underlying planar graph for a diagram K the *universe* for K (see [6]). This terminology distinguishes the underlying planar graph from the link projection and from other graphs that can arise. Thus we speak of the states of a universe.

Since splitting all the vertices of a state results in a configuration of disjoint circles, we see that the states are in one-to-one correspondence with final configurations in the expansion of the bracket. Accordingly, we define $\langle K|S\rangle$ for a diagram K and a state S by the formula

$$\langle K|S\rangle = A^i B^j$$

where i is the number of state markers touching A labels, and j is the number of state markers touching B labels. The total contribution of a given state to the polynomial is then given by the formula.

$$\langle K|S\rangle\, d^{|S|-1}$$

where $|S|$ denotes the number of circles in the splitting of S. View Fig. 2.

These observations are summarized in the statement of the following proposition, whose proof we omit.

PROPOSITION 2.2. $\langle \mathrm{K}\rangle$ *is uniquely determined on diagrams by the rules* (i), (ii), (iii). *It is given by the formula*

$$\langle \mathrm{K}\rangle = \sum_{\mathrm{S}} \langle \mathrm{K}|\mathrm{S}\rangle\, \mathrm{d}^{|\mathrm{S}|-1}$$

where this summation is over all states of the diagram, and S *denotes the number of components in the splitting of a state* S.

We now see how $\langle K\rangle$ behaves under elementary diagram moves, and consequently determine how to adjust A, B, and d to obtain a topological invariant.

LEMMA 2.3. *The following formula holds, where the three diagrams represent the same projection except in the area indicated.*

$$\langle \asymp\!\!\!\!\supset\!\subset \rangle = AB\langle \supset\subset \rangle + (ABd + A^2 + B^2)\langle = \rangle.$$

Hence $\langle \asymp\!\!\!\!\supset\!\subset \rangle = \langle \supset\subset \rangle$ *for all diagrams if*

$$AB = 1 \quad \textit{and} \quad d = -A^2 - A^{-2}.$$

Proof.

$$\langle \; \rangle = A\langle \; \rangle + B\langle \; \rangle.$$

Thus
$$\langle \; \rangle = A\left[B\langle \; \rangle + A\langle \; \rangle\right]$$
$$+ B\left[B\langle \; \rangle + A\langle \; \rangle\right].$$

Hence
$$\langle \; \rangle = (ABd + A^2 + B^2)\langle = \rangle + AB\langle \supset\subset \rangle.$$

This completes the proof.

LEMMA 2.4. *Type II invariance for* $\langle\ \rangle$ *implies type III invariance.*

Proof.

$$\langle \; \rangle = A\langle \; \rangle + B\langle \; \rangle$$
$$= A\langle \; \rangle + B\langle \; \rangle$$

(by II-invariance)

$$= \langle \; \rangle.$$

Hence $\langle \; \rangle = \langle \; \rangle$. This is type III invariance.

Thus we see that by choosing

$$B = A^{-1}, d = -A^2 - A^{-2}$$

$\langle K \rangle$ becomes a Laurent polynomial in A, and it is an invariant of regular isotopy (i.e. invariant under moves of type II and III). It is not invariant under the type I moves, but behaves as follows:

PROPOSITION 2.5. *With*

$$B = A^{-1}, d = -A^2 - A^{-2}$$

then

$$\langle \; \rangle = (-A^3)\langle - \rangle$$

$$\langle \; \rangle = (-A^{-3})\langle - \rangle.$$

Proof. This is a direct calculation, and is omitted.

From now on, unless otherwise specified, we assume that A, B *and* d *are chosen as indicated in Proposition* 2.5. With these choices, the bracket polynomial is an invariant of regular isotopy for unoriented knots and links.

The simplest invariant of regular isotopy for oriented diagrams is the *twist number* (or writhe) $w(K)$. This is the sum of the signs of all the crossings where each crossing is given a sign of plus or minus 1 according to the conventions shown in Fig. 3.

To obtain an invariant of ambient isotopy for oriented knots and links we define a Laurent polynomial $f[K]$ by the formula

$$f[K]=(-A)^{-3w(K)}\langle K\rangle,$$

where $w(K)$ denotes the twist number of the diagram K. The bracket is defined on oriented diagrams by forgetting the orientation.

THEOREM 2.6. *The polynomial* $\mathrm{f[K]}\in \mathrm{Z[A,A^{-1}]}$ *defined above is an ambient isotopy invariant for oriented links* K.

Proof. By combining the behaviour of the twist number under type I Reidemeister moves with the behaviour of the bracket (Proposition 2.5), it follows that $f[K]$ is invariant under type I moves. Thus $f[K]$ is invariant under all three moves, and is therefore an invariant of ambient isotopy.

Both the bracket and the polynomial $f[K]$ behave as follows on taking mirror images:

PROPOSITION 2.7. *Let* K! *denote the mirror image of* K *(obtained by reversing all the crossings of* K*). Then*

$$\langle \mathrm{K!}\rangle\,(\mathrm{A})=\langle \mathrm{K}\rangle\,(\mathrm{A}^{-1})$$

$$\mathrm{f[K!]}\,(\mathrm{A})=\mathrm{f[K]}\,(\mathrm{A}^{-1}).$$

Proof. Just note that switching all crossings results in the replacement of every appearance of A by its inverse in the expansion of the bracket. This proves the first part. Since the twist number of a mirror image is the negative of the twist number of the original, the second part follows as well.

The Jones polynomial

Now recall that the Jones polynomial [5] is defined by the identities:

$$V_{\bigcirc}=1$$

$$t^{-1}V_{\times}-tV_{\times}=\left(\sqrt{t}-\frac{1}{\sqrt{t}}\right)V_{=}\,.$$

The Jones polynomial is an ambient isotopy invariant of oriented knots and links.

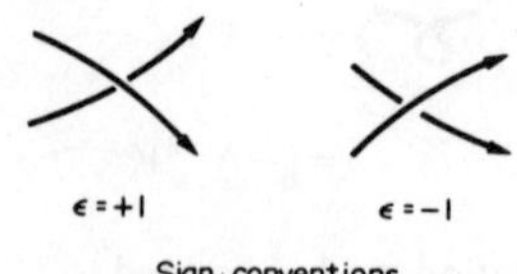

Sign conventions

Fig. 3.

THEOREM 2.8.

$$V_K(t) = f[K](t^{-1/4}).$$

Proof.

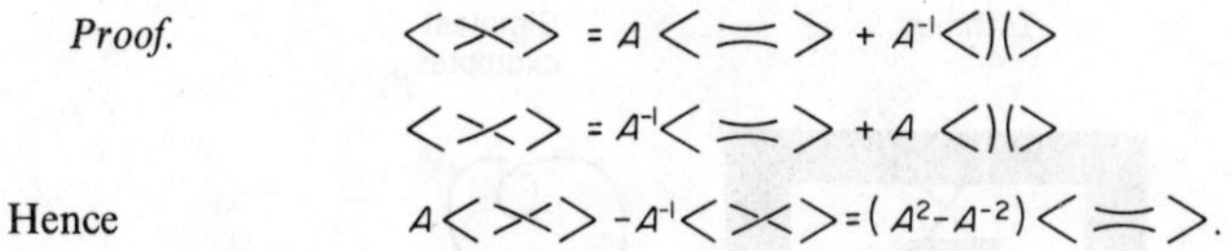

Multiplying this last formula by appropriate writhes, we have

The result follows at once by substituting t raised to the negative one-quarter power for A.

Figure 4 illustrates the calculation of the bracket for the Hopf link and for the trefoil knot. This calculation, in conjunction with Proposition 2.7, produces a short elementary proof of the distinction between the trefoil and its mirror image.

Remark. Some formal results about the Jones polynomial (reversing formula [11], [12], Birman "infinity"-formula [21]) follow immediately and trivially from Theorem 2.8 and the definition of the bracket polynomial. We leave the verification of these relations as an exercise for the interested reader.

Alternating links

We now give applications to alternating links. The first result determines the highest and lowest degree terms in the bracket polynomial for an alternating diagram.

PROPOSITION 2.9. *Let* K *be an alternating knot or link diagram that is connected and reduced. Let* K *be shaded so that all the crossings are of shaded type* A *(the regions labelled* "A" *are shaded). Then the term in* $\langle K \rangle$ *of highest degree in* A *has degree* $V + 2W - 2$, *where* V *is the number of crossings in* K, *and* W *is the number of white regions for this shading. The co-efficient of this power of* A *in* $\langle K \rangle$ *is* $(-1)^{W-1}$.

With the same hypotheses, the lowest degree term has degree $-V - 2(B-1)$, *where* B *denotes the number of black (shaded) regions in the diagram. This term is monic with co-efficient* $(-1)^{B-1}$.

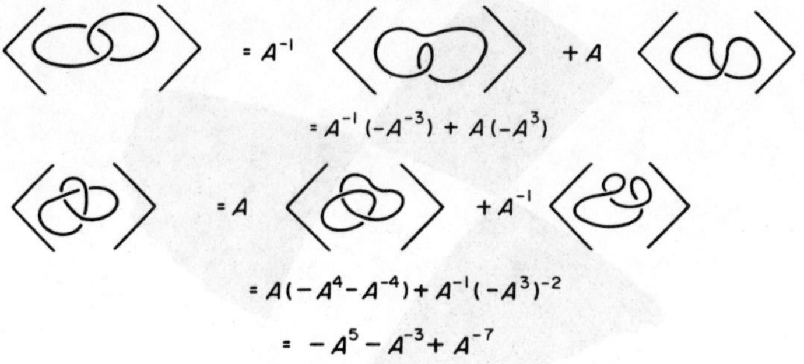

= A^{-1}
+ A
= $A^{-1}(-A^{-3}) + A(-A^{3})$
= A
+ A^{-1}
= $A(-A^{4}-A^{-4})+A^{-1}(-A^{3})^{-2}$
= $-A^{5}-A^{-3}+A^{-7}$

Sample bracket calculations

Fig. 4.

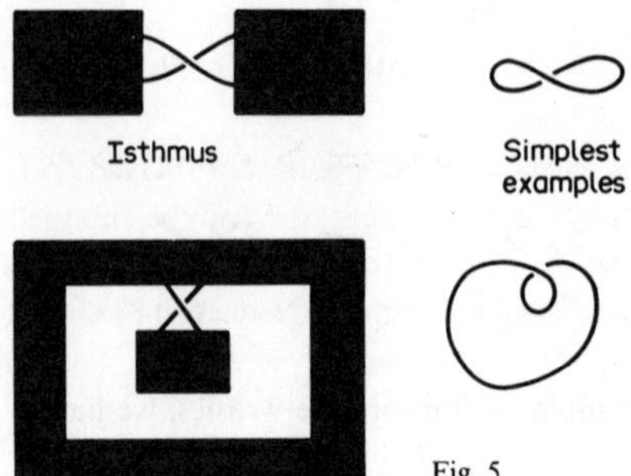

Fig. 5.

Before proving this result, some commentary on terminology is needed. A *reduced* diagram is one that does not contain an isthmus as shown in Fig. 5. An *isthmus* is a crossing in the diagram so that two of the four local regions at the crossing are part of the same region in the larger diagram.

Note that *a connected alternating diagram, when shaded in checkerboard fashion, has all of its crossings of the same shaded type* (see Fig. 6). By flipping the shading, if necessary, we may assume that it is the A-labelled regions that are shaded.

We shall call a diagram satisfying these hypotheses (connected, reduced) a *simple diagram*. Thus Proposition 2.9 says that if K is a simple diagram, then

$$\max\deg\langle K\rangle = V+2W-2$$

$$\min\deg\langle K\rangle = -V-2B+2.$$

The idea behind our pinpointing of the maximal degree is this: by choosing the state obtained by splitting every crossing in the A-direction, we obtain W components (where W is the number of white regions), and hence a degree of $V+2(W-1)$ from the corresponding part of the summation for the bracket polynomial.

Proof of 2.9. Let S be the state obtained by splitting every crossing in the diagram in the A-direction. Then $\langle K|S\rangle = A^V$, and $|S| = W$, where W is the number of white regions in the shading. Thus this state contributes the term

$$\langle K|S\rangle d^{|S|-1} = A^V d^{W-1}$$

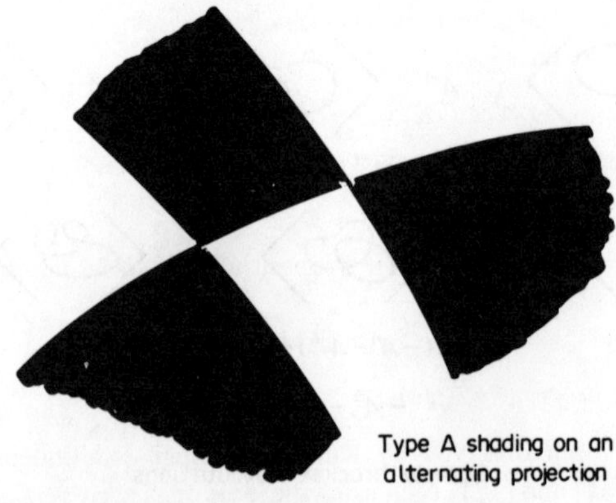

Fig. 6.

to the state expansion of $\langle S \rangle$. Since

$$d = -A^2 - A^{-2}$$

this means that the highest degree contribution of the state S is the degree $V + 2(W-1)$.

Now consider any other state S'. The state S' can be obtained from S by switching some subset of state markers of S. Thus there is a sequence of states $S(0), S(1), S(2), \ldots, S(n)$ so that $S = S(0)$, $S' = S(1)$, and $S(i+1)$ is obtained from $S(i)$ by switching one state marker from type A to type A^{-1}. Since a state marker of type A^{-1} contributes $(1/A)$, we see that $\langle K | S(i+1) \rangle = A^{-2} \langle K | S(i) \rangle$. Also, $|S(i+1)|$ is within 1 of $|S(i)|$, since switching a single state marker can change the component count of the split state by at most one. It follows that the maximal degree contribution of $S(i+1)$ is less than or equal to the maximal degree contribution of $S(i)$.

However, we assert that the maximal degree contribution actually falls from $S(0)$ to $S(1)$. This follows from the above assumption, if it is shown that *switching any state marker in* S *will cause a decrease in the number of components of the corresponding split state.* This follows from the assumption of diagram simplicity. (If the number of split components did not decrease from $S = S(0)$ to $S(1)$, then some white region would touch both sides of a crossing. This can only happen in the presence of an isthmus.)

Thus we have shown that the term of maximal degree in the entire bracket polynomial is contributed by the state S, and is not cancelled by terms from any other state. This completes the proof.

Finally we give the main application.

THEOREM 2.10. *The number of crossings in a simple alternating projection of a link* L *is a topological invariant of* L. *Hence any two simple alternating projections of a given link have the same number of crossings.*

Proof. Let *span* (K) denote the difference

$$\text{span } (K) = \text{maxdeg } \langle K \rangle - \text{mindeg } \langle K \rangle.$$

Then

$$\text{span } (K) = [V + 2W - 2] - [-V - 2B + 2]$$
$$= 2V + 2(W+B) - 4.$$

Since $W + B$ equals the total number of regions in the diagram, and this exceeds the number of crossings by two, we have

$$\text{span } (K) = 2y + 2(V+2) - 4 = 4V.$$

This completes the proof.

Remark. It is worth remarking that Proposition 2.9 can be generalized to imply an inequality:

$$\textit{span } (\text{K}) \leqslant 4\text{V},$$

for an arbitrary (not necessarily alternating) diagram K.

This result has been observed by Kunio Murasugi and (independently) Morwen Thistlethwaite (see [14] and [17]), each using the ideas of an early version of this paper. We give here our short proof via the:

DUAL STATE LEMMA 2.11. *Let* S *be a state for a connected universe* U. *Let* $\hat{S}$ *denote the state obtained from* S *by reversing all the state markers of* S *(call this the dual state of* S). *Then*

$$|S|+|\hat{S}| \leqslant R$$

where R *denotes the number of regions in* U.

Proof. The proof is by induction on the number of vertices V in U. It is easily seen to be true for $V=0, 1, 2$. Therefore suppose the result true for all universes with less than V vertices. Let U be connected, with V vertices. Let U' and U'' be the two universes obtained by splitting U at a given vertex P in the two possible ways. Then, by connectivity of U, one of U' or U'' is also connected. We may suppose that U' is connected. Apply the induction hypothesis to U'.

If S is a state of U, then either S or $\hat{S}$ is split at the vertex P in the same direction that formed U'. We can assume that S is so split. Then, by ignoring the site at P, S can be construed as a state S' of U'.

By induction, $|S'|+|\hat{S}'| \leqslant R'$ where R' denotes the number of regions of U'. By construction, $R'=R-1$ where R is the number of regions of U. And S' and S have the same number of split components: $|S|=|S'|$. On the other hand, it is possible that $\hat{S}$, being obtained from $\hat{S}'$ by splicing at the site p, may have (at most) one less split component than $\hat{S}'$. Thus $|\hat{S}'|+1 \geqslant \hat{S}$. These facts imply the inequality $|S|+|\hat{S}| \leqslant R$, completing the inductive proof of the Lemma.

The inequality mentioned prior to the proof of this lemma now follows by repeating the proof of 2.9, using the state S where all the markers are of type A. No relation with the checkerboard shading is required, nor do we need assume anything other than connectivity of the diagram. We then obtain:

$$\text{maxdeg } \langle K\rangle \leqslant V+2(|S|-1)$$

$$\text{mindeg } \langle K\rangle \geqslant -V-2(|\hat{S}|-1).$$

The lemma, in conjunction with the calculation of 2.10 then gives the inequality: span $(K) \leqslant 4V$.

Finally we note that both Murasugi and Thistlethwaite prove the stronger inequality: span $(K) < 4V$ when K is a connected non-alternating diagram. Wu [18] gives a proof of this inequality by strengthening our dual state Lemma.

§3. CHIRALITY OF ALTERNATING LINKS

Here we apply Proposition 2.9 to obtain a necessary condition for an oriented alternating link to be achiral (ambient isotopic to its mirror image).

THEOREM 3.1. *Let* K *be a simple alternating diagram shaded as in Proposition 2.9. Let* W *and* B *denote the number of white and black regions in this shading. Suppose that* K *has twist number* w(K). *If* K *is ambient isotopic to* K! *then*

$$3w(K)=W-B.$$

It follows from this formula that

$$W=(1/2)\,(V+3w(K))+1$$

$$B=(1/2)\,(V-3w(K))+1$$

where V *denotes the number of crossings in the diagram.*

Proof. We use the Jones polynomial in the form of the ambient isotopy invariant $f[K]$ (A) (see Theorem 2.8). It follows from 2.9 and the definition of $f[K]$ that $f[K]$ has maximal and minimal A-degrees *max* (f) and *min* (f) given by the formulas

$$\max(f) = -3w + V + 2(W-1)$$

$$\min(f) = -3w - V - 2(B-1)$$

where w is the twist number $w = w(K)$.

In order for K to be achiral it is necessary that $-min$ (f) $= max$ (f) (by Proposition 2.6). Thus

$$3w + V + 2(B-1) = -3w + V + 2(W-1).$$

Hence

$$3w = W - B.$$

The remaining formulas follow by using the relation $W + B = V + 2$. This completes the proof.

COROLLARY 3.2. *Let K be a simple alternating diagram. Let* $T = |w(K)|$. *Assume K is not the unknotted circle diagram. Then, if* $T \geqslant V/3$ (*V is the number of crossings in the diagram*) *then K is chiral.*

This corollary is an easy consequence of Theorem 3.1. We omit the proof. It follows at once from Corollary 3.2 that special alternating (simple) links are chiral. This fact was first proved by Murasugi [13], using his signature invariant.

Remark. There exist chiral alternating knots satisfying the condition $3w(K) = W - B$ of Theorem 3.1. Thus the condition of 3.1 is necessary but not sufficient for achirality. Figure 7 depicts such an example. A signature calculation shows that this knot (10_4 in Rolfsen's tables [15]) is chiral.

Note also that it is easy to produce an alternating prime knot of zero twist number that is chiral (remove two crossings from the top line of Fig. 7). Erica Flapan has observed the non-alternating knot 10_{125} to have the same property (chiral, twist number zero, prime).

Remark. It is entirely possible that the venerable conjecture (that the twist number is a topological invariant for reduced alternating diagrams) is true. This would generalize 3.1 except for the case of twist number zero. In this case, 3.1 demands $B = W$ for achirality. More may be true. We make the following conjecture.

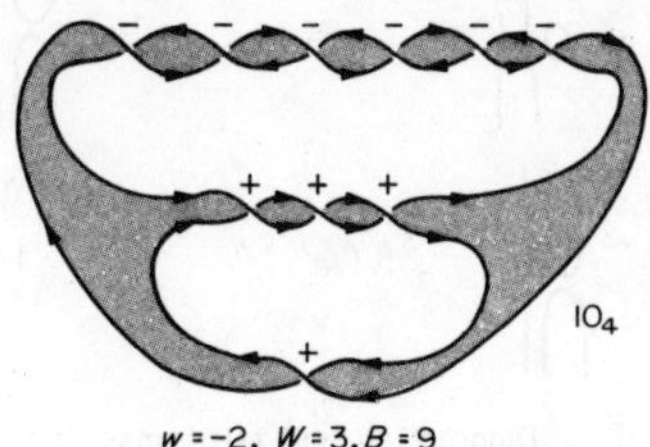

Fig. 7.

CONJECTURE. *Let* K *be a prime reduced alternating diagram, and suppose that* K *is achiral with twist number zero. Then the two planar graphs* B(K) *and* W(K) *are isomorphic as abstract graphs.*

Here $B(K)$ is the graph formed from the black shading: one vertex for each shaded region, one edge for each crossing shared by shaded regions. $W(K)$ is the planar dual to $B(K)$ formed from the white regions. (See Note added in Proof.)

§4. BRAID STATES AND THE DIAGRAM ALGEBRA

The bracket also provides an entry into the representation theory associated with the Jones polynomial. In order to see this we define a *diagram algebra*, $D[n]$, based on the patterns shown in Fig. 8. Here diagrams with free ends are multiplied as braids, while multiplication by the closed loop δ denotes disjoint union. Addition is formal with no imposed relations. As the figure shows, the resulting (multiplicative) monoid has relations

$$\begin{cases} h_i^2 = \delta h_i = h_i \delta \\ h_i h_{i+1} h_i = h_i \\ h_{i+1} h_i h_{i+1} = h_{i+1} \\ h_i h_j = h_j h_i, \quad |i-j| > 1. \end{cases}$$

(We omit the proof here that it has exactly these relations. See [7], [8].)

The original Jones polynomial was defined via a representation into an abstract algebra satisfying (essentially) these multiplicative relations. The diagram algebra forms the beginning of a direct geometric connection between these algebras and the theory of braids. In our terms the relation $\langle \asymp\!\!\!\!\times \rangle = A \langle \asymp \rangle + A^{-1} \langle\,)(\,\rangle$ becomes the pattern for a representation of the braid group (see [5]) into the free additive algebra with the above multiplicative relations and

$$\delta = d = -A^2 - A^{-2}$$

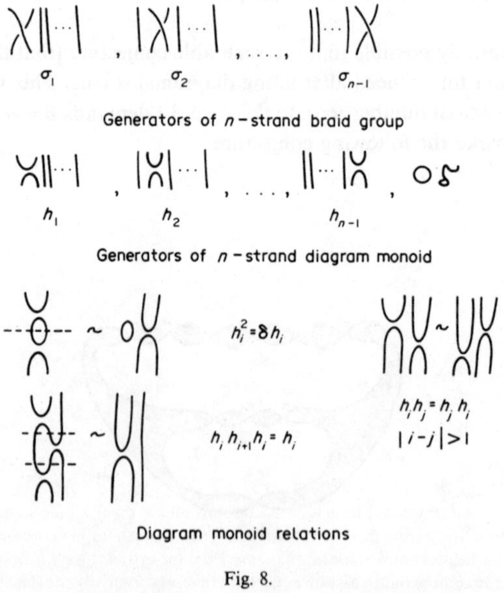

σ_1, σ_2, …, σ_{n-1}

Generators of n-strand braid group

h_1, h_2, …, h_{n-1}, δ

Generators of n-strand diagram monoid

$h_i^2 = \delta h_i$

$h_i h_{i+1} h_i = h_i$

$h_i h_j = h_j h_i$
$|i-j| > 1$

Diagram monoid relations

Fig. 8.

(specializing the loop variable). If σ_i is the ith braid generator then the representation is given by the formula

$$\rho(\sigma_i) = Ah_i + A^{-1}1.$$

Let $B[n]$ denote the n-strand braid group. A braid is a product of the generators

$$\sigma_1^{\pm 1}, \sigma_2^{\pm 1}, \ldots, \sigma_{n-1}^{\pm 1}.$$

(See Fig. 8.) For a braid b, let $\langle b \rangle$ denote the evaluation of the bracket polynomial on the closure of b.

Since the states corresponding to a given braid b are obtained by eliminating crossings horizontally $\langle \asymp \rangle$ or vertically $\langle\ )(\ \rangle$, we see that the generators for the diagram monoid correspond to horizontal splits on the generators of the braid group. (Vertical splits give identity braids.)

The relations in the diagram monoid allow one to algebraically determine the number of components in the closure of h—hence the value of $\langle h \rangle$, for any product h in $D[n]$. This gives a diagrammatic interpretation to trace computations in the representation theory.

Note how the bracket expansion and the form of the representation fit together. By representing each braid generator as a sum of two algebraic terms, the product corresponding to a braid word has a power-of-two number of terms. Each term is a power of A multiplied by a product of generators of the diagram algebra. Each such product corresponds to one of the states in the bracket expansion.

Finally it is worth mentioning that a rich generalization of the braid group is obtained by allowing products of braid generators with elements of the diagram monoid. The resulting system, up to regular isotopy, can be defined so that it has relations corresponding to standard braiding relations, diagram monoid relations, plus extra relations of the type illustrated in Fig. 9. Call this structure the *Braid Monoid.* (See [7] and [8].) Yetter [19] has studied a version of the braid monoid, and Birman and Wenzel [4] use an algebra derived from it to study representations of the braid group and the Kauffman polynomial [8].

This section has been an introduction to braid states and the diagram algebra. For relations with chromatic polynomials and the Potts model ([1], [5]), see [7]. (By associating an appropriate alternating link diagram to any planar graph, and by choosing A, B, d appropriately, dichromatic polynomials and partition functions (for the Potts model) can be computed using the bracket expansion. Thus our formalism provides a relationship among knot theory, graph theory and physics. In the case of braids, the simple topology of the diagram algebra underlies all three aspects.)

$h_{i+1}\sigma_i\sigma_{i+1} = h_{i+1}h_i = \sigma_i\sigma_{i+1}h_i$

A basic braid monoid relation

Fig. 9.

Acknowledgements—The author would like to take this opportunity to thank a number of people: Ken Millett, Ray Lickorish and Bob Brandt for sharing their work and ideas, Joan Birman for her continued intense support of this work, Vaughan Jones for helpful conversations, Massimo Ferri for extraordinary hospitality while this work was coming forth, Sostenes Lins for the right question at the right time, and for many conversations about combinatorics and three-dimensional topology, Mario Rasetti for insightful and patient conversations about knots and physics, Ivan Handler for clarification of Proposition 2.9, Murasugi for a conversation resulting in Theorem 2.10 and finally, M. Kervaire and the mathematics department at Geneva for a most stimulating visit.

Work on this paper was partially supported by CNR (Italy) while the author was Visiting Professor at the Dipartimento Matematica, Universita di Bologona, by ISI (Torino) while the author was visiting at Politecnico di Torino, and by ONR Grant No. N0014-84-K-0099 (Stereochemical Topology Project—University of Iowa).

Note added in Proof

The invariance of the twist number for reduced alternating diagrams has been proved (independently) by Murasugi and Thistlethwaite.

REFERENCES

1. R. J. Baxter: *Exactly solved models in statistical mechanics.* Academic Press, London, 1982.
2. J. S. Birman: Jones braid-plat formulae, and a new surgery triple, to appear.
3. J. S. Birman: *Braids, Links, and Mapping Class Groups.* Annals of Mathematics Studies No. 82, Princeton University Press, 1974.
4. J. S. Birman and H. Wenzel: Link polynomials and a new algebra. Preprint, 1986.
5. V. F. R. Jones: A polynomial invariant for knots via Von Neumann algebras. *Bull. Am. math. Soc.* Vol. 12, Number 1, (1985), 103–111.
6. L. H. Kauffman: *Formal Knot Theory.* Lecture Notes No. 30, Princeton University Press, 1983.
7. L. H. Kauffman: *Statistical Mechanics and the Jones Polynomial.* (to appear in proceedings of July 1986 conference on Artin's braid group. Santa Cruz, California).
8. L. H. Kauffman: *An invariant of Regular Isotopy*, to appear.
9. T. P. Kirkman: The enumeration, description and construction of knots with fewer than 10 crossings. *Trans. R. Soc. Edinb.* **32** (1865), 281–309.
10. C. N. Little: Non-alternate ±knots. *Trans. R. Soc. Edinb.* **35** (1889), 663–664.
11. W. B. R. Lickorish and K. C. Millet: The reversing result for the Jones polynomial. *Pacif. J. Math.*, to appear.
12. H. R. Morton: The Jones polynomial for unoriented links. *Q. Jl Math.* **37** (1986), 55–60.
13. K. Murasugi: On a certain numerical invariant of link types. *Trans. Am. Math. Soc.* **114** (1965), 377–383.
14. K. Murasugi: Jones polynomials and classical conjectures in knot theory. Preprint, 1986.
15. D. Rolfsen: *Knots and Links.* Mathematics Lecture Series No. 7 Publish or Perish Press, 1976.
16. P. G. Tait: *On Knots I, II, III.* Scientific Papers Vol. I, pp. 273–347. Cambridge University Press, London (1898).
17. M. Thistlethwaite: A spanning tree expansion of the Jones polynomial. Preprint, 1986.
18. Y. Q. Wu: Jones polynomial and the crossing number of links. Preprint, 1986.
19. D. N. Yetter: Markov algebras. Preprint, 1986.

Department of Mathematics, Statistics and Computer Science
The University of Illinois at Chicago
Chicago, IL 60680, U.S.A.

Invent. math. 92, 527–553 (1988)

Inventiones mathematicae

The Yang-Baxter equation and invariants of links

V.G. Turaev

Leningrad Department of Steklov Mathematical Institute (LOMI),
Fontanka 27, Leningrad 191011, USSR

§ 1. Introduction

The Yang-Baxter equation first appeared in the independent papers of C.N. Yang and R.J. Baxter in the late 1960's – early 1970's. This equation and its solutions play fundamental role in the theory of completely integrable quantum systems and in the theory of exactly solved models of statistical mechanics (see [1, 9]). A relationship between the Yang-Baxter equation and the new polynomial invariants of links was implicit already in the pioneer paper of Jones [5]. In that paper Jones introduced his famous polynomial of links via a study of certain finite dimensional von Neumann algebras. A remark of D. Evans mentioned in [5] points out that these algebras were earlier discovered by physicists who used them to study the Potts model and the ice-type model of statistical mechanics.

After appearance of [5] several authors introduced two new isotopy invariants of links P and F which are (up to reparametrization) Laurent polynomials of 2 variables (see [3, 7]). Both P and F contain the Jones polynomial but can not be deduced from it. Known constructions of P appeal either to von Neumann algebras, or to Hecke algebras, or to a geometric iterative procedure based on a Conway-type relation. The only known construction of F, due to Kauffman [7], appeal to an analogous geometric procedure.

Recently, Jones [6] has shown that P can be constructed using explicit matrix representations of Hecke algebras, introduced in works on the quantum inverse scattering method and related to the Yang-Baxter equation. It is stressed in [6] that "a consistent general picture of the polynomials is starting to emerge, the relevant mathematical formalism being quantum inverse scattering method and quantum statistical mechanics".

The key observation which underlies the present paper is to the effect that one can directly construct P and F using some solutions of the Yang-Baxter equation. This leads to a general scheme which enables one to introduce these (and other) invariants of links. The resources of this scheme are far from being exhausted.

Note, however, that this approach does not shed light on the conceptual problem of understanding the polynomials from the viewpoint of algebraic topology. In particular, it is by no means clear how to extend the definition of the polynomials P, F given below to links in homology 3-spheres. (It is curious to note that a related invariant – the multivariable Conway polynomial can be defined for links in homology spheres, see [10].)

Though I do not consider von Neumann and Hecke algebras in this paper, it would be of great importance to comprehend the algebraic nature of the invariants.

I am indebted to O.Ja. Viro and W.B.R. Lickorish for helpful remarks. I am especially thankful to N.Yu. Reshetikhin for valuable discussions.

Organization of the paper. In § 2 the Yang-Baxter equation is recalled and the so-called EYB-operators are introduced. In § 3 with each EYB-operator S I associate an isotopy invariant of links T_S. In § 4 some special EYB-operators are considered, and the corresponding link invariants are studied. These invariants are shown to be equivalent to the polynomials P, F mentioned above. In § 5 under some restrictions on the EYB-operator S a state model for the invariant T_S is presented. In § 6 a nonoriented version of this state model is discussed; this model is used to prove Theorem 4.3.4 formulated in § 4.

Notation and agreements. In the whole paper the symbol K denotes a fixed commutative ring with 1 and V denotes a fixed finitely generated free K-module of rank $m \geqq 1$. For a natural n the n-times tensor product $V \otimes_K V \otimes \ldots \otimes_K V$ is denoted by $V^{\otimes n}$. In particular, $V^{\otimes 1} = V$ and $V^{\otimes 2} = V \otimes_K V$. Each basis $v_1, \ldots, v_m$ in V gives rise to a basis in $V^{\otimes n}$ which consists of vectors $v_{i_1} \otimes \ldots \otimes v_{i_n}$ with $i_1, \ldots, i_n \in \{1, 2, \ldots, m\}$. Having this basis, each (K-linear) endomorphism f of $V^{\otimes n}$ determines the multiindexed matrix $(f^{j_1, \ldots, j_n}_{i_1, \ldots, i_n})$, $1 \leqq i_1, j_1, \ldots, i_n, j_n \leqq m$ defined by the equation

$$f(v_{i_1} \otimes \ldots \otimes v_{i_n}) = \sum_{1 \leqq j_1, \ldots, j_n \leqq m} f^{j_1, \ldots, j_n}_{i_1, \ldots, i_n} v_{j_1} \otimes \ldots \otimes v_{j_n}.$$

The symbol K^* will denote the set of invertible elements of K.

By a link we shall mean a tame link in $\mathbf{R}^3$.

§ 2. The Yang-Baxter operators

2.1. Let $R: V^{\otimes 2} \to V^{\otimes 2}$ be a (K-linear) isomorphism. For natural n, i with $n-1 \geqq i \geqq 1$ denote by $R_i(n)$ the isomorphism

$$\mathrm{Id}_V^{\otimes (i-1)} \otimes R \otimes \mathrm{Id}_V^{\otimes (n-i-1)}: V^{\otimes n} \to V^{\otimes n}.$$

Thus for any $v_1, \ldots, v_n \in V$

$$R_i(n)(v_1 \otimes \ldots \otimes v_n) = v_1 \otimes \ldots \otimes v_{i-1} \otimes R(v_i, v_{i+1}) \otimes v_{i+2} \otimes \ldots \otimes v_n.$$

The isomorphism $R: V^{\otimes 2} \to V^{\otimes 2}$ is called a *Yang-Baxter operator* (or, briefly, a YB-operator) if the automorphisms $R_1 = R_1(3)$ and $R_2 = R_2(3)$ of $V^{\otimes 3}$ satisfy the equality

$$R_1 \circ R_2 \circ R_1 = R_2 \circ R_1 \circ R_2. \tag{1}$$

This is the Yang-Baxter equality (with zero spectral parameter). For examples of YB-operators and further information the reader is referred to [2, 4, 8]; see also § 4.

2.2. Recall that for each homomorphism $f\colon V^{\otimes n} \to V^{\otimes n}$ one can define its "operator trace" $\mathrm{Sp}_n(f)$ which is a homomorphism $V^{\otimes(n-1)} \to V^{\otimes(n-1)}$. If $v_1, \ldots, v_m$ is a basis in V then for any $i_1, \ldots, i_{n-1} \in \{1, 2, \ldots, m\}$

$$\mathrm{Sp}_n(f)(v_{i_1} \otimes \ldots \otimes v_{i_{n-1}}) = \sum_{1 \leq j_1, \ldots, j_{n-1}, j \leq m} f^{j_1, \ldots, j_{n-1}, j}_{i_1, \ldots, i_{n-1}, j}\, v_{j_1} \otimes \ldots \otimes v_{j_{n-1}}.$$

($\mathrm{Sp}_n(f)$ does not depend on the choice of basis of V). It is clear that $\mathrm{Sp}(\mathrm{Sp}_n(f)) = \mathrm{Sp}(f) \in K$ where Sp is the ordinary trace of a homomorphism.

2.3 By an *enhanced Yang-Baxter operator* (briefly, EYB-operator) I will understand a collection {a Yang-Baxter operator $R\colon V^{\otimes 2} \to V^{\otimes 2}$; a K-homomorphism $\mu\colon V \to V$; invertible elements α, β of K} which satisfy the following two conditions:

(i) The homomorphism $\mu \otimes \mu\colon V^{\otimes 2} \to V^{\otimes 2}$ commutes with R;

(ii) $\mathrm{Sp}_2(R \circ (\mu \otimes \mu)) = \alpha\beta\mu$; $\mathrm{Sp}_2(R^{-1} \circ (\mu \otimes \mu)) = \alpha^{-1}\beta\mu$.

Note that if μ is an isomorphism then Condition (ii) is equivalent to the following:

$$\mathrm{Sp}_2(R^{\pm 1} \circ (\mathrm{Id}_V \otimes \mu)) = \alpha^{\pm 1}\beta\, \mathrm{Id}_V.$$

We shall mainly consider the case when μ is an isomorphism presented by a diagonal matrix with respect to some basis of V. The following theorem restates Conditions (i), (ii) in this case.

2.3.1. **Theorem.** *Let $R\colon V^{\otimes 2} \to V^{\otimes 2}$ be a YB-operator. Let $v_1, \ldots, v_m$ be a basis of V and μ be an isomorphism $V \to V$ which transforms v_i into $\mu_i v_i$ for $i = 1, \ldots, m$ with $\mu_1, \ldots, \mu_m \in K^*$. The collection $(R, \mu, \alpha \in K^*, \beta \in K^*)$ is a EYB-operator if and only if the following two conditions are satisfied:*

(i)′ *For any $i, j, k, l \in \{1, 2, \ldots, m\}$*

$$(\mu_i \mu_j - \mu_k \mu_l)\, R^{k,l}_{i,j} = 0. \tag{2}$$

(ii)′ *For any $i, k \in \{1, 2, \ldots, m\}$*

$$\sum_{j=1}^{m} R^{k,j}_{i,j}\, \mu_j = \alpha\beta\delta^k_i; \qquad \sum_{j=1}^{m} (R^{-1})^{k,j}_{i,j}\, \mu_j = \alpha^{-1}\beta\delta^k_i$$

(here δ^k_i is the Kronecker symbol: $\delta^i_i = 1$, $\delta^k_i = 0$ for $k \neq i$).

Proof. Obvious.

2.4. *Remarks.* Clearly, $\mu \otimes \mu$ commutes with R iff $\mu \otimes \mu$ commutes with R^{-1}. Therefore, any of the conditions (i), (i)′ implies that for arbitrary i, j, k, l

$$(\mu_i \mu_j - \mu_k \mu_l)(R^{-1})^{k,l}_{i,j} = 0. \tag{3}$$

The condition (ii)′ of Theorem 2.3.1 implies that the product of the square $m\times m$-matrix $[R_{i,j}^{i,j}]$ with the column

$$\begin{bmatrix}\mu_1\\ \vdots\\ \mu_m\end{bmatrix}$$

is equal to the constant column

$$\begin{bmatrix}\alpha\beta\\ \vdots\\ \alpha\beta\end{bmatrix}.$$

The same is true for the matrix $[(R^{-1})_{i,j}^{i,j}]$ if we replace α by α^{-1}. Therefore, if at least one of these two square matrices is invertible over K then there exists at most one sequence $\mu_1, \ldots, \mu_m$ which satisfy (ii)′ for given α, β.

In the general case $\mu_1, \ldots, \mu_m$ (if exist) are not uniquely determined by R, α, β. For example, for any homomorphism $\mu\colon V\to V$ the collection $(\mathrm{Id}_{V^{\otimes 2}}, \mu, \alpha=1, \beta=\mathrm{Sp}\,\mu)$ is a EYB-operator.

§ 3. Invariants of braids and links

3.1 *Invariants of braids.* Every YB-operator $R\colon V^{\otimes 2}\to V^{\otimes 2}$ gives rise to a finite-dimensional representation of the Artin n-string braid group

$$B_n=\langle \sigma_1, \ldots, \sigma_{n-1}: \sigma_i\sigma_j=\sigma_j\sigma_i \quad \text{for } |i-j|\geqq 2;$$
$$\sigma_i\sigma_{i+1}\sigma_i=\sigma_{i+1}\sigma_i\sigma_{i+1} \quad \text{for } i=1, \ldots, n-1\rangle$$

(here $n\geqq 1$). Namely, put $R_i=R_i(n)\colon V^{\otimes n}\to V^{\otimes n}$ and notice that

$$R_i R_j=R_j R_i \quad \text{for } |i-j|\geqq 2$$

and (in view of the Yang-Baxter equality)

$$R_i R_{i+1} R_i=R_{i+1} R_i R_{i+1} \quad \text{for } i=1, \ldots, n-1.$$

Therefore, there is a unique homomorphism $B_n\to \mathrm{Aut}(V^{\otimes n})$ which transforms σ_i into R_i for all i. Denote this homomorphism by b_R. We shall also use the homomorphism w from B_n to the additive group of integers which sends $\sigma_1, \ldots, \sigma_{n-1}$ into 1.

Every EYB-operator $S=(R, \mu, \alpha, \beta)$ determines a mapping $T_S\colon \coprod_{n\geqq 1} B_n\to K$ as follows. For $n\geqq 1$ denote the homomorphism $\mu\otimes\mu\otimes\ldots\otimes\mu\colon V^{\otimes n}\to V^{\otimes n}$ by $\mu^{\otimes n}$. For a braid $\xi\in B_n$ put

$$T_S(\xi)=\alpha^{-w(\xi)}\beta^{-n}\,\mathrm{Sp}(b_R(\xi)\circ\mu^{\otimes n}\colon V^{\otimes n}\to V^{\otimes n}).$$

The most important properties of T_S are given by the following theorem.

3.1.2. **Theorem.** *For any* $\xi, \eta \in B_n$

$$T_S(\eta^{-1}\xi\eta) = T_S(\xi\sigma_n) = T_S(\xi\sigma_n^{-1}) = T_S(\xi).$$

To prove this theorem we need the following lemma which is a direct consequence of definitions.

3.1.3. **Lemma.** *If f, g, h are endomorphisms respectively of* $V^{\otimes(n+1)}$, $V^{\otimes n}$, $V^{\otimes 2}$ *then*

$$\begin{aligned}
&\mathrm{Sp}_{n+1}(f\circ(g\otimes \mathrm{Id}_V)) = \mathrm{Sp}_{n+1}(f)\circ g;\\
&\mathrm{Sp}_{n+1}((g\otimes \mathrm{Id}_V)\circ f)) = g\circ \mathrm{Sp}_{n+1}(f);\\
&\mathrm{Sp}_{n+1}(\mathrm{Id}_V^{\otimes(n-1)}\otimes h) = \mathrm{Id}_V^{\otimes(n-1)}\otimes \mathrm{Sp}_2(h).
\end{aligned}$$

3.1.4. *Proof of Theorem* 3.1.2. It follows from the definition of EYB-operator that $\mu^{\otimes n}$ commutes with $b(\eta)$ for any $\eta\in B_n$ where $b=b_R$: $B_n\to \mathrm{Aut}(V^{\otimes n})$. Thus

$$\mathrm{Sp}(b(\eta^{-1}\xi\eta)\circ\mu^{\otimes n}) = \mathrm{Sp}(b(\eta^{-1})\circ b(\xi)\circ\mu^{\otimes n}\circ b(\eta)) = \mathrm{Sp}(b(\xi)\circ\mu^{\otimes n}).$$

Also, $w(\eta^{-1}\xi\eta)=w(\xi)$. Therefore $T_S(\eta^{-1}\xi\eta)=T_S(\xi)$.

Let us prove that $T_S(\xi\sigma_n)=T_S(\xi)$. Clearly,

$$b(\xi\sigma_n) = (b(\xi)\otimes \mathrm{Id}_V)\circ R_n\colon V^{\otimes(n+1)}\to V^{\otimes(n+1)}.$$

Thus,

$$\begin{aligned}
&\mathrm{Sp}(b(\xi\sigma_n)\circ\mu^{\otimes(n+1)})\\
&\quad = \mathrm{Sp}[(b(\xi)\otimes \mathrm{Id}_V)\circ R_n\circ(\mathrm{Id}_V^{\otimes(n-1)}\otimes\mu\otimes\mu)\circ(\mu^{\otimes(n-1)}\otimes \mathrm{Id}_V^{\otimes 2})]\\
&\quad = \mathrm{Sp}\{\mathrm{Sp}_{n+1}[(b(\xi)\otimes \mathrm{Id}_V)\circ(\mathrm{Id}_V^{\otimes(n-1)}\otimes[R\circ(\mu\otimes\mu)]\circ(\mu^{\otimes(n-1)}\otimes \mathrm{Id}_V^{\otimes 2})]\}.
\end{aligned}$$

Lemma 3.1.3 implies that the expression in the figured brackets is equal to

$$b(\xi)\circ[\mathrm{Id}_V^{\otimes(n-1)}\otimes \mathrm{Sp}_2(R\circ(\mu\otimes\mu))]\circ(\mu^{\otimes(n-1)}\otimes \mathrm{Id}_V).$$

In view of the definition of EYB-operator, this is equal to $\alpha\beta(b(\xi)\circ\mu^{\otimes n})$. Hence

$$\mathrm{Sp}(b(\xi\sigma_n)\circ\mu^{\otimes(n+1)}) = \alpha\beta\,\mathrm{Sp}(b(\xi)\circ\mu^{\otimes n}).$$

Clearly, $w(\xi\sigma_n)=w(\xi)+1$. These equalities imply that $T_S(\xi\sigma_n)=T_S(\xi)$. The equality $T_S(\xi\sigma_n^{-1})=T_S(\xi)$ is proved similarly.

3.2. *Invariants of links.* Recall briefly the well known relationship between braids and links. Each braid gives rise to an oriented link via closing (see Fig. 1). A theorem of J. Alexander asserts that any oriented link is isotopic to the closure of some braid. A theorem of A. Markov asserts that the closures of two braids are isotopic (in the category of oriented links) if and only if these braids are equivalent with respect to the equivalence relation in $\coprod_n B_n$ generated by Markov moves $\xi\to\eta^{-1}\xi\eta$, $\xi\mapsto\xi\sigma_n^{\pm1}$ where $\xi,\eta\in B_n$.

Theorem 3.2 shows that for any EYB-operator $S=(R,\mu,\alpha,\beta)$ the mapping $T_S\colon \coprod_n B_n\to K$ induces a mapping of the set of oriented isotopy classes of links into K. This latter mapping is also denoted by T_S.

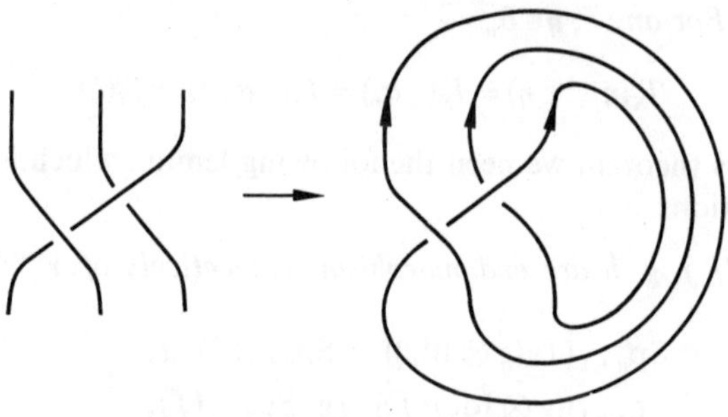

Fig. 1

For the trivial knot O we have

$$T_S(\mathrm{O})=\beta^{-1}\,\mathrm{Sp}(\mu). \tag{4}$$

It is easy to show (using the evident equality $\mathrm{Sp}(f\otimes g)=\mathrm{Sp}(f)\,\mathrm{Sp}(g)$) that T_S is multiplicative: If a link L is the disjoint union of two links L_1 and L_2 then $T_S(L)=T_S(L_1)\cdot T_S(L_2)$. In particular, if L is the trivial n-component link then $T_S(L)=[\beta^{-1}\,\mathrm{Sp}(\mu)]^n$.

To formulate the next property of T_S we will need the following terminology. Let τ be a mapping of the set of oriented isotopy link types into K. Let $f(t)=\sum_{i=p}^{q} k_i t^i$ be a Laurent polynomial over K (i.e. $f(t)\in K[t,t^{-1}]$). Let us say that $f(t)$ *annihilates* τ and write $f(t)*\tau=0$ if for any oriented links $L_p, L_{p+1},\ldots,L_q$, which have diagrams coinciding outside some disk and looking as in Fig. 2 inside this disk, we have $\sum_{i=p}^{q} k_i\tau(L_i)=0$. In particular, when $f(t)=k_-t^{-1}+k_0+k_+t$ the equation $f(t)*\tau=0$ is a Conway-type relation between the invariants of links $L_-=L_{-1}$, L_0, $L_+=L_1$ (see Fig. 3).

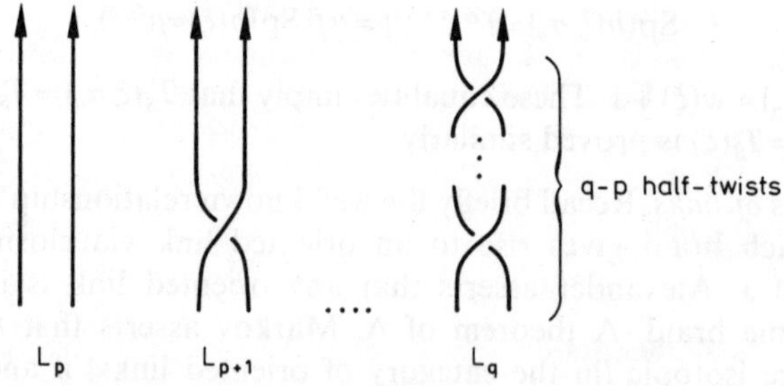

Fig. 2

3.2.1. **Theorem.** *Let $S=(R,\mu,\alpha,\beta)$ be a* EYB*-operator. If the automorphism R of $V^{\otimes 2}$ satisfies the equation $\sum_{i=p}^{q} k_i R^i=0$ with $k_p,\ldots,k_q\in K$ then the polynomial $\sum_{i=p}^{q} k_i\alpha^i t^i$ annihilates T_S.*

Fig. 3

Proof. Let $L_p, \ldots, L_q$ be oriented links which have diagrams as above. Then for some braid η these links are isotopic to the closures of the braids η, $\sigma_1 \eta, \ldots, \sigma_1^{q-p} \eta$. Let n be the number of strings of η. Then

$$T_S(L_i) = T_S(\sigma_1^i \eta) = \alpha^{-i-w(\eta)} \beta^{-n} \operatorname{Sp}[(R_1)^i \circ b_R(\eta) \circ \mu^{\otimes n}].$$

Hence,

$$\sum_{i=p}^{q} k_i \alpha^i T_S(L_i) = \alpha^{-w(\eta)} \beta^{-n} \operatorname{Sp}\left[\sum_{i=p}^{q} k_i R_1^i \circ b_R(\eta) \circ \mu^{\otimes n}\right] = 0.$$

3.2.2. **Corollary.** *For any* EYB*-operator S in V the isotopy invariant T_S is annihilated by a polynomial of degree $\leqq m^2$ (where $m = \operatorname{rk}_K V$).*

Proof. Every endomorphism of $V^{\otimes 2}$ is annihilated by its characteristic polynomial.

3.3. *Remarks.* (i) Without loss of generality we can confine ourselves to EYB-operators (R, μ, α, β) with $\alpha = \beta = 1$. Indeed, if $S = (R, \mu, \alpha, \beta)$ is a EYB-operator then $S' = (\alpha^{-1} R, \beta^{-1} \mu, 1, 1)$ also is a EYB-operator and $T_S = T_{S'}$. However, sometimes it may be convenient to have non-trivial α, β.

(ii) If $S = (R, \mu, \alpha, \beta)$ is a EYB-operator then $S_1 = (-R, -\mu, \alpha, \beta)$ and $S_2 = (R, \mu, -\alpha, -\beta)$ are EYB-operators and for any n-component link L

$$T_{S_1}(L) = T_{S_2}(L) = (-1)^n T_S(L).$$

(iii) It is easy to verify that (in the notation used above) $f(t) * \tau = 0$ implies $t^i f(t) * \tau = 0$ for any integer i. If two polynomials annihilate τ then their sum also annihilates τ. Therefore, the set of polynomials annihilating τ is an ideal of the ring $K[t, t^{-1}]$. Let us call this ideal the annulator of τ. Theorem 3.2.1 shows that for any EYB-operator $S = (\mathrm{R}, \mu, \alpha, \beta)$ the annulator of T_S is contained in the annulator of the endomorphism $\alpha^{-1} R$ of $V^{\otimes 2}$. I do not know if this inclusion can be proper.

§4. Examples and applications

4.1. At present there is a general method which in principle enables one to construct EYB-operators from representations of simple (complex) Lie algebras (see [2, 4]). Each pair (a simple Lie algebra X, an automorphism of the Dynkin diagram of X) determines a "universal" YB-operator acting in an infinite dimensional vector space; with any representation of X in a vector space W one associates an induced YB-operator $W^{\otimes 2} \to W^{\otimes 2}$. These induced operators have been explicitly described for the fundamental representations of the Lie algebras of series A_n^1, B_n^1, C_n^1, D_n^1, A_n^2 and D_n^2 (see [4]; here the upper index denotes the order of the automorphism of the Dynkin diagram: 1 corresponds to the identity and 2 corresponds to the non-trivial involution). The YB-operators which correspond to series A^1, B^1, C^1, D^1 and A^2 can be enhanced to EYB-operators, see Sect. 4.2 and 4.3. The case of D^2 has remained unclear. (In this case one definitely can not enhance the YB-operator in the diagonal way with respect to the natural basis).

Up to the end of § 4, K is the Laurent polynomial ring $\mathbf{Z}[q, q^{-1}]$; V is the free K-module with a fixed basis $v_1, \dots, v_m$. The symbol $E_{i,k}$ denotes the homomorphism $V \to V$ which transforms v_i in v_k and transforms v_r, with $r \neq i$, into 0. The homomorphism $E_{i,k} \otimes E_{j,l}$: $V^{\otimes 2} \to V^{\otimes 2}$ clearly transforms $v_i \otimes v_j$ into $v_k \otimes v_l$ and transforms other basis vectors of type $v_r \otimes v_s$ into 0.

4.2. *The series* A^1. The fundamental vector representation of the simple Lie algebra A_{m-1}^1 gives rise to the following YB-operator $V^{\otimes 2} \to V^{\otimes 2}$ (see [4] and references therein):

$$R = -q \sum_i E_{i,i} \otimes E_{i,i} + \sum_{i \neq j} E_{i,j} \otimes E_{j,i} + (q^{-1} - q) \sum_{i<j} E_{i,i} \otimes E_{j,j}.$$

(Here $i, j = 1, 2, \dots, m$.) Note that our notation differ from that of [4]; in particular, our operator R corresponds to $(k\xi)^{-1} \check{R}(0)$ in [4]. The equality (1) for R can be rather easily checked directly. From [4] one can also extract a formula for R^{-1}:

$$R^{-1} = -q^{-1} \sum_i E_{i,i} \otimes E_{i,i} + \sum_{i \neq j} E_{i,j} \otimes E_{j,i} + (q - q^{-1}) \sum_{i>j} E_{i,i} \otimes E_{j,j}.$$

It is clear that

$$R - R^{-1} = (q^{-1} - q) \operatorname{Id}_V^{\otimes 2}. \tag{5}$$

4.2.1. **Theorem.** *Put* $\mu_i = q^{2i-m-1}$ *for* $i = 1, \dots, m$. *Put* $\alpha = -q^m$, $\beta = 1$. *Then* $S = (R, \mu = \operatorname{diag}(\mu_1, \dots, \mu_m), \alpha, \beta)$ *is a EYB-operator such that for any triple of links* (L_+, L_-, L_0) *as in Fig. 3*

$$q^m T_S(L_+) - q^{-m} T_S(L_-) = (q - q^{-1}) T_S(L_0) \tag{6}$$

and $T_S(\mathrm{O}) = (q^m - q^{-m})/(q - q^{-1})$.

Proof. The matrix of R with respect to the basis $\{v_i \otimes v_j \mid i, j = 1, \dots, m\}$ in $V^{\otimes 2}$ looks as follows:

$$R_{i,j}^{k,l} = \begin{cases} -q & \text{if} \quad i = j = k = l \\ 1 & \text{if} \quad i = l \neq k = j \\ q^{-1} - q & \text{if} \quad i = k < l = j \\ 0 & \text{otherwise} \end{cases}$$

In particular, if $R_{i,j}^{k,l} \neq 0$ then the non-ordered pairs i, j and k, l coincide. The same property holds for the matrix of R^{-1}. Thus, the condition (i)′ of Theorem 2.3.1 is satisfied and the condition (ii)′ is equivalent to the equalities

$$\sum_{j=1}^{m} R_{i,j}^{i,j}\mu_j=\alpha\beta; \qquad \sum_{j=1}^{m}(R^{-1})_{i,j}^{i,j}\mu_j=\alpha^{-1}\beta. \tag{7}$$

We have

$$\sum_{j=1}^{m} R_{i,j}^{i,j}\mu_j=-q\,\mu_i+\sum_{j=i+1}^{m}(q^{-1}-q)\,\mu_j=-q^{2i-m}+(q^{-1}-q)$$
$$\cdot\,[q^{2i-m+1}+q^{2i-m+3}+\ldots+q^{m-1}]=-q^m=\alpha\beta.$$

The second from the formulas (7) is verified analogously. Hence, (R,μ,α,β) is a EYB-operator. Other statements of the theorem follow directly fom the results of Sect. 3.1 and the formula (5).

4.2.2. For an oriented link L *by* $P_m(L)$ I will denote the invariant $T_S(L)$ produced by Theorem 4.2.1 and the constructions of § 3. Using $P_2, P_3, \ldots$ I will give a new proof of the following theorem.

4.2.3. **Theorem** (see [3]). *There exists a unique mapping P from the set of isotopy types of oriented links into the ring $\mathbf{Z}[x,x^{-1},y,y^{-1}]$ such that $P(\mathrm{O})=1$ and for any triple (L_+,L_-,L_0) as above*

$$x\,P(L_+)+x^{-1}\,P(L_-)=y\,P(L_0).$$

4.2.4. **Lemma.** *Let D be a diagram of an oriented n-component link L. Let u be the number of crossing points of D. If $m\geqq 4u+2n+1$ then the Laurent polynomial $(q-q^{-1})^{u+n}\,P_m(L)$ can be uniquely expressed as a (finite) sum*

$$\sum_{a,b\in\mathbf{Z}} r_{a,b}\,q^{a+mb}, \qquad (r_{a,b}\in\mathbf{Z}), \tag{8}$$

so that $r_{a,b}=0$ for $|a|>2u+n$. The coefficients $\{r_{a,b}\}$ do not depend on the choice of $m\geqq 4u+2n+1$.

Proof. The inequality $2u+n<m/2$ implies that if the desirable decomposition (8) exists then it is unique. Let us proof existence. It is well known that trading overcrossings for undercrossings one can transform any link diagram into a diagram of a trivial link. Therefore, applying the formula (6) in the iterative fashion we obtain that $P_m(L)$ is a finite sum of polynomials of type $\pm q^{me}(q-q^{-1})^f$ $P_m(G_d)$ where $e,f\in\mathbf{Z}$; $0\leqq f\leqq u$; G_d is the trivial d-component link, and $d\leqq u+n$. Clearly,

$$(q-q^{-1})^{u+n}[q^{me}(q-q^{-1})^f\,P_m(G_d)]=q^{me}(q^m-q^{-m})^d(q-q^{-1})^{f+u+n-d}.$$

Note that $f+u+n-d\leqq f+u+n\leqq 2u+n$. This immediately implies existence of the decomposition (8).

The last statement of Lemma follows directly from the construction of the decomposition (8).

4.2.5. *Proof of Theorem* 4.2.3. The proof of uniqueness of P is standard and therefore I omit it. Let us prove existence. Let D, L, n, u be the same objects as in the statement of Lemma 4.2.4. Let $m\geqq 4u+2n+1$ and let $\{r_{a,b}\}$ be the coefficients of the sum (8). Put

$$N(L)=(q-q^{-1})^{-u-n}\sum_{a,b\in\mathbf{Z}} r_{a,b}\,q^a t^b.$$

It follows from Lemma 4.2.4 that $N(L)$ is a Laurent polynomial of the variables q, t which does not depend on the choice of m. Since P_m is an isotopy invariant, $N(L)$ is preserved under the Reidemeister moves. (Note that $m=4u+2n+9$ is suitable both for D and for any diagram obtained from D by a single Reidemeister move). Thus $N(L)$ is an isotopy invariant of L. If follows from (6) that

$$tN(L_+)-t^{-1}N(L_-)=(q-q^{-1})\,N(L_0).$$

If G is the trivial n-component link then

$$N(G)=(t-t^{-1})^n/(q-q^{-1})^n.$$

Hence for any link L the function $N(L)$ is a Laurent polynomial of t and $q-q^{-1}$. Substituting $t=\sqrt{-1}\,x$ and $q-q^{-1}=\sqrt{-1}\,y$ and multiplying the resulting polynomial by $(q-q^{-1})/(t-t^{-1})=y/(x+x^{-1})$ we get $P(L)(x, y)$.

4.2.6. *Remark.* For any link L

$$P_m(L)=(q^m-q^{-m})(q-q^{-1})^{-1}\,P(L)(\sqrt{-1}\,q^m, \sqrt{-1}(q-q^{-1})).$$

4.3. *Series B^1, C^1, D^1 and A^2.* Fix $\nu\in\{1,-1\}$. We shall assume that if m is odd then $\nu=-1$. For $i=1, \ldots, m$ put $i'=m+1-i$ and

$$\bar{i}=\begin{cases} i-\nu/2 & \text{if} \quad 1\leq i<(m+1)/2 \\ i & \text{if} \quad i=(m+1)/2, \quad m \text{ being odd} \\ i+\nu/2 & \text{if} \quad (m+1)/2<i\leq m \end{cases}$$

$$\varepsilon(i)=\begin{cases} 1 & \text{if} \quad 1\leq i\leq(m+1)/2 \\ -\nu & \text{if} \quad (m+1)/2\leq i\leq m \end{cases}$$

According to [4] the fundamental representations of simple Lie algebras of series B^1, C^1, D^1, A^2 give rise to the following YB-operator R_ν:

$$\begin{aligned} R_\nu = {} & q\sum_{\substack{i\\ i\neq i'}} E_{i,i}\otimes E_{i,i}+\sum_{\substack{i\\ i=i'}} E_{i,i}\otimes E_{i,i}+\sum_{\substack{i,j\\ i\neq j,j'}} E_{i,j}\otimes E_{j,i} \\ & +q^{-1}\sum_{\substack{i\\ i\neq i'}} E_{i,i'}\otimes E_{i',i}+(q-q^{-1})\sum_{i<j} E_{i,i}\otimes E_{j,j} \\ & +(q^{-1}-q)\sum_{i<j}\varepsilon(i)\,\varepsilon(j)\,q^{\bar{i}-\bar{j}}\,E_{i,j'}\otimes E_{i',j}. \end{aligned}$$

Here for Lie algebras B_n^1, C_n^1, D_n^1, A_n^2 the pair (m, ν) is respectively $(2n+1, -1)$, $(2n, 1)$, $(2n, -1)$, $(n+1, -1)$. It is understood that in the case of odd m the ring $K=\mathbf{Z}[q, q^{-1}]$ is extended to $\mathbf{Z}[q^{1/2}, q^{-1/2}]$.

A direct, purely computational verification of the equality (1) for R_ν seems to be extremely difficult. However, (1) can be verified for R_ν using ideas suggested by a study of link diagrams.

From [4] one can extract a formula for R_ν^{-1}:

$$R_\nu^{-1}=q^{-1}\sum_{\substack{i\\ i\neq i'}} E_{i,i}\otimes E_{i,i}+\sum_{\substack{i\\ i=i'}} E_{i,i}\otimes E_{i,i}+\sum_{\substack{i,j\\ i\neq j,j'}} E_{i,j}\otimes E_{j,i}$$
$$+q\sum_{\substack{i\\ i\neq i'}} E_{i,i'}\otimes E_{i',i}+(q^{-1}-q)\sum_{i>j} E_{i,i}\otimes E_{j,j}$$
$$+(q-q^{-1})\sum_{i>j}\varepsilon(i)\,\varepsilon(j)\,q^{\bar{i}-\bar{j}}\,E_{i,j'}\otimes E_{i',j}.$$

4.3.1. *Remarks.* (i) The case of even m is somewhat easier since $i\neq i'$ for all i in this case.

(ii) The formula for $\bar{i}$ given in [4] contains erroneous signs $\pm$, the correct signs used above were pointed out to me by N. Reshetikhin.

4.3.2. **Theorem.** *Put* $\mu_i=q^{2\bar{i}-m-1}$ *for* $i=1, 2, \ldots, m$. *Put* $\alpha=q^{m+\nu}$ *and* $\beta=1$. *Then* $S_\nu=(R_\nu, \mu=\operatorname{diag}(\mu_1, \ldots, \mu_m), \alpha, \beta)$ *is a* EYB-*operator.*

Proof. The matrices of R_ν and R_ν^{-1} have the following property: If $(R_\nu)_{i,j}^{k,l}\neq 0$ or $(R_\nu^{-1})_{i,j}^{k,l}\neq 0$ then either the non-ordered pairs $\{i,j\}$, $\{k,l\}$ coincide, or $j=i'$ and $l=k'$ (or both). If $j=i'$ and $l=k'$ then

$$\mu_i\,\mu_j=\mu_i\,\mu_{i'}=q^{2(\bar{i}+\bar{i}')-2m-2}=1=\mu_k\,\mu_l.$$

Thus, S_ν satisfies the first condition (i)′ of Theorem 2.3.1. Let us check (ii)′. Put $a(i,j)=(R_\nu)_{i,j}^{i,j}$ and $b(i,j)=(R_\nu^{-1})_{i,j}^{i,j}$ where $i,j=1, \ldots, m$. We have to prove that for any i

$$\sum_{j=1}^{m} a(i,j)\,q^{2\bar{j}-m-1}=q^{m+\nu}, \tag{9}$$

$$\sum_{j=1}^{m} b(i,j)\,q^{2\bar{j}-m-1}=q^{-(m+\nu)}. \tag{10}$$

It is easy to check up that for any i,j,k,l

$$(R_\nu^{-1})_{i,j}^{k,l}=\varphi((R_\nu)_{i',j'}^{k',l'})$$

where φ is the automorphism of $\mathbf{Z}[q^{1/2}, q^{-1/2}]$ sending $q^{1/2}$ into $q^{-1/2}$. In particular, $b(i,j)=\varphi(a(i',j'))$. Thus

$$\sum_{j=1}^{m} b(i,j)\,q^{2\bar{j}-m-1}=\sum_{j=1}^{m}\varphi(a(i',j')\,q^{2\bar{j}'-m-1})=\varphi\left(\sum_{j=1}^{m} a(i',j)\,q^{2\bar{j}-m-1}\right).$$

Therefore, (9) implies (10).

Using the equalities

$$\varepsilon(i)\,\varepsilon(i')=-\nu \quad\text{and}\quad \bar{i}+\bar{i}'=m+1$$

it is easy to compute $a(i,j)$:

$$a(i,j)=\begin{cases} 0 & \text{if} \quad i>j \\ q & \text{if} \quad i=j\neq i' \\ 1 & \text{if} \quad i=j=i' \\ q-q^{-1} & \text{if} \quad i<j\neq i' \\ (q-q^{-1})(1+v\,q^{2\bar{i}-m-1}) & \text{if} \quad i<j=i' \end{cases}$$

To verify (9) we shall consider 3 cases: $i<i'$, $i=i'$, $i>i'$. If $i>i'$, then $i>(m+1)/2$ and

$$\sum_{j=1}^{m} a(i,j)\,q^{2\bar{j}-m-1}=q^{2i-m+v}+(q-q^{-1})\sum_{j=i+1}^{m} q^{2j-m-1+v}=q^{m+v}.$$

If $i=i'$, then $i=(m+1)/2$, m is odd, $v=-1$ and the computation is similar. Let $i<i'$. Then $i<(m+1)/2$ and

$$\sum_{j=1}^{m} a(i,j)\,q^{2\bar{j}-m-1}=q^{2\bar{i}-m}+(q-q^{-1})\sum_{j=i+1}^{m} q^{2\bar{j}-m-1}+v(q-q^{-1}). \tag{11}$$

The sequence

$$q^{2\bar{j}-m-1}, \quad j=i+1, i+2, \ldots, m$$

is the geometric progression

$$q^{2\bar{i}+1-m}, \quad q^{2\bar{i}+3-m}, \ldots, q^{m+v-1}$$

with one superfluous member $q^0=1$ in case $v=-1$ or one omitted member $q^0=1$ in case $v=1$. This excess or omission is exactly compensated by $v(q-q^{-1})$. Therefore the right-hand side of (11) equals

$$q^{2\bar{i}-m}+(q-q^{-1})[q^{2\bar{i}+1-m}+q^{2\bar{i}+3-m}+\ldots+q^{m+v-1}]=q^{m+v}.$$

4.3.3. For an oriented link L the invariant $T_S(L)$ with $S=S_v$ will be denoted by $Q_{m,v}(L)$. It follows from (4) that

$$Q_{m,v}(\bigcirc)=-v+(q^{m+v}-q^{-m-v})/(q-q^{-1}).$$

Apriori, if m is odd then $Q_{m,v}(L)\in\mathbf{Z}[q^{1/2}, q^{-1/2}]$. The next theorem shows among other things that actually $Q_{m,v}(L)\in\mathbf{Z}[q,q^{-1}]$. In the statement of this theorem $\sqrt{-v}=1$, if $v=-1$, and $\sqrt{-v}$ is the complex unit $\sqrt{-1}$ if $v=1$.

4.3.4. **Theorem.** *Let $v\in\{1,-1\}$. For any diagram D of an oriented link L the polynomial*

$$\tilde{Q}_{m,v}(D)=(\sqrt{-v}\,q^{m+v})^{w(D)}\,Q_{m,v}(L)$$

does not depend on the choice of orientation of L. (Here $w(D)$ is the writhe of D – see Sect. 5.1). If D_+, D_-, D_0 and D_∞ are link diagrams coinciding outside some disk and looking as in Fig. 4 inside this disk then

$$\tilde{Q}_{m,\nu}(D_+)+\nu\tilde{Q}_{m,\nu}(D_-)=\sqrt{-\nu}(q-q^{-1})[\tilde{Q}_{m,\nu}(D_0)+\nu\tilde{Q}_{m,\nu}(D_\infty)]. \quad (12)$$

This theorem is proved in § 6 using the results of § 5.

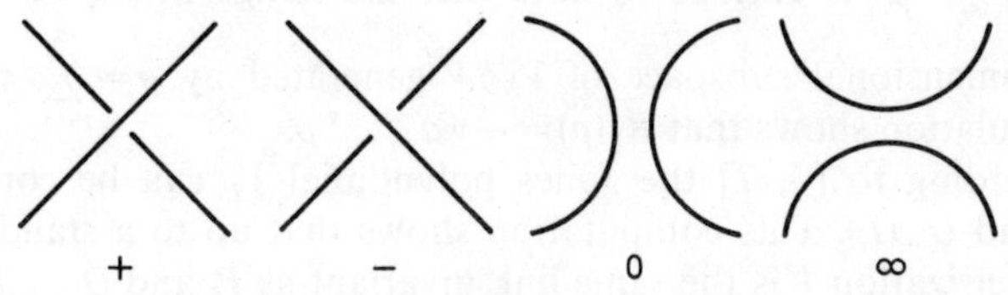

Fig. 4

4.3.5. **Corollary.** *Let* $\nu\in\{1,-1\}$. *There exists a unique mapping* Q_ν *of the set of isotopy classes of oriented links into the ring* $\mathbf{Z}[x,x^{-1},y,y^{-1}]$ *such that* $Q_\nu(\mathrm{O})=1$ *and*

1) *for any diagram D of an oriented link L the polynomial* $\tilde{Q}_\nu(D)=x^{w(D)}\,Q_\nu(L)$ *does not depend on the choice of orientation of L;*

2) *if* D_+, D_-, D_0, D_∞ *are link diagrams as in the statement of Theorem 4.3.4 then*

$$\tilde{Q}_\nu(D_+)+\nu\,\tilde{Q}_\nu(D_-)=y[\tilde{Q}_\nu(D_0)+\nu\,\tilde{Q}_\nu(D_\infty)].$$

This Corollary is deduced from Theorem 4.3.4 exactly in the same fashion as Theorem 4.2.3 was deduced from Theorem 4.2.1. (In particular, Lemma 4.2.4 remains true if one replaces in its statement P_m by $Q_{m,\nu}$ and all (other) entries of m by $m+\nu$.)

The polynomial Q_1 was introduced by Kauffman [7]. (It is denoted by F in [7]). It was pointed out to me by W.B.R. Lickorish that the link invariants $Q_-=Q_{-1}$ and $Q_+=Q_1$ are essentially equivalent: If L is an n-component link then

$$Q_-(L)(x,y)=(-1)^{n-1}\,Q_+(L)(\sqrt{-1}\,x,-\sqrt{-1}\,y).$$

It is clear that

$$Q_{m,\nu}(L)=\left[-\nu+\frac{q^{m+\nu}-q^{-m-\nu}}{q-q^{-1}}\right]Q_\nu(L)(\sqrt{-\nu}\,q^{m+\nu},\sqrt{-\nu}(q-q^{-1})). \quad (13)$$

4.3.6. *Remarks.* (i) Using Q_ν or $\tilde{Q}_\nu$ one can easily define an isotopy invariant of non-oriented links. Namely, if D is a diagram of an oriented link L and if u is the sum of the crossing signs (± 1) over all self-crossing of components of L then the polynomial $x^{-u}\,\tilde{Q}_\nu(D)$ does not depend on the choice of orientation of L. This polynomial is easily seen to be preserved under Reidemeister moves. Hence, it is an isotopy invariant.

(ii) It is easy to deduce from the properties of Q_ν stated in Corollary 4.3.5 that Q_ν is annihilated by the polynomial

$$(x^2\,t-1)(x\,t^2-y\,t+\nu\,x^{-1})\in(\mathbf{Z}[x,x^{-1},y,y^{-1}])[t].$$

This fact is equivalent to the identity

$$(R_\nu + \nu q^{-m-\nu} I)(R_\nu + q^{-1} I)(R_\nu - q I) = 0$$

where $I = \mathrm{Id}_{V\otimes V}$. It is curious to note that the image of $(R_\nu + q^{-1} I)(R_\nu - q I)$ is the one-dimensional subspace of $V\otimes V$ generated by $\rho = \sum_{i=1}^{m} \varepsilon(i)\, q^{\bar{i}}\, v_i \otimes v_{i'}$. A direct calculation shows that $R_\nu(\rho) = -\nu q^{-m-\nu} \rho$.

(iii) According to [3, 7] the Jones polynomial V_L can be computed from both $P(L)$ and $Q_+(L)$. This computation shows that up to a standard multiple and reparametrization V is the same link invariant as P_2 and $Q_{2,+}$. If one introduced $Q_{m,-}$ with $m<0$ by the formula (13) then one would similarly have that V is equivalent to $Q_{-2,-}$. The polynomial $Q_{2,-}$ stays somewhat aside: It can be completely computed from the linking coefficients of the components of the link. In particular, if L is a knot then $Q_{2,-}(L) = 2$. This follows from the equality $\tilde{Q}_{2,-}(D) = \sum_{\omega\in\Omega} q^{w(\omega)}$ where D is an arbitrary link diagram, Ω is the set of orientations of D, $w(\omega)$ is the writhe of the oriented link diagram (D, ω).

§ 5. State models for the invariants of links

L. Kauffman constructed for the Jones polynomial of links a "state model" of striking beauty and simplicity (see [7]). The construction is based on ideas which came from the study of the Potts model in statistical mechanics (see [1]). Jones [6] constructed a state model for the polynomials $P(L)(\sqrt{-1}\, q^{-m}, \sqrt{-1}(q-q^{-1}))$, $m = 2, 3, \ldots$ For $m=2$ the Jones model is related to the Kauffman model via "arrow coverings" (see [1, 6]). In this section under certain conditions on the EYB-operator S I construct a state model for T_S generalizing the Jones construction.

Fix a EYB-operator $S = (R, \mu, \alpha, \beta)$. We shall assume that the K-module V is provided with a basis so that μ is diagonal regarding this basis, $\mu = \mathrm{diag}(\mu_1, \ldots, \mu_m)$, $m = \mathrm{rk}_K V$.

5.1. *States of diagrams.* Let D be a diagram of an oriented link L. The diagram D determines a planar graph Γ_D which is obtained from D by identifying each overcrossing point with the corresponding undercrossing point. (Γ_D is the projection of L in R^2; see Fig. 5). The orientation of L induces orientation of all edges of Γ_D so that Γ_D is an oriented graph. By vertices and edges of D I will mean respectively vertices and (oriented) edges of Γ_D. The sets of vertices and edges of D will be denoted respectively by Vert D and Edg D. The writhe $w(D)$ of D is defined to be $w_+(D) - w_-(D)$ where $w_+(D)$ and $w_-(D)$ are respectively the numbers of positive and negative vertices of D (see Fig. 6).

A *state* of D is an arbitrary mapping f: Edg $D \to \{1, 2, \ldots, m\}$. The set of all states of D is denoted by St(D). With each state f of D and each vertex u of D we associate an element $\pi_u(f)$ of K as follows. If a, b, c, d are edges of D incident to u as in Fig. 7 then

$$\pi_u(f) = \begin{cases} R^{f(c), f(d)}_{f(a), f(b)} & \text{if } u \text{ is positive} \\ (R^{-1})^{f(c), f(d)}_{f(a), f(b)} & \text{if } u \text{ is negative} \end{cases}$$

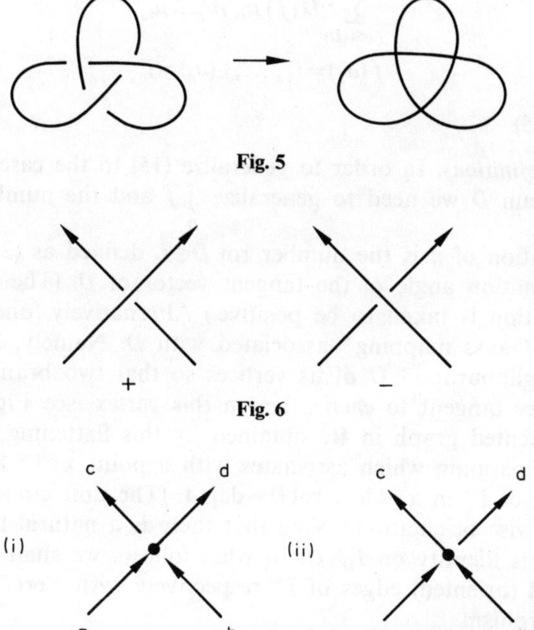

Fig. 5

Fig. 6

Fig. 7

(Here the matrix elements of R, R^{-1} are taken with respect to the basis in $V^{\otimes 2}$ constructed as usual from the fixed basis in V.)

Put

$$\Pi(f) = \prod_{u \in \operatorname{Vert} D} \pi_u(f) \in K. \tag{14}$$

5.2. *State models for special diagrams.* Let D be the diagram of a link L obtained by closing a diagram of a certain n-string braid ξ. Let $a_1, \ldots, a_n$ be the (oriented) edges of D which tie the top and bottom ends of ξ (see Fig. 1). For a state f of D put

$$\int_D f = \prod_{i=1}^{n} \mu_{f(a_i)} \in K.$$

5.2.1. **Theorem.**

$$T_S(L) = T_S(\xi) = a^{-w(D)} \beta^{-n} \sum_{f \in \operatorname{St}(D)} \Pi(f) \int_D f. \tag{15}$$

Proof. Let $v_1, \ldots, v_m$ be the fixed basis of V. Consider the matrix of $b_R(\xi) \circ \mu^{\otimes n}$: $V^{\otimes n} \to V^{\otimes n}$ with respect to the basis $\{v_{i_1} \otimes \ldots \otimes v_{i_n} \mid 1 \leq i_1, \ldots, i_n \leq m\}$. It is easy to see that the diagonal element of this matrix corresponding to the basis vector $v_{i_1} \otimes \ldots \otimes v_{i_n}$ is equal to

$$\sum_{f \in \mathrm{St}(D)} \Pi(f)\, \mu_{i_1} \mu_{i_2} \cdots \mu_{i_n}$$
$$f(a_1)=i_1, \ldots, f(a_n)=i_n.$$

This implies (15).

5.3. *Further definitions.* In order to generalize (15) to the case of an arbitrary oriented diagram D we need to generalize $\int_D f$ and the number of strings n. The generalization of n is the number rot $D \in \mathbf{Z}$ defined as $(2\pi)^{-1}\psi$ where ψ is the total rotation angle of the tangent vector of D. (The direction of the clockwise rotation is taken to be positive.) Alternatively, one can define rot D using the "Gauss mapping" associated with D. Namely, let us flatten Γ_D in a small neighbourhood U of its vertices so that two branches incident to any vertex were tangent to each other in this vertex (see Fig. 8). Denote by $\Gamma^o = \Gamma^o_D$ the oriented graph in $\mathbf{R}^2$ obtained by this flattening. Let $\Delta\colon \Gamma^o \to S^1$ be the Gauss mapping which associates with a point $x \in \Gamma^o$ the unit positive tangent vector of Γ^o in x. Then $\operatorname{rot} D = \deg \Delta$. (The unit circle S^1 is provided with the clockwise orientation.) Note that there is a natural homeomorphism $\Gamma_D \to \Gamma^o$ which is identity on $\Gamma_D \backslash U$. In what follows we shall identify the sets of vertices and (oriented) edges of Γ^o respectively with Vert$D$ and Edg$D$ via this homeomorphism.

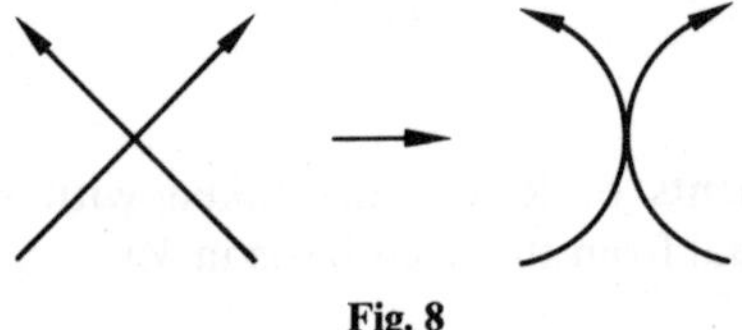

Fig. 8

To define $\int_D f$ I will assume that our EYB-operator $S=(R, \mu, \alpha, \beta)$ satisfies the following two conditions:

(5.3.1) $\mu_1, \ldots, \mu_m \in K^*$;

(5.3.2). If $R^{k,l}_{i,j} \neq 0$ or $(R^{-1})^{k,l}_{i,j} \neq 0$ then $\mu_i \mu_j = \mu_k \mu_l$.

These conditions are not too restrictive. In particular, if the ring K has no zero divisors then 5.3.2 holds for an arbitrary EYB-operator (see § 2).

The integral $\int_D f$ will be defined for the so-called *contributing states* of D. A state f of D is called contributing if $\pi_u(f) \neq 0$ for all $u \in \operatorname{Vert} D$. The set of contributing states of D is denoted by CSt(D).

Let $f \in \mathrm{CSt}(D)$. If $u \in \operatorname{Vert} D$ and if a, b, c, d are edges incident to u as in Fig. 7 then $\pi_u(f) \neq 0$ and in view of (5.3.2) $\mu_{f(a)} \mu_{f(b)} = \mu_{f(c)} \mu_{f(d)}$. Therefore the formal sum $\sum_{a \in \operatorname{Edg} D} \mu_{f(a)}\, a$ is a one-dimensional cycle in $\Gamma_D \approx \Gamma^o_D$ with coefficients in the multiplicative abelian group K^*. Let $[f]$ be the class of this cycle in $H_1(\Gamma^o; K^*) = H_1(\Gamma^o; \mathbf{Z}) \otimes_{\mathbf{Z}} K^*$. Then $\Delta_*([f]) \in H_1(S^1; K^*)$. The chosen orientation of S^1 determines an isomorphism $\Psi\colon H_1(S^1; K^*) \to K^*$. Put $\int_D f$

$=\Psi(\Delta_*([f]))$. It is clear that the integral $\int_D f$ is preserved by ambient isotopies of D in $\mathbf{R}^2$.

5.3.3. *Remarks*. 1. It is convenient to calculate $\int_D f$ as follows. Let $p \in S^1$ be a generic value of the mapping $\Delta: \Gamma^o \to S^1$. Then $\Delta^{-1}(p)$ is a finite set of non-vertex points of Γ^o in which the tangent vector is parallel (and equally directed) to the unit vector p. For $x \in \Delta^{-1}(p)$ denote by $a(x)$ the (oriented) edge of Γ^o containing x. Put $\varepsilon(x)=1$ if when one goes along $a(x)$ through x the tangent vector rotates in the clockwise direction and put $\varepsilon(x)=-1$ in the opposite case. Then

$$\int_D f = \prod_{x \in \Delta^{-1}(p)} \mu_{f(a(x))}^{\varepsilon(x)}. \tag{16}$$

2. If D is the closure of a braid diagram then the definitions of $\int_D f$ given in Sections 5.2, 5.3 are equivalent. The easiest way to see this is to apply the preceding remark to the vector p directed downwards in the plane of Fig. 1.

5.4. **Theorem.** *Let* $S=(R, \mu=\operatorname{diag}(\mu_1, \ldots, \mu_m), \alpha, \beta)$ *be a* EYB-*operator which satisfies Conditions* 5.3.1, 5.3.2 *and the following condition:*

5.4.1. *For any* $i, j, k, l \in \{1, 2, \ldots, m\}$

$$\sum_{1 \leq x, y \leq m} (R^{-1})_{y,i}^{x,j} R_{x,l}^{y,k} \mu_j \mu_y^{-1} = \delta_l^j \delta_k^i. \tag{17}$$

Then for any diagram D of an oriented link L

$$T_S(L) = \alpha^{-w(D)} \beta^{-\operatorname{rot} D} \sum_{f \in \mathrm{CSt}(D)} \Pi(f) \int_D f. \tag{18}$$

Condition 5.4.1 looks rather unpleasant. However, it is necessary for the right-hand side of (18) to be preserved by the Reidemeister move $\Omega 2b$ (see Fig. 9). Anyway, it is easy to check up that the operator S constructed in Theorem

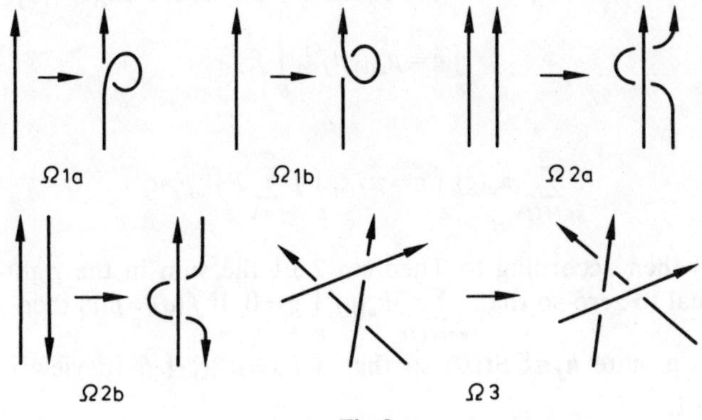

Fig. 9

4.2.1 satisfies 5.4.1. Both operators S_+ and S_- constructed in Theorem 4.3.2 also satisfy 5.4.1; this is verified in § 6. Thus, Theorem 5.4 gives state models for link invariants P_m, $Q_{m,+}$, $Q_{m,-}$ with $m=2, 3, \ldots$

Proof of Theorem 5.4. If D is the closure of a braid diagram then (18) follows straightforwardly from (15). Therefore, it suffices to check up the invariance of the right-hand side of (18) under the Reidemeister moves. It suffices to consider the moves $\Omega 1a$, $\Omega 1b$, $\Omega 2a$, $\Omega 2b$, $\Omega 3$ pictured in Fig. 9. Other Reidemeister moves (obtained from these by a change of orientations of branches) may be presented as compositions of the listed moves and their inverses (see, for example, Fig. 10).

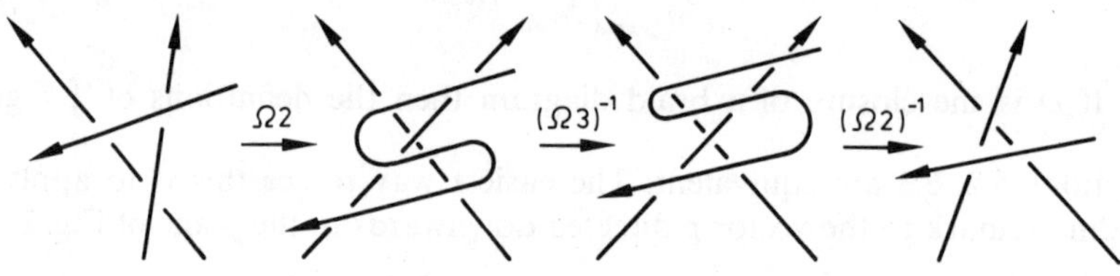

Fig. 10

Let E be a link diagram obtained from D by an application of $\Omega 1a$. Clearly, $w(E)=w(D)+1$ and $\operatorname{rot} E=\operatorname{rot} D+1$. Denote by u the additional vertex of E created by the move. Denote by a, b, c the edges of E incident to u (see Fig. 11).

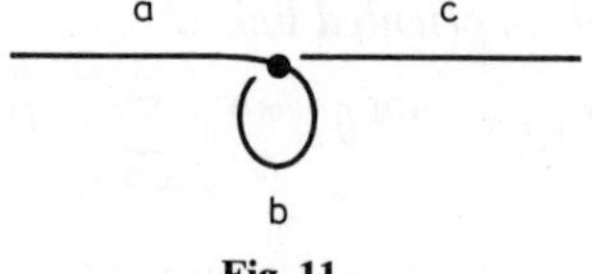

Fig. 11

For a contributing state f of E denote by $M(f)$ the set of contributing states of E $\operatorname{Edg} E \to \{1, 2, \ldots, m\}$ which are equal to f on the set $\operatorname{Edg} E \setminus \{b\}$. If $g \in M(f)$ then

$$\int_E g = \mu_{g(b)} \mu_{f(b)}^{-1} \int_E f.$$

Hence,

$$\sum_{g \in M(f)} \pi_u(g) \int_E g = \mu_{f(b)}^{-1} \int_E f \sum_{j=1}^{m} R_{f(a),j}^{f(c),j} \mu_j. \tag{19}$$

If $f(a) \neq f(c)$ then according to Theorem 2.3.1 the sum in the right-hand side of (19) is equal to zero so that $\sum_{g \in M(f)} \pi_u(g) \int_E g = 0$. If $f(a)=f(c)$ then f evidently gives rise to a state $h_f \in \operatorname{CSt}(D)$ so that $\int_D h_f = \mu_{f(b)}^{-1} \int_E f$. In view of (19) and Theorem 2.3.1,

$$\sum_{g \in M(f)} \Pi(g) \int_E g = \prod_{v \in \operatorname{Vert} D} \pi_v(h_f) \cdot \alpha\beta \int_D h_f = \alpha\beta \Pi(h_f) \int_D h_f.$$

This implies that

$$\alpha^{-w(E)}\beta^{-\operatorname{rot}E}\sum_{g\in \mathrm{CSt}(E)}\Pi(g)\int_E g=\alpha^{-w(D)}\beta^{-\operatorname{rot}D}\sum_{h\in \mathrm{CSt}(D)}\Pi(h)\int_D h.$$

The move $\Omega 1b$ is considered similarly.

The moves $\Omega 2$ and $\Omega 3$ do not change writhe and rot of link diagrams. Therefore it is sufficient to verify that these moves do not change the sum $\sum_{f\in \mathrm{CSt}(D)}\Pi(f)\int_D f$.

Let E be a link diagram obtained from D by an application of $\Omega 2a$. Denote the additional vertices and (oriented) edges of E respectively by u, v and a, b, c, d, r, s (see Fig. 12). For a contributing state f of E denote by $M(f)$ the set of contributing states of E coinciding with f on $\operatorname{Edg} E\setminus\{r,s\}$. If $g\in M(f)$ then $\int_E g=\int_E f$; this easily follows from Remark 5.3.3.1 applied to a vector directed oppositely to the tangent vectors of a, b, c, d. It is clear that

$$\sum_{g\in M(f)}\pi_u(g)\,\pi_v(g)=\sum_{1\leq x,y\leq m}R^{x,y}_{f(a),f(b)}(R^{-1})^{f(c),f(d)}_{x,y}$$

$$=\begin{cases}0, & \text{if } f(a)\neq f(c) \text{ or } f(b)\neq f(d)\\ 1, & \text{if } f(a)=f(c) \text{ and } f(b)=f(d)\end{cases}.$$

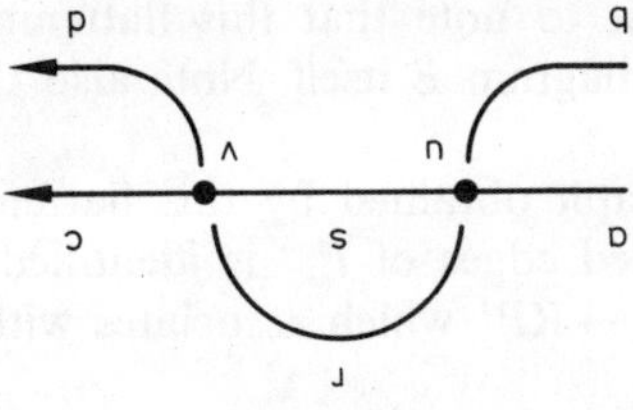

Fig. 12

If $f(a)=f(c)$ and $f(b)=f(d)$ then f gives rise to a state $h_f\in \mathrm{CSt}(D)$ so that $\int_D h_f=\int_E f$. Thus,

$$\sum_{g\in \mathrm{CSt}(E)}\Pi(g)\int_E g=\sum_{h\in \mathrm{CSt}(D)}\Pi(h)\int_D h.$$

The moves $\Omega 2b$ and $\Omega 3$ are considered along the same lines using (instead of the identity $RR^{-1}=1$) respectively formulas (17) and (1).

§ 6. Proof of Theorem 4.3.4

6.1. *Preliminaries.* Let $S=(R\colon V^{\otimes 2}\to V^{\otimes 2},\ \mu\colon V\to V,\ \alpha,\beta)$ be a EYB-operator. Assume that V has a preferred basis and that μ is the diagonal homomorphism diag$(\lambda_1^2, \lambda_2^2, \ldots, \lambda_m^2)$ where $\lambda_1, \ldots, \lambda_m \in K^*$. Assume also that the following holds true:

(6.1.1). $\lambda_{i'} = \lambda_i^{-1}$ for any $i = 1, \ldots, m$ (where $i' = m + 1 - i$);

(6.1.2). For any $i, j, k, l \in \{1, 2, \ldots, m\}$

$$R_{i,j}^{k,l} = R_{l',k'}^{j',i'}. \tag{20}$$

(6.1.3). If $R_{i,j}^{k,l} \neq 0$, then $\lambda_i \lambda_j = \lambda_k \lambda_l$.

In this setting we shall modify definitions of § 5 to make them applicable to non-oriented link diagrams.

Let E be a non-oriented link diagram. The graph $\Gamma_E \subset \mathbf{R}^2$ and the set Vert E are defined as in Sect. 5.1. By an oriented edge of E we shall mean a pair (an edge of E, an orientation of this edge). The set of oriented edges of E is denoted by Edg E. For an edge $a \in \operatorname{Edg} E$ we denote by a' the same edge with the opposite orientation. A state of E (with respect to S) is a mapping f: Edg $E \to \{1, 2, \ldots, m\}$ such that $f(a') = (f(a))'$ for all $a \in \operatorname{Edg} E$. Denote the set of states of E by St(E).

With each state f of E and each vertex u of E we associate $\pi_u(f) \in K$: If a, b, c, d are oriented edges of E incident to u as in Fig. 7 (i) then $\pi_u(f) = R_{f(a),f(b)}^{f(c),f(d)}$. Correctness of this definition follows from (20). For a state f of E define $\Pi(f)$ by the formula (14). A state f of E is called contributing if $\pi_u(f) \neq 0$ for all $u \in \operatorname{Vert} E$. The set of contributing states of E is denoted by CSt(E).

Let us flatten Γ_E in a small neighbourhood of Vert E in the way depicted in Fig. 13. It is important to note that this flattening depends not only on the graph Γ_E but on the diagram E itself. Note also that this flattening differs from the one used in § 5.

Denote the planar graph obtained by this flattening by $\Gamma^\wedge = \Gamma_E^\wedge$. As in Sect. 5.3 the set of oriented edges of $\Gamma^\wedge$ is identified with Edg E. Denote by Δ the Gauss mapping $\Gamma^\wedge \to RP^1$ which associates with a point $x \in \Gamma^\wedge$ the line in $\mathbf{R}^2$ tangent to $\Gamma^\wedge$ in x.

Let f be a contributing state of E. Provide every edge of $\Gamma^\wedge$ with an orientation and consider the sum $\Sigma = \Sigma \lambda_{f(a)}\, a$ where a runs through this family of oriented edges. Conditions 6.1.1 and 6.1.3 imply that Σ is a cycle in $\Gamma^\wedge$ with coefficients in K^* and that its class in $H_1(\Gamma^\wedge; K^*)$ does not depend on the choice of orientation in the edges of E. Denote by $\int_E f$ the image of this class under the homomorphism $\Delta_*: H_1(\Gamma^\wedge; K^*) \to H_1(RP^1; K^*) = K^*$. Here the isomorphism $H_1(RP^1; K^*) \to K^*$ is induced by the orientation of RP^1 corresponding to the clockwise rotation of a line in $\mathbf{R}^2$.

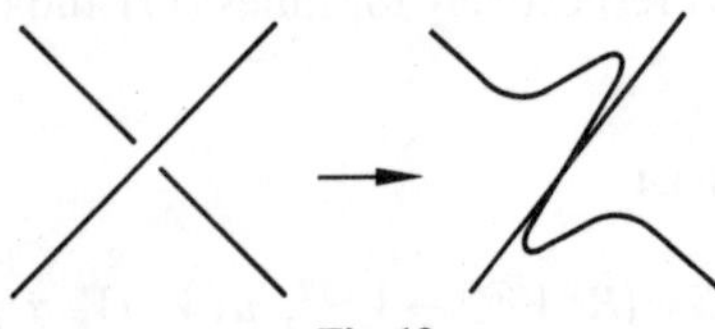

Fig. 13

The next formula is quite analogous to (16): If $p \in RP^1$ is a generic value of $\Delta: \Gamma^\wedge \to RP^1$ then

$$\int_E f = \prod_{x \in \Delta^{-1}(p)} \lambda_{f(a(x))}^{\varepsilon(x, a(x))}. \tag{21}$$

Here for a non-vertex point x of $\Gamma^\wedge$ the symbol $a(x)$ denotes an arbitrarily oriented edge of $\Gamma^\wedge$ containing x. Note that $\lambda_{f(a(x))}^{\varepsilon(x,a(x))}$ does not depend on the choice of orientation in this edge, since $\lambda_{f(a')}^{\varepsilon(x,a')} = \lambda_{f(a)'}^{-\varepsilon(x,a)} = [\lambda_{f(a)}^{-1}]^{-\varepsilon(x,a)} = \lambda_{f(a)}^{\varepsilon(x,a)}$ where $a = a(x)$.

6.2. **Lemma.** *Let $S = (R, \mu = \mathrm{diag}(\lambda_1^2, \ldots, \lambda_m^2), \alpha, \beta)$ be a* EYB-*operator with $\lambda_1, \ldots, \lambda_m \in K^*$. Suppose that Conditions 6.1.1, 6.1.2 and 6.1.3 hold true and that for all $i, j, k, l \in \{1, 2, \ldots, m\}$*

$$(R^{-1})_{i,j}^{k,l} = \lambda_i \lambda_l^{-1} R_{k',i}^{l,j'}. \tag{22}$$

If D is a diagram of an oriented link L and if E is the underlying non-oriented diagram then

$$\alpha^{w(D)} \beta^{\mathrm{rot} D} T_S(L) = \sum_{f \in \mathrm{CSt}(E)} \Pi(f) \int_E f. \tag{23}$$

Proof. Put $\mu_i = \lambda_i^2$ for $i = 1, \ldots, m$. It is easy to deduce from 6.1.1, 6.1.3 and the formula (22) that S satisfies Conditions 5.3.1 and 5.3.2. In view of (22) for any $i, j, k, l \in \{1, 2, \ldots, m\}$ $(R^{-1})_{y,i}^{x,j} = \lambda_y \lambda_j^{-1} R_{x',y}^{j,i'}$ and $R_{x,l}^{y,k} = \lambda_y \lambda_l^{-1} (R^{-1})_{l,k'}^{x',y}$. Therefore the equality (17) is satisfied:

$$\sum_{1 \le x, y \le m} (R^{-1})_{y,i}^{x,j} R_{x,l}^{y,k} \mu_j \mu_y^{-1}$$

$$= \sum_{1 \le x, y \le m} R_{x',y}^{j,i'} (R^{-1})_{l,k'}^{x',y} \mu_j \lambda_j^{-1} \lambda_l^{-1} = \delta_l^j \delta_{k'}^{i'} \lambda_j \lambda_l^{-1} = \delta_l^j \delta_k^i.$$

Now Theorem 5.4 implies that

$$\alpha^{w(D)} \beta^{\mathrm{rot}(D)} T_S(L) = \sum_{f \in \mathrm{CSt}(D)} \Pi(f) \int_D f. \tag{24}$$

We shall prove that the right-hand sides of formulas (23) and (24) are equal. Rewrite first (20, 22) as follows:

$$R_{i,j}^{k,l} = \lambda_k \lambda_j^{-1} (R^{-1})_{j,l'}^{i',k} = \lambda_{j'} \lambda_{k'}^{-1} (R^{-1})_{k',i}^{l,j'} = R_{l',k'}^{j',i'}. \tag{25}$$

Here the first term is equal to the second and fourth ones because of (20, 22); the third and fourth terms are equal because of (22).

Since all edges of D are oriented we have an inclusion $\mathrm{Edg}\, D \hookrightarrow \mathrm{Edg}\, E$. If $g \in \mathrm{St}(E)$ then the restriction of g to $\mathrm{Edg}\, D$ is a state of D denoted by g^*. The mapping $g \mapsto g^*: \mathrm{St}(E) \to \mathrm{St}(D)$ is bijective.

In a neighbourhood of a vertex $u \in \mathrm{Vert}\, E$ the diagram D has one of 4 possible orientations, given in Fig. 14. Let a, b, c, d be edges of E incident to u and oriented as in Fig. 14 (i). Then $\pi_u(g) = R_{g(a),g(b)}^{g(c),g(d)}$ and $\pi_u(g^*)$ is equal to one of the following 4 matrix elements:

$$R_{g(a),g(b)}^{g(c),g(d)}; \quad (R^{-1})_{g(b),g(d')}^{g(a'),g(c)}; \quad (R^{-1})_{g(c'),g(a)}^{g(d),g(b')}; \quad R_{g(d'),g(c')}^{g(b'),g(a')}.$$

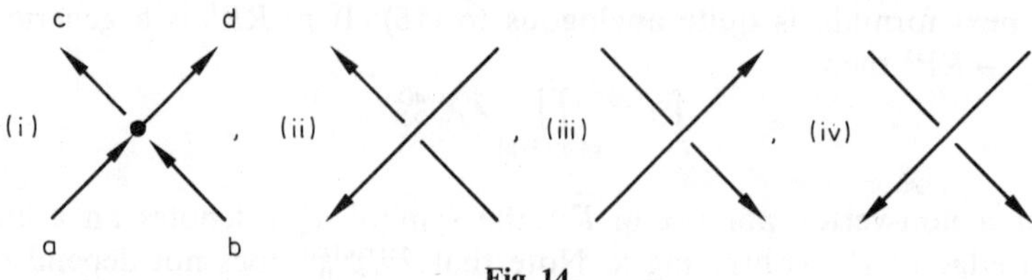

Fig. 14

It follows from (25) and the identity $g(e')=g(e)'$ for $e\in \operatorname{Edg} E$ that these 4 elements of K are non-zero iff at least one of them is non-zero. Thus, $\pi_u(g)\neq 0$ iff $\pi_u(g^*)\neq 0$. Therefore, the mapping $g\mapsto g^*: \mathrm{St}(E)\to \mathrm{St}(D)$ transforms the set $C\,\mathrm{St}(E)$ bijectively onto $C\,\mathrm{St}(D)$. To complete the proof of the lemma we shall show that for any $g\in C\,\mathrm{St}(E)$

$$\Pi(g)\int_E g=\Pi(g^*)\int_D g^*. \tag{26}$$

Let z_1, z_2 be the coordinates in $\mathbf{R}^2$. Applying if necessary an ambient isotopy we may assume that in a small disc neighbourhood W_u of any vertex u of D the diagram D looks as in Fig. 14, i.e. the overcrossing branch is a segment of line parallel to the line $z_1=z_2$, and the undercrossing branch is a segment parallel to the line $z_1=-z_2$. We shall also assume that D lies in a generic position with respect to the projection $(z_1, z_2)\mapsto z_2: \mathbf{R}^2\to\mathbf{R}$. Let X be the (finite) set of points of D in which the tangent to D line is parallel to the horizontal line $z_2=0$. Note that $X\subset D\backslash\cup W_u$. Let us flatten Γ_D as in Sect. 5.3. Let Γ^o be the graph in $\mathbf{R}^2$ obtained by this flattening (see Fig. 8). In each disc W_u may lie points of Γ^o in which the tangent line of Γ^o is parallel to the line $z_2=0$. Denote this set of points by Y_u and put $Y=\bigcup_u Y_u$. Deforming, if necessary, Γ^o assume that $u\notin Y_u$ for all vertices u. Let $a(x)$ and $\varepsilon(x)$ be the same objects as in Sect. 5.3.3. To prove (26) we shall need the following formula: If $f\in C\,\mathrm{St}(D)$ then

$$\int_D f=\prod_{x\in X\cup Y}\lambda_{f(a(x))}^{\varepsilon(x)} \tag{27}$$

(compare with (16) and (21)). In the proof of (27) (and only here) I will use the additive notation for the group operation (multiplication) in K^*. In particular, $\mu_i=2\lambda_i$ for all i. As in Sect. 6.1, the formal sum $\Sigma\lambda_{f(e)}e$, over $e\in \operatorname{Edg} D$ is a 1-cycle in Γ^o with coefficients in K^*. Denote the class of this cycle in $H_1(\Gamma^o; K^*)$ by σ. Clearly, $2\sigma=[f]$ where $[f]$ is the homological class of the sum $\sum_e \mu_{f(e)}e$. Let $r\colon S^1\to S^1$ be the two-sheeted covering. It is evident that r acts in $H_1(S^1; K^*)$ as multiplication by 2. If $\Delta\colon \Gamma^o\to S^1$ is the Gauss mapping (see Sect. 5.3) then

$$\int_D f=\Delta_*([f])=2\Delta_*(\sigma)=p_*(\Delta_*(\sigma))=(p\circ\Delta)_*(\sigma).$$

The equality $\int_D f=(p\circ\Delta)_*(\sigma)$ implies (27).

Flatten Γ_E in $\bigcup_u W_u$ according to the instructions of Sect. 6.1. In each disc W_u lie exactly 4 points of the flattened graph in which its tangent line is parallel to the line $z_2=0$, see Fig. 13. It is evident that the product of the expressions $\lambda_{g(a(x))}^{\varepsilon(x,a(x))}$ corresponding to these 4 points is equal to 1. Thus (21) implies that for $g\in \mathrm{CSt}(E)$

$$\int_E g=\prod_{x\in X} \lambda_{g(a(x))}^{\varepsilon(x,a(x))}.$$

A comparison of this formula with (27), where $f=g^*$, shows that to prove (26) we have to prove the local equality

$$\pi_u(g)=\pi_u(g^*)\prod_{x\in Y_u} \lambda_{g(a(x))}^{\varepsilon(x,a(x))} \tag{28}$$

for every $u\in \operatorname{Vert} E$.

Let a, b, c, d be the edges of E incident to u and oriented as in Fig. 14 (i). Put $i=g(a)$, $j=g(b)$, $k=g(c)$ and $l=g(d)$. Then $\pi_u(g)=R_{i,j}^{k,l}$. In accordance with 4 possible orientations of D in W_u (see Fig. 14) the right-hand side of (28) is easily computed to be one of 4 terms of (25). Therefore, (25) implies (28).

6.3. *Proof of Theorem* 4.3.4. Put $\lambda_i=q^{\bar{i}-(m+1)/2}$ for $i=1, \ldots, m$. It is easy to verify that the EYB-operator (R_ν,μ,α,β) constructed in Sect. 4.3 and the sequence $\lambda_1, \ldots, \lambda_m$ satisfy Conditions 6.1.1, 6.1.2 and 6.1.3. Instead of (22) we have

$$(R_\nu^{-1})_{i,j}^{k,l}=\varepsilon(i)\,\varepsilon(l)\,\lambda_i\,\lambda_l^{-1}(R_\nu)_{k',i}^{l,j'} \tag{29}$$

for any $i, j, k, l\in\{1, 2, \ldots, m\}$. I will first consider the case $\nu=-1$. In this case $\varepsilon(i)\equiv 1$ so that (29) coincides with (22) which enables us to apply Lemma 6.2.

Denote the YB-operator R_{-1} by R and the EYB-operator $(R_{-1}, \mu, \alpha, \beta)$ by S. If D is a diagram of an oriented link L and E is the underlying non-oriented diagram then Lemma 6.2 and equality $\beta=1$ imply

$$\tilde{Q}_{m,-1}(D)=\alpha^{w(D)}\,T_S(L)=\sum_{g\in \mathrm{CSt}(E)} \Pi(g)\int_E g. \tag{30}$$

This directly implies the first statement of the theorem.

Let us prove (12) confining ourselves for simplicity to the case of even m (so that $i'\neq i$ for all i). Denote the sum which stands in the right-hand side of (30) by $Q(E)$. We have to prove that for any non-oriented link diagrams E_+, E_-, E_0, E_∞ coinciding outside some disc and looking as in Fig. 4 inside this disc

$$Q(E_+)-Q(E_-)=(q-q^{-1})(Q(E_0)-Q(E_\infty)).$$

Schematically:

$$Q(\times)-Q(\times=(q-q^{-1})(Q()()-Q(\asymp)).$$

Denote by u the vertex of E_+, E_- pictured in Fig. 4. Let a, b, c, d be edges of E_+, incident to u and oriented as in Fig. 14 (i). Let $g\in\mathrm{CSt}(E_+)$. Put $i=g(a)$, $j=g(b)$, $k=g(c)$ and $l=g(d)$. Since $R_{i,j}^{k,l}=\pi_u(g)\neq 0$ there are 6 mutually exclusive cases: (I) $k=l=j=i$ and then $\pi_u(g)=q$; (II) $k=j$, $l=i\neq j,j'$ and $\pi_u(g)=1$; (III)

$k=l'=j=i'$ and $\pi_u(g)=q^{-1}$; (IV) $k=i<l=j\neq i'$ and $\pi_u(g)=q-q^{-1}$; (V) $k=i<l=j=i'$ and $\pi_u(g)=(q-q^{-1})(1-\lambda_i^2)$; (VI) $i=j'<l=k'$, $i\neq k$ and $\pi_u(g)=(q^{-1}-q)\lambda_i\lambda_l^{-1}$. The set $\mathrm{CSt}(E_+)$ splits in the disjoint union of 6 subsets, say, $A_1, \ldots, A_6$ singled out respectively by these possibilites for $\pi_u(g)$. This splitting induces a splitting of $Q(E_+)$ in a sum of 6 summands. Schematically:

$$Q(E_+)=Q\begin{pmatrix} i & & i \\ & \times & \\ i & & i \end{pmatrix}+\underset{j\neq i,\, i'}{Q\begin{pmatrix} j & & i \\ & \times & \\ i & & j \end{pmatrix}}+Q\begin{pmatrix} i' & & i \\ & \times & \\ i & & i' \end{pmatrix}$$

$$+\underset{i<j\neq i'}{Q\begin{pmatrix} i & & j \\ & \times & \\ i & & j \end{pmatrix}}+\underset{i<i'}{Q\begin{pmatrix} i & & i' \\ & \times & \\ i & & i' \end{pmatrix}}+\underset{k\neq i<k'}{Q\begin{pmatrix} k & & k' \\ & \times & \\ i & & i' \end{pmatrix}}$$

(It is understood that we sum up over all permitted i, j, k.) Each state $g\in A_1$ determines in the evident fashion a (contributing) state g^* of E_0. This gives all contributing states of E_0 whose values on two pictured in Fig. 4 and oriented upwards edges of E_0 are equal. Clearly

$$\int_{E_0} g^*=\int_E g, \qquad \Pi(g)=\pi_u(g)\,\Pi(g^*)=q\,\Pi(g^*).$$

Thus,

$$Q\begin{pmatrix} i & & i \\ & \times & \\ i & & i \end{pmatrix}=q\,Q(i\uparrow\uparrow i).$$

Analogously, using (21),

$$Q\begin{pmatrix} i' & & i \\ & \times & \\ i & & i' \end{pmatrix}=q^{-1}Q\begin{pmatrix} i \\ \rightleftarrows \\ i \end{pmatrix};$$

$$\underset{i<j\neq i'}{Q\begin{pmatrix} i & & i' \\ & \times & \\ i & & i' \end{pmatrix}}=(q-q^{-1})\underset{i<j\neq i'}{Q(i\uparrow\uparrow j)};$$

$$\underset{i<i'}{Q\begin{pmatrix} i & & i' \\ & \times & \\ i & & i' \end{pmatrix}}=(q-q^{-1})\left[\underset{i<i'}{Q(i\uparrow\uparrow i')}-\underset{i<i'}{Q\begin{pmatrix} i' \\ \rightleftarrows \\ i \end{pmatrix}}\right];$$

$$\underset{k\neq i<k'}{Q\begin{pmatrix} k & & k' \\ & \times & \\ i & & i' \end{pmatrix}}=(q^{-1}-q)\underset{i<j\neq i'}{Q\begin{pmatrix} j \\ \rightleftarrows \\ i \end{pmatrix}}$$

Here each expression $Q(\ldots)$ denotes the sum of products $\Pi(g)\int g$ over those contributing stages g of the corresponding non-oriented (!) link diagram, whose values on the pictured oriented edges satisfy the pointed out equalities and inequalities.

Summing it up, we obtain

$$Q(\times)=qQ(i\,\curvearrowleft\!\curvearrowright\, i)+\underset{j\neq i,i'}{Q\begin{pmatrix} j & & i\\ & \times & \\ i & & j\end{pmatrix}}+q^{-1}Q\begin{pmatrix} i\\ \asymp\\ i\end{pmatrix}$$

$$+(q-q^{-1})\underset{i<j}{Q(i\,\curvearrowleft\!\curvearrowright\, j)}+(q^{-1}-q)\underset{i<j}{Q\begin{pmatrix} j\\ \asymp\\ i\end{pmatrix}}. \tag{31}$$

An analogous formula holds for $Q(\times)$. Namely, rotate E_- around u to the angle 90°, apply (31) and then rotate all the diagrams involved in the right-hand side of (31) to $-90°$. After some evident renotation we get

$$Q(\times)=qQ\begin{pmatrix} i\\ \asymp\\ i\end{pmatrix}+\underset{j\neq i,i'}{Q\begin{pmatrix} j & & i\\ & \times & \\ i & & j\end{pmatrix}}+q^{-1}Q(i\,\curvearrowleft\!\curvearrowright\, i)$$

$$+(q-q^{-1})\underset{j<i}{Q\begin{pmatrix} j\\ \asymp\\ i\end{pmatrix}}+(q^{-1}-q)\underset{i>j}{Q(i\,\curvearrowleft\!\curvearrowright\, j)}. \tag{32}$$

Here I used the obvious equalities

$$\underset{i<j}{Q(i\,\curvearrowleft\!\curvearrowright\, j)}=\underset{i<j}{Q(i'\,\curvearrowleft\!\curvearrowright\, j')}=\underset{i>j}{Q(i\,\curvearrowleft\!\curvearrowright\, j)}.$$

It is easy to understand that

$$\underset{j\neq i,i'}{Q\begin{pmatrix} j & & i\\ & \times & \\ i & & j\end{pmatrix}}=\underset{j\neq i,i'}{Q\begin{pmatrix} j & & i\\ & \times & \\ i & & j\end{pmatrix}}.$$

(Here it is important that for $g\in A_2$, $\pi_u(g)=1$). Therefore, subtracting (32) from (31) we get

$$Q(\times)-Q(\times)=(q-q^{-1})\Bigg[Q(i\,\curvearrowleft\!\curvearrowright\, i)+\underset{i<j}{Q(i\,\curvearrowleft\!\curvearrowright\, j)}+\underset{i>j}{Q(i\,\curvearrowleft\!\curvearrowright\, j)}$$

$$-Q\begin{pmatrix} i\\ \asymp\\ i\end{pmatrix}-\underset{i<j}{Q\begin{pmatrix} j\\ \asymp\\ i\end{pmatrix}}-\underset{i>j}{Q\begin{pmatrix} j\\ \asymp\\ i\end{pmatrix}}\Bigg]$$

$$=(q-q^{-1})[Q(\;\;)(\;\;)-Q(\asymp)].$$

Consider now the case $v=1$. We shall slightly modify R_1 to satisfy (22). Put $R=R_1$ and put

$$\tilde{R}=q\sum_i E_{i,i}\otimes E_{i,i}+\sum_{\substack{i,j\\ i\neq j,j'}}\varepsilon(i)\,\varepsilon(j)\,E_{i,j}\otimes E_{j,i}$$

$$-q^{-1}\sum_i E_{i,i'}\otimes E_{i',i}+(q-q^{-1})\sum_{i<j}[E_{i,i}\otimes E_{j,j}+q^{\bar{i}-\bar{j}}\,E_{i,j'}\otimes E_{i',j}].$$

(Here $i, j=1, 2, \ldots, m$). The matrices of R and $\tilde{R}$ are related by the formula

$$\tilde{R}^{k,l}_{i,j}=\varepsilon(i)\,\varepsilon(k)\,R^{k,l}_{i,j}.$$

This implies that $\tilde{R}$ is invertible and

$$(\tilde{R}^{-1})^{k,l}_{i,j}=\varepsilon(i)\,\varepsilon(k)(R^{-1})^{k,l}_{i,j}.$$

If $R^{k,l}_{i,j}\neq 0$ then either the ordered pairs $\varepsilon(k)$, $\varepsilon(l)$ and $\varepsilon(i)$, $\varepsilon(j)$ are equal or $\varepsilon(i)=-\varepsilon(j)$ and $(\varepsilon(k), \varepsilon(l))=(\varepsilon(j), \varepsilon(i))$. In the first case $\tilde{R}^{k,l}_{i,j}=R^{k,l}_{i,j}$, in the second case $\tilde{R}^{k,l}_{i,j}=-R^{k,l}_{i,j}$. An analogous statement is valid for R^{-1}. Therefore, if z: $V^{\otimes n}\to V^{\otimes n}$ is a composition of several homomorphisms $R_i(n)$, $R_i(n)^{-1}$ and if $\tilde{z}$: $V^{\otimes n}\to V^{\otimes n}$ is the corresponding composition of $\tilde{R}_i(n)$, $\tilde{R}_i(n)^{-1}$, then the matrix elements of $z, \tilde{z}$ are related as follows. Let $I=(i_1, \ldots, i_n)$ and $J=(j_1, \ldots, j_n)$ be two sequences of elements of the set $\{1, 2, \ldots, m\}$. If $z^J_I=0$, then $\tilde{z}^J_I=0$. If $z^J_I\neq 0$ then the sequences $(\varepsilon(i_1), \ldots, \varepsilon(i_n))$ and $(\varepsilon(j_1), \ldots, \varepsilon(j_n))$ may be obtained from each other by several, say, $p(I, J)$ transposition $(1, -1)\leftrightarrow(-1, 1)$. Then

$$\tilde{z}^J_I=(-1)^{P(I,J)}\,p^J_I.$$

This easily implies that $\tilde{R}$ is a YB-operator and that the same μ, α, β as in Theorem 4.3.2 (the case $v=1$) enhance $\tilde{R}$ to a EYB-operator. The identity $p(I, I)=0$ implies that (R, μ, α, β) and $(\tilde{R}, \mu, \alpha, \beta)$ give rise to the same invariant of braids and links. The formula (29) gives the equalities

$$(\tilde{R}^{-1})^{k,l}_{i,j}=\varepsilon(k)\,\varepsilon(k')\,\lambda_i\,\lambda_l^{-1}\,\tilde{R}^{l,j'}_{k',i}=-\lambda_i\,\lambda_l^{-1}\,\tilde{R}^{l,j'}_{k',i}.$$

We see, finally, that the EYB-operator $(\sqrt{-1}\,\tilde{R}, \mu, \sqrt{-1}\,\alpha, \beta)$ and the sequence $\lambda_1, \ldots, \lambda_m$ satisfy Conditions 6.1.1, 6.1.2, 6.1.3 and the equality (22). This EYB-operator gives rise to the same link invariant as (R, μ, α, β), namely, to $Q_{m,1}$. Now the same argument as in the case $v=-1$ can be applied to this EYB-operator which proves the theorem for $v=1$.

6.4. *Remark.* In Sect. 6.1 and 6.2 one could use instead of the involution $i\mapsto i'=m+1-i$ an arbitrary involution of the set $\{1, 2, \ldots, m\}$.

References

1. Baxter, E.J.: Exactly solved models in statistical mechanics. New York, London: Academic Press 1982
2. Drinfel'd, V.G.: Quantum groups. Zap. Nauchn. Semin. Leningr. Otd. Mat. Inst. Steklova **155**, 18–49 (1986) (in Russian)
3. Freyd, P., Yetter, D., Hoste, J., Lickorish, W.B.R., Millett, K., Ocneanu, A.: A new polynomial invariant of knots and links. Bull. Am. Math. Soc. **12**, 239–246 (1985)
4. Jimbo, M.: Quantum R matrix for the generalized Toda system. Commun. Math. Phys. **102**, 537–547 (1986)
5. Jones, V.F.R.: A polynomial invariant for knots via von Neumann algebras. Bull. Am. Math. Soc. **12**, 103–111 (1985)
6. Jones, V.F.R.: Notes on a talk in Atiyah's seminar. Handwritten manuscript (November 1986)
7. Kauffman, L.H.: New Invariants in the Theory of Knots. Preprint (1986)
8. Kulish, P.P., Sklyanin, E.K.: On the solutions of the Yang-Baxter equation. Zap. Nauchn. Semin. Leningr. Otd. Mat. Inst. Steklova 95, 129–160 (1980) (in Russian)
9. Tahtadjan, L.A., Faddeev, L.D.: The quantum method for the inverse problem and the XYZ Heisenberg model. Uspekhi Mat. Nauk **34**, (N 5) 13–63 (1979); (English transl. Russian Math. Surv. **34**, (N 5) 11–61 (1979))
10. Turaev, V.G.: The Reidemeister torsion in the knot theory. Uspekhi Mat. Nauk **41**, (N 1) 97–147 (1986); (English transl. Russian Math. Surv. **41** (N 1) 119–182 (1986)

REPRESENTATIONS OF THE ALGEBRA $U_q(sl(2))$, q-ORTHOGONAL POLYNOMIALS AND INVARIANTS OF LINKS

A.N.Kirillov, N.Yu. Reshetikhin

Leningrad Branch of Steklov Mathematical Institute, Fontanka 27, 191011, Leningrad, USSR

INTRODUCTION

This work shows how quantized universal enveloping algebras are connected with other areas of mathematics, using algebra sl(2) as an example. It is shown that in the representation theory of the algebra $U_q(sl(2))$ the q-analogues of $6j$-symbols which were introduced by Askey and Wilson [1] in connection with q-orthogonal polynomials, appear naturally. The connection between the quantized universal algebras and the theory of invariants of links, discovered in [2], is considered in more detail. With the help of q-analogues of $6j$ -symbols we propose a new representation for the invariants of links, related to $U_q(sl(2))$, which is to a great extend similar to SOS models of statistical physics. The representation theory of algebra $U_q(sl(2))$ is closely connected with Temperly-Lieb-Jones algebra, which emerged in statistical mechanics [3] and in the theory of von-Neumann algebras [4]. It happens that the matrix elements of generators in irreducible representations of Jones algebras are special values of q-analogues of $6j$ -symbols.

Let us make some historical comments. Quantum universal enveloping algebras appeared as a result of research into on algebraic aspects of quantum integrable systems [5, 6]. The first example of such an algebra was the algebra $U_q(sl(2))$ found by Kulish and Reshetikhin [7]. The structure of Hopf algebra on $U_q(sl(2))$ was discovered independently by Sklyanin [8], Drinfeld [9], Jumbo [10] who built the q-deformation of the universal enveloping algebra for any simple Lie algebra $\mathfrak{g}$ [9, 10]. A new approach to $U_q(\mathfrak{g})$ algebras, which reflects most adequatly their connection with quantum integrable systems, was proposed by Faddeev, Reshetikhin and Takhtadjan [11], (see also [12]). Finite dimensional representationsoof $U_q(sl(2))$ are described in [7]. First substantial results in the representation theory of algebras $U_q(\mathfrak{g})$ were obtained by Lusztig [13] and Rosso [14] for simple $\mathfrak{g}$. The q-analogues of the Rakah-Fock formulae for $3j$-symbols were obtained by Vaksman [15]. Other representation of the q-analogues of $3j$-symbolds, as well as the properties of their symmetry were found by Kirillov [16].

The connection between universal enveloping algebras with the link theory was established in [9]. We should also mention that the invariants of links, connected with tensor representations of $U_q(\mathfrak{g})$ for classical Lie algebras, can be obtained with the help of cabling (Murakami [17]). Cabling invariants correspond in the terminology of [2] to invariants parametrized by the tensor products of the vector representations of $U_q(\mathfrak{g})$. In particular, the invariants corresponding to finite dimensional representations of $U_q(sl(2))$ are built by Akutzu and Wadati [18] with help of braid representation using the results from vertex models of statistical mechanics. The main results obtained in this direction, are represented in the review by V. Jones [29].

Then studying classical orthogonal polynomials Askey and Wilson [1], proposed q-analogues of $6j$-symbols. As it turns out the q-analogues of $6j$-symbols, that occur in the representation theory of $U_q(sl(2))$ are proportional to the ones defined in [1]. The orthogonality of these polynomials as well as the recurrent relations for them follows from equalities for q- $6j$ -symbols in the representation theory of the algebra $U_q(sl(2))$.

Let us briefly consider the content of this paper. Section 1 contains the description of the algebra $U_q(sl(2))$, of the corresponding universal R-matrix and also presents some useful formulae. The irreducible representations of $U_q(sl(2))$ and the q-analog of the Weyl element are described in Section 2. In the same section an extension of algebra $U_q(sl(2))$ by the q-analog of the Weyl element is introduced.

The decomposition of tensor product of two irreducible finite-dimensional representation of $U_q(sl(2))$ is given in Section 3. It also contains the relations between R-matrices and Clebsh-Gordan coefficients (CGC). We prove that the extension of Uq(sl(2)) by the Weyl element is a Hopf algebra. In Section 4, following [2], a graphical representation of relations between R-matrices and GGC is proposed. In Section 5 the q-analogues of $6j$ -symbols are described. It is shown that they are defined by q-hypergeometrical function ${}_4\varphi_3$ and the symmetries of them are found. Note that the q-analogues of CGC and $6j$ -symbols correspond to the function ${}_3\varphi_2$ and ${}_4\varphi_3$ for such values of arguments when these functions are polynomials. Graphical representations for q- $6j$ -symbols are introduced in Section 6. Relations between q- $6j$-symbols, particularly their orthogonality, are easily obtained with their help.

In Section 7 a new representation for the invariants of links connected with $U_q(sl(2))$ is built with the help of q- $6j$-symbols. This representation is an exact analogue of SOS models in statistical mechanics [19]. The relation of SOS models in statistical mechanics to q-analogue of $6j$-symbols was recently found also by Pasquier [31].

1. Algebra $U_q(sl(2))$

The algebra $U_q(sl(2))$ [7-10] is generated by elements $H, X^{\pm}$ with the commutation relations

$$[X^{\pm}, H] = \mp 2X^{\pm}, \quad [X^+, X^-] = \frac{q^{H/2} - q^{-H/2}}{q^{1/2} - q^{-1/2}} . \qquad (1.1)$$

As a linear algebra $U_q(sl(2))$ space consists of convergent power series in H and of polynomials in $X^{\pm}$. The following formulae for the comultiplication, the antipode and counit on the generators define the structure of a Hopf algebra on $U_q(s\ell(2))$:

$$\Delta(X^{\pm}) = X^{\pm} \otimes q^{H/4} + q^{-H/4} \otimes X^{\pm}, \quad \Delta(H) = H \otimes 1 + 1 \otimes H, \qquad (1.2)$$

$$S(X^{\pm}) = -q^{\pm 1/2} X^{\pm}, \quad S(H) = -H, \qquad (1.3)$$

$$\varepsilon(H) = \varepsilon(X^{\pm}) = 0. \qquad (1.4)$$

The maps $\Delta' = \sigma \circ \Delta$, $S' = S^{-1}$ where σ is the permutation in $U_q(s\ell(2))^{\otimes 2}$, $\sigma(a \otimes b) = b \otimes a$ also define the structure of Hopf algebra on $U_q(s\ell(2))$. Let us denote this Hopf algebra as by $U_q(s\ell(2))'$. It is evident from (1.2), (1.3) that

$$U_q(s\ell(2))' = U_{q^{-1}}(s\ell(2)). \qquad (1.5)$$

Comultiplications Δ and Δ' are connected in $U_q(s\ell(2))^{\otimes 2}$ by the following automorphism [9]:

$$\Delta'(a) = R\,\Delta(a)\,R^{-1}, \qquad (1.6)$$

where $R \in U_q(sl(2))^{\otimes 2}$ and

$$R = \exp\left(\frac{h}{4} H \otimes H\right) \sum_{n \geq 0} \frac{(1-q^{-1})^n}{[n]!} e^n \otimes f^n , \quad q = e^h . \tag{1.7}$$

Here $[n] = (q^{n/2} - q^{-n/2}) / (q^{1/2} - q^{-1/2})$, $e = \exp(\frac{h}{4} H) X^+$, $f = \exp(-\frac{h}{4} H) X^-$.

The element R is called the universal R-matrix. It satisfies the relations:

$$(\Delta \otimes id) R = R_{13} R_{23} , \tag{1.8}$$

$$(id \otimes \Delta) R = R_{13} R_{12} , \tag{1.9}$$

$$(S \otimes id) R = R^{-1} , \tag{1.10}$$

where the indices show the embeddings of R into $U_q(sl(2))^{\otimes 3}$.

Formula (1.8) (or (1.9)) imply the Yang-Baxter equation for R :

$$R_{12} R_{13} R_{23} = R_{23} R_{13} R_{12} . \tag{1.11}$$

The center of the algebra $U_q(sl(2))$ is generated by the q-analog of Casimir's elements [8, 10]:

$$C = \left(\frac{q^{\frac{H+1}{4}} - q^{-\frac{H+1}{4}}}{q^{1/2} - q^{-1/2}} \right)^2 + X^- X^+ . \tag{1.12}$$

For real q one can introduce $*$ -antiinvolution

$$(X^{\pm})^* = X^{\mp} , \quad H^* = H . \tag{1.13}$$

The real form of $U_q(sl(2))$ corresponding to this $*$-antiinvolution is denoted by $U_q(su(2))$. The element (1.12) is invariant under the action of involution (1.13) and therefore all finite-dimensional representations of $U_q(su(2))$ are completely reducible.

Let us give now some useful formulae:

$$\Delta^{(N)}((X^{\pm})^m) = \sum_{0 \le a_1 \le \dots \le a_{N-1} \le m} \begin{bmatrix} m \\ a_1, a_2 - a_1, a_3 - a_2, \dots, m - a_{N-1} \end{bmatrix}_q$$

$$\bigotimes_{i=1}^{N} (X^{\pm})^{a_i - a_{i-1}} q^{-(m - a_i - a_{i-1}) \cdot \frac{H}{4}}, \quad a_N = m, \tag{1.14}$$

$$[X_+, X_-^m] = X_-^{m-1} [m]_1 \frac{q^{\frac{m-1}{2}} q^{H/2} - q^{-\frac{m-1}{2}} q^{-H/2}}{q^{1/2} - q^{-1/2}}, \tag{1.15}$$

where $\Delta^{(N)} : U_q(sl(2)) \longrightarrow U_q(sl(2))^{\otimes N}$ is the composition of comultiplications and

$$\begin{bmatrix} m \\ b_1 \, b_2 \dots b_N \end{bmatrix}_q = \frac{[m]!}{[b_1]! \dots [b_N]!}, \tag{1.16}$$

$b_1 + \dots + b_N = m$.

2. Irreducible Representations Of $U_q(su(2))$ And Finite Dimensional R-matrices

All irreducible representations of $U_q(su(2))$ are finite dimensional. It is useful to describe them in the weight basis. In this basis the element H is a diagonal matrix. The irreducible representations of $U_q(su(2))$ are parametrized by integer (or halfinteger) numbers j. The representation π^j have dimension $2j+1$. The generators $H, X^{\pm}$ act in weight basis e^j_m in the following way:

$$\pi^j(X^{\pm})e^j_m = ([j \mp m][j \pm m+1])^{1/2} e^j_{m\pm 1}, \quad (2.1)$$

$$\pi^j(H)e^j_m = 2m e^j_m , \quad (2.2)$$

where $-j \leq m \leq j$, $2m \equiv 2j \pmod 2$). We shall denote by V^j the representation space of the representation π^j. From (1.1) and (1.2) at $m=j$ we find the value of the central element (1.12) in the irreducible representation:

$$c\, V^j = \left(\frac{q^{\frac{2j+1}{4}} - q^{-\frac{2j+1}{4}}}{q^{1/2} - q^{-1/2}} \right)^2 V^j = [j+\tfrac{1}{2}]^2 V^j. \quad (2.3)$$

The following important formula holds

$$\pi^j((X^{\pm})^a)e^j_n = \left(\frac{[j\mp n]!\,[j \pm n+a]!}{[j \mp n-a]!\,[j\pm n]!} \right)^{1/2} e^j_{n\pm a} . \quad (2.4)$$

Here the r.h.s. is nonzero only for $|n \pm a| \leq j$. Let us consider the matrix $\mathbb{R}^{j_1 j_2} = (\pi^{j_1} \otimes \pi^{j_2})R$ acting in $V^{j_1} \otimes V^{j_2}$. It is not difficult find the matrix elements of $\mathbb{R}^{j_1 j_2}$:

$$\left(\mathbb{R}^{j_1 j_2}\right)^{n_1\, n_2}_{n_1+n,\, n_2-n} = \frac{(1-q^{-1})^n}{[n]!}\, q^{\frac{n_1-n_2+2n}{2}n} .$$

$$\left(\frac{[j_1-n_1]!\,[j_1+n_1+n]!\,[j_2+n_2]!\,[j_2-n_2+n]!}{[j_1-n_1-n]!\,[j_1+n_1]!\,[j_2+n_2-n]!\,[j_2-n_2]!}\right)^{1/2}. \tag{2.5}$$

Define the matrix w^j acting in V^j with the elements:

$$w^j_{mm'} = q^{-\frac{c_j}{2}} (-1)^{j-m} \delta_{m,-m'}\, q^{-\frac{m}{2}}, \tag{2.6}$$

where $c_j = j(j+1)$ and consider the algebra $\widetilde{U_q}(su(2))$ which is the extension of $U_q(su(2))$ by the element with the value (2.6) in any representation. Let τ be the linear antiautomorphism of $U_q(su(2))$ which is the transposition in

$$\tau(X^{\pm}) = X^{\mp},\quad \tau(H) = H. \tag{2.7}$$

PROPOSITION 2.1. In $\widetilde{U_q}(su(2))$ we have:

$$w a w^{-1} = \tau S(a),\quad \forall a \in U_q(su(2)). \tag{2.8}$$

To prove (2.8) it is sufficient to check it in any irreducible $U_q(su(2))$-module.

From (1.10) and (2.8) we obtain the crossing-symmetry of universal R-matrix:

$$(\tau \otimes id) R^{-1} = (w \otimes 1) R (w^{-1} \otimes 1), \tag{2.9}$$

and finite dimensional R-matrices $\mathbb{R}^{j_1 j_2}$,

$$\left((\mathbb{R}^{j_1 j_2})^{-1}\right)^{t_1} = w_1 \mathbb{R}^{j_1 j_2} w_1^{-1}. \tag{2.10}$$

Here t_1 is the transposition in the first space in $V^{j_1} \otimes V^{j_2}$ and $w_1 = w^{j_1} \otimes 1$.

3. q-Analog Of Clebsch-Gordan Coefficients

Let us consider the tensor product $\pi^{j_1} \otimes \pi^{j_2}$ of two irreducible representations of $U_q(su(2))$. In accordance with the complete reducibility of representations of $U_q(su(2))$, $\pi^{j_1} \otimes \pi^{j_2}$ is decomposed into the following sum of irreducible components [10]:

$$V^{j_1} \otimes V^{j_2} = \sum^{\oplus}_{\substack{|j_1 - j_2| \leq j \leq j_1 + j_2 \\ 2j \equiv 2j_1 + 2j_2 \pmod 2}} V^{j} \quad . \tag{3.1}$$

Let $e^{j}_{m}(j_1 j_2)$ be the weight basis in the irreducible component V^{j}. The coordinates of the vectors $e^{j}_{m}(j_1 j_2)$ in the basis $e^{j_1}_{m_1} \otimes e^{j_2}_{m_2}$, by analogy with $q=1$ case, will are called the Clebsch-Gordan coefficients (CGC):

$$e^{j}_{m}(j_1 j_2) = \sum_{m_1, m_2} \begin{bmatrix} j_1 & j_2 & j \\ m_1 & m_2 & m \end{bmatrix}_q e^{j_1}_{m_1} \otimes e^{j_2}_{m_2} \; . \tag{3.2}$$

Since π^{j_1} are $*$ -representations of $U_q(su(2))$ we have also

$$e^{j_1}_{m_1} \otimes e^{j_2}_{m_2} = \sum_{j, m} \begin{bmatrix} j_1 & j_2 & j \\ m_1 & m_2 & m \end{bmatrix}_q e^{j}_{m}(j_1 j_2)\, \delta(j_1 j_2 j). \tag{3.2'}$$

Here $\delta(j_1 j_2 j) = 1$ if $|j_1 - j_2| \leq j \leq j_1 + j_2$, $2j \equiv 2j_1 + 2j_2 (2)$ and $\delta(j_1 j_2 j) = 0$ in other cases.

The coefficients in (3.2), (3.2') can be considered as q -analog of CGC for $su(2)$. The coefficients $\begin{bmatrix} j_1 & j_2 & j \\ m_1 & m_2 & m \end{bmatrix}_q$ can be calculated using the formulae of Section 2. We shall omit the calculations and present only the final formulae:

q-analog of the Majumbar formula [16]:

$$\begin{bmatrix} j_1 & j_2 & j \\ m_1 & m_2 & m \end{bmatrix}_q = \delta_{m_1+m_2,m}\; q^{-\frac{1}{4}(j_1+j_2-j)(j_1-j_2+j+1)-j\frac{m_2}{2}-j_2\frac{m}{2}}$$

$$(-1)^{j_1-m_1}\left\{\frac{[j-m]!\,[j+m]!\,[j_1+m_1]!\,[j_2+m_2]!\,[j_1+j_2-j]!\,[2j+1]}{[j_1-m_1]!\,[j_2-m_2]!\,[j_1-j_2+j]!\,[j_2-j_1+j]!\,[j_1+j_2+j+1]!}\right\}^{1/2}$$

$$\sum_{r\geq 0}(-1)^r\, q^{\frac{r(m-j-1)}{2}}\,\frac{[2j_2-r]!\,[j_1-j_2+j+r]!}{[r]!\,[j_2+m_2-r]!\,[j-j_2+m_1+r]!\,[j_1+j_2-j-r]!}. \tag{3.3}$$

q-analog of the Rakah-Fock formula [15]:

$$\begin{bmatrix} j_1 & j_2 & j \\ m_1 & m_2 & m \end{bmatrix}_q = \delta_{m_1+m_2,m}\,(-1)^{j_1-m_1}\, q^{\frac{1}{4}\left(j_2(j_2+1)-j_1(j_1+1)-j(j+1)\right)+\frac{m_1(m_1+1)}{2}}$$

$$\left\{\frac{[j+m]!\,[j-m]!\,[j_1-m_1]!\,[j_2-m_2]!\,[j_1+j_2-j]!\,[2j+1]}{[j_1+m_1]!\,[j_2+m_2]!\,[j_1-j_2+j]!\,[j_2-j_1+j]!\,[j_1+j_2+j+1]!}\right\}^{1/2} \tag{3.4}$$

$$\sum_{r\geq 0}(-1)^r\, q^{\frac{1}{2}r(m+j+1)}\,\frac{[j_1+m_1+r]!\,[j_2+j-m_1-r]!}{[r]!\,[j-m-r]!\,[j_1-m_1-r]!\,[j_2-j+m_1+r]!}.$$

q-analog of the Van der Waerder formulae[16]:

$$\begin{bmatrix} j_1 & j_2 & j \\ m_1 & m_2 & m \end{bmatrix}_q = \delta_{m_1+m_2,m}\,\Delta(j_1 j_2 j)\, q^{\frac{1}{4}(j_1+j_2-j)(j_1+j_2+j+1)+\frac{j_1 m_2 - j_2 m_1}{2}}$$

$$\{[j_1+m_1]!\,[j_1-m_1]!\,[j_2+m_2]!\,[j_2-m_2]!\,[j+m]!\,[j-m]!\,[2j+1]\}^{1/2}$$

$$\sum_{r\geq 0} (-1)^r\, q^{-\frac{1}{2}r(j_1+j_2+j+1)} \cdot \tag{3.5}$$

$$\cdot \frac{1}{[r]!\,[j_1+j_2-j-r]!\,[j_1-m_1-r]!\,[j_2+m_2-r]!\,[j-j_2+m_1+r]!\,[j-j_1+r-m_2]!},$$

where

$$\Delta(abc) = \left\{ \frac{[-a+b+c]!\,[a-b+c]!\,[a+b-c]!}{[a+b+c+1]!} \right\}^{1/2}. \tag{3.6}$$

The proofs of these formulae are given in [16].

The following relations between CGC reflect the completness and the orthogonality of the basis $e^j_m(j_1 j_2)$ in $V^{j_1} \otimes V^{j_2}$:

$$\sum_{m_1 m_2} \begin{bmatrix} j_1 & j_2 & j \\ m_1 & m_2 & m \end{bmatrix}_q \begin{bmatrix} j_1 & j_2 & j' \\ m_1 & m_2 & m' \end{bmatrix}_q = \delta_{jj'}\,\delta_{mm'}\,\delta(j_1 j_2 j). \tag{3.7}$$

$$\sum_{\substack{|j_1-j_2|\le j\le j_1+j_2\\ |m|\le j}} \begin{bmatrix} j_1 & j_2 & j \\ m_1 & m_2 & m \end{bmatrix}_q \begin{bmatrix} j_1 & j_2 & j \\ m_1' & m_2' & m \end{bmatrix}_q = \delta_{m_1 m_1'}\,\delta_{m_2 m_2'} . \quad (3.8)$$

Applying operators $\Delta(X^{\pm})$ to both sides of (3.2) we obtain the following recurrence relations for $q-3j$ -symbols

$$\{[j\mp m][j\pm m+1]\}^{1/2} \begin{bmatrix} j_1 & j_2 & j \\ m_1 & m_2 & m\pm 1 \end{bmatrix}_q =$$

$$= q^{\frac{m_2}{2}} \{[j_1\pm m_1][j_1\mp m_1+1]\}^{1/2} \begin{bmatrix} j_1 & j_2 & j \\ m_1\mp 1 & m_2 & m \end{bmatrix}_q + \quad (3.6')$$

$$+ q^{-\frac{m_1}{2}} \{[j_2+m_2][j_2\mp m_2+1]\}^{1/2} \begin{bmatrix} j_1 & j_2 & j \\ m_1 & m_2\mp 1 & m \end{bmatrix}_q .$$

It is interesting that there is a simple relation between $q-3j$ -symbols and Hahn q -polynomials. Let us remind that Hahn polynomials $Q_n(x)$ and dual Hahn polynomials $R_n(x)$ are defined by the following expressions $(n, N\in \mathbb{Z}_+ ,\ n\le N)$:

$$Q_n(x) := Q_n(x; a, b, N|q) = {}_3\varphi_2\left(\begin{matrix} q^{-n},\ abq^{n+1},\ x \\ aq,\ q^{-N} \end{matrix} ; q, q\right),$$

$$R_n(\mu(x)):=R_n(\mu(x);a,b,N|q)={}_3\varphi_2\left(\begin{matrix}q^{-n}, abq^{x+1}, q^{-x}\\ aq, q^{-N}\end{matrix};q,q\right),$$

where $\mu(x)=q^{-x}+abq^{x+1}$. For $0\le x, n\le N$ we have:

$$R_n(\mu(x);a,b,N|q)=Q_x(q^{-n};a,b,N|q),$$

and $Q_n(x)$ (or $R_n(\mu(x))$) is the polynomial of degree n of the variable x (or $\mu(x)$).

These polynomials satisfy the following orthogonality relations

$$\sum_{x=0}^{N} Q_m(q^{-x})Q_n(q^{-x})\rho(x)=\delta_{nm}\, d_n^2,$$

$$\sum_{x=0}^{N} R_m(\mu(x))R_n(\mu(x))\rho(x)=\delta_{nm}\, d_n^2,$$

where $\rho(x)$ and d_n are weight and norm of Hahn polinomials.

$$\rho(x):=\rho(x;a,b,N|q)=\frac{(aq;q)_x\,(bq;q)_x}{(q;q)_x\,(q;q)_{N-x}}(aq)^{-x},$$

$$d_n^2:=d_n^2(a,b,N|q)=\frac{(1-abq)\,(abq^2;q)_N\,(aq)^{-N}(-aq)^n}{(1-abq^{2n+1})\,(q;q)_N}\cdot$$

$$\cdot\frac{(q, bq, abq^{N+2};q)_n}{(aq, abq, q^{-N};q)_n}\cdot q^{\binom{n}{2}-nN}$$

for Hahn polynomials $Q_n(x)$, and

$$\rho(x) := \rho(x; a, b, N|q) = \frac{(1-abq^{2x+1})(aq, abq, q^{-N}; q)_x}{(1-abq)(q, bq, abq^{N+2}; q)_x} \cdot (-aq)^{-x} q^{Nx - \binom{x}{2}},$$

$$d_n^2 := d_n^2(a, b, N|q) = \frac{(abq^2; q)_N \, (q, b^{-1}q^{-N}; q)_n}{(bq; q)_N \, (aq, q^{-N}; q)_n} (aq)^{-N} (abq)^n$$

for dual Hahn polynomials $R_n(\mu(x))$.

From the Racah-Fock representation (3.4) we obtain the relation between $q-3j$-symbols and Hahn polynomials:

$$\begin{bmatrix} j_1 & j_2 & j \\ m_1 & m_2 & m \end{bmatrix}_q = (-1)^{j_1-m_1} \frac{\{\rho(x)\}^{1/2}}{d_n} Q_n(q^{-x}; a, b, N|q), \quad (3.8')$$

where $n = j - m$, $a = q^{j_2 - j + m_1}$, $b = q^{j_1 - j_2 + m}$, $x = j_1 - m_1$, $N = j + j_2 - m_1$ $(0 \le x, n \le N)$ and the relation between $q-3j$ -symbols and dual Hahn polynomials

$$\begin{bmatrix} j_1 & j_2 & j \\ m_1 & m_2 & m \end{bmatrix}_q = (-1)^{j_1-m_1} \frac{\{\rho(x)\}^{1/2}}{d_n} R_n(\mu(x); a, b, N|q), \quad (3.8'')$$

where n, x, N, are the same as in (3.8'), $b = q^{j - j_2 + m_1}$, $\mu(x) = q^{m_1 + 1/2}(q^{j_1 + 1/2} - q^{-j_1 - 1/2})$. Here $\rho(x)$ and d_n are weight and norm of Hahn polynomials.

The orthogonality relations for Hahn polynomials and dual Hahn polynomials follows immediately from (3.8), (3.8"). The recurrence relations (3.6′) for $q-3j$ -symbols imply the recurrence relations for Hahn and dual Hahn polynomials.

THEOREM 3.1. There are the following relations between CGC:

$$\sum_{m_1' m_2'} \left(\mathbb{R}^{j_1 j_2}\right)_{m_1 m_2}^{m_1' m_2'} \begin{bmatrix} j_1 & j_2 & j \\ m_1' & m_2' & m \end{bmatrix}_q = \tag{3.9}$$

$$= (-1)^{j_1+j_2-j} q^{\frac{1}{2}(c_j - c_{j_1} - c_{j_2})} \begin{bmatrix} j_2 & j_1 & j \\ m_2 & m_1 & m \end{bmatrix}_q ,$$

$$\begin{bmatrix} j_1 & j_2 & j \\ m_1 & m_2 & m \end{bmatrix}_q = (-1)^{j_1+j_2-j} \begin{bmatrix} j_2 & j_1 & j \\ m_2 & m_1 & m \end{bmatrix}_{q^{-1}} , \tag{3.10}$$

$$\begin{bmatrix} j_1 & j_2 & j \\ m_1 & m_2 & m \end{bmatrix}_q = (-1)^{j_1-m_1} q^{-\frac{m_1}{2}} \left\{ \frac{[2j+1]}{[2j_2+1]} \right\}^{1/2} \begin{bmatrix} j_1 & j & j_2 \\ m_1, & -m, & -m_2 \end{bmatrix}_{q^{-1}} , \tag{3.11}$$

where the numbers c_j and $[n]$ defined in (2.6), (1.7).

The proof of this theorem follows from the representations (3.3) -(3.5) (see also [2]).

THEOREM 3.2.

$$\sum_{m'} \left(\mathbb{R}^{j j_3}\right)_{m' m_3}^{m m_3'} \begin{bmatrix} j_1 & j_2 & j \\ m_1 & m_2 & m' \end{bmatrix}_q = \sum_{m_1' m_2' m_3''} \begin{bmatrix} j_1 & j_2 & j \\ m_1' & m_2' & m \end{bmatrix}_q \cdot$$

$$\left(\mathbb{R}^{j_1 j_3}\right)_{m_1 m_3}^{m_1' m_3''} \left(\mathbb{R}^{j_2 j_3}\right)_{m_2 m_3''}^{m_2' m_3'} . \tag{3.12}$$

To prove this formula it is sufficient to consider relation (1.8) in the representation $\pi^{j_3} \otimes \pi^{j_1} \otimes \pi^{j_2}$ (for more details see [2]).

From (1.11) it follows that the matrices $\mathbb{R}^{j_1 j_2}$ satisfy the Yang-Baxter relation:

$$\mathbb{R}_{12}^{j_1 j_2} \mathbb{R}_{13}^{j_1 j_3} \mathbb{R}_{23}^{j_2 j_3} = \mathbb{R}_{23}^{j_2 j_3} \mathbb{R}_{13}^{j_1 j_3} \mathbb{R}_{12}^{j_1 j_2}. \tag{3.13}$$

Here all matrices act in $V^{j_1} \otimes V^{j_2} \otimes V^{j_3}$ and the indices show how these matrices are acting in the product space.

To conclud this section let us prove that the extension of $U_q(su(2))$ by the Weyl element w is a Hopf algebra.

THEOREM 3.3. Formulae

$$\Delta(w) = R^{-1}(w \otimes w), \quad \varepsilon(w) = 1 \tag{3.14}$$

define the structure of a Hopf algebra on $\widetilde{U_q(su(2))}$.

PROOF. Let us check (3.14) in all irreducible representations. In accordance with the definition of CGC we have:

$$(\pi^{j_1} \otimes \pi^{j_2}) \Delta(w)_{m_1' m_2'}^{m_1 m_2} = \sum_{j, m, m'} \begin{bmatrix} j_1 & j_2 & j \\ m_1 & m_2 & m \end{bmatrix}_q w_{mm'}^{j} \begin{bmatrix} j_1 & j_2 & j \\ m_1' & m_2' & m' \end{bmatrix}_q. \tag{3.15}$$

The symmetries of CGC imply the identity

$$\begin{bmatrix} j_1 & j_2 & j \\ m_1 & m_2 & m \end{bmatrix}_q = \sum_{m_1' m_2'} \left((\mathbb{R}^{j_1 j_2})^{-1} \right)_{m_1' m_2'}^{m_1 m_2} \begin{bmatrix} j_1 & j_2 & j \\ -m_1' & -m_2' & -m \end{bmatrix}_q (-1)^{j_1+j_2-j} q^{\frac{1}{2}(c_j - c_{j_1} - c_{j_2})}.$$

Comparing this identity with (3.15) we obtain the equality (3.14) in the representation

$$(\pi^{j_1} \otimes \pi^{j_2}) \Delta w = (\mathbb{R}^{j_1 j_2})^{-1} w^{j_1} \otimes w^{j_2}.$$

Formulae (1.8) and (1.9) imply thecoassociativity of the action (3.14).

Let $\Delta(a)=\sum_i a^i\otimes a_i$. The Hopf axiom for the antipode is $S(a^i)a_i = a^i S(a_i) = \varepsilon(a)\mathbb{1}$. To define the action of the antipode on w we need the following lemma.

LEMMA. Let A be the quasitriangle Hopf algebra (see [9][12][36] and $R = e_i\otimes e^i$ is the universal R-matrix; then

1. The element $u=\sum_i S(e^i)e_i$ is invertible and $u^{-1}=\sum_j e^j S^2(e_j)$.
2. $S^2(a) = u a u^{-1}$ for any $a\in A$.

We do not give here the proof of this lemma, because it would have demanded the description of many auxiliary constructions. The proof of this lemma for arbitrary quasitriangle Hopf algebra was given by V.G.Drinfeld [36] and for quasitriangle algebras of special structure (for dubles of Hopf algebras) by one of the authors (N.R. u published). Let us check now the property of the antipode in $\widehat{U_q}(su(2))$. From (3.14) we obtain $(S\otimes id)\Delta w = S(w)S^2(e^i)\otimes e_i w$ and therefore we must have

$$S(w)S^2(e_i)e^i w = \varepsilon(w)\cdot\mathbb{1} . \qquad (3.16)$$

Lemma 1 implies that $S(u^{-1}) = S^2(e_i)e^i$. Comparing with (3.16) we get

$$S(w)w = \varepsilon(w)u .$$

Further, the equalities

$$\varepsilon(w)S(u)u = S(w)u w = S(w)^2 w^2 =$$

$$= S(w)w\cdot w S(w) , \quad S(u)u = u S(u)$$

prove the relation

$$w S(w) = \varepsilon(w)S(u) .$$

Using this relation we have:

$$e_i w S(w) e^i = \varepsilon(w) e_i S(u) e^i = \varepsilon(w) S(u) S(u^{-1}) =$$

$$= \varepsilon(w) \cdot \mathbb{1} .$$

So, the conditions of the axiom for an antipode in $\widetilde{U_q}(su(2))$ are satisfied. To calculate $\varepsilon(w)$ we substitute w into (2.8). Since $\tau^2 = id$ we obtain

$$S(w) = \tau(w).$$

Therefore in every irreducible representation we have

$$(w^j)^t w^j = \varepsilon(w) \pi^j(u), \quad w^j (w^j)^t = \varepsilon(w) \pi^j(S(u)).$$

Comparing this formula with (2.6) we get

$$\varepsilon(w) = 1, \quad \pi^j(u) = q^{-c_j + H/2}, \quad \pi^j(S(u)) = q^{-c_j - H/2} .$$

4. Graphical Representation of R-matrices And q-Glebsch-Gordan Coefficients

The relations (3.7)-(3.13) between CGC and R-matrices can be represented grahhically [2].

Let us represent the R-matrices and CGC by graphs with strings colored by numbers j and with states $\{m_i\}$ on the end of strings:

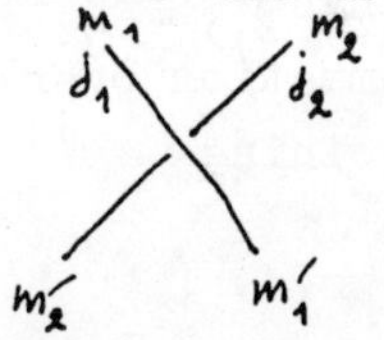

$$\longrightarrow \left(\mathbb{R}^{j_2 j_1}\right)^{m_2' \, m_1'}_{m_2 \, m_1} . \qquad (4.1)$$

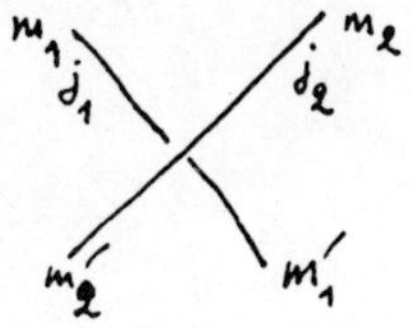

$$\longrightarrow \left(\left(\mathbb{R}^{j_2 j_1}\right)^{-1}\right)^{m_1' \, m_2'}_{m_1 \, m_2} . \qquad (4.2)$$

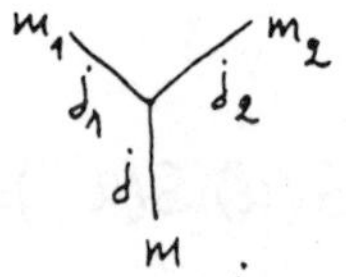

$$\longrightarrow \begin{bmatrix} j_1 & j_2 & j \\ m_1 & m_2 & m \end{bmatrix}_q . \tag{4.3}$$

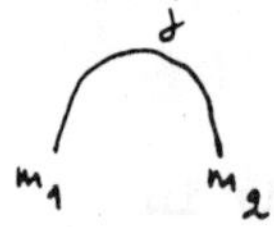

$$\longrightarrow \delta_{m_1,-m_2} (-1)^{-j+m_2} q^{\frac{m_2}{2}} , \tag{4.4}$$

$$\longrightarrow \delta_{m_1,-m_2} (-1)^{j-m_1} q^{-\frac{m_1}{2}} , \quad j \Big|_{m'}^{m} \to \delta_{mm'} . \tag{4.5}$$

The multiplication of matrices correspond to the joining of graphs of these matrices together. For example if the matrix A acts from $V^{j_1'} \otimes \dots \otimes V^{j_L'}$ into $V^{j_1} \otimes \dots \otimes V^{j_N}$, the matrix B acts from $V^{j_1''} \otimes \dots \otimes V^{j_M''}$ to $V^{j_1'} \otimes \dots \otimes V^{j_L'}$ and they are represented by graphs:

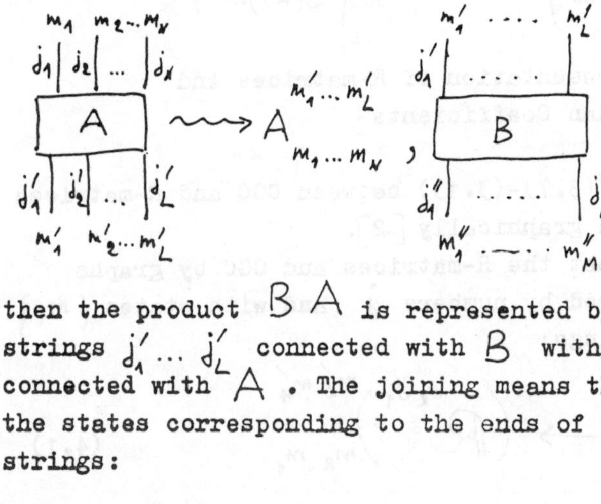

$$\rightsquigarrow A^{m_1' \dots m_L'}_{m_1 \dots m_N} , \qquad \rightsquigarrow B^{m_1'' \dots m_M''}_{m_1' \dots m_L'} ,$$

then the product BA is represented by the joining of strings $j_1' \dots j_L'$ connected with B with strings $j_1', \dots, j_L'$ connected with A. The joining means the summation over the states corresponding to the ends of the joining strings:

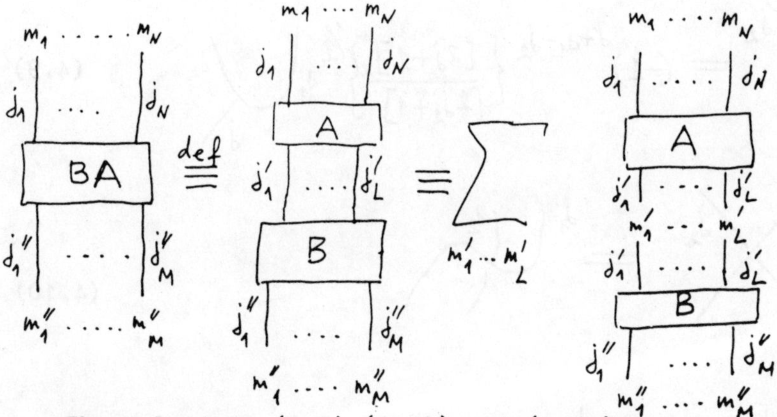

$m_1 \cdots m_N$, $j_1 \cdots j_N$, BA, $j''_1 \cdots j''_M$, $m''_1 \cdots m''_M$ $\stackrel{def}{\equiv}$ $m_1 \cdots m_N$, $j_1 \cdots j_N$, A, $j'_1 \cdots j'_L$, B, $j''_1 \cdots j''_M$, $m''_1 \cdots m''_M$ $\equiv \sum_{m'_1 \cdots m'_L}$ $m_1 \cdots m_N$, $j_1 \cdots j_N$, A, $j'_1 \cdots j'_L$, $m'_1 \cdots m'_L$, $j'_1 \cdots j'_L$, B, $j''_1 \cdots j''_M$, $m''_1 \cdots m''_M$

The relations (3.7)-(3.13) and (2.10) are represented by the following graphical equalities:

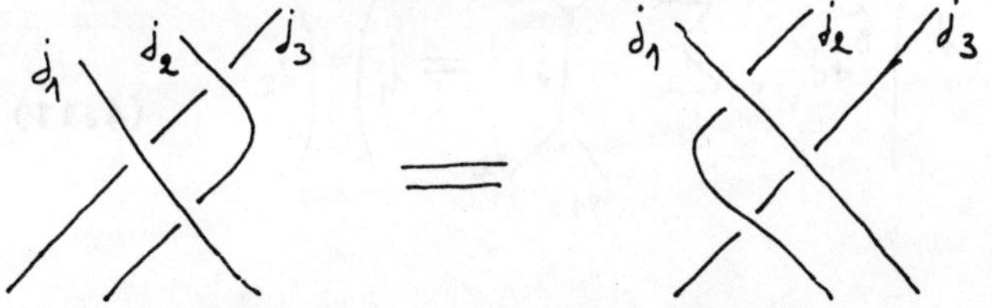

$j_1, j_2, j_3 \quad = \quad j_1, j_2, j_3$ (4.6)

$j_3, j_1, j_2, j \quad = \quad j_3, j_1, j_2, j$ (4.7)

$$j_1, j_2, j \quad = (-1)^{j_1+j_2-j}\, q^{\frac{1}{2}(c_j - c_{j_1} - c_{j_2})} \quad j_1, j_2, j \tag{4.8}$$

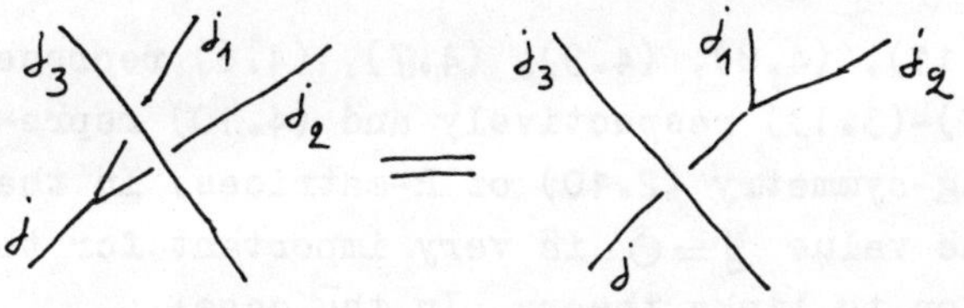

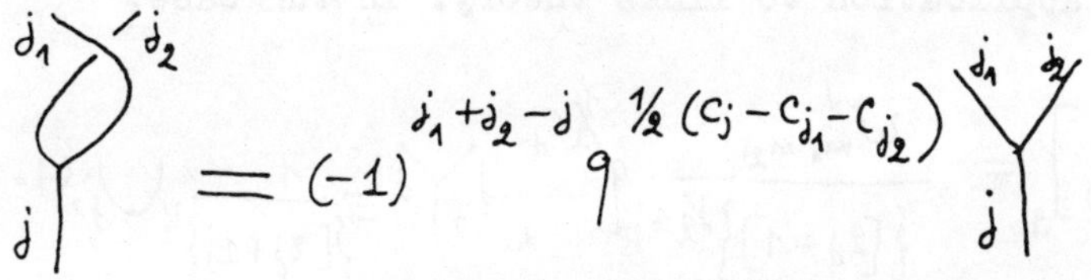

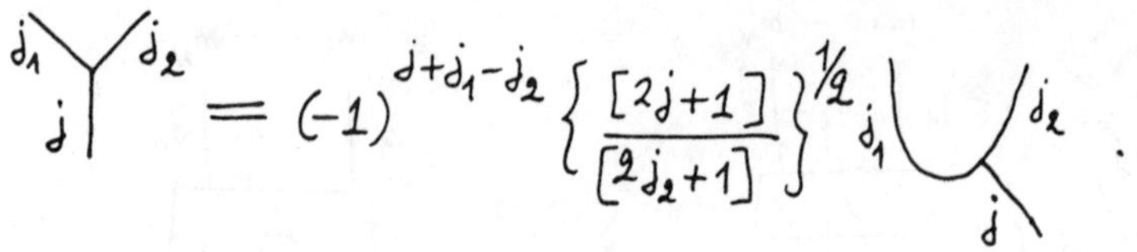

$$j_1 \; j_2 \; j \;=\; (-1)^{j+j_1-j_2}\left\{\frac{[2j+1]}{[2j_2+1]}\right\}^{1/2} \; j_1 \; j_2 \; j \; . \qquad (4.9)$$

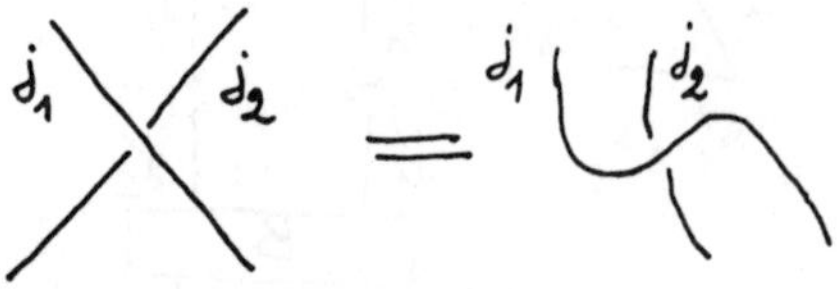

$$j_1 \; j_2 \;=\; j_1 \; j_2 \qquad (4.10)$$

$$j \; j_1 \; j_2 \; j' \;=\; j \; \delta_{jj'} \, , \quad \sum_j \; j_1 \; j_2 \; j \; j_2 \; j_1 \;=\; j_1 \; j_2 \; . \qquad (4.11)$$

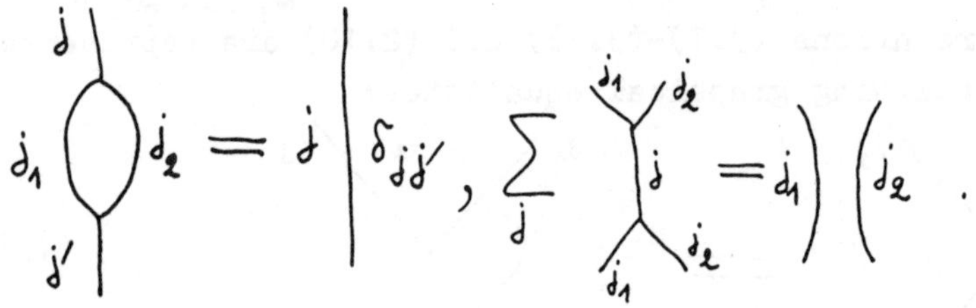

The relations (4.11), (4.8), (4.9), (4.7), (4.6) represent the formulae (3.7)-(3.13) respectively and (4.10) represents the crossing-symmetry (2.10) of R-matrices. In the relation (4.8) the value $j=0$ is very important for the further application to links theory. In this case:

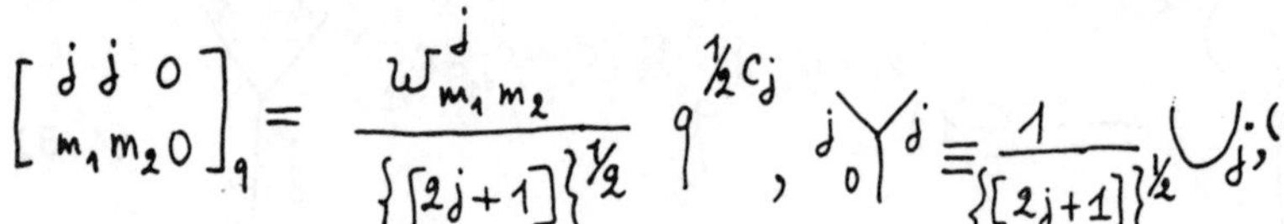

$$\begin{bmatrix} j & j & 0 \\ m_1 & m_2 & 0 \end{bmatrix}_q = \frac{w^j_{m_1 m_2}}{\{[2j+1]\}^{1/2}}\, q^{\frac{1}{2}c_j} \, , \quad j \; 0 \; j \equiv \frac{1}{\{[2j+1]\}^{1/2}} \; j \, , \qquad (4.12)$$

and therefore from (4.8) with $j=0$ it follows that

$$j \;=\; (-1)^{2j} q^{-c_j} \; j \, , \qquad j \;=\; (-1)^{2j} q^{c_j} \; j \; . \qquad (4.13)$$

5. The q-Analogs Of The Wigner-Rakah $6j$ -Symbols

Let us consider now the q-analog of $6j$-symbols and the properties of these $q-6j$ -symbols. For this purpose let us consider tensor product $V^{j_1}\otimes V^{j_2}\otimes V^{j_3}$ of three irreducible representations of $U_q(su(2))$. There are two symplest ways to obtain irreducible components in this representation. One is to decompose first $V^{j_1}\otimes V^{j_2} = \sum_{j_{12}}^{\oplus} V^{j_{12}}$ and then to take irreducible submodules in $V^{j_{12}}\otimes V^{j_3}$. The other is to decompose first $V^{j_2}\otimes V^{j_3} = \sum_{j_{23}}^{\oplus} V^{j_{23}}$ and then $V^{j_1}\otimes V^{j_{23}}$. These two ways give two complete orthogonal bases in $V^{j_1}\otimes V^{j_2}\otimes V^{j_3}$:

$$e_m^{j_{12}j}(j_1 j_2 | j_3) = \sum_{m_1 m_2 m_3} \begin{bmatrix} j_{12} & j_3 & j \\ m_{12} & m_3 & m \end{bmatrix}_q \begin{bmatrix} j_1 & j_2 & j_{12} \\ m_1 & m_2 & m_{12} \end{bmatrix}_q \cdot e_{m_1}^{j_1}\otimes e_{m_2}^{j_2}\otimes e_{m_3}^{j_3}, \qquad (5.1)$$

$$e_m^{j_{23}j}(j_1 | j_2 j_3) = \sum_{m_1 m_2 m_3} \begin{bmatrix} j_1 & j_{23} & j \\ m_1 & m_{23} & m \end{bmatrix}_q \begin{bmatrix} j_2 & j_3 & j_{23} \\ m_2 & m_3 & m_{23} \end{bmatrix} \cdot e_{m_1}^{j_1}\otimes e_{m_2}^{j_2}\otimes e_{m_3}^{j_3}. \qquad (5.2)$$

The matrix elements of the matrix, connecting these bases will be called $q-6j$ -symbols :

$$e_m^{j_{12}j}(j_1,j_2|j_3) = \sum_{j_{23}} \begin{Bmatrix} j_1 & j_2 & j_{12} \\ j_3 & j & j_{23} \end{Bmatrix}_q e_m^{j_{23}j}(j_1|j_2j_3). \qquad (5.3)$$

In the case $q=1$ we have

$$\begin{Bmatrix} j_1 & j_2 & j_{12} \\ j_3 & j & j_{23} \end{Bmatrix}_{q=1} = \left\{ [2j_{12}+1][2j_{23}+1] \right\}^{1/2} (-1)^{j_1+j_2-j-j_3-2j_{12}} \begin{Bmatrix} j_1 & j_2 & j_{12} \\ j_3 & j & j_{23} \end{Bmatrix},$$

where $\begin{Bmatrix} a & b & e \\ d & c & f \end{Bmatrix}$ is Racah-Wigner $6j$-symbol.

For $q \neq 1$ we shall also use Racah-Wigner normalization:

$$\begin{Bmatrix} j_1 & j_2 & j_{12} \\ j_3 & j & j_{23} \end{Bmatrix}_q^{RW} = \left\{ [2j_{12}+1][2j_{23}+1] \right\}^{-1/2} (-1)^{j_1+j_2-j-j_3-2j_{12}} \begin{Bmatrix} j_1 & j_2 & j_{12} \\ j_3 & j & j_{23} \end{Bmatrix}_q.$$

If we use the graphical technique of the previous section the definition (5.3) of $q-6j$-symbols will have the form

j_1 j_2 j_3 j

$$= \sum_{j_{23}} \begin{Bmatrix} j_1 & j_2 & j_{12} \\ j_3 & j & j_{23} \end{Bmatrix}_q \qquad (5.4)$$

j_1 j_2 j_3 j

Using the orthogonality (4.11) of CGC we obtain an expression of $q-6j$-symbols in terms of CGC:

$$\begin{Bmatrix} j_1 & j_2 & j_{12} \\ j_3 & j & j_{23} \end{Bmatrix}_q = \begin{bmatrix} j_1 & j_{23} & j \\ m_1 & m_{23} & m \end{bmatrix}_q^{-1} \sum_{\substack{m_2, m_3 \\ m_2+m_3=m-m_1}} \begin{bmatrix} j_2 & j_3 & j_{23} \\ m_2 & m_3 & m_{23} \end{bmatrix}_q \cdot \qquad (5.5)$$

$$\begin{bmatrix} j_{12} & j_3 & j \\ m_{12} & m_3 & m \end{bmatrix}_q \begin{bmatrix} j_1 & j_2 & j_{12} \\ m_1 & m_2 & m_{12} \end{bmatrix}_q ,$$

or

$$\text{[diagram: } j_1, j_2, j_{23}, j_{12}, j_3, j \text{]} = \begin{Bmatrix} j_1 & j_2 & j_{12} \\ j_3 & j & j_{23} \end{Bmatrix}_q \text{[diagram: } j_1, j_{23}, j \text{]} . \quad (5.6)$$

THEOREM 5.1.

$$\begin{Bmatrix} a & b & e \\ d & c & f \end{Bmatrix}_q^{RW} = \Delta(abe)\Delta(acf)\Delta(ced)\Delta(dbf) \cdot$$

$$\sum_z (-1)^z [z+1]! \Big\{ [z-a-b-e]! [z-a-c-f]! \quad (5.7)$$

$$[z-b-d-f]! [z-d-c-e]! [a+b+c+d-z]!$$

$$[a+d+e+f-z]! [b+c+e+f-z]! \Big\}^{-1}$$

Here the sum is taken only over z with nonnegative arguments in square brackets, $[0]! \equiv 1$.

The proof of this stheorem is given in .

REMARK 1. The sum (5.7) can be expressed through the generalized hypergeometric function ${}_4\varphi_3$ (see [1]):

$$\frac{(-1)^{a+b+c+d} [a+c+b+d+1]!}{[a+b-e]! [a+c-f]! [c+d-e]! [b+d-f]! [e+f-c-b]! [e+f-a-d]!} \cdot$$

$${}_4\varphi_3 \left(\begin{matrix} q^{-a-b+e}, \; q^{-a-c+f}, \; q^{-c-d+e}, \; q^{-b-d+f} \\ q^{-a-b-c-d-1}, \; q^{e+f-c-b+1}, \; q^{e+f-a-d+1} \end{matrix} ; q, q \right)$$

Here we use the following notation

$$ {}_{p+1}\varphi_p\left(\begin{matrix} a_1,\dots,a_{p+1} \\ b_1,\dots,b_p \end{matrix};q,z\right)=\sum_{k=0}^{\infty}\frac{(a_1,\dots,a_{p+1};q)_k}{(b_1,\dots,b_p,q;q)_k}\cdot z^k, $$

where

$$ (a;q)_k=\prod_{i=0}^{k-1}(1-aq^i), $$

$$ (a_1,\dots,a_p;q)_k=\prod_{j=1}^{p}(a_j;q)_k . $$

Let us suppose that one of the numbers qa , qbd or qc is equal to q^{-M}. Then

$$ R_n(\lambda(x);a,b,c,d\,|\,q)={}_4\varphi_3\left(\begin{matrix} q^{-n},\ q^{n+1}ab,\ q^{-x},\ q^{x+1}cd \\ aq\ ,\ bdq\ ,\ cq \end{matrix};q,q\right) $$

is the polynomial of degree n on $\lambda(x)=q^{-x}+q^{x+1}cd$. These polynomials are called the Racah q-polynomials. The Racah-Wilson polynomials correspond to $c=q^{-M-1}$ and are denoted by $W_n(x)$:

$$ W_n(x):=W_n(x;a,b,c,M\,|\,q)={}_4\varphi_3\left(\begin{matrix} q^{-n},\ abq^{n+1},\ q^{-x},\ cq^{x-M} \\ aq\ ,\ q^{-M}\ ,\ bcq \end{matrix};q,q\right), $$

$n=0,1,\dots,M$. This is a polynomial of degree n in $\mu(x)=q^{-x}+cq^{x-M}$, and $W_n(x;a,b,0,M\,|\,q)=$ $=Q_n(q^{-x};a,b,M\,|\,q)$.

From the theorem 5.1 we have

$$ \left\{\begin{matrix} a & b & e \\ d & c & f \end{matrix}\right\}_q=\frac{\{\rho(x)\}^{1/2}}{d_n}W_n(x;\alpha,\beta,\gamma,M\,|\,q), $$

where $n=a+b-e$, $x=c+d-e$, $M=a+b+c+d+1$, $\alpha=q^{-a-d+e+f}$, $\beta=q^{-a-b-c+d-1}$, $\gamma=q^{a+e+f-d+1}$, $\rho(x)$ and d_n are the weight and norm of the Racah-Wilson polynomials:

$$\rho(x):=\rho(x;\alpha,\beta,\gamma,M\,|\,q)=$$

$$=\frac{(1-\gamma q^{2x-M})(\alpha q,\beta\gamma q,\gamma q^{-M},q^{-M};q)_x}{(1-\gamma q^{-M})(q,\beta^{-1}q^{-M},\gamma q,\alpha^{-1}\gamma q^{-M};q)_x}\,(\alpha\beta q)^{-x},$$

$$d_n^2:=d_n^2(\alpha,\beta,\gamma,M\,|\,q)=$$

$$=\frac{(1-\alpha\beta q)(q,\beta q,\alpha\gamma^{-1}q,\alpha\beta q^{M+2};q)_n}{(1-\alpha\beta q^{2n+1})(\alpha\beta q,\alpha q,\beta\gamma q,q^{-M};q)_n}\,\frac{(\alpha^{-1}\gamma,\beta^{-1};q)_M}{(\gamma q,\alpha^{-1}\beta^{-1}q^{-1};q)_M}\cdot$$

$$\cdot(\gamma q^{-M})^n .$$

The symmetries of $q-6j$-symbols follow from (A.18)

1. $$\begin{Bmatrix} b & a & e\\ c & d & f\end{Bmatrix}_q^{RW}=\begin{Bmatrix} a & b & e\\ d & c & f\end{Bmatrix}_q^{RW}$$

2. $$\begin{Bmatrix} a & e & b\\ d & f & c\end{Bmatrix}_q^{RW}=\begin{Bmatrix} a & b & e\\ d & c & f\end{Bmatrix}_q^{RW}$$

3. $$\begin{Bmatrix} a & c & f\\ d & b & e\end{Bmatrix}_q^{RW}=\begin{Bmatrix} a & e & b\\ d & f & c\end{Bmatrix}_q^{RW}$$

4. q-Regge symmetry. Let

$$s_1=\frac{b+c+e+f}{2},\ s_2=\frac{a+d+e+f}{2},\ s_3=\frac{a+b+c+d}{2},$$

then the following equalities hold

$$\begin{Bmatrix} a & s_1-c & s_1-f \\ d & s_1-b & s_1-e \end{Bmatrix}_q^{RW} = \begin{Bmatrix} a & e & b \\ d & f & c \end{Bmatrix}_q^{RW}$$

$$\begin{Bmatrix} s_2-d & b & s_2-f \\ s_2-a & c & s_2-e \end{Bmatrix}_q^{RW} = \begin{Bmatrix} a & b & e \\ d & c & f \end{Bmatrix}_q^{RW}$$

$$\begin{Bmatrix} s_3-d & s_3-c & e \\ s_3-a & s_3-b & f \end{Bmatrix}_q^{RW} = \begin{Bmatrix} a & b & e \\ d & c & f \end{Bmatrix}_q^{RW}$$

$$\begin{Bmatrix} s_2-d & s_3-c & s_1-f \\ s_2-a & s_3-b & s_1-e \end{Bmatrix}_q^{RW} = \begin{Bmatrix} a & b & e \\ d & c & f \end{Bmatrix}_q^{RW}$$

5.

$$\lim_{N\to\infty} \begin{Bmatrix} N-m, & b, & N-m_2 \\ d, & N, & f \end{Bmatrix}_q^{RW} = \left\{\frac{1}{[2f+1]}\right\}^{1/2} \begin{bmatrix} b, & d, & f \\ m_1 & m_2 & m \end{bmatrix}_q,$$

$m_1+m_2=m$.

The symmetries 1-3 of $q-6j$-symbols have a simple graphical interpretation. To describe it let us write $q-6j$-symbol as the following trace:

$$\begin{Bmatrix} j_1 & j_2 & j_{12} \\ j_3 & j & j_{23} \end{Bmatrix}_q = \frac{1}{[2j+1]} \quad (5.8)$$

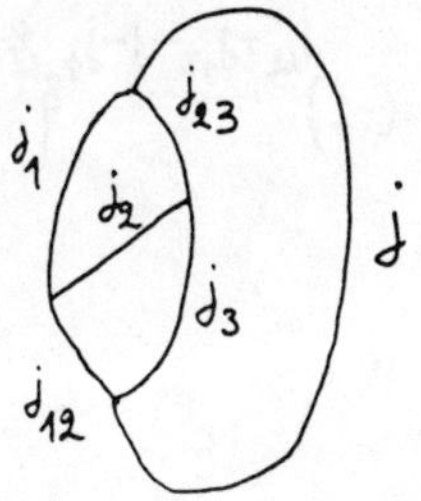

The symmetries 1-3 of $q-6j$-symbols correspond to rotating of thetraedrom formed by the edges (j_1, j_2, j_3, j_{12}, j_{23}, j) in accordance with rules (4.6)-(4.10). For example we have:

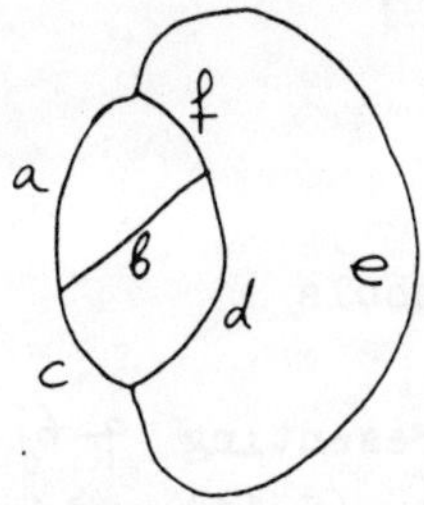

$$= \frac{[2e+1]}{[2a+1]} \quad (5.9)$$

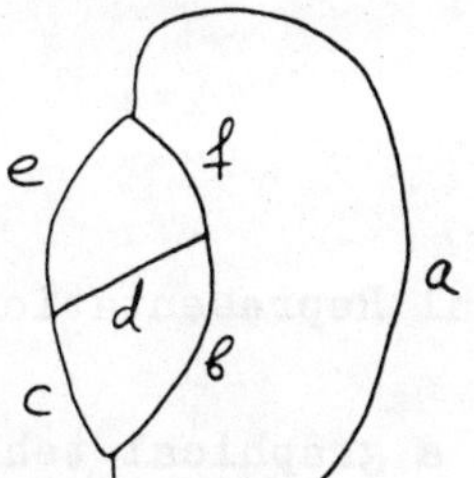

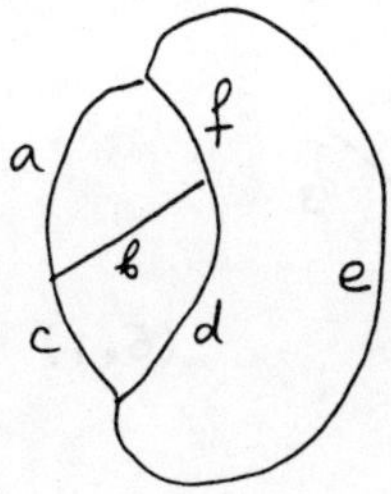

$$= \frac{[2e+1]}{[2c+1]} (-1)^{a+f-c-d} \left\{ \frac{[2a+1][2f+1]}{[2c+1][2d+1]} \right\}^{1/2} \quad (5.10)$$

To conclude this section we give two important formulae with $q-6j$ -symbols

$$\text{[diagram: } j_1, j_2, j_3, j_{13}, j\text{]} = \sum_{j_{12}} (-1)^{j_{12}+j_{13}-j-j_1} q^{\frac{1}{2}(C_j + C_{j_1} - C_{j_{13}} - C_{j_{12}})} \begin{Bmatrix} j_3 & j_1 & j_{13} \\ j_2 & j & j_{12} \end{Bmatrix}_q \text{[diagram: } j_1, j_2, j_{12}, j_3, j\text{]} \qquad (5.11)$$

$$\text{[diagram: } j_1, j_2, j_3, j_{23}, j\text{]} = \sum_{j_{12}} \begin{Bmatrix} j_3 & j_2 & j_{23} \\ j_1 & j & j_{12} \end{Bmatrix}_q \text{[diagram: } j_1, j_2, j_3, j_{12}, j\text{]} \qquad (5.12)$$

6. Graphical Representation Of $q-6j$-Symbols

To give a graphical tehcnique for representing $q-6j$-symbols let us rewrite the relations (5.5), (5.11) and (5.12) in the following form:

$$\text{[diagram: } j_1, j_2, j_3, j_{13}, j\text{]} = \sum_{j_{12}} \text{[diagram: } j_1, j_2, j_3, j_{12}, j_{13}, j\text{]} \qquad (6.1)$$

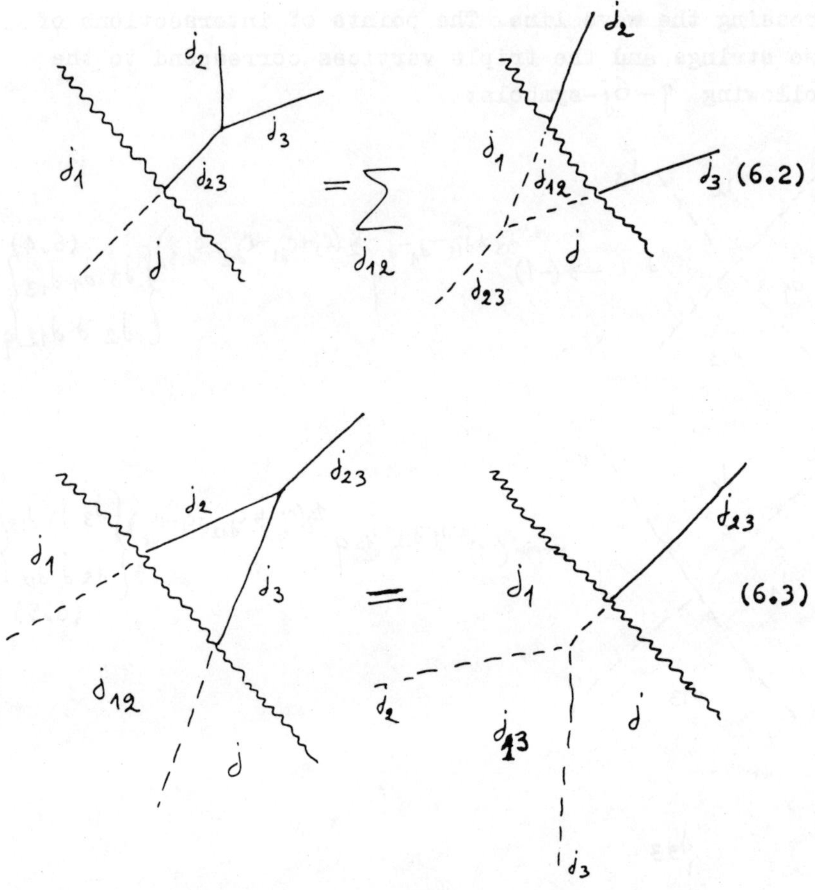

Here a wave line divides the plane into two parts. In the upper part the strings are colored by the numbers j , the ends of these strings are marked by the states $\{m\}$ and vertices represent the R-matrices and CGC according to the rules (4.1)-(4.5). In the lower part the numbers $\dot{j}$ color the strings and the sectors placed between the strings. The colors of the strings are not changed after crossing the wave line. The points of intersections of two strings and the triple vertices correspond to the following $q-6j$-symbols:

$$\rightarrow (-1)^{j_{13}+j_{12}-j_1-j}\, q^{\frac{1}{2}(c_j+c_{j_1}-c_{j_{12}}-c_{j_{13}})} \begin{Bmatrix} j_3 & j_1 & j_{13} \\ j_2 & j & j_{12} \end{Bmatrix}_q \qquad (6.4)$$

$$\rightarrow (-1)^{j+j_1-j_{12}-j_{13}}\, q^{\frac{1}{2}(c_{j_{13}}+c_{j_{12}}-c_j-c_{j_1})} \begin{Bmatrix} j_3 & j_1 & j_{13} \\ j_2 & j & j_{12} \end{Bmatrix}_q \qquad (6.5)$$

$$\rightarrow \begin{Bmatrix} j_1 & j_2 & j_{12} \\ j_3 & j & j_{23} \end{Bmatrix}_q \qquad (6.6)$$

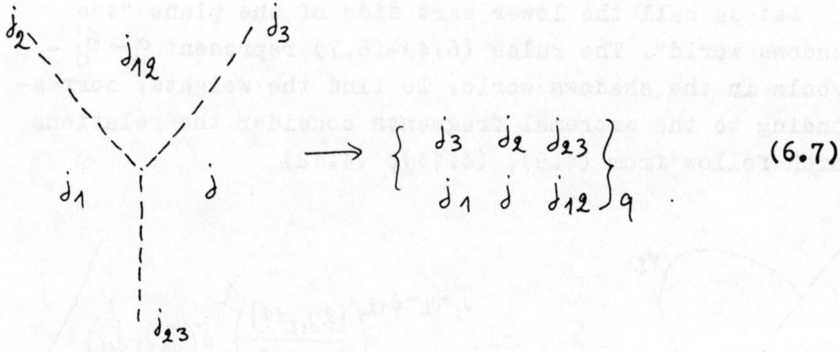

The points of intersections of strings with the wave line correspond to CGC:

$$\begin{bmatrix} j_1 & j_2 & j_{12} \\ m_1 & m_2 & m_{12} \end{bmatrix}_q . \qquad (6.8)$$

The joining of the fragments (6.8) by the wave line correspond to the summation over the states of the joining ends:

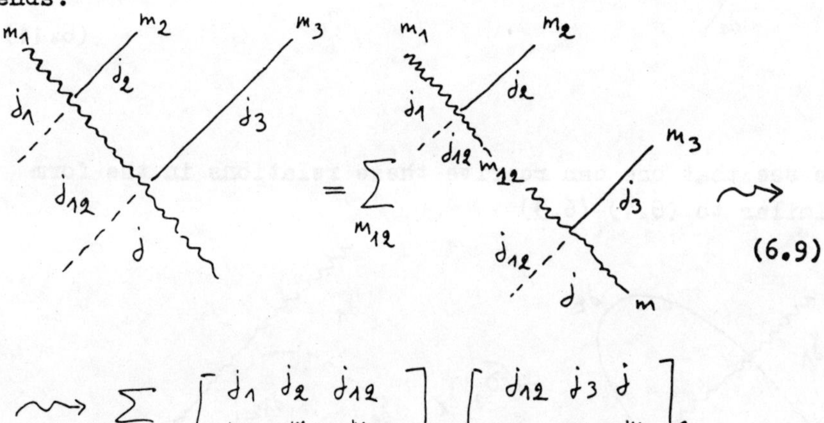

$$\rightsquigarrow \sum_{m_{12}} \begin{bmatrix} j_1 & j_2 & j_{12} \\ m_1 & m_2 & m_{12} \end{bmatrix}_q \begin{bmatrix} j_{12} & j_3 & j \\ m_{12} & m_3 & m \end{bmatrix}_q .$$

Let us call the lower part side of the plane "the shadows world". The rules (6.4)-(6.7) represent $q-6j$-sybols in the shadows world. To find the weights, corresponding to the extremal fragments consider the relations which follow from (4.9), (4.11), (4.12)

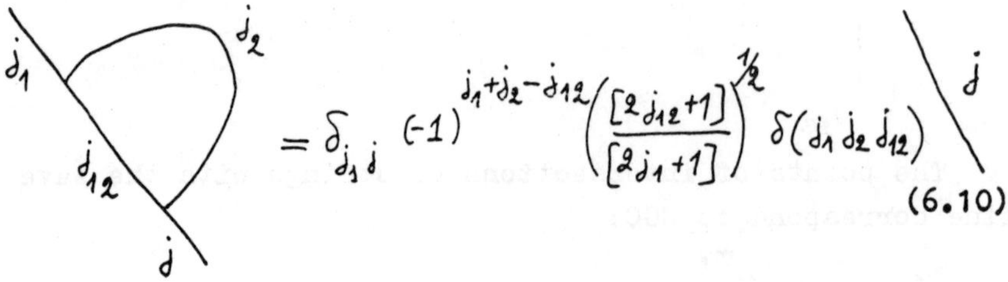

$$= \delta_{j_1 j}\,(-1)^{j_1+j_2-j_{12}}\left(\frac{[2j_{12}+1]}{[2j_1+1]}\right)^{1/2}\delta(j_1 j_2 j_{12}) \tag{6.10}$$

$$\sum_{j_{12}} (-1)^{j_2-j_1-j_{12}}\,\delta(j_1 j_2 j_{12})\left(\frac{[2j_{12}+1]}{[2j_1+1]}\right)^{1/2} = \tag{6.11}$$

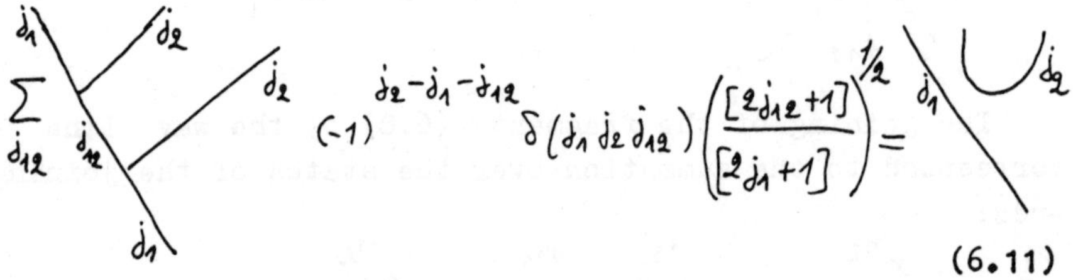

We see that one can rewrite these relations in the form similar to (6.1)-(6.3)

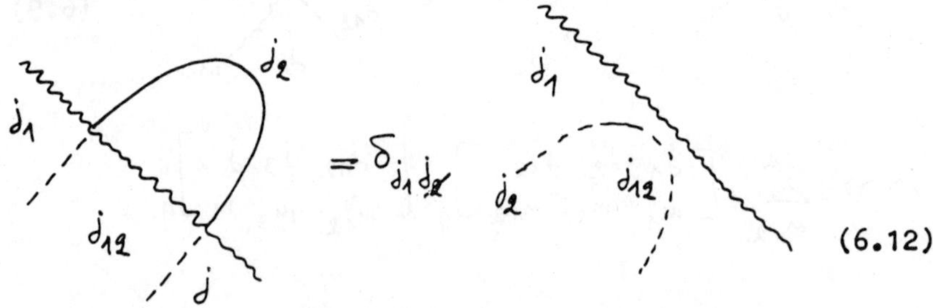

(6.12)

318

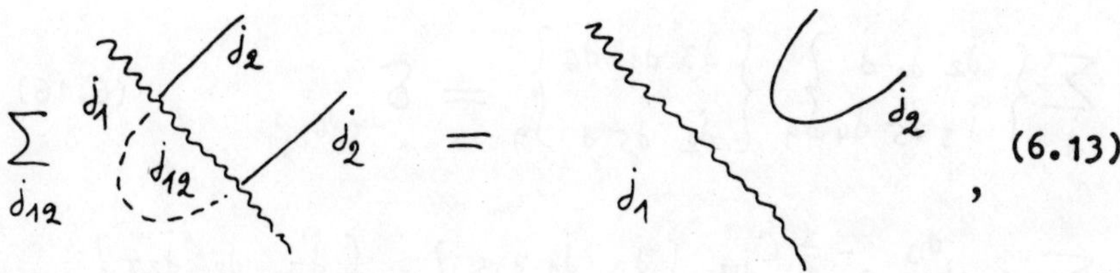

(6.13)

if we associate the following weights with extremal fragments in the shadows world

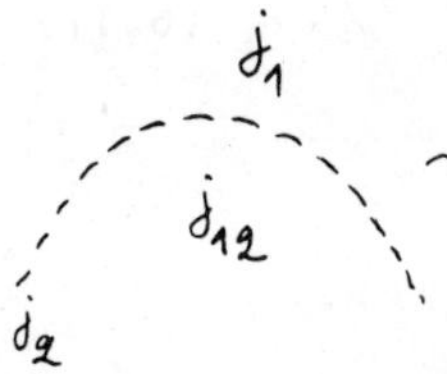

$$\rightsquigarrow (-1)^{j_1+j_2-j_{12}} \left(\frac{[2j_{12}+1]}{[2j_1+1]} \right)^{1/2} \delta(j_1 j_2 j_{12}) \tag{6.14}$$

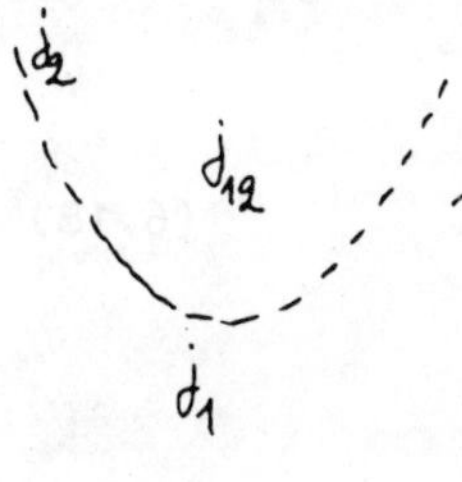

$$\rightsquigarrow (-1)^{j_1+j_2-j_{12}} \left(\frac{[2j_{12}+1]}{[2j_1+1]} \right)^{1/2} \delta(j_1 j_2 j_{12}) \tag{6.15}$$

So, using the relations (6.1)-(6.15) we can transpose every picture representing the combinations of R-matrices and CGC in into the shadows world. Using this process we can obtain the relations between $q-6j$ -symbols from the corresponding relations for R-matrices and CGC.

THEOREM 6.1. The following relations between $q-6j$ - bymbols hold:

$$\sum_{j}\begin{Bmatrix} j_2 & j_1 & j \\ j_3 & j_5 & j_4 \end{Bmatrix}_q \begin{Bmatrix} j_3 & j_1 & j_6 \\ j_2 & j_5 & j \end{Bmatrix}_q = \delta_{j_4 j_6} . \tag{6.16}$$

$$\sum_{j_{13}} (-1)^{j_{13}} q^{-\frac{1}{2}C_{j_{13}}} \begin{Bmatrix} j_3 & j_1 & j_{13} \\ j_2 & j & j_{12} \end{Bmatrix}_q \begin{Bmatrix} j_3 & j_2 & j_{23} \\ j_1 & j & j_{13} \end{Bmatrix}_q =$$

$$= (-1)^{j+j_1+j_2+j_3-j_{12}-j_{23}} q^{\frac{1}{2}\left(C_{j_{23}}+C_{j_{12}}-C_{j_2}-C_{j_3}-C_{j_1}-C_j\right)} \times \tag{6.17}$$

$$\times \begin{Bmatrix} j_3 & j_2 & j_{23} \\ j_1 & j & j_{12} \end{Bmatrix}_q .$$

$$\sum_{d}\begin{Bmatrix} j_2 & a & d \\ j_1 & c & b \end{Bmatrix}_q \begin{Bmatrix} j_3 & d & e \\ j_1 & f & c \end{Bmatrix}_q \begin{Bmatrix} j_3 & j_2 & j_{23} \\ a & e & d \end{Bmatrix}_q =$$

$$= \begin{Bmatrix} j_{23} & a & e \\ j_1 & f & b \end{Bmatrix}_q \begin{Bmatrix} j_3 & j_2 & j_{23} \\ b & f & c \end{Bmatrix}_q . \tag{6.18}$$

$$\sum_{g} (-1)^{a-b-g-f} q^{\frac{1}{2}(C_a-C_b-C_g-C_f)} \begin{Bmatrix} j_2 & a & g \\ j_1 & c & b \end{Bmatrix}_q \begin{Bmatrix} j_3 & g & e \\ j_1 & d & c \end{Bmatrix}_q \begin{Bmatrix} j_3 & a & f \\ j_2 & e & g \end{Bmatrix}_q =$$

$$= \sum_{g} (-1)^{d-c-g-e} q^{\frac{1}{2}(C_d-C_c-C_g-C_e)} \begin{Bmatrix} j_3 & b & g \\ j_2 & d & c \end{Bmatrix}_q \begin{Bmatrix} j_3 & a & f \\ j_1 & g & b \end{Bmatrix}_q \begin{Bmatrix} j_2 & f & e \\ j_1 & d & g \end{Bmatrix}_q . \tag{6.19}$$

$$\sum_{c}(-1)^{c}q^{-\frac{c(c+1)}{2}}\begin{Bmatrix} j & c & b\\ j & a & b\end{Bmatrix}\frac{[2c+1]}{[2b+1]}= \tag{6.20}$$

$$=(-1)^{2j+2b-a}\,q^{-c_j+c_b-\frac{1}{2}c_a}\,.$$

The relation (6.16) is called the orthogonality relation between $q-6j$ -symbols. The relation (6.17) is the q-analog of the Racah identity, the relation (6.18) is the q-analog of the Bidenharn-Elliot identity.

PROOF. Let us rewrite the relations (6.16)-(6.20) in the graphical representation:

$$\sum_{j} \ldots = \ldots\ \delta_{j_4 j_6} \tag{6.16'}$$

$$\sum_{j_{13}} \ldots = \ldots\ (-1)^{j_2+j_3-j_{23}}\,q^{\frac{1}{2}(c_{j_{23}}-c_{j_2}-c_{j_3})} \tag{6.17'}$$

(6.18')

(6.19')

(6.20')

Now we see that these relations follow from (4.11), (4.8), (4.7), (4.13) and (4.6) respectively if we transpose the latter into the shadows world. It seems that this is the simplest proof of the identities (6.18)-(6.20).

REMARK 1. The relations (6.16) are equivalent to the orthogonality relations for the Racah-Wilson polynomials (see [1] formulae (4.1)). The relations (6.17) and (6.18) give identities between the Racah-Wigner polynomials which seems to be new.

REMARK 2. Substituting the special values (6.21) of $q-6j$ -symbols we obtain the recurrence relations:

$$[2c+1][2d][2f+1]\begin{Bmatrix} a & b & e \\ d & c & f \end{Bmatrix}_q^{RW} = \{[b+d-f][d+e-c][b+f-d+1][c+e-d+1][a+c+f+2]\}^{1/2}\begin{Bmatrix} a & b & e \\ d-1/2 & c+1/2 & f+1/2 \end{Bmatrix}_q^{RW} +$$

$$+\{[b+d-f][b+f-d+1][c+d-e][c+d+e+1][a+c-f][a+f-c+1]\}^{1/2}\cdot\begin{Bmatrix} a & b & e \\ d-1/2 & c-1/2 & f+1/2 \end{Bmatrix}_q^{RW} + \{[d+f-b][b+d+f+1]$$

$$[c+d-e][c+d+e+1][c+f-a][a+c+f+1]\}^{1/2}\times\begin{Bmatrix} a & b & e \\ d-1/2 & c-1/2 & f-1/2 \end{Bmatrix}_q^{RW} + \{[d+f-b][b+d+f+1][d+e-c]$$

$$[c+e-d+1][a+f-c][a+c-f+1]\}^{1/2}\begin{Bmatrix} a & b & e \\ d-1/2 & c+1/2 & f-1/2 \end{Bmatrix}_q^{RW},$$

$$\begin{Bmatrix} a & b & e \\ \frac{1}{2}, & e+\frac{1}{2}, & b+\frac{1}{2} \end{Bmatrix}_q^{RW} = (-1)^{a+b+e+1}\left\{\frac{[a+b+e+2][b+e-a+1]}{[2e+2][2b+1]}\right\}^{1/2},$$

$$\begin{Bmatrix} a & b & e \\ \frac{1}{2} & e+\frac{1}{2}, & b-\frac{1}{2} \end{Bmatrix}_q^{RW} = (-1)^{a+b+e}\left\{\frac{[a+b-e][a+e-b+1]}{[2e+2][2b+1]}\right\}^{1/2}, \qquad (6.21)$$

$$\begin{Bmatrix} a & b & e \\ \frac{1}{2} & e-\frac{1}{2}, & b+\frac{1}{2} \end{Bmatrix}_q^{RW} = (-1)^{a+b+e}\left\{\frac{[a+e-b][a+b-e+1]}{[2e][2b+1]}\right\}^{1/2},$$

$$\begin{Bmatrix} a & b & e \\ \frac{1}{2} & e-\frac{1}{2} & b-\frac{1}{2} \end{Bmatrix}_q^{RW} = (-1)^{a+b+e}\left\{\frac{[a+b+e+1][e+b-a]}{[2e][2b+1]}\right\}^{1/2}.$$

From (5.7), (5.7'), (5.7") it follows that these relations are equivalent to the recurrence relation for W_n given in [1](formulae (4.6)).

REMARK 3. Identity (6.19) is the face form of Yang-Baxter equation [20] for constant R-matrices.

REMARK 4. In a similar way one can define $q-9j$ -symbols. These symbols are connected with q -ortogonal polynomials depending on two variables. The details will be given in separate publication.

7. The Invariants Of Links Associated With $U_q(su(2))$

Using the q-analog of $6j$ -symbols we give here a new model for the invariants corresponding to higher representations of $U_q(su(2))$,[2, 17, 18]. This model is based on the graphical representation (6.4), (6.5) for $q-6j$ -symbols and is obtained from the model based on R-matrices [2] by transposing the latter into the shadow world in accordance with the rules of the section 6.

DEFINITION 7.1. Let $\mathcal{D}_L$ be the diagram of the link L.

a) to each component of L we associate a number j_α (α numerates the components of L), which we call the colour of the component.

b) let us paint the plane on which the diagram is located into different colours (numerated by $j \in \frac{1}{2}\mathbb{Z}_+$) following the rules described below:

- to the extremal part of the plane we associate the number $j=0$.
- to those parts of the plane which can be reached from the extremal part by crossing only one string, we associate the colour of this string (these are parts neighbouring to the extremal one).
- other internal parts of the diagram are pointed according to the following inductive rule: let O_K be a part which can be reached from the exteriour by crossing a minimum of K strings and let O^{α}_{K-1} be the neighbours of this part which can be reached by crossing $(K-1)$ -strings; let be the colour of O^{α}_{K-1} parts, then the colour j_K of the part O_K must satisfy the inequalities $|j^{\alpha}_{K-1} - \ell^{\alpha}_{K,K-1}| \leq \leq j_K \leq j^{\alpha}_K + \ell^{\alpha}_{K,K-1}$ for any α ; here $\ell^{\alpha}_{K,K-1}$ is the colour of the string dividing O_K and O_{K-1} . We shall call each set of colours satisfying these inequalities state on the diagram.

c) let the diagram $\mathcal{D}_L$ be in a general position.

d) for each state on $\mathcal{D}_L$ let us associate a weight to each intersecting and each extremal fragment following the rules (6.4), (6.5), (6.14), (6.15). Then we multiply all of these weights and sum up the product over all possible states.

The obtained functional is denoted by $Z_{j_1 \cdots j_K}(\mathcal{D}_L)$ where K is the number of the components of L and $j_1, j_2, \cdots, j_K$ are the colours of these components.

An example of the calculation of the functional $Z_{j_1 \cdots j_K}(\mathcal{D}_L)$.

Let $\mathcal{D}_L$ be the diagram given in Fig.1. The numbers j_1 and j_2 are the colours of the components of L . The states on $\mathcal{D}_L$ are given in Fig.1. For colours K and j

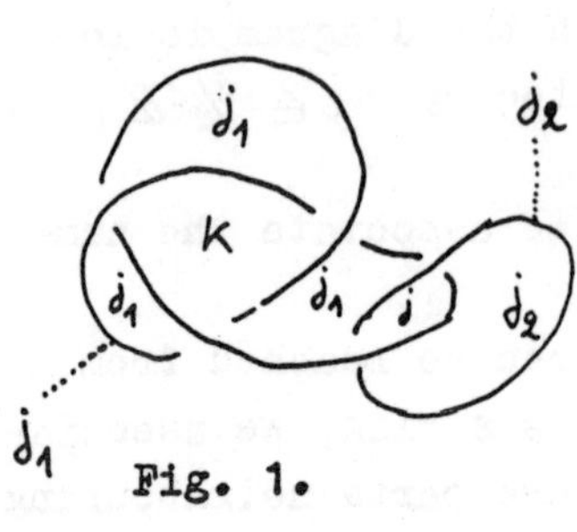

Fig. 1.

we have the following restrictions: $0 \le k \le 2j_1$, $|j_1 - j_2| \le j \le j_1 + j_2$. In accordance with Definition .1 for the functional $\mathcal{D}_{j_1 j_2}(\mathcal{D}_L)$ we obtain the following expression:

$$Z_{j_1 j_2}(\mathcal{D}_L) = \sum_{\substack{0 \le k \le 2j_1 \\ |j_1 - j_2| \le j \le j_1 + j_2}} (-1)^{k+2j_1}$$

$$q^{\frac{-c_k + 2c_{j_1}}{2}} \begin{Bmatrix} j_1 & 0 & j_1 \\ j_1 & k & j_1 \end{Bmatrix}_q (-1)^{k-2j_1} q^{\frac{2c_{j_1} - c_k}{2}} \begin{Bmatrix} j_1 & k & j_1 \\ j_1 & 0 & j_1 \end{Bmatrix}_q$$

$$(-1)^{k-2j_1} q^{\frac{-2c_{j_1} + c_k}{2}} \begin{Bmatrix} j_1 & j_1 & 0 \\ j_1 & j_1 & k \end{Bmatrix}_q (-1)^{j_1+j_2-j} q^{\frac{c_j - c_{j_1} - c_{j_2}}{2}} .$$

$$\begin{Bmatrix} j_2 & j_1 & j \\ j_1 & j_2 & 0 \end{Bmatrix}_q (-1)^{j_1+j_2-j} q^{\frac{c_j - c_{j_1} - c_{j_2}}{2}} \begin{Bmatrix} j_1 & j_1 & 0 \\ j_2 & j_2 & j \end{Bmatrix}_q .$$

$$(-1)^{k-2j_1} \left\{ \frac{[2k+1]}{[2j_1+1]} \right\}^{1/2} [2j_1+1]^{3/2} [2j_2+1] =$$

$$= \sum_{k,j} q^{\frac{1}{2}(4c_{j_1} - 3c_k + 2c_j - 2c_{j_2})} \begin{Bmatrix} j_1 & 0 & j_1 \\ j_1 & k & j_1 \end{Bmatrix} .$$

$$\begin{Bmatrix} j_1 & k & j_1 \\ j_1 & 0 & j_1 \end{Bmatrix}_q \begin{Bmatrix} j_1 & j_1 & 0 \\ j_1 & j_1 & k \end{Bmatrix}_q \begin{Bmatrix} j_2 & j_1 & j \\ j_1 & j_2 & 0 \end{Bmatrix}_q \begin{Bmatrix} j_1 & j_1 & 0 \\ j_2 & j_2 & j \end{Bmatrix}_q \cdot$$

$$\cdot [2k+1]^{1/2} [2j_1+1][2j_2+1] =$$

$$= \frac{1}{[2j_1+1]} \sum_{\substack{0 \le k \le 2j_1 \\ |j_1-j_2| \le j \le j_1+j_2}} q^{1/2(4c_{j_1} - 3c_k + 2c_j - 2c_{j_2})} \cdot$$

$$\cdot (-1)^{2j_1-k} [2k+1][2j+1] \in \mathbb{Z}[q, q^{-1}].$$

In [2] with the help of the matrices (2.4) and (2.6) a functional, analogous to the one described was defined. Let us remind the reader this definition.

DEFINITION 7.2.

a) let the diagram be in the general position

b) to each component of L we associate the number $j_\alpha \in \frac{1}{2}\mathbb{Z}_+$ (colour of the components)

c) divide the diagram into elementary fragments

d) to each edge connecting elementary fragments we associate the states $|m| \le j_\alpha$, $2m \equiv 2j_\alpha \pmod 2$ were j_α is the colour of the component.

e) to each elementary fragments we associate the weights using the rules (4.1), (4.2) (4.4), (4.5) of the section 4.

g) multiplying the matrix elements corresponding to the elementary fragment over all fragments and taking the sum of resulting product over all states on $\mathcal{D}_L$ we obtain the functional $\tilde{Z}_{j_1 \cdots j_K}(\mathcal{D}_L)$.

<u>An example of calculation of the functional</u> $\tilde{Z}(\mathcal{D}_L)$

Let $\mathcal{D}_L$ is the diagram given in Fig.1. In accordance with the definition of $\tilde{Z}_{j_1 j_2}(\mathcal{D}_L)$ we have:

$$\tilde{Z}_{j_1 j_2}(\mathcal{D}_L) = \sum_{\{m\}} w^{j_1}_{m_1 m_2} \left(\left(\mathbb{R}^{j_1 j_1}\right)^{-1}\right)^{m_1' m_3'}_{m_1 m_3} w^{j_2}_{m_3 m_4}$$

$$\left(\left(\mathbb{R}^{j_1 j_1}\right)^{-1}\right)^{m_4' m_2'}_{m_4 m_2} \left(\mathbb{R}^{j_1 j_1}\right)^{m_7 m_5}_{m_1' m_2'} w^{j_1}_{m_6 m_5} w^{j_1}_{m_3' m_7}$$

$$\left(\left(\mathbb{R}^{j_1 j_2}\right)^{-1}\right)^{m_6' m'}_{m_1' m} \left(\left(\mathbb{R}^{j_2 j_1}\right)^{-1}\right)^{m'' m_6}_{m' m_6'} w^{j_2}_{m_8 m} w^{j_2}_{m'' m_8} \cdot$$

$$q^{2C_{j_1} + C_{j_2}}$$

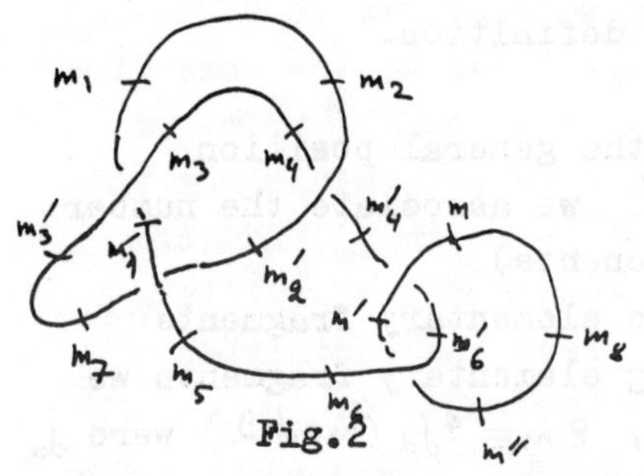

Fig.2

The states $\{m\}$ on the diagram $\mathcal{D}_L$ are given in Fig.2.

Let us introduce an orientation on L and define the numbers

$$w_\alpha(\mathcal{D}_L) = N^+_\alpha(\mathcal{D}_L) - N^-_\alpha(\mathcal{D}_L)$$

where $N^\pm_\alpha(\mathcal{D}_L)$ are the numbers of positive (N^+) and negative (N^-) selfintersections on the component

$$\nearrow\!\!\!\nwarrow \rightsquigarrow N^+ \quad , \quad \nwarrow\!\!\!\nearrow \rightsquigarrow N^- .$$

It is easy to see that the numbers $w_\alpha(\mathcal{D}_L)$ do not depend on the orientation of the component α .

THEOREM 7.1. i) The functionals $Z_{j_1 \cdots j_K}(\mathcal{D}_L)$ and $\tilde{Z}_{j_1 \cdots j_K}(\mathcal{D}_L)$ coincide,

ii) the functional

$$\varphi_{j_1 \cdots j_K}(\mathcal{D}_L) = \prod_{\alpha=1}^{K} \left(q^{c_{j_\alpha}} (-1)^{2j_\alpha} \right)^{w_\alpha(\mathcal{D}_L)} Z_{j_1 \cdots j_K}(\mathcal{D}_L) \quad (7.1)$$

is the invariant of the link L .

PROOF. i) Consider the functional $\tilde{Z}_{j_1 \cdots j_K}(\mathcal{D}_L)$ as the product of the matrices corresponding to the elementary fragments and the matrix $I^j \tilde{Z}_{j_1 \cdots j_K}(\mathcal{D}_L)$ acting in V^j (I^j is a unit matrix). In accordance with the graphical rules this matrix is represented by the diagram $\mathcal{D}_L$ situated in the upper side of the wave line which correspond to the space V^j . Using the rules of section 6 we can transpose (see Fig.3) this diagram to the shadow world. From the rules (6.4)-(6.7) and (6.14), (6.15) we obtain a new representation for calculating the functional $\tilde{Z}_{j_1 \cdots j_K}(\mathcal{D}_L)$. From the definition of this procedure it follows that wthe value of $\tilde{Z}_{j_1 \cdots j_K}(\mathcal{D}_L)$ calculated in the shadow world do not depend on j , and therefore we can put $j = 0$ after that we obtain the rules for calculating $Z_{j_1 \cdots j_K}(\mathcal{D}_L)$.

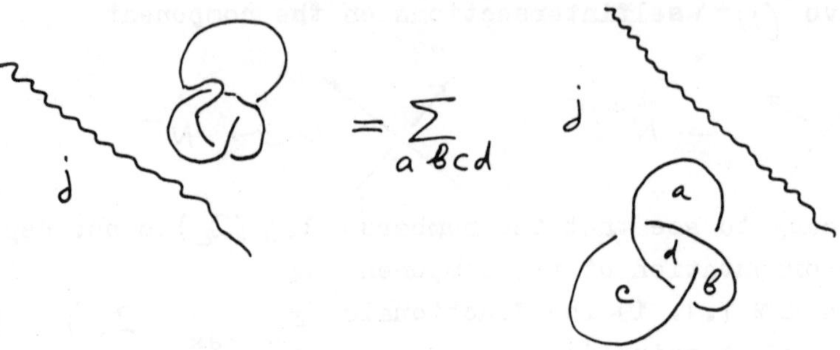

Fig. 3

The part ii) of the theorem was proved in [2] (see also [18, 29]). The main idea is that the invariance of $\widetilde{Z}(\mathcal{D}_L)$ under regular isotopies follows from the relations (4.6), (4.10) and the invariance under singular isotopies follows from (4.12) and from the structure of the multiplier before in (7.1).

Let us give a representation for the invariant φ in the case when L is given in the closed braid form. Consider the case $j_1 = j_2 = \cdots = j_K = j$.

We should remind that the braid group B_N [35] is generated by the elements s_i, $i=1,\ldots,N-1$ with the following relations

$$s_i\, s_{i+1}\, s_i = s_{i+1}\, s_i\, s_{i+1} \;,$$

$$s_i\, s_j = s_j\, s_i \;, \quad |i-j|>1 \,.$$

One can define a representation π of the braid group in the space $(V^j)^{\otimes N}$

$$\pi(s_i) := g_i = 1 \otimes \cdots \otimes P_{i\,i+1} R^{jj}_{i\,i+1} \otimes \ldots \otimes 1 \tag{7.2}$$

where $P_{i\,i+1}$ is the permutation matrix.

It is well known that for $j=1/2$ this representation is the representation of the braid group in the Temperley-Lieb-Jones algebra. Indeed, in this case

$$\pi(s_i)=1\otimes\cdots\otimes q^{-\frac{1}{2}}\left(e_i(q+q^{-1})-q^{-\frac{1}{2}}\right)\otimes\cdots\otimes 1 \tag{7.3}$$

where the elements e_i form T.L.J. algebra:

$$e_i e_{i+1} e_i = \tau e_i\,,\quad e_i^2 = e_i$$

$$e_{i+1} e_i e_{i+1} = \tau e_{i+1} \tag{7.4}$$

$$\tau=(q+q^{-1})^{-2}$$

Representations (7.2) for $j>\frac{1}{2}$ can be obtained from one for $j=\frac{1}{2}$ by the fusion-procedure [2]. Let us define the following matrices acting in $(V^{1/2})^{\otimes 2jN}$:

$$\mathcal{P}_j^{(n+1,\dots,n+2j)} = \prod_{\ell=1}^{2j}\left(q^{2j-2\ell+1}-q^{-(2j-2\ell+1)}\right)^{-1}\cdot$$

$$\sum_{\varepsilon_1\cdots\varepsilon_{2j}=\pm 1}\ \prod_{\ell=1}^{2j}\varepsilon_\ell\, q^{(2j+2-\ell)\varepsilon_\ell/2}\left(R_{n+1}^{\varepsilon_1}\cdots R_{n+2j}^{\varepsilon_{2j}}\right)$$

$$\left(R_{n+1}^{\varepsilon_2}\cdots R_{n+2j-1}^{\varepsilon_2}\right)\cdots R_{n+1}^{\varepsilon_{2j}}$$

$$\mathbb{R}^{(j)}_{n} = \left(R_{n+2j}\cdots R_{n+4j-1}\right)\left(R_{n+2j-1}\cdots\right.$$

$$\left.\cdots R_{n+4j-2}\right)\cdots\left(R_{n+1}\cdots R_{n+2j}\right) \tag{7.6}$$

$$R_i = 1\otimes\cdots\otimes P_{ii+1}R^{\frac{1}{2}\frac{1}{2}}_{ii+1}\otimes\cdots\otimes 1$$

THEOREM 7.2. The map $\tilde{\pi} : B_N \longrightarrow End\left(\left(V^{1/2}\right)^{\otimes 2jN}\right)$. defined by the following formula

$$\tilde{\pi}(g_i) = R^{(j)}_{i} = \varphi_j^{(2ji+1,\dots,2j(i+1))}\,\varphi_j^{(2j(i+1)+1,\dots,2j(i+2))}\,\mathbb{R}^{(j)}_{2ji} \tag{7.7}$$

is the representation of the braid-group B_N and this representation is isomorphic to the representation (7.2).

PROOF. Let us give graphical prove of this theorem. The representation $\tilde{\pi}$ can be written as the following picture:

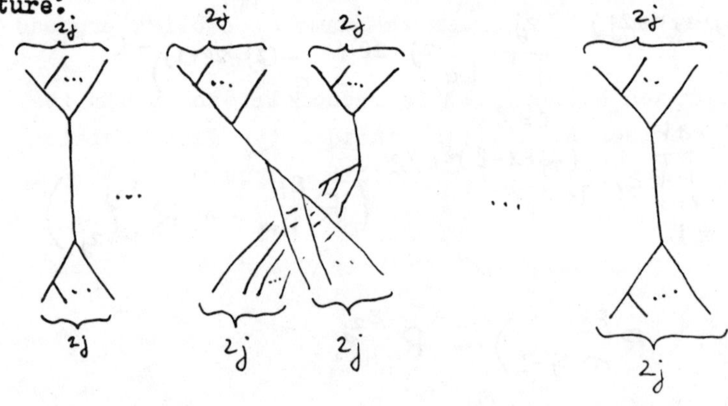

Using the relations (4.6) and (4.7) we can transform this picture in the form

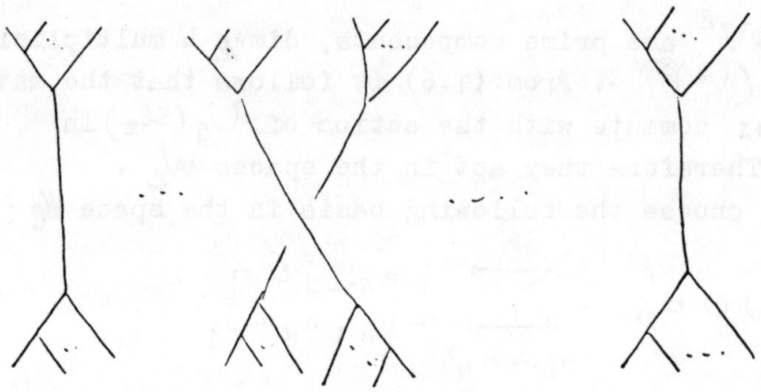

This picture represent the action (7.2) of braid group. So, the theorem is proved.

The fusion-procedure can be applied to the construction of links invariants corresponding to higher representations starting from the invariants that correspond to the vector representation. This procedure was independently considered in [17] under the name of cabling procedure.

The representation (7.2) is reducible. To obtain its irreducible components let us consider the decomposition of $(V^j)^{\otimes N}$ into $U_q(sl_2)$-irreducible components:

$$(V^j)^{\otimes N} = \sum_{\ell}^{\oplus} (W_\ell \otimes V^\ell)$$

where $W_\ell \otimes V^\ell$ are prime components, $\dim W_\ell$ = multiplicity of V^ℓ in $(V^j)^{\otimes N}$. From (1.6) it follows that the matrices $R_{i\,i+1}$ commute with the action of $U_q(sl_2)$ in $(V^j)^{\otimes N}$. Therefore they act in the spaces W_ℓ.

Let us choose the following basis in the space W_ℓ:

$$E^\ell(a) \otimes e^\ell_m = \sum_{n_1 \dots n_N} \begin{bmatrix} a_{N-1} & j & \ell \\ n_{N-1} & n_N & m \end{bmatrix}_q \cdots$$

$$\cdots \begin{bmatrix} a_1 & j & a_2 \\ n_1 & n_3 & n_2 \end{bmatrix}_q \begin{bmatrix} j & j & a_1 \\ n_1 & n_2 & n_1 \end{bmatrix}_q e^j_{n_1} \otimes \cdots \otimes e^j_{n_N} \qquad (7.8)$$

The elements of this basis are numerated by the sequences $(a) \equiv (j, a_2, \dots, a_{N-1}, \ell)$ of numbers $a_i \in \frac{1}{2}\mathbb{Z}_+$ satisfying the following conditions $0 \le a_2 \le 2j_1, \dots, |a_k - j| \le a_{k+1} \le a_k + j, \dots, a_N = \ell$.

Let us define the following action of the braid group in the space W_ℓ:

$$\pi_j(s_i)\left(E^\ell(a) \otimes e^\ell_m\right) = \left(\pi^\ell_j(s_i) E^\ell(a)\right) \otimes e^\ell_m \qquad (7.9)$$

PROPOSITION.

$$\pi^\ell_j(s_i) E^\ell(a) = \sum_{a'_i} (-1)^{a'_i + a_i - a_{i+1} - a_{i-1}} \begin{Bmatrix} j & a_{i-1} & a_i \\ j & a_{i+1} & a'_i \end{Bmatrix}_q$$

$$q^{\pm \frac{c_{a_{i-1}} + c_{a_{i+1}} - c_{a_i} - c_{a'_i}}{2}} E^\ell(j, a_2, \dots, a'_i, \dots, a_{N-1}, \ell)$$

$$q^{\pm\frac{c_{a_{i-1}}+c_{a_{i+1}}-c_{a_i'}-c_{a_i}}{2}} E^{\ell}(j, a_2, \ldots, a_i', \ldots, a_{N-1}, \ell) \qquad (7.10)$$

PROOF. According to the graphical rules of section 4 the vector $E^{\ell}(a) \otimes e^{\ell}_m$ is represented by the following picture:

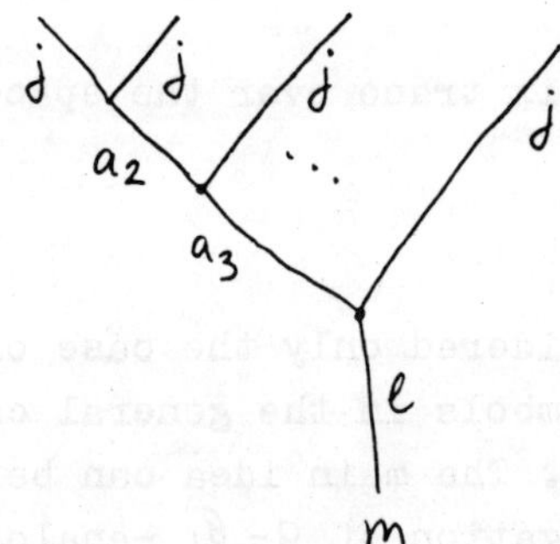

The left hand side of (7.9) is written as

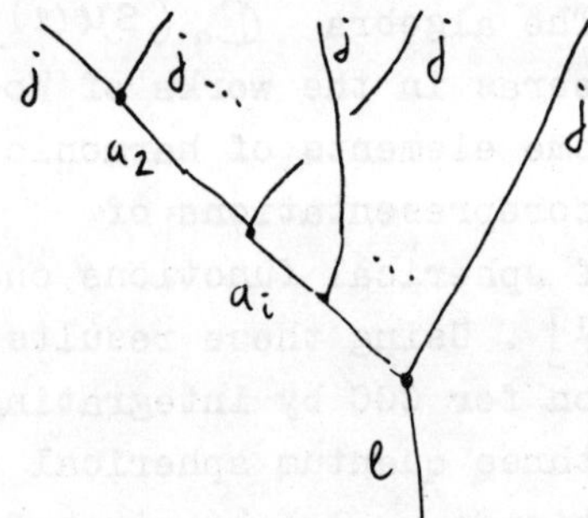

The formula (7.9) follows immediately from (5.11).

Now let the link L be a closure of the braid α. In this case from the definition $\tilde{Z}(L)$ and $Z(L)$ we have:

$$\tilde{Z}_j(\alpha) = tr_{(V^j)^{\otimes N}}\left(\left(q^{H/2} \otimes \dots \otimes q^{H/2}\right)\pi_j(\alpha)\right) =$$
$$= \sum_{\ell \in j^{\otimes N}} [2\ell+1]_q \, tr_{W_\ell}\left(\pi_j^\ell(\alpha)\right), \qquad (7.11)$$

where $tr_V(\cdot)$ means the matrix trace over the space V.

CONCLUSION

In the present work we considered only the case of the algebra $\mathfrak{g} = s\ell(2)$. $q-6j$ symbols in the general case may be defined in a similar way. The main idea can be extracted from [2]. The investigation of $q-6j$ -analogous of CGC for $U_q(s\ell(n))$ will be given in a separate publication.

Here we don't consider the dual algebra $U_q(s\ell(2))^* = \mathbb{C}_q(SL(2))$. The algebra $\mathbb{C}_q(SL(2))$ appeared in different contexts in [9, 25]. The algebra $\mathbb{C}_q(SU(2))$ appeared in the theory C^* algebras in the works of Woronovich [26],who constructed some elements of harmonic analysis for this algebra. The corepresentations of $\mathbb{C}_q(SU(2))$ and q-analogue of spherical functions on $SU(2)$ were studied in [27][37]. Using these results one can obtain the representation for CGC by integrating over $\mathbb{C}_q(SU(2))$ the product of three quantum spherical functions. A similar representation can also be given for $q-6j$ -symbols.

As in the case of $q=1$ the relations between $q-6j$.-symbols can be organized in the Wigner-Recah algebra [23].

We do not consider in details the case when q is a root of unity. In this case one can introduce the calculus of restricted $q-6j$ -symbols. These $q-6j$ -symbols are defined by the formulas of § 5, but the values of their arguments are restricted by the additional condition $a, b, c, d, e, f \leq r-2$. As in the case of general q the arguments of restricted $q-6j$ -symbols satisfy also standard inequalities $|b-e| \leq a \leq e+c$, $|c-f| \leq a \leq c+f$, $|d-f| \leq b \leq d+f$, $|e-c| \leq d \leq e+c$. From the symmetries of $q-6j$ -symbols it follows that for $q^{\frac{r}{2}}=1$

$$\begin{Bmatrix} a & b & e \\ d & c & f \end{Bmatrix}_q = (-1)^{b+f-c-e+1+2a} \begin{Bmatrix} \bar{a} & b & e \\ d & c & f \end{Bmatrix}_q =$$

$$= \begin{Bmatrix} \bar{a} & b & e \\ \bar{d} & c & f \end{Bmatrix}_q ,$$

$$\begin{Bmatrix} a & b & e \\ d & c & f \end{Bmatrix}_q = (-1)^{f-b-d} \begin{Bmatrix} a & \bar{b} & e \\ \bar{d} & c & \bar{f} \end{Bmatrix}_q$$

where $\bar{a} = \frac{r}{2} - 1 - a$.

It is easy to check that restricted $q-6j$ -symbols satisfy all relations (6.16)-(6.19).

There is an interesting application of restricted $q-6j$ -symbols in conformal field theory. Moore and Seiberg found the polynomial equations describing operator algebras of conformal field theories. It is not hard to check that the relations (6.16)-(6.19) for restricted $q-6j$ -symbols coincide with polynomial equation of Moore and Seiberg, and therefore restricted $q-6j$ -symbols define some operator algebra. It follows from [32] that it is the operator algebra of the Wess–Zumino model of level $r-2$ with central charge

$$c = \frac{3(r-2)}{r}$$

and with the anomaleous dimensions $h_j = \frac{j(j+1)}{r}$,
$j = 0, 1/2, \dots, \frac{r-2}{2}$.

We thank L.Faddeev, V.Bazhanov, M.Semenov-Tian-Shanksy, J.Soybelman, L.Takhtajan and L.Vaksman for interesting discussions.

References

1. R.Askey, J.Wilson, SIAM J.Math.Anal., v.10, N 5, 1979.
2. N.Yu.Reshetikhin, Quantized universal enveloping algebras, the Yang-Baxter equation and invariants of links. I. II. LOMI-preprints, E-4-87, E-17-87, Leningrad, 1988.
3. H.N.Temperly, E.Lieb, Prob.Roy.Soc. (London),(1971), 252-280.
4. V.Jones, Invent.Math., 72 (1983), 1-25.
5. L.D.Faddeev, Integrable models in 1+1-Dimensional Quantum Field Theory. In Les Houches Lectures 1982, Elsevier, Amsterdam, 1984.
6. P.P.Kulish, E.K.Sklyanin, Lecture Notes in Physics, 1982, v.151, p.61.
7. P.P.Kulish, N.Yu.Reshetikhin, Zap.nauch.semin.LOMI, 1980, v.101, p.112.
8. E.K.Sklyanin, Uspehi Mat.Nauk, 1985, 40, N 2, p.214.
9. V.G.Drinfeld, Dokl.Akad.Nauk SSSR, 1985, v.283, N 5, p.1060-1064.
10. M.Jimbo, Lett.Math.Phys., 1985, 10, 63-69.
11. L.D.Faddeev, N.Yu.Reshetikhin, L.A.Takhtajan, LOMI - preprint, E- 87, 1987.

12. N.Yu.Reshetikhin, M.Semenov-Tian-Shansky. Quantum R-matrices and factorization problem in quantum groups. Imperial College-preprint, April 1988.
13. G.Luztig, Quantum deformations of certain simple modules over enveloping algebra n.I.T.preprint, December 1987.
14. M.Rosso. C.R.Acad.Sci.Paris t.305, Série I, p.587-590, 1987.
15. L.L.Vaksman, Dokl.Akad.Nauk SSSR, 1988, v.288.
16. A.N.Kirillov, Zap.nauch.semin.LOMI, 1988, v.168.
17. J.Murakami, The parallel version of polinomial invariants of links. Osaka University preprint 1988.
18. Y.Akutsu, M.Wadati, J.Phys.Soc.Jpn. 56 (1987) 839; 56 (1987) 3039; 59 (1987) 3464.
19. E.Date, M.Jimbo, T.Miwa, M.Okado. Solvable lattice models. RIMS - 590 - preprint, 1987.
20. M.Jimbo, T.Miwa, M.Okado, An $A_n^{(1)}$ family of solvable lattice models. RIMS - 579 - prepint, 1987.
21. E.Abe, Hopf algebras. Cambridge tracts in mathematics 74. Cambridge University Press, 1980.
22. H.Wenzl, Representations of Hecke algebras and subfactors. Ph.D.Thesis. University of Pennsilvania (1985).
23. L.C.Biedenharn, J.D.Louck, Angular momentum in Quantum Physics, Encidopedia of Math.and Appl. volume 8, Addison-Wesley Publ.Comp.1981.
24. D.A.Varchalowich, A.N.Moskalov, V.K.Hersonski, The theory of Quantum angular momentum. - Leningrad, Nauka, 1975.
25. L.Faddeev, L.Takhtajan. Lect.Notes Phys. 246 (1986), 166-179.
26. S.Woronowich Publ.RIMS, Kyoto Univ., 23, 1987, 117-181.
27. T.Masuda, K.Mimachi, Y.Nakagami, M.Noumi, K.Ueno. Representations of quantum groups and q -analoge of or-

thogonal polynomials, RIMS 613, preprint, 1988.

28. P.Podles, Lett.Math.Phys., 304 (1987) 323-326.
29. V.Jones. On knot invariants related to some statistical mechanical models. California University preprint, Berkley, 1988.
30. A.P.Jueys, A.A.Bandzaitis, Angular Momentum Theory in Quantum Physics. - Vilnius, Mokslas, 1977.
31. V.Pasquier. Etiology of IRF models, Sacley-preprint 1988.
32. Y.Kanie, A.Tsuchiya, Advanced Studies in Pure Mathematics. Vol.16, p.297-372, Kinokuniay, Tokyo Japan.
33. T.Deguchi, M.Wadati, Y.Akutsu.Link Polinomials constructed from Solvable Models in Statistical Mechanics. Tokio University Preprint, 1988.
34. Freyd P., Yetter D., Lickorish W.B.R., Millett K., Ocneanu A., Hoste J., Bull.Math.Soc., 1985, v.12, N 2, p.239-246.
35. Turaev V.G., The Yang-Baxter equation and invariants of links. - LOMI-preprint E-3-87, 1987.
36. Drinfeld V.G. Algebra and Analis. v.1, N2, 1989.
37. Soybelman Ya.,Vaksman L. Funk. Anal. apl. v.22,N 4,p. 75, 1988

Commun. Math. Phys. 117, 243–259 (1988)

Communications in Mathematical Physics

Knots, Links, Braids and Exactly Solvable Models in Statistical Mechanics

Yasuhiro Akutsu[1] and Miki Wadati[2]

[1] Institute of Physics, Kanagawa University, Rokkakubashi, Kanagawa-ku, Yokohama 221, Japan
[2] Institute of Physics, College of Arts and Sciences, University of Tokyo, Komaba, Meguro-ku, Tokyo 153, Japan

Abstract. We present a general method to construct the sequence of new link polynomials and its two variable extension from exactly solvable models in statistical mechanics. First, we find representations of the braid group from the Boltzmann weights of the exactly solvable models. Second, we give the Markov traces associated with new braid group representations and using them construct new link polynomials. Third, we extend the theory into a two-variable version of the new link polynomials. Throughout the paper, we emphasize the essential roles played by the exactly solvable models and the underlying Yang-Baxter relation.

1. Introduction

In physics we often deal with the configuration problem of one-dimensional objects, for instance, polymers, magnetic fluxes, dislocation lines and trajectories of particles. We generally call a one-dimensional object a string. A knot is a closed string which does not cross with itself. As a more generalized object, an assembly of knots with mutual entanglements is called a link. Classification of knots and links is known to be a longstanding problem in mathematics [1, 2]. In this paper we report an unexpected close connection between physics and mathematics. Namely, we present a general method to construct topological invariants for knots and links by using the theory of exactly solvable models in statistical mechanics.

We begin with the braid and the braid group. Braids are formed when n points on a horizontal line are connected by n strings to n points on another horizontal line directly below the first n points. A trivial n-braid is a configuration where no intersection between the strings is present. A general n-braid is constructed from the trivial n-braid by successive applications of the operation b_i, $i=1,2,\ldots,n-1$. The operation b_i and its inverse b_i^{-1} are best understood by the graphs (Fig. 1). A the set of generators, $b_1, b_2, \ldots, b_{n-1}$, define the braid group B_n [3]. By regarding the trivial n-braid as the identity operation in B_n, we can identify any element in B_n as an n-braid. To guarantee the topological equivalence between different

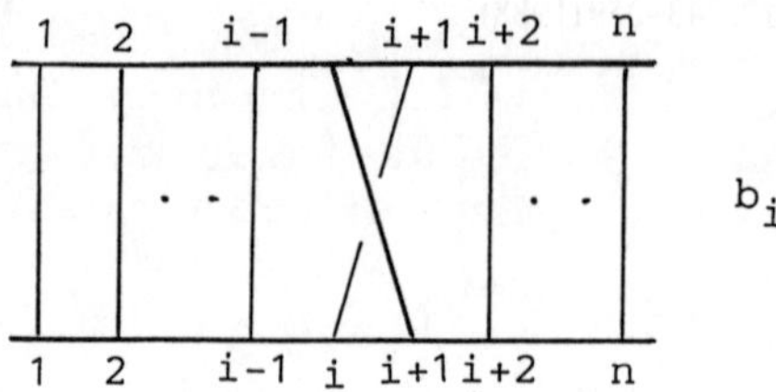

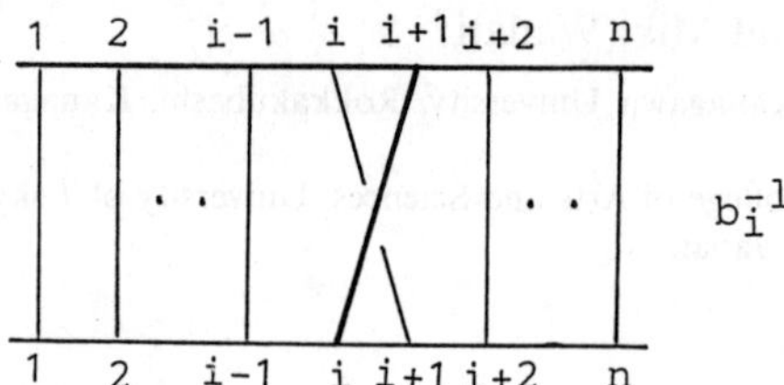

Fig. 1. Operations b_i and b_i^{-1}

(a)

(b)

Fig. 2. Defining relations of the braid group

expressions of a braid in terms of braid group elements, Artin proved that the following conditions are necessary and sufficient (Fig. 2):

$$b_i b_j = b_j b_i, \quad |i-j| \geqq 2, \tag{1.1a}$$

$$b_i b_{i+1} b_i = b_{i+1} b_i b_{i+1}. \tag{1.1b}$$

We call them the defining relation of B_n. Then, each topologically equivalent class of the braids is identified with an element in B_n.

Given a braid, one may form a link by tying opposite ends. Conversely, according to Alexander's theorem [4], any link is represented by a closed braid. This fact gives the braid group a fundamental role in the knot theory. However, the representation of a link as a closed braid is highly non-unique. Therefore, the following theorem due to Markov [5] is important. The equivalent braids expressing the same link are mutually transformed by successive applications of two types of operations, type I and type II Markov moves:

$$\text{I. } AB \to BA\,(A, B \in B_n),$$
$$\text{II. } A \to Ab_n, A \to Ab_n^{-1}, \quad (A \in B_n, b_n \in B_{n+1}). \tag{1.2}$$

Hence, to obtain a link polynomial which is a topological invariant, we first construct a suitable representation B_n' of the braid group B_n and then find a Markov move invariant quantity defined on B_n'.

Let us denote the representation of b_i by g_i and a link polynomial by $\alpha(\cdot)$. The link polynomial $\alpha(\cdot)$ must satisfy the conditions:

$$\text{I. } \alpha(AB) = \alpha(BA)\,(A, B \in B_n'), \tag{1.3a}$$
$$\text{II. } \alpha(Ag_n) = \alpha(Ag_n^{-1}) = \alpha(A) \quad (A \in B_n', g_n \in B_{n+1}'). \tag{1.3b}$$

This quantity is readily obtained if we can find a linear functional $\phi(\cdot)$ on B_n', called the Markov trace [6, 7], which have the following properties (the Markov properties):

$$\text{I. } \phi(AB) = \phi(BA)\,(A, B \in B_n'), \tag{1.4a}$$
$$\text{II. } \phi(Ag_n) = \tau\phi(A), \phi(Ag_n^{-1}) = \bar{\tau}\phi(A) \quad (A \in B_n', g_n \in B_{n+1}'), \tag{1.4b}$$

with the parameters τ and $\bar{\tau}$ being given by

$$\tau = \phi(g_i), \quad \bar{\tau} = \phi(g_i^{-1}) \quad \text{for all } i. \tag{1.5}$$

With the Markov trace $\phi(\cdot)$, the link polynomial $\alpha(\cdot)$ is given by

$$\alpha(A) = (\tau\bar{\tau})^{-(n-1)/2} \left(\frac{\bar{\tau}}{\tau}\right)^{e(A)/2} \phi(A) \quad (A \in B_n'). \tag{1.6}$$

Here $e(A)$ is the exponent sum of g_i's appearing in the braid A. For instance, if $A = g_2^3 g_1^{-2}$, then $e(A) = 3 - 2 = 1$. The properties (1.3a) and (1.3b) are easily verified from (1.4a), (1.4b) and the definition of $e(A)$.

In the recent discovery of a new polynomial invariant (Jones polynomial) for knots and links, Jones [7] utilized a C^*-algebra $A_{q,n}$ generated by $\{1, e_1, e_2, \ldots, e_n\}$. The generators $\{e_j\}$ are essentially identical with the Temperley-Lieb operators [8] and satisfy the relations:

$$e_i^* = e_i, \quad e_i^2 = e_i,$$
$$e_i e_j = e_j e_i \quad \text{for } |i-j| \geq 2,$$
$$e_i e_{i\pm 1} e_i = q^{-1} e_i. \tag{1.7}$$

The Temperley-Lieb algebra is known to describe the transfer matrices of the ferroelectric model (6-vertex model) and the self-dual Potts model (critical Potts model) in statistical mechanics [9, 10]. An ingenious idea by Jones is to use the algebra $A_{q,n}$ for a construction of a representation (Hecke algebra [11] representation) of the braid group. To observe this, we introduce a parameter t and a generator $\hat{g}_i$ by

$$q=2+t^{-1}+t, \tag{1.8}$$

$$\hat{g}_i=(t+1)e_i-1 \qquad (e_i \in A_{q,n-1}). \tag{1.9}$$

Then the Hecke algebra $H(t,n)$ is generated by $\{\hat{g}_i\}$. The generators satisfy both the defining relation (1.1) of the braid group and the quadratic relation (the reduction relation):

$$\hat{g}_i^2=(t-1)\hat{g}_i+t. \tag{1.10}$$

In most cases, the Jones polynomial is more powerful than the classic Alexander polynomial [12] in the sense that it detects properties of a link which could not be detected by the latter. The Jones polynomial and the Alexander polynomial are special cases of the two-variable extension [13] of the Jones polynomial. It is known that the Jones polynomial and its two-variable extension are still not complete. There exist infinitely many different links which have the same polynomial [14, 15]. The aim of the present paper is clear. We further pursue a close relation between the exactly solvable models and the link polynomials, and present a general method which leads to a sequence of new link polynomials and its two-variable extension.

The paper is organized as follows. In Sect. 2, we give a general prescription to have a braid group representation from the Boltzmann weights which satisfy the Yang-Baxter relation. In Sect. 3, we apply the prescription to the N-state vertex model and obtain a sequence of braid group representations. Further, defining the Markov traces associated with the new braid group representation, we construct a sequence of new link polynomials. In Sect. 4, we give an alternative method to obtain the braid group representation by introducing "composite" string with symmetrizers. In Sect. 5, we use the composite string representation of the braid group to obtain the two-variable extension of the new link polynomials. The last section is devoted to a summary of the paper.

2. Construction of Braid Group Representations from Solvable Models

Recent development in the theory of quantum completely integrable systems provides us a unified treatment of various exactly solvable models in 1+1 dimensional field theory and in 2-dimensional classical statistical mechanics [16–20]. The central idea is that to each solvable model we can associate a family of commuting transfer matrices which are the generators of an infinite number of conserved quantities. The commutability condition is called the Yang-Baxter relation.

The Yang-Baxter relation takes different expressions depending on the types of models. For the 1+1 dimensional field theory, the Yang-Baxter relation is the

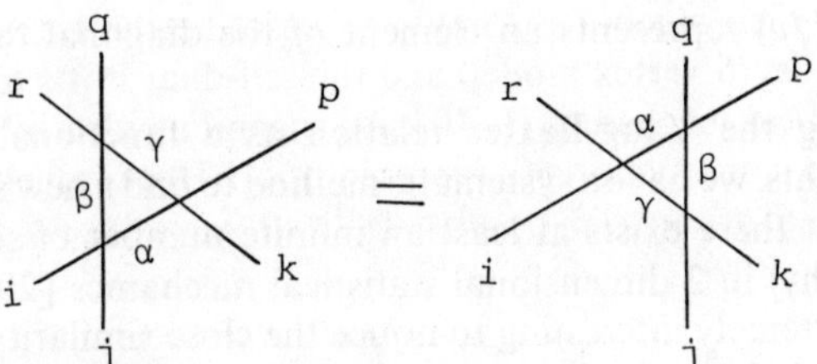

Fig. 3. The factorization equation

factorization condition [21] for the many-body *S*-matrices and is known as the factorization equation. We denote the scattering amplitude for the process $(i,j) \to (k,l)$ by $S^{ik}_{jl}(u)$, where u is the rapidity (we also refer to it as the spectral parameter). Then, the factorization equation reads as (Fig. 3)

$$\sum_{\alpha,\beta,\gamma} S^{\beta q}_{\gamma r}(v)S^{\alpha p}_{k\gamma}(u+v)S^{i\alpha}_{j\beta}(u) = \sum_{\alpha,\beta,\gamma} S^{\alpha p}_{\beta q}(u)S^{i\alpha}_{\gamma r}(u+v)S^{j\beta}_{k\gamma}(v). \tag{2.1}$$

In 2-dimensional statistical mechanics, we have two types of models, the vertex models and the IRF (interaction round a face) models [22]. Some of the IRF models are equivalent to the vertex models through the Wu-Kadanoff-Wegner transformation [23]. It is known that any factorized *S*-matrix can be interpreted as the Boltzmann weight of a solvable vertex model and that the factorization equation is the commutability condition of the transfer matrices of the vertex model [24]. For the IRF models, the Yang-Baxter relation is called the star-triangle relation which reads as

$$\begin{aligned}&\sum_c w(b,d,c,a;u)w(a,c,f,g;u+v)w(c,d,e,f;v)\\&\quad=\sum_c w(a,b,c,g;v)w(b,d,e,c;u+v)w(c,e,f,g;u).\end{aligned} \tag{2.2}$$

Here $w(a,b,c,d;u)$ denotes the Boltzmann weight for the spin configuration (a,b,c,d) round a face.

The Boltzmann weights satisfying (2.1) or (2.2) define a set of operators [22], $\{X_i(u)\}$, which satisfy the relations:

$$X_i(u)X_j(v) = X_j(v)X_i(u) \quad \text{for} \quad |i-j| \geqq 2, \tag{2.3a}$$

$$X_i(u)X_{i+1}(u+v)X_i(v) = X_{i+1}(v)X_i(u+v)X_{i+1}(u). \tag{2.3b}$$

In the vertex model we have the following expression of the operator $X_i(u)$:

$$X_i(u) = \sum_{klmn} S^{km}_{ln}(u) \cdot I^{(1)} \otimes I^{(2)} \otimes \ldots \otimes E^{(i)}_{nk} \otimes E^{(i+1)}_{ml} \otimes I^{(i+2)} \otimes \ldots, \tag{2.4a}$$

where $I^{(j)}$ is the identity acting on the jth position and E_{nk} is a matrix whose elements are $(E_{nk})_{pq} = \delta_{np}\delta_{kq}$. In the IRF model, matrix elements of $X_i(u)$ are given by

$$\begin{aligned}[X_i(u)]_{\{l'\},\{l\}} = \;&\delta_{l'_0 l_0}\delta_{l'_1 l_1}\ldots\delta_{l'_{i-1} l_{i-1}}\\&\times w(l_i, l_{i+1}, l'_i, l_{i-1}; u) \cdot \delta_{l'_{i+1} l_{i+1}} \ldots \delta_{l'_n l_n}.\end{aligned} \tag{2.4b}$$

The operator $X_i(u)$ represents an element of the diagonal to diagonal transfer matrix [22].

By regarding the Yang-Baxter relation as a functional equation for the Boltzmann weights, we have a systematic method to find a new solvable model. We have shown that there exists at least an infinite number of solvable hierarchies (Grand Hierarchy) in 2-dimensional statistical mechanics [20, 25–29].

Now it is extremely interesting to notice the close similarity between (1.1) and (2.3). A graphical similarity between Figs. 2 and 3 is also intriguing. The only difference is that in (2.3) we have spectral parameters as the arguments of the operators. In other words, the Yang-Baxter relation contains superfluous information which might be important in future studies. For the present purpose, to have a braid group representation, we eliminate the spectral parameters. We have found [30, 31] that an interesting identification is, after suitable normalization if necessary, given by

$$b_i \to g_i = \lim_{u\to\infty} X_i(u). \tag{2.5}$$

Depending on the function of parametrization, all the known solutions to the Yang-Baxter relations are classified into three cases; (1) elliptic (2) trigonometric or hyperbolic (3) rational. Any model in the case (1) [respectively the case (2)] at criticality corresponds to a model in the case (2) [respectively case (3)]. Then, to obtain an interesting braid group representation as a well-defined limit (2.5), we should use a model at criticality which is parametrized by the trigonometric or hyperbolic functions. This explains an observation by Jones that the Temperley-Lieb algebra for the transfer matrices of the 6-vertex model (8-vertex model at criticality) and the self-dual (critical) Potts model appear in the construction of the braid group representation.

In what follows, we consider only the solvable models whose Boltzmann weights are parametrized by the hyperbolic (or equivalently trigonometric) functions. We assume that the unitarity condition holds:

$$\sum_{p,q} S^{ql}_{pk}(-u)S^{ip}_{jq}(u)=\delta_{ik}\delta_{jl}, \quad \text{for the vertex model}, \tag{2.6a}$$

$$\sum_{e} w(e,c,d,a;-u)w(b,c,e,a;u)=\delta_{bd}, \quad \text{for the IRF model}. \tag{2.6b}$$

Due to the unitarity (2.6) which amounts to

$$X_i(-u)=X_i(u)^{-1}, \tag{2.7}$$

we have the identification

$$g_i^{-1} = \lim_{u\to\infty} X_i(-u). \tag{2.8}$$

3. The N-State Vertex Model and New Link Polynomials

Among many solvable models with hyperbolic (or equivalently trigonometric) parametrization, we consider a series of vertex models (the N-state vertex model) proposed by Sogo, Akutsu and Abe [32]. This series includes the 6-vertex model as $N=2$ case and the 19-vertex model [33] as $N=3$ case.

The edge variables i, j, k, and l of the Boltzmann weights $\{S_{jl}^{ik}(u)\}$ for the N-state vertex model take the following values:

$$i, j, k, l = -s, -s+1, \ldots, s-1, s, \tag{3.1}$$

where "spin" s is related to the state number N by

$$N = 2s+1. \tag{3.2}$$

The model has the properties:

a) charge conservation condition;

$$S_{jl}^{ik}(u) = 0 \text{ unless } i+j = k+l, \tag{3.3}$$

b) CPT invariances;

$$S_{jl}^{ik}(u) = S_{-j-l}^{-i-k}(u) = S_{ik}^{jl}(u) = S_{lj}^{ki}(u), \tag{3.4}$$

c) crossing symmetry;

$$S_{jl}^{ik}(u) = F(u) \cdot S_{ji}^{-k-l}(\lambda - u). \tag{3.5}$$

Here the parameter λ is called the crossing point of the spectral parameter and $F(u)$ is some function. For instance, parametrization of the weights which satisfy the unitarity condition (2.8) are shown below, for $N=2$, 3, and 4 cases.

1) $N=2$ $(s=1/2)$ case

$$S_{1/2\ 1/2}^{1/2\ 1/2}(u) = 1, \quad S_{-1/2\ -1/2}^{1/2\ 1/2}(u) = \frac{\sinh u}{\sinh(\lambda - u)},$$

$$S_{-1/2\ 1/2}^{1/2\ -1/2}(u) = \frac{\sinh \lambda}{\sinh(\lambda - u)}. \tag{3.6}$$

2) $N=3$ $(s=1)$ case

$$S_{1\ 1}^{1\ 1}(u) = 1, \quad S_{-1\ -1}^{1\ 1}(u) = \frac{\sinh u \sinh(\lambda+u)}{\sinh(\lambda-u)\sinh(2\lambda-u)},$$

$$S_{-1\ 1}^{1\ -1}(u) = \frac{\sinh\lambda \sinh 2\lambda}{\sinh(\lambda-u)\sinh(2\lambda-u)},$$

$$S_{0\ 0}^{1\ 1}(u) = \frac{\sinh u}{\sinh(2\lambda-u)}, \quad S_{0\ 1}^{1\ 0}(u) = \frac{\sinh 2\lambda}{\sinh(2\lambda-u)},$$

$$S_{0\ -1}^{0\ 1}(u) = \frac{\sinh 2\lambda \sinh u}{\sinh(\lambda-u)\sinh(2\lambda-u)},$$

$$S_{0\ 0}^{0\ 0}(u) = \frac{\sinh\lambda \sinh 2\lambda - \sinh u \sinh(\lambda-u)}{\sinh(\lambda-u)\sinh(2\lambda-u)}. \tag{3.7}$$

3) $N=4$ $(s=3/2)$ case

$$S^{3/2\ 3/2}_{3/2\ 3/2}(u)=1\,,$$

$$S^{3/2\ 3/2}_{-3/2\ -3/2}(u)=\frac{\sinh u\sinh(\lambda+u)\sinh(2\lambda+u)}{\sinh(\lambda-u)\sinh(2\lambda-u)\sinh(3\lambda-u)},$$

$$S^{3/2\ -3/2}_{-3/2\ 3/2}(u)=\frac{\sinh\lambda\sinh 2\lambda\sinh 3\lambda}{\sinh(\lambda-u)\sinh(2\lambda-u)\sinh(3\lambda-u)},$$

$$S^{3/2\ 3/2}_{1/2\ 1/2}(u)=\frac{\sinh u}{\sinh(3\lambda-u)},$$

$$S^{3/2\ 3/2}_{-1/2\ -1/2}(u)=\frac{\sinh u\sinh(\lambda+u)}{\sinh(2\lambda-u)\sinh(3\lambda-u)},$$

$$S^{3/2\ 1/2}_{1/2\ 3/2}(u)=\frac{\sinh 3\lambda}{\sinh(3\lambda-u)},$$

$$S^{3/2\ 1/2}_{-1/2\ 1/2}(u)=\frac{2\sqrt{\sinh\lambda\sinh 3\lambda}\cosh\lambda\sinh u}{\sinh(2\lambda-u)\sinh(3\lambda-u)},$$

$$S^{3/2\ 1/2}_{-3/2\ -1/2}(u)=\frac{\sinh 3\lambda\sinh u\sinh(\lambda+u)}{\sinh(\lambda-u)\sinh(2\lambda-u)\sinh(3\lambda-u)},$$

$$S^{3/2\ -1/2}_{-1/2\ 3/2}(u)=\frac{\sinh 2\lambda\sinh 3\lambda}{\sinh(2\lambda-u)\sinh(3\lambda-u)},$$

$$S^{3/2\ -1/2}_{-3/2\ 1/2}(u)=\frac{\sinh 2\lambda\sinh 3\lambda\sinh u}{\sinh(\lambda-u)\sinh(2\lambda-u)\sinh(3\lambda-u)},$$

$$S^{1/2\ 1/2}_{1/2\ 1/2}(u)=\frac{\sinh 2\lambda\sinh 3\lambda-\sinh u\sinh(\lambda-u)}{\sinh(2\lambda-u)\sinh(3\lambda-u)},$$

$$S^{1/2\ 1/2}_{-1/2\ -1/2}(u)=\frac{\sinh u[\sinh 2\lambda\sinh 3\lambda-\sinh u\sinh(\lambda-u)]}{\sinh(\lambda-u)\sinh(2\lambda-u)\sinh(3\lambda-u)},$$

$$S^{1/2\ -1/2}_{-1/2\ 1/2}(u)=\frac{\sinh 2\lambda[\sinh\lambda\sinh 3\lambda-2\cosh\lambda\sinh u\sinh(\lambda-u)]}{\sinh(\lambda-u)\sinh(2\lambda-u)\sinh(3\lambda-u)}.\tag{3.8}$$

The S-matrices for general N have been given in [32].

In order to construct an interesting braid group representation, we further introduce the symmetry-breaking transformation [32],

$$\tilde{S}^{ik}_{jl}(u)=\exp[(k-i-l+j)u/2]\cdot S^{ik}_{jl}(u)\,.\tag{3.9}$$

Due to the charge conservation condition (3.3), the resulting asymmetrized Boltzmann weights $\{\tilde{S}^{ik}_{jl}(u)\}$ also satisfy the factorization equation. We note that the transformation does not spoil the unitarity condition. Substituting $\tilde{S}^{ik}_{jl}(u)$ into (2.4a), we obtain the generator g_i of the braid group representation. We find that g_i satisfies the reduction relation

$$(g_i-c_1)(g_i-c_2)\dots(g_i-c_N)=0\,,\tag{3.10}$$

where

$$c_i=(-1)^{i+1}t^{[N(N-1)-(N-i+1)(N-i)]/2}, \tag{3.11a}$$

$$t=e^{2\lambda}. \tag{3.11b}$$

The reduction relation (3.10) for $N=2$ case is, by an identification $g_i \to -\hat{g}_i$, that of the the Hecke algebra (1.10).

A remaining task is to introduce the Markov trace which should satisfy the property (1.4). We have found that the Markov trace $\phi(\cdot)$ associated with our braid group representation is given in a form [30, 31]

$$\phi(A)=\mathrm{Tr}(HA). \tag{3.12}$$

Here Tr stands for the ordinary trace, that is, the sum of diagonal elements of the matrix. And, the matrix H is given by a tensor product of a diagonal matrix h:

$$H=h\otimes h\otimes \ldots \otimes h\otimes \ldots, \tag{3.13}$$

with

$$(h)_{pq}=t^{-p}\delta_{pq}\Bigg/\left(\sum_{k=-s}^{s} t^{-k}\right). \tag{3.14}$$

The denominator in (3.14) gives the proper normalization for the identity I_n in B_n':

$$\phi(I_n)=1. \tag{3.15}$$

The Markov trace defined by (3.12)–(3.15) is a generalization of the Powers state [34, 35]. The parameters τ and $\bar{\tau}$ in (1.5) are given by

$$\tau(t)=1/(1+t+t^2+\ldots+t^{N-1}), \tag{3.16a}$$

$$\bar{\tau}(t)=t^{N-1}/(1+t+t^2+\ldots+t^{N-1})=\tau(1/t). \tag{3.16b}$$

Thus, we arrive at a conclusion that the link polynomial $\alpha(A)$ for elements A in B_n' is given by

$$\alpha(A)=[t^{-(N-1)/2}(t+t+t^2+\ldots+t^{N-1})]^{n-1}[t^{(N-1)/2}]^{e(A)}\phi(A), \tag{3.17}$$

where $e(A)$ is the exponent sum of g_i's appearing in the braid A.

From the reduction relation (3.10), the Alexander-Conway relation for the Jones polynomial (the $N=2$ case) [7, 12, 36] is derived:

$$\alpha(L_+)=(1-t)t^{1/2}\alpha(L_0)+t^2\alpha(L_-). \tag{3.18}$$

In the above, by L_+, L_0, and L_-, we have denoted links which have the configurations of g_i, g_i^0, and g_i^{-1}, respectively, at an intersection. Similarly, we can show that

$$\alpha(L_{2+})=t(1-t^2+t^3)\alpha(L_+)+t^2(t^2-t^3+t^5)\alpha(L_0)-t^8\alpha(L_-) \tag{3.19}$$

for the $N=3$ case and

$$\alpha(L_{3+})=t^{3/2}(1-t^3+t^5-t^6)\alpha(L_{2+})+t^6(1-t^2+t^3+t^5-t^6+t^8)\alpha(L_+)$$
$$-t^{9/2}t^8(1-t+t^3-t^6)\alpha(L_0)-t^{20}\alpha(L_-), \tag{3.20}$$

for the $N=4$ case. In the above expressions, the meanings of L_{2+} and L_{3+} should be clear.

Since we have the explicit forms of the braid group representations and the Markov traces, the evaluation of link polynomials, $\alpha(A)$ defined by (3.17), is automatic. In fact, we have given new link polynomials up to closed 3-braids by applying the $N=3$ theory [37].

There exists a link polynomial corresponding to the exactly solvable model. Our theory can be extended into any solvable vertex models and IRF models. The braid group representation and the Markov trace are readily found for the IRF models.

4. Composite String Representations

The N-state vertex model can be considered to describe the scattering for spin $(N-1)/2$ particles which has the factorization property. In particular, the 6-vertex model ($N=2$) corresponds to the spin 1/2 factorized S-matrix. We recall that a multiplet of spin 1/2 particles contains higher spin particles. For example, from a pair of spin 1/2 particles, we can make two "composite particles"; one with spin 1 and the other with spin 0. Kulish and Sklyanin [18] pointed out that the $N=3$ vertex model can be interpreted as the factorized S-matrix for the composite spin 1 particles. The spin 1 factorized S-matrix can be constructed through the following two steps:

Step 1: prepare the four-body S-matrix expressing the collision of two pairs of spin 1/2 particles.

Step 2: multiply projectors (symmetrizers) to extract the spin 1 component for each pair.

This process of making higher spin S-matrices from lower spin S-matrices is sometimes termed as the fusion procedure. By applying the formula (2.5) to the resulting S-matrix, the fusion procedure is directly translated into the braid group representations.

Let us denote the spin s representation of B_n by $B_n^{[s]}$. For instance, $B_n^{[1/2]}$ is equivalent to $H(t,n)$ by an identification $g_i \to -\hat{g}_i$. Note that we always regard the $B_n^{[s]}$ as a group algebra. For definiteness, let us explain the $N=3$ case. To have a spin 1 representation of B_n, we prepare n pairs of strings (totally $2n$ strings). From the generators $\{g_j;\ j=1,2,\ldots,2n-1\}$ of $B_n^{[1/2]}$, we define the following operator $G_i (i=1,2,\ldots,n-1)$:

$$G_i = P_i^{(+)} P_{i+1}^{(+)} g_{2i} g_{2i-1} g_{2i+1} g_{2i} P_i^{(+)} P_{i+1}^{(+)}, \tag{4.1}$$

where the symmetrizer $P_i^{(+)} ([P_i^{(+)}]^2 = P_i^{(+)})$ acting the ith pair is explicitly given by

$$P_i^{(+)} = \frac{1}{1+t}(t + g_{2i-1}). \tag{4.2}$$

Noting that the projector can be identified as a sum of elements in B_2 (or $B_2^{[1/2]}$ as a group algebra), we immediately see that the generators $\{G_i;\ i=1,2,\ldots,n-1\}$ satisfy the defining relations of B_n. We call this representation of B_n the composite string representation.

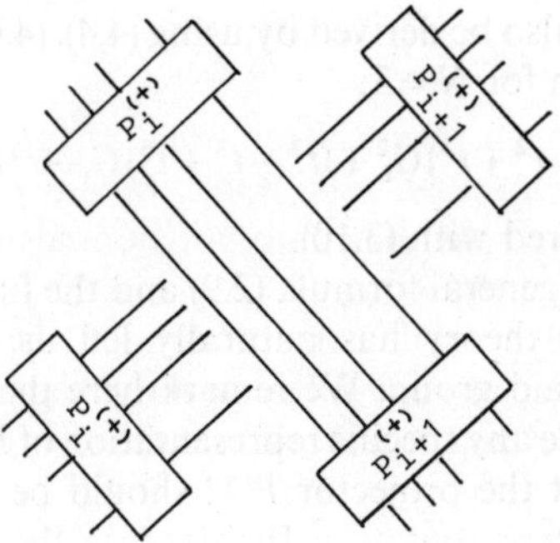

Fig. 4. Generator G_i defined in (4.4)

The above procedure is readily extended into higher spin representations. The general prescription of the composite string representation of B_n for the N-state case may be summarized as follows. For notational simplicity, we write k for $N-1$. Prepare n sets of k strings (totally $k\times n$ strings). Define an operator $g_j^{(p)}$ (p: positive integer) by

$$g_j^{(p)}=\prod_{m=1}^{p} g_{j-p-1+2m}\,. \tag{4.3}$$

In the above, we are free from the operator ordering problem due to the property (1.1). Then we have the expression of the generator G_i for the N-state case (Fig. 4):

$$G_i=P_i^{(+)}P_{i+1}^{(+)}\bar{G}_iP_i^{(+)}P_{i+1}^{(+)} \tag{4.4a}$$

with

$$\bar{G}_i=g_{ki}^{(1)}g_{ki}^{(2)}\ldots g_{ki}^{(k-1)}g_{ki}^{(k)}g_{ki}^{(k-1)}\ldots g_{ki}^{(2)}g_{ki}^{(1)}, \tag{4.4b}$$

where the symmetrizer $P_i^{(+)}$ acts on the ith set of k strings. We can verify that the symmetrizer $P^{(+)}$ is an element in $B_k^{[1/2]}$, which guarantees $\{G_i;\ i=1,2,\ldots,n\}$ to satisfy the defining relation of B_n. For example, $P_i^{(+)}$ for $N=4$ case is given by

$$\begin{aligned}P_i^{(+)}=&\frac{1}{(1+t)(1+t+t^2)}\,[t^3+t^2(g_{3i-2}+g_{3i-1})\\&+t(g_{3i-2}g_{3i-1}+g_{3i-1}g_{3i-2})+g_{3i-2}g_{3i-1}g_{3i-2}]\,.\end{aligned} \tag{4.5}$$

The projector $P_i^{(+)}$ for general N can be constructed, for instance, recursively. Thus we have constructed the representation of braid group generator b_i. Further, we have the following identifications of the remaining operators:

$$I\to P_1^{(+)}P_2^{(+)}\ldots P_n^{(+)}, \tag{4.6}$$

$$G_i^{-1}\to P_i^{(+)}P_{i+1}^{(+)}(\bar{G}_i)^{-1}P_i^{(+)}P_{i+1}^{(+)}\,. \tag{4.7}$$

The appearance of $P_i^{(+)}$'s in (4.6) and (4.7) means that we are considering only the highest spin space for each set of strings. We have confirmed that the expression (4.4) indeed reproduces the results in previous sections for any $N>2$. The reduction relations can also be derived by using (4.4), (4.6), and (4.7). For instance, we have a cubic relation for $N=3$:

$$G_i^3=(1-t^2+t^3)G_i^2+(t^2-t^3+t^5)G_i-t^5P_i^{(+)}P_{i+1}^{(+)}, \tag{4.8}$$

which should be compared with (3.10).

Combination of our general formula (2.9) and the fusion procedure known in the factorized S-matrix theory has naturally led us to the composite string representation of the braid group. We remark here that the above construction procedure does not utilize any specific representation of $B_n^{[1/2]}$ we started from. The only requirement is that the projector $P^{(+)}$ should be an element (or a sum of elements) of the starting representation. In this sense, the projector itself should not necessarily be the symmetrizer. In fact, for the $N=3$ case, we can construct a different braid group representation from that in (4.2) by using the antisymmetrizer

$$P_i^{(-)}=\frac{1}{1+t}(1-g_{2i-1}). \tag{4.9}$$

5. Two-Variable Extension

In a previous section we have introduced the composite string representation of the braid group. We use the composite string representation to extend our new link polynomials to those with two variables. In the case of the Jones polynomial, the two-variable extension [13] has been made both in a combinatorial way and in a C^*-algebraic way. As the latter, Ocneanu introduced a trace functional $\psi(\cdot)$ defined on $B_n^{[1/2]}$ which has the proper normalization

$$\psi(I)=1, \tag{5.1}$$

and has the Markov properties

$$\psi(AB)=\psi(BA)\,(A,B\in B_n^{[1/2]}), \tag{5.2}$$

$$\psi(Ag_n)=z\cdot\psi(A)\,(A\in B_n^{[1/2]},\,g_n\in B_{n+1}^{[1/2]}), \tag{5.3a}$$

$$\psi(Ag_n^{-1})=\bar{z}\cdot\psi(A)\,(A\in B_n^{[1/2]},\,g_n\in B_{n+1}^{[1/2]}), \tag{5.3b}$$

with

$$z=\psi(g_n), \tag{5.4a}$$

$$\bar{z}=\psi(g_n^{-1}). \tag{5.4b}$$

An important point is that the quantity

$$z=\psi(g_i)\quad\text{for all } i \tag{5.5}$$

is independent of the parameter t and is a free parameter. This contrasts to the Jones's trace $\phi(\cdot)$ [see (3.17), (3.18), and (3.19) with $N=2s+1=2$]. The pair of variables (t,z) enters into the two-variable Jones polynomial. Hence, we may attain the two-variable extension of the link polynomials $N\geqq 3$ by generalizing the Ocneanu's trace. Interpretation of our higher spin representation of B_n in terms of composite strings presented in the previous section is helpful for this purpose.

Let us explain the simplest case $N=3$. Since the spin 1 representation $B_n^{[1]}$ can be interpreted as a sub group-algebra in $B_{2n}^{[1/2]}$, the Ocneanu's trace $\psi(\cdot)$ naturally acts on $B_n^{[1]}$. Then the induced functional $\psi^{[1]}(\cdot)$ defined by

$$\psi^{[1]}(A)=\psi(A)/[\psi(P_i^{(+)})]^n \qquad (A\in B_n^{[1]}) \tag{5.6}$$

is a natural candidate of the generalized Ocneanu's trace. The quantity appearing in the denominator is given by

$$\psi(P_i^{(+)})=\frac{1}{1+t}(t+z), \tag{5.7}$$

where we have combined (4.2) and (5.3). The denominator in (5.6) is important to guarantee the proper normalization. A task left for us is to check the Markov properties

$$\psi^{[1]}(AB)=\psi^{[1]}(BA)\,(A,B\in B_n^{[1]}), \tag{5.8}$$

$$\psi^{[1]}(AG_n)=Z\cdot\psi^{[1]}(A)\,(A\in B_n^{[1]},G_n\in B_{n+1}^{[1]}), \tag{5.9a}$$

$$\psi^{[1]}(AG_n^{-1})=\bar{Z}\cdot\psi^{[1]}(A)\,(A\in B_n^{[1]},G_n\in B_{n+1}^{[1]}), \tag{5.9b}$$

with

$$Z=\psi^{[1]}(G_i)=\frac{(1+t)z^2}{t+z}, \tag{5.10a}$$

$$\bar{Z}=\psi^{[1]}(G_i^{-1})=\frac{(1+t)\bar{z}^2}{1+t\bar{z}}. \tag{5.10b}$$

Using (5.2)–(5.4) in the definition (5.6), we can confirm that (5.8) and (5.9) hold. Hence the functional $\psi^{[1]}(\cdot)$ is indeed the generalized Ocneanu's trace.

We apply the formula (5.6) to obtain the two-variable link polynomial. Before doing this, we introduce a variable ω by

$$\omega=\bar{z}/z=\psi(g_i^{-1})/\psi(g_i)=\frac{z+t-1}{tz} \tag{5.11}$$

and change variables from (t,z) to (t,ω). We then have

$$Z=\frac{1-t}{1-\omega t}\cdot\frac{1-t^2}{1-\omega t^2}, \tag{5.12a}$$

$$\bar{Z}=\omega^2 Z. \tag{5.12b}$$

Therefore we obtain the two-variable extension $\alpha_\omega^{[1]}(\cdot)$ of the $N=3$ polynomial as

$$\alpha_\omega^{[1]}(A)=\left(\frac{1-t}{1-\omega t}\,\frac{1-t^2}{1-\omega t^2}\cdot\omega\right)^{-(n-1)}\omega^{e(A)}\psi^{[1]}(A)\qquad(A\in B_n^{[1]}). \tag{5.13}$$

It is interesting to compare this expression with that of the $N=2$ case,

$$\alpha_\omega^{[1/2]}(A)=\left(\frac{1-t}{1-\omega t}\cdot\sqrt{\omega}\right)^{-(n-1)}(\sqrt{\omega})^{e(A)}\psi^{[1/2]}(A). \tag{5.14}$$

The one variable polynomials are reproduced by setting

$$\omega = t. \tag{5.15}$$

We know that for general N the projector $P_i^{(+)}$ is expressed as sum of elements in $B_n^{[1/2]}$. Then the construction of two-variable link polynomials for general N can be done in a similar fashion [38].

To close this section, we want to mention the relationship between our $N=3$ polynomial and the Kauffman polynomial [39]. Kauffman constructed a two-variable link polynomial which has a combinatorial definition and is different from the two-variable Jones polynomial. Very recently, Birman and Wenzl [40], and independently Murakami [41] devised an algebra corresponding to the Kauffman polynomial, which we call Birman-Wenzl-Murakami (BWM) algebra. The BWM algebra denoted by $C_n(l, m)$ is generated by two kinds of operators: one is $\{\tilde{G}_i\}_{i=1}^{n-1}$ satisfying the defining relation of the braid group and the other is $\{\tilde{E}_i\}_{i=1}^{n-1}$. They satisfy the following relations

$$\tilde{G}_i + \tilde{G}_i^{-1} = m(1 + \tilde{E}_i), \tag{5.16a}$$

$$\tilde{E}_i \tilde{E}_{i\pm1} \tilde{E}_i = \tilde{E}_i, \tag{5.16b}$$

$$\tilde{G}_{i\pm1} \tilde{G}_i \tilde{E}_{i\pm1} = \tilde{E}_i \tilde{G}_{i\pm1} \tilde{G}_i = \tilde{E}_i \tilde{E}_{i\pm1}, \tag{5.16c}$$

$$\tilde{G}_{i\pm1} \tilde{E}_i \tilde{G}_{i\pm1} = \tilde{G}_i^{-1} \tilde{E}_{i\pm1} \tilde{G}_i^{-1}, \tag{5.16d}$$

$$\tilde{G}_{i\pm1} \tilde{E}_i \tilde{E}_{i\pm1} = \tilde{G}_i^{-1} \tilde{E}_{i\pm1}, \tag{5.16e}$$

$$\tilde{E}_{i\pm1} \tilde{E}_i \tilde{G}_{i\pm1} = \tilde{E}_{i\pm1} \tilde{G}_i^{-1}, \tag{5.16f}$$

$$\tilde{G}_i \tilde{E}_i = \tilde{E}_i \tilde{G}_i = l^{-1} \tilde{E}_i, \tag{5.16g}$$

$$\tilde{E}_i \tilde{G}_{i\pm1} \tilde{E}_i = l \tilde{E}_i, \tag{5.16h}$$

$$\tilde{E}_i \tilde{E}_j = \tilde{E}_j \tilde{E}_i \quad \text{if } |i-j| \geq 2, \tag{5.16i}$$

$$\tilde{E}_i^2 = [m^{-1}(l + l^{-1}) - 1] \tilde{E}_i, \tag{5.16j}$$

$$\tilde{G}_i^2 = m(\tilde{G}_i + l^{-1} \tilde{E}_i) - 1. \tag{5.16k}$$

Combining (5.16a) and (5.16k) we see that the generator $\tilde{G}_i$ satisfies a cubic relation

$$\tilde{G}_i^3 = \left(m + \frac{1}{l}\right) \tilde{G}_i^2 - \left(1 + \frac{m}{l}\right) \tilde{G}_i + \frac{1}{l}. \tag{5.17}$$

Associated with the algebra $C_n(l, m)$, Birman and Wenzl defined a Markov trace by using the Kauffman polynomial. Hence the Kauffman polynomial is a trace type invariant. Since ours is also a trace type invariant with a braid group generator having a cubic relation, the Kauffman polynomial should be related to our $N=3$ polynomial. In fact, by renormalizing our braid group generator G_i as

$$\tilde{G}_i = \frac{G_i}{t\sqrt{-1}} \tag{5.18}$$

and *defining* the operator $\tilde{E}_i$ through the relation

$$\tilde{G}_i+\tilde{G}_i^{-1}=\frac{1-t^2}{t\sqrt{-1}}(1+\tilde{E}_i), \tag{5.19}$$

we can directly verify that the operators $\{\tilde{G}_i\}$ and $\{\tilde{E}_i\}$ indeed satisfy the relation (5.16) with

$$m=\frac{1-t^2}{t\sqrt{-1}}, \tag{5.20}$$

$$l=\frac{\sqrt{-1}}{t^2}. \tag{5.21}$$

Hence we see that our $N=3$ braid group representation $B_n^{[1]}$ corresponds to a special case of $C_n(l,m)$. But this does not mean that our $N=3$ polynomial is a special case of the Kauffman polynomial. The reason is the following. Both of the two variables in the Kauffman polynomial enter into the algebra. In our $N=3$ two-variable polynomial, roles of two variables are different; one for the algebra and the other for the trace functional. Then our two-variable polynomial is different from the Kauffman polynomial. It is quite interesting to observe that our two-variable polynomial at the point $\omega=t$ coincides with the Kauffman polynomial with (5.20) and (5.21).

We expect that the analog of BWM algebra structure should also exist for $N\geqq 4$ polynomials [38]. This will give us combinatorial or "graphical" interpretation of our hierarchy of link polynomials.

6. Summary

In this paper, we have discussed a close relation between the knot theory and the exactly solvable models in statistical mechanics. We have presented a general prescription to construct a braid group representation from the Boltzmann weights of a solvable model satisfying the Yang-Baxter relation (factorization equation, star-triangle relation). Through the discussion, we have shown that the starting solvable model must be critical. This gives a natural explanation for the appearance of the Temperley-Lieb algebra in the Jones polynomial. We have applied the formula to a sequence of solvable vertex models to have a sequence of new braid group representations. With the Markov traces associated with these new representations, we have successfully constructed new link polynomials. Further, by utilizing the "fusion" method which is known to give higher spin factorized S-matrices, we have been led to the composite string representation of our hierarchy of braid group representations. This representation enables us to construct the two-variable extensions of our new polynomials.

Recently Murakami [42] presented braid group representations also by using multiple strings. His representation is seemingly similar, but it is different from ours in that the symmetrizer is not used. At present, we have not found an interpretation of his result from a viewpoint of the factorized S-matrix theory. Detailed account on this subject will be given in a separate paper [38].

In conclusion, we like to emphasize that the sequence of new link polynomials in this paper is a specialization of our theory. The method of constructing the braid group representation is quite general and can be used for any vertex model and IRF model. For example, we can apply our method to an IRF model called the eight-vertex SOS model [25]. Then, an important question is how many independent link polynomials can be made from solvable models. This problem will be the subject of our future study.

Acknowledgements. We wish to thank Prof. A. Jaffe for his encouragement, Prof. C. A. Tracy for constant interest in our recent works, Prof. J. S. Birman, Prof. L. H. Kauffman and Prof. J. Murakami for sending their preprints prior to publication. We also thank Mr. T. Deguchi for helpful discussions.

References

1. Birman, J.S.: Braids, links, and mapping class groups. Princeton, NJ: Princeton University Press 1974
2. Rolfsen, D.: Knots and links. Berkeley, CA: Publish or Perish 1976
3. Artin, E.: Ann. Math. **48**, 101 (1947)
4. Alexander, J.W.: Proc. Natl. Acad. Sci. USA **9**, 93 (1923)
5. Markov, A.A.: Recueil Math. Moscov 73 (1935)
6. Jones, V.F.R.: Invent. Math. **72**, 1 (1983)
7. Jones, V.F.R.: Bull. Am. Math. Soc. **12**, 103 (1985)
8. Temperley, H.N.V., Lieb, E.H.: Relations between the "percolation" and "colouring" problem and other graph-theoretical problems associated with regular planar lattices: Some exact results for the "percolation" problem. Proc. Roy. Soc. Lond. A **322**, 251 (1971)
9. Lieb, E.H., Wu, F.Y.: In: Phase transitions and critical phenomena, Vol. 1, p. 331. Domb, C., Green, M.S. (eds.). London: Academic Press 1972
10. Baxter, R.J., Kelland, S.B., Wu, F.Y.: Equivalence of the Potts' model or Whitney polynomial with an ice-type model. J. Phys. A **9**, 397 (1976)
11. Bourbaki, N.: Groupes et algebres de Lie. Paris: Hermann 1968, Chap. 4
12. Alexander, J.W.: Trans. Am. Math. Soc. **30**, 275 (1928)
13. Freyd, P., Yetter, D., Hoste, J., Lickorish, W.B.R., Millett, K., Ocneanu, A.: Bull. Am. Math. Soc. **12**, 239 (1985)
14. Birman, J.S.: Invent. Math. **81**, 287 (1985)
15. Kanenobu, T.: Math. Ann. **275**, 555 (1986)
16. Faddeev, L.D.: Sov. Sci. Rev. Math. Phys. C **1**, 107 (1981)
17. Thacker, H.B.: Exact integrability in quantum field theory and statistical mechanics. Rev. Mod. Phys. **53**, 253 (1981)
18. Kulish, P.P., Sklyanin, E.K.: Lecture Notes in Physics, Vol. 151, p. 61. Berlin, Heidelberg, New York: Springer 1982
19. Wadati, M.: In: Dynamical problems in soliton systems, p. 68. Takeno, S. (ed.). Berlin, Heidelberg, New York: Springer 1985
20. Wadati, M., Akutsu, Y.: Exactly solvable models in statistical mechanics. In: Springer Series in Nonlinear Dynamics. Lakshmanan, M. (ed.). Berlin, Heidelberg, New York: Springer 1988
21. Karowski, M., Thun, H.J., Truong, T.T., Weisz, P.H.: On the uniqueness of a purely elastic *S*-matrix in (1+1) dimensions. Phys. Lett. **67**B, 321 (1977)
    Zamolodchikov, A.B., Zamolodchikov, A.B.: Factorized *S*-Matrices in two dimensions as the exact solutions of certain relativistic quantum field theory models. Ann. Phys. (NY) **120**, 253 (1979)
    Sogo, K., Uchinami, M., Nakamura, A., Wadati, M.: Nonrelativistic theory of factorized *S*-matrix. Prog. Theor. Phys. **66**, 1284 (1981)

Sogo, K., Uchinami, M., Akutsu, Y., Wadati, M.: Classification of exactly solvable two-component models. Prog. Theor. Phys. **68**, 508 (1981)
22. Baxter, R.J.: Exactly solved models in statistical mechanics. London: Academic Press 1982
23. Wu, F.Y.: Ising model with four-spin interactions. Phys. Rev. B **4**, 2312 (1971)
Kadanoff, L.P., Wegner, F.J.: Some critical properties of the eight vertex model. Phys. Rev. B **4**, 3989 (1981)
24. Zamolodchikov, A.B.: Z_4-symmetric factorized S-matrix in two space-time dimensions. Commun. Math. Phys. **69**, 165 (1979)
25. Andrews, G.E., Baxter, R.J., Forrester, P.J.: Eight-vertex SOS model and generalized Rogers-Ramanujan-type identities. J. Stat. Phys. **35**, 193 (1984)
26. Kuniba, A., Akutsu, Y., Wadati, M.: J. Phys. Soc. Jpn **55**, 1092, 2170, and 3338 (1986)
Kuniba, A., Akutsu, Y., Wadati, M.: An exactly solvable 4-state IRF model. Phys. Lett. **116**A, 382 (1986) and An exactly solvable 5-state IRF model. **117**A, 358 (1986)
Baxter, R.J., Andrews, G.E.: Lattice gas generalization of the hard hexagon model. I. Star-triangle relation and local densities. J. Stat. Phys. **44**, 249 (1986)
Andrews, G.E., Baxter, R.J.: Lattice gas generalization of the hard hexagon model. II. The local densities as elliptic functions. J. Stat. Phys. **44**, 713 (1986)
27. Akutsu, Y., Kuniba, A., Wadati, M.: J. Phys. Soc. Jpn. **55**, 1466 and 1880 (1986)
28. Kuniba, A., Akutsu, Y., Wadati, M.: J. Phys. Soc. Jpn. **55**, 2605 (1986)
29. Akutsu, Y., Kuniba, A., Wadati, M.: J. Phys. Soc. Jpn. **55**, 290 (1986)
Date, E., Jimbo, M., Miwa, T., Okado, M.: Fusion of the eight vertex SOS model. Lett. Math. Phys. **12**, 209 (1986)
30. Akutsu, Y., Wadati, M.: J. Phys. Soc. Jpn. **56**, 839 (1987)
31. Akutsu, Y., Wadati, M.: J. Phys. Soc. Jpn. **56**, 3039 (1987)
32. Sogo, K., Akutsu, Y., Abe, T.: New factorized S-matrix and its application to exactly solvable q-state model. I and II. Theor. Phys. **70**, 730 and 739 (1983)
33. Zamolodchikov, A.B., Fateev, V.A.: A model of facterized S-matrix and an integrable spin-1 Heisenberg chain. Sov. J. Nucl. Phys. **32**, 298 (1980)
34. Powers, R.T.: Ann. Math. **86**, 138 (1967)
35. Pimsner, M., Popa, S.: Preprint
36. Conway, J.H.: In: Computational problems in abstract algebra, p. 329. Leach, J. (ed.). London: Pergamon Press 1969
37. Akutsu, Y., Deguchi, T., Wadati, M.: J. Phys. Soc. Jpn. **56**, 3464 (1987)
38. Deguchi, T., Akutsu, Y., Wadati, M.: J. Phys. Soc. Jpn. **57**, No. 3 (1988)
39. Kauffman, L.H.: Preprint
40. Birman, J.S., Wenzl, H.: Preprint
41. Murakami, J.: Preprint
42. Murakami, J.: Preprint

Communicated by A. Jaffe

Received August 6, 1987; in revised form January 15, 1988

C. R. Acad. Sci. Paris, t. 307, Série I, p. 207-210, 1988 207

Algèbre/*Algebra*

Groupes quantiques et modèles à vertex de V. Jones en théorie des nœuds

Marc Rosso

Résumé – On montre qu'à toute représentation de dimension finie d'une algèbre enveloppante universelle quantifiée, on peut associer un modèle à vertex tel que défini par Vaughan Jones dans ses constructions d'invariants d'isotopie d'entrelacs.

Quantum groups and V. Jones's vertex models for knots

Abstract – *It is shown that, to each finite dimensional representation of a quantized universal enveloping algebra, one can associate a vertex model as defined by Vaughan Jones in his constructions of isotopy link invariants.*

Introduction. – Dans [5], V. Jones introduit des modèles à vertex (semblables à ceux rencontrés en Mécanique statistique), à partir desquels il construit des invariants d'isotopie régulière pour les entrelacs orientés dans $\mathbf{R}^3$; l'invariant est donné par une certaine fonction de partition associée à la donnée du modèle et d'une projection plane générique de l'entrelac. Il suggère alors que le formalisme des groupes quantiques devrait permettre d'associer un tel modèle à toute représentation de dimension finie d'une algèbre de Lie complexe simple.

Nous allons montrer que c'est effectivement le cas, en travaillant au niveau universel du groupe quantique : ceci repose sur une propriété très simple de la structure quasitriangulaire de l'algèbre enveloppante universelle quantifiée.

I. Modèles à vertex en théorie des nœuds [5].

Définition. – *Un modèle à vertex est donné par un espace vectoriel* V *muni d'une base privilégiée* **e**, *par deux familles d'applications linéaires* $R_\pm(\lambda)\in \mathrm{End}(V\otimes V)$, $\lambda\in(0,\pi)$, *ainsi qu'une famille à un paramètre d'endomorphismes inversibles* $L(\lambda)\in\mathrm{End}(V)$, *diagonaux dans la base* **e**, *le tout étant soumis aux conditions suivantes* [*on note* $P\in\mathrm{End}(V\otimes V)$ *l'opérateur* $a\otimes b\mapsto b\otimes a$ *et* $\check{R}_\pm(\lambda)=P.R_\pm(\lambda)$]:

(1) $R_\pm(\lambda+\delta)=(L(-\delta)\otimes 1)R_\pm(\lambda)(L(-\delta)\otimes 1)^{-1}=(1\otimes L(\delta))R_\pm(\lambda)(1\otimes L(\delta))^{-1}$, [*en particulier*, $\check{R}_\pm(\lambda)$ *commute à* $L(\delta)\otimes L(\delta)$ *et* $R_\pm(\lambda)$ *est complètement déterminé par* $R_\pm(0)$].

(2) $\check{R}_-(0)\,\check{R}_+(0)=\mathrm{id}$.

(3) *Pour* $A\in\mathrm{End}(V\otimes V)$, *posons* $A^{t_1}=(\mathrm{Transposition}\otimes\mathrm{id})(A)$, $A^{t_2}=(\mathrm{id}\otimes\mathrm{Transposition})(A)$. *On demande alors que :* $[PR_+(\pi)^{t_1}][PR_-(\pi)^{t_2}]=\mathrm{id}$.

(4) (*Équation de Yang-Baxter*) $R_{12}(\lambda)R_{13}(\lambda+\mu)R_{23}(\mu)=R_{23}(\mu)R_{13}(\lambda+\mu)R_{12}(\lambda)$ $(R=R_+)$ [*ce qui équivaut, grâce à* (1), *à:* $R_{12}(0)R_{13}(0)R_{23}(0)=R_{23}(0)R_{13}(0)R_{12}(0)$].

Des variantes de tels modèles ont aussi été étudiées par Turaev [7]. *Voir* aussi [8] pour une autre approche à la construction d'invariants en utilisant les R-matrices.

Soit $U_h\mathcal{G}$ le groupe quantique associé à l'algèbre de Lie simple $\mathcal{G}$ (Drinfeld [1], Jimbo [3], [4]). Nous allons voir que, en prenant pour $R_+(0)$ l'image de la R-matrice quantique dans une représentation de dimension finie, en interprétant la transposition comme un certain antiautomorphisme de $U_h\mathcal{G}$ et en choisissant correctement $L(\delta)$, on obtient un modèle à vertex.

Note présentée par Alain Connes.

208 C. R. Acad. Sci. Paris, t. 307, Série I, p. 207-210, 1988

II. LE GROUPE QUANTIQUE $U_h\mathcal{G}$ ET SA STRUCTURE QUASITRIANGULAIRE.

1. DÉFINITION [1]. — *Soient $\mathcal{G}$ une algèbre de Lie complexe simple de dimension finie, $\mathcal{H}$ une sous-algèbre de Cartan, $(\alpha_i)_{1\leq i\leq N}$ une base de racines simples, (,) un produit scalaire invariant sur $\mathcal{H}^*$, H_i les vecteurs de racine correspondant à α_i par l'isomorphisme $\mathcal{H}^*\to\mathcal{H}$ et $A=(a_{ij})$ la matrice de Cartan.*

$U_h\mathcal{G}$ est la $\mathbf{C}[[h]]$-algèbre engendrée au sens h-adique par $\mathcal{H}$ et des générateurs X_i^+, X_i^- avec les relations :

$$[a_1, a_2]=0,\quad a_1, a_2\in\mathcal{H},\qquad [a, X_i^\pm]=\pm\alpha_i(a)X_i^\pm,\quad a\in\mathcal{H}$$

$$[X_i^+, X_j^-]=\delta_{ij}\frac{\operatorname{sh}((h/2)H_i)}{\operatorname{sh}(h/2)}$$

et, pour $i\neq j$, posant $q_i=\exp[h/2(\alpha_i,\alpha_i)]$:

$$\sum_{k=0}^{1-a_{ij}}(-1)^k\binom{1-a_{ij}}{k}_{q_i} q_i^{-k(1-a_{ij}-k)/2}(X_i^\pm)^k X_j^\pm (X_i^\pm)^{1-a_{ij}-k}=0$$

où $\binom{n}{k}_q=(q^n-1)\ldots(q^{n-k+1}-1)/(q^k-1)\ldots(q-1)$.

$U_h\mathcal{G}$ est une algèbre de Hopf : la comultiplication Δ et l'antipode S sont données par :

$$\Delta(a)=a\otimes 1+1\otimes a,\quad a\in\mathcal{H},\qquad \Delta(X_i^\pm)=X_i^\pm\otimes\exp\frac{h}{4}H_i+\exp\frac{-h}{4}H_i\otimes X_i^\pm$$

$$S(a)=-a,\qquad S(X_i^\pm)=-e^{\pm h/4(\alpha_i,\alpha_i)}X_i^\pm$$

Il sera utile de considérer $Y_i=(\operatorname{sh}h/2).(\operatorname{sh}h/2(\alpha_i,\alpha_i))^{-1}X_i$ au lieu de X_i^-.

2. LA STRUCTURE QUASITRIANGULAIRE DE $U_h\mathcal{G}$. — Par définition, elle est donnée par un élément inversible R de $U_h\mathcal{G}\otimes U_h\mathcal{G}$ tel que :

(*a*) $R\,\Delta(x)R^{-1}=\Delta'(x)$, $\forall x\in U_h\mathcal{G}$.

(*b*) $(\Delta\otimes\mathrm{id})(R)=R_{13}R_{23}$, $(\mathrm{id}\otimes\Delta)(R)=R_{13}R_{12}$.

[(*b*) implique l'équation de Yang-Baxter].

Selon Drinfeld [1], une telle structure quasitriangulaire peut être obtenue à partir de la structure canonique sur le double quantique $\mathcal{D}(U_h b_+)$ de $U_h b_+$ (sous-algèbre de Hopf engendrée par $\mathcal{H}$ et les X_i^+). Rappelons simplement que si $(U_h b_+)^0$ est l'algèbre enveloppante quantifiée duale de $U_h b_+$, avec la comultiplication opposée, alors :

(i) $\mathcal{D}(U_h b_+)$ contient $U_h b_+$ et $(U_h b_+)^0$ comme sous-algèbres de Hopf.

(ii) R est l'image de l'élément canonique de $U_h b_+\otimes(U_h b_+)^0$ par l'inclusion

$$U_h b_+\otimes(U_h b_+)^0\subsetneq\mathcal{D}(U_h b_+)\otimes\mathcal{D}(U_h b_+).$$

(iii) L'application linéaire : $U_h b_+\otimes(U_h b_+)^0\to\mathcal{D}(U_h b_+)$, $a\otimes b\mapsto ab$, est bijective.

De plus, on a un isomorphisme $\varphi:(U_h b_+)^0\xrightarrow{\sim}U_h b_-$, si bien que :

$$\mathcal{D}(U_h b_+)\simeq U_h\mathcal{G}\otimes U\mathcal{H}.$$

Soit $\varepsilon: U\mathcal{H}\to\mathbf{C}$ l'augmentation canonique. La structure quasitriangulaire $\bar{R}$ de $U_h\mathcal{G}$ est l'image de $R\in\mathcal{D}(U_h b_+)\otimes\mathcal{D}(U_h b_+)$ par la composée :

$$\mathcal{D}(U_h b_+)\otimes\mathcal{D}(U_h b_+)\xrightarrow{\sim}(U_h\mathcal{G}\otimes U\mathcal{H})\otimes(U_h\mathcal{G}\otimes U\mathcal{H})\xrightarrow{(\mathrm{Id}\otimes\varepsilon)^{\otimes 2}}U_h\mathcal{G}\otimes U_h\mathcal{G}.$$

(Cette application, restreinte à $U_h b_+\otimes 1$ ou à $1\otimes U_h b_-$, n'est autre que l'inclusion naturelle.)

C. R. Acad. Sci. Paris, t. 307, Série I, p. 207-210, 1988

PROPOSITION. — $(S\otimes Id)(R)=R^{-1}$ *dans* $U_h b_+\otimes(U_h b_+)^0$, *si bien que* $(S\otimes Id)(\bar{R})=\bar{R}^{-1}$ *dans* $U_h \mathcal{G}\otimes U_h \mathcal{G}$.

Démonstration. — Le produit de convolution confère à $End(U_h b_+)$ une structure d'algèbre qui, restreinte à $U_h b_+\otimes(U_h b_+)^0$, est la structure produit tensoriel d'algèbres. Or, R, vu dans $End(U_h b_+)$ est l'application identité, dont l'inverse pour le produit de convolution est S; dans $U_h b_+\otimes(U_h b_+)^0$, S est représenté par $(S\otimes Id)(R)$.

III. INVOLUTION DE CARTAN, ANTIAUTOMORPHISME CANONIQUE ET ANTIPODE. — 1. Il y a, sur $U_h\mathcal{G}$, un automorphisme d'algèbre (et antiautomorphisme de cogèbre) Θ tel que : $\Theta(X_i^+)=-Y_i$, $\Theta(Y_i)=-X_i^+$, $\Theta(H_i)=-H_i$.

2. On peut aussi définir un antiautomorphisme T par : $T(X_i^+)=Y_i$, $T(Y_i)=X_i^+$, $T(H_i)=H_i$.

3. L'antipode S est un antiautomorphisme d'algèbre et de cogèbre et S^2 est l'automorphisme intérieur défini par $\exp(hH_\delta)$, où δ est la demi-somme des racines positives. On le note $Int(\exp(hH_\delta))$.

De plus $S(X_i^+)=-Int(\exp(h/2)H_\delta)(X_i^+)$, $S(Y_i)=-Int(\exp(h/2)H_\delta)Y_i)$.

Indiquons quels sont les objets de $U_h\mathcal{G}$ dont les images vont définir les modèles à vertex, et montrons qu'ils vérifient déjà les propriétés (1) à (4) au niveau universel : posons $R_+(0)=\bar{R}$, $L(\lambda)=\exp(-(h/2)(\lambda/\pi)H_\delta)$.

Soit $\sigma\in End(U_h\mathcal{G}\otimes U_h\mathcal{G})$ définie par $\sigma(x\otimes y)=y\otimes x$; bien que P n'existe pas dans le contexte universel, on peut tout de même considérer $P.x\otimes y.P=\sigma(x\otimes y)=y\otimes x$ pour tout $x, y\in U_h\mathcal{G}$.

On pose alors : $R_-(0)=\sigma(R_+(0)^{-1})=\sigma((Id\otimes S)(\bar{R}))$.

Du fait que $\bar{R}\Delta(x)=\Delta'(x)\bar{R}$, $\forall x\in U_h\mathcal{G}$, $R_+(0)$ et $R_-(0)$ commutent à $L(\lambda)\otimes L(\lambda)$ d'où (1) et (2).

(4) est immédiate.

Pour vérifier (3), on interprète la transposition des matrices comme étant donnée, au niveau universel, par l'antiautomorphisme T (ce sera justifié plus loin). Alors,

$$\begin{aligned} R_+(\pi)^{t_1} &= \left(T\circ Int\left(\exp\frac{h}{2}H_\delta\right)\otimes Id\right)(\bar{R}) \\ &= \left(Int\left(\exp-\frac{h}{2}H_\delta\right)\circ T\otimes Id\right)(\bar{R}) \\ &= ((Int(\exp-hH_\delta)\circ\Theta\circ S')\otimes Id)(\bar{R}) \\ &= (\Theta\circ S\otimes Id)(\bar{R}) \\ &= (\Theta\otimes Id)(\bar{R}^{-1}). \end{aligned}$$

Par ailleurs,

$$\begin{aligned} R(\pi)^{t_2} &= \left[Id\otimes T\circ Int\left(\exp-\frac{h}{2}H_\delta\right)\circ S\right](\sigma\bar{R}) \\ &= \left[Id\otimes Int\left(\exp\frac{h}{2}H_\delta\right)Int\left(\exp\left(\frac{-h}{2}H_\delta\right)\right)\circ\Theta\right](\sigma\bar{R}) \\ &= (Id\otimes\Theta)\,\sigma(\bar{R})=\sigma((\Theta\otimes Id)(\bar{R})) \\ &= \sigma[(R_+(\pi)^{t_1})^{-1}] \end{aligned}$$

d'où la propriété (3).

210 C. R. Acad. Sci. Paris, t. 307, Série I, p. 207-210, 1988

IV. CONSTRUCTION D'UN MODÈLE A VERTEX À PARTIR D'UNE REPRÉSENTATION DE DIMENSION FINIE D'UN GROUPE QUANTIQUE. — Notons que dans la présentation de $U_h\mathscr{G}$, le paramètre de déformation h n'intervient que par $e^{h/4}$: on peut travailler sur **C** en introduisant un paramètre non nul $t(=e^{-h/4})$ fixé et considérant la **C**-algèbre $U_t\mathscr{G}$ engendrée par X_i^+, Y_i et $k_i=\exp((h/4)H_i)$, avec les relations qui se déduisent immédiatement de celles énoncées pour $U_h\mathscr{G}$. (C'est l'approche de Jimbo.)

Nous supposons désormais que t n'est pas une racine de l'unité.

Soit (ρ, V) une représentation de dimension finie de $U_t\mathscr{G}$; comme une telle représentation est complètement réductible [6], on peut supposer ρ irréductible. V sera l'espace de définition du modèle à vertex. On va déterminer la base privilégiée **e** de façon à ce que la transposition des matrices représentant les endomorphismes de V dans **e** corresponde à l'antiautomorphisme T, et pour cela on introduit la forme bilinéaire contravariante sur V. Celle-ci est définie de façon générale, comme dans le cas classique (*cf.* Humphreys [2], p. 36), sur chaque module de plus haut poids, en remplaçant $\lambda\in\mathscr{H}^*$ par le caractère de la sous-algèbre engendrée par les k_i noté t^λ dans [6] : $t^\lambda\ k_i\mapsto t^{\lambda(H_i)}$. C'est une forme bilinéaire symétrique (,) sur V telle que $(\rho(u)v, w)=(v, \rho(T(u))w)$, $\forall v, w\in V$, $\forall u\in U_t\mathscr{G}$, et elle est non dégénérée si V est irréductible.

Elle détermine donc ici un isomorphisme $J: V\to V^*$ tel que : $J(v)(w)=(v, w)$ et ${}^t\rho(u)(J(v))=J(\rho(T(u))v)$, $\forall v\in V$, $\forall u\in U_t\mathscr{G}$.

Comme V est un espace vectoriel sur **C**, il y a une base $\mathbf{e}=(\mathbf{e}_1, \ldots, \mathbf{e}_n)$ de V telle que $(e_i, e_j)=\delta_{ij}$. Alors $J(e_i)=e_i^*$ définit la base duale $\mathbf{e}^*$ et on vérifie immédiatement l'égalité entre matrices : $\mathrm{Mat}(T(u), \mathbf{e})=\mathrm{Mat}({}^tu, \mathbf{e}^*)={}^t\mathrm{Mat}(u, \mathbf{e})$. On choisit **e** comme base privilégiée.

On pose $L(\lambda)=\rho[k_{-2\delta}]^{\lambda/\pi}$ où, pour $\alpha=\sum_{1}^{N} n_i\alpha_i$, $k_\alpha=k_1^{n_1}\ldots k_N^{n_N}$.

Bien que $\bar{R}\in U_h\mathscr{G}\otimes U_h\mathscr{G}$ n'existe pas dans $U_t\mathscr{G}\otimes U_t\mathscr{G}$, on peut néanmoins considérer $(\rho\otimes\rho)(\bar{R})=R_+(0)$: en effet, $\bar{R}$ est donné par une série formelle, dont seuls un nombre fini de termes sont non nuls dans une représentation de dimension finie (car X_i^+, X_i^- sont nilpotents), et dont les coefficients sont des fonctions rationnelles de puissances de t (*cf.* Drinfeld [1], p. 816). Ainsi $R_+(0)$ a un sens et, comme ρ est un morphisme d'algèbres, les propriétés montrées au niveau universel dans le paragraphe précédent se transportent immédiatement dans la représentation : les relations (1) à (4) sont vraies et on a un modèle à vertex.

Ce travail a été effectué alors que je visitais l'Université de Californie à Berkeley. Je tiens à remercier V. Jones pour son invitation et pour m'avoir suggéré cette question.

Note reçue et acceptée le 6 juin 1988.

RÉFÉRENCES BIBLIOGRAPHIQUES

[1] V. G. DRINFELD, Quantum groups, *Proceedings I.C.M.*, Berkeley, 1, 1986.
[2] J. HUMPHREYS, *in* Lie theories and their applications, *Queen's papers in Pure and Applied Math.*, 48.
[3] M. JIMBO, A q-analogue of $U(gl(n+1))$, Hecke algebra, and the Yang-Baxter equation, *Lett. in Math. Phys.*, 11, 1986, p. 247-252.
[4] M. JIMBO, A q-difference analogue of $U(\mathscr{G})$ and the Yang-Baxter equation, *Lett. in Math. Phys.*, 10, 1985, p. 63-69.
[5] V. F. R. JONES, *On knot invariants related to some statistical mechanical models.*
[6] M. ROSSO, Finite dimensional representations of the quantum analog of the enveloping algebra of a complex simple Lie algebra, *Comm. Math. Phys.*
[7] V. G. TURAEV, *The Yang-Baxter equation and invariants of links.*
[8] H. WENZL, *Braid group representations and the quantum Yang-Baxter equation.*

Centre de Mathématiques pures, École polytechnique, 91128 *Palaiseau Cedex.*

Contemporary Mathematics
Volume 78, 1988

STATISTICAL MECHANICS AND THE JONES POLYNOMIAL

Louis H. Kauffman

ABSTRACT. This paper studies a relationship between the formalism of knot theory and certain models in statistical mechanics. It is shown how the partition function for the Potts model may be computed from an associated link diagram, and how this provides a common algorithmic model with the Jones polynomial. Certain features of both Jones polynomial and the Potts model can be treated in common, such as the appearance of the Temperley-Lieb algebra for braid diagrams and the geometry of the ice-model for piecewise linear diagrams.

I. INTRODUCTION

In his paper [3] on the Jones polynomial, Vaughan Jones described a connection between his new polynomial invariant of links and the Potts model in statistical mechanics. Both the Potts model and the Jones polynomial involve traces defined on a von Neumann algebra A[n]. For the knot theory, Jones implicated this algebra by constructing a representation to it with domain the Artin braid group B[n]. In statistical mechanics the same algebra is used to calculate the partition function for the Potts model (also by a trace), and the algebra is known to the physicists as the Temperly-Lieb Algebra [1].

The purpose of this paper is to exhibit a direct diagrammatic connection between these two subjects. In particular, I show that the partition function of the Potts model for a planar lattice can be calculated from a link diagram that is canonically associated with the lattice. (A link diagram is a schematic planar picture of a knot or link embedded in three-dimensional space.) The algorithm for computing the partition function from the link diagram is a special case of a three-variable polynomial defined on such diagrams that I call the bracket polynomial (denoted [L] for a diagram L. See [5], [6].). The Jones polynomial, up to a normalization, is also a special case of the bracket polynomial. In this way, the Jones polynomial and the Potts partition function are both aspects of a single algorithm defined on diagrams of knots and links.

1980 Mathematics Subject Classification (1985 Revision) - 57M25, 05C15, 82A05, 82A25.

The actual relation of the partition function of the Potts model with topological properties of knots and links is an open question. Remarkably, the Potts partition function for a planar lattice G is computed by bracket for an alternating link K(G). (An alternating link has a weaving pattern so that a given thread is seen to pass alternately under and over successive strands.) If K(G) is reduced (see section 2), then the long-standing conjectures of Tait and Little imply that [K(G)] is an invariant of the topological type of K(G). Thus, if this conjecture is true, then the Potts partition function of the lattice G is actually a topological invariant of the associated link K(G). This leads to many questions regarding the meaning of this connection between physics and topology.

The paper is organized as follows. Section II constructs the general bracket, and shows how it specializes to the Jones polynomial and to the dichromatic polynomial for planar graphs. Section III gives the background for the Potts model, shows that the Potts partition function is a dichromatic polynomial, and relates these results to the discussion of the dichromate in section II. Section IV discusses the diagram monoid of states for the bracket, and its relation to computing the dichromatic polynomial and the Potts partition function. This formalism is related to the Temperly-Lieb algebra via a diagrammatic tensor formalism due to Roger Penrose [10]. Section V returns to the Potts model, and shows how our viewpoint illuminates the discussion of the critical point for the ferromagnetic case. The link K(G) and its mirror image $K(G)^*$ correspond to the graph G and its dual G^*. This clarifies the structure of an argument [16] locating the conjectured critical point for the model.

Finally, in section VI, I show how a translation of the 6-vertex ice-model [1] of statistical mechanics into our knot-theoretic formalism gives rise to a different state-expansion for the bracket in terms of arrow coverings and local angular data for piecewise-linear diagrams. This gives a new formalism for the Jones polynomial and lets us raise further questions about the topology and the physics.

II. THE GENERAL BRACKET

I first define a three-variable polynomial $[K](A,B,d) \in Z[A,B,d]$ (Z denotes the integers) defined for unoriented link diagram K. The polynomial [K] will be referred to as the general bracket polynomial or the square bracket. Only by specializing its variables does the bracket become a topological invariant, but it is well-defined on link diagrams as combinatorial entities.

First recall a few basic facts about link diagrams and their relationship with planar graphs: A universe (or link shadow) is a planar graph with four edges locally incident to each of its vertices. See Figure 1. A link diagram is a universe endowed with extra structure at each vertex, indicating a crossing. Again see Figure 1. A link diagram can be seen as a schematic drawing of a knot or link and it can be regarded as a special species of planar graph.

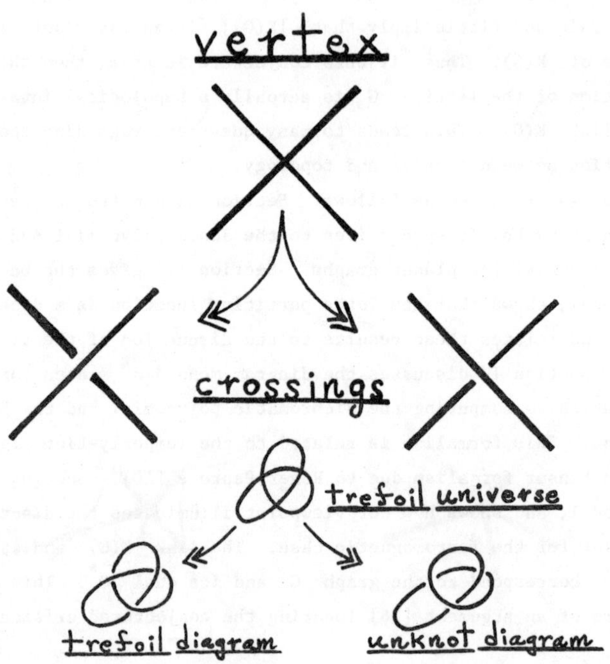

Figure 1

There is another planar graph associated with a link diagram K: This graph, G(K), is obtained by first checkerboard shading the diagram K as in Figure 2 (let the unbounded region be unshaded). The vertices of G(K) are in one-to-one correspondence with the shaded regions of K. Two vertices are joined by an edge if the two regions touch at a crossing in the knot diagram.

The graph G(K) depends only upon the underlying universe for K. If we want a complete graphical translation of the link diagram, this can be done by assigning signs (+1 or -1) to the edges of G(K) to indicate the unoriented crossing types (See [5].).

The association U $\longrightarrow\!\!\!>$ G(U) of a planar universe to a planar graph is invertible. The inverse process takes a planar graph, G, and produces a universe, M(G). M(G) is sometimes called the <u>medial graph</u> for G. The medial graph is produced by placing a crossing on each edge of G, and then connecting M(G) as shown in Figure 3.

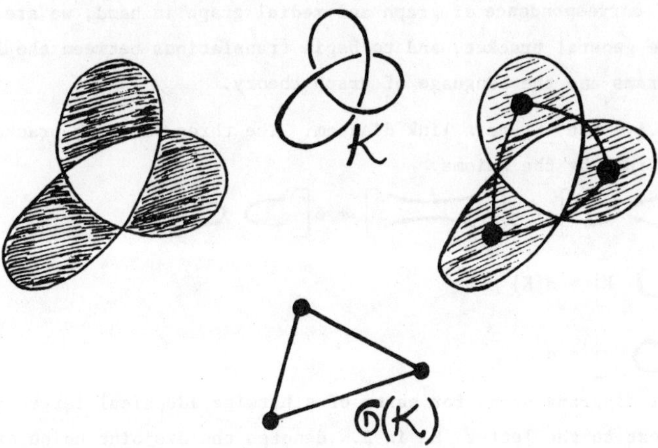

Checkerboard Shading and Associated Graph

Figure 2

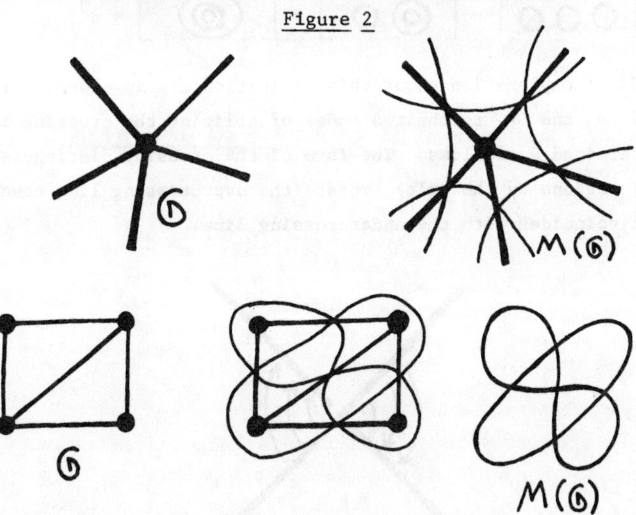

Figure 3

Note that in producing the medial graph M(G) from the planar graph G, we extend each crossing segment (of the crossings placed at each edge of G) toward the corresponding vertex, connecting it to a crossing segment from the next edge that is incident to this vertex - in clockwise or counterclockwise order. By this construction, planar graphs and universes (shadows of link diagrams) are in one-to-one correspondence.

This is the basic connection between link diagrams and graph theory.

With the correspondence of graph and medial graph in hand, we are prepared to define the general bracket, and to begin translations between the language of link diagrams and the language of graph theory.

DEFINITION 2.1. Let K be a link diagram. The three-variable bracket polynomial is defined by the axioms:

1. $[\times] = A[\asymp] + B[\supset\subset]$

2. $[\bigcirc\ K] = d[K]$

3. $[\bigcirc] = d$

The small diagrams stand for parts of otherwise identical larger diagrams; the circle next to the letter K in 2. denotes the disjoint union of K with a Jordan curve. Thus

$$[\bigcirc\bigcirc\bigcirc] = [\circledcirc\ \bigcirc] = [\circledcirc] = d^3.$$

Some words of explanation about this definition are in order. First, the assignment of A and B to the two modes of splicing the crossing in equation 1 is determined as follows. The form of the crossing distinguishes a pair of local regions by the rule: rotate the overcrossing line counterclockwise until it coincides with the undercrossing line.

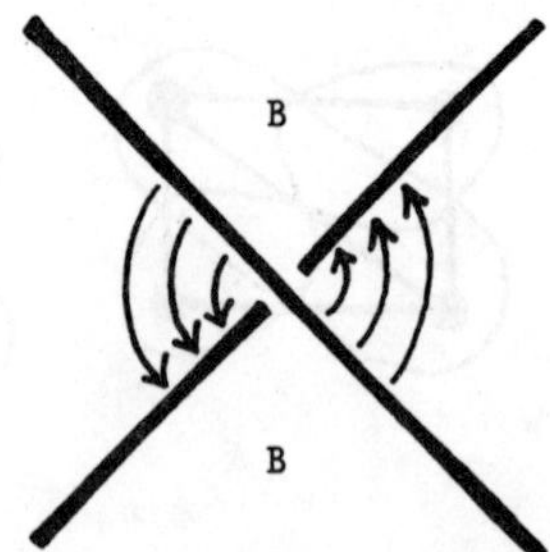

The two regions swept out by this rotation are labelled A. The other two are labelled B. When splitting the vertex fuses the two A-regions, I say that this *splitting opens an A-channel*. This is the part labelled A in formula 1. The other split opens the B-channel. See Figure 4.

The bracket is then well-defined recursively. Note that the order of applications of formula 1. does not affect the final result, since the choice of A or B is entirely local. Note also that switching a crossing reverses the roles of A and B. Thus

[ ] = B[ ] + A[ ].

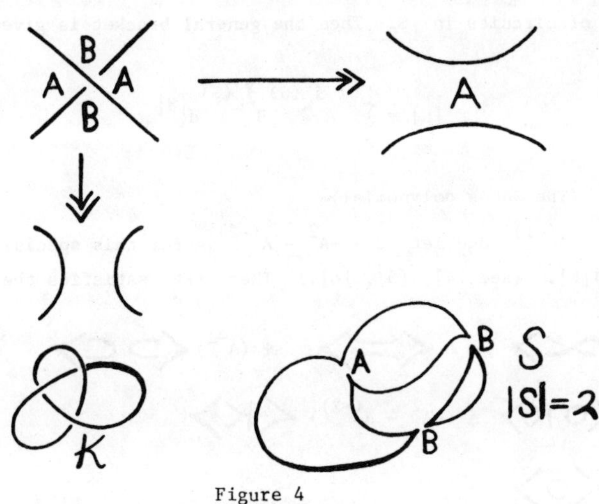

Figure 4

Here are a few sample bracket calculations:

(i) [ ] = B[ ] + A[ ]

$= Bd + Ad^2$

(ii) [ ] = B[ ] + A[ ]

$= Bd^2 + Ad$

(iii) [ ] = A[ ] + B[ ]

$= A(Bd + Ad^2) + B(Bd^2 + Ad)$

$= A^2d^2 + B^2d^2 + 2ABd$

In general, we can give an explicit formula for the bracket by considering the full recursion. I define a state S of K as the configuration of simple closed curves in the plane obtained by splitting each crossing in one of the two possible ways. There are 2^V states where V denotes the number of crossings in the diagram K. See Figure 4 for an example of a state for the trefoil diagram. The formula for [K] will sum over the contributions of each state.

Let S be a state of K. Let $i_K(S)$ denote the number of A-channels in S, and let $j_K(S)$ denote the number of B-channels. Let $|S|$ denote the number of circuits in S. Then the general bracket is given by the formula:

$$[K] = \sum_S A^{i_K(S)} B^{j_K(S)} d^{|S|}.$$

EXAMPLE 1. (The Jones polynomial)

Let $B = A^{(-1)}$ and let $d = -A^2 - A^{(-2)}$. For this specialization, let $\langle K\rangle = (d^{-1})[K]$. (See [4], [5], [6].) Then $\langle K\rangle$ satisfies the axioms:

1. $\langle \asymp \rangle = A \langle = \rangle + (A^{-1}) \langle \supset \subset \rangle$

2. $\langle O K \rangle = (-A^2 - A^{(-2)}) \langle K \rangle$

3. $\langle O \rangle = 1$

This special bracket is invariant under the Reidemeister moves of type II and type III. Under type I moves $\langle K\rangle$ is multiplied by $-A^3$ or its inverse.

For K an oriented link, let w(K) denote the writhe of K, where this number is the sum of the signs of crossings. Crossing signs are +1 or -1 as shown below.

Now define a normalized polynomial by the formula

$$f_K = ((-A)^{(-3w(K))})\langle K\rangle.$$

In [5] I show that f_K <u>is an ambient isotopy invariant of the oriented link</u> K. Furthermore, <u>if</u> $V_K(t)$ denotes the original Jones polynomial [3], then

$$f_K(t^{(-1/4)}) = V_K(t).$$

This gives a model for the Jones polynomial as a (normalized) specialization of the general bracket.

The Jones polynomial $V_K(t)$ is an invariant of ambient isotopy for oriented knots and links, and it is the first polynomial invariant capable of detecting chirality for many knots. A knot is said to be chiral if it is not ambient isotopic to its mirror image (obtained by switching all crossings). The bracket provides an elementary route to the Jones polynomial. For the reader interested in pursuing this direction, I record here the three types of Reidemeister moves:

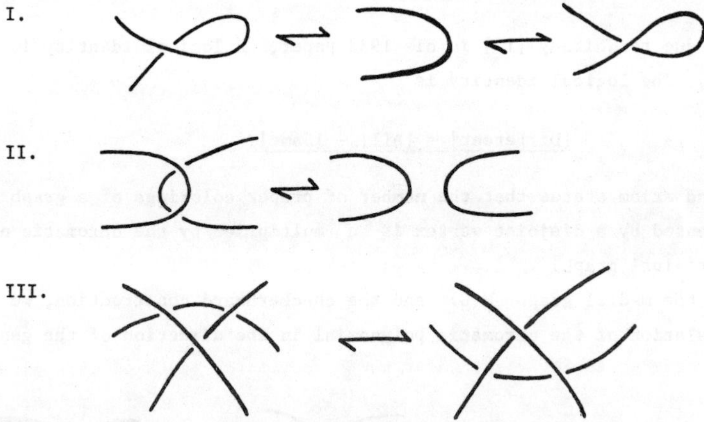

Ambient isotopy is generated diagramatically by these three types of local move. Each small diagram represents a corresponding situation in a larger diagram.

The existence of combinatorial invariants such as the Jones polynomial poses extraordinary problems for classical knot theory. For example, it is not known as of this writing whether $V_K(t)$ detects knottedness in all cases. In other words, does $V_K(t) = 1$ imply that a knot K is unknotted (ambient isotopic to an unknotted circle)?

EXAMPLE 2. (Chromatic Polynomial)

Let G be a planar graph. Let $C[G](q)$ denote the number of ways to properly color G with q colors. A proper coloring of G is an assignment of colors to the vertices of G so that two vertices sharing an edge are colored differently. The coloring polynomial satisfies the following graphical axioms:

1. C G = C G′ − C G″

2. $C_{\bullet G} = qC_G$

The first axiom states that if three graphs G, G', G" are related so that G' is obtained from G by deleting an edge, and G" is obtained from G by collapsing the same edge, then

$$C[G] = C[G'] - C[G''].$$

The proof is due to Whitney [15] in his 1932 paper, "A logical identity in mathematics". The logical identity is

<u>[Different] = [All] - [Same]</u>.

The second axiom states that the number of proper colorings of a graph that is augmented by a disjoint vertex is q multiplied by the chromatic number of the original graph.

By using the medial graph M(G) and the checkerboard construction, we can begin a translation of the chromatic polynomial in the direction of the general bracket.

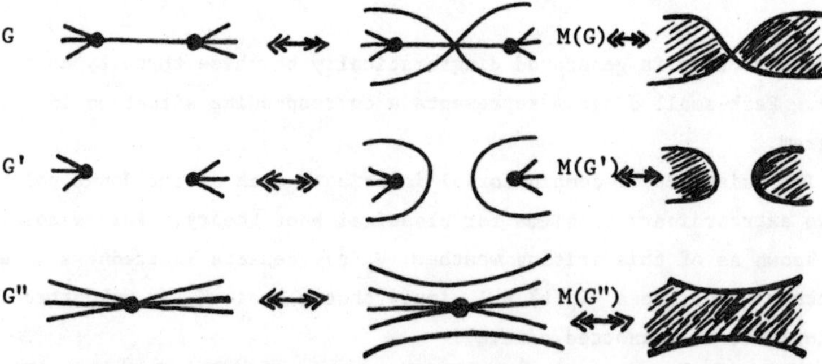

We rewrite the axioms for the chromatic polynomial in terms of shaded universes as follows:

1. C[ ] = C[ ] - C[ ]

2. $C[B \cup M] = qC[M]$ where B is any shaded component that is free of crossings.

For example,

G = •———•, M(G) =

C = C − C

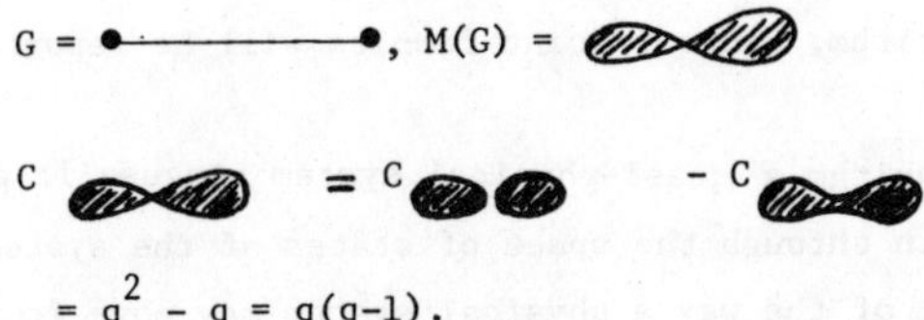

$= q^2 - q = q(q-1)$.

In general, for a shaded universe M(G), we can expand C[M(G)] via the same states as the bracket expansion, except that we keep track of the number of shaded regions, and it is necessary to classify split vertices as internal or external:

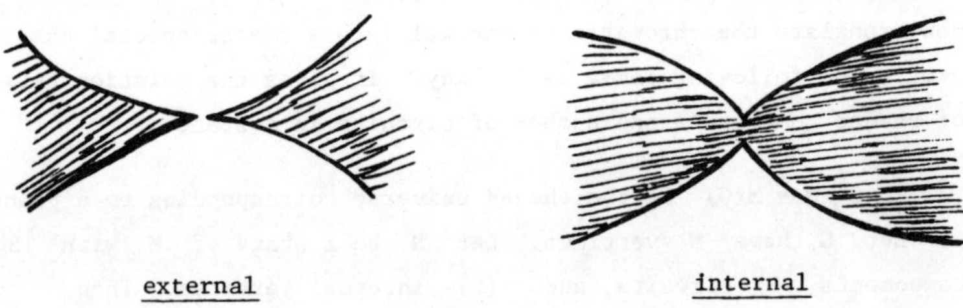

The open channel for an external vertex is unshaded, while the channel for an internal vertex is shaded.

Let M be a shaded universe, and S(M) its collection of shaded states. Let I(S) denote the number of internal vertices in the state S. Let $\|S\|$ denote the number of shaded components in the state S. Then our axioms imply that

$$C[M] = \sum_{S(M)} (-1)^{I(S)} q^{\|S\|} .$$

This gives a specific algorithm for computing chromatic polynomials. The algorithm at this level is of independent interest. I have been investigating its behaviour as a quasi-physical system in collaboration with Mario Rasetti, Corrado Agnes and Amelia Sparavigna of the Politecnico di Torino, Torino, Italy. We compute C[G] by listing the states, starting with the state with maximal number of shaded components (all external vertices). The algorithm moves through the "space of states" changing only one vertex at a time (via a gray code enumeration). We watch the partial sums for C[G]. On a log-log plot of partial sum versus number of iterations (in the listing of states) the picture suggests a remarkable linearity of relationship for the chromatic numbers of subgraphs. These numbers arise after every 2^n steps. In other words, it seems to be possible to estimate the chromatic number rather long before the computation is completed! See Figure 5 for an example of one of our plots. This algorithm, and our joint results will be reported in more detail elsewhere.

I called this algorithm a quasi-physical system because it performs an artificial ergodic path through the space of states of the system. Thus it is a very idealized model of the way a physical system may move from state to state. Obviously, there is much experimental-mathematical and theoretical work to be done here. As we shall see in the next section, the chromatic polynomial (and its generalization to the dichromatic polynomial) is a central feature in certain models in statistical mechanics.

EXAMPLE 3. (More translation between link diagrams and graphs).

We now translate the chromatic polynomial into a direct special case of the bracket. The following lemma is the key. It gives the relation between number of shaded components and number of circuits in a state.

LEMMA 2.2. Let $M = M(G)$ be the shaded universe corresponding to a planar graph G. Let G have N vertices. Let S be a state of M with $\|S\|$ shaded components, $|S|$ circuits, and $I(S)$ internal vertices. Then $\|S\| = (1/2)(N - I(S) + |S|)$.

PROOF: Associate to each state S a graph $G(S)$: The vertices of $G(S)$ are the vertices of G (our given graph G), one for each shaded region of $M = M(G)$. Two vertices of $G(S)$ are connected by an edge exactly when this edge corresponds to an interior vertex of S. Thus $G(S)$ has $I(S)$ edges. I assert that $G(S)$ has $|S| - \|S\| + 1$ faces. (Count faces of $G(S)$ by counting circuits of S, but discard circuits forming outer boundaries of shaded components of S; add 1 to count the unbounded face.) By construction, $G(S)$ has $\|S\|$ components. According to Euler

$$(\text{vertices}) - (\text{edges}) + (\text{faces}) = (\text{components}) + 1$$

for a planar graph. Hence

$$N - I(S) + (|S| - \|S\| + 1) = \|S\| + 1.$$

This completes the proof.

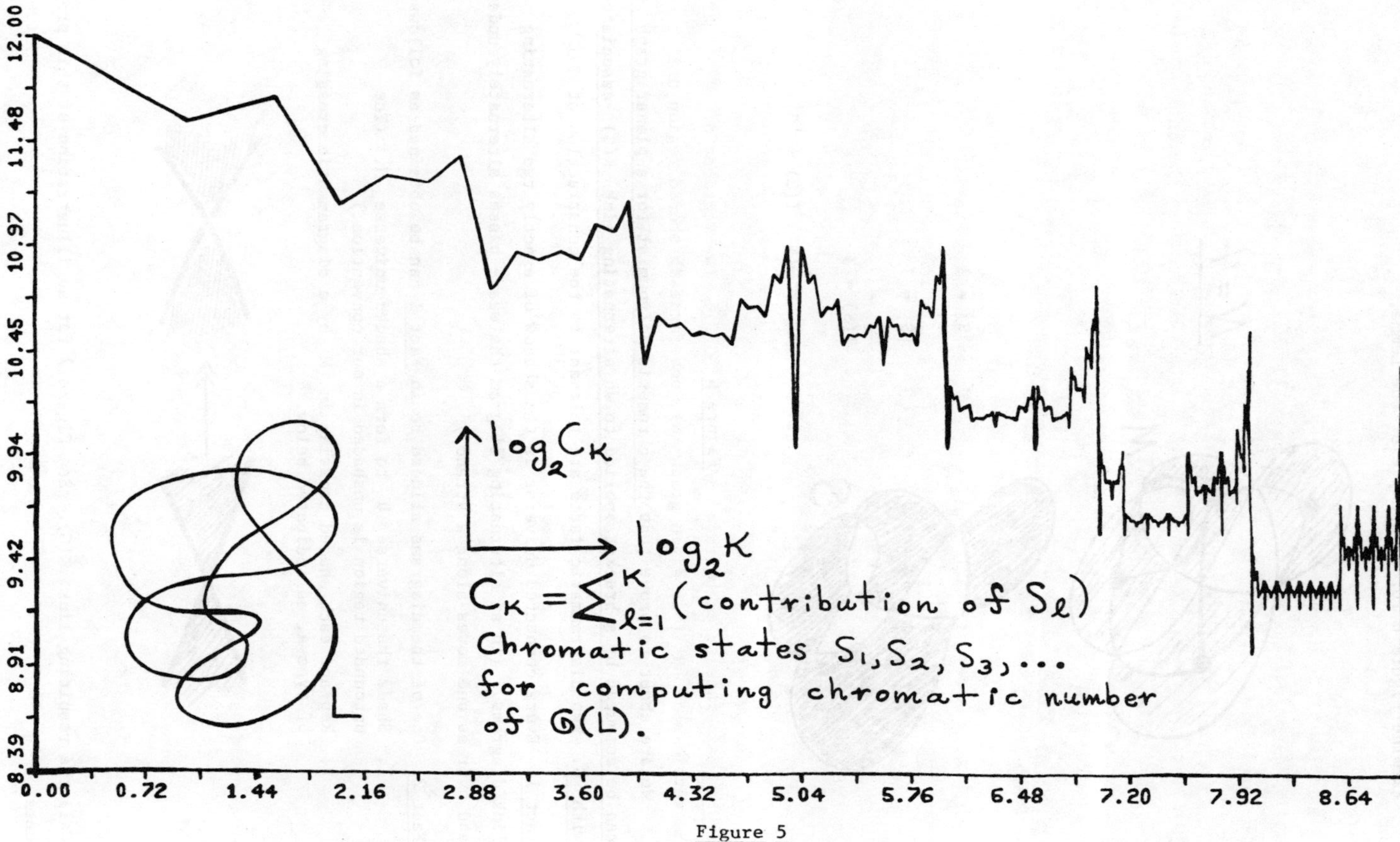

$\log_2 C_K$
$\log_2 K$
$C_K = \sum_{\ell=1}^{K}$ (contribution of S_ℓ)
Chromatic states $S_1, S_2, S_3, \ldots$
for computing chromatic number
of $G(L)$.
L
12.00
11.48
10.97
10.45
9.94
9.42
8.91
8.39
0.00
0.72
1.44
2.16
2.88
3.60
4.32
5.04
5.76
6.48
7.20
7.92
8.64

Figure 5

Examine Figure 6 for an example of the Lemma.

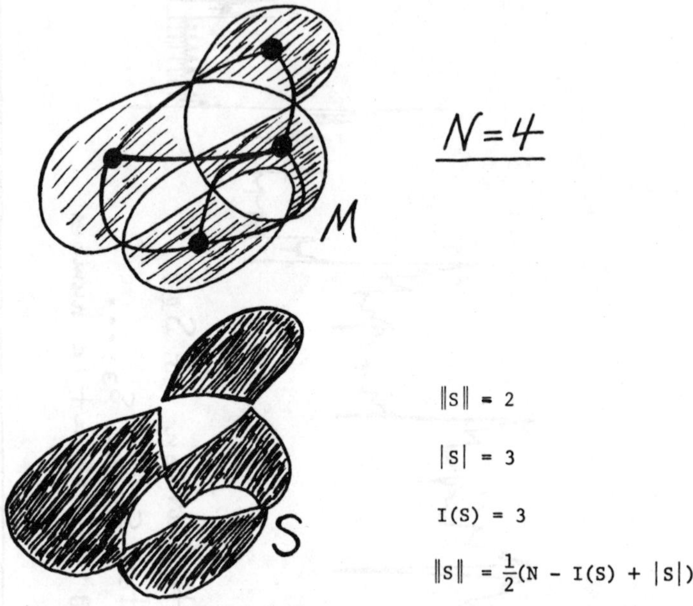

Figure 6

We are about to prove that the chromatic polynomial for a planar graph G can be computed by a bracket applied to an alternating link K(G) associated with G. Two diagrammatic facts are relevant to the construction of K(G):

Fact 1. Every connected universe is the shadow of exactly two alternating link diagrams. (In an alternating diagram the weave passes alternately under and over as one moves along a strand.)

Fact 2. One of the diagrams alluded to in Fact 1 can be obtained as follows:

1. Shade the universe U to form a shaded universe M. (The unbounded region is unshaded in our convention.)
2. Replace each shaded crossing in M by a diagrammatic crossing of type-A, as indicated below.

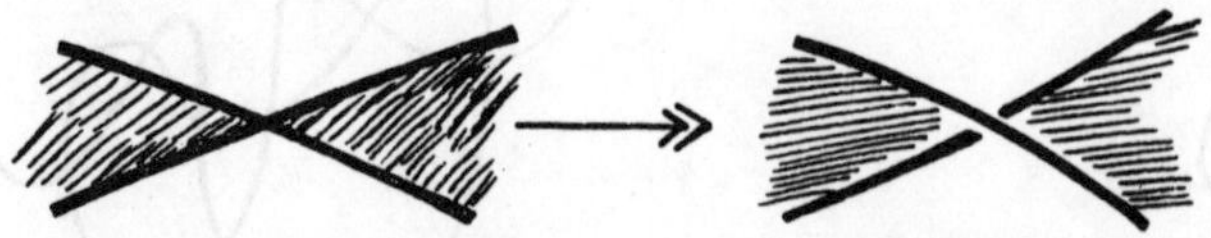

Call the resulting link K(G). See Figure 7 for an illustration of this process.

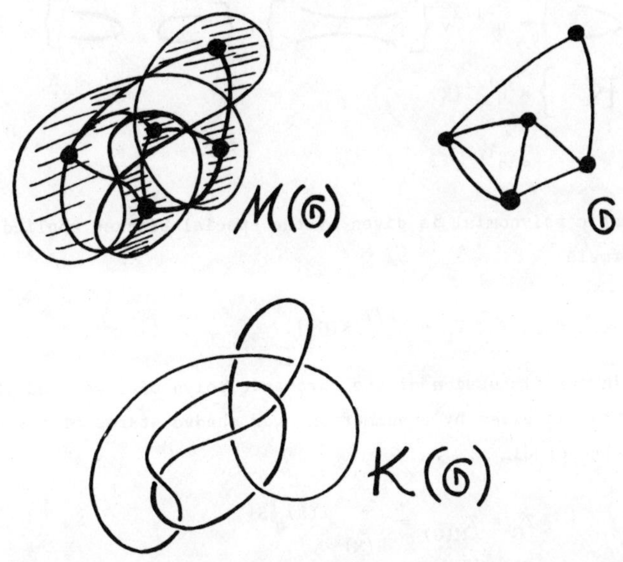

Figure 7

I shall make the translation for a generalization of the chromatic polynomial -- the dichromatic polynomial, Z[G](q,v). The dichromatic polynomial has axioms:

1. $Z_{\succ\!\!-\!\!\prec} = Z_{\succ\ \prec} + vZ_{\succ\!\prec}$

2. $Z_{\bullet G} = qZ_G$

Thus $Z[G](q,-1) = C[G](q)$, giving the chromatic polynomial as a special case when the variable v in the dichromatic polynomial is equal to minus one.

EXAMPLE:

$$Z_{\bullet\!\!-\!\!-\!\!\bullet} = Z_{\bullet\bullet} + vZ_{\bullet} = q^2 + vq$$

$$Z_{\circ} = Z_{\bullet} + vZ_{\bullet} = q(1+v)$$

PROPOSITION 2.3. Let G be a planar graph with N vertices. Let $K(G)$ be the alternating link diagram associated with G by the process described above. For any diagram K let $\{K\}$ be the special bracket defined by $\{K\} = [K](q^{-1/2}v,1,q^{1/2})$. That is, K satisfies the recursion

1. $\{ \times \} = q^{-1/2} v \{ \asymp \} + \{ \supset \subset \}$

2. $\{ \bigcirc K \} = q^{1/2} \{K\}$

$\{ \bigcirc \} = q^{1/2}$.

Then the dichromatic polynomial is given by the special bracket applied to $K(G)$ via the formula

$$Z_G = q^{N/2} \{K(G)\}.$$

PROOF: Just as in our discussion of the chromatic polynomial we find the dichromatic polynomial is given by a summation over shaded states of the universe $M(G)$ by the formula

$$Z_G = Z_{M(G)} = \sum_{S(M)} v^{I(S)} q^{\|S\|}.$$

Now use the formula from Lemma 2.2.

$$Z_M = \sum_{S(M)} v^{I(S)} q^{1/2(N-I(S) + |S|)}$$

(*)
$$Z_M = q^{N/2} \sum_{S(M)} (q^{-1/2} v)^{I(S)} (q^{1/2})^{|S|}.$$

This gives a formula for the dichromatic polynomial in terms of internal vertices and circuits in the states S.

Note that internal vertices in states of $M(G)$ are in one-to-one correspondence with type-A splits in the states of $K(G)$.

The theorem follows from this remark, formula (*), and the form of the state expansion of the general bracket.

EXAMPLE.  G  M(G)  K(G)

$\{K(G)\} = \{$  $\} = q^{-1/2} v \{$ 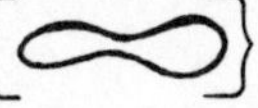 $\} + \{$  $\} = q^{-1/2} v q^{1/2} + q$

$\therefore \{K(G)\} = v + q$

$q^{N/2}\{K(G)\} = qv + q^2.$

SUMMARY. We have defined a general bracket polynomial on link diagrams and have shown how it specializes to the Jones polynomial for links, and to the dichromatic polynomial for planar graphs via the use of the alternating link diagram associated to a planar graph G.

III. THE POTTS MODEL AND THE DICHROMATIC POLYNOMIAL.

Here we discuss a model from statistical mechanics. The framework consists in a lattice G, taken to be any planar graph. (It is also of interest to consider lattices in three dimensional space.) The states of the physical system associated with G consist in assignments of "spins" to the vertices of G. The spins are assumed to be available in q discrete values, where q is a positive integer. Thus we may take the neutral term color in place of spin, and consider a state to be an assignment of colors to the vertices of G (not necessarily a proper coloring). The colors may correspond to spins of particles located at these vertices, or with other localizable and discrete physical states.

Hence we shall speak of states σ of a graph G where $\sigma = \{\sigma_i\}$ is an assignment σ_i to each vertex i of G where σ_i has q possible values. These states should not be confused with the states associated with a link diagram of section 2. If necessary, I shall refer to the latter as diagram states and the former as graph states. A graph state is simply any assignment of q colors to the vertices of a graph G. A diagram state is a mode of splitting the crossings of a link diagram.

To each state σ of a graph G there is an associated energy, $E(\sigma)$, and for the ensemble of all the system's states there is the partition function

$$Z = \sum_{\sigma} e^{-\frac{1}{kT}E(\sigma)}$$

where T denotes temperature, and k is a constant (Boltzman's constant). From the partition function can be deduced many physical properties of the system. For example, the probability p(E) of the system being in a state of energy E is given by the formula

$$p(E) = e^{-\frac{1}{kT}E}/Z.$$

When the partition function itself is written in exponential form

$$Z = e^{-F/kT},$$

then F is the so-called <u>free energy</u> of the system. The <u>average energy</u> U is the expectation value

$$U = \sum_{\sigma} E(\sigma)e^{-E(\sigma)/kT}/Z$$

and one can show that

$$U = F + ST$$

where T denotes temperature, as above, and $S = -\partial F/\partial T$ is the <u>entropy</u> of the system.

It should be mentioned that the intent of this type of model is to create a mathematical situation that embodies the characteristics of a system of interacting particles and its changes under changes of temperature. Thus, on physical grounds, one would expect the system to exhibit <u>phase transition</u>. This means, for example, that as the temperature is started at a high value and then lowered, the distributions of spins will go from very random (at high temperature) to clumping into domains of alignment (at low temperature). The corresponding transition in the distribution of probable states should then be exhibited as a discontinuity or sharp change in the partition function (for very large lattices) at certain critical temperatures.

Historically, very simple rules for the energy of a state have been considered. The <u>Potts model</u> [1] assigns energy by the formula

$$E(\sigma) = \varepsilon \sum_{\langle i,j\rangle} \delta(\sigma_i,\sigma_j)$$

where i and j are vertices of G, ε is ± 1, $\langle i,j\rangle$ denotes an edge connecting i and j, and δ is the Kronecker delta:

$$\delta(x,y) = \begin{cases} 1 & \text{if } x = y \\ 0 & \text{if } x \neq y. \end{cases}$$

In the case $q = 2$, the Potts model is equivalent to the <u>Ising Model</u>. The Ising model was solved exactly by Onsager in the 1940's. It is the first case of an exactly solved model, and was shown to have a phase transition for the limit of a square $N \times N$ planar lattice as N goes to infinity. No exact expression is known for the limit of the Potts partition function on this lattice for general q.

The epsilon ($\varepsilon = \pm 1$) in this expression for the energy divides the Potts models into two cases: anti-ferromagnetic ($\varepsilon = +1$) and ferromagnetic ($\varepsilon = -1$). Consider the ferromagnetic case. Here it is higher energy for neighboring spins to be different. Thus at low energies (low temperatures) one expects neighboring spins to be aligned as happens in magnetic domains in iron or other magnetic materials. In the anti-ferromagnetic case the low temperature domains will correspond to regions of the graph that are properly colored (neighboring spins different). Physically distinct, these two cases have the same mathematical formalism.

We shall now show that the partition function for the Potts model for G planar is a special case of the general bracket. In fact, I will give a proof of the well-known [12] fact that the Potts partition function is a dichromatic polynomial. Then we are in a position to apply our Proposition 2.3, expressing the dichromatic polynomial as a bracket.

PROPOSITION 3.1. Let G be a planar graph. Let Z be the partition function for the q-state Potts model on G. Let

$$v = e^{-\varepsilon/kT} - 1.$$

Then the partition function is the dichromatic polynomial in q and v.

$$Z = Z[G](q,v)$$

PROOF: By the definition of the partition function, we have the following sequence of equalities

$$Z = \sum_{\sigma} e^{-\frac{1}{kT}E(\sigma)} = \sum_{\sigma} e^{-\frac{\varepsilon}{kT}\sum_{\langle i,j\rangle}\delta(\sigma_i,\sigma_j)}$$

$$= \sum_{\sigma}\prod_{\langle i,j\rangle} e^{-\frac{\varepsilon}{kT}\delta(\sigma_i,\sigma_j)}$$

$$= \sum_{\sigma}\prod_{\langle i,j\rangle}(1 + v\delta(\sigma_i,\sigma_j)),\ v = e^{-\frac{\varepsilon}{kT}} - 1.$$

It is easy to see that this last formula is the dichromatic polynomial in q and v. Just verify the recursive axioms (section II) directly. This completes the proof.

Since we know how to translate the dichromatic polynomial into a bracket polynomial, we can finally state

<u>COROLLARY 3.2</u>. Let $Z[G]$ be the q-state Potts partition function for a planar graph G. Let $v = e^{-\frac{\varepsilon}{kT}} - 1$ where k is Boltzman's constant and T is the temperature. Then

$$Z[G] = q^{N/2}\{K(G)\}$$

where N is the number of vertices of G, and $K(G)$ is the alternating link associated with G (as in section II). The symbol $\{K\}$ denotes the special bracket defined by

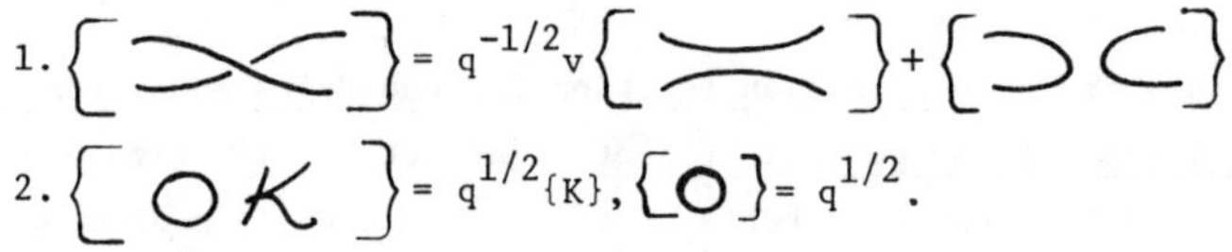

1. $\{\times\} = q^{-1/2} v \{\asymp\} + \{\supset\subset\}$
2. $\{O\,K\} = q^{1/2}\{K\}, \; \{O\} = q^{1/2}.$

PROOF: Apply 2.3 to 3.1.

In [9] we shall discuss specific results of this translation for the Potts model. It is remarkable that both the Jones polynomial and the Potts model fit into exactly the same combinatorial framework.

One may wonder whether there is a direct relation between the topology of the link $K(G)$ and the Potts model. This may be the case. It follows from the <u>Tait flyping conjecture</u> that $[K(G)](A,B,d)$ (the general bracket) is a topological invariant of $K(G)$ whenever G is a connected planar graph without isthmus. The Tait conjecture states that topological equivalences of such alternating knots are generated by special moves (flypes) that each preserve the alternating structure. These moves preserve the general bracket, and hence they preserve the Potts partition function.

This means that (modulo the Tait conjecture) we can take a planar lattice G, form the alternating link $K(G)$, transform this link topologically to any other <u>alternating</u> link K' (in reduced form), take the planar graph G' of K'. Then G and G' will have the same partition function. In [9] we shall explore examples of this process.

The full Tait conjecture is not yet proved, but some of its corollaries have been verified (minimality and invariance of the number of crossings and of the writhe for reduced alternating diagrams). These verifications depend crucially on the Jones polynomial, the bracket model, and on a two-variable generalization of the Jones polynomial due to the present author (see [5], [6], [10], [13], [14]).

Much remains to be explored in this conduit between physics and topology.

<u>REMARK</u>: It is interesting to reformulate the general bracket as a generalization of the Potts model. To do this, we can regard any link diagram as corresponding to the medial graph of a <u>signed graph</u> G (G embedded in the plane). By a signed graph G I mean that each edge of G is assigned ± 1. This encodes the crossing type for the medial graph according to the convention

Thus

while

Changing all the signs creates the mirror image.

Letting K(G) denote the knot or link diagram corresponding to a signed graph G, we then find (as in Proposition 2.3) that $[K] = d^{-N} W[G]$ where N is the number of vertices in G, and W satisfies recursions:

1. $W[\bullet \overset{+}{—} \bullet] = AdW[\,—\bullet—\,] + BW[\,—\bullet \quad \bullet—\,]$

2. $W[\bullet \overset{-}{—} \bullet] = BdW[\,—\bullet—\,] + AW[\,—\bullet \quad \bullet—\,]$

3. $W[\bullet \sqcup G] = d^2 W[G]$

This generalized contraction-deletion algorithm can then be translated into a corresponding energy model for signed graphs (with d^2 spins per site, and the interaction energy dependent on the sign of the bond). This model will be investigated elsewhere.

IV. BRACKET, BRAID DIAGRAMS AND THE TEMPERLEY-LIEB ALGEBRA.

The n-strand braid group B[n] is generated by elementary braids

$$\sigma_1, \bar{\sigma}_1, \sigma_2, \bar{\sigma}_2, \ldots, \sigma_{n-1}, \bar{\sigma}_{n-1}$$

as shown in Figure 8.

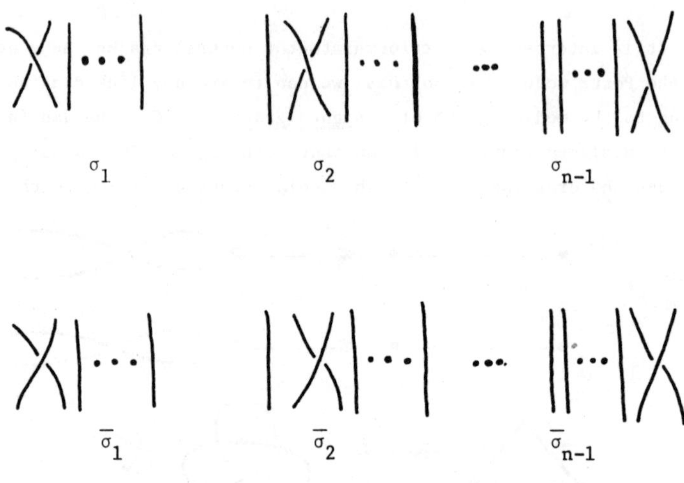

Figure 8

A braid diagram is any specific diagram obtained by taking a word in these generators. Thus $\sigma_1\bar{\sigma}_1$ is the diagram

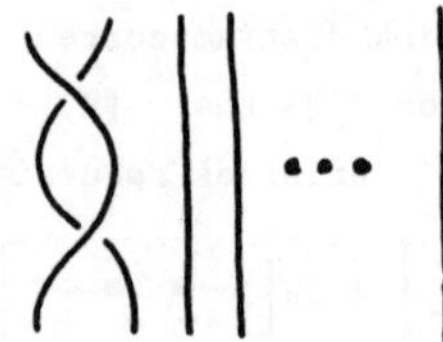

taken in this form (even though it is isotopic to the identity braid).

If b is a braid diagram, let $\bar{b}$ denote its closure, obtained by connecting top to bottom strands as shown in Figure 9. Let P(b) denote the plat closure for braids with an even number of strands as also shown in Figure 9.

As explained in [5], [6], [8], we can represent any state in the expansion for the general bracket for a braid diagram b as a product of generators $h_1, h_2, \ldots, h_{n-1}$ of local splittings. These generators satisfy diagrammatic relations as shown in Figure 10, and form a multiplicative monoid. We call the diagram algebra the free additive algebra over this monoid, with coefficients in the polynomial ring in the variables A, B. The evaluation of the bracket is invariant under the relations in the diagram algebra.

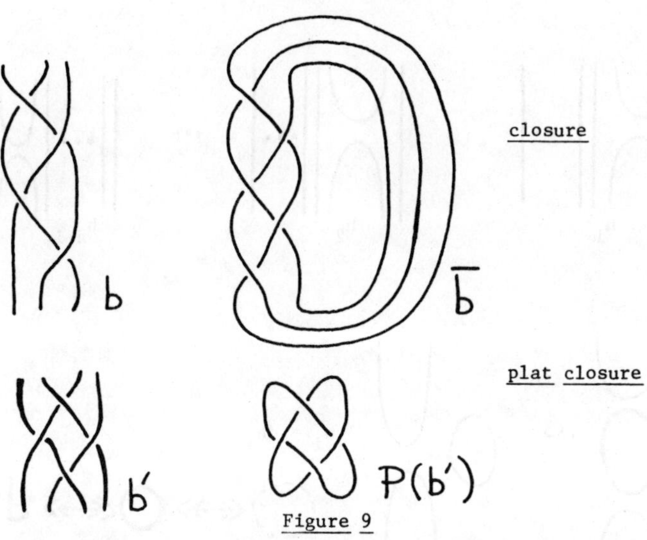

Figure 9

Since, in the diagram algebra, $h_i^2 = dh_i$, it may seem appropriate to regard d as a special element of the algebra. Since d commutes with all elements it has the same status as the polynomial generators A and B. Thus it is sometimes useful to regard the ring Z[A,B,d] as a coefficient ring for the algebra. Note however, that as we have defined it, the diagram algebra is the free additive algebra over Z[A,B] that is generated multiplicatively by the diagram monoid.

In the expression of the relations for Figure 10 the letter d stands for the closed loop. By using these relations we can expand the general bracket, needing only to know the value of the bracket on a product of the monoid generators. This value depends upon the type of closure, since the value of the bracket on a disjoint collection of Jordan curves is d raised to the number of curves. The case of plat closure is particularly interesting.

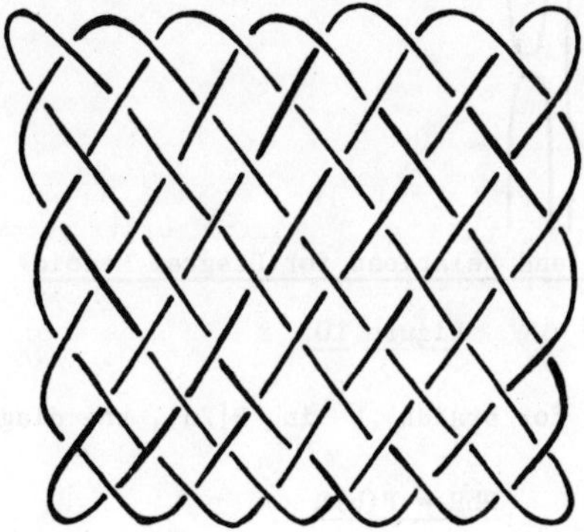

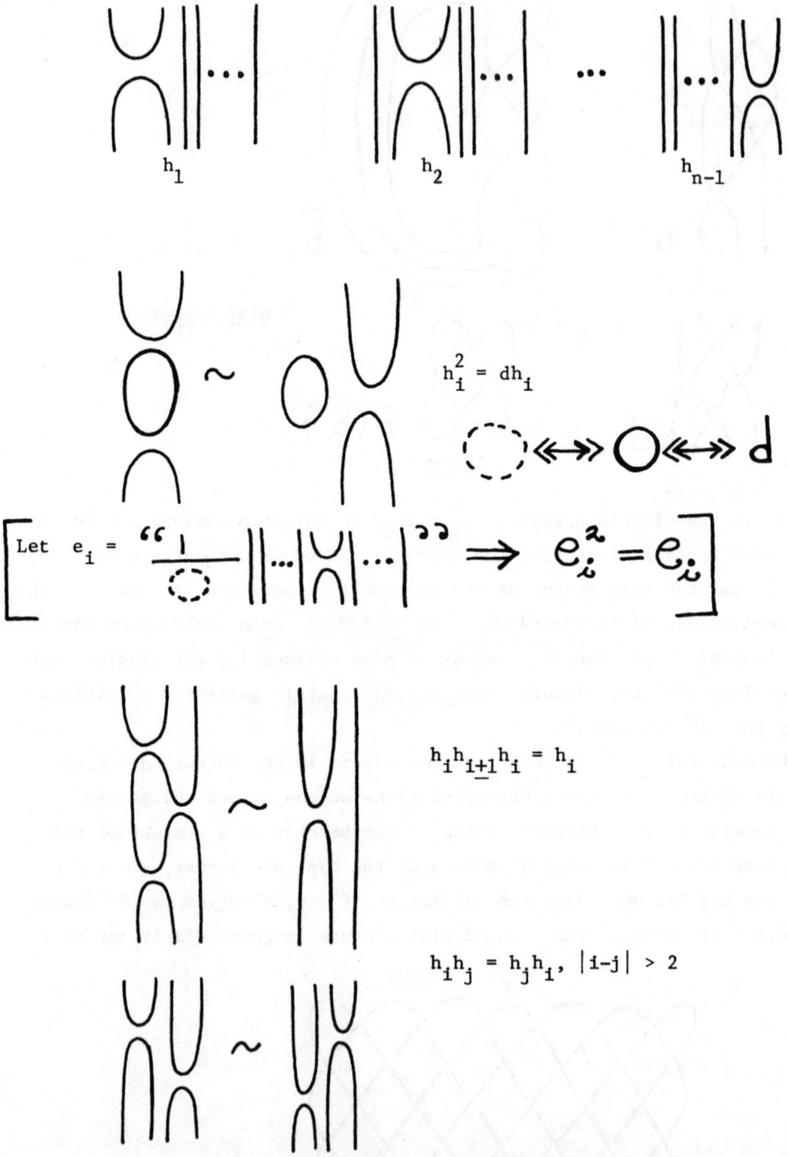

h_1 h_2 h_{n-1}

$h_i^2 = dh_i$

$\Longleftrightarrow\ \bigcirc \Longleftrightarrow\ d$

[Let e_i = "$\frac{1}{\bigcirc}$ ||…|X|…|" $\Rightarrow e_i^2 = e_i$]

$h_i h_{i\pm1} h_i = h_i$

$h_i h_j = h_j h_i$, $|i-j| > 2$

<u>Generators and Relations for Diagram Monoid</u>

<u>Figure 10</u>

As Figure 11 shows we have, for braids b in $B[2n]$, the diagrammatic formula

$$\underline{HbH = P(b)H}$$

where H denotes the product $h_1h_3h_5 \ldots h_{2n-1} = H$.

$H = h_1h_3h_5 \ldots h_{2n-1}$

(here $n = 3$)

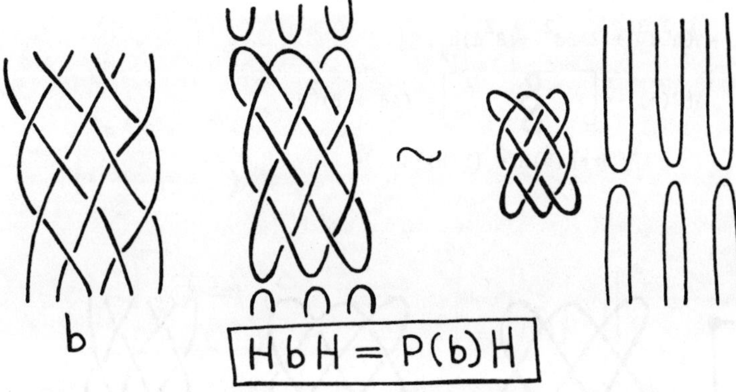

$\text{H}\,\text{rep}(b)\,\text{H} = [b]\text{H}$

where

$\text{rep}(\sigma_i) = Ah_i + B$

[╳] = A[⌣⌢] + B[ )( ]

$\text{rep}(\overline{\sigma}_i) = Bh_i + A$

[╳] = B[⌣⌢] + A[ )( ]

Figure 11

A consequence of this is the following description for the general bracket on the plat closure P(b):

1. The braid diagram b is given as a word in

$$\sigma_1, \overline{\sigma}_1, \ldots, \sigma_{n-1}\, \overline{\sigma}_{n-1}.$$

Replace each instance of σ_i by $Ah_i + B$, and each instance of $\overline{\sigma}_i$ by $Bh_i + A$.

Call the resulting element of the diagram algebra rep(b).

2. Then $Hrep(b)H = [P(b)]H$ is a valid identity in the diagram algebra. By using the relations in the algebra systematically, this formula provides an algorithm for computing $[P(B)]$.

EXAMPLE 1. $b = \sigma_1\sigma_1 =$, $P(b) =$, $rep(b) = (Ah_1 + B)^2$

$$Hrep(b)H = h_1(Ah_1 + B)^2h_1 = h_1(A^2h_1^2 + 2ABh_1 + B^2)h_1$$

$$= (A^2d^3 + 2ABd^2 + B^2d)h_1.$$

$$[P(b)] = [\quad] = (Ad + B)^2d$$

$$\therefore Hrep(b)H = [P(b)]H$$

EXAMPLE 2.

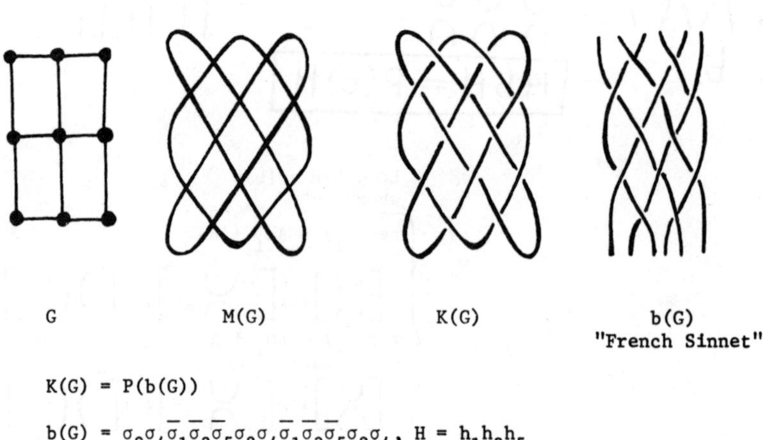

$$K(G) = P(b(G))$$

$$b(G) = \sigma_2\sigma_4\bar{\sigma}_1\bar{\sigma}_3\bar{\sigma}_5\sigma_2\sigma_4\bar{\sigma}_1\bar{\sigma}_3\bar{\sigma}_5\sigma_2\sigma_4, \ H = h_1h_3h_5$$

For $\{K\}$: $rep(\sigma_i) = q^{-1/2}v\, h_i + 1$

$$rep(\bar{\sigma}_i) = h_i + q^{-1/2}v$$

$$\left\{\begin{array}{l} h_i^2 = q^{1/2}h_i \\ h_ih_{i\pm1}h_i = h_i \\ h_ih_j = h_jh_i, \ |i-j| > 1 \end{array}\right\}$$

$$Z_G = q^{N/2}\{K(G)\}, \ \{K(G)\}H = Hrep(b(G))H$$

The formalism of this diagram algebra is identical to the Temperley-Lieb Algebra [1] used in evaluating the Potts model. As the example above indicates, the plat closure of the "French Sinnet" braid is the alternating link associated with a rectangular lattice in the plane. The calculation we have indicated for the dichromatic polynomial in this example is exactly paralleled in Baxter [1] in his chapter on the Potts model - but without mention of braids. The entire formalism was invented in the lattice context.

Finally, it is worth indicating that the diagram algebra is actually a "skeleton" of a specific matrix representation for the algebra. Following Penrose [11], et al, let

denote a Kronecker delta δ^{β}_{α}.

In general, let a diagram in the form

$\beta_1\ \beta_2\ \beta_3\ \cdots\ \beta_n$

T

$\alpha_1\ \alpha_2\ \alpha_2\ \cdots\ \alpha_n$

$$= T^{\beta_1\beta_2\cdots\beta_n}_{\alpha_1\alpha_2\cdots\alpha_n}$$

denote a matrix or tensor element. Depict contraction of indices (summation over a repeated index) by connecting corresponding lines. Thus

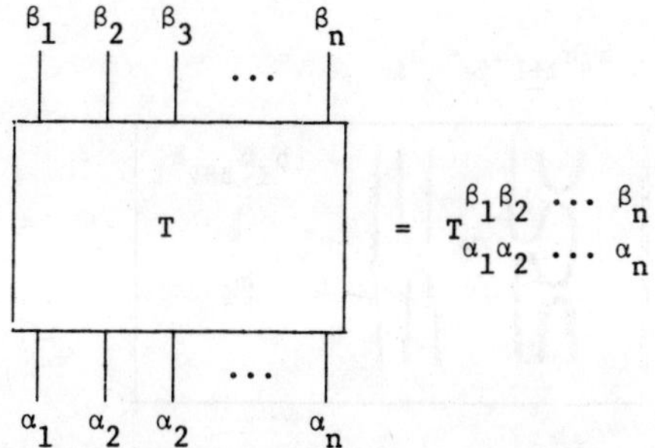

$$A^{\alpha}_{\beta}B^{\beta}_{\delta} = $$

Let juxtaposition of these forms correspond to tensor product.

We may let

$$= M^{\alpha\beta}$$

and

$= M_{\alpha\beta}$

Then

$h \Longleftrightarrow$ $= M_{\alpha\beta}M^{\varepsilon\nu}$

and

$h^2 \Longleftrightarrow$ $\Longleftrightarrow M_{\alpha\beta}M^{\varepsilon\nu}M_{\varepsilon\nu}M^{\lambda\mu} = d(M_{\alpha\beta}M^{\lambda\mu})$

$= dh$

where

$d = M^{\varepsilon\nu}M_{\varepsilon\nu} \Longleftrightarrow$.

In order to have the relations

$h_i h_{i\pm 1} h_i = h_i$

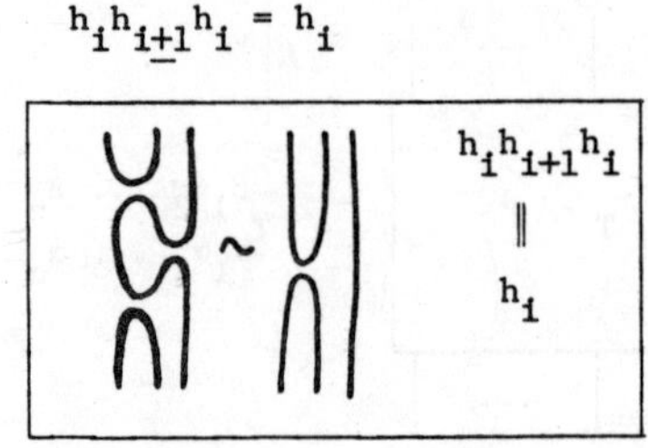

we need that

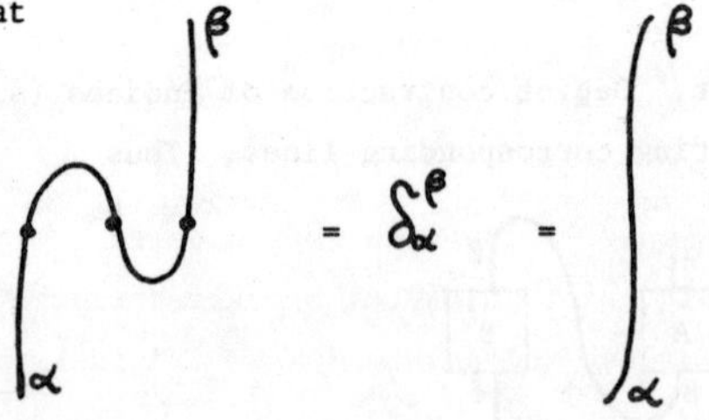

$= \delta_\alpha^\beta =$

which is equivalent to the demand that

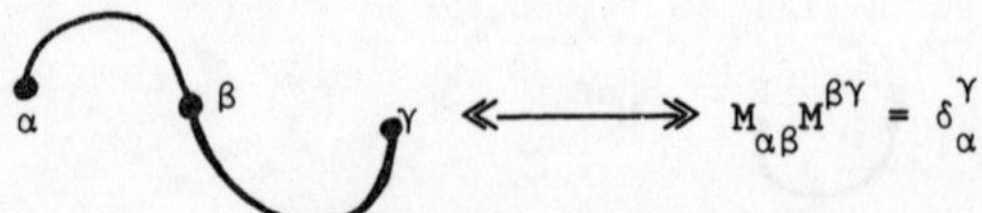

$\Longleftrightarrow M_{\alpha\beta}M^{\beta\gamma} = \delta_\alpha^\gamma$

If $M_{\alpha\beta} = M^{\alpha\beta}$, then this requirement is equivalent to the demand that $MM = I$. Thus we can take M to be a 2×2 Pauli matrix such as

$$M = \begin{pmatrix} 0 & \sqrt{-1}\,A \\ -\sqrt{-1}\,A^{-1} & 0 \end{pmatrix}$$

and set

$$h_1 = (M \otimes M) \otimes I \otimes I \otimes \ldots \otimes I \Longleftrightarrow$$

$$h_2 = I \otimes (M \otimes M) \otimes I \otimes \ldots \otimes I \Longleftrightarrow$$

$$\vdots$$

$$h_{n-1} = I \otimes I \otimes I \otimes \ldots \otimes (M \otimes M) \Longleftrightarrow$$

This particular representation is specifically the algebra of Temperly and Lieb. In this form note that

$$M \otimes M = \begin{array}{|c|c|c|c|} \hline 0 & 0 & 0 & 0 \\ \hline 0 & \lambda & 1 & 0 \\ \hline 0 & 1 & \lambda^{-1} & 0 \\ \hline 0 & 0 & 0 & 0 \\ \hline \end{array} \qquad \lambda = -A^2$$

$$\begin{pmatrix} \lambda & 1 \\ 1 & \lambda^{-1} \end{pmatrix}^2 = (\lambda + \lambda^{-1}) \begin{pmatrix} \lambda & 1 \\ 1 & \lambda \end{pmatrix} = (-A^2 - A^{-2}) \begin{pmatrix} \lambda & 1 \\ 1 & \lambda \end{pmatrix}$$

and recall that

$$\langle O K \rangle = (-A^2 - A^{-2})\langle K \rangle \ .$$

Here $\underline{rep}: B_n \to$ (Temperley-Lieb) is a representation of the braid group. $\sigma_i \to Ah_i + A^{-1}$, $\sigma_i \to A^{-1}h_i + A$. Thus the formalism of the topological bracket $\langle K \rangle$ (see section II, Example 1) lives in this algebra. It is a nice exercise to see how to compute the topological bracket through this representation. The result is equivalent to Vaughan Jones' original construction of his polynomial for braids.

For the topological bracket we define $\langle b \rangle = \langle \overline{b} \rangle$ where $\overline{b}$ denotes the closure of the braid b, as in Figure 9. Then for an element H in the diagram monoid, $\langle H \rangle = d^{|H| - 1}$ where H denotes the number of disjoint closed curves in $\overline{H}$, and $d = -A^2 - A^{-2}$.

In order to detect this evaluation in the matrix algebra, define

$$\eta = (\eta_a^b) = \begin{pmatrix} -A^2 & 0 \\ 0 & -A^{-2} \end{pmatrix}$$

and $\eta_n = \eta \otimes \ldots \otimes \eta$ (n-copies of η).

Let M_n denote the matrix algebra we have constructed, representing the n-strand diagram algebra. Let $tr: M_n \to Z[A,A^{-1}]$ be the standard matrix trace. Then for any $H \in M_n$, we have the following trace formula for the bracket:

$$\langle H \rangle = d^{-1} tr(\eta H).$$

This is our version of Vaughan Jones' trace formula for his polynomial. The bracket "is" the trace.

V. THE FERROMAGNETIC CRITICAL POINT

First an exercise about the general bracket. Let K be any link diagram, and let K^* be its mirror image obtained by reversing all the crossings of K. Suppose that K has $c(K)$ crossings.

LEMMA 5.1. Under the above assumptions,

$$[K^*](A,B) = (A/B)^{c(K)} [K](B^2/A,B).$$

PROOF: Let $g[K] = (A/B)^{c(K)}[K]$. Then it is easy to check that

$$g\left[\times\right] = (A^2/B)g\left[\asymp\right] + Ag\left[\,)(\,\right]$$

from the expansion formula for the general bracket. It then follows by induction that

$$g[K](A,B) = [K](A^2/B,A).$$

Now use $[K^*](A,B) = [K](A,B)$.

Specialize this lemma to the bracket $\{K^*\}$ where $A = q^{-1/2}v^*$, $B = 1$ and we find

$$\{K^*\}(q^{-1/2}v^*,1) = (q^{-1/2}v)^{c(K)}\{K\}(q^{1/2}v^{-1*},1).$$

In the case where K is the alternating link diagram associated with a rectangular lattice, then K^* is the diagram associated with its planar dual. If $\{K\} = \{K\}(q^{-1/2}v,1)$ then we see that we should make the identification

$$q^{1/2}v^{*-1} = q^{-1/2}v$$

in order to compare these models. (v and v^* being the modified temperature variables in each case). Thus

$$vv^* = q.$$

A heuristic argument [16] then suggests that unless the critical point occurs at $v = v^*$ there will, by duality, be at least <u>two</u> critical points. (And this does not make sense physically.) Therefore, we expect the criticality to be at

$$v^2 = q.$$

Here $e^{\frac{1}{kT}} - 1 = v$ (ferromagnetic case), so that

$$T_c = T_{critical} = \frac{1}{k\,\ell n(1 + \sqrt{q})} .$$

Note that at this temperature $\{\asymp\!\!\!\times\} = \{\asymp\} + \{\,)\,(\,\}$. Hence, at this critcality the recursive expansion for the partition function is particularly simple, and it is independent of crossing types, depending only on the medial graph.

It would be very interesting to see this ferromagnetic critical point verified by an exact calculation of the Potts model.

VI. ICE AND ANGLES.

In this section I explain a method for calculating the general bracket on piecewise-linear link diagrams. The method depends on "arrow-coverings", a state-notion derived from the ice model [1] in statistical mechanics. In the ice-model (6-vertex model) the underlying graph M is 4-valent, hence it is the shadow of a link diagram. An <u>arrow-covering</u> of M is a choice of orientation for each M so that <u>at each 4-valent vertex two arrows point into the vertex and two arrows point out of it</u>.

We shall also allow 2-valent vertices, but here the arrow covering must give a consistent orientation across the vertex (one in and one out).

Figure 12 illustrates the possible arrow configuration at a vertex (four-valent).

Arrow Configuration Split

(a)

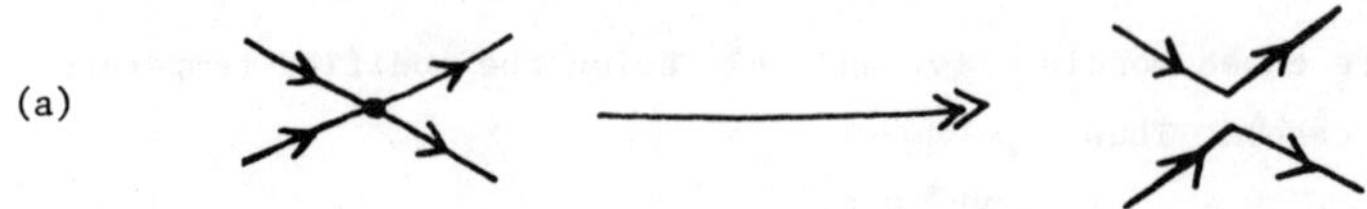

(a')

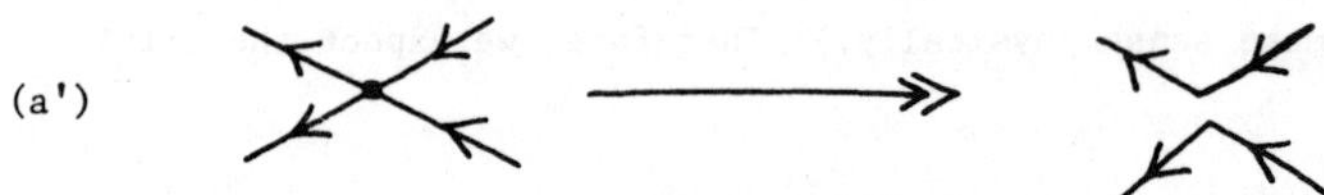

(b)

(b')

(c)

(d)

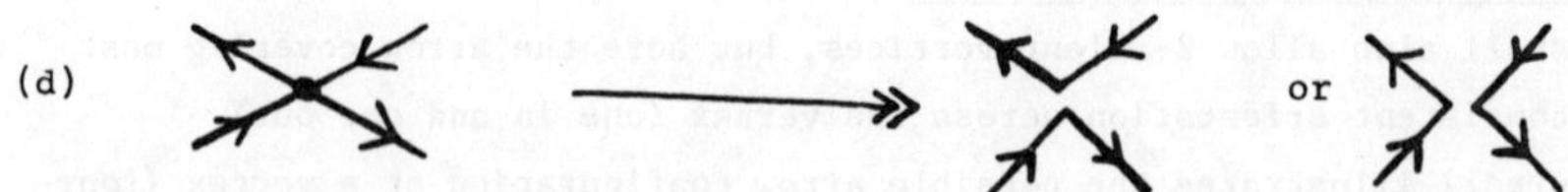

Arrow Configurations and Their Corresponding Splittings

Figure 12

Each arrow configuration gives rise to one or two (see cases (c) and (d) of Figure 12) modes of splitting the diagram at the vertex. We insist that the splits be oriented so that the split diagrams inherit arrow coverings.

Let K be an unoriented link diagram, and let S denote the collection of link diagrams states S, obtained by splitting vertices as in section 1. Let $\overline{S}$ denote the collection of oriented states $\overline{S}$ where each simple closed curve in $\overline{S}$ has been assigned an orientation. It is then easy to see that each $\overline{S}$ in $\overline{S}$ closes to a unique arrow covering $\overline{A} \in A$ where A is the set of arrow coverings of U (U is the shadow of K).

Since any simple closed curve μ in a diagram state S will appear with two orientations in the states S we shall write

$$d = Z^{2\pi} + Z^{-2\pi}$$

where Z is a new variable. Then each 2-valent vertex will contribute a power of Z, Z^{θ}, according to the rule

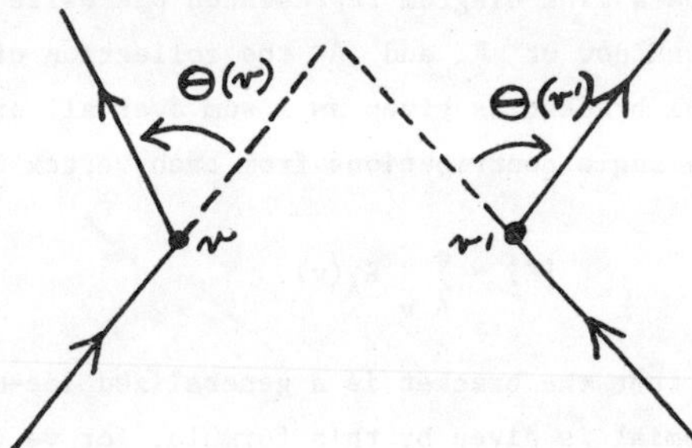

Counterclockwise rotations give positive angles; clockwise rotations give negative angles.

For a piecewise linear oriented simple closed curve the product $Z^{\theta_1} Z^{\theta_2} \ldots Z^{\theta_n}$ over its vertices will equal $Z^{2\pi}$ or $Z^{-2\pi}$ according as the curve is positively or negatively oriented.

By taking into account both orientations on a simple closed curve, and summing these two products we retrieve $d = Z^{2\pi} + Z^{-2\pi}$.

Now let $\overline{A} \in A$ be an arrow covering of a piecewise-linear (i.e., the graph is composed of straight line segments) link shadow M. Let M be the shadow of the link K. For each vertex v of M, define K(v) as follows:

(i) If v is 2-valent then $K_A(v) = Z^{\theta(v)}$ where $\theta(v)$ is the angle corresponding to this vertex and arrow covering.

(ii) If v is 4-valent, let v', v'' the two local vertices obtained by splitting the diagram according to the arrow covering. Let $K_A(v)$ = the sum over allowed splittings (there may be two) of $A^{i(v)}B^{j(v)}Z^{\theta(v') + \theta(v'')}$ where A or B appear according to the usual rules of section 1.

For example: if K = [crossing] locally then K_A ([split]) $= AZ^{\theta_1 + \theta_2}$ where

θ_1 θ_2 .

Each contribution $K_A(v)$ is strictly local, once the arrow covering has been chosen. We then have

THEOREM 6.1. Let K be a link diagram represented piecewise linearly in the plane. Let M be the shadow of K, and $\mathcal{A}$ the collection of arrow coverings of M. Then the general bracket is given as a sum over all arrow coverings of the products of the ice-angle contributions from each vertex of M:

$$[K] = \sum_{\mathcal{A}} \prod_{v} K_A(v).$$

This theorem shows that the bracket is a generalized ice-model. In particular, the Jones polynomial is given by this formula, for we can write the topological bracket in the form above, taking $Z = (\sqrt{-1}\,A)^{1/\pi}$. This reformulation of the bracket calculation in terms of strictly local data may pave the way towards new (infinitesimal or differential geometric) interpretations of the new invariants of knots and links.

In this section we have carried an idea from statistical mechanics into the knot-theoretic context.

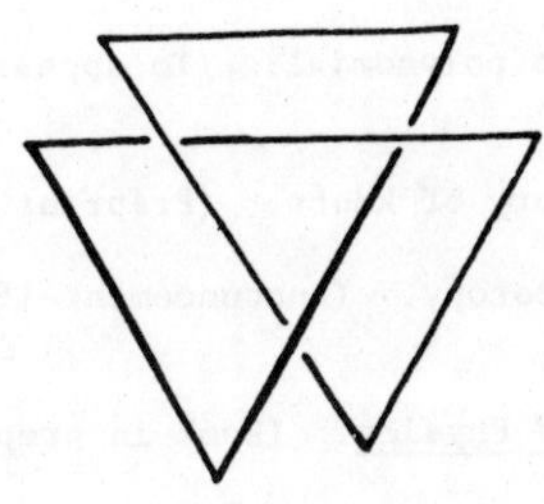

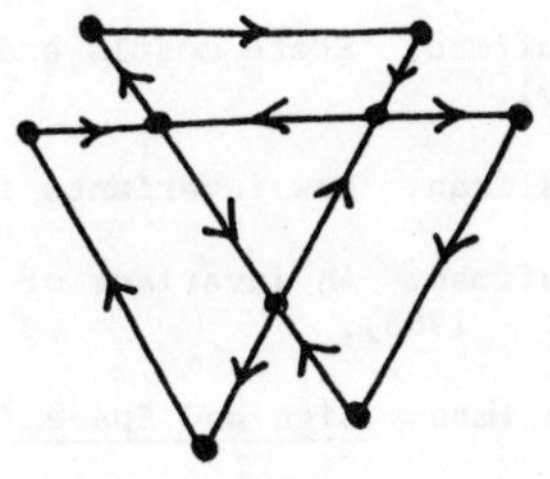

K piecewise linear.

An arrow covering of the shadow M.

Figure 13

As pointed out by Baxter [1], in the case of the Potts model this change of variables gives $q^{1/2} = Z^{2\pi} + Z^{-2\pi} = 2\cosh(\lambda)$ when $Z = e^{\lambda/2\pi}$. For q taking values in the Beraha numbers

$$q = 4\cos^2(\pi/n),\ n = 2,3,4,\ \ldots$$

we then have the λ-values

$$\lambda = i\pi/2,\ i\pi/3,\ i\pi/4,\ \ldots$$

It is possible that this reformulation of the bracket will shed geometric light on the mysterious appearances of the Beraha numbers in chromatic problems $(q = 4,\ v = -1)$, the Jones algebra [3] and other recent work [2] about the meaning of these models at these special values.

REFERENCES

1. R.J. Baxter. Exactly Solved Models in Statistical Mechanics. Academic Press, 1982. (Chapter 12).

2. D. Friedan, Z. Qiu, and S. Shenker. Conformal Invariance, Unitarity and two dimensional critical exponents. In Vertex Operators in Mathematics and Physics - Proceedings of a Conference November 10-17, 1983. Pub. of MSRI#3, Springer-Verlag 1984.

3. V.F.R. Jones. A polynomial invariant for links via von Neumann algebras. Bull. Am. Math. Soc., 12(1985), 103-112.

4. V.F.R. Jones. Hecke algebra representations of braid groups and link polynomials. (preprint 1986).

5. L.H. Kauffman. State models and the Jones polynomial. (To appear in Topology).

6. L.H. Kauffman. New invariants in the theory of knots. (Preprint 1986).

7. L.H. Kauffman. An invariant of regular isotopy. (Announcement-1985, preprint - 1986).

8. L.H. Kauffman. Sign and Space / Knots and Physics. (Book in preparation).

9. L.H. Kauffman, M. Rasetti, C. Agnes, and A. Sparavigna. Knot theory and the Potts model. (In preparation).

10. K. Murasugi. Jones polynomials and classical conjectures in knot theory I and II. (To appear in Topology).

11. R. Penrose. Applications of negative dimensional tensors. Combinatorial Mathematics and Its Applications. Edited by D.J.A. Welsh, Academic Press (1971).

12. H.N.V. Temperley. Graph Theory and Applications. Ellis Horwood Ltd. (1981).

13. M. Thistlethwaite. A spanning tree expansion of the Jones polynomial. (To appear in Topology).

14. M. Thisthethwaite. Kauffman's polynomial and alternating links. (Preprint 1986).

15. H. Whitney. A logical expansion in mathematics. Bull. Am. Math. Soc., 38(1932).

16. F.Y. Wu. The Potts Model. Review of Modern Physics, Vol. 54, No. 1, January 1982.

DEPARTMENT OF MATHEMATICS, STATISTICS AND COMPUTER SCIENCE
UNIVERSITY OF ILLINOIS AT CHICAGO
BOX 4348
CHICAGO, ILLINOIS 60680

March 23, 1987

PACIFIC JOURNAL OF MATHEMATICS
Vol. 137, No. 2, 1989

ON KNOT INVARIANTS RELATED TO SOME STATISTICAL MECHANICAL MODELS

V. F. R. JONES

Dedicated to the memory of Henry Dye

We use three different kinds of statistical mechanical models to construct link invariants. The vertex models emerge as the most general. Our treatment of them is essentially the same as Turaev's. Using the work of Goldschmidt we are able to define models whose invariants are homology invariants for branched covers. Thus the statistical mechanical framework embraces both the "classical" and the "new" link invariants.

0. Introduction. In this paper we shall discuss three types of statistical mechanical models—vertex models, Potts type models, and IRF models. In all cases we shall see that the models may be defined on a knot diagram (replacing the lattice of the model), and that a suitable variation on the partition function of the system is often a knot invariant, i.e. depends only on the knot as a three-dimensional entity and not on the chosen diagram.

The connection between knot theory and statistical mechanics was first established, indirectly, in **[J1]** where it was observed that the Temperley-Lieb algebra of the Potts and ice-type models (see **[Ba]** and **[TL]**) can be used to define a knot invariant using the theory of braids and a certain trace on the Temperley-Lieb algebra, discovered in the course of investigations into type II_1 factors (see **[J2]**). (This invariant is a Laurent polynomial in $\sqrt{t}$ which we shall write $V_L(t)$, where L is some oriented link.) But it was Kauffman who first began to understand this connection in a direct way with his "states model" for V_L, which freed the understanding of V_L from the use of braids or inductive methods (as in **[F+]**). Kauffman's model seems very special to V_L, but another approach to such explicit formulae was suggested by the braid formalism. The author succeeded in "unbraiding" a trace formula for a series of specializations of the two variable polynomial of **[F+]**. The relevant braid group representations were discovered by Jimbo **[Ji]**, Drinfeld **[D]**, and Wenzl **[W1]**. This unbraiding was reported in a letter to Kauffman and we give the details of it in this paper. It was immediately generalized by Turaev **[Tu]** to embrace the Kauffman polynomial. We present our own version of Turaev's formalism, which has the interesting feature that it uses the angles formed at the crossings of a knot projection.

All the formulae referred to in the preceding paragraph are analogues of the partition function for a statistical mechanical system on a two-dimensional diagram of the link. Turaev's formulae correspond to a certain class of models known as "vertex models". We shall see that for Potts-type and IRF models it is possible to define non-trivial link invariants as analogues of the partition function. Vertex models turn out to be the most general, though in a perhaps artificial way.

The formalism is interesting for its generality. Using the work of Goldschmidt and the author in **[GJ]** we recover models which calculate the homology of 2-fold branched covers, and in view of **[Go]** it seems likely that all of the information in a Seifert matrix is accessible to our method. We are also able to situate Kauffman's model quite clearly as an intermediary between the Potts model and a vertex model (called "ice-type") via a piece of combinatorics that seems unique to the Potts model and the ice-type model.

Although the formalism is quite general and not tied to braid presentations or induction, it is still hampered by the need for a two-dimensional projection (shadow) of a three-dimensional object. Our main reason for doing this work was as a step towards a useful and genuinely three-dimensional understanding of the invariants. So far we have not succeeded. The situation is the same as that of the poor prisoners in Plato's allegory of the cave **[P1**, pp. 253–254].

1. Vertex models. In statistical mechanics a vertex model is defined on a graph with vertices $V = \{v\}$ and edges $\mathscr{E} = \{e\}$. A finite set Θ (or more generally, a measure space) of "states per edge" is given and a state of the system is a function $\sigma\colon \mathscr{E} \to \Theta$. Let $\mathscr{H}$ be the set of all states. For each vertex v there is an energy function $E_v\colon \mathscr{H} \to \mathbf{R}$ such that $E_v(\sigma)$ is determined by the restriction of σ to the set of edges incident to v (nearest neighbor interaction). The partition function of the vertex model is the sum $Z = \sum_{\sigma\in\mathscr{H}} \exp(-(1/kT)(\sum_v E_v(\sigma)))$. If the system is in interaction with its surroundings a term needs to be added to $\sum_{v\in V} E_v(\sigma)$ (see **[Hi]**). This should be the origin of the "angle" term in our formalism below. The *Boltzmann weights* are the terms $w_v(\sigma) = \exp(-(1/kT)E_v(\sigma))$ so that $Z = \sum_{\sigma\in\mathscr{H}} \prod_{v\in V} w_v(\sigma)$. The $w_v(\sigma)$ are in general functions of certain variables.

We shall be interested in four-valent graphs only and the vertices will be divided into two classes, "+" and "−", so that $w_v(\sigma) = w_\pm(a, b|x, y)$ where $\{a, b, x, y\}$ are the elements of Θ assigned by σ to the edges incident to v, ordered according to some convention. Thus we make the following formal definition (which is intended for knot theory, the equations (1.2)–(1.5) are too restrictive for statistical mechanics).

DEFINITION 1.1. A *vertex model* is a finite set Θ together with two functions $w_\pm(a,b|x,y)(\lambda) = w_\pm(a,b|x,y)(\lambda,h)$ where $a,b,x,y\in\Theta$, $\lambda\in(0,\pi)$ (and we are deliberately vague about the values of w; they should at least lie in a commutative ring), a "random variable" $f:\Theta\to\mathbf{R}$, $f(a)=f_a$, and a constant $\hbar = h/2\pi$ such that

$$(1.2)\qquad w_\pm(a,b|x,y)(\lambda+\delta) = w_\pm(a,b|x,y)(\lambda)e^{\hbar(f_a-f_b)\delta} = w_\pm(a,b|x,y)e^{-\hbar(f_x-f_y)\delta}$$

$$(1.3)\qquad \sum_{b,y} w_+(a,b|x,y)(\lambda)w_-(y,z|b,c)(\lambda)\exp(\lambda\hbar(f_y-f_b)) = \delta(a,c)\delta(x,z)\exp(\lambda\hbar(f_a-f_x)),$$

$$(1.4)\qquad \sum_{b,y} w_+(a,b|y,x)(\lambda)w_-(z,y|b,c)(\lambda)\exp((\lambda-\pi)\hbar(f_b+f_y)) = \delta(a,c)\delta(x,z)\exp((\lambda-\pi)\hbar(f_a+f_z)),$$

$$(1.5)\qquad \sum_{b,y,s} w_+(a,b|x,y)(\lambda)w_+(b,c|r,s)(\lambda+\mu)w_+(y,z|s,t)(\mu) = \sum_{b,y,s} w_+(x,y|r,s)(\mu)w_+(a,b|s,t)(\lambda+\mu)w_+(b,c|y,z)(\lambda).$$

Comments on Definition 1.1. We let V be a vector space with basis Θ and define linear maps $R_\pm(\lambda)\in\operatorname{End}(V\otimes V)$, $\check{R}_\pm(\lambda)\in\operatorname{End}(V\otimes V)$ and $L(\delta)\in\operatorname{End}(V)$ by

$$(1.6)\qquad R_\pm(\lambda)(a\otimes x) = \sum_{b,y} w(a,b|x,y)b\otimes y,$$

$$(1.7)\qquad \check{R}_\pm(\lambda)(a\otimes x) = \sum_{b,y} w(a,b|x,y)y\otimes b,$$

$$(1.8)\qquad L(\delta)a = \exp(\delta\hbar f_a)a.$$

Also let $P\in\operatorname{End}(V\otimes V)$ be given by $P(a\otimes b) = b\otimes a$. Then note: $\check{R}_\pm(\lambda) = PR_\pm(\lambda)$.

Equation (1.2) shows that $w_\pm(a,b|x,y)(\lambda)$ may be extended to all $\lambda \in \mathbf{R}$ and that $w_\pm(a,b|x,y)(0)$ determines $w_\pm(a,b|x,y)(\lambda)$. Provided the image of exp contains no zero divisors, it follows that $w_\pm(a,b|x,y)(\lambda) = 0$ unless $f_a + f_x = f_b + f_y$. Thus our framework is no more general than Turaev's [**Tu**]. We may rewrite (1.2) as

$$(1.2') \qquad \begin{aligned} R_\pm(\lambda+\delta) &= (L(-\delta)\otimes 1)R_\pm(\lambda)(L(-\delta)\otimes 1)^{-1} \\ &= (1\otimes L(\delta))R_\pm(\lambda)(1\otimes L(\delta))^{-1} \end{aligned}$$

so that in particular $\check{R}_\pm(\lambda)$ commutes with $L(\delta)\otimes L(\delta)$ and $\check{R}_\pm(\lambda) = (L(\lambda)\otimes 1)\check{R}_\pm(0)(1\otimes L(-\lambda))$. We suppose that (1.2) holds for the rest of the discussion.

Equation (1.3) is the same as

$$(1.3') \qquad \check{R}_-(0)\check{R}_+(0) = \mathrm{id}.$$

Thus the values of $w_-(a,b|x,y)(\lambda)$ are determined by those of $w_+(a,b|x,y)(\lambda)$ and + and − can be interchanged in (1.2).

Equation (1.4) may be recast as follows. For $A \in \mathrm{End}(V\otimes V)$ let A^{t_1} and A^{t_2} be (transpose $\otimes$ id)(A) and (id $\otimes$ transpose) (A) respectively. Then (1.4) is the same as

$$(1.4') \qquad [P(R_+(\pi))^{t_1}][P(R_-(\pi))^{t_2}] = \mathrm{id}.$$

Equation (1.5) is the well-known Yang Baxter equation, to be found often in [**Ba**]. Using an obvious notation for the operations of elements of $\mathrm{End}(V\otimes V)$ on $V\otimes V\otimes V$ we can rewrite (1.5), with $R = R_+$, as

$$(1.5') \qquad R_{12}(\lambda)R_{23}(\lambda+\mu)R_{13}(\mu) = R_{13}(\mu)R_{23}(\lambda+\mu)R_{12}(\lambda).$$

Because we are supposing (1.2), this is equivalent to the braid relation with $\check{R} = \check{R}(0)$

$$(1.5'') \qquad \check{R}_{12}\check{R}_{23}\check{R}_{12} = \check{R}_{23}\check{R}_{12}\check{R}_{23}.$$

As we noted earlier, the conditions (1.2), (1.3), (1.4) are quite strong and the physically interesting solutions of (1.5) in [**Ba**] do not satisfy them.

Now suppose that we are given a generic planar projection of a smooth oriented link in $\mathbf{R}^3$, so that the only singularities are double points and at these the intersections are transversal. Then the link L

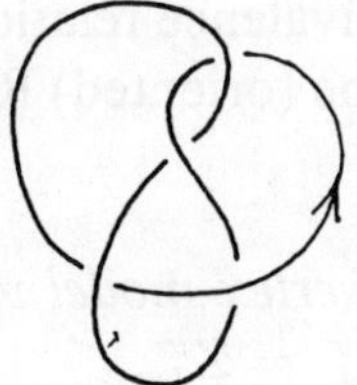

FIGURE 1.9
Diagram of an oriented link

may be adequately represented by the planar projection together with crossing data as shown in Figure 1.9.

DEFINITION 1.10. Given a vertex model $\nu = (\Theta, w_\pm, f, \hbar)$ and an oriented link diagram L we define the *partition function* as

$$Z_L^\nu = \sum_{\text{states } \sigma} \left(\prod_{\substack{\text{crossings} \\ \text{of } L}} w_\pm(a, b|x, y)(\lambda) \right) \exp\left(\hbar \int_L f_\sigma \, d\Theta\right)$$

(an empty product is equal to 1) where a state σ is a function from the edges of the planar graph subjacent to L to Θ. So a state defines a configuration around each crossing represented by Fig. 1.11 which also establishes the convention as to the order in which a, b, x, y appear as arguments of $w_\pm$

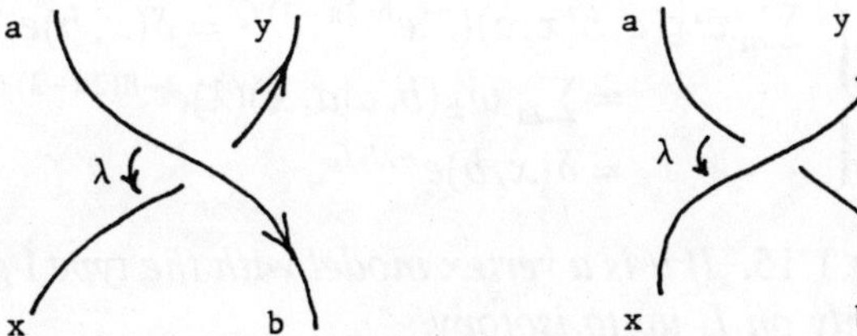

FIGURE 1.11

We take λ to be the "ingoing" angle measured in radians, in the sense of Fig. 1.11. Obviously $0 < \theta < \pi$.

Finally, f_σ is the locally constant function on L defined as being $f_{\sigma(e)}$ along any edge e, and $d\theta$ is the "change of angle" or curvature 1-form on L, defined as the pull-back to L of the angle form on the circle via the mapping $L \to S^1$ given by a unit tangent vector pointing in the direction of the orientation of L. This ends Definition 1.10.

Regular isotopy is the equivalence relation on link diagrams (generated by planar isotopy and the (oriented) Reidemeister moves of types II and III (see [**BZ**]).

THEOREM 1.12. *Given a vertex model* ν *and two regularly isotopic link diagrams* L_1 *and* L_2, $Z^\nu_{L_1} = Z^\nu_{L_2}$.

Proof. Equation (1.2) implies that each term in Z^ν_L is unchanged by changing the angles made at the crossings. It follows fairly easily that each term in Z^ν_L is invariant under planar isotopy. Once this is established, (1.3), (1.4) and (1.5) are clearly sufficient conditions for Z^ν_L to be invariant under all type II Reidemeister moves and a special configuration of type III move. Invariance under the other oriented type III moves follows by manipulating (1.2′) and (1.5″), and a simple topological argument in one case. We refer to [**Tu**] for details. □

One is most interested in isotopy of links in $\mathbf{R}^3$ which is expressed by invariance under all the Reidemeister moves, but, as emphasized by Kauffman, any regular isotopy invariant provided one pays attention to the Tait number Tait(L) (=number of crossings, counted according to their sign) and the rotation number rot(L). But all the examples we shall give behave in a particularly simple way under the type I Reidemeister moves which we now formalize.

DEFINITION 1.13. A vertex model $\nu = (\Theta, w, f, \hbar)$ *has the type* I *property* if

$$(1.14)\qquad \begin{cases} \sum_a w_\pm(a,b|x,a)(\lambda)e^{\hbar(2\pi-\lambda)f_a} = \delta(x,b)e^{-\lambda\hbar f_b} \\ \qquad = \sum_a w_\pm(b,a|a,x)(\lambda)e^{-\hbar(2\pi-\lambda)f_a} \\ \qquad = \delta(x,b)e^{+\lambda\hbar f_b}. \end{cases}$$

COROLLARY 1.15. *If* ν *is a vertex model with the type* I *property then* Z^ν_L *depends only on* L *up to isotopy.*

(Often a model will not have the type I property but can be changed into one with it simply by multiplying w_+ by a constant factor and w_- by its inverse.)

Let us write tr_1 and tr_2 for the partial traces from $\mathrm{End}(V\otimes V)\to \mathrm{End}\,V$ (e.g. if $A(a\otimes x) = \sum A(a,b|x,y)(b\otimes y)$, then $\mathrm{tr}_1(A)(x) = \sum_{y,a} A(a,a|x,y)y$).

Equation (1.14) is equivalent to

$$(1.14')\qquad \mathrm{tr}_1(\check{R}_\pm(2\pi)) = L(-2\pi), \quad \mathrm{tr}_2(\check{R}_\pm(2\pi)) = L(2\pi)$$

or

$$(1.14'') \quad \operatorname{tr}_1((L(2\pi)\otimes 1)\check{R}_\pm(0)) = \mathrm{id}, \quad \operatorname{tr}_2(\check{R}_\pm(0)(1\otimes L(-2\pi))) = \mathrm{id}.$$

We see that, in the notation of Turaev, $R_{ij}^{kl} = w(i,l|j,k)(0)$ and $\mu_i = e^{hf_i}$, $\hbar = h/2\pi$, when the set Θ is the set $\{1,2,\dots,n\}$.

Examples of vertex models.

EXAMPLE 1.16. We begin with the example reported in a letter to Kauffman. For $n = 1,2,\dots$, let P_n be the vertex model defined by $\Theta = \{-n,-n+2,\dots,n-2,n\}$, $f_a = a$ and

$$w_+(a,b|x,y)(\lambda,h) = \begin{cases} 0 & \text{if } \{a,x\}\neq\{y,b\} \text{ or } a<b, \\ \exp\left(\frac{h}{2}\right) & \text{if } a=b=x=y, \\ 1 & \text{if } a=b,\ x=y,\ a\neq x, \\ 2\sinh\left(\frac{h}{2}\right)\exp(\lambda\hbar(b-a)) & \text{if } a>b,\ a=y,\ b=x, \end{cases}$$

$$w_-(a,b|x,y)(\lambda,h) = w_+(x,y|a,b)(\lambda,-h).$$

Then (1.2)–(1.5) are satisfied. It is easy to show that if σ is given, the term $T = (\prod_\nu \exp(\hbar(b-a)))\exp(\hbar\int_L \sigma\, d\theta)$ may be calculated as follows: replace any crossing with $a>b$, $a=y$, $b=x$ by the "smoothed" picture as in Fig. 1.17 (similarly for negative crossings).

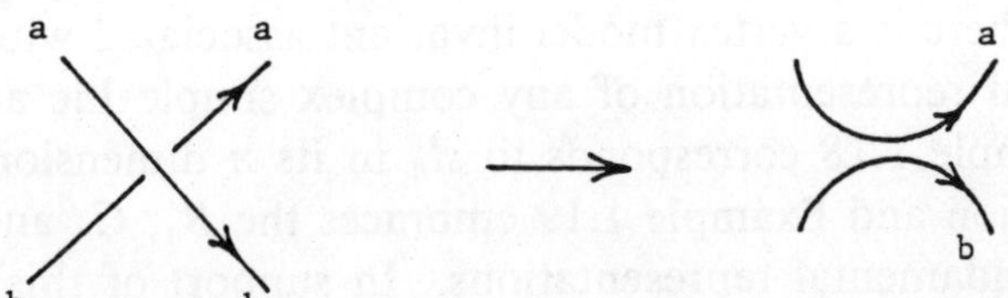

FIGURE 1.17

In the resulting diagram L', σ is constant along connected components and $T = \exp(h\sum_{a=-n}^{+n} a\,\mathrm{rot}(L'_a))$ where L'_a is the subdiagram of L' consisting of those connected components for which $\sigma = a$.

The vertex models P_n can be multiplied by a factor so that they have the type I property. Let $P_L(n,h)$ denote the partition function of this normalized model. One may show

(a) that $P_L(n,h)$ is a Laurent polynomial in $e^{h/2}$ divisible by $P_{\text{unknot}}(n,h)$

(b) that if L_+, L_- and L_0 are link diagrams identical except in the neighborhood of one crossing, where they are as in Fig. 1.18, then

$$\exp\left((n+1)\frac{h}{2}\right) P_{L_+}(n,h) - \exp\left(-\frac{(n+1)h}{2}\right) P_{L_-}(n,h)$$
$$= \sinh\left(\frac{h}{2}\right) P_{L_0}(n,h).$$

Thus as n varies, the $P_L(n,h)$ define a family of specializations of the two-variable invariant of [F+], sufficient to determine it (and thus prove its existence).

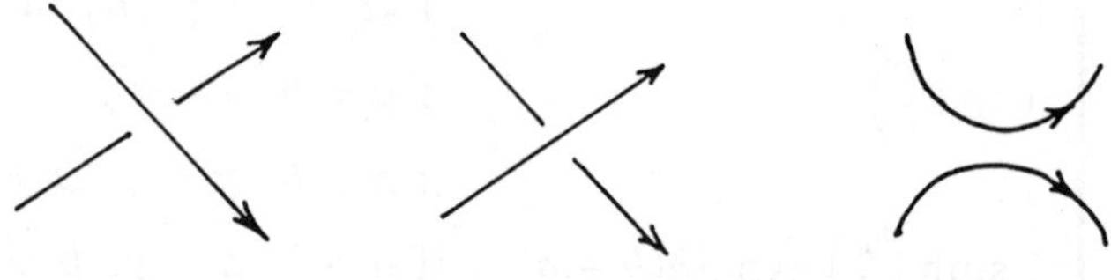

FIGURE 1.18

EXAMPLE 1.19. Turaev's examples in [Tu] defining the Kauffman polynomial satisfy our equations with the obvious choice of $w_\pm$, f, $q = e^{h/2}$.

EXAMPLE 1.20. The quantum group formalism of [Ji], [Dr] suggests that there is a vertex model invariant associated with any finite dimensional representation of any complex simple Lie algebra. Indeed, Example 1.18 corresponds to sl_n in its n dimensional identity representation and Example 1.19 embraces the B_n, C_n and D_n series in their fundamental representations. In support of this conjecture we give another example, corresponding to the N-dimensional irreducible representation of sl_2. The matrix $R(0)$ can be deduced from [Dr] and [Ji2]. These examples have apparently also been discovered using braids by Akutsu and Wadati [AW] and Wenzl [W2] although it is difficult to be absolutely sure, as only the first three cases are given in [AW] and only an existence result occurs in [W2] (which also gives many other invariants in support of the conjecture). Our models are as follows. For each $N = 2, 3, 4, \ldots$, let X_{2n} be the vertex model with

$$\Theta = \left\{-\frac{N-1}{2}, -\frac{N-1}{2}+1, \ldots, \frac{N-3}{2}, \frac{N-1}{2}\right\},$$

$f_a = a$ and $w(a, b|x, y)(\lambda, h)$ defined by

$$w_+(a, b|x, y)(\lambda, h) = \begin{cases} \exp\left\{\frac{h}{2}\left[ax + by + \frac{k(k-1)}{2} + \frac{\lambda}{\pi}(a-b)\right]\right\}\Gamma^N_{x,k}(h) \\ \quad \text{where } k = a - b \text{ if } a + x = b + y \text{ and } a \geq b, \\ 0 \quad \text{otherwise} \end{cases}$$

and $w_-(a, b|x, y)(\lambda, h) = w_+(x, y|a, b)(\lambda, -h)$, where

$$\Gamma^N_{x,k}(h) = 2^k \prod_{j=0}^{k-1} \frac{\sinh\left(\frac{h}{2}\left(x + \frac{N+1}{2} + j\right)\right)\sinh\left(\frac{h}{2}\left(x - \frac{N-1}{2} + j\right)\right)}{\sinh\left(-\frac{h}{2}(1+j)\right)}$$

(empty products are one).

I would like to thank M. Rosso for some help with w_-. One sees immediately that (20) and (22) of **[Tu]** are satisfied, with $\mu_a = \exp(ha)$, $\nu_a = -\exp(ha/2)$ and $a' = -a$, so that (17) of **[Tu]**, which is a version of our (1.4), is satisfied. This also proves that the invariants of this example are, up to powers of e^h, invariants of unoriented links. Equation (1.2) is obvious and equation (1.3) follows from direction computation. Equation (1.5) follows from **[Dr]** and **[Ji2]**.

Using the notation of **[An]**, the identity $\sum_{\alpha=0}^{\beta}(q^{-\beta})_\alpha(z)_\alpha q^\alpha/(q)_\alpha = z^\beta$ is easily proven by induction (or follows from the proof of (3.3.12) of **[An]** as pointed out to the author by Andrews). It shows that $w_\pm$ may be multiplied by powers of e^h so that the normalized vertex model has the type I property. The resulting link invariant is a Laurent polynomial in $e^{h/2}$. The case $N = 2$ is the same as the case $n = 1$ of Example 1.18, namely, the polynomial V_L. W. Baxter has written a program for calculating these invariants using braids. The method is effective for $N \leq 5$ for braids on three strings and for $N \leq 4$ for braids on five strings.

EXAMPLE 1.21. This example was discovered by A. Lipson **[Li]**. It has the feature that the Boltzmann weights are λ-independent. We show how to deduce these weights from invariant theory, as pointed out by R. Howe.

If E is n-dimensional Euclidean space with inner product $\langle\, , \rangle$, there is a privileged element e of $\mathrm{End}(E \otimes E)$ which is orthogonal projection onto the one-dimensional subspace spanned by $\sum v_i \otimes v_i$, where $\{v_i\}$ is an orthonormal basis. It follows immediately that $e_1e_2e_1 = (1/n^2)e_1$, $e_2e_1e_2 = (1/n^2)e_2$ in $\mathrm{End}(E \otimes E \otimes E)$ where $e_1 = e_{12}$, $e_2 = e_{23}$ in the notation of (1.5′). By the calculations of **[Jo4]**, if we put $\check{R}_{\pm}(\lambda) = (t^{\pm 1}+1)e - 1$ where $4\cosh^2(h/2) = n^2$, $t = e^h$, then the coefficients of the matrix $\check{R}$ give Boltzmann weights satisfying (1.3) and (1.5). Explicitly we define the model with $\Theta = \{1, 2, \dots, n\}$, $f_a = 0$ and

$$w_+(a, b|x, y)(\lambda, h) = \begin{cases} 0 & \text{if } a \notin \{x, y\} \text{ or } x \notin \{a, b\}, \\ \exp(-h/4) & \text{if } a = y,\ x = b,\ a \neq b, \\ \exp(h/4) & \text{if } a = x,\ y = b,\ a \neq b, \\ 2\cosh(h/4) & \text{if } a = x = y = b, \end{cases}$$

$$w_-(a, b|x, y)(\lambda, h) = w_+(a, b|x, y)(\lambda, -h).$$

By inspection the assignment of the Boltzmann weights at a crossing is independent of the directions of the arrows. Thus (1.4) is the same as (1.3). The type I property is also satisfied after normalization. This example obviously calculates $V_L(e^h)$.

EXAMPLE 1.22. Another unoriented example with no angle dependence due to Lipson is defined in **[Li]**, **[LM]**. It is trivial on knots and determined in general by linking numbers of sublinks with their complements.

We see that even vertex models with no angle dependence can give highly non-trivial information. It would be nice to know if such models are also sufficient to determine the invariants of Examples 1.18, 1.19 and 1.20.

We add the final comment that (1.2)–(1.5) do not mix $\{a, b\}$ and $\{x, y\}$ so one should be able to define invariants that treat the different components of a link differently.

2. Spin models, the Potts model, IRF models. We use the terminology "spin model" to denote a statistical mechanical model, such as the Potts model, defined on a graph with vertices $\{v\}$ and edges $\{e\}$. The vertices may be in a (finite) set Θ of "spins" and a state is a function $\sigma\colon \{v\} \to \Theta$. The model is defined by an energy function $E\colon \Theta \times \Theta \to \mathbf{R}$ so that the energy of an edge joining vertices v_1 and v_2 is $E(\sigma(v_1), \sigma(v_2)) = E(a, b)$. As before write $w(a, b) = \exp(-\beta E(a, b))$. The partition function is then

$$\sum_\sigma \exp(-\beta \sum_e E(a, b)) = \sum_\sigma \prod_e w(a, b).$$

Although it is possible to develop a more precise theory for oriented links, all our examples will be unoriented, so we now outline a theory, analogous to that of §1, which allows us to construct invariants of unoriented links as partition functions of spin models.

DEFINITION 2.1. A *spin model* $S = \{\Theta, w_\pm\}$ will be a set of n "spins" Θ and functions $w_\pm(a,b)$, $a, b \in \Theta$ such that (for all $a, b, c \in \Theta$)

$$w_\pm(a,b) = w_\pm(b,a), \tag{2.2}$$

$$w_+(a,b)w_-(a,b) = 1, \tag{2.3}$$

$$\sum_{x\in\Theta} w_-(a,x)w_+(x,c) = n\delta(a,c), \tag{2.4}$$

$$\sum_{x\in\Theta} w_+(a,x)w_+(b,x)w_-(c,x) = \sqrt{n}w_+(a,b)w_-(b,c)w_-(c,a). \tag{2.5}$$

(Equation (2.5) is known as the "star-triangle relation" and occurs essentially in **[On]**.)

DEFINITION 2.6. Let $S = (\Theta, w_\pm)$ be a spin model and L be an *unoriented* link diagram. We define the partition function Z_L^S as follows.

First we shade the regions of L black and white so that adjacent regions have different colours and the unbounded region is white as in Figure 2.7:

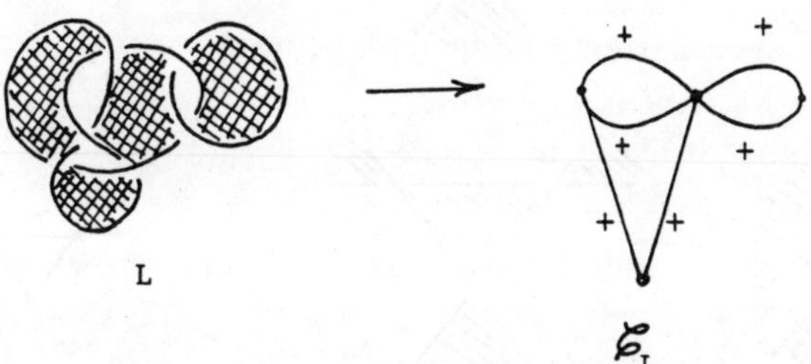

FIGURE 2.7

We then form the planar graph $\mathscr{C}_L$, with signed edges, with V vertices which are the black regions of L and whose edges are the crossings of L. One may distinguish two types of crossings of L, equivalently, edges of $\mathscr{C}_L$ according to a convention established by Fig. 2.9.

A state σ will be a function from the vertices of $\mathscr{H}_L$ to Θ, and given a state σ and an edge e between vertices v_1 and v_2 with sign + or −, we will write $w_\pm(a,b)$ for $w_\pm(\sigma(v_1), \sigma(v_2))$. Then we set

$$Z_L^S = \left(\frac{1}{\sqrt{n}}\right)^{V-1} \sum_{\text{states}} \prod_{\text{edges}} w_\pm(a,b),$$

where as usual an empty product is one. This ends Definition 2.6.

THEOREM 2.8. *If S is a spin model and L_1 and L_2 are unoriented connected link diagrams then $Z^S_{L_1} = Z^S_{L_2}$ if L_1 and L_2 are regularly isotopic.*

Proof. Clearly Z^S does not change under planar isotopy. Invariance under type II Reidemeister moves is guaranteed by (2.3) and (2.4) corresponding to the type II moves

↔ and ↔ .

Note that in the second case the number of shaded regions changes which necessitates a normalization of the form $(1/\sqrt{n})^V$. The number of shaded regions may be assumed to change by two in this move since the L_1 and L_2 are connected.

There are two kinds of type III moves to be considered as shown in Fig. 2.9.

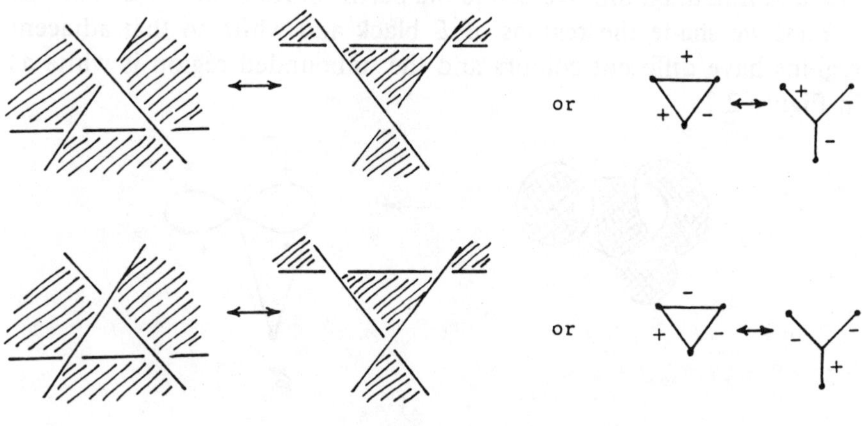

FIGURE 2.9

Invariance under the first kind is assured by (2.5). Invariance under the second kind is given by the following:

$$(2.10) \qquad \sum_{x\in\Theta} w_+(a,x)w_-(b,x)w_-(c,x)$$
$$= \sqrt{n}w_+(a,b)w_-(b,c)w_+(c,a).$$

Now consider an n-dimensional vector space V with basis Θ and on $V\otimes V$ define the operators X and Δ by $X(a\otimes b) = \sum_x w_+(a,b)x\otimes b$ and $\Delta(a\otimes b) = w_-(a,b)a\otimes b$. Then (2.5) is the same as

$$(2.5') \qquad X\Delta X = \sqrt{n}\Delta X\Delta$$

and by (2.2)–(2.4), (2.10) is the same as

(2.10′) $$X\Delta X^{-1} = \frac{1}{\sqrt{n}}\Delta^{-1}X\Delta.$$ □

Perhaps surprisingly, all spin models have the type I property of (1.14) in the following sense. (Note that type I Reidemeister moves are positive or negative according to the following convention:

$+1 \to$ [positive kink diagram] , $-1 \to$ [negative kink diagram] .)

PROPOSITION 2.11. *Let (S, Θ) be a spin model. Then there is a constant A, called the modulus of S such that if L′ is obtained from L by adding a loop via a ±1 type* I *Reidemeister move then* $Z_{L'} = A^{\mp 1} Z_L$.

Proof. Putting $b = c$ in (2.5) and using (2.3) we have $\sum_{x\in\Theta} w_+(a, x) = \sqrt{n}w_-(b, b)$ for all a, b. If we set

$$A = w_-(b, b) = (1/\sqrt{n}) \sum_{x\in\Theta} w_+(a, x)$$

then a careful analysis of the shading possibilities proves the result. □

Thus if $\vec{L}$ is an *oriented* link with associated unoriented link L, and S is a spin system of modulus A, we may define the quantity

(2.12) $$Z^S_{\vec{L}} = A^{\mathrm{Tait}(\vec{L})} Z^S_L.$$

PROPOSITION 2.13. $Z^S_{\vec{L}}$ *is an invariant of oriented links.*

Proof. Just note that both possible orientations give the sign +1 to the crossing involved in a positive type I Reidemeister move. □

Note that the reasoning of Theorem 2.8 shows that we would have obtained a link invariant $\check{Z}^S_L$ for the spin model S by agreeing to colour the unbounded region black. The next result shows that nothing new happens. Note that the result is not entirely trivial as the two graphs corresponding to the different colourings may look very different.

PROPOSITION 2.14 (*Duality*). *If S is a spin model and L is an oriented link projection, then* $Z_L^S = \check{Z}_L^S$.

Proof. The sequence of Reidemeister moves of Fig. 2.15 converts L into L' such that the white-unbounded colouring of L' is the same as the black-unbounded colouring of L. □

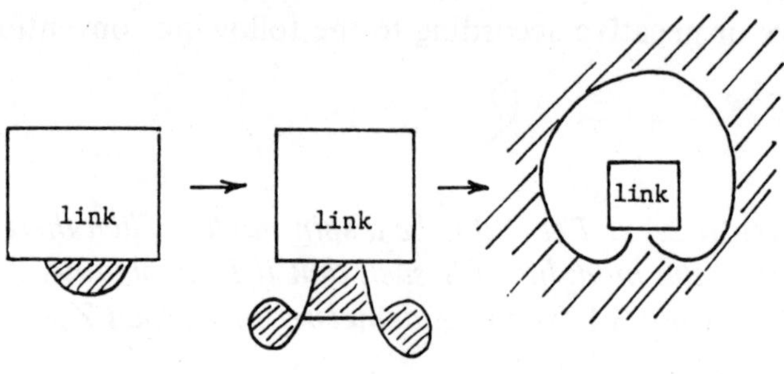

FIGURE 2.15

Note that equations (2.2)–(2.5) impose many restrictions on $w_\pm(a, b)$. The following is quite useful.

PROPOSITION 2.16. *For each z let k_z be the number of ordered pairs (a, b) for which $w_+(a, b) = z$. Then k_z is a multiple of n (and* $\sum_z k_z = n^2$).

Proof. By (2.13′) the matrices Δ and $1/\sqrt{n}X$ are conjugate, but Δ is diagonal with eigenvalues $w_-(a, b)$ and X is of the form $\pi \otimes \mathrm{id}$ on $V \otimes V$ so that the multiplicities of all of its eigenvalues are multiples of n. □

EXAMPLES. With Proposition 2.16 it is easy to find all solutions (2.2)–(2.5) when $n = 2$ or 3. We leave it to the reader to check that one obtains the invariants $V_L(i)$, $V_L(e^{i\pi/3})$ and nothing more. When $n = 4$ one also obtains the invariant of (1.22) and some others of no great significance. We shall be especially interested in the following two examples for both of which the set Θ can be given the structure of a group so that $w_+(a, b)$ depends only on ab^{-1}.

EXAMPLE 2.17 (The Potts model). In statistical mechanics the Potts model is the spin model for which the Boltzmann weights depend only on whether the two atoms are in the same state or not. Correct choice of the parameters leads, for each n, to the following choice of $w_\pm$ (by

(2.3) it suffices to give w_+):

$$w_+(a,b) = \begin{cases} 1 & \text{if } a = b, \\ -t^{-1} & \text{otherwise,} \end{cases}$$

where $2+t+t^{-1} = n$. It is easy to check that these Boltzmann weights satisfy (2.2)–(2.6). One may also check directly that the invariant of unoriented links is an unoriented version of $V_L(t)$.

EXAMPLE 2.18 (Metaplectic invariants) **[GJ]**. If the set Θ is a group G, we may impose the condition that $w_+(a,b)$ depend only on ab^{-1}, say $w_+(a,b) = f(ab^{-1})$. Then (2.2), (2.4) and (2.5) become

$$(2.19) \quad \begin{cases} \text{(a)} & f(g) = f(g^{-1}), \\ \text{(b)} & \frac{1}{|G|}\sum_{h\in G} f(h)[f(h^{-1}g)]^{-1} = \delta_{g,1}, \\ \text{(c)} & f(xy)f(x)^{-1}f(y)^{-1} \\ & \quad = \frac{1}{\sqrt{|G|}}\sum_{g\in G} f(y^{-1}g)f(xg)f(g)^{-1}. \end{cases}$$

If $G = \mathbf{Z}/n\mathbf{Z}$ we see that the Boltzmann weights of Example 2.17 satisfy (2.19). Another example, with $G = \mathbf{Z}/n\mathbf{Z}$, n odd, which coincides with Example 2.17 when $n = 3$ but differs for $n > 3$, was discovered in **[GJ]**. The function $f(a)$ is a Gaussian: $f(a) = K\, e^{\pi i a^2/n}$ where $K^{-2} = \sum_{a\in\mathbf{Z}/n\mathbf{Z}} e^{2\pi i a^2/n}$. The knot invariant was shown in **[GJ]** to be determined by a Seifert matrix and to be essentially given by the rank of the homology of the twofold branched cover of S^3, branched over the knot, with mod n coefficients.

A lot of "classical" knot invariants were obtained in **[GJ]**. It seems likely that they may all be accessible to the partition function method using slightly more elaborate spin models. Thus the formalism is able to embrace both the "classical" (Seifert matrix) and "new" (V_L, P_L, Kauffman polynomial) invariants as well.

I would like to thank D. Goldschmidt and T. Ziman for many stimulating discussions about spin models.

The spin models present many approachable problems. In general they seem more susceptible to brute force calculations than other models because the whole data is just an $n \times n$ matrix. It would be quite feasible to determine all models for $n = 5, 6$. It would also be nice to discover a machine, analogous to the quantum group machine, for creating solutions to (2.2)–(2.5).

IRF Models. IRF (interaction round a face) models are similar to spin models in that a state is defined by a function σ: vertices of a lattice $\to \Theta$. They differ in that the energy of a state is the sum of the energies of each face, the energy of a given face being determined by the configuration of states on the vertices of its boundary. Thus if the lattice is a square lattice in two dimensions one is given a function $w(a,b,c,d)$, $a,b,c,d \in \Theta$, and the partition function for a finite rectangular sublattice will be $\sum_{\text{states}} \prod_{\text{faces}} w(a,b,c,d)$ where a, b, c, d are taken in some fixed order around each face. IRF models as such do not generalize conveniently to an arbitrary graph, but if L is an oriented link projection, its underlying graph is 4-valent so all faces of the planar dual are quadrilaterals and we may proceed in a manner analogous to spin models.

DEFINITION 2.19. An IRF model $I = \{\Theta, w\}$ will be a set on n "spins" Θ and functions $w_\pm(a,b,c,d)$ with $a,b,c,d \in \Theta$ such that (for all $a,b,c,d,e,f \in \Theta$)

$$\sum_{x\in\Theta} w_\pm(a,b,a,x) = 1, \tag{2.20}$$

$$\sum_x w_+(a,b,x,d)w_-(x,b,e,d) = \delta_{a,e}, \tag{2.21}$$

$$\sum_x w_+(d,a,b,x)w_-(b,e,d,x) = \delta_{a,e}, \tag{2.22}$$

$$\sum_x w_+(a,b,x,f)w_+(b,c,d,x)w_+(x,d,e,f) = \sum_x w_+(b,c,x,a)w_+(a,x,e,f)w_+(x,c,d,e). \tag{2.23}$$

Although we shall not construct any IRF models not derived from spin models, we define the link invariants associated with them.

DEFINITION 2.24. Let $I = \{\Theta, w_\pm\}$ be an IRF model and L an oriented link projection. We define the *partition function* as

$$Z_L^I = \sum_{\text{states}} \left(\prod_{\substack{\text{faces}\\ \text{of } \mathscr{H}}} w_\pm(a,b,c,d) \right)$$

where $\mathscr{H}$ is the planar dual of the graph subjacent to L, with signed faces which are + or − according to the sign of the unique crossing of L which they contain. A state σ is a function from the vertices of $\mathscr{H}$ to Θ. For a given face in the product we choose $w_\pm(a,b,c,d)$ according to the sign of the face and the configuration of spins at its four corners, ordered according to Fig. 2.25.

This ends Definition 2.24.

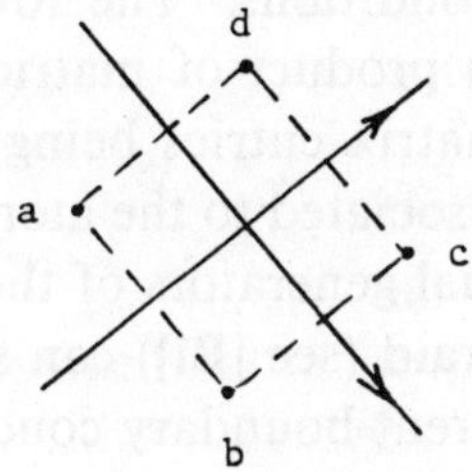

FIGURE 2.25

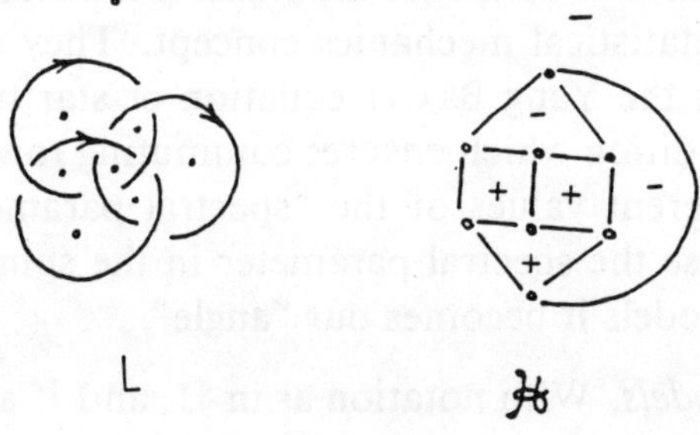

FIGURE 2.26

An example of the passage from L to $\mathscr{H}$ is given in Fig. 2.26. The proof of the next result is trivial.

THEOREM 2.27. *If I is an IRF model and L_1 and L_2 isotopic link models then* $Z^I_{L_1} = Z^I_{L_2}$.

IRF models have also been considered from the braid point of view by Akutsu and Wadati [AW2]. Note that we could have dropped (2.20) if we had only been interested in regular isotopy.

3. Braids and transfer matrices. The role of braid group representations in knot theory is closely analogous to that of transfer matrices in statistical mechanics. Any oriented link can be represented as a closed braid, which can thus be thought of as a particularly ordered way to represent a link. To each of the partition function invariants presented above there is a braid group representation such that the trace of the braid is equal to the partition function. (In the case of vertex models with non-trivial angle dependence the trace must be suitably weighted.) In statistical mechanics the system is typically on a lattice, so is already well ordered. The technique of transfer matrices associates to each row of the lattice a (large) matrix, the trace of some power of which is the partition function for a (finite) square lattice with periodic boundary conditions. The row-to-row transfer matrix can often be written as a product of matrices corresponding to the atoms in that row, the matrix entries being the Boltzmann weights. These transfer matrices associated to the atoms correspond to the matrices representing the usual generators of the braid group.

The plat closure of a braid (see [**Bi**]) can sometimes be used. This corresponds to using different boundary conditions (non-periodic) on the square lattice (see [**Jo3**]).

We now outline the braid group representations in the three cases. It will be apparent how to construct the transfer matrices corresponding to atoms in the statistical mechanics concept. They can be found in [**Ba**]. In all cases the Yang Baxter equation or star triangle equation is a sufficient condition which ensures commuting row-to-row transfer matrices for different values of the "spectral parameter". We have been unable to use the spectral parameter in the spin and IRF cases. For the vertex models it becomes our "angle".

3.1. *Vertex models.* With notation as in §1, and V as usual, a vector space with basis Θ, we represent the braid group B_k on $\bigotimes^k V$ using the $\check{R}(0)$ matrix which acts on $V \otimes V$ by

$$\check{R}(0)(a \otimes x) = \sum_{b,y} w_+(a, b|x, y)\, y \otimes b.$$

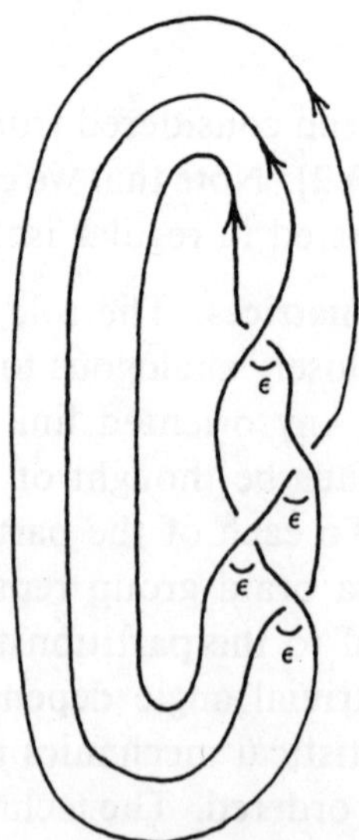

FIGURE 3.2

We define $\check{R}_i$ on $\bigotimes^k V$ by

$$\check{R}_i((v_1 \otimes \cdots \otimes v_{i-1}) \otimes v_i \otimes v_{i+1} \otimes (v_{i+1} \otimes \cdots \otimes v_k))$$
$$= (v_1 \otimes \cdots \otimes v_{i-1}) \otimes \check{R}(0)(v_i \otimes v_{i+1}) \otimes (v_{i+1} \otimes \cdots \otimes v_k).$$

Then $\check{R}_i^{-1}$ is obtained by applying the same prescription to $\check{R}^{-1}(0)$. Thus we get a representation of B_k by sending σ_i to $\check{R}_i$. We define the linear functional ϕ on $\mathrm{End}(\bigotimes^k V)$ by $\phi(A) = \mathrm{trace}(DA)$ where $D = L(2\pi) \otimes L(2\pi) \otimes \cdots \otimes L(2\pi)$. If α is a braid in B_k with closure $\hat{\alpha}$ then we may arrange the braid picture of $\hat{\alpha}$ to look as in Fig. 3.2.

The angle ε may be made arbitrarily small so that the only angle contribution to the partition function is "around the back" of the braid. We see that in the limit, $\varepsilon \to 0$, we have $\phi(\alpha) = Z_{\hat{\alpha}}$ (see 5.2 of [**Tu**]). This is in fact how the partition function approach to P_L was discovered, in an attempt to reverse the procedure just outlined.

3.3. *Spin models.* Given the Boltzmann weights $w_\pm(a, b)$ and a vector space V with basis Θ we consider three ways to form elements of $\mathrm{End}(V \otimes V)$, already used in Theorem 2.11. Let

$$\Delta(a \otimes b) = w_-(a, b) a \otimes b,$$
$$X_1(a \otimes b) = \frac{1}{\sqrt{n}} \sum_x w_+(a, x) v_x \otimes v_b \quad \text{and}$$
$$X_2(a \otimes b) = \frac{1}{\sqrt{n}} \sum_x w_+(b, x) v_a \otimes v_x.$$

Obviously, X_1 and X_2 commute and (2.5) implies $\Delta X_1 \Delta = X_1 \Delta X_1$ and $\Delta X_2 \Delta = X_2 \Delta X_2$. Moreover Δ^{-1}, X_1^{-1} and X_2^{-1} exist by (2.3) and (2.4). Thus we may define a representation of B_{2k} on $\bigotimes_{i=1}^k V$ by sending σ_{2i-1} to Δ_i, which is Δ on the ith tensor factor, tensored with the identity on the others, and σ_{2i} to X_i where

$$X_i(u_1 \otimes u_2 \otimes u_3) = u_1 \otimes X(u_2) \otimes u_3 \quad \text{where}$$
$$u_1 \in \bigotimes_{j=1}^{i-1} V, \quad u_2 \in V, \quad u_3 \in \bigotimes_{j=k+1}^{k} V \quad \text{and} \quad X_1 = X \otimes \mathrm{id}.$$

We leave it to the reader to contemplate Fig. 3.4 and see that, for a braid $\alpha \in B_{2k}$, the trace of the matrix representing α on $\otimes^k V$ is the same (up to normalization) as the partition function invariant of $\hat{\alpha}$. Notice how the matrix Δ, using w_-, corresponds to horizontal bonds, whereas X, corresponding to vertical ones, uses w_+. Note also that duality occurs in the braid picture as invariance of the trace under a shift "half a step" to the right.

FIGURE 3.4

3.5. *IRF models.* Given the Boltzmann weights $w_{\pm}(a, b, c, d)$ of an IRF model and V with basis Θ, we define an operator R on $V \otimes V \otimes V$ by

$$R(b \otimes a \otimes d) = \sum_c w_+(a, b, c, d) b \otimes c \otimes d.$$

Then by (2.21) R^{-1} exists and is given by the same formula with w_+ replaced by w_-. If we define R_1 and R_2 on $V \otimes V \otimes V \otimes V$ by $R_1 = R \otimes \mathrm{id}$ and R_2 by $\mathrm{id} \otimes R$ then (2.23) is the same as $R_1 R_2 R_1 = R_2 R_1 R_2$. So we may define representations of B_k on $\otimes^{k+1} V$ in what should by now be an obvious manner, shifting R to the right by one for each succeeding generator. We leave it to the reader to contemplate Fig. 3.4, without the shading and with $\times$'s in all the regions, to establish that the trace of a braid α in this representation is the same as the partition function invariant of $\hat{\alpha}$.

We end §3 by pointing out a puzzling feature common to all the models, related to Markov's theorem [**Ma**]. It is that, from the braid point of view, all that is required for a link invariant is invariance under the Markov moves (see [**J1**] or [**Tu**]). This is guaranteed by the equations for the model expressing invariance under the types I and III Reidemeister moves. The existence of an inverse matrix is also required, invoking one kind of type II equation ((1.3) or (2.21)) but *not* the other. Thus one may find R matrices defining invariants via the braid representation picture, but not defining models as above. Indeed here is an example. Let X and Y be defined on the vector space V with basis $v_1, v_2, \ldots, v_n$ by $Xv_k = \omega^k v_k$, $Yv_k = v_{(k+1) \bmod n}$ ($\omega = e^{2\pi i/n}$). Then defining $\check{R} \in \mathrm{End}(V \otimes V)$ by $\check{R} = \sum_{k \in \mathbf{Z}/n\mathbf{Z}} \omega^{k^2} X^k \otimes Y^k$

it is easy to check that all the conditions for a vertex model *except* (1.4) are satisfied (with $f_a = 0$). In fact, the trace of the braid group representation calculates the invariants of **[GJ]** outlined in (2.18).

On the other hand, if we adopt the regular isotopy point of view we may forget about type I Reidemeister moves and one could imagine models for which the braid/trace picture did not behave in any simple way under the type II Markov moves. The situation on these points seems unclear at this juncture.

4. Equivalences. The author would like to thank H. Au-Yang and J. Perk for pointing out (in statistical mechanics) the existence for every spin model of an equivalent IRF model, and for every IRF model of an equivalent vertex model. These equivalences persist in the knot theory context but are more subtle because of positive and negative crossings. It is also true that, in statistical mechanics, to every vertex model on a square lattice there is an equivalent IRF model (see **[PW]**) but we have been unable to make such an equivalence in the knot theory version. For these reasons we state our equivalences as formal results.

PROPOSITION 4.1. *Let $S = \{\Theta, w_\pm\}$ be a proper spin model. Then there is an IRF model $I = \{\Theta, \tilde{w}_\pm\}$ such that $Z_L^I = (Z_L^S)^2$ for all oriented connected link diagrams L, where we forget the orientation of L for Z_L^S.*

Proof. Let Θ have n elements. Set

$$\tilde{w}_\pm(a, b, c, d) = (1/\sqrt{n}) w_\pm(a, c) w_\mp(b, d).$$

It is easy to check (2.20) $\to$ (2.23). Note that for a given link diagram L, the set of vertices $\mathscr{C}$ of the graph $\mathscr{H}$ is the union of the set of vertices for $\mathscr{C}$ and its dual $\check{\mathscr{C}}$. The faces of $\mathscr{H}$ are quadrilaterals whose diagonals are the edges of $\mathscr{C}$ and $\check{\mathscr{C}}$. So to each state of $\mathscr{H}$ there are unique states of $\mathscr{C}$ and $\check{\mathscr{C}}$. Thus by (2.14) we have

$$\begin{aligned}(Z_L^S)^2 &= \check{Z}_L^S Z_L^S \\ &= (\sqrt{n})^{\#(\text{faces of } L)+2} \sum_{\substack{\text{states} \\ \text{of } \mathscr{H}}} \left(\prod_{\substack{\text{edges} \\ \text{of } \mathscr{C}}} w_\varepsilon(a, c) \right) \left(\prod_{\substack{\text{edges} \\ \text{of } \check{\mathscr{C}}}} w_{\varepsilon'}(b, d) \right)\end{aligned}$$

where ε, ε' are $+$ and $-$. Pairing up the products according to the faces of $\mathscr{H}$, the proof is completed by observing that the choice of ε

FIGURE 4.2

according to Fig. 2.10 corresponds to our definition of $\tilde{w}_{\pm}(a,b,c,d)$, regardless of the orientation at a crossing. See Figures 4.2 and 2.25.

Note also that since L is connected, #(faces of L)=#(faces of $\mathscr{K}$)+2.

PROPOSITION 4.3. *Let* $I = \{\Theta, w_{\pm}\}$ *be an IRF model. Then there exists a vertex model* $v = \{\Theta \times \Theta, \tilde{w}_{\pm}, f\}$ *(with* $f \equiv 0$*) for which* $nZ_L^I = Z_L^v$ *for any oriented link. Moreover* v *has the type* I *property.*

Proof. Define

$$\tilde{w}_{\pm}((d,a),(b,c)|(a',b')(c',d')) = \begin{cases} w_{\pm}(a,b,c,d) & \text{if } (a,b,c,d) = (a',b',c',d'), \\ 0 & \text{otherwise.} \end{cases}$$

It is easily checked that this defines a vertex model and

$$\sum_{\substack{\text{states}\\ \text{of } v}} \prod_{\substack{\text{crossings}\\ \text{of } L}} \tilde{w}_{\pm}(\bar{a},\bar{b}|\bar{x},\bar{y}) = \sum_{\substack{\text{states}\\ \text{of } I}} \prod_{\substack{\text{faces}\\ \text{of } \mathscr{K}}} w_{\pm}(a,b,c,d) = Z_L^I. \qquad \square$$

Thus up to a choice of square root, vertex models are more general than spin models and IRF models. We would add, however, that this should not be taken as a reason to ignore the spin and IRF models as the above equivalences are a little trivial and the natural context for the spin model is not the vertex model. For instance, the number of sites per edge is the square of the number of spins per site, which immediately makes the vertex model approach rather cumbersome.

For particular models there may be non-trivial equivalences between spin and vertex models. For instance, the vertex model defining V_L ($n = 1$ in Example 1.16) is equivalent to the collection of all Potts models (Example 2.17) by what is known to physicists as Temperley-Lieb equivalence. This was discovered in [TL] using the transfer matrix picture but a direct proof for arbitrary planar graphs is given in [Ba]. The idea is to represent the two states per edge of the vertex model by arrows and given a state of the model, to eliminate all crossings according to the arrow configuration at the crossing. A clever counting argument shows that one obtains the partition function for the Potts model. The idea of eliminating the crossing according to the configuration around a vertex gives precisely the states model of Kauffman for V_L (see **[K1]**, **[K2]**). Thus the Kauffman states model is an intermediate step in the proof of Temperley-Lieb equivalence.

REFERENCES

[An] G. Andrews, *The Theory of Partitions*, Addison-Wesley, 1976.

[AW1] Y. Akutsu and M. Wadati, *Exactly solvable models and new link polynomials, I: N-state vertex models*, preprint 1987.

[Ba] R. Baxter, *Exactly Solved Models in Statistical Mechanics*, Academic Press, London, 1982.

[Bi] J. Birman, *Braids, links and mapping class groups*, Ann. Math. Studies, **82** (1974).

[BZ] G. Burde and H. Zieschang, *Knots*, DeGruyter, 1985.

[Dr] V. Drinfeld, *Quantum groups* in *Proceedings of the International Congress of Mathematicians*, Berkeley 1986, vol. 1, 798–820.

[F+] P. Freyd, D. Yetter, J. Hoste, W. Lickorish, K. Millet, and A. Ocneanu, *A new polynomial invariant of knots and links*, Bull. Amer. Math. Soc., **12** (1985), 183–312.

[GJ] D. Goldschmidt and V. Jones, *Metaplectic link invariants*, Univ. of California, Berkeley preprint (1987).

[Go] D. Goldschmidt, *Classical link invariants via the Burau representation*, Univ. of California, Berkeley preprint (1987).

[Hi] T. Hill, *Statistical Mechanics*, Dover Publications, Inc., 1956.

[Ji1] M. Jimbo, *A q-analogue of $U(\mathrm{gl}(n+1))$, Hecke algebra, and the Yang-Baxter equation*, Letters in Math. Phys., (1986), 247–252.

[Ji2] ———, *A q-difference analogue of $U(\mathscr{H})$ and the Yang-Baxter equation*, Letters in Math. Phys., **10** (1985), 63–69.

[Jo1] V. Jones, *A polynomial invariant for knots via von Neumann algebras*, Bull. Amer. Math. Soc., **12** (1985), 103–111.

[Jo2] ———, *Index for subfactors*, Invent. Math., **72** (1983), 1–25.

[Jo3] ———, *Hecke algebra representations of braid groups and link polynomials*, Ann. Math., **126** (1987), 335–388.

[Jo4] ———, *Braid groups, Hecke algebras and type II_1 factors*, from *Geometric Methods in Operator Algebras*, Longman Sci. & Tech., 1986, 242–273.

[K1] L. Kauffman, *An invariant of regular isotopy*, preprint.

[K2] ———, *State models and the Jones polynomial*, Topology, **26** (1987), 395–407.

[LM] W. B. R. Lickorish and K. Millet, *An evaluation of the F polynomial of a link*, preprint.

[Li] A. S. Lipson, *Some more states models for link invariants*, to appear in Pacific J. Math.

[Ma] A. A. Markov, *Uber die freie Aquivalenz geschlossener Zopfe*, Recevil Mathematique Moscou, **1** (1935), 73–78.

[On] L. Onsager, *Crystal statistics* I. *A two dimensional model with an order-disorder transition*, Phys. Rev., **65** (1944), 117–149.
[P1] Plato, *The Republic* (transl. B. Jowett), Prometheus Books, 367 B.C.
[PW] J. Perk and F. Wu, *Graphical approach to the non-intersecting string model*, Physica, **138A** (1986), 100–124.
[TL] H. N. V. Temperley and E. Lieb, *Relations between the "percolation" and "colouring" problem and other graph-theoretical problems associated with regular planar lattices: some exact results for the "percolation" problem*, Proc. Roy. Soc. London, **322** (1971), 251–280.
[Tu] V. Turaev, *The Yang-Baxter equation and invariants of links*, LOMI preprint (Leningrad, 1987).
[W1] H. Wenzl, private communication.
[W2] ———, *Braid group representations and the quantum Yang-Baxter equation*, (to appear).

Received April 15, 1988.

UNIVERSITY OF CALIFORNIA
BERKELEY, CA 94720

Marukami, J.
Osaka J. Math.
26 (1989), 1–55

THE PARALLEL VERSION OF POLYNOMIAL INVARIANTS OF LINKS

Dedicated to Professor Nagayoshi Iwahori on his sixtieth birthday

JUN MURAKAMI

(Received April 10, 1987)

Introduction

Let $\mathcal{L}$ be the set of link isotopy types and X: $\mathcal{L} \rightarrow \boldsymbol{C}$ a $\boldsymbol{C}$-valued invariant of links. For any positive integer r, one can define the *r-parallel version* $X^{(r)}$: $\mathcal{L} \rightarrow \boldsymbol{C}$ of X by putting

$$X^{(r)}(K) = X(K^{(r)}), \qquad K \in \mathcal{L},$$

where $K^{(r)}$ is the r-parallel link of K obtained by applying the operation $\phi^{(r)}$ (see Figure 1) to each crossing point of K.

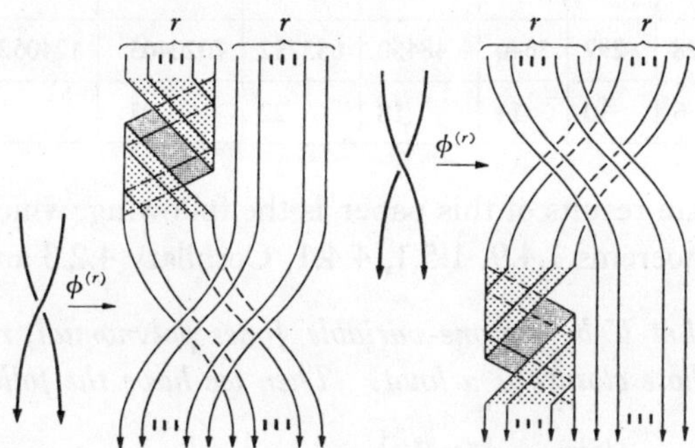

Figure 1

Recently, Morton-Short [24] and Yamada [34] independently noticed the existence of a pair of knots which is indistinguishable by the Jones polynomial [14], but is distinguishable by its 2-parallel version. Moreover there exist mutant knots distinguishable by the 3-parallel version of the two-variable Jones polynomial $P^{(3)}$ ([25] and Section 6.2 of this paper). Thus it seems to be worth while studying the r-parallel version of a link invariant from a general

* This research was supported in part by Grant-in-Aid for Scientific Research, The Ministry of Education, Science and Culture.

view-point.

Throughout this paper we assume that all the invariants of links are of 'trace' type. A link invariant X is called of trace type iff X satisfies the following conditions: (i) Let $b^\wedge$ be the closure of an n-braid b. Then $X(b^\wedge)$ can be written as a linear combination of characters of representations of the braid group B_n. (ii) The characters in (i) satisfy some compatibility conditions with respect to the natural inclusion $B_n \to B_{n+1}$ ($n=1, 2, \cdots$) (see Definition 1.1.4). For example, the (two-variable) Jones polynomial [7], [14], [29] and the Kauffman polynomial [17], [22] fulfill this assumption. (See also [2], [10], [15], [26].) The purpose of this paper is to give an efficient method to calculate the r-parallel version $X^{(r)}$ of the invariant X. A direct calculation of $X^{(r)}(K)$ is, in principle, possible, but the degrees of the characters of the representations involved soon become enormous even for relatively small r. We show that the characters can be reduced to a sum of those of much smaller degrees, and we have only to deal with representation matrices of much smaller sizes. See Table 1.

Table 1 The maximal degree of characters needed to get the r-parallel version of the Jones polynomial of the closures of n-braids.

braid \ index r		1	2	3	4	5	6	7	8	9
3	direct method	2	9	48	297	2002	13260	90440	653752	4601610
	our method	2	3	4	5	6	7	8	9	10
4	direct method	3	28	297	3640	48450	653752	7020405	124062000	1739969550
	our method	3	6	10	14	18	22	27	32	38

One of the main results of this paper is the following, which is an immediate consequence of Thoerems 1.4.9, 1.5.1, 4.4.1, Corollary 4.2.3 and (4.1.10).

Theorem. *Let V be the one-variable Jones polynomial, r a positive integer and b a 3-braid whose closure is a knot. Then we have the following.*

$$V^{(r)}(b^\wedge) = \sum_{j=0}^{[r/2]} c_{r,j} \Big(\sum_{i=0}^{[3(r-j)/2]} a_{3(r-j),i}(V) \operatorname{Trace}(\pi_{3(r-j),i}(b)) \Big),$$

where

$$c_{r,j} = \frac{r-2j+1}{r+1} \binom{r+1}{j},$$

$$a_{k,i}(V) = (-1)^{k+1} \frac{t^{k/2+1/2-i} - t^{-k/2-1/2+i}}{t-t^{-1}},$$

and $\pi_{3r,i}$'s, are representations of B_3 given in Theorem 4.4.1.

The degree of the representation $\pi_{3r,i}$ is equal to $i+1$ if $0 \le i \le r$ and

$3r-2i+1$ if otherwise. Applications of the above theorem are given in (4.4.4)–(4.4.10).

In the last section of this paper, we shall give a necessary condition for the existence of mutant knots distinguishable by the r-parallel version $X^{(r)}$ of the link invariant X (Theorem 6.2.4). We construct mutant knots K_1, K_2, K_3, K_4 for which the 3-parallel version $P^{(3)}(K_i)$ $(1 \leq i \leq 4)$ of the two-variable Jones polynomial are all distinct (Section 6.2).

We shall give a formula for an invariant of cable links. Let X be a link invariant of trace type. Then we shall construct several link invariants $X^{(r,1)}$, $X^{(r,2)}$, $\cdots$ by 'decomposing' the r-parallel version $X^{(r)}$ of X (Theorems 1.5.1 and 2.2.1). Let K be a knot, L a link in the solid torus and K_L the satellite link [4] coming from K and L. Let X_L be a knot invariant denfied by $X_L(K)=X(K_L)$. Then X_L can be written as a linear combinations of invariants $X^{(r,1)}$, $X^{(r,2)}$, $\cdots$ if L is a 'cable' (Definition 1.6.3), or if X is the Jones polynomial or the Kauffman polynomial (Theorem 1.6.4, 2.2.1, 4.3.2, 5.3.1).

We prove, in Sections 1–2, fundamental formulas for $X^{(r)}$ and show some related results of general nature. In Sections 3–5 we apply the results of Sections 1–2 to the r-parallel version of the (two-avriable) Jones polynomial and the Kauffman polynomial, and investigate them more closely. It is discussed in Section 6 how $X^{(r)}$ works at mutant knots.

As a conclusion of this paper, the r-parallel version of link invariants seems to be quite promising in attacking the classification problem of link types.

Acknowledgement. I would like to thank S. Yamada for the useful discussions including the results of [34], which is the motivation of this study. I would like to express my appreciation to N. Iwahori, who suggested that the matrix entries given in Theorem 3.3.1 might be expressed by Gauss' polynomial, and to T. Kanenobu, who informed me about pairs of closed 3-braids with the same Kauffman polynomial. I am profoundly indebted to A. Gyoja, who gave me a lot of useful information about the Iwahori algebras (or Hecke algebras), for example, the result of [10]. I am also indebted to T. Kobayashi, who introduced me to the theory of the (two-variable) Jones polynomial. I am very grateful to N. Kawanaka, H. Nagao, K. Nagatomo, M. Nakaoka and M. Ochiai for their useful advice and kind encouragement. Last of all, I must thank muSIMP/muMATH (Symbolic Mathematics Package), TURBO Pascal for MS-DOS on the personal computer PC-9801 (NEC) and AOS/VS Pascal on Eclipse MV/2000DC (Deta General), which are used for the calculations in the proof of Proposition 4.4.10 and in Example 6.2.7.

1. The parallel version of link invariants, I (the case of knots). We shall discuss in this and next sections general properties of a 'parallel version' of a link invariant of 'trace' type. In this section we first state and prove the results in the case of knots, and, in Section 2, generalize them to the case of links.

1.1. Link invariants of trace type. Let B_n be Artin's braid group on n-strings with standard generators $\sigma_1, \sigma_2, \cdots, \sigma_{n-1}$ as in Figure 2,

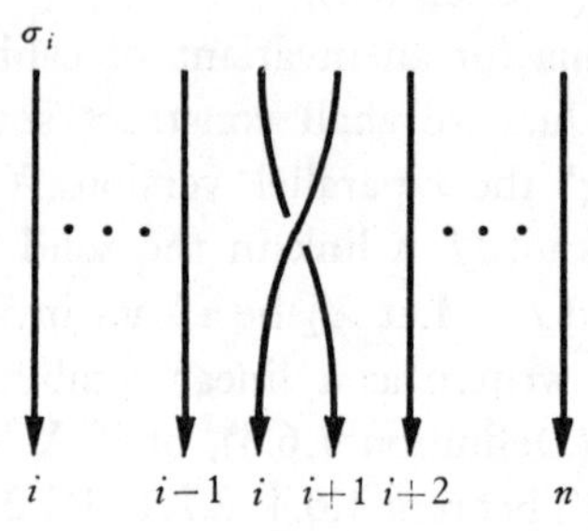

Figure 2

i.e.

$$B_n = \langle \sigma_1, \sigma_2, \cdots, \sigma_{n-1} | \sigma_i\sigma_{i+1}\sigma_i\sigma_{i+1}\sigma_i\sigma_{i+1} \ (1 \le i \le n-2), \\ \sigma_i\sigma_j = \sigma_j\sigma_i \ (1 \le i < j-1 \le n-2) \rangle.$$

Let $\xi_n^{(r)}: B_n \to B_{n+r}$ be the group homomorphism defined by $\xi_n^{(r)}(\sigma_i) = \sigma_i$ for $n, r \in \boldsymbol{N} = \{1, 2, \cdots\}$ and $1 \le i \le n-1$. We regard B_n as a subgroup of B_{n+r} with respect to the inclusion $\xi_n^{(r)}$. Let $B = \{(b, n) | b \in B_n \text{ for } n = 1, 2, \cdots\}$. A *closure* of a braid (b, n), written $(b, n)^\wedge$ or simply $b^\wedge$, is the link formed by joining the n points at the top of (b, n) to those at the bottom without further crossings. Two links K_1, K_2 are called *equivalent* if K_1 is ambient isotopic to K_2 in $\boldsymbol{R}^3$ ([3], Chapt. 1, B).

Theorem 1.1.1 (Alexander [1] p. 42). *Every link is equivalent to the closure of a braid.*

DEFINITION 1.1.2 (Markov class). Let $\sim$ be the equivalence relation on B generated by the following:

(i) $(bb', n) \sim (b'b, n)$ for $b, b' \in B_n$,

(ii) $(b, n) \sim (\sigma b_n^{\pm 1}, n+1)$ for $b \in B_n$.

The equivalence classes of B by the above relation are called *Markov classes.*

Theorem 1.1.3 (Markov [1] p. 51). *The closures of two braids b_1, b_2 are equivalent if and only if b_1 and b_2 are contained in the same Markov class.*

DEFINITION 1.1.4 (link invariant). A mapping X from B to a set S is called a *link invariant* if X is constant on each Markov class of B.

Let $\boldsymbol{C}$ be the field of complex numbers and $B_n^\wedge$ the set of equivalence classes of finite dimensional irreducible representations of B_n over $\boldsymbol{C}$. Let f be a $\boldsymbol{C}$-valued function on B. We assume that f can be written as a finite linear combination of characters of the braid group, i.e. for $b \in B_n$,

$$f(b, n) = \sum_{\mu \in B_n^\wedge} a_\mu(f) \chi_\mu(b) , \tag{1.1.5}$$

where, for each $\mu \in B_n^\wedge$, χ_μ is the corresponding character and $a_\mu(f)$ is an element of $\boldsymbol{C}$ independent of b. Let (ρ_μ, V_μ) be the representation of B_n corresponding to $\mu \in B_n^\wedge$, i.e. V_μ is the representation space and ρ_μ is the group homomorphism of B_n to $GL(V_\mu)$. Then ρ_μ is extended uniquely to a $\boldsymbol{C}$-algebra homomorphism of the group ring $\boldsymbol{C}B_n$ to $\mathrm{End}(V_\mu)$ and so its character χ_μ is extended to a linear function on $\boldsymbol{C}B_n$, which are also denoted by ρ_μ and χ_μ respectively. Hence the function f is extended to a function on $\boldsymbol{C}B = \{(b, n) \mid b \in \boldsymbol{C}B_n \text{ for } n \in \boldsymbol{N}\}$ by (1.1.5), which is also denoted by f.

DEFINITION 1.1.6 (associated algebra). Let $A_n(f)^\wedge = \{\mu \in B_n^\wedge \mid a_\mu(f) \neq 0\}$ for $n \in \boldsymbol{N}$ and $A_n(f) = \bigoplus_{\mu \in A_n(f)^\wedge} \rho_\mu(\boldsymbol{C}B_n)$. We call $A_n(f)$ $(n=1, 2, \cdots)$ the *associated algebra* of f. Let $p_n \colon \boldsymbol{C}B_n \to A_n(f)$ be the algebra homomorphism defined by $p_n = \bigoplus_{\mu \in A_n(f)^\wedge} \rho_\mu$.

DEFINITION 1.1.7 (trace type). A $\boldsymbol{C}$-valued function f on B is called of *trace type* if f satisfies the following conditions:

(i) The function f can be written as a linear combination of characters of B_n as in (1.1.5) for each $n \in \boldsymbol{N}$.

(ii) Let $A_1(f), A_2(f), \cdots$ be the associated algebras of f. Then there are algebra homomorphisms $\zeta_n \colon A_n(f) \to A_{n+1}(f)$ for which $p_{n+1} \circ \xi_n^{(1)} = \xi_n \circ p_n$.

REMARK 1.1.8. Both the one (two)-variable Jones polynomial and the Kauffman polynomial are of trace type for the generic values of their parameter(s).

To specify the notations, we review the definitions of the polynomial invariants of links mentioned in Remark 1.1.8. We also recall the definitions of the regular isotopy invariants of unoriented link diagrams called the bracket polynomial and the L-polynomial [16], [17] which will be needed in Sections 4 and 5.

DEFINITION 1.1.9 (writhe). The *writhe* of a link diagram K is the sum of the signatures of all crossing points of K (Figure 3) and is denoted by $w(K)$.

DEFINITION 1.1.10. Let t be a non-zero complex number. The *bracket polynomial* $\langle \cdot \rangle = \langle \cdot \rangle (t)$ with values in $\boldsymbol{C}$ is uniquely defined for regular isotopy classes of the unoriented link diagrams by the following formulas:

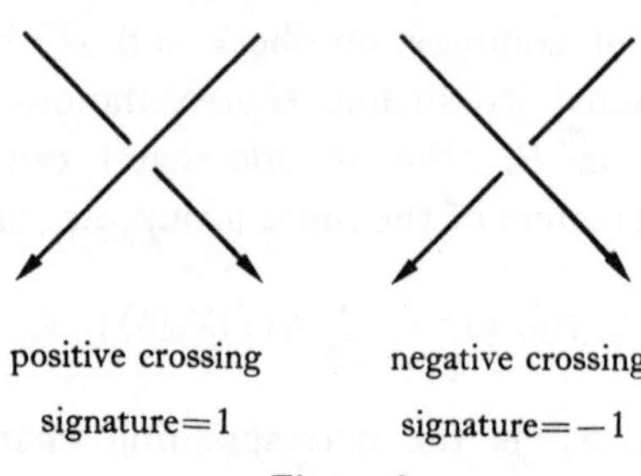

Figure 3

$$\langle K_+\rangle = t^{-1/4}\langle K_0\rangle + t^{1/4}\langle K_\infty\rangle, \qquad \langle O\rangle = 1 \text{ for the unknot } O,$$

where the K_* are identical except within a ball where they are as in Figure 4.

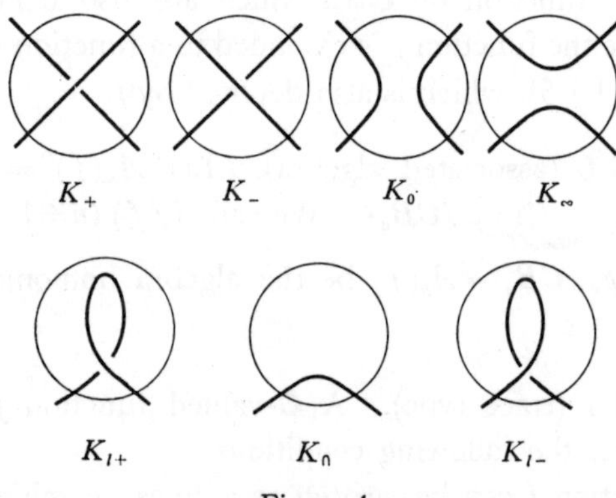

Figure 4

Let a and x be non-zero complex numbers. The *D-polynomial* $D(\cdot)=D(\cdot)(a, x)$ with values in $\boldsymbol{C}$ is uniquely defined for regular isotopy classes of the unoriented link diagrams by the following formulas:

$$D(K_+)-D(K_-) = x(D(K_0)-D(K_\infty)),$$
$$D(K_{l^+}) = aD(K_\cap), \quad D(K_{l^-}) = a^{-1}D(K_\cap), \qquad D(O) = 1,$$

where the K_* are identical except within a ball where they are as in Figure 4.

In the following, we define isotopy invariants of oriented links. Let t be a non-zero complex number. The *one-variable Jones polynomial* $V(\cdot)=V(\cdot)(t)$ with values in $\boldsymbol{C}$ is defined by

$$\text{(1.1.11)} \qquad t^{-1}V(K_+)-tV(K_-) = (t^{1/2}-t^{-1/2})V(K_0), \qquad V(O) = 1,$$

where the K_* are identical except within a ball where they are as in Figure 5. To specify the parameter t, we also denote $V(K)$ by $V(K)(t)$ for a link K. Let K be a link diagram and $|K|$ its unoriented diagram. Then it is known [16] that

$$\text{(1.1.12)} \qquad V(K)(t) = (-t^{3/4})^{w(K)}\langle |K| \rangle(t).$$

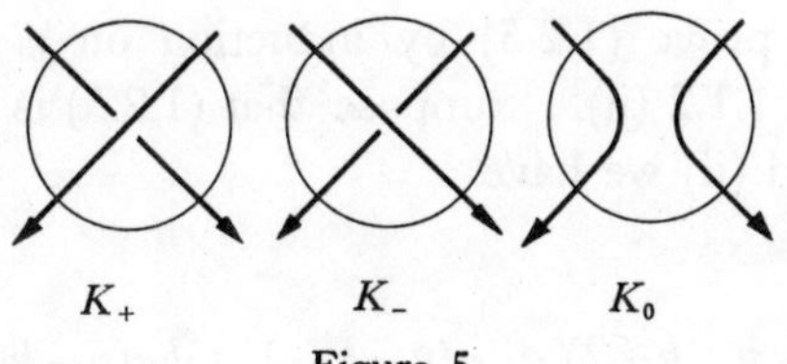

Figure 5

Let l and m be non-zreo complex numbers. The *two-variable Jones polynomial* $P(\cdot)=P(\cdot)(l, m)$ with values in $\boldsymbol{C}$ is defined by

$$l^{-1}P(K_+)+lP(K_-)+mP(K_0)=0\,, \qquad P(O)=1\,,$$

where we use the notation in (1.1.11).

Let a and x be non-zero complex numbers and K be a link diagram. The *Kauffman polynomial* $F(\cdot)=F(\cdot)(a, x)$ with values in $\boldsymbol{C}$ is defined by $F(K)=a^{-w(K)}D(|K|)$.

1.2. Parallel versions. Fix a positive integer r. Let $\phi_n^{(r)}\colon B_n\to B_{rn}$ be the group homomorphism defined by Figure 1, or, equivalently, by

$$\text{(1.2.1)} \quad \phi_n^{(r)}(\sigma_i)=\sigma(ri-r+1, ri-1)^{-r}\sigma(ri, ri+r-1)\sigma(ri-1, ri+r-2)\cdots \sigma(ri-r+1, ri) \qquad (1\leq i\leq n-1)\,,$$

where $\sigma(i, j)=\sigma_i\sigma_{i+1}\cdots\sigma_j$. Let $\phi^{(r)}\colon B\to B$ be the map defined by $\phi^{(r)}((b, n))=(\phi_n^{(r)}(b), rn)$.

Theorem 1.2.2. *Let (b_1, n_1) and (b_2, n_2) be two braids. Then the links $(\phi_{n_1}^{(r)}(b_1), rn_1)^\wedge$ and $(\phi_{n_2}^{(r)}(b_2), rn_2)^\wedge$ are equivalent if $(b_1, n_1)^\wedge$ and $(b_2,n_2)^\wedge$ are equivalent.*

The following lemma and Theorem 1.1.3 yield the above theorem.

Lemma 1.2.3. *Let $\sim$ denote the Markov equivalence relation (Definition 1.1.2). Then we have the following:*

(a) *For $b_1, b_2\in B_n$, $(\phi_n^{(r)}(b_1b_2), rn)\sim(\phi_n^{(r)}(b_2b_1), rn)$,*

(b) *For $b\in B_n$, $(\phi_n^{(r)}(b), rn)\sim(\phi_{n+1}^{(r)}(b\sigma_n^{\pm1}), rn+r)$.*

Proof. The part (a) holds since $\phi_n^{(r)}$ is a group homomorphism. The following Lemma 1.2.4 shows the part (b). □

Lemma 1.2.4. *For $b\in B_n$ $(n\geq r)$,*

$$\text{(1.2.5)} \qquad (b, n)\sim(b\sigma_n^{(r)}, n+r)\,,$$

$$\text{(1.2.6)} \qquad (b, n)\sim(b(\sigma_n^{(r)})^{-1}, n+r)\,.$$

where $\sigma_n^{(r)}=\sigma(n-r+1, n-1)^{-r}\sigma(n, n+r-1)\sigma(n-1, n+r-2)\cdots\sigma(n-r+1, ri)$.

Proof. We first prove (1.2.5) by induction on r. For $r=1$, (1.2.5) is identical to Definition 1.1.2 (ii). Suppose that (1.2.5) is true for $r=k-1$. By Definition 1.1.2 (i) and (ii) we have

$$\begin{aligned}(1.2.7)\quad & (b\sigma_n^{(k)}, n+k) \\ & \sim(\sigma(n-1, n+k-2)\cdots\sigma(n-k+1, n)b\sigma(n-k+1, n-1)^{-k} \\ & \quad\times\sigma(n, n+k-2)\sigma_{n+k-1}, n+k) \\ & \sim(\sigma(n-1, n+k-2)\cdots\sigma(n-k+1, n)b\sigma(n-k+1, n-1)^{-k} \\ & \quad\times\sigma(n, n+k-2), n+k-1) \\ & \sim(b\sigma(n-k+1, n-1)^{-k}\sigma(n, n+k-2)\sigma(n-1, n+k-2)\cdots \\ & \quad\sigma(n-k+1, n), n+k-1).\end{aligned}$$

By using the relation of the braid group, we have $\sigma_k^{-1}\sigma(i, j)^{-1}=\sigma(i, j)^{-1}\sigma_{k+1}^{-1}$, and so we get

$$\begin{aligned}(1.2.8)\quad & \sigma(n-k+1, n-1)^{-k+1} \\ & =\sigma(n-k+2, n-1)^{-1}\sigma_{n-k+1}^{-1}\sigma(n-k+1, n-1)^{-k+2} \\ & =\sigma(n-k+2, n-1)^{-1}\sigma(n-k+1, n-1)^{-k+2}\sigma_{n-1}^{-1} \\ & =\sigma(n-k+2, n-1)^{-2}\sigma_{n-k+1}^{-1}\sigma(n-k+1, n-1)^{-k+3}\sigma_{n-1}^{-1} \\ & =\sigma(n-k+2, n-1)^{-2}\sigma(n-k+1, n-1)^{-k+3}\sigma_{n-2}^{-1}\sigma_{u-1}^{-1} \\ & =\cdots=\sigma(n-k+2, n-1)^{-k+1}\sigma_{n-k+1}^{-1}\sigma_{u-k+2}^{-1}\cdots\sigma_{n-1}^{-1}.\end{aligned}$$

By using the relation of the braid group, we have $\sigma(i, j)\sigma(i-1, j)=\sigma(i-1, j)\sigma(i-1, j-1)=\sigma_{i-1}\sigma(i, j)\sigma(i-1, j-1)$. Hence we get

$$\begin{aligned}(1.2.9)\quad & \sigma(n, n+k-2)\sigma(n-1, n+k-2)\sigma(n-2, n+k-3)\cdots\sigma(n-k+1, n) \\ & =\sigma_{n-1}\sigma(n, n+k-2)\sigma(n-1, n+k-3)\sigma(n-2, n+k-3)\cdots \\ & \quad\sigma(n-k+1, n) \\ & =\sigma_{n-1}\sigma_{n-2}\sigma(n, n+k-2)\cdots\sigma(n-2, n+k-4)\sigma(n-3, n+k-4)\cdots \\ & \quad\sigma(n-k+1, n) \\ & =\cdots=\sigma_{n-1}\sigma_{n-2}\cdots\sigma_{n-k+1}\sigma(n, n+k-2)\cdots\sigma(n-k+1, n-1).\end{aligned}$$

By substituting (1.2.8) and (1.2.9) into (1.2.7), we obtain

$$\begin{aligned} & (b\sigma_n^{(r)}, n+k) \\ & \sim(b\sigma(n-k+1, n-1)^{-k}\sigma(n, n+k-2)\sigma(n-1, n+k-2)\cdots \\ & \quad\sigma(n-k+1, n), n+k-1) \\ & =(b\sigma(n-k+1, n-1)^{-1}\sigma(n-k+2, n-1)^{-k+1}\sigma_{n-k+1}^{-1}\sigma_{n-k+2}^{-1}\cdots \\ & \quad\sigma_{n-1}^{-1}\sigma_{n-1}\sigma_{n-2}\cdots\sigma_{n-k+1}\sigma(n, n+k-2)\cdots \\ & \quad\sigma(n-k+2, n-1)\sigma(n-k+1, n-2), n+k-1)\end{aligned}$$

$$= (b\sigma(n-k+1, n-1)^{-1}\sigma(n-k+2, n-1)^{-k+1}\sigma(n, n+k-2)\cdots$$
$$\sigma(n-k+2, n-1)\sigma(n-k+1, n-1), n+k-1)$$
$$= (b\sigma(n-k+1, n-1)^{-1}\sigma_n^{(k-1)}\sigma(n-k+1, n-1), n+k-1)\,.$$

Moreover, Lemma 1.2.3 (a) and the induction hypothesis imply that

$$(b\sigma(n-k+1, n-1)^{-1}\sigma_n^{(k-1)}\sigma(n-k+1, n-1), n+k-1)\sim(b, n).$$

Hence (1.2.5) is proved. An analogous argument yields (1.2.6). □

DEFINITION 1.2.10 (r-parallel version). Let K be a link and $(b, n)\in B$ the braid whose closure is equivalent to K. The link $(\phi_n^{(r)}(b), n)^\wedge$ is called the *r-parallel version* of K and denoted by $K^{(r)}$. The *r-parallel version* $f^{(r)}$ of a function f on B is defined by $f^{(r)}(b,n)=f(\phi_n^{(r)}(b), rn)$.

The r-parallel version of a link is well-defined by Theorem 1.2.2. Hence the parallel version $X^{(r)}$ of a link invariant X is again a link invariant.

1.3. Wreath products. We review representation theory of wreath products. The references are [19] and [6]. Throughout this paper, a semigroup has a unit, a semigroup homomorphism preserves the unit and a linear representation of a group or a semigroup sends the unit to the identity transformation. Let $\Omega=\{1, 2, \cdots, n\}$ and H a semigroup with a semigroup homomorphism $\theta: H\to S_n$, where S_n denotes the symmetric group of degree n actnig on Ω naturally. For $h\in H$, let $\langle\theta(h)\rangle$ denote the subgroup of S_n generated by $\theta(h)$. Fix a group G and an element h_0 of H.

(1.3.1) $\quad H(h_0)=\{h\in H\,|\,\text{for every } \langle\theta(h_0)\rangle\text{-orbit } O \text{ of } \Omega,\ \theta(h)\cdot O=O\}$,

which is a subsemigroup of H. The semigroup $H(h_0)$ acts on G^n by the following. For $h\in H(h_0)$ and $g=(g_1, g_2, \cdots, g_n)\in G^n$, ${}^hg=(g_{\theta(h)^{-1}(1)}, \cdots, g_{\theta(h)^{-1}(n)})$.

DEFINITION 1.3.2. The set $G^n\times H(h_0)$ together with the composition law $(g, h)(g', h')=(g({}^hg'), hh')$ is a semigroup called the *wreath product* $G^n\rtimes H(h_0)$ of G with $H(h_0)$.

For a group G let $G^\wedge$ denote the set of equivalence classes of finite dimensional irreducible representations of G over $\boldsymbol{C}$, and for $\nu\in G^\wedge$, (ρ_ν, V_ν) denotes the corresponding representation. For $\nu=(\nu_1, \cdots, \nu_n)$ $(\nu_j\in G^\wedge)$, let $\rho_\nu=\rho_{\nu_1}\otimes\cdots\otimes\rho_{\nu_n}$, which is an element of $(G^n)^\wedge$. In the following, we assume that $\nu_i=\nu_j$ if $\langle\theta(h_0)\rangle\cdot i=\langle\theta(h_0)\rangle\cdot j$. We define an action of the semi-group $G^n\rtimes H(h_0)$ on V_ν by the following: For $(g, h)\in G^n\rtimes H(h_0)$ and $(v_1, \cdots, v_n)$ $\in V_{\nu_1}\otimes\cdots\otimes V_{\nu_n}$,

$$(g, h)(v_1, \cdots, v_n)=\rho_\nu(g)(v_{\theta(h)^{-1}(1)}, \cdots, v_{\theta(h)^{-1}(n)})\,.$$

We denote by $\rho_\nu^{\sim}$ the representation of the semigroup $G^n \rtimes H(h_0)$ given by the above action on V_ν. If $\rho_\nu(g)=(\rho_{\nu_1}(g_1)_{\alpha_1\beta_1}\cdots\rho_{\nu_n}(g_n)_{\alpha_n\beta_n})$ $(1\leq\alpha_1, \beta_1\leq\dim V_{\nu_1},\cdots,1\leq\alpha_n, \beta_n\leq\dim V_{\nu_n})$ is a matrix representing $g=(g_1,\cdots,g_n)\in G^n$, then we get for $h\in H(h_0)$:

$$(1.3.3) \qquad \rho_\nu^{\sim}(g,h)=(\rho_{\nu_1}(g)_{1\alpha_1\beta_{\theta(h)^{-1}(1)}}\cdots\rho_{\nu_n}(g_n)_{\alpha_n\beta_{\theta(n)^{-1}(h)}}).$$

Let π be a representation of the semigroup $H(h_0)$. By composing the canonical projection $G^n \rtimes H(h_0)\to H(h_0)$, π is naturally extended to a representation of the semigroup $G^n \rtimes H(h_0)$, which will also be denoted by π.

Proposition 1.3.4. *Let (ρ, V) be a finite dimensional linear representation of the semigroup $G^n \rtimes H(h_0)$ such that the restriction $\rho|_{G^n}$ is a completely reducible group representation. We assume that the irreducible components of $\rho|_{G^n}$ are all equivalent to ρ_ν. Then there is a representation π of the semigroup $H(h_0)$ such that ρ is equivalent to $\rho_\nu^{\sim}\otimes\pi$.*

This is a version of a standard result found, e.g., in [6], Theorem 5.1.7. Although G^n may not be a finite group or $H(h_0)$ may not be a group, an argument analogous to the proof of [loc. cit.] works because of our assumptions, i.e. the finite dimensionality of V and the complete reducibility of $\rho|_{G^n}$.

1.4. The characters associated with the parallel version. Fix a $\boldsymbol{C}$-valued function f on B of trace type and r a positive integer. The r-parallel version $f^{(r)}$ of f is also of trace type since we have the following from (1.1.5).

$$(1.4.1) \qquad f^{(r)}((b,n)^{\wedge})=\sum_{\mu\in A_{rn}(f)^{\wedge}} a_\mu(f)\chi_\mu(\phi_n^{(r)}(b)).$$

We decompose the characters $\chi_\mu\circ\phi_n^{(r)}$ $(\mu\in A_{rn}(f)^{\wedge})$ into sums of characters of B_n of smaller degrees. We need some preparations. Let ι_j: $\boldsymbol{C}B_r\to\boldsymbol{C}B_{rn}$ $(1\leq j\leq n)$ be the homomorphism defined by

$$(1.4.2) \qquad \iota_j(\sigma_i)=\sigma_{i+rj-r} \qquad (1\leq i\leq r-1)$$

and ι: $\boldsymbol{C}B_r^{\otimes n}\to\boldsymbol{C}B_{rn}$ the homomorphism defined by $\iota(b_1\otimes\cdots\otimes b_n)=\iota_1(b_1)\cdots\iota_n(b_n)$ for $b_j\in\boldsymbol{C}B_r$, where $\boldsymbol{C}B_r^{\otimes n}$ denotes the n-fold tensor product $\boldsymbol{C}B_r\otimes\cdots\otimes\boldsymbol{C}B_r$ of $\boldsymbol{C}B_r$. We regard $\boldsymbol{C}B_r^{\otimes n}$ as a subalgebra of $\boldsymbol{C}B_{rn}$ by the inclusion ι. Let θ be the group homomorphism from B_n to the symmetric group S_n of degree n defined by $\theta(\rho_i)=(i\ i+1)$ (the transposition of i and $i+1$). We define an action of B_n on $\Omega=\{1,\cdots,n\}$ by $b(i)=\theta(b)(i)$ for $i\in\Omega$. Then we have the following.

Lemma 1.4.3. *For $b\in B_n$ and $b'\in B_r$,*

$$\phi_n^{(r)}(b)\iota_k(b')=\iota_{b(k)}(b')\phi_n^{(r)}(b) \qquad (1\leq k\leq n).$$

Proof. The following formulas imply the statement of the lemma.

(1.4.4. a) $\phi_n^{(r)}(\sigma_i)\iota_k(\sigma_j)=\iota_k(\sigma_j)\phi_n^{(r)}(\sigma_i)$ $(1\leq i\leq n-1,\ 1\leq j\leq r-1,\ k\neq i,\ i+1)$,

(1.4.4. b) $\phi_n^{(r)}(\sigma_i)\iota_{i+1}(\sigma_j)=\iota_i(\sigma_j)\phi_n^{(r)}(\sigma_i)$ $(1\leq i\leq n-1,\ 1\leq j\leq r-1)$,

(1.4.4. c) $\phi_n^{(r)}(\sigma_i)\iota_i(\sigma_j)=\iota_{i+1}(\sigma_j)\phi_n^{(r)}(\sigma_i)$ $(1\leq i\leq n-1,\ 1\leq j\leq r-1)$.

The formula (1.4.4a) is a consequence of (1.2.1) and the relations $\sigma_i\sigma_j=\sigma_j\sigma_i$ $(|i-j|\geq 2)$. We prove (1.4.4b). Let $\sigma'(i,j)=\sigma_i\sigma_{i-1}\cdots\sigma_j$ $(i\geq j)$. Then

$$\phi_n^{(r)}(\sigma_i)=\sigma(ri-r+1,\ ri-1)^{-r}\sigma'(ri,\ ri-r+1)\sigma'(ri+1,\ ri-r+2)\cdots \sigma'(ri+r-1,\ ri)\,.$$

Since the relations of the braid group imply $\sigma'(i,j)\sigma_k=\sigma_{k-1}\sigma'(i,j)$ $(i>k\geq j)$, we have

$$\begin{aligned}&\phi_n^{(r)}(\sigma_i)\iota_{i+1}(\sigma_j)\\ &\quad=\sigma(ri-r+1,\ ri-1)^{-r}\sigma'(ri,\ ri-r+1)\sigma'(ri+1,\ ri-r+2)\cdots\\ &\qquad\sigma'(ri+r-1,\ ri)\sigma_{ri+j}\\ &\quad=\sigma(ri-r+1,\ ri-1)^{-r}\sigma_{ri-r+j}\sigma'(ri,\ ri-r+1)\sigma'(ri+1,\ ri-r+2)\cdots\\ &\qquad\sigma'(ri+r-1,\ ri)\,.\end{aligned}$$

We also know ([1], p. 28, Corollary 1.8.4) that the element $\sigma(ri-r+1, ri-1)^{-r}$ is contained in the center of the subgroup $\langle\sigma_{ri-r+1}, \sigma_{ri-r+2}, \cdots, \sigma_{ri-1}\rangle$ of B_{rn}. Hence we have

$$\phi_n^{(r)}(\sigma_i)\iota_{i+1}(\sigma_j)=\sigma_{ri-r+j}\,\phi_n^{(r)}(\sigma_i)=\iota_i(\sigma_j)\phi_n^{(r)}(\sigma_i)\,.$$

This proves (1.4.4.b). The formula (1.4.4.c) is proved analogously by using

$$\phi_n^{(r)}(\sigma_i)=\sigma(ri-r+1,\ ri-1)^{-r}\sigma(ri,\ ri+r-1)\sigma(ri-1,\ ri+r-2)\cdots\sigma(ri-r+1,\ ri)$$

and the relations $\sigma(i,j)\sigma_k=\sigma_{k+1}\sigma(i,j)$ $(i\leq k<j)$. □

Lemma 1.4.3 implies that the subgroup $\iota(B_r^n)\phi_n^{(r)}(B_n)$ of B_{rn} is isomorphic to the wreath product $B_r^n\rtimes B_n$ with respect to $\theta_n\colon B\to S_n$.

DEFINITION 1.4.5 (isotypic subspace). Let A be a semisimple algebra over $\boldsymbol{C}$, U an A-module and ρ an irreducible representation of A. An A-submodule W of U is called the *ρ-isotypic subspace* if W is the maximal subspace on which the action of A is isomorphic to $\rho\oplus\cdots\oplus\rho$ (n-times) for some $n=0, 1, 2, \cdots$.

For $\nu(1), \cdots, \nu(n)\in A_r(f)^\wedge$ and $\mu\in A_{rn}(f)^\wedge$, let $V_{\mu,\nu(1),\cdots,\nu(n)}$ be the $\rho_{\nu(1)}\otimes\cdots\otimes\rho_{\nu(n)}$-isotypic subspace of V_μ as a B_r^n-module and $Y=V_{\nu(1)}\otimes\cdots\otimes V_{\nu(n)}$. Then $V_{\mu,\nu(1),\cdots,\nu(n)}\cong Y\oplus\cdots\oplus Y\cong Y\otimes\boldsymbol{C}^{d(\mu,\nu(1),\cdots,\nu(n))}$, where $d(\mu, \nu(1), \cdots, \nu(n))$ denote the multiplicity of $\rho_{\nu(1)}\otimes\cdots\otimes\rho_{\nu(n)}$ in $\rho_\mu|_{B_r^n}$. For $b\in B_n$, $(b_1', \cdots, b_n')\in B_r^n$ and $v_1\otimes\cdots\otimes v_n\otimes w\in V_{\nu(1)}\otimes\cdots\otimes V_{\nu(n)}\otimes\boldsymbol{C}^{d(\mu\nu(1),\cdots,\nu(n))}=V_{\mu,\nu(1),\cdots,\nu(n)}$, we have

$$(1.4.6)\quad \begin{aligned}&\rho_\mu(\iota(b'_1\otimes\cdots\otimes b'_n)\phi^{(r)}_n(b))(v_1\otimes\cdots\otimes v_n\otimes w)\\ &=\rho_\mu(\phi^{(r)}_n(b)\iota(b'_{b(1)}\otimes\cdots\otimes b'_{b(n)}))(v_1\otimes\cdots\otimes v_n\otimes w)\\ &=\rho_\mu(\phi^{(r)}_n(b))(\rho_{\nu(1)}(b'_{b(1)})v_1\otimes\cdots\otimes\rho_{\nu(n)}(b'_{b(n)})v_\nu\otimes w)\,,\end{aligned}$$

from (1.4.3). The above formula implies the following:

Lemma 1.4.7. *For $b\in B_n$, the subspace $\rho_\mu(\phi^{(r)}_n(b))V_{\mu,\nu(1),\cdots,\nu(n)}$ is equal to the $\rho_{\nu(b^{-1}(1))}\otimes\cdots\otimes\rho_{\nu(b^{-1}(n))}$-isotypic subspace $V_{\mu,\nu(b^{-1}(1)),\cdots,\nu(b^{-1}(n))}$ of V_μ.*

In particular, $V_{\mu,\nu,\cdots,\nu}$ is invariant relative to the actiono of $\iota(B^n_r)\phi^{(r)}_n(B_n)$. Let $\rho_{\mu,\nu}$ be the representation of $\iota(B^n_r)\phi^{(r)}_n(B_n)$ obtained by restricting ρ_μ on $V_{\mu,\nu,\cdots,\nu}$ and $\chi_{\mu,\nu}$ its character. Then we have the following.

Proposition 1.4.8. *For $b'\in B_r$ and $b\in B_n$ such that $b^\wedge$ is a knot, we have*

$$\chi_\mu(\iota_1(b')\phi^{(r)}_n(b))=\sum_{\mu\in A_r(f)^\wedge}\chi_{\mu,\nu}(\iota_1(b')\phi^{(r)}_n(b))\,.$$

Proof. Recall that f is of trace type. The condition (ii) of Definition 1.1.7 implies that there are algb algebra homomorphisms $\iota_i{}^\sim: A_r(f)\to A_{rn}(f)$ $(1\leq i\leq n)$ such that $\iota_i{}^\sim\circ p_r=p_{rn}\circ\iota_i$. Hence the action of B^n_r on V_μ is factored by p^n_r: $B^n_r\to A_r(f)^n$. Since $A_r(f)$ is the associated algebra of f, $A_r(f)$ is completey reducible and so we have $V_\mu=\bigoplus_{\nu\in(A_r(f)^n)^\wedge}V_{\mu,\nu(1),\cdots,\nu(n)}$. Lemma 1.4.7 implies that

$$\begin{aligned}\rho_\mu(\iota_1(b')\phi^{(r)}_n(b))V_{\mu,\nu(1),\cdots,\nu(n)}&=\rho_\mu(\iota_1(b'))V_{\mu,\nu(b^{-1}(1)),\cdots,\nu(b^{-1}(n))}\\ &=V_{\mu,\nu(b^{-1}(1)),\cdots,\mu(b^{-1}(n))}\,.\end{aligned}$$

Hence if $V_{\mu,\nu(1),\cdots,\nu(n)}\in V_{\mu,\nu(b^{-1}(()),\cdots,\nu(b^{-1}(n))}$, then the diagonal part of the matrix $\iota_1(b')\phi^{(r)}_n(b)$ corresponding to the subspace $V_{\mu,\nu(1),\cdots,\nu(n)}$ is equal to 0. Thus $\chi_\mu(\iota_1(b')\phi^{(r)}_n(b))$ is equal to the sum of the diagonal elements of $\rho_\mu(\iota_1(b')\phi^{(r)}_n(b))$ corresponding to the subspaces $V_{\mu,\nu(1),\cdots,\nu(n)}$ with $V_{\mu,\nu(n),\mu,\nu(n)}=V_{\mu,\nu(b^{-1}(1)),\cdots,\nu(b^{-1}(n))}$. Since the closure of b is a knot, $\theta(b)$ is an n-cycle. Thus $\nu(i)=\nu(b^{-1}(i))$ $(1\leq i\leq n)$ imply $\nu(1)=\cdots=\nu(n)$. But the sum of diagonals of $\rho_\mu(\iota_1(b')\phi^{(r)}_n(b))$ on $V_{\mu,\nu,\cdots,\nu}$ is equal to $\chi_{\mu,\nu}(\iota_1(b')\phi^{(r)}_n(b))$. This proves the proposition. □

Since $\iota(B^n_r)\phi^{(r)}_n(B_n)$ is isomorphic to $B^n_r\rtimes B_n$, we see, from Proposition 1.3.4, that ther is a representation $\pi_{\mu,\nu}$ of B_n such that

$$(1.4.9)\qquad \rho_{\mu,\nu}\cong(\rho^{\otimes n}_\nu)^\sim\otimes\pi_{\mu,\nu}\,,$$

where $(\rho^{\otimes n}_\nu)^\sim$ is the representation of $\iota(B^n_r)\phi^{(r)}_n(B_n)$ coming from the representation $\rho^{\otimes n}_\nu$ of B^n_r as in Proposition 1.3.4. Let $\chi\nu^\sim$, $\omega_{\mu,\nu}$ be the characters of $(\rho^{\otimes n}_\nu)^\sim$, $\pi_{\mu,\nu}$ respectively. Now we can state our main theorem.

Theorem 1.4.10. *Let b be an n-braid whose closure is a knot. Then we have the following.*

(i) $\chi_\mu(\phi_n^{(r)}(b)) = \sum_{\nu\in A_r(f)^\wedge} \chi_\nu(1)\omega_{\mu,\nu}(b)$,

(ii) $f^{(r)}(b, n) = \sum_{\mu\in A_{rn}(f)^\wedge} (\sum_{\nu\in A_r(f)^\wedge} a_\mu(f)\chi_\nu(1)\omega_{\mu,\nu}(b))$.

Proof. Proposition 1.4.8 and (1.4.9) yield that

$$\chi_\mu(\phi_n^{(r)}(b)) = \sum_{\nu\in A_r(f)^\wedge} \chi_{\mu,\nu}(\phi_n^{(r)}(b)) = \sum_{\nu\in A_r(f)^\wedge} \chi_\nu^{\sim}(\phi_n^{(r)}(b))\omega_{\mu,\nu}(b)\,.$$

Hence the first formula is an immediate consequence of the following lemma with $b'=1$. The second one is obtained from (i) and (1.4.1). □

Lemma 1.4.11. *For $b'\in B_r$ and $b\in B_n$ such that $b^\wedge$ is a knot, we have*

$$\chi_\nu^{\sim}(\iota_1(b')\phi_n^{(r)}(b)) = \chi_\nu(b')\,.$$

Proof. By using 1.3.3, we have

$$\begin{aligned}
&\chi_\nu^{\sim}(\iota_1(b')\phi_n^{(r)}(b)) \\
&= \sum_{\alpha_1,\beta_1,\cdots,\alpha_n,\beta_n} \delta_{\alpha_1\beta_1}\cdots\delta_{\alpha_n\beta_n}(\rho_\nu(b')_{\alpha_1\beta_{b^{-1}(1)}}\rho_\nu(1)_{\alpha_2\beta_{b^{-1}(2)}}\cdots\rho_\nu(1)_{\alpha_n\beta_{b^{-1}(n)}}) \\
&= \sum_{\alpha_1,\beta_1,\cdots,\alpha_n,\alpha_n} \delta_{\alpha_1\beta_1}\cdots\delta_{\alpha_n\beta_n}(\rho_\nu(b')_{\alpha_1\beta_{b^{-1}(1)}}\delta_{\alpha_2\beta_{b^{-1}(2)}}\cdots\delta_{\alpha_n\beta_{b^{-1}(n)}})\,.
\end{aligned}$$

Since $b^\wedge$ is a knot and so $\theta(b)$ is an n-cycle, each term of the above summention is 0 except when $\alpha_1=\beta_{b^{-1}(1)}=\alpha_{b^{-1}(1)}=\cdots=\alpha_{b^{-n+1}(1)}=\beta_1$, i.e. $\alpha_1,\cdots,\alpha_n$, $\beta_1,\cdots,\beta_n$ are all equal. Hence we have

$$\chi_\nu^{\sim}(\iota_1(b')\phi_n^{(r)}(b)) = \sum_{\alpha_1}\rho_\nu(b')_{\alpha_1\alpha_1} = \chi_\nu(b')\,.$$

This completes the proof. □

1.5. A decomposition of $X^{(r)}$ into invariants. For a $\boldsymbol{C}$-valued function f on B of trace type, let

$$f^{(r,\nu)}(b, n) = \sum_{\mu\in A_{rn}(f)^\wedge} a_\mu(f)\,\omega_{\mu,\nu}(b)\,.$$

The r-parallel version $X^{(r)}$ of a link invariant X of trace type is a sum of invariants $X^{(r,\nu)}$ parametrized by $A_r(X)^\wedge$:

Theorem 1.5.1. *For a $\boldsymbol{C}$-valued function f on B of trace type and $b\in B_n$, we have the following*

(i) *$f^{(r,\nu)}$ is a link invariant if f is a link invariant.*

(ii) *$f^{(r)}(K)= \sum_{\nu\in A_r(f)} \chi_\nu(1)f^{(r,\nu)}(K)$ for a knot K.*

Proof. (ii) is an immediate consequence of the definition of $f^{(r,\nu)}$ and Theorem 1.4.10 (ii). To show (i), we check the invariance of $f^{(r,\nu)}$ relative to the relations (i), (ii) of Definition 1.1.2. Let b and b' be elements of B_n. Since

$\omega_{\mu,\nu}(bb')=\omega_{\mu,\nu}(b'b)$, we have $f^{(r,\nu)}(bb', n)=f^{(r,\nu)}(b'b, n)$. It remains to show that

$$f^{(r,\nu)}(b, n) = f^{(r,\nu)}(b\sigma_n^{\pm 1}, n+1)\,. \tag{1.5.2}$$

Let f_n: $\boldsymbol{C}B_n \to \boldsymbol{C}$ be a $\boldsymbol{C}$-linear function defined by $f_n(b)=f(b, n)$ for $b \in B_n$. For $\nu \in A(f)^\wedge$, let η_ν be an element of $\boldsymbol{C}B_r$ such that $\rho_\nu(\eta_\nu)=\mathrm{id}$ and $\rho_{\nu'}(\eta_\nu)=0$ for $\nu \neq \nu' \in A_r(f)^\wedge$. Then $p_r(\eta_\nu)$ is contained in the center of $A_r(f)$, $p_r(\eta_\nu)^2=p_r(\eta_\nu)$, and $p_r(\eta_\nu)p_r(\eta_{\nu'})=0$ if $\nu \neq \nu'$. Since $\rho_\mu(\iota_1(\eta_\nu)\iota_2(\eta_\nu)\cdots\iota_n(\eta_\nu))$ is a projection from V_μ to the $\rho_\nu^{\otimes n}$-isotypic subspace $V_{\mu,\nu,\dots,\nu}$ of V_μ, we have $\rho_\mu(\iota_1(\eta_\nu)\iota_2(\eta_\nu)\cdots\iota_n(\eta_\nu)\phi_n^{(r)}(b))$ $=\rho_{\mu,\nu}(\phi_n^{(r)}(b))$. Hence by (1.4.9) and Lemma 1.4.11, we get

$$\begin{aligned}\omega_{\mu,\nu}(b) &= \chi_\nu(1)^{-1}\chi_\mu(\iota_1(\eta_\nu)\iota_2(\eta_\nu)\cdots\iota_n(\eta_\nu)\phi_n^{(r)}(b))\\ &\text{for } \mu \in A_{rn}(f)^\wedge \text{ and } \nu \in A_r(f)^\wedge.\end{aligned} \tag{1.5.3}$$

This formula implies that

$$\begin{aligned}&f^{(r,\nu)}(b\sigma_n^{\pm 1}, n+1)\\ &= \chi_\nu(1)^{-1} f_{rn+r}(\iota_1(\eta_\nu)\cdots\iota_n(\eta_\nu)\iota_{n+1}(\eta_\nu)\phi_{n+1}^{(r)}(b\sigma_n^{\pm 1}))\,.\end{aligned}$$

But we know, from Lemma 1.4.3, that $\iota_{n+1}(\eta_\nu)\phi_n^{(r)}(b\sigma_n^{\pm 1})=\phi_{u+1}^{(r)}(b\sigma_r^{\pm 1})\iota_n(\eta_\nu)$ and so we have

$$\begin{aligned}&f^{(r,\nu)}(b\sigma_n^{\pm 1}, n+1)\\ &= \chi_\nu(1)^{-1} f_{rn+r}(\iota_1(\eta_\nu)\cdots\iota_n(\eta_\nu)\phi_{n+1}^{(r)}(b\sigma_n^{\pm 1})\iota_n(\eta_\nu))\\ &= \chi_\nu(1)^{-1} f_{rn+r}(\iota_n(\eta_\nu)\iota_1(\eta_\nu)\cdots\iota_n(\eta_\nu)\phi_{n+1}^{(r)}(b\sigma_n^{\pm 1}))\\ &= \chi_\nu(1)^{-1} f_{rn+r}(\iota_1(\eta_\nu)\cdots\iota_{n-1}(\eta_\nu)\iota_n(\eta_\nu)\phi_{n+1}^{(r)}(b\sigma_n^{\pm 1}))\,.\\ &= \chi_\nu(1)^{-1} f_{rn+r}(\iota_1(\eta_\nu)\cdots\iota_{n-1}(\eta_\nu)\iota_n(\eta_\nu)\phi_n^{(r)}(b\sigma_n^{\pm 1}))\,.\end{aligned}$$

In the last step of the above calculation, we use $p_r(\eta_\nu)^2=p_r(\eta_\nu)$. Since $\phi_n^{(r)}(\sigma_n^{\pm 1})=$ $(\sigma_{nr}^{(r)})^{\pm 1}$, Lemma 1.2.4 implies that the last term of the above is equal to $\chi_\nu(1)^{-1} f_{rn+r}(\iota_1(\eta_\nu)\cdots\iota_{n-1}(\eta_\nu)\iota_n(\eta_\nu)\phi_n^{(r)}(b))$, which is equal to $f^{(r,\nu)}(b, n)$ by (1.5.3). Hence we have (1.5.2). □

1.6. Invariants of cable links. Let ι_i: $B_r \to B_{rn}$ ($1 \leq i \leq n$) be the group homomorphism defined in Section 1.4.

Proposition 1.6.1. *Let $b' \in B_r$ and $\sim$ denote the Markov equivalence relation (Definition 1.1.2). Then we have the following:*

(a) *For $b_1, b_2 \in B_n$ such that the closure of b_1b_2 is a knot, $(\iota_1(b')\phi_n^{(r)}(b_1b_2), rn) \sim$ $(\iota_1(b')\phi_n^{(r)}(b_2b_1), rn)$,*

(b) *For $b \in B_n$, $(\iota_1(b')\phi_n^{(r)}(b), rn) \sim (\iota_1(b')\phi_{n+1}^{(r)}(b\sigma_n^{\pm 1}), rn+r)$.*

Proof. Lemma 1.2.4 shows the part (b). By using Definition 1.1.2 (ii) and Lemma 1.4.3, we have

$$\begin{aligned}(\iota_1(b')\phi_n^{(r)}(b_1b_2), rn) &\sim (\phi_n^{(r)}(b_1b_2)(b'), rn) \\ &= (\iota_{b_1b_2(1)}(b')\phi_n^{(r)}(b_1b_2), rn) \sim \cdots \\ &\sim (\iota_{(b_1b_2)^k(1)}(b')\phi_n^{(r)}(b_1b_2), rn) \qquad (k=1, 2, \cdots) \\ &\sim (\iota_{b_2(b_1b_2)^k(1)}(b')\phi_n^{(r)}(b_2b_1), rn) .\end{aligned}$$

Since the closure of b_1b_2 is a knot, we have $\{(b_1b_2)^k(1)\,|\,k=0, 1, \cdots, n-1\}=\{1, \cdots, n\}$ and so there is k' with $(b_1b_2)^{k'}(1)=b_2^{-1}(1)$. Hence we have $(\iota_{b_2(b_1b_2)^{k'}(1)}(b')\phi_n^{(r)}(b_2b_1), rn)=(\iota_1(b')\phi_n^{(r)}(b_2b_1), rn)$. This proves the part (a). □

The above proposition and Theorem 1.1.3 yield the following.

Corollary 1.6.2. *Let $b'\in B_r$ and (b_1, n_1), (b_2, n_2) be two braids such that the closures of b_1 and b_2 are knots. Then the closures of $(\iota_1(b')\phi_{n_1}^{(r)}(b_1), rn_1)$ and $(\iota_1(b')\phi_{n_2}^{(r)}(b_2), rn_2)$ are equivalent if $(b_1, n_1)^\wedge$ and $(b_2, n_2)^\wedge$ are equivalent.*

DEFINITION 1.6.3 (cable links). For a braid (b, n) and $b'\in B_r$, we call $(\iota_1(b')\phi_n^{(r)}(b), rn)^\wedge$ the *(r-strand) cable link* of $(b, n)^\wedge$ associated with b'. For a function f on B, let $f_{b'}^{(r)}(b, n)=f(\iota_1(b')\phi_n^{(r)}(b), rn)$.

The r-strand cable links is well-defined by Corollary 1.6.2, and so $X_{b'}^{(r)}$ is an invariant of link isotopy types for a link invariant X and $b'\in \boldsymbol{C}B_r$.

Theorem 1.6.4. *Let $b'\in \boldsymbol{C}B_r$. For a braid $b\in B_n$ whose closure is a knot and a $\boldsymbol{C}$-valued function f on B of trace type, $f_{b'}^{(r)}$ satisfies*

$$f_{b'}^{(r)}(b, n) = \sum_{\nu\in A_r(f)^\wedge} \chi_\nu(b')f^{(r,\nu)}(b, n) .$$

Proof. Since $f_{b'}^{(r)}(b, n)=\sum_{\mu\in A_{rn}(f)^\wedge} \chi_\mu(\iota_1(b')\phi_n^{(r)}(b))$, we get the stateent of the theorem by (1.4.9) and Lemma 1.4.11. □

2. The parallel version of link invariants, II (the case of general links). In this section, we give a generalization of Theorems 1.4.10 and 1.6.4 for braids whose closures are links.

2.1. The characters associated with the parallel version. Fix a $\boldsymbol{C}$-valued function f on B of trace type and fix an element $b_0\in B_n$. Let $B_n(b_0)=\{b\in B_n\,|$ every $\langle\theta(b)\rangle$-orbit of Ω is contained in a single $\langle\theta(b_0)\rangle$-orbit of $\Omega\}$, where $\Omega=\{1, \cdots, n\}$. We use the notations in Sections 1.3 and 1.4. Lemma 1.4.3 implies that the subgroup $\iota(B_r^n)\phi_n^{(r)}(B_n(b_0))$ of B_{rn} is isomorphic to the wreath product $B_r^n \rtimes B_n(b_0)$ with respect to θ: $B_n\to S_n$. For $\nu=(\nu_1, \cdots, \nu_n)$ $(\nu_i\in A_r(f)^\wedge)$, let $\rho_\nu=\rho_{\nu_1}\otimes\cdots\otimes\rho_{\nu_n}$ be the representation of B_r^n. For $\mu\in A_{rn}(f)^\wedge$, let $V_{\mu,\nu}$ be the ρ_ν-isotypic subspace of V_μ as a $\boldsymbol{C}B_{rn}$-module Let $\delta(j)=\min\langle\theta(b_0)\rangle\cdot j$ for $j\in\Omega$. If $\nu_j=\nu_{\delta(j)}$ for $1\le j\le n$, then by Lemma 1.4.3, $V_{\mu,\nu}$ is invariant relative to the action of $B_r^n \rtimes B_n(b_0)$. In this case, let $\rho_{\mu,\nu}$ denote the representation of $\iota(B_{rn})\phi_n^{(r)}(B_n(b_0))$ obtained by restricting ρ_μ on $V_{\mu,\nu}$ and $\chi_{\mu,\nu}$ its character. Then we have the following.

Proposition 2.1.1. *Let k be the number of $\langle\theta(b_0)\rangle$-orbits of Ω and $\{\delta(j) \mid j\in\Omega\}=\{m_1, \cdots, m_k\}$. Then we have*

$$\chi_\mu(\phi_n^{(r)}(b_0)) = \sum_{\nu_{m_1}\in A_r(f)^\wedge} \sum_{\nu_{m_k}\in A_r(f)^\wedge} \chi_{\mu,(\nu_{\delta(1)}\cdots,\nu_{\delta(n)})}\phi_n^{(r)}((b_0)) .$$

Proof. The proof of Proposition 1.4.8 shows that $\chi_\mu(\phi_n^{(r)}(b_0))$ is equal to the sum of the diagonal elements of $\rho_\mu(\phi_n^{(r)}(b_0))$ corresponding to the subspaces $V_{\mu,(\nu_1,\cdots,\nu_n)}$ such that $V_{\mu,(\nu_1,\cdots,\nu_n)}=V_{\mu,(\nu_{\theta(b_0)^{-1}(1)},\cdots,\nu_{\theta(b_0)^{-1}(n)})}$. If $V_{\mu,(\nu_1,\cdots,\nu_n)}=V_{\mu,(\nu_{\theta(b_0)^{-1}(1)},\cdots,\nu_{\theta(b_0)^{-1}(n)})}$, then we have $\nu_j=\nu_{\delta(j)}$ and the sum of diagonal elements of $\rho_\mu(\phi_n^{(r)}(b_0))$ corresponding to the subspace $V_{\mu,(\nu_1,\cdots,\nu_n)}$ is equal to $\chi_{\mu,(\nu_{\delta(1)},\cdots,\nu_{\delta(n)})}\phi_n^{(r)}((b_0))$. Hence we get conclusion. □

Let $\nu=(\nu_1, \cdots, \nu_n)\in(A_r(f)^\wedge)^n$ with $\nu_j=\nu_{\delta(j)}$. Since $\iota(B_r^n)\phi_n^{(r)}(B_n(b_0))$ is isomorphic to $B_r^n \rtimes B_n(b_0)$, we see, from Proposition 1.3.4, that there is a representation $\pi_{\mu,\nu}$ of B_n such that

$$\rho_{\mu,\nu}\cong\rho_\nu^\sim\otimes\pi_{\mu,\nu}\,, \tag{2.1.2}$$

where $\rho_\nu^\sim$ is the representation of $\iota(B_r^n)\phi_n^{(r)}(B_n(b_0))$ coming from the representation ρ_ν of B_r^n as in Proposition 1.3.4. Let $\chi_\nu^\sim$, $\omega_{\mu,\nu}$ be the characters of $\rho_\nu^\sim$, $\pi_{\mu,\nu}$ respectively. Now we can state a generalized version of Theorem 1.4.10.

Theorem 2.1.3. (i) *For $\mu\in A_{rn}(f)^\wedge$,*

$$\chi_\mu(\phi_n^{(r)}(b_0)) = \sum_{\nu_{m_1}\in A_r(f)^\wedge} \sum_{\nu_{m_k}\in A_r(f)^\wedge} \Big(\prod_{j=1}^{k} \chi_{\nu_{m_j}}(1)\Big)\omega_{\mu,(\nu_{\delta(1)},\cdots,\nu_{\delta(n)})}(b_0) .$$

(ii)

$$f^{(r)}(b_0) = \sum_{\mu\in A_{rn}(f)^\wedge} \sum_{\nu_{m_1}\in A_r(f)^\wedge} \cdots \sum_{\nu_{m_k}\in A_r(f)^\wedge} a_\mu(f)\Big(\prod_{j=1}^{k} \chi_{\nu_{m_j}}(1)\Big)\omega_{\mu,(\nu_{\delta(1)},\cdots,\nu_{\delta(n)})}(b_0) .$$

Proof. Proposition 2.1.1 and (2.1.2) yield that

$$\begin{aligned}\chi_\mu(\phi_n^{(r)}(b_0)) &= \sum_{\nu_{m_1}\in A_r(f)^\wedge} \cdots \sum_{\nu_{m_k}\in A_r(f)^\wedge} \chi_{\mu,(\nu_{\delta(1)},\cdots,\nu_{\delta(n)})}(b_0) \\ &= \sum_{\nu_{m_1}\in A_r(t)^\wedge} \cdots \sum_{\nu_{m_k}\in A_r(f)^\wedge} (\chi_\nu^\sim(\phi_n^{(r)}(b_0)))\omega_{\mu,(\nu_{\delta(1)},\cdots,\nu_{\delta(n)})}(b_0) .\end{aligned}$$

Hence the first formula is an immediate consequence of following Lemma 2.1.4. The second one is obtained from (i) and (1.4.1). □

Lemma 2.1.4. *We have*

$$\chi_\nu{}^{\sim}(b_0)=\prod_{j=1}^{k}\chi_{\nu m_j}(1)\,.$$

Proof. By using (1.3.3), we have

$$\chi_\nu{}^{\sim}\phi_n^{(r)}((b_0))$$
$$=\sum_{\alpha_1,\beta_1,\cdots,\alpha_n,\beta_n}\delta_{\alpha_1\beta_1}\cdots\delta_{\alpha_n\beta_n}(\rho_{\nu_1}(1)_{\alpha_{\theta(b_0)^{-1}(1)}}\rho_{\nu_2}(1)_{\alpha_2\beta_{\theta(b_0)^{-1}(2)}}\cdots\rho_{\nu_n}(1)_{\alpha_n\beta_{\theta(b_0)^{-1}(n)}})\,.$$

Let $|S|$ denote the cardinality of a set S. Each term of the above summention is zero unless

$$\alpha_{m_1}=\beta_{\theta(b_0)^{-1}(m_j)}=\alpha_{\theta(b_0)^{-1}(m_j)}=\cdots=\alpha_{\theta(b_0)^{-|\psi^{-1}(j)|+1}(m_j)}=\beta_{m_j}\qquad(1\leq j\leq k)\,.$$

But the above equalities implies that $\alpha_i=\beta_i=\alpha_j=\beta_j$ for i and j such that $\langle\theta(b_0)\rangle\cdot i=\langle\theta(b/)_0\rangle\cdot j$. Hence we have

$$\chi_\nu{}^{\sim}(b_0)=\prod_{j=1}^{k}\left(\sum_{\alpha_{m_j}}\rho_{\nu_j}(1)_{\alpha_{m_j},\alpha_{m_j}}\right)=\prod_{j=1}^{k}\chi_{\nu_j}(1)\,.$$

This proves Lemma 2.1.4. □

2.2. Further generalizations. In this section, we state a generalized version of Theorem 1.6.4 for cable links of multi-component links. We omit proofs of the results in this section, which are anologous to those of corresponding facts in Section 1.

Let K be a k-component link. A bijection Ψ from the set of connected components of K to $\{1, 2, \cdots, k\}$ is called a *marking* of K and the pair (K, Ψ) is called a *(k-component) marked link*. The connected component C of K with $\Psi(C)=i$ is called the *i-th component* of K. Let $R=(r_1, r_2, \cdots, r_k)\in\boldsymbol{N}^k$. Let $(K, \Psi)^{(R)}$ denote the link diagram obtained by replacing each crossing point of a link diagram of K as in Figure 6. The link type of $(K, \Psi)^{(R)}$ is not depend on the choice of link diagrams of K. We call $(K, \Psi)^{(R)}$ the *R-parallel version* of (K, Ψ). For a link invariant X, let $X^{(R)}(K, \Psi)=X((K, \Psi)^{(R)})$. Then $X^{(R)}$ is an invariant of k-component marked links and is called the *R-parallel version* of X. We can show a generalized version of Theorem 2.1.3 (ii) for $X^{(R)}$. We omit the details.

Let (K, Ψ) be a k-component marked link, $R=(r_1, r_2, \cdots, r_k)\in\boldsymbol{N}^k$ and $b=(b_1, b_2, \cdots, b_k)\in B_{r_1}\times\cdots\times B_{r_k}$. Let $(K, \Psi)_b$ denote the link obtained by inserting the braid b_i $(1\leq i\leq k)$ to the bunch of the components of $(K, \Psi)^{(R)}$ corresponding to the component of K sent to i by Ψ as in Figure 7. We call $(K, \Psi)_b$ the *cable link* of (K, Ψ) associated with b. For a link invariant X, let $X_b(K, \Psi)=X((K, \Psi)_b)$. Then X_b is an invariant of k-component marked links. Then we have:

Theorem 2.2.1. *Let X be a link invariant of trace type, $R=(r_1, r_2, \cdots, r_k)\in$*

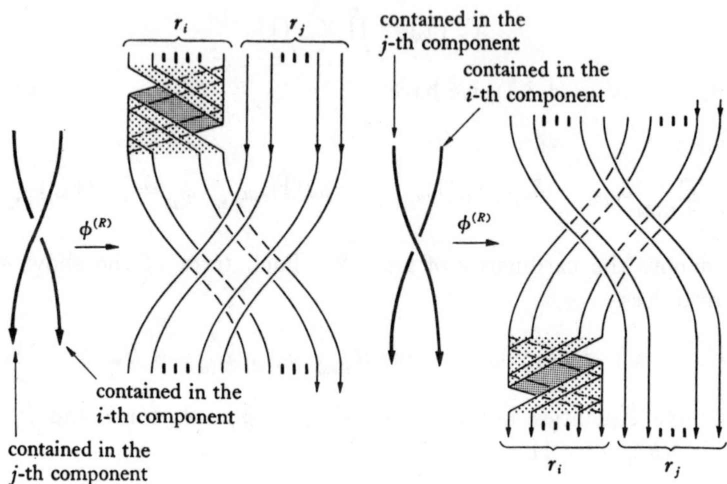

Figure 6

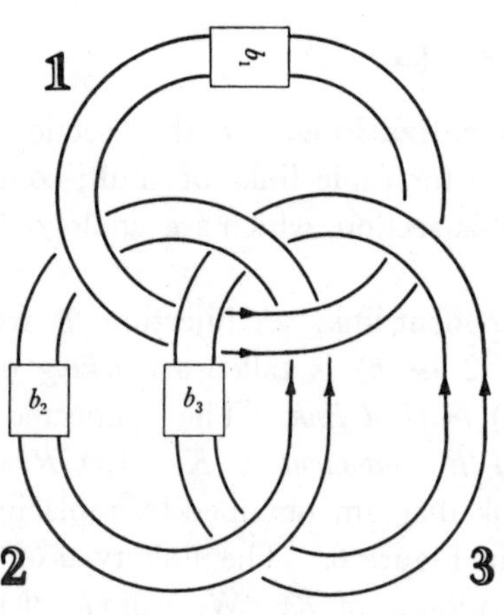

Figure 7

N^k and $b=(b_1, b_2, \cdots, b_k)\in B_{r_1}\times\cdots\times B_{r_k}$. Then there are invariants $X^{(R,\nu)}$ of k-component marked links parametrized by $\nu=(\nu_1, \nu_2, \cdots, \nu_k)\in A_{r_1}(X)^\wedge\times\cdots\times A_{r_k}(X)^\wedge$ which satisfy the following. For a k-component marked link (K, Ψ),

$$X_b(K, \Psi) = \sum_{(\nu_1,\cdots,\nu_k)\in A_{r_1}(X)^\wedge\times\cdots\times A_{r_k}(X)^\wedge} \left(\prod_{j=1}^{k} \chi_{\nu_j}(b_j)\right) X^{(R,\nu)}(K, \Psi)\,.$$

This theorem is a generalization of Theorems 1.5.1 and 1.6.4.

3. The r-parallel version of the two-variable Jones polynomial.

In this section we review representation theory of Iwahori algebras and the two-varaible Jones polynomial. At the end of this section, we shall construct representations of B_3 assoiated with the two-parallel version of the two-variable Jones polynomial. Fix non-zero complex numbers α and s such that s is not equal to any root of unity. Let $q^{k/2}=s^k$ for an integer k.

DEFINITION 3.1 ([3], p. 55, ex. 23; [18]). Let $H_n(q)$ be a $\boldsymbol{C}$-algebra with a unit defined by the following relations:

$$(3.2)\quad \begin{aligned} H_n(q) = \langle T_1, T_2, \cdots, T_{n-1} \mid\, & T_i^2+(1-q)T_i-q = 0 && (1\leq i\leq n-1), \\ & T_iT_{i+1}T_i = T_{i+1}T_iT_{i+1} && (1\leq i\leq n-2), \\ & T_iT_j = T_jT_i && (1\leq i<j-1\leq n-2)\rangle . \end{aligned}$$

We call $H_n(q)$ the *Iwahori's Hecke algebra of type* A_{n-1}. For simplicity, we denote $H_n(q)$ by H_n if there is no fear of confusion.

It is known that H_n is semisimple (see, e.g. [11]) and isomorphic to the group algebra $\boldsymbol{C}S_n$ of the symmetric group S_n of degree n.

DEFINITION 3.3. Let $\Lambda(n)$ be the set of partitions of n, i.e.

$$\Lambda(n) = \{(\lambda_1, \lambda_2, \cdots) \mid \sum_i \lambda_i = n,\ \lambda_i \in \boldsymbol{N} \cup \{0\},\ \lambda_1 \geq \lambda_{i+1}\ (i \in \boldsymbol{N})\} .$$

For $\lambda=(\lambda_1, \lambda_2, \cdots)\in\Lambda(n)$, let $\lambda_i^*=\max\{j \mid \lambda_j\geq i\}$ if $i\leq\lambda_1$ and $\lambda_i^*=0$ if $i>\lambda_1$. We call $\lambda^*=(\lambda_1^*, \lambda_2^*, \cdots)$ the *dual partition of* λ.

As is well-known, the irreducible representations of S_n are a parametrized by $\Lambda(n)$. Hence $\Lambda(n)$ parametrizes the irreducible representations of H_n, too. Let $(\rho_\lambda, V_\lambda)$ be the representation of H_n parametrized by a partition $\lambda\in\Lambda(n)$ and χ_λ its character. For each $\lambda\in\Lambda(n)$, we know [12] the representation matrices of $\rho_\lambda(T_i)$ $(1\leq i\leq n-1)$ with respect to a basis parametrized by the standard tableaus corresponding to the partition λ.

REMARK 3.4. Fix $\lambda\in\Lambda(n)$, and let $\mathcal{S}_\lambda$ be the set of standard tableaus corresponding to the partition λ. For $S\in\mathcal{S}_\lambda$, S^* denote the transpose of S, which is a standard tableau corresponding to λ^*. Let $\{e_S \mid S\in\mathcal{S}_\lambda\}$ and $\{e_{S^*} \mid S\in\mathcal{S}_\lambda\}$ be the basis of V_λ and V_{λ^*} given in [12] respectively. Let η: $V_\lambda\to V_{\lambda^*}$ be the linear isomorphism defined by $\eta(e_S)=e_{S^*}$ $(S\in\mathcal{S}_\lambda)$. Then there is a diagonal matrix D with respect to the basis $\{e_S \mid S\in\mathcal{S}_\lambda\}$ such that

$$\rho_\lambda(T_i)(q) = D^{-1}\eta^{-1}(-q\rho_{\lambda^*}(T_i)(q^{-1}))\eta D .$$

By the defining relations of B_n and H_n, there is an algebra homomorphism $p_n^{(\alpha)}$: $\boldsymbol{C}B_n\to H_n$ defined by $p_n^{(\alpha)}(\sigma_i)=\alpha T_i$.

Theorem 3.5 ([7]). *Let* $l=(-q)^{1/2}\alpha$, $m=(-1)^{1/2}(q^{-1/2}-q^{1/2})$ *and* $P(\cdot)(l, m)$ *be the two-variable Jones polynomial. Then, for* $b\in B_n$, *we have*

$$P(b^\wedge)=\sum_{\lambda\in\Lambda(n)} a_\lambda(P)\chi_\lambda(p_n^{(\alpha)}(b)),$$

where the coefficients $a_\lambda(P)\in \boldsymbol{C}$ $(\lambda\in\Lambda(n))$ *are given in* [10] *or* [15].

The coefficients $a_\lambda(P)$ $(\lambda\in\Lambda(n))$ satisfies

$$a_\lambda(P)(\alpha, q)=a_{\lambda^*}(P)(-\alpha q, q^{-1})\,. \tag{3.6}$$

Let K be a link. For a positive integer r and $\nu\in\Lambda(r)$, let $P^{(r,\nu)}$ be the invariant parametrized by (r, ρ_ν) as in Section 1.5. Then Remark 3.4 and (3.6) implies the following.

$$P^{(r,\nu)}(K)(\alpha, q)=P^{(r,\nu^*)}(K)(-\alpha q, q^{-1})\,. \tag{3.7}$$

Let I be a subset of $\{1, \cdots, n-1\}$, H_I the subalgebra of H_n generated by $\{T_i\,|\,i\in I\}$, and S_I the subgroup of S_n generated by $\{(i\,i+1)\,|\,i\in I\}$. Let ρ'_λ be the representation of S_n parametrized by $\lambda\in\Lambda(n)$. Let $\{\lambda_1, \lambda_1+\lambda_2, \cdots, \lambda_1+\lambda_2+\cdots+\lambda_p\}=\{1, 2, \cdots, n\}\backslash I$ and $\Lambda_I=\Lambda(\lambda_1)\times\Lambda(\lambda_2)\times\cdots\times\Lambda(\lambda_p)$. Then the irreducible representations of H_I and S_I are parametrized by Λ_I. The restriction $\rho_\lambda|_{H_I}$and $\rho'_\lambda|_{S_I}$ are sums of irreducible representations of H_I and S_I respectively, e.g. $\rho_\lambda|_{H_I}=\bigoplus_{\mu\in\Lambda_I} m_{\lambda,\mu}\rho_\mu$ and $\rho'_\lambda|_{H_I}=\bigoplus_{\mu\in\Lambda_I} m'_{\lambda,\mu}\rho'_\mu$, where ρ_μ and ρ'_μ are representation of H_I and S_I parametrized by $\mu\in\Lambda_I$.

Proposition 3.8. *Let* $\mu_0\in\Lambda_I$ *such that the corresponding representation* ρ'_{μ_0} *of* S_b *is trivial. Then the above multiplicities* m_{λ,μ_0} *and* m'_{λ,μ_0} *are equal.*

Proof. The construction of each irreducible representations of H_n and H_I in [12] satisfies the following: The entries of the matrices of the generators T_i are all rational functions of the parameter q with poles at roots of unity. On the other hand, the above proposition is proved for infinitely many integer values of q by using [5] Theorem 7.2. But we know that $\boldsymbol{C}S_I$ and H_I are semi-simple if q is not equal to any roots of unity. Hence m_{λ,μ_0} and m'_{λ,μ_0} are equal if q is not equal to any roots of unity. □

REMARK 3.9. Because of the above proposition, we can calculate m_{λ,μ_0} by the Littlewood-Richardson rule ([13], 2.8.13).

Let ρ_λ be the representation of H_{rn} parametrized by $\lambda\in\Lambda(rn)$ and ρ_ν that of H_r parametrized by $\nu\in\Lambda(r)$. Let $\pi_{\lambda,\nu}$ be the representation of B_{rn} parametrized by ρ_λ and ρ_ν as in Section 1.4. In Table 2, the representation matrices of $\pi_{\lambda,\nu}$ are given in the case of $n=3$, $r=2$ and $\nu=(2)$. By using Theorems 1.4.10, 1.5.1, 3.5 and (3.7), we may calculate $P^{(2,(2))}$, $p^{(2,(11))}$ and $P^{(2)}$ for knots equivalent to the closures of 3-braids. To obtain these matrices, we use the representation matrices of the generators of H_6 given by W-graphs introduced in [18]. The W-graphs correspond,ng to the irreducible representations of H_6 are actually

Table 2

$\nu_0=(2)$

$\lambda=(6)$ $\quad \pi_{\lambda,\nu_0}(\sigma_1)=\pi_{\lambda,\nu_0}(\sigma_2)=\alpha^2q^2,$

$\lambda=(51)$ $\quad \pi_{\lambda,\nu_0}(\sigma_1)=\alpha^2\begin{pmatrix}-1 & 0\\ b & q^2\end{pmatrix},\ \pi_{\lambda,\nu_0}(\sigma_2)=\alpha^2\begin{pmatrix}-1 & q\\ 0 & q^2\end{pmatrix},$

$\lambda=(42)$ $\quad \pi_{\lambda,\nu_0}(\sigma_1)=\alpha^2\begin{pmatrix}1/q & 0 & 0\\ -1/\sqrt{q} & -1 & 0\\ 1 & q\sqrt{q} & q^2\end{pmatrix},\ \pi_{\lambda,\nu_0}(\sigma_2)=\alpha^2\begin{pmatrix}q^2 & q\sqrt{q} & 1\\ 0 & -1 & -1/\sqrt{q}\\ 0 & 0 & 1/q\end{pmatrix},$

$\lambda=(411)$ $\quad \pi_{\lambda,\nu_0}(\sigma_1)=\pi_{\lambda,\nu_0}(\sigma_2)=-\alpha^2,$

$\lambda=(33)$ $\quad \pi_{\lambda,\nu_0}(\sigma_1)=\pi_{\lambda,\nu_0}(\sigma_2)=-\alpha^2,$

$\lambda=(321)$ $\quad \pi_{\lambda,\nu_0}(\sigma_1)=\alpha^2\begin{pmatrix}1/q & 0\\ -1/\sqrt{q} & -1\end{pmatrix},\ \pi_{\lambda,\nu_0}(\sigma_1)=\alpha^2\begin{pmatrix}-1 & -1/\sqrt{q}\\ 0 & 1/q\end{pmatrix},$

$\lambda=(222)$ $\quad \pi_{\lambda,\nu_0}(\sigma_1)=\pi_{\lambda,\nu_0}(\sigma_2)=\alpha^2/q,$

$\lambda=(3111),\ (2211),\ (21111),\ (111111)$ $\quad \dim V_{\lambda,\nu_0}=0.$

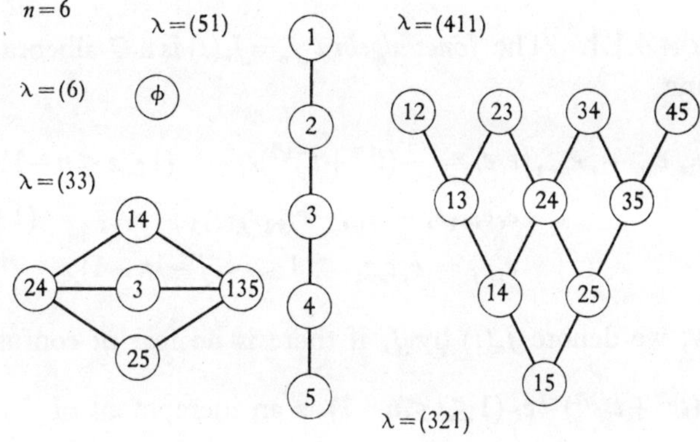

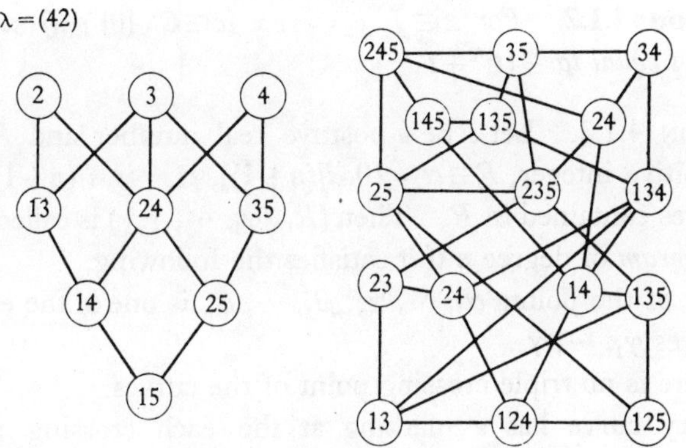

Figure 8

constructed by Naruse and Gyoja (Figure 8).

4. The r-parallel version of the one-variable jones polynomial.

In this section we discuss the r-parallel version $V^{(r)}$ of the one-variable Jones polynomial V in detail. We give a formula for the one-variable Jones polynomials of satellite links (Theorem 4.3.2). Let $V^{(r,\nu)}$ denote the invariant associated with $\nu\in A_r(V)^\wedge$. The representation matrices of $\pi_{\mu,\nu}\in B_3{}^\wedge$ ($\mu\in A_{3r}(V)^\wedge$, $\nu\in A_r(V)^\wedge$) associated with $V^{(r,\nu)}$ are gives given explicitly (Theorem 4.4.1). By using these matrices we can easily compute $V^{(r)}(b^\wedge)$ for a knot $b^\wedge$ which is the closure of a 3-braid b. The cases $r=2$ and $r=3$ are discussed closely in (4.4.3)–(4.4.10).

4.1. The jones algebra. In this section we review the definition of the Jones algebra J_n, which is the associated algebra $A_n(V)$ (Definition 1.1.6) of the one-variable Jones polynomial V. We also review the construction of the irreducible representations of J_n [15] in terms of rectangular diagrams for our later convenience. Fix a non-zero complex number s which is not equal to any root of unity. Let $t^{k/4}=s^k$ for an integer k.

DEFINITION 4.1.1. The *Jones algebra* $J_n=J_n(t)$ is a $\boldsymbol{C}$-albebra with 1 defined by the following.

$$\begin{aligned} J_n(t) = \langle e_1, e_2, \cdots, e_{n-1} \mid e_ie_i &= -(t^{1/2}+t^{-1/2})e_i \qquad (1\leq i\leq n-1)\,, \\ e_ie_{i+1}e_i &= e_i\,, \quad e_{i+1}e_ie_{i+1} = e_{i+1} \qquad (1\leq i\leq n-2)\,, \\ e_ie_j &= e_je_i \qquad (1\leq i<j-1n-1)\rangle\,. \end{aligned}$$

For simplicity, we denote $J_n(t)$ by J_n if there is no fear of confusion.

Since $-(t^{1/2}+t^{-1/2})^{-1}e_i$ $(1\leq i\leq n-1)$ is an idempotent of J_n, we have

Proposition 4.1.2. *For* $x\in J$, $e_ix=cx$ $(c\in\boldsymbol{C}\backslash\{0\})$ *iff* $x\in e_iJ_n$. *In this case, c is always equal to* $-(t^{1/2}+t^{-1/2})$.

DEFINITION 4.1.3. Let s be a positive real number and $R=[0, s]\times[0, 1]$. Let n be a positive integer, $R\ni\alpha_i=(0, i/(n+1))$, $\beta_i=(s, i/(n+1))$ $(1\leq i\leq n)$ and $\gamma_1, \cdots, \gamma_n$ curves contained in R. Then $(R, \{\gamma_1, \cdots, \gamma_n\})$ is called an (unoriented) *rectangular diagram* of degree n if it satisfies the following.

(i) Any of the points $\alpha_1, \cdots, \alpha_n, \beta_1, \cdots, \beta_n$ is one of the end points of the curves $\gamma_1, \cdots, \gamma_n$.

(ii) There is no triple crossing point of the curves.

(iii) Every curve has a marking at the each crossing point indicating whether the curve under consideration is the over path or the under path at the crossing.

(iv) The intersection of the curves $\gamma_1, \cdots, \gamma_n$ and the boundary of R is equal to $\{\alpha_1, \cdots, \alpha_n, \beta_1, \cdots, \beta_n\}$.

A diagram $(R, \{\gamma_1, \cdots, \gamma_n\})$ is called a *rectangualr diagram without crossing points* of degree n if the curves $\gamma_1, \cdots, \gamma_n$ do not intersect themselves. If orientations are given to all the curves in R, then R is called an *oriented rectangular diagram* of degree n.

We call $\alpha_1, \cdots, \alpha_n$ (respectively $\beta_1, \cdots, \beta_n$) the points at the *top* (respectively *bottom*) of R. Two rectangular diagrams $([0, s]\times[0, 1], \{\gamma_1, \cdots, \gamma_n\})$ and $([0, s']\times[0, 1], \{\gamma'_1, \cdots, \gamma'_n\})$ of degree n are called *equivalent* if there is a homeomorphism f: $[0, s]\times[0, 1]\rightarrow[0, s']\times[0, 1]$ such that $f((0, t))=(0, t)$, $f((s, t))=(s', t)$ $(t\in[0, 1])$ and $f(\gamma_i)=\gamma'_i (1\leq i\leq n)$.

We define algebras D_n $(n\in \boldsymbol{N})$ over $\boldsymbol{C}$. Let E'_n denote the set of the equivalence classes of rectangular diagrams aithout crossing points of degree n. With the convention $\#E'_0=1$, the number $\#E'_n$ of elements of E'_n $(n\in \boldsymbol{N})$ satisfy the recursive relations $\#E'_n=\sum_{k=0}^{n-1}\#E'_k\#E'_{n-k-1}$. Hence $\#E'_n$ is equal to the Catalan number $\binom{2n}{n}/(n+1)$ ([20], § 2.3.4.4). As a $\boldsymbol{C}$-vector space, we put $D_n=\bigotimes_{e\in E'_n}\boldsymbol{C}e$; in particular,

$$\dim_{\boldsymbol{C}} D_n = \binom{2n}{n}/(n+1)\,. \tag{4.1.4}$$

To define the multiplication in D_n, it is enough to define the product ab for two equivalence classes a and b of rectangular diagrams $([0, s[\times[0, 1], \{\gamma_1, \cdots, \gamma_n\})$ and $([0, s']\times[0, 1], \{\gamma'_1, \cdots, \gamma'_n\})$. This is done similarly as in the case of braids by the following rule.

(a) Let $R=[0, s+s']\times[0, 1]$. Let g: $[0, s]\times[0, 1]\rightarrow R$ and g': $[0, s']\times[0, 1]\rightarrow R$ be the mappings defined by $g((x, y))=(x, y)$ and $g'((x, y))=(x+s, y)$ respectively.

(b) Let $(\bigcup_{i=1}^{m}\delta_i)\cup(\bigcup_{i=1}^{n}\gamma''_i)$ be the decomposition of $(\bigcup_{i=1}^{n}g(\gamma_i))\cup(\bigcup_{i=1}^{n}g'(\gamma'_i))$ into connected components, where δ_i $(1\leq i\leq m)$ are closed curves and γ''_i $(1\leq i\leq n)$ are curves with end points.

(c) Let d be the equivalence class of the rectangular diagram $(R, \{\gamma''_1, \cdots, \gamma''_n\})$. Then $ab=(-t^{1/2}-t^{-1/2})^m d$.

Let e'_i $(1\leq i\leq n-1)$ denote the rectangular diagrams given in Figure 9. Then we have

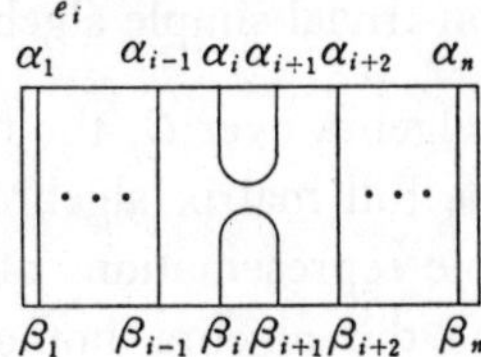

Figure 9

$$\text{(4.1.5)}\qquad \begin{aligned} &e_i'^2 = (-t^{1/2}-t^{-1/2})e_i' && (1\leq i\leq n-1)\,. \\ &e_i' e_{i+1}' e_i' = e_i'\,, \quad e_{i+1}' e_i' e_{i+1}' = e_{i+1}' && (1\leq i\leq n-2)\,, \end{aligned}$$

Hence there is a homomorphism η from J_n to D_n sending e_i to e_i'. But, by [15] and (4.1.4), we have

$$\dim_{C} J_n = \binom{2n}{n}/(n+1) = \dim_{C} D_n\,.$$

Hence η is an siomorphism. Let $E_n=\eta^{-1}(E_n')$. Then E_n is a basis of J_n. The definition of the multiplication in D_n implies that, for e_i' and $x\in E_n'$, $e_i'x$ is a non-zero scalar multiple of an element of E_n'. Hence e_iJ_n is spanned by a subset of E_n. Combining this observation with Proposition 4.1.2, we get the following:

Proposition 4.1.6. *The subspace e_iJ_n of J_n is spanned by $E_n^{(i)}=\{\mu^{-1}(x)\mid x\in E_n'$, the i-th and $(i+1)$-th points at the top of x are connected by a curve of $x\}$.*

For $y\in E_n'$, let $T_D(y)$ denote the number of strings of y connecting two points at the top of y. For example,

$$\text{(4.1.7)}\qquad T_D(\eta(e_1e_3\cdots e_{2j-1}))=j\,.$$

Let $E_{n,i}=\{\eta^{-1}(y)\mid y\in E_n',\ T_D(y)\geq i\}$ and $J_{n,1}^{\bullet}$ the two-sided ideal of J_n generated by the elements of $E_{n,i}$. Since $T_D(yy')\geq\max(T_D(y),\ T_D(y'))$ for $y,\ y'\in E_n$, $J_{n,i}^{\bullet}$ is equal to the $\boldsymbol{C}$-linear span of $E_{n,i}$ in J_n. For $x\in J_n$, we have

$$\text{(4.1.8)}\qquad e_{i(1)}e_{i(2)}\cdots e_{i(k)}x\in J_{n,k}^{\bullet}\qquad (i(j)<i(j+1)-1 \text{ for } 1\leq j\leq k)\,.$$

Let $J_{n,i}=J_{n,i}^{\bullet}/J_{n,i+i}^{\bullet}$. Since J_n is known to be semisimple, the canonical projection $J_{n,i}^{\bullet}\to J_{n,i}$ splits. Hence $J_{n,i}$ can be regarded as a two-sided ideal of J_n.

Proposition 4.1.9 *The two-sided ideals $J_{n,i}$ $(0\leq i\leq[n/2])$ of J_n are simple algebras and $J_n\simeq\bigoplus_{i=0}^{[n/2]}J_{n,i}$.*

Proof. Let $e(1,\ 2i-1)=e_1e_3\cdots e_{2i-1}$ $(0\leq i\leq[n/2])$. Since $e(1,\ 2i-1)\in J_{n,i}^{\bullet}$ and $e(1,\ 2i-1)\notin J_{n,i+1}^{\bullet}$, $J_{n,i}\neq\{0\}$ for $0\leq i\leq[n/2]$. We also know ([15], § 11) that J_n is a direct sum of $[n/2]+1$ two-sided ideals which are simple algebras. Hence, by using Jordan-Holder's Theorem, we conclude that the two-sided ideals $J_{n,i}$ $(0\leq i\leq[n/2])$ are non-trivial simple algebras. □

Since J_n is a semisimple algebra over $\boldsymbol{C}$, the two-sided ideals J_{ni} of J_n $(0\leq i\leq[n/2])$ are isomorphic to the full matrix algebras over $\boldsymbol{C}$. Let $\rho_{n,i}\colon J_n\to J_{n,i}$ denote the associated irreducible representations of J_n, and $\chi_{n,i}$ their charactews.

Let $p_n,\ p_n'\colon \boldsymbol{C}B_n\to J_n(t)$ be the algebra homomorphisms defined by $\sigma_i\to -t^{1/2}-te_i$ and $\sigma_i\to t^{-1/4}+t^{1/4}e_i$ respectively. Then the Jones polynomial V and the bracket polynomial $\langle\cdot\rangle$ for the closure of an n-braid b are written as linear combinations of these characters, *i.e.*

$$V(b\hat{}) = \sum_{i=0}^{[r/2]} a_{n,i}(V)\chi_{n,i}(p_n(b)),$$

(4.1.10)

$$\langle|b\hat{}|\rangle = \sum_{i=0}^{[r/2]} a_{n,i}(V)\chi_{n,i}(p'_n(b)).$$

The coefficients $a_{n,i}(V)$ are given by

(4.1.11) $$a_{n,i}(V) = (-1)^{n+1}(t^{n/2+1/2-i} - t^{-n/2-1/2+i})(t-t^{-1}),$$

(see [10], [15], [16]).

Note that none of the coefficients $a_{n,i}(V)$ are equal to zero because t is not a root of unity.

REMARK 4.1.12. Let $\varphi_n: H_n(t) \to J_n(t)$ be the algebra homomorphism defined by $\varphi_n(T_i) = -1 - t^{1/2}e_i$. The compositions $\rho_{n,i}\circ\varphi_n$ for $0 \leq i \leq [n/2]$ give irreducible representations of $H_n(t)$ parametrized by the partitions $(\underbrace{22\cdots2}_{i\text{-time}}\underbrace{11\cdots1}_{(n-2i)\text{-time}})$.

DEFINITION 4.1.13. Let G_n denote the free semigriup generated by 1, f_i, σ_i, σ_i^{-1} $(1 \leq i \leq n-1)$, $G = \{(g, n) \mid g \in G_n\}$ and for $(g, n) \in G$, let $d(g, n)$ denote the rectangular diagram (Definition 4.1.3) of (g, n) defined as in Figure 10.

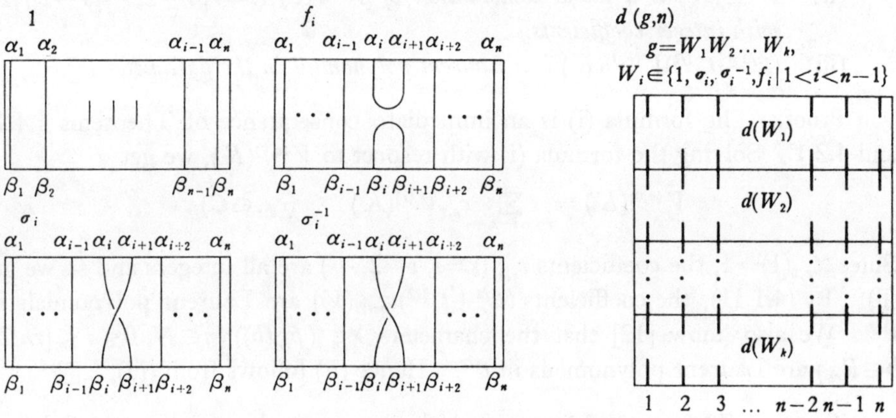

Figure 10

By the analogy with braids we use the term *closure* of a rectangular diagram $d(g, n)$, written $(g, n)\hat{}$, to mean the link formed by joining the n points at the top of $d(g, n)$ to those at the bottom without further crossings. Let $\Psi_n: CG_n \to J_n$ denote the algebra homomorphism defined by

$$\Psi_n(f_i)=e_i\,,\quad \Psi_n(\sigma_i)=t^{-1/4}+t^{1/4}e_i\,,\quad \Psi_n(\sigma_i^{-1})=t^{1/4}+t^{-1/4}e_i\,.$$

By using the original definition of $\langle\cdot\rangle$ in [16], we can show that

$$\langle (g,n)^\wedge\rangle=\sum_{i=0}^{[r/2]}a_{n,i}(V)\chi_{n,i}(\Psi_n(g))\,, \tag{4.1.14}$$

where the coefficients $a_{n,i}(V)$ are given by (4.1.11).

4.2. Relations among the invariants associated with the decomposition (1.5.1) of $V^{(r)}$. Let p_r be the homomorphism from $\boldsymbol{C}B_r$ to J_r defined just before (4.1.10). Let $V^{(r,i)}$ denote the invariant associated with the irreducible representation $\rho_{r,i}\circ p_r$ of B_r (see Sections 1.4 and 4.1). In this section, we give relations among $V^{(r,i)}$. In the following we use the conventions that $V^{(0)}(K)=V^{(0,0)}(K)=\langle|K|\rangle^{(0)}=\langle|K|\rangle^{(0,0)}(K)=(-t^{-1/2}-t^{1/2})^{-1}$ for any link diagram K.

Theorem 4.2.1. *Let K be a knot. Then we have*

$$V^{(r,i)}(K)=V^{(r-2k,i-k)}(K)\qquad \text{for}\quad r\geq 0,\ 0\leq i\leq [r/2],\ 0\leq k\leq i\,.$$

REMARK 4.2.2. The same is also true in the case when K is a general link. We restrict our attention to the present case for simplicity.

Corollary 4.2.3. *Let K be a knot. Then we have*

(i) $V^{(r)}(K)=\sum_{i=0}^{[r/2]}\chi_{r,i}(1)V^{(r-2i,0)}(K)$ *for* $r\geq 0$,

(ii) $V^{(r,0)}(K)$ *is a linear combination of* $V^{(s)}(K)$ $(s=r, r-2, \cdots, r-2[r/2])$ *with integer coefficients.*

(iii) $(t^{1/2}+t^{-1/2})V^{(r,j)}(K)$ *is a Laurent polynomial in the parameter* $t^{1/2}$.

Proof. The formula (i) is an immediate consequence of Theorems 1.4.10 and 4.2.1. Solving the formula (i) with respect to $V^{(s,0)}(K)$, we get

$$V^{(s,0)}(K)=\sum_{r=s,s-2,\cdots}c_{s,r}V^{(r)}(K)\qquad (c_{s,r}\in\boldsymbol{C})\,.$$

Since $\chi_{r,0}(1)=1$, the coefficients $c_{s,r}$ $(r=s, s-2, \cdots)$ are all integers and so we get (ii). By (4.1.11), the coefficients $(t^{1/2}+t^{-1/2})a_{rn,i}(V)$ are Laurent polynomials in $t^{1/2}$. We also know [12] that the characters $\chi_{rn,i}(p_{rn}(b))$ $(n\in\boldsymbol{N},\ 0\leq i\leq [rn/2],\ b\in B_{rn})$ are Laurent polynomials in $t^{1/2}$. Hence (ii) follows from (iii). □

To prove Theorem 4.2.1, we need the bracket polynomial $\langle\cdot\rangle$ defined on G (see the last paragraph of Section 4.1). By (4.1.10), the bracket polynomial is of trace type. Applying Theorem 1.6.4 to $\langle\cdot\rangle$, we get the following.

Proposition 4.2.4. *For $b'\in\boldsymbol{C}B_r$ and a braid (b,n) we have*

$$\langle (b,n)^\wedge\rangle_{b'}^{(r)}=\sum_{i=0}^{[r/2]}\chi_{r,i}(p_r'(b'))\langle (b,n)\rangle^{(r,i)}\qquad \text{for}\quad r\geq 0\,.$$

Let $f'_i \in CB_{rn}$ *such that* $p'_{rn}(f'_i)=e_i$ $(1\leq i\leq rn-1)$. *Then we have the following:*

Proposition 4.2.5. *For* $b\in B_n$, *we have*

$$\langle (b, n)\rangle^{(r)}_{f'_1 f'_3\cdots f'_{2j-1}} = (-t^{3/4})^{-2jw(b^\wedge)}(-t^{1/2}-t^{-1/2})^j \langle (b, n)\rangle^{r-2j}.$$

Proof. For $g\in G_{rn}$, we put $\langle g\rangle=\langle d(g, n)^\wedge\rangle\in \boldsymbol{C}$. By Definition 1.1.10 of the bracket polynomial, the mapping $\langle\cdot\rangle$: $G_{rn}\to \boldsymbol{C}$ is factored by the projection Ψ_{rn}: $G_{rn}\to J_{rn}$ (see the last paragraph of Section 4.1). Since $p'_{rn}(f'_1 f'_3, \cdots, f'_{2j-1})= \Psi_{rn}(f_1 f_3 \cdots f'_{2j-1})$, we have

$$\begin{aligned}\langle (b, n)\rangle^{(r)}_{f'_1 f'_3\cdots f'_{2j-1}} &= \langle (f'_1 f'_3 \cdots f'_{2j-1}\phi^{(r)}_n(b), rn)\rangle \\ &= \langle (f_1 f_3 \cdots f_{2j-1}\phi^{(r)}_n(b), rn)\rangle .\end{aligned}$$

The following lemma shows that the link diagram $(f_1 f_3\cdots f_{2j-1}\phi^{(r)}_n(b), rn)^\wedge$ is regular isotopic to

$$(\phi^{(r-j)}_n(b), (r-2j)n)^\wedge \cup (\bigcup_{i=1}^{j}(f_1\sigma_1^{-2w(b^\wedge)}, 2)^\wedge) \qquad \text{(disjoint union)}.$$

For a link diagram K which is a disjoint union of two link diagrams K_1 and K_2, it is known that $\langle K\rangle=(-t^{1/2}-t^{-1/2})\langle K_1\rangle\langle K_2\rangle$. It is also known ([16], Proposition 2.5) that $\langle (f_1\sigma_1^{-2w(b^\wedge)}, 2)\rangle=(-t^{3/4})^{-2w(b^\wedge)}$. Hence we have

$$\begin{aligned}&\langle (f_1 f_3 \cdots f_{2j-1}\phi^{(r)}_n)(b), rn)^\wedge\rangle \\ &= \langle \phi^{(r-j)}_n((b), (r-2j)n)^\wedge \cup (\bigcup_{i=1,\cdots,j} (f_1\sigma_1^{-2w(b^\wedge)}, 2)^\wedge)\rangle \\ &= (-t^{3/4})^{-2jw(b^\wedge)}(-t^{1/2}-t^{-1/2})^j \langle (\phi^{(r-j)}_n(b), (r-2j)n)\rangle ,\end{aligned}$$

as required. □

Lemma 4.2.6. *For* $b\in B_n$, *the link diagram* $(f_1 f_3 \cdots f_{2j-1}\phi^{(\pm)}_n(b), rn)^\wedge$ *is regular isotopic to the disjoint union of the link diagrams* $(\phi^{(r-j)}_n(b), (r-2j)n)^\wedge$ *and copies* j *of* $(f_1\sigma_1^{-2w(b^\wedge)}, 2^\wedge)$.

Proof. Let $\beta=s^{(\varepsilon(1))}_{i(1)} s^{(\varepsilon(2))}_{i(2)} \cdots s^{(\varepsilon(k))}_{i(k)}$ where $3\leq i(j)\leq r-1$, $-1\leq \varepsilon\leq (j)\leq 1$ $(1\leq j\leq k)$, $s^{(0)}_{i(j)}=f_{i(j)}$, $s^{(1)}_{i(j)}=\sigma_{i(j)}$ and $s^{(-1)}_{i(j)}=\sigma^{-1}_{i(j)}$. Let $\beta^*=s^{(\varepsilon(1))}_{i(1)-2} s^{(\varepsilon(2))}_{i(2)-2} \cdots s^{(\varepsilon(k))}_{i(k)-2}$. Then Figure 11 demonstrates that the link diagram $(f_2\beta\phi^{(r)}_n(b), rn)^\wedge$ is regular isotopic to the disjoint union of the link diagrams $(\beta^*\phi^{(r-2)}_n(b), (r-2)n)^\wedge$ and $(f_1, \sigma_1^{-2w(b^\wedge)}, 2)^\wedge$. Hence an induction on j proves the statement of the lemma. □

Lemma 4.2.7. *Let* $e(j)=e_1 e_3 \cdots e_{2j-1}\in J_r$ $(1\leq j\leq [r/2])$. *Let* $\chi_{r,i}$ *be the character of the representation* $\rho_{r,i}$ *of* J_r, *defined in Section* 4.1. *We have*

(a) $\chi_{r,i}(e(j))=0$ *if* $i<j$,

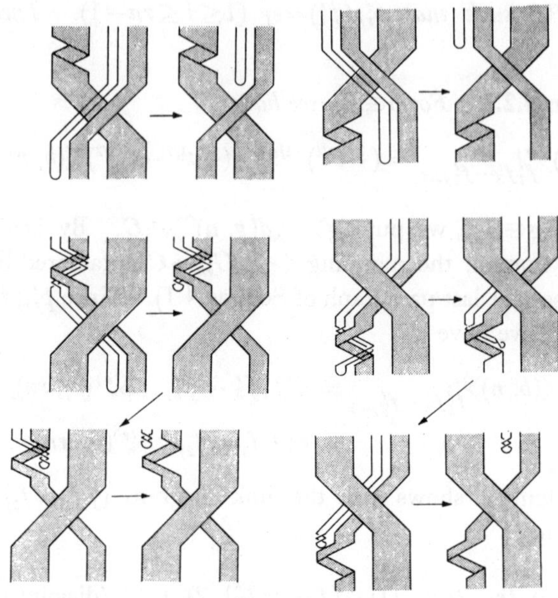

Figure 11

(b) $\chi_{r,i}(e(j))=(-t^{1/2}-t^{-1/2})^j\chi_{r-2j,i-j}(1)$ *if* $0\leq j\leq i$,
where we used the convention that $\chi_{0,0}(1)=1$.

Proof. Since $T_D(e(j))=j$ by (4.1.7), $\rho_{r,i}(e(j))=0$ if $j<i$ and so we get the first formula of the above lemma. We regard J_{r-1} as a subalgebra of J_r by the inclusion homomorphism sending $e_i\in J_{r-1}$ to $e_i\in J_r$ $(1\leq i\leq r-2)$. For an element $x\in J_r$ contained in the subalgebra J_{r-1}, we know [15] that

$$\begin{aligned}&\chi_{r,0}(x)=\chi_{r-1,0}(x)\,,\\ (4.2.8)\qquad&\chi_{r,i}(x)=\chi_{r-1,i}(x)+\chi_{r-1,i-1}(x)\qquad\text{if}\quad 0<i<r/2\,,\\ &\chi_{2r,r}(x)=\chi_{2r-1,r-1}(x)\,.\end{aligned}$$

Using (4.2.8) and Lemma 4.2.9 below, we get the part (b) by induction on r. The details are omitted. □

Lemma 4.2.9. $\chi_{2r,r}(e(r))=(-t^{1/2}-t^{-1/2})^r$.

Proof. By Lemma 4.2.7 (a) and (4.2.8), it is enough to show:

$$\chi_{2r,r}(e(r))=(-t^{1/2}-t^{-1/2})\chi_{2r-1,r-1}(e(r-1))\,. \tag{4.2.10}$$

Let $e(i)'$ be an element of $\boldsymbol{CB}_{2r}$ with $p_{2r}(e(i)')=e(i)$ $(1\leq i\leq r)$. By Lemma 4.2.7 (a), $\chi_{2r,i}(e(r))=0$ if $i<r$ and $\chi_{2r-1,i}(e(r-1))=0$ if $i<r-1$. Hence we have

$$V_{2r}(e(r)')=a_{2r,r}(V)\chi_{2r,r}(e(r))$$

and

$$V_{2r-1}(e(r-1)') = a_{2r-1,r-1}(V)\chi_{2r-1,r-1}(e(r-1)).$$

Since V is a link invariant, we have

$$V_{2r}(p_{2r}(\sigma_{2r-1})e(r-1)') = V_{2r}(p_{2r}(\sigma_{2r-1}^{-1})e(r-1)') = V_{2r-1}(e(r-1)')$$
(Definition 1.1.2 (ii)).

Hence, by using

$$e_i = t^{1/2}(t^{1/2}-t^{-1/2})^{-1}p_{2r}(\sigma_i) - t^{-1/2}(t^{1/2}-t^{-1/2})^{-1}p_{2r}(\sigma_i^{-1}),$$

we have

$$\begin{aligned} a_{2r,r}(V)\,\chi_{2r,r}(e(r)) &= V(e(r)', 2r) \\ &= V((t^{1/2}(t^{1/2}-t^{-1/2})^{-1}\,p_{2r}(\sigma_{2r-1}) - t^{-1/2}(t^{1/2}-t^{-1/2})^{-1}\,p_{2r}(\sigma_{2r-1}^{-1}))\,e(r-1)', 2r) \\ &= V(e(r-1)', 2r-1) = a_{2r-1,r-1}(V)\,\chi_{2r-1,r-1}(e(r-1)). \end{aligned}$$

This and (4.1.11) imply (4.2.10). □

Proof of Theorem 4.2.1. From Proposition 4.2.4 and Lemma 4.2.7 we have

$$\begin{aligned} &\langle(b,n)\rangle^{(r)}_{f_1' f_3' \cdots f_{2j-1}'} \\ &= (-t^{1/2}-t^{-1/2})^j \Big(\sum_{i=j}^{[r/2]} \chi_{r-2j,i-j}(1)\langle(b,n)\rangle^{(r,i)}\Big) \quad (0 \le r,\ 0 \le j \le [r/2]). \end{aligned}$$

By interchanging r with $r+2j$ and i with $i+j$, we get

$$\begin{aligned} &\langle(b,n)\rangle^{(r+2j)}_{f_1' f_3' \cdots f_{2j-1}'} \\ &= (-t^{1/2}-t^{-1/2})^j \Big(\sum_{i=0}^{[r/2]} \chi_{r,i}(1)\langle(b,n)\rangle^{(r+2j,i+j)}\Big) \quad (0 \le r,\ 0 \le j \le [r/2]). \end{aligned}$$

Hence Proposition 4.2.5 implies that

$$\langle(b,n)\rangle^{(r)} = (-t^{3/4})^{2jw(b^\wedge)} \Big(\sum_{i=0}^{[r/2]} \chi_{r,i}(1)\langle(b,n)\rangle^{(r+2j,i+j)}.$$

By using (1.1.12), we get $V^{(r)}(b,n) = (-t^{3/4})^{rw(b^\wedge)}\langle(b,n)\rangle^{(r)}$ and $V^{(r,i)}(b,n) = (-t^{3/4})^{rw(b^\wedge)}\langle(b,n)\rangle^{(r,i)}$. Therefore we have

$$V^{(r)}(b,n) = \sum_{i=0}^{[r/2]} \chi_{r,i}(1)\, V^{(r+2j,i+j)}(b,n).$$

But the left hand side of the above formula does not depend on j, and so we obtain inductively that

$$\begin{aligned} V^{(0)}(b,n) &= V^{(0,0)}(b,n) = V^{(2,1)}(b,n) = \cdots = V^{(2i,i)}(b,n) = \cdots, \\ V^{(1)}(b,n) &= V^{(1,0)}(b,n) = V^{(3,1)}(b,n) = \cdots = V^{(1+2i,i)}(b,n) = \cdots, \\ &\cdots \end{aligned}$$

$$V^{(r)}(b, n) - \sum_{i=1}^{[r/2]} \chi_{r,i}(1)\, V^{(r,i)}(b, n)$$
$$= V^{(r,0)}(b, n) = V^{(r+2,1)}(b, n) = \cdots = V^{(r+2j,j)}(b, n) = \cdots ,$$
$$\cdots .$$

These are the equalities we wanted to show. □

4.3. The Jones polynomial of satellite links. We give a satellite link version of Theorem 1.6.4 for the one-variable Jones polynomial V.

DEFINITION 4.3.1. Let K and L be diagrams of links in the 3-sphere and the solid torus respectively. Let $N(K)$ the tubular neighborhood of K. Then there is a bijection f, called the faithful embedding, from the solid torus to $N(K)$. This mapping is determined canonically up to ambient isotopy. Let $K_L = f(L)$. Then K_L is a link in the 3-sphere and is called the *satellite link* of K with respect to L. For a link invariant X, put $X_L(K) = X(K_L)$. Then X_L is a link invariant.

For a link diagram L, let $|L|$ denote the unoriented link diagram obtained by forgetting the orientation of L. For $(g, n) \in G$ (Definition 4.1.13), let $(g, n)^{\sim}$ be the unoriented link diagram in the annulus obtained by joining the points at the top of $d(g, n)$ and those at the bottom without further crossings as in Figure 12. Conversely, for any link diagram L in the annulus, there is an element $(g, n) \in G$ such that $|L|$ is regular isotopic to $(g, n)^{\sim}$ in the annulus. Let Ψ_r: $G_r \to J_r$ be the projection defined in Section 4.1.

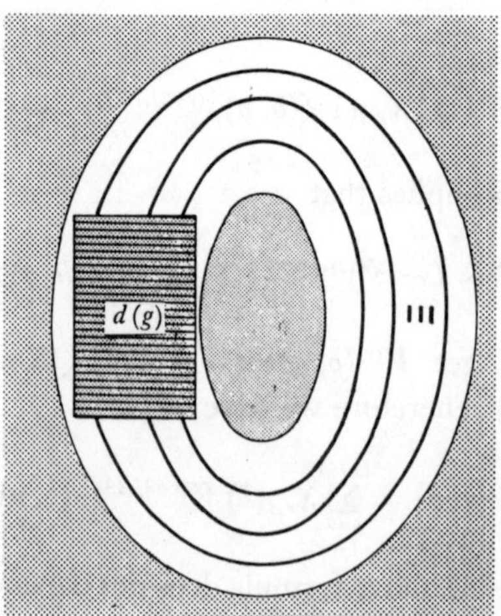

Figure 12

Theorem 4.3.2. *Let K be a knot, L an oriented link diagram in the annulus, $w(L)$ the writhe of L, and (g, r) an element of G whose closure is regular isotopic to $|L|$ in the annulus. Then we have*

$$V_L(K)=(-t^{3/4})^{w(L)}\sum_{i=0}^{[r/2]}\chi_{r,i}(\Psi_r(g))\,V^{(r,i)}(K)\,.$$

Proof. Let $(b, n)\in B$ be a braid whose closure is equivalent to K. Then K_L is equivalent to $(g\phi_n^{(r)}(b), rn)^\wedge$. We regard G_r as a subsemigroup of G_{rn} by the inclusion homomorphism sending $g\in G_r$ to the isotopy class of g' in G_{rn} as in Figure 13.

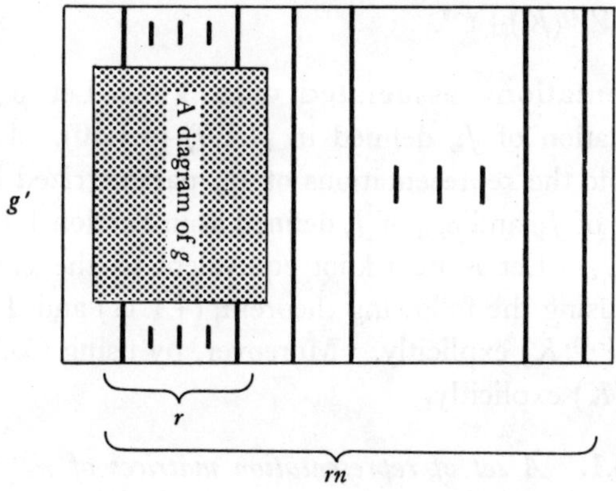

Figure 13

By applying the results of Section 1.3 to the wreath product $G_n^r \rtimes G_n$, we get

$$\langle (g\phi_n^{(r)}(b), rn)\rangle=\sum_{i=0}^{[r/2]}\chi_{r,i}(\Psi_r(g))\langle (b, n)\rangle^{(r,i)}\,.$$

The proof of this formula is analogous to that of Theorem 1.6.4. We know that $V^{(r,i)}(b, n)=(-t^{3/4})^{w(b^\wedge)}\langle (b, n)\rangle^{(r,i)}$ and $w((g\phi_n^{(r)}(b), rn)^\wedge)=w((g, r)^\wedge)+rw(b^\wedge)$. By using these formulas, the above statement can be proved. □

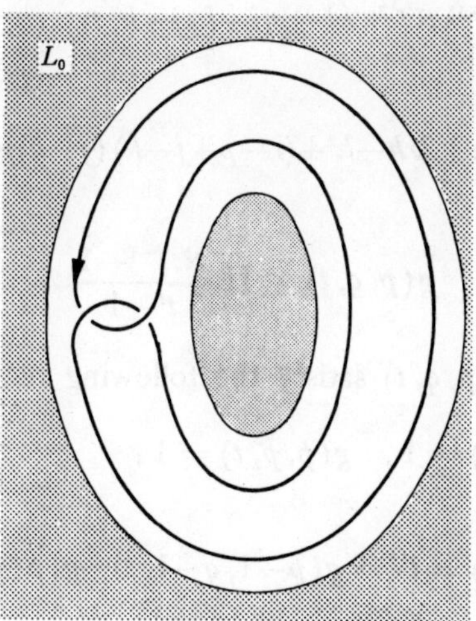

Figure 14

Theorem 4.3.2 can be extended for marked links as Theorem 2.5.4.

EXAMPLE 4.3.3. (doubled knot [33]). Let L_0 be the link diagram in the annulus as in Figure 14. Then L_0 is regular isotopic to $(g_0, 4)^{\sim}$, where $g_0 = f_1 f_2 f_3 \sigma_2^2$. For a knot K, we have $V_{L_0}(K) = t^{-3/2}((1-t)V^{(2,0)}(K) - (t^{-1}+t)V^{(0,0)}(K))$ since $\chi_{4,0}(\Psi_4(g_0)) = 0$, $\chi_{4,1}(\Psi_4(g_0)) = 1-t$ and $\chi_{4,2}(\Psi_4(g_0)) = -(t^{-1}+t)$. But we know that $V^{(2,0)}(K) = V^{(2)}(K) + (t^{-1/2}+t^{1/2})^{-1}$ from Corollary 4.2.3 and so we have $V_{L_0}(K) = t^{-3/2}(1-t)V^{(2)}(K) + t^{-2}$.

4.4. Representations associated with $V^{(r)}$. Let $\rho_{m,k}$ denote the irreducible representation of J_m defined in (4.1.9)–(4.1.10). In this section, $\pi_{3r,i}$ $(0 \leq i \leq [3r/2])$ denote the representations of B_{3r} parametrized by two irreducible representations $\rho_{3r,i}$ of J_{3r} and $\rho_{r,0}$ of J_r defined as in Section 1.4. Let $\omega_{3r\ i}$ denote the character of $\pi_{3r,i}$. Let K be a knot equivalent to the closure of a 3-braid $b \in B_3$. Then, by using the following theorem, (4.1.11) and Theorem 1.4.10 (ii), we can calculate $V^{(r,0)}(K)$ explicitly. Moreover, by using Corollary 4.2.3 (i), we may calculate $V^{(r)}(K)$ explicitly.

Theorem 4.4.1. *A set of representation matrices of* $\pi_{3r,i}(\sigma_j)$ $(0 \leq i \leq [3r/2]$, $j=1, 2)$ *is given by the following.*

$$\pi_{3r,i}(\sigma_1) = (\alpha_{i,j,k}(t))_{0 \leq j,k \leq d(r,i)} ,$$
$$\pi_{3r,i}(\sigma_2) = (\alpha_{i,d(r,i)-j,d(r,i)-k}(t))_{0 \leq j,k \leq d(r,i)} ,$$

where $d(r, i) = \deg(\pi_{3r,i}) - 1 = i$ *(if* $0 \leq i \leq r$*) or* $3r - 2i$ *(if* $r < i \leq [3r/2]$*) and*

$$\begin{aligned}
\alpha_{i,j,k}(t) &= (-1)^{(r+k)}\, t^{\gamma(r,i,k,j)}\, g(i-k, j-k, t) && (k \leq j,\ 0 \leq i \leq r)\,, \\
\alpha_{i,j,k}(t) &= 0 && (j < k,\ 0 \leq i \leq r)\,, \\
\alpha_{i,j,k}(t) &= (-1)^{(r+k)}\, t^{\gamma(r,2r-i,k,j)}\, g(i-k, j-k, t) && (k \leq j,\ r < i \leq [3r/2])\,, \\
\alpha_{i,j,k}(t) &= 0 && (j < k,\ r < i \leq [3r/2])\,,
\end{aligned}$$

where $\gamma(r, p, k, j) = r + k + 2rk - k^2 + (r - p + j - k)(j - k))/2$ *and* $g(p, q, t)$ *is the Gauss' polynomial, i.e.*

$$g(p, q, t) = \prod_{i=1}^{q} \frac{t^{p+i-1} - 1}{t^i - 1} .$$

The polynomial $g(p, q, t)$ satisfy the following recursive relations:

$$g(p, 0, t) = 1\,, \quad g(p, p, t) = 1\,,$$

(4.4.2)

$$g(p, q, t) = t^{(p-q)}\, g(p-1, q-1, t) + g(p-1, q, t)\,.$$

Before proving the above theorem, we give examples and applications.

Table 3

$r=2$:

$$\pi_{(6,0),(2,0)}(\sigma_1)=t,$$

$$\pi_{(6,1),(2,0)}(\sigma_1)=\begin{pmatrix} t & 0 \\ t^2 & -t^3 \end{pmatrix},$$

$$\pi_{(6,2),(2,0)}(\sigma_1)=\begin{pmatrix} t & 0 & 0 \\ t^{3/2}(1+t) & -t^3 & 0 \\ t^3 & -t^{7/2} & t^4 \end{pmatrix},$$

$$\pi_{(6,3),(2,0)}(\sigma_1)=-t^3,$$

$r=3$:

$$\pi_{(9,0),(3,0)}(\sigma_1)=-t^{3/2},$$

$$\pi_{(9,1),(3,0)}(\sigma_1)=\begin{pmatrix} -t^{3/2} & 0 \\ -t^3 & t^{9/2} \end{pmatrix},$$

$$\pi_{(9,2),(3,0)}(\sigma_1)=\begin{pmatrix} -t^{3/2} & 0 & 0 \\ -t^{5/2}(1+t) & t^{9/2} & 0 \\ -t^3 & t^{11/2} & -t^{13/2} \end{pmatrix},$$

$$\pi_{(9,3),(3,0)}(\sigma_1)=\begin{pmatrix} -t^{3/2} & 0 & 0 & 0 \\ -t^2(1+t+t^2) & t^{9/2} & 0 & 0 \\ -t^{7/2}(1+t+t^2) & t^5(1+t) & -t^{13/2} & 0 \\ -t^6 & t^{13/2} & -t^7 & t^{15/2} \end{pmatrix},$$

$$\pi_{(9,4),(3,0)}(\sigma_1)=\begin{pmatrix} t^{9/2} & 0 \\ t^{11/2} & -t^{13/2} \end{pmatrix}.$$

EXAMPLE 4.4.3. The representation matrices of $\sigma_1\in B_3$ for the case $r=2$ and $r=3$ are given in Table 3. The matrices of σ_2 are obtained as follows.

$$\pi_{3r,i}(\sigma_2)=K\,\pi_{3r,i}(\sigma_1)\,K, \quad \text{where} \quad K=\begin{pmatrix} 0 & & 1 \\ & 1 & \\ & \therefore & \\ 1 & & 0 \end{pmatrix}$$

Let $C^{\cdot}=\{c\in C\,|\,c\neq 0,\ c^m\neq 1 \text{ for any } m\in N\}$.

Proposition 4.4.4. *Let (b, b') be a pair of 3-braids such that $w(b^{\wedge})=w(b'^{\wedge})$ and $V(b^{\wedge})(s)=V(b'^{\wedge})(s)$ for all $s\in C^{\cdot}$. Then, for $t\in C^{\cdot}$, $V^{(0,2)}(b^{\wedge})(t)=V^{(2,0)}(b'^{\wedge})(t)$ iff $\omega_{6,2}(b)(t)=\omega_{6,2}(b')(t)$.*

Proof. Recall that $a_{n,i}(V)(t)\neq 0$ for $0\leq i\leq [n/2]$ by (4.1.11). From (4.1.10) and Theorem 1.4.10, it is enough to show that

(4.4.5) $$\omega_{6,i}(b)(t)=\omega_{6,i}(b')(t) \qquad \text{for} \quad i=0,1,3\,.$$

We have $\omega_{6,i}(b)(t)=\omega_{6,i}(b')(t)$ for $i=1, 3$ and $\chi_{3,0}(b)(t)=\chi_{3,0}(b')(t)$ since they are linear characters of B_3 and $w(b^{\wedge})=w(b'^{\wedge})$. The assumptions $w(b^{\wedge})=w(b'^{\wedge})$

and $V(b^{\wedge})(s)=V(b'^{\wedge})(s)$ for all $s\in\boldsymbol{C}^{\bullet}$ yields $\chi_{3,1}(b)(s)=\chi_{3,1}(b')(s)$ since $V(\cdot)(s)=a_{3,0}(V)(s)\,\chi_{3,0}(\cdot)(s)+a_{3,1}(V)(s)\,\chi_{3,1}(\cdot)(s)$ for 3-braids. But we know that $\omega_{6,1}(b)(t)=(-1)^{w(b^{\wedge})}\chi_{3,1}(b)(t^2)$, and so we have $\omega_{6,1}(b)(t)=\omega_{6,1}(b')(t)$. Hence (4.4.5) is proved. □

From the above Proposition and Corollary 4.2.3, we have the following.

Corollary 4.4.6. *Let (b, b') be a pair of 3-braids such that $w(b^{\wedge})=w(b'^{\wedge})$ and $V(b^{\wedge})(s)=V(b'^{\wedge})(s)$ for all $s\in\boldsymbol{C}^{\bullet}$. Then, for $t\in\boldsymbol{C}^{\bullet}$, $V^{(2)}(b^{\wedge})(t)=V^{(2)}(b'^{\wedge})(t)$ iff $\omega_{6,2}(b)(t)=\omega_{6,2}(b')(t)$.*

Let $P=P(\cdot)(\alpha, q)$ denote the two-variable Jones polynomial with non-zero complex parameters α and q where q is not equal to any roots of unity. Then we have the following.

Corollary 4.4.7. *Let (b, b') be a pair of 3-braids such that $w(b^{\wedge})=w(b'^{\wedge})$, $V(b^{\wedge})(s)=V(b'^{\wedge})(s)$ and $V^{(2)}(b^{\wedge})(s)=V^{(2)}(b'^{\wedge})(s)$ for all $s\in\boldsymbol{C}^{\bullet}$. Then we have*

(a) $P^{(2,(2))}(b^{\wedge})(\alpha, q)=P^{(2,(2))}(b'^{\wedge})(\alpha, q)$,
(b) $P^{(2,(11))}(b^{\wedge})(\alpha, q)=P^{(2,(11))}(b'^{\wedge})(\alpha, q)$,
(c) $P^{(2)}(b^{\wedge})(\alpha, q)=P^{(2)}(b'^{\wedge})(\alpha, q)$.

Proof. The representations $\pi_{\lambda,\nu}(\lambda\in\Lambda(6),\ \nu=(2)\in\Lambda(2))$ of B_3 associated with $P^{(2)}$ are given in Table 3. With these representation matrices, an analogous argument of the proof of Proposition 4.4.4 shows the part (a). The part (a) and (3.6) implies the part (b). The part (c) follows from the parts (a), (b) and Theorem 1.5.1. □

Proposition 4.4.8. *Let (b, b') be a pair of 3-braids whose closures are knot with $w(b^{\wedge})=w(b'^{\wedge})$ and $F(b^{\wedge})(a, x)=F(b'^{\wedge})(a, x)$ for all $a, x\in\boldsymbol{C}^{\times}$, where F denotes the Kauffman polynomial. Then, for $t\in\boldsymbol{C}^{\bullet}$, $V^{(3,0)}(b^{\wedge})(t)=V^{(3,0)}(b'^{\wedge})(t)$ iff $\omega_{9,3}(b)(t)=\omega_{9,3}(b')(t)$.*

Proof. Because $a_{9,3}(V)(t)\neq 0$, it is enough to show that $\omega_{9,i}(b)(t)=\omega_{9,i}(b')(t)$ for $i=0, 1, 2, 4$. We have $V(b^{\wedge})(s)=V(b'^{\wedge})(s)\,(s\in\boldsymbol{C}^{\bullet})$ from $F(b^{\wedge})=F(b'^{\wedge})$ [22]. Thus, as the proof of Proposition 4.4.4, we have $\omega_{9,i}(b)(t)=\omega_{9,i}(b')(t)$ for $i=0, 1, 4$. Let χ be the character of the representation ρ of B_3 defined by the following.

$$\rho(\sigma_1)=\begin{pmatrix} a^{-1} & 0 & x \\ 0 & 0 & -1 \\ 0 & 1 & x \end{pmatrix}, \quad \rho(\sigma_2)=\begin{pmatrix} 0 & -1 & 0 \\ 1 & x & 0 \\ 0 & a^{-1}x & a^{-1} \end{pmatrix}.$$

From Theorem 12.2 of [26] we have $\chi(b)(a, x)=\chi(b')(a, x)$ if the values of a and x are generic. This yields $\chi(b)(a, x)=\chi(b')(a, x)$ for all $a, x\in\boldsymbol{C}^{\times}$ since $\chi(\beta)(a, x)$ are Laurent polynomials of the parameters a and x for all $\beta\in B_3$. Let χ'

be the character obtained from χ by substituting $x=-t^{-3/2}+t^{3/2}$ and $a=t^{-7/2}$. Then $\omega_{9,2}(\cdot)(t)=t^3\chi'(\cdot)(t)$ and so we have $\omega_{9,2}(b)(t)=\omega_{9,2}(b')(t)$. This proves the proposition. □

From the above proposition and Corollary 4.2.3, we have the following.

Corollary 4.4.9. *Let (b, b') be a pair of 3-braids whose closures are knot with $w(b^\wedge)=w(b'^\wedge)$ and $F(b^\wedge)(a, x)=F(b'^\wedge)(a,x)$ for all $a, x\in \mathbf{C}^\times$, where F denotes the Kauffman polynomial. Then, for $t\in \mathbf{C}^\cdot$, $V^{(3)}(b^\wedge)(t)=V^{(3)}(b')(t)$ iff $\omega_{9,3}(b)(t)=\omega_{9,3}(b')(t)$.*

Proposition 4.4.10. *The invariant $V^{(3)}$ is independent from V, $V^{(2)}$, and the Kauffman polynomial F.*

Proof. We give a pair of braids (b, b') such that $V(b^\wedge)=V(b'^\wedge)$, $V^{(2)}(b^\wedge)=V^{(2)}(b'^\wedge)$, $F(b^\wedge)=F(b'^\wedge)$ but $V^{(3)}(b^\wedge)\neq V^{(3)}(b'^\wedge)$. Let

$$b(m, n)=\sigma_1^2\sigma_2^{-2}\sigma_1\sigma_2^2\sigma_1^{-2}\sigma_2^m\sigma_1^2\sigma_2^{-2}\sigma_1^{-1}\sigma_2^2\sigma_1^{-2}\sigma_2^n,$$

for odd integers m, n. Then T. Kanenobu noted that the knots $b(m, n)^\wedge$ and $b(n, m)^\wedge$ have the same Kauffman polynomial. Therefore they have the same Jones polyr omial [22] and the same 2-parallel version of the Jones polynomial [35]. But T. Kanenobu conjectured that $b(m, n)^\wedge$ and $b(n, m)^\wedge$ are not equivalent except a finite number of pairs (m, n). From a calculation with a computer we have $\omega_{9,3}(b(3, 1))\neq\omega_{9,3}(b(1, 3))$ by substituting 4 to the parameter t. Thus we have $V^{(3)}(b(1, 3)^\wedge)\neq V^{(3)}(b(3, 1)^\wedge)$ from Corollary 4.4.9. □

From now on, we prepare some notations which is needed in the proof of Theorem 4.1.1. We use the notations in Section 4.1. Fix an integer i with $0\leq i\leq[3r/2]$. Let

$$V_{3r,i}=J_{3r,i}\,e_1\,e_3\cdots e_{2i-1}/(J_{3r,i}\,e_1\,e_3\cdots e_{2i-1}\cap J^{\cdot}_{3r,i+1})\,.$$

From Proposition 4.1.9, $V_{3r,i}$ is an irreducible left J_{3r}-module and the left action of J_{3r} on $V_{3r,i}$ are equivalent to $\rho_{r,i}$. By Proposition 4.1.6, there is a subset $G_{3r,i}$ of E_{3r} such that $\{x \bmod J^{\cdot}_{3r,i+1}\,|\,x\in G_{3r,i}\}$ is a basis of $V_{3r,i}$. More precisely, $G_{3r,i}=E_{r,i}\cap J_{3r}\,e_1\,e_3\cdots e_{2i-1}$. Let

$$G_{3r,i,1}=\{x\in G_{3r,i}\,|\,\text{there is } j\neq 0 \pmod r \text{ such that the } j\text{-th point at the top of } \eta(x) \text{ is connected to the } (j+1)\text{-th point at the top of } \eta(x)\}\,,$$

and $V_{3r,i,1}$ the subspace of $V_{3r,i}$ spanned by the image of $G_{3r,i,1}$. For $0\leq j_1, j_2, j_3\leq [r/2]$, let V_{3r,i,j_1,j_2,j_3} be the $\rho_{i,j_1}\otimes\rho_{r,j_2}\otimes\rho_{r,j_3}$-isotypic subspace with respect to $J_r^{\otimes 3}$ naturally embedded in J_{3r}. Let $I=\{k\,|\,1\leq k\leq 3r-1,\ k\neq 0 \bmod r\}$ and J_I the subalgebra of J_{3r} generated by $e_k(k\in I)$. Let λ_0 denote the one-dimensional representation of J_I defined by $\lambda_0(e_k)=0\ (k\in I)$. For a J_I-module U, let $U_+=\sum_{k\in I} e_k U$. Then U_+ is a J_I-submodule of U. Let U_0 denote the λ_0-isotypic subspace of U. Since $-(t^{1/2}+t^{-1/2})^{-1}\,e_k(k\in I)$ is a projection and $U_0=\bigcap_{k\in I}(1+(t^{1/2}+t^{-1/2})^{-1}\,e_k)\,U$, we have the following.

Proposition 4.4.11. *The composition of natural homomorphisms $U_0\to U\to U/U_+$ is a J_I-module isomorphism.*

From Proposition 4.1.8, we have $V_{3r,i,1}=\sum_{k\in I} e_k\,V_{3r,i}$. Since J_I is isomorphic to $J_r^{\otimes 3}$, we get the following.

Corollary 4.4.12. *The $J_r^{\otimes 3}$-submodule $V_{3,i,0,0,0}$ of $V_{r,i}$ is isomorphic to $V_{3r,i}/V_{3r,i,1}$ as a $J_r^{\otimes 3}$-module.*

Let $V_{3r,i,0}=V_{3r,i}/V_{3r,i,1}$. Then the above corollary implies that the representation $\pi_{3r,i}$ of B_3 in Theorem 4.4.1 is equivalent to the action of B_3 on $V_{3r,i,0}$ induced by restricting the representation $\rho_{3r,i}\circ\phi_3^{(r)}$. We give a basis of $V_{3r,i,0}$ and representation matrices of $\pi_{3r,i}$ with respect to the above basis. For $0\le k,\ m<j$, let

$$e^{(d)}(j-k,j+k,j-m,j+m)=\begin{cases} e(j-k,j+k)\,e(j-k-1,j+k+1)\cdots e(j-m,j+m) & \text{if } m\ge k\,,\\ e(j-k,j+k)\,e(j-k+1,j+k-1)\cdots e(j-m,j+m) & \text{if } m<k\,,\end{cases}$$

and

$$e^{(h)}(j-k,j+m,j-m,j+k)=e(j-k,j+m)\,e(j-k-1,j+m-1)\cdots e(j-m,j+k)\,.$$

Let $H_{r,i}=\{h_{r,i,j}\,|\,0\le j\le i \text{ if } i\le r \text{ and } i-r\le j\le 2r-i \text{ if } r<i\}$, where the elements $h_{r,i,j}$ are given as follows. In the case of $0\le i\le r$, let

$$\begin{aligned} h_{r,i,j}{}^{(1)}&=e^{(d)}(r,r,r-j+1,r+j-1)\,,\\ h_{r,i,j}{}^{(2)}&=e^{(d)}(2r,2r,2r-i+j+1,2r+i-j-1)\,,\\ h_{r,i,j}{}^{(3)}&=e^{(h)}(2r-i+j,2r+i-j-2,r+j+1,r-j+2i-1)\,,\\ h_{r,i,j}{}^{(4)}&=e^{(h)}(r+j,r-j+2i-2,1,2i-2j-1)\,,\\ h_{r,i,j}&=h_{r,i,j}{}^{(1)}\,h_{r,i,j}{}^{(2)}\,h_{r,i,j}{}^{(3)}\,h_{r,i,j}{}^{(4)}\,,\end{aligned}$$

for $0\le j\le i$. In the case of $r<i\le[3r/2]$, let

$$\begin{aligned} h_{r,i,j}{}^{(1)}&=e^{(d)}(r,r,r-j+1,r+j-1)\,,\\ h_{r,i,j}{}^{(2)}&=e^{(d)}(2r,2r,r+j+1,3r-j-1)\,,\\ h_{r,i,j}{}^{(3)}&=e^{(d)}(r-j,3r-j,2r-i-j+1,2r+i-j-1)\,,\\ h_{r,i,j}{}^{(4)}&=e^{(h)}(2r-i-j,2r+i-j-2,1,2i-1)\,,\end{aligned}$$

$$h_{r,i,j} = h_{r,i,j}^{(1)}\, h_{r,i,j}^{(2)}\, h_{r,i,j}^{(3)}\, h_{r,i,j}^{(4)}\,.$$

for $i-r \leq j \leq 2r-i$. The rectangular diagrams of $\eta(h_{r,i,j}) \in D_{3r}$ are given as in Figure 15.

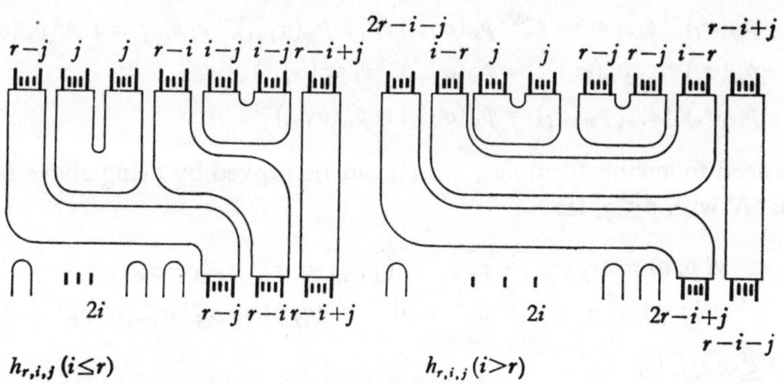

Figure 15

Let $W_{3r,i}$ denote the inverse image of $V_{3r,i,1}$ with respect to the canonical projection

$$J_{3r}\, e_1\, e_3 \cdots e_{2i-1} \rightarrow J_{3r}\, e_1\, e_3 \cdots e_{2i-1} / (J_{3r,i+1} \cap J_{3r}\, e_1\, e_3 \cdots e_{2i-1})\,.$$

Proposition 4.4.13. *The set* $H_{r,i}^{\cdot} = \{h \bmod W_{3r,i} \mid h \in H_{r,i}\}$ *is a basis of* $V_{3r,i,0}$.

Proof. For $h \in H_{r,i}$, we know that $h \in E_{3r,i}$, $h \in J_{3r}\, e_1\, e_3 \cdots e_{2i-1}$, $h \notin E_{3r,i+1}$ and $h \notin E_{3r,i,1}$. Hence $H_{r,i}^{\cdot}$ is linearly independent since $W_{3r,i}$ is a subspace of $J_{r,i}^{\cdot}$ spanned by $(E_{r,i+1} \cap J_{3r}\, e_1\, e_3 \cdots e_{2i-1}) \cup E_{r,i,1}$. Since the dimension of the representation $\rho_{r,0}^{\otimes 3}$ of $J_r^{\otimes 3}$ is equal to one, the dimension of the $\rho_{r,0}^{\otimes 3}$-isotypic subspace of $V_{3r,i}$ with respec to $J_r^{\otimes 3}$ is equal to the multiplicity of $\rho_{r,0}^{\otimes 3}$ in $\rho_{3r,i} \mid_{J_r^{\otimes 3}}$. From Remarks 4.1.12 and 3.8, we may use the Littlewood-Richardson rule (Remark 3.8) for computing $\dim_C V_{3r\ i,0}$. The result is

$$\dim_C V_{3r,i,0} = i+1 \quad \text{if} \quad 0 \leq i \leq r \quad \text{and} \quad 2r-i+1 \quad \text{if} \quad r < i \leq [3r/2]\,,$$

which is equal to $\# H_{r,i}$. This shows that $H_{3r,i}^{\cdot}$ is a basis of $V_{3r,i,0}$. □

Let $p_{3r}\colon B_{3r} \rightarrow J_{3r}$ be the algebra homomorphism introduced in Section 4.1. We use the following formulas in the proof of Theorem 4.4.1, which can be proved immediately from the definitions of J_{3r}, p_{3r} and $W_{3r,i}$. For $v \in J_r\, e_1\, e_3 \cdots e_{2i-1}$ and $1 \leq k \leq 3r-1$ with $k \not\equiv 0 \pmod r$, we have

(4.4.14) $$p_{3r}(\sigma_i)\, v = -t^{1/2}\, v \pmod{W_{3r,i}}\,.$$

For $1 \leq k \leq 3r-1$, we have

$$p_{3r}(\sigma_i)\, e_i = e_i\, p_{3r}(\sigma_i) = t^{3/2}\, e_i\,. \tag{4.4.15}$$

For $1 \leq k \leq 3r-2$, we have

$$\begin{aligned}
&p_{3r}(\sigma_1)\, e_{i+1}\, e_i = t^{3/2}\, p_{3r}(\sigma_{i+1})^{-1}\, e_i\,, \quad p_{3r}(\sigma_{i+1})\, e_i\, e_{i+1} = t^{3/2}\, p_{3r}(\sigma_i)^{-1}\, e_{i+1}\,, \\
&p_{3r}(\sigma_i)^{-1}\, e_{i+1}\, e_i = t^{-3/2}\, p_{3r}(\sigma_{i+1})\, e_i\,, \quad p_{3r}(\sigma_{i+1})^{-1}\, e_i\, e_{i+1} = t^{-3/2}\, p_{3r}(\sigma_i)\, e_{i+1}\,, \\
&p_{3r}(\sigma_i)\, e_{i+1}\, p_{3r}(\sigma_i)^{-1} = p_{3r}(\sigma_{i+1})^{-1}\, e_i\, p_{3r}(\sigma_{i+1})\,, \\
&p_{3r}(\sigma_i)^{-1}\, e_{i+1}\, p_{3r}(\sigma_i) = p_{3r}(\sigma_{i+1})\, e_i\, p_{3r}(\sigma_{i-1})^{-1}\,.
\end{aligned} \tag{4.4.16}$$

We also need following formulas, which can be proved by using above formulas. For $p, q \in \boldsymbol{N}$ with $p \leq q$, let

$$\begin{aligned}
&\sigma(p, q) = \sigma_p\, \sigma_{p+1} \cdots \sigma_q\,, \quad \sigma^-(p, q) = \sigma_p^{-1}\, \sigma_{p+1}^{-1} \cdots \sigma_q^{-1}\,, \\
&\sigma'(q, p) = \sigma_q\, \sigma_{q-1} \cdots \sigma_p \quad \text{and} \quad \sigma'^-(q, p) = \sigma_q^{-1}\, \sigma_{q-1}^{-1} \cdots \sigma_p^{-1}\,.
\end{aligned}$$

For $0 \leq k \leq m < j$, let

$$\sigma(j \pm k, j+m, j-m, j \mp k) = \sigma(j \pm k, j+m)\, \sigma(j \pm k-1, j+m-1) \cdots \sigma(j-m, j \mp k)\,.$$

and

$$\begin{aligned}
\sigma^-(j \pm k, j+m, j-m, j \pm k) = {} & \sigma^-(j \pm k, j+m) \\
& \sigma^-(j \pm k-1, j+m-1) \cdots \sigma^-(j-m, j \pm k)\,.
\end{aligned}$$

For $j, k, m, n \in \boldsymbol{N}$ with $j > m-k+n$, we have

$$\begin{aligned}
&\sigma(j-k+m, j+m+n-1, j-k+1, j+n)\, \sigma(j-k, j-k-n+1, j-1, j-n) \\
&e^{(d)}(j, j, j-n+1, j+n-1) \\
&\quad = \sigma^-(j-k, j-k+1, j+k+m-1, j-k-n+m) \\
&\qquad \sigma(j-k+m, j+m-1, j-k-n+1, j-n) \\
&\qquad\quad e^{(d)}(j+m, j+m, j+m-n+1, j+m+n-1) \\
&\qquad\quad e^{(h)}(j+m-n, j+m+n-2, j-n+1, j+n-1)
\end{aligned} \tag{4.4.17}$$

For $j, k \in \boldsymbol{N}$, we have

$$\begin{aligned}
&\sigma_i^{\pm 1}\, e(i+2, i+2k)\, e(i+1, i+2k-1)\, e(i, i+2k-2) \\
&\quad = e(i+2, i+2k)\, e(i+1, i+2k-1)\, e(i, i+2k-2)\, \sigma_{i+2k}^{\pm 1}\,.
\end{aligned} \tag{4.4.18}$$

Formula (4.4.17) with $m=1$, (4.4.15), (4.4.16) and (4.4.18) implies that

$$\begin{aligned}
&\sigma(j, j+k-1)\, \sigma'(j-1, j-k+1)\, e^{(d)}(j, j, j-k+1, j+k-1) \\
&\quad = t^{3/2}\, t^{3(k-1)}\, \sigma'^-(j-1, j-k+1)\, \sigma'(j-1, j-k+1)^{-1} \\
&\qquad e^{(d)}(j, j, j-k+1, j+k-1)\,.
\end{aligned} \tag{4.4.19}$$

An induction on m and (4.4.19) shows that

(4.4.20) $\sigma(j, j-m+1, j+m-1, j)\, e^{(d)}(j, j, j-m+1, j+m-1)$
$= (t^{3/2})^{m2}\, \sigma'^{-}(j-1, j-m+1)\, \sigma'(j-1, j-m+1)^{-1}$
$\sigma'^{-}(j-2, j-m+1)\, \sigma'(j-2, j-m+1)^{-1}$
$\cdots\sigma'^{-}(j-m, j-m)\, \sigma'(j-m, j-m)^{-1}\, e^{(d)}(j, j, j-m+1, j+m-1)\,.$

Proof of Theorem 4.4.1. We calculate the representation matrices of $\pi_{3r,i}(\sigma_j)$ $(j=1, 2)$ with respect to the basis $H_{3r,i}^{\bullet}$. This can be done by writing down $p_{3r}(\phi_3^{(r)}(\sigma_j))\, h_{r,i,k}$ mod $W_{3r,i}(j=1, 2)$ as $\boldsymbol{C}$-linear combinations of elements of $H_{3r,i}^{\bullet}$. Steps 1–7 can be proved by using formulas (4.4.14)–(4.4.20).

STEP 1. $p_{3r}(\phi_3^{(r)}(\sigma_j))\, h_{r,i\,k}$
$\equiv (-t^{1/2})^{r(r-1)}\, \sigma(r, 2r-1, 1, r)\, h_{r,i,k} \pmod{W_{3r,i}}\,.$

STEP 2. $\sigma(r, 2r-1, 1, r)\, h_{r,i,k}$
$\equiv (t^{3/2})^{k^2}\, \sigma(r, 2r-1, k+1, r+k)\, \sigma(k, r-1, 1, r-k)$
$\sigma'^{-}(r-1, r-k+1)\, \sigma'(r-1, r-k+1)^{-1}\, \sigma'^{-}(r-2, r-k+2)$
$\sigma'(r-2, r-k+2)^{-1}\cdots\sigma'^{-}(r-k+1, r-k+1)$
$\sigma'(r-k+1, r-k+1)^{-1}\, h_{r,i,k} \pmod{W_{3r,i}}\,.$

STEP 3. $\sigma(r, 2r-1, k+1, r+k)\, \sigma(k, r-1, 1, r-k)$
$\sigma'^{-}(r-1, r-k+1)\, \sigma'(r-1, r-k+1)^{-1}\, \sigma'^{-}(r-2, r-k+2)$
$\sigma'(r-2, r-k+2)^{-1}\cdots\sigma'^{-}(r-k+1, r-k+1)$
$\sigma'(r-k+1, r-k+1)^{-1}\, h_{r,i,k}$
$\equiv (-t^{-1/2})^{k(k-1)}\, \sigma(r, 2r-1, k+1, r+k)\, \sigma(k, r-1, 1, r-k)\, h_{r,i,k}$
$\pmod{W_{3r,i}}\,.$

STEP 4. $\sigma(r, 2r-1, k+1, r+k)\, \sigma(k, r-1, 1, r-k)\, h_{r,i,k}{}^{(2)}$
$= t^{3k(r-k)/2}\, \sigma^{-}(k, r-1, 1, r-k)\, \sigma(r, 2r-k-1, r-k+1, 2r-2k)$
$\sigma(r-k, 2r-2k-1, 1, r-k)$
$e^{(d)}(2er-k, 2r-k, 2r-2k+1, 2r-1)$
$e^{(h)}(2r-2k, 2r-2, r-k+1, r+k-1)\,.$

STEP 5. $\sigma^{-}(k, r-1, 1, r-k)\, \sigma(r, 2r-k-1, 1, r-k)$
$e^{(d)}(2r-k, 2r-k, 2r-2k+1, 2r-1)$
$e^{(h)}(2r-2k, 2r-2, r-k+1, r+k-1)$
$e^{(d)}(2r, 2r, 2r-i+k+1, 2r+i-k-1)$
$e^{(h)}(2r-i+k, 2r+i-k-2, r+k+1, r-k+2i-1)$
$e^{(h)}(r-k, r-k+2i-2, 1, 2i-1)$
$\equiv (-t^{-1/2})^{k(r-k)}\, \sigma(r-k, 2r-2k-1, 1, r-k)$

$$e^{(d)}(2r-k, 2r-k, 2r-i-k+1, 2r+i-k-1)$$
$$e^{(h)}(2r-i-k, 2r+i-k-2, 1, 2i-1) \pmod{W_{3r,i}}.$$

Step 6. $\sigma(r-k, 2r-2k-1, 1, r-k)$
$$e^{(d)}(2r-k, 2r-k, 2r-i-k+1, 2r+i-k-1)$$
$$e^{(h)}(2r-i-k, 2r+i-k-2, 1,2i-1)$$
$$\equiv t^{3i(r-k)/2}(-t^{1/2})^{(r-i)(r-k)-k(r-k)}\, \sigma^{-}(2r, r+k+1, 2r+i-k-1, r+i)$$
$$e^{(d)}(r, r, r-i+1, r+i-1)\, e^{(h)}(r-i, r+i-2, 1,2i-1)$$
$$\pmod{W_{3r,i}}.$$

Step 7. *For non-negative integers i, k and m such that $k<i$ and $m<i-k$, we have*

$$e^{(d)}(2r, 2r, 2r+m-2, 2r-m+2)$$
$$e^{(h)}(2r-m+1, 2r+m-1, r+k+1, r+k+2m-1)$$
$$\sigma^{-}(2r+m, r+k+2m+1, 2r+i-k-1, r+i+m)$$
$$e^{(d)}(r, r, r-k+1, r+k-1)$$
$$e^{(d)}(r-k, r+k+2m, r-i+m+1, r+i+m-1)$$
$$e^{(h)}(r-i+m, r+i+m-2, 1, 2i-1)$$
$$\equiv (-t^{-1/2})^{(r-2k-2m+i-1)}\, e^{(d)}(2r, 2r, 2r+m-2, ,2r-m+2)$$
$$e^{(h)}(2r-m+1, 2r+m-1, r+k+2, r+k+2m)$$
$$\sigma^{-}(2r+m, r+k+2m+2, 2r+i-k-2, r+i+m)$$
$$e^{(d)}(r, r, r-k, r+k)$$
$$e^{(d)}(r-k-1, r+k+2m+1, r-i+m+1, r+i+m-1)$$
$$e^{(h)}(r-i+m, r+i+m-2, 1, 2i-1)$$
$$+(-t^{-1})^{(r-2k-2m+i-1)}\, e^{(d)}(2r, 2r, 2r+m-1, 2r-m+1)$$
$$e^{(h)}(2r-m, 2r+m, r+k+1, r+k+2m+1)$$
$$\sigma^{-}(2r+m+1, r+k+2m+3, 2r+i-k-1, r+i+m+1)$$
$$e^{(d)}(r, r, r-k+1, r+k-1)$$
$$e^{(d)}(r-k, r+k+2m+2, r-i+m+2, r+i+m)$$
$$e^{(h)}(r-i+m+1, r+i+m-1, 1, 2i-1) \pmod{W_{3r,i}}.$$

Step 7 implies the following.

Step 8. $e^{(d)}(2r, 2r, 2r+m-2, 2r-m+2)$
$$e^{(h)}(2r-m+1, 2t+m-1, r+k+1, r+k+2m-1)$$
$$\sigma^{-}(2r+m, r+k+2m+1, 2r+i-k-1, r+i+m)$$
$$e^{(d)}(r, r, r-k+1, r+k-1)$$

$$e^{(d)}(r-k, r+k+2m, r-i+m+1, r+i+m-1)$$
$$e^{(h)}(r-i+m, r+i+m-2, 1, 2i-1)$$
$$\equiv \sum_{j=k}^{i} \beta_{k\,m,j}\, h_{r,i,j}\,,$$

where $\beta_{k,m,j}(0\leq m\leq i-k)$, *satisfy the following relations.*

(4.4.21) $\beta_{k,i-k,k}=1\ (0\leq k\leq i)\,,\quad \beta_{k,i-k,j}=0\ (0\leq j\leq i, j\neq k)\,,$

$$\beta_{k,m,j}=(-t^{-1/2})^{(r-2k-2m+i-1)}\,\beta_{k+1,m,j}+(-t^{-1})^{(r-2k-2m+i-1)}\,\beta_{k,m+1,j}$$
$$(0\leq m<i-k)\,.$$

STEP 9. *For* $0\leq k\leq j$ *and* $0\leq m\leq i-j$, *we have*

$$\beta_{k,m,j}$$
$$=(-1)^{(r-i+1)(i-k-m)}\,(t^{-1/2})^{(r-i+j-k)(j-k)+2(r+j-2k-m)(i-j-m)}g_{1-k-m,j-k}(t)\,.$$

Especially, we have

$$\beta_{k,0,j}=(-1)^{(r-i+1)(i-k)}\,(t^{-1/2})^{(r-i+j-k)(j-k)+2(r+j-2k)(i-j)}\,g_{i-k,j-k}(t)\,.$$

Proof. Since

$$(-1)^{(r-i+1)(i-k-m)}\,(t^{-1/2})^{(r-i+j-k)(j-k)+2(r+j-2k-m)(i-j-m)}\,g_{i-k-m,j-k}(t)$$

satisfy the relation (4.4.21), these are identical to $\beta_{k,m,j}$. □

STEP 10. Combining the results of Steps 1–9, we get the statement of Theorem 4.4.1 for the case $0\leq i\leq r$. An analogous argument proves the statement of the theorem for the case $r<i\leq[3r/2]$.

5. The parallel version of the Kauffman polynomial.

The associated algebras C_1, C_2, $\cdots$ of the Kauffman polynomial F (Definition 1.1.10) are obtained in [2] and [26]. The invariant F is known to be of trace type if the values of the parameters a and x are generic. Therefore we can apply our theory in Section 1 to F with generic parameters. We give a formula for the Kauffman polynomials of satellite links (Theorem 5.3.1).

5.1. The associated algebra C_n.

Let $a, \beta\in \boldsymbol{C}\setminus\{0\}$ such that β is not equal to any root of unity. Let $x=\beta-1/\beta$. The associated algebra C_n of the Kauffman polynomial is defined as a $\boldsymbol{C}$-algebra with 1 by the following.

(5.1.1) $C_n=$

$$\langle \tau_i, \tau_i^{-1}, \varepsilon_i\ (1\leq i\leq n-1)\,|\, \tau_i\,\tau_{i+1}\,\tau_i=\tau_{i+1}\,\tau_i\,\tau_{i+1}\,,$$
$$\varepsilon_i\,\varepsilon_{i+1}\,\varepsilon_i=\varepsilon_i\,,\ \varepsilon_{i+1}\,\varepsilon_i\,\varepsilon_{i+1}=\varepsilon_{i+1}\,,$$
$$\tau_i^{\pm1}\,\varepsilon_{i+1}\,\varepsilon_i=\tau_{i+1}^{\mp1}\,\varepsilon_i\,,\ \tau_{i+1}^{\pm1}\,\varepsilon_i\,\varepsilon_{i+1}=\tau_i^{\mp1}\,\varepsilon_{i+1}\,,$$

$$\varepsilon_i \varepsilon_{i+1} \tau_i^{\pm 1} = \varepsilon_i \tau_{i+1}^{\mp 1}, \ \varepsilon_{i+1} \varepsilon_i \tau_{i+1}^{\pm 1} = \varepsilon_{i+1} \tau_i^{\mp 1} \quad (1 \leq i \leq n-2),$$
$$\tau_i \tau_j = \tau_j \tau_i, \ \varepsilon_i \tau_j = \tau_j \varepsilon_i, \ \varepsilon_i \varepsilon_j = \varepsilon_j \varepsilon_i \quad (1 \leq i \leq j-1 \leq n-2),$$
$$\tau_i - \tau_j^{-1} = x(1-\varepsilon_i), \ \tau_i \varepsilon_i = \varepsilon_i \tau_i = a^{-1} \varepsilon_i, \ \tau_i \tau_i^{-1} = \tau_i^{-1} \tau_i = 1$$
$$(1 \leq i \leq n-1)\rangle.$$

This definition is equivalent to the one in [2]. Let $C_{n,i} = C_n e_1 e_3 \cdots e_{2i-1} C_n$. Then $C_{n,i}$ is a two sided ideal and satisfies $C_n = C_{n,0} \supset C_{n,1} \supset \cdots \supset C_{n,[n/2]}$. Let $D_{n,i} = C_{n,i}/C_{n,i+1} (0 \leq i \leq [n/2]-1)$ and $D_{n,[n/2]} = C_{n,[n/2]}$. Note that

(5.1.2) $D_{n,0} \cong H_n((-1)^{-1/2}\beta)$ (the Iwahori's Hecke algebra of type A_{n-1}).

Let $c_{n,i} = \binom{n}{2}\binom{n-2}{2} \cdots \binom{n-2i}{2}/(i!)$. For $\mu \in \Lambda(n)$, let $d(\mu)$ denote the degree of the character χ_μ of H_n parametrized by the partition μ. Then we have ([2], Theorem 3.7)

$$(5.1.3) \qquad D_{n,i} \equiv \bigoplus_{\mu \in \Lambda(n-2i)} M_{d(\mu)c_{n,i}}(\boldsymbol{C}) \quad (0 \leq i \leq [n/2]).$$

Hence the irreducible representations of $D_{n,i}$ are parametrized by $\Lambda(n-2i)$ and so

$$(5.1.4) \qquad C_n^\wedge = \{(i, \mu) \mid 0 \leq i \leq [n/2], \ \mu \in \Lambda(n-2i)\}.$$

Let $\rho_{i,\mu}$ denote the corresponding irreducible representation of (i, μ) and $\chi_{i,\mu}$ its character. Let $p'_n : \boldsymbol{C}G_n \to C_n$ be the homomorphism defined by $p'_n(\sigma_i) = \tau_i$, $p'_n(\sigma_i^{-1}) = \tau_i^{-1}$, $p'_n(f_i) = \varepsilon_i$, and $p_n : \boldsymbol{C}B_n \to C_n$ the homomorphism defined by $p_n(\sigma_i) = a\tau_i$ and $p_n(\sigma_i^{-1}) = a^{-1}\tau_i^{-1}$ for $1 \leq i \leq n-1$. Then the main result of [2] can be reformulated as follows.

Theorem 5.1.5. *For an n-braid b, there are* $a_{i,\mu}(F) \in \boldsymbol{C}(0 \leq i \leq [n/2], \mu \in \Lambda(n-2i))$ *such that*

$$F((b, n)^\wedge) = \sum_{(i,\mu) \in C_n^\wedge} a_{i,\mu}(F) \chi_{i,\mu}(p_n(b)).$$

The coefficients $a_{i,\mu}(F)$ are given explicitly in [28].

5.2. Relations among the invariants associated with the decomposition of $F^{(r)}$. Let $F^{(r,(i,\mu))}$ denote the invariant associated with the decomposition of the r-parallel version $F^{(r)}$ of the Kauffman polynomial F parametrized by $(i, \mu) \in C_n^\wedge$ (5.1.4) as in Section 1.3. In this section we show that some of the above invariants are the same one. In the following we use the conventions that $F^{(0)}(K) = F^{(0,(0,\phi))}(K) = (-1+(a+a^{-1})/x)^{-1}$ for any link diagram K.

Theorem 5.2.1. *Let K be a knot. Then we have*

$$F^{(r,(i,\mu))}(K)=F^{(r-2k,(i-k,\mu))}(K) \quad for \quad 0\leq r, 0\leq i\leq [r/2], 0\leq k\leq i .$$

Corollary 5.2.2. *Let K be a knot. Then we have*

$$F^{(r)}(K)=\sum_{i=0}^{[r/2]}\sum_{\mu\in\Lambda(r-2i)} \chi_{i,\mu}(1)\, F^{(r-2i,(0,\mu))}(K) .$$

The above results may be generalized to the case of marked links.

We may prove Theorem 5.2.1 by an anologous argument used in the proof of Theorem 4.2.1. We use the D-polynomial instead of the bracket polynomial. In this case, the following lemma takes the role of Lemma 4.2.7 in the proof of Theorem 4.2.1.

Lemma 5.2.3. *Let* $e(j)=e_1 e_3\cdots e_{2j-1}\in C_r(1\leq j\leq [r/2])$. *For the character* $\chi_{i,\mu}$ *of the representation* $\rho_{i,\mu}$ *of* C_r, *we have*

(a) $\chi_{i,\mu}(e(j))=0$ *if* $i<j$,

(b) $\chi_{i,\mu}(e(j))=(1-(a-a^{-1})/x)^j\, \chi_{i-j,\mu}(1)$ *if* $1\leq j\leq i$,

where $\chi_{i-j,\mu}\in C_{r-2j}\hat{}$ *and we use the convention that* $\chi_{0,\phi}(1)=1$.

To prove this lemma, we use the folloaing formula in [2]. For $\mu=(\mu_1, \mu_2, \cdots)\in\Lambda(n)$ and $\mu'=(\mu_1', \mu_2', \cdots)\in\Lambda(n-1)$, we denote $\mu'<\mu$ iff $\mu_i'\leq\mu_i$ for all $i\in\boldsymbol{N}$. For $y\in C_{r-1}$, $0\leq i\leq [r/2]$ and $\mu\in\Lambda(r-2i)$, we have

$$\chi_{i,\mu}(y)=\sum_{\substack{\mu'\in\Lambda(r-2i-1),\\ \mu'<\mu}} \chi_{i,\mu'}((y))+(\sum_{\substack{\mu''\in\Lambda(r-2i+1),\\ \mu<\mu''}} \chi_{i-1,\mu''}(y)) . \tag{5.2.4}$$

The details of proofs of Theorem 5.2.1 and Lemma 5.2.3 are omitted.

5.3. The Kauffman polynomial of satellite links. In the case of the Kauffman polynomial, we have a formula for srtellite links as in the case of the one-variable Jones polynomial (discussed in Section 4.3).

Theorem 5.3.1. *Let K be a knot, L a link diagram in the annulus, and* (g, r) *an element of G whose closure is regular isotopic to* $|L|$. *Let* $F_L(K)=F(K_L)$, *where* K_L *is the satellite link of K with respect to L. Then we have*

$$F_L(K)=a^{-w(L)}\,(\sum_{(i,\mu)\in C_r^{\wedge}} \chi_{i,\mu}(p_r'(g))\, F^{(i,\mu)}(K)) .$$

The proof is similar to that of Theorem 4.3.2 and we omit it. This theorem may be generalized to the case of marked links.

EXAMPLE 5.3.2 (doubled knot [33]). Let L_0 be the link diagram in the annulus as in Figure 14. Then as in Example 4.3.3, $|L_0|$ is regular isotopic to $(g_0, 4)^{\sim}$, where $g_0=f_1 f_2 f_3 \sigma_2^2$. For a knot K, we have $F_{L_0}(K)=-a^{-2}x(\beta^{-1}+a^{-1})F^{(2,(0,(2)))}(K)+a^{-2}x(\beta-a^{-1})F^{(2,(0,(11)))}(K)+a^{-2}(x^3(a-a^{-1}-x)^{-1}+x^2+1)$ by calculating $\chi_{j,\mu}(g_0)$ for $(i,\mu)\in C_4\hat{}$.

6. Mutation and the r-parallel version of a link invariant. Let K and K' be two distinct mutant links (Definition 6.2.3). We are interested in comparing the r-parallel versions $X^{(r)}(K)$ and $X^{(r)}(K')$ of a link nivariant X. In the case of the (two-variable) Jones polynomial V, P and the Kauffman polynomial F, it is already known [23], [25] that none of $V^{(r)}$, $P^{(2)}$, and $F^{(2)}$ can distinguish two mutant knots. But it is announced [25] that the Kinoshita-Terasaka knot and the Conway's 11–corssing knot, which are mutant, have distinct $P^{(3)}$. We can show that there are four mutant knots having distinct $P^{(3)}$.

6.1. Tangles.

DEFINITION 6.1.1. Let n be a positive integer. An oriented rectangular diagram T which consists of oriented curves is called a *n-tangle* if T contains some or no closed curves, and n non-closed curves starting at the top of the rectangle and terminating at the bottom of it (Figure 16). Two n-tangles T and S are called *isotopic* if there is a sequence of Reidemeister moves from T to S. We denote R_n the set of isotopy classes of n-tangles. A *closure* of an n-tangle T, written $T^\wedge$, is the link diagram formed by joining the n points at the top of T to those at the bottom without further crossings. For an m-braid b $(m \geq n)$, the *n-closure* of b is the n-tangle formed by joinig the i-th poont $(n+1 \leq i \leq m)$ at the top of b to that at the bottom without further crossings (Figure 16).

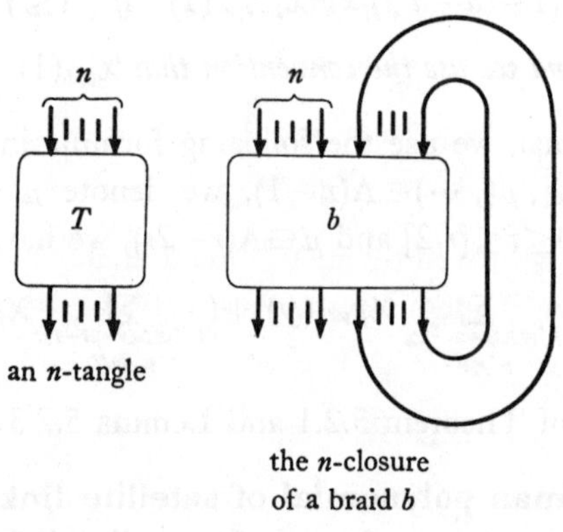

Figure 16

As in Theorem 1.1.1, we have the following.

Theorem 6.1.2. *Every n-tangle is isotopic to the n-closure of a braid.*

For two n-tangles T and S, we define the composite tangle TS by connecting the points at the bottom of T to those at the top of S as in Figure 17. The set R_n together with the above composition law is a semigroup called the *n-tangle semigroup*. An n-braid is an n-tangle and we regard B_n as a subsemigroup of

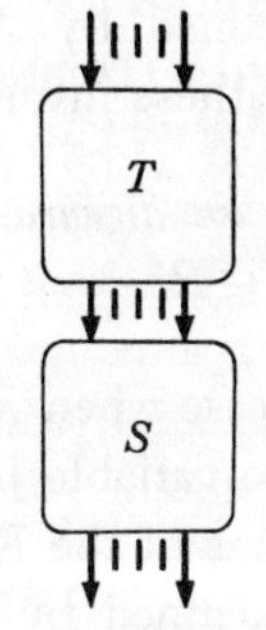

a product tangle

Figure 17

R_n. Let CR_n denote the semigroup algebra of R_n called the *n-tangle algebra*.

DEFINITION 6.1.3. For a positive integer r, the *r-parallel version* $T^{(r)}$ of an n-tangle T is obtained by rep'acing each crossings of T as in Figure 1.

Proposition 6.1.4. *Let T and S be n-tangles. Their r-parallel versions $T^{(r)}$ and $S^{(r)}$ are isotopic if T and S are isotopic.*

This proposition is proved by an argument analogous to the proof of Theorem 1.2.2. We use the Reidemeister moves instead of the relations in Definition 1.1.2.

Let $\phi_n^{(r)}: CR_n \to CR_{rn}$ be the algebra homomorphism defined by $\phi_n^{(r)}(T) = T^{(r)}$. Let $\iota_k^{\sim}: B_r \to CB_{rn}$ be the composition of ι_k defined by (1.4.2) and the natural inclusion $B_{rn} \to CR_{rn}$. Let $\theta: R_n \to S_n$ be the mapping such that, for $T \in R_n$, the i-th point at the top of T is joined to the $\theta(T)(i)$-th point at the bottom by a curve of T. Then θ is a semigroup homomorphism.

Lemma 6.1.5. *For $T \in R_n$ and $b \in B_r$,*

$$\phi_n^{(r)}(T)\,\iota_k^{\sim}(b') = \iota_{\theta(T)(K)}^{\sim}(b')\,\phi_n^{(r)}(T) \quad (1 \le k \le n)\,.$$

Proof. Let $(b, m) \in B$ such that the n-closure of b is isotopic to T. Then $\phi_n^{(r)}(T)$ is isotopic to the rn-closures of $\phi_m^{(r)}(b)$. Hence the statement of the lemma follows from Lemma 1.4.3 and the fact that the rn-closures of $(\phi_m^{(r)}(b)\,\iota_k(c))$ and $(\iota_k(c)\,\phi_m^{(r)}(b))$ are isotopic for $c \in B_r$ if $n+1 \le k \le m$. □

In the following of this section let X be the one-variable Jones polynomial (two-variable Jones polynomial, the Kauffman polynomial respectively). We fix the complex parameters of the invariant X so that X is of trace type and the corresponding coefficients $a_{n,i}(V)$ in (4.1.9) ($a_\lambda(P)$ in Theorem 3.4, $a_{i,\mu}(F)$ in Theorem 5.1.5 respectively) are all non-zero. Let X_n be the linear function on the associated algebra $A_n(X)$ defined by $X((b, n)^\wedge)=X_n(p_n(b))$ for $b\in B_n$. Then from the defining relation of these invariants, we have the following.

Proposition 6.1.6. *There is an algebra homomorphism.* $\Xi_n: \boldsymbol{C}R_n\to A_n(X)$ *such that* $X(T^\wedge)=X_n(\Xi_n(T))$ *for* $T\in R_n$.

Proof. We prove this in the case when X is the Kauffman polynomial $F(\cdot)(a, x)$. The proofs for one and two-variable Jones polynomials are similar. Let $C_m(m\in \boldsymbol{N})$ be the associated algebras of the Kauffman polynomial (Section 5.1). Let $F_m: C_m\to \boldsymbol{C}$ be the mapping defined by $F_m(p_m(b))=F(b^\wedge)$ for $b\in B_m$. We first define a linear mapping $\Phi_m: C_m\to C_{m-1}$. Let $(\cdot, \cdot)_m$ be the symmetric bilinear form on C_m defined by using $F_m: C_m\to \boldsymbol{C}$ as follows:

$$(6.1.7)\qquad (y_1, y_2)_m = F_m(y_1 y_2) \qquad \text{for} \quad y_1, y_2\in C_m\,.$$

Since $(\cdot, \cdot)$ is non-degenerate by the assumption of the parameters (a, x) of F, we can uniquely define Φ_m so that Φ_m satisfies $(y_1, \Phi_m(y_2))_{m-1}=(y_1, y_2)_m$ for all $y_1\in C_{m-1}$ and $y_2\in C_m$. By putting $y_1=1$, we have $F_m(y_2)=F_{m-1}(\Phi_m(y_2))$. By using the defining relations (5.1.1) of C_m, any element $y\in C_m$ can be written as $y=z_1+z_2\, z_{m-1}\, z_3+z_4\, \tau_{m-1}\, z_5$ for some $z_1, \cdots, z_5\in C_{m-1}$. By using Definition 1.1.10, we know that $\Phi_m(y)=(1-(a-a^{-1})/x)\, y_1+y_2\, z_2+a\, y_3\, z_3$. Let $\lambda_r: C_m\to C_{m+r}$ and $\lambda'_r: B_m\to B_{m+r}$ be the algebra homomorphisms defined by $\lambda_r(\varepsilon_i)=\varepsilon_{i+r}$, $\lambda_r(\tau_i^{\pm1})=\tau_{i+r}^{\pm1}$ and $\lambda'_r(\sigma_i^{\pm1})=\sigma_{i+r}^{\pm1}$ respectively. Then

$$(6.1.8)\qquad \Phi_{m+r}(\lambda_r(y)) = \lambda_r(\Phi_m(y)) \qquad \text{for} \quad y\in C_m\,,$$

and

$$(6.1.9)\qquad p_{m+r}(\lambda'_r(b)) = \lambda_r(p_m(b)) \qquad \text{for} \quad b\in B_m\,.$$

Hence we have

$$(6.1.10)\qquad \Phi_m(y_1 y_2) = \Phi_m(y_1)\, y_2 \qquad \text{for} \quad y_1\in C_m \quad \text{and} \quad y_2\in C_{m-1}\,,$$

where $p_m: \boldsymbol{C}B_m\to C_m$ is the projection defined in Section 5.1. Let T be an n-angle. Then there is $(b, m)\in B$ whose n-closure is isotopic to T. Let

$$(6.1.11)\qquad \Xi_n(T) = \Phi_{n+1}(\cdots(\Phi_{m-1}(\Phi_m(p_m(b))))\cdots)\,.$$

From now on we prove that $\Xi_n(TS)=\Xi_n(T)\Xi_n(S)$ for two n-tangles T and S. Let (b, m), $(b', m')\in B$ whose n-closures are isotopic to T and S respectively. Then the product tangle TS is isotopic to the n-closure of $(\delta^{-1}\lambda_{m'-n}(b)\,\delta b', m+m'-n)$ where $\delta=\sigma(m'-n, 1, m'-1, n)$ (see (4.4.17)). By using (6.1.8)–(6.1.11), for $y\in C_n$, we have

$$
\begin{aligned}
&F_n(y\Xi_n(TS))\\
&= F_n(y\Phi_{n+1}(\cdots(\Phi_{m+m'-n}(p_{m+m'-n}(\delta^{-1}\lambda'_{m'-n}(b)\,\delta b'))))\cdots))\\
&= F_n(y\Phi_{n+1}(\cdots(\Phi_{m'}(\delta^{-1}\Phi_{m'+1}(\cdots(\Phi_{m+m'-n}(p_{m+m'-n}(\lambda'_{m'-n}(b))))\cdots)\,\delta p_{m'}(b')))\cdots))\\
&= F_n(y\Phi_{n+1}(\cdots(\Phi_{m'}(\delta^{-1}\lambda_{m'-n}(\Phi_{n+1}(\cdots(\Phi_m(p_m(b)))\cdots))\,\delta p_{m'}(b')))\cdots))\\
&= F_n(y\Phi_{n+1}(\cdots(\Phi_{m'}(\Phi_{n+1}(\cdots(\Phi_m(p_m(b)))\cdots)\,p_{m'}(b')))\cdots))\\
&= F_n(y\Phi_{n+1}(\cdots(\Phi_{m'}(\Phi_{n+1}(\cdots(\Phi_m(p_m(b)))\cdots)\,\Phi_{n+1}(\cdots(\Phi_{m'}(p_{m'}(b')))\cdots))\\
&= F_n(y\Xi_n(T)\,\Xi_n(S)).
\end{aligned}
$$

Hence we get $\Xi_n(TS)=\Xi_n(T)\,\Xi_n(S)$. □

For $\mu\in A_{rn}(X)^{\wedge}$ (respectively $\nu\in A_r(X)^{\wedge}$), let (ρ_μ, V_μ) (respectively (ρ_ν, V_ν)) be the corresponding representation of $A_{rn}(X)$ (respectively $A_r(X)$) and χ_μ (respectively χ_ν) its character. We regard B_r^n as a subgroup of B_{rn} as in Section 1.4. Then V_μ can be regarded as a B_r^n-module. For $\nu\in A_r(X)^{\wedge}$, let $V_{\mu,\nu,\dots,\nu}$ be the $(\rho_\nu\circ p_r)^{\otimes n}$-isotypic subspace of V_μ. Then $V_{\mu,\nu,\dots,\nu}$ is invariant relative to the action of $\phi_n^{(r)}(R_n)$. Lct $(\rho_{\mu,\nu}, V_{\mu,\nu,\dots,\nu})$ denote the representation $\rho_\mu\circ\Xi_{rn}\phi_n^{(r)}|_{V_{\mu,\nu}}$ of R_{rn}. From Proposition 1.3.4, we know that there is a representation $\pi_{\mu,\nu}$ of R_n such that $\rho_{\mu,\nu}\cong(\rho_\nu^{\otimes n})^{\sim}\otimes\pi_{\mu,\nu}$, where $(\rho_\nu^{\otimes n})^{\sim}$ is the representation of $B_r^n\rtimes R_n$ coming from $\rho_\nu^{\otimes n}$ as in Proposition 1.3.4. Let $\omega_{\mu,\nu}$ be the character of $\pi_{\mu,\nu}$. Then the argument proving Theorem 1.4.10 implies the next one.

Theorem 6.1.12. *Let T be an n-tangle whose closure is a knot. Then we have the following.*

(i) $\chi_\mu(\Xi_{rn}(\phi_n^{(r)}(T)))=\sum_{\nu\in A_r(X)^{\wedge}}\chi_\nu(1)\,\omega_{\mu,\nu}(T)$,

(ii) $X^{(r)}(T^{\wedge})=\sum_{\mu\in A_{rn}(X)^{\wedge}}\Big(\sum_{\nu\in A_r(X)^{\wedge}}a_\mu(X)\,\chi_\nu(1)\,\omega_{\mu,\nu}(T)\Big)$.

6.2. Mutation.

In this section, let X be the one-variable Jones polynomial V (the two-variable Jones polynomial P, the Kauffman polynomial F respectively).

DEFINITION 6.2.1. For T in the n-tangle algebra R_n, we define three non-trivial involutions γ_1, γ_2 and γ_3 (see Figure 18). Let γ_1 be the involution defined by $\gamma_1 T=gTg^{-1}$ where $g=\sigma(1,n)\,\sigma(1,n-2)\cdots\sigma(1,1)$ and $\sigma(i,j)=\sigma_i\,\sigma_{i-1}\cdots\sigma_j$ $(i\le j)$. Let D_T be a rectangular diagram of T and H the half-turn of D_T about the center of D_T. Put $T'=H(D_T)$. Let $\gamma_2 T$ be the element of R_n which is the isotopy class of the tangle obtained by inverting the orientations of all the strings of T'. Let γ_3 denote the composition of γ_1 and γ_2.

The involution γ_1 induces an automorphism of the 2-tangle algebra $\boldsymbol{CR}_2$, and γ_2 and γ_3 induce anti-automorphisms of $\boldsymbol{CR}_2$. Hence they induce (anti-) automorphisms of the subalgebra $\Xi_{rn}(\phi_n^{(r)}(\boldsymbol{CR}_n))$ of $A_{rn}(X)$, where $\Xi_{rn}\colon \boldsymbol{CR}_{rn}\to A_{rn}(X)$ is the algebra homomorphism introduced in Proposition 6.1.6.

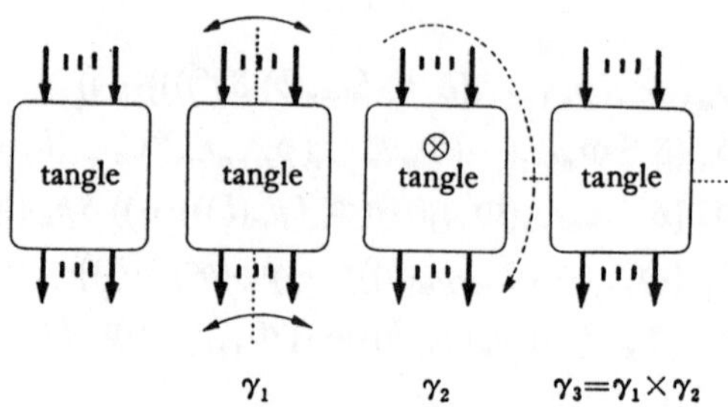

involutions of an n-tangle

Figure 18

Proposition 6.2.2. *Let $\mu\in A_{rn}(X)^\wedge$ and $\nu\in A_r(X)^\wedge$. If the degree (i.e. the dimension of the representation space) of the representation $\pi_{\mu,\nu}$ (introduced at the end of the last section) is equal to zero or one, then we have*

$$\pi_{\mu,\nu}(T)=\pi_{\mu,\nu}(\gamma_i T) \qquad \text{for} \quad T\in R_n,\ 1\leq i\leq 3\,.$$

Proof. Lemma 6.2.9, which will be gieven at the end of this section, shows that the actions of $\gamma_i(1\leq i\leq 3)$ induce $\boldsymbol{C}$-algebra (anti-)automorphisms of $\pi_{\mu,\nu}(\boldsymbol{C}R_2)$. But $\pi_{\mu,\nu}(\boldsymbol{C}R_2)$ is isomorphic to $\boldsymbol{C}$ or $\{0\}$ by the assumption. Hence their actions must be trivial ones. □

DEFINITION 6.2.3 (see, e.g. [8]). Two links K and K' are called *mutant* if there are two 2-atngles T, S and an involution γ_i (i=1, 2 or 3) of S such that K and K' are equivalent to $TS^\wedge$ and $(T\gamma_i S)^\wedge$ respectively.

The following theorem is the main result of this section.

Theorem 6.2.4. *Let K and K' be mutant knots. Then, for a positive integer r and $\nu\in A_r(X)^\wedge$, $X^{(r,\nu)}(K)=X^{(r,\nu)}(K')$ if the degrees of the representations $\pi_{\mu,\nu}$ ($\mu\in A_{2r}(X)$) are all equal to zero or one.*

Proof. Let T and S be the tangles and γ an involution of S such that K and K' are equivalent to $TS^\wedge$ and $(T\gamma S)^\wedge$ respectively. Theorem 6.1.9 implies that

$$X^{(r,\nu)}(K)=\chi_\nu(1)\sum_{\mu\in A_{2r}(X)^\wedge} a_\mu(X)\,\mathrm{trace}\,(\pi_{\mu,\nu}(T)\,\pi_{\mu,\nu}(S))\,,$$

$$X^{(r,\nu)}(K')=\chi_\nu(1)\sum_{\mu\in A_{r2}(X)^\wedge} a_\mu(X)\,\mathrm{trace}\,(\pi_{\mu,\nu}(T)\,\pi_{\mu,\nu}(\gamma S))\,.$$

Hence the statement of the theorem follows from Proposition 6.2.2. □

Corollary 6.2.5. *Let K and K' be mutant knots. Then we have*

(i) $$V^{(r,\nu)}(K)=V^{(r,\nu)}(K')\,,$$

(ii) ([23], [25]) $V^{(r)}(K)=V^{(r)}(K')$,

(iii) $P^{(r,\mu)}(K)=P^{(r,\mu)}(K')$ *if the partition* μ *of* r *is equal to* (r) *or* (1^r),

(iv) ([23]) $P^{(2)}(K)=P^{(2)}(K')$,

(v) ([23]) $F^{(2)}(K)=F^{(2)}(K')$.

Proof. Recall that the degree of $\pi_{\mu,\nu}$ is equal to the multiplicities of $\rho_\nu^{\otimes 2}$ in ρ_μ for $\mu\in A_{2r}(X)^\wedge$ and $\nu\in A_r(X)$. By Remark 3.8, the above multiplicities may be calculated by the Littlewood-Richardson rule in the case that X is the Jones polynomial V or the two-variable Jones polynomial P. This rule shows that the degree of the representations $\pi_{\mu,\nu}$ for the case (i)-(iv) are all equal to 1 or 0. To prove (v), we show that the degree of $\pi_{\mu,\nu}$ is equal to 1 or 0 for $\mu\in A_4(F)^\wedge$ and $\nu\in A_2(F)^\wedge$. This can be checked by using the realizations of irreducible representations of $A_4(F)$ given in [2]. □

Theorem 6.2.6. *Let K and K' be mutant knots, i.e. there are tangles T and S and an involution* γ *of S such that K and K' are equivalent to* $TS^\wedge$ *and* $(T\gamma S)^\wedge$. *Put* $\lambda_0=(321)\in\Lambda(6)$ *and* $V_0=(21)\in\Lambda(3)$. *Then, with the notations in Theorem* 6.1.12 *we have* $P^{(3)}(K)\neq P^{(3)}(K')$ *iff* $a_{\lambda_0}(P)\neq 0$ *and* $\omega_{\lambda_0,\nu_0}(TS)\neq\omega_{\lambda_0,\nu_0}(T\gamma S)$.

Proof. From Theorem 1.5.1 and Corollary 6.2.5 we have $P^{(3)}(K)\neq P^{(3)}(K')$ iff $P^{(3,\nu_0)}(K)\neq P^{(3,\nu_0)}(K')$. But from Littlewood-Richardson rule (Remark 3.8), the degree of π_{λ_0,ν_0} is equal to 2 and those of π_{λ,ν_0} are equal to 1 or 0 if $\lambda\neq\lambda_0$. Hence Proposition 6.2.2 implies the statement of the theorem. □

EXAMPLE 6.2.7. We give four mutant knots K_1, K_2, K_3 and K_4 for which the invariants $P^{(3)}(K_i)$ $(1\leq i\leq 4)$ are all distinct. Let S_1, S_2, σ and $\sigma^\sim$ be the tangles as in Figure 19, $S_3=S_1\,\sigma S_3$, $S_4=S_1\,S_2$ and $K_i=(S_3\,\gamma_i(S_4))^\wedge$ $(1\leq i\leq 4$, Figure 19). Let q be an indeterminate and $H_6^\bullet$ the $\boldsymbol{Z}[q^{1/2},q^{-1/2}]$-algebra with unit defined by the relations (3.2). Let $\Phi_m: H_m^\bullet\to H_{m-1}^\bullet$ be the mapping defined as in the proof of Proposition 6.1.6. Then, from the definition of the two-variable Jones polynomial, we know that the image $\Phi_m(H_m^\bullet)$ is contained in $H_{m-1}^\bullet\otimes_{\boldsymbol{Z}}\boldsymbol{Z}[\alpha,\alpha^{-1},(\alpha q^{1/2}-\alpha^{-1}q^{-1/2})^{-1}]$. Hence $\Xi_n(R_n)$ $(n=3,6)$ are contained in $H_n^\bullet\otimes_{\boldsymbol{Z}}\boldsymbol{Z}[\alpha,\alpha^{-1},(\alpha q^{-1}-\alpha^{-1}q^{-1/2})^{-1}]$ by (6.1.11). On the other hand, Naruse and Gyoja constructed W-graphs [18] of all the irreducible repre sentations of H_6 (Figure 8). This means that, for each irreducible representation, they give a basis of the representation space and representation matrices of generators of H_6 with respect to the basis. The entries of the above representation matrices of the generators $T_i\in H_6^\bullet$ $(1\leq i\leq 5)$ are all contained in $\boldsymbol{Z}[q^{1/2},q^{-1/2}]$. This implies the following two facts. Let $\mathfrak{R}=\boldsymbol{Z}[q^{1/2},q^{-1/2},\alpha,\alpha^{-1},(\alpha q^{1/2}-\alpha^{-1}q^{-1/2})^{-1}]$. The first fact is that the composition of Ξ_6 and the above representations of $H_6^\bullet$ define irreducible representations of R_6 over $\mathfrak{R}$. The second fact is that the ν_0-isotypic subspace V_{λ_0,ν_0} of the representation space

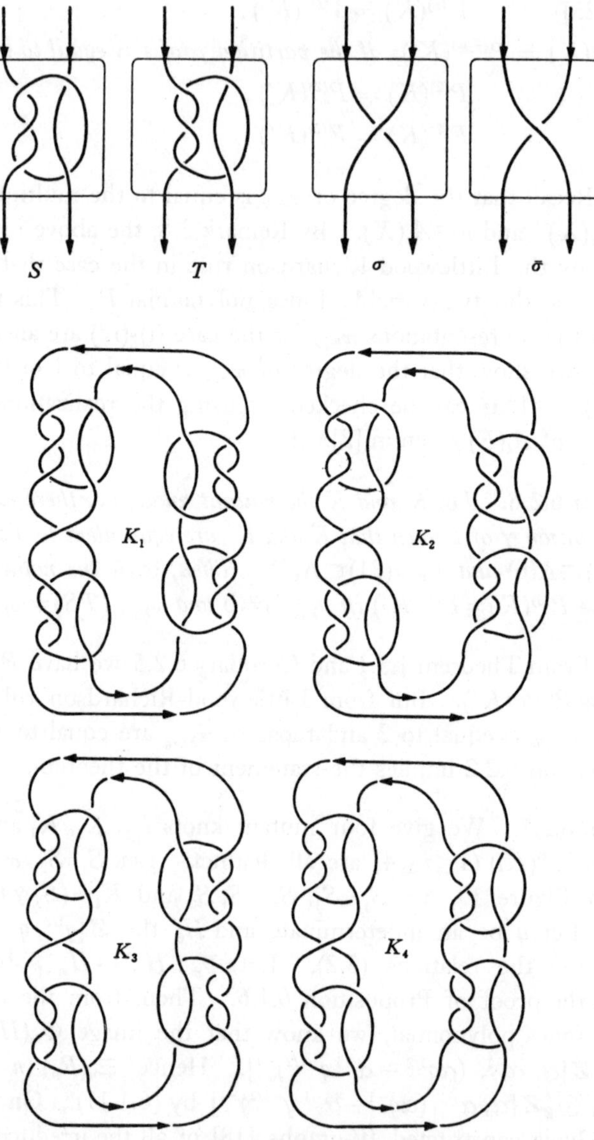

Figure 19

V_{λ_0} of ρ_{λ_0} is defined over $\mathcal{R}$. Hence we can define the reduction modulo p of the representation π_{λ_0,ν_0} of R_2. This is defined over $\boldsymbol{F}_p[q^{1/2}, q^{-1/2}, \alpha, \alpha^{-1}, (\alpha q^{1/2}-\alpha^{-1}q^{-1/2})^{-1}]$, where $\boldsymbol{F}_p$ is the finite field of prime order p. In the following, we put $p=23$, $\alpha=1$ and $q=2$. With these parameters, we know that $\alpha q^{1/2}-\alpha^{-1}q^{-1/2}\neq 0$ and $a_{\lambda_0}(P)\neq 0$ (mod 23) from the explicit formula for $a_{\lambda_0}(P)$ given in [10] or [15]. Therefore we may use Theorem 6.2.7. We calculate the elements $\Xi_6(\phi_2^{(3)}(S_1))$, $\Xi_6(\phi_2^{(3)}(S_2))$, $\Xi_6(\phi_2^{(3)}(\sigma))$ and $\Xi_6(\phi_2^{(3)}(\sigma^{\sim}))$ of $(H_6^{\cdot}\otimes_{\boldsymbol{Z}}$

$\boldsymbol{F}_{23})_{\alpha=1,q=2}$ with the aid of computers (Turbo PASCAL on MS-DOS for PC–9801 (NEC) and AOS/VS PASCAL on Eclipse MV–2000DC (DG)). Their images under π_{λ_0,ν_0} with respect to the explicit basis mentioned above are

$$\pi_{\lambda_0,\nu_0}(\phi_2^{(y)}(S_1))\equiv\begin{bmatrix}9 & 1\\ 7 & 16\end{bmatrix},\quad \pi_{\lambda_0,\nu_0}(\phi_2^{(3)}(S_2))\equiv\begin{bmatrix}8 & 19\\ 0 & 5\end{bmatrix},$$

$$\pi_{\lambda_0,\nu_0}(\phi_2^{(3)}(\sigma))\equiv\begin{bmatrix}0 & 19\\ 19 & 0\end{bmatrix},\quad \pi_{\lambda_0,\nu_0}(\phi_2^{(3)}(\sigma^{\sim}))\equiv\begin{bmatrix}0 & 21\\ 21 & 0\end{bmatrix}\pmod{23}.$$

Hence we have

$$\pi_{\lambda_0,\nu_0}(\phi_2^{(3)}(S_3)) = \pi_{\lambda_0,\nu_0}(\phi_2^{(3)}(S_1\,\sigma S_2))\equiv\begin{bmatrix}14 & 20\\ 17 & 1\end{bmatrix},$$

$$\pi_{\lambda_0,\nu_0}(\phi_2^{(3)}(S_4)) = \pi_{\lambda_0,\nu_0}(\phi_2^{(3)}(S_1\,S_2))\equiv\begin{bmatrix}3 & 15\\ 10 & 6\end{bmatrix},$$

$$\pi_{\lambda_0,\nu_0}(\phi_2^{(3)}(\gamma_1\,S_4)) = \pi_{\lambda_0,\nu_0}(\phi_2^{(3)}(\sigma S_1\,S_2\,\sigma^{\sim}))\equiv\begin{bmatrix}6 & 10\\ 15 & 3\end{bmatrix},$$

$$\pi_{\lambda_0\nu_0}(\phi_2^{(3)}(\gamma_2\,S_4)) = \pi_{\lambda_0,\nu_0}(\phi_2^{(3)}(S_2\,S_1))\equiv\begin{bmatrix}21 & 13\\ 12 & 11\end{bmatrix},$$

$$\pi_{\lambda_0,\nu_0}(\phi_2^{(3)}(\gamma_3\,S_4)) = \pi_{\lambda_0,\nu_0}(\phi_2^{(3)}(\sigma S_2\,S_1\,\sigma^{\sim}))\equiv\begin{bmatrix}11 & 12\\ 13 & 21\end{bmatrix}\pmod{23}.$$

Thus we have $\omega_{\lambda_0,\nu_0}(\phi_2^{(3)}(S_3\,S_4))\equiv 20$, $\omega_{\lambda_0,\nu_0}(\phi_2^{(3)}(S_3\,\gamma_1\,S_4))\equiv 5$, $\omega_{\lambda_0,\nu_0}(\phi_2^{(3)}(S_3\,\gamma_2\,S_4))\equiv 7$ and $\omega_{\lambda_0,\nu_0}(\phi_2^{(3)}(S_3\,\gamma_3\,S_4))\equiv 18 \pmod{23}$. Hence the invariants $P^{(3)}(K_i)$ $(1\leq i\leq 4)$ are all distinct by Theorem 6.2.6 and so any two of the knots K_1, K_2, K_3 and K_4 are not equivalent.

In the rest of this paper, we shall prove the remaining part of the proof of Proposition 6.2.2.

Lemma 6.2.8. *Let (ρ, U) be an irreducible representation of $A_{rn}(X)$. Then $\gamma_i (i=1, 2$ or $3)$ induces an automorphism (respectively anti-automorphism) of* End(U) *defined by ${}^{\gamma_i}(\rho(y))=\rho(\gamma_i\,y)$ $(y\in A_{rn}(X))$.*

Proof. 1°) The case $i=1$. For $y\in A_{rn}(X)$, we have $\gamma_1\,y=p_{rn}(\phi_n^{(r)}(g))\,y p_{rn}(\phi_n^{(r)}(g))^{-1}$ where $g=\sigma(1, n)\,\sigma(1, n-2)\cdots\sigma(1, 1)$, $\sigma(i, j)=\sigma_i\,\sigma_{i+1}\cdots\sigma_j (i\leq j)$ and p_{rn} be the projection $\boldsymbol{C}B_{rn}\to A_{rn}(X)$. Hence γ_1 acts on End(U) by ${}^{\gamma_1}(\rho(y))=\rho(p_{rn}(\phi_n^{(r)}(g))\,\rho(y)\,\rho(p_{rn}(\phi_n^{(r)}(g)))^{-1}$.

2°) The case $\gamma=\gamma_2$. In this case, the statement of the lemma follows from 1°) and 3°) since $\gamma_2=\gamma_1\circ\gamma_3$.

3°) The case $\gamma=\gamma_3$. We know realizations of every irreducible representations of $A_{rn}(X)$ such that $p_{rn}(\sigma_i)$ $(1 \leq i \leq rn-1)$ are represented by symmetric matrices. When $X=V$ (P, F respectively), such constructions are given in [31] ([31], [28] respectively). For a matrix M, let ${}^tM$ denote the transposed matrix of M. We may assume that ${}^t\rho(p_{rn}(\sigma_i))=\rho(p_{rn}(\sigma_i))$. Since γ_3 satisfies $\gamma_3(p_{rn}(\sigma_i))=p_{rn}(\sigma_i)$ $(1 \leq i \leq rn-1)$ and $\gamma_3(hh')=\gamma_3(h')\,\gamma_3(h)$ for $h, h' \in A_{rn}$, we have ${}^{\gamma_3}\rho(h)={}^t\rho(h)$ $(h \in A_{rn}(X))$. Hence γ_3 acts on End(U) by the matrix transposition. □

Lemma 6.2.9. *Let $(\pi_{\mu,\nu}, W_{\mu,\nu})$ $(\mu \in A_{2r}(X)^{\wedge}, \nu \in A_r(X)^{\wedge})$ be the representation of $\boldsymbol{CR}_2$ given at tne end of Section 6.1. Then $\gamma_i (i=1, 2, 3)$ induces an (anti-) autooautomorphism of $\pi_{\mu,\nu}(\boldsymbol{CR}_2)$ defined by ${}^{\gamma_i}(\pi_{\mu,\nu}(y))=\pi_{\mu,\nu}(\gamma_i\, y))$ $(y \in A_{2r}(X))$.*

Proof. 1°) The case $\gamma=\gamma_1$. The statement of the lemma follows from ${}^{\gamma_1}(\pi_{\mu,\nu}(y))=\pi_{\mu,\nu}(g)\,\pi_{\mu,\nu}(y)\,\pi_{\mu,\nu}(g)^{-1}$ where $g=\sigma(1, n-1)\,\sigma(1, n-2)\cdots\sigma(1, 1)$ and $\sigma(i, j)=\sigma_i\,\sigma_{i+1}\cdots\sigma_j$ $(i \leq j)$.

2°) The case $\gamma=\gamma_2$. In this case, the statement of the lemma follows from 1°) and 3°) since $\gamma_2=\gamma_1\circ\gamma_3$.

3°) The case $\gamma=\gamma_3$. Let (ρ_μ, V_μ) be the irreducible representation of $A_{rn}(X)$ parametrized by μ. We shall show that $\pi_{\mu,\nu}(\gamma_3\, y)={}^t(\pi_{\mu,\nu}(y))$ with respect to a basis of $W_{\mu\ \nu}$. In this proof, we use the following fact.

Lemma 6.2.10. *Let s be an indeterminate. Let $y_1(s)$, $y_2(s) \in M_m(\mathcal{K})$ with ${}^ty_1(s)=y_2(s)$ where $\mathcal{K}$ is the algebraic closure of $\boldsymbol{C}(s)$. Let $U(s)$ be a subspace of $\mathcal{K}^m$ such that $y_1(s)\,U(s) \subseteq U(s)$ and ${}^ty_1(s)\,U(s) \subseteq U(s)$. We assume that there is a open set $\mathcal{U}$ of $\boldsymbol{R}$ (the set of real numbers) and a basis $\{u_1(s), \cdots, u_m(s)\}$ of $\mathcal{K}^m$ such that $\{u_1(s), \cdots, u_k(s)\}$ is a basis of $U(s)$ which satisfy the following conditions.*

(i) *The coordinates of $u_i(s)$ $(1 \leq i \leq m)$ have no branch point in $\mathcal{U}$ and they are contained in $\boldsymbol{R}$ for $s \in \mathcal{U}$.*

(ii) *The vectors $u_i(s)$ $(1 \leq i \leq m)$ are linearly independent for $s \in \mathcal{U}$.*

(iii) *The elements of the matrix of $y_1(s)$ with respect to the above basis have no branch point in $\mathcal{U}$.*

Then there is a basis of $U(s)$ such that the transpose of the matrix of $y_1(s)|_{U(s)}$ with respect to this basis is equal to the matrix of $y_2(s)|_{U(s)}$.

Proof. At first, fix $s \in \mathcal{U}$. By applying the Schmidt's method to $\{u_1(s), \cdots, u_m(s)\}$, we get an orthonormal basis $\{v_1(s), \cdots, v_m(s)\}$ of $\boldsymbol{R}^m$ with respect to the standard bilinear form such that $\{v_1(s), \cdots, v_k(s)\}$ is an orthonormal basis of $U(s)$. With respect to the basis $\{v_1(s), \cdots, v_k(s)\}$, the transpose of the matrix of $y_1(s)$ is equal to $y_2(s)$. Now we discuss for general s. From the above argument, we know that $(y_1(s))_{ij}=(y_2(s))_{ji}$ $(1 \leq i, j \leq m)$ for $s \in \mathcal{U}$. But the assumption of the lemma implies that the elements of the matrix of $y_1(s)$ with respect to the basis $\{v_1(s), \cdots, v_k(s)\}$ are algebraic functions in s. Hence the equalities $(y_1(s))_{ij}=(y_2(s))_{ji}$ $(1 \leq i, j \leq m)$ also hold when we regard$s$ as an indeterminate. □

Now, we return to the proof of Lemma 6.2.9. Let $(\rho_{\mu,\nu}, V_{\mu,\nu})$ and $((\rho_\nu^{\otimes n})^\sim, V_\nu^{\otimes n})$ be the representations of $B_r^n \rtimes R_n$ given at the end of Section 6.1. From the definition of $\phi_n^{(r)}$, we have

$$\phi_n^{(r)}(\gamma_3\,\sigma_i) = (\sigma_{ri-r+1}\,\sigma_{ri-r+2}\cdots\sigma_{ri-1})^{-r}\,(\sigma_{ri+1}\,\sigma_{ri+2}\cdots\sigma_{ri+r-1})^r\,\gamma_3\,\phi_n^{(r)}(\sigma_i)\,.$$

We also have $\phi_n^{(r)}(\gamma_3\,y)=c_y\,\gamma_3\,\phi_n^{(r)}(y)$ where c_y is a product of some $(\sigma_{ri+1}\,\sigma_{ri+2}\cdots\sigma_{ri+r-1})^{\pm r}(1\leq i\leq n-1)$. Note that the exponent sum of c_y is equal to zero. The elements $\rho_\mu((\sigma_{ri+1}\,\sigma_{ri+2}\cdots\sigma_{ri+r-1})^{-r})\,(1\leq i\leq n-1)$ act on $V_{\mu,\nu}$ by the same scalar $\rho_\nu((\sigma_1\,\sigma_2\cdots\sigma_{r-1})^{-r})$ since $(\sigma_1\,\sigma_2\cdots\sigma_{r-1})^{-r}$ are contained in the center of the subgroup B_r. Since the exponent sum of c_y is equal to zero, we have $\rho_{\mu,\nu}(c_y)=1$. Recall that $V_{\mu,\nu}$ is the $\rho_\nu^{\otimes n}$-isotypic subspace of V_μ as a $\boldsymbol{C}B_r^n$-module and the representation $\rho_{\mu,\nu}$ of R_n is given by restricting $\rho_\mu\circ\Xi_{rn}\circ\phi_n^{(r)}$ to $V_{\mu,\nu}$. The constructions of the irreducible representations of $A_k(X)$ $(k\in\boldsymbol{N})$ refered in the proof of Lemma 6.2.8 imply that V_μ, $V_{\mu,\nu}$ and $\rho_{\mu,\nu}(y)\,(y\in R_n)$ satisfy the assumption of Lemma 6.2.10 for $s=q^{1/2}$. Therefore we have $\rho_{\mu,\nu}(\gamma_3\,y)={}^t(\rho_{\mu,\nu}(y))$ $(1\leq i\leq n-1)$ with respect to some basis of $V_{\mu,\nu}$. Let $\theta\colon R_n\to S_n$ be the mapping defined just before Lemma 6.1.5. The representation $(\rho_\nu^{\otimes n})^\sim(y)\,(y\in R_n)$ gives a parmutation of a basis of $V_\nu^{\otimes n}$. Hence $(\rho_\nu^{\otimes n})^\sim$ containes the trivial representation of R_n with positive multiplicities. This implies that $\rho_{\mu,\nu}$ containes the representation $\pi_{\mu,\nu}$ with positive multiplicities since $\rho_{\mu,\nu}=(\rho_\nu^{\otimes n})^\sim\otimes\pi_{\mu,\nu}$. Let $U_{\mu,\nu}$ be a R_n-invariant subspace of V_μ where the representation $\pi_{\mu,\nu}{}^{\cdot}$ of R_n on $U_{\mu,\nu}$ is equivalent to $\pi_{\mu,\nu}$. We identify $(\pi_{\mu,\nu}{}^{\cdot}, U_{\mu,\nu})$ and $(\pi_{\mu,\nu}, W_{\mu,\nu})$. As in the case of $\rho_{\mu,\nu}$, we can show that there is a basis of $W_{\mu,\nu}$ such that $\pi_{\mu,\nu}(\gamma_3\,y)={}^t(\pi_{\mu,\nu}(y))$ for $y\in\boldsymbol{C}R_n$. This is what we wanted to show. □

References

[1] J.S. Birman: Braids, Links, and Mapping Class Groups, Annals of Math. Studies 82, Princeton, 1974.

[2] J.S. Birman and H. Wenzl: *Braids, link polynomials and a new algebra,* preprint.

[3] N. Bourbaki: Groupes et algebre de Lie, Chapt. IV, V, VI, Hermann, Paris, 1968.

[4] G. Burde and H. Zieschang: Knots, Walter de Gruyter, 1985.

[5] C.W. Curtis, N. Iwahori and R. Kilmoyer: *Hecke algebras and characters of parabolic type of finite groups with (B,N)-pairs,* I.H.E.S. Publ. Math. **40** (1972), 81–116.

[6] C.W. Curtis and I. Reiner: Representation theory of finite groups and associative algebras, Interscience Publishers, New York 1962.

[7] P. Freyd, D. Yetter, J. Hoste, W.B.R. Lickorish, K. Millet and A. Ocneanu: *A new polynomial invariant of knots and links,* Bull. Amer. Math. Soc. **12** (1985), 239–246.

[8] C.A. Giller: *A family of links and the Conway calculus,* Trans. Amer. Math. Soc. **270** (1982), 75–109.

[9] A. Gyoja: *On the existence of a W-graph for an irreducible representation of a Coxeter group,* Journal of Algebra **86** (1984), 422–438.

[10] A. Gyoja: *Topological invariants of links and representations of Hecke algebras II,* preprint, Osaka University.

[11] A. Gyoja and K. Uno: *On the semisimplicity of Hecke algebras,* J. Math. Soc. Japan **41** (1989), 75–79.

[12] P.N. Hoefsmit: *Representations of Hecke algebras of finite groups with BN-pairs of classical type,* thesis, The University of British Columbia.

[13] G. James and A. Kerber: The representation theory of the symmetric group, Encyclopedia of math. and its applications 16, Addison-Wesley, 1981.

[14] V.F.R. Jones: *A polynomial invariant for knots via von Neumann algebras,* Bull. Amer. Math. Soc. **12** (1985), 103–111.

[15] V.F.R. Jones: *Hecke algebra representations of braid groups and link polynomials,* Ann. Math. **126** (1987), 335–388.

[16] L.H. Kauffman: *State models and the Jones polynomial,* Topology **26** (1987), 395–407.

[17] L.H. Kauffman: *An invariant of regular isotopy,* to appear.

[18] D. Kazhdan and G. Lusztig: *Representations of Coxeter groups and Hecke algebras,* Invent. Math. **53** (1979), 165–184.

[19] A. Kerber: Representations of permutation groups I, Lecture Note in Math. 240, Springer, Berlin, 1971.

[20] D.E. Knuth: The Art of Computer Programming (2nd ed.), Vol. 1, Fundamental Algorithms, Addison-Wesley, 1973.

[21] A. Lascoux and M.P. Schutzenberger: *Polynomes de Kazhdan & Lusztig pour les grassmanniennes,* Young tableaux and Schur functions in algebra and geometry (Torun, 1980), pp. 249–266, Asterisque, **87–88**, Soc. Math. France, Paris, 1981.

[22] W.B.R. Lickorish: *A relationship between link polynomials,* Math. Proc. Camb. Philos. Soc. **100** (1986), 109–112.

[23] W.B.R. Lickorish and A.S. Lipson: *Polynomials of 2-cable-like links,* preprint, Camblidge University, 1986.

[24] H.R. Morton and H.B. Short: *The 2-variable polynomial of cable knots,* Math. Proc. Camb. Philos. Soc., to appear.

[25] H.R. Morton and P. Traczyk: *The Jones polynomial of satellite links around mutants,* preprint.

[26] J. Murakami: *The Kauffman polynomial of links and representation theory,* Osaka J. Math. **24** (1987), 745–758.

[27] J. Murakami: *On the Jones invariant of parallel links and linear representations of braid groups,* unpublished manuscript.

[28] J. Murakami: *The representations of Brauer's centralizer algebra and the Kauffman polynomial of links,* in preparation.

[29] J.H. Przytycki and P. Traczyk: *Invariants of links of Conway type,* Kobe J. Math. **4** (1988), 115–139.

[30] D. Rolfsen: Knots and links. Publish or Perish Inc., 1976.

[31] H. Wenzl: *Representations of Hecke algebras and subfactors,* Ph.D. Thesis, University of Pensylvania.

[32] H. Wenzl: *On the structure of Brauer's centralizer algebras,* preprint, University of Calfornia, Berkeley.

[33] J.H.C. Whitehead: *On doubled knots,* J. London Math. Soc. **12** (1937), 63–71.
[34] S. Yamada: *On the 2-variable Jones polynomial of satellite links,* Topology and Computer Science, 295–300, Kinokuniya, Tokyo 1987.
[35] S. Yamada: *Some regular isotopy invariants and a link invariant operator,* to appear in Topology.

Department of Mathematics
Osaka University
Toyonaka, Osaka 560
Japan

ALGEBRAIC ASPECTS, BACKGROUND FROM C* ALGEBRAS

Invent. math. 72, 1-25 (1983)

Inventiones mathematicae

Index for Subfactors

V.F.R. Jones

Department of Mathematics, University of Pennsylvania, Philadelphia, PA 19104, USA

§1. Introduction

One of the first things Murray and von Neumann did with their theory of continuous dimension for subspaces affiliated with a type II_1 factor was to define an invariant for its action on a Hilbert space. This invariant has come to be known as the coupling constant and it measures the relative mobility of the factor and its commutant. To be precise, if M is a II_1 factor on $\mathscr{H}$ and M' is its commutant, the coupling constant C_M is infinite if M' is an infinite factor and if M' is finite, one takes any non-zero vector ξ in $\mathscr{H}$ and one considers the closed subspaces $\overline{M\xi}$ and $\overline{M'\xi}$ affiliated with M' and M, respectively. It is a nontrivial result of Murray and von Neumann that the ratio $C_M = \dim_{M'}(\overline{M\xi})/\dim_M(\overline{M'\xi})$ does not depend on ξ. This real number between 0 and ∞ is called the coupling constant.

From a more modern point of view, the coupling constant measures the dimension of a Hilbert space on which M acts and may be defined in terms of intertwining maps in the category of normal M-modules. This leads to the notation $C_M = \dim_M(\mathscr{H})$. This notation is more natural and has been useful in the study of the von Neumann algebra of a foliation where the dimensions of certain geometric Hilbert spaces are measured by a von Neumann algebra (see [8]). We will use this more suggestive notation while keeping the term "coupling constant." Two normal representations of M are unitarily equivalent iff they have the same coupling constant.

It is an easy observation that the coupling constant can be used to define a conjugacy invariant for subfactors of II_1 factors. We call this invariant the index since if the subfactor comes from a subgroup in the group constructions of II_1 factors, the conjugacy invariant is the index of the subgroup. The index is defined in general as $\dim_N(L^2(M, \mathrm{tr}))$ where N is the subfactor and tr is the trace on M. This definition was probably noticed by Murray and von Neumann and appears more or less explicitly in works by Goldman [14], Suzuki

Supported in part by NSF Grant # MCS 79-03041

[26] and others. In fact Goldman proves that if a subfactor has index 2 then the whole factor may be expressed as the crossed product of the subfactor by a $\mathbb{Z}_2$ action. This is analogous to the fact that any subgroup of index 2 of a group is normal. It may be combined with Connes' classification of periodic automorphisms of the hyperfinite II_1 factor R ([6]) to yield the pleasing positive result that there is, up to conjugacy, only one subfactor of index 2 of R.

The question immediately arises: what possible values can the index take? The difficulty in answering this question is that there is very little to play with given an arbitrary subfactor of a II_1 factor. The only general result available is the existence of a conditional expectation onto the subfactor shown by Umegaki in [30]. In fact this tool will allow us to determine completely the possible values of the index for subfactors of R. Experience with dimension in II_1 factors suggests that the index will take on a continuum of values. This is indeed true but surprisingly one cannot turn on this effect until index 4 and even then it involves the fundamental group. It seems entirely plausible that for II_1 factors without full fundamental group, whose existence was shown by Connes in [9], the index may take only countably many values.

What, then, are the possible values of the index for subfactors of R? The answer is $\{4\cos^2\pi/n|n=3,4,\ldots\}\cup\{r\in\mathbb{R}|r\geqq 4\}\cup\{\infty\}$. The value ∞ and the real values $\geqq 4$ are easily obtained (2.1.19 and 2.2.5). The situation between 1 and 4 is more difficult. The basic idea of the analysis for index <4 is as follows: Let $N\subseteq M$ be II_1 factors. One represents M on $L^2(M,\mathrm{tr})$ and considers the extension e_N to $L^2(M,\mathrm{tr})$ of the conditional expectation onto N. One defines $\langle M, e_N\rangle$ to be the II_1 factor generated by M and e_N on $L^2(M,\mathrm{tr})$. (This construction appears also in [5], [24]).) The crucial observation at this point is that the index of M in $\langle M, e_N\rangle$ is the same as that of N in M. Thus one may iterate this extension process and one obtains a sequence of II_1 factors, each one obtained from the previous one by adding a projection. The inductive limit gives a II_1 factor and if the projections in the construction are numbered $e_1, e_2, \ldots$, then they satisfy $e_i e_{i\pm1} e_i=\tau e_i$, $e_i e_j=e_j e_i$ if $|i-j|\geqq 2$ and $\mathrm{tr}(e_{i_1} e_{i_2}\ldots e_{i_n})=\tau^n$ if $|i_j-i_k|\geqq 2$ for $j\neq k$. Here τ is the reciprocal of the index of N in M and tr denotes the trace on the inductive limit. Analysis of the algebra generated by the e_i's yields that if $\tau>1/4$ it can only be $\frac{1}{4}\sec^2\pi/n$ for $n=3,4,\ldots$. A large bonus of the analysis is that it shows how to construct subfactors with these values as τ.

It seems strange that the set of values contains a discrete and a continuous part, but this may yet be understood by the fact that, if the index is less than 4, the relative commutant is trivial while the constructions available for the continuous part all have nontrivial relative commutant. We have very little information on what happens to the values of the index if the relative commutant is required to be trivial, but note that a result of Popa in [23], together with Connes' result on injective factors [7] shows that any II_1 factor has a subfactor of infinite index with trivial relative commutant.

Before closing the introduction, I would like to pose 4 problems which I hope will lead to a more profound understanding of subfactors.

Problem 1 (due to Connes). What are the possible values of the index for subfactors of R with trivial relative commutant? Or, what is $\mathscr{C}_R$?

Problem 2. For each $n=3, 4, 5, \ldots$ are there only finitely many subfactors of R up to conjugacy with index $4\cos^2 \pi/n$?

Problem 3. If N is a subfactor of R, is N conjugate to $N \otimes R$ in a decomposition $R \cong R \otimes R$? ($[R:N] < \infty$)

Problem 4. If N is a subfactor of M which is regular and has trivial relative commutant, is M the crossed product of N by a group action? (True if $M=R$ by a result of Ocneanu [22], see [17].)

It would be impossible to thank everyone who helped me with this paper. I have tried to mention individual contributions in the text, but let me also thank especially B. Baker, A. Connes, F. Goodman, R. Powers, M. Takesaki, and A. Wassermann for many fruitful conversations. This paper is an extended version of the Comptes Rendus note [18].

Contents

§2. Generalities

§2.1. *The Global Index*

If M is a finite factor acting on a Hilbert space $\mathscr{H}$ with finite commutant M', the coupling constant $\dim_M(\mathscr{H})$ of M is defined as $\mathrm{tr}_M(E_\xi^{M'})/\mathrm{tr}_{M'}(E_\xi^M)$ where ξ is a non-zero vector in $\mathscr{H}$, tr_A denotes the normalized trace and E_ξ^A is the projection onto the closure of the subspace $A\xi$. This definition, due to Murray and von Neumann in [20], is independent of ξ.

We recall some rules of calculation associated with $\dim_M$. (See [10, p. 263].)

$$\dim_M(\mathscr{H})>0 \tag{2.1.1}$$

$$\dim_M(\mathscr{H})=(\dim_{M'}(\mathscr{H}))^{-1} \tag{2.1.2}$$

$$\text{If } e \text{ is a projection in } M',\ \dim_{M_e}(e\mathscr{H})=\mathrm{tr}_{M'}(e)\dim_M(\mathscr{H}) \tag{2.1.3}$$

$$\text{If } E \text{ is a projection in } M,\ \dim_{M_e}(e\mathscr{H})=(\mathrm{tr}_M(e))^{-1}\dim_M(\mathscr{H}) \tag{2.1.4}$$

If $M\otimes 1$ is the amplification of M on $\mathscr{H}\otimes\mathscr{K}$,

$$\dim_M(\mathscr{H}\otimes\mathscr{K})=\dim_{\mathbb{C}}(\mathscr{K})\dim_M(\mathscr{H}) \tag{2.1.5}$$

$\dim_M(\mathscr{H})=1$ iff M is standard on $\mathscr{H}$, i.e. there is a cyclic trace vector for M. (2.1.6)

Agree to put $\dim_M(\mathscr{H})=\infty$ if M' is infinite.

Proposition 2.1.7. *Let M be as above and N be a subfactor. The number $\dim_N(\mathscr{H})/\dim_M(\mathscr{H})$ is independent of $\mathscr{H}$ provided M' is finite.*

Proof. Any two such representations differ by an amplification and an induction. By (2.1.3) and (2.1.5), both $\dim_M$ and $\dim_N$ are multiplied by the same constants in this process. Q.E.D.

Definition. If N is a subfactor of M, the number $\dim_N(\mathscr{H})/\dim_M(\mathscr{H})$ defined in 2.1.7 is called the (global) *index* of N in M and written $[M:N]$. Note that $[M:N]=\infty$ means that N' is infinite for any normal representation of M.

By (2.1.7) and (2.1.6), $[M:N]=\dim_N(L^2(M,\mathrm{tr}))$. Thus $[M:N]$ is a conjugacy invariant for N as a subfactor of M.

The rules of calculation for $\dim_M$ give some rules for $[M:N]$.

Proposition 2.1.8. *If $P\subseteq Q\subseteq M$ are II_1 factors then*

$$[M:M]=1 \tag{2.1.9}$$

$$[M:P]\geqq 1 \tag{2.1.10}$$

$$[M:P]=[M:Q][Q:P] \tag{2.1.11}$$

$$[M:P]\geqq[M:Q] \tag{2.1.12}$$

$$[M:P]=[M:Q] \quad \textit{implies } P=Q \tag{2.1.13}$$

$$[M:P]=[P':M'] \quad \textit{if } P' \textit{ is finite.} \tag{2.1.14}$$

Proof. The only nontrivial property is (2.1.13). To prove it first note that by (2.1.11) and (2.1.9) we may suppose that $M=Q$ and that M acts on $L^2(M,\mathrm{tr})$ with cyclic trace vector ξ. Then $P\xi$ is dense in $\mathscr{H}$ by hypothesis. But then for any $a\in M$ there is a net b_n of elements of P with $b_n\xi\to a\xi$ in $\mathscr{H}$, i.e. $\|b_n-a_n\|_2\to 0$. By [10, Lemma 1, p. 270], this implies $a\in P$. Q.E.D.

We next examine how the index behaves under tensor products.

Proposition 2.1.15. *Let N_1 and N_2 be subfactors of the finite factors M_1 and M_2, respectively. Then $N_1\otimes N_2$ is a subfactor of $M_1\otimes M_2$ and $[M_1\otimes M_2:N_1\otimes N_2]=[M_1:N_1][M_2:N_2]$.*

Proof. Let M_1 and M_2 act with cyclic trace vectors $\xi_1 \in \mathscr{H}_1$ and $\xi_2 \in \mathscr{H}_2$ respectively. Then if e_1 and e_2 are the projections onto $\overline{N_1 \xi_1}$ and $\overline{N_2 \xi_2}$, $e_1 \otimes e_2$ is the projection onto $\overline{N_1 \otimes N_2(\xi_1 \otimes \xi_2)}$. Moreover $M_1 \otimes M_2$ is standard on $\mathscr{H}_1 \otimes \mathscr{H}_2$ and $(N_1 \otimes N_2)' = N_1' \otimes N_2'$. Thus $\mathrm{tr}_{(N_1 \otimes N_2)'}(e_1 \otimes e_2) = \mathrm{tr}_{N_1'}(e_1)\,\mathrm{tr}_{N_2'}(e_2)$ which gives the desired result. Q.E.D.

Proposition 2.1.16. *Let N_i, M_i be as above and suppose $N_i' \cap M_i = \mathbb{C}$, $i = 1, 2$. Then $(N_1 \otimes N_2)' \cap (M_1 \otimes M_2) = \mathbb{C}$.*

Proof. $(N_1 \otimes N_2)' \cap (M_1 \otimes M_2) = (N_1' \otimes N_2') \cap M_1 \otimes M_2$ (see [28, p. 227]) and $(N_1' \otimes N_2') \cap (M_1 \otimes M_2) = (N_1' \cap M_1) \otimes (N_2' \cap M_2)$. Q.E.D.

We now introduce two isomorphism invariants for II_1 factors.

Definition. If M is a finite factor let

$$\mathscr{I}_M = \{r \in \mathbb{R} \cup \{\infty\} \mid \text{there is a } \mathrm{II}_1 \text{ subfactor } N \text{ of } M \text{ with } [M:N] = r\}$$

$$\mathscr{C}_M = \{r \in \mathbb{R} \cup \{\infty\} \mid \text{there is a } \mathrm{II}_1 \text{ subfactor } N \text{ of } M \text{ with } [M:N] = r \text{ and } N' \cap M = \mathbb{C}\}.$$

The determination of $\mathscr{I}_M$ and $\mathscr{C}_M$ is in general rather difficult. We shall gather some immediate results about them.

Proposition 2.1.17. *If $M \cong M \otimes M$ then both $\mathscr{I}_M$ and $\mathscr{C}_M$ are subsemigroups (with 1) of $\{r \in \mathbb{R} \mid r \geqq 1\} \cup \{\infty\}$ under multiplication.*

Proof. This follows from 2.1.15 and 2.1.16. Q.E.D.

Lemma 2.1.18. *If N is a hyperfinite subfactor of M then $[M:N] < \infty$ implies that M is hyperfinite.*

Proof. If M acts in such a way that N′ is finite, then N′ is hyperfinite (e.g. [29]) and by [7], M' and so M is hyperfinite. Q.E.D.

Corollary 2.1.19. *For any II_1 factor M, $\infty \in \mathscr{I}_M$.*

Proof. If $M \cong R$ then $R \otimes 1 \subseteq R \otimes R$ is of infinite index. Otherwise by [21] there is a hyperfinite subfactor of M which is of infinite index by 2.1.18. Q.E.D.

Proposition 2.1.20. *For any separable II_1 factor M, $\infty \in \mathscr{C}_M$.*

Proof. A result of Popa in [23] says that we can find a maximal abelian subalgebra A and a unitary u with $uAu^* = A$ such that u and A generate a subfactor N of M, isomorphic to R. Since N contains a maximal abelian subalgebra, $N' \cap M = \mathbb{C}$.

So if M is not hyperfinite, combining this result with 2.1.18 gives the required subfactor. If $M = R$, see [17] or §2.3. Q.E.D.

In this paper we will show that

$$\mathscr{I}_M \cap [1, 4) = \mathscr{C}_M \cap [1, 4) \subseteq \{4 \cos^2 \pi/n \mid n = 3, 4, \ldots\}$$

and that

$$\mathscr{I}_R = \{4 \cos^2 \pi/n \mid n = 3, 4, \ldots\} \cup \{r \in \mathbb{R} \mid r \geqq 4\} \cup \{\infty\}.$$

§ 2.2. The Local Index

If the global index of a subfactor is finite, one may define a finer invariant, which I call the local index, obtained by restricting the trace on N' to $N'\cap M$. I would like to thank A. Connes for suggesting this approach.

Definition. Let $N\subseteq M$ be II_1 factors and let $p\in N'\cap M$ be a projection. The *index of N at p* will be $[M_p:N_p]=[M:N]_p$.

Lemma 2.2.1. *The index at p and the global index are related by the formula*

$$[M:N]_p=[M:N]\operatorname{tr}_M(p)\operatorname{tr}_{N'}(p).$$

Proof. If M begins in standard form on $\mathscr{H}$ then by 2.1.4, $\dim_{M_p}(p\mathscr{H})=\operatorname{tr}_M(p)^{-1}$. Also $\dim_N(\mathscr{H})=[M:N]$ so by 2.1.3, $\dim_{N_p}(p\mathscr{H})=[M:N]\operatorname{tr}_{N'}(p)$. Thus $[M:N]_p=\dim_{N_p}(p\mathscr{H})/\dim_{M_p}(p\mathscr{H})=[M:N]\operatorname{tr}_M(p)\operatorname{tr}_{N'}(p)$. Q.E.D.

Lemma 2.2.2. *If $\{p_i\}$ is a partition of unity in $N'\cap M$ then*

$$[M:N]=\sum_i \operatorname{tr}_M(p_i)^{-1}[M:N]_{p_i}.$$

Proof. For each i, $[M:N]\operatorname{tr}_{N'}(p_i)=\operatorname{tr}_M(p_i)^{-1}[M:N]_{p_i}$ and summing over i gives the result.

Corollary 2.2.3. *If $[M:N]<\infty$ then $N'\cap M$ is finite dimensional.*

Proof. If $N'\cap M$ were infinite dimensional we could find arbitrarily large partitions of unity and by 2.2.2 $[M:N]=\infty$. Q.E.D.

In fact with a little more care one may obtain the bound $[M:N]\geqq\dim_{\mathbb{C}}(N'\cap M)$.

Corollary 2.2.4. *If $[M:N]<4$, $N'\cap M=\mathbb{C}$.*

Proof. If $N'\cap M\neq\mathbb{C}$, then it contains two mutually orthogonal non-zero projections and since $[M:N]_{p_i}\geqq 1$, $[M:N]\geqq\operatorname{tr}(p_1)^{-1}+\operatorname{tr}(p_2)^{-1}\geqq 4$. Q.E.D.

Corollary 2.2.5. *If M has fundamental group* $=\mathbb{R}$, $\mathscr{I}_M$ *contains* $\{r\in\mathbb{R}\,|\,r\geqq 4\}$.

Proof. We must exhibit for any $r\geqq 4$ a subfactor of index r. Choose $d\in(0,1)$ with $1/d+1/(1-d)=r$ and choose a projection p with $\operatorname{tr}_M(p)=d$. Then M_p and M_{1-p} are isomorphic so choose some isomorphism $\theta\colon M_p\to M_{1-p}$ and let N be the subfactor $\{x+\theta(x)\,|\,x\in M_p\}$. Then $N_p=M_p$ and $N_{1-p}=M_{1-p}$ so by 2.2.2, $[M:N]=1/d+1/(1-d)=r$. Q.E.D.

§ 2.3. Examples

We give three examples of subfactors and their indices.

Example 2.3.1. Suppose $M=N\otimes P$ where P is a type I_n factor. Then $[M:N\otimes 1]=n^2$. This follows from 2.1.5.

Thus for any II_1 factor M, $\mathscr{I}_M$ always contains $\{n^2 | n \in \mathbb{Z} - \{0\}\} \cup \{\infty\}$. It is not inconceivable that there are II_1 factors for which $\mathscr{I}_M$ is no larger than this set.

Example 2.3.2. Let A be a von Neumann algebra and G a countable discrete group of automorphisms for which the crossed product $A \rtimes G$ is a finite factor. If H is a subgroup of G such that $A \rtimes H$ is a finite factor then $[A \rtimes G : A \rtimes H] = [G:H]$.

Proof. Write G as a disjoint union of cosets, $G = \coprod_{i \in I} H g_i$. Then let $V_i = \overline{(A \rtimes H)\, u_{g_i}}$. The V_i are subspaces of $L^2(A \rtimes G, \mathrm{tr})$ affiliated with $(A \rtimes H)'$ which are mutually orthogonal, where tr is the extension to $A \rtimes G$ of a faithful normal G-invariant trace on A and the u_g's are the implementing unitaries of the crossed product. Moreover if p_i is the projection onto V_i then $\sum_{i \in I} p_i = 1$ and the p_i are mutually equivalent in $(A \rtimes H)'$ since $V_i = J u_{g_i} J V_0$ where $V_0 = \overline{A \rtimes H}$ and J is the involution on $L^2(A \rtimes G, \mathrm{tr})$. This is because $J u_{g_i} J \in (A \rtimes G)' \subseteq (A \rtimes H)'$.

Thus if $[G:H] = \infty$, $(A \rtimes H)'$ is infinite so $[A \rtimes G : A \rtimes H] = \infty$. If $[G:H] < \infty$, $(A \rtimes H)'$ is finite since reduction by p_0 puts $A \rtimes H$ in standard form. So $[G:H]\, \mathrm{tr}_{(A \rtimes H)'}(p_0) = 1$ which establishes that $[A \rtimes G : A \rtimes H] = [G:H]$. Q.E.D.

This example is the justification for the name "index".

Example 2.3.3. If M is a II_1 factor and G is a finite group of outer automorphisms of M with fixed point algebra M^G, $[M:M^G] = |G|$.

Proof. This result could be established using 2.3.2 but I shall give a different proof which brings in the basic construction of Chapter 3.

Let M act on $L^2(M, \mathrm{tr})$ and let u_g be the unitaries extending the action of G on M. Then the u_g's act also on M' and it is established in [1] that $(M^G)'$ is isomorphic in the obvious way to $M' \rtimes G$. The projection onto $\overline{M^G}$ is $|G|^{-1} \sum_{g \in G} u_g$ and by the isomorphism with the crossed product its trace is $|G|^{-1}$. Thus $[M:M^G] = |G|$. Q.E.D.

§ 3. The Basic Construction

§ 3.1. *Extending Finite von Neumann Algebras by Subalgebras*

Let M be a finite von Neumann algebra with faithful normal normalized trace tr and let N be a von Neumann subalgebra. By [30] there is a conditional expectation $E_N \colon M \to N$ defined by the relation $\mathrm{tr}(E_N(x)y) = \mathrm{tr}(xy)$ for $x \in M$, $y \in N$. The map E_N is normal and has the following properties:

$$E_N(axb) = aE_N(x)b \quad \text{for } x \in M,\ a, b \in N \text{ (the bimodule property)} \quad (3.1.1)$$

$$E_N(x^*) = E_N(x)^* \quad \text{for all } x \in M \tag{3.1.2}$$

$$E_N(x^*)E_N(x) \leqq E_N(x^* x) \quad \text{and} \quad E_N(x^* x) = 0 \quad \text{implies } x = 0. \tag{3.1.3}$$

Let ξ be the canonical cyclic trace vector in $L^2(M, \mathrm{tr})$. Identify M with the algebra of left multiplication operators on $L^2(M, \mathrm{tr})$. The conditional expectation E_N extends to a projection e_N on $\mathscr{H}$ via $e_N(x\xi) = E_N(x)\xi$. Let J be the involution $x\xi \mapsto x^*\xi$.

Proposition 3.1.4.

(i) *For* $x \in M$, $e_N x e_N = E_N(x) e_N$.

(ii) *If* $x \in M$ *then* $x \in N$ *iff* $e_N x = x e_N$.

(iii) $N' = \{M' \cup \{e_N\}\}''$.

(iv) J *commutes with* e_N.

Proof. (i) If y is an arbitrary element of M then $e_N x e_N(y\xi) = e_N x(E_N(y)\xi) = E_N(x)E_N(y)\xi$ by 3.1.1, and $E_N(x)e_N(y\xi) = E_N(x)E_N(y)\xi$. But the vectors $y\xi$ are dense in $L^2(M, \mathrm{tr})$.

(ii) Relation 3.1.1 shows that e_N commutes with N as in (i). Moreover if $x \in M$ and $e_N x = x e_N$ then $(e_N x)\xi = E_N(x)\xi = (xe)\xi = x\xi$. Since ξ is separating, $x = E_N(x)$.

(iii) It suffices to show that $\{M' \cup \{e_N\}\}' = N$. This follows from (i).

(iv) This follows from 3.1.2. Q.E.D.

These calculations lead to the following.

Definition. Let $\langle M, e_N \rangle$ be the von Neumann algebra on $L^2(M, \mathrm{tr})$ generated by M and e_N. This is the basic construction.

Proposition 3.1.5. (i) $\langle M, e_N \rangle = JN'J$.

(ii) *Operators of the form* $a_0 + \sum_{i=1}^{n} a_i e_N b_i$ *with* a_i, $b_i \in M$, *give a dense* $*$-*subalgebra of* $\langle M, e_N \rangle$.

(iii) $x \mapsto x e_N$ *is an isomorphism of* N *onto* $e_N \langle M, e_N \rangle e_N$.

(iv) *The central support of* e_N *in* $\langle M, e_N \rangle$ *is* 1.

(v) $\langle M, e_N \rangle$ *is a factor iff* N *is.*

(vi) $\langle M, e_N \rangle$ *is finite iff* N' *is.*

Proof. (i) and (ii) follow immediately from 3.1.4. For (iii), to show that $e_N \langle M, e_N \rangle e_N \subseteq N e_N$ it suffices by (ii) to show that $e_N(a e_N b) e_N \in N e_N$. This follows from (i) of 3.1.4. Moreover if $x e_N = 0$ then $x e_N \xi = x\xi = 0$ and ξ is separating. Thus $x \mapsto x e_N$ is injective. Affirmations (iv), (v) and (vi) are now easy. Q.E.D.

We want to consider special traces on $\langle M, e_N \rangle$.

Definition. If P is a subalgebra of $\langle M, e_N \rangle$, a trace Tr on $\langle M, e_N \rangle$ is called a (τ, P) trace if Tr extends tr and $\mathrm{Tr}(e_N x) = \tau\, \mathrm{tr}(x)$ for $x \in P$.

Lemma 3.1.6. *A* (τ, N) *trace is a* (τ, M) *trace.*

Proof. If $x\in M$, $\mathrm{Tr}(xe_N)=\mathrm{Tr}(e_Nxe_N)=\mathrm{Tr}(E_N(x)e_N)=\tau\,\mathrm{tr}(E_N(x))=\tau\,\mathrm{tr}(x)$. Q.E.D.

We shall now concentrate on the case where M and N are factors.

Proposition 3.1.7. *If M and N are factors then $[M:N]<\infty$ iff $\langle M,e_N\rangle$ is finite and in this case the canonical trace* Tr *on $\langle M,e_N\rangle$ is a (τ,M) trace where $\tau=[M:N]^{-1}$. In particular $\mathrm{Tr}(e_N)=[M:N]^{-1}$. Also $[\langle M,e_N\rangle:N]=[M:N]$.*

Proof. By 3.1.6 it suffices to show that Tr is a (τ,N) trace. But consider the map $y\to\mathrm{Tr}(e_Ny)$ defined on N. This is a trace by (ii) of 3.1.4 so since N is a factor there is a constant K such that $\mathrm{Tr}(e_Ny)=K\,\mathrm{tr}(y)$. Moreover the trace of e_N in N' is by definition τ so by (iv) of 3.1.4, (i) of 3.1.5 and uniqueness of the trace, this is the same as $\mathrm{Tr}(e_N)$. Thus $K=\tau$, and Tr is a (τ,N) trace.

To prove this last assertion note that

$$[\langle M,e_N\rangle:M]=\dim_{M'}(L^2(M,\mathrm{tr}))/\dim_{N'}(L^2(M,\mathrm{tr}))=[\dim_{N'}(L^2(M,\mathrm{tr}))]^{-1}$$
$$=\dim_N(L^2(M,\mathrm{tr}))=[M:N].\quad \text{Q.E.D.}$$

For some general results about projections onto finite subalgebras see Skau's paper [24].

Finally in this section I show that the basic construction is generic for subfactors of finite index in the sense that all such subfactors are of the form $M\subseteq\langle M,e_N\rangle$, although not canonically.

Lemma 3.1.8. *Let N be a II_1 factor acting on $L^2(N,\mathrm{tr})$ and let M be a II_1 factor containing N. Then $[M:N]$ is finite and there is a subfactor P of N such that $M=\langle N,e_P\rangle$.*

Proof. Since M' is a II_1 factor, $[M:N]<\infty$. Let $P=JM'J$. Then $[N:P]=[M:P]$ as in 3.1.7 and $\langle N,e_P\rangle$ is a subfactor of M with $[M:\langle N,e_P\rangle]=1$. Thus by 2.1.13, $M=\langle N,e_P\rangle$. Q.E.D.

Corollary 3.1.9. *Let $N\subseteq M$ be II_1 factors with $[M:N]<\infty$ then there is a subfactor P of N and an isomorphism θ: $M\to\langle N,e_P\rangle$ with $\theta|_N=\mathrm{id}$.*

Proof. Represent M on $L^2(M,\mathrm{tr})$. Choose a projection $p\in M'$ with $\mathrm{tr}(p)=[M:N]^{-1}$. Then by [20], on $pL^2(M,\mathrm{tr})$, M and N act with N in standard form. By 3.1.8 we are through. Q.E.D.

§3.2. Inclusions of Complex Semisimple Algebras

Let $N\subseteq M$ be finite dimensional complex semisimple algebras and $N=\bigoplus_{i=1}^{n}N_i$, $M=\bigoplus_{j=1}^{m}M_j$ be their canonical decompositions as direct sums of simple algebras, $N_i\cong M_{n_i}(C)$, $M_j\cong M_{m_j}(C)$. The inclusion of N in M is specified up to conjugacy by an $n\times m$ matrix $A_N^M=(a_{ij})$ where (a_{ij}) is the number of simple components of a simple M_j module viewed as an N_i module. This may be zero or any positive integer. If $\{p_i\}$ are the central idempotents for the N_i and q_j those for the M_j, $a_{ij}=0$ iff $p_iq_j=0$. We will call the matrix A_N^M the inclusion matrix.

The inclusion can also be described diagramatically as follows:

$n_1 \quad n_2 \quad \ldots \quad n_i \quad \ldots \quad n_n$

(a_{ij})

$m_1 \quad m_2 \quad \ldots \quad m_j \quad \ldots \quad m_m$

Here there are a_{ij} lines between n_i and m_j. This diagram will be called the Bratteli diagram after [4].

If the identity of M is the same as that of N we have the obvious relation $m_j = \sum_{i=1}^{n} a_{ij} n_i$ which we shall write as

$$\vec{m} = \vec{n} A_N^M. \tag{3.2.1}$$

A concrete example is

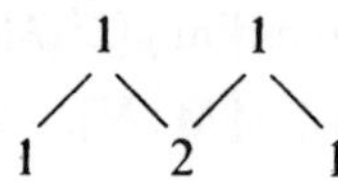

which is the diagram for the inclusion of $\mathbb{C}S_2$ in $\mathbb{C}S_3$.

If $N \subseteq M \subseteq P$, note the formula

$$A_N^P = A_N^M A_M^P. \tag{3.2.2}$$

If V is a faithful M-module, the centres of M and M'' are identical so that (if 3.2.1 holds) in the decompositions of M' and N' as simple algebras we may write $N' = \bigoplus_{i=1}^{n} N_i'$ and $M' = \bigoplus_{j=1}^{m} M_j'$. The following formula is in [3, §5, ex. 17].

$$A_{M'}^{N'} = (A_N^M)^T. \tag{3.2.3}$$

Since there is only one normalized trace on $M_n(\mathbb{C})$, a trace on $M = \bigoplus_j M_j$ may be specified by a column vector $\vec{t} = \begin{pmatrix} t_1 \\ t_2 \\ \vdots \\ t_m \end{pmatrix}$ where t_j is the trace of a minimal idempotent in M_j. The trace of the identity is the product $\vec{m} \cdot \vec{t}$, where $\vec{m} = (m_1, m_2, \ldots, m_m)$.

If $\vec{t}$ defines a trace on M whose restriction to N is defined by the vector $\vec{s}$ then the following relation is immediate.

$$\vec{s} = A_N^M \vec{t}. \tag{3.2.4}$$

Conversely if $\vec{s}$ and $\vec{t}$ define traces on N and M, respectively, then they agree on N if 3.2.4 holds.

§3.3. Finite Dimensional C-Algebras*

A finite dimensional C^*-algebra is of course semisimple so the discussion and notation of §3.2 applies. The presence of the $*$-operation allows us to perform the basic construction of §3.1. The main question to be answered in this section is: given a faithful trace tr on the finite dimensional C^*-algebra M, when does there exist a (faithful) (τ, M) trace on $\langle M, e_N\rangle$? All traces in this section will be positive, i.e. $\vec{t}$ is a positive vector in the natural ordering of $\mathbb{R}^n$.

We begin with a lemma giving more information on the basic construction in finite dimensions.

Lemma 3.3.1. *Let $N \subseteq M$ be finite dimensional C^*-algebras and let* tr *be a faithful positive normalized trace on M. Let e_N and $\langle M, e_N\rangle$ be as in §3.1. Suppose $\{p_i | i=1,2,\ldots,n\}$ are the minimal central projections of N. Then*

(i) *Jp_iJ are the minimal central projections of $\langle M, e_N\rangle$*

(ii) *$A_M^{\langle M, e_N\rangle} = (A_N^M)^T$ (with the obvious identification of the indices, $p_i \leftrightarrow Jp_iJ$)*

(iii) *$e_N J p_i J = e_N p_i$*

(iv) *$x \to e_N x J p_i J$ is an isomorphism from $p_i N$ onto $(e_N J p_i J)\langle M, e_N\rangle(e_N J p_i J)$.*

Proof. (i) The p_i are the minimal central projections of N' and $\langle M, e_N\rangle = JN'J$.

(ii) This follows from 3.2.3 and (i).

(iii) If $x \in M$, $(e_N J p_i J)(x\xi) = e_N(x p_i \xi) = E_N(x) p_i \xi$, and

$$(e_N p_i)(x\xi) = e_N(p_i x \xi) = p_i E_N(x)\xi = E_N(x) p_i \xi.$$

(iv) Injectivity follows from (iii), and (iii) of 3.1.5. If $y \in p_i e_N \langle M, e_N\rangle p_i e_N$ then $y = e_N z$ for $z \in N$ and $p_i z = z$ by (iii) of 3.1.5 so $y = e_N p_i z$. Q.E.D.

Theorem 3.3.2. *Let M, N and* tr *be as above and let* tr *be given on M by the vector $\vec{t}$ and on N by the vector $\vec{s}$. Then there is a (τ, M) trace* Tr *on $\langle M, e_N\rangle$ iff*

(i) $A^T A \vec{t} = (1/\tau)\vec{t}$

(ii) $A A^T \vec{s} = (1/\tau)\vec{s}$

where $A = A_N^M$.

Proof. ($\Rightarrow$) Let Tr be given on $\langle M, e_N\rangle$ by $\vec{r}$, r_i being the trace of a minimal projection in $Jp_iJ\langle M, e_N\rangle$. By (iv) of 3.3.1, such a minimal projection may be chosen of the form $e_N q$, q being a minimal projection in $p_i N$. Since Tr is a (τ, M) trace, $r_i = \tau s_i$ so $\vec{r} = \tau \vec{s}$. But by (ii) of 3.3.1, 3.2.2 and 3.2.4, $\vec{s} = AA^T\vec{r}$ $= \tau AA^T\vec{s}$, i.e. $AA^T\vec{s} = (1/\tau)\vec{s}$. Also $\vec{t} = A^T\vec{r} = \tau A^T AA^T\vec{r} = \tau A^T A\vec{t}$.

($\Leftarrow$) Define a trace Tr on $\langle M, e_N\rangle$ by the vector $\vec{r} = \tau\vec{s}$. It is a faithful positive trace since tr is. By 3.1.6 it suffices to show that it extends tr and that it is a (τ, N) trace. For the former, by 3.2.4 we need $A^T\vec{r} = \vec{t}$. But $A\vec{t} = \vec{s}$ and by (i), $A^T\vec{s} = (1/\tau)\vec{t}$ so $A^T\vec{r} = \vec{t}$. For the latter let q be a minimal projection in $p_i N$. Then $e_N q$ is a minimal projection in $Jp_iJ\langle M, e_N\rangle$ and by definition $\mathrm{Tr}(e_N q)$ $= \tau\,\mathrm{tr}(q)$. The map $x \to \mathrm{Tr}(e_N x)$ is a trace on $p_i N$ so by uniqueness and linearity, $\mathrm{Tr}(e_N x) = \tau\,\mathrm{tr}(x)$ for all $x \in N$. Q.E.D.

§3.4. Two Extensions; Goldman's Theorem

Let N be a proper von Neumann subalgebra of the finite von Neumann algebra M with faithful normal normalized trace tr. Suppose there is a faithful normal (τ, M) trace Tr on $\langle M, e_N\rangle$. Then we may form the extension $\langle\langle M, e_N\rangle, e_M\rangle$.

Proposition 3.4.1.

(i) $e_M e_N e_M = \tau e_M$

(ii) $e_N e_M e_N = \tau e_N$

(iii) $e_M \wedge e_N = e_M \wedge e_N^\perp = e_M^\perp \wedge e_N = 0$.

Proof. (i) By (i) of 3.1.4, $e_M e_N e_M = E_M(e_N) e_M$ and since Tr is a (τ, M) trace, $E_M(e_N) = \tau$.

(ii) By (ii) of 3.1.5 it suffices to verify the relation on vectors of the form $a\xi$ and $a e_N b \xi$ where $a, b \in M$. But

$$e_N e_M e_N(a\xi) = e_N(E_M(e_N a)\xi) = \tau e_N(a\xi),$$

and

$$e_N e_M e_N(a e_N b\xi) = e_N e_M(E_N(a) e_N b\xi) = e_N(\tau E_N(a) b\xi)$$

while $\tau e_N(a e_N b\xi) = \tau e_N E_N(\tau) b\xi$.

(iii) $e_N \wedge e_M = s - \lim_{n\to\infty} (e_N e_M e_N)^n = 0$ since $\tau < 1$.

The other relations follow from $(1-\tau) < 1$. Q.E.D.

Now suppose there is a faithful $(\tau, \langle M, e_N\rangle)$ trace Tr on $\langle\langle M, e_N\rangle, e_M\rangle$.

Corollary 3.4.2. *If $\tau \neq 1/2$, the von Neumann algebra generated by e_N and e_M is isomorphic to $M_2(\mathbb{C}) \oplus \mathbb{C}$. If $\tau = 1/2$ it is isomorphic to $M_2(\mathbb{C})$ and we have the relation $e_M + e_N - e_M e_N - e_N e_M = 1/2$.*

Proof. The relations of 3.4.1 show immediately that the von Neumann algebra generated by e_N and e_M has dimension at most 5 and is not abelian. But $\mathrm{Tr}(e_M \vee e_N) = 2\tau - \mathrm{Tr}(e_M \wedge e_N) = 2\tau$. This is enough to prove the affirmations about the structure of the algebra. An easy calculation shows that $p = (1-\tau)^{-1}(e_M + e_N - e_M e_N - e_N e_M)$ is a projection with $p e_N = e_N$ and $p e_M = e_M$. Thus $p = e_1 \vee e_2$ and if $\tau = 1/2$, $p = 1$. Q.E.D.

Corollary 3.4.3. (Goldman's theorem [14]). *Let N be a subfactor of the II_1 factor M with $[M:N] = 2$. Then M decomposes as the crossed product of N by an outer action of $\mathbb{Z}_2$.*

Proof. By 3.1.9 we know that M is of the form $\langle N, e_P\rangle$ for a subfactor of index 2 of N. Thus N is generated by N and $u = 2e_P - 1$. Moreover since tr is a $(1/2, N)$ trace on M, $\mathrm{tr}(ux) = 0$ for $x \in N$, and the relationship of 3.4.2 implies that, if $v = 2e_N - 1$, $uv = -vu$. Thus for $x \in N$, $v u x u^* = u x u^* v$. So $u x u^*$ commutes with e_N and by (ii) of 3.1.4, $u x u^* \in N$. Thus by [25], M is the crossed product of N by a $\mathbb{Z}_2$ action which is necessarily outer since M is a factor. Q.E.D.

§4. Possible Values of the Index

§4.1. Certain Algebras Generated by Projections

Chapter 4 will be largely devoted to proving the following result.

Theorem 4.1.1. *Let M be a von Neumann algebra with faithful normal normalized trace* tr. *Let $\{e_i | i = 1, 2, \ldots\}$ be projections in M satisfying*

a) $e_i e_{i\pm 1} e_i = \tau e_i$ *for some* $\tau \leqq 1$

b) $e_i e_j = e_j e_i$ *for* $|i-j| \geqq 2$

c) $\mathrm{tr}(w e_i) = \tau\, \mathrm{tr}(w)$ *if* w *is a word on* $1, e_1, e_2, \ldots, e_{i-1}$

Then if P *denotes the von Neumann algebra generated by the* e_i*'s,*

(i) $P \cong R$ *(the hyperfinite* II_1 *factor)*

(ii) $P_\tau = \{e_2, e_3, \ldots\}''$ *is a subfactor of* P *with* $[P:P_\tau] = \tau^{-1}$.

(iii) $\tau \leqq 1/4$ *or* $\tau = \frac{1}{4}\sec^2 \pi/n$, $n = 3, 4, \ldots$.

In this section we will prove (i) and (ii). We begin the proof with some notation.

Definition. Let $A_{m,n}$ be the $*$-algebra generated by $1, e_m, e_{m+1}, \ldots, e_n$ for $1 \leqq m \leqq n \leqq \infty$. Let A_n be $A_{1,n}$, $A_0 = \mathbb{C}$. Thus $A_\infty = A_{1,\infty}$ is the $*$-algebra generated by 1 and the e_i's.

Next some combinatorial results. If w is an (associative) word on the e_i's, call it *reduced* if it is of minimal length for the grammatical rules $e_i e_{i\pm 1} e_i \leftrightarrow e_i$, $e_i e_j \leftrightarrow e_j e_i$ for $|i-j| \geqq 2$, $e_i^2 \leftrightarrow e_i$.

Lemma 4.1.2. *Let* $e_{i_1} e_{i_2} \ldots e_{i_k}$ *be a reduced word. Then if* $m = \max\{i_1, i_2, \ldots, i_k\}$, m *occurs only once in the list* $i_1, i_2, \ldots, i_k$.

Proof. By induction on the length of a reduced word. It is trivial for words of length $\leqq 1$. Suppose true for words of length $\leqq n$ and let w be a reduced word of length $n+1$. Suppose $w = w_1 e_m w_2 e_m w_3$ where m is the maximum index and w_2 does not contain e_m. Then there are 2 possibilities.

a) w_2 does not contain e_{m-1}. In this case e_m commutes with all the e_i's in w_2 so the length of w may be shortened using $e_m^2 \to e_m$.

b) w_2 contains e_{m-1}. Then since w is reduced, so is w_2 and by induction $w_2 = v_1 e_{m-1} v_2$ where v_1 and v_2 are words on $e_1, e_2, \ldots, e_{m-2}$. But then e_m commutes with v_1 and v_2 so that the length of w may be reduced using $e_m e_{m-1} e_m \to e_m$. Q.E.D.

It is clear that in the algebra A, any word on the e_i's is proportional to a reduced word.

Corollary 4.1.3. (i) A_n *is finite dimensional.*

(ii) *For* $x \in A_n$, $e_{n+1} x e_{n+1} = E_{A_{n-1}}(x) e_{n+1}$ *(here* $E_{A_{n-1}}$: $A_n \to A_{n-1}$ *is with respect to the restriction of* tr).

(iii) $x \mapsto x e_{n+1}$ *is an isomorphism of* A_{n-1} *onto* $e_{n+1} A_{n+1} e_{n+1}$.

Proof. (i) If there are only finitely many reduced words, A_n is finite dimensional. But this follows immediately by induction from 4.1.2.

(ii) First of all $\mathrm{tr}(x e_{n+1}) = \tau\, \mathrm{tr}(x)$ for $x \in A_n$ follows immediately from 4.1.1(c) so that $E_{A_{n-1}}(e_n) = \tau$. By 4.1.2 and linearity it suffices to consider x of the form $w e_n w'$ with w and w' in A_{n-1}. But then $e_{n+1} x e_{n+1} = \tau w w' e_{n+1}$ by 4.1.1(a) and (b), and $E_{A_{n-1}}(x) e_{n+1} = \tau w w' e_{n+1}$ by the bimodule property of $E_{A_{n-1}}$.

(iii) If w is a reduced word on $e_1, e_2, \ldots, e_{n+1}$ write $w = x e_{n+1} y$ with $x, y \in A_n$. Then by (ii), $e_{n+1} w e_{n+1} = t e_{n+1}$ for $t \in A_{n-1}$. Thus $e_{n+1} A_{n+1} e_{n+1} \subseteq A_{n-1} e_{n+1}$. To show that the map is an isomorphism, suppose $x e_{n+1} = 0$, $x \in A_{n-1}$. Then $x x^* e_{n+1} = 0$ so $\mathrm{tr}(x x^*) = 0$, i.e. $x = 0$. Q.E.D.

Aside 4.1.4. In fact it is possible to uniquely order reduced words by pushing $e_{\max}$ to the right as far as possible. It is easy to show that such an ordered reduced word is of the form

$$(e_{j_1}e_{j_1-1}\dots e_{k_1})(e_{j_2}e_{j_2-1}\dots e_{k_2})\dots(e_{j_p}e_{j_p-1}\dots e_{k_p})$$

where j_p is the maximum index, $j_i \geqq k_i$ and $j_{i+1} > j_i$, $k_{i+1} > k_i$. To each such word we may associate an increasing path on the integer lattice between $(0,0)$ and $(n+1, n+1)$, which does not cross the diagonal. For instance $(e_3e_2e_1)(e_4e_3)(e_5e_4)$ in A_5 would correspond to the path (I owe this observation to H. Wilf):

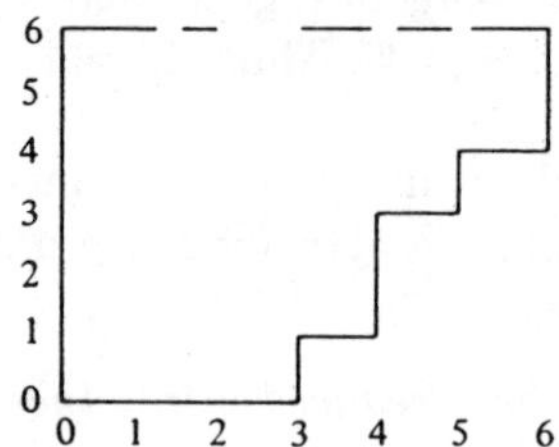

It is well known that such paths are counted by the Catalan numbers $1/(n+2)\binom{2(n+1)}{n+1}$ so we obtain $\dim A_n \leqq 1/(n+2)\binom{2(n+1)}{n+1}$. Uniqueness and linear independence of ordered reduced words would follow from $\dim A_n = 1/(n+2)\binom{2(n+1)}{n+1}$ which we will prove in §5.1 for $\tau \leqq 1/4$. See Aside 5.1.1.

We now want to study traces on A_∞. For this define a *totally reduced word* to be a reduced word on the e_i's where we also allow cyclic permutations. The following result is obvious.

Remark 4.1.5. Any trace on A_n is determined by its effect on totally reduced words.

Lemma 4.1.6. Any totally reduced word is of the form $w = e_{i_1}e_{i_2}\dots e_{i_n}$ with $|i_j - i_k| \geqq 2$, $j \neq k$, and $\mathrm{tr}(w) = \tau^n$.

Proof. The last assertion is immediate from 4.1.1(c). We prove the first assertion by induction on the length of a totally reduced word. It suffices to prove that if $m = \max\{i_1, i_2, \dots, i_n\}$ in a totally reduced word $e_{i_1}e_{i_2}\dots e_{i_n}$, then e_{m-1} does not occur. For this note that by a cyclic permutation we may suppose the word is of the form $e_m w$ and that e_{m-1} occurs at most once in w (since a totally reduced word is reduced). But then we may proceed to $e_m w e_m$ using $e_m^2 = e_m$ and then eliminate e_{m-1} from w using (a) and (b) of 4.1.1. Q.E.D.

We shall now show that any normal normalized trace on P is equal to tr. For this it suffices to show that it equals tr on completely reduced words. For each subset $I \subseteq N$ with $|i-j| \geqq 2$ whenever $i, j \in I$, $i \neq j$, define A_I to be the algebra generated by $\{e_i | i \in I\}$.

Lemma 4.1.7. *For any finite permutation σ of I, there is a unitary $u \in A_\infty$ such that $u e_k u^* \doteq e_{\sigma(k)}$ for all $k \in I$.*

Proof. It suffices to show that any transposition $e_i \leftrightarrow e_j$ $(j \geqq i)$ can be effected by a unitary. We may even suppose that no k strictly between i and j is in I. If we can find a unitary $u \in A_{i,j}$ with $u e_i u^* = e_j$ and $u e_j u^* = e_i$ then this u will do since it commutes with all the other e_k's, $k \in I$. But for this it suffices to show that e_i and e_j are equivalent in $A_{i,j}$. And $e_i e_{i+1} \dots e_j$ is a multiple of a partial isometry v with $v v^* = e_i$, $v^* v = e_j$. Q.E.D.

Corollary 4.1.8. *Any normal normalized trace on P is equal to* tr *on A_I''.*

Proof. Since all the e_i's are independent for tr, A_I'' may be identified with an infinite tensor product Bernoulli shift algebra. By 4.1.7 the normalizer induces the obvious action of S_∞ on A_I''. This action is well known to be ergodic. Hence any invariant measure which is absolutely continuous with respect to tr is proportional to tr. Q.E.D.

Corollary 4.1.9. *P is a II_1 factor isomorphic to R.*

Proof. By 4.1.4, 4.1.5 and 4.1.8 there is only one normal normalized trace on P. Thus P is a factor. By [21] and (a) of 4.1.3, $P \cong R$. Q.E.D.

Note that our proof that P is a factor follows similar lines to the scheme laid out in [27].

Corollary 4.1.10. *P_τ is a subfactor of P.*

Proof. Writing $f_i = e_{i+1}$, the f_i satisfy the same relations as the e_i. Q.E.D.

Lemma 4.1.11. *For each n, the map $e_i \to e_{n-i}$ extends to a* tr*-preserving $*$-automorphism σ_n of A_n.*

Proof. The map obviously extends to a $*$-automorphism of the free involutive monoid on the self-adjoint e_i. It thus suffices to know that if w is a word on e_1, $e_2, \dots, e_n$ then $\mathrm{tr}(\sigma_n(w)) = \mathrm{tr}(w)$. For then if $x = \sum_i c_i w_i$ for $c_i \in \mathbb{C}$,

$$\mathrm{tr}(x x^*) = \mathrm{tr}\Big(\sum_{i,j} c_i \bar{c}_j w_i w_j^*\Big) = \mathrm{tr}\Big(\sum_{i,j} c_i \bar{c}_j \sigma(w_i) \sigma(w_j^*)\Big)$$

so that $\sum_i c_i w_i \to \sum_i c_i \sigma(w_i)$ is a well defined isometry for the definite hermitian scalar product defined by tr. But the trace of w is determined by 4.1.1(a) and (b) and the formula of 4.1.6, all of which are invariant under the interchange $e_i \leftrightarrow e_{n-i}$. Q.E.D.

Corollary 4.1.12.

(i) $E_{A_{2,n}}(e_1) = \tau$

(ii) *for* $x \in A_{2,n}$, $e_1 x e_1 = E_{A_{3,n}}(x) e_1$

(iii) $E_{P_\tau}(e_1) = \tau$, $e_1 x e_1 = E_{A_{3,\infty}''}(x) e_1$ *for* $x \in P_\tau$.

Proof. (i) and (ii) follow from σ_n applied to (ii) of 4.1.3, and (iii) is just the limit as $n \to \infty$ of (i) and (ii). Q.E.D.

Proof of (ii) of 4.1.1 (calculation of $[P:P_\tau]$). Do the basic construction to obtain $\langle P, e_{P_\tau}\rangle$. Then $e_{P_\tau} e_1 e_{P_\tau} = \tau e_{P_\tau}$ follows from (i) of 4.1.12. We further claim that $e_1 e_{P_\tau} e_1 = \tau e_1$. By (iii) of 4.1.12, elements of the form $a_0 + \sum_{i=1}^{n} a_i e_1 b_i$ with a_i, $b_i \in P_\tau$ are dense in P so it suffices to verify $e_1 e_{P_\tau} e_1 = \tau e_1$ on $x\xi$ and $x e_1 y \xi$ with $x, y \in P_\tau$. But $e_1 e_{P_\tau} e_1 (x\xi) = \tau e_1 x \xi = (\tau e_1)(x\xi)$, and

$$e_1 e_{P_\tau} e_1 (x e_1 y \xi) = e_1 (E_{P_\tau}(e_1 E_{A''_{3,\infty}}(x) y)\xi) = \tau e_1 E_{A'_{3,\infty}}(x) y \xi = \tau e_1 (x e_1 y \xi).$$

These two relations imply that e_1 and e_{P_τ} are equivalent in $\langle P, e_{P_\tau}\rangle$. Now by (iii) of 3.1.5, e_{P_τ} is a finite projection and e_1 is in a II_1 factor so that $\langle P, e_{P_\tau}\rangle$ is necessarily a finite factor, i.e. $[P:P_\tau] < \infty$. But we know that $\mathrm{tr}(e_1) = \tau$ so $[P:P_\tau] = \mathrm{tr}(e_{P_\tau})^{-1} = \tau^{-1}$. Q.E.D.

§4.2. Restrictions on τ

We shall now prove (iii) of 4.1.1. We keep the notation of §4.1. Also define $s_n = e_1 \vee e_2 \vee \ldots \vee e_n$.

Lemma 4.2.1. *If $1 - s_n \neq 0$ then it is a minimal projection in A_n which belongs to $Z(A_n)$.*

Proof. If w is a word on $e_1, e_2, \ldots, e_n$ then

$$(1 - e_1 \vee e_2 \vee \ldots \vee e_n) w = 0. \quad \text{Q.E.D.}$$

Lemma 4.2.2. *If $s_n \neq 1$ then $e_{n+1} \wedge s_n = e_{n+1} s_{n-1}$.*

Proof. By 4.1.3(ii), $e_{n+1} s_n e_{n+1} = E_{A_{n-1}}(s_n) e_{n+1}$. But by 4.2.1 and the bimodule property of $E_{A_{n-1}}$, $E_{A_{n-1}}(s_n) \in Z(A_{n-1})$. Let $p_0, p_1, \ldots, p_k$ be the minimal projections in $Z(A_{n-1})$ with $p_0 = 1 - s_{n-1}$, and write $E_{A_{n-1}}(s_n) = \sum_{i=0}^{k} \lambda_i p_i$. Then since $e_{n+1} \wedge s_n \geqq e_{n+1} s_{n-1}$ and $\lim_{m\to\infty} (e_{n+1} s_n e_{n+1})^m = e_{n+1} \wedge s_n$, and by (iii) of 4.1.3, $\lambda_1 = \lambda_2 = \ldots = \lambda_k = 1$. Since $s_n \neq 1$, $E_{A_n}(s_n) \neq 1$ so by (iii) of 4.1.3, $\lambda_k < 1$. Thus $e_{n+1} \wedge s_n = e_{n+1} s_{n-1}$. Q.E.D.

Corollary 4.2.3. *If* $\mathrm{tr}(s_k) > 0$, $\mathrm{tr}(1 - s_{k+1}) = \mathrm{tr}(1 - s_k) - \tau\,\mathrm{tr}(1 - s_{k-1})$. *And* $\mathrm{tr}(1 - s_1) = 1 - \tau$, $\mathrm{tr}(1 - s_2) = 1 - 2\tau$.

Proof. The trace tr satisfies $\mathrm{tr}(p \vee q) = \mathrm{tr}(p) + \mathrm{tr}(q) - \mathrm{tr}(p \wedge q)$. So by 4.2.2 and (c) of 4.1.1, $\mathrm{tr}(1 - s_{k+1}) = \mathrm{tr}(1 - s_k) - \tau\,\mathrm{tr}(1 - s_{k-1})$. Q.E.D.

For this reason we define the polynomials $P_n(x)$ by $P_0(x) = 0$, $P_1(x) = 1$ and $P_{n+1} = P_n - x P_{n-1}$, so that if $P_{k+2}(\tau) > 0\ \forall k \leq n$ then $P_{k+2}(\tau) = \mathrm{tr}(1 - s_k)$.

Lemma 4.2.4. *Let* $\sigma = \dfrac{1 + \sqrt{1 - 4x}}{2}$, $\tilde{\sigma} = \dfrac{1 - \sqrt{1 - 4x}}{2}$. *Then*

(i) $P_n(x) = \dfrac{\sigma^n - \tilde{\sigma}^n}{\sigma - \tilde{\sigma}}$

(ii) $P_n(\frac{1}{4}\sec^2\theta)=\sin n\theta/(2^{n-1}\cos^{n-1}\theta\sin\theta)$

(iii) $\deg P_n=\left[\frac{n-1}{2}\right]$.

Proof. (i) The general solution to the difference equation is $P_n=A\sigma^n+B\tilde{\sigma}^n$. The initial conditions give $A+B=0$, $A(\sigma-\tilde{\sigma})=1$.

(ii) Putting $\sigma=re^{i\theta}$, $\tilde{\sigma}=re^{-i\theta}$, $x=\frac{1}{4}\sec^2\theta$, $r=\frac{1}{2}\sec\theta$, $\sigma^n-\tilde{\sigma}^n=2ir^n\sin n\theta$, $\sigma-\tilde{\sigma}=2ir\sin\theta$.

(iii) Follows easily by induction from the difference equation. Q.E.D.

Corollary 4.2.5. (i) *The smallest root of* P_n *is* $\frac{1}{4}\sec^2\frac{\pi}{n}$.

(ii) $P_n(\tau)>0$ *for* $\tau<\frac{1}{4}\sec^2\frac{\pi}{n}$.

(iii) $P_{n+1}(\tau)<0$ *for* τ *between* $\frac{1}{4}\sec^2\frac{\pi}{n+1}$ *and* $\frac{1}{4}\sec^2\frac{\pi}{n}$.

Proof. (i) By counting the number of distinct values of $\frac{1}{4}\sec^2\frac{m\pi}{n}$ (which are roots of P_n by 4.2.4) we find that all roots of P_n are real and they are the numbers $\frac{1}{4}\sec^2\frac{m\pi}{n}$ with $\frac{m\pi}{n}<\frac{\pi}{2}$. The smallest is $\frac{1}{4}\sec^2\frac{\pi}{n}$.

(ii) By induction the coefficient of $x^{[(n-1)/2]}$ in $P_n(x)$ is positive when $\left[\frac{n-1}{2}\right]$ is even and negative when $\left[\frac{n-1}{2}\right]$ is odd.

(iii) $P_{n+1}(\tau)$ must be negative between its first and second real roots, and $\sec^2\pi/(n+1)<\sec^2\pi/n<\sec^2 2\pi/(n+1)$. Q.E.D.

Proof of (iii) of 4.1.1. Suppose $\tau>1/4$ and $\tau\neq\frac{1}{4}\sec^2\frac{\pi}{n}$, $n=3,4,5,\ldots$. Then there is a $k\geqq 3$ with $\frac{1}{4}\sec^2\pi/(k+1)<\tau<\frac{1}{4}\sec^2\pi/k$. But then $P_n(\tau)>0$ for all $n\leqq k$ so $P_{n+1}(\tau)=\mathrm{tr}(1-s_{n-1})$ by 4.2.3. But by (iii) of 4.2.5, $P_{n+1}(\tau)<0$ which is impossible since $1-s_{n-1}$ is a projection. Q.E.D.

§4.3. Values of the Index

Theorem 4.3.1. *If N is a subfactor of the* II_1 *factor M then either* $[M:N]\geqq 4$ *or* $[M:N]=4\cos^2\pi/n$ *for some* $n\geqq 3$.

Proof. If $[M:N]<\infty$, define the increasing sequence M_i, $i=0,1,2,\ldots$ of II_1 factors by the relations $M_0=M$, $M_1=\langle M,e_N\rangle$, $M_{i+1}=\langle M_i,e_{M_{i-1}}\rangle$ for $i\geqq 1$. The inductive limit becomes a II_1 factor with faithful normal normalized trace tr (by uniqueness of the trace - see also [19]). Moreover if $\tau=[M:N]^{-1}$ and $e_i=e_{M_i}$ then the e_i satisfy the conditions of 4.1.1 by 3.4.2 and 3.1.7. By Theorem 4.1.1 either $[M:N]\geqq 4$ or $[M:N]=4\cos^2\pi/n$ for some $n\in\mathbb{Z}$, $n\geqq 3$. Q.E.D.

Theorem 4.3.2. *For each* $n=3,4,\ldots$ *there is a subfactor* P_τ *of* R *with* $[R:P_\tau]=4\cos^2\pi/n$ *and* $P'\cap R=\mathbb{C}$. *For each* $r\geqq 4$, $r\in\mathbb{R}$, *there is a subfactor* P *of* R *with* $[R:P]=r$.

Proof. The existence of subfactors with index $r\geqq 4$ was shown in 2.2.5 and the assertion about the relative commutant when $[R:P]=4\cos^2\pi/n$ follows from 2.2.4. Thus we only need to construct subfactors with index $4\cos^2\pi/n$.

To do this note that the conditions of 3.3.2 for finite dimensional C^*-algebras $N\subseteq M$ together with (ii) of 3.3.1 show, by interchanging A and A^T, $\vec{s}$ and $\vec{t}$, that if there is a (τ, M) trace on $\langle M, e_N\rangle$ then there is a $(\tau,\langle M, e_N\rangle)$ trace on $\langle\langle M, e_N\rangle, e_M\rangle$ and so on. Thus we may iterate the basic construction once we have started it. Once the construction has been iterated, the inductive limit has a faithful normal normalized trace on it and the e_i's resulting from the iteration satisfy the conditions of 4.1.1 so by the result of 4.1.1 we may choose P_τ as the subfactor of R.

Thus it suffices to find $N\subseteq M$, finite dimensional C^*-algebras with inclusion matrix A and positive vectors $\vec{s}$ and $\vec{t}$ with $A^TA\vec{t}=(4\cos^2\pi/n)\vec{t}$ and $AA^T\vec{s}=(4\cos^2\pi/n)\vec{s}$. In fact the matrix A is enough since the subalgebra N can then be taken as the direct sum of as many copies of $\mathbb{C}$ as there are rows in the matrix.

Let A be the square $n\times n$ matrix (a_{ij}) with $a_{ij}=1$ if $|i-j|=1$ and 0 otherwise, e.g. $\begin{pmatrix}0&1&0\\1&0&1\\0&1&0\end{pmatrix}$. We leave it to the reader to check the linear algebra. This choice of A was suggested by F. Goodman. Q.E.D.

Remark 4.3.3. If we had started with any $m\times n$ non-negative integer valued matrix we could have made the construction of 4.3.2. Thus 4.1.1 implies that for such a matrix, if $\|A\|\leqq 2$ then $\|A\|=2\cos\pi/n$, $n=3,4,\ldots$. This can also be proved using the well-known result of Kronecker which asserts that if z is an algebraic integer all of whose conjugates have absolute value equal to 1, then z is a root of unity.

§5. The Bratteli Diagram; the Relative Commutant

§5.1. The Bratteli Diagram when $\tau\leqq 1/4$

Let $\{e_1,e_2,\ldots\}$, M and tr be as in §4.1. Let $A_n=\{e_1,e_2,\ldots,e_n\}''$ and B_n be the algebra generated by $\{e_1,e_2,\ldots,e_n\}$ (without 1). We know that A_n and B_n are finite dimensional C^*-algebras. We want to determine the Bratteli diagram for them and the value of tr on minimal projections. In this section we suppose $\tau\leqq 1/4$. This means that

$$P_{n+2}(\tau)=\mathrm{tr}(1-e_1\vee e_2\vee\ldots\vee e_n)>0 \quad \text{for all } n\geqq 1$$

(where P_n is as in §4.2).

Let $\begin{Bmatrix} n \\ b \end{Bmatrix} = \binom{n}{b} - \binom{n}{b-1}$ (ordinary binomial symbols with the convention $\binom{n}{-1} = 0$). We shall show by induction that

(a) $A_n = \bigoplus_{k=0}^{[\frac{n+1}{2}]} Q_k^n$ where $Q_k^n \cong M_{\{ {n+1 \atop k} \}}(\mathbb{C})$.

(b) $Q_0^n = (1 - e_1 \vee e_2 \vee \ldots \vee e_n)\mathbb{C}$ so that

(c) $B_n = \bigoplus_{k=1}^{[\frac{n+1}{2}]} Q_k^n$

(d) The trace of a minimal projection in Q_k^n is $\tau^k P_{n+2-2k}(\tau)$ for $k = 0, 1, \ldots, \left[\frac{n+1}{2}\right]$.

(e) The inclusion matrix of A_{n-1} in A_n is

(i) When n is even

$A = (a_{ij})$ with $a_{ij} = \begin{cases} 1 & \text{if } j = i \text{ or } i+1 \\ 0 & \text{otherwise,} \end{cases}$

$i, j = 0, 1, \ldots, \left[\frac{n+1}{2}\right]$

(here the indices i and j refer to the subscript of Q).

(ii) when n is odd

$A = (a_{ij})$ $a_{ij} = \begin{cases} 1 & \text{if } j = i \text{ or } i+1 \\ 0 & \text{otherwise} \end{cases}$

$i = 0, ., \ldots, (n+1)/2; j = 0, 1, \ldots, (n+3)/2.$

All this information is summed up by the following diagram (which also appears on p. 118 of [31]).

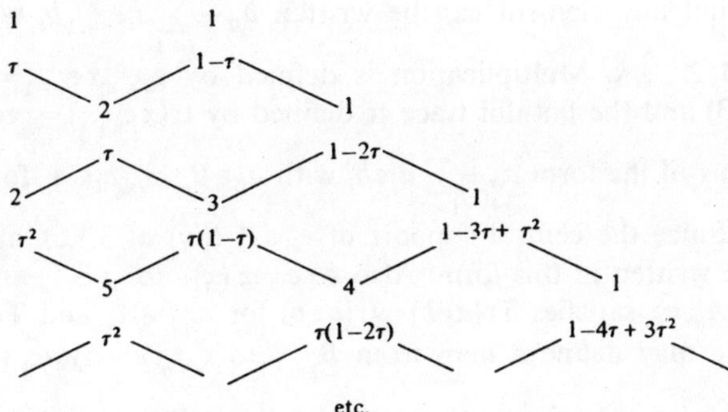

The numbers are the $\begin{Bmatrix} n+1 \\ k \end{Bmatrix}$ and the polynomials, which give the trace on minimal projections in the corresponding matrix algebra, change by multiplication by τ each step down a vertical column.

Proof. The proof will use the basic construction for the inclusion $A_{n-1} \subseteq A_n$ to obtain an almost faithful representation of A_{n+1}.

The truth of assertions (a)→(e) for $n=2$ follows immediately from §3.4. For the inductive step we shall treat only the case $n=2m$, the odd case being essentially identical. Suppose (a)→(e) are true for all $k \leq n$ and apply the basic construction of §3.1 with $M=A_n$, $N=A_{n-1}$ with respect to tr. Let $E=E_{A_{n-1}}$, $e_{A_{n-1}}=e$. By induction and (ii) of 3.3.1 we know that we have the following Bratteli diagram:

$$\begin{array}{lccccccc}
A_{n-1} & \cdots & \left\{\begin{matrix}n\\m\end{matrix}\right\} & & \left\{\begin{matrix}n\\m-1\end{matrix}\right\} & & \cdots & \left\{\begin{matrix}n\\0\end{matrix}\right\} \\
\cap| & & & \searrow & & \searrow & \swarrow & \searrow \\
A_n & \cdots & & \left\{\begin{matrix}n+1\\m\end{matrix}\right\} & & \left\{\begin{matrix}n+1\\m-1\end{matrix}\right\} \cdots \left\{\begin{matrix}n+1\\1\end{matrix}\right\} & & \left\{\begin{matrix}n+1\\0\end{matrix}\right\} \\
\cap| & & \swarrow & \searrow & \swarrow & \searrow \quad \swarrow & \searrow & \swarrow \\
\langle A_n, e\rangle & \cdots & \left\{\begin{matrix}n+2\\m+1\end{matrix}\right\} & & \left\{\begin{matrix}n+2\\m\end{matrix}\right\} & \cdots & \left\{\begin{matrix}n+2\\1\end{matrix}\right\} &
\end{array}$$

The numbers on the bottom line follow from 3.7.1, (ii) of 3.3.1 and the identities

$$\left\{\begin{matrix}a\\b\end{matrix}\right\}+\left\{\begin{matrix}a\\b-1\end{matrix}\right\}=\left\{\begin{matrix}a+1\\b\end{matrix}\right\}, \quad \left\{\begin{matrix}2m+2\\m+1\end{matrix}\right\}=\left\{\begin{matrix}2m+1\\m\end{matrix}\right\}.$$

By assertion (c), the algebra generated by $\{e_1, e_2, \ldots, e_n\}$ is the direct sum of the first m terms on the A_n line so if we define the faithful (non-normalized) trace Tr on $\langle A_n, e\rangle$ by the rule $\mathrm{Tr}(ep_k)=\tau\,\mathrm{tr}(p_k)$ where p_k is a minimal projection in Q^{n-1}_{k-1} $(k=1,2,\ldots,m+1)$, the identity $P_j=P_{j+1}+\tau P_{j-1}$ ensures that Tr agrees with tr on B_n. Moreover as in 3.3.2, uniqueness of the trace and linearity show that $\mathrm{Tr}(ex)=\tau\,\mathrm{tr}(x)$ for all x in A_{n-1} and hence all x in A_n. Also Tr is a faithful positive trace since all the polynomials $P_n(\tau)$ are positive.

But now consider the algebras B_{n+1} and $\langle A_n, e\rangle$. B_{n+1} is generated by B_n and e_{n+1} so that any element can be written $a_0+\sum_{i=1}^{i} a_i e_{n+1} b_i$ with $a_0 \in B_n$, a_i, $b_i \in A_n$ for $i=1,2,\ldots,k$. Multiplication is defined by $e_{n+1} x e_{n+1}=E(x)e_{n+1}$ for $x \in A_n$ (see 4.1.3) and the faithful trace tr defined by $\mathrm{tr}(xe_{n+1})=\tau\,\mathrm{tr}(x)$ for $x \in A_n$. In $\langle A_n, e\rangle$, sums of the form $a_0+\sum_{i=1}^{j} a_i e b_i$ with $a_0 \in B_n$, a_i, $b_i \in A_n$ for $i \geq 1$, form a 2-sided ideal. Since the central support of e is 1 ((iv) of 3.1.5), any element of $\langle A_n, e\rangle$ can be written in this form. Also $exe=E(x)e$ for $x \in A_n$ and the faithful trace Tr on $\langle A_n, e\rangle$ satisfies $\mathrm{Tr}(aeb)=\mathrm{tr}(aeb)$ for a, $b \in A_n$ and $\mathrm{Tr}(b)=\mathrm{tr}(b)$ for $b \in B_n$. Thus we may define a map from B_{n+1} to $\langle A_n, e\rangle$ by $a_0+\sum_i a_i e_{n+1} b_i \mapsto a_0+\sum_i a_i e b_i$ which is a surjective isometry for the definite hermitian scalar products defined by tr and Tr (and hence is well defined).

At this stage (†) we have obtained assertion (c) for $n+1$ and the values of tr on the minimal projections in Q^{n+1}_1, Q^{n+1}_2, ..., Q^{n+1}_{m+1}. But A_{n+1} is just $\{B_{n+1} \cup \{1\}\}''$ and since $\mathrm{tr}(1-e_1 \vee \ldots \vee e_{n+1})=P_{n+3}(\tau)>0$, A_{n+1} is $B_{n+1} \oplus (1-e_1 \vee e_2 \vee \ldots \vee e_{n+1})\mathbb{C}$. Moreover $x(1-e_1 \vee e_2 \vee \ldots \vee e_{n+1})=0$ for any $x \in B_n$

and $(1-e_1 \vee \ldots \vee e_{n+1})(1-e_1 \vee \ldots \vee e_n) \neq 0$ so the Bratteli diagram for $A_n \subseteq A_{n+1}$ is forced to be

$$
\begin{array}{lcccccccc}
A_n \cdots & & \left\{\begin{matrix} n+1 \\ m \end{matrix}\right\} & & \left\{\begin{matrix} n+1 \\ m+1 \end{matrix}\right\} & \cdots & \left\{\begin{matrix} n+1 \\ 1 \end{matrix}\right\} & & \left\{\begin{matrix} n+1 \\ 0 \end{matrix}\right\} \\
A_{n+1} \cdots & \left\{\begin{matrix} n+2 \\ m+1 \end{matrix}\right\} & & \left\{\begin{matrix} n+2 \\ m \end{matrix}\right\} & & \cdots & \left\{\begin{matrix} n+2 \\ 1 \end{matrix}\right\} & & \left\{\begin{matrix} n+2 \\ 0 \end{matrix}\right\}
\end{array}
$$

This proves assertions (a) and (e) for $n+1$, and assertion (d) follows from $\operatorname{tr}(1-e_1 \vee \ldots \vee e_{n+1}) = P_{n+3}(\tau)$. This ends the proof. Q.E.D.

Aside 5.1.1. The binomial identity

$$\sum_{i=0}^{[\frac{n+1}{2}]} \binom{n+1}{i}^2 = \frac{1}{n+2}\binom{2(n+1)}{n+1}$$

follows from [13, p. 63]. This shows that $\dim A_n$ is the same as the number of ordered reduced words which are thus linearly independent. See 4.1.4.

§5.2. *The Bratteli Diagrams* $\tau = \frac{1}{4}\sec^2\frac{\pi}{n}$

If $\tau = \frac{1}{4}\sec^2 \pi/(n+2)$ then $P_k(\tau) > 0$ for all $k \leqq n+1$ so the inductive argument of §5.1 goes through until the point marked (†) for assertions (a)→(e) up to step n. At this stage we find that $e_1 \vee e_2 \vee \ldots \vee e_n = 1$ so $B_n = A_n$. From this point on the basic construction for the pair $A_k \subseteq A_{k+1}$ will give an isometric (so faithful) surjective representation of A_{k+2} and the Bratteli diagram will not grow any wider.

To convince the reader of these assertions without boring him with the details, we treat the case $n=4$. The argument of §5.1 shows that we have the following diagram

$$
\begin{array}{lcccccc}
A_1 \cdots & 1 & & & 1 & & \\
 & \tau & & & 1-\tau & & \\
A_2 \cdots & & & 2 & & 1 & \\
 & & & \tau & & 1-2\tau & \\
A_3 \cdots & 2 & & & 3 & & 1 \\
 & \tau^2 & & & \tau(1-\tau) & & 1-3\tau+\tau^2 \\
A_4 \cdots & & & 5 & & 4 & \\
 & & & \tau^2 & & \tau(1-2\tau) &
\end{array}
$$

Now if $\vec{s} = \begin{pmatrix} \tau \\ 1-2\tau \end{pmatrix}$, $\vec{t} = \begin{pmatrix} \tau^2 \\ \tau(1-\tau) \\ 1-3\tau+\tau^2 \end{pmatrix}$, and $A = \begin{pmatrix} 1 & 0 \\ 1 & 1 \\ 0 & 1 \end{pmatrix}$, $\vec{s}$ is an eigenvector for $A^T A$ with eigenvalue $3 = \tau^{-1} = 4\cos^2\frac{\pi}{6}$ and $\vec{t}$ is an eigenvector for AA^T with the same eigenvalue. They give normalized traces on A_2 and A_3 so by §3.3 the basic construction will continue to give faithful representations of the A_i by the same argument as §5.1, since $A_n = B_n$ for $n \geqq 4$.

The Bratteli diagrams will be:

For $n=2$ ($\tau = 1/2$)

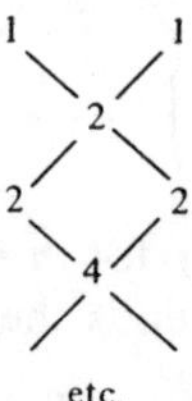

1 1
2
2 2
4
etc.

Note that this is the same as the complex Clifford algebras (see [16, p. 148]).

For $n=3$ ($\tau = 1/\varphi^2$, φ = golden ratio)

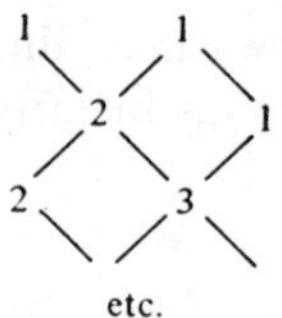

This diagram already appears in [4], [11].

For $n=4$

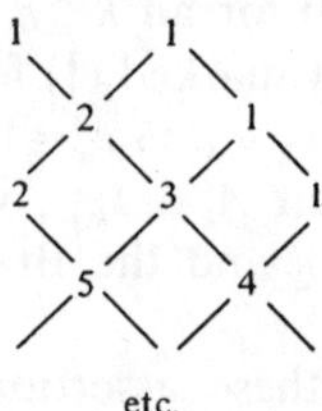

The pattern is now clear. The traces on minimal projections are the same as for the case $\tau < 1/4$ where this makes sense.

§5.3. The Relative Commutant when $\tau \leqq 1/4$

One of the main motivations of this paper was to decide whether there is a continuum of values of the index realized by subfactors with trivial relative commutant (see problem 1). It was originally thought that the subfactors P_τ, $\tau \leqq 1/4$, had trivial relative commutant. In this section we shall show that this is only true when $\tau = 1/4$. The calculations were first finished by A. Wassermann to whom the author is grateful. The difficulty is the absence of an orthonormal basis which makes it impossible to exhibit an element of the relative commutant.

The case $\tau < 1/4$. Let us adopt the notation of §4.1 so R is presented as the algebra generated by the e_i's. We intend to show that $E_{R \cap P'_\tau}(e_1)$ is not a scalar, i.e. $E_{R \cap P'_\tau}(e_1) \neq \tau$. For this we will show that $\|E_{R \cap P'_\tau}(e_1) - \tau\|_1 \neq 0$ (remember $\|x\|_1 = \mathrm{tr}(|x|)$). We shall need some lemmas.

Lemma 5.3.1. $E_{R\cap P'_\tau}(x)=s-\lim_{n\to\infty} E_{R\cap A'_{2,n}}(x)$ *for* $x\in R$.

Proof. Since $P_\tau=\left(\bigcup_{n=1}^{\infty} A_{2,n}\right)''$, the algebras $R\cap A'_{2,n}$ are a decreasing sequence of von Neumann algebras with intersection $P'_\tau\cap R$. The rest is well known (e.g. it is true in $L^2(R,\mathrm{tr})$). Q.E.D.

Lemma 5.3.2. $\|E_{A'_{2,n+1}\cap R}(e_1)-\tau\|_1=\|E_{A'_n\cap A_{n+1}}(e_{n+1})-\tau\|_1$.

Proof. Let du denote Haar measure on the unitary group of a finite dimensional C^*-algebra. Then

$$\|E_{A'_{2,n+1}\cap R}(e_1-\tau)\|_1=\|\int_{U(A_{2,n+1})}(ue_1u^*-\tau)\,du\|_1.$$

Applying the isomorphism σ_{n+1} of 4.1.11 we find that this expression equals

$$\|\int_{U(A_n)}(ue_{n+1}u^*-\tau)\,du\|_1$$

which equals $\|E_{A'_n\cap A_{n+1}}(e_{n+1})-\tau\|_1$. Q.E.D.

Thus to show that $P'_\tau\cap R\neq C$, it suffices to show that

$$\lim_{n\to\infty}\|E_{A'_n\cap A_{n+1}}(e_{n+1})-\tau\|_1\neq 0.$$

Since we know that the limit exists, it suffices to consider n odd; say $n=2m-1$. From §5.1 let $p_0, p_1, \dots, p_{m-1}$ be the minimal central projections in A_{n-1} corresponding to $Q_0^{n-1}, Q_1^{n-1}, \dots, Q_{m-1}^{n-1}$, similarly $q_0, q_1, \dots, q_m$ for A_n and $r_0, r_1, \dots, r_m$ for A_{n+1}. So $p_0=1-e_1\vee e_2\vee\dots\vee e_{n-1}$, $q_0=1-e_1\vee e_2\vee\dots\vee e_n$ and $r_0=1-e_1\vee e_2\vee\dots\vee e_{n+1}$. We also know from §5.1 and (iii) of 3.3.1 that

$$e_{n+1}r_i=e_{n+1}p_{i-1}\quad \text{for } i\geqq 1. \tag{5.3.3}$$

Since all the embeddings on the Bratteli diagram are of multiplicity at most one, the relative commutant of A_n in A_{n+1} is the abelian algebra generated by the mutually orthogonal projections $q_0r_0, q_0r_1, q_1r_0, \dots, q_mr_m$. Thus

$$E_{A'_n\cap A_{n+1}}(e_1)=\frac{\mathrm{tr}(q_0r_0e_{n+1})}{\mathrm{tr}(q_0r_0)}q_0r_0+\frac{\mathrm{tr}(q_0r_1e_{n+1})}{\mathrm{tr}(q_0r_1)}q_0r_0+\dots$$
$$+\frac{\mathrm{tr}(q_mr_me_{n+1})}{\mathrm{tr}(q_mr_m)}q_mr_m$$

so that

$$\|E_{A'_n\cap A_{n+1}}(e_{n+1})-\tau\|_1=\sum_{i=0}^{m}|\mathrm{tr}(q_ir_ie_{n+1})-\tau\,\mathrm{tr}(q_ir_i)|$$
$$+\sum_{i=0}^{m-1}|\mathrm{tr}(q_ir_{i+1}e_{n+1}-\tau\,\mathrm{tr}(q_ir_i)|$$
$$\geqq\sum_{i=1}^{m}\tau|\mathrm{tr}(q_ip_{i-1})-\mathrm{tr}(q_ir_i)|\quad\text{by } 5.3.3$$
$$\geqq\tau\left|\sum_{i=1}^{m}\mathrm{tr}(q_ip_{i-1})-\mathrm{tr}(q_ir_i)\right|.$$

But from the Bratteli diagram, $q_i p_{i-1}$ is a projection of rank $\begin{Bmatrix} 2m-1 \\ i-1 \end{Bmatrix}$ in the matrix algebra Q_i^{2m-1} $(=q_i A_n)$ and $q_i r_i$ is of rank $\begin{Bmatrix} 2m \\ i \end{Bmatrix}$ in $Q_i^{2m}(=r_i A_{n+1})$. So the last sum may be written

$$L=\tau\left|\sum_{i=1}^{m}\left(\begin{Bmatrix} 2m-1 \\ i-1 \end{Bmatrix}\tau^i P_{2(m-i)+1}(\tau)-\begin{Bmatrix} 2m \\ i \end{Bmatrix}\tau^i P_{2(m-i)+2}(\tau)\right)\right|.$$

We saw in 4.2.4 that $P_n(\tau)=(\sigma^n-\tilde{\sigma}^n)/(\sigma-\tilde{\sigma})$ where $\sigma=(1+\sqrt{1-4\tau})/2$ and $\tilde{\sigma}=(1-\sqrt{1-4\tau})/2$. Note that for $\tau<1/4$, $\sigma>1/2$, $\tilde{\sigma}<1/2$ and $\tilde{\sigma}+(\tau/\tilde{\sigma})=1$, $\sigma+(\tau/\sigma)=1$. Let us now calculate the relevant limits.

Lemma 5.3.6. (a) $\lim\limits_{m\to\infty}\left(\sum\limits_{i=1}^{m}\begin{Bmatrix} 2m-1 \\ i-1 \end{Bmatrix}\tilde{\sigma}^{2(m-i)+1}\tau^i\right)=0$

(b) $\lim\limits_{m\to\infty}\left(\sum\limits_{i=1}^{m}\begin{Bmatrix} 2m \\ i \end{Bmatrix}\tilde{\sigma}^{2(m-i)+2}\tau^i\right)=0$

(c) $\lim\limits_{m\to\infty}\left(\sum\limits_{i=1}^{m}\begin{Bmatrix} 2m-1 \\ i-1 \end{Bmatrix}\sigma^{2(m-i)+1}\tau^i\right)=\tau(1-\tau/\sigma^2)$

(d) $\lim\limits_{m\to\infty}\left(\sum\limits_{i=1}^{m}\begin{Bmatrix} 2m \\ i \end{Bmatrix}\sigma^{2(m-i)+2}\tau^i\right)=\sigma^2(1-\tau/\sigma^2)$.

Proof. (a) and (c). Note that $\sigma^{2(m-i)+1}\tau^i=\tau(\tau/\sigma)^{i-1}\sigma^{2m-i}$ so if $x=\sigma$ or $\tilde{\sigma}$, in both cases we have to evaluate

$$\lim_{m\to\infty}\tau\left(\sum_{i=1}^{m}\binom{2m-1}{i-1}\left(\frac{\tau}{x}\right)^{i-1}x^{2m-i}-\frac{\tau}{x^2}\sum_{i=1}^{m}\binom{2m-1}{i-2}\left(\frac{\tau}{x}\right)^{i-2}x^{2m-i+1}\right).$$

Since $x+\tau/x=1$, we recognize the probability of $\geqq m$ successes in $2m-1$ Bernoulli trials. If $x=\sigma$, the probability of success is $>1/2$ so by the de Moivre-Laplace central limit theorem the limit is $\tau(1-\tau/\sigma^2)$ and if $x=\tilde{\sigma}$, the probability of success is $<1/2$ so the limit is 0.

(b) and (d) are proved in the same way with the substitution $x^{2(m-i)+2}\tau^i=x^2(\tau/x)^i x^{2m-i}$. Q.E.D.

Expanding L and using 4.3.5 we see that

$$(\sigma-\tilde{\sigma})/L=\tau|(1-\tau/\sigma^2)(\tau-\sigma^2)|\neq 0 \qquad \text{for } \tau\neq 1/4.$$

This shows that $P_\tau'\cap R\neq\mathbb{C}$ for $\tau<1/4$.

The case $\tau=1/4$. In this case we contend that $P_\tau'\cap R=\mathbb{C}$. Luckily we can use another model of $P_{1/4}$. Let R be realized as the closure of the Fermion algebra $\bigotimes_{i=1}^{\infty}(M_2(\mathbb{C}))_i$ with respect to the trace tr. The fixed point algebra of the obvious infinite product action of $U(2)$ is generated by the representation of S_∞ coming from interchanging the tensor product components. The transpositions between successive components may be written $2e_i-1$ with $\mathrm{tr}(e_i)=1/4$ and it is a matter of calculation to show that the e_i's satisfy $e_ie_{i\pm1}e_i=\frac{1}{4}e_i$, $e_ie_j=e_je_i$ for $|i-j|\geqq 2$. Thus the e_i algebra is $R^{U(2)}$ and the subfactor $P_{1/4}$ is $M_2(C)_1'\cap R^{U(2)}$.

But it is shown in [31] that $(R^{U(2)})' \cap R = \mathbb{C}$ so that $P'_{1/4} \cap R = M_2(\mathbb{C})_1$ and $M_2(\mathbb{C})_1 \cap R^{U(2)} = \mathbb{C}$. Thus $P'_{1/4} \cap R^{SU(2)} = \mathbb{C}$. For $R^{U(2)}$ see also [12], [15], [31], [2]. Q.E.D.

References

1. Aubert, P.L.: Théorie de Galois pour une W^*-algèbre. Comment. Math. Helv. **39**, 411–433 (1976)
2. Baker, B., Powers, R.: Product states and C^* dynamical systems of product type. Preprint, University of Pennsylvania
3. Bourbaki, N.: Algèbre, Ch. 8. Hermann, 1959
4. Bratteli, O.: Inductive limits of finite dimensional C^*-algebras. Trans. A.M.S. **171**, 195–234 (1972)
5. Christensen, E.: Subalgebras of a finite algebra. Math. Ann. **243**, 17–29 (1979)
6. Connes, A.: Periodic automorphisms of the hyperfinite factor of type II_1. Acta Sci. Math. **39**, 39–66 (1977)
7. Connes, A.: Classification of injective factors. Ann. Math. **104**, 73–115 (1976)
8. Connes, A.: Sur la théorie non-commutative de l'intégration. In: Algèbres d'opérateurs. Lecture Notes in Mathematics, Vol. 725. Berlin-Heidelberg-New York: Springer 1978
9. Connes, A.: A II_1 factor with countable fundamental group. J. Operator Theory **4**, 151–153 (1980)
10. Dixmier, J.: Les algèbres d'opérateurs dans l'espace Hilbertien. Deuxieme édition. Ganthier Villars, 1969
11. Effros, E., Shen, C.-L.: $A-F$ C^*-algebras and continued fractions. Indiana Math. J. **29**, 191–204 (1980)
12. Enomoto, M., Watatani, Y.: $C^*(\Pi)$ and Cuntz algebras from Young diagrams. Preprint
13. Feller, W.: Probability theory and its applications. I. Wiley and Sons, Inc. 1957
14. Goldman, M.: On subfactors of factors of type II_1. Mich. Math. J. **7**, 167–172 (1960)
15. Handelman, D.: Notes on ordered K-theory. Vancouver conference on operator algebras (1982)
16. Husemoller, D.: Fibre bundles. Grad. Texts in Math. Vol. 20. Berlin-Heidelberg-New York: Springer 1975
17. Jones, V.: Sur la conjugaison des sous-facteurs des facteur de type II_1. C.R. Ac. Sci. Paris t. **286**, 597–598 (1977)
18. Jones, V.: L'indice des sous-facteurs de facteurs de type II_1. C.R. Ac. Sci. Paris t. 286 (1977), 597–598
19. Kadison, R., Fugledge, B.: On a conjecture of Murray and von Neumann. Proc. Natl. Acad. Sci. U.S. **37**, 420–425 (1951)
20. Murray, F., von Neumann, J.: On rings of operators, II. Trans. A.M.S. **41**, 208–248 (1937)
21. Murray, F., von Neumann, J.: On rings of operators. IV. Ann. Math. **44**, 716–808 (1943)
22. Ocneanu, A.: Actions of discrete amenable groups on factors. To appear, Springer Lecture notes in Math. Berlin-Heidelberg-New York: Springer
23. Popa, S.: On a problem of R.V. Kadison on maximal abelian $*$-subalgebras in factors. Invent. Math. **65**, 269–281 (1981)
24. Skau, C.: Finite subalgebras of a von Neumann algebra. J. Funct. Anal. **25**, 211–235 (1977)
25. Suzuki, N.: Crossed products of rings of operators. Tohoku Math. J. 7, 167–172 (1960)
26. Suzuki, N.: Extensions of rings of operators in Hilbert spaces. Tohoku Math. J. **14**, 217–232 (1962)
27. Stratila, S., Voiculescu, D.: Representations of AF algebras and of the group $U(\infty)$. Springer Lecture notes in Math.
28. Takesaki, M.: Theory of operator algebras I. Berlin-Heidelberg-New York: Springer 1979
29. Topping, D.: Lectures on von Neumann algebras. Nostrand Reinhold Math. Studies, 1969
30. Umegaki, H.: Conditional expectation in an operator algebra I. Tohoku Math. J. **6**, 358–362 (1954)
31. Wasserman, A.: Thesis, University of Pennsylvania, 1981

Invent. math. 92, 349–383 (1988)

Inventiones mathematicae

Hecke algebras of type A_n and subfactors

Hans Wenzl*

Department of Mathematics, University of California, Berkeley, CA 94720, USA

In his paper [J-1] V. Jones introduced an index, which 'measures' the size of a subfactor in a II_1 factor. The main result of that paper is that the index of a subfactor has to be either greater or equal than 4 or it has to be equal to $4\cos^2(\pi/l)$ for some $l \in \mathbf{N}$, $l \geqq 3$ and that there exist subfactors for all these index values. Similarly as for subgroups, the index alone does not characterize the subfactor up to conjugacy by automorphisms. The fact that there are only countably many possible index values <4 seems to be related to another invariant. Subfactors with index less than 4 always have trivial centralizers, (or relative commutants), i.e. the only elements of the factor which commute with every element of the subfactor are multiples of the identity. On the other hand, the examples given in [J-1] for subfactors with index greater than 4 all have nontrivial centralizers. Furthermore, all known examples of subfactors with trivial relative commutants have as index an algebraic integer. At the current state of knowledge, it is still unknown whether there are only countably many values possible for the index of subfactors with trivial centralizers. Note however, that the set of all possible index values of a subfactor with trivial centralizer in an arbitrary II_1 factor has to be a closed subset of $\mathbf{R}$ (see [HW]).

Our original motivation for this paper was to study how subfactors of the hyperfinite II_1 factor can be constructed via *AF* algebras. We provide a method of computing the index and we give an upper bound for the size of the centralizer of the constructed subfactor. Our general results will then be applied to the series of complex Hecke algebras $H_n(q)$, $n \in \mathbf{N}$ of type A_{n-1}. Their standard generators $g_1, g_2, \ldots, g_{n-1}$ satisfy the same relations as a set of simple reflections of the symmetric group S_n except that the reflection property $g_i^2=1$ is replaced by $g_i^2=(q-1)g_i+q$. It is well-known that $H_n(q)$ is isomorphic to $\mathbf{C}S_n$ if q is not a root of unity. If the parameter is a root of unity, $H_n(q)$ may no longer be semisimple and its structure is not known in general. This is, however, the most interesting case as far as subfactors are concerned. We define representations ρ of $H_\infty(q)$ such that $\rho(H_n(q))$ is semisimple for all $n \in \mathbf{N}$. Together with

* Supported in part by NSF grant # DMS 85-13467

a suitable trace, we obtain for $q=e^{2\pi i/l}$, $l=4, 5, \ldots$ a subfactor $\rho(\langle g_2, g_3, \ldots\rangle)''$ of $\rho(H_\infty(q))''$. Refining this method we construct for each irreducible representation χ of $SU(k)$, $k\geqq 2$ a sequence of subfactors with trivial centralizers such that if $l\to\infty$, its index values converge to the square of the dimension of χ. Jones' examples of subfactors in [J-1] correspond to $SU(2)$ in its standard representation.

In more detail, our results are as follows. In the first section, we determine the structure of an extension which is a generalization of Jones' basic construction. We obtain from this general statements about the relative commutants of the subfactor in extensions of the factor coming from iterations of Jones' construction. To construct examples of subfactors, we consider ascending sequences of finite dimensional C^* algebras (A_n) and (B_n) with $A_n\subset B_n$. We assume that there is a factor trace on the inductive limit B_∞ of these algebras such that also the weak limit of the union of the A_n's gives a subfactor A of the resulting factor B. To be able to compute the index of A, we need the following crucial condition:

Let ε_{A_n}, ε_{B_n} and $\varepsilon_{A_{n+1}}$ be the conditional expectations onto A_n, B_n and A_{n+1} respectively. Then we require that

$$\varepsilon_{A_{n+1}}\varepsilon_{B_n}=\varepsilon_{A_n}.$$

This implies in the limit that $\varepsilon_A\varepsilon_{B_n}=\varepsilon_{A_n}$ (see also [PP1]). The main difficulty of this approach consists of finding examples which satisfy the condition.

We will furthermore assume that the inclusion matrices for $A_n\subset B_n$, $A_n\subset A_{n+1}$ and for $B_n\subset B_{n+1}$ become periodic for all n greater some n_0. In this case, we show that the index $[B:A]$ of the resulting subfactor is the square of the norm of the inclusion matrix for $A_n\subset B_n$ for any $n\geqq n_0$. Furthermore, we prove that the dimension of the relative commutant $A'\cap B$ is less or equal than the dimension of $pA'_n\cap pB_np$ for all $n\geqq n_0$ and all nonzero projections $p\in A_n$. This upper bound for the size of the relative commutant has proven to be sharp in all known (periodic) cases.

In the second section, we construct representations of Hecke algebras of type A. If q is not a root of unity, they already appear in Hoefsmit's thesis ([H], unpublished), and were later rediscovered by Ocneanu [O] and by this author. If $q=1$, they specialize to Young's orthogonal representations of the symmetric groups. In this paper, we will use modifications of these representations which can also be defined for q a primitive l-th root of unity. We single out for every $k\in\mathbf{N}$, $1\leqq k\leqq l-1$ a special class of Young diagrams, which will be called (k, l) diagrams. These are all Young diagrams with at most k rows for which the numbers of boxes in the first and in the k-th row differs by at most $l-k$. Using the above mentioned modifications, we define a representation $\pi^{(k,l)}$ of $H_\infty(q)$ such that the irreducible components of $\pi^{(k,l)}(H_n(q))$ are in 1-1 correspondence with (k, l) diagrams with n boxes. If $q=e^{2\pi i/l}$, we obtain from this an AF algebra with periodic Bratteli diagram for the sequence $\ldots\subset\pi^{(k,l)}(H_n(q))\subset\pi^{(k,l)}(H_{n+1}(q))\subset\ldots$. We conclude this section with a complete (except for $q=e^{\pm 2\pi i/l}$ and $q=0$) classification of all C^* representations of $H_n(q)$, i.e. representations on a Hilbert space such that the images of the generators $g_1, g_2\ldots g_{n-1}$ of $H_n(q)$ are normal.

In the following section we will consider a special class of traces tr on $H_\infty(q)$ which we will call Markov traces. They are defined inductively by $\mathrm{tr}(1)=1$ and by $\mathrm{tr}(g_n w)=z\,\mathrm{tr}(w)$, where $z\in\mathbf{C}$ is a fixed number and $w\in H_n(q)$. They were first studied for the full Hecke algebra by A. Ocneanu who used them to generalize Jones' link polynomial (see [J-3] and [O]). Markov traces are important for our purposes, because whenever they factor over a C^* representation ρ of $H_\infty(q)$ on which they are faithful and positive, the condition of the first section is satisfied for $A_n=\rho(\langle g_2, g_3, \ldots g_{n-1}\rangle)$ and $B_n=\rho(H_n(q))$. We will give complete proofs of the classification of positive Markov traces which was anounced in [O]. If $q=e^{\pm 2\pi i/l}$, we have exactly one positive Markov trace for each k, $k=1, 2, \ldots, l-1$. The representations coming from the GNS construction with respect to these traces correspond to the representations $\pi^{(k,l)}$ mentioned in the last paragraph. Having all these data available, the results of the first section can be applied for the following situation:

Let $m\in\mathbf{N}$ and let A be the subfactor $\pi^{(k,l)}(\langle g_{m+1}, g_{m+2}\ldots\rangle)''$ of the factor $B=\pi^{(k,l)}(H_\infty(q))''$. Then the relative commutant $A'\cap B$ is equal to $\pi^{(k,l)}(H_m(q))$. So if $m=1$, the relative commutant is trivial. In this case the index is equal to $\sin^2(k\pi/l)/\sin^2(\pi/l)$. For $m>1$, we get subfactors with trivial relative commutant by reducing by a minimal projection $p\in A'\cap B$. So we obtain for each (k, l) diagram λ with m boxes a subfactor $A_p\subset B_p$ with trivial relative commutant and index

$$[B_p\colon A_p]=\prod_{1\leqq r<s\leqq k}\frac{\sin^2((\lambda_r-\lambda_s+s-r)\,\pi/l)}{\sin^2((s-r)\,n/l)}.$$

If $l\to\infty$ the index values tend to the square of the dimension of the irreducible representation of $SU(k)$ corresponding to λ.

This relationship to irreducible representations of $SU(k)$ will be studied again in a subsequent paper using M. Jimbo's solutions of the quantum Yang-Baxter equations. This approach will also enable us to compute the higher relative commutants of our examples of subfactors. Moreover, we will also show that the subfactors coming from the Hecke algebra approach for real and positive parameters have nontrivial relative commutants except for $q=1$. The corresponding subfactors for $q=1$ have been studied by A. Wassermann, who computed their indices and higher relative commutants.

Acknowledgements. This paper is a slightly expanded version of my Ph.D. thesis at University of Pennsylvania. I would like to thank my adviser Vaughan Jones for useful suggestions and for making it possible for me to stay at MSRI, Berkeley during the special year for operator algebras. Special thanks are going to Adrian Ocneanu for telling me his formula for the weights of the Markov traces and for useful discussions about the representations of the Hecke algebras. Part of the work on this paper was done at University of Ottawa. I would like to thank David Handelman for his invitation and hospitality. Finally, it is a pleasure to thank Dan Voiculescu and Antony Wassermann for useful comments and references.

§1. Construction of subfactors by AF algebras

The goal of this section is to provide tools for the construction of subfactors of the hyperfinite II_1 factor by AF algebras and for computing their invariants. We will first recall some basic results about subfactors (see [J-1], §3 and §4).

In all of this paper A and B will be finite von Neumann algebras. We will always assume that there exists a finite faithful normal trace tr on B with $\mathrm{tr}(1)=1$. We can then define a map ε_A: $B \to A$, the so-called trace preserving conditional expectation by the equation

$$\mathrm{tr}(a\varepsilon_A(b)) = \mathrm{tr}(ab) \quad \text{for all } a \in A, \quad b \in B \tag{1.1}$$

Let $L^2(B, \mathrm{tr})$ be the Hilbert space generated by B and the inner product given by the trace. To be more precise we denote the element of $L^2(B, \mathrm{tr})$ coming from the element $b \in B$ by b_ξ. Let B be represented in standard form on $L^2(B, \mathrm{tr})$ via left multiplaction. We define a projection e_A by

$$e_A(b_\xi) = \varepsilon_A(b)_\xi \quad \text{for } b \in B \quad \text{and} \quad b_\xi \in L^2(B, \mathrm{tr})$$

(see [J-1], §3). Let (B, e_A) be the von Neumann algebra generated on $L^2(B, \mathrm{tr})$ by B and e_A. This is called the basic construction for $A \subset B$ and plays an important role in the theory of subfactors. e_A has the following algebraic properties:

(a) $e_A b e_A = \varepsilon_A(b) e_A$ for $b \in B$ and $e_A A \cong A$.

(b) Finite sums of elements of the form $a e_n b$ or c *with* $a, b, c \in B$ are dense in $\langle B, e_A \rangle$.

(c) $e_A \langle B, e_A \rangle e_A = A e_A$. (1.2)

Let J: $L^2(B, \mathrm{tr}) \to L^2(B, \mathrm{tr})$ be defined by $J(b_\xi) = b_\xi^*$. Then $\langle B, e_A \rangle \cong JA'J$ where A' is the commutant of A on $L^2(B, \mathrm{tr})$. We obtain as an immediate consequence that the centers of A and $\langle B, e_A \rangle$ are isomorphic. In particular, if A is a factor, then so is $\langle B, e_A \rangle$. If it is a finite factor, we define the index $[B:A]$ of A in B to be the number $1/\mathrm{Tr}(e_A)$, where Tr denotes the unique normalized trace on $\langle B, e_A \rangle$. If $\langle B, e_A \rangle$ is not finite, the index is defined to be infinte. If the index is finite, we also have

$$\mathrm{Tr}(e_A b) = 1/[B:A]\, \mathrm{Tr}(b) \quad \text{for all } b \in B. \tag{1.3}$$

Let M_n be the C^* algebra of all complex $n \times n$ matrices and let A and B be finite dimensional C^* algebras. Then $A \cong \bigoplus_{i=1}^{m} A_i$ and $B \cong \bigoplus_{j=1}^{m'} B_j$ with $A_i \cong M_{a_i}$, $B_j \cong M_{b_j}$ and $a_i, b_j \in \mathbf{N}$. The vectors $\vec{a} = (a_i)$ and $\vec{b} = (b_j)$ are called the dimension vectors of A and B. As a simple B_j module is also an A module, it can be decomposed into a direct sum of simple A modules. Let g_{ij} be the number of simple A_i modules in this decomposition. The matrix $G = (g_{ij})$ is called the inclusion matrix for $A \subset B$. By the above mentioned isomorphism between the centers of A and $\langle B, e_A \rangle$, we can write $\langle B, e_A \rangle = \bigoplus_{i=1}^{m} \langle B, e_A \rangle_i$. With these notations, we have

If p is a minimal projection of A_i, $e_A p$ is a minimal projection of $\langle B, e_A \rangle_i$. (1.4)

A trace tr on B is completely determined by a weight vector $\vec{t}=(t_j)$ with $t_j=\mathrm{tr}(p_j)$ where p_j is a minimal projection of B_j. We can similarly characterize a trace on A by a vector $\vec{s}=(s_i)$. The weight vectors and dimension vectors of A and B are related by the following equations:

$$G\vec{t}=\vec{s} \quad \text{and, if } A \text{ and } B \text{ have the same 1,} \quad G^t\vec{a}=\vec{b}. \tag{1.5}$$

If tr is faithful, the inclusion matrix for $B\subset\langle B, e_A\rangle$ is the transposed G^t of G. So similar equations, with G replaced by G^t, relate trace and dimension vectors of B and $\langle B, e_A\rangle$. It holds in general that if B is contained in a finite dimensional C^* algebra C with inclusion matrix H then $A\subset C$ has inclusion matrix GH. In particular we have that $A\subset\langle B, e_A\rangle$ has inclusion matrix GG^t.

For the computation of the indices in our examples we will need to know how the projection obtained from the basic construction acts on usually finite dimensional subalgebras of our given algebras.

Theorem 1.1. *Let $A\subset B$ be finite dimensional C^* algebras acting on a Hilbert space H with a faithful trace* tr *on B. Let ε_A be the trace preserving conditional expectation onto A and let e be a projection on H such that*

(∗) *$ebe=e\varepsilon_A(b)$ for $b\in B$ and*

(∗∗) *$eA\cong A$.*

Let $\langle B, e\rangle$ be the C^* algebra generated by B and e.

(i) *$\langle B, e\rangle\cong Q\oplus K$, where Q is isomorphic to $\langle B, e_A\rangle$, the basic construction for $A\subset B$ and K is isomorphic to a subalgebra of B.*

(ii) *Let z be the central projection onto Q. Then z is equal to the central support of e in $\langle B, e\rangle$.*

(iii) *Let* Tr *be a trace on $\langle B, e\rangle$ extending* tr. *Then $\mathrm{Tr}(e)\geq d\,\mathrm{Tr}(z)$, where $d=\min\{a_i/(GG^t\,\vec{a})_i,\ i=1, \dots, m\}$.*

(iv) *If $\vec{s}$ is the weight vector of $\mathrm{Tr}_{|Q}$, then $GG^t\vec{s}\leq\vec{s}$ (where $\vec{v}\leq\vec{w}$ for $\vec{v}$, $\vec{w}\in\mathbf{R}^m$ if $v_i\leq w_i$ for $i=1, 2, \dots m$).*

Proof. (i) Let us denote by $\bar{b}$ the image of $b\in B$ in the standard representation of B on $L^2(B, \mathrm{tr})$ and let e_A be the projection of the basic construction for $A\subset B$. It follows from (∗) that elements of the form $a_0+\sum a_i e b_i$ with a_0, a_i, $b_i\in B$ generate $\langle B, e\rangle$ (see also [J-1], (3.1.5), (ii)). Using (1.2), it is easy to check that $\Phi:\langle B, e\rangle\rightarrow\langle B, e_A\rangle$ with

$$\Phi(a_0+\textstyle\sum a_i e b_i)=\bar{a}_0+\sum\bar{a}_i\, e_A\,\bar{b}_i$$

defines a surjective homomorphism, provided it is well-defined.

So let $a_0+\sum a_i e b_i=0$, $b\in B$ and $\tilde{b}=a_0 b+\sum a_i\,\varepsilon_A(b_i b)$. Then

$$0=(a_0+\textstyle\sum a_i\, e b_i)\, b e=\tilde{b} e$$

and

$$0=e\tilde{b}^*\,\tilde{b}e=\varepsilon_A(\tilde{b}^*\tilde{b})\, e.$$

Hence $\tilde{b}=0$ by (**) and the faithfulness of the conditional expectation. It follows

$$\Phi(a_0+\sum a_i e b_i)(b_\xi)=\tilde{b}_\xi=0.$$

The statement about K will follow from (ii).

(ii) The central support of e contains z by (i) and (1.4). On the other hand, we can write $1-z$ as an expression of the form $a_0+\sum a_i e b_i$. Obviously $\Phi(1-z)=0$. Hence, with $\tilde{b}=a_0+\sum a_i \varepsilon_A(b_i)$, $\tilde{b} e_A=\Phi((1-z)\,\tilde{b} e)=0$. It follows, as in the proof of (i) that $\tilde{b}=0$, hence $\tilde{\tilde{b}}=0$. But then $(1-z)\,e=0$ and $(1-z)\langle B, e\rangle=(1-z)B$, which shows the rest of (i).

(iii) Let p_i be a minimal central projection of A and let $\tilde{p}_i$ be the corresponding central projection of $Q\cong\langle B, e_A\rangle$. If the weight vector Tr for Q is given by $\vec{\tilde{s}}$, we have by (**) and (1.5)

$$\mathrm{Tr}(e\tilde{p}_i)/\mathrm{Tr}(\tilde{p}_i)=a_i\,\tilde{s}_i/(GG^t\vec{a})_i\,\tilde{s}_i\geqq d.$$

Hence $\mathrm{Tr}(e)=\sum \mathrm{Tr}(e\tilde{p}_i)\geqq d\,\mathrm{Tr}(z)$.

(iv) Let $\vec{r}$ be the weight vector of Tr restricted to K. By (i) and the remarks above, the inclusion matrix for $B\subset\langle B, e\rangle$ is of the form (G^t, R). So

$$\vec{s}=G\vec{t}=G(G^t\vec{\tilde{s}}+R\vec{r})\geqq GG^t\vec{\tilde{s}},$$

as all coordinates of $\vec{r}$ and R are nonnegative. □

Let now A and B be finite von Neumann algebras on a Hilbert space H with a finite trace tr on B. Let for a projection p on H A_p and B_p be the reduced algebras, i.e. $B_p=\{pbp,\, b\in B\}''$ and $A_p=\{pap,\, a\in A\}''$. If A is a factor and if $p\in A'\cap B$, $p\neq 0$, we have

$$pA\cong A \quad\text{and}\quad \mathrm{tr}(ap)=\mathrm{tr}(a)\,\mathrm{tr}(p), \tag{1.6}$$

where the second equality follows from the uniqueness of the trace on a factor up to scalar multiples.

By the same argument we have for $0\neq p\in A$ that $pA'\cong A'$ and, in particular

$$p(A'\cap B)\cong A'\cap B. \tag{1.7}$$

If A is a finite dimensional algebra and z is a minimal central projection in A, we obtain from this

$$p(A'\cap B)\cong z(A'\cap B)\quad \text{for any nonzero projection } p\in A \text{ such that } pz=p. \tag{1.8}$$

We shall use our results so far for the computation of additional, finer invariants of subfactors. Let $B_1=\langle B, e_A\rangle$. Then $[B_1{:}B]=[B{:}A]$ by [J-1] and we can again apply the basic construction for $B\subset B_1$. We thus obtain an extension $B\subset B_1\subset B_2\subset\ldots$. Let $C_i=A'\cap B_i$ be the relative commutant (or centralizer) of A in B_i. It is easy to check that an automorphism $\Phi\in\mathrm{Aut}(B)$ can be extended to an isomorphism $\tilde{\Phi}\colon\langle B, e_A\rangle\to\langle B, e_{\Phi(A)}\rangle$ by defining $\tilde{\Phi}(e_A)=e_{\Phi(A)}$. If follows by induction that the structure of the algebras $C_1, C_2, \ldots$ is an invariant of subfactors of B. We will refer to them as higher relative commutants of $A\subset B$. A. Ocneanu has proposed that these higher relative commutants with some additional structure are a complete invariant for a large class of subfactors of the hyperfinite II_1 factor up to conjugacy by automorphisms. We will apply theorem (1.1) to determine the structure of a direct summand of C_n from C_{n-1} and C_{n-2}. The following result was first mentioned by Ocneanu.

Corollary 1.2. *C_n contains a direct summand which is isomorphic to $\langle C_{n-1}, e_{C_{n-2}}\rangle$.*

Proof. Let e_n be the projection of the basic construction for $B_{n-2}\subset B_{n-1}$. Let $a\in A$, $c\in C_{n-1}$ and let $b\in B_{n-2}$. Then we have

$$\mathrm{tr}(a\varepsilon_{B_{n-2}}(c)\,b)=\mathrm{tr}(acb)=\mathrm{tr}(cab)=\mathrm{tr}(\varepsilon_{B_{n-2}}(c)\,ab).$$

Hence it follows from (1.1) that $\varepsilon_{B_{n-2}}(c)\in C_{n-2}$ for $c\in C_{n-1}$. So $\varepsilon_{B_{n-2}}$ and $\varepsilon_{C_{n-2}}$ coincide on C_{n-1}. Using (1.2), it is easy to check that the hypotheses of theorem (1.1) hold for $A=C_{n-2}$, $B=C_{n-1}$ and $e=e_n$. Hence it follows from that theorem that $\langle C_{n-1}, e_n\rangle$ contains a direct summand S which is isomorphic to the basic construction for $C_{n-2}\subset C_{n-1}$. It remains to show that S is a direct summand in C_n.

Let z be the central projection (in $\langle C_{n-1}, e_n\rangle$) onto S. Assume that z is not central in C_n. Then, by Theorem (1.1), (ii) there exists a partial isometry v such that $v^*v\leqq e_n$ and $vv^*\leqq 1-z$. By [PP1], Lemma 1.2, we can write v as $v=we_n$ with $w\in B_{n-1}$. We can show as before that $\varepsilon_{B_{n-1}}(v)=\varepsilon_{C_{n-1}}(v)$. On the other hand, it follows from (1.1) and (1.3)

$$\varepsilon_{B_{n-2}}(v)=1/[B\colon A]\,w\in C_{n-1}.$$

This would imply $w\in C_{n-1}$ and $v\in S$, a contradiction to $vv^*\leqq 1-z$. Hence z is also central in C_n.

Observe that by (1.4) any minimal projection p in S is equivalent to a projection of the form $e_n p$, where p is a minimal projection of C_{n-2}. As $e_n B_n e_n = e_n B_{n-2}\cong B_{n-2}$ (see (1.2)), we obtain

$$A'_{e_n p}\cap(B_n)_{e_n p}=((A_p)'\cap(B_{n-2})_p)_{e_n}\cong(A_p)'\cap(B_{n-2})_p\cong\mathbf{C}.$$

Hence $e_n p$ is a minimal projection of C_n, which implies that any minimal projection in S is also a minimal projection in C_n.

Next, let p_1, p_2 be minimal projections in C_{n-2} with $p_1 \nsim p_2$ and let $b\in C_n$. By (1.2), (c) $e_n b e_n = a e_n$ for a suitable $a\in C_{n-2}$. It follows

$$(e_n p_1)\,b(e_n p_2)=p_1\,a p_2\,e_n=0.$$

Hence also $e_n p_1 \nsim e_n p_2$ and any 2 projections which are not equivalent in S, are also not equivalent in C_n.

The last 2 paragraphs show that the Bratteli diagram for $S\subset zC_n$ is the trivial one, i.e. $S=zC_n$ is a direct summand of C_n. □

One way of constructing subfactors with trivial relative commutants is to reduce a given factor and a subfactor of it by a minimal projection of the relative commutant. The following proposition relates the basic constructions of the reduced and of the original factors.

Proposition 1.3. *Let p be a projection in $A'\cap B$. Then*

$$\langle B_p, e_{A_p}\rangle \cong \langle B, e_A\rangle_{pJpJ}$$

Proof. Note that $JpJp$ projects onto the subspace of $L^2(B, \mathrm{tr})$ which is generated by the elements of B_p. Moreover, if tr_p is the normalized trace on B_p, we have

$$\mathrm{tr}_p(b)=1/\mathrm{tr}(p)\,\mathrm{tr}(b) \quad \text{for all} \quad b\in B_p. \tag{*}$$

Hence $pJpJ\; L^2(B, \mathrm{tr})\cong L^2(B_p, \mathrm{tr}_p)$, as their inner products only differ by a fixed scalar multiple. It follows immediately from (1.1) and (*) that

$$\varepsilon_{A_p}(b)=1/\mathrm{tr}(p)\,\varepsilon_A(b)\,p \quad \text{for all} \quad b\in B_p.$$

We obtain from this that

$$1/\mathrm{tr}(p)(pJpJ\,e_A\,pJpJ)\,b_\xi=1/\mathrm{tr}(p)(pJpJ\,\varepsilon_A(pbp))_\xi=\varepsilon_{A_p}(pbp)_\xi.$$

Hence the map induced by

$$b\in B_p\mapsto pJpJ\,bpJpJ$$

and by

$$e_{A_p}\mapsto 1/\mathrm{tr}(p)\,pJpJ\,e_A\,pJpJ$$

extends to a spatial isomorphism from $\langle B_p, e_{A_p}\rangle$, represented on $L^2(B_p, \mathrm{tr})$ into $\langle B, e_A\rangle_{pJpJ}$, represented on $pJpJ\,L^2(B, \mathrm{tr})$. It remains to show the surjectivity. Note that

$$\mathrm{tr}(pb(1-p)\,a)=\mathrm{tr}(b(1-p)\,ap)=0 \quad \text{for all } a\in A, b\in B.$$

Hence $\varepsilon_A(pb(1-p))=\varepsilon_A((1-p)\,bp)=0$ for all $b\in B$ by (1.1) or, equivalently

$$e_A(pJ(1-p)\,J))=e_A((1-p)\,JpJ)=0.$$

Recall from (1.2) that linear combinations of elements of B or elements of the form $b_1 e_A b_2$ with $b_1, b_2\in B$ form a dense subset of $\langle B, e_A\rangle$. As JpJ commutes with any element of B, we have

$$JpJ\,b_1\,e_A\,b_2\,JpJ=b_1\,JpJ(p+1-p)\,e_A(p+1-p)\,JpJ\,b_2=b_1\,JpJp\,e_A\,pJ\,pJ\,b_2.$$

So our spatial homomorphism is onto by this and (1.2). □

For the construction of subfactors we will have the following set-up (see also [PP1]):

(i) Let $B_1\subset B_2\subset\ldots$ be an ascending sequence of finite dimensional C^* algebras with B_n a proper subalgebra of B_{n+1} for all $n\in\mathbf{N}$. Let furthermore tr be a positive finite extremal trace on its inductive limit B_∞ and let π_{tr} be the GNS construction with respect to tr. Then it is well-known that the weak closure B of $\pi_{\mathrm{tr}}(B_\infty)$ is isomorphic to R, the hyperfinite II_1 factor (see [SV]).

(ii) Similarly, let (A_n) be an ascending sequence of subalgebras $A_n\subset B_n$ such that the weak closure A of $\pi_{\mathrm{tr}}(A_\infty)$ is a subfactor of B.

We are, of course, mainly interested in constructing proper subfactors of B. This can not be expected in general even if A_n is a proper subalgebra of B_n for all $n\in\mathbf{N}$. We therefore need the following additional condition (see also [PP1]):

(iii) Let $\varepsilon_{A_{n+1}}$, ε_{A_n} and ε_{B_n} be the trace preserving conditional expectations onto A_{n+1}, A_n and B_n respectively. Then we will require

$$\varepsilon_{A_{n+1}}\,\varepsilon_{B_n}=\varepsilon_{A_n}\quad \text{for all}\quad n\in\mathbf{N}.$$

It can be shown easily by induction on k (see also the proof of [PP1], Prop. (2.6)), that this implies $\varepsilon_{A_{n+k}}\,\varepsilon_{B_n}=\varepsilon_{A_n}$ for all $k\in\mathbf{N}$ and in the limit

$$\varepsilon_A\,\varepsilon_{B_n}=\varepsilon_{A_n}.$$

So whenever we write $(A_n)\subset(B_n)$, we will mean sequences of finite dimensional C^* algebras with a trace tr for which the just described properties (i)–(iii) hold. The inclusion matrix for $A_n\subset B_n$ will be denoted by G_n, the weight vectors of the restrictions of tr to A_n and B_n by $\vec{s}_n$ and $\vec{t}_n$ and their dimension vectors by $\vec{a}_n$ and $\vec{b}_n$ respectively.

Let e_A be the projection coming from the basic construction for $A\subset B$. Then $e_A b e_A=\varepsilon_A(b)\,e_A=\varepsilon_{A_n}(b)\,e_A$ for any $b\in B_n$. Hence it follows from Theorem 1.1 that

$$\langle B_n, e_A\rangle=Q_n\oplus K_n,\quad \text{with}\ Q_n\cong\langle B_n, e_{A_n}\rangle.$$

K_n is isomorphic to a subalgebra of B_n.

Let z_n be the central projection for Q_n. Then, again by Theorem 1.1, z_n is the central support of e_A in $\langle B_n, e_A\rangle$. Hence, as the union of the $\langle B_n, e_A\rangle$ is weakly dense in the factor $\langle B, e_A\rangle$, the z_n's are an increasing sequence converging to 1 weakly. By (1.2) and (1.3), the weight vector of Tr on Q_n is equal to $1/[B:A]\,\vec{s}_n$. So, with $\langle .,. \rangle$ the usual scalar product, we have

$$\mathrm{Tr}(z_n)=\langle G_n G_n^t \vec{a}_n, 1/[B:A]\,\vec{s}_n\rangle. \tag{1.9}$$

We will be mainly concerned with the case when the sequences $(A_n)\subset(B_n)$ have periodic Bratteli diagrams. Let us recall first that a real quadratic matrix D with nonnegative entries is called *primitive* if D^l has only positive entries for some $l\in\mathbf{N}$. We will say that the sequence (A_n) is *periodic* with period k if there is an $n_0\in\mathbf{N}$ such that for all $n>n_0$ the inclusion matrix for $A_{n+k}\subset A_{n+k+1}$ is the same (after a possible relabeling of the central projections) as the one for $A_n\subset A_{n+1}$. We include in our definition here the additional condition that the inclusion matrix for $A_n\subset A_{n+k}$ is primitive, as this is the only interesting case for us.

We will say that $(A_n)\subset(B_n)$ is periodic if both (A_n) and (B_n) are periodic with same period k and if also the inclusion matrices for $A_{n+k}\subset B_{n+k}$ and $A_n\subset B_n$ are the same. Hereby we assume the same identification of central projections of A_n and A_{n+k} and of B_n and B_{n+k} as for the periodicity of (A_n) and (B_n).

It is easy to see that we can always make a periodic sequence with period k a periodic sequence with period 1 by omitting algebras in between and by a suitable relabeling. By omitting sufficiently many algebras at the beginning of the sequence we can also assume $n_0=1$. In this case the inclusion matrix for $A_n \subset A_{n+1}$ is a fixed primitive matrix D with nonnegative integer entries for all $n \in \mathbf{N}$. Similarly, also the inclusion matrix for $A_n \subset B_n$ is a constant matrix G.

It follows from the Perron-Frobenius theorem (see for instance [G], p. 53 and 63) that the largest eigenvalue μ of D is positive and that all the coefficients of the corresponding eigenvector are positive. It is well-known that in this case $\bigcap_{n\in\mathbf{N}} D^n(\mathbf{R}^m_+)$ consists of all positive multiples of this eigenvector. Hence it follows from the remarks after (1.5)

The weight vector $\vec{s}_1$ of tr for A_1 is a multiple of the Perron vector of D and

$$\vec{s}_{n+1}=(1/\mu)^n\,\vec{s}_1. \tag{1.10}$$

Similarly, the vectors $(1/\mu)^n\,\vec{a}_{n+1}=(1/\mu\,D^t)^n\,\vec{a}_1$ converge to a positive multiple of the Perron vector $\vec{v}$ of D^t. So if $z_{n,i}$ is the i-th minimal central projection of A_n with dimension $a_{n,i}$ and weight $s_{n,i}$, we have

$$\lim_{n\to\infty} \operatorname{tr}(z_{n,i})=\lim_{n\to\infty} s_{n,i}((1/\mu\,D^t)^{n+1}\,\vec{a}_1)_i=\lambda s_{1,i}\,v_i>0 \quad \text{with } \lambda>0.$$

In particular, there exists a constant $c>0$ such that

$$(1/\mu)^{n-1}\,a_{n,i}\geqq c \quad \text{and} \quad \operatorname{tr}(z_{n,i})\geqq c \qquad \text{for all } n\in\mathbf{N}, \qquad i=1,\dots m. \tag{1.11}$$

We will derive formulas for the index of subfactors constructed by periodic sequences $(A_n)\subset(B_n)$. We first show that such subfactors always have finite index (see also [J-1], proof of Theorem (4.1.1), (ii)).

Lemma 1.4. *Let $(A_n)\subset(B_n)$ be periodic. Then $[B:A]<\infty$.*

Proof. Let $d_{n,i}=a_{n,i}/(GG^t\,\vec{a}_n)_i>0$. Then $d_{n,i}\to v_i/(GG^t\,\vec{v})_i$ for $n\to\infty$, where $\vec{v}$ is the Perron vector of D^t. Hence $d=\inf\{d_{n,i},\, n\in\mathbf{N},\, i=1,\dots m\}>0$.

Let Tr be a (possibly only semifinite) trace on $\langle B, e_A\rangle$. As $e_A\langle B, e_A\rangle e_A = A\,e_A\cong A$ is finite, e_A is a finite projection and $\operatorname{Tr}(e_A)<\infty$. By Theorem 1.1 (iii), $\operatorname{Tr}(e_A)\geqq d\operatorname{Tr}(z_n)$ for all $n\in\mathbf{N}$. As $z_n\to 1$, $\operatorname{Tr}(1)<\infty$. Hence $\langle B, e_A\rangle$ and $[B:A]$ are finite. □

The following theorem expresses the index in terms of the weight vectors of the trace. A similar result is proven in [PP1], Theorem (5.1) by different methods. A formula for local indices is given in [J-1], (2.2.1). The difficulty is that in general it is not true that $\operatorname{tr}_{A'}(p)=\operatorname{tr}_B(p)$ for $p\in A'\cap B$.

Theorem 1.5. *Let the factors $A\subset B$ be constructed from sequences of finite dimensional C^* algebras $(A_n)\subset(B_n)$ with properties* (i)–(iii). *Let with the notations as above $\vec{g}_i^{(n)}$ be the i-th column vector of G_n^t. Then*

(i) $[B:A]\geqq \sup_{i,n} \|\vec{g}_i^{(n)}\|^2$.

(ii) *If* (A_n) *is periodic and if* $[B:A]<\infty$, *there exists an* $n_1\in\mathbf{N}$ *such that for* $n>n_1$

$$G_n G_n^t \vec{s}_n=[B:A]\,\vec{s}_n.$$

If e_A *is the projection coming from the basic construction for* $A\subset B$, *we also have for* $n>n_1$

$$\langle B_n, e_A\rangle\cong\langle B_n, e_{A_n}\rangle.$$

(iii) *If* $(A_n)\subset(B_n)$ *is periodic, there exists an* $n_1\in\mathbf{N}$ *such that for* $n>n_1$

$$[B:A]=\|\vec{s}_n\|^2/\|\vec{t}_n\|^2.$$

If $p\in A'\cap B_n$ *for some* $n\in\mathbf{N}$, *then*

$$[B_p:A_p]=\mathrm{tr}(p)^2[B:A].$$

Proof. (i) Let $s_{n,i}$ be the i-th coordinate of $\vec{s}_n$. By theorem (1.1), (iv)

$$[B:A]\,s_{n,i}\geqq(G_n G_n^t \vec{s}_n)_i\geqq\langle\vec{g}_i^{(n)},\vec{g}_i^{(n)}\rangle s_{n,i}$$

for all $n\in\mathbf{N}$ and $i=1,2,\dots m$.

(ii) We can assume without restriction of generality the special conditions of periodicity with $n_0=1$ and $k=1$. So the inclusion matrix for $A_n\subset B_n$ is a fixed matrix G for all $n\in\mathbf{N}$.

Recall from (1.9) that the weight vector for Q_1 is equal to $1/[B:A]\,\vec{s}_1$. If $GG^t\vec{s}_1\neq[B:A]\,\vec{s}_1$, the vector $\vec{s}_1-GG^t\,1/[B:A]\,\vec{s}_1$ would be nonnegative by Theorem 1.1 (iv). Let d be its largest coordinate and let c be as in (1.11). Then we have, using the periodicity of the Bratteli diagrams and (1.9)

$$\begin{aligned}\mathrm{Tr}(1-z_n)&=\langle\vec{a}_n,\vec{s}_n\rangle-\langle GG^t\vec{a}_n, 1/[B:A]\,\vec{s}_n\rangle\\&=\langle(1/\mu)^{n-1}\,\vec{a}_n,\vec{s}_1-GG^t\,1/[B:A]\,\vec{s}_1\rangle\geqq cd\end{aligned}$$

for all $n\in\mathbf{N}$. But this is a contradiction to $z_n\to 1$ for $n\to\infty$. Hence $GG^t\vec{s}_1=[B:A]\,\vec{s}_1$. This entails $z_n=1$ and $\langle B_n, e_A\rangle=Q_n$. As $G\vec{t}_1=\vec{s}_1$ by (1.5), we also have $[B:A]\;\vec{t}_1=G^tG\;\vec{t}_1$. Hence $\|\vec{s}_1\|^2=\langle\vec{t}_1,\;GG^t\vec{t}_1\rangle=[B:A]\,\|\vec{t}_1\|^2$. As $\vec{s}_n=(1/\mu)_1^{n-1}\,\vec{s}_1$ and $\vec{t}_n=(1/\mu)^{n-1}\,\vec{t}_1$, the formula also holds in general.

(iii) By Lemma 1.4, we can use the formula in (ii) to compute the index $[B:A]$ if $(A_n)\subset(B_n)$ is periodic. Let $p\in A'\cap B$ and let tr_p be the normalized trace on B_p. By the uniqueness of the trace on $A\cong pA$, the weight vector of tr_p on pA_n is also $\vec{s}_n$. As the inclusion matrix for $B_n\subset B_{n+1}$ is primitive, there exists an l such that $z_{n,j}\,p\neq 0$ for any $n>l$ and any minimal central projection $z_{n,j}\in B_n$. As $\mathrm{tr}_p(b)=1/\mathrm{tr}(p)\;\mathrm{tr}(b)$ for $b\in B_p$, the weight vector of tr_p on $(B_n)_p$ is equal to $1/\mathrm{tr}(p)\;\vec{t}_n$. So by (ii) and by $[B_p:A_p]\leqq[B:A]<\infty$ (see [J-1], (2.2.1)), we have

$$[B_p:A_p]=\mathrm{tr}(p)^2\,\|\vec{s}_n\|^2/\|\vec{t}_n\|^2=\mathrm{tr}(p)^2[B:A].$$

Remark. We will construct examples with period k where only after k steps we will have the same inclusion matrix. However for all of these matrices we have $\|G_n^t G_n\| = \|G_{n+1}^t G_{n+1}\| = \ldots = [B:A]$ for $n > n_0$.

In the rest of this section we will give an upper bound for the dimension of the relative commutant of a subfactor constructed by periodic Bratteli diagrams. We will need the following simple observation:

Let V_1 and V_2 be finite dimensional subspaces of a Hilbert space H with $\dim V_2 < \dim V_1$. If e_1 and e_2 are the corresponding orthogonal projections onto these spaces, the linear map $e_1 e_2 e_1 : V_1 \to V_1$ has rank $< \dim V_1$. Hence there exists a nonzero $\xi \in \ker(e_1 e_2 e_1) \cap V_1$ such that

$$\|e_2 \xi - \xi\|^2 = \|(1-e_1) e_2 \xi - \xi\|^2 \geqq \|\xi\|^2. \tag{1.12}$$

Theorem 1.6. *Let $(A_n) \subset (B_n)$ be periodic and let n_0 be as in the definition of periodic diagrams. Then we have for all $n > n_0$ and any nonzero projection $p \in A_n$*

$$\dim A' \cap B \leqq \dim p(A'_n \cap B_n).$$

Proof. We know by Lemma 1.4 that $[B:A] < \infty$, hence $A' \cap B$ has to be finite dimensional (see [J-1], (2.2.3)). Let $x \in A' \cap B$. Observe that $\varepsilon_{B_n}(x)$ is in $A'_n \cap B_n$ (which can be checked easily using (1.1)) hence it is equal to $\varepsilon_{A'_n \cap B_n}(x)$. Moreover, if $n \to \infty$, $\varepsilon_{B_n}(x)$ converges to x in trace norm as $\bigcup B_n$ is dense in B. As the unit ball of $A' \cap B$ is compact, we have uniform convergence i.e. there exists for all $\varepsilon > 0$ an $n_1 \in \mathbf{N}$ such that

$$\|\varepsilon_{B_n}(x) - x\|_2 \leqq \varepsilon \|x\|_2 \quad \text{for all } x \in A' \cap B \quad \text{and} \quad n > n_1. \tag{$*$}$$

Assume now that there exists for some $n > n_1$ a nonzero projection $p \in A_n$ such that $\dim p(A'_n \cap B_n) < \dim p(A' \cap B) \leqq \dim A' \cap B$. We can assume p to be a minimal projection of A_n. But then the same inequality also holds for the minimal central projection z majorizing p instead of p by (1.7). Applying (1.12) for the finite dimensional subspaces $z(A'_n \cap B_n)$ and $z(A' \cap B)$ of $L^2(B, \mathrm{tr})$, we find $x \in A' \cap B$ such that

$$\|\varepsilon_{B_n}(xz) - xz\|_2 \geqq \|xz\|_2 = c^{1/2} \|x\|_2,$$

where c is the lower bound for $\mathrm{tr}(z)$ as in (1.11). Hence also $\|\varepsilon_{B_n}(x) - x\|_2 \geqq c^{1/2} \|x\|_2$, which would contradict $(*)$. □

Let us finally remark that we could already apply our results so far to construct new subfactors from Jones' subfactors in the following way:

Let $e_1, e_2, \ldots$ be the projections and let tr be the trace as in [J-1], (4.1.1), (a)–(c). Using Theorem 1.6, we can show that

$$\langle e_{m+1}, e_{m+2}, \ldots \rangle' \cap \langle e_1, e_2, \ldots \rangle'' = \langle e_1, \ldots, e_{m-1} \rangle$$

for the special values $\tau = 1/(4\cos^2(\pi/l))$ and $l = 3, 4, \ldots$. By reducing by minimal projections of these relative commutants we obtain subfactors with trivial relative commutants and index $\sin^2(k\pi/l)/\sin^2(\pi/l)$ for $l = 3, 4, \ldots$ and $k = 1, 2, \ldots l-1$. We will use similar ideas for a more general situation in Theorem 3.7.

§2. Orthogonal representations of Hecke algebras of type A_n

In this context, we will mean by the Hecke algebra $H_n(q)$ of type A_{n-1} the free complex algebra with generators $1, g_1, \ldots, g_{n-1}$ and parameter $q \in \mathbf{C}$ with the defining relations

$$(H1)' \; g_i g_{i+1} g_i = g_{i+1} g_i g_{i+1} \quad \text{for } i=1,2,\ldots,n-2,$$
$$(H2)' \; g_i g_j = g_j g_i, \quad \text{whenever } |i-j| \geqq 2,$$
$$(H3)' \; g_i^2 = (q-1) g_i + q \quad \text{for } i=1,2,\ldots,n-1.$$

Note that for $q=1$ the last relation becomes $g_i^2=1$. In this case the g_i's satisfy exactly the same relations as a set of simple reflections of the symmetric group S_n. It is known that $H_n(q) \cong \mathbf{C} S_n$ if q is not a root of unity (see [Bou], p. 54–56). We will give an independent proof of this result in Theorem 2.2. Similarly as for $\mathbf{C} S_n$, we can show that $H_n(q)$ is spanned linearly by elements of the form

$$\beta_1 \beta_2 \ldots \beta_n, \quad \text{where either } \beta_i = g_{j_i} g_{j_i+1} \ldots g_{i-1} \quad \text{for } 1 \leqq j_i \leqq i-1 \quad \text{or} \quad \beta_i = 1 \tag{2.1}$$

(see for instance [W-2], Lemma (2.1)). Note that $(H1)'$ and $(H2)'$ are the defining relations for the braid group B_n (see [Bi], p. 11). So representations of $H_n(q)$ yield in particular also representations of B_n (see [J-2]).

Observe that $(H3)'$ implies that g_i has at most 2 different spectral values, hence also at most 2 spectral projections. It will be more convenient for us to define the representations for these projections.

So let for $q \neq -1$

$$e_i = \frac{(q-g_i)}{(q+1)}$$

be the spectral projection belonging to the eigenvalue -1. Then $g_i = q(1-e_i) - e_i = q-(1+q)\, e_i$. So $\langle 1, g_1, \ldots, g_{n-1} \rangle = \langle 1, e_1, e_1, \ldots, e_{n-1} \rangle$ and the relations $(H1)' - (H3)'$ translate to

$$(H1) \; e_i e_{i+1} e_i - q/(1+q)^2 e_i = e_{i+1} e_i e_{i+1} - q/(1+q)^2 e_{i+1} \quad \text{for } i=1,\ldots,n-2,$$
$$(H2) \; e_i e_j = e_j e_i \quad \text{whenever } |i-j| \geqq 2,$$
$$(H3) \; e_i^2 = e_i \quad \text{for } i=1,\ldots,n-1.$$

We obtain as an easy consequence of (2.1)

$$H_n(q) \text{ is the linear span of elements of the form } a e_{n-1} b \text{ or } c \text{ with } a,b,c \in H_{n-1}(q). \tag{2.2}$$

For the definition of representations of Hecke algebras we will need rational functions a_d, $d \in \mathbf{Z} \setminus \{0\}$ defined by

$$a_d(q) = \frac{(1+q+\ldots+q^d)}{(1+q)(1+q+\ldots+q^{d-1})} = \frac{(1-q^{d+1})}{(1+q)(1-q^d)} \quad \text{for } q \neq 1.$$

Note that these functions, in the first notation, are well-defined for all $q \in \mathbf{C}$ except for $q=1$ or q a d-th root of unity. As we are mainly interested in the case $q \neq 1$, we will usually use the slightly more convenient second notation.

Lemma 2.1. (i) $a_d + a_{-d} = 1$.

(ii) *Let* $k, l, m \in \mathbf{N}$ *with* $k+m=l$. *Then*

$$q/(1+q)^2 = a_k\, a_m + a_l\, a_{-m} - a_k\, a_l = a_k\, a_m + a_l\, a - k - a_m\, a_l$$

and $a_d\, a_{-d-1} = q/(1+q)^2$.

(iii) $a_d(q) = a_d(q^{-1})$ *for all* $q \in \mathbf{C}$.

(iv) *If* $q \geqq 1$, $(a_n(q))$ *is a strictly decreasing sequence converging to* $q/(1+q)$ *and* $(a_{-n}(q))$ *is a strictly increasing sequence converging to* $1/(1+q)$. *In particular,* $a_d(q) \geqq 0$ *for* $d \in \mathbf{Z} \setminus \{0\}$.

(v) *If* $q = e^{2\pi i x}$, $x \in [0, 1/2)$ *such that* $1/x \notin \mathbf{N}$, *then* $a_1(q) > a_2(q) > \ldots > a_{[1/x]-1}(q) > 0 > a_{[1/x]}(q)$, *where* $[1/x]$ *is the largest integer less than* $1/x$.

If $x = 1/n$, $n \in \mathbf{N}$, *we have the same inequalities until* $a_{n-1}(q) = 0$.

(vi) $a_m(q) = a_n(q)$ *if and only if* $q^{m-n} = 1$.

Proof. The proofs of (i), (ii), (iii) and (vi) are straightforward computations. For (iv) note that

$$a_d(q) = \frac{q}{(1+q)} + \frac{1}{(1+q)(1+q+\ldots+q^{d-1})}$$

and that $a_{-d}(q) = 1 - a_d(q)$ by (i). For (v) we write $a_d(q) = (\sin(d+1)\, x\pi)/2\cos(x\pi)\, \sin(dx\pi)$. The function $t \mapsto (\sin(t + x\pi)/\sin(t)$ has the derivative $-(\sin(x\pi))/\sin^2(t)$, which is negative for $x \in [0, 1/2]$ and $t \in (0, \pi)$. Hence $a_d(q)$ is decreasing. The other statements follow easily from properties of the sine function. □

We are now going to define the above mentioned representations. Similarly as for the symmetric group, we can label them by Young diagrams. We mean by a Young diagram $\lambda = [\lambda_1, \lambda_2, \ldots, \lambda_k]$ an array of n boxes with λ_1 boxes in the first row, λ_2 boxes in the second row, and so on. The set of all Young diagrams with n boxes is denoted by Λ_n. As in [L], we will abbreviate s rows with the same number l of boxes by l^s (e.g. the Young diagram with all of its n boxes in the first column would be $[1^n]$). We shall write $\mu < \lambda$ if μ is obtained from λ by removing appropriate boxes. If μ has only one or 2 boxes less than λ, it will usually be denoted by λ' or λ'' respectively.

Let T_λ be the set of all standard tableaux belonging to λ and let T_n be the set of all standard tableaux with n boxes. If $\lambda \in \Lambda_n$ and $t \in T_\lambda$, we define t' to be the tableau t without the box containing n. Similarly, t'' is t without the boxes containing n and $n+1$.

There is an obvious action of S_n on tableaux, which permutes the numbers in the boxes. We extend this action to products of the generators g_i by identifying g_i with the transposition $(i, i+1)$. So $g_i(t)$ would be the tableau obtained from t by interchanging the numbers i and $i+1$.

Let V_λ be the complex vector space with basis $\{\vec{v}_t, t \in T_\lambda\}$. To simplify the proofs, we will embed V_λ in the larger vector space $\tilde{V}_\lambda$, whose basis is labelled by arbitrary tableaux of shape λ. Our representations will depend on the relative position of boxes containing successive numbers. If the numbers l and m are contained in the c_l-th box of the r_l-th row and in the c_m-th box of the r_m-th row of a tableau t respectively, we define

$$d_{t,l,m}=c_l-c_m+r_m-r_l.$$

If follows immediately from the definition that

$$d_{t,m,l}=-d_{t,l,m}.$$

For brevity, we will write $a_{t,i,j}$ for $a_{d_{t,i,j}}$. We define the linear map $\pi_\lambda(e_i)$ on the subspace of $\tilde{V}_\lambda$ spanned by vectors $\vec{v}_t$ for which $a_{t,i,i+1}(q)$ is well-defined by

$$\pi_\lambda(e_i)\,\vec{v}_t=a_{t,i,i+1}(q)\,\vec{v}_t+(a_{t,i,i+1}(q)\,a_{t,i+1,i}(q))^{1/2}\,\vec{v}_{g_i(t)}. \tag{2.3}$$

Observe that for any tableau t with at most n boxes, we have $0\leqq|d_{t,i,i+1}|<n$ for $1\leqq i<n$. So if q is not a k-th root of unity with $2\leqq k\leqq n-1$, $a_{t,i,i+1}(q)$ is always well-defined except if $d_{t,i,i+1}=0$. This can never happen if t is a standard tableau (see lemma (2.11) at the end of this section). So in this case $\pi_\lambda(e_i)$ is well-defined on V_λ for all $\lambda\in\Lambda_n$ and $i=1,2,\dots n-1$. To show semisimplicity of these representations, we will need a slightly stronger restriction. We will say that a complex number q is *n-regular*, if $q\neq 0$ and if q is not a k-th root of unity for $k=2,3,\dots n$. It follows from the remarks above that in this case $a_{t,i,i+1}(q)=0$ if and only if $d_{t,i,i+1}=-1$.

Next, we show that V_λ is a subspace of $\tilde{V}_\lambda$ which is invariant under all $\pi_\lambda(e_i)$. Indeed, it is easy to see that if t is a standard tableau, $g_i(t)$ is not a standard tableau only if i and $i+1$ are in the same row or in the same column. In these cases either $d_{t,i,i+1}=-1$ or $d_{t,i+1,i}=-d_{t,i,i+1}=-1$ which implies $a_{t,i,i+}(q)\,a_{t,i+1,i}(q)=0$ for any value of q. So the claim directly follows from the definition of $\pi_\lambda(e_i)$ in (2.3).

Let t be any arbitrary tableau for which $\pi_\lambda(e_i)$ is defined and let V be the span of $\vec{v}_t$ and $\vec{v}_{g_i(t)}$. Obviously, V is $\pi_\lambda(e_i)$-invariant. If $d=d_{t,i,i+1}$, then $d_{g_i(t),i,i+1}=-d$ and $\pi_\lambda(e_i)|_V$ is given by the matrix

$$\begin{pmatrix} a_d & (a_d\,a_{-d})^{1/2} \\ (a_d\,a_{-d})^{1/2} & a_{-d} \end{pmatrix} \tag{2.4}$$

Using Lemma 2.1 (i) we see that this matrix is an idempotent. This shows that $\pi_\lambda(e_i)$ satisfies relation $(H3)$.

If $|i-j|\geqq 2$, we have $g_i\,g_j(t)=g_j\,g_i(t)$ for any tableau t with n boxes. With this observation it follows from (2.3) that $\pi_\lambda(e_i)\,\pi_\lambda(e_j)\,\vec{v}_t=\pi_\lambda(e_j)\,\pi_\lambda(e_i)\,\vec{v}_t$ whenever all occurring coefficients are well-defined. So in this case π_λ also preserves relation $(H2)$.

$(H1)$ is more difficult to show. Our proof here is slightly different to the one in [W-2] (which is essentially the same as the ones in [O] and [H]). Let

$t \in T_n$ and let $i \in \mathbf{N}$, $i \leqq n-2$. Let also

$$k = d_{t,i,i+1}, \quad l = d_{t,i,i+2} \quad \text{and} \quad m = d_{t,i+1,i+2}.$$

We assume that $a_k(q)$, $a_l(q)$ and $a_m(q)$ are well-defined. If $p = g_{i_1} g_{i_2} \dots g_{i_j}$, we define $p^* = g_{i_j} g_{i_{j-1}} \dots g_1$. It is easy to check by induction on j that

$$d_{p(t),r,s} = d_{t,p^*(r),p^*(s)}.$$

We then obtain

$$\begin{aligned}
&\pi_\lambda(e_i)\,\pi_\lambda(e_{i+1})\,\pi_\lambda(e_i)\,\vec{v}_t \\
&= (a_k^2\, a_m + a_k\, a_{-k}\, a_l)\,\vec{v}_t + (a_k\, a_{-k})^{1/2}(a_k\, a_m + a_{-k}\, a_l)\,\vec{v}_{g_i(t)} \\
&\quad + a_k(a_m\, a_{-m})^{1/2}\, a_l\,\vec{v}_{g_{i+1}(t)} + a_m(a_k\, a_{-k})^{1/2}(a_l\, a_{-l})^{1/2}\,\vec{v}_{g_{i+1}g_i(t)} \\
&\quad + a_k(a_m\, a_{-m})^{1/2}(a_l\, a_{-l})^{1/2}\,\vec{v}_{g_i g_{i+1}(t)} + (a_k\, a_{-k})^{1/2}(a_m\, a_{-m})^{1/2}(a_l\, a_{-l})^{1/2}\,\vec{v}_{g_i g_{i+1} g_i(t)}.
\end{aligned}$$

For $\pi_\lambda(e_{i+1})\ \pi_\lambda(e_i)\ \pi_\lambda(e_{i+1})\,\vec{v}_t$, we obtain the same expression with i and $i+1$ and with k and m interchanged. It follows

$$\begin{aligned}
&(\pi_\lambda(e_i)\,\pi_\lambda(e_{i+1})\,\pi_\lambda(e_i) - \pi_\lambda(e_{i+1})\,\pi_\lambda(e_i)\,\pi_\lambda(e_{i+1}))\,\vec{v}_t \\
&= (a_k^2\, a_m + a_k\, a_{-k}\, a_l - a_m^2\, a_k - a_m\, a_{-m}\, a_l)\,\vec{v}_t \\
&\quad + (a_k\, a_{-k})^{1/2}(a_k\, a_m + a_{-k}\, a_l - a_m\, a_l)\,\vec{v}_{g_i(t)} \\
&\quad + (a_m\, a_{-m})^{1/2}(a_k\, a_l - a_m\, a_k - a_l\, a_{-m})\,\vec{v}_{g_{i+1}(t)}.
\end{aligned} \tag{$*$}$$

On the other hand we have

$$\begin{aligned}
&q/(1+q)^2(\pi_\lambda(e_i) - \pi_\lambda(e_{i+1}))\,\vec{v}_t \\
&= q/(1+q)^2((a_k - a_m)\,\vec{v}_t + (a_k\, a_{-k})^{1/2}\,\vec{v}_{g_i(t)} - (a_m\, a_{-m})^{1/2}\,\vec{v}_{g_{i+1}(t)}).
\end{aligned} \tag{$**$}$$

It follows from Lemma 2.1, (ii) that ($*$) and ($**$) are equal.

Note that if t is a standard tableau, $d_{t,i,j} \neq 0$ for all $1 \leqq i < j \leqq n$ with $|i-j| \leqq 2$ by Lemma 2.11. So if q is n-regular, our computation above always makes sense even if some of the tableaux occurring there should no longer be standard tableaux. Recall that V_λ is an invariant subspace of $\tilde{V}_\lambda$ for any value of q. So if $\pi_\lambda(e_i)$ just denotes the restriction to V_λ of the map defined in (2.3), it is well-defined everywhere and the map $e_i \to \pi_\lambda(e_i)$ for $i = 1, 2, \dots n-1$ preserves relations $(H1)-(H3)$. So π_λ is a representation of $H_n(q)$.

The following simple facts will be important in the proof of the next theorem. Observe that the map $t \mapsto t'$ defines a bijection between T_λ and $\bigcup_{\lambda' < \lambda} T_{\lambda'}$. So we obtain in particular

$$V_\lambda \cong \bigoplus_{\lambda' < \lambda} V_{\lambda'}. \tag{2.5}$$

A brief look at the definitions of π_λ and $\pi_{\lambda'}$ shows that the above equation gives us the decomposition of V_λ as an $H_{n-1}(q)$ module. So we have

$$\pi_\lambda|_{H_{n-1}(q)} \cong \bigoplus_{\lambda' < \lambda} \pi_{\lambda'}. \tag{2.6}$$

We can now determine the structure of $H_n(q)$ for all but finitely many values of q. For similar results see [Bou], p. 54, 55 and 56, [H] and [O].

Theorem 2.2. *Let $q \in \mathbf{C}$ be n-regular, i.e. $q \neq 0$ and q is not an l-th root of unity for $l=2, 3, \ldots, n$. Then π_λ, as defined in (2.3) is an irreducible representation of $H_n(q)$ on V_λ for all $\lambda \in \Lambda_n$. If $\mu \in \Lambda_n$, π_λ is isomorphic to π_μ if and only if $\lambda=\mu$. The map*

$$\pi_n \colon x \in H_n(q) \mapsto \bigoplus_{\lambda \in \Lambda_n} \pi_\lambda(x)$$

defines a faithful representation of $H_n(q)$. In particular, $H_n(q) \cong \mathbf{C}\, S_n$.

Proof. By induction assumption, $H_{n-1}(q)$ is a semisimple algebra with minimal central idempotents $z_{\lambda'}$, labeled by Young diagrams $\lambda' \in \Lambda_{n-1}$. Let $0 \neq W \subset V_\lambda$ be an $H_n(q)$ module. As V_λ decomposes as an $H_{n-1}(q)$ module into the direct sum of irreducible modules $V_{\lambda'}$ by (2.5), there exists a $\lambda' \in \Lambda_{n-1}$, $\lambda' < \lambda$ such that $V_{\lambda'} \subset W$. Let $\tilde{\lambda}' \neq \lambda'$ be another Young diagram with $n-1$ boxes such that $\tilde{\lambda}' < \lambda$. There is exactly one $\lambda'' \in \Lambda_{n-2}$ contained in both λ' and $\tilde{\lambda}'$. Let $t \in T_\lambda$ be such that $t' \in T_{\lambda'}$ and $t'' \in T_{\lambda''}$. Then $(g_{n-1}(t))' \in T(\tilde{\lambda}')$ and therefore

$$\pi_\lambda(z_{\tilde{\lambda}'})\, \pi_\lambda(e_{n-1})\, \vec{v}_t = (a_d\, a_{-d})^{1/2}\, \vec{v}_{g_{n-1}(t)} \in V_{\tilde{\lambda}'},$$

where $d = d_{t,n-1,n}$. As $1 \notin \{q^{d-1}, q^d, q^{d+1}\}$, $a_d\, a_{-d}$ is well-defined and nonzero. Hence the irreducible $H_{n-1}(q)$ module $V_{\tilde{\lambda}'}$, is contained in W. As $\tilde{\lambda}'$ was arbitrary, $W \supset \bigoplus_{\lambda' < \lambda} V_{\lambda'} = V_\lambda$.

Next we show that the V_λ's are mutually nonisomorphic $H_n(q)$ modules. This is obvious for $n=1$ and follows for $n=2$ from $\pi_{[2]}(e_1) = 1 \neq 0 = \pi_{[1^2]}(e_1)$. If $n > 2$ and λ, $\tilde{\lambda} \in \Lambda_n$ with $\lambda \neq \tilde{\lambda}$, one of these diagrams has a subdiagram which is not a subdiagram of the other one (see Lemma 2.11 at the end of this section). Hence, by (2.5) V_λ and $V_{\tilde{\lambda}}$ differ already as $H_{n-1}(q)$ modules.

Let ρ_λ be the representation of the symmetric group S_n over $\mathbf{C}$ corresponding to λ. It is well-known that $\rho_\lambda|_{S_{n-1}} \cong \bigoplus_{\lambda' < \lambda} \rho_{\lambda'}$. So $\dim(\rho_\lambda) = \dim(\pi_\lambda)$ by (2.6) and induction. Hence

$$\dim\Big(\bigoplus_{\lambda \in \Lambda_n} \pi_\lambda(H_n(q)\Big) = \dim \mathbf{C} S_n = n!.$$

This shows that π_n is faithful. □

Remarks. 1. If $q=1$, the representations defined above coincide with Young's orthogonal representations. So we can refer to them as the orthogonal representations for $H_n(q)$. The disadvantage of having square roots in this representation can be avoided by replacing them by *semi-orthogonal representations* in analogy to Young (see [H]). This can be used to show that the Hecke algebra $H_n(q)$ of type A_{n-1} over an arbitrary field k has the same matrix decomposition as the group algebra of S_n in characteristic 0 provided that $1+q+q^2+\ldots+q^r \neq 0$ for $r=1, 2, \ldots, n$.

2. It can be checked by straightforward computations that relations $(H1)-(H3)$ also hold for $1-e_i$ instead of e_i for $i=1,2,\ldots,n-1$. Using the same inductive procedure as in the proof of the last theorem, we can show that the representation $e_i \mapsto \pi_\lambda(1-e_i)$ is equivalent to π_{λ^*}, where λ^* is obtained from λ by interchanging rows with columns (e.g. $[1^n]^*=[n]$).

As a first consequence of theorem (2.2) we can inductively compute special minimal idempotents of $H_n(q)$. Let us denote the diagram belonging to the tableau t by $\lambda(t)$. Then with z_λ the minimal central idempotent belonging to $\pi_\lambda(H_n(q))$, we define for each standard tableau $t\in T_n$ elements p_t inductively by

$$p_{[1]}=1 \quad \text{and} \quad p_t=z_{\lambda(t)}\, p_{t'}. \tag{2.7}$$

It follows from (2.6) and induction that if $t_0\in T_n$, $\pi_n(p_{t_0})$ is a rank 1 projection with

$$\pi_n(p_{t_0})\Big(\sum_{t\in T_n} \alpha_t \vec{v}_t\Big)=\alpha_{t_0}\vec{v}_{t_0}.$$

Let $r\in T_{n-1}$. It follows similarly from this and (2.6) that

$$\pi_n(p_r)\Big(\sum_{t\in T_n} \alpha_t \vec{v}_t\Big)=\sum_{t\in T_n,\, t'=r} \alpha_t \vec{v}_t.$$

Let us abbreviate $d_{t,n-1,n}$ by d_t. We obtain from (2.3) and (2.6) that

$$\pi_n(p_r\, e_{n-1}\, p_r)\,\vec{v}_t=a_{d_t}\vec{v}_t \quad \text{if } t'=r. \tag{2.8}$$

To apply this formula we will first show that if $t_1, t_2\in T_n$ such that $t_1'=r=t_2'$ and if q is n-regular, then $a_{d_{t_1}}(q)=a_{d_{t_2}}(q)$ only if $t_1=t_2$. Assume that r belongs to the Young diagram $\lambda=[\lambda_1,\ldots\lambda_k]\in\Lambda_{n-1}$ and let t_1, t_2 be obtained by adding the box with n in the i_1-th and in the i_2-th row of r respectively. Then we have

$$|d_{t_1}-d_{t_2}|=|\lambda_{i_1}-\lambda_{i_2}+i_2-i_1|\leqq\lambda_1+k-2<n.$$

Hence $a_{t_1}(q)\neq a_{t_2}(q)$ by lemma (2.1), (vi) for q n-regular.

So we can compute $\pi_n(p_t)$ as the spectral projection of $\pi_n(p_r\, e_{n-1}\, p_r)$ corresponding to the eigenvalue a_{d_t}. This already shows part (d) of the following corollary by the faithfulness of π_n. Part (e) follows similarly from (2.8) while part (a) is a consequence of the formula after (2.7). The rest of the corollary can be shown by induction using (2.7). It follows

Corollary 2.3. *Let q be n-regular and let for each standard tableau $t\in T_n$ the minimal idempotent p_t be defined as in (2.7). Then we have*

(a) $p_t\, p_{\tilde{t}}=p_{\tilde{t}}\, p_t=0 \quad$ if $t\neq\tilde{t}$.

(b) $\sum_{t\in T_\lambda} p_t=z_\lambda$.

(c) $\sum_{t\in T_n} p_t=1$.

(d) *If* $t_0 \in T_n$ *and* $r = t'_0$, p_{t_0} *can be computed inductively by*

$$p_{t_0} = \prod_{t'=r,\, t \neq t_0} \frac{p_r e_{n-1} p_r - a_{d_t} p_r}{a_{d_{t_0}} - a_{d_t}}.$$

(e) $p_{t_0} e_n p_{t_0} = a_{d_{t_0}} p_{t_0}$.

We remark here that for $q=1$, these idempotents do not specialize to the well known Young idempotents but to a different class of minimal idempotents belonging to Young's seminormal (or orthogonal) representations (see [KJ], Theorem 3.2.14).

If q is a primitive l-th root of unity, $a_l(q)$ is not well-defined. We can, however, still define representations of the same type as above for special classes of Young diagrams.

Definition 2.4. Let $k, l \in \mathbf{N}$ with $k \leqq l-1$. A Young diagram λ is called a (k, l) *diagram* if it has at most k rows and $\lambda_1 - \lambda_k \leqq l-k$. Let $\Lambda_n^{(k,l)}$ denote the set of all (k, l) diagrams with n nodes.

Similarly as in the general case, we define for a (k, l) diagram λ the set $T_\lambda^{(k,l)} \subset T_\lambda$, consisting of tableaux $t \in T_\lambda$ such that $t' \in T_{\lambda'}^{(k,l)}$ for a (k, l) diagram $\lambda' \in \Lambda_{n-1}^{(k,l)}$.

Let $V_\lambda^{(k,l)}$ be the subspace of V_λ with basis $\{\vec{v}_t,\ t \in T_\lambda^{(k,l)}\}$. If q is a primitive l-th root of unity, we define the linear map $\pi_\lambda(e_i)$ as in (2.4) for all vectors $\vec{v}_t$ for which $d_{t,i,i+1}$ is not divisible by l. The restriction of this map to $V_\lambda^{(k,l)}$ will be denoted by $\pi_\lambda^{(k,l)}(e_i)$, i.e. we have for $t \in T_\lambda^{(k,l)}$ as in (2.3)

$$\pi_\lambda^{(k,l)}(e_i)\, \vec{v}_t = a_{t,i,i+1}\, \vec{v}_t + (a_{t,i,i+1}\, a_{t,i+1,i})^{1/2}\, \vec{v}_{g_i(t)}. \tag{2.9}$$

It remains to show that these linear maps are well-defined for each $t \in T_\lambda^{(k,l)}$. For $i < n-1$ note that as in (2.6)

$$\pi_\lambda^{(k,l)}(e_i) = \bigoplus_{\lambda' < \lambda} \pi_{\lambda'}^{(k,l)}(e_i). \tag{2.10}$$

So by induction assumption $\pi_\lambda^{(k,l)}(e_i)$ is well-defined.

For $i = n-1$ observe first that for any (k, l) tableau $t \in T_n$ belonging to the (k, l) diagram λ

$$|d_{t,n-1,n}| \leqq \lambda_1 - \lambda_k + k - 1 < l. \tag{2.11}$$

Hence the coefficients occuring in the definition of $\pi_\lambda^{(k,l)}(e_{n-1})\, \vec{v}_t$ are well-defined for each $t \in T(\lambda)^{(k,l)}$.

To show that $\pi_\lambda^{(k,l)}$ is a representation, we first prove that $V_\lambda^{(k,l)}$ is an invariant subspace of V_λ. So let t be a (k, l) tableau such that $g_{n-1}(t)$ is standard but not a (k, l) tableau. By (2.11) this can only happen if the 2 boxes containing $n-1$ and n are in the first and in the last row respectively and if $d_{t,n-1,n} = 1-l$. But in this case $a_{g_{n-1}(t),n-1,n} = a_{l-1} = 0$ and $\pi_\lambda^{(k,l)}(e_{n-1})\, \vec{v}_t = a_{1-l}\, \vec{v}_t = \vec{v}_t$.

To be able to use the computation for checking $(H1)$ before Theorem 2.2, we have to make sure that all coefficients there are well-defined even if some of the tableaux occurring there are not (k, l) tableaux. Again by (2.10), we only need to consider the case with e_{n-2} and e_{n-1}. By (2.11) the only problem could occur if $|d_{t,n-2,n}|=l$. But as $\lambda(t')$ is also a (k,l) diagram, this could only happen if the box containing $n-2$ is in the last row next to the last box there and the box containing n is at the end of the first row. But then it is easy to see that t''' would no longer be a (k,l) tableau.

With this observation the proof that $\pi_\lambda^{(k,l)}$ is a representation is the same as for the regular case. We thus obtain

Corollary 2.5. *Let q be a primitive l-th root of unity with $l \geqq 4$ and let $\Lambda_n^{(l)}$ denote the collection of all (k,l) diagrams for $k=1,2,\ldots l$-1. Then there exists for every $\lambda \in \Lambda_n^{(l)}$ a semisimple irreducible representation $\pi_\lambda^{(l)}$ of $H_n(q)$. If λ is a (k,l) diagram for some $k \in \mathbf{N}, k \leqq l-1$, $\pi_\lambda^{(l)}$ coincides with $\pi_\lambda^{(k,l)}$ as defined in (2.9). So we obtain a in general not faithful, semisimple representation*

$$\pi_n^{(l)}: x \in H_n(q) \to \bigoplus_{\lambda \in \Lambda_n^{(l)}} \pi_\lambda^{(l)}(x).$$

Representations belonging to different (l) diagrams are not equivalent.

Proof. We have already shown that $\pi_n^{(l)}$ is a representation. Assume that a Young diagram $\lambda \in \Lambda_n$ is both a (k_1,l) and a (k_2,l) diagram with $k_1 > k_2$. Then $\lambda_{k_1}=0$ by definition of (k_2,l) diagram. It follows that $\lambda_1 < l-k_1 < l-k_2$. Hence in particular, any subdiagram of λ is both a (k_1,l) and a (k_2,l) diagram. But then $\pi_\lambda^{(k_1,l)} = \pi_\lambda^{(k_2,l)}$ by (2.10) and induction assumption for $H_{n-1}(q)$.

The irreducibility can be shown as in Theorem 2.2 by observing that $a_{t,i,i+1}(q)\, a_{t,i+1,i}(q) \neq 0$ for all $t \in T_\lambda^{(k,l)}$ such that also $g_i(t) \in T_\lambda^{(k,l)}$. The fact that representations belonging to different (k,l) diagrams are not equivalent is also shown as in Theorem 2.2 using Lemma 2.11 at the end of this section. □

Let us denote the minimal central projections of $\pi_n^{(l)}(H_n(q))$ by $z_\lambda^{(l)}$ with $\lambda \in \Lambda_n^{(l)}$. Similarly as in (2.7), we define minimal idempotents $p_t^{(l)}$ of $\pi_n^{(l)}(H_n(q))$, labelled by the elements of $T_n^{(l)}$ by

$$p_{[1]}^{(l)} = 1 \quad \text{and} \quad p_t^{(l)} = z_{\lambda(t)}^{(l)} p_{t'}^{(l)}.$$

Note that these idempotents are only defined for the image of $H_n(q)$ under the representation $\pi_n^{(l)}$. As this representation is not faithful in general, there is no obvious interpretation for them in $H_n(q)$ itsself. Otherwise we can prove for these idempotents the same properties (with exponent $^{(l)}$ added to p_t, T_n and z_λ) as stated for the p_t's in corollary (2.3).

Let us fix $k, l \in \mathbf{N}$ *with* $l \geqq 4$ and $1 \leqq l-1$ and let for $q=e^{2\pi i/l}$

$$B_n = \bigoplus_{\lambda \in \Lambda_n^{(k,l)}} \pi_\lambda^{(k,l)}(H_n(q)). \tag{2.12}$$

By definition of the $\pi_\lambda^{(k,l)}$'s, the restriction of this representation to $H_{n-1}(q)$ is isomorphic to B_{n-1}. With this identification we can define the representation $\pi^{(k,l)}: H_\infty(q) \to B_\infty$ of the corresponding inductive limites by

$$\pi^{(k,l)}(x) = \bigoplus_{\lambda \in \Lambda_n^{(k,l)}} \pi_\lambda^{(k,l)}(x) \quad \text{for } x \in H_n(q). \tag{2.13}$$

Let $n<m$ and let $G_{n,m}$ be the inclusion matrix for $B_n \subset B_m$. By (2.10), the inclusion matrix $G_{n-1,n}=(g_{\lambda',\lambda})$, where $\lambda' \in \Lambda_n^{(k,l)}$, is given by

$$g_{\lambda',\lambda}=\begin{cases}1 & \text{if } \lambda'<\lambda \\ 0 & \text{if } \lambda' \nless \lambda\end{cases} \tag{2.14}$$

It is easy to show by induction on $m-n$ that for $\mu \in \Lambda_n$ and $\lambda \in \Lambda_m$ the entry $g_{\mu,\lambda}$ is positive if and only if $\mu<\lambda$. Our aim is to show that the inclusion diagrams for the B_n's eventually become periodic with period k. To do so, we first need some combinatorial results.

Lemma 2.6. *Let* $F\colon \bigcup_{n\in\mathbf{N}} \Lambda_n^{(k,l)} \to \bigcup_{n\in\mathbf{N}} \Lambda_n^{(k,l)}$ *be the map that adds a column of* k *nodes on the left hand side of a* (k,l) *diagram* (e.g. *for* $k=3$ *we have* $F([21])=[321]$).

(i) *If* $\lambda \in \Lambda_n^{(k,l)}$ *and* $n>(l-1)(k-1)+sk$, *then* $\lambda_k \geqq s+1$.
(ii) *If* $n \geqq n_0=(l-k-1)(k-1)$, *then* F *restricted to* $\Lambda_n^{(k,l)}$ *is a bijection between* $\Lambda_n^{(k,l)}$ *and* $\Lambda_{n+k}^{(k,l)}$.
(iii) F *preserves the inclusion pattern, i.e.* $F(\lambda')<F(\lambda)$ *if and only if* $\lambda'<\lambda$.
(iv) *If* $s>2l$, $\mu \in \Lambda_n^{(k,l)}$ *and* $\lambda \in \Lambda_{n+sk}^{(k,l)}$, *then* $\mu<\lambda$.

Proof. (i) Assume $\lambda_k \leqq s$. As λ is a (k,l) diagram, $\lambda_1 \leqq s+l-k$ and hence also $\lambda_r \leqq s+l-k$ for $r=2, 3, \ldots k-1$. But then $n \leqq (s+l-k)(k-1)+s=(l-k)(k-1)+sk$.

(ii) Obviously F is injective. Moreover, as $n+k \geqq (l-k-1)(k-1)+k=(l-k)(k-1)+1$, each $\lambda \in \Lambda_{n+k}^{(k,l)}$ has at least one node in its last row, i.e. $\lambda \in F(\Lambda_n^{(k,l)})$.

(iii) This is clear.

(iv) It is enough to show that $\lambda_k \geqq \mu_1$. If this were not the case, $\lambda_r \leqq \lambda_k+l-1<\mu_1+l-1$ and

$$\lambda_r \leqq \lambda_1<\mu_1+l-1 \leqq \mu_r+2(l-1) \quad \text{for } r=1,2,3,\ldots,k.$$

But then $sk<2k(l-1)$.

Corollary 2.7. *Let* B_n *be as in* (2.12) *and let* n_0 *be as in Lemma* 2.6 (i). *Then there exists a primitive matrix* D *such that if* $n>n_0$, $G_{n,n+rk}=D^r$ *for all* $r\in\mathbf{N}$.

Proof. By Lemma 2.6 (ii), F induces a bijection between the minimal central projections of B_n and B_{n+k} and between the minimal central projections of B_{n+1} and B_{n+k+1}. By (2.14) and (iii), $g_{\lambda',\lambda}=g_{F(\lambda'),F(\lambda)}$ for $\lambda' \in \Lambda_n^{(k,l)}$ and $\lambda \in \Lambda_{n+1}^{(k,l)}$. Hence $G_{n,n+1}$ can be identified with $G_{n+k,n+k+1}$ under F. As $G_{n,n+k}=G_{n,n+1}\,G_{n+1,n+2}\cdots G_{n+k-1,n+k}$ (see remarks after (1.4)) and as a similar computation holds for $G_{n+k,n+2k}$, we can also identify $G_{n,n+k}$ and $G_{n+k,n+2k}$ in the same way. As $n \geqq n_0$, $D=G_{n,n+k}$ is a square matrix which has only nonnegative entries and $G_{n,n+rk}=D^r$. By the remark after (2.14) and (iv) $D^s=G_{n,n+sk}$ only has positive entries for $s>2l$. Hence D is primitive. □

For our considerations so far we could have used the semi-orthogonal representations as well. This would have had the advantage of not having to use square roots. We will need the orthogonal representations in the following when positivity will play an important role. So from now on V_λ is assumed to be a complex Hilbert space with orthonormal basis $\{\vec{v}_t,\ t\in T_\lambda\}$. It will be essential for the construction of subfactors that for a representation ρ of $H_n(q)$ also subalgebras of the form $\rho(\langle e_m, e_{m+1}, \ldots, e_{n-1}\rangle)$ are C^* algebras.

Definition 2.8. A representation ρ of $H_n(q)$ or $H_\infty(q)$ on a Hilbert space is called a C^* representation if $\rho(e_i)$ is a selfadjoint projection for $i=1, 2, \ldots, n-1$ or for all $i\in\mathbf{N}$ respectively.

Note that $\pi_{[n]}(e_i)=0$ and $\pi_{[1^n]}(e_i)=1$ for $n\in\mathbf{N}$ and $i=1, 2, \ldots, n-1$. Obviously these are C^* representations. Any C^* representation isomorphic to a direct sum of such representations will be called a trivial C^* representation. To characterize nontrivial C^* representations, we say that a Young diagram $\lambda\in\Lambda_n$ is an m-diagram if $\lambda_1+k<m+2$, where k is the number of rows of λ. Note that this equivalent to $|d_{t,i,j}|<m$ for all $t\in T_\lambda$ and $1\leqq i,j\leqq n$.

For the proof of the next proposition we will use the following property of $H_\infty(q)$. Let

$$\Delta_{m,n}=(g_m g_{m+1}\cdots g_{n-1})(g_m g_{m+1}\cdots g_{n-2})\cdots(g_m g_{m+1})\, g_m.$$

By [B], Lemma (2.5.1) we have for $m\leqq i\leqq n-1$ that $\Delta_{m,n}\, g_i\, \Delta_{m,n}^{-1}=g_{m+n-i-1}$. Hence conjugation by $\Delta_{m,n}$ yields an inner automorphism $\nu_{m,n}$ of $H_\infty(q)$ or $H_k(q)$, $k\geqq n$ with $\nu_{m,n}(e_i)=e_{m+n-i-1}$ for $m\leqq i<n$ and with $\nu_{m,n}(e_i)=e_i$ for $i\leqq m-2$ or $i\geqq n+1$. We use this to show by induction on i that if ρ is a representation of $H_n(q)$ with $\rho(e_1)=\rho(e_2)$, then

$$\rho(e_i)=\rho(e_1) \qquad \text{for } i=1, 2, \ldots n-1. \tag{2.15}$$

Indeed, this is trivially true for $i=2$. Assume it is true for i and $i-1$. Using the fact that $\eta_{i-1,i+2}$ is inner we have

$$\rho(e_{i+1})=\rho(\nu_{i-1,i+2}(e_{i-1}))=\rho(\nu_{i-1,i+2}(e_i))=\rho(e_i).$$

We can now use the positivity trick of [W-1] to find necessary conditions for the existence of C^* representations of $H_n(q)$ (see also [O]).

Proposition 2.9. *There are nontrivial C^* representations of $H_n(q)$ only if q is real and not negative or if $|q|=1$. Let in the second case $q=e^{2\pi i x}$ with $0<|x|<1/2$ and assume that there is an $m\in\mathbf{N}$ such that $m-1<|1/x|<m$. Then any irreducible C^* representation of $H_n(q)$ is equivalent to π_λ for some m-diagram in Λ_n. In particular, there are no nontrivial C^* representations of $H_n(q)$ for $n>((m+1)/2)^2$.*

Proof. Let ρ be an irreducible C^* representation of $H_n(q)$. Note that $e_1=p_{[1^2]}$. So it follows from corollary (2.3) that the only possible spectral values of the positive element $\rho(e_1 e_2 e_1)$ are 0, 1 and $q/(1+q)^2$. If $q/(1+q)^2$ is not a spectral value, $\rho_{|H_3(q)}$ would only contain the trivial representations $\pi_{[2]}$ and $\pi_{[1^2]}$. But then $\rho(e_1)=\rho(e_2)$ by definition of these representations. It follows from (2.15) that $\rho(e_i)=\rho(e_1)$ for $i=1, 2, \ldots n-1$. As ρ is irreducible, it has to be one dimensional, i.e. it has to be a trivial C^* representation. So if ρ is a nontrivial C^* representation, $q/(1+q)^2$ has to be a nonnegative spectral value of the positive element $\rho(e_1 e_2 e_1)$. This implies that q is real and not negative or that $|q|=1$.

To show the statement for the second case, we proceed by induction on n. By induction assumption, $\rho(H_{n-1}(q))$ decomposes into a direct sum of irreducible C^* representations of $H_{n-1}(q)$ which are isomorphic to representations labeled by m-diagrams with $n-1$ boxes. Let $\lambda\in\Lambda_{n-1}$ be such a diagram and let $r\in T_\lambda$ with p_r as in (2.7). Again by corollary (2.3), the spectral values of the positive element $\rho(p_r e_{n-1} p_r)$ are contained in the set $\{a_{t,n-1,n},\ t\in T_n,\ t'=r\}\cup\{0\}$.

Now every tableau t with $t'=r$ belongs to an $(m+1)$-diagram μ. As $m-1<x<m$, $q=e^{2\pi ix}$ can not be an l-th root of unity for $|l|\leqq m+1$. Hence p_t is well-defined for all $t\in T_\mu$ by Lemma 2.1 (vi) and Corollary 2.3 and hence also z_μ. If μ is not an m-diagram, there is a tableau $t\in T_\mu$ such that $d_{t,n-1,n}=m$. But then $\rho(p_t e_{n-1} p_t)=a_m(q)\,\rho(p_t)$ and $\rho((p_t e_{n-1} p_t))^2=a_m(q)\,\rho(p_t e_{n-1} p_t)$. By lemma (2.1), (v) $a_m(q)<0$. Hence $\rho(p_t)$ can only be positive if $\rho(p_t)=0$. But then also the well-defined central idempotent z_λ is in the kernel of ρ.

It is not hard to see that for any $n>((m+1)/2)^2$ and any nontrivial Young diagram $\lambda\in\Lambda_n$ with $k\geqq 2$ rows we have $\lambda_1+k\geqq m+2$. □

Together with the results of the last proposition the following proposition gives a complete classification of all irreducible C^* representations of $H_n(q)$ except for $q=e^{\pm 2\pi i/l}$ with $l\in\mathbf{N}$, $l\geqq 4$ and for $q=0$.

Proposition 2.10. *If q is real and positive, there is a faithful C^* representation of $H_n(q)$ for all $n\in\mathbf{N}$. If $q=e^{2\pi ix}$ with $|x|<1/2$ and $m\in\mathbf{N}$ with $m<|1/x|<m+1$, π_λ is a C^* representation for all m-diagrams $\lambda\in\Lambda_n$. If $q=e^{\pm 2\pi i/l}$ for $l\in\mathbf{N}$ with $l\geqq 4$, $\pi_\lambda^{(l)}$ is a C^* representation for all (l) diagrams λ.*

In particular, there exist nontrivial C^* representations of $H_\infty(q)$ if q is real and positive of if $q=e^{2\pi i/l}$ with $l\in\mathbf{N}$, $l\geqq 4$.

Proof. It is enough to check that the matrix blocks given in (2.4) are selfadjoint. It is easy to see that this happens if and only if both $a_d\geqq 0$ and $a_{-d}\geqq 0$. This is always the case if q is real and positive. If $q=e^{\pm 2\pi ix}$ with $|x|<1/2$ and $m\in\mathbf{N}$ with $m<|1/x|<m+1$, the positivity condition holds if $|d|\leqq m$ by Lemma 2.1. If λ is an m-diagram, this is the case for any $t\in T_\lambda$ and $d=d_{t,i,i+1}$ for $i=1,2,\ldots,n-1$.

The same argument goes through for $q=e^{\pm 2\pi i/l}$ and t belonging to a (k,l) tableau by definition of the representations $\pi_\lambda^{(l)}$. If $t\in T_\lambda^{(k,l)}$ and $g_i(t)$ is not in $T_\lambda^{(k,l)}$, $\pi_\lambda^{(k,l)}(e_i)\,\vec{v}_t=v\,\vec{v}_t$ with $v\in\{0,1\}$. So also in this case we obtain a self-adjoint matrix. □

The following lemma proves several simple combinatorial results which are needed in the proofs of Theorem 2.2 and Corollary 2.5.

Lemma 2.11. (a) *Let $t\in T_n$ be a standard tableau and let $1\leqq i,j\leqq n$. If $0<|i-j|\leqq 2$, then $d_{t,i,j}\neq 0$.*

(b) *Let λ and $\tilde{\lambda}$ be 2 distinct Young diagrams with n boxes, $n\geqq 3$. Then there is for at least one of them a subdiagram with $n-1$ boxes which is not a subdiagram of the other one. The same statement is true if the given diagrams are (k,l) diagrams with $l\geqq 5$ and if we also require the subdiagram to be a (k,l) diagram.*

(c) *Representations belonging to different* (4) *diagrams are not isomorphic.*

Proof. (a) As $d_{t,j,i}=-d_{t,i,j}$, we can assume $i<j$. If $d_{t,i,j}=0$, the box containing j has to be both to the right of and below the box containing i. But if t is standard, there have to be at least 2 boxes in t which contain numbers between i and j (see Fig. 1).

(b) Let $\lambda=[\lambda_1, \dots \lambda_r]$ and let $\tilde{\lambda}=[\tilde{\lambda}_1, \dots \tilde{\lambda}_r]$. Let s be the smallest index for which $\lambda_s \neq \tilde{\lambda}_s$ and let without restriction of generality $\lambda_s<\tilde{\lambda}_s$. If $s \neq 1$, take the subdiagram of λ obtained by taking away the last box of the $(s-1)$-th row.

If $s=1$ and $\tilde{\lambda}$ has more than 1 row, take any subdiagram of $\tilde{\lambda}$ which can be obtained by taking away a box from a row other than the first one. Otherwise $\tilde{\lambda}=[n]$, $n \geqq 3$ and $\lambda=[n-1, 1]$. In this case we take the subdiagram $[n-2, 1]$ of λ.

For the (k,l) case, the same strategy works except if $\tilde{\lambda}=[s, r^{k-1}]$, $\lambda=[s-1, r+1, r^{k-2}]$ and $s-r=l-k$. If $k>2$, we can choose the diagram $[s-1, r+1, r^{k-3}, r-1]$, while if $k=2$ and $s \geqq r+3$, we can take $[s-2, r+1]$. Otherwise, it follows from $s-r \leqq l-k$ that $l \leqq 4$.

(c) If $l=4$, there is only one (1, 4) and one (3, 4) diagram with n boxes for $n \geqq 2$, which label the trivial representations of $H_n(i)$. It is easy to check that if $n=2r+1$ is odd, the only (2, 4) diagram with n boxes is $[r+1, r]$ and if $n=2r$ is even, the only (2, 4) diagrams are $[r, r]$ and $[r+1, r-1]$. It follows as in theorem (2.2) that $\pi_{[1,1]}^{(2,4)} \not\cong \pi_{[2]}^{(2,4)}$. It is easy to check by (2.9) that the representation $x \in H_{n-2}(i) \mapsto \pi_{[r,r]}^{(2,4)}(e_{n-1}x)$ is isomorphic to $\pi_{[r-1,r-1]}^{(2,4)}$ and the representation $x \in H_{n-2}(i) \mapsto \pi_{[r+1,r-1]}^{(2,4)}(e_{n-1}x)$ is isomorphic to $\pi_{[r,r-2]}^{(2,4)}$. So $\pi_{[r,r-2]}^{(2,4)} \not\cong \pi_{[r+1,r-1]}^{(2,4)}$ by induction assumption. □

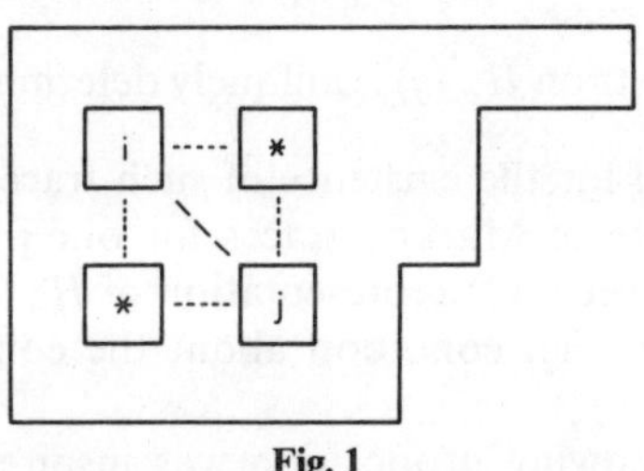

Fig. 1

§3. Markov traces for Hecke algebras

In the last section, we constructed C^* algebras from the sequence $(H_n(q))$ of Hecke algebras. In this section we compute special traces which will make these C^* algebras II_1 factors. Moreover, they will also satisfy the crucial condition of §1 for the construction of subfactors which involves the conditional expectations of various subalgebras.

These traces were found by A. Ocneanu in connection with his study of link invariants (see [O]). He also found explicit formulas for the weight vectors of these traces in terms of Schur functions if $H_n(q)$ is semisimple and determined the values of the parameters for which there is a positive Markov trace on a C^* representation of $H_\infty(q)$. We will give complete proofs of the results announced in [O] as far as necessary for our purposes.

In the following a trace tr on $H_\infty(q)$ will always mean a linear functional $\mathrm{tr}: H_\infty(q) \to \mathbf{C}$ such that $\mathrm{tr}(ab) = \mathrm{tr}(ba)$ for all $a, b \in H_\infty(q)$ and $\mathrm{tr}(1) = 1$.

Definition 3.1. A trace tr on $H_\infty(q)$ is said to be a Markov trace if there is an $\eta \in \mathbf{C}$ such that $\mathrm{tr}(xe_n) = \eta\, \mathrm{tr}(x)$ for all $n \in \mathbf{N}$ and $x \in H_n(q)$.

Recall that the first 2 defining relations of the Hecke algebra $H_n(q)$ are identical with the ones for the braid group B_n. The name for traces as in definition (3.1) comes from the fact that suitably normalized Markov traces stay invariant under the Markov moves for closed braids (see [J-4]).

Let $m < n$ and let $H_{m,n}(q)$ be the subalgebra of $H_n(q)$ generated by 1, e_m, $e_{m+1}, \ldots e_{n-1}$. We will show for any Markov trace tr on $H_\infty(q)$ that

$$\mathrm{tr}(xy) = \mathrm{tr}(x)\,\mathrm{tr}(y) \quad \text{for } x \in H_m(q) \text{ and } y \in H_{m,n}(q). \tag{3.1}$$

Indeed, this is trivially true for $m = n$. Assume that (3.1) is true for a given $n \geqq m$. It follows from (2.2) that every element $y \in H_{m,n+1}(q)$ can be written as a sum of elements of the form ae_nb or c with a, b, $c \in H_{m,n}(q)$. So the claim follows from

$$\mathrm{tr}(xae_nb) = \eta\, \mathrm{tr}(xab) = \mathrm{tr}(x)\, \eta\, \mathrm{tr}(ab) = \mathrm{tr}(x)\, \mathrm{tr}(ae_nb).$$

It follows immediately from (2.1) that there exists a linear basis of $H_n(q)$ consisting either of elements of $H_{n-1}(q)$ or of elements of the form $ae_{n-1}b$ with $a, b \in H_{n-1}(q)$. Hence it follows by induction on n that

$$\text{A Markov trace tr on } H_\infty(q) \text{ is uniquely determined by } \eta = \mathrm{tr}(e_1). \tag{3.2}$$

An elementary proof for the existence of such traces for every $\eta \in \mathbf{C}$ is given in [J-4]. The importance of Markov traces for our purposes comes from the fact that if they factor over a C^* representation of $H_\infty(q)$ they are factor traces, which also satisfy the crucial condition about the conditional expectations in §1.

The proof of the following proposition was inspired by the proof of [J-1], Corollary 4.1.8. We will also need the inner automorphisms $v_{n,m}$ of $H_\infty(q)$, defined in the last section. Recall that $v_{n,m}(e_i) = e_{m+n-i-1}$ for $n \leqq i < m$ and $v_{n,m}(e_i) = e_i$ for $i \leqq n-2$ or $i \geqq m+1$.

Proposition 3.2. *Any positive Markov trace* tr *which factors over a C^* representation ρ of $H_\infty(q)$, is a factor trace. Let $B_n = \rho(H_n(q))$ and let $A_n = \mathbf{C}$ for $n \leqq m$ and $A_n = \rho(H_{m,n}(q)) = \rho(\langle e_{m+1}, e_{m+2}, \ldots e_{n-1}\rangle)$ for $n > m$. We also assume that* tr *is faithful. Then we have for the conditional expectations onto these subalgebras*

$$\varepsilon_{A_{n+1}}\varepsilon_{B_n} = \varepsilon_{A_n}.$$

Proof. Let ρ be a C^* representation of $H_\infty(q)$. It is enough to show that any normalized trace $\widetilde{\mathrm{tr}}$ on $\rho(H_\infty(q))$ which is absolutely continuous with respect to tr is equal to tr. Let $x \in \rho(H_n(q))$ for some $n \in \mathbf{N}$. We regard x as an element of the subalgebra of $\rho(H_\infty(q))$ generated by $\{\rho(H_{(k-1)n+1,kn}(q)),\ k \in \mathbf{N}\}$

$=\{\rho(v_{kn}(H_n(q))), k\in\mathbf{N}\}$. Note that this algebra is isomorphic to $\otimes^\infty \rho(H_n(q))$ by $(H2)$. As v_{nk} is an inner automorphism for all $k\in\mathbf{N}$ the restriction of tr to this subalgebra can be regarded as a symmetric trace, i.e. $\operatorname{tr}(\otimes a_j)=\Pi\operatorname{tr}(a_j)$ for $a_j\in H_n(q)$ and $j\in\mathbf{Z}$.

Hence tr is an extremal trace on $\otimes^\infty \rho(H_n(q))$ by [St], Theorem (2.8) and therefore tr and $\widetilde{\operatorname{tr}}$ coincide on this subalgebra. In particular, $\operatorname{tr}(x)=\widetilde{\operatorname{tr}}(x)$.

To prove the second statement, recall that for $n>m$ any element $x\in A_{n+1}=\rho(H_{m,n+1}(q))$ can be written as a linear combination of elements of the form $a\rho(e_n)b$ or c with $a, b, c\in A_n$ (see (2.2)). If $y\in B_n$ and $x\in A_{n+1}$ such that $x=a\rho(e_n)\,b$ with $a, b\in A_n$ we obtain

$$\begin{aligned}\operatorname{tr}(x\varepsilon_{A_{n+1}}(y))&=\operatorname{tr}(xy)=\operatorname{tr}(a\rho(e_n)\,by)=\eta\operatorname{tr}(aby)=\eta\operatorname{tr}(ab\,\varepsilon_{A_n}(y))\\&=\operatorname{tr}(a\rho(e_n)\,b\varepsilon_{A_n}(y))=\operatorname{tr}(x\varepsilon_{A_n}(y)).\end{aligned}$$

This shows the second statement by (1.1). □

We will determine all positive Markov traces on $H_\infty(q)$ and we will compute their weight vectors.

Proposition 3.3. (a) *Let q not be a root of unity and let* tr *be a trace on $H_\infty(q)$ such that its restriction to $H_n(q)$ is given by the weight vector $(w_\lambda)_{\lambda\in\Lambda_n}$. Then* tr *is a Markov trace with* $\operatorname{tr}(e_1)=\eta$ *if and only if*

$$\sum_{t\in T_{n+1},\, t'=r} a_{t,n,n+1}\, w_{\lambda(t)}=\eta\, w_\mu$$

for all $r\in T_\mu$ and $\mu\in\Lambda_n$ (where again $\lambda(t)$ denotes the Young diagram belonging to t).

(b) *Let q be a primitive l-th root of unity and let* tr *be a trace factoring over $\pi^{(l)}$ (as defined in Corollary 2.5) such that its restriction to $\pi^{(l)}(H_n(q))$ is given by the weight vector $(w_\lambda)_{\lambda\in\Lambda_n^{(l)}}$. Then* tr *is a Markov trace with* $\operatorname{tr}(e_1)=\eta$ *if and only if*

$$\sum_{t\in T^{(l)}_{n+1},\, t'=r} a_{t,n,n+1}\, w_{\lambda(t)}=\eta\, w_\mu$$

for all $r\in T^{(l)}_\mu$ and $\mu\in\Lambda^{(l)}_n$

Proof. (a) We will use the set $\{p_r, r\in T_n\}$ of minimal idempotents as defined in (2.7). As $\sum_{r\in T_n} p_r=1$ by Corollary 2.3, tr being a Markov trace is equivalent to $\operatorname{tr}(e_n xp_r)=\eta\operatorname{tr}(xp_r)$ for all $r\in T_n$ and $x\in H_n(q)$. As p_r is a minimal projection in $H_n(q)$, there is an $\alpha(x)\in\mathbf{C}$ such that $p_r xp_r=\alpha(x)\,p_r$. So if $r\in T_\mu$, we have

$$\operatorname{tr}(xp_r)=\alpha(x)\,w_\mu. \tag{$*$}$$

If $t\in T_{n+1}$ such that $t'=r$, it follows from (2.3) and (2.7) that $e_n\,p_t=(p_t+p_{g_n(t)})\,e_n\,p_t$. As $g_n(t)'\in T_{\tilde\mu}$ with $\tilde\mu\neq\mu$, we have by (2.6) $z_\mu\, p_{g_n(t)}=z_\mu\, z_{\tilde\mu}\, p_{g_n(t)}=0$. So we obtain

$$p_r\, x e_n\, p_r=p_r\, xz_\mu\, e_n\, p_r=\sum_{t\in T_{n+1},\, t'=r} p_r\, xp_r\, p_t\, e_n\, p_t=\sum_{t\in T_{n+1},\, t'=r} a_{t,n,n+1}\,\alpha(x)\, p_t.$$

It follows

$$\mathrm{tr}(e_n x p_r) = \sum_{t \in T_{n+1}, t'=r} a_{t,n,n+1}\, \alpha(x)\, w_{\lambda t)}.$$

Comparing this with (*) we see that $\mathrm{tr}(e_n x p_r) = \eta\, \mathrm{tr}(x p_r)$ is equivalent to the formula in the statement.

For the proof of (b) we use the set of idempotents $p_t^{(l)}$ as defined after Corollary 2.5 in the same way as we used the idempotents p_t in (a). The rest of the proof goes exactly as in (a). □

Similarly as with class functions on the infinite symmetric group (see [Th] and [Wa]), the trace vectors for the Hecke algebras are expressed by Schur functions.

For our purposes, Schur functions are defined conveniently in the following way (for the general background see [Mc]): Let $h_1, h_2, \ldots$ be algebraically independent variables. If $\lambda = [\lambda_1, \ldots \lambda_m]$ is a Young diagram, we define $h_\lambda = \prod_{i=1}^{m} h_{\lambda_i}$.

Let moreover for an m-tupel $(a_1, \ldots a_m)$ and for $1 \leqq i < j \leqq m$ the raising operator R_{ij} be defined by

$$R_{ij}(a_1, \ldots a_m) = (\ldots a_i + 1, \ldots a_j - 1, \ldots a_m).$$

Then we define for each Young diagram $\lambda = [\lambda_1, \ldots \lambda_m]$ the function $s_\lambda(h_1, h_2, \ldots)$ by

$$s_\lambda = \prod_{1 \leqq i < j \leqq m} h_\lambda - h_{R_{ij}(\lambda)}. \tag{3.3}$$

In this context we will call any set $\{s_\lambda\}$ of functions which are obtained from the ones defined above for any special choice of the variables $h_1, h_2, \ldots$ a set of *Schur function*. Obviously, any algebraic relation between the functions defined in (3.3) will also hold in general for any set of Schur functions. In particular, if $\mu \in \Lambda_{n-1}$, we have as a special case of the Littlewood-Richardson rule (see [Mc], (9.2))

$$s_{[1]}\, s_\mu = \sum_{\lambda \in \Lambda_n, \lambda > \mu} s_\lambda. \tag{3.4}$$

We will apply this for the following case: Let

$$H(t) = \prod_{i=0}^{\infty} \frac{1 - b q^i t}{1 - a q^i t}.$$

We obtain functions h_r, $r \in \mathbf{N}$, depending on a, b and q by writing $H(t) = \sum_{r=0}^{\infty} h_r t^r$.

To express the corresponding Schur functions we need the following notations: We denote a box of a Young diagram λ by a pair (i, j), indicating the row and column of that box. Recall that λ_j^* denotes the number of boxes in the

j-th column of λ. It follows from [Mc], I. §3, Ex. 3 (following [L], p. 125) that

$$s_\lambda(a,b,q)=q^{n(\lambda)}\prod_{(i,j)\in\lambda}\frac{a-bq^{j-i}}{1-q^{h(i,j)}}, \tag{3.5}$$

where $n(\lambda)=\sum_{i\geq 1}(i-1)\lambda_i$ and $h(i,j)=\lambda_i-i+\lambda_j^*-j+1$ is the length of the hook through (i,j). For the special choice $a=1-\eta(1+q)$ and $b=q-\eta(1+q)$ we have

$$s_\lambda(\eta,q)=q^{n(\lambda)}\prod_{(i,j)\in\lambda}\frac{(q-q^{j-i})-(1+q)(1-q^{j-i})\eta}{1-q^{h(i,j)}}. \tag{3.6}$$

Let μ be the Young diagram obtained from λ by removing the box (i,λ_i). Observe that $\mu_i=\lambda_i-1$, $\mu_{\lambda_i}^*=i-1=\lambda_{\lambda_i}^*-1$ and otherwise $\lambda_j=\mu_j$ and $\lambda_j^*=\mu_j^*$. So the hooklengths of a box (j_1,j_2) are the same in λ and in μ except if $j_1=i$ or $j_2=\lambda_i+1$. Using this we obtain from (3.5)

$$\begin{aligned}\frac{s_\lambda}{s_\mu}&=q^{i-1}\frac{a-bq^{\lambda_i-i}}{1-q}\prod_{j=1}^{\lambda_i}\frac{1-q^{h(i,j)-1}}{1-q^{h(i,j)}}\prod_{j=1}^{i-1}\frac{1-q^{h(j,\lambda_i)-1}}{1-q^{h(j,\lambda_i)}}\\&=\frac{a-bq^{\lambda_i-i}}{1-q}\prod_{j=1}^{\lambda_i}\frac{1-q^{h(i,j)-1}}{1-q^{h(i,j)}}\prod_{j=1}^{i-1}\frac{1-q^{-h(j,\lambda_i)+1}}{1-q^{-h(j,\lambda_i)}}.\end{aligned} \tag{3.7}$$

Lemma 3.4. *Let $\mu\in\Lambda_{n-1}$ and let $s_\mu(\eta,q)$ be as in (3.5). Moreover let $r\in T_\mu$. Then*

$$\sum_{t\in T_n,\,t'=r}(1+q)\,a_{t,n-1,n}\,s_{\lambda(t)}=\frac{qa-b}{1-q}s_\mu,$$

where $\lambda(t)$ is the Young diagram belonging to t. For the special choice of parameters in (3.4) we have

$$\sum_{t\in T_n,\,t'=r}a_{t,n-1,n}\,s_{\lambda(t)}(\eta,q)=\eta\,s_\mu(\eta,q).$$

Proof. Let $\lambda=[\lambda_1,\dots\lambda_m]$ and let $i_0\in\mathbf{N}$, $1\leqq i_0\leqq m$. We will modify λ such that our claim can be reduced to the special case of the Littlewood-Richardson rule in (3.4). Let $\lambda^{(i_0)}$ be the Young diagram defined by

$$\lambda_i^{(i_0)}=\lambda_i+1\qquad\text{for } i<i_0$$

and

$$\lambda_i^{(i_0)}=\lambda_{i+1}\qquad\text{for } i\geqq i_0.$$

Note that this implies

$$(\lambda^{(i_0)})_i^*=\lambda_i^*-1\qquad\text{for } i\leqq\lambda_{i_0}$$

and

$$(\lambda^{(i_0)})_i^*=\lambda_{i-1}^*\qquad\text{for } i>\lambda_{i_0}.$$

Let $h_\lambda(i, j)$ and $h_{\lambda^{(i_0)}}(i, j)$ denote the hook lengths corresponding to the box (i, j) of λ and $\lambda^{(i_0)}$ respectively. It follows from the previous remarks that

$$h_\lambda(i,j) = h_{\lambda^{(i_0)}}(i, j+1) \quad \text{for } i < i_0 \quad \text{and} \quad j > \lambda_{i_0}$$

and

$$h_\lambda(i,j) = h_{\lambda^{(i_0)}}(i-1, j) \quad \text{for } i > i_0.$$

If $i < i_0$ and $j \leqq \lambda_{i_0}$, $h_\lambda(i,j) = h_{\lambda^{(i_0)}}$. We define $\tilde{s}_\lambda(a, b, q) = s_\lambda(qa, b, q)$. Let $\mu \in \Lambda_{n-1}$ and $r \in T_\mu$ such that the box containing $n-1$ is in the i_0-th row. Let $t \in T_n$ be obtained from r by adding a box in the i-th row and let $\lambda = \lambda(t)$ be the corresponding Young diagram. Using (3.7) for s_λ/s_μ and for $\tilde{s}_{\lambda^{(i_0)}}/\tilde{s}_{\mu^{(i_0)}}$, it follows from the remarks above that these 2 quotients differ only by the factor belonging to either $(i, \lambda_{i_0}+1)$ in $\tilde{s}_{\lambda^{(i_0)}}/\tilde{s}_{\mu^{(i_0)}}$ (if $i < i_0$) or to (i_0, λ_i) in s_λ/s_μ (if $i_0 < i$). It follows

$$\frac{\tilde{s}_{\lambda^{(i_0)}}}{\tilde{s}_{\mu^{(i_0)}}} = (1+q)\, a_{t,n-1,n} \frac{s_\lambda}{s_\mu}.$$

Now observe that $\tilde{s}_{[1]} = (qa-b)/(1-q)$. Hence it follows from (3.4)

$$\sum_{t \in T_n, t' = r} (q+1)\, a_{t,n-1,n}\, s_{\lambda(t)}/s_\mu = \sum_{t \in T_n, t' = r} \tilde{s}_{\lambda^{(i_0)}}/\tilde{s}_{\mu^{(i_0)}} = \frac{qa-b}{1-q}.$$

For the last statement just observe that if $a = 1 - \eta(1+q)$ and $b = q - \eta(1+q)$, the expression $(qa-b)/(1+q)(1-q)$ is equal to η. □

The following special cases of s_λ will be particularly interesting for us. Let $k \in \mathbf{Z} \setminus \{0\}$ and let $s_{\lambda,k}(q)$ be the specialization of s_λ obtained by setting $a = (1-q)/(1-q^k)$ and $b = q^k(1-q)/(1-q^k)$. Note that we can also obtain these functions from (3.6) by setting $\eta = a_{-k}(q) = (1-q^{-k+1})/(1+q)(1-q^{-k})$. Using (3.5) one can write these functions as

$$s_{\lambda,k}(q) = q^{n(\lambda)} \prod_{(i,j) \in \lambda} \frac{(1-q)(1-q^{j-i+k})}{(1-q^k)(1-q^{h(i,j)})}. \tag{3.8}$$

Note also that if $\chi^{(\lambda)}(q)$ denotes the character of a matrix in $Gl(k)$ with eigenvalues $q^{(-k+1)/2}, q^{(-k+3)/2} \ldots q^{(k-3)/2}, q^{(k-1)/2}$ and if the number of boxes of λ is n, $s_{\lambda,k}(q) = \chi^{(\lambda)}(q)/(\chi^{([1])}(q))^n$.

Lemma 3.5. *Let $k \in \mathbf{N}$.*

(a) *If λ is a diagram with more than k rows, $s_{\lambda,k} = 0$.*

(b) *Let F be as in Lemma 2.6. If λ has at most k rows, $s_{F(\lambda),k} = (q^{(k-1)/2}(1-q)/(1-q^k))^k s_{\lambda,k}$.*

(c) *Let λ be a Young diagram with at most k rows and let $s_{\lambda,k}$ be the rational function in q obtained from (3.8) after carrying out all possible cancellations. Then $s_{\lambda,k}(q)$ is well-defined for q a primitive l-th root of unity with $l > k$ if*

$$\lambda_1 - \lambda_k + k - 1 \leqq l \tag{$*$}$$

and for such diagrams $s_{\lambda,k}(q)=0$ *iff equality holds in* $(*)$. *In particular,* $s_{\lambda,k}(q)\neq 0$ *for all* (k,l) *diagrams.*

(d) *If* $q=e^{\pm 2\pi i/l}$ *and if* λ *is a* (k,l) *diagram,* $s_{\lambda,k}(q)>0$.

Proof. For (a) note that the factor belonging to the box $(k+1,1)$ is equal to 0. For (b), observe that $h_{F(\lambda)}(i,j+1)=h_\lambda(i,j)$ and that $n(F(\lambda))-n(\lambda)=k(k-1)/2$. Hence we obtain for $s_{F(\lambda),k}(q)$ the expression

$$q^{n(F(\lambda))}\frac{\prod_{(i,j)\in F(\lambda),\,j\neq\lambda_i+1}(1-q)(1-q^{j-i+k})\prod_{i=1}^{k}(1-q)(1-q^{\lambda_i+1-i-k})}{\prod_{(i,j)\in F(\lambda),\,j\neq 1}(1-q^k)(1-q^{h(i,j)})\prod_{i=1}^{k}(1-q^k)(1-q^{\lambda_i+1-i-k})}$$

$$=q^{k(k-1)/2}\,s_{\lambda,k}(q)\frac{(1-q)^k}{(1-q^k)^k}.$$

By (b), it is enough to show (c) for Young diagrams λ with $\lambda_k=0$. As $0<h(i,j)\leqq h(1,1)<\lambda_1-1+k$, $(*)$ implies that $1-q^{h(i,j)}\neq 0$ for all $(i,j)\in\lambda$. So $s_{\lambda,k}(q)$ is well-defined for q a primitive l-th root of unity. If we have equality in $(*)$, the numerator in (3.8) becomes 0 if (and only if) $(i,j)=(1,\lambda_1)$. If we have a strict inequality in $(*)$, the numerators are not equal to 0 for all $(i,j)\in\lambda$.

(d) Observe that $(q^{(k-1)/2}(1-q)/(1-q^k))^k=\sin(\pi/l)/\sin(k\pi/l)>0$. Hence it is enough to show the claim for (k,l) diagrams λ with $\lambda_k=0$ by (b). In this case we have for all $(i,j)\in\lambda$

$$0<h(i,j)\leqq h(1,1)\leqq\lambda_1-\lambda_k+k-1<l$$

and

$$0<j+k-i\leqq\lambda_1+k-i<l.$$

It follows from (3.8)

$$s_{\lambda,k}(q)=\prod_{(i,j)\in\lambda}\frac{\sin(\pi/l)\sin((j-i+k)\pi/l)}{\sin(k\pi/l)\sin(h(i,j)\pi/l)}>0.\quad\square$$

The following theorem determines all values of η and q for which we have a positive Markov trace factoring over a C^* representation of $H_\infty(q)$. We also describe the image of the representations coming from the *GNS* construction with respect to these traces. Except of this last statement for q a root of unity all the other statements can be found in [O].

Recall that we have nontrivial C^* representations if and only if either q is real and positive or if $q=e^{\pm 2\pi i/l}$, $l=4,5,\ldots$.

Theorem 3.6. (a) *Let* q *be not a root of unity and let* $\eta\in\mathbf{C}$. *Then* tr *is a trace with Markov property on* $H_\infty(q)$ *with* $\mathrm{tr}(e_1)=\eta$ *if and only if* $\mathrm{tr}_{|H_n(q)}$ *has the weight vector* $(w_\lambda)_{\lambda\in\Lambda_n}$ *with* $w_\lambda=s_\lambda(\eta,q)$ (see (3.6)). *If* ρ *is a* C^* *representation of* $H_\infty(q)$, *there exists a positive trace with Markov property on* $\rho(H_\infty(q))$ *for* $q\in\mathbf{R}$, $q\geqq 1$ *if and only if either*

(α) $1/(1+q) \leqq \eta \leqq q/(1+q)$ or
(β) $\eta = a_k(q)$, $\quad k \in \mathbf{Z} \setminus \{0\}$.

If $0<q<1$, *the same statements hold except that one has to reverse the* "$\leqq$" *sign in* (α). *The traces are faithful in case* (α). *In case* (β), *they are nonzero exactly on those representations of* $H_n(q)$ *which belong to Young diagrams with at most k columns for* $k>0$ *and with at most* $|k|$ *rows for* $k<0$.

(b) *If* $q = e^{\pm 2\pi i/l}$, *there is a Markov trace factoring over a* C^* *representation of* $H_\infty(q)$ *if and only if* $k \in \{1, 2, \ldots l-1\}$ and $\eta = a_{-k}(q) = (1-q^{-k+1})/(1+q)(1-q^{-k})$. *In this case, the representation corresponding to the GNS construction with respect to this trace is equivalent to* $\pi^{(k,l)}$ *as defined in* (2.13). *The weight vector for the restriction of* tr *to* $H_n(q)$ *is given by the vector* $(w_\lambda)_{\lambda \in \Lambda_n^{(k,l)}}$ *with* $w_\lambda = s_{\lambda,k}(q)$ (see (3.8)).

Proof. (a) Let the weight vector of the restriction of tr to $H_n(q)$ be defined as in the statement. Note that $s_{[1]}(\eta, q) = 1$ for any choice of η and q. So tr is defined consistently on $H_\infty(q)$ by (3.4) and (1.5) and it is a Markov trace with $\mathrm{tr}(e_1) = \eta$ by Proposition 3.3 and Lemma 3.4.

For positivity, we only need to check for which values of η $w_\lambda \geqq 0$ for all Young diagrams λ. Observe that all numerators in the product in (3.6) are of the form $(1+q)(1-q^k)\eta - (q-q^k)$ and if $q>1$, all denominators are negative. So $w_\lambda > 0$ for all Young diagrams if

$$\eta > (q-q^k)/(1+q)(1-q^k) = a_{-k}(q) \quad \text{for } k \in \mathbf{N}$$

and

$$\eta < (q-q^k)/(1+q)(1-q^k) = a_{-k}(q) \quad \text{for } -k \in \mathbf{N}.$$

By Lemma 2.1 (vi), this happens if $1/(1+q) \leqq \eta \leqq q/(1+q)$.

If $\eta = (q-q^k)/(1+q)(1-q^k)$, $s_\lambda(\eta, q) = 0$ for any Young diagram λ containing the box $(1, k+1)$. If a Young diagram λ has at most k columns, $j-i<k$ for all boxes $(i,j) \in \lambda$. So $s_\lambda(\eta, q) > 0$ as all factors in (3.6) are so.

On the other hand let $\eta > q/(1+q)$ (the case $\eta < 1/(1+q)$ can be treated similarly). Then again by Lemma 2.1, there is a smallest $k \in \mathbf{N}$ such that $\eta \geqq (q-q^k)/(1+q)(1-q^k)$. If we have a proper inequality, it is easy to see that all factors of $s_{[k+1]}(\eta, q)$ in (3.6) are nonnegative except the one belonging to the box $(1, k+1)$. So, as $\pi_{[k+1]}$ is a one dimensional representation, we obtain $\mathrm{tr}(z_{[k+1]}) = w_{[k+1]} = s_{[k+1]}(\eta, q) < 0$. Hence tr can not be positive, which shows the only if part.

(b) Using Proposition 3.3 (b), (3.4) and Lemma 3.5 (c), it can be checked easily as in (a) that the traces defined by the given weight vectors are well-defined and have Markov property. The positivity statement (and faithfulness on $\pi^{(k,l)}(H_n(q))$) follows from Lemma 3.5 (d).

On the other hand, assume that tr is a Markov trace factoring over a C^* representation of $H_\infty(q)$. Observe that all Young diagrams with $\leqq l-1$ boxes are (l) diagrams. Hence there exists a faithful C^* representation of $H_{l-1}(q)$ and the idempotents p_t as in Corollary 2.3 are well-defined for all $t \in T_{l-1}$. In particular, it follows as in (a) that the weights of a Markov trace, restricted to $H_{l-1}(q)$,

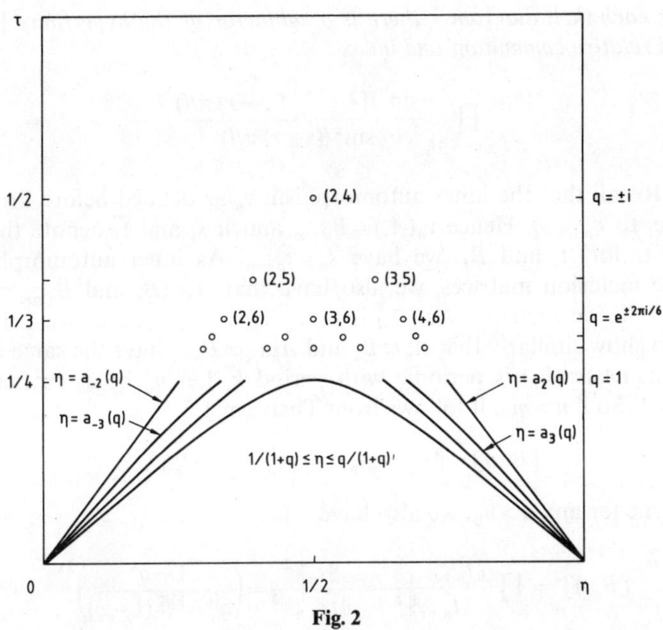

Fig. 2

are given by the Schur functions $s_\lambda(q, \eta)$ with $\lambda \in \Lambda_{l-1}$. Let $p_{[l-1]}$ be the minimal projection corresponding to the only standard tableau belonging to $[l-1]$. By (2.8), the eigenvalues of $p_{[l-1]} e_{l-1} p_{[l-1]}$ are 0 and $a_{l-1}(q)=0$. So if ρ is a C^* representation, the self-adjoint element $\rho(p_{[l-1]} e_{l-1} p_{[l-1]})$ is equal to 0. If the Markov trace tr factors over ρ, it follows from that

$$0 = \mathrm{tr}(e_{l-1}\, p_{[l-1]}) = \eta\, \mathrm{tr}(p_{[l-1]}) = \eta\, s_{[l-1]}(\eta, q).$$

This can only be true if either $\eta = 0 = a_{l-1}(q)$ or if $s_{[l-1]}(\eta, q) = 0$. It follows from (3.6) that in this case $\eta = a_k(q)$ for some $k \in \mathbf{N}$, $1 \leqq k \leqq l-1$. □

The values of η and q for which the Markov trace is positive are shown in Fig. 2, which is due to Ocneanu. For simplicity, we use instead of q the parameter $\tau = q/(1+q)^2$. Note that $\tau \leqq 1/4$ corresponds to q real and positive, while $\tau = 1/(4\cos^2(\pi/l))$ corresponds to $q = e^{\pm 2\pi i/l}$.

We are now in the position to use this information to produce examples of subfactors with trivial relative commutants. Let $q = e^{\pm 2\pi i/l}$ and let $\rho = \pi^{(k,l)}$ be the C^* representation of $H_\infty(q)$ as defined in (2.13). Let us define for a fixed $m \in \mathbf{N}$ the finite dimensional algebras A_n and B_n by $B_n = \pi^{(k,l)}(H_n(q))$ and $A_n = \pi^{(k,l)}(\langle e_{m+1}, e_{m+2}, \ldots e_{n-1} \rangle)$. It follows from Proposition 3.2 and Theorem 3.6 that the Markov trace with $\eta = a_k(q)$ is a (positive) factor trace.

Theorem 3.7. *Let $q = e^{\pm \pi i/l}$ and let $k \in \mathbf{N}$, $1 \leqq k \leqq l-1$. Let (A_n), (B_n) and* tr *be as above and let A and B be the resulting factors* (see § 1). *Then*

(a) $[B:A] = \sin^{2m}(k\pi/l)/\sin^2(\pi/l)$.

(b) $A' \cap B = B_m$.

(c) *For each* (k, l) *diagram* λ *there is a subfactor of the hyperfinite* II_1 *factor with trivial relative commutant and index*

$$\prod_{1 \leq r < s \leq k} \frac{\sin^2((\lambda_r - \lambda_s + s - r)\,\pi/l)}{\sin^2((s-r)\,\pi/l)}.$$

Proof. (a) Recall that the inner automorphism v_n as defined before Proposition 2.9 maps e_i to e_{n+1-i}. Hence $v_n(A_n) = B_{n-m}$ and if $\vec{s}_i$ and $\vec{t}_j$ denote the weight vectors of tr for A_i and B_j, we have $\vec{s}_n = \vec{t}_{n-m}$. As inner automorphisms do not change inclusion matrices, we also have that $A_n \subset B_n$ and $B_{n-m} \subset B_n$ have the same inclusion matrices. As (B_n) is periodic with period k (see Lemma 2.6) we can also show similarly that $A_n \subset B_n$ and $A_{n+k} \subset B_{n+k}$ have the same inclusion matrices, i.e. $(A_n) \subset (B_n)$ is periodic with period k. Let n_0 be as in Lemma 2.6 and let $m = 1$. So if $n > m_0$, it follows from Theorem 1.5

$$[B:A] = \|\vec{s}_n\|^2 / \|\vec{t}_n\|^2 = \|\vec{t}_{n-1}\|^2 / \|\vec{t}_n\|^2.$$

As this is true for any $n > n_0$, we also have

$$[B:A]^k = \prod_{i=1}^{k} \frac{\|\vec{t}_{n+i-1}\|^2}{\|\vec{t}_{n+i}\|^2} = \frac{\|\vec{t}_n\|^2}{\|\vec{t}_{n+k}\|^2} = \left(\frac{1-q^k}{q^{(k-1)/2}(1-q)}\right)^{2k}$$

by Theorem 3.6 (b) and Lemma 3.5 (b). We obtain

$$[B:A] = \frac{\sin^2(k\pi/l)}{\sin^2(\pi/l)}.$$

If $m > 1$, we compute by the same method

$$[B:A] = \frac{\|\vec{t}_{n-m}\|^2}{\|\vec{t}_n\|^2} = \frac{\sin^{2m}(k\pi/l)}{\sin^{2m}(\pi/l)}.$$

(b) It follows from $(H2)$ that $B_m \subset A' \cap B$. On the other hand choose $s \in \mathbf{N}$ such that $ks > n_0$. Choose an arbitrary (k, l) tableau $t \in T([s^k])^{(k,l)}$ and let $p_t^{(l)}$ be the corresponding minimal idempotent in B_{sk}. Now observe that the Bratteli diagram for

$$p_t^{(l)} B_{sk+1} p_t^{(l)} \subset p_t^{(l)} B_{sk+2} p_t^{(l)} \subset \ldots \subset p_t^{(l)} B_{sk+m} p_t^{(l)}$$

is the same as the Bratteli diagram for

$$B_1 \subset B_2 \subset \ldots \subset B_m.$$

Hence $p_t^{(l)} B_{sk+m} p_t^{(l)} \cong B_m$. Applying the inner automorphism v_{sk+m} we see that $v_{sk+m}(p_t^{(l)})$ is a minimal projection of A_{sk+m} and $v_{sk+m}(p_t^{(l)}) B_{sk+m} v_{sk+m}(p_t^{(l)}) \cong B_m$. Hence $\dim B_m \geq \dim(A' \cap B)$ by Theorem 1.6 which implies $B_m = A' \cap B$.

(c) If p is a minimal projection of $A' \cap B$, A_p is a subfactor of B_p with $A_p' \cap B_p = (A' \cap B)_p = \mathbf{C}\,p$. Recall that the weight vectors of tr are given by the Schur functions $s_{\lambda,k}$ (see Theorem 3.6). In view of (a), (b) and Theorem 1.5

(iii) it is enough to show for every $\lambda \in \Lambda_m^{(k,l)}$ that

$$s_{\lambda,k}^2(q)\left(\frac{1-q^k}{q^{(k-1)/2}(1-q)}\right)^{2m} = \prod_{1 \leqq r \leqq s \leqq k} \frac{\sin^2((\lambda_r-\lambda_s+s-r)\pi/l)}{\sin^2((s-r)\pi/l)}. \qquad (*)$$

As usual we prove this by induction on m. If $m=0$, we have $A=B$ and therefore $[B:A]=1$, while on the other hand $s_{[0],k}(q)=1$.

Let us denote the right hand side of (*) by $\operatorname{ind}(\lambda)$. Assume that the formula holds for all (k,l) diagrams with $m-1$ boxes. Let μ be the diagram obtained from λ by removing the box (i,λ_i). Then

$$\frac{\operatorname{ind}(\lambda)}{\operatorname{ind}(\mu)} = \prod_{j=1}^{i-1} \frac{\sin^2((\lambda_j-\lambda_i+i-j)\pi/l)}{\sin^2((\lambda_j-\lambda_i+i-j+1)\pi/l)} \prod_{j=i+1}^{k} \frac{\sin^2((\lambda_j-\lambda_i+i-j)\pi/l)}{\sin^2((\lambda_j-\lambda_i+i-j-l)\pi/l)}.$$

Now observe that if $\lambda_p^* = \lambda_{p+1}^* = \ldots = \lambda_k^*$, we have

$$\prod_{j=p}^{r} \frac{1-q^{h(j,i)}}{1-q^{h(i,j)-1}} = \frac{1-q^{h(j,r)-1}}{1-q^{h(j,p)}}.$$

Moreover note that the hooklength for the box (i,j) of λ is equal to $h(i,j) = \lambda_i - \lambda_{\lambda_j^*} + \lambda_j^* - i + 1$. With these observations and with (3.6) it is now easy to check that

$$\frac{\operatorname{ind}(\lambda)}{\operatorname{ind}(\mu)} = \frac{s_{\lambda,k}^2(q)(1-q^k)^2}{s_{\mu,k}^2(q)(1-q)^2\, q^{k-1}}.$$

From this follows our claim by induction on m. □

Remarks. 1. If $m=1$, $A' \cap B$ is trivial and $[B:A] = \sin^2(k\pi/l)/\sin^2(\pi/l)$. These values converge to k^2 if $l \to \infty$. If $k=2$, $\sin^2(2\pi/l)/\sin^2(\pi/l) = 4\cos^2(\pi/l)$, which are the values obtained by Jones in [J-1]. In fact it is not hard to show that the C^* algebras obtained from the $(2,l)$ representations coincide with the AF algebras in [J-1].

2. Observe that if a diagram λ is a (k,l) diagram as well as a (k',l) diagram with $k \neq k'$, we usually obtain different index values for λ depending on whether λ is viewed as a (k,l) or a (k',l) diagram. For example if $\lambda=[1]$ is viewed as a (k,l) diagram for $k=1,2,\ldots$, the corresponding index values are $\sin^2(k\pi/l)/\sin^2(\pi/l)$.

3. The formulas for the indices show a remarkable similarity with Weyl's dimension formulas for the characters of $SU(k)$. Indeed, if $l \to \infty$ (or equivalently if $q \to 1$), the value of the index corresponding to the diagram λ converges to the square of the dimension of the character of $SU(k)$ belonging to λ. In fact, theorem (3.7), (c) can also be stated in terms of characters of $SU(k)$ in the following way:

For any character λ of $SU(k)$ with weights $\lambda_1, \ldots \lambda_{k-1}$ such that $\lambda_1 \leqq l-k$ there exists a subfactor of R with trivial relative commutant and index

$$\prod_{1 \leqq r < s \leqq k} \frac{\sin^2((\lambda_r-\lambda_s+s-r)\pi/l)}{\sin^2((s-r)\pi/l)},$$

where we set $\lambda_k=0$. The index is equal to the character of the element with eigenvalues $q^{(-k+1)/2}, q^{(-k+3)/2}, \dots q^{(k-3)/2}, q^{(k-1)/2}$ with $q=e^{2\pi i/l}$.

The subfactors in the limiting case $q=1$ have been studied by Wassermann in [Wa], who computed their indices and higher relative commutants using techniques different from our periodicity arguments.

4. We will treat the subfactors coming from the Hecke algebras with real and positive parameters in a future paper using their connection with solutions of quantum Yang-Baxter equations. This approach will also enable us to compute higher relative commutants of the examples given in Theorem 3.7.

References

[B] Birman, J.: Braids, links and mapping class groups. Ann. Math. Stud. **82** (1974)

[F..] [FYHLMO] Freyd, P., Yetter, D., Hoste, J., Lickorish, W.B.R., Millett, K., Ocneanu, A.: A new polynomial invariant of knots and links. Bull. Am. Math. Soc. **12**, 239–246 (1985)

[G] Gantmacher, F.R.: Matrix theory. Vol. II, Chelsea Publishing Company: New York, 1974

[HW] Handelman, D., Wenzl, H.: Closedness of index values for subfactors. Proc. Am Math. Soc. **101**, 277–282 (1987)

[H] Hoefsmit, P.N.: Representations of Hecke algebras of finite groups with BN pairs of classical type. Thesis, University of British Columbia, 1974

[JK] James, G., Kerber, A.: The representations of the symmetric group. Addison-Wesley: Reading, Mass

[J-1] Jones, V.F.R.: Index for subfactors. Invent. Math. **72**, 1–25 (1983)

[J-2] Jones, V.F.R.: Braid groups, Hecke algebras and type II_1 factors. Japan-US Conference proceedings, 1983

[J-3] Jones, V.F.R.: A polynomial invariant for knots via von Neumann algebras. Bull. Am. Math. Soc. **12** 103–111 (1985)

[J-4] Jones, V.F.R.: Hecke algebra representations of braid groups. Ann. Math. **126**, 335–388 (1987)

[L] Littlewood, D.: The theory of group characters. Oxford University Press, 1940

[Mc] McDonald, I.: Symmetric functions and Hall polynomials. Clarendon Press: Oxford, 1979

[M] Murnaghan, F.: The theory of group representations. Dover, New York, 1939

[O] Ocneanu, A.: A polynomial invariant for knots: A combinatorial and an algebraic approach (Preprint, see [FYHLMO] for a shortened version)

[PP1] Pimsner, M., Popa, S.: Entropy and index for subfactors. Ann. Sci. Ec. Norm. Super, 4[e] serie, **19**, 57–106 (1986)

[PP2] Pimsner, M., Popa, S.: Iterating the basic construction (Preprint, INCREST)

[St] Størmer, E.: Symmetric states of infinite tensor products of C^* algebras. J. Funct. Anal. **3**, 48–68 (1969)

[SV] Stratila, S., Voiculescu, D.: Representations of AF-algebras and of the group $U(\infty)$. Lect. Notes Math., Vol. 486. Berlin Heidelberg New York: Springer 1975

[T] Thoma E.: Die unzerlegbaren, positiv-definiten Klassenfunktionen der abzaehlbaren, unendlichen, symmetrischen Gruppe. Math. Z. **85**, 40–61 (1964)

[Wa] Wassermann, A.: Automorphic actions of compact groups on operator algebras. Thesis, University of Pennsylvania 1981

[W-1] Wenzl, H.: On sequences of projections. C.R. Math. Rep. Acad. Sci. Canada IX, **1**, 5–9 (1987)

[W-2] Wenzl, H.: Representations of Hecke algebras and subfactors. Thesis, University of Pennsylvania 1985

TRANSACTIONS OF THE
AMERICAN MATHEMATICAL SOCIETY
Volume 313, Number 1, May 1989

BRAIDS, LINK POLYNOMIALS AND A NEW ALGEBRA

JOAN S. BIRMAN AND HANS WENZL

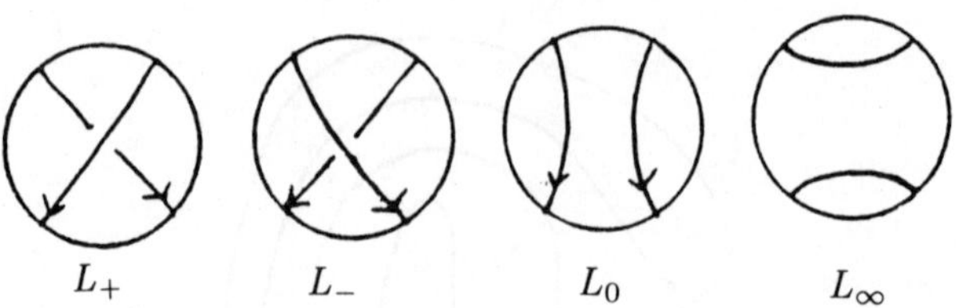

ABSTRACT. A class function on the braid group is derived from the Kauffman link invariant. This function is used to construct representations of the braid groups depending on 2 parameters. The decomposition of the corresponding algebras into irreducible components is given and it is shown how they are related to Jones' algebras and to Brauer's centralizer algebras.

In [J,3] Vaughan Jones announced the discovery of a new polynomial invariant of knots and links, which bore many similarities to the classical Alexander polynomial, but was seen to detect properties of a link which could not be detected by the Alexander invariants. The discovery was a real surprise, one of those exciting moments in mathematics when two seemingly unrelated disciplines turn out to have deep interconnections. The discovery came about in the following way. Jones' earlier contributions in the area of Operator Algebras had produced, in [J,1], a family of algebras $A_n(t)$, $t \in \mathbf{C}$, indexed by the natural numbers $n = 1, 2, 3, \ldots$, and equipped with a trace function $\tau: A_n(t) \to \mathbf{C}$. His algebra $A_n(t)$ was a quotient of the well-known Hecke algebra of the symmetric group, which we denote by $\mathscr{H}_n(l, m)$ to delineate our particular 2-parameter version of it. Jones had discovered, in [J,2], that there were representations of Artin's braid group B_n in the algebra $A_n(t)$, in fact there were maps

$$B_n \xrightarrow{\phi} \mathscr{H}_n(l, m) \xrightarrow{\rho_2} A_n(t)$$

from B_n into the multiplicative group of $A_n(t)$ which factored through $\mathscr{H}_n(l, m)$.

Links enter the picture via braids. Each oriented link L in oriented S^3 can be represented by a (nonunique) element β in some braid group B_n. There is an equivalence relation on $B_\infty = \coprod_{n=1}^{\infty} B_n$, known as Markov equivalence, which determines a 1-1 correspondence between equivalence classes $[\beta] \in B$ and isotopy types of the associated oriented links L_β. Jones' discovery was that with a small renormalization his trace function on $A_\infty(t) = \coprod_{n=1}^{\infty} A_n(t)$ could be made into a function which lifted to an invariant on Markov classes

Received by the editors December 9, 1987.

1980 *Mathematics Subject Classification* (1985 *Revision*). Primary 57M25; Secondary 20F29, 20C07.

The work of the first author was supported in part by NSF grant #DMS-8503758.

The work of the second author was supported in part by NSF grant #DMS-8510816.

in B_∞. That modified trace, described in [J,3] and in more detail in [J,4], is the Jones polynomial $V_L(t)$. (It becomes a polynomial when the parameter t is regarded as an indeterminate.)

The polynomial $V_L(t)$ was quickly generalized in a six-author paper [FYHLMO], to a 2-variable polynomial $P_L(l,m)$. One of the authors was A. Ocneanu. Ocneanu's interpretation of $P_L(l,m)$, as described in [FYHLMO] and [J,4], is via a lift of the Jones trace to the Hecke algebra $\mathscr{H}_n(\mathbf{1},m)$. Others, notably Lickorish and Millett [L-M] and Hoste [H], had discovered the identical polynomial $P_L(l,m)$ by combinatorial methods which had little to do with braids, traces or algebras.

On the heels of $P_L(l,m)$ came two additional polynomials, the Kauffman polynomial $K_L(l,m)$ and a precurser later identified as $K_L(l,m)$. References are [B-L-M and K,1]. These were proved to be well defined by various combinatorial methods, including not only generalizations of the methods used in [L-M and H], but also new techniques in [K,1], which again bypassed braids. The polynomials $P_L(l,m)$ and $K_L(l,m)$ were shown to be independent, with each distinguishing links the other could not distinguish. On the other hand (see [L]), $V_L(t) = K_L(t^{-3/4}, -(t^{1/4}+t^{-1/4})) = P_L(it^{-1}, -i(t^{1/2}-t^{-1/2}))$. (We shall have more to say about this curious fact in §4 below.)

The purpose of this note is to reverse the process begun by Jones. We will use the existence of $K_L(l,m)$, and apply the methods used to construct it in [K,1] to construct a new two-parameter family of finite-dimensional algebras, $\{\mathscr{C}_n(l,m);\ n = 1,2,3,\dots\}$, complete with trace, such that $K_L(l,m)$ is, after appropriate renormalization, that trace, just as $P_L(l,m)$, renormalized, was shown by Ocneanu to be the trace on $\mathscr{H}_n(l,m)$.

In §1 below, we will review the background. In §2 we define our algebra by generators and relations, and explain our motivation, which is based upon Kauffman's work. In §3 we study the algebra $\mathscr{C}_n(l,m)$. We prove that $\mathscr{C}_\infty(l,m) = \coprod_{n\in\mathbf{Z}}^{\infty} \mathscr{C}_n(l,m)$ supports a nondegenerate trace. The existence of this trace is a direct consequence of the fact that $K_L(l,m)$ is a link type invariant. We use the existence of the trace to uncover the structure of $\mathscr{C}_n(l,m)$, by techniques which derive from the "basic construction" of Jones in [J,1]. We prove that $\mathscr{C}_n(l,m)$ is semisimple and is a direct sum $\mathscr{H}_n \oplus \mathscr{H}_n'$, where $\mathscr{H}_n = \mathscr{H}_n(l,m)$. The structure of $\mathscr{H}_n'$ as an algebra over $\mathbf{C}[l,\alpha]$, where $m = \alpha+\alpha^{-1}$, is determined inductively from the inclusion pattern $\mathscr{C}_{n-2} \subset \mathscr{C}_{n-1}$. The dimension of $\mathscr{H}_n(l,m)$ is $n!$ (see [J,4] for a proof of this well-know fact). It follows from our structure theorem that the dimension of $\mathscr{C}_n(l,m)$ is $1\cdot 3\cdot 5\cdots(2n-1)$.

The relationship between $\mathscr{C}_n(l,m)$ and the 1-parameter Jones algebra $A_n(t)$ is very interesting. We prove in §4 that there are two distinct homomorphisms from $\mathscr{C}_n(l,m)$ onto $A_n(t)$. One of them factors through $\mathscr{H}_n(l,m)$, and is not unexpected in view of the fact that $\mathscr{C}_n(l,m) = \mathscr{H}_n \oplus \mathscr{H}_n'$, with $\mathscr{H}_n(l,m)$ mapping in a known way onto $A_n(t)$. The other does *not* factor through $\mathscr{H}_n(l,m)$. The manner in which the irreducible representations of $\mathscr{C}_n(l,m)$ go over to irreducible representations of $A_n(t)$ in the second homomorphism seems quite remarkable. We discovered that there are two homomorphsims in our attempts to understand the curious fact that $V_L(t)$ occurs both as $P_L(it^{-1}, i(t^{1/2}-t^{-1/2}))$ and as $K_L(t^{-3/4}, -t^{1/4}-t^{-1/4})$.

The existence of these two homomorphisms is not the only curious fact about $\mathscr{C}_n(l, m)$. In §5 we discuss an algebra which was studied by R. Brauer in 1937 [Br]. Like $\mathscr{C}_n(l, m)$, Brauer's algebra has dimension $1 \cdot 3 \cdot 5 \cdots (2n-1)$. It is defined by pictures which bear a striking similarity to our pictures, indeed his pictures give a very easy way to find a basis for our algebra. We will show that our algebras can be regarded as a deformation of Brauer's similarily as $\mathscr{H}_n(l, n)$ is a deformation of $\mathbf{C}S_n$. This was also observed independently by H. Morton and P. Traczyk (see [MT]).

The final section of the paper, §6, discusses a possible application of our work to the question of whether the braid group is a linear group.

ACKNOWLEDGEMENT

The definition of our algebra $\mathscr{C}_n(l, m)$ in §2 below was motivated by pictures of a certain monoid which generalizes Artin's braid group B_n. The monoid has been used by various authors, although to our knowledge it remains a bit mysterious at this writing. The first author first learned of this monoid years ago from Claude Bourin [Bo], and again more recently had discussions about it with David Yetter [Y]. It has figured most recently in the work of Kauffman [K,1 and K,2].

We first learned about Brauer's algebra from Vaughan Jones. We thank him for telling us about it, and for his continuing lively interest in this work. We also wish to thank Phillip Hanlon for telling us many interesting things about the Brauer algebra.

Related results have been obtained simultaneously and independently by J. Murakami [M], who also constructs an algebra from Kauffman's polynomial duplicating our §2, but not our §§3–6.

1. The braid group B_n, the algebra $\mathscr{H}_n$, and the link polynomial $P_L(l, m)$

Let $\vec{p} = (p_1, \dots, p_n)$ be an n-tuple of distinct points p_i on the complex plane, which for convenience may be assumed to lie along the x-axis. For present purposes, a *braid* β on n-strands is an equivalence class $[e]$ of level-preserving embeddings of n disjoint copies of the unit interval $I^{(n)} = \coprod_{j=1}^{n} I_j$ in $\mathbf{C} \times I$, where e sends the n copies of $\{0\}$ (respectively $\{1\}$) to $\vec{p} \times \{0\}$ (respectively $\vec{p} \times \{1\}$). The equivalence relation is $e \approx e'$ if e is isotopic to e' rel $\partial I^{(n)}$. Multiplication is by concatenation and rescaling, and the identity element is the constant embedding $e_j(t) = p_j \times t$ for each $t \in I_j$, where e_j is the jth coordinate function, $j = 1, \dots, n$. This makes the set of all n-braids into a group B_n, *Artin's braid group.* A typical element of B_4 is illustrated in

Figure 1, via its projection onto the $x-t$ plane, but with double points replaced by overpasses and underpasses. From such pictures it is intuitively clear that the elementary braids $\sigma_1, \dots, \sigma_{n-1}$ shown in Figure 2 generate B_n. This was proved by Artin in [Ar], who also proved that defining relations are:

$$\sigma_i\sigma_j = \sigma_j\sigma_i \quad \text{if } |i-j| \geq 2,\ 1 \leq i, j \leq n-1; \tag{1}$$

$$\sigma_i\sigma_{i+1}\sigma_i = \sigma_{i+1}\sigma_i\sigma_{i+1}, \quad 1 \leq i \leq n-2. \tag{2}$$

Note that there is a natural way to orient a braid, determined by the orientation on the unit interval. We orient our braids from top to bottom.

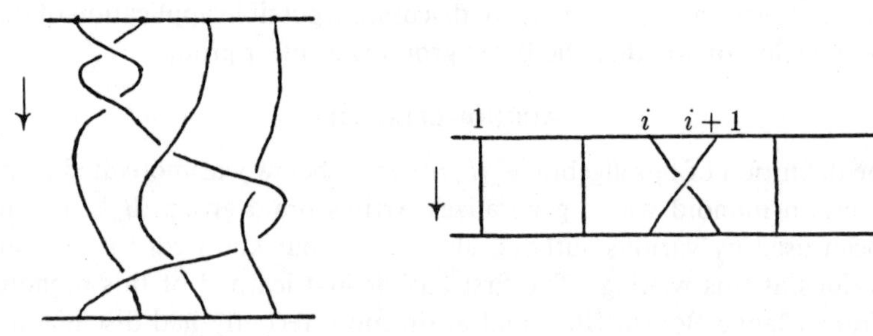

FIGURE 1 FIGURE 2

Links are obtained from braids when the free ends at the top and bottom of a braid are joined up by n disjoint arcs as in Figure 3 to form a *closed braid.* If $\beta \in B_n$, we denote the link so-obtained by L_β. It carries a natural orientation, determined by the orientation on β.

Two oriented links L, L' in oriented 3-space are *equivalent* if L is isotopic to L'. The equivalence class is a *link type.* Alexander proved in [A1] that every oriented link is isotopic to L_β for some (nonunique) element $\beta \in B_\infty$, $B_\infty = \coprod_{n=1}^{\infty} B_n$. Markov's theorem (see [B]) asserts that the equivalence relation on B_∞ which is generated by the following two moves:

(i) $\beta \leftrightarrow \alpha\beta\alpha^{-1}$, $\alpha, \beta \in B_n$;

(ii) $B_n \leftrightarrow B_{n+1}$, by $\beta \leftrightarrow \beta\sigma_n^{\pm 1}$

determines the equivalence class of all elements $\beta \in B$ which close to determine a given link type. Thus the problem of classifying link types is equivalent to the algebraic problem of classifying *Markov classes* $[\beta]$ in B_∞.

In view of Markov's first move ((i) above) a necessary condition for a function with domain B_∞ to be invariant on Markov classes is that it be a class function on each group B_n. We now review how Markov class invariants have been constructed out of class functions on representations of the braid groups.

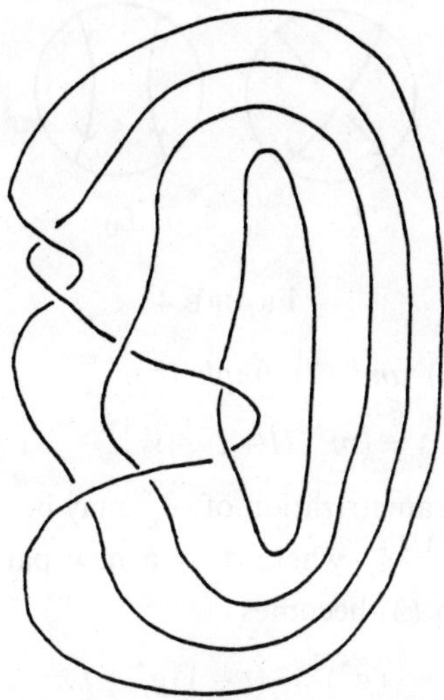

FIGURE 3

The Hecke algebra $\mathcal{H}_n(l, m)$ of the symmetric group is a 2-parameter family of algebras, with parameters $l, m \in \mathbf{C}$, and with $n \in \mathbf{N}$ a fixed natural number. For our purposes the most convenient definition of $\mathcal{H}_n(l, m)$ will be by generators and relations. Our generators will be denoted $\tilde{g}_1, \ldots, \tilde{g}_{n-1}$.

Notation. Instead of writing the relation $\tilde{g}_i\tilde{g}_j = \tilde{g}_j\tilde{g}_i$, $|i-j| \geq 2, \ldots$, we will write $(1)_{\tilde{g}}$, and similarly $(2)_{\tilde{g}}$ for relation (2) in the variables $\tilde{g}_1, \ldots, \tilde{g}_{n-1}$. With this convention, $\mathcal{H}_n(l, m)$ has defining relations $(1)_{\tilde{g}}$, $(2)_{\tilde{g}}$ and

$$l\tilde{g}_i + l^{-1}\tilde{g}_i^{-1} = m. \tag{3}$$

The algebra $\mathcal{H}_n = \mathcal{H}_m(l, m)$ has been studied by Ocneanu [O] with a full account of the relevant features of his work given by Jones in [J,4]. The facts which we will need here are summarized by

Proposition 1.1 [J,4]. *The algebras $\{\mathcal{H}_n ; n \in \mathbf{N}\}$ are ordered by inclusion $\mathcal{H}_1 \subset \mathcal{H}_2 \subset \mathcal{H}_3 \subset \cdots$, with $\mathcal{H}_n$ the subalgebra of $\mathcal{H}_{n+1}$ generated by $\tilde{g}_1, \ldots, \tilde{g}_{n-1}$. The algebra $\mathcal{H}_n$ has dimension $n!$. Each $\mathcal{H}_n$ supports a trace function $\tau : \mathcal{H}_n \to \mathbf{C}$, which is characterized uniquely by the conditions:*

$$\tau(a+b) = \tau(a) + \tau(b); \tag{4}$$

$$\tau(ab) = \tau(ba); \tag{5}$$

$$\tau(1) = 1; \tag{6}$$

$$\tau(w\tilde{g}_n) = (m^{-1}(l + l^{-1}))^{-1}\tau(w) \quad \textit{whenever } w \in \mathcal{H}_n \subset \mathcal{H}_{n+1}. \tag{7}$$

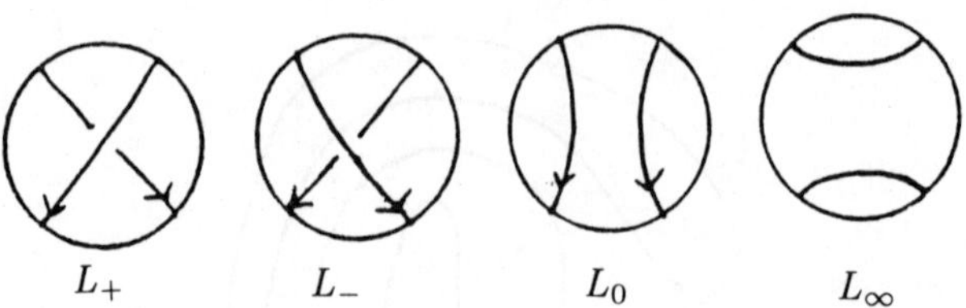

FIGURE 4

Note that (3), (4), (6) *and* (7) *imply that*:

(8) $$\tau(\tilde{g}_i) = \tau(\tilde{g}_i^{-1}) = (m^{-1}(l + l^{-1}))^{-1}, \qquad 1 \le i \le n-1.$$

The more common parametrization of $\mathscr{H}_n$ may be recovered by introducing new generators $g_i^* = t^{-1/2}\tilde{g}_i$ where t is a new parameter, with $l = it^{-1}$, $m = i(t^{1/2} - t^{-1/2})$. Then (3) becomes

(3)* $$(g_i^*)^2 = (t-1)g_i^* + t.$$

This gives a 1-parameter version of the algebra $\mathscr{H}_n(l, m)$ which we will denote by the symbol $\mathscr{H}_n(t)$. A second parameter z can then be introduced via the trace, if one replaces (7) by

(7)* $$\tau(w g_n^*) = z\tau(w) \quad \text{whenever } w \in \mathscr{H}_n(t) \subset \mathscr{H}_{n+1}(t).$$

Equations (3)*, (4), (6) and (7)* imply:

(8)* $$\tau(g_i^*) = z, \qquad \tau(g_i^{*-1}) = t^{-1}z + t^{-1} - 1$$

which is less convenient than (8) because of Markov's second move. This explains our choice of the generators $\tilde{g}_i$.

Since the trace is a class invariant in $\phi(B_n)$, and since (by (7) and (8)) it behaves nicely under the mapping $B_n \leftrightarrow B_{n+1}$ defined by Markov's second move ((ii) above) one sees immediately that the function

(9) $$P_L(l, m) = (m^{-1}(l + l^{-1}))^{n-1}\tau(\phi(\beta))$$

is invariant on the Markov class of each $\beta \in B_\infty$. It is the six-author polynomial of [FYHLMO].

Since the unknot is the closure of the 1-braid $1 \in B_1$, one sees from (9) that for an unknotted circle:

(10) $$P_{\text{unknot}}(l, m) = 1.$$

Let L_+, L_-, L_0 be links which are defined as the closures of n-braids β_+, β_-, β_0, where the defining braids are products of the elementary braids which are identical except for a single letter, with $\beta_+ = \alpha\sigma_i$, $\beta_- = \alpha\sigma_i^{-1}$, $\beta_0 = \alpha$, $\alpha \in B_n$, $1 \le i \le n-1$. This means that L_+, L_-, L_0 have link diagrams which are identical everywhere except in a small disc, where they differ in the manner indicated in the first three pictures in Figure 4. It follows immediately from formula (3) and the linearity of the trace function that their polynomials are related by the *crossing change formula*

$$lP_{L_+}(l,m) + l^{-1}P_{L_-}(l,m) = mP_{L_0}(l,m). \tag{11}$$

This is Proposition 6.2 of [J,4].

The key idea in the papers of Lickorish and Millett [L-M] and Hoste [H] was to use (10) and (11) as the basis of a *definition* of $P_L(l,m)$, thereby bypassing algebras, traces and all of the attendent machinery and focusing on the combinatorics of link diagrams. Their proof that $P_L(l,m)$ is a well-defined link invariant is equivalent to Jones' proof of Proposition 1.1. Our idea in the next section will be to reverse the procedure, using combinatorics to motivate a way to define a new algebra.

2. The definition of $\mathscr{C}_n(l,m)$

For each natural number n we define a 2-parameter family of algebras $\mathscr{C}_n(l,m)$ with generators $G_1,\dots,G_{n-1}$ and relations $(1)_G$, $(2)_G$, and others which involve elements $E_1,\dots,E_{n-1}$ defined by

$$G_i + G_i^{-1} = m(1+E_i). \tag{12}$$

The additional relations are:

$$E_iE_{i\pm1}E_i = E_i, \tag{13}$$

$$G_{i\pm1}G_iE_{i\pm1} = E_iG_{i\pm1}G_i = E_iE_{i\pm1}, \tag{14}$$

$$G_{i\pm1}E_iG_{i\pm1} = G_i^{-1}E_{i\pm1}G_i^{-1}, \tag{15}$$

$$G_{i\pm1}E_iE_{i\pm1} = G_i^{-1}E_{i\pm1}, \tag{16}$$

$$E_{i\pm1}E_iG_{i\pm1} = E_{i\pm1}G_i^{-1}, \tag{17}$$

$$G_iE_i = E_iG_i = l^{-1}E_i, \tag{18}$$

$$E_iG_{i\pm1}E_i = lE_i. \tag{19}$$

These imply the further relations:

$$E_iE_j = E_jE_i \quad \text{if } |i-j| \geq 2, \tag{20}$$

$$E_i^2 = (m^{-1}(l+l^{-1})-1)E_i, \tag{21}$$

$$G_i^2 = m(G_i + l^{-1}E_i) - 1. \tag{22}$$

In every case the indices are chosen from $1 \leq i, j \leq n-1$ to be all for which the relation in question makes sense.

Remark. Our choice of l and m as parameters suggests (as we will prove in §3) that $\mathscr{H}_n(l,m)$ is a quotient of $\mathscr{C}_n(l,m)$. The homomorphism $\phi : B_n \to \mathscr{H}_n(l,m)$ defined earlier will be seen to factor through $\mathscr{C}_n(l,m)$, via $\sigma_i \to l^{-1}G_i \to \tilde{g}_i$.

The rest of this section will be devoted to explaining our motivation in writing down these particular defining relations for $\mathscr{C}_n(l, m)$, using ideas from [K,1]. The reader who has not studied that paper may wish to omit the rest of this section on a first reading. The underlying idea is that we want to define our algebra in such a way as to force it to admit a trace, with the Kauffman polynomial a renormalization of that trace.

Recall that in §1 we showed that $P_L(l, m)$ satisfied the crossing-change formula (11). A similar situation exists for $K_L(l, m)$, but it requires a digression. Let L be an oriented link which is defined by a link diagram D, i.e. a regular planar projection. Let ε be the algebraic crossing number of the diagram, where crossings are counted as being positive (or negative) to correspond to the first (or second) picture in Figure 4. Kauffman defines in [K,1] a precurser of the polynomial $K_L(l, m)$, which he calls the " L-polynomial". We denote it by $\widehat{K}_D(l, m)$, defining it by:

(23.1) $$\widehat{K}_D(l, m) = l^{\varepsilon} K_L(l, m).$$

Note that $\widehat{K}_D(l, m)$ cannot be a link type invariant if $K_L(l, m)$ is, because the addition of a trivial loop in a link diagram, for example using Markov's second move, must leave $K_L(l, m)$ invariant but must then change $\widehat{K}_D(l, m)$. In fact, it must be true that:

(23.2) If D and D' are diagrams which are identical except for the addition of a trivial loop in D', which increases the algebraic crossing number, then $\widehat{K}_{D'}(l, m) = l\widehat{K}_D(l, m)$.

Continuing, let D_+, D_-, D_0, D_∞ be 4 link diagrams which are identical except inside a small disc, where they differ in the manner indicated in the four pictures in Figure 4. The $\widehat{K}_D$-polynomials $\widehat{K}_{D_+}, \widehat{K}_{D_-}, \widehat{K}_{D_0}, \widehat{K}_{D_\infty}$ which are associated to these diagrams are related by a crossing change formula, vis:

(23.3) $$\widehat{K}_{D_+}(l, m) + \widehat{K}_{D_-}(l, m) = m(\widehat{K}_{D_0}(l, m) + \widehat{K}_{D_\infty}(l, m)).$$

Two other properties of $\widehat{K}_D(l, m)$ are:

(23.4) If D is a planar circle, then $\widehat{K}_D(l, m) = 1$;

(23.5) If D, D' are diagrams which are related by regular isotopy in the plane (see [K,1]), then $\widehat{K}_D(l, m) = \widehat{K}_{D'}(l, m)$.

The 5 axioms (23.2)–(23.5) suffice to determine $\widehat{K}_D(l, m)$ on all diagrams, and (adding (23.1)) to define $K_L(l, m)$ on all links. It is proved in [K,1] that $K_L(l, m)$ is well defined and a link type invariant, and that $\widehat{K}_D(l, m)$ is well defined on *diagrams*, and is an invariant of equivalence classes of diagrams, under "regular isotopy" of diagrams in the plane (see [K,1] for the definition of regular isotopy).

Our idea is as follows. Let $\beta = \sigma_{\mu_1}^{\varepsilon_1} \cdots \sigma_{\mu_r}^{\varepsilon_r}$ be a braid which closes to an oriented link $L = L_\beta$. The braid determines a diagram $D = D_\beta$ for the link

L, and associated to this diagram is the polynomial $\widehat{K}_D(l,m)$. Eventually, we will define the trace (see §3) so that

$$\operatorname{trace}(G_{\mu_1}^{\varepsilon_1}\cdots G_{\mu_r}^{\varepsilon_r}) = (m^{-1}(l+l^{-1})-1)^{1-n}\widehat{K}_D(l,m).$$

Since $\widehat{K}_D(l,m)$ is well defined on diagrams, the trace will be seen to be well defined on monomials in the algebra. Then, if $g_i = l^{-1}G_i$, $1 \le i \le n-1$, and if $\varepsilon = \varepsilon_1 + \varepsilon_2 + \cdots + \varepsilon_{n-1}$, this is equivalent to

$$\operatorname{trace}(g_{\mu_1}^{\varepsilon_1}\cdots g_{\mu_r}^{\varepsilon_r}) = (m^{-1}(l+l^{-1})-1)^{1-n}K_L(l,m),$$

or (cf equation (9) above):

$$K_L(l,m) = (m^{-1}(l+l^{-1})-1)^{n-1}\operatorname{trace}(g_{\mu_1}^{\varepsilon_1}\cdots g_{\mu_r}^{\varepsilon_r}).$$

We now return to our algebra $\mathscr{C}_n(l,m)$, and to our attempts to motivate the relations in $\mathscr{C}_n(l,m)$. Recall from the discussion in §1 that if L_+, L_-, L_0 are as defined there, and if L_+ is the closure of a braid $\beta_+ \in B_n$, then it must be true that $\beta_+ = \alpha\sigma_i$, $\beta_- = \alpha\sigma_i^{-1}$, $\beta_0 = \alpha$, for some $\alpha \in B_n$, up to cyclic permutation of the defining braid. There seems to be no such braid interpretations for L_∞, indeed L_∞ cannot be defined by a braid in the usual sense.

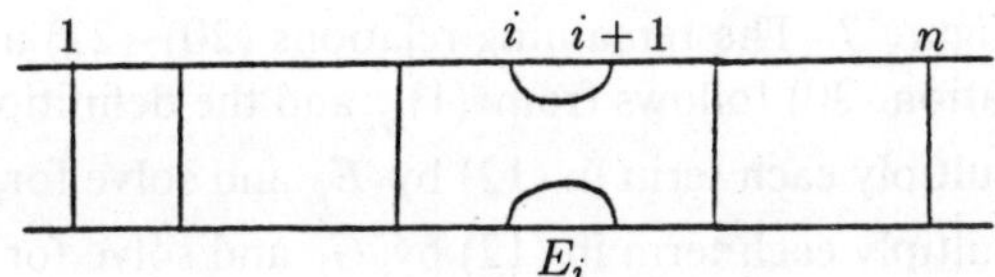

FIGURE 5

We introduce $(n-1)$ new elements $E_1, \ldots, E_{n-1}$, using (12) as our *definition* of E_i. Clearly E_i ought to have come from a braid-like object in a "braid monoid", namely the embedding on $I^{(n)} = \coprod_{j=1}^{n} I_j$ in $\mathbf{C} \times I$ which goes with the picture in Figure 5, and it will be helpful to think of it in this way [Bo, Y, K,2 and K,3].

There is no difficulty in composing such generalized braids with one another and with ordinary braids by concatenation. This allows us to interpret relations such as $E_iE_{i+1}E_i = E_i$ (relation (13)) and $G_{i+1}G_iE_{i+1} \approx E_iE_{i+1}$ (relation (14)) by pictures (Figure 6). Note that relations (13), (14), (15), (16), (17) are all associated to pictures which involve only regular isotopy. On the other hand, relations (18) and (19) require something more, because on the passage from the left side of (18) or (19) to the right we must delete a trivial loop, and axiom (23.2) asserts that this must be accounted for by the addition of a multiplicative

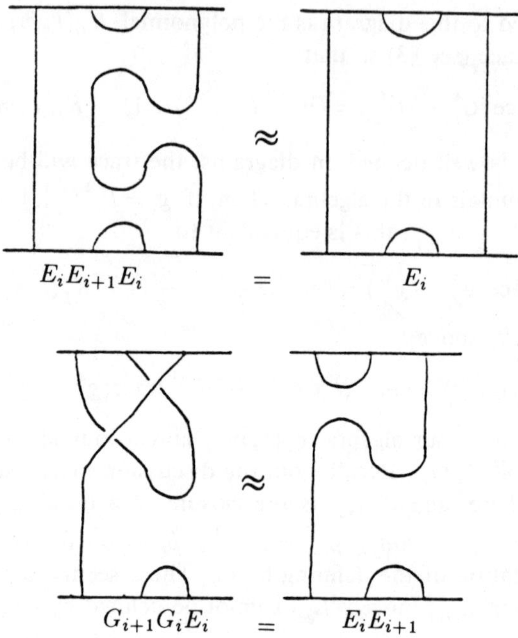

FIGURE 6

factor $l^{\pm 1}$. See Figure 7. The remaining relations (20)–(22) are consequences of the others. Relation (20) follows from $(1)_G$ and the definition of E_i in (12). To obtain (21), multiply each term in (12) by E_i and solve for E_i^2, using (18). To obtain (22), multiply each term in (12) by G_i and solve for G_i^2, using (18). This accounts for all of the relations in $\mathscr{C}_n(l,m)$, and explains how we were led to our definition. In the next section we will study the algebra.

3. PROPERTIES OF $\mathscr{C}_n(l,m)$

Lemma 3.1. *$\mathscr{C}_n(l,m)$ is finite dimensional. Its dimension is bounded below by $n!$. Each element of $\mathscr{C}_n(l,m)$ can be written as a linear combination of elements of the form $w_1\gamma w_2$ with $\gamma \in \{G_{n-1}, E_{n-1}, 1\}$ and w_1, w_2 monomials in $1, G_1, E_1, \ldots, G_{n-2}, E_{n-2}$.*

Proof. The mapping $\psi : \mathscr{C}_n(l,m) \to \mathscr{H}_n(l,m)$ defined by $G_i \to lg_i$, $E_i \to 0$ is a homomorphism, so $\dim \mathscr{C}_n(l,m) > \dim \mathscr{H}_n(l,m) = n!$.

To prove that $\dim \mathscr{C}_n(l,m) < \infty$, it suffices to prove that each $w \in \mathscr{C}_n = \mathscr{C}_n(l,m)$ is a sum of monomials of the form $w'\gamma w''$ where w', w'' belong to the subalgebra $\widetilde{\mathscr{C}}_{n-1}$ of C_n generated by $1, G_1, \ldots, G_{n-1}$ and where $\gamma = G_{n-1}$ or G_{n-1}^{-1} or E_{n-1} or 1. The statement of the lemma then follows, because we can use (12) to eliminate $G_1^{-1}, \ldots, G_{n-1}^{-1}$.

If $n = 2$ the assertion is trivially true, so assume $n > 2$. Let w be a monomial in $\mathscr{C}_n$. Then $w = w_0\gamma_0 w_1\gamma_1 \cdots w_r\gamma_r$, where each $w_j \in \widetilde{\mathscr{C}}_{n-1}$ and each $\gamma_j = G_{n-1}$, E_{n-1} or 1. If $r = 0$ we are done, so assume $r \geq 1$. Then $w = w_0\gamma_0 w_1\gamma_1 z$, where $z = w_2\gamma_2 \cdots w_r\gamma_r$. Since $w_1 \in \widetilde{\mathscr{C}}_{n-1}$, induction on r yields $w_1 = v_0\alpha v_1$, where $v_0, v_1 \in \widetilde{C}_{n-2}$ and $\alpha = G_{n-2}$ or E_{n-2} or 1. Then v_0, v_1 commute with γ_0, γ_1 by $(1)_G$ so $w = (w_0 v_0)(\gamma_0\alpha\gamma_1)(v_1 z)$, and it suffices to prove that the product $\gamma_0\alpha\gamma_1$ is a sum of monomials which involve G_{n-1} or E_{n-1} once, for all $\gamma_0, \gamma_1 \in \{G_{n-1}, E_{n-1}\}$ and for $\alpha \in \{1, G_{n-2}, E_{n-2}\}$.

If $\alpha = 1$, then relations (18), (21), (22) suffice to reduce each of these as claimed. Assume $\alpha \neq 1$. Then

$$\begin{aligned}\gamma_0\alpha\gamma_1 \in \{&G_{n-1}G_{n-2}G_{n-1},\ G_{n-1}G_{n-2}E_{n-1},\\ &G_{n-1}E_{n-2}G_{n-1},\ G_{n-1}E_{n-2}E_{n-1},\ E_{n-1}G_{n-2}G_{n-1},\\ &E_{n-1}G_{n-2}E_{n-1},\ E_{n-1}E_{n-2}G_{n-1},\ E_{n-1}E_{n-2}E_{n-1}\}.\end{aligned}$$

Using relations $(2)_G$, (12), (13), (14), (15), (16), (17), and (19) each of these can be reduced to a sum of monomials, each of which contains E_{n-1} or G_{n-1} once. Induction on r completes the proof. □

Remark. See Theorem 3.7 for a precise formula for $\dim \mathscr{C}_n(l, m)$.

We now show that our algebras support a trace function. The trace is most conveniently defined if we rescale the generators, so let $g_i = l^{-1}G_i$, $1 \leq i \leq n-1$.

Theorem 3.2. *Each $\mathscr{C}_n(l, m)$ supports a trace function $\tau' : \mathscr{C}_n \to \mathbf{C}$ which is characterized by the properties*:

$$(4)' \qquad \tau'(a+b) = \tau'(a) + \tau'(b),$$

$$(5)' \qquad \tau'(ab) = \tau'(ba),$$

$$(6)' \qquad \tau'(1) = 1 \quad \textit{where } 1 \in \mathscr{C}_n \textit{ for any } n \in N,$$

$$(7)' \qquad \tau'(wg_n) = \tau'(wg_n^{-1}) = \tau'(wE_n) = z\tau(w) \textit{ if } w \in \textit{subalgebra of } \mathscr{C}_n \textit{ generated by } g_1, \ldots, g_{n-1}, \textit{ where } z^{-1} = m^{-1}(l + l^{-1}) - 1.$$

Proof. There is an algebra homomorphism $\Delta : \mathbf{C}B_n \to \mathscr{C}_n = \mathscr{C}_n(l, m)$ which sends σ_i to g_i. For any monomial $w = g_{\mu_1}^{\varepsilon_1} g_{\mu_2}^{\varepsilon_2} \cdots g_{\mu_r}^{\varepsilon_r} \in \mathscr{C}_n$, let $W = \sigma_{\mu_1}^{\varepsilon_1}\sigma_{\mu_2}^{\varepsilon_2} \cdots \sigma_{\mu_r}^{\varepsilon_r} \in \Delta^{-1}(w)$. The braid W determines a closed braid $\widehat{W}$ and so a link which we also refer to as $\widehat{W}$, with Kauffman polynomial $K(\widehat{W})$. Let

$$(24) \qquad \tau'(w) = z^{n-1}K(\widehat{W}), \quad \text{where } z^{-1} = m^{-1}(l + l^{-1}) - 1,$$

and extend linearly to all of $\mathscr{C}_n(l, m)$.

For later use, note that by (23.1) the link $\widehat{W}$ has diagram polynomial $\widehat{K}(\widehat{W}) = l^{\varepsilon_1+\varepsilon_2+\cdots+\varepsilon_r}K(\widehat{W})$.

Assuming for the moment that τ' is well defined, independently of the choice of $W \in \Delta^{-1}(w)$, we note that τ' satisfies properties $(4)'$–$(7)'$: Property $(4)'$ is satisfied by definition. Property $(5)'$ holds because the closed braid $\widehat{W}$ depends only on the *cyclic* word W. Property $(6)'$ follows from the fact that $1 \in \mathscr{C}_n$ lifts to $1 \in B_n$, and $\hat{1}$ is the unlink of n-components, with Kauffman polynomial z^{1-n}. Thus $\tau'(1) = z^{n-1}z^{1-n} = 1$. Property $(7')$ follows from the fact that the links defined by closing the braids $W \in B_n$ and $W\sigma_n^{\pm 1} \in B_{n+1}$ are equivalent, and so have the same Kauffman polynomial. It is immediate that these four properties determine τ'.

It remains to prove that τ' is well defined. For this it suffices to show that K takes the value 0 on the kernel of Δ. Recall that $\mathbf{C}B_n$ is generated by $\sigma_1, \dots, \sigma_{n-1}$, with defining relations $(1)_\sigma$ and $(2)_\sigma$, while $\mathscr{C}_n$ is generated by $G_1, \dots, G_{n-1}$, with defining relations (12)–(19). Set $\xi_1 = m^{-1}(l\sigma_i + l^{-1}\sigma_i^{-1}) - 1 \in B_n$. Then $\Delta(\xi_i) = E_i$. Lift G_i to $l\sigma_i$. Then $\Delta(\sigma_i) = l^{-1}G_i$.

Each of the relations in (13)–(19) lifts to an element in the kernel of Δ, say R_j $(j = 1, \dots, q)$, and the kernel of Δ is the smallest two-sided ideal in $\mathbf{C}B_n$ spanned by the elements $R_1, \dots, R_q$. Each R_j is a linear combination of monomials in $\mathbf{C}B_n$. For example, one of the relations in (18) yields:

$$R_1 = l^2\sigma_i\xi_i - \xi_i.$$

If we use the symbol $\widehat{WR_j}$ for the sum of the closed braids determined by left-multiplying R_j by an arbitrary monomial $W \in \mathbf{C}B_n$, we must prove that $K(\widehat{WR_j}) = 0$. Note that we need only use left multiplication because closed braids belong to cyclic words in B_n.

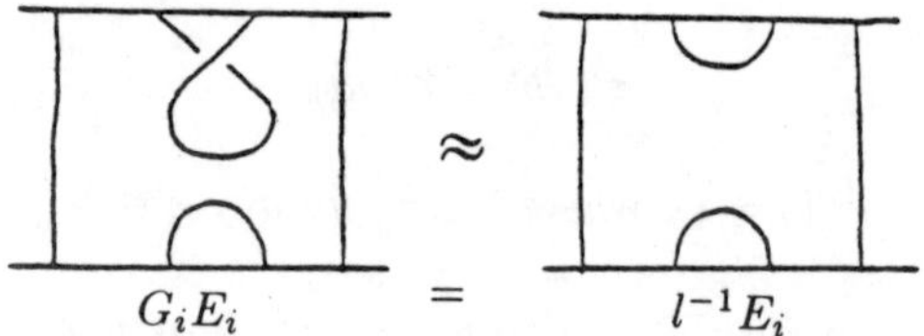

FIGURE 7

We do the calculation in the case of R_1, defined above. Recall that R_1 came from the relation (18), which was motivated by the pictures in Figure 7. We will use Figure 7 in a roundabout way, to be described. First, if we replace ξ_1 in R_1 by its expression as a sum of elementary braids σ_i and σ_i^{-1} we obtain:

$$mR_1 = l^2[l\sigma_i^2 + l^{-1} - m\sigma_i] - [l\sigma_i + l^{-1}\sigma_i^{-1} - m].$$

Therefore, we must prove that

$$l[l^2K(\widehat{W\sigma_i^2}) + K(\widehat{W}) - mlK(\widehat{W\sigma_i})] - [lK(\widehat{W\sigma_i}) + l^{-1}K(\widehat{W\sigma_i^{-1}}) - mK(\widehat{W})] = 0.$$

Equivalently, via (23.1), we must prove that:

$$l[\widehat{K}(\widehat{W\sigma_i^2}) + \widehat{K}(\widehat{W}) - m\widehat{K}(\widehat{W\sigma_i})] - [\widehat{K}(\widehat{W\sigma_i}) + \widehat{K}(\widehat{W\sigma_i^{-1}}) - m\widehat{K}(\widehat{W})] = 0.$$

To do so, we first turn to Figure 4, and recall the crossing-change formula (23.3) which relates the diagram polynomials of four links L_+, L_-, L_0, L_∞ which are defined by diagrams which differ only in a small region, where the difference is as specified in Figure 4. Note that the closed braid diagrams for $\widehat{W\sigma_i^2}, \widehat{W}, \widehat{W\sigma_i}$ make up one such triplet L_+, L_-, L_0 and that $\widehat{W\sigma_i}, \widehat{W\sigma_i^{-1}}, \widehat{W}$ make up another L'_+, L'_-, L'_0. Therefore if we bring the links L_∞ and L'_∞ (with obvious notation) into the picture, we must show that

$$l\widehat{K}(L_\infty) - \widehat{K}(L'_\infty) = 0$$

where our links L_∞ and L'_∞ are defined by diagrams which are identical everywhere except inside a region where they differ in the manner indicated in Figure 7, the picture on the left corresponding to L_∞ and that on the right to L'_∞. Since Kauffman's axiom (23.2) asserts that $\widehat{K}(L'_\infty) = l\widehat{K}(L_\infty)$ as required, and the proof is complete for R_1.

The other relations can be checked in the same way as the previous example. Note that it is enough to show that $\tau'(R_j) = 0$. Indeed, if $W \in B_n$ and R_j is one of our relators, the links corresponding to $\widehat{WR_j}$ are the same everywhere except in a small area in which they differ from each other in the same way as the ones corresponding to $\widehat{R}_j$. So we can show that $K(\widehat{WR_j}) = 0$ using the same crossing change rules coming from the definition of Kauffman's polynomial as we use for showing $K(\widehat{R}_j) = 0$. This implies $\tau'(WR_j) = 0$.

The fact that $\tau'(R_j) = 0$ follows essentially from the definition of our relations which were motivated by pictures coming from the definition of Kauffman's link invariant (see the end of §2). □

We now go on to determine the algebraic structure of $\mathscr{C}_n(l, m)$. Our main tool will be a generalized version of Jones' basic construction in [J,1]. For this, we will first have to extend methods already developed in [J,1 and W]. Our main observation is that the positivity assumptions in those papers are not necessary for our purposes. The only assumption will be that our traces are nondegenerate. This is made precise below.

We will assume throughout this section that $A < B$ are finite dimensional algebras over a field S of characteristic zero. Let $M_k(S)$ be the algebra of all $k \times k$ matrices with entries from S. Assume that $A = \bigoplus A^{(i)}$, $B = \bigoplus B^{(j)}$ with $A^{(i)} \cong M_{a_i}(S)$, $B^{(j)} \cong M_{b_j}(S)$ with $a_i, b_j \in \mathbf{N}$. The vectors $\vec{a} = (a_i)$ and $\vec{b} = (b_j)$ are called the *dimension vectors* of A resp. B.

Let tr be a nondegenerate trace on B, i.e.

$$\text{(25)} \qquad \operatorname{tr}(xy) = 0 \quad \forall y \in B \quad \text{implies } x = 0.$$

It is well known and easy to check that, in our case, this is equivalent to $\operatorname{tr}(p) \neq 0$ for every minimal idempotent $p \in B$. Let us also recall that if tr is nondegenerate, the map $B \ni b \mapsto \operatorname{tr}(b\cdot) \in B^*$, the dual of B, is an isomorphism between

B and B^* (as usual, $\mathrm{tr}(b\cdot)$, denotes the map $x \mapsto \mathrm{tr}(bx)$), i.e. for any linear map $\varphi: B \to S$ there is a unique $b \in B$ such that $\varphi(x) = \mathrm{tr}(bx)$ for all $x \in B$.

Let us also assume that the restriction of tr onto A is nondegenerate. Then, using the isomorphism above for A and A^*, there exists for every $b \in B$ a necessarily unique $\varepsilon_A(b) \in A$ such that

$$\mathrm{tr}(b\cdot)|_A = \mathrm{tr}(\varepsilon_A(b)\cdot)|_A.$$

It can be checked easily that the map $B \ni b \to \varepsilon_A(b) \in A$ is linear and that

$$\mathrm{tr}(b) = \mathrm{tr}(\varepsilon_A(b)), \tag{26}$$

$$\varepsilon_A(a_1 b a_2) = a_1 \varepsilon_A(b) a_2, \qquad a_1, a_2 \in A,\ b \in B. \tag{27}$$

We call ε_A the trace preserving conditional expectation from B to A. Let us represent B via the left regular representation onto itself. To avoid confusion we write B_ξ if we regard B as the representation space and b_ξ for an element of B_ξ. Let $\mathscr{L}(B_\xi)$ be the set of all linear maps on B_ξ. For any algebra C on B_ξ let

$$C' = \{x \in \mathscr{L}(B_\xi),\ xc = cx,\ \forall c \in C\}.$$

It is well known (and can be checked by explicit matrix multiplications) that C is isomorphic to a direct sum of full matrix algebras iff C' is isomorphic to a direct sum of full matrix algebras. If B (and B_ξ) is finite dimensional it follows that $(B')' = B$ (see for instance [La, XVIII, 3, Theorem 1]).

By our assumptions on the structure of B, we can regard B as a concrete matrix algebra such that with $b \in B$, its transpose, denoted here by b^*, is in B. Moreover, we can choose this matrix representation such that with a also a^* is in $A \subset B$. Note that $^*: B \to B$, $b \mapsto b^*$ is an involution with $(ab)^* = b^* a^*$. We can then define a map $J: B_\xi \to B_\xi$ by $J(b_\xi) = b_\xi^*$. A straightforward computation shows that $JBJ \subseteq B'$. On the other hand, if $b' \in B'$ such that $b'(1_\xi) = b_\xi$, we have $b'(a_\xi) = ab_\xi = (Jb^*J)(a_\xi)$ for any $a \in B$. Hence $JBJ = B'$. As in [J,1], we define $e_A \in L(B_\xi)$ by $e_A(b_\xi) = (\varepsilon_A(b))_\xi$.

Lemma 3.3. *With the above notation, we have*

(i) $e_A x e_A = \varepsilon_A(x) e_A$ *for* $x \in B$.
(ii) *Let* $x \in B$. *Then* $x \in A$ *iff* $e_A x = x e_A$.
(iii) $A' = \{B' \cup \{e_A\}\}''$.
(iv) J *commutes with* e_A.

Proof. See [J, (3.1.4)]. For (ii), replace the statement "ξ is separating" by "the map $B \ni b \to b_\xi \in B_\xi$ is injective".

Obviously, a simple $B^{(j)}$ module is also an A module. Let g_{ij} be the number of simple $A^{(i)}$ modules in its decomposition into simple A modules. The matrix $G = (g_{ij})$ is called the *inclusion matrix* for $A \subset B$. If A and B have the same identity element, we have

$$\vec{b} = G^t \vec{a}. \tag{28}$$

Under these assumptions we can give a precise description of $\langle B, e_A \rangle$, the algebra generated by B and e_A. As in [J,1], we call $\langle B, e_A \rangle$ the *basic construction* for $A \subset B$.

Proposition 3.4. (i) $\langle B, e_A \rangle \cong A'$. *In particular,* $\langle B, e_A \rangle$ *is isomorphic to a direct sum of full matrix algebras.*

(ii) *There exists a canonical* 1-1 *correspondence between the direct summands of* A *and* $\langle B, e_A \rangle$ *such that if* q *is a minimal idempotent of* $A^{(i)}$, $e_A q$ *is a minimal idempotent of* $\langle B, e_A \rangle^{(i)}$.

(iii) *The inclusion matrix for* $B \subset \langle B, e_A \rangle$ *is* G^t.

(iv) *Every element* $x \in \langle B, e_A \rangle$ *can be written in the form* $x = \sum a_i e_A b_i + c$, *where* $a_i, b_i c \in B$.

Proof. (i) As $A^* = A$, it follows as in [J,1, (3.1.5)] that $\langle B, e_A \rangle = JA'J$.

(ii)–(iv) follow from [J,1, (3.3.1)]. □

The following theorem is an extension of [W, Proposition (1.2)] to semisimple algebras.

Theorem 3.5. *Let* A, B, tr *and* ε_A *be as above. Assume that* B *is contained in an algebra* C *and there is an element* $e \in C$ *such that*

(i) $e^2 = e$,

(ii) $exe = \varepsilon_A(x)e = e\varepsilon_A(x)$ *for all* $x \in B$,

(iii) *The map* $a \in A \mapsto ae \in \langle B, e \rangle$ *is an injective homomorphism with* $e1 = e$.

Then the 2*-sided ideal* $\langle e \rangle \subset \langle B, e \rangle$ *generated by* e *is isomorphic to* $\langle B, e_A \rangle$ *as an algebra. The quotient* $\langle B, e \rangle / \langle e \rangle$ *is isomorphic to a subalgebra of* B *and splits as a direct summand in* $\langle B, e \rangle$.

Proof. For convenience, let us not distinguish between $B \subset C$ and the isomorphic image of B in the representation on B_ξ. Let us define $\varphi : B \cup \{e\} \to \mathscr{L}(B_\xi)$ by $b \mapsto b. \in \mathscr{L}(B_\xi)$ and $e \mapsto e_A$. We claim that φ extends to a well-defined homomorphsim from $\langle B, e \rangle$ onto $\langle B, e_A \rangle$. The only nontrivial part is to show that φ is well defined. Let $x = \sum a_i e b_i + c = 0$, $a_i, b_i, c \in B$. (It follows from (i)–(iii) that this is the most general form of an element of $\langle B, e \rangle$; see also [J,1].) Let $\tilde{y}. \in B$ and let $\tilde{x} = \sum a_i \varepsilon_A(b_i \tilde{y}) + c\tilde{y}$. Then $0 = eyxe = \varepsilon_A(y\tilde{x})e$ $\forall y. \in B$. By (iii), $\varepsilon_A(y\tilde{x}) = 0$ and therefore also $\mathrm{tr}(y\tilde{x}) = 0$ for all $y. \in B$. Hence $\tilde{x} = 0$ and we obtain

$$\varphi(x)(\tilde{y}\xi) = \tilde{x}\xi = 0.$$

As $\tilde{y}$ was arbitrary, $\varphi(x) = 0$.

On the other hand, let $x = \sum a_i e b_i + c. \in ke\varphi$. Then $0 = \varphi(x)(\tilde{y}\xi) = (\sum a_i \varepsilon_A(b_i \tilde{y}) + c\tilde{y})\xi$ $\forall \tilde{y} = B$. Hence $\sum a_i \varepsilon_A(b_i \tilde{y}) + c\tilde{y} = 0$ and in particular,

$$(*) \qquad x\tilde{y}e = 0 \qquad \forall \tilde{y} \in B.$$

But then we also have

$$\begin{aligned}\operatorname{tr}\left(y\left(\sum a_i\varepsilon_A(b_i\tilde{y})+c\tilde{y}\right)\right) &= \operatorname{tr}\left(\sum \varepsilon_A(ya_i)\varepsilon_A(b_i\tilde{y})+yc\tilde{y}\right)\\ &= \operatorname{tr}\left(\sum \varepsilon_A(ya_i)b_i+yc)\tilde{y}\right)=0\end{aligned}$$

for all $y, \tilde{y}. \in B$.

So, as tr is nondegenerate, we obtain $\sum \varepsilon_A(ya_i)b_i + yc = 0$ and in particular

$$(**) \qquad eyx = 0 \qquad \forall y. \in B.$$

Hence if $x \in \ker\varphi$ and $y \in BeB$, we have $xy = yx = 0$ by $(*)$ and $(**)$. In particular, if $z \in \ker\varphi \cap BeB$, z is annihilated both by BeB and $\ker\varphi$.

As $\varphi(BeB) = \langle B, e_A\rangle$ (see Proposition 3.4, (iv)), BeB and $\ker\varphi$ generate $\langle B, e\rangle$. So $z = 1z = 0$, as by (iii) $1 \in B$ is also the multiplicative identity in $\langle B, e\rangle$. It follows from this that $BeB \cong \langle B, e_A\rangle$, which is a direct sum of full matrix algebras.

By (i)–(iii), $\langle e\rangle \cong BeB$. Obviously, the quotient $\langle B, e\rangle/\langle e\rangle$ is already generated by the image of B, hence it is isomorphic to a subalgebra of the semisimple algebra B. It is well known that in this case

$$\langle B, e\rangle \cong \langle e\rangle \oplus \langle B, e\rangle/\langle e\rangle.$$

Lemma 3.6. *Let $e_n = zE_n$, where $z = 1/(m^{-1}(l+l^{-1})-1)$ and assume that τ' is nondegenerate on C_{n-1} and C_n. Then e_n has properties* (i)–(iii) *of Theorem* 3.5 *for $C_{n-1} \subset C_n$ and C_{n-1} is isomorphic to the subalgebra of C_n generated by $G_1, \dots, G_{n-2}, E_1, \dots, E_{n-2}$.*

Proof. Using Theorem 3.2, $(7)'$ and the fact that τ' is nondegenerate we can show easily that $\mathscr{C}_{n-1} \ni x \mapsto xe_n \in \mathscr{C}_{n+1}$ is injective. From this follows the last statement of the lemma and property (iii). By Lemma 3.1, any element of $\mathscr{C}_n(l, m)$ can be written as a linear combination of elements of the form $a\chi b$ with $a, b \in \mathscr{C}_{n-1}(l, m)$ and $\chi \in \{1, G_{n-1}, E_{n-1}\}$.

By Theorem 3.2, $(7)'$ we obtain for χ as above and $c \in \mathscr{C}_{n-1}(l, m)$

$$\tau'((a\chi b)c) = \tau'(\chi)\tau'(abc).$$

Hence $\varepsilon(a\chi b) = \tau'(\chi)ab$ where

$$\varepsilon : \mathscr{C}_n(l, m) \to \mathscr{C}_{n-1}(l, m)$$

is the trace preserving conditional expectation. On the other hand

$$\begin{aligned} e_n(a\chi b)e_n = z^2aE_n\chi E_nb &= \begin{cases} z^2aE_nb & \text{if } \chi = E_{n-1},\\ lz^2aE_nb & \text{if } \chi = G_{n-1},\end{cases}\\ &= \tau'(\chi)abe_n = \varepsilon(a\chi b)e_n.\end{aligned}$$

This shows property (ii). Property (i) follows from (21). □

We can now determine the structure of C_n as an algebra over an appropriate field.

Theorem 3.7. *Let us regard* $C_{n+1} = C_{n+1}(l, m)$ *as an algebra over* $\mathbf{C}(l, \alpha)$ *where* $m = \alpha + \alpha^{-1}$. *Then* $C_1 \cong \mathbf{C}(l, \alpha)$, $C_2 \cong \mathbf{C}(l, \alpha)^3$ *and*

$$C_{n+1} \cong H_{n+1} \oplus \langle C_n, e_{C_{n-1}} \rangle,$$

where $H_{n+1} = H_{n+1}(l, \alpha + \alpha^{-1})$ *and the second summand is Jones' basic construction for* $C_{n-1} \subset C_n$, *which is semisimple. In particular* C_{n+1} *is semisimple, it has dimension* $1 \cdot 3 \cdot 5 \cdots (2n+1)$ *and* τ', *the trace as defined in Theorem 3.2, is nondegenerate.*

Proof. Clearly, $C_1 \cong \mathbf{C}(l, \alpha)$. Let z be the rational function in l and $m = \alpha + \alpha^{-1}$ as defined in the previous lemma and let

$$p_1 = zE_1, \qquad p_2 = \frac{G_1 - \alpha}{\alpha^{-1} - \alpha}(1 - p_1) \quad \text{and} \quad p_3 = \frac{G_1 - \alpha^{-1}}{\alpha - \alpha^{-1}}(1 - p_1).$$

Then it follows from (12), (18) and (21) that $p_i^2 = p_i$, $p_i p_j = 0$ for $i, j = 1, 2, 3$, and $i \neq j$ and that $p_1 + p_2 + p_3 = 1$. Using property $(7)'$ of Theorem (3.2) one sees that $\tau'(p_i) \neq 0$ for $i = 1, 2, 3$ which shows the assumptions for C_2.

It has already been pointed out earlier that the additional relation $E_i = 0$ reduces the generating relations of C_n to the defining relations of the Hecke algebra H_n. We thus obtain a homomorphism ψ from C_n onto H_n. By definition of ψ, its kernel is equal to the ideal I generated by the E_i's. By a theorem of Benson and Curtis (see [Lu, Theorem (3.1)])

$$(*) \qquad C_n / \ker \psi \cong H_n \cong \mathbf{C}(l, \alpha) S_n,$$

where S_n denotes the symmetric group on n letters.

Using relation (13) repeatedly, we obtain

$$E_i = E_i E_{i+1} \cdots E_{n-1} E_n E_{n-1} \cdots E_{i+1} E_i.$$

Hence I coincides with the ideal $\langle E_n \rangle \subset C_{n+1}$ generated by E_n. Using relations (13), (16) and (17), we show as in Lemma 3.1 that $\langle E_n \rangle$ is the linear span of products of generators of the form $w_1 E_n w_2$ with $w_1, w_2 \in C_n$. It follows from this, Theorem 3.5 and Lemma 3.6 that

$$\ker \psi = \langle E_n \rangle = C_n E_n C_n \cong \langle C_n, e_{C_{n-1}} \rangle.$$

We shall show at the end of this section by a simple dimension argument that $\langle C_n, e_{C_{n-1}} \rangle = \langle C_n, e_{C_{n-1}} \rangle''$. So by Proposition (3.4), (i) $\ker \psi$ is semisimple which is also true for the quotient $C_{n+1} / \ker \psi$ by $(*)$. Hence

$$C_{n+1} \cong H_{n+1} \oplus C_n E_n C_n.$$

It remains to show that τ' is nondegenerate. Let p be a minimal projection of C_{n-1}. Then $\tau'(e_n p) = z^2 \tau'(p) \neq 0$ by induction assumption. So τ' is nondegenerate on $C_n E_n C_n$ by Proposition (3.4), (ii) and the remark below (25).

Note that if $l = 1$ and $\alpha = i$, $z(1,i) = 0$ and $\tau'(p_2)(1,i) = \tau'(p_3)(1,i) = 1/2$. By the induction assumption $\tau'(p)(1,i) = 0$ for any minimal projection in $C_{n-2}E_{n-2}C_{n-2}$ and $\tau'(p)(1,i) \neq 0$ for any minimal projection in the H_{n-1} summand of C_{n-1}. It follows as in the previous paragraph that $\tau'(p)(1,i) = 0$ for any minimal projection in $C_nE_nC_n$. So τ' can be considered as a trace on H_{n+1} having Markov property $(7)'$. These traces are uniquely determined by $q = -\alpha^2$ and $\eta = (\mathrm{tr}(\tilde{g}_i) + 1)/(1+q)$. For our special choice of values we obtain $q = 1$ and $\eta = 1/4$. But then it follows from Ocneanu's classification of positive traces on H_∞ that $\tau'(p)(1,i) > 0$ for all minimal projections in H_{n+1} (see [W, proof of Theorem (3.5), (i), α]). Hence $\tau'(p) \neq 0$ for any minimal projection p in the H_{n+1}-summand of C_{n+1}. This shows that τ' is nondegenerate on C_{n+1}.

We next describe how to use "Bratteli diagrams" to determine the structure of $C_n(l,\alpha)$. These diagrams allow one to compute the irreducible representations of these algebras by an iterative procedure. We will first assume that $\langle C_n, e_{C_{n-1}}\rangle = \langle C_n, e_{C_{n-1}}\rangle''$, where the right-hand side is semisimple by Proposition 3.4. We will then show by computing dimensions that the left-hand side cannot be a proper subalgebra of the right-hand side. An example will be given after we describe the basic setup. Related constructions are to be found in [J,2] (for the algebra we called $A_n(t)$ in the introduction) and in [W] (for $\mathscr{H}_n(t)$). See also [J,4], where the underlying construction for the group algebra $\mathbf{C}S_n$ of the symmetric group by "Young diagrams" is described in a particularly lucid manner.

The Bratteli diagram for $\mathscr{C}_n(l,m)$ is a graph which we now describe. See Figure 8. The vertices are arranged in horizontal rows, those in the nth row from the top being in 1-1 correspondence with the irreducible representations of $\mathscr{C}_n(l,m)$, $n = 1,2,3,\ldots$. A vertex in row n is joined to one in row $n-1$ by k edges if the restriction of the corresponding representation of $\mathscr{C}_n$ to $\mathscr{C}_{n-1}$, under the natural inclusion $\mathscr{C}_{n-1} \hookrightarrow \mathscr{C}_n$, contains that irreducible representation of $\mathscr{C}_{n-1}$ k times. The vertices are labeled by integers which denote the dimension; the dimension of a representation of $\mathscr{C}_n$ is the sum of the dimensions of all those representations of $\mathscr{C}_{n-1}$ which are joined to it by edges. Note that $\mathscr{C}_n = \mathscr{H}'_n \oplus \mathscr{H}_n$ by Theorem 3.7, where $\mathscr{H}'_n$ is $\mathscr{C}_nE_n\mathscr{C}_n$. We now describe how to construct row n of the diagram from row $n-1$, $n \geq 3$. One first reflects that part of the diagram which consists of the vertices in rows $n-2$ and $n-1$ and all edges joining them about the horizontal line corresponding to the representations of $\mathscr{C}_{n-1}$. One then inserts the dimension of the representations of $\mathscr{C}_{n-1}$, and assigns to each representation of $\mathscr{C}_n$ the sum of the dimensions of the representations lying above it. See Figure 8. This determines that part of $\mathscr{C}_n$ which we have called $\mathscr{H}'_n$. One then augments it by adding more vertices, one for each irreducible representation of $\mathscr{H}_n$, using the conventions in [W] (or in [J,2]; see especially §3 of [J,2]). This makes sense because each $\mathscr{C}_j$, $j < n$, has already been constructed as $\mathscr{H}'_j \oplus \mathscr{H}_j$, so that the Bratteli diagram for $\{\mathscr{H}_n; n = 1, 2, 3, \ldots\}$ is embedded as a subdiagram of that for $\{\mathscr{C}_n; n = 1, 2, 3, \ldots\}$. Finally, we note that the inductive construction begins with the unique 1-dimensional trivial representation of $\mathscr{C}_1$, and three 1-dimensional representations of $\mathscr{C}_2$, one belonging to $\mathscr{H}'_2$ and the other two to $\mathscr{H}_2$.

Figure 8 illustrates the Bratteli diagram through $n = 4$, and shows the construction of $\mathscr{H}'_5$ from that of $\mathscr{C}_4$ and $\mathscr{C}_3$. Later, in §6, we will use this diagram to obtain explicit matrix representations of B_3 and B_4.

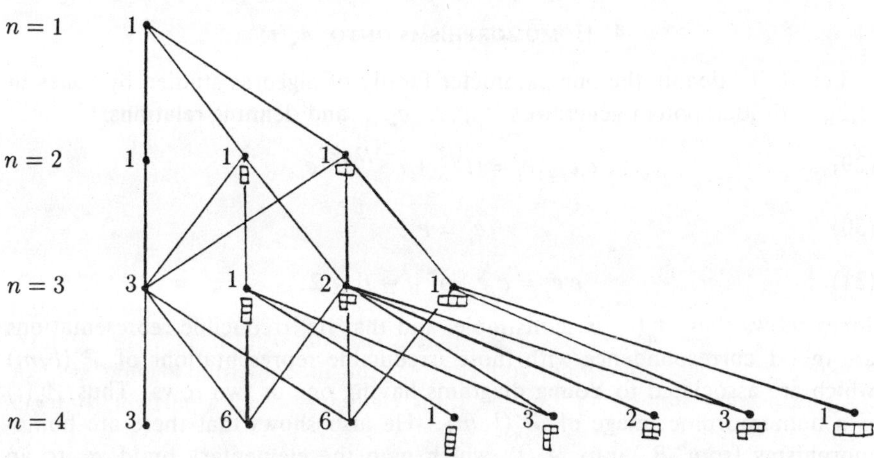

(a) The construction of $H'_n \oplus H_n$, $1 \leq n \leq 4$

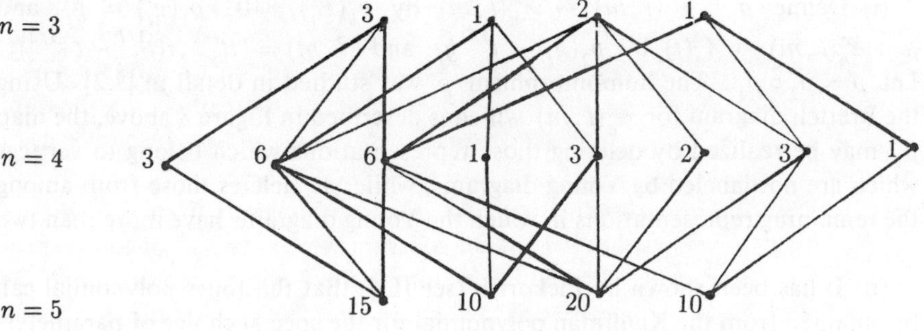

(b) The construction of H'_n, $n = 5$

FIGURE 8

We now show the missing part in the proof of Theorem 3.7, namely that $\langle C_n, e_{C_{n-1}}\rangle = \langle C_n, e_{C_{n-1}}\rangle''$. It is a well-known combinatorial fact that the algebras constructed by this procedure (i.e. with $\langle C_n, e_{C_{n-1}}\rangle''$ instead of $\langle C_n, e_{C_{n-1}}\rangle$) have dimension $1\cdot 3\cdots(2n+1)$ (see for instance [S, Lemma 8.7]). On the other hand, we shall see in §5 that for a special choice of (slightly modified) parameters $C_{n+1}(l,\alpha)$ specializes to Brauer's algebra $D_{n+1}(x)$ which has dimension $1\cdot 3\cdots(2n+1)$. So also the dimension of $C_{n+1}(l,\alpha)$ has to be at least that number. This shows again by induction on n that $\langle C_n, e_{C_{n-1}}\rangle = \langle C_n, e_{C_{n-1}}\rangle''$. □

4. Homomorphisms onto $A_n(t)$

Let $A_n(t)$ denote the one-parameter family of algebras studied by Jones in [J,2], with idempotent generators $e_1, \dots, e_{n-1}$ and defining relations:

$$e_i e_{i\pm 1} e_i = (t^{1/2} + t^{-1/2})^{-2} e_i, \tag{29}$$

$$e_i^2 = e_i, \tag{30}$$

$$e_i e_j = e_j e_i \quad \text{if } |i-j| \geq 2. \tag{31}$$

Jones shows that $A_n(t)$ is semisimple, and that its irreducible representations are in 1-1 correspondence with those irreducible representations of $\mathscr{H}_n(l,m)$ which are associated to Young diagrams having one or two rows. Thus $A_n(t)$ is a homomorphic image of $\mathscr{H}_n(l,m)$. He also shows that there are homomorphisms from B_n into $A_n(t)$ which map the elementary braid σ_i to an appropriate linear combination of e_i and 1, which we will denote by $\hat{g}_i$ here.

We now show that there are two inequivalent homomorphisms ρ and ψ from $\mathscr{C}_n(l,m) \to A_n(t)$. Inequivalent means that there cannot be an isomorphism i of $\mathscr{C}_n(l,m)$ with $\rho = \psi \circ i$.

(i) Define $\rho_1 : \mathscr{C}_n(l,m) \to \mathscr{H}_n(l,m)$ by $\rho_1(E_i) = 0$, $\rho_1(g_i) = \tilde{g}_i$, and $\rho_2 : \mathscr{H}_n(l,m) \to A_n(t)$ by $\rho_2(\tilde{g}_i) = t^{1/2}\hat{g}_i$, and $(l,m) = (it^{-1}, i(t^{1/2} - t^{-1/2}))$. Let $\rho = \rho_2 \circ \rho_1$. The homomorphism ρ was studied in detail in [J,2]. Using the Bratteli diagram for $\mathscr{C}_n(l,m)$ which is described in Figure 8 above, the map ρ_1 may be realized by deleting those representations which belong to vertices which are not labeled by Young diagrams, while ρ_2 deletes those from among the remaining representations in which the Young diagrams have more than two rows.

(ii) It has been shown by Lickorish (see [Li]) that the Jones polynomial can be obtained from the Kauffman polynomial for the special choice of parameters $(l,m) = (t^{-3/4}, -t^{1/4} - t^{-1/4})$. Hence, by factoring over the ideal I_n defined by

$$I_n = \{a \in C_n(t^{-3/4}, -t^{1/4} - t^{-1/4}),\ \tau'(ab) = 0 \quad \text{for all } b \in C_n(t^{-3/4}, -t^{1/4} - t^{-1/4})\},$$

we obtain a well-defined surjective homomorphism

$$\Psi : C_n(t^{-3/4}, -t^{1/4} - t^{-1/4}) \mapsto C_n(t^{-3/4}, -t^{1/4} - t^{-1/4})/I_n \cong A_n(t).$$

This homomorphism can be described explicitly by (see also [J,2, §3])

$$\Psi(G_i) = t^{-1/4}((1+t)e_i - 1),$$

which implies

$$\Psi(E_i) = -(t^{1/2} + t^{-1/2})e_i.$$

The last equality shows that Ψ does not factor through H_n. An explicit computation shows that τ' annihilates the idempotent p_3 as defined in the proof of Theorem 3.7 for the special choice $\alpha = -t^{1/4}$ and $l = t^{-3/4}$. Hence the hypotheses of Theorem 3.5 are no longer satisfied. In particular, we can no longer construct Bratteli diagrams as in Theorem 3.7. Similar phenomena with Brauer's centralizer algebras (see next section and [W,2]) suggest that C_3 (and also higher C_n's) will no longer be semisimple for this special choice of l and m. Thus the kernel of Ψ sits in a complicated way in $C_n(l, m)$ completely different from our first kernel.

5. Brauer's centralizer algebras

In 1937 R. Brauer introduced, in [Br], a family of algebras which we denote by $\{\mathscr{D}_n(x)\}$, where n is a positive integer and x a parameter or indeterminate. His algebras are ordered by inclusion, $\mathscr{D}_1(x) \subset \mathscr{D}_2(x) \subset \cdots$. We will show that our algebras are perturbations or deformations of $\mathscr{D}_n(x)$ in much the same way as the Hecke algebra $\mathscr{H}_n(t)$ of type A_{n-1} is a deformation or perturbation of $\mathbf{C}S_n$.

First, we explain what we mean about $\mathscr{H}_n(t)$ and $\mathbf{C}S_n$. The group algebra $\mathbf{C}S_n$ is generated by elements $s_1, \dots, s_{n-1}$, where s_i denotes the transposition $(i, i+1)$. It has defining relations $(1)_s, (2)_s$ and $s_i^2 = 1$. On the other hand, the algebra $\mathscr{H}_n(t)$ is generated by $g_1^*, \dots, g_{n-1}^*$, with defining relations $(1)_{g^*}, (2)_{g^*}$ and $(g_i^*)^2 = (t-1)g_i^* + t$. In particular, this last relation tells us that $\mathscr{H}_n(1)$ is isomorphic to $\mathbf{C}S_n$. Even more, we can use similar pictures to describe monomials in $\mathscr{H}_n(t)$ and in $\mathbf{C}S_n$. Recall the geometric definition of a braid in §1 above as an equivalence class of embeddings of $I^{(n)}$ in $\mathbf{C} \times I$. Similarly, a permutation $\alpha \in S_n$ may be regarded as an equivalence class of immersions of $I^{(n)}$ in $\mathbf{R} \times I \subset \mathbf{C} \times I$. See Figure 9. Multiplication is by concatenation and rescaling, as for braids. There are $n!$ possible equivalence classes of embeddings, and $\mathbf{C}S_n$ is the free algebra on these $n!$ possible patterns.

Pictures similar to those we have just given in $\mathbf{C}S_n$ were used by R. Brauer in [Br] to describe another algebra $\mathscr{D}_n(x)$ which contains $\mathbf{C}S_n$ as a subalgebra. One has the same $2n$ base points, i.e. n in $\mathbf{R} \times \{0\}$ and n in $\mathbf{R} \times \{1\}$, only now the immersions of $I^{(n)}$ in $\mathbf{R} \times I$ are allowed to join these $2n$ points in pairs in an arbitrary fashion, that is we no longer require that the arcs join a point on $\mathbf{R}\times\{0\}$ to a point on $\mathbf{R}\times\{1\}$. One then obtains a set of $1\cdot 3\cdot 5\cdots(2n-1)$ pictures because the first point can be joined to $2n-1$ others, the second to $2n-3$ others, etc. Multiplication is as before by concatenation, erasure of the middle line and rescaling, but there is a new phenomenon. If S, T are two such graphs, let U be the graph obtained by concatenation and erasure as above. Note that on concatenation there may be closed loops which arise in the middle. Let $m(S,T)$ be the number of such loops. Then $ST = (x)^{m(S,T)}U$, where x is a parameter. The free algebra on the $1\cdot 3\cdot 5\cdots(2n-1)$ generalized patterns with this product rule, is Brauer's algebra $\mathscr{D}_n(x)$. It is easy to see that it is generated by $2n-2$ elements $s_1, \dots, s_{n-1}, e_1, \dots, e_{n-1}$ where s_i and e_i are the patterns shown in Figure 9.

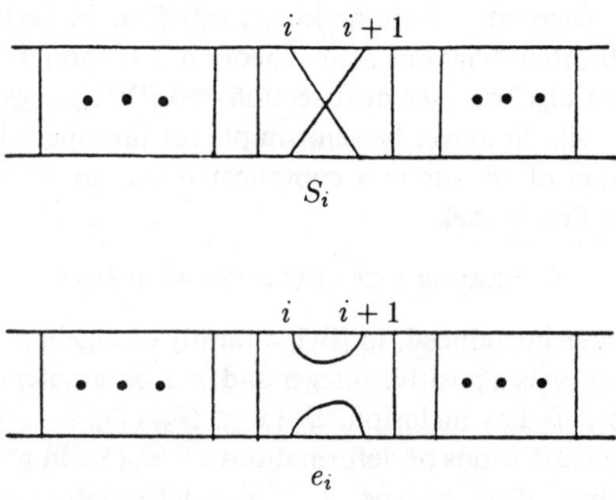

FIGURE 9

Note that for the point $l = i$, $m = 0$ in the parameter space of our algebra E^2 cannot be determined by relation (21). We will avoid this singularity by choosing different parameters (which results in "blowing up" the point $(i,0)$ to a line). After these substitutions it will be easy to see how our algebras are connected with Brauer's. So let us define elements $\widehat{G}_j$, $\widehat{E}_j$ and parameters $\hat{l}$ and x by

$$G_j = i\widehat{G}_j, \quad E_j = -\widehat{E}_j, \quad l = -i\hat{l} \quad \text{and} \quad m = i\frac{\hat{l}^{-1}-\hat{l}}{x+1}$$

or, if we solve for x

$$x = m^{-1}(l + l^{-1}) - 1.$$

It is immediate that whenever x and m are well defined, the algebra $\widehat{\mathscr{C}}_n(\hat{l}, x)$ generated by $\widehat{G}_1, \widehat{E}_1, \dots, \widehat{G}_{n-1}, \widehat{E}_{n-1}$ is equal to $\mathscr{C}_n(l, m)$. Note that relations (13)–(20) also hold for $\widehat{G}_j$ and $\widehat{E}_j$ instead of G_j and E_j, while (21) becomes $\widehat{E}_j^2 = x\widehat{E}_j$. If we set $\hat{l} = 1$, we obtain from (22) $\widehat{G}_j^2 = 1$. One may now check (by pictures) that $\widehat{G}_j \to s_j$ and $\widehat{E}_j \to e_j$ can be extended to a homomorphism from $\widehat{\mathscr{C}}_n(1, x)$ onto $\mathscr{D}_n(x)$. Comparing the dimensions, it follows that

$\widehat{\mathscr{C}}_n(1, x)$ is isomorphic to $\mathscr{D}_n(x)$. So Brauer's algebra can be considered as a special case of our algebra. We finally remark that this does not determine the structure of $\mathscr{D}_n(x)$. As we have seen in §4, the algebras at special values of the parameters can have quite a different structure. The above mentioned problem is studied in a separate paper [W,2] which uses the connections of Brauer's algebra with the representation theory of orthogonal groups.

6. Matrix representations of B_n

A classical open problem is whether the braid groups $\{B_n ; n = 1, 2, 3, \ldots\}$ are linear groups. For $n = 1, 2, 3$ the answer is known to be "yes", but for $n > 3$ the problem is unsolved. Since our work gives new linear representations of B_n it seems worthwhile to give some of the new representations explicitly.

There are three mutually nonisomorphic 1-dimensional representations of C_2 belonging to the idempotents p_1, p_2 and p_3 as defined in the proof of Theorem 3.7.

We now turn to B_3. It follows directly from the defining relations that E_1, E_2E_1 and G_2E_1 form a basis of a minimal left ideal of $C_2E_2C_2$. Using this basis, we obtain the following new matrix representations of B_3:

$$\sigma_1 \mapsto \begin{bmatrix} l^{-1} & m & 0 \\ 0 & m & 1 \\ 0 & -1 & 0 \end{bmatrix}, \qquad \sigma_2 \mapsto \begin{bmatrix} 0 & 0 & -1 \\ 0 & l^{-1} & l^{-1}m \\ 1 & 0 & m \end{bmatrix}.$$

Theorem 3.7 determines the structure of $\mathscr{C}_3(l, m)$ inductively. See Figure 8, or proceed as follows. We know that

$$\mathscr{C}_2 \cong \mathscr{H}_2 \oplus \mathscr{C}_1 E_1 \mathscr{C}_1 \cong \mathbf{C}(l, \alpha)^3.$$

The inclusion matrix G for $\mathscr{C}_1 \subset \mathscr{C}_2$ is a 3×1 matrix given by

$$G = \begin{pmatrix} 1 \\ 1 \\ 1 \end{pmatrix}.$$

Using (28) and Proposition 3.4, (iii) we obtain that

$$\mathscr{C}_2 E_2 \mathscr{C}_2 \cong M_3(\mathbf{C}(l, \alpha)).$$

This shows that the matrices above generate the full matrix algebra of 3×3 matrices. On the other hand $\mathscr{H}_3 \cong \mathbf{C}[l, \alpha]S_3$, the group algebra of the symmetric group on 3 letters. Hence

$$\mathscr{H}_3 \cong \mathbf{C}(l, \alpha)^2 \oplus M_2(\mathbf{C}(l, \alpha)).$$

Similarly, we can determine

$$\mathscr{C}_3 E_3 \mathscr{C}_3 \cong M_6(\mathbf{C}(l, \alpha)) \oplus M_6(\mathbf{C}(l, \alpha)) \oplus M_3(\mathbf{C}(l, \alpha)).$$

Then $\mathscr{H}_4'$ is also $\mathscr{C}_3 E_3 \mathscr{C}_3$, a direct sum of two six-dimensional representations and one 3-dimensional representation. Since the representations of B_4 in $\mathscr{H}_4$

are known to be determined by the Burau representation of B_4 (see [J,2]), we focus on $\mathscr{H}_4'$. The 3-dimensional representation of B_4 factors through B_3, and so it cannot be faithful, but the 6-dimensional ones stand a chance. We therefore describe one of them in detail.

As above, let $m = \alpha + \alpha^{-1}$, and let

$$e_i = \frac{1}{m^{-1}(l + l^{-1}) - 1} E_i .$$

Let $v = (G_1 - \alpha)(1 - e_1)$. Then, a basis for the 6-dimensional part which we investigate is vE_3, E_2vE_3, G_2vE_3, $E_1E_2vE_3$, $G_1E_2vE_3$, and $G_1G_2vE_3$. Allowing the generators G_1, G_2, G_3 of $\psi(B_4)$ to act on this basis on the left, and using the defining relations for C_4, we obtain the following two parameter matrix representations of B_4:

$$\sigma_1 \to \begin{bmatrix} \alpha^{-1} & 0 & 0 & 0 & 0 & 0 \\ 0 & 0 & 0 & 0 & -1 & 0 \\ 0 & 0 & 0 & 0 & 0 & -1 \\ 0 & 0 & 0 & l^{-1} & ml^{-1} & \alpha ml^{-1} \\ 0 & 1 & 0 & 0 & m & 0 \\ 0 & 0 & 1 & 0 & 0 & m \end{bmatrix},$$

$$\sigma_2 \to \begin{bmatrix} 0 & 0 & -1 & 0 & 0 & 0 \\ 0 & l^{-1} & ml^{-1} & m & 0 & 0 \\ 1 & 0 & m & 0 & 0 & 0 \\ 0 & 0 & 0 & m & 1 & 0 \\ 0 & 0 & 0 & -1 & 0 & 0 \\ 0 & 0 & 0 & 0 & 0 & \alpha^{-1} \end{bmatrix},$$

$$\sigma_3 \to \begin{bmatrix} l^{-1} & m & 0 & 0 & \alpha^{-1}m & 0 \\ 0 & m & 1 & 0 & 0 & 0 \\ 0 & -1 & 0 & 0 & 0 & 0 \\ 0 & 0 & 0 & \alpha^{-1} & 0 & 0 \\ 0 & 0 & 0 & 0 & m & 1 \\ 0 & 0 & 0 & 0 & -1 & 0 \end{bmatrix}.$$

We ask the question: Is this representation faithful? By the methods described in [Bi] it would probably suffice to prove that the matrix group generated by $\sigma_1\sigma_3^{-1}$ and $\sigma_2\sigma_1\sigma_3^{-1}\sigma_2^{-1}$ is free of rank 2.

REFERENCES

[Al] J. W. Alexander, *A lemma on systems of knotted curves*, Proc. Nat. Acad. Sci. U.S.A. **9** (1923), 93–95.

[Ar] E. Artin, *Theorie der Zopfe*, Hamburg Abh. **4** (1925), 47–72.

[Bi] J. S. Birman, *Braids, links and mapping class groups*, Ann. of Math. Studies, no. 82, Princeton Univ. Press, 1974.

[B-L-M] R. Brandt, W. B. R. Lickorish and K. Millett, *A polynomial invariant for unoriented knots and links*, preprint.

[Bo] Claude Bourin, Unpublished notes.

[Br] R. Brauer, *On algebras which are connected with the semisimple continuous groups*, Ann. of Math. **38** (1937), 857–872.

[FYHLMO] P. Freyd, D. Yetter, J. Hoste, W.B.R. Lickorish, K. Millett, and A. Ocneanu, *A new polynomial invariant of knots and links*, Bull. Amer. Math. Soc. (N.S.) **12** (1985), 239–246.

[H] J. Hoste, *A polynomial invariant of knots and links*, Pacific J. Math. **124** (1986), 295–320.

[J,1] V. F. R. Jones, *Index for subfactors*, Invent. Math. **72** (1983), 1–25.

[J,2] ——, *Braid groups, Hecke algebras and type* II *factors*, Geometric Methods in Abstract Algebras, Proc. U.S.-Japan Symposium, Wiley, 1986, pp. 242–273.

[J,3] ——, *A polynomial invariant for knots via Von Neumann algebras*, Bull. Amer. Math. Soc. (N.S.) **12** (1985), 103–111.

[J,4] ——, *Hecke algebra representations of braid groups and link polynomials*, Ann. of Math. **126** (1987), 335–388.

[K,1] L. Kauffman, *An invariant of regular isotopy*, preprint.

[K,2] ——, *A states model for the Jones polynomial*, Topology **26** (1987), 395–407.

[K,3] ——, *Sign and space: knots and physics*, Lecture Notes, Torino.

[Ki] M. Kidwell, *On the dimension of the Birman-Wenzl algebra*, Unpublished notes.

[La] S. Lang, *Algebra*, Addison-Wesley, 1965.

[L-M] W. B. R. Lickorish and K. Millett, *A polynomial invariant of oriented links*, Topology **26** (1987), 107–141.

[Li] W. B. R. Lickorish, *A relationship between link polynomials*, Math. Proc. Cambridge Philos. Soc. **100** (1986), 109–112.

[Lu] George Lusztig, *On a theorem of Benson and Curtis*, J. Algebra **71** (1981).

[M] J. Murakami, *The Kauffman polynomial of links and representation theory*, Osaka J. Math. **24** (1987), 745–758.

[M-S] H. Morton and H. B. Short, *Calculating the 2-variable polynomial for knots presented as closed braids*, preprint.

[M-T] H. Morton and P. Traczyk, *Knots, skeins and algebras*, preprint.

[O] A. Ocneanu, *A polynomial invariant for knots; a combinatorial and algebraic approach*, preprint.

[S] S. Sundaram, *On the combinatorics of representations of* $S_p(2n;\mathbf{C})$, Mem. Amer. Math. Soc. (to appear).

[W,1] H. Wenzl, *Representations of Hecke algebras and subfactors*, Invent. Math. **24** (1988), 349–383.

[W,2] ——, *On the structure of Brauer's centralizer algebras*, Ann. of Math. **128** (1988), 179–193.

[Y] David Yetter, Private correspondence with the first author.

DEPARTMENT OF MATHEMATICS, COLUMBIA UNIVERSITY, NEW YORK, NEW YORK 10027

DEPARTMENT OF MATHEMATICS, UNIVERSITY OF CALIFORNIA, SAN DIEGO, LA JOLLA, CALIFORNIA 92093

Murakami J.
Osaka J. Math.
24 (1987), 745–758

THE KAUFFMAN POLYNOMIAL OF LINKS AND REPRESENTATION THEORY*

JUN MURAKAMI

(Received July 3, 1986)

1. Introduction. In 1984, V. Jones [8] introduced a new polynomial invariant of link isotopy types which is now called the (one variable) Jones polynomial, which was subsequently generalized to the two variable Jones polynomial [4], [13]. These invariants are closely related to the traces of irreducible representations of Iwahori algebras (or Hecke algebras) [2], [7] associated with the symmetric groups (see [5], [9]).

The purpose of this paper is to show that the Kauffman polynomial [12] can also be interpreted as a function F on a certain associative algebra. We define a *knit semi-group* D_n of degree n which is generated by the generators of the braid group B_n on n strings and elements $e_1, e_2, \cdots, e_{n-1}$ in Figure 4. We call an element of D_n an *n-knit*. We get a link $d^\wedge$ in the 3-sphere by closing an n-knit d. We call $d^\wedge$ a *closed n-knit* coming from d. We also define an algebra $E_n(\alpha, \beta)$ for non-zero complex numbers $\alpha, \beta \in \boldsymbol{C}-\{0\}$ as a quotient of a semi-group algebra $\boldsymbol{C}[D_n]$ of D_n over $\boldsymbol{C}$. Then the Kauffman polynomial of a closed n-knit is obtained through $E_n(\alpha, \beta)$.

From Section 7 on, we treat the case $n=3$. Then we can show that the function F is a sum of traces of irreducible representations of the algebra $E_3(\alpha, \beta)$ (Theorem 10.1). The author expects the same is true for general n. In Sections 12–16, we apply our formula to closed 3-braids and 2-bridge links, since they are special types of closed 3-knits. For example, if two closed 3-braids have the same writhe (or twist number) and the same Jones polynomial, they also have the same Q-polynomial (Theorem 13.1). Thus for the closed 3-braid, the Alexnader polynomial and the writhe determine the two variable Jones, Jones, and Q-polynomials. In the actual claculation of the examples in Sections 14 b), c) and 16, the author used a personal computer (NEC PC-9801) with muMATH-83 (Symbolic Mathematics Package) for MS-DOS.

Acknowledgments. I am profoundly indebted to A. Gyoja who gave me many informations about Iwahori algebras. As well, the notion of knit semi-

* This research was supported in part by the Grant in Aid for Scientific Research, the Mintisry of Education, Science and Culture.

groups arose in discussions with him. I would like to thank T. Kanenobu who introduced me the statement of Theorem 13.1 holds for every example he calculated.

2. The Kauffman polynomial and the L-polynomial. Fix non-zero complex numbers α, β. Let $L(\alpha, \beta, \cdot)$ be the L-polynomial [11] which is a regular isotopy invariant of unoriented link diagrams defined by following relations (2.1), (2.2) and (2.3):

(2.1) $$L(\alpha, \beta, O) = 1 \quad \text{for a planar circle } O,$$

(2.2) $$L(\alpha, \beta, |K_+|)+L(\alpha, \beta, |K_-|) = (\beta+\beta^{-1})\,(L(\alpha, \beta, |K_0|)+L(\alpha, \beta, |K_\infty|)),$$

(2.3) $$L(\alpha, \beta, |K_{l+}|) = \alpha\, L(\alpha, \beta, |K_\sim|), \qquad L(\alpha, \beta, |K_{l-}|) = \alpha^{-1}\, L(\alpha, \beta, |K_\sim|),$$

where $|K_*|$'s are unoriented link diagrams, identical without a circle where they are as in Figures 1 and 2.

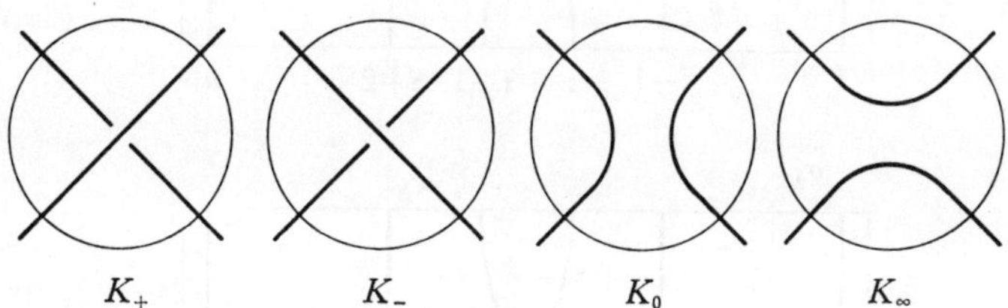

Figure 1

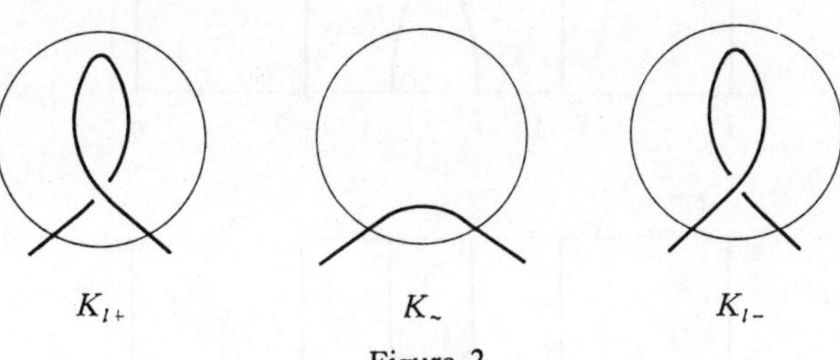

Figure 2

For an oriented link K, the writhe (or twist number) $w(K)$ is the sum of the signs of all crossings. The sign of a crossing point is defined as in Figure 3. The Kauffman polynomial $F(\alpha, \beta, K)$ [12] for an oriented link diagram K is defined by:

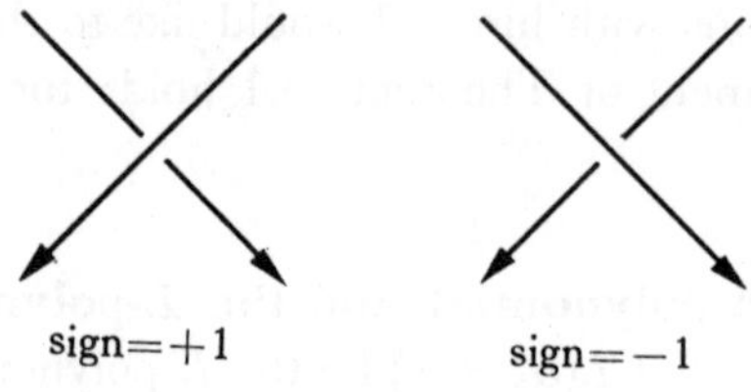

Figure 3

$$F(\alpha, \beta, K) = \alpha^{-w(K)} L(\alpha, \beta, |K|), \tag{2.4}$$

where $|K|$ is an unoriented link coming from K in disregard of its orientation. Then it is an ambient isotopy invariant and is an invariant of link isotopy types.

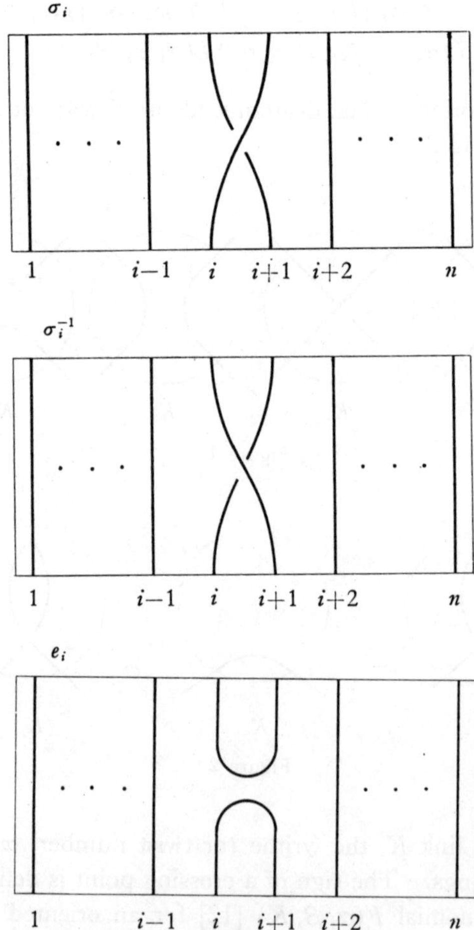

Figure 4

3. Rectangular diagrams. Fix a positive integer n. Let $G_n=\{\sigma_i, \sigma_i^{-1}, e_i \mid i=1, 2, \cdots, n-1\}$. Let F_n be a free semi-group with identity element 1 generated by G_n. Then $F_n=\{w_1 w_2 \cdots w_r \mid r\in N, w_i\in G_n\} \cup \{1\}$. A *rectangular diagram* $R(w)$ corresponding to $w\in G_n$ are defined as in Figure 4. A rectangular diagram $R(w)$ corresponding to $w\in F_n$ is defined as in Figure 5.

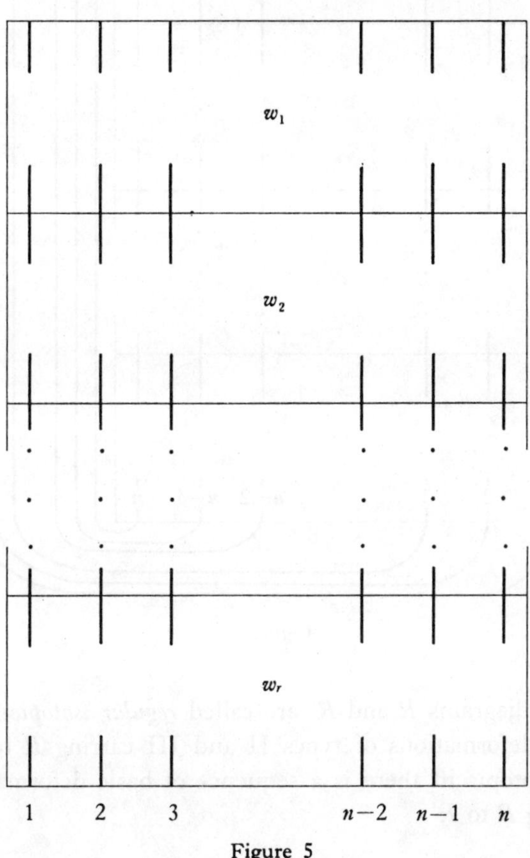

Figure 5

As in Figure 6, we get an unoriented link diagram $w^\wedge$ associated with $w\in F_n$, connecting upper end points and lower end points of a rectangular diagram $R(w)$.

4. Basic deformations. The three basic deformations of rectangular diagrams are as shown in Figure 7. It shows representative situations for each deformation. If a rectangular diagram has the local forms as shown in this figure, the deformation is performed without disturbing the rest of the diagram.

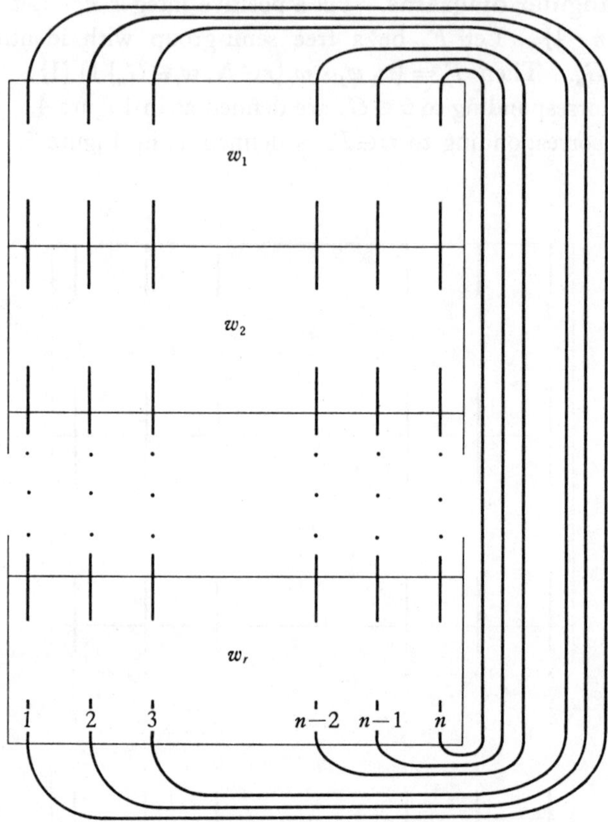

Figure 6

Two rectangular diagrams R and R' are called *regular isotopic* iff there is a sequence of basic deformations of types II and III carring R to R'. They are called ambient isotopic iff there is a sequence of basic deformations of types I, II and III carring R to R'.

5. Knit semi-groups. Let

(5.1) $\quad D_n = F_n/(w = w'$, if $R(w)$ and $R(w')$ are regular isotopic).

We call D_n a *knit semi-group* of degree n and call its element an n-knit or simply a knit. For an n-knit b, we denote by $R(b)$ the regular isotopy class of the rectangular diagram associated to b. We denote by $b^\wedge$ the set of link diagrams obtained by closing the rectangular diagrams associated with the elements of F_n contained in the class of b. We call $b^\wedge$ the *closed n-knit* coming from b.

Proposition 5.2. *The L-polynomial may be considered as a map from* D_n *to*

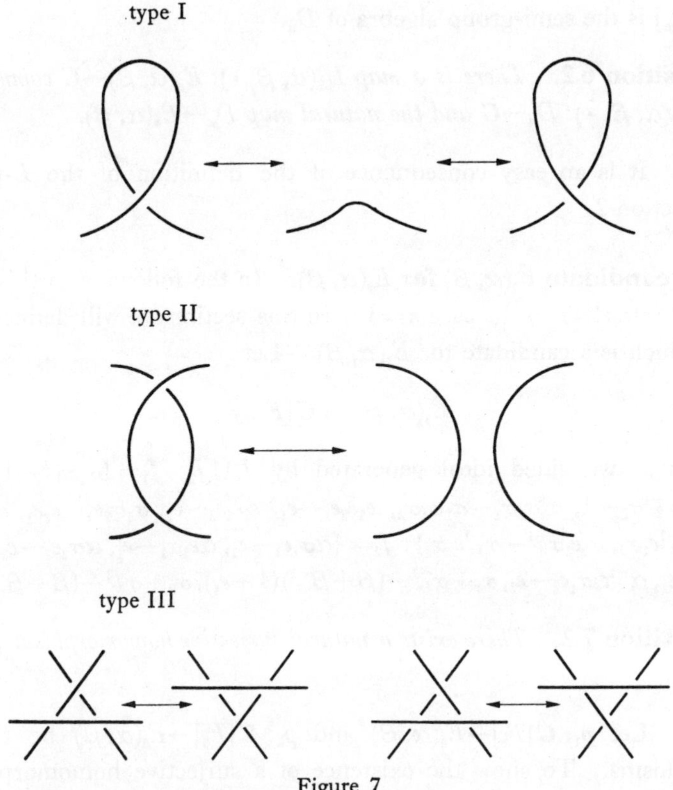

Figure 7

C. We denote it by $L_D(\alpha, \beta, \cdot)$.

Proof. For $x \in D_n$, let $L_D(\alpha, \beta, x) = L(\alpha, \beta, \hat{x}^{\wedge})$ where $\hat{x}$ is a representative of x in F_n. Because the L-polynomial is a regular isotopy invariant, $L_D(\alpha, \beta, x)$ does not depend on the choice of $\hat{x}$.

6. An algebra associated with the *L*-polynomial. The *writhe* (or twist number) of a string (or component) of a rectangular diagram is the sum of the signs of all the crossings of this string with itself. There are two ways to give an orientation to the string. But the writhe of the string does not depend on the choice of it. The *writhe* of a rectangular diagram R is the sum of the writhe of all strings of it. We denote it $w(R)$. It is invariant under the regular isotopy. Hence we may define the *writhe* of a knit $x \in D_n$ by $w(x) = w(R(x))$. For $x, y \in D_n$, we denote $x \sim y$ if $R(x)$ and $R(y)$ are ambient isosopic. Let

$$(6.1)\quad E_n(\alpha, \beta) = \boldsymbol{C}[D_n]/(\sigma_i + \sigma_i^{-1} - (\beta + \beta^{-1})(1 + e_i) \quad \text{for} \quad 1 \leqq i \leqq n-1\,, \\ \alpha^{-w(x)}x - \alpha^{-w(y)}y \quad \text{for} \quad x, y \in D_n \text{ s.t. } x \sim y)\,,$$

where $\boldsymbol{C}[D_n]$ is the semi-group algebra of D_n.

Proposition 6.2. *There is a map $L_E(\alpha, \beta, \cdot)\colon E_n(\alpha, \beta) \to C$ commuting with the map $L_D(\alpha, \beta, \cdot)\colon D_n \to \boldsymbol{C}$ and the natural map $D_n \to E_n(\alpha, \beta)$.*

Proof. It is an easy consequence of the definition of the L-polynomial given in Section 2.

7. A candidate $\bar{E}_3(\alpha, \beta)$ for $E_3(\alpha, \beta)$. In the following in this paper, we will restrict ourselves to the case $n=3$. In this section we will define an algebra $\bar{E}_3(\alpha, \beta)$ which is a candidate for $E_3(\alpha, \beta)$. Let

$$\bar{E}_3(\alpha, \beta) = \boldsymbol{C}[F_3]/\bar{J} \tag{7.1}$$

where $\bar{J}$ is a two sided ideal generated by $J_1 \cup J_2$; $J_1 = \{\sigma_1\sigma_1^{-1}-1,\ \sigma_1^{-1}\sigma_1-1,\ \sigma_2\sigma_2^{-1}-1,\ \sigma_2^{-1}\sigma_2-1,\ \sigma_1\sigma_2\sigma_1-\sigma_2\sigma_1\sigma_2,\ e_1e_2e_1-e_1,\ e_2e_1e_2-e_2,\ \sigma_1\sigma_2e_1-e_2e_1,\ \sigma_2\sigma_1e_2-e_1e_2,\ \sigma_1e_2\sigma_1^{-1}-\sigma_2^{-1}e_1\sigma_2,\ \sigma_2e_1\sigma_2^{-1}-\sigma_1^{-1}e_2\sigma_1\}$, $J_2 = \{\alpha\sigma_1e_1-e_1,\ \alpha e_1\sigma_1-e_1,\ \alpha\sigma_2e_2-e_2,\ \alpha e_2\sigma_2-e_2,\ \alpha^{-1}e_1\sigma_2e_1-e_1,\ \alpha^{-1}e_2\sigma_1e_2-e_2,\ \sigma_1+\sigma_1^{-1}-(\beta+\beta^{-1})(1+e_1),\ \sigma_2+\sigma_2^{-1}-(\beta+\beta^{-1})(1+e_2)\}$.

Proposition 7.2. *There exists a natural surjective homomorphism $\varphi\colon \bar{E}_3(\alpha, \beta) \to E_3(\alpha, \beta)$.*

Proof. Let $p_1\colon \boldsymbol{C}[F_3] \to \bar{E}_3(\alpha, \beta)$ and $p_2\colon \boldsymbol{C}[F_3] \to E_3(\alpha, \beta)$ be the natural homomorphisms. To show the existence of a surjective homomorphism from $\bar{E}_3(\alpha, \beta)$ to $E_3(\alpha, \beta)$ commuting with p_1 and p_2, we must check that $\ker p_2 \supset \ker p_1 = \bar{J}$. Because the two terms of elements of J_1 are regular isotopic, they are contained in $\ker p_2$ according to the definition (5.1) of D_n. Because of the definition (6.1) of $E_3(\alpha, \beta)$, the elements of J_2 are all contained in $\ker p_2$. Hence any element of $\bar{J}$ is contained in $\ker p_2$ and Proposition 7.2 is proved.

8. Structure of $\bar{E}_3(\alpha, \beta)$

Proposition 8.1. $\bar{E}_3(\alpha, \beta) \cong M_3(\boldsymbol{C}) + \boldsymbol{C} + M_2(\boldsymbol{C}) + \boldsymbol{C}$ *for generic α, β.*

Proof. A basis of $\bar{E}_3(\alpha, \beta)$ over $\boldsymbol{C}$ is $V_0 = \{1, \sigma_1, \sigma_2, \sigma_1\sigma_2, \sigma_2\sigma_1, \sigma_1\sigma_2\sigma_1, e_1, \sigma_2e_1, e_2e_1, e_2, \sigma_1e_2, e_1e_2, \sigma_1e_2\sigma_1, e_1\sigma_2, e_2\sigma_1\}$. Let $I(\alpha, \beta)$ be the two sided ideal of $\bar{E}_3(\alpha, \beta)$ generated by e_1. Then a bisis of $I(\alpha, \beta)$ is $\{e_1, \sigma_2e_1, e_2e_1, e_2, \sigma_1e_2, e_1e_2, e_1\sigma_2, e_2\sigma_1, \sigma_1e_2\sigma_1\}$ and $I(\alpha, \beta)$ is isomorphic to $M_3(C)$ as an algebra for generic α, β. We also have $\bar{E}_3(\alpha, \beta)/I(\alpha, \beta) \cong H_3(\beta)$ where $H_3(\beta) = \boldsymbol{C}\langle\sigma_1, \sigma_2\rangle/(\sigma_i+\sigma_i^{-1}-(\beta+\beta^{-1})\cdot 1 (i=1, 2),\ \sigma_1\sigma_2\sigma_1-\sigma_2\sigma_1\sigma_2)$, which is the Iwahori algebra ([2], Chapt. IV, §2, Ex. 22–25) associated with the symmetric group S_3 of degree 3. Except for finite numbers of β, $H_3(\beta)$ is known to be completely reducible and isomorphic to the group ring of S_3: $H_3(\beta) \cong \boldsymbol{C}[S_3] \cong \boldsymbol{C} + M_2(\boldsymbol{C}) + \boldsymbol{C}$. We also know that $\dim_{\boldsymbol{C}} H_3(\beta)$

$+\dim_{\boldsymbol{C}} I(\alpha, \beta)=\dim_{\boldsymbol{C}} \bar{E}_3(\alpha, \beta)$ because $\dim_{\boldsymbol{C}} H_3(\beta)=6$, $\dim_{\boldsymbol{C}} I(\alpha, \beta)=9$ and $\dim_{\boldsymbol{C}} \bar{E}_3(\alpha, \beta)=15$. Hence the cannonical projection $\bar{E}_3(\alpha, \beta)\to H_3(\beta)$ splits and we get the proposition.

In the following of this paper, fix generic $\alpha, \beta\in \boldsymbol{C}-\{0\}$ so that Proposition 8.1 holds.

9. Irreducible representations of D_3 through $\bar{E}_3(\alpha, \beta)$. As shown in Proposition 8.1, $\bar{E}_3(\alpha, \beta)$ consists of four simple algebras. Let p_0, p_1, p_2, p_3, be the cannonical projections from $\bar{E}_3(\alpha, \beta)$ to its irreducible components $M_3(\boldsymbol{C})$, $\boldsymbol{C}$, $M_2(\boldsymbol{C})$, $\boldsymbol{C}$, respectively. Then a composition of the cannonical projection $\boldsymbol{C}[D_3]\to\bar{E}_3(\alpha, \beta)$ and $p_i(i=0, 1, 2, 3)$ induces an irreducible linear representations of D_3, say ρ_i. Representation matrices of $\rho_0, \rho_1, \rho_2, \rho_3$ are given as follows:

$$\rho_0(\sigma_1)=\begin{bmatrix}\alpha^{-1} & 0 & z\\ 0 & 0 & -1\\ 0 & 1 & z\end{bmatrix},\quad \rho_0(\sigma_2)=\begin{bmatrix}0 & -1 & \alpha^{-1}\\ 1 & z & 0\\ 0 & \alpha^{-1}z & 0\end{bmatrix},$$

$$\rho_0(e_1)=\begin{bmatrix}\mu & \alpha & 1\\ 0 & 0 & 0\\ 0 & 0 & 0\end{bmatrix},\qquad \rho_0(e_2)=\begin{bmatrix}0 & 0 & 0\\ 0 & 0 & 0\\ 1 & \alpha & \mu\end{bmatrix},$$

$$\rho_1(\sigma_1)=\beta\,,\quad \rho_1(\sigma_2)=\beta\,,\quad \rho_1(e_1)=0\,,\quad \rho_1(e_2)=0\,,$$

$$\rho_2(\sigma_1)=\begin{bmatrix}\beta & \sqrt{-1}\\ 0 & \beta^{-1}\end{bmatrix},\quad \rho_2(\sigma_2)=\begin{bmatrix}\beta^{-1} & 0\\ \sqrt{-1} & \beta\end{bmatrix},\quad \rho_2(e_1)=\rho_2(e_2)=\begin{bmatrix}0 & 0\\ 0 & 0\end{bmatrix},$$

$$\rho_3(\sigma_1)=\beta^{-1}\,,\quad \rho_3(\sigma_2)=\beta^{-1}\,,\quad \rho_3(e_1)=0\,,\quad \rho_3(e_2)=0\,,$$

where $z=\beta+\beta^{-1}$ and $\mu=(\alpha+\alpha^{-1})/z-1$. The matrices of ρ_1, ρ_2, ρ_3 are obtained from that of irreducible representations of the Iwahori algebra associated with S_3. Let W be a subspace of $\bar{E}_3(\alpha, \beta)$ spanned by e_1, $\sigma_2 e_1$, $e_2 e_1$ then W is invariant under the left action of $\bar{E}_3(\alpha, \beta)$. The matrix representations of the restrictions of $\rho_0(\sigma_i)$, $\sigma_0(e_i)$ $(i=1, 2)$ on W with respect to the basis $\{e_1, \sigma_2 e_1, e_2 e_1\}$ are as above. For example, $\sigma_1 e_1=\alpha^{-1}e_1$, $\sigma_1 e_2 e_1=\sigma_2 e_1$, $\sigma_1 e_2 e_1=\sigma_2^{-1}e_1=-\sigma_2 e_1+z e_2 e_1+z e_1$ and so we have the matrix of $\rho_0(\sigma_1)$.

10. Formulas for the L-polynomial, the Kauffman polynomial and the Q-polynomial of a closed 3-knit

Theorem 10.1. *Let K be an oriented link diagram such that the unoriented link diagram $|K|$ coming from K in disregard of its orientation is euqal to a closed*

3-knit $x^\wedge$ associated with $x \in D_3$. Let $\chi_0, \chi_1, \chi_2, \chi_3$ be the traces of irreducible representations $\rho_1, \rho_2, \rho_3, \rho_4$ of D_3 defined in Section 9. *Then we have*

$$L(\alpha, \beta, |K|) = \chi_0(x)+a_1\chi_1(x)+a_2\chi_2(x)+a_3\chi_3(x)\,, \tag{10.2}$$

$$F(\alpha, \beta, K) = \alpha^{-w(K)}(\chi_0(x)+a_1\chi_1(x)+a_2\chi_2(x)+a_3\chi_3(x))\,, \tag{10.3}$$

where $w(K)$ is the writhe of K defined in Section 2 *and a_1, a_2, a_3 are given as follows:*

$$\begin{aligned} a_1 &= P(\beta^{-3}(\mu^2-3)+(1-2\beta^{-2})(\alpha\mu-z-\alpha^{-1})+(\beta^{-1}-\beta)\alpha^2)\,, \\ a_2 &= P((\beta-\beta^{-1})(\mu^2-3)+(\beta^{-2}-\beta^2)(\alpha\mu-z-\alpha^{-1})+(\beta-\beta^{-1})\alpha^2)\,, \\ a_3 &= P(-\beta^3(\mu^2-3)+(2\beta^2-1)(\alpha\mu-z-\alpha^{-1})+(\beta^{-1}-\beta)\alpha^2)\,, \end{aligned} \tag{10.4}$$

where $z=\beta+\beta^{-1}$, $\mu=(\alpha+\alpha^{-1})/z-1$ and $P=1/(\beta^{-1}-\beta)(\beta^2-1+\beta^{-2})$.

The *Q-polynomial* (or absolute polynomial) of an unoriented link $|K|$ is defined by $Q(\beta, |K|)=L(1, \beta, |K|)$, which is an invariant of isotopy types [3], [6].

Corollary 10.5. *Let K and x be as in Theorem* 10.1. *Then we have*

$$\begin{aligned} Q(\beta, |K|) &= \chi_0(x)+\frac{2z^2+2z-2}{z^2(z^2-3)}(\chi_1(x)+\chi_3(x)) \\ &\quad +\frac{-z^4-2z^3+3z^2+4z-4}{z^2(z^2-3)}\chi_2(x)\,. \end{aligned} \tag{10.6}$$

Proof of Theorem 10.1. According to Propositions 5.2 and 6.2, we already know that $L(\alpha, \beta, |K|)=L_D(\alpha, \beta, x)=L_E(\alpha, \beta, \varphi(p(x)))$. Because of the definition of the L-polynomial, we have $L_E(\alpha, \beta, \varphi(p(x)))=L_{\bar{E}}(\alpha, \beta, \varphi(p(gxg^{-1})))$ for $x\in D_3$ and $g\in B_3\subset D_3$. But $p(B_3)$ generates $\bar{E}_3(\alpha, \beta)$ as a $\boldsymbol{C}$-algebra and $\bar{E}_3(\alpha, \beta)$ is semi-simple, and so $L_E(\alpha, \beta, \varphi(p(x)))$ is the sum of traces of irreducible representations of D_3 coming from the irreducible components of $\bar{E}_3(\alpha, \beta)$ treated in Section 9, i.e. $L_E(\alpha, \beta, \varphi(p(x)))=a_0\chi_0(x)+a_1\chi_1(x)+a_2\chi_2(x)+a_3\chi_3(x)$. But the values of $\chi_0, \chi_1, \chi_2, \chi_3$ and $L_D(\alpha, \beta, \cdot)$ for $1, \sigma_1, \sigma_1\sigma_2, e_1$ in D_3 are given as follows:

(10.7)

D_3	χ_0	χ_1	χ_2	χ_3	$L_D(\alpha, \beta, \cdot)$
1	3	1	2	1	μ^2
σ_1	$\beta+\beta^{-1}+\alpha^{-1}$	β	$\beta+\beta^{-1}$	β^{-1}	$\alpha\mu$
$\sigma_1\sigma_2$	0	β^2	1	β^{-2}	α^2
e_1	μ	0	0	0	μ

Hence we have $a_0=1$ and a_1, a_2, a_3 as in (10.4).

11. $E_3(\alpha, \beta)$ is isomorphic to $\bar{E}_3(\alpha, \beta)$

Corollary 11.1. *The cannonical projection φ: $\bar{E}_3(\alpha, \beta) \to E_3(\alpha, \beta)$ defined in Proposition 7.2 is an isomorphism for generic α, β.*

Proof. Assume that $\ker \varphi \supsetneqq \{0\}$. Then $\ker \varphi$ contains at least one irreducible component of $\bar{E}_3(\alpha, \beta)$. But in this situation, the coefficient a_i in (10.5) corresponding ot above irreducible component must be zero. But $a_0=1$ and a_1, a_2, a_3 are all non-zero for generic α, β according to (10.4). It contradicts the assumption and we get $\ker \varphi = \{0\}$. Hence φ is an isomorphism.

12. The Kauffman polynomials of closed 3-braids.

Let B_3 be a braid group on three strings. B_3 is contained in D_3 as a sub-semigroup. Let b be a 3-braid and $b^\wedge$ be its closed braid, then the writhe $w(b^\wedge)$ is equal to the exponent sum $\mathcal{E}(b)$ of b. Let

$$C = \{(b, b') \in B_3 \times B_3 \mid V(\hat{b}) = V(\hat{b}'),\ \mathcal{E}(b) = \mathcal{E}(b')\} \tag{12.1}$$

where V: $\{\text{link}\} \to \boldsymbol{C}(t^{1/2})$ is the Jones invariant of links.

Theorem 12.2. *Let (b_1, b_2) be a pair of 3-braids in C. Then $F(\alpha, \beta, b_1^\wedge) = F(\alpha, \beta, b_2^\wedge)$ iff $\chi_0(b_1) = \chi_0(b_2)$.*

Proof. For (b_1, b_2) in C, $w(b_1^\wedge) = w(b_2^\wedge)$ and $\chi_i(b_1) = \chi_i(b_1)$ for $i=1, 3$ since $\mathcal{E}(b_1) = \mathcal{E}(b_2)$. Because

$$\beta \cdot P^{-1} \rho_2(\sigma_1) P = \begin{bmatrix} -t & 1 \\ 0 & 1 \end{bmatrix}, \quad \beta \cdot P^{-1} \rho_2(\sigma_2) P = \begin{bmatrix} 1 & 0 \\ t & -t \end{bmatrix}, \tag{12.3}$$

where $t = -\beta^2$ and $P = \begin{bmatrix} \sqrt{-1} \cdot \beta & 0 \\ 0 & 1 \end{bmatrix}$, we have $\chi_2(b_1) = \chi_2(b_2)$ as shown in [1]. Using (10.3), we get $F(\alpha, \beta, b_1^\wedge) - F(\alpha, \beta, b_2^\wedge) = \alpha^{-w(b_1)}(\chi_0(b_1) - \chi_0(b_2))$. This proves Theorem 12.2.

13. The Q-polynomial of closed 3-braids

Theorem 13.1. *Let C be the subset of $B_3 \times B_3$ defined in Section 12. Then for $(b, b') \in C$, we have $Q(\beta, \hat{b}) = Q(\beta, \hat{b}')$.*

Therefore, we have, from [1], [8] and [9],

Corollary 13.2. *For the closed 3-braid, the two variable Jones, Jones, and Q-polynomials are determined by the Alexander polynomial and the writhe.*

Proof of Theorem 13.1. Let ρ_0, ρ_2 be the representations of D_3 and χ_0, χ_2 be their traces defined in Section 9. We also denote $\rho_0(y)=\rho_0(\alpha, \beta, y)$, $\chi_0(y)=\chi_0(\alpha, \beta, y)$, $\rho_2(y)=\rho_2(\beta, y)$ and $\chi_2(y)=\chi_2(\beta, y)$ for a 3-knit y if the parameters α, β are needed. Because of Theorem 12.2 and $Q(\beta, \hat{b})=F(1, \beta, \hat{b})$, we get $Q(\beta, \hat{b})-Q(\beta, \hat{b}')=\chi_0(1, \beta, b)-\chi_0(1, \beta, b')$. Let

$$P=\begin{bmatrix} 1 & -1/\beta & 0 \\ 0 & 1 & 0 \\ 0 & -\beta & 1 \end{bmatrix}\begin{bmatrix} \sqrt{-1} & 0 & 0 \\ -\sqrt{-1}\cdot B_1B_3 & B_3 & -\sqrt{-1}\cdot B_1B_3 \\ 0 & 0 & -\sqrt{-1} \end{bmatrix} \tag{13.3}$$

where $B_1=\beta^{1/2}-\beta^{-1/2}$ and $B_3=1/(\beta^{3/2}+\beta^{-3/2})$. Then we have

$$\begin{aligned} P^{-1}\rho_0(1, \beta, \sigma_1)\, P &= \begin{bmatrix} \beta & \sqrt{-1}\cdot\beta^{1/2} & -1 \\ 0 & 1 & 2\sqrt{-1}\cdot\beta^{-1/2} \\ 0 & 0 & \beta^{-1} \end{bmatrix}, \\ P^{-1}\rho_0(1, \beta, \sigma_2)\, P &= \begin{bmatrix} \beta^{-1} & 0 & 0 \\ 2\sqrt{-1}\cdot\beta^{-1/2} & 1 & 0 \\ -1 & \sqrt{-1}\cdot\beta^{1/2} & \beta \end{bmatrix}. \end{aligned} \tag{13.4}$$

The representation matrices of σ_1 and σ_2 of the symmetric tensor of degree 2 of $\rho_2(\beta, \cdot)$, say $\rho_2^{(2)}(\beta, \cdot)$, are equal to those of $P^{-1}\rho_0(1, \beta^2, \cdot)\, P$, where the matrix of symmetric tensor of degree 2 of a 2 by 2 matrix is given as follows:

$$S^{(2)}\colon A=\begin{bmatrix} a & b \\ c & d \end{bmatrix} \to \begin{bmatrix} a^2 & ab & b^2 \\ 2ac & ad+bc & 2bd \\ c^2 & cd & d^2 \end{bmatrix}. \tag{13.5}$$

An elementary calculation shows that $S^{(2)}(AB)=S^{(2)}(A)\,S^{(2)}(B)$ for $A, B\in M_2(\boldsymbol{C})$ and the trace of $S^{(2)}(A)$ is equal to $\mathrm{trace}(A)^2-\det(A)$. Hence we get $\chi_0(1, \beta^2, \cdot)=\chi_2(\beta, \cdot)^2-1$ and $Q(\beta^2, \hat{b})=Q(\beta^2, \hat{b}')$. But this equality holds for every generic complex number β and so we have $Q(\beta, \hat{b})=Q(\beta, \hat{b}')$.

14. Examples for Theorem 12.2. In [1], Birman gave examples of pairs of knots coming from closed 3-braids for whihc the Jones invariant does not work well. According to Theorem 13.1, the Q-polynomial is not good for them, too. Here we show the Kauffman invariant is good enough for some of them.

a) Let $x=\sigma_1^{p_1}\sigma_2^{q_1}\cdots\sigma_1^{p_s}\sigma_2^{q_s}(p_i>0,\ q_i<0$ for $1\leqq i\leqq s$, $\varepsilon(x)=p_1+\cdots+p_s+q_1+\cdots+q_s=6r\neq 0)$ and $y=(\sigma_1\sigma_2\sigma_1)^{4r}x^{-1}$. Then $V(\hat{x})=V(\hat{y})$ and $Q(\beta, \hat{x})=Q(\beta, \hat{y})$ but $F(\alpha, \beta, \hat{x})\neq F(\alpha, \beta, \hat{y})$ because the highest degree of α in $\chi_0(x)$ is equal to $-(q_1+q_2+\cdots+q_s)$ but that of α in $\chi_0(y)$ is equal to $(p_1+p_2+\cdots+p_s)/3-2(q_1+q_2+\cdots+q_s)/3$ and they are different since $p_1+\cdots+p_s+q_1+\cdots+q_s\neq 0$. These pairs are examples treated in Proposition 2 of [1].

b) $\delta_1=\sigma_1^{-3}\sigma_2^6\sigma_1^{-1}\sigma_2^9$, $\delta_2=\sigma_1^{-3}\sigma_2^9$, $\delta_3=\sigma_1^{-2}\sigma_2^3$, $\gamma=\delta_1\,\delta_2\,\delta_3$ and $\mu=\delta_1\,\delta_3\,\delta_2$. Then $V(\hat{\gamma})=V(\hat{\mu})$ and $Q(\beta,\hat{\gamma})=Q(\beta,\hat{\mu})$ but $F(\alpha,\beta,\hat{\gamma})\neq F(\alpha,\beta,\hat{\mu})$.

c) $\delta_1'=\sigma_1^{-7}\sigma_2^8\sigma_1^{-1}\sigma_2^7$, $\delta_2'=\sigma_1^{-7}\sigma_2^7$, $\delta_3'=\sigma_1^{-6}\sigma_2^{-1}$, $\gamma'=\delta_1'\,\delta_2'\,\delta_3'$ and $\mu'=\delta_1'\,\delta_3'\,\delta_2'$. Then $V(\hat{\gamma}')=V(\hat{\mu}')$ and $Q(\beta,\hat{\gamma}')=Q(\beta,\hat{\mu}')$ but $F(\alpha,\beta,\hat{\gamma}')\neq F(\alpha,\beta,\hat{\mu}')$.

15. The Kauffman polynomials of 2-bridge links.

Let m be a positive odd integer. A 2-bridge link $K(a_1, a_2, \cdots, a_m)$ is an oriented link associated to a closed 3-knit $(e_1\,\sigma_2^{a_1}\,\sigma_1^{a_2}\cdots\sigma_2^{a_m})^\wedge$ with an orientation as in Figure 8.

2-bridge link

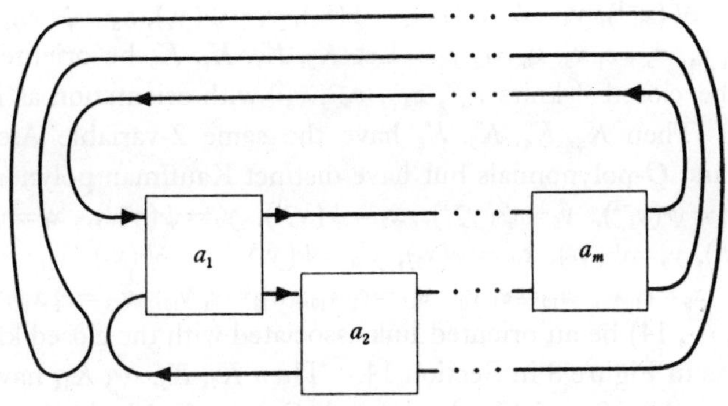

Where

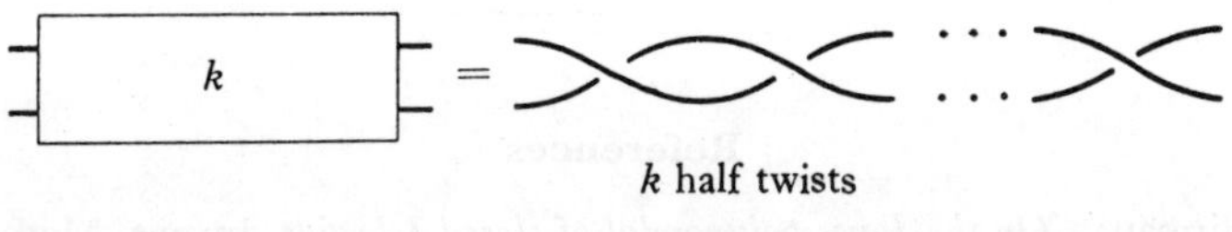

$-k$ =

$(k>0)$ k half twists

Figure 8

Theorem 15.1. *For an oriented 2-bridge link $K=K(a_1, a_2, \cdots, a_m)$, we have*

$$F(\alpha, \beta, K) = \alpha^{-w(K)}\,\chi_0(e_1\,\sigma_2^{a_1}\,\sigma_1^{a_2}\cdots\sigma_2^{a_m})\,. \tag{15.2}$$

Proof. Becaue $\rho_1(e_1)=0$, $\rho_2(e_1)=\begin{bmatrix} 0 & 0 \\ 0 & 0 \end{bmatrix}$, $\rho_3(e_1)=0$, we get (15.2) from (10.3).

16. Examples for Theorem 15.1. In Theorem 7 of [10], arbitrarily many 2-bridge links with the same 2-variable Alexander, 2-variable Jones and Q-polynomials are given. The Kauffman polynomials with specialized values of α and β of some of above links are calculated and it turned out that they can be distinguished by the Kauffman polynomials.

Let ψ be a map from invertible elements of $\bar{E}_3(\alpha, \beta)$ to $\bar{E}_3(\alpha, \beta)$ defined by $\psi(x)=x\sigma_1^2 x^{-1}$.

a) Let $x_1=\sigma_2^2\sigma_1^2\sigma_2^2\sigma_1^{-2}\sigma_2^{-2}$, $y_1=\sigma_2^{-2}\sigma_1^{-2}\sigma_2^2\sigma_1^2\sigma_2^2$, $x_2=\psi(x_1)$, $y_2=\psi(y_1)$, $z_1=e_1x_2$, $z_2=e_1y_2$. Let K_1, K_2 be oriented links associated with the closed 3-knits $z_1\hat{}$ and $z_2\hat{}$ with orientations as in Figure 8 in Section 15. Then K_1 and K_2 have the same 2-variable Alexander, 2-variable Jones and Q-polynomials but have distinct Kauffman polynomials.

b) Let $x_3=\psi(x_1^{-1})$, $y_3=\psi(y_1^{-1})$, $x_4=\psi(x_2)$, $y_4=\psi(y_2)$, $x_5=\psi(x_3)$, $y_5=\psi(y_3)$, $z_3=e_1x_4$, $z_4=e_1y_4$, $z_5=e_1x_5$, $z_6=e_1y_5$. Let K_3, K_4, K_5, K_6 be oriented links associated with the closed 3-knits $z_3\hat{}$, z_4 , $z_5\hat{}$, $z_6\hat{}$ with orientation as in Figure 8 in Section 15. Then K_3, K_4, K_5, K_6 have the same 2-variable Alexander, 2-variable Jones and Q-polynomials but have distinct Kauffman polynomials.

c) Let $x_6=\psi(x_2^{-1})$, $y_6=\psi(y_2^{-1})$, $x_7=\psi(x_3^{-1})$, $y_7=\psi(y_3^{-1})$, $x_8=\psi(x_4)$, $y_8=\psi(y_4)$, $x_9=\psi(x_5)$, $y_9=\psi(y_5)$, $x_{10}=\psi(x_6)$, $y_{10}=\psi(y_6)$, $x_{11}=\psi(x_7)$, $y_{11}=\psi(y_7)$, $z_7=e_1x_8$, $z_8=e_1y_8$, $z_9=e_1x_9$, $z_{10}=e_1y_9$, $z_{11}=e_1x_{10}$, $z_{12}=e_1y_{10}$, $z_{13}=e_1x_{11}$, $z_{14}=e_1y_{11}$. Let $K_i(i=7, 8, \cdots, 14)$ be an oriented link associated with the closed knit $z_i\hat{}$ with an orientation as in Figure 8 in Section 14. Then $K_7, K_8, \cdots, K_{14}$ have the same 2-variable Alexander, 2-variable Jones and Q-polynomials but have distinct Kauffman polynomials.

References

[1] J.S. Birman: *On the Jones polynomial of closed 3-braids,* Invent. Math. **81** (1985), 287–294.

[2] N. Bourbaki: Eléments de mathématique, groupes et algèbres de Lie, Chapitres 4, 5 et 6. Hermann, Paris, 1968.

[3] R.D. Brandt, W.B.R. Lickorish, K.C. Millett: *A polynomial invariant for unoriented knots and links,* Invent. Math. **84** (1986), 563–573.

[4] P. Freyd and D. Yetter, J. Hoste, W.B.R. Lickorish and K.C. Millett, A. Ocneanu: *A new polynomial invariant of knots and lindks,* Bull. Amer. Math. Soc. **12** (1985), 239–246.

[5] A. Gyoja: *Topological invariants of links and representations of Hecke algebras* II, preprint, Osaka University.

[6] C.F. Ho: *A new polynomial invariant for knots and links,* preliminary report, Abstracts Amer. Math. Soc. **6** (1985), 300.

[7] N. Iwahori: *On the structure of the Hecke ring of a Chevalley group over a finite field,* J. Fac. Sci. Univ. Tokyo **10** (part 2), (1964), 215–236.

[8] V.F.R. Jones: *A polynomial invariant for knots via von Neumann algebras,* Bull. Amer. Math. Soc. **12** (1985), 103–111.

[9] V.F.R. Jones: *Hecke algebra representations of braid groups and link polynomials,* Ann. of Math. **126** (1987), 335–388.

[10] T. Kanenobu: *Examples on polynomial invariants of knots and links,* preprint.

[11] L.H. Kauffman: *State models for knot polynomials,* preprint.

[12] W.B.R. Lickolish: *A relationship between link polynomials,* Math. Proc. Cambridge Philos. Soc. **100** (1986), 109–112.

[13] J.H. Przytycki and P. Traczyk: *Invariants of links of Conway type,* to appear in Kobe J. Math.

Department of Mathematics
Osaka University
Toyonaka, Osaka 560, Japan

BACKGROUND FROM TOPOLOGY, CLASSICAL WORKS

Topology Vol. 20, pp. 101–108
Pergamon Press Ltd., 1981.

THE CONWAY POLYNOMIAL†

LOUIS H. KAUFFMAN

(*Received* 11 *June* 1979)

§1. INTRODUCTION

THE CLASSICAL Alexander polynomial[1] $\Delta(x) = \Delta_K(x)$ of a knot or link $K \subset S^3$ is an invariant of ambient isotopy that is well-defined up to sign and multiplication by powers of the variable x. The Conway polynomial $\nabla(z) = \nabla_K(z)$ is a direct invariant of ambient isotopy. This polynomial, first introduced by Conway in[2], has remarkable properties that allow its computation from a knot diagram without recourse to matrices or determinants. It is related to the Alexander polynomial via the potential function $\nabla(x - x^{-1})$ which is (up to sign and powers of x) equivalent to $\Delta(x^2)$.

The purpose of this paper is to give an exposition of the Conway polynomial, and to explain the source of its properties by modelling it in analogy to the Alexander polynomial. There is a good geometric explanation for the potential function; this proceeds as follows:

Given an oriented knot or link $K \subset S^3$, let $F \subset S^3$ be a connected, oriented spanning surface for K. Let θ: $H_1(F) \times H_1(F) \to Z$ be the Seifert pairing (see §3). Let θ also denote any matrix of this pairing with respect to a basis for $H_1(F)$. Define the *potential function* $\Omega_K(x)$ by the formula $\Omega_K(x) = D(x\theta - x^{-1}\theta')$ where D denotes determinant and θ' *is the transpose of* θ. (If $H_1(F) = 0$, Let $\Omega_K(x) = 1$).

Our main result is the following:

THEOREM 3.4. *Let $K \subset S^3$ be an oriented knot or link, $F \subset S^3$ a connected, oriented spanning surface for K, and let $\Omega_K(x) = D(x\theta - x^{-1}\theta)$.*

(i) $\Omega_K(x)$ is an invariant of ambient isotopy of K. That is, if K is ambient isotopic to K' then $\Omega_K(x) = \Omega_{K'}(x)$.

(ii) If K is an unknot, then $\Omega_K(x) = 1$. If K is an unlink (with two or more components), then $\Omega_K(x) = 0$.

(iii) If K_+, K_-, $K_\circ$ are related by single crossing changes as indicated below

then

$$\Omega_{K_+} - \Omega_{K_-} = (x - x^{-1})\Omega_{K_\circ}.$$

These properties suffice for a recursive computation of Ω_K, independent of its definition as a determinant. The Alexander polynomial may be defined by the equation $\Delta(x) = D(x\theta - \theta')$. Hence $\Omega(x)$ is, up to powers of x, a version of $\Delta(x^2)$.

Comparison of Theorem 3.5. with the descriptions of Conway's potential function $\nabla(x - x^{-1})$ shows that $\Omega(x) = \nabla(x - x^{-1})$. Thus $\Omega(x)$ *is* the Conway potential function. Since $\Omega(x)$ is a polynomial in $z = x - x^{-1}$, we denote this polynomial by $\nabla(z)$.

This paper is organized as follows: In §2, we give an axiomatic exposition of the Conway polynomial. Theorem 3.4 is proved in §3. Section 4 discusses applications of these methods to the Arf invariant, and to knots of polynomial equal to 1.

†Research partially supported by the National Science Foundation.

§2. AXIOMS AND COMPUTATIONS

To each knot or link $K \subset S^3$ there is associated a polynomial (the Conway polynomial) $\nabla_K(z)$ satisfying the following axioms: (a) $\nabla_K(z)$ is an invariant of ambient isotopy of $K \subset S^3$. (b) If K is the unknot, then $\nabla_K(z) = 1$. (c) If three knots (links) have diagrams identical in all respects except for the changes indicated below, then

$$\nabla_K - \nabla_{\bar{K}} = z\nabla_L.$$

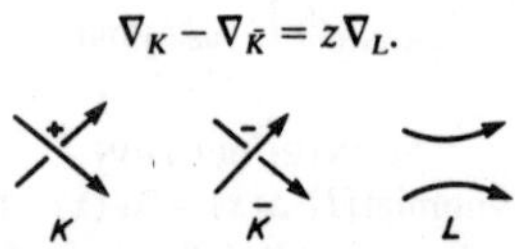

Definition 2.1 A link $L \subset S^3$ is *split* (a split link) if $L = L_1 \cup L_2$, a union of non-empty sublinks such that there are two disjoint embedded three-balls in S^3 containing L_1 and L_2 respectively.

LEMMA 2.2. *If L is a split link then* $\nabla_L(z) = 0$.

Proof. Refer to Fig. 1. Since K and $\bar{K}$ are ambient isotopic (Rotate the bottom part of K by 360°) we see that $0 = \nabla_K - \nabla_{\bar{K}} = z\nabla_L$. Hence $\nabla_L = 0$.

It is a consequence of Theorem 3.4 that these axioms are consistent. It is an easy consequence of the axioms that they alone (coupled with observations of unknots and unlinks) suffice to compute $\nabla(z)$.

Here is a computation for the trefoil knot (refer to Fig. 2.): If K is the trefoil and K, $\bar{K}$, L, $\bar{L}$ and U are as in Fig. 2, then $\nabla_K - \nabla_{\bar{K}} = z\nabla_L$, $\nabla_L - \nabla_{\bar{L}} = z\nabla_U$, $\nabla_{\bar{L}} = 0$, $\nabla_U = 1$, $\nabla_{\bar{K}} = 1$. Hence $\nabla_L = z$ and $\nabla_K = 1 + z^2$.

In general, this axiomatic approach shows the surprising simplicity and recursive nature of the invariant $\nabla_K(z)$. It seems nothing short of miraculous that such a scheme should produce good invariants! The remainder of this section contains more examples of ways of calculation.

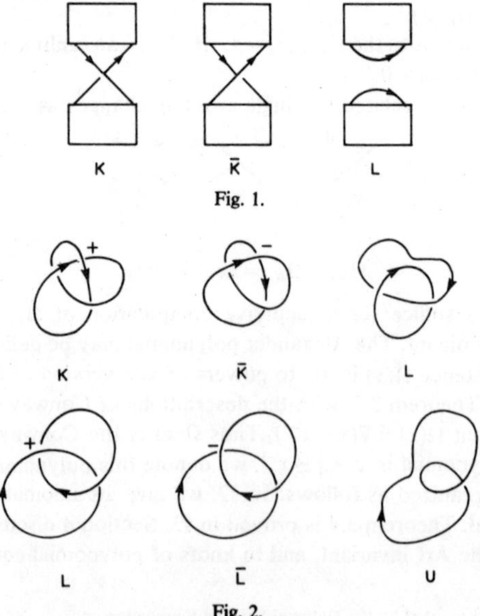

Fig. 1.

Fig. 2.

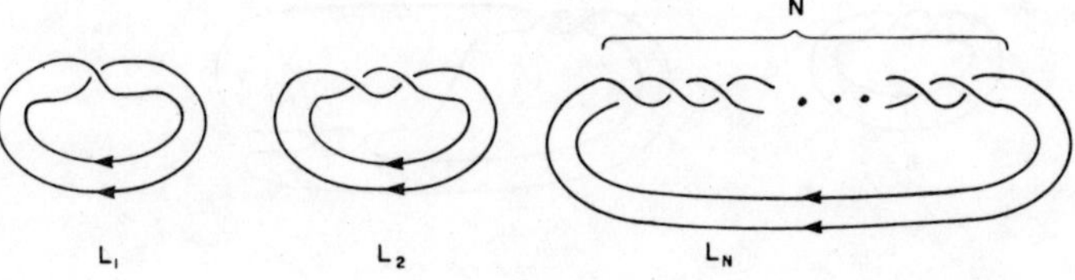

Fig. 3.

Let L_n be a $(2, n)$ torus link as depicted in Fig. 3. Let $\nabla_n = \nabla_{L_n}$. Then the axioms show that $\nabla_n = z\nabla_{n-1} + \nabla_{n-2}$ and $\nabla_1 = 1$, $\nabla_2 = z$. Thus $\nabla_3 = 1 + z^2$, $\nabla_4 = 2z + z^3, \ldots$ Since $\nabla_n(1) = \nabla_{n-1}(1) + \nabla_{n-2}(1)$, $\nabla_n(1)$ is the nth Fibonacci number.

Along with this direct recursive procedure Conway has also devised some more efficient methods of calculations: A *tangle* (see [2]) is a piece of a knot diagram with two input strings and two output strings, oriented as in Fig. 4. Each input is connected to one output, and there may be any other knotting or linking (without free ends) inside the tangle box. Given tangles A and B the tangle $A + B$ is defined as in Fig. 5. Also, there are two ways to form a knot or link from a given tangle A. These are denoted $N(A)$ (numerator) and $D(A)$ (denominator) as in Fig. 5.

Given a tangle A, let $A^N = \nabla_{N(A)}(z)$ and $A^D = \nabla_{D(A)}(z)$. The *fraction* of the tangle A is then defined by the formula $F(A) = A^N/A^D$.

THEOREM 2.3. (*Conway*). *Let A and B be tangles. Then $F(A + B) = F(A) + F(B)$. That is, $(A + B)^N = A^N B^D + A^D B^N$ and $(A + B)^D = A^D B^D$.*

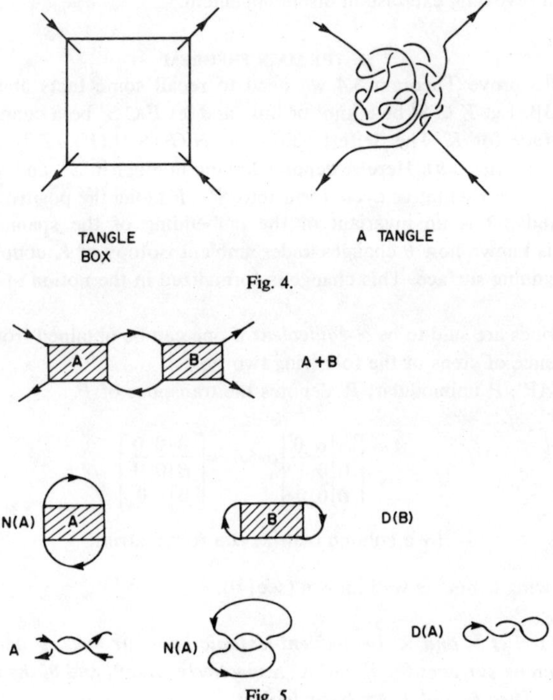

Fig. 4.

Fig. 5.

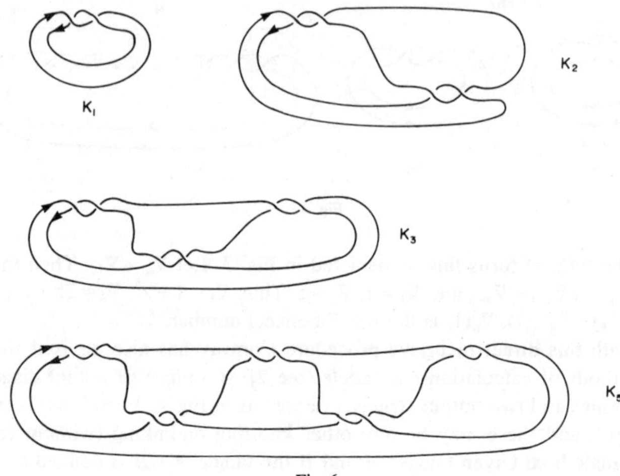

Fig. 6.

As an application of this addition theorem, let K_n denote the knot (link) illustrated in Fig. 6. It then follows from 2.3 that $\nabla_{K_n}(z) = \text{num}\ (z = \frac{1}{z+}\frac{1}{z+}\cdots+\frac{1}{z})$ where this formula denotes the numerator of a continued fraction with n terms. Note that polynomial fractions are added formally with no cancellation between numerators and denominators. It then follows that $\nabla_{K_n} = \nabla_{L_{n+1}}$ where L_n is the torus link of Fig. 3. This leads to the guess that *K_n and L_{n+1} are ambient isotopic.* This guess is indeed correct; the proof is a revealing exercise in disentanglement.

§3. THE MAIN THEOREM

In order to prove Theorem 3.4 we need to recall some facts about the Seifert pairing (see [3]). Let $K \subset S^3$ be a knot or link, and let $F \subset S^3$ be a connected oriented spanning surface for K. The Seifert pairing θ: $H_1(F) \times H_1(F) \to Z$ is defined by the formula $\theta(a, b) = \Lambda(a^+, b)$. Here Λ denotes linking number in S^3, and a^+ is the result of translating a representative cycle for a into $S^3 - F$ along the positive normals to F.

As it stands, θ is an invariant of the embedding of the spanning surface F. However, it is known how θ changes under ambient isotopy of K coupled with a new choice of spanning surface. This change is formalized in the notion of S-equivalence of matrices.

Two matrices are said to be *S-equivalent* if one can be obtained from the other by a finite sequence of steps of the following two types:

(i) $A \leftrightarrow PAP'$, P unimodular, P' denotes the transpose of P.

(ii) $$A \leftrightarrow \left[\begin{array}{c|cc} A & \alpha & 0 \\ \hline 0 & 0 & 1 \\ 0 & 0 & 0 \end{array}\right] \text{ or } A \leftrightarrow \left[\begin{array}{c|cc} A & 0 & 0 \\ \hline \beta & 0 & 0 \\ 0 & 1 & 0 \end{array}\right]$$

(α a column matrix; β a row matrix).

The following lemma is well known (see [4]).

LEMMA 3.1. *Let K and K' be ambient isotopic knots or links, F and F' connected oriented spanning surfaces for K and K' respectively, and θ_1 and θ_2 the Seifert pairings for F and F'. Then θ_1 and θ_2 are S-equivalent.*

We shall be using the *Seifert surface* for a link diagram. This is an orientable spanning surface that is obtained by an algorithm due to Seifert (see [5]). The Seifert surface is constructed by first finding the set of Seifert circles. A given Seifert circle is obtained by choosing a point on the link diagram and traversing the diagram, always declining to pass through a crossing (by jumping to a nearby strand). The surface is obtained by attaching disks to the Seifert circles, and filling in each crossing with a twisted band. Figure 7 depicts twisted bands, and illustrates how the Seifert surface will appear near each of the crossing configurations for the Conway polynomial.

Now recall that when θ: $H_1(F)\times H_1(F)\to Z$ is the Seifert pairing for a connected, oriented spanning surface (F) for a knot or link K, then we have defined the potential function $\Omega_K(x)=D(x\theta-x^{-1}\theta')$.

LEMMA 3.2. *$\Omega_K(x)$ is independent of the choice of spanning surface. In particular, if K is an unknotted circle, then $\Omega_K(x)=1$.*

Proof. Independence follows from Lemma 3.1 and a simple determinant computation. The key point is that

$$D\left(x\begin{bmatrix}0&1\\0&0\end{bmatrix}-x^{-1}\begin{bmatrix}0&0\\1&0\end{bmatrix}\right)=D\begin{bmatrix}0&x\\-x^{-1}&0\end{bmatrix}=1.$$

This same observation takes care of the case of the unknot.

LEMMA 3.3. *If L is a split link, then $\Omega_L(x)=0$.*

Proof. Since L is split (see Definition 2.1.), there are disjoint oriented connected spanning surfaces F_1, $F_2\subset S^3$ such that $L=L_1\cup L_2$, $\partial F_1=L_1$, $\partial F_2=L_2$. Excise a disk from the interior of each surface and connect the surfaces by a tube whose ends fit these holes. Otherwise the tube is disjoint from either surface. This procedure constructs a connected surface F whose boundary is the link L. Let $\mathcal{M}\in H_1(F)$ denote the homology class of a meridean on the tube. Since L_1 and L_2 are non-empty, $\mathcal{M}\neq 0$. If θ: $H_1(F)\times H_1(F)\to Z$ is the Seifert pairing, then $\theta(\mathcal{M},\alpha)=\theta(\alpha,\mathcal{M})=0$ for all $\alpha\in H_1(F)$. Hence $D(x\theta-x^{-1}\theta')=0$. This fact coupled with Lemma 3.2, completes the proof.

Proof of Theorem 3.4 Part (i) follows at once from Lemmas 3.1 and 3.2. Part (ii) follows from Lemma 3.3 (since an unlink is split). In order to prove (iii) it suffices to

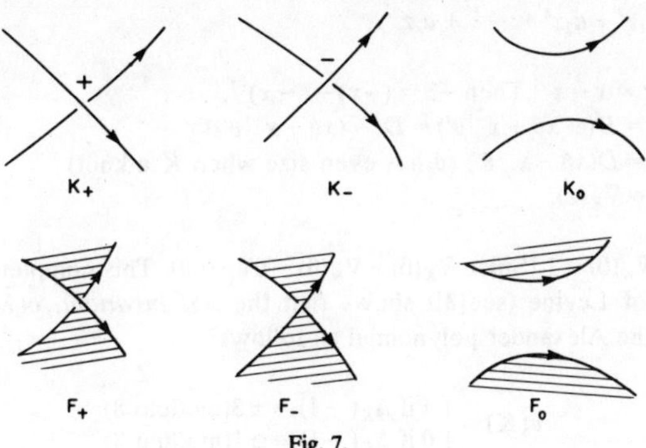

Fig. 7.

assume that each knot or link come equipped with a specific diagram. The potential function will then be calculated from the Seifert surface for this diagram. On such a Seifert surface F, the three cases of Fig. 7 are related by changing the twist on a band, or by eliminating it entirely. The notation will be that of Fig. 7 with links K_+, K_-, K_0 and corresponding surfaces F_+, F_-, F_0.

In the remainder of this proof the word split will refer to a *split diagram.* A link diagram is split if it can be divided into two disjoint non-empty portions each contained within disjoint disks in the plane of the diagram. Thus a link with a split diagram is certainly split in the previous sense. Note that the Seifert surface for a non-split link diagram is connected (and hence may be used to compute a potential function).

Note that if K_+ is split, then both K_- and K_0 are also split. Hence, if K_+ or K_- is split then there is nothing to prove (since by 3.3 split links have vanishing potential function). It may happen that K_0 is split while K_+ and K_- are not split. However, in this case the upper and lower strands of K_0 must be in the split halves. Thus the situation is as in Fig. 1, and K_+ and K_- are ambient isotopic. Therefore the potential functions of K_+ and K_- are equal, and theorem is verified for this case.

Finally, assume that K_+, K_-, and K_0 are all non-split. Then $H_1(F_+) \cong H_1(F_-) \cong H_1(F_0) \oplus Z$ where the extra factor of Z is generated by $\alpha_+ \in H_1(F_+)$ and by $\alpha_- \in H_1(F_-)$, each represented by a curve passing once through the corresponding twisted band of Fig. 7. By extending a basis from $H_1(F_0)$ to $H_1(F_+)$ and to $H_1(F_-)$ it is now clear that Seifert matrices θ_+ and θ_- for F_+ and F_- take the form indicated below. Here θ_0 denotes a Seifert matrix for F_0, and $a = \theta_-(\alpha_-, \alpha_-) = \theta_+(\alpha_+, \alpha_+) - 1$.

$$\theta_- = \left[\begin{array}{c|c} a & \mu \\ \hline \eta & \theta_0 \end{array}\right], \quad \theta_+ = \left[\begin{array}{c|c} a+1 & \mu \\ \hline \eta & \theta_0 \end{array}\right].$$

(η = column matrix, μ = row matrix).
A determinant computation now shows that

$$\Omega_{K_+}(x) - \Omega_{K_-}(x) = (x - x^{-1})\Omega_{K_0}(x).$$

This completes the proof of the theorem.

§4. APPLICATIONS

(a) Arf invariant

PROPOSITION 4.1. *If K is a knot then $\nabla_K(z)$ is a polynomial in even powers of z: $\nabla_K(z) = 1 + a_1z^2 + a_2z^4 + \cdots + a_nz^{2n}$.*

Proof. Let $z = x - x^{-1}$. Then $-z = (-x) - (-x)^{-1}$.
Hence $\nabla_K(-z) = D(-x\theta + x^{-1}\theta') = D(-(x\theta - x^{-1}\theta'))$
$= D(x\theta - x^{-1}\theta')$ (θ has even size when K a knot)
$= \nabla_K(z)$.

It is clear that $\nabla_K(0) = 1$ (Since $\nabla_K(0) - \nabla_{\bar{K}}(0) = 0\nabla_L = 0$). This completes the proof.

A theorem of Levine (see [3]) shows that the *Arf invariant*, $c(K)$, of a knot is obtained from the Alexander polynomial as follows:

$$c(K) = \begin{cases} 1 \text{ if } \Delta_K(-1) \equiv \pm 3 (\text{modulo } 8) \\ 0 \text{ if } \Delta_K(-1) \equiv \pm 1 (\text{modulo } 8) \end{cases}.$$

Since $\Delta(x) = \nabla(\sqrt{x} - \sqrt{x^{-1}})$, we see that $\Delta(-1) = \nabla(i - i^{-1}) = \nabla(2i)$. Hence by 4.1, $\Delta(-1) \equiv \nabla(2)$ (modulo 8), and therefore

$$c(K) = \begin{cases} 1 \text{ if } \nabla_K(2) \equiv \pm 3 (\text{modulo } 8) \\ 0 \text{ if } \nabla_K(2) \equiv \pm 1 (\text{modulo } 8) \end{cases}.$$

With ∇_K as in 4.1, $\nabla_K(2) \equiv 1 + 4a_1$ (modulo 8). Thus the first coefficient of the Conway polynomial determines the Arf invariant.

(b) Knots of polynomial equal to one

Let $K \subset S^3$ be a knot, and associate to K the tangle $T(K)$ obtained by taking oppositely oriented parallel strands as in Fig. 8. This tangle has trivial denominator. Its numerator is a (twisted) double of K. Let λ denote the linking number of the two components of $\hat{K} = N(T(K))$. Then it is easy to see that $\hat{K}$ has a Seifert matrix $\theta = (\lambda)$. Hence $\nabla_{\hat{K}} = \lambda z$, and therefore the tangle $T(K)$ has fraction $F(T(K)) = \lambda z/1$.

We can now use the Conway fraction Theorem (2.1) to show that certain knots have $\nabla_K = 1$. Refer to Fig. 9. The knot illustrated is the numerator of the tangle sum $A + B + C$ where $F(A) = 1/z$, $F(B) = 3z/1$, and (using the remarks above) $F(C) = -3z/1$. Hence the knot has polynomial equal to the numerator of the fraction $1/z + 3z/1 + -3z/1$, and this is equal to one.

This result could be obtained directly from a Seifert pairing. It is nevertheless illuminating to see its relation to fraction and tangle addition.

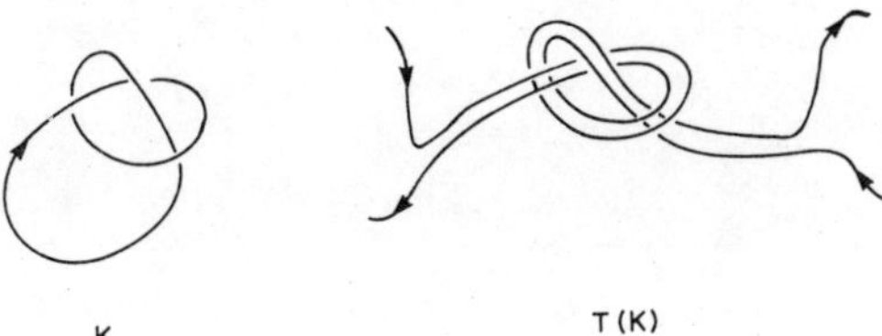

Fig. 8.

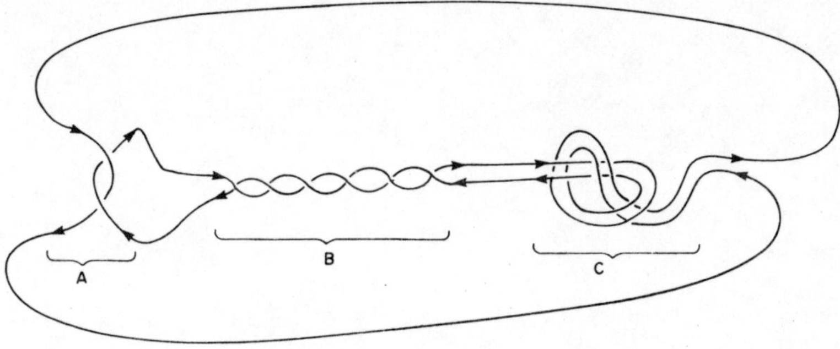

Fig. 9.

REFERENCES

1. J. W. ALEXANDER: Topological invariants of knots and links. *TAMS* **30** (1928), 275–306.
2. J. H. CONWAY: An enumeration of knots and links, and some of their algebraic properties. *Comput. Problems in Abstract Algebra*, Pergamon Press, New York (1970), 329–358.
3. J. LEVINE: Knot corbordism groups in codimension two. *Comm. Math. Helv.* **44** (1969), 229–244.
4. J. LEVINE: An algebraic classification of some knots of codimension two. *Comm. Math. Helv.* **45** (1970), 185–198.
5. H. SEIFERT: Über das Geschlecht von Knotcn. *Math. Ann.* **110** (1934), 571–592.
6. R. H. FOX: A quick trip through know theory. *Topology of 3-Manifolds*, pp. 120–167. Prentice Hall (1962).
7. L. H. KAUFFMAN and F. VARELA: *Form Dynamics. J. Social and Biological Structures.* To appear.
8. J. W. MILNOR: Infinite cyclic coverings. *Topology of Manifolds* (Michigan State University, 1967). Prindle, Weber and Schmidt, Boston (1968).

The University of Illinois at Chicago Circle
Chicago, IL 60680, U.S.A.

The minimal number of Seifert circles equals the braid index of a link

Shuji YAMADA

Department of Mathematics, Ehime University,

Matsuyama, Ehime, 790, Japan.

Introduction

Alexander proved that any oriented link diagram can be transformed into a closed braid by an ambient isotopy ([1],[2]). But we note that his transformation does not keep invariant the number of Seifert circles and the writhe between the original diagram and the obtaind closed braid.

In this paper we will give an alternative method which keeps invariant the number of Seifert circles and the writhe (Theorem 1).

The existense of such a transformation gives that $s(L) \geq b(L)$ for any oriented link L (Theorem 2), where $s(L)$ denotes the minimal number of Seifert circles of all diagrams for L and $b(L)$ denotes the braid index of L.

It is obvious that $s(L) \leq b(L)$, therefore we have the following theorem.

Theorem 3. *For any oriented link* L, $s(L) = b(L)$.

Two estimates for the degree of a variable of the two variable Jones polynomial are given in [3], [5] and [6]. We will show that

Invent. Math. 88, 347–356 (1987).

they are essentially equivalent to each other.

1. Notations and definitions

Let D be an oriented link diagram. We define the writhe $wr(D)$ of D by $wr(D) = \sum_{c} sign(c)$ where c ranges all of the crossings of D and $sign(c)$ is defined as in Figure 1. Let $s(D)$ be the number of Seifert circles of D, where Seifert circles of D are the circles obtained by smoothing all the crossings of D as in Figure 2.

Figure 1 Figure 2

For a braid b, let $\hat{b}$ be the closed braid diagram got by closing b, i.e. tying the top ends to the bottom ends of b as in Figure 3, $e(b)$ be the exponent sum of b, and $n(b)$ be the number of strings of b.

Figure 3

Let C and C' be oriented circles on the 2-sphere S^2, we say that C and C' are coherent (anti-coherent resp.) iff $[C] = [C']$ ($-[C']$ resp.)$\in H_1(A)$, where A is the annulus bounded by C and C' on S^2. Let $C_1,\ldots,C_n$ be mutually disjoint oriented circles on S^2, and suppose that each circle has a positive integer weight and let $w(C_i)$ denote the weight of C_i. Let $a_1,\ldots,a_r$ be mutually disjoint oriented simple arcs on S^2 such that for all a_j, if

$x \in (C_i \cap a_j) \cup \partial a_j$ then a neighbourhood of x is diffeomorphic to one of the situations in Figure 4.

Figure 4

Then we say that $\{C_1, \ldots, C_n ; a_1, \ldots, a_r\}$ is a system of weighted Seifert circles. See an example of a system in Figure 5.

Figure 5

Let $\mathcal{S}$ denote the set of all systems of weighted Seifert circles.

Let $S = \{C_1, \ldots, C_n ; a_1, \ldots, a_r\}$ be an element of $\mathcal{S}$. Then we will give a method of producing a number of link diagrams from S.

First we replace every C_i with an arbitrary $w(C_i)$-string closed braid. Secondly we replace every a_j with an arbitrary $[w(C_{i_1})+\ldots+w(C_{i_m})]$-string braid so that, it wedges into each $w(C_{i_k})$-string braid C_{i_k}, where $C_{i_1}, \ldots, C_{i_m}$ are the circles of S which intersect a_j. So, we obtain a number of link diagrams from S.

If a link diagram D is obtained from S by using the method as the above, then we say that D is derived from S. There is an example of a diagram in Figure 6 which is derived from the system in Figure 5.

Figure 6

Let $\mathscr{D}(S)$ denote the set of link diagrams which are derived from S. For each integer t, let $\mathscr{D}(S,t) = \{D\in\mathscr{D}(S) \mid wr(D)=t\}$. Let $\mathscr{L}(S,t)$ ($\mathscr{L}(S)$ resp.) denote the set of all oriented link types which are presented by a diagram in $\mathscr{D}(S,t)$ ($\mathscr{D}(S)$ resp.).

It is shown easily that for any oriented link diagram D, there exists a system $S\in\mathscr{S}$ such that $D\in\mathscr{D}(S)$. We can choose $S = \{C_1,\dots,C_n\ ;\ a_1,\dots,a_r\}$ for such a system, where $C_1,\dots,C_n$ are the Seifert circles of D and $a_1,\dots,a_r$ are placed at each crossing of the diagram and $w(C_i) = 1$ for all C_i.

Let $S = \{C_1,\dots,C_n\ ;\ a_1,\dots,a_r\} \in\mathscr{S}$. We define the total weight $w(S)$ of S by $w(S) = w(C_1)+\dots+w(C_n)$.

For each positive integer p, let $\mathscr{S}_p = \{S\in\mathscr{S} \mid w(S)=p\}$, and let $T_p = \{C;\ \}\in\mathscr{S}_p$ where $w(C) = p$. Clearly $\mathscr{D}(T_p)$ is the set of all p-string closed braid diagrams. We call T_p the trivial system of total weight p.

2. Bunching operations

We define two operations, the bunching operations of type I and type II, by the following.

Let $S = \{C_1,\dots,C_n\ ;\ a_1,\dots,a_r\}$ be an element of $\mathscr{S}$.

If there are two coherent circles C_i and C_j such that $Int(A)\cap\{C_1\cup\dots\cup C_n\} = \phi$, where A is the annulus boundedby C_i and C_j on S^2, then we define a new system S' as follows.

Let C be an abstract circle and $f:A\rightarrow C$ be a continuous map such that $f|_{C_i}$ and $f|_{C_j}$ are homeomorphisms, if $A\cap a_k \neq \phi$ then

$f(A\cap a_k)$ = {one point} and if $C_i\cap a_k \neq \phi$, $C_j\cap a_l \neq \phi$ and $a_k \neq a_l$ then $f(C_i\cap a_k) \neq f(C_j\cap a_l)$. Then the quotient space $(S^2\cup C)/f(x)\sim x$ is a 2-sphere, let $S' = (S\cup C)/f(x)\sim x$ with $w(C) = w(C_i)+w(C_j)$.

We say that S' is derived from S by applying the bunching operation of type I to C_i and C_j. See Figure 7.

Figure 7

If there are two anti-coherent circles C_i and C_j of S and a band b on S^2 such that $b\cap S = \partial b\cap(C_i\cup C_j) = d_i\cup d_j$ and $b\cap\{a_1\cup\ldots\cup a_r\} = \phi$, where d_i and d_j are subarcs of C_i and C_j respectively, then let $S' = \{C_1,\ldots,\check{C}_i,\ldots,\check{C}_j,\ldots,C_n,C\ ;\ a_1,\ldots,a_r\}$. Here $\check{C}_i$, $\check{C}_j$ means the deleting of these circles, $C = (C_i\cup C_j\cup\partial b)-Int(d_i\cup d_j)$, and the orientation of C is determined from those of C_i and C_j naturally with $w(C) = w(C_i)+w(C_j)$.

We say that S' is derived from S by applying the bunching operation of type II to C_i and C_j. See Figure 8.

Figure 8

3. Lemma and Theorem 1

Lemma. *Let S and S' be elements of $\mathcal{S}$. If S' is derived from S by the bunching operation of* type I *or* type II *then $w(S) = w(S')$ and $\mathcal{L}(S,t)\subset\mathcal{L}(S',t)$ for any integer t.*

Proof. It is obvious that $w(S) = w(S')$ by the definition of the bunching operations. We will show the latter claim.

In the case of type I, we can show easily that $\mathcal{D}(S,t) \subset \mathcal{D}(S',t)$. Therefore $\mathcal{L}(S,t) \subset \mathcal{L}(S',t)$.

In the case of type II, let C_i and C_j be the circles of S to which the bunching operation of type II is applied. Let b, d_i, d_j be as in the above description. Let N_k be a regular neighbourhood of C_k such that $N_k \cap (C_1 \cup \ldots \cup C_n) = C_k$ for $k = i, j$. Let C_i^+ be the component of ∂N_i which intersects b. Let C_j^- be the component of ∂N_j which does not intersect b. Let b_1 be a thin band which joins C_i and C_j^-, Let b_2 be a thick band which joins C_i^+ and C_j, as in Figure 9. Let $C'_i = (C_i \cup C_j^- \cup \partial b_1) - Int((C_i \cup C_j^-) \cap \partial b_1)$, $C'_j = (C_j \cup C_i^+ \cup \partial b_2) - Int((C_j \cup C_i^+) \cap \partial b_2)$, $d'_i = C_i \cap b_1$, $d'_j = C_j \cap b_2$, $e_i = C'_i - C_i$ and $e_j = C'_j - C_j$.

Let $S'' = \{C_1, \ldots, \check{C}_i, \ldots, \check{C}_j, \ldots, C_n, C'_i, C'_j \,;\, a_1, \ldots, a_r\}$, where the orientations of C'_i and C'_j are defined from the those of C_i and C_j naturally with $w(C'_i) = w(C_i)$ and $w(C'_j) = w(C_j)$. See Figure 10.

Figure 9 Figure 10

For any element D of $\mathcal{D}(S,t)$, we can deform D to a element D' of $\mathcal{D}(S'',t)$ under a regular isotopy as in Figure 11. We call this deformation the bunching deformation.

Figure 11

Figure 11 is understood as the following.

We can assume that d'_i and d'_j are replaced by trivial braids when we produce D from S. Let τ_i and τ_j denote those trivial braids. First we stretch out τ_j to τ'_j and set it on e_j. Secondly we stretch out τ_i to τ'_i and set it on e_i. Let e be e_i (e_j resp.) and τ be τ_i (τ_j resp.). If an arc a of S intersects e then the braid β placed at a is changed by the following map φ,

$$\varphi: B_{p+q} \longrightarrow B_{p+q+r},$$

$$\sigma_k \longmapsto \sigma_k \qquad \text{if } 1 \le k \le p-1,$$

$$\sigma_p \longmapsto \rho^{-1}\sigma_p\rho \qquad \text{where } \rho = \prod_{l=p+1}^{p+r} \sigma_l \quad \text{(Figure 13)},$$

$$\sigma_k \longmapsto \sigma_{k+r} \qquad \text{if } p+1 \le k \le p+q-1,$$

where σ_k is the generator of the braid group which exchanges the k-th string and $(k+1)$-th string by a right hand twist (Figure 12), $p = \sum_{\substack{C\cap a' \neq \phi \\ C \in S}} w(C)$, $q = \sum_{\substack{C\cap a'' \neq \phi \\ C \in S}} w(C)$, where a' (a'' resp.) is the component of $a-e$ which contains the starting point (terminal point resp.) of the oriented arc a; and $r = n(\tau)$.

Figure 12 Figure 13

S' is derived from S'' by the bunching operation of type I, therefore $D' \in \mathfrak{D}(S', t)$.

We shall note that φ keeps invariant the exponent sum of the braids. Therefore $wr(D) = wr(D')$.

This completes the proof of the lemma. □

We note that the bunching operations decrease the number of circles of a system.

The next Theorem asserts that any oriented link diagram can be deformed to a closed braid by the bunching deformations.

Theorem 1. *If* S *is a non trivial system then there is a pair of circles of* S *to which the bunching operation of* type I *or* type II *can be applied.*

Proof. Let $S = \{C_1,\ldots,C_n\ ;\ a_1,\ldots,a_r\}$. We cut the S^2 along $C_1,\ldots,C_n$, then there is a piece P which is not a disk because S is not trivial.

In the case that P is an annulus, let $C_i \cup C_j = \partial P$. If C_i and C_j are coherent then the bunching operation of type I can be applied to them. If C_i and C_j are anti-coherent then the bunching operation of type II can be applied to them.

In the case thet P is not an annulus, we cut P along $a_1,\ldots,a_r$. We can assume that every piece is a disk, because if there is a piece which is not a disk then we can add oriented arcs to S to cut the piece into disks. Then there is a piece Q such that $\#(\partial Q \cap (C_1 \cup \ldots \cup C_n)) \geq 4$ because P is neither an annulus nor a disk. We note that $\#(\partial Q \cap (C_1 \cup \ldots \cup C_n))$ will always be even because of the orientation of arcs.

Let $\partial Q \cap (C_1 \cup \ldots \cup C_n) = E_1 \cup \ldots \cup E_m$, where $E_1,\ldots,E_m$ are on ∂Q in this order, $m \geq 4$, and m is even. Let $E_j \subset C_{i_j}$. Then C_{i_j}

and $C_{i_{j+1}}$ are coherent ($1 \leq j \leq m-1$) therefore C_{i_j} and $C_{i_{j+2}}$ are anti-coherent ($1 \leq j \leq m-2$). We claim that $C_{i_1} \neq C_{i_3}$ or $C_{i_2} \neq C_{i_4}$. Assume that $C_{i_1} = C_{i_3}$ and $C_{i_2} = C_{i_4}$. Let D_j be the disk bounded by C_{i_j} which does not contain Q ($j=1,2$). Then $Q \cup D_1 \cup D_2$ contains a surface of genus 1, a contradiction.

Hence the bunching operation of type II can be applied to one of the pairs (C_{i_1}, C_{i_3}) and (C_{i_2}, C_{i_4}).

This completes the proof of Theorem 1. □

4. Applications

By the lemma and Theorem 1, we get the following Theorem.

Theorem 2. *For any* $S \in \mathcal{S}_p$ *and integer* t, $\mathcal{L}(S,t) \subset \mathcal{L}(T_p,t)$.

This Theorem means that $b(L) \leq s(L)$ for any oriented link L. Then we get Theorem 3 in the introduction.

Let $\underline{d}(L)$ ($\overline{d}(L)$ resp.) be the lowest (highest resp.) degree of $P_L(l,m)$ about the variable l, where $P_L(l,m)$ is the two vatiable Jones polynomial of L ([4]). It is shown in [3], [5] and [6] that for any oriented link L, the following inequalities hold.

(1) $m(L) \leq \underline{d}(L) \leq \overline{d}(L) \leq M(L)$,

(2) $m'(L) \leq \underline{d}(L) \leq \overline{d}(L) \leq M'(L)$,

where m, M, m' and M' are defined by the following.

$$m(L) = \max\{\, wr(D)-(s(D)-1) \mid D \text{ is a diagram for } L \,\},$$
$$M(L) = \min\{\, wr(D)+(s(D)-1) \mid D \text{ is a diagram for } L \,\},$$
$$m'(L) = \max\{\, e(b)-(n(b)-1) \mid \hat{b} \text{ is a closed braid for } L \,\},$$
$$M'(L) = \min\{\, e(b)+(n(b)-1) \mid \hat{b} \text{ is a closed braid for } L \,\}.$$

The next Corollary means that the two inequalities (1) and (2) are essentially equivalent to each other.

Corollary. *For any oriented link* L, $m(L) = m'(L)$ *and* $M(L) = M'(L)$.

Proof. It is obvious that $m'(L) \leq m(L)$ and $M(L) \leq M'(L)$ because $wr(\hat{b}) = e(b)$ and $s(\hat{b}) = n(b)$ for any braid b.

To show the reverse inequalities of the above, we assume that D_1 and D_2 are diagrams for L such that $m(L) = wr(D_1)-(s(D_1)-1)$ and $M(L) = wr(D_2)+(s(D_2)-1)$. By Theorem 2 we can transform D_1 and D_2 to a closed braid $\hat{b}_1$ and $\hat{b}_2$ respectively such that $wr(D_1) = wr(\hat{b}_1)$, $s(D_1) = s(\hat{b}_1)$, $wr(D_2) = wr(\hat{b}_2)$ and $s(D_2) = s(\hat{b}_2)$. Then we get that $m'(L) \geq m(L)$ and $M(L) \geq M'(L)$. □

Acknowledgement

I wish to thank Professor H. Murakami, Professor Y. Nakanishi and Professor T. Kobayashi for their suggestions for this problem and helpful advices. I wish also to thank Professor M. Ochiai for his constant encouragement.

References

1. Alexander, J. W. : A lemmma on system of knotted curves. Proc. Nat. Academ. Science USA, 9, 93-95 (1923).

2. Birman, J. S. : Braids, links, and mapping class groups. Ann. Math. Studies; no 82, Princeton, New Jersey, Princeton Univ. Press (1974).

3. Franks, J., Williams, R. F. : Braids and the Jones-Conway polynomial. (Preprint 1985).

4. Freyd, P., Yetter, D.; Hoste, J.; Lickorish, W. B. R., Millett, K.; Ocneanu, A. : A new polynomial invariant of knots and links. Bull. Amer. Math. Soc. 12, 239-246 (1985).

5. Morton, H. R. : Seifert circles and knot polynomials. Math. Proc. Camb. Phil. Soc. 99, 107-109 (1986).

6. Morton, H. R. : Closed braid representations for a link, and its 2-variable polynomial. (Preprint Liverpool 1985).

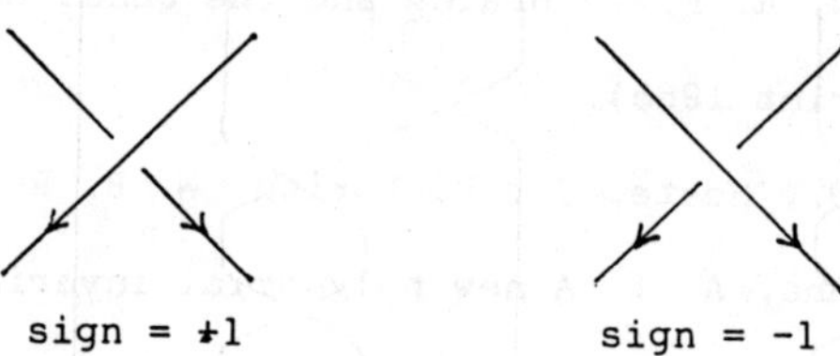

Figure 1

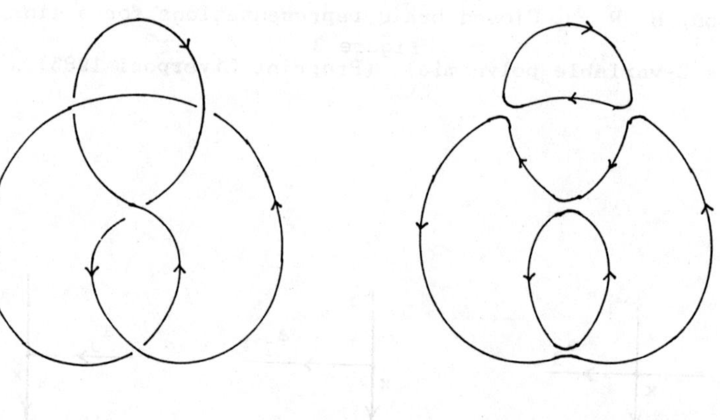

Figure 2

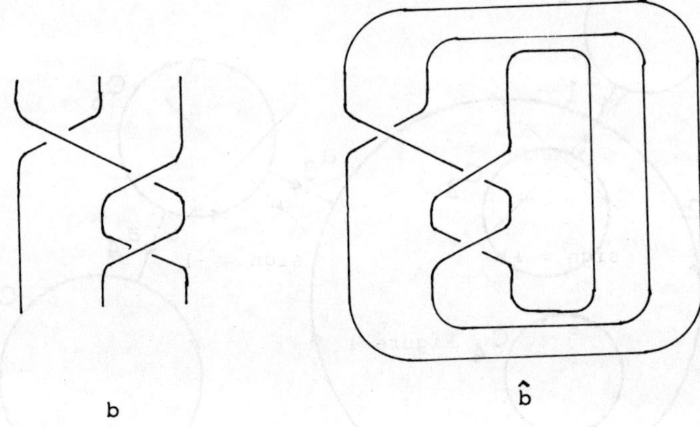

Figure 3

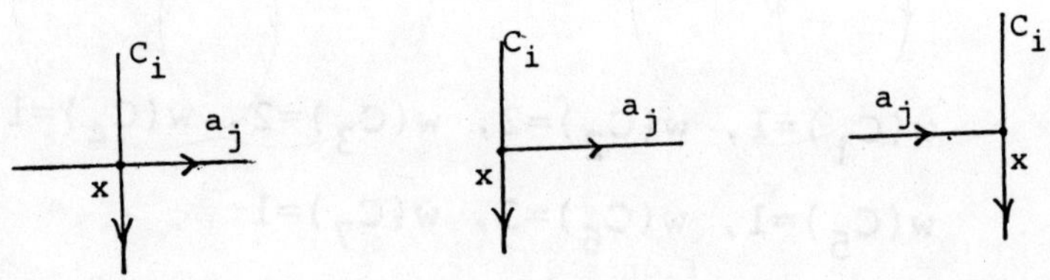

Figure 4

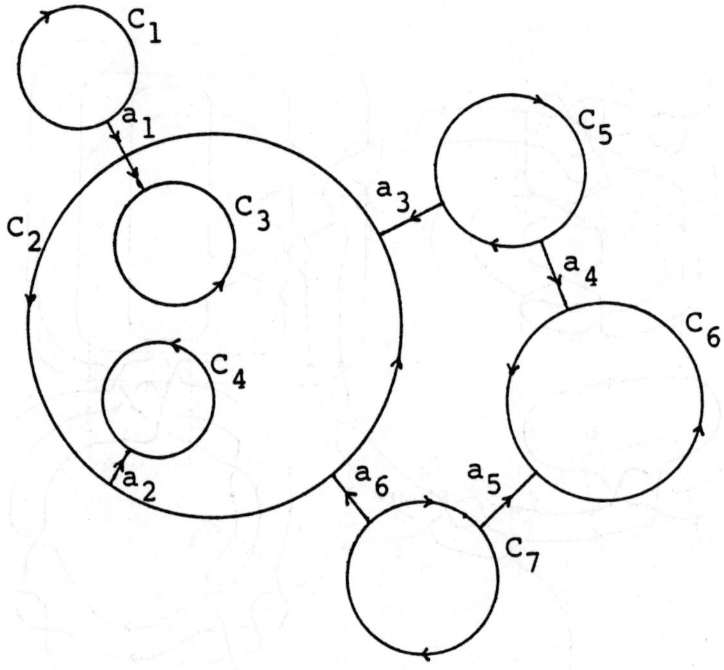

$w(C_1)=1,\ w(C_2)=2,\ w(C_3)=2,\ w(C_4)=1,$

$w(C_5)=1,\ w(C_6)=3,\ w(C_7)=1$

Figure 5

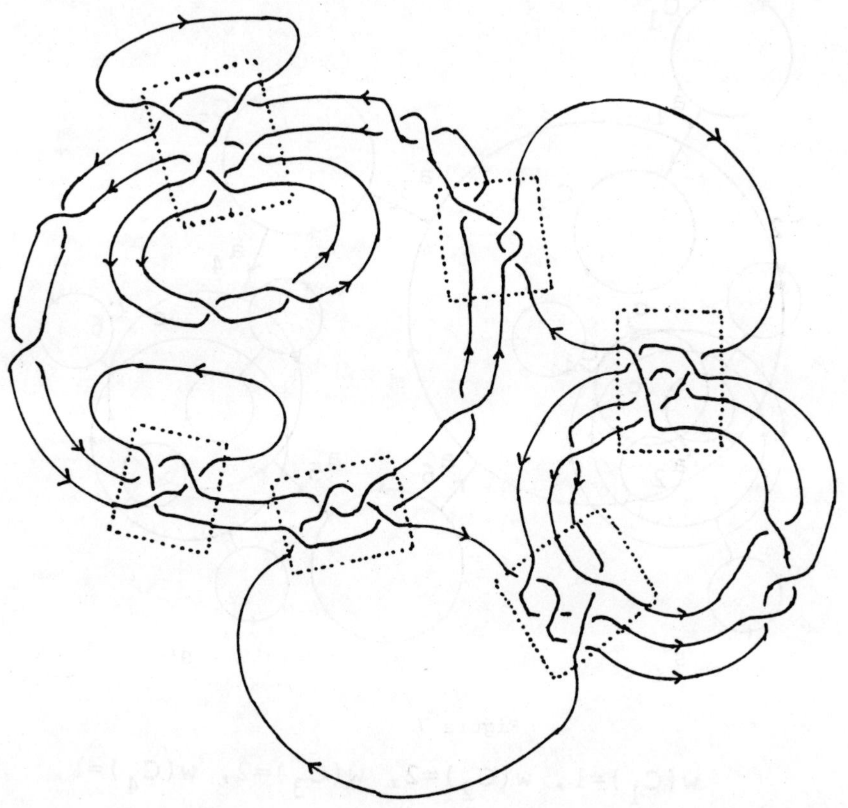

Figure 6

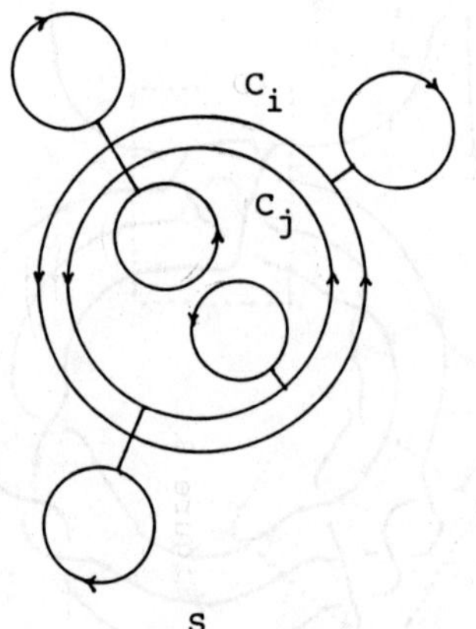

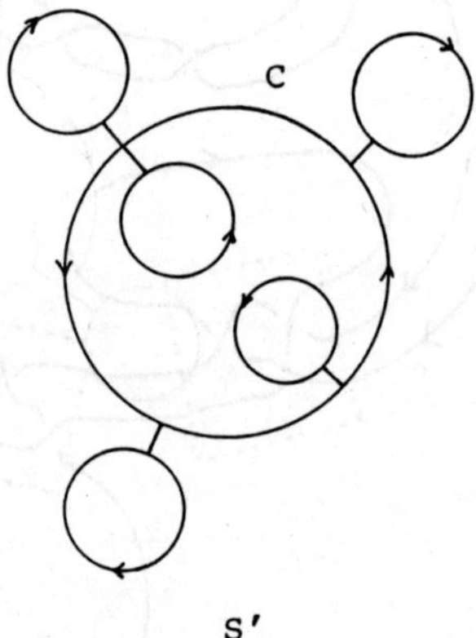

Figure 7

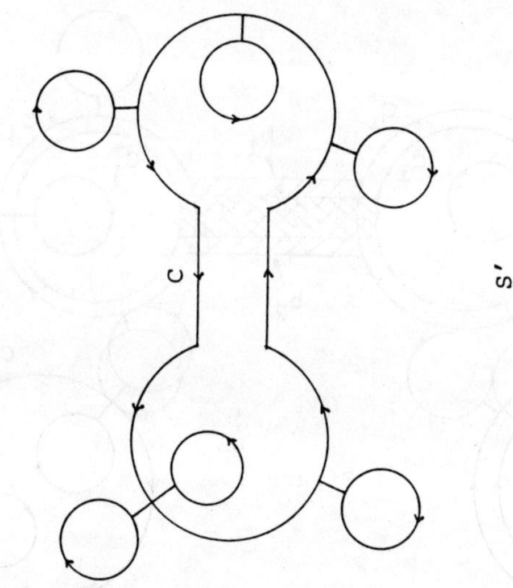

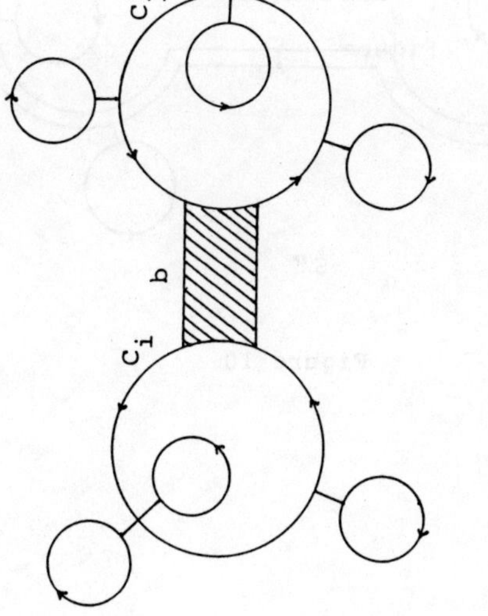

Figure 8

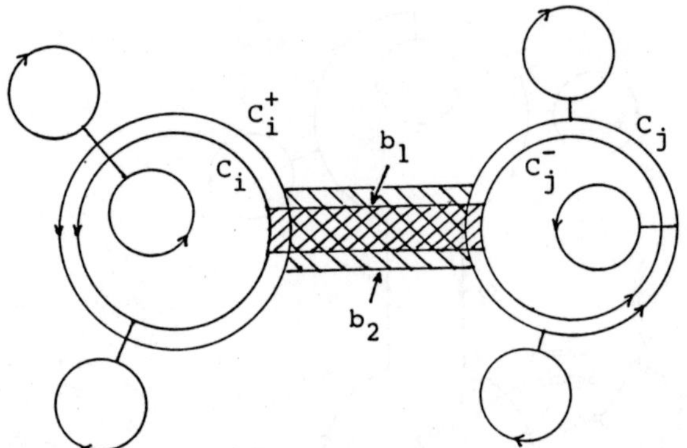

Figure 9

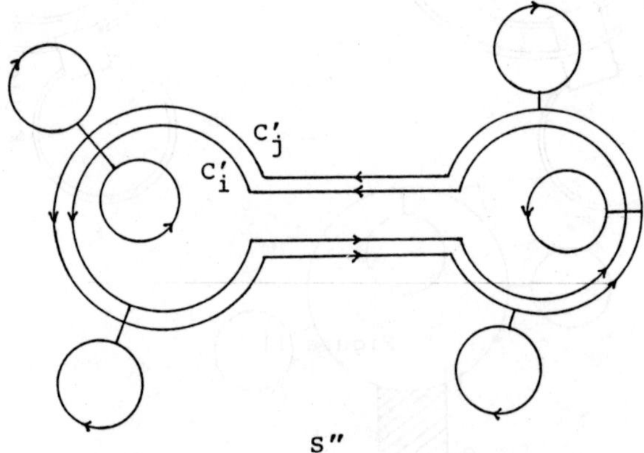

Figure 10

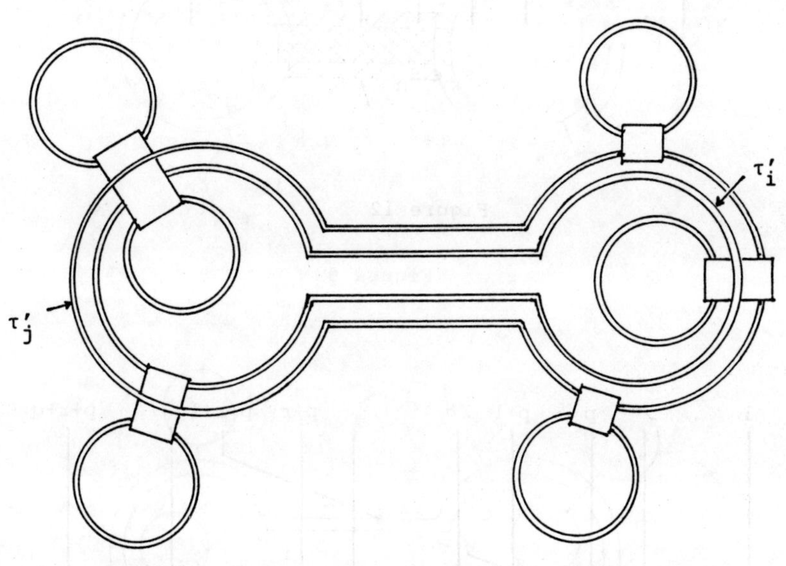

Figure 11

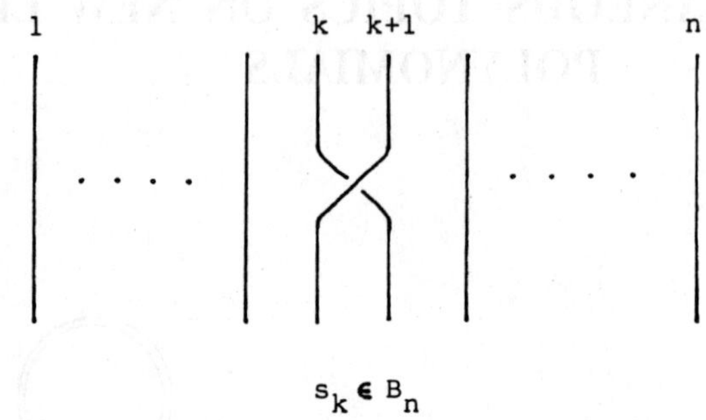

$s_k \in B_n$

Figure 12

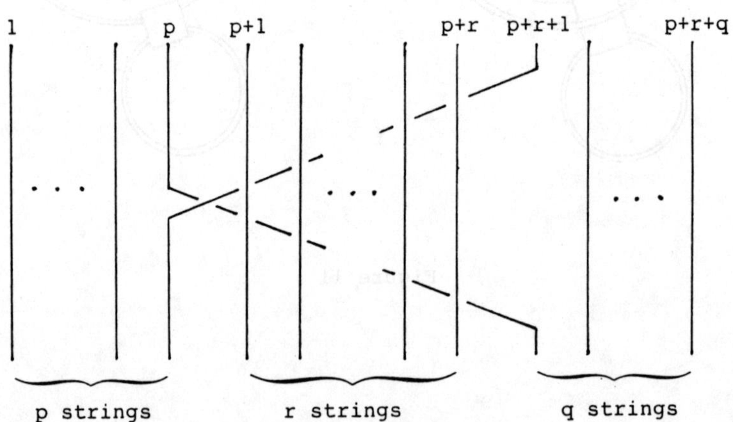

Figure 13

MISCELLANEOUS TOPICS ON NEW LINK POLYNOMIALS

On the Jones polynomial of closed 3-braids

Joan S. Birman*

Columbia University, Department of Mathematics, New York, NY 10027, USA

Introduction

In [J 2] Vaughan Jones introduced a new polynomial $V_L(t)$ which is an invariant of the isotopy type of an oriented knot or link L. The polynomial can be computed from an arbitrary representation of L as a closed braid, i.e. from an element in one of the Artin braid groups B_n, $n=1, 2, 3, \ldots$. It is very powerful, distinguishing the trefoil and its mirror image, the unknot and the Kinoshita-Terasaka knot, and any two members of the infinite sequence of Whitehead links, all of which have homeomorphic complements. In an early version of [J 2] (before the results reported here were complete) Jones had conjectured that his polynomial was injective on closed 3-braids. In this note we use trace identities in the Burau matrix representation of B_3 to construct myriad counterexamples to that conjecture. At the same time, we also disprove a second conjecture of Jones from [J 2], that a link L is amphicheiral iff $V_L(t)=V_L(t^{-1})$. The 2-variable generalized Jones polynomial introduced in [FYHLMO] also fails to distinguish any of our link pairs. Finally, our examples answer in the negative a question of Morton, who asked whether the Alexander polynomial is a complete invariant of the link obtained by adjoining the braid axis to a closed 3-braid link.

In the monograph [Mu] Murasugi began a classification of closed 3-braid links, work which was extended by Hartley in [H]. From the partial results in [Mu] and [H] it seemed reasonable to conjecture that, omitting various obvious exceptional cases (links with braid index < 3, composite links and torus links) the link type of a closed 3-braid is determined by its conjugacy class in B_3, up to the equivalences which correspond to orientation changes. This conjecture remains open. When we began this work we had hoped to settle it, expecting that the conjugacy class of a braid was determined by the Jones or Alexander polynomial of the associated closed braid. However, this is far from the truth. The problem of characterizing all of the 3-braids whose closures have a given Jones or Alexander or 2-variable polynomial seems to be very non-trivial and possibly even to be too complicated to have an interesting solution.

* Supported in part by NSF grant # MCS 79 04715

Invent. math. 81, 287–294 (1985)

Inventiones mathematicae

On the Jones polynomial of closed 3-braids*

Joan S. Birman
Columbia University, Department of Mathematics, New York, NY 10027, USA

Introduction

In [J, 2] Vaughan Jones introduced a new polynomial $V_L(t)$ which is an invariant of the isotopy type of an oriented knot or link $L \subset S^3$. The polynomial can be computed from an arbitrary representation of L as a closed braid, i.e. from an element in one of the Artin braid groups B_n, $n=1, 2, 3, \ldots$. It is very powerful, distinguishing the trefoil and its mirror image, the unknot and the Kinoshita-Terasaka knot, and any two members of the infinite sequence of Whitehead links, all of which have homeomorphic complements. In an early version of [J, 2] (before the results reported here were complete) Jones had conjectured that his polynomial was injective on closed 3-braids. In this note we use trace identities in the Burau matrix representation of B_3 to construct myriad counterexamples to that conjecture. At the same time, we also disprove a second conjecture of Jones from [J, 2], that a link L is amphicheiral if $V_L(t)=V_L(t^{-1})$. The 2-variable generalized Jones polynomial introduced in [FYHLMO] also fails to distinguish any of our link pairs. Finally, our examples answer in the negative a question of Morton, who asked whether the Alexander polynomial is a complete invariant of the link obtained by adjoining the braid axis to a closed 3-braid link.

In the monograph [Mu] Murasugi began a classification of closed 3-braid links, work which was extended by Hartley in [H]. From the partial results in [Mu] and [H] it seemed reasonable to conjecture that, omitting various obvious exceptional cases (links with braid index <3, composite links and torus links) the link type of a closed 3-braid is determined by its conjugacy class in B_3, up to the equivalences which correspond to orientation changes. This conjecture remains open. When we began this work we had hoped to settle it, expecting that the conjugacy class of a braid was determined by the Jones or Alexander polynomial of the associated closed braid. However, this is far from the truth. The problem of characterizing all of the 3-braids whose closures have a given Jones or Alexander or 2-variable polynomial seems to be very non-trivial and possibly even to be too complicate to have an interesting solution.

* Supported in part by NSF grant # MCS 79-04715

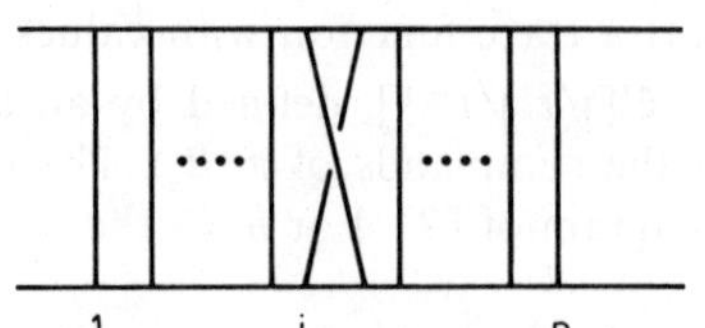

Fig. 1. The elementary braid s_i

1. The Jones polynomial

Let B_n be the n-string Artin braid group [A] and let $s_1, \ldots, s_{n-1}$ be the elementary braids which generate B_n (Fig. 1). Let $\{A_n(\tau);\ \tau \in \mathbb{R}^+,\ n=1, 2, 3, \ldots\}$ be the family of algebras which are described by Jones in [J, 1] and [J, 2]. Each $A_n = A_n(\tau)$ is a finite-dimensional algebra of operators on a Hilbert space, generated by projections $\{1, e_1, e_2, \ldots, e_{n-1}\}$, and is a direct sum of matrix algebras. For fixed n there are $[(n+2)/2]$ summands, which are indexed by the entries in the nth row of a truncated Pascal's triangle, the dimensions of the summands being described by the entries in that row (see the first diagram in §2 of [J, 1]). It was proved in [J, 1] that for each t satisfying $\tau = t/(1+t)^2$, where now $0 < \tau \leqq \frac{1}{4}$, there is a representation π_t: $B_n \to A_n$, defined by $\pi_t(s_i) = \sqrt{t}(te_i - (1-e_i))$, $1 \leqq i \leqq n-1$. Since the A_n's are matrix algebras, the homomorphisms π_t determine, for each admissible t and each positive integer n a finite-dimensional representation of B_n over $\mathbb{C}$, or (regarding t as an indeterminate rather than a parameter) over $\mathbb{C}[\sqrt{t}, \sqrt{t}^{-1}]$. Since A_n is a direct sum of $[(n+2)/2]$ matrix algebras, one obtains in this way $[(n+2)/2]$ matrix representations of B_n over $\mathbb{C}[\sqrt{t}, \sqrt{t}^{-1}]$.*

Two summands have been identified explicitly in [J, 1], namely the ones corresponding to the last two entries in the nth row of the aforementioned Pascal's triangle. To describe the first summand, let ϕ: $B_n \to S_n$ be the homomorphism onto the symmetric group defined by $\phi(s_i) = (i, i+1)$, $1 \leqq i \leqq n-1$. If $\alpha \in B_n$, let e_α denote the exponent sum of α as a word in the elementary braids $s_i, \ldots, s_{n-1}$. Then, for each $n \geqq 1$ the representation π_t contains as one of its summands a 1-dimensional representation which is defined by mapping $\alpha \in B_n$ to $(-\sqrt{t})^{e_\alpha}$. Theorem 3.3 of [J, 1] identifies a second summand of $\pi_t(B_n)$, $n \geqq 2$, as the Kronecker product of the Burau representation of B_n (a representation ψ: $B_n \to GL(n-1, \mathbb{Z}[t, t^{-1}])$ introduced in [Bu]) with the normalized parity representation $s_j \to -\sqrt{t}$, $j = 1, \ldots, n-1$. For $n=3$ we have:

$$\pi_t(\alpha) = (-\sqrt{t})^{e_\alpha} \oplus (-\sqrt{t})^{e_\alpha} \psi(\alpha), \qquad \alpha \in B_3, \tag{1}$$

where ψ is defined explicitly by:

$$\psi(s_i) = \begin{bmatrix} -t & 1 \\ 0 & 1 \end{bmatrix}, \quad \psi(s_2) = \begin{bmatrix} 1 & 0 \\ t & -t \end{bmatrix}. \tag{2}$$

* In [J, 1] the representation π_t is defined by $s_i \to te_i - (1-e_i)$, and in J, 2] as given. This discrepency leads to some changes in the formulas in the passage from [J, 1] to [J, 2]

Jones' algebras support a trace function with values in $\mathbb{C}$, which determines a function trace: $\pi_t(B_n)\to\mathbb{C}[\sqrt{t},\sqrt{t}^{-1}]$, defined by an appropriate linear combination of the traces of the summands of $\pi_t(B_n)$. The coefficients are specified in the Pascal's triangle diagram of [2]. For $n=3$ the formula is:

$$\text{trace}\,\pi_t(\alpha)=(1-2\tau)(-\sqrt{t})^{e_\alpha}+\tau(-\sqrt{t})^{e_\alpha}\,\text{trace}\,\psi(\alpha), \tag{3}$$

where as before $\tau=t/(1+t)^2$.

For $\alpha\in B_n$, let $L=L_\alpha$ denote the link determined by the closed braid $\hat{\alpha}$. The Jones polynomial $V_L(t)$ is defined in [J, 2] to be:

$$V_L(t)=\left(-\frac{t+1}{\sqrt{t}}\right)^{n-1}\text{trace}\,\pi_t(\alpha). \tag{4}$$

For the special case $n=3$ formulas (3) and (4) combine to give:

$$V_L(t)=(-\sqrt{t})^{e_\alpha}(t+t^{-1}+\text{trace}\,\psi(\alpha)). \tag{5}$$

Thus $V_L(t)$ is determined by e_α and trace $\psi(\alpha)$.

The 2-variable Jones polynomial was introduced in [FYHLMO]. There are several ways to define it, and for our purposes the approach of Ocneanu is most relevant. Ocneanu defined it as a weighted sum of traces in a matrix representation of B_n which is much like the one just described, only the Pascal's triangle is not truncated (so that there are new summands), also the weights depend upon two parameters. For the special case $n=3$ the only new summand which enters into the two-variable version is a new trivial representation of B_3, hence the two-variable polynomial is determined by e_α and trace $\psi(\alpha)$, just as the one-variable version is.

Jones' conjectures are:

Conjecture 1. Let α, $\beta\in B_3$, and let L_α, L_β be the links determined by the closed braids $\hat{\alpha},\hat{\beta}$. Assume that L_α, L_β have the same number of components. Then $V_{L_\alpha}(t)=V_{L_\beta}(t)$ implies $L_\alpha\approx L_\beta$.

Conjecture 2. Let $\alpha\in B_n$, with L_α a knot. Then $V_{L_\alpha}(t)=V_{L_\alpha}(t^{-1})$ implies that L is amphicheiral.

In view of Eq. (5) and the remarks which follow, it is clear that if $\alpha,\alpha'\in B_3$, with $L=\hat{\alpha}$, $L'=\hat{\alpha}'$, and if $e_\alpha=e'_\alpha$, trace $\psi(\alpha)=$ trace $\psi(\alpha')$, then L and L' will have the same Jones and two-variable Jones polynomials.

2. The Alexander polynomial

Let $\alpha\in B_n$. Let $A_\alpha(t)$ denote the Alexander polynomial of the closed braid α, where if $\hat{\alpha}$ is a link the polynomial is the reduced Alexander polynomial, with orientation determined by the orientations on the braid. Let $\mathscr{A}_\alpha(t,x)$ denote the Alexander polynomial of the link $\hat{\alpha}\cup\hat{\lambda}$ obtained by adjoining the braid axis $\hat{\lambda}$ to the link $\hat{\alpha}$.

Assume that $\mathscr{A}_\alpha(t,x)$ is normalized so that $\mathscr{A}_\alpha(t,x)=\mathscr{A}_\alpha(\bar{t},\bar{x})$, where $\bar{t}=t^{-1}$, $\bar{x}=x^{-1}$, and $A_\alpha(1)=1$, when $\hat{\alpha}$ is a knot. Let $e=e_\alpha$. We claim:

$$\begin{aligned} &\text{(i)} \quad A_\alpha(t,x)=(-1)^e A_\alpha(t,1)\,x^{\frac{1-n}{2}}\,t^{-\frac{e}{2}}|\psi(\alpha)-xI|, \\ &\text{(ii)} \quad t^{\frac{1-n}{2}}(1+t+\ldots+t^{n-1})\,A_\alpha(t)=(-1)^{\frac{e}{2}}A_\alpha(t,1). \end{aligned} \tag{6}$$

To see this, note first that Eq. (6) include a formula for $A_\alpha(t)$ discovered by Burau (see Theorem 3.11 of [Bu]), and also its subsequent generalization by Morton [Mo] to a formula for $A_\alpha(t,x)$, however there is something new. In Burau's formula, $A_\alpha(t)$ was only determined up to a multiplicative constant $\pm t^k$ whereas in our formula this constant has been fixed by normalizing $A_\alpha(t)$ so that it is symmetric and so that $A_\alpha(1)=1$. Therefore the only thing we must verify is the normalization. Let $M(t)$ denote the matrix which represents $\psi(\alpha)$, $\alpha\in B_n$, where $M(t)$ is as defined by Eq. (2), so that $\det M(t)=(-t)^e$. By [Sq], $M(\bar{t})$ is conjugate in the general linear group to $(M(t))^{-1}$. Note that $M(t)$ has dimension $n-1$. We have:

$$\begin{aligned} A_\alpha(\bar{t},\bar{x}) &= x^{\frac{n-1}{2}}\,t^{\frac{e}{2}}\,|M(\bar{t})-\bar{x}I| \\ &= x^{\frac{n-1}{2}}\,t^{\frac{e}{2}}\,|(M(t))^{-1}-\bar{x}I| \\ &= \frac{x^{\frac{n-1}{2}}\,t^{\frac{e}{2}}\,|xI-M(t)|}{(-t)^e x^{n-1}} \\ &= (-1)^{n-1-e}A_\alpha(t,x). \end{aligned}$$

Since $n-1\pm e$ is even when $\hat{\alpha}$ is a knot, this gives the desired symmetry. To see that $A_\alpha(1)=1$ for $\hat{\alpha}$ a knot, we use the calculations in the proof of Corollary 3.11.2 of [Bu], which show that if the permutation associated to α is an n-cycle, then $|M(1)-I|=n(-1)^{n-1}$. Formulas (6) therefore give $A_\alpha(1)=1$.

Morton's conjecture is:

Conjecture 3. Let α, $\beta\in B_3$. Then $A_\alpha(t,x)=A_\beta(t,x)$ implies $\hat{\alpha}\cup\hat{\lambda}=\hat{\beta}\cup\hat{\lambda}$.

In the special case $n=3$ equations (6) specialize to

$$\begin{aligned} &\text{(i)} \quad A_\alpha(t,x)=x^{-1}\,t^{\frac{e}{2}}-t^{-\frac{e}{2}}\,\text{trace}\,\psi(\alpha)+x\,t^{-\frac{e}{2}}, \\ &\text{(ii)} \quad (t^{-1}+1+t)\,A_\alpha(t)=(-1)^e A_\alpha(t,1). \end{aligned} \tag{7}$$

Since $A_\alpha(t,x)$ is determined by $e=e_\alpha$ and trace $\psi(\alpha)$, it will follow that our counterexamples to Conjecture 1 are also counterexamples to Conjecture 3.

3. The examples

Recall [G] that an arbitrary element $\alpha\in B_3$ can be written in the form

$$a=\Delta^{2k}s_1^{p_1}s_2^{q_1}\ldots s_1^{p_u}s_2^{q_u}, \quad k=0,\pm1,\pm2,\ldots \tag{8}$$

where $\Delta = s_1 s_2 s_1$ and where either all the p_i's and q_j's are positive, or all the p_i's are negative and all the q_j's are positive.

The integer $2k$ is the *power* of α, and $s_1^{p_1} s_2^{q_1} \dots s_1^{p_u} s_2^{q_u}$ is the principal part. The braid α is a *principal braid* if $k=0$. A principal braid is a *positive* braid if all the p_i's, q_j's are positive and an *alternating* braid if the p_i's are negative and the q_j's are positive. In the alternating case the power and the cyclic array of exponents in the principal part determine the conjugacy class in B_3 [Sc]. In the positive case a similar result holds [G].

We begin with a family of counterexamples to Conjecture 2:

Proposition 1. *Let δ be an alternating principal braid, chosen so that $\phi(\delta)$ is a 3-cycle and $e_\delta = 6r$ for some even integer $r \neq 0$. Let $\alpha = \Delta^{-2r}\delta$. Then $V_{L_\alpha}(t) = V_{L_\alpha}(t^{-1})$ but L_α is not amphicheiral.*

Proof. In formulas (7) the polynomials $1+t+t^{-1}$, $A_\alpha(t)$ and $\sqrt{t^{e_\alpha}} + \sqrt{t^{-e_\alpha}}$ are all symmetric, therefore $\sqrt{t^{-2_\alpha}}$ trace $\psi(\alpha)$ is too. From this it follows that if $e_\alpha = 0$, then trace $\psi(\alpha)$ is symmetric. In our examples, $e_\alpha = 0$, therefore from (5) we see that $V_{L_\alpha}(t)$ is symmetric. To see that $L_\alpha \not\approx L_{\alpha^{-1}}$, it suffices to prove that the signature $\sigma(L_\alpha) \neq 0$, because the signature changes sign when L_α is replaced by its mirror image. To compute $\sigma(L_\alpha)$, we use the formulas given in [Mu], in the remarks following the statements of Propositions 10.1 and 11.1, and in the first part of Proposition 11.1 We obtain $\sigma(L_\alpha) = 2r \neq 0$. □

Very similar ideas give an initial set of counter examples to Conjectures 1 and 3.

Proposition 2. *Let α be an alternating principal braid, with $e_\alpha = 6r \neq 0$. Let $\beta = \Delta^{4r}\alpha^{-1}$. Then L_α and L_β have the same connectivity and $e_\alpha = e_\beta$, trace $\psi(\alpha)$ = trace $\psi(\beta)$, but $L_\alpha \not\approx L_\beta$.*

Proof. Since $\phi(\Delta^2) =$ identity, $\phi(\beta) = \phi(\alpha^{-1})$, so L_α, L_β have the same connectivity. Since Δ^{4r} has exponent sum $12r$, we have $e_\beta = 6r = e_\alpha$. Define:

$$\rho: B_3 \to SL(2, \mathbb{C}[\sqrt{t}, \sqrt{t^{-1}}])$$

$$s_j \to \frac{1}{i\sqrt{t}} \psi(s_j), \quad j=1,2.$$

Then

$$\text{trace } \psi(\alpha) = (i\sqrt{t})^{e_\alpha} \text{ trace } \rho(\alpha), \quad \alpha \in B_3. \tag{9}$$

Since $\rho(\Delta^4) = I$, it follows that trace $\rho(\beta) =$ trace $\rho(\alpha^{-1}) =$ trace $\rho(\alpha)$. Therefore, by (5), (7) and (9) we see that L_α and L_β have the same Jones and Alexander polynomials.

To see that $L_\alpha \not\approx L_\beta$, we calculate their signatures as in the proof of Proposition 1. To compute $\sigma(L_\beta)$, note that the relation $s_1 s_2 s_1 = s_2 s_1 s_2$ in the group B_3 implies that $\Delta s_1 \Delta^{-1} = s_2$, $\Delta s_2 \Delta^{-1} = s_1$, $\Delta^2 \in$ Center B_3. Hence the formulas of [M] can be applied to $\Delta\beta\Delta^{-1}$. Since conjugate braids determine the same link, $L_{\Delta\beta\Delta^{-1}} = L_\beta$, $\sigma(L_\beta) = 2r \neq 6r = \sigma(L_\alpha)$, so $L_\alpha \not\approx L_\beta$. □

The examples of Proposition 2 are somewhat special since α and β have mutually inverse principal parts, however we will see that there are many other examples. First, we note some well-known facts about the traces of matrices in $SL(2, R)$, where R is a commutative ring with 1. A direct computation shows that if $X, Y \in SL(2, R)$, then

(10.1) trace X = trace X^{-1}

(10.2) trace XY = trace YX (10)

(10.3) trace XY = (trace X) (trace Y) − trace XY^{-1}

Induction on word length, using (10), then establishes the well-known fact that if Z is an arbitrary element in the group generated by X and Y, then

(10.4) trace Z is a polynomial with integer coefficients in trace X, trace Y and trace XY.

These facts then yield, almost immediately, the following lemma of Troels Jorgenson (unpublished):

Lemma 3. (Jorgenson). *Let* $W(X, Y) \in gp\{X, Y\}$, $X, Y \in SL(2, R)$. *Then trace* X = *trace* Y *implies trace* $W(X, Y)$ = *trace* $W(Y, X)$.

Proof. By (10.4), trace $W(X, Y)$ is a polynomial in trace X, trace Y and trace XY = trace YX. The result follows. □

To apply Lemma 3 to the construction of new examples, replace the braid α in Proposition 2 by α', where a' is a conjugate of α which has been chosen so that α' does not commute with β. Then apply Lemma 3 to $X = \alpha'$, $Y = \beta$ to get examples of braids with distinct principal parts which determine links having the same Jones and Alexander polynomials, using the signature invariant to distinguish them as before.

We have still not settled the question of whether V_L might be injective on closed braids with power zero (again with the exceptional cases omitted). These are key examples, since by [Mu] the classification of closed 3-braid knots rests on them. Hence the following lemma is of interest:

Lemma 4. *Let* p_1, q_1, p_2, q_2 *be artibrary integers. Let* $\delta_1, \delta_2, \delta_3 \in B_3$ *be defined by* $\delta_1 = s_1^{p_1} s_2^{q_1} s_1^{p_2} s_2^{q_2}$, $\delta_2 = s_1^{p_1} s_2^{q_2}$, $\delta_3 = s_1^{p_1 - p_2} s_2^{q_2 - q_1}$. *Let* $\gamma = \delta_1 \delta_2 \delta_3$, $\mu = \delta_1 \delta_3 \delta_2$. *Then* $e_\gamma = e_\mu$ *and trace* $\rho(\gamma)$ = *trace* $\rho(\mu)$.

Proof. Trivially, $e_\gamma = e_\mu$. To relate the traces of $\rho(\gamma)$, $\rho(\mu)$, apply (10.3) with $X = \delta_1$ and $Y = \delta_2 \delta_3$ (for γ) and $\delta_3 \delta_2$ (for μ). The result follows by (10.2). □

In order to see if we could distinguish the link types of the links constructed by the last procedure we computed some examples. Note that we want to be sure that γ isn't conjugate to $\mu^{\pm 1}$ or to μ read backwards, and for that it suffices to show that the cyclic array of exponents $(p_1, q_1, p_2, q_2, p_1, q_2, p_1 - p_2, q_2 - q_1)$ for γ is not equivalent to that for μ, read forwards or backward [G, Sc]. With these restrictions, a simple example is $\delta = s_1^{-2} s_2^3 s_1^{-1} s_2^4$, $\delta_2 = s_1^{-2} s_2^4$, $\delta_3 = s_1^{-1} s_2$, giving a pair of 3-component links γ, μ, defined as in Lemma 4, with distinct linking matrices. The case of knots is more subtle. While Lemma 4 gives plenty of candidates, it proved to be a very tricky matter to distinguish them. After a number of failures, we consulted Kenneth Perko and with his advice and help studied the examples $\delta_1 = s_1^{-3}s_2^6 s_1^{-1}s_2^9$, $\delta_2 = s_1^{-3}s_2^9$, $\delta_3 = s_1^{-2}s_2^3$. With these choices the groups $\pi_1(S^3 - \hat{\gamma})$ and $\pi_1(S^3 - \hat{\mu})$ each have two non-trivial, inequivalent homomorphisms onto the symmetric group Σ_4 on 4 symbols, such that meridians map to transpositions. These homomorphisms define two 4-sheeted covering spaces with $\hat{\gamma}$ and $\hat{\mu}$ as branch sets. These homomorphisms onto Σ_4 determine two distinguished homomorphisms onto Σ_3, yielding associated 3-sheeted coverings branched over $\hat{\gamma}$ and $\hat{\mu}$. In the 3-sheeted coverings the branch sets $\hat{\gamma}$ and $\hat{\mu}$ are covered by two curves, one of branching index 2 and the other of branching index. The linking numbers between these two curves are 12 and 0 for both $\hat{\gamma}$ and $\hat{\mu}$. Focusing on the case where the linking numbers are 0, we then have unique 4-sheeted covering spaces to investigate. Both 4-sheeted coverings are simply-connected. There are 3 curves covering the branch set in each case, one of branching index 2 and the other two of branching index 1. Since the linking numbers of the distinguished curve of branching index 2 with the other two curves are 9 and 9 in the case of $\hat{\gamma}$, but 5 and 13 in the case of $\hat{\mu}$, the knots are distinct. □

Remark. Natural generalizations of (10.1)–(10.4) to the groups $\pi_t(B_n)$, $n > 3$ are formulas IV–VI of [J, 2] for computing traces in $A_n(\tau)$. One wonders whether analogues of Lemmas 3 and 4 exist in $\pi_t(B_n)$, $n > 3$, producing non-trivial knots with $V_L(t) = 1$ or non-trivial links with the polynomial of the unlink. Such examples do not occur when $n = 3$.

4. A conjecture

The examples in § 3 leave open the possibility that the conjugacy class of a closed 3-braid determines its link type, up to orientation changes, if one omits the exceptional cases when the braid index is <3 or the link is composite or a torus link (see Theorem 13.1 and § 16 of [Mu] for a careful listing of these exceptional cases). This possibility was implicit in the work in [Mu] and [H]. Since the conjugacy problem in B_n and in particular in B_3 has been solved, it would imply that there is at hand a complete invariant for closed 3-braid links. More generally, one might ask for conditions on n-braids such that a link which is a closed n-braid is characterized by its conjugacy class in B_n (again up to orientation changes). Call such a link a *nice link*. Here are some conjectures.

(i) Let $\alpha \in B_n$, and let $\beta = \Delta^{2k}\alpha$, where Δ is Garside's fundamental braid (see [Bu] or [G]). Then for sufficiently large k the link $\hat{\beta}$ is nice.

(ii) Let $\alpha = s_{\mu_1}^{m_1} s_{\mu_2}^{m_2} \dots s_{\mu_r}^{m_r}$ be a cyclic word. Assume each $\mu_i \neq \mu_{i+1}$, each $m_i \geq 3$, and $r \geq 2$. Assume also that $\hat{\alpha}$ is prime. Then $\hat{\alpha}$ is nice.

(iii) Let α be as in (ii), but now assume that r is even, that $|\mu_{i+1} - \mu_i| = 1$ for each $i = 1, \dots, r$; that each $m_i \neq 0$, and that $m_{i+1} < 0$ if and only if $m_i > 0$. Assume $\hat{\alpha}$ is prime. Then $\hat{\alpha}$ is nice.

Acknowledgement. We thank Vaughan Jones for many informative discussions, about his new polynomial. Also, we thank Kenneth Perko for his expert help in the computation of the dihedral linking numbers which we used to distinguish the final set of examples.

The referee has informed us that some of the counter examples which we constructed in Proposition 2 had also been discovered simultaneously and independently by Maité Lozano and Hugh Morton (unpublished work).

References

[A] Artin, E.: Theorie der Zopfe. Abh. Math. Semin. Univ. Hamb. **4**, 47–72 (1925)

[Bu] Burau, W.: Über Zopfgruppen und gleichsinnig verdrillte Verkettungen. Abh. Math. Semin. Hans. Univ. **11**, 171–178 (1936)

[G] Garside, F.: The braid group and other groups. Q. J. Math., Oxf. **20**, 235–254 (1969)

[H] Hartley, R.: On the classification of 3-braid links. Abh. Math. Semin. Univ. Hamb. **50**, 108–117 (1980)

[J, 1] Jones, V.: Braid groups. Hecke algebras and type II_1 factors. (MSRI, Berkeley) preprint

[J, 2] Jones, V.: A polynomial invariant for knots via von Neumann algebras. Bull. AMS **12**, 103–111 (1985)

[Mo] Morton, H.: Infinitely many fibered knots with the same Alexander polynomial. Topology **17**, 101–104 (1978)

[Mu] Murasugi, K.: On closed 3-braids. Memoirs AMS No. 151 (1974). Am. Math. Soc., Providence, R.I.

[Sc] Schreier, O.: Uber die Gruppen $A^a B^b = 1$. Abh. Math. Semin. Univ. Hamb. **3**, 167–169 (1923)

[Sq] Squier, C.: The Burau representation is unitary. Proc. AMS **90**, (2) 199–202 (1984)

[FYHLMO] Freyd and Yetter, Hoste, Lickorish and Millett, Ocneanu: A new polynomial invariant of knots and links. Ball AMS **12**, 239–246 (April 1985)

Errata, "On the Jones Polynomial of closed 3-braids":

page 290, equation (6) should read:

(i) $A_\alpha(t,x) = x^{(1-n)/2} t^{-e} |\psi(\alpha) - x\mathbf{I}|$,

(ii) $t^{(1-n)/2}(1+t+\ldots+t^{n-1})A_\alpha(t) = (-1)^e A_\alpha(t,1)$

page 291, line 14:

... therefore $\sqrt{t}^{\,-e_\alpha}$ trace $\psi(\alpha)$ is too......

TRANSACTIONS OF THE
AMERICAN MATHEMATICAL SOCIETY
Volume 295, Number 1, May 1986

JONES POLYNOMIALS OF ALTERNATING LINKS

KUNIO MURASUGI

ABSTRACT. Let $J_K(t) = a_r t^r + \cdots + a_s t^s$, $r > s$, be the Jones polynomial of a knot K in S^3. For an alternating knot, it is proved that $r - s$ is bounded by the number of double points in any alternating projection of K. This upper bound is attained by many alternating knots, including 2-bridge knots, and therefore, for these knots, $r - s$ gives the minimum number of double points among all alternating projections of K. If K is a special alternating knot, it is also proved that $a_s = 1$ and s is equal to the genus of K. Similar results hold for links.

1. Introduction. Let K be an oriented knot or link in S^3 and let $J_K(t)$ be the polynomial defined by V. Jones [**5**] which is now called the *Jones polynomial* of K. $J_K(t)$ is an invariant of a knot or link type. It is not clear, however, to what extent $J_K(t)$ is related to known algebraic or topological invariants in knot theory. Although it is shown [**10**] that $J_K(t)$ determines the Arf invariant of a knot or a link, it is also known that $J_K(t)$ does not determine the genus or the signature of K [**8**].

In this paper, we prove that for an alternating knot, $J_K(t)$ provides some information that has never been obtained from other algebraic invariants. To be more precise, let K be an alternating knot in S^3 and $\tilde{K}$ an alternating projection of K. Let $J_K(t)$ be the Jones polynomial of K.[1] Write $J_K(t) = a_r t^r + \cdots + a_s t^s$, where $r > s$ and $a_r \neq 0 \neq a_s$. Let $h(\tilde{K})$ be the number of double points in $\tilde{K}$. Then we can prove

THEOREM A. $r - s \leq h(\tilde{K})$.

It is probably not surprising that the "reduced" degree $r - s$ is bounded by the number of double points of $\tilde{K}$, but what is surprising is the fact that equality in Theorem A holds for many alternating knots including alternating algebraic knots [Theorem 11.2], alternating pretzel knots and alternating closed 3-braids $\beta = \sigma_1^{p_1}\sigma_2^{-q_1}\cdots\sigma_1^{p_r}\sigma_2^{-q_r}$, where $r \geq 2$ and p_i, $q_i > 0$ [**5**]. It means that for these knots, $h(\tilde{K})$ is a knot type invariant and $h(\tilde{K})$ gives the minimum number of double points among all alternating projections of K. This proves

THEOREM B (SEE COROLLARY 9.3). *Let K be an alternating knot and $\tilde{K}$ an alternating projection. If $r - s = h(\tilde{K})$, then for any alternating projection K^* of K, the number of double points of K^* is at least $h(\tilde{K})$. Therefore, $\tilde{K}$ is an*

Received by the editors April 9, 1985 and, in revised form, May 25, 1985.
1980 *Mathematics Subject Classification*. Primary 57M25.
Key words and phrases. Knot, link, Jones polynomial, alternating knot, reduced Alexander polynomial.

[1] The original Jones polynomial $V_K(t)$ in [**5**] is given by $J_K(t^{-1})$.

alternating projection of K with a minimum number of double points. (*$\tilde{K}$ will be called a* minimum *alternating projection.*)

Using Theorem B, for instance, we are able to determine the minimum number of double points an alternating projection of a 2-bridge knot can have (see Corollary 11.3).

Until the present the minimum number of double points of a knot projection has been determined only for very limited knot types [**1**, **3**].

Since it is unlikely that a nonalternating projection of an alternating knot has fewer double points, it is plausible that the minimum number of double points of an alternating knot will be given by $r - s$.

Now, since an alternating knot is "formed" from special alternating knots or links, a proof of Theorem A requires more information on Jones polynomials on special alternating knots. Let K be a special alternating knot in S^3. That is, K has a connected alternating projection in which one of the chessboard surfaces is orientable. Let $\tilde{K}$ be such a special alternating projection. For simplicity, we assume that at each double point of $\tilde{K}$, an oriented segment crosses over the other from left to right

We say that $\tilde{K}$ is of positive type. (Otherwise, $\tilde{K}$ is of negative type.) A special alternating knot has a special alternating projection of either positive or negative type, but it cannot have both types. For these knots, we can prove slightly more precise results.

THEOREM C (SEE THEOREM 2.1). *Let K be a special alternating knot which has a special alternating projection $\tilde{K}$ of positive type. Let $J_K(t)$ be the Jones polynomial of K. Write $J_K(t) = a_r t^r + \cdots + a_s t^s$, where $r > s$ and $a_r \neq 0 \neq a_s$. Then*

(I) $a_s = 1$.

(II) $s \geq 0$ *and s is equal to the genus of K.*

(III) *Let $h(\tilde{K})$ be the number of double points in $\tilde{K}$. Then*

$$r - s \leq h(\tilde{K}). \tag{1.1}$$

Note that if K is of negative type, then K is the mirror image of a special alternating knot K^* of positive type, and, hence, $J_K(t) = J_{K^*}(t^{-1})$, and the rôles of r and s should be reversed.

Theorem A, in principal, follows from Theorem C. It should be noted here that Theorems A and C are, in fact, proved for alternative links introduced by L. H. Kauffman [**7**].

One of the outstanding problems still unsolved is whether *any* alternating projection of an alternating link without removable points

is minimum. The answer is not known even for 2-bridge links, for which we now have the minimum alternating projection.

This paper is organized as follows. In §2, we state the first main theorem (Theorem 2.1) that is a slight generalization of Theorem C. The proof of this theorem will be completed at the end of §8. In §§9–10, we will prove the second main theorem (Theorem 9.1) that implies Theorem A. In §11, we will prove that equality in Theorem A holds for alternating algebraic links considered in [**13**].

2. Main Theorem I. In order to prove Theorem C, we need to consider links rather than knots, and to this purpose, the polynomials defined in [**8**] will be more convenient than the original Jones polynomials.

Let L be an oriented link in S^3. Let $P_L(l,m)$ be the polynomial in two variables l and m introduced in [**8**]. $P_L(l,m)$ is defined recursively by using the following fundamental identities (2.1)[2] and (2.2).

(I) If $L+$, L_- and L_0 are completely identical projections except at one double point c, where they are related by Figure 1, then

$$lP_{L_-}(l,m) + l^{-1}P_{L_+}(l,m) + mP_{L_0}(l,m) = 0. \tag{2.1}$$

(II) If L is a trivial link of n components, then

$$P_L(l,m) = (-(l+l^{-1})/m)^{n-1}. \tag{2.2}$$

Now for an n-component link L in S^3, we define the polynomial $Q_L(l,m)$ by

$$Q_L(l,m) = (m/l)^{n-1}P_L(l,m). \tag{2.3}$$

$Q_L(l,m)$ is an integer polynomial in $l^{\pm 2}$ and m^2; i.e.,

$$Q_L(l,m) \in \mathbf{Z}[m^2, l^2, l^{-2}].$$

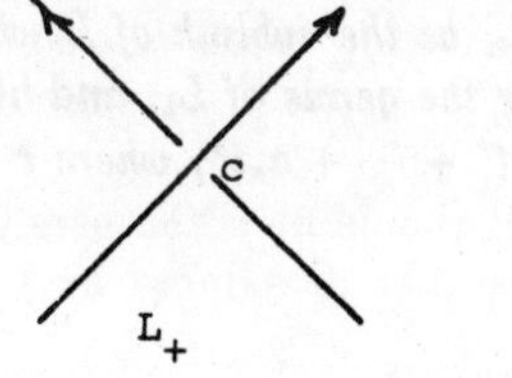

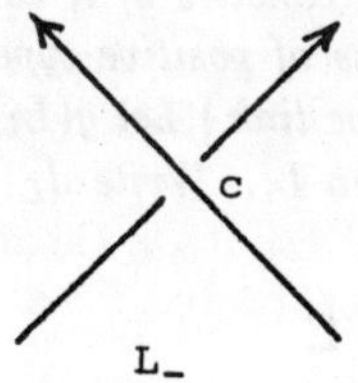

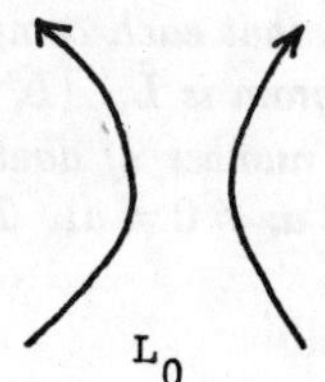

FIGURE 1

[2]Formula (2.1) is slightly different from the original formula given in [**8**]. $P_{L_+}(l,m)$ is actually $K_{L_+}(l^{-1},m)$ in [**8**].

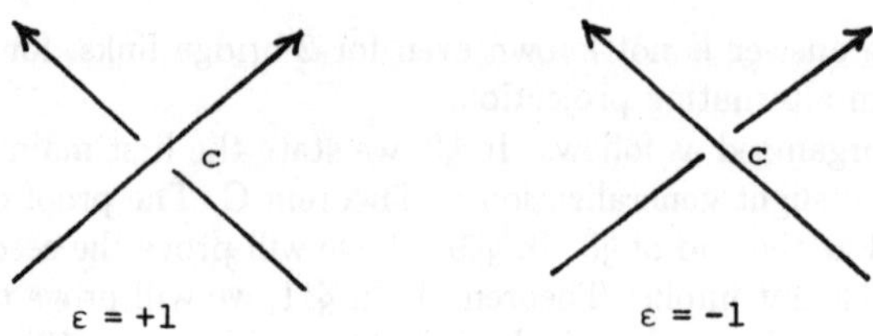

FIGURE 2

Finally we define $J_L(t)$ as

$$J_L(t) = Q_L(ti, -(\sqrt{t} - 1/\sqrt{t})i), \tag{2.4}$$

where $i^2 = -1$.

$J_L(t)$ is an integer polynomial in $t^{\pm 1}$. If K is a knot, then $J_K(t^{-1})$ coincides with the original Jones polynomial. However, if L is a link of n (> 1) components, $J_L(t)$ will be called the *reduced* Jones polynomial.

EXAMPLE 2.1. For an n-component trivial link L,

$$\begin{aligned} Q_L(l,m) &= \left(\frac{m}{l}\right)^{n-1} P_L(l,m) = \left(\frac{m}{l}\right)^{n-1} \left(-\frac{l+l^{-1}}{m}\right)^{n-1} \\ &= (-1)^{n-1}(1+l^{-2})^{n-1}, \end{aligned}$$

and, hence, $J_L(t) = (-1)^{n-1}(1-t^{-2})^{n-1}$.

Let $\tilde{L}$ be a link diagram of a link L; i.e., $\tilde{L}$ is a projection of L on S^2. To each double point c of $\tilde{L}$, we assign $\varepsilon(c) = +1$ or -1 as shown in Figure 2. $\tilde{L}$ is said to have *positive* (*negative*) *type* if $\varepsilon(c) = +1$ ($\varepsilon(c) = -1$) for all double points c in $\tilde{L}$. An oriented link L is called a link of positive (negative) type, or simply a positive (negative) link if L has a diagram of positive (negative) type. A special alternating link is a link that has an alternating and positive (negative) link diagram.

The family of positive or negative (not necessarily alternating) links includes positive or negative closed braids; in particular, it includes any torus links.

The main theorem I (Theorem 2.1) of this paper holds not only for special alternating links but also for positive or negative links.

THEOREM 2.1. *Let L be an oriented n-component positive link. Let $\tilde{L}$ be a link diagram. Suppose that $\tilde{L}$ consists of λ connected components $\tilde{L}_1, \tilde{L}_2, \ldots, \tilde{L}_\lambda$, and that each component $\tilde{L}_i$ is of positive type. Let L_i be the sublink of L whose diagram is $\tilde{L}_i$. (L_i is a positive link.) Let $g(L_i)$ denote the genus of L_i, and $h(\tilde{L}_i)$ the number of double points on $\tilde{L}_i$. Write $J_L(t) = a_r t^r + \cdots + a_s t^s$, where $r > s$ and $a_r \neq 0 \neq a_s$. Then*

$$\begin{aligned} &(1)\quad a_s = 1, \\ &(2)\quad s = \sum_{i=1}^{\lambda} g(L_i) - (n-1) - (\lambda - 1), \\ &(3)\quad r - s \leq \sum_{i=1}^{\lambda} h(\tilde{L}_i) + (n-1) + (\lambda - 1). \end{aligned} \tag{2.5}$$

Note that for a general positive link L, equality in (3) may not hold. The simplest example is the closure $\hat{\beta}$ of a 3-braid $\beta = (\sigma_1\sigma_2)^2$. $\hat{\beta}$ is a trefoil knot, but the natural diagram of the closed braid contains four double points.

COROLLARY 2.2. *In Theorem 2.1, if*

$$r - s = h(\tilde{L}) + (n-1) + (\lambda - 1),$$

then for any positive diagram of L, the number of double points is at least $h(\tilde{L})$. Therefore, $\tilde{L}$ is a minimum positive diagram of L.

Since $h(\tilde{L})$, n and λ are independent from the orientation, if equality holds in (2.5)(3), Corollary 2.2 yields a curious conclusion that s depends on the orientation of L but $r - s$ does not.

3. Proof of Theorem 2.1 (I). First we fix notation that will be used throughout this paper. All links are oriented.

Given a link L, a diagram of L is usually denoted by $\tilde{L}$, and from the context, it is generally easy to understand which diagram is being referred to.

For a link L:

(3.1)
(1) $\lambda(L)$ is the number of sublinks of L which split each other ,
(2) $\mu(L)$ is the number of components of L ,
(3) $\delta(L)$ is the degree of the reduced Alexander polynomial of L .

For a link diagram $\tilde{L}$,

(4) $\lambda(\tilde{L})$ is the number of connected components of $\tilde{L}$,
(5) $h(\tilde{L})$ is the number of double points on $\tilde{L}$.

For a Laurent polynomial $f(t) = \sum_{j=s}^{r} a_j t^j$, $a_r \neq 0 \neq a_s$, $r > s$,

$$\begin{aligned}(6)\quad & l\text{-deg}\, f(t) = s, \quad h\text{-deg}\, f(t) = r, \quad r\text{-deg}\, f(t) = r - s, \\ & l\text{-coef}\, f(t) = a_s, \quad h\text{-coef}\, f(t) = a_r.\end{aligned}$$

It is well known [**2, 11**] that if $\tilde{L}$ is a positive (negative) diagram of L, then $\lambda(\tilde{L}) = \lambda(L)$. In other words, a positive (negative) link L is not split iff its diagram is connected.

Since a diagram $\tilde{L}$ uniquely determines a link L, we may write $\mu(\tilde{L})$ or $\delta(\tilde{L})$ for $\mu(L)$ or $\delta(L)$. To emphasize a link L, we frequently write $P(L; l, m)$ or $Q(L; l, m)$ for $P_L(l, m)$ or $Q_L(l, m)$.

Now let L, L_i, $\tilde{L}$, $\tilde{L}_i$ and $J_L(t)$ be defined as in Theorem 2.1. Let $J_i(t)$ denote the reduced Jones polynomial of L_i defined by (2.4). First we prove

PROPOSITION 3.1. *Let $\lambda = \lambda(L)$.*
(1) *If l-coef $J_i(t) = 1$, $1 \leq i \leq \lambda$, then l-coef $J(t) = 1$.*
(2) $l\text{-deg}\, J(t) = \sum_{i=1}^{\lambda} l\text{-deg}\, J_i(t) - 2(\lambda - 1)$.
(3) $h\text{-deg}\, J(t) = \sum_{i=1}^{\lambda} h\text{-deg}\, J_i(t)$.

PROOF. Since L consists of exactly λ split sublinks $L_1, L_2, \dots, L_\lambda$, it follows from [8, (3)] that

$$P(L;l,m) = (-1)^{\lambda-1}\left(\frac{l+l^{-1}}{m}\right)^{\lambda-1}\prod_{i=1}^{\lambda} P(L_i;l,m). \tag{3.2}$$

Since

$$n-1 = \sum_{i=1}^{\lambda}\mu(L_i) - 1 = \sum_{i=1}^{\lambda}\{\mu(L_i)-1\} + (\lambda-1),$$

it follows from (3.2) that

$$\begin{aligned} Q(L;l,m) &= \left(\frac{m}{l}\right)^{n-1} P(L;l,m) \\ &= (-1)^{\lambda-1}\left(\frac{m}{l}\right)^{n-1}\left(\frac{l+l^{-1}}{m}\right)^{\lambda-1}\prod_{i=1}^{\lambda} P(L_i;l,m) \\ &= (-1)^{\lambda-1}\left(\frac{m}{l}\right)^{\lambda-1}\left(\frac{l+l^{-1}}{m}\right)^{\lambda-1}\prod_{i=1}^{\lambda} Q(L_i;l,m) \\ &= (-1)^{\lambda-1}(1+l^{-2})^{\lambda-1}\prod_{i=1}^{\lambda} Q(L_i;l,m). \end{aligned}$$

Therefore, we have

$$J_L(t) = (-1)^{\lambda-1}(1-t^{-2})^{\lambda-1}\prod_{i=1}^{\lambda} J_i(t),$$

from which Proposition 3.1 follows immediately.

Now the proof of Theorem 2.1 will be by induction on $h(\tilde{L}) = \sum_{i=1}^{\lambda} h(\tilde{L}_i)$, the total number of double points on $\tilde{L}$.

As we have seen in Example 2.1, Theorem 2.1 is true for $h(\tilde{L}) = 0$. (Furthermore, r-deg $J_L(t) = (n-1)+(\lambda-1)$.) Therefore, we may assume that Theorem 2.1 holds for a positive link L^* with $h(\tilde{L}^*) < h(\tilde{L})$.

First, we will show that it is sufficient to prove Theorem 2.1 for a nonsplit positive link.

Assume that Theorem 2.1 is proved for $\lambda = 1$. Suppose $\lambda > 1$. Then two cases can occur.

Case 1. $h(\tilde{L}_i) < h(\tilde{L})$ for all $i = 1, 2, \dots, \lambda$.

Case 2. $h(\tilde{L}_i) = h(\tilde{L})$ for some i, and $h(\tilde{L}_j) = 0$ for $j \neq i$, and hence $\tilde{L}_j$ is a circle on S^2.

For Case 1, we apply the induction hypothesis on each L_i, and we have

$$l\text{-deg}\, J_i(t) = g(L_i) - \{\mu(L_i) - 1\}$$

and

$$r\text{-deg}\, J_i(t) \leq h(\tilde{L}_i) + \{\mu(L_i) - 1\}.$$

Therefore, Proposition 3.1 and the induction hypothesis imply that

$$\begin{aligned}
&(1)\quad l\text{-coef}\, J_L(t) = 1,\\
&(2)\quad l\text{-deg}\, J_L(t) = \sum_{i=1}^{\lambda} l\text{-deg}\, J_i(t) - 2(\lambda-1)\\
&\qquad = \sum_{i=1}^{\lambda}\{g(L_i) - \mu(L_i) + 1\} - 2(\lambda-1)\\
&\qquad = \sum_{i=1}^{\lambda} g(L_i) - \sum_{i=1}^{\lambda}\mu(L_i) + \lambda - 2(\lambda-1)\\
&\qquad = \sum_{i=1}^{\lambda} g(L_i) - (n-1) - (\lambda-1),\\
(3.3)\quad &(3)\quad r\text{-deg}\, J_L(t) = \sum_{i=1}^{\lambda} h\text{-deg}\, J_i(t) - \sum_{i=1}^{\lambda} l\text{-deg}\, J_i(t) + 2(\lambda-1)\\
&\qquad = \sum_{i=1}^{\lambda} r\text{-deg}\, J_i(t) + 2(\lambda-1)\\
&\qquad \le \sum_{i=1}^{\lambda}\{h(\tilde{L}_i) + \mu(L_i) - 1\} + 2(\lambda-1)\\
&\qquad = \sum_{i=1}^{\lambda} h(\tilde{L}_i) + n - \lambda + 2(\lambda-1)\\
&\qquad = \sum_{i=1}^{\lambda} h(\tilde{L}_i) + (n-1) + (\lambda-1).
\end{aligned}$$

If Case 2 occurs, then L splits into one link L_i and a trivial $(\lambda-1)$-component link, and hence,

$$P(L;l,m) = P(L_i;l,m)(-(l+l^{-1})/m)^{\lambda-1}.$$

Therefore, we have

$$(3.4)\qquad Q(L;l,m) = \left(\frac{m}{l}\right)^{n-1} P(L_i;l,m)(-1)^{\lambda-1}\left(\frac{l+l^{-1}}{m}\right)^{\lambda-1}.$$

Since $n = \mu_i + \lambda - 1$, $\mu_i = \mu(L_i)$, we see from (3.4) that

$$\begin{aligned}
Q(L;l,m) &= (m/l)^{\mu_i-1}P(L_i;l,m)(-1)^{\lambda-1}(1+l^{-2})^{\lambda-1}\\
&= Q(L_i;l,m)(-1)^{\lambda-1}(1+l^{-2})^{\lambda-1},
\end{aligned}$$

and hence

$$J_L(t) = (-1)^{\lambda-1}(1-t^{-2})^{\lambda-1}J_i(t).$$

Since $l\text{-deg}\, J_i(t) = g(L_i) - (\mu_i - 1)$, it follows from Proposition 3.1(2) that

$$\begin{aligned}
l\text{-deg}\, J_L(t) &= g(L_i) - (\mu_i - 1) - 2(\lambda-1)\\
&= g(L_i) - (\mu_i + \lambda - 2) - (\lambda-1) = g(L_i) - (n-1) - (\lambda-1).
\end{aligned}$$

Furthermore, since h-deg $J_L(t) = h$-deg $J_i(t)$, it follows from Proposition **3.1(2)**, (3) that

$$\begin{aligned} r\text{-deg}\, J_L(t) &= h\text{-deg}\, J_{L_i}(t) - l\text{-deg}\, J_{L_i}(t) + 2(\lambda - 1) \\ &= r\text{-deg}\, J_{L_i}(t) + 2(\lambda - 1) \le h(\tilde{L}_i) + (\mu_i - 1) + 2(\lambda - 1) \\ &= h(\tilde{L}_i) + (\mu_i + \lambda - 2) + (\lambda - 1) = h(\tilde{L}) + (n - 1) + (\lambda - 1). \end{aligned}$$

This proves Theorem 2.1. Therefore, in the following proof, we may assume that L is not split.

4. Proof of Theorem 2.1 (II). Let L be a nonsplit positive link and $\tilde{L}$ its positive diagram. To compute $P(L; l, m)$, we apply the fundamental identities at several double points on $\tilde{L}$ to make $\tilde{L}$ "simpler".

Let L_+, L_- and L_0 be link diagrams described in §2(I). For convenience, we say that L_- (or L_+) is obtained from L_+ (or L_-) by *changing* the incident sign ε at c, and L_0 is obtained from L_- (or L_+) by *eliminating* the double point c.

Now we study the effect of an elimination of a double point c from a link diagram.

First we recall a relation between $g(L)$ and $\delta(L)$ for a positive (negative) link K. Let $\tilde{L}$ be a connected positive diagram. $\tilde{L}$ divides S^2 into finitely many domains.

It is known that $\tilde{L}$, considered as a 1-complex in S^2, is divided into finitely many oriented circles in S^2, called Seifert circles in [**7**, p. 57] or Seifert circuits in [**12**, p. 390].

PROPOSITION 4.1. *Let L be a nonsplit positive (negative) link in S^3, and $\tilde{L}$ a positive (negative) diagram of L. Let $m(\tilde{L})$ be the number of Seifert circles in $\tilde{L}$ and $\alpha(\tilde{L})$ the number of domains in S^2 divided by $\tilde{L}$. Then*

$$(4.1) \qquad \begin{aligned} &(1) \quad g(L) = \tfrac{1}{2}\{\delta(L) - \mu(L) + 1\}, \\ &(2) \quad \delta(L) = \alpha(\tilde{L}) - m(\tilde{L}) - 1. \end{aligned}$$

This proposition was essentially proved in [**11**], but for the details, see [**2**].

Now to each double point c, four (not necessarily distinct) domains are incident. If all domains are distinct, c will be called *proper*. Otherwise, c is called ***removable***. Note that if c is removable, then we can "remove" c from $\tilde{L}$ without changing its link type.

PROPOSITION 4.2. *Let $\tilde{L}$ be a positive link diagram with λ $(= \lambda(\tilde{L}))$ connected components $\tilde{L}_1, \tilde{L}_2, \ldots, \tilde{L}_\lambda$. Let $\tilde{L}^*$ be the diagram obtained from $\tilde{L}$ by eliminating a double point c on $\tilde{L}$. Then*

(1) *$\tilde{L}^*$ is a positive diagram. Let $\tilde{L}_1^*, \tilde{L}_2^*, \ldots, \tilde{L}_{\lambda^*}^*$ be connected components of $\tilde{L}^*$.*

(2) (i) *If c is proper, then $\lambda^* = \lambda$, and*

$$(4.2) \qquad \sum_{i=1}^{\lambda} \delta(\tilde{L}_i) = \sum_{i=1}^{\lambda} \delta(\tilde{L}_i^*) + 1.$$

(ii) *If c is removable, then $\lambda^* = \lambda + 1$, and*

$$(4.3) \qquad \sum_{i=1}^{\lambda} \delta(\tilde{L}_i) = \sum_{i=1}^{\lambda+1} \delta(L_i^*).$$

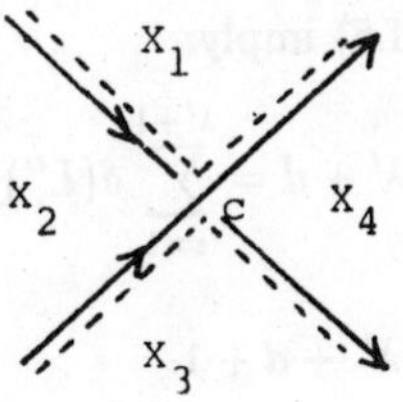

FIGURE 3

PROOF. (1) is obvious. To prove (2), let X_1, X_2, X_3, X_4 be four domains meeting at c (see Figure 3).

Suppose c is proper. Then X_2 and X_4 are distinct domains and $\lambda(\tilde{L}^*) = \lambda(\tilde{L})$. But the elimination of c makes X_2 and X_4 amalgamate, and hence $\alpha(\tilde{L}^*) = \alpha(\tilde{L}) - 1$, from which (4.2) follows by using (4.1)(2). If c is a removable double point on $\tilde{L}_\lambda$, say, then $\lambda(L^*) = \lambda(\tilde{L}) + 1$, and $\tilde{L}_\lambda$ splits into two diagrams $\tilde{L}^*_\lambda$ and $\tilde{L}^*_{\lambda+1}$. Obviously $\tilde{L}_\lambda$ is a (Schubert) product of two links L^*_λ and $L^*_{\lambda+1}$, and hence $\delta(L_\lambda) = \delta(L^*_\lambda) + \delta(L^*_{\lambda+1})$. Since $\delta(L_i) = \delta(L^*_i)$ for $i = 1, 2, \ldots, \lambda - 1$, (4.3) follows immediately.

We generalize Proposition 4.2 to the following.

PROPOSITION 4.3. *Let $\tilde{L}$ be a connected positive link diagram. Suppose we obtain λ connected diagrams $\tilde{L}_1, \tilde{L}_2, \ldots, \tilde{L}_\lambda$ by eliminating d double points on $\tilde{L}$. Then*

$$\sum_{i=1}^{\lambda} \delta(\tilde{L}_i) - \lambda + d + 1 = \delta(\tilde{L}). \tag{4.4}$$

PROOF. A proof will be by induction on d.

If $d = 1$, then $\lambda = 1$ or 2 and Proposition 4.3 follows from Proposition 4.2.

Inductively, suppose (4.4) is true for $j < d$. Let $\tilde{L}'$ be the link diagram obtained from $\tilde{L}$ by eliminating $d - 1$ double points $c_1, c_2, \ldots, c_{d-1}$ on $\tilde{L}$. Suppose $\tilde{L}'$ has λ' connected components $\tilde{L}'_1, \ldots, \tilde{L}'_{\lambda'}$. We may assume inductively

$$\sum_{i=1}^{\lambda'} \delta(\tilde{L}'_i) - \lambda' + (d-1) + 1 = \delta(\tilde{L}). \tag{4.5}$$

Now we eliminate c_d so that we have a link diagram $\tilde{L}''$ of λ'' connected components $\tilde{L}''_1, \ldots, \tilde{L}''_{\lambda''}$.

If $\lambda'' = \lambda'$, then $\sum_{i=1}^{\lambda''} \delta(L''_i) + 1 = \sum_{i=1}^{\lambda'} \delta(L'_i)$, and hence, by (4.5), we have

$$\begin{aligned}\delta(\tilde{L}) &= \sum_{i=1}^{\lambda'} \delta(L'_i) - \lambda' + d = \sum_{i=1}^{\lambda''} \delta(L''_i) + 1 - \lambda' + d \\ &= \sum_{i=1}^{\lambda''} \delta(L''_i) - \lambda'' + d + 1.\end{aligned}$$

If $\lambda'' = \lambda' + 1$, then (4.3) and (4.5) imply

$$\delta(\tilde{L}) = \sum_{i=1}^{\lambda'} \delta(L_i') - \lambda' + d = \sum_{i=1}^{\lambda'+1} \delta(L_i'') - (\lambda' + 1) + (d+1)$$
$$= \sum_{i=1}^{\lambda''} \delta(L_i'') - \lambda'' + d + 1.$$

The induction is now completed.

5. Proof of Theorem 2.1 (III). Let $\tilde{L}$ be a connected positive link diagram of a positive link L. Let $c_1, c_2, \ldots, c_k$ be double points on $\tilde{L}$ to which we apply the fundamental identity (2.1).

In this section, we will prove a reduction formula (5.5) to compute $J_L(t)$.

Now first we change the incident sign at c_1 to get $\tilde{L}_1$. Since $\varepsilon(c_1) = +1$, the fundamental identity (2.1) gives us

(5.1) $$(-1)P(L; l, m) = l^2 P(L_1; l, m) + lmP(L_1^*; l, m),$$

where L_1^* is the link diagram obtained from $\tilde{L}$ by eliminating c_1 from $\tilde{L}$. Note that $\tilde{L}_1^*$ is again a positive diagram. Next, change the incident sign at c_2 on $\tilde{L}_1$. Then the fundamental identity (2.1) tells us

(5.2) $$(-1)P(L_1; l, m) = l^2 P(L_{1,2}; l, m) + lmP(\hat{L}_{1,2}; l, m),$$

where $L_{1,2}$ is the link obtained from L_1 by changing the incident sign at c_2 and $\hat{L}_{1,2}$ is the link obtained from L_1 by eliminating c_2. Since L_1 is not a positive diagram, $\hat{L}_{1,2}$ is not a positive diagram.

To evaluate $P(\hat{L}_{1,2}; l, m)$, we change the incident sign at c_1 on $\hat{L}_{1,2}$ again, and apply the fundamental formula (2.1) to obtain

(5.3) $$(-1)P(\hat{L}_{1,2}; l, m) = l^{-2} P(L_2^*; l, m) + l^{-1} mP(L_{1,2}^*; l, m),$$

where $\tilde{L}_2^*$ is the diagram obtained from $\tilde{L}$ by eliminating c_2, and $\tilde{L}_{1,2}^*$ is the diagram obtained from $\tilde{L}$ by elimimating c_1 and c_2. Both $\tilde{L}_2^*$ and $\tilde{L}_{1,2}^*$ are positive diagrams. Substituting (5.3) into (5.2), we obtain (5.4) from (5.1):

(5.4)

$$\begin{aligned}(-1)^2 P(L; l, m) &= (-1)l^2 P(L_1; l, m) + (-1)lmP(L_1^*; l, m)\\ &= l^4 P(L_{1,2}; l, m) + l^3 mP(\hat{L}_{1,2}; l, m) + (-1)lmP(L_1^*; l, m)\\ &= l^4 P(L_{1,2}; l, m) + (-1)l^3 m\{l^{-2} P(L_2^*; l, m) + l^{-1} mP(L_{1,2}^*; l, m)\}\\ &\quad + (-1)lmP(L_1^*; l, m)\\ &= l^4 P(L_{1,2}; l, m) + (-1)lm\{P(L_1^*; l, m) + P(L_2^*; l, m)\}\\ &\quad + (-1)l^2 m^2 P(L_{1,2}^*; l, m).\end{aligned}$$

Now an easy induction argument proves the basic reduction formula

(5.5) $$\begin{aligned}P(L; l, m) &= (-1)^k l^{2k} P(L_{1,2,\ldots,k}; l, m)\\ &\quad + (-1)\sum_{d=1}^{k} (lm)^d \sum_{1 \le j_1 < \cdots < j_d \le k} P(L_{j_1,\ldots,j_d}^*; l, m),\end{aligned}$$

where $\tilde{L}^*_{j_1,\dots,j_d}$ is a positive diagram obtained from $\tilde{L}$ by eliminating d double points $c_{j_1},\dots,c_{j_d}$, and $\tilde{L}_{1,2,\dots,k}$ is a diagram obtained from $\tilde{L}$ by changing the incident signs at $c_1, c_2, \dots, c_k$.

Since L has n-components, it follows from (5.5) that

$$\begin{aligned} Q(L;l,m) &= \left(\frac{m}{l}\right)^{n-1} P(L;l,m) \\ &= (-1)^k \left(\frac{m}{l}\right)^{n-1} l^{2k} P(L_{1,2,\dots,k};l,m) \\ &\quad + (-1)\sum_{d=1}^{k}(lm)^d \left(\frac{m}{l}\right)^{n-1} \sum_{1\le j_1<\cdots<j_d\le k} P(L^*_{j_1,\dots,j_d};l,m) \\ &= (-1)^k l^{2k} Q(L_{1,2,\dots,k};l,m) \\ &\quad + (-1)\sum_{d=1}^{k}(lm)^d \left(\frac{m}{l}\right)^{n-1} \sum_{1\le j_1<\cdots<j_d\le k} P(L^*_{j_1,\dots,j_d};l,m). \end{aligned} \tag{5.6}$$

In the next section, we will evaluate l-deg, l-coef and r-deg of each term in the second summation in (5.6).

6. Proof of Theorem 2.1 (IV). Let

$$R(j_1,\dots,j_d) = (lm)^d (m/l)^{n-1} P(L^*_{j_1,\dots,j_d};l,m).$$

By the substitutions $l = t\sqrt{-1}$ and $m = -(\sqrt{t} - 1/\sqrt{t})\sqrt{-1}$, $R(j_1,\dots,j_d)$ becomes an integer polynomial $R^*(j_1,\dots,j_d)$ in $t^{\pm 1}$. In this section we will show that l-deg $R^*(j_1,\dots,j_d)$ depends on d, not a sequence $j_1, j_2, \dots, j_d$. In fact, we will prove the following lemma.

LEMMA 6.1. *For any $1 \le d \le k$ and $1 \le j_1 < \cdots < j_d \le k$,*

$$\begin{aligned} &(1)\quad l\text{-coef}\, R^*(j_1,\dots,j_d) = (-1)^d; \\ &(2)\quad l\text{-deg}\, R^*(j_1,\dots,j_d) = \tfrac{1}{2}\{\delta(L) - n + 1\} - (n-1) \\ &\qquad\qquad\qquad\qquad\quad = g(L) - (n-1); \end{aligned} \tag{6.1}$$

(3) *if $L^*_{j_1,\dots,j_d}$ is not split, then*

$$r\text{-deg}\, R^*(j_1,\dots,j_d) \le h(\tilde{L}) + (n-1).$$

Note that $n = \mu(L)$.

PROOF. Suppose $\tilde{L}^* = \tilde{L}^*_{j_1,\dots,j_d}$ consists of λ connected components $\tilde{L}_1,\dots,\tilde{L}_\lambda$. Since the number of components of a link increases or decreases by one, each time we eliminate a double point of a link diagram, we see that $\mu(L^*) \equiv n - d \pmod 2$, and hence we can write $\mu(L^*) = n - d + 2q$ for some integer q. Therefore,

$$\sum_{i=1}^{\lambda}\{\mu(L_i) - 1\} = \mu(L^*) - \lambda = n - d + 2q - \lambda.$$

Since

$$P(L^*;l,m) = \left(-\frac{l+l^{-1}}{m}\right)^{\lambda-1} \prod_{p=1}^{\lambda} P(L_p;l,m)$$

and
$$Q(L_p;l,m) - (m/l)^{n_p-1}P(L_p;l,m), \qquad n_p = \mu(L_p),$$
for $p = 1, 2, \ldots, \lambda$, we have

$$R(j_1,\ldots,j_d) = (lm)^d \left(\frac{m}{l}\right)^{n-1} \left(-\frac{l+l^{-1}}{m}\right)^{\lambda-1} \prod_{p=1}^{\lambda} P(L_p;l,m)$$
$$= (lm)^d \left(\frac{m}{l}\right)^{n-1} \left(-\frac{l+l^{-1}}{m}\right)^{\lambda-1}$$
$$\times \prod_{p=1}^{\lambda} \left\{ \left(\frac{m}{l}\right)^{n_p-1} P(L_p;l,m) \right\} \prod_{p=1}^{\lambda} \left(\frac{l}{m}\right)^{n_p-1}.$$

Noting $\sum_{p=1}^{\lambda}(n_p - 1) = n - d + 2q - \lambda$, we see
(6.2)
$$R(j_1,\ldots,j_d) = (lm)^d \left(\frac{m}{l}\right)^{n-1} \left(\frac{l}{m}\right)^{n-d+2q-\lambda} \left(-\frac{l+l^{-1}}{m}\right)^{\lambda-1} \prod_{p=1}^{\lambda} Q(L_p;l,m)$$
$$= (lm)^d \left(\frac{m}{l}\right)^{d-2q+\lambda-1} (-1)^{\lambda-1} \left(\frac{l+l^{-1}}{m}\right)^{\lambda-1} \prod_{p=1}^{\lambda} Q(L_p;l,m)$$
$$= (-1)^{\lambda-1}(lm)^d \left(\frac{m}{l}\right)^{d-2q} (1+l^{-2})^{\lambda-1} \prod_{p=1}^{\lambda} Q(L_p;l,m)$$
$$= (-1)^{\lambda-1} m^{2(d-q)} l^{2q} (1+l^{-2})^{\lambda-1} \prod_{p=1}^{\lambda} Q(L_p;l,m).$$

Therefore, we obtain
(6.3)
$$R^*(j_1,\ldots,j_d) = (-1)^{\lambda-1}(-1)^{d-q}(-1)^q(t-2+t^{-1})^{d-q}t^{2q}(1-t^{-2})^{\lambda-1} \prod_{p=1}^{\lambda} J(L_p;t)$$
$$= (-1)^{\lambda+d-1}(t-2+t^{-1})^{d-q}t^{2q}(1-t^{-2})^{\lambda-1} \prod_{p=1}^{\lambda} J(L_p;t).$$

Since $h(\tilde{L}_p) \le h(\tilde{L}) - d < h(\tilde{L})$ and $\tilde{L}_p$ is connected, we can apply the induction hypothesis on each reduced Jones polynomial $J(L_p;t)$, and we have

For each $p = 1, 2, \ldots, \lambda$,

(6.4)
(1) l-coef $J(L_p;t) = 1$;
(2) l-deg $J(L_p;t) = \frac{1}{2}\{\delta(L_p) - n_p + 1\} - (n_p - 1)$;
(3) r-deg $J(L_p;t) \le h(\hat{L}_p) + (n_p - 1)$.

Therefore

$$l\text{-coef}\, R^*(j_1,\ldots,j_d) = (-1)^{\lambda+d-1}(-1)^{\lambda-1} \prod_{p=1}^{\lambda} l\text{-coef}\, J(L_p;t) = (-1)^d.$$

This proves (6.1)(1).

To prove (6.1)(2), first we see from (6.3) that

$$\begin{aligned} l\text{-deg}\, R^*(j_1,\ldots,j_d) &= -(d-q)+2q-2(\lambda-1)+\sum_{p=1}^{\lambda} l\text{-deg}\, J(L_p;t) \\ &= -d+3q-2(\lambda-1)+\sum_{p=1}^{\lambda}\left[\frac{1}{2}\{\delta(L_p)-n_p+1\}-(n_p-1)\right] \\ &= \frac{1}{2}\left\{\sum_{p=1}^{\lambda}\delta(L_p)-3\sum_{p=1}^{\lambda}n_p+3\lambda-2d+6q-4\lambda+4\right\} \\ &= \frac{1}{2}\left\{\sum_{p=1}^{\lambda}\delta(L_p)-3(n-d+2q)-2d+6q-\lambda+4\right\} \\ &= \frac{1}{2}\left\{\sum_{p=1}^{\lambda}\delta(L_p)-3n+d-\lambda+4\right\}. \end{aligned}$$

Now Proposition 4.3 (4.4) shows that $\sum_{p=1}^{\lambda}\delta(L_p)-\lambda+d+1=\delta(L)$ and hence

$$l\text{-deg}\, R^*(j_1,\ldots,j_d)=\tfrac{1}{2}\{\delta(L)-3n+3\}=\tfrac{1}{2}\{\delta(L)-n+1\}-(n-1).$$

This proves (6.1)(2).

Finally, suppose $L^*=L^*_{j_1,\ldots,j_d}$ is connected, and we will prove (6.1)(3). Putting $\lambda=1$ in (6.3), we have

$$h\text{-deg}\, R^*(j_1,\ldots,j_d)=d-q+2q+h\text{-deg}\, J(L^*;t)=d+q+h\text{-deg}\, J(L^*;t).$$

Since $l\text{-deg}\, R^*(j_1,\ldots,j_d)=l\text{-deg}\, J(L^*;t)+3q-d$, it follows that

$$r\text{-deg}\, R^*(j_1,\ldots,j_d)=r\text{-deg}\, J(L^*;t)+2d-2q.$$

Since $h(\tilde{L}^*)=h(\tilde{L})-d<h(\tilde{L})$, we can apply the induction hypothesis to obtain

$$r\text{-deg}\, J(L^*;t)\le h(\tilde{L}^*)+(\mu(L^*)-1) \tag{6.5}$$

and hence

$$\begin{aligned} r\text{-deg}\, R^*(j_1,\ldots,j_d) &\le h(\tilde{L}^*)+\mu(L^*)-1+2d-2q \\ &= h(\tilde{L}^*)+n-d+2q-1+2d-2q \\ &= h(\tilde{L}^*)+d+(n-1)=h(\tilde{L})+(n-1). \end{aligned}$$

This proves (6.1)(3).

7. Proof of Theorem 2.1 (V). In order to apply the induction hypothesis on the reduction formula (5.5), we are required to show that

(7.1) $\quad$ *$L_{1,2,\ldots,k}$ is isotropic to a positive link L' such that $h(\tilde{L}')<h(\tilde{L})$.*

For an arbitrary choice of double points $c_1,c_2,\ldots,c_k$, (7.1), in general, is not true. In this section, however, we will show that there exist double points $c_1,\ldots,c_k$

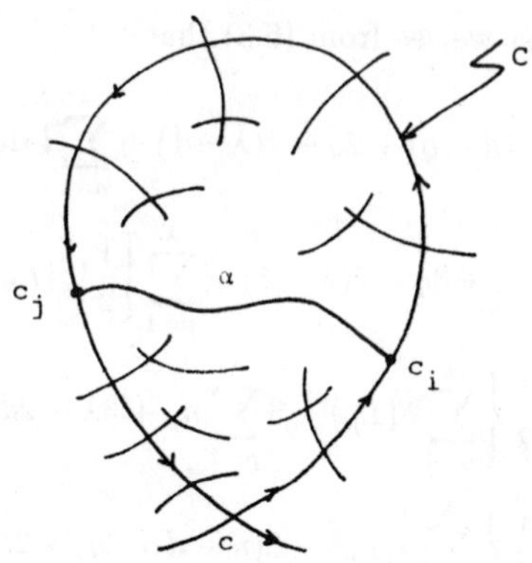

FIGURE 4

for which (7.1) holds. In fact we will prove

LEMMA 7.1. *Let L be a nonsplit positive link and $\tilde{L}$ a connected positive link diagram of L. Then there are double points $c_1, c_2, \ldots, c_k$ on $\tilde{L}$ such that*

(1) $L_{1,2,\ldots,k}$ is isotopic to a link L' such that $h(\tilde{L}') = h(\tilde{L}) - 2k$, and

(2) for any sequence $j_1, \ldots, j_d$, $1 \le d \le k$, $1 \le j_1 < j_2 < \cdots < j_d \le k$, $L^*_{j_1,\ldots,j_d}$ is connected.

PROOF. A proof is based upon a standard geometric argument.

Take a component L_1, say, of a link L. It is then easy to see that L_1 has two points P and P' (P may be P') such that the part of L_1 joining P and P' projects on a circle C in $\tilde{L}$ and such that P and P' project on a double point c of $\tilde{L}$. Since $\tilde{L}$ is connected, such P and P' exist. If $P = P'$, then L_1 is unknotted.

Now the circle C divides S^2 into two domains. The left side domain of C (with respect to its direction) is called the *interior* of C, denoted by $\overline{C}$.

First we consider the case $P \ne P'$.

Let c be the double point of $\tilde{L}$ on which P and P' project. If C has no other double points, then c is removable, and hence we can remove c from $\tilde{L}$ without changing the link type L, and there is nothing to prove here.

Suppose C contains m double points $c_1, c_2, \ldots, c_m$ besides c. m is obviously even. A simple arc α joining two double points c_i and c_j $(1 \le i < j \le m)$ in $\overline{C}$ is called a *separating* arc of C if (1) α does not intersect with $\tilde{L}$ except its end points, and (2) each of two domains into which α divides $\overline{C}$ has the nonempty intersection with $\tilde{L}$ except its end points (see Figure 4).

Let α be a separating arc of C and α joins c_i and c_j, $1 \le i < j \le m$. α divides C into two arcs α_+ and α_-, where α_+ contains $c_i, c_{i+1}, \ldots, c_j$ and α_- contains c_j, $c_{j+1}, \ldots, c_m, c, c_1, \ldots, c_{i-1}, c_i$.

Now α is called the *innermost* separating arc of C if the subdomain of $\overline{C}$ bounded by α_+ and α does not have any other separating arcs of C.

If C has separating arcs, C must have at least one innermost separating arc. Let β be an innermost separating arc and suppose β joins c_i and c_j, $1 \le i < j \le m$. Then $j - i$ must be even; i.e., β_+ contains an even number of double points. Since $\tilde{L}$ is positive, half of these double points are overcrossing points on C including

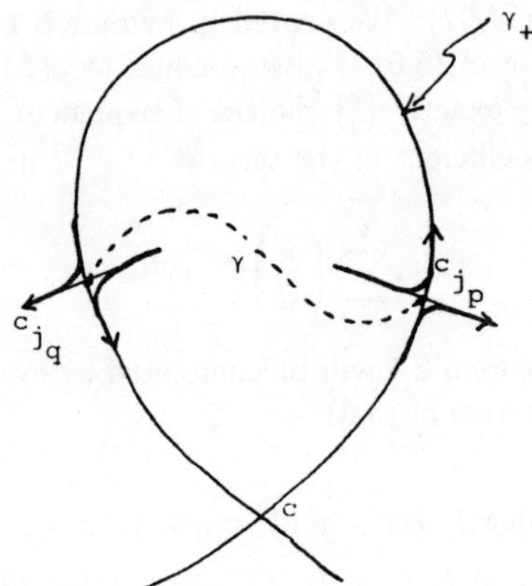

FIGURE 5

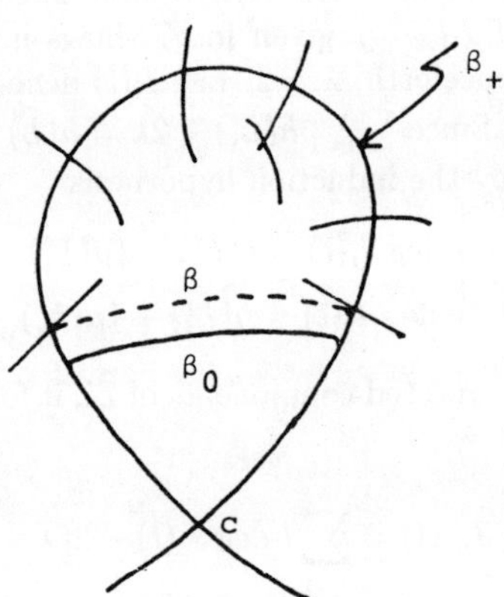

FIGURE 6

possibly c_i and c_j. These double points are exactly what we are looking for. If C has no separating arcs, then all the overcrossing double points on C are what we sought. In fact, if $L^*_{j_1,\ldots,j_d}$ is disconnected, then there is a separating arc γ in C which joins overcrossing double points c_{j_p} and c_{j_q}, say. Since β is innermost, γ must be β. However, γ_+ contains an odd number of double points as is seen in Figure 5, while β_+ contains an even number of double points, a contradiction.

When $P = P'$, then we include c in the sequence of double points $c_1, c_2, \ldots, c_m$, and apply a similar argument.

Obviously, $L_{1,2,\ldots,k}$ is isotopic to L' with $h(\tilde{L}') = h(\tilde{L}) - 2k$ by pulling β_+ down to β_0 (Figure 6).

This proves Lemma 7.1.

8. Proof of Theorem 2.1 (VI). We return to the proof of Theorem 2.1.

We may assume now that $L^*_{j_1,\ldots,j_d}$ is connected and $L_{1,2,\ldots,k}$ is isotopic to L' such that $h(\tilde{L}') \leq h(\tilde{L}) - 2k$, $k \geq 1$.

Consider $Q_L(l,m)$ or $J_L(t)$. We proved in Lemma 6.1 that l-deg of each term in the second summation of (5.6) is always equal to $g(L)-(n-1)$ and its l-coef is $(-1)^d$. Since there are exactly $\binom{k}{d}$ choices of sequences $j_1,\ldots,j_d$ from k distinct indices $1,2,\ldots,k$, the coefficient of the term $t^{g(L)-n+1}$ in the second term of (5.6) is

$$(-1)\sum_{d=1}^{k}\binom{k}{d}(-1)^d=+1.$$

Now the proof of Theorem 2.1 will be completed by evaluating l-deg and h-deg of the first and second terms of (5.6).

In order to prove Theorem 2.1, therefore, it is enough to show

$$\text{(8.1)}\quad \begin{aligned}&(1)\quad 2k+l\text{-deg}\,J_{L'}(t)>g(L)-(n-1);\\ &(2)\quad 2k+h\text{-deg}\,J_{L'}(t)-\{g(L)-(n-1)\}\le h(\tilde{L})+(n-1).\end{aligned}$$

Note that $\tilde{L}$ is connected.

Suppose $\tilde{L}'$ consists of λ connected positive link diagrams $\tilde{L}'_1,\tilde{L}'_2,\ldots,\tilde{L}'_\lambda$. However, the construction of $L_{1,2,\ldots,k}$ given in §7 shows implicitly that $\lambda=1$ or 2. Therefore, we assume, henceforth, $\lambda\le 2$. Let $J_i(t)$ denote the reduced Jones polynomial of L'_i, $1\le i\le\lambda$. Since $\sum_{i=1}^{\lambda}h(\tilde{L}'_i)+2k=h(\tilde{L})$ and $k\ge 1$, it follows that $h(\tilde{L}'_i)<h(\tilde{L})$; therefore by the induction hypothesis

$$\text{(8.2)}\quad \begin{aligned}&(1)\quad l\text{-deg}\,J_i(t)=g(L'_i)-\{\mu(L'_i)-1\};\\ &(2)\quad r\text{-deg}\,J_i(t)\le h(\tilde{L}'_i)+\{\mu(L'_i)-1\}.\end{aligned}$$

Since $L'_1,\ldots,\tilde{L}'_\lambda$ are connected components of $\tilde{L}'$, it follows from Proposition 3.1 that

$$\text{(8.3)}\quad \begin{aligned}(1)\quad l\text{-deg}\,J_{L'}(t)&=\sum_{i=1}^{\lambda}l\text{-deg}\,J_i(t)-2(\lambda-1)\\ &=\sum_{i=1}^{\lambda}\{g(L'_i)-\mu(L'_i)+1\}-2(\lambda-1)\\ &=\sum_{i=1}^{\lambda}g(L'_i)-n+\lambda-2(\lambda-1)\\ &=\sum_{i=1}^{\lambda}g(L'_i)-(n-1)-(\lambda-1),\\ (2)\quad r\text{-deg}\,J_{L'}(t)&\le h(\tilde{L}')+(n-1)+(\lambda-1).\end{aligned}$$

Therefore, (8.1)(1) is equivalent to

$$\text{(8.4)}\qquad \sum_{i=1}^{\lambda}g(L'_i)-(\lambda-1)+2k>g(L).$$

Now let $\alpha(\tilde{X})$ and $s(\tilde{X})$ denote, respectively, the number of domains and Seifert circles in a link diagram $\tilde{X}$. Then Proposition 4.1 and a simple computation show that (8.4) is equivalent to

$$4k + s(\tilde{L}) - \sum_{i=1}^{\lambda} s(\tilde{L}'_i) > \alpha(\tilde{L}) - \sum_{i=1}^{\lambda} \alpha(\tilde{L}'_i) + 2(\lambda - 1). \tag{8.5}$$

Furthermore, (8.3)(2) implies that

$$\begin{aligned} &2k + h\text{-deg}\, J_{L'}(t) - \{g(L) - n + 1\} \\ &\quad = 2k + r\text{-deg}\, J_{L'}(t) + l\text{-deg}\, J_{L'}(t) - \{g(L) - n + 1\} \\ &\quad \le 2k + h(\tilde{L}') + (n - 1) + (\lambda - 1) + l\text{-deg}\, J_{L'}(t) - g(L) + (n - 1) \\ &\quad = h(\tilde{L}) + 2(n - 1) + (\lambda - 1) + l\text{-deg}\, J_{L'}(t) - g(L). \end{aligned}$$

Therefore, in order to prove (8.1)(2), it suffices to show that

$$h(\tilde{L}) + 2(n - 1) + (\lambda - 1) + l\text{-deg}\, J_{L'}(t) - g(L) \le h(\tilde{L}) + (n - 1),$$

or equivalently,

$$(\lambda - 1) + l\text{-deg}\, J_{L'}(t) \le g(L) - n + 1. \tag{8.6}$$

Using Proposition 4.1 and (8.3)(1), we see that (8.6) is equivalent to

$$(\lambda - 1) + \sum_{i=1}^{\lambda} g(L'_i) - (n - 1) - (\lambda - 1) \le g(L) - (n - 1).$$

That is,

$$\sum_{i=1}^{\lambda} g(L'_i) \le g(L). \tag{8.7}$$

Now a simple calculation shows that (8.7) is equivalent to

$$s(\tilde{L}) - \sum_{i=1}^{\lambda} s(\tilde{L}'_i) \le \alpha(\tilde{L}) - \sum_{i=1}^{\lambda} \alpha(\tilde{L}'_i). \tag{8.8}$$

Therefore the final lemma needed to complete our proof is

LEMMA 8.1.

$$\text{(8.9)} \quad \begin{aligned} &(1) \quad 4k + s(\tilde{L}) - \sum_{i=1}^{\lambda} s(\tilde{L}'_i) > \alpha(\tilde{L}) - \sum_{i=1}^{\lambda} \alpha(\tilde{L}'_i) + 2(\lambda - 1); \\ &(2) \quad \alpha(\tilde{L}) - \sum_{i=1}^{\lambda} \alpha(\tilde{L}'_i) \ge s(\tilde{L}) - \sum_{i=1}^{\lambda} s(\tilde{L}'_i). \end{aligned}$$

PROOF. We will use the same notation as in the previous section.

For simplicity, we let $s(\tilde{L}') = \sum_{i=1}^{\lambda} s(\tilde{L}'_i)$ and $\alpha(\tilde{L}') = \sum_{i=1}^{\lambda} \alpha(\tilde{L}'_i)$. Since we are only interested in $s(\tilde{L}) - s(\tilde{L}')$ and $\alpha(\tilde{L}) - \alpha(\tilde{L}')$ and since these numbers are affected only when a Seifert circle intersects β_+, we consider only these Seifert circles.

First we suppose $P \neq P'$, where P and P' are points on one component of L considered in §7. (For the case $P = P'$, the proof is analogous and therefore is omitted.)

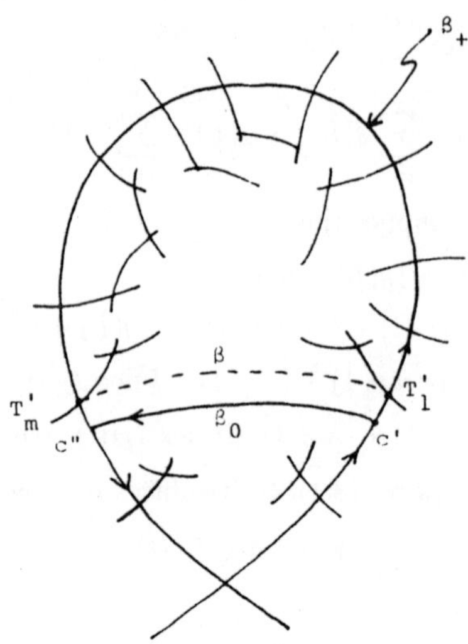

FIGURE 7

Suppose there are m Seifert circles $T_1', T_2', \ldots, T_m'$ in $\tilde{L}'$ that intersect β_+ in $\tilde{L}$. (Beware that T_i' is a Seifert circle in $\tilde{L}'$ but not in $\tilde{L}$.) See Figure 7.

Let c' and c'' be the end points of β_0. c' and c'' are not double points of $\tilde{L}$. We extend β_+ to the arc $\hat{\beta}_+$ in $\tilde{L}$ which joins c' to c''. Therefore, $\hat{\beta}_+$ and β_0 form a simple closed curve in S^2.

Now when β_0 is replaced by $\hat{\beta}_+$, each domain in $\tilde{L}'$ is cut into two domains in $\tilde{L}$. Therefore we have

(8.10) $$\alpha(\tilde{L}) - \alpha(\tilde{L}') = 2k.$$

Suppose there are r Seifert circles in $\tilde{L}$, each of which contains at least one edge in β_+. We will show

(8.11) $$2k \geq r - m \geq 0.$$

The first inequality is obvious. To prove the second inequality, note that each Seifert circle T_i' in $\tilde{L}'$ is cut into two distinct Seifert circles $T_{i,1}$ and $T_{i,2}$ in $\tilde{L}$. However, one of these Seifert circles, say $T_{i,1}$, may form one Seifert circle in $\tilde{L}$ with, say, $T_{i+1,1}$ that is obtained from T_{i+1}'. In any case, we have $r - m \geq 0$. This proves (8.11). See Figure 8, where the broken lines indicate Seifert circles in $\tilde{L}$.

Now, if $\lambda = 2$ and $k = 1$, then L is a product of three links, L_1, L_2 and the Hopf link L_3 (see Figure 9). Then Theorem 2.1 follows from the induction assumption. Therefore we may assume that $\lambda = 1$, or $\lambda = 2$ and $k \geq 2$. Then Lemma 8.1 will follow from (8.10) and (8.11).

Since $\lambda - 1 < k$ and $\alpha(\tilde{L}) - \alpha(\tilde{L}') = 2k$, it follows that

$$\begin{aligned} &4k + s(\tilde{L}) - s(\tilde{L}') - \{\alpha(L) - \alpha(\tilde{L}') + 2(\lambda - 1)\} \\ &= 4k + r - m - 2k - 2(\lambda - 1) \geq 2k - 2(\lambda - 1) > 0. \end{aligned}$$

This proves (8.9)(1). Furthermore,

$$\alpha(\tilde{L}) - \alpha(\tilde{L}') - \{s(\tilde{L}) - s(\tilde{L}')\} = 2k - (r - m) \geq 0.$$

This proves (8.9)(2).

The proof of Theorem 2.1 is now completed.

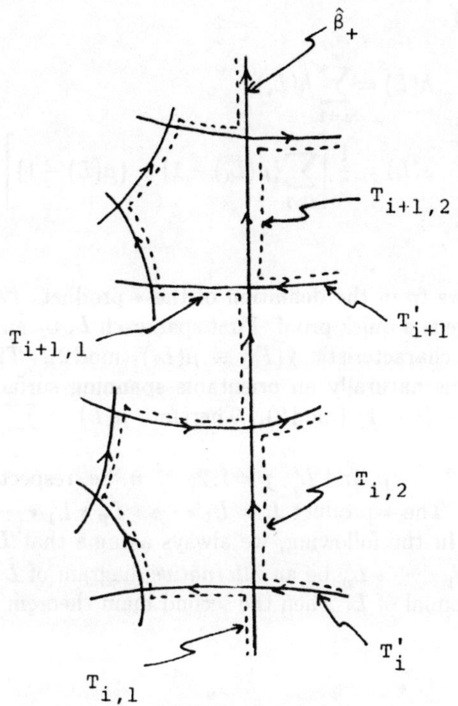

FIGURE 8

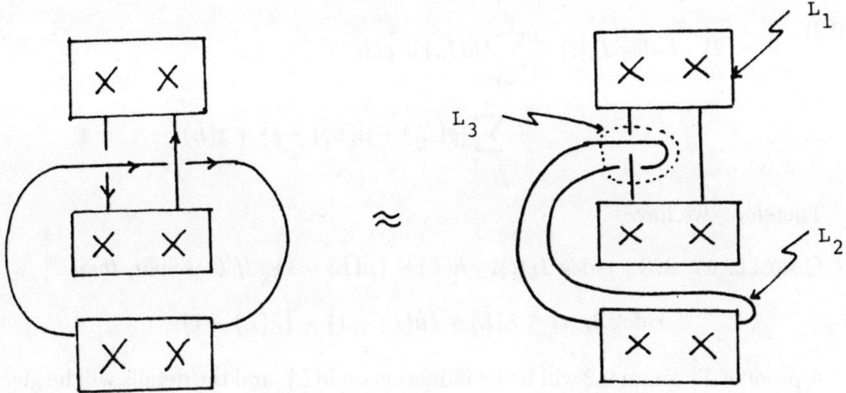

FIGURE 9

9. Alternative links. Let $L_1, L_2, \ldots, L_r$ be r links in S^3. Denote by $L = L_1 * L_2 * \cdots * L_r$ the $*$-product (or Murasugi sum) of these links. (For definition, see [**4** or **12**].) Since L is constructed from L_i, a diagram $\tilde{L}$ of L is naturally obtained from those of the L_i's, and hence we can write $\tilde{L} = \tilde{L}_1 * \tilde{L}_2 * \cdots * \tilde{L}_r$.

PROPOSITION 9.1.

$$(9.1)\quad \begin{aligned} &(1)\quad h(\tilde{L}) = \sum_{i=1}^{r} h(\tilde{L}_i); \\ &(2)\quad s(L) = \frac{1}{2}\left[\sum_{i=1}^{r}\{\mu(L_i) - 1\} - \{\mu(L) - 1\}\right] \end{aligned}$$

is an integer.

PROOF. (1) follows from the definition of the $*$-product. (2) is probably well known. However, here is a quick proof. First span each L_i by an orientable surface F_i. Then the Euler characteristic $\chi(F_i) \equiv \mu(L_i) \pmod 2$. The construction of the $*$-product gives us naturally an orientable spanning surface F of L so that $\chi(F) \equiv \sum_{i=1}^{r} \chi(F_i) - (r-1) \pmod 2$. Therefore $\mu(L) \equiv \sum_{i=1}^{r} \mu(L_i) - (r-1) \pmod 2$ Q.E.D.

Now let L_i, $i = 1, 2, \ldots, p$, and L'_j, $j = 1, 2, \ldots, n$, be, respectively, positive and negative links in S^3. The $*$-product $L = L_1 * \cdots * L_p * L'_1 * \cdots * L'_n$ is called an *alternative* link [**7**]. In the following, we always assume that L is not split. Let $\tilde{L} = \tilde{L}_1 * \cdots * \tilde{L}_p * \tilde{L}'_1 * \cdots * \tilde{L}'_n$ be an *alternative* diagram of $\tilde{L}$. Let $J_L(t)$ be the reduced Jones polynomial of L. Then the second main theorem of this paper is

THEOREM 9.2.

$$(9.2)\quad \begin{aligned} &(1)\quad l\text{-deg}\, J_L(t) \geq \sum_{i=1}^{p} g(L_i) - \sum_{j=1}^{n}\{h(\tilde{L}'_j) + g(L'_j) + \mu(L'_j) - 1\} \\ &\qquad\qquad - \{\mu(L) - 1\} + s(L). \\ &(2)\quad h\text{-deg}\, J_L(t) \leq \sum_{i=1}^{p}\{h(\tilde{L}_i) + g(L_i)\} \\ &\qquad\qquad - \sum_{j=1}^{n}\{g(L'_j) + \mu(L'_j) - 1\} + s(L). \end{aligned}$$

Therefore, we have

COROLLARY 9.3. $r\text{-deg}\, J_L(t) \leq h(\tilde{L}) + \{\mu(L) - 1\}$. *If L is split, then*

$$r\text{-deg}\, J_L(t) \leq h(\tilde{L}) + \{\mu(L) - 1\} + \{\lambda(L) - 1\}.$$

A proof of Theorem 9.2 will be by induction on $h(\tilde{L})$, and the details will be given in the next section. However, since our proof is quite similar to that of Theorem 2.1, most simple or straightforward calculations will be omitted.

10. Proof of Theorem 9.2. First we deduce a reduction formula similar to (5.5) or (5.6).

Let $c_1, \ldots, c_k$ be the double points on $\tilde{L}$ to which we apply the fundamental identity (2.1). Let v and w be, respectively, the number of positive and negative double points among these double points ($k = v + w$). Then the same argument used in §5 gives us the reduction formula

$$\begin{aligned}P(L;l,m) &= (-1)^k l^{2(v-w)} P(L_{1,2,\ldots,k};l,m)\\ &\quad +(-1)\sum_{d=1}^{k}\frac{(lm)^d}{l^{2u}}\sum_{1\le j_1<\cdots<j_d\le k} P(L^*_{j_1,\ldots,j_d};l,m),\end{aligned} \tag{10.1}$$

where u is the number of negative double points among $c_{j_1},\ldots,c_{j_d}$. (10.1) now implies

$$\begin{aligned}Q(L;l,m) &= (-1)^k l^{2(v-w)} Q(L_{1,2,\ldots,k};l,m)\\ &\quad +(-1)\sum_{d=1}^{k}\frac{(lm)^d}{l^{2u}}\left(\frac{m}{l}\right)^{\mu-1}\sum_{1\le j_1<\cdots<j_d\le k} P(L^*_{j_1,\ldots,j_d};l,m),\end{aligned} \tag{10.2}$$

where $\mu=\mu(L)$.

Set $L^*=L^*_{j_1,\ldots,j_d}$ and

$$R(j_1,\ldots,j_d)=\frac{(lm)^d}{l^{2u}}\left(\frac{m}{l}\right)^{\mu-1}P(L^*;l,m).$$

Let R^* be the polynomial on $t^{\pm 1}$ which is obtained from $R(j_1,\ldots,j_d)$ by the substitutions $l=t\sqrt{-1}$ and $m=-(\sqrt{t}-1/\sqrt{t})\sqrt{-1}$. We will prove

PROPOSITION 10.1.

$$\begin{aligned}&(1)\quad l\text{-deg}\, R^* \ge \sum_{i=1}^{p} g(L_i)-\sum_{j=1}^{n}\{h(\tilde{L}'_j)+g(L'_j)+\mu(L'_j)-1\}\\ &\qquad\qquad -\{\mu(L)-1\}+s(L).\\ &(2)\quad h\text{-deg}\, R^* \le \sum_{i=1}^{p}\{h(\tilde{L}_i)+g(L_i)\}-\sum_{j=1}^{n}\{g(L'_j)+\mu(L'_j)-1\}+s(L).\end{aligned} \tag{10.3}$$

PROOF. Since a simple calculation shows that

$$R^*=(-1)^d(t-2+t^{-1})^{d-q}t^{2q-2u}J(L^*;t),$$

where $2q=\mu(L^*)-\mu(L)+d$, we have

$$\begin{aligned}&(1)\quad l\text{-deg}\, R^* = 3q-2u-d+l\text{-deg}\, J(L^*;t),\\ &(2)\quad h\text{-deg}\, R^* = d+q-2u+h\text{-deg}\, J(L^*;t).\end{aligned} \tag{10.4}$$

Write $L^*=L_1^* * \cdots * L_{p^*}^* * L_1'^* * \cdots * L_{n^*}'^*$, where L_i^* are positive links and $L_j'^*$ are negative links. Then the induction assumption on L^* gives us

$$\begin{aligned}&(1)\quad l\text{-deg}\, J(L^*;t) \ge \sum_{i=1}^{p^*} g(L_i^*)-\sum_{j=1}^{n^*}\{h(\tilde{L}_j'^*)+g(L_j'^*)+\mu(L_j'^*)-1\}\\ &\qquad\qquad + s(L^*)-\{\mu(L^*)-1\},\\ &(2)\quad h\text{-deg}\, J(L^*;t) \le \sum_{i=1}^{p^*}\{h(\tilde{L}_i^*)+g(L_i^*)\}\\ &\qquad\qquad -\sum_{j=1}^{n^*}\{g(L_j'^*)+\mu(L_j'^*)-1\}+s(L^*),\end{aligned} \tag{10.5}$$

where

$$s(L^*) = \frac{1}{2}\left[\sum_{i=1}^{p^*}\{\mu(L_i^*) - 1\} + \sum_{j=1}^{n^*}\{\mu(L_j'^*) - 1\} - \{\mu(L^*) - 1\}\right].$$

Since $\sum_{i=1}^{p^*} h(\tilde{L}_i^*) = \sum_{i=1}^{p} h(\tilde{L}_i) - (d-u)$ and $\sum_{j=1}^{n^*} h(\tilde{L}_j'^*) = \sum_{j=1}^{n} h(\tilde{L}_j') - u$, we have

(10.6)

$$(1)\quad \sum_{i=1}^{p^*}\delta(L_i^*) = \sum_{i=1}^{p}\delta(L_i) - (d-u),$$

$$(2)\quad \sum_{j=1}^{n^*}\delta(L_j'^*) = \sum_{j=1}^{n}\delta(L_j) - u.$$

Using (4.1)(1) and (10.6), we obtain

(10.7)

$$(1)\quad \sum_{i=1}^{p^*} g(L_i^*) = \sum_{i=1}^{p} g(L_i) - \frac{1}{2}(d-u) + \frac{1}{2}\sum_{i=1}^{p}\{\mu(L_i) - 1\} - \frac{1}{2}\sum_{i=1}^{p^*}\{\mu(L_i^*) - 1\},$$

$$(2)\quad \sum_{j=1}^{n^*} g(L_j'^*) = \sum_{j=1}^{n} g(L_j') - \frac{1}{2}u + \frac{1}{2}\sum_{j=1}^{n}\{\mu(L_j') - 1\} - \frac{1}{2}\sum_{j=1}^{n^*}\{\mu(L_j'^*) - 1\}.$$

(10.3) is now obtained from (10.4) using (10.5)–(10.7).

Now, to complete the proof of Theorem 9.2, it only remains to show the following

PROPOSITION 10.2. *Let* $\hat{L} = L_{1,2,\dots,k}$.

(10.8)

$$(1)\quad l\text{-deg}\, J(\hat{L};t) + 2(v-w) \geq \sum_{i=1}^{p} g(L_i) - \sum_{j=1}^{n}\{h(L_j') + g(L_j') + \mu(L_j') - 1\} - \{\mu(L) - 1\} + s(L),$$

$$(2)\quad h\text{-deg}\, J(\hat{L};t) + 2(v-w) \leq \sum_{i=1}^{p}\{h(L_i) + g(L_i)\} - \sum_{j=1}^{n}\{g(L_j') + \mu(L_j') - 1\} + s(L).$$

PROOF. Write $\hat{L} = \hat{L}_1 * \cdots * \hat{L}_{\hat{p}} * \hat{L}_1' * \cdots * \hat{L}_{\hat{n}}'$, where $\hat{L}_i$ are positive links and $\hat{L}_j'$ are negative links. Since $h(\hat{L}) = h(\tilde{L}) - 2k < h(\tilde{L})$, the induction hypothesis yields

$$\text{(10.9)}\quad \begin{aligned} &(1)\quad l\text{-deg}\, J(\hat{L};t) \geq \sum_{i=1}^{\hat{p}} g(\hat{L}_i) - \sum_{j=1}^{\hat{n}} \{h(\hat{L}'_j) + g(\hat{L}'_j) + \mu(\hat{L}'_j) - 1\} \\ &\qquad\qquad + s(\hat{L}) - \{\mu(\hat{L}) - 1\}, \\ &(2)\quad h\text{-deg}\, J(\hat{L};t) \leq \sum_{i=1}^{\hat{p}} \{h(\hat{L}_i) + g(\hat{L}_i)\} \\ &\qquad\qquad - \sum_{j=1}^{\hat{n}} \{g(\hat{L}'_j) + \mu(\hat{L}'_j) - 1\} + s(\hat{L}), \end{aligned}$$

where

$$2s(\hat{L}) = \sum_{i=1}^{\hat{p}} \{\mu(\hat{L}_i) - 1\} + \sum_{j=1}^{\hat{n}} \{\mu(\hat{L}'_j) - 1\} - \{\mu(\hat{L}) - 1\}.$$

From (10.9), we will see that it is enough to show (10.10). Note that $\mu(\hat{L}) = \mu(L)$.

$$\text{(10.10)}\quad \begin{aligned} &(1)\quad \sum_{i=1}^{\hat{p}} g(\hat{L}_i) - \sum_{j=1}^{\hat{n}} \{h(\hat{L}'_j) + g(\hat{L}'_j) + \mu(\hat{L}_j) - 1\} + s(\hat{L}) + 2(v - w) \\ &\qquad \geq \sum_{i=1}^{p} g(L_i) - \sum_{j=1}^{n} \{h(L'_j) + g(L'_j) + \mu(L'_j) - 1\} + s(L), \\ &(2)\quad \sum_{i=1}^{\hat{p}} \{h(\hat{L}_i) + g(\hat{L}_i)\} - \sum_{j=1}^{\hat{n}} \{g(\hat{L}'_j) + \mu(\hat{L}'_j) - 1\} + s(\hat{L}) + 2(v - w) \\ &\qquad \leq \sum_{i=1}^{p} \{h(L_i) + g(L_i)\} - \sum_{j=1}^{n} \{g(L'_j) + \mu(L'_j) - 1\} + s(L). \end{aligned}$$

Since $\sum_{i=1}^{\hat{p}} h(\hat{L}_i) = \sum_{i=1}^{p} h(L_i) - 2v$ and $\sum_{j=1}^{\hat{n}} h(\hat{L}'_j) = \sum_{j=1}^{n} h(L'_j) - 2w$, using (4.1)(1), (10.10) can be simplified to the following:

$$\text{(10.11)}\quad \begin{aligned} &(1)\quad \frac{1}{2}\sum_{i=1}^{\hat{p}} \delta(\hat{L}_i) - \frac{1}{2}\sum_{j=1}^{\hat{n}} \delta(\hat{L}'_j) + 2v \geq \frac{1}{2}\sum_{i=1}^{p} \delta(L_i) - \frac{1}{2}\sum_{j=1}^{n} \delta(L'_j), \\ &(2)\quad \frac{1}{2}\sum_{i=1}^{\hat{p}} \delta(\hat{L}_i) - \frac{1}{2}\sum_{j=1}^{\hat{n}} \delta(L'_j) - 2w \leq \frac{1}{2}\sum_{i=1}^{p} \delta(L_i) - \frac{1}{2}\sum_{j=1}^{n} \delta(L'_j). \end{aligned}$$

Therefore, it is sufficient to show, finally, that

$$\text{(10.12)}\quad \begin{aligned} &(1)\quad \sum_{i=1}^{\hat{p}} \delta(\hat{L}_i) + 4v \geq \sum_{i=1}^{p} \delta(L_i) \geq \sum_{i=1}^{\hat{p}} \delta(\hat{L}_i), \\ &(2)\quad \sum_{j=1}^{\hat{n}} \delta(\hat{L}'_j) + 4w \geq \sum_{j=1}^{n} \delta(L'_j) \geq \sum_{j=1}^{\hat{n}} \delta(\hat{L}'_j). \end{aligned}$$

Since $\hat{L}_i$ and L_i are positive links, and $\hat{L}'_j$ and L'_j are negative links, (10.11) is exactly the same as (8.9) in Lemma 8.1.

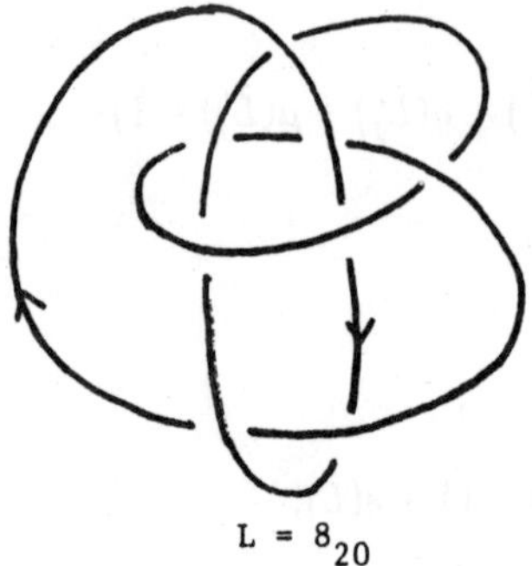

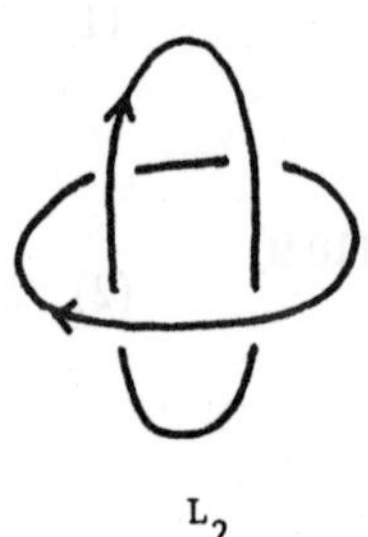

FIGURE 10

The proof of Proposition 10.2 and hence the proof of Theorem 9.2 are now completed.

REMARK. What we have actually proved in Theorem 9.2 is that if $L = L_1 * \cdots * L_m$ is an alternative link, then

$$\text{(10.13)} \qquad r\text{-deg}\, J(L;t) - \{\mu(L) - 1\} \leq \sum_{i=1}^{m} [r\text{-deg}\, J(L_i;t) - \{\mu(L_i) - 1\}].$$

This formula is not true for nonalternative links (see Example 10.1). Nevertheless it seems that $r\text{-deg}\, J(L;t) - \{\mu(L) - 1\}$ is bounded by $h(\tilde{L})$ for any link diagram $\tilde{L}$ of L.

EXAMPLE 10.1. A knot 8_{20} is the $*$-product of two links L_1 and L_2. (See Figure 10.)

It is known that

$$r\text{-deg}\, J(L;t) - \{\mu(L) - 1\} = 6,$$
$$r\text{-deg}\, J(L_1;t) - \{\mu(L_1) - 1\} = 2,$$
$$r\text{-deg}\, J(L_2;t) - \{\mu(L_2) - 1\} = 1.$$

11. Alternating algebraic links. In this section, we prove that equality in Theorem A holds for alternating algebraic links considered in [**13**].

Let T be a weighted finite connected tree, and let w be a weight function. Denote by $V(T)$ the set of all vertices in V and by $v(T)$ the number of vertices of T and $w(T) = \sum_{v \in V(T)} |w(v)|$. T is called *even* and *excessive* if for any $v \in V(T)$, $w(v)$ is even and $|w(v)| \geq \max\{\text{val}(v), 2\}$, where $\text{val}(v)$ denotes the valency of v. $v \in V(T)$ is called *positive* (*negative*) if $w(v) > 0$ ($w(v) < 0$). We assume that $w(v) \neq 0$ for any $v \in V(T)$. A tree is called *positive* (*negative*) if all the vertices are positive (negative).

For a weighted tree, let $A(T)$ be the set of those edges in T which join positive vertices and negative vertices. If all the edges in $A(T)$ are removed from T, T will split into finitely many subtrees $T_1, \ldots, T_k$, each of which is either strictly positive or negative. A collection $\{T_1, T_2, \ldots, T_k\}$ is called the *uniform decomposition* of T.

Now, given an even connected tree T, we can construct an orientable surface F by plumbing as specified by T. The boundary of F is an oriented link, denoted by $l(T)$. The orientation of $l(T)$ is induced from that of F.

It is proved in [**13**].

PROPOSITION 11.1. *If T is a positive (negative) even excessive tree, then $l(T)$ is a positive (negative) alternating link. More generally, for an even tree T, if each component of the uniform decomposition of T is excessive, then $l(T)$ is an alternating link.*

For a proof, see Proposition 4.2 in [**13**]. The main theorem of this section is

THEOREM 11.2. *Let T be an even tree and let $J_L(t)$ be the reduced Jones polynomial of the algebraic link $L = l(T)$ associated with T. Let $\{T_1, T_2, \dots, T_k\}$ be the uniform decomposition of T. If each T_i is excessive, then*

$$\text{r-deg}\, J_L(t) = w(T) - v(T) + k + \mu(L) - 1. \tag{11.1}$$

Furthermore, there is an alternating diagram $\tilde{L}$ of L such that $h(\tilde{L}) = w(T) - v(T) + k$. Therefore, $w(T) - v(T) + k$ is the minimum number of double points any alternating diagram of L can have.

Since a 2-bridge knot or link is an algebraic link associated with an even excessive tree, the next corollary follows immediately.

COROLLARY 11.3. *Let $L(\alpha, \beta)$ be the 2-bridge link of type (α, β), where β is odd and $0 < \beta < \alpha$. Let $\{a_{11}, a_{12}, \dots, a_{1,n_1}, -a_{21}, \dots, -a_{2,n_2}, \dots, (-1)^{k-1}a_{k,1}, \dots, (-1)^{k-1}a_{k,n_k}\}$ be a continued fraction form of $\alpha/(\alpha-\beta)$, where a_{ij} are positive and even. Then*

$$\text{r-deg}\, J_L(t) = \sum_{i=1}^{k} \left[\sum_{j=1}^{n_i} a_{i,j} - (n_i - 1) \right] + \mu(L) - 1, \tag{11.2}$$

and there is an alternating diagram $\tilde{L}$ of L such that

$$h(\tilde{L}) = \sum_{i=1}^{k} \left[\sum_{j=1}^{n_i} a_{i,j} - (n_i - 1) \right].$$

PROOF OF THEOREM 11.2. We will prove equalities in (9.2) by induction on $v(T)$. If $v(T) = 1$, then $k = 1$ and $l(T)$ is a torus link of type $(v(T), 2)$. A direct computation then proves Theorem 11.2.

Now we assume that equalities hold in (9.2) for any algebraic link $l(T')$ with $v(T') < v(T)$.

Let $\{T_1, \dots, T_k\}$ be the uniform decomposition of T. Assume that $T_1, \dots, T_p$ are positive (excessive) trees and $T_{p+1}, \dots, T_k$ are negative (excessive) trees. First we note

For any i,

$$\text{(11.3)} \qquad \begin{aligned} &(1)\quad \delta(l(T_i)) = v(T_i) \qquad [\mathbf{13}, \text{ Proposition 7.1(1)}], \\ &(2)\quad h(l(T_i)) = w(T_i) - v(T_i) + 1. \end{aligned}$$

Now using (4.1)(1) and (11.3), we compute the right-hand sides of (9.2) and find that it is sufficient to prove the following lemma.

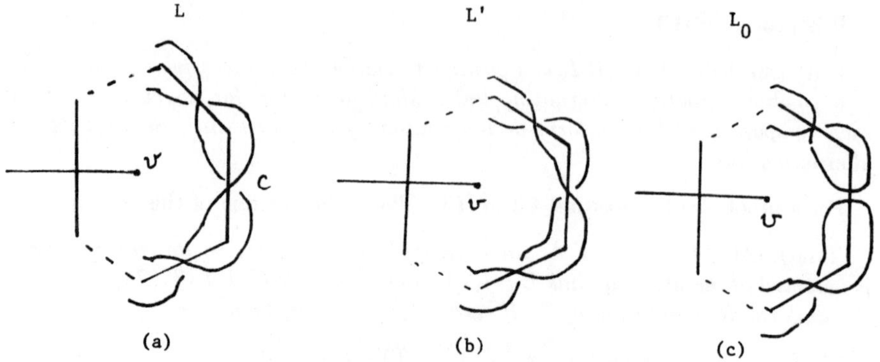

FIGURE 11

LEMMA 11.4. *Let* $L = l(T)$. *Let* $w_P(T) = \sum_{i=1}^{p} w(T_i)$ *and* $w_N(T) = \sum_{j=p+1}^{k} |w(T_j)|$. *Then*

$$(11.4)\quad \begin{aligned} &(1)\quad l\text{-deg}\, J(L;t) = \frac{1}{2}v(T) - w_N(T) - \tfrac{3}{2}\{\mu(L) - 1\} - (k - p), \\ &(2)\quad h\text{-deg}\, J(L;t) = w_P(T) - \frac{1}{2}v(T) - \tfrac{1}{2}\{\mu(L) - 1\} + p. \end{aligned}$$

PROOF Take a stump v; i.e., $\text{val}(v) = 1$. We may assume w.l.o.g. that $v \in T_k$. Let $w(v) = \nu < 0$. Applying the fundamental identity at a double point c, illustrated in Figure 11(a), we have

$$(11.5)\quad \begin{aligned} (-1)Q(L;l,m) &= l^{-2}Q(L';l,m) + (m/l)^{\mu}P(L_0;l,m) \\ &= l^{-2}Q(L';l,m) + (m/l)^{\mu-\mu_0+1}Q(L_0;l,m), \end{aligned}$$

where $\mu = \mu(L)$ and $\mu_0 = \mu(L_0)$.

Let T' and T_0 be weighted trees with which algebraic links L' and L_0 are associated.

Case 1. $|w(v)| = |\nu| = 2$.

Let v' be a vertex of T_k incident to v. (We assume $v(T_k) \geq 2$. For the case $V(T_k) = \{v\}$, the proof is similar and omitted.) Then $\{T_1, \ldots, T_{k-1}, \hat{T}_k(v, v')\}$ is the uniform decomposition of T', where $\hat{T}_k(H)$ denotes the *complement* of a subtree H of T_k. ($\hat{T}_k(H)$ is the tree obtained from T_k by removing all vertices of H and all edges of T_k incident with at least one vertex in H. $\hat{T}_k(H)$ is, in general, a collection of trees.) Furthermore, $\{T_1, \ldots, T_{k-1}, \hat{T}_k(v)\}$ is the uniform decomposition of T_0.

Case 1(i). $\mu - \mu_0 = 1$.

(11.5) yields

$$(11.6)\quad (-1)J(L;t) = (-1)t^{-2}J(L';t) + (t - 2 + t^{-1})t^{-2}J(L_0;t).$$

We then prove

$$(11.7)\quad \begin{aligned} &(1)\quad l\text{-deg}\, J(L';t) - 2 > l\text{-deg}\, J(L_0;t) - 3, \\ &(2)\quad h\text{-deg}\, J(L';t) - 2 < h\text{-deg}\, J(L_0;t) - 1. \end{aligned}$$

Since the proof of (2) is similar to that of (1), we only prove (1).

Since T_k is an even excessive tree, $\hat{T}_k(v, v')$ consists of *at most* $|w(v')|/2$ connected negative trees, and hence the number, ξ, of negative trees in the uniform decomposition of T' is at most $(k-p-1) + |w(v')|/2$. Note that $|w(v')|$ is even and positive. A simple computation then shows that

$$\begin{aligned} l\text{-deg}\, J(L';t) - 2 &= \tfrac{1}{2}\{v(T) - 2\} - \{w_N(T) - 2 - |w(v')|\} - \tfrac{3}{2}\{\mu(L') - 1\} - \xi - 2 \\ &\geq l\text{-deg}\, J(L_0;t) - 3 + \frac{|w(v')|}{2} > l\text{-deg}\, J(L_0;t) - 3. \end{aligned}$$

This proves (11.7).

Now we see from (11.6) that

$$\begin{aligned} l\text{-deg}\, J(L;t) &= l\text{-deg}\, J(L_0;t) - 3 \\ &= \tfrac{1}{2}v(T) - w_N(T) - \tfrac{3}{2}\{\mu(L) - 1\} - (k-p). \end{aligned}$$

This proves (11.4)(1). A proof of (2) is omitted.

Case 1(ii). $\mu - \mu_0 = -1$.

(11.5) yields

$$(-1)J(L;t) = (-1)t^{-2}J(L';t) + J(L_0;t).$$

Then we can prove

$$\text{(11.8)} \qquad \begin{aligned} &(1) \quad l\text{-deg}\, J(L';t) - 2 > l\text{-deg}\, J(L_0;t), \\ &(2) \quad h\text{-deg}\, J(L';t) - 2 < h\text{-deg}\, J(L_0;t). \end{aligned}$$

Therefore, $l\text{-deg}\, J(L;t) = l\text{-deg}\, J(L_0;t)$ and $h\text{-deg}\, J(L;t) = h\text{-deg}\, J(L_0;t)$, and we obtain (11.4). The details will be omitted.

Case 2. $|w(v)| = |\nu| > 2$.

Let T_k' be the same weighted tree as T_k except for one weight at v, where $w(v) = -(\nu - 2)$. Then $\{T_1, \ldots, T_{k-1}, T_k'\}$ is the uniform decomposition of T'. T_0 is the same as before. Then we can prove

$$\text{(11.9)} \qquad \begin{aligned} &\text{(I)} \quad If\, \mu - \mu_0 = 1, then \\ &(1) \quad l\text{-deg}\, J(L';t) - 2 < l\text{-deg}\, J(L_0;t) - 3, \\ &(2) \quad h\text{-deg}\, J(L';t) - 2 < h\text{-deg}\, J(L_0;t) - 1. \\ &\text{(II)} \quad If\, \mu - \mu_0 = -1, then \\ &(1) \quad l\text{-deg}\, J(L';t) - 2 < l\text{-deg}\, J(L_0;t), \\ &(2) \quad h\text{-deg}\, J(L';t) - 2 < h\text{-deg}\, J(L_0;t). \end{aligned}$$

These formulas imply (11.4). Since the proofs are straightforward computations, they are omitted.

The proof of (11.1) is now completed.

A construction of an alternating minimum diagram $\tilde{L}$ can be found in the proof of Proposition 4.2 in [**13**]. Q.E.D.

EXAMPLE 11.1. See Figure 12.

For any positive (or negative) excessive (not necessarily even) connected tree T, $l(T)$ is an alternating link, and an alternating (nonoriented) diagram $\widetilde{l(T)}$ of $l(T)$ is naturally obtained from T. (For details, see [**13**].) Since $r\text{-deg}\, J(l(T);t)$ does not depend on an orientation of each component of $l(T)$ [**6** or **9**], $r\text{-deg}\, J(l(T);t)$ is well defined for each tree T. It seems that Theorem 11.2 is true for these links.

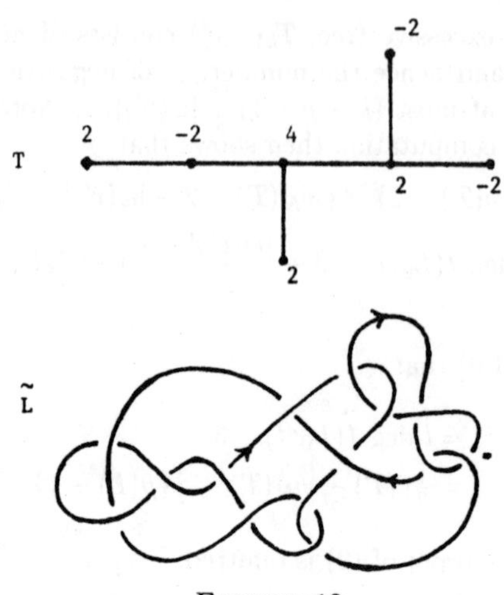

FIGURE 12

Therefore

CONJECTURE. *For a positive excessive connected tree T,*

$$r\text{-deg}\, J(l(T);t) = w(T) - v(T) + \mu(l(T)),$$
$$h(l(T)) = w(T) - v(T) + 1.$$

Therefore, $w(T) - v(T) + 1$ *is the minimum number of double points any alternating diagram of* $l(T)$ *can have.*

ADDED IN PROOF. The author proves the above conjecture in the forthcoming paper [**14**].

REFERENCES

1. C Bankwitz, *Über die Torsionszahlen der alternierenden Knoten*, Math. Ann. **103** (1930), 145–161.
2. R. H. Crowell, *Genus of alternating link types*, Ann. of Math. (2) **69** (1959), 258–275.
3. ——, *Non-alternating links*, Illinois J. Math. **3** (1959), 101–120.
4. D. Gabai, *The Murasugi sum is a natural geometric operation*, Contemp. Math. **20** (1983), 131–143.
5. V. Jones, *A polynomial invariant for knots via von Neumann algebras*, Bull. Amer. Math. Soc. (N.S.) **12** (1985), 103–111.
6. V. F. R. Jones and J. Birman, Seminar notes.
7. L.H. Kauffman, *Formal knot theory*, Math. Notes, no. 30, Princeton Univ. Press, Princeton, N.J., 1983.
8. W. B. R. Lickorish and K. C. Millett, *Polynomial invariant of oriented links* (to appear).
9. H. R. Morton, *The Jones polynomial for unoriented links* (to appear).
10. H. Murakami, *A recursive calculation of the* Arf *invariant of a link* (to appear).
11. K. Murasugi, *On the genus of the alternating knot.* I. II, J. Math. Soc. Japan **10** (1958), 94–105; ibid. **10** (1958), 235–248.
12. ——, *On a certain numerical invariant of link types*, Trans. Amer. Math. Soc. **117** (1965), 387–422.
13. ——, *On the Alexander polynomial of alternating algebraic knots*, J. Austral. Math. Soc. (A) **39** (1985), 317–333.
14. ——, *Jones polynomials and classical conjectures in knot theory* (to appear).

DEPARTMENT OF MATHEMATICS, UNIVERSITY OF TORONTO, TORONTO, CANADA, M5S 1A1

Topology Vol. 26, No. 2, pp. 187–194, 1987

JONES POLYNOMIALS AND CLASSICAL CONJECTURES IN KNOT THEORY

KUNIO MURASUGI

(*Received* 28 *January* 1986)

§1. INTRODUCTION AND MAIN THEOREMS

LET L be a tame link in S^3 and $V_L(t)$ the Jones polynomial of L defined in [2]. For a projection $\tilde{L}$ of L, $c(\tilde{L})$ denotes the number of double points in $\tilde{L}$ and $c(L)$ the minimum number of double points among all projections of L.

A link projection $\tilde{L}$ is called *proper* if $\tilde{L}$ does not contain "removable" double points like ⤫◌ or ⤫◌ .

In this paper, we will prove some of the outstanding classical conjectures due to P.G. Tait [7].

THEOREM A. (*P. G. Tait Conjecture*) *Two (connected and proper) alternating projections of an alternating link have the same number of double points.*

THEOREM B. *The minimal projection of an alternating link is alternating. In other words, an alternating link always has an alternating projection that has the minimum number of double points among all projections. Moreover, a non-alternating projection of a prime alternating link cannot be minimal.*

The primeness is necessary in the last statement of Theorem B, since the connected sum of two figure eight knots is alternating, but it has a minimal non-alternating projection. Note that the figure eight knot is amphicheiral.

Theorems A and B follow easily from Theorems 1–4 (stated below) which show strong connections between $c(\tilde{L})$ and the Jones polynomial $V_L(t)$.

Let $d_{\max} V_L(t)$ and $d_{\min} V_L(t)$ denote the maximal and minimal degrees of $V_L(t)$, respectively, and span $V_L(t) = d_{\max} V_L(t) - d_{\min} V_L(t)$.

THEOREM 1. *For any projection* $\tilde{L}$ *of a link* L,

$$\text{span } V_L(t) \le c(\tilde{L}) + \lambda - 1, \tag{1}$$

where λ *is the number of connected components of* $\tilde{L}$, *and therefore, if* L *has* λ *split components, then*

$$\text{span } V_L(t) \le c(L) + \lambda - 1. \tag{2}$$

If L is an alternating link, then we are able to prove the following:

* The work was done while the author was visiting at the University of Geneva, Switzerland. This research was partially supported by NSERC-No. A4034.

THEOREM 2. *If* $\tilde{L}$ *is a connected proper alternating projection of an alternating link* L, *then*

$$\text{span } V_L(t) = c(\tilde{L}). \tag{3}$$

(1) and (3) now yield that for any alternating link with λ split components,

$$\text{span } V_L(t) = c(L) + \lambda - 1. \tag{4}$$

If L is prime, we can prove the converse of Theorem 2. In fact, we have

THEOREM 3. *Let* L *be a prime link. Then for any non-alternating projection* $\tilde{L}$ *of* L,

$$\text{span } V_L(t) < c(\tilde{L}). \tag{5}$$

We should note that the primeness is necessary in Theorem 3, since the equality in (5) holds for a non-alternating projection of the square knot.

Using Theorems 2 and 3 we are able to give the complete characterization of links for which (4) holds.

THEOREM 4. *Let* L *be a non-split link. Then* (4) *holds for* L *if and only if* L *is the connected sum of alternating links.*

Besides Theorems A and B, these Theorems 1–4 yield several other consequences.

COROLLARY 5. *If a knot* K *is alternating and amphicheiral, then any proper alternating projection has an even number of double points.*

Proof. Let K^* be the mirror image of K. Then $V_{K^*}(t) = V_K(t^{-1})$, [2]. Since $K = K^*$, $V_K(t)$ is symmetric and hence, span $V_K(t)$ is even and Corollary 5 follows from Theorem 2.

COROLLARY 6. *If* L_1 *and* L_2 *are alternating links, then*

$$c(L_1 \# L_2) = c(L_1) + c(L_2),$$

where # *means the connected sum.*

This solves Problem 1 in [1] for alternating links.

Corollary 6 follows from (4) and Theorem 4 since

$$\text{span } V_{L_1 \# L_2}(t) = \text{span } V_{L_1}(t) + \text{span } V_{L_2}(t).$$

To state the final corollary, we define the twist number w at each double point v in a projection $\tilde{L}$ as indicated in Fig. 1.

Define $w(\tilde{L}) = \sum_{v \in L} w(v)$, where the summation is taken over all double points in $\tilde{L}$.

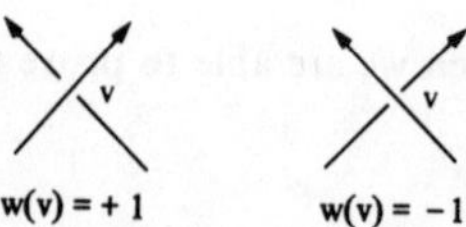

Fig. 1.

COROLLARY 7. *Let* $\tilde{L}_1$ *and* $\tilde{L}_2$ *be proper alternating projections of a special alternating link* L. *Then* $w(\tilde{L}_1) = w(\tilde{L}_2)$.

Proof. Since L is special alternating, any proper alternating projection must be proper special alternating [5]. Therefore, $w(\tilde{L}_i) = c(\tilde{L}_i)$ or $-c(\tilde{L}_i)$, $i = 1, 2$, and hence, $w(\tilde{L}_1) = w(\tilde{L}_2)$ or $-w(\tilde{L}_2)$. However, the sign of the signature of L is determined by the sign of $w(\tilde{L})$ of any special alternating projection $\tilde{L}$ of L, [5]. Therefore, $w(\tilde{L}_1) = w(\tilde{L}_2)$, q.e.d.

Corollary 7 shows that $w(\tilde{K})$ is a knot type invariant as long as we consider a special alternating knot K and its proper alternating projections. It is still not known whether or not $w(\tilde{K})$ is a knot type invariant when K is an alternating knot and $\tilde{K}$ a proper alternating projection.

Proofs of Theorems 1–4 will be given in the next two sections.

This work is inspired by the work of L. Kauffman [4] and conversations with C. Weber of the University of Geneva, to whom I would like to express my gratitude.

After submitting the paper, I learned that M. B. Thistlethwaite also obtained the same results using a completely different method [8].

Finally, I would like to thank the referee for his invaluable suggestions.

§2. PROOF OF THEOREM 1

We use the bracket polynomial $P_{\tilde{L}}(A)$ defined by L. Kauffman [4] rather than the original Jones polynomial.

For a projection $\tilde{L}$ of a link L, $P_{\tilde{L}}(A)$ is defined recursively by using the following fundamental identities (6)–(8):

(6) If $\tilde{L} = 0$, then $P_{\tilde{L}}(A) = 1$.

(7) If $\tilde{L}$ is the disjoint union of two link projections $\tilde{L}_1$ and $\tilde{L}_2$ then

$$P_{\tilde{L}}(A) = -(A^2 + A^{-2})P_{\tilde{L}_1}(A)P_{\tilde{L}_2}(A).$$

(8) If $\tilde{L}$, $\tilde{L}_0$ and $\tilde{L}_\infty$ are completely identical projections except at a small neighborhood of one crossing in $\tilde{L}$, where they are related by the diagrams below, then

$$P_{\tilde{L}}(A) = AP_{\tilde{L}_0}(A) + A^{-1}P_{\tilde{L}_\infty}(A).$$

It is shown that $P_{\tilde{L}}(A)$ is uniquely determined by these identities. However, $P_{\tilde{L}}(A)$ may be different from $P_{\tilde{L}'}(A)$ for another projection $\tilde{L}'$ of the same link. Nevertheless, $P_{\tilde{L}}(A)$ and the Jones polynomial $V_L(t)$ are related by the following formula [4]:

$$V_L(t) = (-t^{3/4})^{w(\tilde{L})}P_{\tilde{L}}(t^{-1/4}), \tag{9}$$

where $w(\tilde{L})$ is the integer defined in Section 1.

Since span $V_L(t) =$ span $P_{\tilde{L}}(t^{-1/4}) =$ span $P_{\tilde{L}}(t^{1/4}) = 1/4$ span $P_{\tilde{L}}(t) = 1/4$ span $P_{\tilde{L}}(A)$, to prove Theorem 1 it is sufficient to show the following:

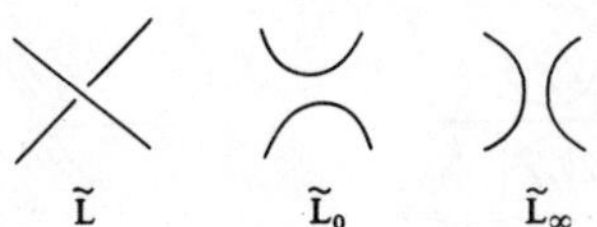

Fig. 2.

For any connected projection $\tilde{L}$ of a link L,

$$\text{span } P_{\tilde{L}}(A) \leq 4c(\tilde{L}). \tag{10}$$

Proof of (10). Let $\tilde{L}$ be a proper connected projection of L in S^2. $\tilde{L}$ divides S^2 into finitely many domains, which we will classify as shaded or unshaded. Let Γ be the graph of a link projection $\tilde{L}$ such that each vertex of Γ corresponds to an unshaded domain and each edge of Γ corresponds to a double point of $\tilde{L}$. We call an edge e of Γ positive or negative according to the diagrams shown in Fig. 3.

We should note that $\tilde{L}$ is alternating if and only if either all the edges of Γ are positive or all the edges are negative. A graph Γ is called *oriented* if every edge is either positive or negative.

Let p and n be the number of positive and negative edges in Γ, respectively, and hence $p+n = c(\tilde{L})$.

Now to evaluate $P_{\tilde{L}}(A)$, we have to *smooth* a double point on an edge e. For convenience, we call these smoothings *parallel* or *transverse* according to the diagrams shown in Fig. 4.

When we apply either a parallel or a transverse smoothing on every edge in Γ, we obtain a trivial link of several components, and $P_{\tilde{L}}(A)$ is the sum of the bracket polynomials of these links multiplied by A^k with some integer k. In order to obtain a more precise formula of $P_{\tilde{L}}(A)$, we will use the folllowing notation in the rest of the paper.

For an oriented graph Γ, we denote by Γ^* the dual graph of Γ which is oriented in such a way that an edge e^* in Γ^* is positive (or negative) if and only if e^* intersects a positive (or negative) edge in Γ.

Γ_+ and Γ_- denote, respectively, the subgraphs of Γ that consist of all positive edges and their end vertices, and of all negative edges and their end vertices.

For integers a and $b, 0 \leq a \leq p$ and $0 \leq b \leq n$, $\tilde{\mathscr{L}}_\Gamma(a, b)$ denotes the link projection obtained from the link projection with Γ as its oriented graph, by applying parallel smoothings on exactly a positive edges and b negative edges, and by applying transverse smoothings on $p-a$ positive edges and $n-b$ negative edges in Γ. ($\tilde{\mathscr{L}}_\Gamma(a, b)$ is called a *state* in [4].) $\tilde{\mathscr{L}}_\Gamma(a, b)$ is a trivial link. For each pair (a, b), there exist $\binom{p}{a}\binom{n}{b}$ such trivial links. Let $S(a, b)$ denote the collection of these links.

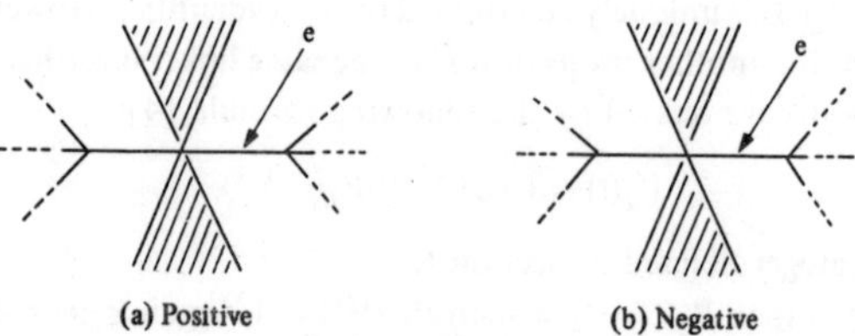

(a) Positive (b) Negative

Fig. 3.

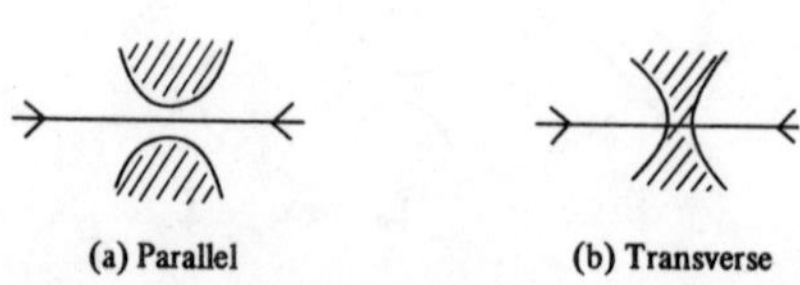

(a) Parallel (b) Transverse

Fig. 4.

Let $\mu_\Gamma(a, b)$ (or $\mu(a, b)$ when no confusion may occur) be the maximal number of components a link in $S(a, b)$ can have. For each of the following pairs $(a, b) = (0, 0)$, $(p, 0)$, $(0, n)$ and (p, n), $S(a, b)$ consists of only one link and we know that $\mu(0, 0)$ and $\mu(p, n)$ are, respectively, the number of vertices of Γ and that of Γ^*, and hence $\mu(0, 0) + \mu(p, n) - 2$ is the number of edges in Γ.

First we prove the following:

Lemma 1. *For any integers a and b, $0 \leq a \leq p$ and $0 \leq b \leq n$, $\mu(a, b) + a + b$ is an increasing function of a and b, and $\mu(a, b) - a - b$ is a decreasing function of a and b.*

Proof. If we change a transverse smoothing to a parallel smoothing, or vice versa, the number of components of the resulting trivial link increases or decreases by one. Therefore,

$$\begin{aligned} &|\mu(a+1, b) - \mu(a, b)| \leq 1, \quad \text{and} \\ &|\mu(a, b+1) - \mu(a, b)| \leq 1, \end{aligned} \tag{11}$$

and hence, we have

$$\begin{aligned} &(1)\ \mu(a+1, b) + 1 \geq \mu(a, b) \geq \mu(a+1, b) - 1, \\ &(2)\ \mu(a, b+1) + 1 \geq \mu(a, b) \geq \mu(a, b+1) - 1. \end{aligned} \tag{12}$$

An easy induction now gives us a proof of Lemma 1.

Using Lemma 1, we can see that

$$\begin{aligned} &(1)\ \mu(0, n) + n - b \geq \mu(0, b) \geq \mu(a, b) - a, \\ &(2)\ \mu(p, 0) + p - a \geq \mu(a, 0) \geq \mu(a, b) - b. \end{aligned} \tag{13}$$

Now let $b_i(\Gamma)$ denote the ith Betti number of a graph Γ as a 1-complex.

Lemma 2. *If Γ is the graph of $\tilde{L}$, then*

$$\begin{aligned} &(1)\ \mu(p, 0) = b_1(\Gamma_+) + b_1(\Gamma^*_-) + 1, \\ &(2)\ \mu(0, n) = b_1(\Gamma_-) + b_1(\Gamma^*_+) + 1, \end{aligned} \tag{14}$$

and hence

$$(3)\ \mu(p, 0) + \mu(0, n) \leq c(\tilde{L}) + 2.$$

Proof. Since $\mu(p, 0)$ and $\mu(0, n)$ are dual, it may suffice to prove (14) (1).

To compute $\mu(p, 0)$, we first remove all negative edges (but no vertices) from Γ, and denote by Γ_0 the resulting (possibly disconnected) planar subgraph of Γ. Then the boundary of a regular neighborhood of Γ_0 in S^2 is exactly the trivial link $\tilde{\mathscr{L}}_\Gamma(p, 0)$. Therefore, $\mu(p, 0)$, the number of components of $\tilde{\mathscr{L}}_\Gamma(p, 0)$ is given by $b_1(\Gamma_0) + b_0(\Gamma_0)$. Since $b_i(\Gamma_0) = \text{rank } H_i(\Gamma_0)$, $i = 0,1$, we have $b_1(\Gamma_0) = b_1(\Gamma_+)$, and $b_0(\Gamma_0) = b_1(S^2 - \Gamma_0) + 1$ yields $b_0(\Gamma_0) = b_1(\Gamma^*_-) + 1$. This proves (14) (1).

To show (14) (3), note that $\mu(p, n)$ is the number of boundary components of a regular neighborhood of $\Gamma(= \Gamma_+ \cup \Gamma_-)$ in S^2. Since $b_1(\Gamma_+ \cap \Gamma_-) = 0$, we have $b_1(\Gamma_+) + b_1(\Gamma_-) \leq b_1(\Gamma)$. Furthermore, since Γ is connected and $\mu(p, n) = b_1(\Gamma) + b_0(\Gamma)$, it follows that $b_1(\Gamma_+) + b_1(\Gamma_-) + 1 \leq \mu(p, n)$. Similarly, $b_1(\Gamma^*_+) + b_1(\Gamma^*_-) + 1 \leq \mu(0, 0)$, and therefore, $\mu(p, 0) + \mu(0, n) \leq \mu(p, n) + \mu(0, 0) = c(\tilde{L}) + 2$. This proves Lemma 2.

We now return to the proof of (10). Repeated applications of (6)–(8) give us the following:

$$P_{\tilde{L}}(A) = \sum_{\tilde{\mathscr{L}}} A^{-a+(p-a)} A^{b-(n-b)} \{-(A^2+A^{-2})\}^{|\tilde{\mathscr{L}}|-1}, \tag{15}$$

where the summation runs over all trivial link diagrams $\tilde{\mathscr{L}}$ in $S(a, b)$, and $0 \leq a \leq p$ and $0 \leq b \leq n$, and $|\tilde{\mathscr{L}}|$ denotes the number of components of $\tilde{\mathscr{L}}$. Since $|\tilde{\mathscr{L}}| \leq \mu(a, b)$, by definition, in (15), we see that

$$d_{\max} P_{\tilde{L}}(A) \leq \max_{a,b} \{p-n-2a+2b+2\mu(a, b)-2\} \quad \text{and}$$

$$d_{\min} P_{\tilde{L}}(A) \geq \min_{a,b} \{p-n-2a+2b-2\mu(a, b)+2\}. \tag{16}$$

However, (13) shows that

$$\begin{aligned} 2\mu(a, b)-2a+2b-2n &\leq 2\mu(0, n), \quad \text{and} \\ -2\mu(a, b)-2a+2b &\geq -2\mu(p, 0)-2p, \end{aligned}$$

and hence

$$\begin{aligned} d_{\max} P_{\tilde{L}}(A) &\leq p+n+2\mu(0, n)-2, \quad \text{and} \\ d_{\min} P_{\tilde{L}}(A) &\geq -p-n-2\mu(p, 0)+2. \end{aligned} \tag{17}$$

Using (14) (3), we obtain finally

$$\begin{aligned} \text{span } P_{\tilde{L}}(A) &\leq 2(p+n)+2\mu(0, n)+2\mu(p, 0)-4 \\ &\leq 2c(\tilde{L})+2\{c(\tilde{L})+2\}-4 \\ &= 4c(\tilde{L}). \end{aligned} \tag{18}$$

This proves (10) and the proof of Theorem 1 is now complete.

§3. PROOFS OF THEOREMS 2–4

We will use the same notation used in Section 2.

Proof of Theorem 2. Let L be a non-split alternating link and $\tilde{L}$ a proper connected alternating projection. We may assume without loss of generality that all edges of the graph Γ of $\tilde{L}$ are positive. Therefore, $\Gamma_+ = \Gamma$ and $\Gamma_- = \phi$, and hence, $p = c(\tilde{L})$ and $n = 0$ (and $b = 0$). Now to prove Theorem 2 we must show that

$$\begin{aligned} &(1)\ d_{\max} P_{\tilde{L}}(A) = p+2\mu(0, 0)-2, \quad \text{and} \\ &(2)\ d_{\min} P_{\tilde{L}}(A) = -p-2\mu(p, 0)+2. \end{aligned} \tag{19}$$

However, to prove (19), it suffices to show that each of $A^{p+2\mu(0,0)-2}$ and $A^{-p-2\mu(p,0)+2}$ appears exactly once in the summation (15). [See (17).] In other words, it is enough to show that

$$\begin{aligned} &(1)\ \mu(0, 0) > \mu(a, 0)-a, && \text{for} \quad 0 < a \leq p, \quad \text{and} \\ &(2)\ \mu(p, 0)+p > \mu(a, 0)+a, && \text{for} \quad 0 \leq a < p. \end{aligned} \tag{20}$$

Now we know that $\mu(0, 0) = \nu(\Gamma)$, the number of vertices in Γ, and $\mu(p, 0) = \nu(\Gamma^*)$ the number of vertices in Γ^*, and further since $\tilde{L}$ is proper, we see easily from the definition that $\mu(1, 0) = \nu(\Gamma)-1$ and hence trivially $\mu(0, 0) > \mu(1, 0)-1$. Since $\mu(a, 0)-a$ is a decreasing function of a (Lemma 1), it follows that $\mu(0,0) > \mu(1,0)-1 \geq \mu(a,0)-a$, for $0 < a \leq p$. This proves (20) (1).

Similarly, since $\tilde{L}$ is proper, $\mu(p,0) = \mu(p-1,0)+1$ and trivially $\mu(p,0)+1 > \mu(p-1,0)$. Since $\mu(a,0)+a$ is an increasing function of a, it follows that $\mu(p,0)+p > \mu(p-1,0)+p-1 \geqq \mu(a,0)+a$, for $0 \leq a < p$. This proves (20) (2), and the proof of Theorem 2 is complete.

Remark. We actually proved that for an alternating link L, the coefficients of the terms of $V_L(t)$ of maximal and minimal degrees are $+1$ or -1. (See [8].)

Another proof of Theorem 2 is also given in [5].

Proof of Theorem 3. We may assume that $\tilde{L}$ is connected and proper. Let Γ be the graph of $\tilde{L}$. If Γ has a cut vertex, then Γ is the one-point union of two subgraphs Γ_1 and Γ_2. Let $\tilde{L}_1$ and $\tilde{L}_2$ be the link projections whose graphs are Γ_1 and Γ_2, respectively. Then L is the connected sum of two links L_1 and L_2 whose projections are $\tilde{L}_1$ and $\tilde{L}_2$, respectively. Since L is prime, one of L_i, say L_1, is unknotted, and hence, $V_L(t) = V_{L_2}(t)$. Therefore, Theorem 1 yields that span $V_L(t) =$ span $V_{L_2}(t) \leq c(\tilde{L}_2) < c(\tilde{L}_1)+c(\tilde{L}_2) = c(\tilde{L})$, since $c(\tilde{L}_1) \neq 0$. Therefore, we may assume that Γ has no cut vertices.

Now suppose span $V_L(t) = c(\tilde{L})$. Then span $P_{\tilde{L}}(A) = 4c(\tilde{L})$, and hence, as we have seen in the proof of Theorem 1, we must have the equality in (14) (3). Therefore, $b_1(\Gamma_+)+b_1(\Gamma_-)+1 = \mu(p,n)$ and $b_1(\Gamma^*_+)+b_1(\Gamma^*_-)+1 = \mu(0,0)$. However, the first equality (and hence the second equality as well) holds if and only if $b_1(\Gamma_+)+b_1(\Gamma_-) = b_1(\Gamma)$. This is possible only if Γ is a positive or negative graph. Therefore, $\tilde{L}$ must be an alternating projection. This proves Theorem 3.

Proof of Theorem 4. If a non-split link L is the connected sum of alternating links L_i, $i = 1, 2, \ldots, k$, then L has a connected proper projection $\tilde{L}$ and each L_i has a connected proper alternating projection $\tilde{L}_i$ such that

$$c(\tilde{L}) = \sum_{i=1}^{k} c(\tilde{L}_i).$$

Since

$$\text{span } P_{\tilde{L}}(A) = \sum_{i=1}^{k} \text{span } P_{\tilde{L}_i}(A),$$

it follows from (3) that

$$\text{span } V_L(t) = \sum_{i=1}^{k} \text{span } V_{L_i}(t) = \sum_{i=1}^{k} c(\tilde{L}_i) = c(\tilde{L}).$$

Conversely, if span $P_{\tilde{L}}(A) = 4c(\tilde{L})$ for some connected proper projection $\tilde{L}$ of a link L, then, as we have seen in the proof of Theorem 3, the equality in (14) (3) must hold; therefore, Γ is either a positive or negative graph, or Γ has cut vertices $v_1, \ldots, v_r$ which separate Γ into positive and/or negative graphs. Therefore, L is the connected sum of (positive or negative) alternating links. This proves Theorem 4.

REFERENCES

1. J. C. HAUSMANN (ed.): Knot theory, *Proceedings Plans-sur-Bex*, Switzerland, *Springer Lecture Notes in Mathematics* **685** (1977).
2. V. F. R. JONES: A polynomial invariant for knots via von Neumann algebras, *Bull. AMS.* **89** (1985), 103–111.

3. L. H. KAUFFMAN: Formal knot theory, Lecture Notes **30**, Princeton University Press, Princeton, 1983.
4. L. H. KAUFFMAN: State models for knot polynomials, to appear.
5. K. MURASUGI: On a certain numerical invariant of link types, *Trans. AMS.* **117** (1965), 387–422.
6. K. MURASUGI: Jones polynomials of alternating links, *Trans. AMS.* **295** (1986), 147–174.
7. P. G. TAIT: On knots, *Scientific paper I*, 273–347, Cambridge University Press, London, 1898.
8. M. B. THISTLETHWAITE: A spanning tree expansion of the Jones polynomial, to appear.

University of Toronto,
Toronto, Canada,
M5S1A1.

Invent. math. 84, 563–573 (1986)

Inventiones mathematicae

A polynomial invariant for unoriented knots and links

Robert D. Brandt[1], W.B.R. Lickorish[3], and Kenneth C. Millett[1,2]*

[1] University of California, Santa Barbara, CA-93106, USA
[2] MSRI, Berkeley, CA 94720, USA
[3] University of Cambridge, 16 Mill Lane, Cambridge CB2 1SB, GB

§ 1. Introduction

The recent discovery by V.F.R. Jones [J] of a link polynomial, which complements the classical Alexander polynomial, has been generalized by the discovery of a new two-variable Laurent polynomial $P(L)$ associated with an oriented classical link L (see [L-M] and [F-Y-H-L-M-O]). This polynomial is calculated by a recursive formula involving changing crossings in a presentation of the link until the unlink is obtained. Here yet another (Laurent one-variable) polynomial $Q(L)$ will be associated to an *unoriented* link L in a manner that is spiritually very close to the way of defining the above mentioned two-variable polynomial. These polynomials are, however, different in the sense that there are pairs of knots distinguished by one polynomial but not the other. The theorem to be proved in this paper is as follows:

Theorem. *There is a unique function Q from the set of unoriented links of S^1's in S^3 to $\mathbf{Z}[x^{\pm 1}]$ such that:*

(i) *$Q(L)$ depends only on the isotopy class of L;*

(ii) *If U is the unknot, then $Q(U)=1$;*

(iii) *If L_+, L_-, L_0, and L_∞ are links that are identical except near one point where they are as in Fig. 1,*

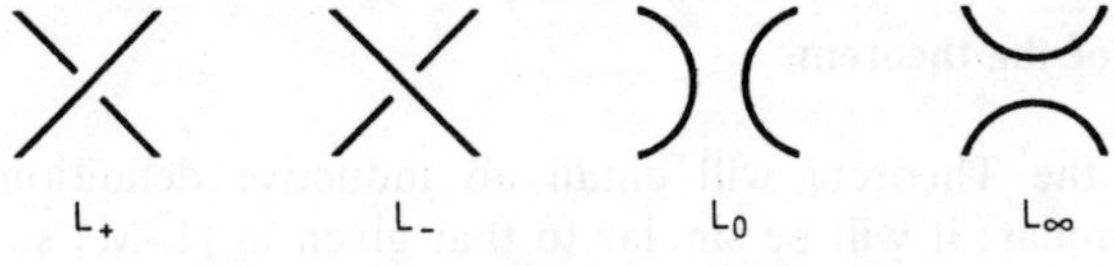

Fig. 1

then

$$Q(L_+)+Q(L_-)=x(Q(L_0)+Q(L_\infty)). \qquad (*)$$

* Partially supported by national science grant No. DMS-8503733

The lack of orientation makes it unclear which diagram in Fig. 1 should be labeled L_+ and which L_-, with similar confusion existing between L_0 and L_∞. However the symmetry of the formula neutralises this ambiguity. As in [L-M] it is easy to show that if such a function exists it is unique, so the proof of the theorem, which constitutes the next section, concentrates entirely on existence. Similarly it is easy to deduce that, if U^c denotes the unlink of c components, then $Q(U^c)=\mu^{c-1}$ where $\mu=2x^{-1}-1$.

Some properties of Q will be proved in §3. In particular, for any link L, $Q(L)(1)=1$, $Q(L)(-1)=(-3)^d$, $Q(L)(-2)=(-2)^{c-1}$, and $Q(L)(2)=(\delta_L)^2$, where c is the number of components of L, d is the dimension of the mod 3 homology of the double cover of S^3 branched along L, and δ_L is the determinant of L. These last results do at least correlate $Q(L)$ with other known invariants of L.

The basic structure of linear skein theory carries over to $Q(L)$ with some obvious modifications. Many of the examples in [L-M] involved 2×2 matrices whereas their replacement in the new theory requires 3×3 matrices, the results on the numerators and denominators and on rational knots are thus more complicated. It is intended to give a brief survey of these formulae and a more comprehensive table of polynomials in another paper. The following results are, however, elementary consequences of the theorem or of linear skein theory.

Property 1. (a) $Q(L_1 \# L_2)=Q(L_1)Q(L_2)$ *where* $L_1 \# L_2$ *denotes any 'connected' sum of* L_1 *and* L_2.

(b) $Q(L_1 \cup L_2)=\mu Q(L_1)Q(L_2)$ *where* $L_1 \cup L_2$ *is the 'distant' union of* L_1 *and* L_2.

(c) $Q(L)=Q(\bar{L})$ *where* $\bar{L}$ *is the mirror image of* L.

(d) *If* L_2 *is a mutant of* L_1 (see [L-M]), *then* $Q(L_1)=Q(L_2)$.

This new polynomial *seems* to be the only new polynomial that can be defined by a perturbation of the ideas and methods of [L-M]. Other ideas for a new polynomial have usually turned out to give the original two-variable polynomial modified in some way by knowledge of the number of components of the link. The fact that 'L_∞' occurs in the definition of Q does suggest that Q might be independent of the two-variable polynomial, and an example will be given to show that that is indeed so.

An announcement has been made by C.F. Ho of his independent discovery of the new polynomial [H].

§2. The proof of the theorem

The proof of the Theorem will entail an inductive definition of Q and a sequence of lemmas; it will be similar to that given in [L-M] so emphasis will be given to points at which the proofs differ.

Let $\mathscr{L}_n$ denote the set of all (generic) projections, with at most n crossings, of links equipped with an ordering $(c_1, c_2, c_3, \ldots)$ of their components and with a basepoint b_i and an orientation for each component c_i. This information induces an ordering on all the points of $L \in \mathscr{L}_n$; the ordering begins at the basepoint b_1 of c_1, proceeds along all of c_1 in the given direction and then transfers to the basepoint b_2 of c_2, and so on. If $L \in \mathscr{L}_n$, the *standard ascending projection* $\alpha L \in \mathscr{L}_n$ associated to L is the projection formed by switching the crossings of L (from over to under or vice versa) so that, proceeding along L in the above ordering of points, each crossing is first encountered as an under-crossing.

If a link is regarded as a subset of $\mathbf{R}^2 \times \mathbf{R}$, the projection $\mathbf{R}^2 \times \mathbf{R} \to \mathbf{R}$ defines a height function on the link. Here links are being regarded as subsets of the plane that are images of immersed 1-manifolds with over-crossing information recorded. Height functions will now be used on such link projections, these being two-valued functions at the cross-over points with the obvious definition of continuity.

Definition. Let $L \in \mathscr{L}_n$. A continuous function $h: L \to \mathbf{R}$ is an *untying function* if:

(i) $y_i \in c_i$, $y_j \in c_j$ and $i<j$ implies that $h(y_i)<h(y_j)$;

(ii) On each c_i the function h is monotonically increasing from the basepoint b_i to some (top)point t_i, and is monotonically decreasing from t_i to b_i;

(iii) At a cross-over point the value of h at the over-pass exceeds that at the under-pass.

Note. (a) Any standard ascending projection has an untying function.

(b) If L has an untying function, it represents an unlink.

Inductive Hypothesis $(n-1)$. Suppose that a function

$$Q: \mathscr{L}_{n-1} \to \mathbf{Z}[x^{\pm 1}]$$

has been defined such that:

(i) Q is independent of choice of basepoints, of ordering of components and of orientations;

(ii) Q is invariant under Reidemeister moves that do not increase the number of crossings beyond $(n-1)$;

(iii) If $L_+ \in \mathscr{L}_{n-1}$, then with the usual notation,

$$Q(L_+)+Q(L_-)=x(Q(L_0)+Q(L_\infty));$$

(iv) If $L \in \mathscr{L}_{n-1}$ and L has an untying function, then $Q(L)=\mu^{c-1}$, where c is the number of components of L and $\mu=2x^{-1}-1$.

Of course any element of $\mathscr{L}_0$ has an untying function, so that Q is *defined* on $\mathscr{L}_0$ by (iv), and this starts the induction.

Recursive Definition (n). If $L \in \mathscr{L}_n$ define $Q(\alpha L)$ to be μ^{c-1} where L has c components. If L and αL differ at at most $(r-1)$ crossings assume that $Q(L)$ has been defined. If now they differ at r crossings let $Q(L)$ be the polynomial calculated by applying the formula $(*)$,

$$Q(L_+)+Q(L_-)=x(Q(L_0)+Q(L_\infty))$$

to the first such crossing encountered in proceeding along L from the basepoint b_1. Note that although here L_0 and L_∞ are not well defined as elements of $\mathscr{L}_{n-1}$ since order, orientations, and basepoints are missing, $Q(L_0)$ and $Q(L_\infty)$ are well defined (by induction on n). By convention, L_0 will be chosen to denote the link obtained by nullifying the relevant crossing of L_+ (or L_-) in the manner *consistent* with orientations.

This now defines Q on $\mathscr{L}_n$, and it will now be checked, in a sequence of lemmas, that Q satisfies the induction hypothesis (n).

Lemma 1 (n). *Suppose $L \in \mathscr{L}_n$. Suppose that the crossings where L and αL differ are labelled in any sequence. The polynomial corresponding to L, calculated from $Q(\alpha L)$ by applying formula $(*)$ to the crossings in the sequence, is $Q(L)$.*

Proof. By induction on the number of crossing differences between L and αL, it is only necessary to consider the effect of exchanging the order of crossings labelled "i" and "j". Let $\sigma_i L$, $\eta_i^0 L$, and $\eta_i^\infty L$ be "L" with the crossing labelled "i" switched, and nullified in the two possible ways. Considering i followed by j, the polynomial calculated for L is $-Q(\sigma_i L)+x\{Q(\eta_i^0 L)+Q(\eta_i^\infty L)\}$, which, by changing crossing j, becomes equal to

$$Q(\sigma_j\sigma_i L)-x[Q(\eta_j^0\sigma_i L)+Q(\eta_j^\infty\sigma_i L)]+x\{-Q(\sigma_j\eta_i^0 L)+x[Q(\eta_j^0\eta_i^0 L)$$
$$+Q(\eta_j^\infty\eta_i^0 L)]-Q(\sigma_j\eta_i^\infty L)+x[Q(\eta_j^0\eta_i^\infty L)+Q(\eta_j^\infty\eta_i^\infty L)]\}.$$

As the σ, η^0, and the η^∞ operations all commute this formula is, by inspection, symmetric in i and j. Hence considering the two crossings in the other order gives the same polynomial for L. □

Lemma 2 (n). *$Q|\mathscr{L}_n$ is independent of choice of basepoints.*

Proof. We need only show that if a basepoint of a component lies on a segment of the projection it can be moved to an adjacent segment without changing the polynomial. Suppose the basepoint on component c_i is to be changed from position b_1 to position b_2, passing a crossing of c_i with c_j. Let L_1 and L_2 denote the relevant elements of $\mathscr{L}_n$ that have basepoints on c_i at b_1 and b_2, respectively, and are otherwise exactly the same.

Case $i \neq j$. Here αL_1 and αL_2 represent exactly the same projection though with different basepoints. Thus $Q(L_1)=Q(L_2)$ as, by Lemma 1(n), the choice of the order in which the sequence of crossing changes is accomplished does not change the polynomial.

Case $i=j$. In this case αL_1 and αL_2 differ only at the crossing under consideration where the associated crossing changes are labelled σ, η^0, and η^∞. By Lemma 1(n), $Q(L_1)$ can be calculated by first changing all other relevant crossings and thereby giving $Q(L_1)=f(Q(\sigma\alpha L_1))$, where f is some linear function. Similarly, $Q(L_2)=f(Q(\alpha L_2))$, with exactly the same function f occuring because the calculation involves only projections of fewer crossings for which (by induction) the position of the basepoints is irrelevant. By definition, $Q(\alpha L_1)$ and $Q(\alpha L_2)$ are equal to μ^{c-1} where c is the number of components. Furthermore $\eta^0\alpha L_1 \in \mathscr{L}_{n-1}$ is ascending and has $c+1$ components and therefore (by induction) has polynomial μ^c.

Now $\eta^\infty \alpha L_1$, the projection obtained from αL_1 by nullifying the crossing with a miss-match of orientations, has $(n-1)$ crossings and c components. But, removal of a neighborhood of the crossing from c_i splits c_i into two parts, and the untying function associated (by ascendingness) to αL_1 restricts to monotone functions on these two parts. These monotone functions combine to give an untying function for $\eta^\infty \alpha L_1$. Thus the induction implies that $Q(\eta^\infty \alpha L_1) = \mu^{c-1}$. Therefore

$$\begin{aligned} Q(\sigma\alpha L_1) &= -Q(\alpha L_1) + x(Q(\eta^0 \alpha L_1) + Q(\eta^\infty \alpha L_1)) \\ &= -\mu^{c-1} + x(\mu^c + \mu^{c-1}) = \mu^{c-1}. \end{aligned}$$

Direct substitution in "f" shows that $Q(L_1) = Q(L_2)$. □

Lemma 3(n). *$Q|\mathscr{L}_n$ satisfies formula* (∗).

Proof. Suppose, with the usual notation, L_+, L_-, L_0, and L_∞ are in $\mathscr{L}_n$. The formula

$$Q(L_+) + Q(L_-) = x(Q(L_0) + Q(L_\infty))$$

is the first step in a calculation (permitted by Lemma 1(n)) of either $Q(L_+)$ from $Q(\alpha L_+)$ or of $Q(L_-)$ from $Q(\alpha L_-)$, depending upon which of $L_\pm$ differs from $\alpha L_\pm$ at the crossing under consideration. □

Lemma 4(n). *Suppose that L is an element of $\mathscr{L}_n$ with c components that has an untying function h: $L \to \mathbf{R}$. Then $Q(L) = \mu^{c-1}$.*

Proof. The link projection L has components $\{c_i\}$ each with an orientation, a basepoint b_i and a top-point t_i. Suppose $v = \sum_1^c v_i$, where v_i is the number of self-crossings of c_i on the segment from t_i to b_i. If $v=0$ the result is true as L is then a standard ascending projection. Thus, inductively, assume the result true for any $L \in \mathscr{L}_n$ with an untying function and a lower value of v than the one given.

Let c_i be a component of L for which $v_i > 0$. Consider the first self-crossing X of c_i after t_i (in the direction specified by the orientation). If that crossing is an over-pass (as encountered proceeding from t_i) the function h may be changed to be increasing on the segment from t_i to X and just beyond X. This produces a smaller value of v, so that $Q(L) = \mu^{c-1}$ by induction on v.

Thus assume the segment from t_i encounters X as an under-pass. The fact that h decreases from t_i to b_i implies that this will be an underpassing of the segment from b_i to t_i. Let σL, $\eta^0 L$, and $\eta^\infty L$ be L with crossing X switched or nullified in each of the two possible ways. The situation is depicted in Fig. 2(a) where L is shown with an unbroken line where h is increasing and a broken line where h is decreasing. In Fig. 2(b) σL is shown and, reasoning as in the preceding paragraph shows that $Q(\sigma L) = \mu^{c-1}$. As shown in Fig. 2(c), $\eta^0 L$ has components c_i' and c_i'' (containing b_i and t_i, respectively) in place of c_i, and c_i'' always crosses c_i' as an over-pass. The function h can be adjusted, as indicated by the broken lines, to be an untying function for $\eta^0 L$ so that, by induction on n, $Q(\eta^0 L) = \mu^c$. Finally, h can be adjusted, as shown in Fig. 2(d), to be an

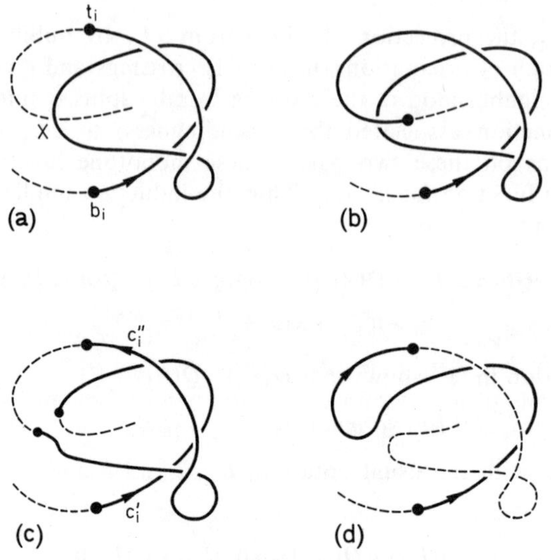

Fig. 2

untying function of $\eta^\infty L$ so that $Q(\eta^\infty L) = \mu^{c-1}$. Thus, using Lemma 3($n$),

$$\begin{aligned} Q(L) &= -Q(\sigma L) + x[Q(\eta^0 L) + Q(\eta^\infty L)] \\ &= -\mu^{c-1} + x[\mu^c + \mu^{c-1}] = \mu^{c-1}. \end{aligned}$$

This completes the induction on v, and hence establishes the lemma. □

Corollary 4.1(n). *$Q|\mathscr{L}_n$ is independent of choice of orientations of components.*

Proof. Let $L \in \mathscr{L}_n$ and let L' be L with the orientation of one component c_i reversed. Let βL be $\alpha L'$ with the ith orientation changed back again. But βL clearly has an untying function, so $Q(\beta L) \equiv \mu^{c-1}$.

The calculation of $Q(L')$ from $Q(\alpha L') \equiv \mu^{c-1}$ is the same as that for $Q(L)$ from $Q(\beta L) \equiv \mu^{c-1}$, so that $Q(L') = Q(L)$. □

Lemma 5(n). *$Q(L)$ is invariant under Reidemeister moves that do not increase the number of crossings beyond n.*

Proof. The proof of Proposition 4(n) of [L-M] translates immediately to give a proof of this lemma. □

Proof of the Theorem. The theorem is proved once the induction hypothesis (n) is established. It only remains to prove that $Q(L)$ is independent of choice of ordering for $L \in \mathscr{L}_n$. This is not obvious. The proof is exactly Propositions 5(n) and 6(n) of [L-M] with the symbols "Q" and "L" replacing "P" and "K" throughout. □

Remark (J.H. Conway). Propositions 5(n) and 6(n) of [L-M] can be thought of as proving that any ascending element of $\mathscr{L}_n$ can be changed to an element of $\mathscr{L}_0$ by a sequence of basepoint changes and Reidemeister moves that do not increase the number of crossings beyond n, provided the move of Fig. 3 is included.

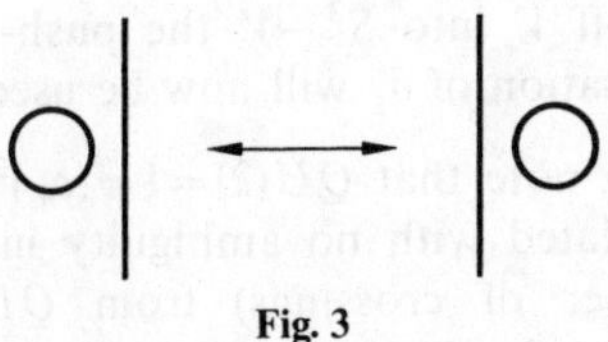

Fig. 3

§ 3. Basic properties of the polynomials

The statement of Property 1 in the introduction recorded the behaviour of the new polynomial under the link operations of connected sum, distant union, reflection and mutation. It follows that the polynomial of the c-component unlink is μ^{c-1} where $\mu = 2x^{-1} - 1$. Further properties of an elementary nature are listed below.

Property 2. *For any link* L, $Q(L) - 1$ *is divisible by* $2(x-1)$.

Proof. Using induction, on the number of crossings in a presentation and the number of crossings that must be changed to achieve the ascending configuration, this follows from

$$Q(L_+) - 1 + Q(L_-) - 1 = x\{Q(L_0) - 1 + Q(L_\infty) - 1\} + 2(x-1). \quad \square$$

Corollary. (i) $QL(1) = 1$.
(ii) *In* $Q(L)$ *the constant term is odd, and all the other coefficients are even.*

Property 3. *If* L *has* c *components,* $QL(-2) = (-2)^{c-1}$.

Proof. The method of proof is a similar double induction using the formula (*). $\square$

Property 4. *For any link* L, $QL(2) = (\delta_L)^2$ *where* δ_L *is the determinant of* L.

A little discussion is in order before proving this result. The determinant of L, δ_L, is a classical invariant of L which may be defined to be the modulus of the Alexander polynomial of L evaluated at -1. Then δ_L is the order of the first homology group of the double cover of S^3 branched over L (that group being infinite when δ_L is zero). This is attractively independent of any choice of orientation.

Because the theory of Q is allergic to orientation it is desirable to work with a definition of δ_L that is comparitively free of orientations. Such an approach was classically given by the fact that δ_L is the modulus of the determinant of the Goeritz matrix associated to any presentation of L, [G]. A new interpretation of the Goeritz method was given by C.McA. Gordon and R.A. Litherland [G-L]. To any (maybe unorientable) connected surface V spanning L they associate a symmetric bilinear form $\mathcal{G}_V\colon H_1(V) \times H_1(V) \to \mathbf{Z}$. The absolute value of the determinant of this bilinear form is an invariant of the link; it is δ_L, for, by judicious choices of V and a base of $H_1(V)$, $\mathcal{G}_V$ is represented by a Goeritz matrix. Suppose that α, $\beta \in H_1(V)$ are represented by 1-cycles a, b. Then $\mathcal{G}_V(\alpha, \beta)$ is defined to be the linking number of a and τb, where τb is $2b$ pushed off V into $S^3 - V$, the push-off being locally to both sides of V. This characterisation of δ_L will now be used.

Proof of Property 4. First note that $QU(2)=1=(\delta_U)^2$ where U is the unknot. Now $QL(2)$ can be calculated with no ambiguity in the usual recursive way (inducting on the number of crossings) from $QL_+(2)+QL_-(2)=2(QL_0(2)+QL_\infty(2))$, where $QU(2)=1$ for U the unknot. Here $L_\pm$, L_0 and L_∞ are links with projections related in the usual way. Thus the property follows at once if it can be established that $(\delta_{L_+})^2+(\delta_{L_-})^2=2((\delta_{L_0})^2+(\delta_{L_\infty})^2)$. Figure 4 shows the links L_i for $i=+, -, 0, \infty$, each with a connected spanning surface V_i shaded in.

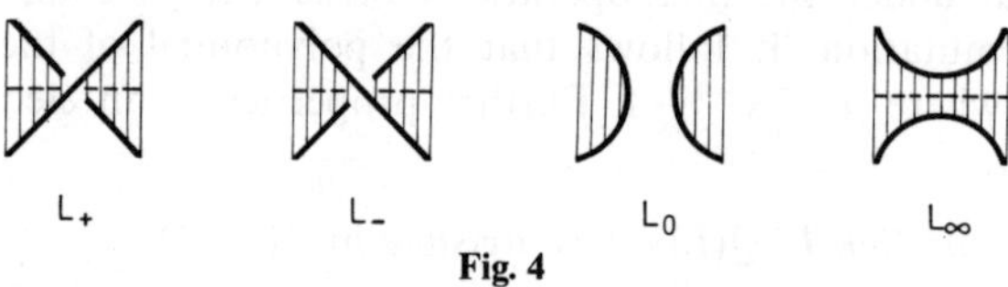

Fig. 4

These surfaces can be constructed by using Seifert's circuit method for V_0, connecting up components with hollow handles away from the area depicted, and modifying as shown to obtain V_+, V_-, and V_∞. Thus all but V_∞ could be taken to be orientable. The four surfaces are identical outside the area shown. Take a base for $H_1(V_0)$ represented by oriented (simple) closed curves on V_0 and for bases of $H_1(V_i)$, $i=+, -, \infty$, take the homology classes of the dotted curves in the diagrams in Fig. 4 (the ends of which join up outside the diagram) and the set of curves already chosen on V_0. The matrices M_i that represent the $\mathscr{G}_{V_i}$ with respect to these bases have the form

$$M_\infty = \begin{bmatrix} n & \rho \\ \rho^\tau & M_0 \end{bmatrix}, \quad M_\pm = \begin{bmatrix} n\mp 1 & \rho \\ \rho^\tau & M_0 \end{bmatrix}.$$

Thus $\det M_\pm = \det M_\infty \mp \det M_0$, and so squaring and adding one obtains

$$(\det M_+)^2 + (\det M_-)^2 = 2((\det M_\infty)^2 + (\det M_0)^2).$$

Recollection that $\delta_{L_i} = |\det M_i|$ yields the required formula. □

The next result was conjectured by J.H. Conway.

Property 5. *$QL(-1)=(-3)^d$, where $d=\dim H_1(D;\mathbf{Z}_3)$ and D is the double cover of S^3 branched over L.*

Proof. In the notation of the previous proof, let D_i be the double cover of S^3 branched over L_i and let $d_i=\dim H_1(D_i;\mathbf{Z}_3)$. $H_1(D_i;\mathbf{Z}_3)$ is presented by $\hat{M}_i$, the matrix M_i reduced modulo 3, so that d_i is the nullity of $\hat{M}_i$. The form of the matrices $\hat{M}_i$ for $i=+, -, 0$, and ∞ shows that three of their nullities take a common value a, the fourth is $a+1$. Hence $(-3)^{d_+}+(-3)^{d_-}+(-3)^{d_0}+(-3)^{d_\infty}=0$, and the usual induction argument completes the proof. □

Remark. The determinant δ_L can be calculated, up to sign, as $\det(A+A^\tau)$ where A is a Seifert matrix for L. The Alexander polynomial is $\det(t^{1/2}A - t^{-1/2}A^\tau)$. It is natural to wonder whether, in the light of the above, $Q(L)$ could be defined from some slight generalisation of $\mathcal{G}_V$.

Property 6. *If L has c components the lowest power of x in $Q(L)$ is precisely $(1-c)$.*

Proof. The usual double induction procedure shows that $(1-c)$ is a lower bound for the powers of x in $Q(L)$. Thus, by the Corollary to Property 2, this result is true when $c=1$. Now induction, on c and on the number of crossing changes necessary to change L to the distant union of its components, produces the result in general. □

Property 7. *If L has c components,*

$$[x^{c-1}Q(L)]_{x=0}=[(-m)^{c-1}P(L)]_{(l,m)=(1,0)}$$

where $P(L)$ is the two-variable polynomial in l and m defined in [L-M].

Proof. The result is trivial if L is the unlink. The substitution $(l,m)=(1,-y)$ changes $P(L)$ to a Laurent polynomial in y defined in the usual way by $P(L_+)+P(L_-)=yP(L_0)$ and $P(U)=1$. Now $x^{c-1}Q(L)$ and $y^{c-1}P(L)$ are genuine polynomials with no negative powers of x and y. Consider the defining formulae for Q and P when $L_\pm$ has c components:

Case (i). L_0 has $c+1$ components and L_∞ has c components.

$$x^{c-1}QL_+ + x^{c-1}QL_- = x^cQL_0 + x(x^{c-1}QL_\infty)$$

$$y^{c-1}PL_+ + y^{c-1}PL_- = y^cPL_0.$$

Case (ii). L_0 and L_∞ each have $(c-1)$ components.

$$x^{c-1}QL_+ + x^{c-1}QL_- = x^2(x^{c-2}QL_0 + x^{c-2}QL_\infty)$$

$$y^{c-1}PL_+ + y^{c-1}PL_- = y^2(y^{c-2}PL_0).$$

The result now follows, substituting $x=0=y$, from the usual induction on the number of crossings and the number of crossing differences from the standard ascending projection. □

Property 8. *The degree of $Q(L)$ is less than the crossing number of L.*

Proof. This follows from the usual double induction using formula (*). □

Examples. Employing the standard notation for knots and links having presentations with few crossings;

$$Q(8_8)=1+4x+6x^2-10x^3-14x^4+4x^5+8x^6+2x^7$$

$$Q(10_{129}\text{ reflected})=1-12x-2x^2+26x^3+4x^4-20x^5-4x^6+6x^7+2x^8$$

$$Q(13_{6714})=1+20x+14x^2-62x^3-40x^4+64x^5+38x^6$$
$$-26x^7-14x^8+4x^9+2x^{10}.$$

Calculations for such examples are best performed by computer; in these examples, however, human verification was also employed. The interest of the above example is that the three given knots have the same two-variable polynomial (see [L-M] Example 16). Thus the Q polynomial is *not* a function of the two-variable polynomial (and so certainly not just a function of the polynomials of Alexander and Jones).

Table. Values of the Q-polynomial

3_1	$-3+2+2$
4_1	$-3-2+4+2$
5_1	$5-2-6+2+2$
5_2	$1-4-2+4+2$
6_1	$1+4-6-4+4+2$
6_2	$5-2-10+0+6+2$
6_3	$5-6-12+4+8+2$
7_1	$-7+4+16-6-10+2+2$
7_2	$-3+6+8-10-6+4+2$
7_3	$-3+2+6-6-4+4+2$
7_4	$1+8-4-12+0+6+2$
7_5	$1+0-4-6+2+6+2$
7_6	$5+2-12-10+6+8+2$
7_7	$5+6-18-14+10+10+2$
8_1	$-3-6+14+12-14-8+4+2$
8_2	$-7+0+22+2-20-4+6+2$
8_3	$1-8+4+12-8-6+4+2$
8_4	$-3+2+14-2-16-2+6+2$
8_5	$-11+14+26-16-24+2+8+2$
8_6	$1-4+2+2-8+0+6+2$
8_7	$-7+4+20-8-20+2+8+2$
8_8	$1+4+6-10-14+4+8+2$
8_9	$-7+4+16-10-16+4+8+2$
8_{10}	$-11+14+22-22-22+8+10+2$
8_{11}	$-3+6+4-12-10+6+8+2$
8_{12}	$5+2-8-12-4+8+8+2$
8_{13}	$-3+10+10-22-16+10+10+2$
8_{14}	$1+8+0-22-10+12+10+2$
8_{15}	$-7+16+10-32-16+16+12+2$
8_{16}	$-3+10+18-22-30+8+16+4$
8_{17}	$-3+6+12-20-24+10+16+4$
8_{18}	$5+2+12-26-36+14+24+6$
8_{19}	$-11+10+20-10-12+2+2$
8_{20}	$-7+12+12-14-8+4+2$
8_{21}	$-7+8+6-12-2+6+2$
Conway-Kinoshita-Terasaka	$17-24-52+54+76-28-48-4+8+2$
$(-3,5,7)$-pretzel	$1+48-72-172+234+256-286-206+162+94-42-22+4+2$
untwisted double of trefoil	$17+8-180+134+556-618-872+978+818-736-470+284$ $+158-54-28+4+2$
2_1^2	$-2x^{-1}+1+2$
4_1^2	$2x^{-1}-1-4+2+2$
5_1^2	$2x^{-1}-1-8+0+6+2$
6_2^3	$4x^{-2}-4+1+0-16+0+12+4$

To give a feel for the nature of this new polynomial, a small table is given above in which the Laurent polynomial $\sum_{r}^{s} a_i x^i$ is written $a_{-r}x^{-r}+a_{-r+1}$ $+a_{-r+2}+\ldots+a_s$, with $a_0 x^0$ written a_0. The usual name of a knot or link preceeds the polynomial.

Direct calculation has also shown that the two knots of [L-M] Example 17, which have distinct signatures (and so are not mutants) have the same Q-polynomial, i.e. $-7+4x+16x^2-6x^3-10x^4+2x^5+2x^6$.

References

[F-Y-H-L-M-O] Freyd, P., Yetter, D., Hoste, J., Lickorish, W.B.R., Millett, K.C., Ocneanu, A.: A new polynomial invariant of knots and links. Bull. Am. Math. Soc. **12**, 239–246 (1985)

[G] Goeritz, L.: Knoten und quadratische Formen. Math. Z. **36**, 647–654 (1933)

[G-L] Gordon, C.McA., Litherland, R.A.: On the signature of a link. Invent. Math. **47**, 53–69 (1978)

[H] Ho, C.F.: A new polynomial for knots and links – preliminary report. Abstracts Am. Math. Soc. **6**, 4 (1985), 300, abstract 821-57-16

[J] Jones, V.F.R.: A polynomial invariant for knots via von Neumann algebras. Bull. Am. Math. Soc. **12**, 103–111 (1985)

[L-M] Lickorish, W.B.R., Millett, K.C.: A polynomial invariant of oriented links. Topology (to appear)

Math. Proc. Camb. Phil. Soc. (1987), **101**, 267

The 2-variable polynomial of cable knots

BY H. R. MORTON AND H. B. SHORT
Department of Pure Mathematics, University of Liverpool, Liverpool L69 3BX

(*Received* 24 *February* 1986)

Abstract

The 2-variable polynomial P_K of a satellite K is shown not to satisfy any formula, relating it to the polynomial of its companion and of the pattern, which is at all similar to the formulae for Alexander polynomials. Examples are given of various pairs of knots which can be distinguished by calculating P for 2-strand cables about them even though the knots themselves share the same P. Properties of a given knot such as braid index and amphicheirality, which may not be apparent from the knot's polynomial P, are shown in certain cases to be detectable from the polynomial of a 2-cable about the knot.

1. *Introduction*

With the development of the 2-variable polynomial P_K of an oriented knot K (of one or more components) [**5**], has come the desire to relate it directly to the geometry of the knot exterior, as Seifert, and later Fox, were able to do with the Alexander polynomial. One obvious place where the geometry should show up is when the knot is a satellite. In this case the exterior is made up of the union of two link exteriors, that of a *companion* and that of a *pattern link*, forming part of a natural decomposition of the exterior in the context of 3-manifolds.

The multi-variable Alexander polynomial of such a union bears a simple relation to those of the two constituent link exteriors. This takes the form of an equation $S_K = S_C S_R$, where K is a satellite with companion C and pattern link R consisting of an unknotted component defining a complementary solid torus V which contains the other component(s) of R. The knot K is formed from the image of these component(s) when a solid torus neighbourhood of C is replaced by V using a faithful (longitude-preserving) homeomorphism. In the equation S is either the multi-variable Alexander polynomial, or, in the case of a 1-component knot, a modified version of it; the variables, which correspond to homology classes in the appropriate link exterior, are each replaced by the element which they represent in the exterior of K. See [**2**] for a general, but not very readable, account.

Possible satellite formula

On the analogy of the Alexander polynomial, we expected some such satellite formula for the other polynomials. For each K a conjectural function S_K of several variables might exist, specializing to P_K in some way and satisfying an equation $S_K = S_C S_R \phi$ for a satellite K of C with pattern R, where the variables in S_C and S_R are substituted by others in a way which depends in some suitable sense only on the gluing of these two exteriors, and ϕ is a possible normalizing factor, again depending only on the gluing homeomorphism.

Under such a framework, if two knots C_1 and C_2 with $S_{C_1} = S_{C_2}$ were used with the same pattern and method of gluing to construct satellites K_1 and K_2 we would then have $S_{K_1} = S_{K_2}$.

Conceivably there may be an S_K which specialises to P_K and satisfies such a satallite formula (possibly specializing also to Kauffman's recently announced polynomial F_K [**8**]). In this paper we show however that no such function S_K can be found with the additional property that $S_K = P_K$ for a 1-component knot K, even if it differs from P_K when K has more components.

The examples given are pairs of knots C_1 and C_2 with $P_{C_1} = P_{C_2}$ whose (2, 1) cables K_1 and K_2 (satellites constructed with the same pattern $R =$ ( ) have $P_{K_1} \neq P_{K_2}$. These are described in a later section under the heading 'Birman's pairs of 3-braids' and the polynomials are displayed in Table 1.

2. *Scope of the calculations*

To compute the polynomial P for these examples we developed a program to calculate P_K from a presentation of K as a closed braid, based on the construction of Ocneanu and Jones [**13**, **6**]. Details of the theoretical basis for the calculations, and their practical implementation are given in [**12**].

Our program can handle interactive calculations for 7-string braids of over 100 crossings and will deal with 8-string braids of up to 150 crossings on the Liverpool University IBM 3083 computer using less than 16 megabytes storage and under 200 seconds of computer time. The time required grows relatively slowly (quadratically) with the number of crossings for braids of a given string index, so within the constraints of presentation as a closed braid on at most 8 strings the method provides an efficient way of handling knots with many crossings. It may be contrasted with Thistlethwaite's encyclopaedic work in producing tables of P for all knots up to 13 crossings based on the Conway recurrence relation [**14**]. His method works well, given information about all knots with fewer crossings, and is not particularly sensitive to braid index, but in general it would face exponential time growth with the number of crossings if calculation of P for an individual knot beyond the range of the table was required.

Annotated copies of the Pascal program to calculate the polynomial P and also the Alexander and Jones polynomials of a knot presented as a closed braid on at most eight strings are available on request.

3. *Further consequences*

While the failure of the satellite formula proves disappointing from the point of view of understanding the general structure of the polynomial P_K, it leaves the way open to using P in a second attempt at distinguishing two knots C_1 and C_2 with $P_{C_1} = P_{C_2}$ by comparing P on satellites of C_1 and C_2. From our limits on the computing power available, we have been restricted to considering 2-string cables where C_1 and

C_2 can themselves be presented as closed braids on at most four strings, so that the cables can be presented as 8-string braids. Other satellites such as doubles or higher string cables could in principle be used.

The only examples so far observed where the satellites are still indistinguishable using P have come when C_2 is a mutant of C_1. In all cases tried so far, if C_1 and C_2 are mutants then their 2-cables, although not apparently mutants themselves, have the same P. Examples are discussed in a later section.

Burau polynomials

The possibility of calculating P_K from the 'Burau polynomial' of a closed braid representative β of K, i.e. from the characteristic polynomial $\det(xI-\beta(t))$ of the Burau matrix $\beta(t)$ for β, was raised tentatively by Jones[**7**]. Our examples used in disproving the existence of a simple satellite formula can also serve to discount this possibility.

To see why this is so, suppose that C is a knot presented as the closure $\hat{\beta}$ of some $\beta \in B_n$. The *complete closure*, $\hat{\beta} \cup L_\beta$, consisting of $\hat{\beta}$ together with the braid axis L_β is then a link whose 2-variable Alexander polynomial is the Burau polynomial of β [**11**]. The $(2, r)$ cable K about C can be presented naturally as the closure of a $2n$-string braid, by doubling all strings in β, with twists as required. The complete closure of this braid is then a satellite constructed using the $(2, r)$ cable pattern R on the string C from the complete closure $C \cup L_\beta$ of β. Using the satellite formula for Alexander polynomials the Burau polynomial of this representation for K can then be calculated from that of $C \cup L_\beta$ and the pattern.

If this construction is applied to two braids β_1 and β_2 with the same Burau polynomial, the resulting braid presentations for the $(2, r)$ cables K_1 and K_2 will then have the same Burau polynomial. The examples of Birman given in tables 1 and 2 are $(2, r)$ cables about 3-braids with the same Burau polynomial. The resulting 6-braids have then the same Burau polynomials, but their closures K_1 and K_2 have $P_{K_1} \neq P_{K_2}$.

With the failure of the satellite formula, features of a knot K which do not show up directly from P_K may nevertheless become apparent from the polynomial of a 2-cable about K. These features include the *braid index* and the question of *amphicheirality* of K, and examples are discussed in the course of the next section.

4. *Discussion of examples*

Braids in the accompanying tables are listed as elements of B_8, using $\pm i$ to stand for the generator $\sigma_i^{\pm 1}$. The polynomial $P_K(v, z)$ is given as a matrix of coefficients (p_{ij}), where $P_K(v, z) = \Sigma p_{ij} z^i v^j$, with the range of i and j indicated at the side. We use the convention that $v^{-1}P_{K^+} - vP_{K^-} = zP_{K^0}$, where K^+, K^- and K^0 differ only as shown:

Putting $v = 1$ gives $\nabla_K(z)$, the Conway polynomial. The substitutions ($z = x - x^{-1}$, $v = 1$) and ($z = x - x^{-1}$, $v = x^2$) give respectively the Alexander and Jones polynomials, $\Delta(x^2)$ and $V(x^2)$. In the tables the negative powers of x in the Alexander polynomial have been omitted, because of its invariance when x is replaced by $-x^{-1}$.

Table 1

Braid representing the 2-cable K_1

*** -2-1-3-2(4354)^{7}167 ***

The polynomial P for K_1

24	26	28	30		Conway polynomial	Alexander polynomial	Jones polynomial
50	−90	45	−4	0	1 (0)	1 (0)	1 (24)
750	−1140	480	−45	2	45 (2)	−1 (6)	1 (28)
4445	−5383	1709	−111	4	660 (4)	1 (8)	1 (32)
14105	−13320	3002	−113	6	3674 (6)	−1 (14)	−1 (34)
27092	−19723	3003	−54	8	10318 (8)	1 (16)	1 (36)
33566	−18654	1820	−12	10	16720 (10)	−1 (22)	−1 (38)
27798	−11643	680	−1	12	16834 (12)	1 (24)	−1 (54)
15656	−4846	153		14	10963 (14)		−1 (58)
6004	−1330	19		16	4693 (16)		1 (60)
1541	−231	1		18	1311 (18)		−1 (62)
253	−23			20	230 (20)		1 (64)
24	−1			22	23 (22)		−1 (66)
1				24	1 (24)		1 (68)

Braid representing the 2-cable K_2

*** $(4354)^2(2132)^2(4354)^2(2132)^2 4354(2132)^{-3}167$ ***

The polynomial P for K_2

20	22	24	26	28	30		Conway polynomial	Alexander polynomial	Jones polynomial
−16	80	−110	70	−35	12	0	1 (0)	1 (0)	−1 (18)
−300	1220	−1210	420	−140	55	2	45 (2)	−1 (6)	1 (20)
−1874	7076	−5467	1029	−217	113	4	660 (4)	1 (8)	1 (22)
−5806	21558	−13365	1340	−166	113	6	3674 (6)	−1 (14)	1 (28)
−10363	39414	−19734	1013	−66	54	8	10318 (8)	1 (16)	1 (30)
−11452	46372	−18655	456	−13	12	10	16720 (10)	−1 (22)	−1 (34)
−8127	36484	−11643	120	−1	1	12	16834 (12)	1 (24)	1 (36)
−3757	19549	−4846	17			14	10963 (14)		−1 (38)
−1123	7145	−1330	1			16	4693 (16)		−1 (42)
−209	1751	−231				18	1311 (18)		1 (44)
−22	275	−23				20	230 (20)		−1 (46)
−1	25	−1				22	23 (22)		1 (48)
	1					24	1 (24)		−1 (54)
									−1 (62)
									1 (64)
									−1 (70)
									1 (72)

Table 2

Braid representing a knot C_1
*** 11122221211-2-2-2111234567 ***
Braid representing a knot C_2
*** 111222212111211-2-2-234567 ***

The polynomial P for C_1 and C_2

10	12	14	
5	−3	−1	0
31	−11	−4	2
52	−9	−4	4
35	−2	−1	6
10			8
1			10

Braid representing the (2, 1)-cable about C_1
*** $(2132)^3(4354)^4 21324354(2132)^2(4354)^{-3}(2132)^3 43541^{-23}67$ ***

The polynomial P for this cable

20	22	24	26	28	30		Conway polynomial	Alexander polynomial	Jones polynomial	
−149	407	−319	46	−3	19	0	1 (0)	11 (0)	−1 (4)	12 (58)
−4382	12124	−9880	2415	−1076	863	2	64 (2)	−10 (4)	1 (6)	−5 (60)
−50472	140079	−114785	34243	−21388	12963	4	640 (4)	7 (8)	−1 (8)	−5 (62)
−317466	886465	−716356	234730	−183608	98595	6	2360 (6)	−3 (12)	1 (10)	3 (64)
−1251252	3544185	−2787908	963172	−918054	453992	8	4135 (8)	1 (20)	−1 (12)	5 (66)
−3336727	9681064	−7332716	2624764	−3019960	1387543	10	3968 (10)		1 (14)	−4 (68)
−6321130	18982322	−13699429	5041481	−6971998	2971026	12	2272 (12)		−1 (16)	−10 (70)
−8788244	27628767	−18789643	7085256	−11748376	4613040	14	800 (14)		1 (18)	14 (72)
−9165353	30555075	−19345884	7466144	−14823467	5313655	16	170 (16)		1 (22)	−1 (74)
−7272714	26086910	−15171599	5991977	−14243411	4608857	18	20 (18)		2 (30)	−11 (76)
−4426196	17366982	−9138087	3695028	−10534332	3036606	20	1 (20)		−2 (34)	7 (78)
−2071246	9059325	−4238499	1755890	−6030207	1524737	22			−1 (36)	4 (80)
−742861	3702454	−1509149	641028	−2673527	582055	24			5 (38)	−8 (82)
−202189	1179084	−408411	178007	−913881	167390	26			−9 (42)	4 (84)
−40980	289138	−82424	36889	−238235	35612	28			5 (44)	1 (86)
−5986	53481	−12004	5520	−46439	5428	30			8 (46)	−3 (88)
−595	7210	−1191	563	−6547	560	32			−11 (48)	2 (90)
−36	668	−72	35	−630	35	34			−1 (50)	1 (92)
−1	38	−2	1	−37	1	36			13 (52)	−2 (94)
	1			−1		38			−9 (54)	1 (96)
									−5 (56)	

Braid representing the (2, 1)-cable about C_2

*** $(2132)^3(4354)^4 21324354 (2132)^3 4354 (2132)^2 (4354)^{-3} 1^{-23} 67$ ***

The polynomial P for this cable							Conway polynomial	Alexander polynomial	Jones polynomial	
20	22	24	26	28	30					
−165	487	−479	206	−83	35	0	1 (0)	11 (0)	−1 (4)	12 (58)
−4682	14144	−15040	8775	−4896	1763	2	64 (2)	−10 (4)	1 (6)	−5 (60)
−52346	156355	−164697	104655	−68114	24787	4	640 (4)	7 (8)	−1 (8)	−6 (62)
−323272	953193	−962106	626410	−464256	172391	6	2360 (6)	−3 (12)	1 (10)	4 (64)
−1261615	3708909	−3525854	2288638	−1930063	724120	8	4135 (8)	1 (20)	−1 (12)	4 (66)
−3348179	9947173	−8810295	5620926	−5434061	2028404	10	3968 (10)		1 (14)	−3 (68)
−6329257	19277235	−15776844	9817955	−11000795	4013978	12	2272 (12)		−1 (16)	−10 (70)
−8792001	27859220	−20907214	12641994	−16620167	5818968	14	800 (14)		1 (18)	14 (72)
−9166476	30683856	−20941807	12287546	−19189381	6326432	16	170 (16)		1 (22)	−1 (74)
−7272923	26138520	−16070255	9153211	−17183061	5234528	18	20 (18)		2 (30)	−11 (76)
−4426218	17381680	−9517127	5270758	−12031334	3322242	20	1 (20)		−2 (34)	8 (78)
−2071247	9062226	−4357579	2352590	−6606731	1620741	22			−1 (36)	3 (80)
−742861	3702831	−1536581	811317	−2840188	605482	24			4 (38)	−8 (82)
−202189	1179113	−412907	213996	−949436	171423	26			1 (40)	4 (84)
−40980	289139	−82920	42346	−243661	36076	28			−8 (42)	2 (86)
−5986	53481	−12037	6081	−46999	5460	30			4 (44)	−4 (88)
−595	7210	−1192	598	−6582	561	32			8 (46)	1 (90)
−36	668	−72	36	−631	35	34			−11 (48)	2 (92)
−1	38	−2	1	−37	1	36			12 (52)	−2 (94)
	1			−1		38			−9 (54)	1 (96)
									−5 (56)	

The Alexander polynomial in particular serves as a useful check for the calculations, since it can be quickly found for a cable using the satellite formula.

It is noticeable in these examples that the coefficients of the Alexander and Jones polynomials are considerably smaller than those in the Conway polynomial. This cannot be true in general, since all integer polynomials in z^2 with constant term 1 are possible as Conway polynomials of some knot, and so some Conway polynomials must correspond to Alexander polynomials with larger coefficients. Most probably it is a result of using fairly positive braids on a relatively small number of strings.

To reduce the size of the coefficients in these examples it would be tempting to rewrite P itself in terms of x rather than z, but this would only be possible for a 1-component knot because of the negative powers of z which occur otherwise.

Birman's pairs of 3-braids

The original motivation for these calculations arose from the simple examples of pairs of 3-braids discovered by M. T. Lozano and H. R. Morton, and simultaneously in greater variety by J. Birman [**1**]. Each example consists of two 3-braids β_1, β_2, with the same exponent sum and the same trace for their Burau matrix, consequently the same Burau polynomial, while closing to inequivalent knots. Since the polynomial P for the closure of a 3-braid is also determined by exponent sum and trace of Burau matrix these pairs give examples of inequivalent knots with the same polynomial P.

One of the simplest such pairs is $\beta_1 = \sigma_1^{-1}\sigma_2^7$, $\beta_2 = \Delta_3^4\sigma_1\sigma_2^{-7}$, conjugate to $\sigma_2^2\sigma_1^2\sigma_2^2\sigma_1^2\sigma_2\sigma_1^{-3}$, where Δ_n is the half twist on n strings, so that $\Delta_3 = \sigma_1\sigma_2\sigma_1$. Their closures give inequivalent knots C_1 (the $(2,7)$ torus knot) and C_2. A 2-cable about each can be presented as a 6-braid, by replacing σ_1 with $\sigma_2\sigma_3\sigma_1\sigma_2$ and σ_2 with $\sigma_4\sigma_5\sigma_3\sigma_4$. The linking number of the two strands in the resulting cable is then the exponent sum of the original 3-braid, so that the cables produced in this way will in this example be the $(2,12)$ cables about C_1 and C_2 respectively. Addition of one extra σ_1 to each 6-braid will give 1-component knots K_1, K_2, the $(2,13)$ cables about C_1, C_2. In table 1 the polynomials P_{K_1} and P_{K_2} are exhibited; they can be seen to differ considerably, as do the Jones polynomials of K_1 and K_2, while their Alexander polynomials, which satisfy a satellite formula, do not.

One further pair is given in table 2 of Birman's more general type. These pairs are given by $\beta_1 = \delta_1\delta_2\delta_3$, $\beta_2 = \delta_1\delta_3\delta_2$, where

$$\delta_1 = \sigma_1^{p_1}\sigma_2^{q_1}\sigma_1^{p_2}\sigma_2^{q_2}, \quad \delta_2 = \sigma_1^{p_1}\sigma_2^{q_2}, \quad \delta_3 = \sigma_1^{p_1-p_2}\sigma_2^{q_2-q_1}.$$

All other Birman pairs which we have tried can be distinguished by their 2-cable polynomials. This may be compared with Birman's difficulties in distinguishing the closed 3-braids using other methods.

Any of these examples will show also that the Burau polynomial is insufficient for calculating P, as noted earlier.

Symmetry and amphicheirality

The polynomial P_K of an amphicheiral knot K is unchanged when v is replaced by $-v^{-1}$. The knot 9_{42} has

$$P(v,z) = (2+z^2)v^{-2} - (3+4z^2+z^4) + (2+z^2)v^2,$$

Table 3

Braid representing the knot 9_{42}
*** -21332-1-32-12-34567 ***

−2	0	2	
2	−3	2	0
1	−4	1	2
	−1		4

Braid representing the (2, 0)-cable about 9_{42}
*** $(4354)^{-1}2132(6576)^2 4354(2132)^{-1}(6576)^{-1}4354(2132)^{-1}4354(6576)^{-1}1^{-2}$ ***

The polynomial P for this cable

−5	−3	−1	1	3	5	7		Jones polynomial
4	−16	29	−29	16	−4		−1	−1 (−21)
30	−109	159	−124	46	5	−7	1	1 (−19)
76	−274	341	−201	43	29	−14	3	−1 (−5)
85	−338	376	−159	16	27	−7	5	1 (−3)
45	−221	231	−65	2	9	−1	7	−1 (−1)
11	−78	79	−13		1		9	−1 (1)
1	−14	14	−1				11	1 (3)
	−1	1					13	−1 (5)
								1 (19)
								−1 (21)

Alexander Polynomial = Conway polynomial = 0

which has this symmetry although 9_{42} is not amphicheiral. The (2, 0) cable about an amphicheiral knot will again be amphicheiral, so its polynomial will be symmetric. Calculation of the polynomial for the (2, 0) cable about 9_{42} is displayed in Table 3. This polynomial is not symmetric, so giving a proof that 9_{42} is not amphicheiral. Notice that the Jones polynomial in table 3 still exhibits symmetry, so that it does not detect the lack of amphicheirality, in this case, nor apparently does Kauffman's new polynomial F.

Braid index

It was shown in [**10**], and also in [**3**], that a lower bound for the braid index of K can be found from the polynomial P_K.

Explicitly, $n \geqslant \frac{1}{2}(e_{\max}-e_{\min})+1$, where n is the braid index of K, i.e. the smallest number of strings needed to present K as a closed braid, $e_{\max}$ and $e_{\min}$ are the largest and smallest degree respectively of the non-Alexander variable v in $P_K(v,z)$. This same inequality is shown in [**10**] to apply also where n is the number of Seifert circles arising from any diagram of K.

In a number of instances the braid index inequality can be shown to be strict, and the braid index calculated exactly, by applying the inequality to a cable about K. For example the (2, 7) cable K about a trefoil can be presented as the closure of the 4-braid $\sigma_1(\sigma_2\sigma_3\sigma_1\sigma_2)^3$, but it has $\frac{1}{2}(e_{\max}-e_{\min})+1 = 3$. This was noted by Franks and Williams [**4**], who asked whether its braid index was actually 3. Our calculations presented in Table 4 exclude this possibility. For if K has a presentation as a 3-braid, then every 2-cable about K has a presentation as a 6-braid, and so the polynomial

Table 4

Braid representing the (2, 27)-cable about the (2, 7)-cable about the trefoil
***** 2132(4354657621324354)^{3}1 *****

The polynomial P for this knot

46	48	50	52	54	56	58	
893	−2935	3740	−2295	675	−79	2	0
24030	−66203	68350	−32340	6775	−485	4	2
310359	−713763	591303	−211414	30170	−1161	1	4
2503447	−4798408	3170667	−844269	79190	−1461		6
13881384	−22194447	11658704	−2284522	136147	−1068		8
55537145	−74293024	30971785	−4416455	161397	−468		10
165324695	−185800514	61398463	−6296288	135570	−121		12
374554353	−355168252	92877922	−6755584	81723	−17		14
657426753	−527978406	108944297	−5525984	35398	−1		16
906832669	−618538005	100234894	−3470613	10901			18
994140130	−576814659	72894389	−1676680	2325			20
873633502	−431235362	42078814	−620805	326			22
619135093	−259594689	19295646	−174377	27			24
355091052	−126031335	7007766	−36455	1			26
164964919	−49280982	2000681	−5489				28
61960207	−15447015	443071	−562				30
18717410	−3846817	74482	−35				32
4505528	−750139	9177	−1				34
851448	−111970	781					36
123451	−12342	41					38
13245	−946	1					40
990	−45						42
46	−1						44
1							46

Table 5

Braid representing a 2-cable about Conway's 11-crossing knot
***** (4354)^{3}2132(6576)$^{-1}$(4354)$^{-2}$2132(4354)$^{-1}$2132(6576)$^{-1}$1 *****

Braid representing a similar 2-cable about the Kinoshita-Teresaka knot
***** (2132)^3(6576)^{2}4354(6576)$^{-1}$(2132)$^{-2}$4354(6576)$^{-1}$(2132)$^{-1}$(4354)$^{-1}$1 *****

The polynomial P for both these knots

−2	0	2	4	6	8	10	12	
15	−97	233	−252	101	33	−43	11	0
146	−861	1917	−1926	646	341	−344	82	2
688	−3533	7068	−6430	1815	1251	−1115	256	4
1831	−8531	15171	−12175	2879	2352	−1982	455	6
2921	−13081	20828	−14371	2800	2547	−2115	471	8
2870	−13145	19014	−10997	1714	1656	−1389	277	10
1757	−8781	11703	−5514	656	651	−562	90	12
667	−3908	4850	−1790	151	151	−136	15	14
152	−1142	1330	−361	19	19	−18	1	16
19	−210	231	−41	1	1	−1		18
1	−22	23	−2					20
	−1	1						22

for each cable would satisfy $\frac{1}{2}(e_{max} - e_{min}) \leqslant 6$, as would also be the case if any diagram for K had just 3 Seifert circles.

The 8-braid used in Table 4 represents the $(2, 27)$ cable about K. The seven non-zero columns of coefficients show that $\frac{1}{2}(e_{max} - e_{min}) + 1 = 7$, so that at least seven strings are needed to present this cable. The knot K is then an example of the closure of a positive braid with braid index strictly larger than $\frac{1}{2}(e_{max} - e_{min}) + 1$.

Mutants

A knot K whose diagram is made up of tangles R and S as in figure 1 is converted into a *mutant* of K by replacing R with $\tau(R)$, where τ is the operation of rotation through π about one of three axes.

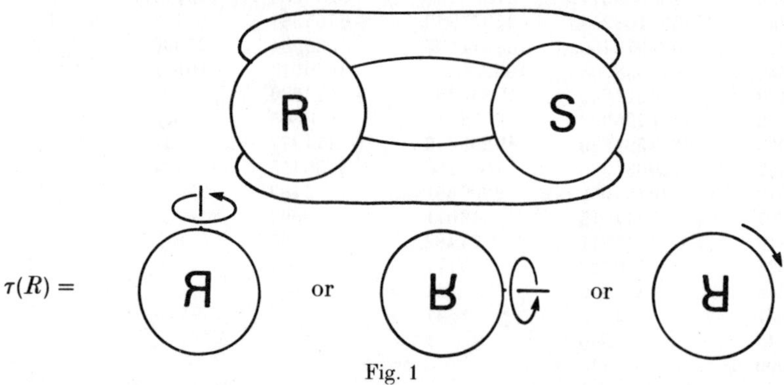

Fig. 1

Mutants have long been known to have the same Alexander polynomial, and more recently to have the same polynomial P [**9**]. Possibly the best known mutant pair are the inequivalent knots of Conway and Kinoshita-Terasaka which both have trivial Alexander polynomial. Although their 2-cables are not obviously mutants they do still share the same polynomial P shown in Table 5. This coincidence of polynomials has given us a measure of confidence in the accuracy of the computer calculations, as it would appear highly unlikely that the coefficients would all agree if there was an error in the algorithm or its implementation.

Other pairs of mutants within the range of our computations have been tried with similar results. A variety of mutants can be tackled using the fact that the 4-braids $w(\sigma_1, \sigma_2)\, v(\sigma_2, \sigma_3)$, $w(\sigma_1, \sigma_2)\, \sigma_2^{-1}\, v(\sigma_2, \sigma_3)\, \sigma_2$, $w(\sigma_1, \sigma_2)$ rev $v(\sigma_2, \sigma_3)$ and $w(\sigma_1, \sigma_2)\, \sigma_2^{-1}$ rev $v(\sigma_2, \sigma_3)\, \sigma_2$ close to mutants, where rev v is the braid v in reverse.

The result suggested by these calculations, that 2-cables of any pair of mutant knots have the same polynomial, has been subsequently proved by Lickorish and Lipson [**15**], and also by Przytycki and Traczyk.

The second author was supported as a Senior Research Assistant by SERC grant no: GR/C/48974.

REFERENCES

[1] J. S. Birman, On the Jones polynomial of closed 3-braids. *Invent. Math* **81** (1985), 287–294.
[2] R. H. Fox. Free differential calculus V. The Alexander matrices re-examined. *Ann. Math.* **71** (1960), 408–422.

[3] J. FRANKS and R. F. WILLIAMS. Braids and the Jones–Conway polynomial. Preprint, Northwestern University, 1985.

[4] J. FRANKS and R. F. WILLIAMS. Braids and the Jones–Conway polynomial. *Abstracts Amer. Math. Soc.* **6** (1985), 355.

[5] P. FREYD, D. YETTER, J. HOSTE, W. B. R. LICKORISH, K. C. MILLETT and A. OCNEANU, A new polynomial invariant of knots and links. *Bull. Amer. Math. Soc.* **12** (1985), 239–246.

[6] V. F. R. JONES. An introduction to Hecke algebras. Seminar notes, M.S.R.I. 1984.

[7] V. F. R. JONES. Private communication.

[8] L. KAUFFMAN. Two two-variable polynomials. Preprint, 1985.

[9] W. B. R. LICKORISH and K. C. MILLETT. A polynomial invariant of oriented links. Preprint, 1985. To appear in *Topology*.

[10] H. R. MORTON. Seifert circles and knot polynomials. *Math. Proc. Cambridge Philos. Soc.* **99** (1986), 107–109.

[11] H. R. MORTON. Exchangeable braids. In *Low dimensional topology*, LMS lecture notes 95 (Cambridge University Press, 1985), 86–105.

[12] H. R. MORTON and H. B. SHORT. Calculating the 2-variable polynomial for knots presented as closed braids. Preprint, Liverpool University 1986.

[13] A. OCNEANU. A polynomial invariant for knots; a combinatorial and an algebraic approach. Preprint, M.S.R.I. 1985.

[14] M. B. THISTLETHWAITE. Tables of knot polynomials. Polytechnic of the South Bank, London, 1985.

[15] W. B. R. LICKORISH and A. S. LIPSON. Polynomials of 2-cable-like links. Preprint, Cambridge University, 1986.

MURAKAMI, H.
KOBE J. MATH.,
3 (1986), 61–64

ON DERIVATIVES OF THE JONES POLYNOMIAL

By Hitoshi MURAKAMI

(Received April 2, 1985)
(Revised January 11, 1986)

In [1] V. F. R. Jones introduces a polynomial invariant $V_L(t)$ for an oriented link L and there he states that $d/dtV_K(1)=0$ for any knot K. In this note we generalize this to links. Moreover we give a formula which expresses $d^2/dt^2V_L(1)$ in terms of the z^2-terms of the Conway polynomials [2, § 5] and linking numbers of the components of L.

The Jones polynomial $V_L(t)$ has the following two properties [1] and one can compute it using these properties. (See for example [2, Lemma 5.2].)

(a) Let O_n be the trivial link with n components. Then

$$V_{O_n}(t)=(-t^{\frac{1}{2}}-t^{-\frac{1}{2}})^{n-1}.$$

(b) If three links L, L', and l are identical except near a crossing point as in Figure 1, then

$$-tV_L(t)+t^{-1}V_{L'}(t)=(t^{\frac{1}{2}}-t^{-\frac{1}{2}})V_l(t).$$

For an oriented link $L=K_1\cup K_2\cup\cdots\cup K_n$, let $\lambda_{ij}(L)$ denote the linking number of K_i and K_j $(i\neq j)$. Then we have

THEOREM 1. $d/dtV_L(1)=3(-2)^{n-2}\{\sum_{i<j}\lambda_{ij}(L)\}$.

THEOREM 2.

$$\begin{aligned}d^2/dt^2V_L(1) &= 3(-2)^n\{\sum_{i=1}^{n}a_2(K_i)\}\\ &+3(-2)^{n-1}\{\sum_{i<j}\lambda_{ij}(L)^2\}\\ &+9(-2)^{n-3}\{\sum_{\substack{i<j,\,s<t\\(i,j)\neq(s,t)}}\lambda_{ij}(L)\lambda_{st}(L)\}\\ &-3(-2)^{n-2}\{\sum_{i<j}\lambda_{ij}(L)\}\\ &+(n-1)(-2)^{n-3},\end{aligned}$$

where $a_2(K_i)$ is the z^2-term of the Conway polynomial of K_i. Note that for a knot

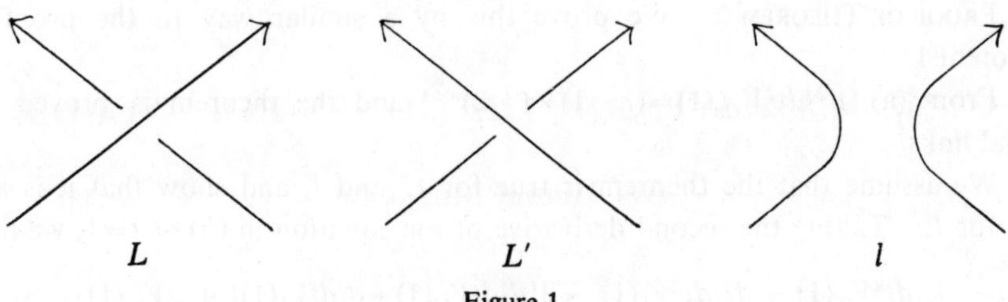

Figure 1

K, $d^2/dt^2 V_K(1) = -6a_2(K)$.

PROOF OF THEOREM 1. From (a), $d/dt V_{O_n}(1)=0$ and the theorem is proved in this case.

We assume that the theorem is true for L' and l, and show that it is also true for L. (L', l, and L are the links as in Figure 1.) Differentiating both sides in (b) at $t=1$, we have

$$-\{V_{L'}(1)+V_L(1)\} + \{d/dt V_{L'}(1) - d/dt V_L(1)\} = V_l(1).$$

Since $V_L(1)=(-2)^{\#(L)-1}$ for any link with $\#(L)$ components [1, Theorem 15] (Proof: $V_L(1)=V_{L'}(1)$ from (b). Thus $V_L(1)$ is invariant under crossing changes and is equal to that of a trivial link $(=(-2)^{\#(L)-1})$.), the above equality turns out to be

$$(-2)^{\#(L)} + \{d/dt V_{L'}(1) - d/dt V_L(1)\} = (-2)^{\#(l)-1}.$$

If $\#(L)=\#(l)-1$, then

$$d/dt V_L(1) = d/dt V_{L'}(1)$$

and the theorem is true since $\lambda(L)=\lambda(L')$. If $\#(L)=\#(l)+1$, then

$$d/dt V_L(1) = d/dt V_{L'}(1) + 3(-2)^{\#(L)-2}$$

and the theorem is true since $\lambda(L)=\lambda(L')+1$.

Next we assume that the theorem is true for L and l, and show that it is also true for L'. The same argument as above shows that $d/dt V_{L'}(1)=d/dt V_L(1)$ if $\#(L')=\#(l)-1$, and that $d/dt V_{L'}(1)=d/dt V_L(1)-3(-2)^{\#(L)-2}$ if $\#(L')=\#(l)+1$. Since $\lambda(L')=\lambda(L)$ in the former case and $\lambda(L')=\lambda(L)-1$ in the latter case, the theorem is true for L'.

Because $V_L(t)$ can be computed recursively as described in [2, Lemma 5.2], Theorem 1 is now proved for any link.

REMARK. Theorem 14 in [1] can be also generalized to links. Namely $V_L(\exp(2\pi i/3))=1$ for any link L if one chooses $(\exp(2\pi i/3))^{\frac{1}{2}}$ to be $\exp\left(\frac{2}{3} \times 2\pi i\right)$. This can be easily seen since $-t+t^{-1}=t^{\frac{1}{2}}-t^{-\frac{1}{2}}$ in this case. (See Property (b).)

PROOF OF THEOREM 2. We prove this by a similar way to the proof of Theorem 1.

From (a), $d^2/dt^2 V_{On}(1)=(n-1)\times(-2)^{n-3}$ and the theorem is proved for trivial links.

We assume that the theorem is true for L' and l, and show that it is also true for L. Taking the second derivative of the equation in (b) at $t=1$, we have

$$d^2/dt^2 V_{L'}(1) - d^2/dt^2 V_L(1) - 2\{d/dt V_{L'}(1)+d/dt V_L(1)\} + 2V_{L'}(1)$$
$$= 2 \times d/dt V_l(1) - V_l(1).$$

From Theorem 1, this equation becomes

$$d^2/dt^2 V_{L'}(1) - d^2/dt^2 V_L(1) + 3 \times (-2)^{\#(L)-1}\{\sum_{i<j} \lambda_{ij}(L)+\sum_{j<i} \lambda_{ij}(L')\}$$
$$-(-2)^{\#(L)} = -3 \times (-2)^{\#(l)-1}\{\sum_{i<j} \lambda_{ij}(l)\} - (-2)^{\#(l)-1}.$$

First suppose that $n=\#(L)=\#(l)-1$. Let $L=K_1 \cup K_2 \cup \cdots \cup K_n$, $L'=K'_1 \cup K'_2 \cup \cdots \cup K'_n$, and $l=k_1 \cup k_2 \cup \cdots \cup k_{n+1}$. We assume that K_n, K'_n, k_n, and k_{n+1} are the knots as indicated in Figure 2.

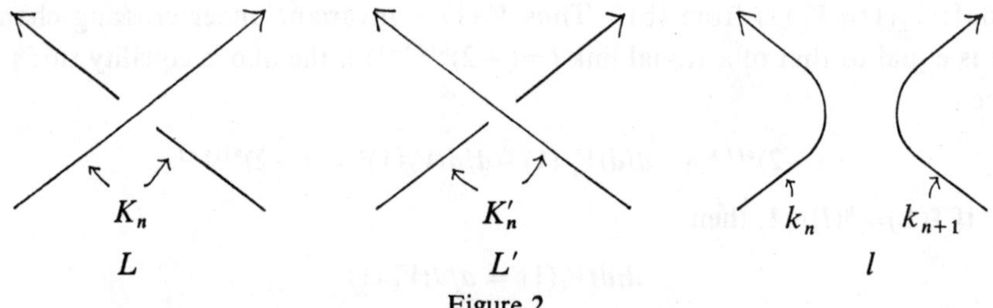

Figure 2

Then from the hypothesis on $d^2/dt^2 V_{L'}(1)$ and the fact:

$$\lambda_{ij}(L) = \lambda_{ij}(L') \quad \text{for any } i \text{ and } j,$$
$$\lambda_{ij}(L) = \lambda_{ij}(l) \quad \text{for } i, j < n,$$
$$\lambda_{in}(L) = \lambda_{in}(l) + \lambda_{i,n+1}(l) \quad \text{for } i < n, \text{ and}$$
$$a_2(K_i) = a_2(K'_i) \quad \text{for } i < n,$$

we have

$$d^2/dt^2 V_L(1) = 3(-2)^n\{\sum_{i=1}^{n-1} a_2(K_i)\}$$
$$+ 3(-2)^{n-1}\{\sum_{i<j} \lambda_{ij}(L)^2\}$$
$$+ 9(-2)^{n-3}\{\sum_{\substack{i<j,s<t\\(i,j)\neq(s,t)}} \lambda_{ij}(L)\lambda_{st}(L)\}$$

$$- 3(-2)^{n-2}\{\sum_{i<j}\lambda_{ij}(L)\}$$
$$+ (n-1)(-2)^{n-3}$$
$$+ 3(-2)^{n}\{a_2(K'_n)+\lambda_{n,n+1}(l)\}.$$

Now from the axiom of the Conway polynomial and [2, Lemma 5.6], $a_2(K_n)-a_2(K'_n)=\lambda_{n,n+1}(l)$. Thus we finally obtain the required formula for $d^2/dt^2V_L(1)$.

Next suppose that $n=\sharp(L)=\sharp(l)+1$. Let $L=K_1\cup K_2\cup\cdots\cup K_n$, $L'=K'_1\cup K'_2\cup\cdots\cup K'_n$, and $l=k_1\cup k_2\cup\cdots\cup k_{n-1}$. We assume that K_{n-1}, K_n, K'_{n-1}, K'_n, and k_{n-1} are the knots as indicated in Figure 3.

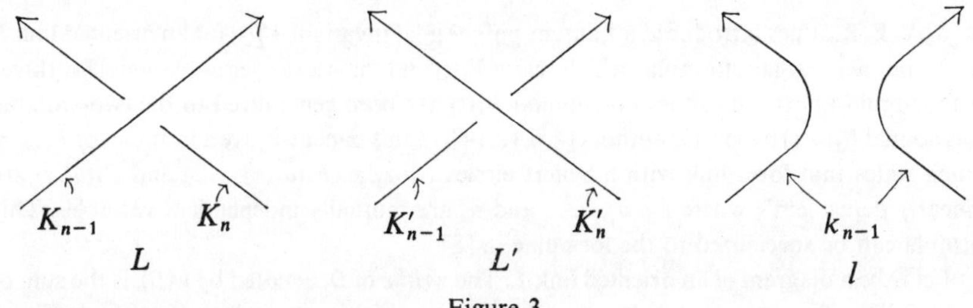

Figure 3

Since

$$\lambda_{ij}(L)=\lambda_{ij}(L') \quad \text{for} \quad (i,j)\neq(n-1,n),$$
$$\lambda_{n-1,n}(L)=\lambda_{n-1,n}(L')+1,$$
$$\lambda_{ij}(l)=\lambda_{ij}(L) \quad \text{for} \quad i,j<n-1,$$
$$\lambda_{i,n-1}(l)=\lambda_{i,n-1}(L)+\lambda_{in}(L) \quad \text{for} \quad i<n-1, \text{ and}$$
$$a_2(K_i)=a_2(K'_i) \quad \text{for any} \quad i,$$

we can also obtain the required formula for $d^2/dt^2V_L(1)$

If we assume that the theorem is true for L and l, we can also prove as above that it is true for L'. Thus the proof is complete.

References

[1] V. F. R. Jones: A polynomial invariant for knots via von Neumann algebras, Bull. Amer. Math. Soc. **12** (1985), 103–112.

[2] L. H. Kauffman: Formal Knot Theory, Math. Notes 30, Princeton Univ. Press, 1983.

Department of Mathematics
Osaka City University
Sugimoto, Sumiyoshi-ku
Osaka, 558, Japan

Topology Vol. 26, No. 4, pp. 409–412, 1987

A FORMULA FOR THE TWO-VARIABLE LINK POLYNOMIAL

HITOSHI MURAKAMI

(*Received in revised form* 29 *January* 1987)

IN [8] V. F. R. Jones introduced a Laurent polynomial invariant $V_L(t)$ for an oriented link L in S^3, and there he gave formulae which relate $V_L(t)$ and the Alexander polynomial for three- or four-braid knots. The Jones polynomial $V_L(t)$ has been generalized to the two-variable polynomial $P_L(a, z)$ by several authors [4, 6, 12, 14]. In this paper we give a formula for $P_L(a, z)$ which states that for a link with n Seifert circles $P_L(a_0, z)$, $P_L(a_1, z)$, ..., and $P_L(a_n, z)$ are "linearly dependent", where a_0, a_1, ..., and a_n are mutually independent variables. This formula can be specialized to the formulae in [8].

Let D be a diagram of an oriented link L. The *writhe* of D, denoted by $w(D)$, is the sum of signs of all the crossing points in D, where the sign is $+1$ if the crossing point is as L_+ in Fig. 1, and -1 if it is as L_- there [10, 11]. (This is equal to $\tilde{c}(D)$ in [13].) Note that if L is represented as a closed braid, then $w(D)$ coincides with the exponent sum of its braid representation.

If we remove all the crossing points in D (i.e. change them as L_0 in Fig. 1), then the resulting diagram splits into some simple closed curves. These circles are called *Seifert circles* of D.

The two-variable polynomial $P_L(a, z) \in \mathbb{Z}[a^{\pm 1}, z^{\pm 1}]$ is defined so that it satisfies the following two axioms [4, 6, 12, 14].

(I) $P_O(a, z) = 1$, where O is the trivial knot.

(II) $a^{-1}P_{L_+}(a, z) - aP_{L_-}(a, z) + zP_{L_0}(a, z) = 0$, where L_+, L_-, and L_0 have the same diagram except near a crossing point where they are as in Fig. 1.

The Jones polynomial $V_L(t) \in \mathbb{Z}[t^{\pm 1/2}]$ is given by $P_L(t, t^{1/2} - t^{-1/2})$ [8] and the (normalized) Alexander polynomial $\Delta_L(t)$ is given by $P_L(1, t^{1/2} - t^{-1/2})$ [1, 3, 5, 9].

Now we will show

THEOREM 1. *Let L be a link with a diagram D containing n Seifert circles. Then we have*

$$(*)\qquad a_0^{-w(D)}P_L(a_0, z) = \sum_{i=1}^{n} a_i^{-w(D)}P_L(a_i, z) \cdot \prod_{\substack{1 \leq j \leq n \\ j \neq i}} \frac{(a_0 a_j^{-1} - a_0^{-1} a_j)}{(a_i a_j^{-1} - a_i^{-1} a_j)},$$

where a_0, a_1, ..., and a_n are mutually independent variables.

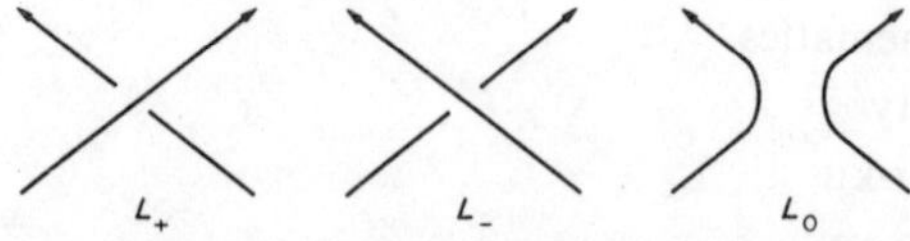

Fig. 1.

Remark 1. The referee pointed out that the formula becomes simpler as above than the original one:

$$\sum_{i=0}^{n}(-1)^i\left\{\prod_{\substack{0\leq j<k\leq n\\ j\neq i,k\neq i}}(a_ja_k^{-1}-a_j^{-1}a_k)\right\}a_i^{-w(D)}P_L(a_i,z)=0.$$

Remark 2. If L is a closed n-braid, then the above formula holds if we put $w(D)=e(L)$, where $e(L)$ is the exponent sum of its representation as a closed n-braid.

Recalling that $P_L(t^{1/2},t^{1/2}-t^{-1/2})=1$ [12], $P_L(t^{-1/2},\ t^{1/2}-t^{-1/2})=P_L(t^{1/2},\ -(t^{1/2}-t^{-1/2}))=(-1)^{\#(L)-1}=(-1)^{n+w(D)-1}$ ($\#(L)$ denotes the number of components of L), $P_L(1,t^{1/2}-t^{-1/2})=\Delta_L(t)$, $P_L(t,t^{1/2}-t^{-1/2})=V_L(t)$, and $P_L(t^{-1},\ t^{1/2}-t^{-1/2})=V_L(-t^{-1})=(-1)^{n+w(D)-1}\ V_L(t^{-1})$, we have the following corollary.

Corollary. *Let L be a link with diagram D.*

(i) *If D has 3 Seifert circles, then*

$$\begin{aligned}P_L(a,t^{1/2}-t^{-1/2})&=G_w(a,t^{1/2})+G'_w(a,t^{1/2})\cdot\Delta_L(t)\\&=H_w(a,t^{1/2})+H'_w(a,t^{1/2})\cdot V_L(t),\end{aligned}$$

where G_w, G'_w, H_w, and H'_w are Laurent polynomials in a and $t^{1/2}$ depending only on $w=w(D)$, given explicitly by Theorem 1.

(ii) *If D has 4 Seifert circles, then $P_L(a,t^{1/2}-t^{-1/2})$ can similarly be expressed in terms of $\Delta_L(t)$ and $V_L(t)$, or in terms of $V_L(t)$ and $V_L(t^{-1})$. It follows that $\Delta_L(t)$ can be expressed in terms of $V_L(t)$ and $V_L(t^{-1})$.*

(iii) *If D has 5 Seifert circles, then $P_L(a,t^{1/2}-t^{-1/2})$ can be expressed in terms of $\Delta_L(t)$, $V_L(t)$ and $V_L(t^{-1})$.*

Remark 3. V. F. R. Jones pointed out to the author in a letter that the two-variable polynomial can be expressed in terms of the Alexander polynomial and the exponent sum for three-braid links, since the corresponding Hecke algebra is the direct sum of the trivial representation, the Burau representation, and the parity representation. See [2, 7, 8] for detail.

To prove Theorem 1 we prepare some lemmas. It is easy to prove the following lemma but it is essential for the proof of Theorem 1.

Lemma 1. *Let L be a link with a diagram D. Consider a Laurent polynomial $(a_0/a_i)^{w(D)}$ $P_L(a_i,z)\in\mathbb{Z}[a_0^{\pm1},\ a_i^{\pm1},\ z^{\pm1}]$. Then it satisfies the axiom (II) for $P_L(a_0,z)$.*

Proof. Let $w=w(D_0)$. Since $w(D_+)=w+1$ and $w(D_-)=w-1$,

$$\begin{aligned}&a_0^{-1}(a_0/a_i)^{w+1}P_{L_+}(a_i,z)-a_0(a_0/a_i)^{w-1}P_{L_-}(a_i,z)+z(a_0/a_i)^{w}P_{L_0}(a_i,z)\\&\qquad=(a_0/a_i)^{w}\{a_i^{-1}P_{L_+}(a_i,z)-a_iP_{L_-}(a_i,z)+zP_{L_0}(a_i,z)\}=0.\end{aligned}$$

Here D_0, D_+, and D_- are diagrams corresponding to L_0, L_+, and L_- respectively. This completes the proof.

Lemma 2. *Every link diagram D can be deformed to a diagram of the trivial link with writhe zero or one after some crossing changings. We note that the writhe is zero if $n+\#(L)$ is* *even and one if $n+\#(L)$ is odd, where n is the number of Seifert circles in D and L is a link represented by D.*

Proof. We prove the lemma by induction on the number of crossing points in a diagram. If there are no crossing points, the lemma is trivially true. So suppose that it is true for any diagram with crossing points fewer than c. Let D be a diagram with c crossing points ($c>0$).

First suppose that there is a component B with a self-crossing point x. Let E be a diagram obtained from D after removing x as L_0 in Fig. 1. Then B splits into two components B_1 and B_2. After some crossing changings we have a diagram E' where B_1 crosses over all the other components in E'. By the inductive hypothesis we assume that B_1 and its complement $E'-B_1$ are diagrams of trivial links with writhe zero or one. Now consider the corresponding diagram D' obtained from D after crossing changings as above. Changing the crossing at x if necessary, we obtain the required diagram since it represents a trivial link for any choice of the sign of x.

If there is no such B, then the conclusion follows immediately since we can change crossings so that D_i crosses over D_j for $i<j$, where D_i's are the components of $D(i=1,2,\ldots,\#(L))$. The proof is complete.

LEMMA 3. $a_0^p=\sum_{i=1}^{n} a_i^p\cdot \prod_{\substack{1\leqq j\leqq n\\ j\neq i}} \frac{(a_0a_j^{-1}-a_0^{-1}a_j)}{(a_ia_j^{-1}-a_i^{-1}a_j)}$ *for* $|p|\leqq n-1$ *and* $p\equiv n-1 \pmod 2$.

Proof. Put $F_p=\sum_{i=0}^{n}(-1)^i\left\{\prod_{\substack{0\leqq j<k\leqq n\\ j\neq i,\,k\neq i}}(a_ja_k^{-1}-a_j^{-1}a_k)\right\}a_i^p.$

We will show that $F_p=0$ for $|p|\leqq n-1$ and $p\equiv n-1 \pmod 2$. We regard F_p as a Laurent polynomial in a_0 and write it as $F_p(a_0)$. Then one can easily see that $\prod_{s=0}^{n} a_s^{n-1}\cdot F_p(a_0)$ is a polynomial (without negative powers) and has degree $2(n-1)$. Since $F_p(\pm a_s)=0$ for $s=1,2,\ldots n$, the equation $F_p(a_0)=0$ has $2n$ solutions. Thus $F_p(a_0)=0$.

The lemma follows from $F_p=0$ after some calculation.

Proof of Theorem 1. We will show

$$(**)\qquad P_L(a_0,z)=\sum_{i=1}^{n}(a_0/a_i)^{w(D)}P_L(a_i,z)\cdot\prod_{\substack{1\leqq j\leqq n\\ j\neq i}}\frac{(a_0a_j^{-1}-a_0^{-1}a_j)}{(a_ia_j^{-1}-a_i^{-1}a_j)}.$$

From Lemma 1 the right hand side of (**) satisfies the axiom (II) for $P_L(a_0,z)$. From Lemma 2 and a standard argument using a resolution tree as in [6], (**) holds for any link with a diagram containing n Seifert circles if we prove it holds for trivial links with writhe zero or one and with components fewer than or equal to n. Thus we must only check that (*) holds when $a_i^{-w(D)}P_L(a_i,z)=1, a_i^{-1}(a_i-a_i^{-1}), (a_i-a_i^{-1})^2,\ldots, a_i^{-1}(a_i-a_i^{-1})^{n-2}$, and $(a_i-a_i^{-1})^{n-1}$ if n is odd and $a_i^{-w(D)}P_L(a_i,z)=a_i^{-1}, (a_i-a_i^{-1}), a_i^{-1}(a_i-a_i^{-1})^2,\ldots, a_i^{-1}(a_i-a_i^{-1})^{n-2}$, and $(a_i-a_i^{-1})^{n-1}$ if n is even. But this follows from Lemma 3, completing the proof.

From the proof of Lemma 3, we see that $F_p\neq 0$ for any p with $|p|\geqq n$. Thus $a^{-w(D)}P_L(a,z)$ contains only a^q with $|q|\leqq n-1$, if L is a link with a diagram containing n Seifert circles. So we have the following theorem due to H. R. Morton, J. Franks and R. F. Williams.

THEOREM 2. [13, 15] *For any diagram D of a link L, we have*

$$w(D)-(n-1)\leqq e\leqq E\leqq w(D)+(n-1),$$

where n is the number of Seifert circles in D, e is the lowest degree of a in $P_L(a,z)$, and E is the highest degree of a in $P_L(a,z)$.

Remark 3. Theorem 1 also follows from Theorem 2 using Lemma 3. So Theorem 1 cannot give better estimate concerning the number of Seifert circles than the above inequality. We also remark that if $E-e=2(n-1)$, then $w(D)=\frac{1}{2}(E+e)$. Thus the writhe is a link type invariant in this case.

Acknowledgements—The author would like to thank Prof. V. F. R. Jones, Prof. Y. Nakanishi and the referee for their helpful advice.

REFERENCES

1. J. W. ALEXANDER: Topological invariants of knots and links, *Trans. Am. Math. Soc.* **30** (1928), 275–306.
2. J. S. BIRMAN: On the Jones polynomial of closed 3-braids, *Invent. Math.* **81** (1985), 287–294.
3. J. H. CONWAY: *An Enumeration of Knots and Links, and some of their Algebraic properties, Computational Problems in Abstract Algebra*, Pergamon Press, Oxford (1969), 329–358.
4. P. FREYD, D. YETTER, J. HOSTE, W. B. R. LICKORISH, K. MILLETT, and A. OCNEANU: A new polynomial invariant of knots and links, *Bull. A.M.S.* **12** (1985), 239–246.
5. C. GILLER: A family of links and the Conway calculus, *Trans. Am. Math. Soc.* **270** (1982), 75–109.
6. J. HOSTE: A polynomial invariant of knots and links, *Pacific J. Math.* **124** (1986), 295–320.
7. V. F. R. JONES: Braid groups, Hecke algebras and type II_1 factors, *Jap.–U.S. Conf. Proc.* (1983).
8. V. F. R. JONES: A polynomial invariant for knots via von Neumann algebras, *Bull. A.M.S.* **12** (1985), 103–111.
9. L. H. KAUFFMAN: The Conway polynomial, *Topology* **20** (1981), 101–108.
10. L. H. KAUFFMAN: A geometric interpretation of the generalized polynomial, preprint.
11. L. H. KAUFFMAN: An invariant of regular isotopy, preprint.
12. W. B. R. LICKORISH and K. C. MILLETT: A polynomial invariant of oriented links, *Topology* **26** (1987), 107–141.
13. H. R. MORTON: Seifert circles and knot polynomials, *Math. Proc. Camb. Phil. Soc.* **99** (1986), 107–109.
14. J. H. PRZYTYCKI and P. TRACZYK: Invariants of links of Conway type, to appear *Kobe J. Math* (1987).
15. J. FRANKS and R. F. WILLIAMS: Braids and the Jones polynomials, preprint.

Department of Mathematics,
Osaka City University,
Sugimoto, Sumiyoshi-ku,
Osaka, 558, *Japan*

L'Enseignement Mathématique, t. 33 (1987), p. 203-225

A SIMPLE PROOF OF THE MURASUGI AND KAUFFMAN THEOREMS ON ALTERNATING LINKS

by V. G. Turaev

The aim of the present paper is to give simplified proofs of several theorems recently obtained by Murasugi and Kauffman with the help of Jones polynomials for links. These theorems settle several old conjectures of Tait on alternating link diagrams. The proofs given here follow the main lines of the proofs given in [3], [6]; however some steps are considerably simplified, including the crucial "extended dual state Lemma".

I thank Claude Weber for careful reading of a preliminary version of this paper and for valuable suggestions. I am also indebted to Pierre de la Harpe for encouraging remarks.

§ 1. Introduction

For the definition of (smooth) links in the 3-sphere, link diagrams, alternating diagrams and alternating links, the reader is referred to [3].

A link diagram is called *reduced* if there is no (smooth) circle $S^1 \subset R^2$ intersecting the diagram in exactly two points which lie near a crossing point, as in the following picture.

Figure 1

A link diagram is called *splittable* if there is a circle $S^1 \subset R^2$ which does not intersect the diagram and such that both components of $R^2 - S^1$ intersect the diagram. A link diagram K is said to be a *connected sum* of link diagrams $K_1, \ldots, K_m$ if $K_1, \ldots, K_m$ lie in disjoint discs in R^2 and if K can be obtained from $K_1, \ldots, K_m$ by band summation (the bands are supposed to lie in R^2 and to have no crossing point with each other and with $\bigcup_i K_i$). Finally, a link diagram is called *weakly alternating* if each of its split components is either a reduced alternating diagram or a connected sum of reduced alternating diagrams. Here is an example of a weakly alternating diagram which is not alternating.

FIGURE 2

For a link diagram K we denote by $c(K)$ the *number of crossing points* of K and by $r(K)$ the *number of split components* of K.

Recall that with each oriented link $L \subset S^3$, V. Jones [4] has associated a polynomial $V_L(t) \in \mathbf{Z}[t^{1/2}, t^{-1/2}]$. If

$$V_L(t) = \sum_{n \leqslant i \leqslant m} a_i t^i \quad \text{with} \quad n, m, i \in \frac{1}{2}\mathbf{Z} \quad \text{and} \quad a_n \neq 0 \neq a_m,$$

then one defines $\operatorname{span}(L) = m - n$.

According to [4], if L has an odd number of components, then $V_L(t) \in \mathbf{Z}[t, t^{-1}]$; if L has an even number of components, then $t^{1/2} V_L(t) \in \mathbf{Z}[t, t^{-1}]$. Therefore, in all cases $\operatorname{span}(L) \in \mathbf{Z}$. Note also that $\operatorname{span}(L)$ is not changed if we invert the orientations of some components of L (thanks to the Jones reversing result, see § 8 of [3]). Thus the integer $\operatorname{span}(L)$ is an invariant of non-oriented links.

This invariant has the following additive properties. If L splits into links $L_1, \ldots, L_r$ then

$$\operatorname{span}(L) = r - 1 + \sum_{i=1}^{r} \operatorname{span}(L_i). \tag{1}$$

This follows from the formula

$$V_L(L) = (-t^{1/2}-t^{-1/2})^{r-1} \prod_{i=1}^{r} V_{L_i}(t)$$

of Jones [4]. If L is a connected sum of two links L' and L'' (performed from the unlinked union on any choice of components), then

$$V_L(L) = V_{L'}(t)V_{L''}(t)$$

so that

$$\text{span}\,(L) = \text{span}\,(L') + \text{span}\,(L'')\,.$$

THEOREM 1 (Murasugi, Kauffman). *Let K be a diagram of a link L. Then:*

(i) $c(K) + r(K) - 1 \geqslant \text{span}\,(L)$,

(ii) $c(K) + r(K) - 1 = \text{span}\,(L)$ *if and only if K is a weakly alternating diagram.*

In particular, as $r(K) = 1$ if L is unsplittable:

COROLLARY 1. *Let K be a diagram of an unsplittable link L. Then* $c(K) \geqslant \text{span}\,(L)$, *with equality if and only if K is a connected sum of reduced alternating diagrams.*

Let us observe that, if K and K' are alternating projections, one can always make connected sums K_1 and K_2 of K and K' in order that K_1 be alternating and K_2 be non-alternating. In particular, it follows that a link which has a weakly alternating projection is indeed an alternating link. See figure 3.

COROLLARY 2. *Two weakly alternating diagrams of the same alternating link L have the same number of split components. This number is equal to the number of split components of L.*

Proof. It is enough to note that every unsplittable weakly alternating diagram represents an unsplittable link. This fact is well known (at least for unsplittable alternating diagrams: see Crowell [1] and references therein). However, for the reader's convenience, we shall give here a proof of this fact which depends only on Theorem 1 and on a few elementary observations.

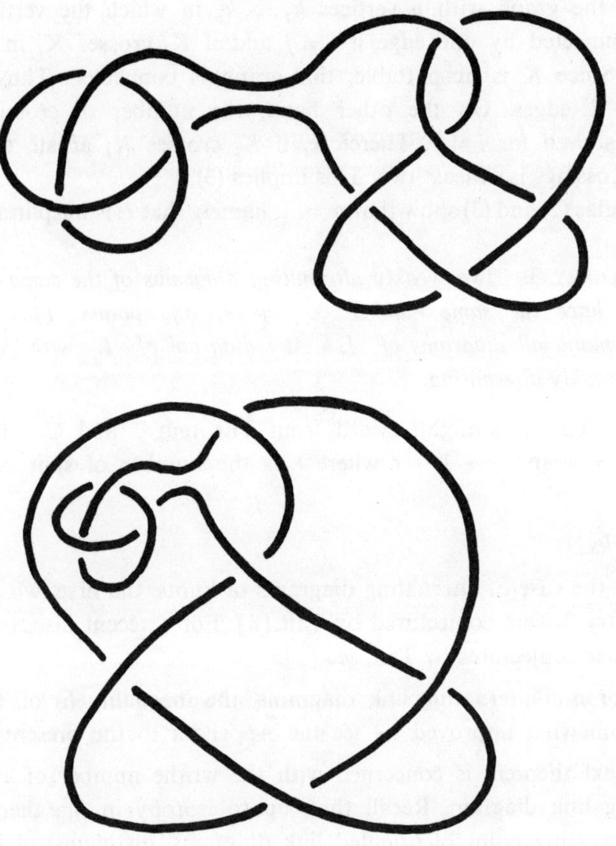

FIGURE 3

Let L be a link presented by an unsplittable weakly alternating diagram K, so that $r(K) = 1$. Suppose that L splits into unsplittable links $L_1, ..., L_p$. Then K is a "union" of subdiagrams $K_1, ..., K_p$ where K_i represents L_i for $i = 1, ..., p$. Since L_i is unsplittable, K_i is also unsplittable. In view of Corollary 1

$$(2) \qquad \sum_{i=1}^{p} c(K_i) \geqslant \sum_{i=1}^{p} \operatorname{span}(L_i) = \operatorname{span}(L) - (p-1) = c(K) - (p-1).$$

Let us prove that

$$(3) \qquad c(K) \geqslant c(K_1) + c(K_2) + \ldots + c(K_p) + 2(p-1).$$

Consider the graph with p vertices $k_1, \dots, k_p$ in which the vertices k_i and k_j are connected by one edge if $i \neq j$ and if K_i crosses K_j in one point at least. Since K is unsplittable, this graph is connected. Thus it has at least $p-1$ edges. On the other hand, the number of crossings of K_i and K_j is even for $i \neq j$. Therefore, if K_i crosses K_j at all, the number of such crossings is at least two. This implies (3).

Formulas (2) and (3) show that $p = 1$, namely that L is unsplittable. □

COROLLARY 3. *Two weakly alternating diagrams of the same alternating link L have the same number c of crossing points. This number is minimal among all diagrams of L. Any diagram of L with c crossing points is weakly alternating.*

Proof. This is straightforward from Theorem 1 and Corollary 2. Of course, $c = \operatorname{span}(L) - 1 + r$ where r is the number of split components of L. □

Remarks.

1. In the case of alternating diagrams of knots, the first two statements of Corollary 3 were conjectured by Tait [8]. For a recent discussion of this and of other conjectures by Tait, see [3].

2. For non-alternating link diagrams, the inequality (i) of Theorem 1 can be somewhat improved — see the Appendix to the present paper.

The next theorem is concerned with the writhe number of an oriented alternating link diagram. Recall that, up to isotopy in R^2, there are two types of crossing point of oriented link diagrams, distinguished by a sign:

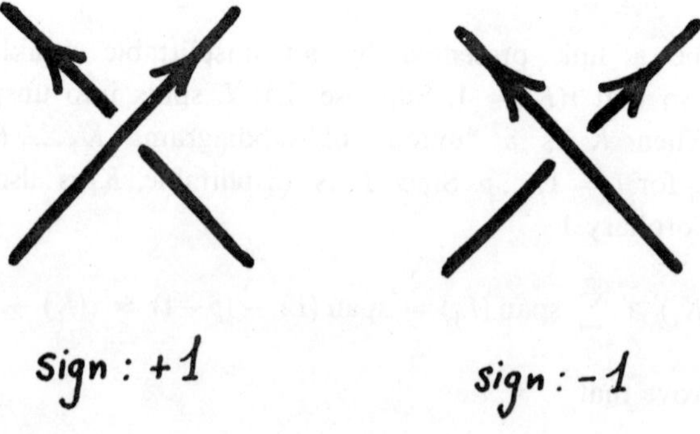

FIGURE 4

The *writhe number* $w(K)$ of an oriented link diagram K is the sum of the signs over all crossing points of K. Little believed that the writhe number of an oriented reduced alternating diagram is a link type invariant. This conjecture has been recently proved independently by Murasugi [6] and Thistlethwaite [9]. It follows directly from the following Theorem.

THEOREM 2 (Murasugi [6]). *If* K *is an oriented weakly alternating diagram, then*

$$w(K) = \sigma(L) - d_{\max}(V_L(t)) - d_{\min}(V_L(t))$$

where the oriented link presented by K *is denoted by* L, *its signature by* $\sigma(L)$, *and where* $d_{\max}$ *and* $d_{\min}$ *denote the maximal and minimal degrees of a polynomial. (Note that Murasugi uses the polynomial* $V = V_L(t^{-1})$, *so that his formula has two plus signs.)*

Theorems 1 and 2 imply that, for oriented weakly alternating diagrams, both the number of positive crossing points and the number of negative crossing points are link type invariants.

It is worth realizing that, if $K^\times$ is the mirror image of an oriented link diagram K, then $w(K^\times) = -w(K)$. Therefore, if K is weakly alternating and represents an amphicheiral link, then Theorem 2 implies that $w(K) = 0$.

*
* *

The remaining part of this paper is organized as follows. In § 2 the extended dual state Lemma, due to Kauffman and Murasugi, is stated and proved. In § 3 I quickly recall the Kauffman state model for the Jones polynomial. Theorem 1 is proved in § 4 and Theorem 2 is proved in § 5. In the Appendix, the inequality (i) of Theorem 1 is somewhat improved.

§ 2. THE EXTENDED DUAL STATE LEMMA

Let Γ be the image of a generic immersion of a finite number of circles into R^2. Note that self-crossing points of Γ are exclusively double points. For each double point x of Γ a small disc in R^2 centered in x is divided by Γ into four parts. These parts appear in two pairs of opposite sectors. Each of these pairs is called a *marker* of Γ at x. In pictures these markers are indicated like that:

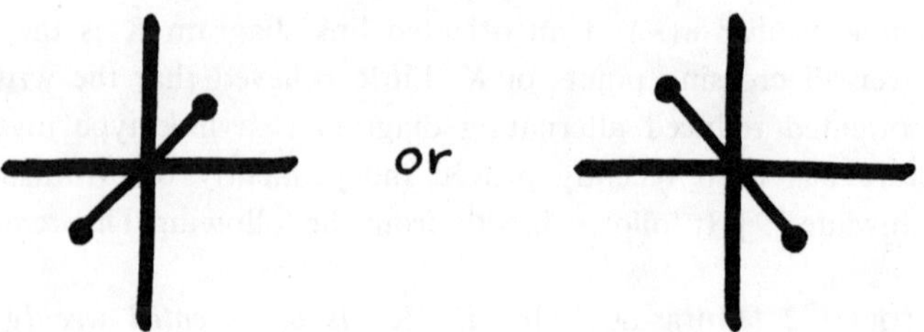

FIGURE 5

One can smooth (or surger) Γ along the markers:

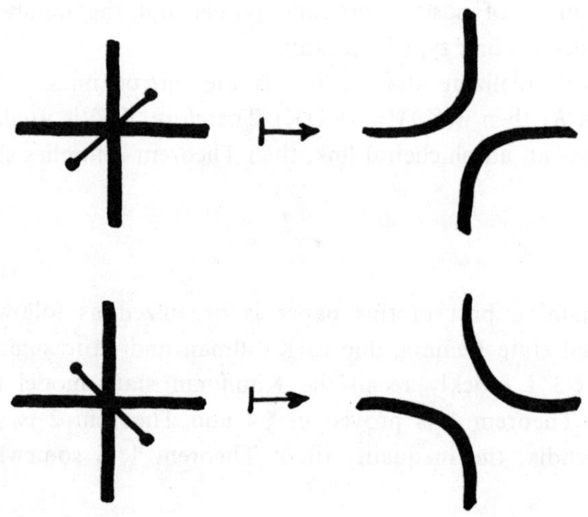

FIGURE 6

A *state* S for Γ is a choice of one marker at each double point of Γ. The opposite choice of marker at each double point defines the *dual state* of S, denoted by $\check{S}$. The dual state of $\check{S}$ is obviously S. If we surger Γ along the markers of a state S we obtain a closed imbedded 1-manifold $\Gamma_S \subset R^2$ as in the following picture.

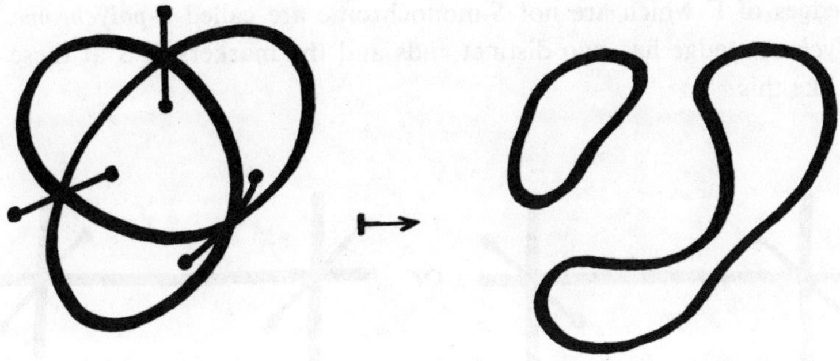

FIGURE 7

Let $|S|$ denote the number of connected components of Γ_S.

Denote by $r = r(\Gamma)$ the number of connected components of the set Γ in R^2, and by $c = c(\Gamma)$ the number of double points of Γ. It is clear that Γ has 2^c states. (If $c = 0$, then, by definition, Γ has one state S with $\Gamma_S = \Gamma$.)

LEMMA 1 (the dual state Lemma [5]). *For any state S of Γ, one has*

$$|S| + |\check{S}| \leqslant c + 2r. \tag{4}$$

To prove this Lemma and to study the case of equality in (4), we need the following definitions.

By an *edge* of Γ, we shall mean an arc in Γ whose interior does not contain any double point, and whose two ends are double points of Γ. The case of coinciding ends is not excluded, and such an edge is called a *loop*.

Let S be a state of Γ. An edge e of Γ is called *S-monochrome* if either e is a loop, or e has distinct ends and the markers of S at these ends look like this:

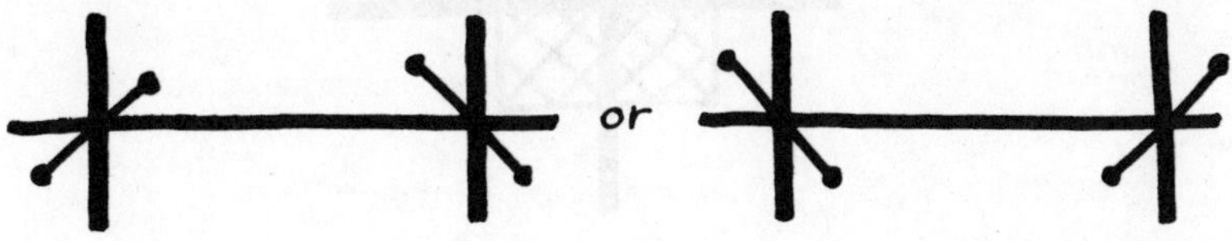

FIGURE 8

The edges of Γ which are not S-monochrome are called *S-polychrome*. Any *S-polychrome* edge has two distinct ends and the markers of S at these ends look like this:

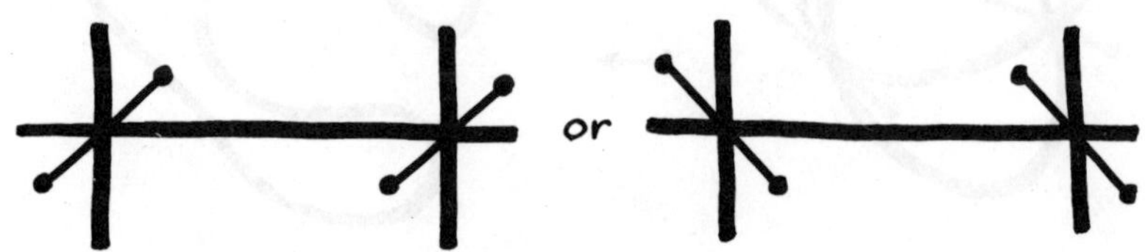

FIGURE 9

The state S of Γ is called *monochrome* if all edges of Γ are S-monochrome. It is clear that S is monochrome if and only if $\check{S}$ is monochrome.

We shall say that Γ is *prime* if each circle $S^1 \subset R^2$ which intersects Γ in exactly two points and transversally bounds a disc in $S^2 = R^2 \bigcup \{\infty\}$ which intersects Γ in a simple arc.

LEMMA 2. *Suppose that* Γ *is prime and connected. Let* S *be a state of* Γ. *Then the equality*

$$|S| + |\check{S}| = c + 2$$

holds if and only if S *is monochrome.*

Proof of lemmas 1 and 2. Let S be a state of Γ. To each double point x of Γ we associate a small square in R^2:

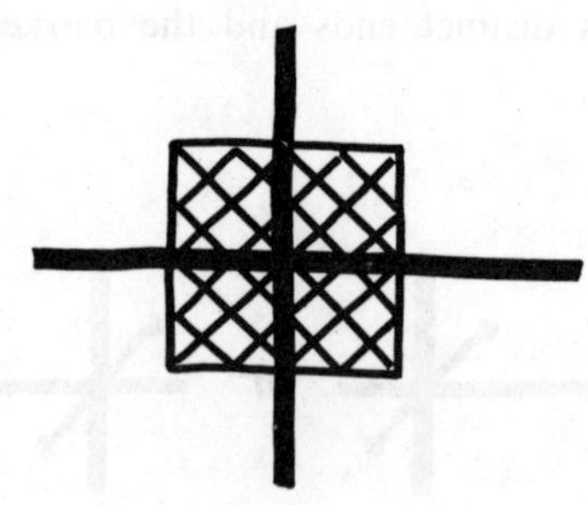

FIGURE 10

To each S-monochrome edge e of Γ we associate a plane band with core e:

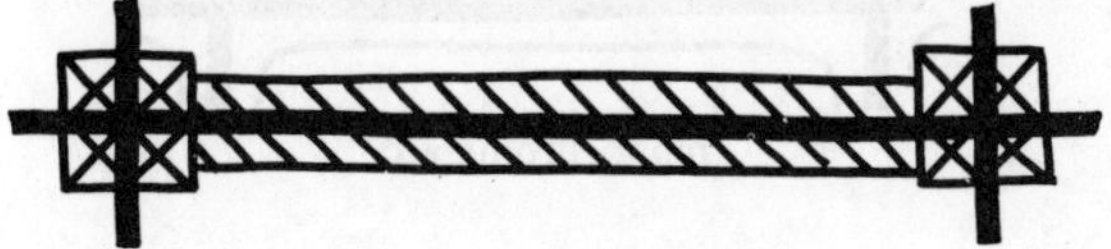

FIGURE 11

If e is a loop, the band looks like this:

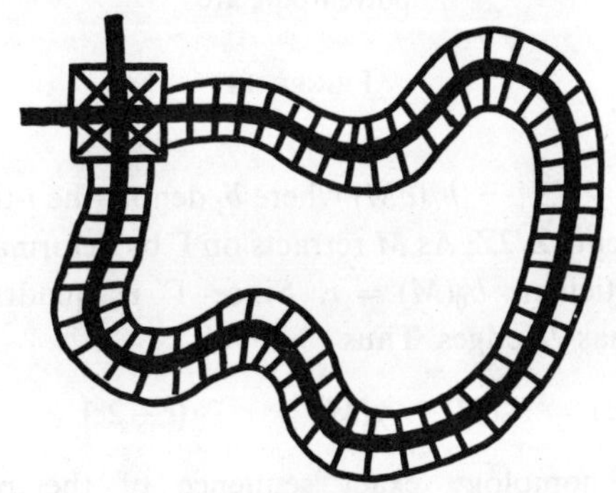

FIGURE 12

To each S-polychrome edge e we associate a 1-twisted band in R^3 with core e:

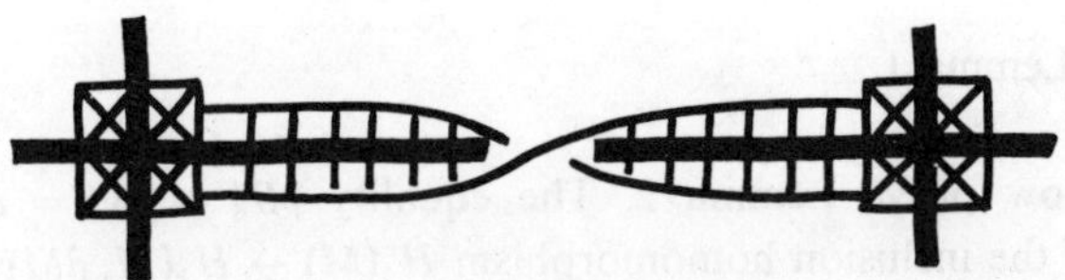

FIGURE 13

Denote by $M = M(S)$ the union of all these squares and bands. It is clear that M is a compact surface in R^3.

It is easy to check that the boundary ∂M of M is the disjoint union $\Gamma_S \amalg \Gamma_{\bar{S}}$, where it is understood that Γ_S and $\Gamma_{\bar{S}}$ are slightly moved away in R^3 to avoid intersections. See the following picture:

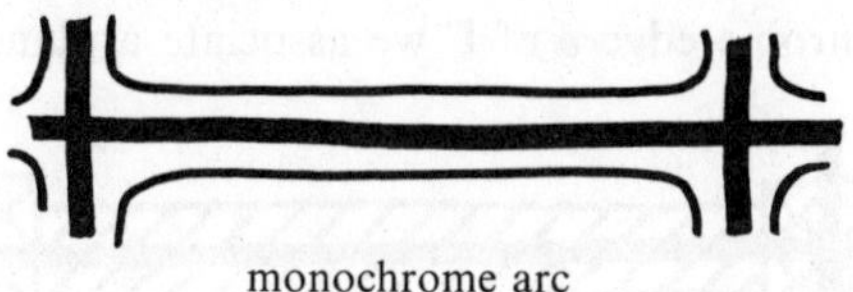
monochrome arc

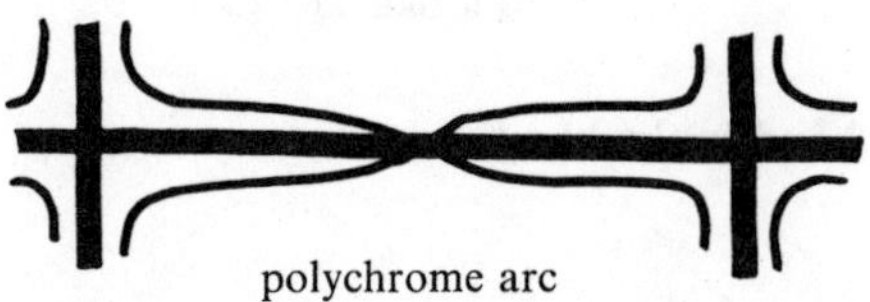
polychrome arc

FIGURE 14

Therefore $|S| + |\check{S}| = b_0(\partial M)$ where b_i denote the i-th Betti number of a space with coefficients $\mathbf{Z}/2\mathbf{Z}$. As M retracts on Γ by deformation, $b_i(M) = b_i(\Gamma)$ for all i. In particular, $b_0(M) = r$. Since Γ is quadrivalent and has c double points, Γ has $2c$ edges. Thus

$$b_1(M) = b_0(M) - \chi(M) = r - (c-2c) = r + c .$$

Consider the homology exact sequence of the pair $(M, \partial M)$ with coefficients $\mathbf{Z}/2\mathbf{Z}$:

$$\dots \to H_1(M) \to H_1(M, \partial M) \to H_0(\partial M) \to H_0(M) \to \{0\} .$$

As $b_1(M, \partial M) = b_1(M) = r + c$ by Poincaré duality, one has

$$|S| + |\check{S}| = b_0(\partial M) \leqslant b_0(M) + b_1(M, \partial M) = 2r + c .$$

This proves Lemma 1.

Let us now prove Lemma 2. The equality $|S| + |\check{S}| = c + 2$ holds if and only if the inclusion homomorphism $H_1(M) \to H_1(M, \partial M)$ in the exact sequence above is equal to zero. This happens if and only if the intersection form

$$H_1(M) \times H_1(M) \to \mathbf{Z}/2\mathbf{Z} \tag{5}$$

is zero. If S is monochrome then $M(S)$ is a planar surface, so that the form (5) is indeed zero.

Suppose that S is not monochrome. We shall prove that the form (5) is non zero. This will imply the strict inequality $|S| + |\check{S}| < c + 2$.

Let e be a S-polychrome edge of Γ. Consider the connected components of $R^2 - \Gamma$ which are adjacent to e. These components are distinct: Otherwise there would exist a simple loop in R^2 intersecting Γ in exactly one regular point, which is impossible. Denote these two components by a and b. It is clear that $\bar{a} \cap \bar{b}$ is a union of edges and double points of Γ, with in particular $e \in \bar{a} \cap \bar{b}$. If $\bar{a} \cap \bar{b}$ were to contain an edge of Γ distinct from e, then the dotted circle in the following picture would intersect Γ in two points.

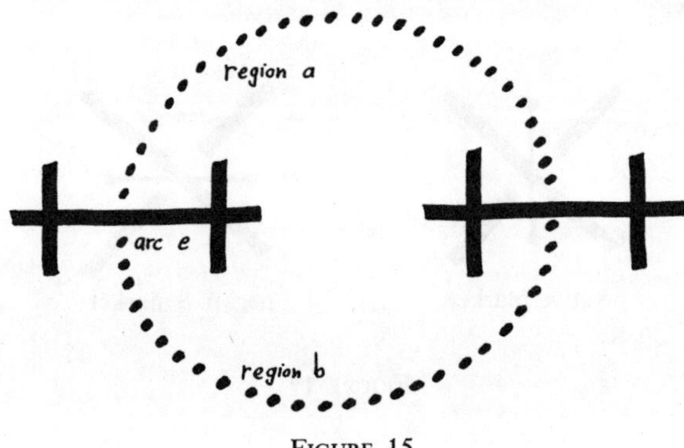

FIGURE 15

But this is impossible because Γ is prime. Thus $\bar{a} \cap \bar{b}$ is equal to the union of e and some double points.

Since e is S-polychrome, the intersection of the homology classes $[\partial \bar{a}]$ and $[\partial \bar{b}]$ in $H_1(M) \approx H_1(\Gamma)$ is equal to 1 (modulo 2):

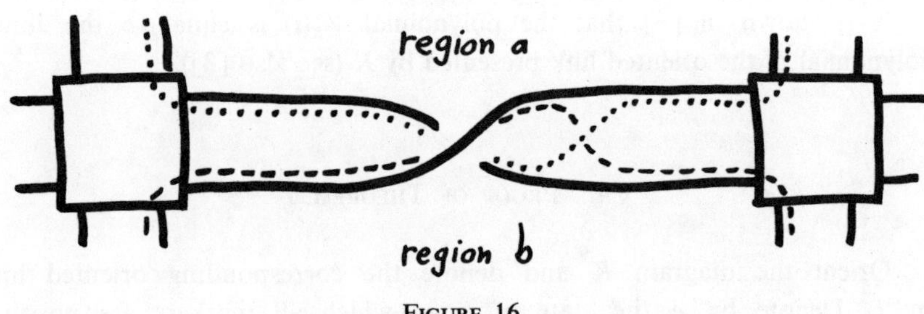

FIGURE 16

Thus (5) is a non-zero form, and the proof is complete. □

Remark. It is not important for us but curious to observe that $M(S)$ is always an orientable surface.

§ 3. Kauffman's state model for the Jones polynomial

Let K be a link diagram. By a state or a marker of K, we mean respectively a state or a marker of the corresponding link projection in R^2 (which is obtained from K by forgetting the overcrossing-undercrossing data). The markers of K are divided into two classes — positive and negative. By definition, if the over-line is rotated counterclockwise around the double point, then the first marker it meets is the positive one and the second one is negative:

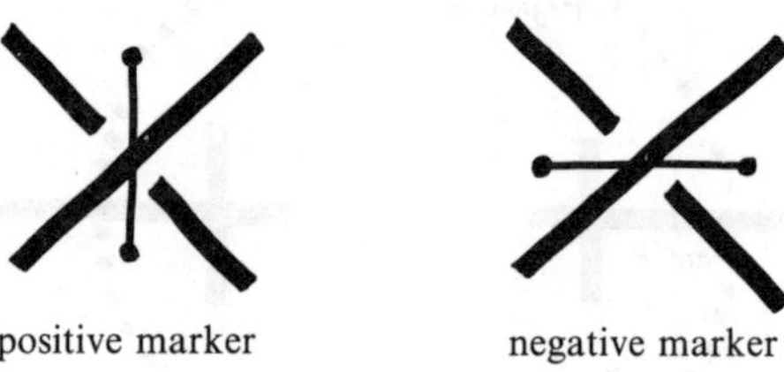

Figure 17

Let the diagram K be oriented. Consider the polynomial

$$V_K(t) = (-t)^{-3w(K)/4} \sum t^{(a_S - b_S)/4}(-t^{1/2} - t^{-1/2})^{|S|-1}$$

where $w(K)$ is the writhe number of K. The summation is over all the states S of K; the number of positive [respectively negative] markers of the state S is denoted by a_S [respectively b_S], and the number $|S|$ is defined in § 2.

It is shown in [5] that the polynomial $V_K(t)$ is equal to the Jones polynomial of the oriented link presented by K (see also [3]).

§ 4. Proof of Theorem 1

Orient the diagram K and denote the corresponding oriented link by L. Denote by A the state of K in which all markers are positive, and by $B = \check{A}$ the dual state in which all markers are negative. For any state S of K, denote by D_S and d_S respectively the maximal and minimal degrees in t in the expression

$$t^{(a_S - b_S)/4}(-t^{1/2} - t^{-1/2})^{|S|-1}$$

(see § 3), namely

$$D_S = (a_S - b_S + 2|S| - 2)/4$$
$$d_S = (a_S - b_S - 2|S| + 2)/4\,.$$

In particular

$$D_A = (c + 2|A| - 2)/4$$

(6)

$$d_B = (-c - 2|B| + 2)/4\,.$$

Proof of (i). If a state S^2 is obtained from a state S by replacing one positive marker by a negative one (at some crossing point), then $a_{S^2} = a_S - 1$, $b_{S^2} = b_S + 1$ and $|S^2| \leqslant |S| + 1$. Thus

$$D_{S^2} - D_S = -\frac{1}{2} + (|S^2| - |S|)/2 \leqslant 0$$

so that $D_{S^2} \leqslant D_S$. This implies that $D_S \leqslant D_A$ for any state S of K. Therefore

$$d_{\max}(V_L(t)) \leqslant -\frac{3}{4} w(K) + D_A$$

$$d_{\min}(V_L(t)) \geqslant -\frac{3}{4} w(K) + d_B\,.$$

Thus in view of equalities (6) and of Lemma 1 of § 2, one has

(7) $$\operatorname{span}(L) \leqslant D_A - d_B = (c + |A| + |B| - 2)/2$$
$$\leqslant (2c + 2r - 2)/2 = c + r - 1\,. \quad \square$$

Proof of (ii). Let $K_1, \ldots, K_r$ be the unsplittable components of K, with $r = r(K)$. Denote by L_i the oriented link represented by K_i. It follows from part (i) of the Theorem and from formula (1) that

$$c(K) = \sum_{i=1}^{r} c(K_i) \geqslant \sum_{i=1}^{r} \operatorname{span}(L_i) = \operatorname{span}(L) - (r-1)\,.$$

Thus the equality $c(K) + r - 1 = \operatorname{span}(L)$ holds if and only if $c(K_i) = \operatorname{span}(L_i)$ for each i. Therefore, to prove (ii), it suffices to consider the unsplittable case $r = 1$.

It is evident that the numbers $c(K)$ and span (L) are both additive under connected sum of diagrams. Therefore it is enough to prove the following assertion (*).

(*) $\left\{\begin{array}{l}\text{For a prime unsplittable diagram } K \text{ of an oriented link } L, \text{ the} \\ \text{equality } c(K) = \operatorname{span}(L) \text{ holds if and only if } K \text{ is a reduced and} \\ \text{alternating diagram.}\end{array}\right.$

In (*), note that, formally, the link L is not supposed to be prime or even unsplittable.

Suppose first that $c(K) = \operatorname{span}(L)$. Then all inequalities above are in fact equalities. As $r = 1$, one has in particular

$$|A| + |B| = c + 2r = c + 2 .$$

Lemma 2 of § 2 shows that the state A is monochrome. This implies that K is alternating, because of the easy but essential lemma:

LEMMA. *Let* K *be an oriented connected link diagram. Then* K *is alternating if and only if the state* A *is monochrome.*

Moreover the diagram K is reduced, since all prime diagrams are reduced except the two diagrams

FIGURE 18

which are excluded by the assumption $c(K) = \operatorname{span}(L)$.

Suppose conversely that K is reduced and alternating. The preceeding Lemma shows that the state A is monochrome. According to Lemma 2 of § 2: $|A| + |B| = c + 2$. We prove below that

$$d_{\max}(V_L(t)) = -\frac{3}{4}w(K) + D_A \tag{8}$$

$$d_{\min}(V_L(t)) = -\frac{3}{4}w(K) + d_B . \tag{9}$$

Thus the inequalities (7) are in fact equalities, so that $\operatorname{span}(L) = c + r - 1 = c$.

By region, we mean hereafter a connected component of $S^2 - K$. (Here $S^2 = R^2 \cup \{\infty\}$.) Since K is alternating, each region intersects either markers which are all positive or markers which are all negative. Shade the regions of the first type:

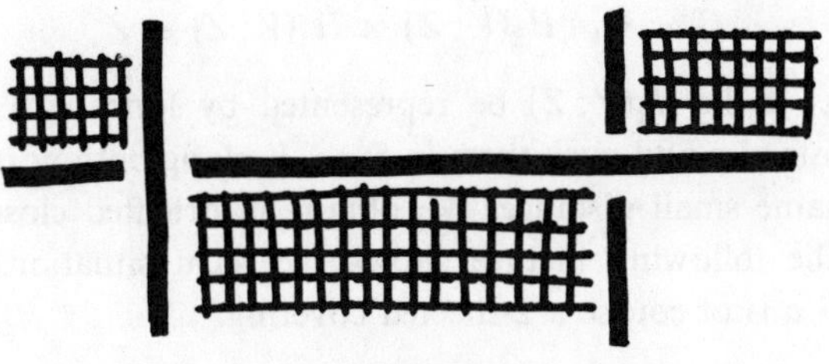

FIGURE 19

Observe that two unshaded regions near one crossing point are necessarily distinct, otherwise the diagram K would not be reduced:

FIGURE 20

It is evident that A is equal to the number of unshaded regions. Let a state S^2 be obtained from A by replacing one positive marker by the negative marker. Under this operation two distinct unshaded regions are connected by a band, and therefore $|S^2| = |A| - 1$. In view of the arguments given in the proof of part (i) of the Theorem, this implies that $D_S < D_A$ for any state S of K. This implies (8). Analogous arguments imply (9), and the proof of (ii) in Theorem 1 is complete.

§ 5. PROOF OF THEOREM 2

Let me first recall the definition of the *signature* of an oriented link L in terms of a (not necessarily orientable) surface V bounded by L (see [2]). One defines a bilinear form

$$Q = Q_V : H_1(V; Z) \times H_1(V; Z) \to Z$$

as follows. Let $\alpha, \beta \in H_1(V; Z)$ be represented by loops a, b in V. Let us double all points of a and push them in $S^3 - V$ along both normal directions to V, at the same small distance. We obtain an oriented closed 1-manifold $\tilde{a} \in S^3 - V$; the following picture shows the local situation. The natural projection $\tilde{a} \to a$ is of course a 2-sheeted covering.

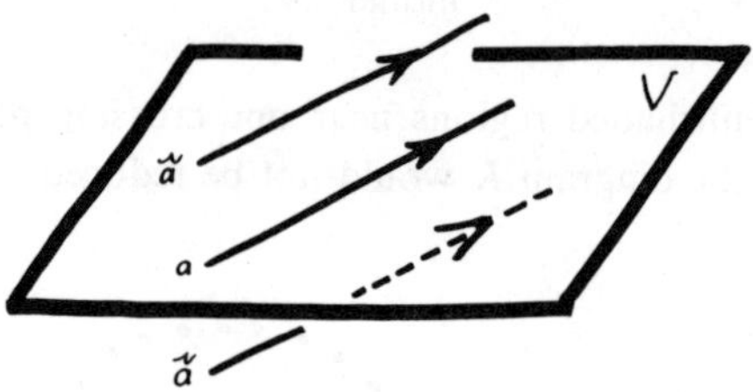

FIGURE 21

Denote by $Q(\alpha, \beta)$ the linking coefficient $Lk(\tilde{a}, b)$ of $\tilde{a}$ and b. It turns out that Q is a well defined symmetric bilinear form. Let L^V be a parallel copy of L in $S^3 - V$. Define

$$\sigma(L) = \text{sign}\,(Q) - \frac{1}{2} Lk(L, L^V)\,.$$

Here sign (Q) denotes the signature of the symmetric bilinear form obtained by factorizing out the annihilator of Q. According to [2], $\sigma(L)$ does not depend on the choice of the spanning surface V. In case V is orientable, $Lk(L, L^V) = 0$ and we get the classical definition of the signature of L due to Murasugi.

All diagrams and links being oriented, it is easy to check that the writhe number of a link diagram, the signature of a link, and the number $d_{\max}(V_L(t)) + d_{\min}(V_L(t))$ are additive with respect to both disjoint unions and connected sums of diagrams. Therefore it is enough to prove Theorem 2 for a diagram K which is connected, prime, alternating and reduced.

Let c_+ and c_- denote the numbers of positive and negative crossing points of such a K.

CLAIM (Murasugi). *One has* $\sigma(L) = |A| - 1 - c_+$.

This claim implies Theorem 2. Indeed, formulas (8), (9) and (6) show that

$$d_{\max}(V_L(t)) + d_{\min}(V_L(t)) + w(K)$$
$$= -w(K)/2 + D_A + d_B = -w(K)/2 + (|A|-|B|)/2\,.$$

Substituting in the last expression

$$w(K) = c_+ - c_-$$
$$|B| = c + 2 - |A|$$
$$c = c_+ + c_-$$

we obtain

$$d_{\max}(V_L(t)) + d_{\min}(V_L(t)) + w(K)$$
$$= |A| - 1 - c_+ = \sigma(L)\,.$$

This implies Theorem 2.

Proof of the Claim. There is a spanning surface V of L associated with the diagram K. It is built up from shaded regions of $S^2 - K$ (see § 4) and small bands connecting these regions which enter one crossing point. In a neighbourhood of a crossing point, V looks like this:

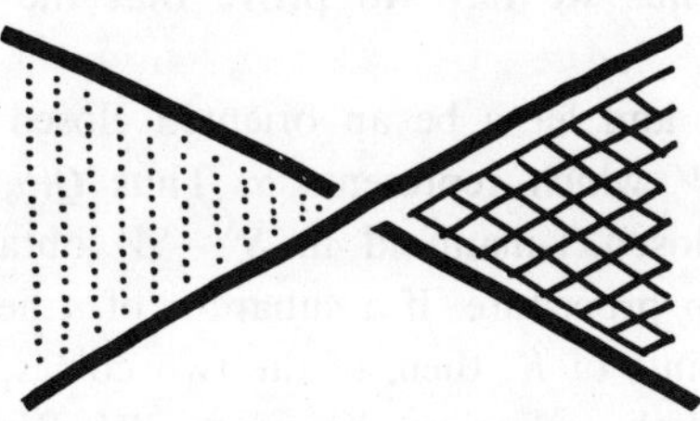

FIGURE 22

We shall prove the claim by using this surface V.

We prove first that the number $-\frac{1}{2}Lk(L, L^V)$ is equal to $-c_+$. We may assume that the push-off L^V of L in $S^3 - V$ lies in the unshaded regions of R^2 except in a neighbourhood of the crossing points. The following picture shows L^V near a crossing point (the orientations of L and L^V are not shown).

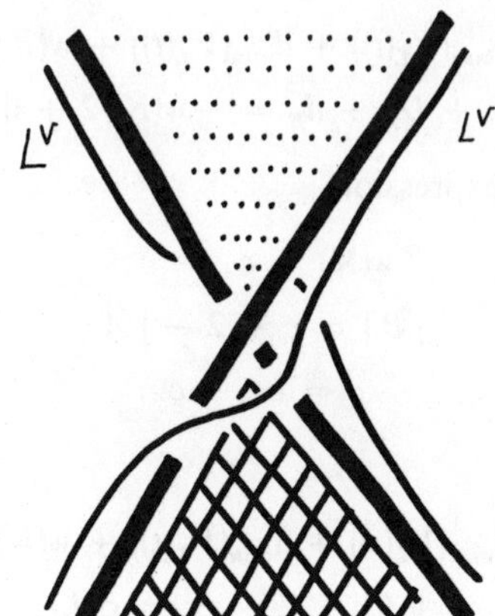

FIGURE 23

We compute $Lk(L, L^V)$, by counting the algebraic number of times L^V passes under L. It is easy to check that each crossing point of L contributes with a 2 if it is positive and with a 0 if it is negative. Thus $Lk(L, L^V) = 2c_+$.

Now, we prove that $\text{sign}(Q_V) = |A| - 1$. The surface V retracts by deformation onto the complement on the unshaded regions in S^2. As the diagram is alternating, the number of unshaded regions is $|A|$, so that $b_1(V) = |A| - 1$. Thus we have to prove that the form Q_V is positive definite.

Let $\alpha \in H_1(V; Z)$ and let a be an oriented closed 1-manifold (possibly non connected) in V which represents α. Thus $Q(\alpha, \alpha) = Lk(\tilde{a}, a)$, where $\tilde{a}$ is the oriented closed 1-manifold in $S^3 - V$ obtained from a by the 2-sheeted blowing up procedure. If a subarc x of a lies in a shaded region far from crossing points of K, then, of the two corresponding subarcs of $\tilde{a}$, one lies over R^2 and the other one lies under R^2. We shall always picture the first (higher) subarc of $\tilde{a}$ on the right side of x (looking from above along a) and the second (lower) subarc of $\tilde{a}$ on the left side of x; see the following picture.

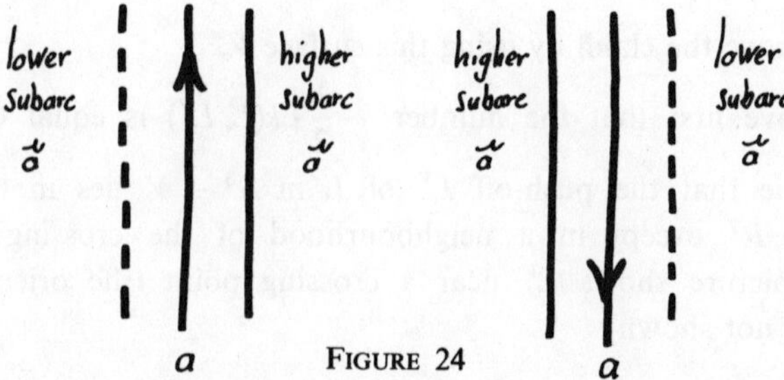

FIGURE 24

Note that the diagram of $\tilde{a}$ misses the diagram of a except in a neighborhood of the crossing points. Surgering if necessary a in V, we may assume that all components of a go through any band of V in one direction. Positions of a like those in the following picture may easily be removed by surgery.

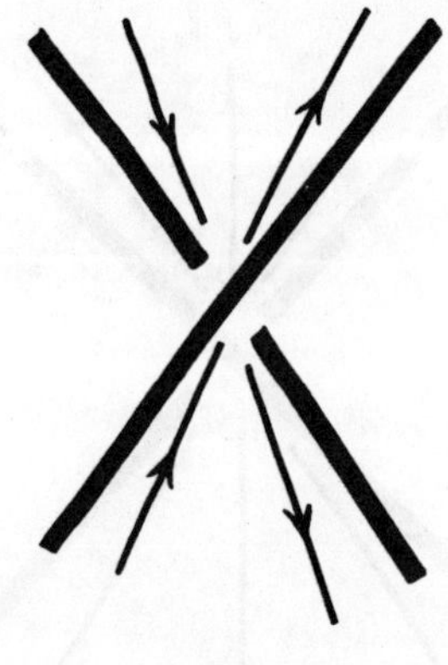

FIGURE 25

For simplicity, consider first a neighbourhood of a crossing point through which a goes only once:

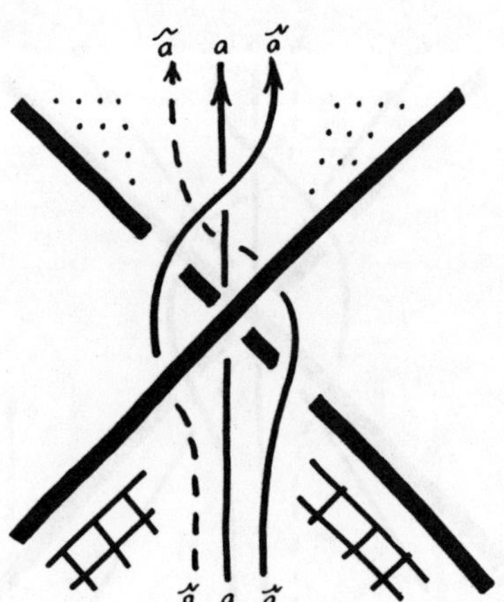

FIGURE 26

It is clear that $\tilde{a}$ passes under a in this neighbourhood one time from right to left.

If a goes through a neighbourhood $\mathscr{U}$ of a crossing point n times, then the relative positions of the corresponding n arcs of a, say $x_1, \ldots, x_n$, are represented as follows:

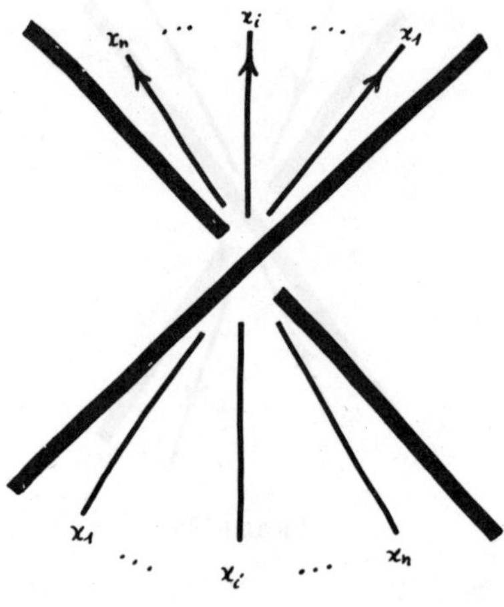

FIGURE 27

In the next picture, we show the two arcs of $\tilde{a}$ which correspond to x_i:

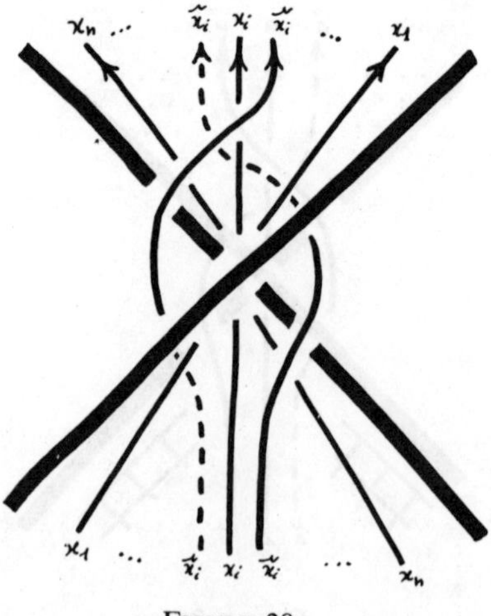

FIGURE 28

It is clear that these two arcs of $\tilde{a}$ pass $2i-1$ times from right to left under a. Thus the contribution of the neighbourhood $\mathcal{U}$ to $Q(\alpha, \alpha)$ is given by

$$\sum_{i=1}^{n} (2i-1) = -n + 2\sum_{i=1}^{n} i = n^2 .$$

This shows that $Q(\alpha, \alpha) > 0$ if a crosses at least one band of V. If not, then $\alpha = 0$.

Thus Q is positive definite. This completes the proof of Theorem 2.

Appendix: An improvement of the inequality of Theorem 1

Though the inequality

$$c(K) + r(K) - 1 \geqslant \operatorname{span}(L) \tag{10}$$

of Theorem 1 becomes an equality for weakly alternating diagrams, it may be sharpened a little for other cases. Let K be a link diagram in R^2 and let $\Gamma \subset R^2$ be the associated link projection. For $P \in S^2 - \Gamma$ (where $S^2 = R^2 \cup \{\infty\}$), let $i(P)$ be the intersection number modulo 2 of Γ with a generic 1-chain connecting P to ∞. Shade the regions of $S^2 - \Gamma$ for which $i \equiv 1 \pmod 2$, so that S^2 is painted like a chessboard. Let $b_1, \ldots, b_m$ be the shaded regions of $S^2 - \Gamma$ and let $w_1, \ldots, w_n$ be the unshaded regions of $S^2 - \Gamma$.

An edge e of Γ is called *K-good* either if e is a loop or if one of the end points of e corresponds to an overcrossing point of K and the other end point of e corresponds to an undercrossing point of K. An edge of Γ which is not K-good is called *K-bad.* For any $i \in \{1, \ldots, m\}$ and for any $j \in \{1, \ldots, n\}$, it is clear that the set $\overline{b_i} \cap \overline{w_j}$ consists of several edges and double points of Γ. Denote by $a(i, j)$ the number modulo 2 of K-bad edges in $\overline{b_i} \cap \overline{w_j}$. Denote by $u(K)$ the rank of the $m-by-n$ matrix $(a(i, j))$.

THEOREM. *If K is a diagram of a link L, then*

$$c(K) + r(K) - 1 \geqslant \operatorname{span}(L) + u(K) . \tag{11}$$

COROLLARY. *If K is a diagram of an unsplittable link L, then*

$$c(K) \geqslant \operatorname{span}(L) + u(K) .$$

Of course, if K is a weakly alternating diagram, then $u(K) = 0$.

The inequalities of the Theorem and of the Corollary may be strict. For example, if we take the diagram $K = 8_{19}$ in Rolfsen's book, then $\operatorname{span}(8_{19}) = 5$ and $u(K) = 2$, so that the inequality (11) amounts to $8 > 7$. Unfortunately, even in the case where (11) is an equality, it does not mean that K is a minimal diagram of L, since $u(K)$ depends on K and is not an invariant of L.

The proof of the Theorem goes along the same lines as the proof of Theorem 1 of § 1. Indeed the proof of Lemma 1 of § 2 shows in fact that $|S| + |\check{S}| \leqslant c + 2r - R$, where R is the rank of the intersection form (5). For the state A, it is easy to show that $R = 2u(K)$, and this gives the desired result.

REFERENCES

[1] Crowell, R. H. Nonalternating links. *Ill. J. Math. 3* (1959), 101-120.
[2] Gordon, C. McA and R. A. Litherland. On the signature of a link. *Invent. math. 47* (1978), 53-69.
[3] de la Harpe, P., M. Kervaire and C. Weber. On the Jones polynomial. *L'Enseignement math. 32* (1986), 271-335.
[4] Jones, V. A polynomial invariant for knots via von Neumann algebras. *Bull. Amer. Math. Soc. 12* (1985), 103-111.
[5] Kauffman, L. State models and the Jones polynomial. Preprint, 1986.
[6] Murasugi, K. Jones polynomials and classical conjectures in knot theory. *Topology 26* (1987), 187-194.
[7] Rolfsen, D. *Knots and links.* Publish or Perish 1976.
[8] Tait, P. G. On Knots I, II, III. *Scientific papers, Vol. 1* (1898), 273-347.
[9] Thistlethwaite, M. B. Kauffman's polynomial and alternating links. Preprint, 1986.

(Reçu le 7 avril 1987)

V. G. Turaev

Steklov Inst. of Math.
Fontaka 27
Leningrad (191011)
USSR

Contemporary Mathematics
Volume 78, 1988

THE JONES POLYNOMIAL OF SATELLITE LINKS AROUND MUTANTS

H. R. Morton and P. Traczyk

ABSTRACT. We prove that the (n,k)-cables around a mutant pair of knots C_1 and C_2 cannot be distinguished by the Jones polynomial V, for any (n,k). We prove further that the same result holds for any other satellite around C_1 and C_2.

1. INTRODUCTION.

Lickorish and Lipson, [LL], showed that if C_1 and C_2 are a mutant pair of knots then their 2-cables cannot be distinguished by the 2-variable polynomial P.

Here we prove that the (n,k)-cables around a mutant pair of knots C_1 and C_2 cannot be distinguished by the Jones polynomial V, for any (n,k). We prove further that the same result holds for any other satellite around C_1 and C_2.

The Jones polynomial for oriented links, originally defined via von Neumann algebras in [J], may be introduced in a simple combinatorial way using Kauffman's approach of regular isotopy, [K]. We recall his definition here.

The <u>bracket polynomial</u>, $\langle\ \rangle$, taking values in $Z[A^{\pm 1}]$, is an invariant of unoriented link diagrams which is defined up to regular isotopy (Reidemeister moves 2 and 3) by the following properties:

(1) $\langle \times \rangle = A \langle \asymp \rangle + A^{-1} \langle)(\rangle$

where the diagrams differ only as shown.

(2) $\langle 0 \sqcup K \rangle = (-A^{-2} - A^2) \langle K \rangle$

where the diagram $0 \sqcup K$ is the distant union of a circle without crossings and the diagram K.

(3) $\langle 0 \rangle = 1$

From a diagram of an unoriented link K we set $f[K] = (-A)^{-3w(K)} \langle K \rangle$, where $w(K)$ is the twist number of the diagram used, i.e. the sum of the crossings with sign. Then $f[K]$ is invariant under all three Reidemeister moves, and the Jones polynomial $V_K(t)$ is given by $f[K](t^{-1/4})$.

1980 <u>Mathematics Subject Classification</u> (1985 <u>Revision</u>). 57M25.
Second author supported by SERC grant no. GR/C/48974

2. TANGLES AND MUTANTS.

We shall refer to an n-tangle constructed from two n-tangles S and R as in figure 1 as their product, SR.

Figure 1.

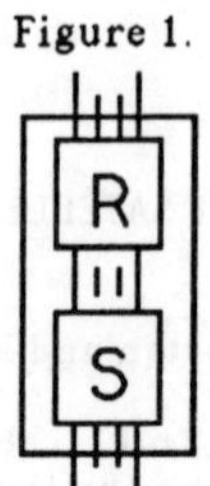

By analogy with braids we use the term closure of a tangle T, written T^, to mean the link formed by joining the n points at the top of T to those at the bottom without further crossings as shown in figure 2.

Figure 2.

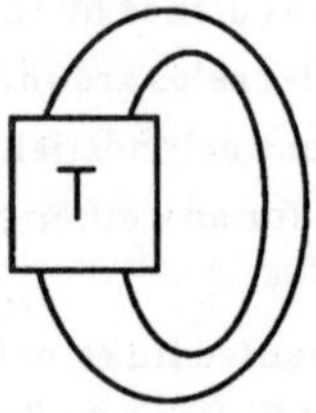

In what follows we shall mainly be concerned with unoriented tangles whose diagrams are defined up to regular isotopy, so that we may refer unambiguously to the bracket polynomial of their closure.

Figure 3.

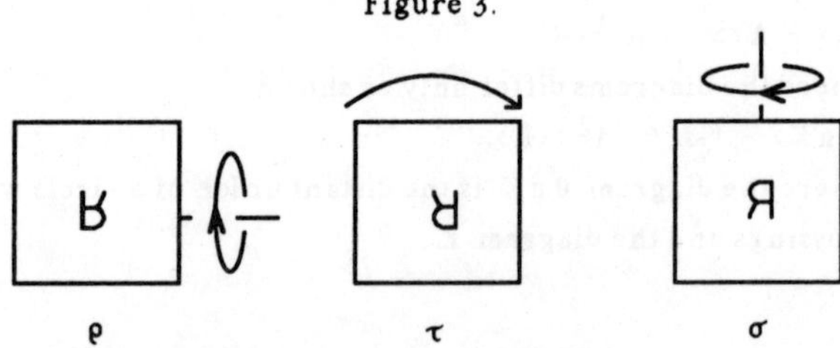

For an n-tangle R we define three rotated tangles ρR, σR and τR, by rotating R through π as shown in figure 3. If a link L has a diagram which can be written as QT^ where Q and T are 2-tangles then any link formed by rotating one of these tangles is called a mutant of L. Using the same rotation on both tangles yields a link isotopic to L, and so, since $\sigma = \rho\tau$, we can write the three possible mutants which arise from this decomposition as $(\rho Q)T^\wedge$, $Q(\tau T)^\wedge$ and $(\rho Q)(\tau T)^\wedge$.

3. BANDED TANGLES.

From a 2-tangle T in which the strings join the top points to the bottom points we shall construct an n-tangle S by banding about T. This consists of replacing each string by a band of m or k parallel strings, with n = m + k, so that crossings are replaced by band crossings as in figure 4. We allow the extreme cases m = 0 or k = 0.

Figure 4.

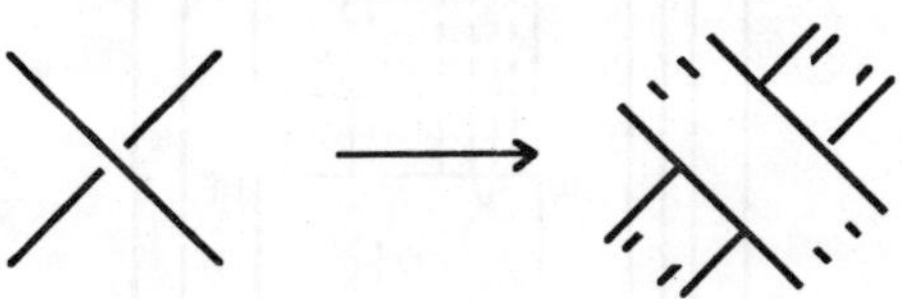

We distinguish two types of 2-tangle T, depending on whether the top points in T are eventually joined to the corresponding bottom points, giving parallel connection, or to the diagonally opposite points, giving diagonal connection.

In considering diagrams for a link L and any mutants we may assume, after a simple redrawing, if necessary, that neither of the 2-tangles from which L has been constructed contains a string joining its two top points, so that the tangles each have parallel or diagonal connections and may be banded as above.

THEOREM 1. Let S be an n-tangle given by banding a 2-tangle T, and let R be any n-tangle. Write L = SR^.

(a) Suppose that T has parallel connections. Write $\rho L = S(\rho R)^\wedge$, and suppose that L and ρL are oriented so that the orientation on S in ρL is the same as that in L, while the orientation on ρR is the reverse of that induced by rotating R. Then $V(\rho L) = V(L)$.

(b) Suppose that T has diagonal connections. Under the assumptions of (a) with ρ replaced by τ we have $V(\tau L) = V(L)$.

PROOF. We shall prove (a). The proof of (b) is analogous.

It is immediate that the diagrams for L and ρL have the same twist number, $w(L) = w(\rho L)$. It is then enough to forget the orientation and prove that $\langle L\rangle = \langle \rho L\rangle$.

We can begin to construct a binary tree for computing $\langle L\rangle$ by applying the formula ‹✕› = A‹≍› + A^{-1}‹)(› to a crossing c in the tangle R. Simultaneously we compute $\langle \rho L\rangle$ by applying this formula to the crossing ρc in ρR. Induction on the number of crossings in R reduces the problem to the case when R is an n-tangle with no crossing points. We may assume further that R has no free circles.

We now proceed by induction on n. The end points of R lie in one of four quadrants, distinguished by which of the bands in S is attached to them. Since T has parallel connections the band attached to the NW quadrant of R will eventually emerge from S and attach to the SW quadrant, as in figure 5.

Figure 5.

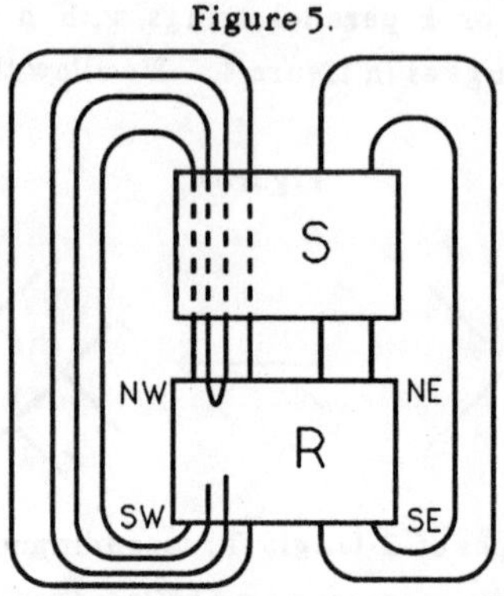

(i) Suppose that there is an arc in R whose endpoints are adjacent and lie in the same quadrant, say the NW quadrant. We may isotop this arc by regular isotopy through the band in S until it reappears in the SW quadrant, altering the diagram SR to S_1R_1, where S_1 is banded from T using two string less on one of the bands than for S. Then $\langle SR^\wedge \rangle = \langle S_1R_1{}^\wedge \rangle$. In R the corresponding arc has endpoints in the SW quadrant. Isotopy through the same band of S will replace $S(\rho R)$ by $S_1(\rho R_1)$. The required result, that $\langle SR^\wedge \rangle = \langle S(\rho R)^\wedge \rangle$, then follows by induction on n.

(ii) If no arcs in R have adjacent endpoints in the same quadrant it follows readily that R has the form shown in figure 6. Then $\rho R = R$ and there is nothing further to prove.

Figure 6.

The following result, proved in a more general setting by Lickorish, [L], is an immediate corollary.

COROLLARY. Let S be given by banding a 1-tangle, and let L be represented by $SR^\wedge$, for some tangle R. Then R may be replaced by ρR or τR without altering the Jones polynomial, so long as the orientations for ρR or τR can be chosen consistently as above.

PROOF. Regard S as given by banding a 2-tangle, using no strings on one of the bands, and apply Theorem 1.

4. SATELLITES.

We now apply Theorem 1 to satellites around mutant knots. A diagram for an r-strand satellite around a knot C, with pattern P, can be easily described, starting from a diagram of C and an r-tangle P. To compare this with other descriptions of satellites, e.g. [MS], we must view P^ as lying in an unknotted solid torus, with the top of the tangle forming a meridian disk. Then we band the diagram of C with an r-band, and modify the band by inserting P, as shown in figure 7, together with -w(C) full twists, where w(C) is the twist number of the given diagram for C.

Figure 7

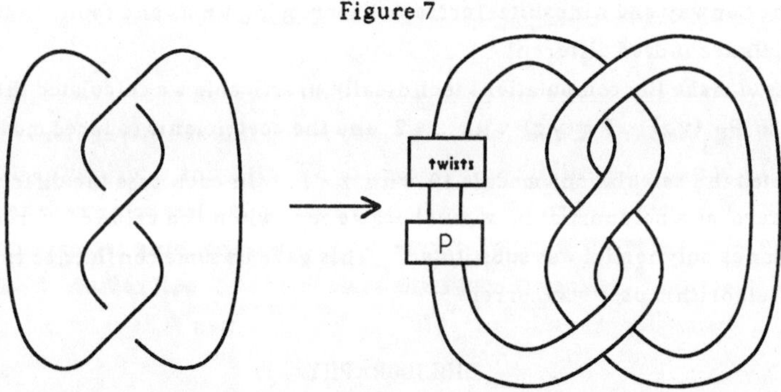

THEOREM 2. Let C_1 be a knot with diagram QT^, where Q and T are 2-tangles, and let C_2 be any mutant of C_1. Let K_1 and K_2 be the satellites of C_1 and C_2 respectively, constructed with the same pattern P. Then $V(K_1) = V(K_2)$.

PROOF. Since C_1 is a knot we may assume that one of Q and T has parallel connections, T say, and that the other has diagonal connections. Then a diagram for C_2 can be chosen as either $(\rho Q)T^\wedge$, $Q(\tau T)^\wedge$ or $(\rho Q)(\tau T)^\wedge$. Let $T^{(r)}$ (and similarly $Q^{(r)}$) be the 2r-tangle formed by banding each string of T with an r-band. Then the satellite K_1 has a diagram $\overline{P}Q^{(r)}T^{(r)\wedge}$, where $\overline{P}$ is the 2r-tangle constructed using P and a number of full twists on the first r strands only. The number of twists required will be the same for C_1 as for any of the possible diagrams for C_2, since the diagrams have the same twist number.

Apply Theorem 1, with $S = T^{(r)}$ and $R = \overline{P}Q^{(r)}$ to show that K_1 and $(\rho R)S^\wedge$ have the same polynomial. Now $\rho R = (\rho Q^{(r)})(\rho\overline{P})$ so the new link is the satellite of $C_2 = (\rho Q)T^\wedge$ constructed with pattern ρP, since the twists in the band may be inserted at any point. By the corollary to Theorem 1 we may replace P by ρP (or

τP) in forming a satellite, without altering V. Hence the satellite K_2 of C_2 with pattern P has $V(K_1) = V(K_2)$.

A similar argument, taking $S = Q^{(r)}$, with τ in place of ρ, will work for the second possible diagram for C_2. This same argument, applied now to $(\rho R)T^{(r)}$ above, taking $S = \rho Q^{(r)}$, deals with the remaining possibility.

5. FURTHER CALCULATIONS.

The result presented here was suggested by an earlier algebraic approach to the problem, which in a natural sense represents a dual attack, [MT]. This appraoch also suggested that the corresponding stronger result for the 2-variable polynomial P, proved in [LL] for 2-cables around mutants, was unlikely to hold for 3-cables around mutants.

We have subsequently compared P for corresponding 3-cables K_1 and K_2 around the Conway and Kinoshita-Teresaka 11-crossing knots and found that the polynomials are indeed different.

So as to make the computations technically practicable we calculated the difference $P_{K_1}(v,z) - P_{K_2}(v,z)$ with $z = 2$, and the coefficients reduced modulo 23, and repeated the calculations modulo 19 with $z = 1$. In each case the difference was non-zero, as a polynomial in v, but became zero when the value of v required to give the Jones polynomial was substituted. This gave us some confidence that the computer algorithm used was correct.

BIBLIOGRAPHY

[J] Jones, V.F.R. "A polynomial invariant for knots via von Neumann algebras," Bull. Amer. Math. Soc. 12 (1985), 103-111.

[K] Kauffman, L.H. "State models for knot polynomials," Preprint, Chicago University, (1985).

[L] Lickorish, W.B.R. "Linear skein theory and link polynomials," Preprint, Cambridge University, (1986).

[LL] Lickorish, W.B.R. and Lipson, A.S. "Polynomials of 2-cable-like links," Preprint, Cambridge University, (1986).

[MS] Morton, H.R. and Short, H.B. "The 2-variable polynomial of cable knots," Math. Proc. Camb. Philos. Soc. 101 (1987), 267-278.

[MT] Morton, H.R. and Traczyk, P. "Knots, skeins and algebras," Preprint, Liverpool University, (1987).

DEPARTMENT OF PURE MATHEMATICS
THE UNIVERSITY, PO BOX 147,
LIVERPOOL L69 3BX ENGLAND.

INSTYTUT MATEMATYKI
PKiN IXp
00-901 WARSZAWA POLAND

PROCEEDINGS OF THE
AMERICAN MATHEMATICAL SOCIETY
Volume 102, Number 3, March 1988

JONES' BRAID-PLAT FORMULA AND A NEW SURGERY TRIPLE

JOAN S. BIRMAN[1] AND TAIZO KANENOBU[2]

(Communicated by Haynes R. Miller)

ABSTRACT. A link $L_\beta(2k, n-2k)$ is defined by a type $(2k, n-2k)$ pairing of an n-braid β if the first $2k$ strands are joined up as in a plat and the remaining $n-2k$ as in a closed braid. The main result is a formula for the Jones polynomials of $L_\beta(2k, n-2k)$, valid for all k, $0 \leqslant 2k \leqslant n$, which generalizes and relates earlier results of Jones for the cases $n = 0$ and $2k$.

1. Introduction. Let B_n denote Artin's n-string braid group **[Ar]**, $n = 1, 2, 3, \ldots$. Thinking of elements of B_n as geometric braids, there are two well-known ways, dating back to Alexander **[Al]** and Reidemeister **[R]**, to construct a link from an element α of B_n. The first identifies the n free ends at the beginning of α with the n free ends at the end of α to form an (oriented) *closed braid*, denoted $\tilde{\alpha}$. The second, assuming n to be even, identifies adjacent pairs of strands at each end of α to form an (unoriented) *plat* $\hat{\alpha}$.

Let A_n, $n = 1, 2, \ldots$, be the sequence of von Neumann algebras described in **[J1]** generated by projections $e_1, \ldots, e_{n-1}$. Let r_t: $B_n \to A_n$, $t \in \mathbf{C}$, be the 1-parameter family of representations of B_n in A_n, and let tr: $A_n \to \mathbf{C}$ be the Jones trace. For each $\alpha \in B_n$, Jones defines a polynomial

$$V_{\tilde{\alpha}}(t) = \delta^{n-1} \operatorname{tr}(r_t(\beta)), \tag{1}$$

where $\delta = -(1+t)/\sqrt{t}$, and shows that it is an invariant of the oriented link type of $\tilde{\alpha}$.

Starting with a fixed braid $\alpha \in B_{2k}$ it seems on the face of it that the link types of $\tilde{\alpha}$ and $\hat{\alpha}$ are quite unrelated; for example, one may alter α so that $\tilde{\alpha}$ remains fixed, while $\hat{\alpha}$ is changed from a knot to a link. It therefore seemed remarkable when Jones discovered in early 1985 (see **[J2]**) that their polynomial invariants were given by closely related formulas, viz (1) and

$$V_{\hat{\alpha}}(t) \equiv \delta^{3k-1} \operatorname{tr}(r_t(\alpha) e_1 e_3 \cdots e_{2k-1}), \tag{2}$$

Received by the editors February 13, 1986.
1980 *Mathematics Subject Classification* (1985 *Revision*). Primary 55A25.
Key words and phrases. Jones polynomial, link, link diagram, braid, plat, orientation.
[1]Supported in part by NSF grant DMS-8503758.
[2]Supported in part by Grant-in-Aid for Encouragement of Young Scientists (No. 6174-0048), Ministry of Education, Science and Culture.

where $\equiv$ means up to a multiplicative power of $\sqrt{t}$. Note that the discovery of (2) as reported in **[J2]** was inspired by work on the Potts model in statistical mechanics, and so involved a chain of ideas which seemed very far from the studies which led to the discovery of (1).

One can also define a link from a closed braid by joining up the $2n$ free ends of the braid strings in pairs in still other ways. Assuming that the braid is defined by a regular projection onto a plane P, we restrict our attention to pairings in which the free ends of α are joined up by n disjoint arcs in P (since if the arcs were to cross, the crossings could be subsumed into the braid). Call such a pairing *type* $(2k, n-2k)$ if the first k pairs of strings at both ends of the braid are joined up as in a plat and the remaining $n-2k$ as in a closed braid (see Figure 1a) and let $L_\alpha(2k, n-2k)$ denote the link so obtained. Up to canonical modifications in the defining braid, all noncrossed pairings may be obtained in this way.

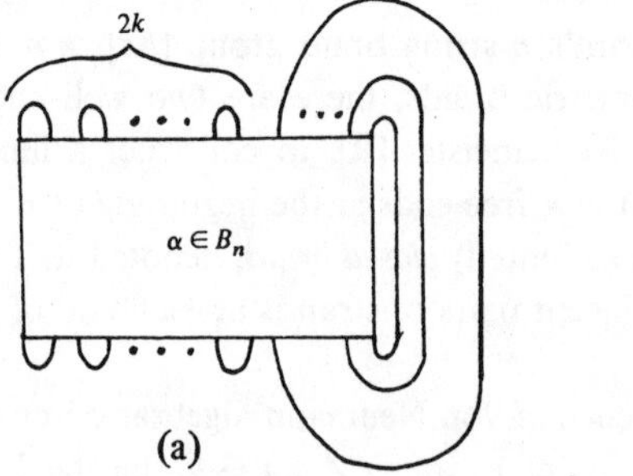

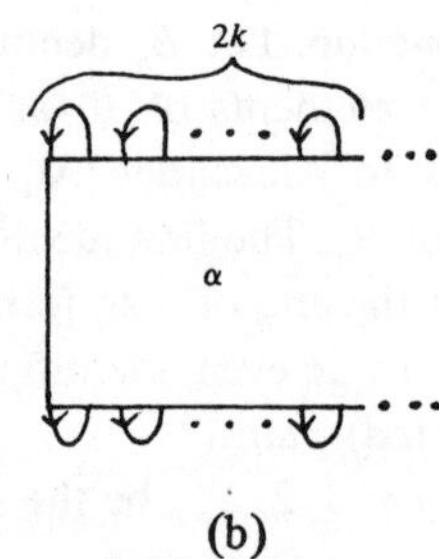

FIGURE 1

The first result in this paper (Theorem 1) generalizes formulas (1) and (2) to a single formula for the Jones polynomial of $L_\alpha(2k, n-2k)$, $0 \leqslant 2k \leqslant n$. At the same time we will show that if one places a mild restriction on the defining braid α, then there is a well-defined "standard" way to orient the plat $L_\alpha(2k, n-2k)$, and with this choice of orientation we can identify the multiplicative power of t in (2) precisely. Our proof is elementary and different from that in **[J2]**, and it explains the relationship between (1) and (2).

The second result is an application of Theorem 1 to prove that there is a "surgery triple" which relates the polynomials V_{L_1}, $V_{L_{-1}}$, V_{L_∞} of three links L_1, L_{-1}, L_∞. Let L_0, L_1, L_{-1}, L_∞ be four links which are defined by diagrams (no longer necessarily coming from braids) which are identical everywhere outside a small disc D, and are as illustrated in Figure 2 in D. Note that L_0, L_1, L_{-1} are all oriented, but the component of L_∞ which is associated to the strands inside D has no natural

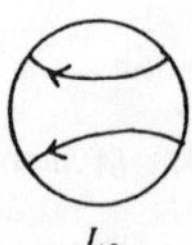

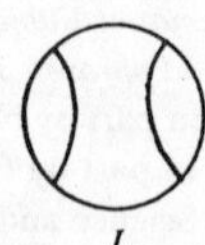

FIGURE 2

orientation. By Theorem 12 of [**J1**] the Jones polynomials of L_0, L_1, L_{-1} are related by

$$(3) \qquad V_{L_0} = \frac{1}{\mu}\left(t^{-1}V_{L_1} - tV_{L_{-1}}\right),$$

where $\mu = (t-1)/\sqrt{t}$. (Caution: see Comment 4.2 at the end of this paper.) We will prove (Theorem 2) that there is a similar formula which relates V_{L_∞}, V_{L_1}, and $V_{L_{-1}}$. The most interesting case (the Corollary) occurs when L_1, and hence also L_{-1} and L_∞, are knots.

REMARKS. This manuscript subsumes two earlier versions, [**B**] and [**K**]. Our results in those two papers were related, with [**B**] proving a more general result than [**K**], but by a more complicated method than that in [**K**]. We decided to combine our two papers.

Theorem 2 was originally proved simultaneously by Jones (private letter) and by the first author in [**B**]. It was then used by Lickorish in [**L**] where a different proof is given. The proof which we give here is different from both that in [**B**] and that in [**L**].

2. The Jones polynomial of $L_\alpha(2k, n-2k)$. The key to pinning down the multiplicative power of t in (2) and in its generalization (below) will be seen to be related to having control over orientations in $L_\alpha(2k, n-2k)$. With this in mind we begin by placing a (mild) restriction on the defining braid $\alpha \in B_n$. The braid α will be said to be *admissible* for $L_\alpha(2k, n-2k)$ if the partial orientations on $L_\alpha(2k, n-2k)$ are those indicated in Figure 1b.

LEMMA 1. *Any braid $\beta \in B_n$ may be altered to an admissible braid α, with $L_\alpha(2k, n-2k) = L_\beta(2k, n-2k)$, by adding appropriate half-twists at the top and bottom of β.*

PROOF. Clear. □

Assume from now on that $\alpha \in B_n$ is admissible for $L_\alpha(2k, n-2k)$, and that the plat part has been oriented as in Figure 1b. If this does not determine orientations on all components of $L_\alpha(2k, n-2k)$, then the unoriented components may be oriented as in a closed braid, say from top to bottom in the braid part. We call this the *standard orientation* for $L_\alpha(2k, n-2k)$. Observe that the standard orientation on $L_\alpha(2k, n-2k)$ determines orientations on the individual braid strands of α. When we need to specify these, we will write $\overleftrightarrow{\alpha}$, to distinguish $\overleftrightarrow{\alpha}$ from $\vec{\alpha}$, the same braid with each strand oriented top-to-bottom.

LEMMA 2. *Let $\overleftrightarrow{\alpha}$ be defined as above. If the strands of $\overleftrightarrow{\alpha}$ are identified by a type $(0, n)$ pairing, as in a closed braid, then the resulting link $\overleftrightarrow{L}_\alpha(0, n)$ can be given a consistent orientation, using the orientations on the strands of $\overleftrightarrow{\alpha}$.*

PROOF. Modify the diagram for $L_\alpha(2k, n-2k)$ by pulling the upper plat parts around the braid until they are next to the corresponding lower plat parts, as in Figure 3. Now do "surgery" on the link, changing the tangles inside the dotted discs from the tangle labeled 0 in Figure 2 to the tangle labeled 1. The result will be $\overleftrightarrow{L}_\alpha(0, n)$ oriented consistently to agree with the orientation on $\overleftrightarrow{\alpha}$. □

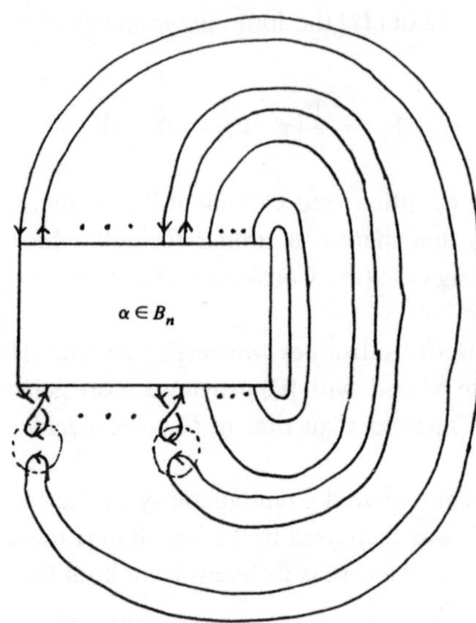

FIGURE 3

We are now ready to state and prove our first result. Let $\vec{L}_\alpha(0,n)$ be the closed braid link determined by $\vec{\alpha}$. Then $\overleftarrow{L}_\alpha(0,n)$ is obtained from $\vec{L}_\alpha(0,n)$ by reversing the orientation of the "wrongly ordered" sublink $\overleftarrow{K}$. Let λ be the linking number $\operatorname{lk}(\overleftarrow{K}, \overleftarrow{L}_\alpha(0,n) - \overleftarrow{K})$.

THEOREM 1. *Let $\alpha \in B_n$ be admissible for $L_\alpha(2k, n-2k)$. Assume that the link $L_\alpha(2k, n-2k)$ has the standard orientation. Then*

$$V_{L_\alpha(2k,n-2k)} = t^{3\lambda}\delta^{n+k-1}\operatorname{tr}\bigl(r_t(\alpha)e_1e_3\cdots e_{2k-1}\bigr). \tag{4}$$

PROOF. See Figure 3. Let $\{L_{\varepsilon_1\varepsilon_2\cdots\varepsilon_K};\ \varepsilon_i \in \{0,1,-1\}\}$ be the collection of 3^k links which are obtained from the link in Figure 3 by inserting in each of the k discs the tangles from Figure 2 which correspond to the choices of $\varepsilon_1, \ldots, \varepsilon_K$. Note that each triplet $(L_{\varepsilon_1\cdots 0\cdots\varepsilon_K}, L_{\varepsilon_1\cdots 1\cdots\varepsilon_K}, L_{\varepsilon_1\cdots -1\cdots\varepsilon_K})$ is a 0, 1, -1 surgery triple, so that the corresponding Jones polynomials are related by equation (3).

Consider the special cases when each $\varepsilon_i = +1$. Examples are given in Figure 4, when $2k = n = 4$. Each $\overleftarrow{L}_{\varepsilon_1\varepsilon_2\cdots\varepsilon_K}$ is the closure of a braid $\tilde{\alpha}_{\varepsilon_1\varepsilon_2\cdots\varepsilon_K} = \alpha\sigma_1^{1-\varepsilon_1}\sigma_3^{1-\varepsilon_2}\cdots\sigma_{2K-1}^{1-\varepsilon_K}$, and so can be obtained from the closed braid $\vec{L}_{\varepsilon_1\varepsilon_2\cdots\varepsilon_K}$ by reversing the orientation on a sublink $\vec{K}_{\varepsilon_1\varepsilon_2\cdots\varepsilon_K}$. We study how the linking numbers $\lambda_{\varepsilon_1\varepsilon_2\cdots\varepsilon_K} = \operatorname{lk}(\vec{K}_{\varepsilon_1\varepsilon_2\cdots\varepsilon_K}, \vec{L}_{\varepsilon_1\varepsilon_2\cdots\varepsilon_K} - \vec{K}_{\varepsilon_1\varepsilon_2\cdots\varepsilon_K})$ vary for different choices of the ε_i's in the set $\{\pm 1\}$. Note that each appearance of -1 in the array $\varepsilon_1\varepsilon_2\cdots\varepsilon_K$ corresponds to a square σ_{2i-1}^2 in $\alpha\sigma_1^{1-\varepsilon_1}\sigma_3^{1-\varepsilon_2}\cdots\sigma_{2K-1}^{1-\varepsilon_K}$. Now, the portion of the link diagram which is associated to this square always involves distinct components. The reason is that the two strands are oppositely oriented, and there is no way for an upward-oriented braid strand to be joined to a downward-oriented braid strand in a link of type $(0,n)$. Let $m_{\varepsilon_1\varepsilon_2\cdots\varepsilon_K}$ be the number of occurrences of -1 in the array $\varepsilon_1\varepsilon_2\cdots\varepsilon_K$.

Then

$$(5) \qquad \lambda_{\varepsilon_1\varepsilon_2\cdots\varepsilon_K} = -\lambda + m_{\varepsilon_1\varepsilon_2\cdots\varepsilon_K}.$$

Now Jones' reversing result (see **[L-M]** or **[M]**) tells us that

$$(6) \qquad \overleftarrow{V}_{\varepsilon_1\varepsilon_2\cdots\varepsilon_K} = t^{-3\lambda_{\varepsilon_1\varepsilon_2\cdots\varepsilon_K}} \overrightarrow{V}_{\varepsilon_1\varepsilon_2\cdots\varepsilon_K}.$$

Formulas (1), (5), and (6) then imply that the polynomial $\overleftarrow{V}_{\varepsilon_1\varepsilon_2\cdots\varepsilon_K}$ is given by

$$(7) \qquad \overleftarrow{V}_{\varepsilon_1\varepsilon_2\cdots\varepsilon_K} = \delta^{n-1}t^{3\lambda-3m_{\varepsilon_1\varepsilon_2\cdots\varepsilon_K}}\operatorname{tr}\left(ag_1^{1-\varepsilon_1}g_3^{1-\varepsilon_2}\cdots g_{2K-1}^{1-\varepsilon_K}\right)$$

where $a = r_t(\alpha)$, $g_i = r_t(\sigma_i) = \sqrt{t}\,(te_i - (1-e_i))$.

We will have no more use for $\vec{L}_{\varepsilon_1\varepsilon_2\cdots\varepsilon_K}$, so from now on we drop the arrow, writing α, L, V for $\tilde{\alpha}$, $\overleftarrow{L}$, $\overleftarrow{V}$. We are ready to compute $V_{00\cdots 0} = V_{L_\alpha(2k,n-2k)}(t)$. The reader may find it helpful to follow the pictures in Figure 4, which illustrate the case $2k = n = 4$. Construct a tree of $2^{k+1}-1$ links, arranged in $k+1$ rows, with $L_{00\cdots 0}$ at the base. Place the two links $\{L_{\varepsilon_1 0\cdots 0};\ \varepsilon_1 = \pm 1\}$ in the second row from the bottom, the four links $\{L_{\varepsilon_1\varepsilon_2 0\cdots 0};\ \varepsilon_i = \pm 1\}$ above these, and so forth, ending with the 2^k links $\{L_{\varepsilon_1\varepsilon_2\cdots\varepsilon_K};\ \varepsilon_i = \pm 1\}$ in the top row. There is a natural order in each row, dictated by working up from the bottom and requiring that the links in the tree group themselves into $(0,+,-)$ surgery triples $(L_{\varepsilon_1\cdots 0\cdots\varepsilon_K}, L_{\varepsilon_1\cdots 1\cdots\varepsilon_K}, L_{\varepsilon_1\cdots -1\cdots\varepsilon_K})$, with the "+" and "−" member of each triple directly above the "0" member. Applying (3) repeatedly, we can then work our way up the tree, to express $V_{00\cdots 0}$ as a sum of the polynomials of the links which lie above $L_{00\cdots 0}$ in the tree, viz:

$$\begin{aligned} V_{00\cdots 0} &= \frac{1}{\mu}\left(t^{-1}V_{10\cdots 0} - tV_{-10\cdots 0}\right) \\ &= \frac{1}{\mu^2}\left(t^{-1}\left(t^{-1}V_{110\cdots 0} - tV_{1-10\cdots 0}\right) - t\left(t^{-1}V_{-110\cdots 0} - tV_{-1-10\cdots 0}\right)\right) \\ &= \cdots = \frac{1}{\mu^k}\sum_{\varepsilon_i=\pm 1}(-1)^{m_{\varepsilon_1\varepsilon_2\cdots\varepsilon_k}}t^{-k-2m_{\varepsilon_1\varepsilon_2\cdots\varepsilon_k}}V_{\varepsilon_1\varepsilon_2\cdots\varepsilon_k} \end{aligned}$$

which by virtue of (7) becomes

$$V_{00\cdots 0} = \frac{\delta^{n-1}t^{3\lambda}}{\mu^k}\sum_{\varepsilon_i=\pm 1}(-1)^{m_{\varepsilon_1\varepsilon_2\cdots\varepsilon_k}}t^{-k-m_{\varepsilon_1\varepsilon_2\cdots\varepsilon_k}}\operatorname{tr}\left(ag_1^{1-\varepsilon_1}g_3^{1-\varepsilon_2}\cdots g_{2k-1}^{1-\varepsilon_k}\right).$$

Magically, one recognizes that the 2^k terms in the sum combine into a single formula:

$$(8) \qquad V_{00\cdots 0} = \frac{\delta^{n-1}t^{3\lambda}(-1)^k t^{-2k}}{\mu^k}\operatorname{tr}\left(a\left(g_1^2 - t\right)\left(g_3^2 - t\right)\cdots\left(g_{2k-1}^2 - t\right)\right).$$

Then $g_i = \sqrt{t}\,((t+1)e_i - 1)$ (see **[J1]**) implies that

$$(9) \qquad g_i^2 - t = -t^2\mu\delta e_i, \qquad 1 \leqslant i \leqslant n-1.$$

Substituting (9) into (8) we obtain (4), and our proof is complete. □

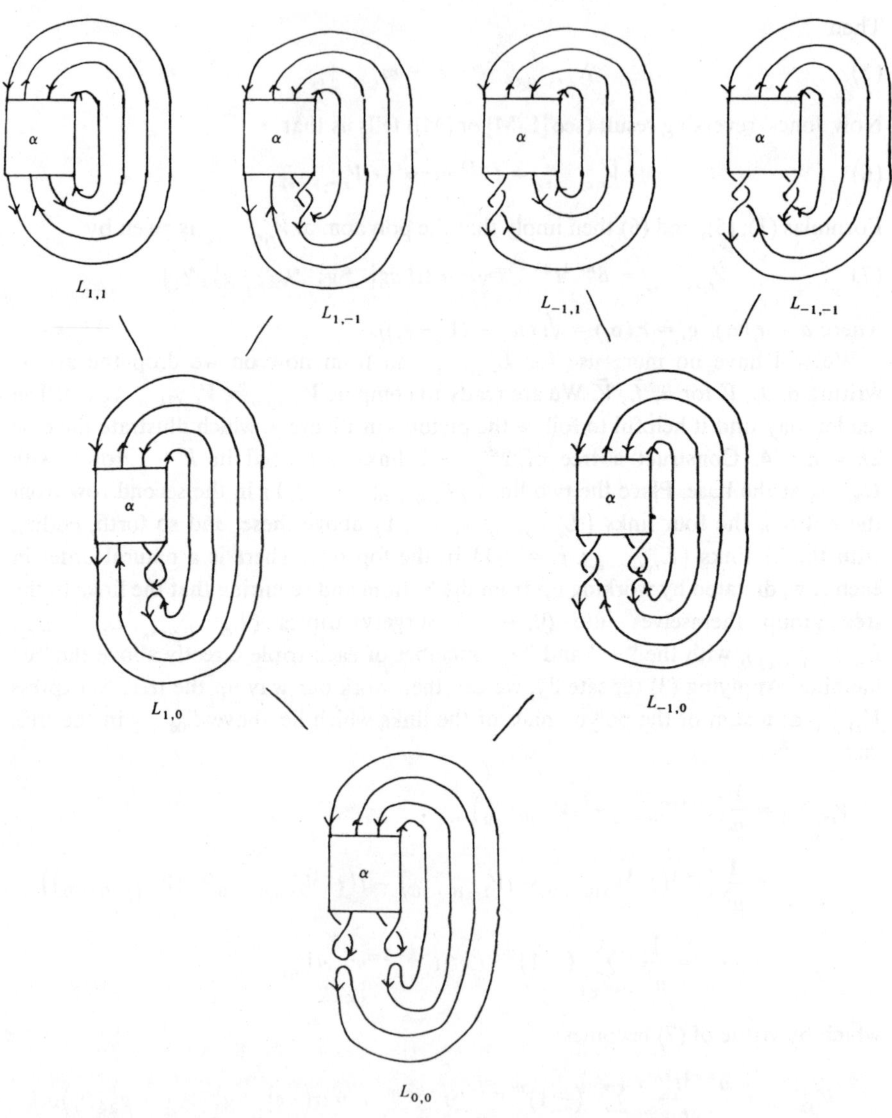

FIGURE 4

3. Surgery triples. In this section we apply Theorem 1 to give the promised formula which relates the polynomials V_1, V_{-1}, V_∞ of three links L_1, L_{-1}, L_∞ which are defined by the surgeries in Figure 2.

Recall that if K is a sublink of L, then $\lambda(K)$ means $\operatorname{lk}(K, L-K)$. Note that there is no canonical way to orient the components of L_∞ which extend the surgered arcs. The reader is referred to Figure 5 for the description of the various possible components K_j^i of L_j, $j = 1, -1, \infty$.

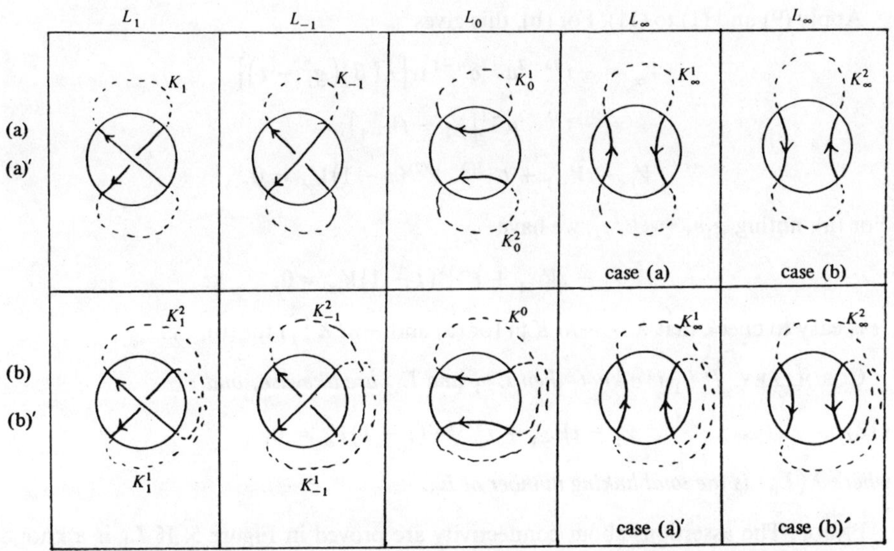

FIGURE 5

THEOREM 2. *The Jones polynomials of* L_1, L_{-1}, L_∞ *are related by*

(10) $$V_1 - tV_{-1} + t^{3q}(t-1)V_\infty = 0$$

where q is given by

case	q
a	$\lambda(K_0^2)$
b	$\lambda(K_{-1}^1) + \frac{1}{2}$
a'	$\lambda(K_0^2) - \lambda(K_\infty^2)$
b'	$\lambda(K_{-1}^1) + \frac{1}{2} - \lambda(K_\infty^2)$

PROOF. We consider the cases (a) and (b), since (a′) and (b′) follow from (a) and (b) by Jones' reversing result. For the purposes of the proof, we may assume that L_1 is a closed braid, for any link may be so represented, and the deformations which take it to a closed braid can be taken to avoid the surgered disc. By isotopy and conjugation, the crossing of interest can then be moved to the lower left part of the braid. Finally, inserting $\sigma_1^{-1}\sigma_1$ in the braid, if necessary, we can assume the braid has the form $\beta\sigma_1^2$. Thus we assume: $L_1 = (\widehat{\beta\sigma_1^2})$, $L_{-1} = \hat{\beta}$, $L_\infty = L_{\beta\sigma_1}(2, n-2)$ for (a) and $L_\beta(2, n-2)$ for (b). By Theorem 1 we then have

(11) $$V_\infty = \begin{cases} t^{3\lambda}\delta^n \operatorname{tr}(r_t(\beta\sigma_1)e_1) & \text{for (a)}, \\ t^{3\lambda}\delta^n \operatorname{tr}(r_t(\beta)e_1) & \text{for (b)}. \end{cases}$$

The integer λ in equation (11) is defined as in Theorem 1, viz: orient the plat part of L_∞ as in Figure 1b, and then construct the associated closed braid (in our case it will be $(\widehat{\beta\sigma_1})$ or $(\hat{\beta})$); then λ is the linking number of the "wrongly ordered" sublink of that closed braid.

Apply (9) and (1) to (11). For (b), this gives

$$V_\infty = -t^{3\lambda-2}\mu^{-1}\delta^{n-1}\operatorname{tr}\left[r_t(\beta)\left(g_1^2 - t\right)\right]$$
$$= -t^{3\lambda-2}\mu^{-1}[V_1 - tV_{-1}],$$
$$V_1 - tV_{-1} + t^{-3(\lambda-1/2)}(t-1)V_\infty = 0.$$

For (a), noting $g_1 e_1 = t^{3/2}e_1$, we have

$$V_1 - tV_{-1} + t^{-3\lambda}(t-1)V_\infty = 0.$$

It is easy to check that $\lambda = -\lambda(K_0^2)$ for (a) and $-\lambda(K_{-1}^1)$ for (b). □

COROLLARY. *If L_1 is a knot, then L_{-1} and L_∞ are likewise, and*

$$V_1 - tV_{-1} + t^{3\lambda(L_0)}(t-1)V_\infty = 0, \tag{12}$$

where $\lambda(L_0)$ is the total linking number of L_0.

PROOF. The assertions about connectivity are proved in Figure 5. If L_1 is a knot, then L_0 has two components, and the result follows from Theorem 2. □

4. Comments.

4.1. The Corollary allows one to compute $V_k(t)$ for an arbitrary knot k by choosing a minimal set of crossing changes which unknot k, and then constructing a tree of knots in which each interior node corresponds to a single $(1, -1, \infty)$ surgery and each braid ends in the unknot, with polynomial 1. Orientations will not matter, because we are working with knots. In principle, this ought to give a very simple tool for investigating $V_k(t)$ for knots, were it not for the fact that each time one does the $(1, -1, \infty)$ surgery one has to compute $\lambda(L_0)$, and the numbers so obtained seem unpredictable. They appear to play a crucial role in $V_k(t)$.

4.2. In this paper we use conventions in our definition of V_L which are consistent with formula (3), correcting an error in Theorem 12 of **[J1]**, which is inconsistent with earlier conventions in that same paper.

4.3. Very new results of Louis Kauffman (private letter) show that e_i's in Jones' algebra A_n may be interpreted geometrically as braid-like objects (see Figure 6) which compose in much the same way as braids, by concatenation and rescaling. With this interpretation the formula (2) becomes transparent, because on adding the "weaving pattern" $e_1e_3 \cdots e_{2k-1}$ to the end of a geometric braid α and then identifying top and bottom as in a closed braid, one obtains immediately the link $L_\alpha(2k, n-2k)$.

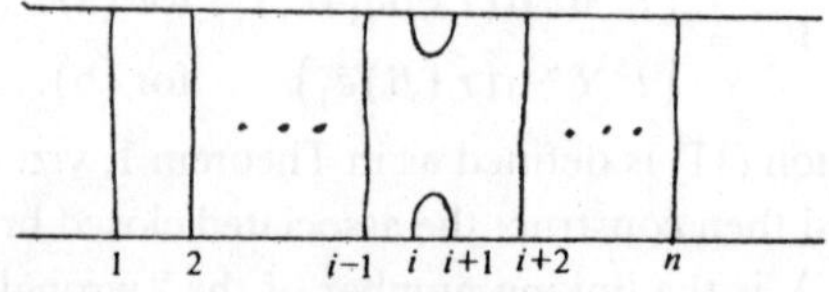

FIGURE 6

REFERENCES

[A1] J. W. Alexander, *A lemma on systems of knotted curves*, Proc. Nat. Acad. Sci. U.S.A. **9** (1923), 93–95.

[Ar] E. Artin, *Theorie der Zöpfe*, Abh. Math. Sem. Univ. Hamburg **4** (1925), 47–72.

[B] J. Birman, *Jones' braid-plat formula and a new surgery triple*, preprint.

[J1] V. F. R. Jones, *A polynomial invariant for knots via von Neumann algebras*, Bull. Amer. Math. Soc. **12** (1985), 103–111.

[J2] ______, Notes from lectures on the "Potts Model", MSRI, 1984–1985.

[K] T. Kanenobu, *The Jones polynomial of a plat*, preprint.

[L] W. B. R. Lickorish, *A relationship between link polynomials*, Math. Proc. Cambridge Philos. Soc. **100** (1986), 109–112.

[L-M] W. B. R. Lickorish and K. C. Millett, *The reversing result for the Jones polynomial*, Pacific J. Math. **124** (1986), 173–176.

[M] H. Morton, *The Jones polynomial for unoriented links*, Quart. J. Math. Oxford **37** (1986), 55–60.

[R] K. Reidemeister, *Knotentheorie*, Springer, 1932 (reprinted 1974, Springer-Verlag).

DEPARTMENT OF MATHEMATICS, COLUMBIA UNIVERSITY, NEW YORK, NEW YORK 10027.

DEPARTMENT OF MATHEMATICS, KYUSHU UNIVERSITY 33, FUKUOKA 812, JAPAN

MONODROMY OF BRAID GROUPS, A RELATION WITH CONFORMAL FIELD THEORY

Advanced Studies in Pure Mathematics 16, 1988
Conformal Field Theory and Solvable Lattice Models
pp. 297–372

Vertex Operators in Conformal Field Theory on $\mathbb{P}^1$ and Monodromy Representations of Braid Group

Dedicated to Professor Hirosi Toda on his 60th birthday

Akihiro Tsuchiya and Yukihiro Kanie

Contents

§ 0. Introduction

The 2-dimensional conformal field theory was initiated by A.A. Belavin, A.N. Polyakov and A.B. Zamolodchikov [BPZ] and was developed by many physicists, e.g. [DF], [ZF] etc. In the paper [BPZ], the significance of the primary fields for this theory is pointed out. V.G. Knizhnik and A.B. Zamolodchikov [KZ] developed the theory with current algebra symmetry, proposed the notion of primary fields with gauge symmetry, and gave the differential equations of multipoint correlation functions.

Our aim in this paper is to give rigorous foundations to the work of [KZ], and to reformulate and develop the operator formalism in the conformal field theory on the complex projective line $\mathbb{P}^1$. The space $\mathscr{H}$ of operands is taken to be a sum $\mathscr{H}=\sum_{j=0}^{\ell/2}\mathscr{H}_j$ of the integrable highest weight modules $\mathscr{H}_j$ of the affine Lie algebra $\hat{\mathfrak{g}}=\mathfrak{sl}(2, \mathbb{C})\otimes\mathbb{C}[t, t^{-1}]\oplus\mathbb{C}c$ of type $A_1^{(1)}$. We fix the value ℓ (positive integer) of the central element c of $\hat{\mathfrak{g}}$ on the space $\mathscr{H}$. The Virasoro algebra $\mathscr{L}$ acts on each $\mathscr{H}_j$ through

Received March 4, 1987.

the Sugawara forms $L(m)$, $m \in \mathbb{Z}$. For each $X \in \mathfrak{sl}(2, \mathbb{C})$, the field operator $X(z)=\sum_{m\in\mathbb{Z}} X(m)z^{-m-1}$ obeys the equations of motion:

$$[L(m), X(z)]=z^m\left(z\frac{d}{dz}+m+1\right)X(z).$$

The currents $X(z)$, $X \in \mathfrak{sl}(2, \mathbb{C})$ and the energy-momentum tensor $T(z)=\sum_{m\in\mathbb{Z}} L(m)z^{-m-2}$ preserve each $\hat{\mathfrak{g}}$-module $\mathscr{H}_j$. Thus each space $\mathscr{H}_j$ may be considered as a free theory. In order to introduce operators describing the interactions in the theory, we define the vertex operators due to V.G. Knizhnik and A.B. Zamolodchikov [KZ].

The vertex operators play a central role in this paper. In Section 2, we show the existence and the uniqueness theorem of the vertex operators. In Section 3 we get the differential equations satisfied by N-point functions, which have only regular singularities. The properties of vertex operators are derived from these differential equations (called the fundamental equations). First, we get the convergence of compositions of vertex operators. The commutation relation of vertex operators is equivalently rephrased in terms of the connection matrix of the fundamental equations, and is calculated explicilty in a special case. The monodromies of the fundamental equations give rise to representations of the braid group B_N. We determine explicitly this monodromy representation in a more special case. In fact, it gives an irreducible representation of the Hecke algebra $H_N(q)$ of type A_{N-1}, where $q=\exp(2\pi\sqrt{-1}/(\ell+2))$. Here it is remarkable that the vacuum expectation values of the products of vertex operators provide canonical bases of these representation spaces and the commutation relations of vertex operators give a 'factorization' of the monodromy representations.

Fix a positive integer ℓ for the value of the central element c, and a half integer j with $0\leq 2j\leq \ell$, then there is a unique (up to isomorphisms) irreducible highest weight left $\hat{\mathfrak{g}}$-module $\mathscr{H}_j$ with a highest weight vector $u_j(j)$. The Lie algebra $\hat{\mathfrak{g}}$ has a decomposition $\hat{\mathfrak{g}}=\mathfrak{m}_+\oplus\mathfrak{g}\oplus\mathbb{C}c\oplus\mathfrak{m}_-$, where $\mathfrak{g}=\mathfrak{sl}(2, \mathbb{C})=\mathbb{C}F\oplus\mathbb{C}H\oplus\mathbb{C}E$ and $\mathfrak{m}_\pm=\mathfrak{g}\otimes\mathbb{C}[t^{\pm 1}]t^{\pm 1}$ (see Section 1.1) The subspace $V_j=\{v \in \mathscr{H}_j;\ \mathfrak{m}_+v=0\}$ is an irreducible $\mathfrak{g}$-module of highest weight $2j$, i.e. of dimension $2j+1$.

We can define the corresponding irreducible highest weight right $\hat{\mathfrak{g}}$ (or $\mathfrak{g}$)-module $\mathscr{H}_j^\dagger$ (or $V_j^\dagger$) (and fix a highest weight vector $u_j^\dagger(j)$), and the nondegenerate bilinear pairing (called *vacuum expectation value*) $\langle\,|\,\rangle\colon \mathscr{H}_j^\dagger \times\mathscr{H}_j\to\mathbb{C}$ such that $\langle u_j^\dagger(j)\,|\,u_j(j)\rangle=1$ and $\langle va\,|\,w\rangle=\langle v\,|\,aw\rangle$ for any $v \in \mathscr{H}_j^\dagger$, $a \in \hat{\mathfrak{g}}$, $w \in \mathscr{H}_j$. Its restriction on $V_j^\dagger\times V_j$ is also nondegenerate.

Let $\mathscr{H}=\sum_{j=0}^{\ell/2}\mathscr{H}_j$ and $\mathscr{H}^\dagger=\sum_{j=0}^{\ell/2}\mathscr{H}_j^\dagger$. By an *operator* we mean a linear mapping $\Phi\colon \mathscr{H}\to\hat{\mathscr{H}}$, where $\hat{\mathscr{H}}$ is a completion of $\mathscr{H}$. Note that an operator Φ is characterized by a bilinear mapping $\hat{\Phi}: \mathscr{H}^\dagger \times \mathscr{H} \to \mathbb{C}$ defined by $\langle v|\hat{\Phi}|w\rangle = \langle v|\Phi(w)\rangle$ for any $v \in \mathscr{H}^\dagger$ and $w \in \mathscr{H}$. Two operators may not always be composable (see Section 2.1 for the definition of the composability).

For a positive half-integer j, a multi-valued, holomorphic, operator-valued function $\Phi(z)$ on the manifold $M_1 = \mathbb{C}^*$ is called a *vertex operator of spin j* if for any $u \in V_j$ and $z \in M_1$, $\Phi(z)$: $V_j \otimes \mathscr{H} \to \hat{\mathscr{H}}$ satisfies the following:

(Gauge Condition) $\quad [X(m), \Phi(u; z)] = z^m \Phi(Xu; z) \qquad (X \in \mathfrak{g}, m \in \mathbb{Z})$;

(Equation of Motion) $\quad [L(m), \Phi(u; z)] = z^m \left\{ z\dfrac{d}{dz} + (m+1)\Delta_j \right\} \Phi(u; z) \qquad (m \in \mathbb{Z})$,

for any $u \in V_j$ and $z \in M_1$, where the number $\Delta_j = (j^2 + j)/(\ell+2)$ is called the *conformal dimension* of the vertex operator $\Phi(z)$ and $\Phi(u; z)$: $\mathscr{H} \to \hat{\mathscr{H}}$ is an operator defined by $\Phi(u; z)(w) = \Phi(z)(u \otimes w)$ for $w \in \mathscr{H}$.

Remark (Proposition 2.4) that there are no vertex operators of spin j for $j > \ell/2$.

A triple $\mathrm{v} = \begin{pmatrix} j \\ j_2 j_1 \end{pmatrix}$ of nonnegative half integers j_2, j_1 and j is called a *vertex*. Put $\hat{\Delta}(\mathrm{v}) = \Delta_j + \Delta_{j_1} - \Delta_{j_2}$. Then the *Clebsch-Gordan condition*

$$|j_1 - j_2| \leq j \leq j_1 + j_2 \quad \text{and} \quad j_1 + j_2 + j \in \mathbb{Z}$$

for a vertex v is a condition for $\mathrm{Hom}_{\mathfrak{g}}(V_j \otimes V_{j_1}, V_{j_2}) \neq 0$. In this case $\mathrm{Hom}_{\mathfrak{g}}(V_j \otimes V_{j_1}, V_{j_2}) = \mathbb{C}$ and v is called a CG-vertex.

For a vertex $\mathrm{v} = \begin{pmatrix} j \\ j_2 j_1 \end{pmatrix}$ with $j_2, j_1 \leq \ell/2$, a vertex operator $\Phi(z)$ of spin j is called *of type* v, if $\Phi(u; z) = \Pi_{j_2} \Phi(u; z) \Pi_{j_1}$ for any $u \in V_j$, where Π_i is the projection of $\mathscr{H}$ (or $\hat{\mathscr{H}}$) onto $\mathscr{H}_i$ (or $\hat{\mathscr{H}}_i$ respectively). Then we get the condition for the existence of vertex operators:

Theorem 1 (Proposition 2.1 and Theorem 2.2).

i) *A vertex operator* $\Phi(z)$ *of type* v *is uniquely determined by the form* (*initial term*) $\Phi_0 \in \mathrm{Hom}_{\mathfrak{g}}(V_{j_2}^\dagger \otimes V_j \otimes V_{j_1}, \mathbb{C})$ *defined by*

$$\Phi_0(v, u, w) = (z^{\hat{\Delta}(\mathrm{v})} \langle v|\Phi(u; z)|w\rangle)|_{z=0} \qquad (v \in V_{j_2}^\dagger, u \in V_j, w \in V_{j_1}).$$

ii) *There exists a nonzero vertex operator* Φ *of type* $\mathrm{v} = \begin{pmatrix} j \\ j_2 j_1 \end{pmatrix}$ *on* $\mathscr{H}$, *if and only if the vertex* v *is an* ℓCG-*vertex, that is, it satisfies the* ℓ-*constrained Clebsch-Gordan condition*:

$$|j_1 - j_2| \leq j \leq j_1 + j_2, \quad j_1 + j_2 + j \in \mathbb{Z} \quad \textit{and} \quad j_1 + j_2 + j \leq \ell.$$

Remark. i) The inequalities $j_1+j_2+j\leq \ell$ and $|j_1-j_2|\leq j\leq j_1+j_2$ imply the conditions $j_1, j_2, j\leq \ell/2$.

ii) Nonzero vertex operators of a fixed type v are unique up to a constant multiple. For each ℓCG-vertex $\mathrm{v}=\begin{pmatrix} j \\ j_2 j_1 \end{pmatrix}$, we choose and fix a nonzero element $\varphi_{\mathrm{v}} \in \mathrm{Hom}_{\mathfrak{g}}(V_j\otimes V_{j_1}, V_{j_2})=\mathrm{Hom}_{\mathfrak{g}}(V_{j_2}^{\dagger}\otimes V_j\otimes V_{j_1}, \mathbb{C})\,(\cong\mathbb{C})$ and denote by $\Phi_{\mathrm{v}}(z)$ the associated vertex operator of type v with the initial term $\Phi_{\mathrm{v},0}=\varphi_{\mathrm{v}}$.

For each ℓCG-vertex $\mathrm{v}=\begin{pmatrix} j \\ j_2 j_1 \end{pmatrix}$, introduce the $\mathfrak{g}$-module $\mathscr{P}(\mathrm{v})$ defined by $\mathscr{P}(\mathrm{v})=\{\Phi_{\mathrm{v}}(u; z); u \in V_j\}$: $X\Phi_{\mathrm{v}}(u; z)=\Phi_{\mathrm{v}}(Xu; z)$ $(X \in \mathfrak{g})$.

We can show that any operators of the form $X(\zeta)$, $X \in \mathfrak{g}$, $T(\zeta)$ and vertex operators are composable. The composability of vertex operators is obtained by using the fact that the differential equations of N-point functions have only regular singular points.

Introduce the space $\mathcal{O}(\mathrm{v})$ of operators on $\mathscr{H}$ as the $\mathbb{C}$-vector space spanned by the set

$$\left\{\frac{1}{(2\pi\sqrt{-1})^N}\int_{C_N}\cdots\int_{C_1} d\zeta_N\cdots d\zeta_1(\zeta_N-z)^{m_N}\cdots(\zeta_1-z)^{m_1}X_N(\zeta_N)\cdots\right.$$
$$\left.\cdots X_1(\zeta_1)\Phi(u; z);\ N \in \mathbb{Z}_{\geq 0},\ X_i \in \mathfrak{g},\ m_i \in \mathbb{Z}\ (1\leq i\leq N),\ u \in V_j\right\},$$

where C_i's are contours around C_{i-1} such that 0 is outside C_N and z is inside C_1.

Introduce a $\hat{\mathfrak{g}}$-module structure and an $\mathscr{L}$-module structure in $\mathcal{O}(\mathrm{v})$ defined by

$$\hat{X}(m)A(z)=\frac{1}{2\pi\sqrt{-1}}\int_C d\zeta(\zeta-z)^m X(\zeta)A(z) \in \mathcal{O}(\mathrm{v})$$

and

$$\hat{L}(m)A(z)=\frac{1}{2\pi\sqrt{-1}}\int_C d\zeta(\zeta-z)^{m+1}T(\zeta)A(z) \in \mathcal{O}(\mathrm{v})$$

for $A(z) \in \mathcal{O}(\mathrm{v})$, $X \in \mathfrak{g}$, $m \in \mathbb{Z}$, and some contour C around z such that 0 is outside C.

Theorem 2 (Theorem 2.9). *For each ℓCG-vertex* v, *the* $\mathfrak{g}$*-module mapping* Φ: $V_j \ni u\mapsto\Phi_{\mathrm{v}}(u; z) \in \mathscr{P}(\mathrm{v})$ *is extended to the* $\hat{\mathfrak{g}}$*-isomorphism of* $\mathscr{H}_j$ *onto* $\mathcal{O}(\mathrm{v})$.

Here we summarize the relations satisfied by vertex operators:

Fundamental relations for vertex operators

Let $\Phi(z)$ be a vertex operator of spin j. Then

$$\hat{X}(m)\Phi(u; z)=0 \qquad (m\geq 1, X \in \mathfrak{g}, u \in V_j);$$

$$\hat{X}(0)\Phi(u; z)=[X(0), \Phi(u; z)]=\Phi(Xu; z) \qquad (X \in \mathfrak{g}, u \in V_j);$$

$$\hat{L}(m)\Phi(u; z)=0 \qquad (m\geq 1, u \in V_j);$$

$$\hat{L}(0)\Phi(u; z)=\Delta_j\Phi(u; z) \qquad (u \in V_j);$$

$$\hat{L}(-1)\Phi(u; z)=\frac{\partial}{\partial z}\Phi(u; z) \qquad (u \in V_j);$$

$$\hat{E}(-1)^{\ell-2j+1}\Phi(u_j(j); z)=0.$$

Remark that the last equation is derived from the structure of the irreducible $\hat{\mathfrak{g}}$-module $\mathcal{H}_j$ by using Theorem 2.

Now we call the vectors $|\mathrm{vac}\rangle=u_0(0) \in \mathcal{H}_0$ and $\langle\mathrm{vac}|=u_0^\dagger(0) \in \mathcal{H}_0^\dagger$ the *Virasoro vacuum*. They satisfies the equalities

$$X(m)|\mathrm{vac}\rangle=L(n)|\mathrm{vac}\rangle=0 \qquad (X \in \mathfrak{g}, m\geq 0, n\geq -1);$$

$$\langle\mathrm{vac}|X(m)=\langle\mathrm{vac}|L(n)=0 \qquad (X \in \mathfrak{g}, m\leq 0, n\leq 1).$$

For an N-ple $\mathbb{J}=(j_N, \cdots, j_1)$ of half integers with $0\leq 2j_i\leq \ell$, let $V^\vee(\mathbb{J})=V_{j_N}^\vee\otimes\cdots\otimes V_{j_1}^\vee$, and let $V_0^\vee(\mathbb{J})$ denote the invariant subspace of $V^\vee(\mathbb{J})$ under the diagonal $\mathfrak{g}$-action, where $V_j^\vee$ denotes the dual $\mathfrak{g}$-module of V_j. Let $\Phi_i(z_i)$ be a vertex operator of spin j_i $(1\leq i\leq N)$, then the vacuum expectation value of the composed operator

$$\langle\Phi_N(z_N)\cdots\Phi_1(z_1)\rangle=\langle\mathrm{vac}|\Phi_N(z_N)\cdots\Phi_1(z_1)|\mathrm{vac}\rangle$$

is considered as a $V^\vee(\mathbb{J})$-valued, formal Laurent series on $(z_N, \cdots, z_1)$ and is called an *N-point function* (of spin $\mathbb{J}$): If $\Phi_i(z_i)$ is of type $\mathbb{v}_i$ $(1\leq i\leq N)$,

$$\langle\Phi_N(z_N)\cdots\Phi_1(z_1)\rangle=\prod_{i=1}^N z_i^{-\Delta(\mathbb{v}_i)}\sum C_{m_N\cdots m_1}z_N^{-m_N}\cdots z_1^{-m_1},$$

where $C_{m_N\cdots m_1} \in V^\vee(\mathbb{J})$ and the sum is taken over integers $m_k \in \mathbb{Z}$ $(1\leq k \leq N)$ with $m_N\geq 0$ and $m_1\leq 0$.

Let π_i be the $\mathfrak{g}$-action on the i-th component of $V^\vee(\mathbb{J})$ and introduce the operator Ω_{ik} defined by

$$\Omega_{ik}=\frac{1}{2}\pi_i(H)\pi_k(H)+\pi_i(E)\pi_k(F)+\pi_i(F)\pi_k(E),$$

and $\Omega_i=\Omega_{ii}$ is the action of the Casimir element $\Omega=\frac{1}{2}HH+EF+FE$ on the i-th component of $V^{\vee}(\mathbb{J})$. Then $\Omega_{ik}=\frac{1}{2}\{(\pi_i+\pi_k)(\Omega)-\Omega_i-\Omega_k\}$ $(i\neq k)$ and $\Omega_i=2(j_i^2+j_i)$ id on $V^{\vee}(\mathbb{J})$.

Then we get a system of differential equations and a system of algebraic equations for N-point functions:

Theorem 3 (Theorem 3.1). *Let $\Phi_i(z_i)$ be a vertex operator of spin j_i $(1\leq i\leq N)$, then the N-point function $\langle\Phi_N(z_N)\cdots\Phi_1(z_1)\rangle$ satisfies the following equations:*

(I) *(projective invariance) For $m=-1, 0$ and 1,*

$$\sum_{i=1}^{N} z_i^m\left(z_i\frac{\partial}{\partial z_i}+(m+1)\Delta_{j_i}\right)\langle\Phi_N(z_N)\cdots\Phi_1(z_1)\rangle=0.$$

(II) *(gauge invariance) For any $X\in\mathfrak{g}$,*

$$\sum_{i=1}^{N}\pi_i(X)\langle\Phi_N(z_N)\cdots\Phi_1(z_1)\rangle=0.$$

(III) *For each $i=1,\cdots,N$,*

$$\left((\ell+2)\frac{\partial}{\partial z_i}-\sum_{\substack{k=1\\k\neq i}}^{N}\frac{\Omega_{ik}}{z_i-z_k}\right)\langle\Phi_N(z_N)\cdots\Phi_1(z_1)\rangle=0.$$

(IV) *For each i $(1\leq i\leq N)$ and any $u_k\in V_{j_k}$ $(k\neq i)$,*

$$\sum_{\mathfrak{m}_i}\binom{L_i}{\mathfrak{m}_i}\sum_{k\neq i}(z_k-z_i)^{-m_k}\langle\Phi_N(E^{m_N}u_N; z_N)\cdots\Phi_i(u_{j_i}(j_i); z_i)\cdots\Phi_1(E^{m_1}u_1; z_1)\rangle$$
$$=0,$$

where $\mathfrak{m}_i=(m_N,\cdots,\hat{m}_i,\cdots,m_1)\in(\mathbb{Z}_{\geq 0})^{N-1}$ with $\sum_{k\neq i}m_k=L_i=\ell-2j_i+1$ and $\binom{L_i}{\mathfrak{m}_i}$ is the multinomial coefficient.

Consider the systems E(J) of differential equations and B(J) of algebraic equations for $V_0^{\vee}(\mathbb{J})$-valued functions $\Phi(z_N,\cdots,z_1)$ on the manifold $X_N=\{z=(z_N,\cdots,z_1)\in\mathbb{C}^N;\ z_i\neq z_k\ (i\neq k)\}$:

$$\mathrm{E}(\mathbb{J}):\quad\left((\ell+2)\frac{\partial}{\partial z_i}-\sum_{\substack{k=1\\k\neq i}}^{N}\frac{\Omega_{ik}}{z_i-z_k}\right)\Phi(z_N,\cdots,z_1)=0\qquad(1\leq i\leq N),$$

$$\mathrm{B}(\mathbb{J}):\ \sum_{\mathfrak{m}_i}\binom{L_i}{\mathfrak{m}_i}\prod_{k\neq i}(z_k-z_i)^{-m_k}\Phi(z_N,\cdots,z_1)\,(E^{m_N}u_N,\cdots,u_{j_i}(j_i),\cdots,E^{m_1}u_1)$$
$$=0,$$

for each i $(1\leq i\leq N)$ and any $u_k\in V_{j_k}$ $(k\neq i)$, where $\mathfrak{m}_i=(m_N,\cdots,\hat{m}_i,$

$\cdots, m_1) \in (\mathbb{Z}_{\geq 0})^{N-1}$ with $\sum_{k \neq i} m_k = L_i = \ell - 2j_i + 1$.

Introduce the set $\mathscr{P}_\ell(\mathbb{J})$ defined by

$$\mathscr{P}_\ell(\mathbb{J}) = \Big\{ \mathfrak{p} = (p_N, \cdots, p_1, p_0); p_i \in \frac{1}{2}\mathbb{Z}_{\geq 0} \ \ \mathrm{v}_i = \begin{pmatrix} & j_i & \\ p_i & & p_{i-1} \end{pmatrix} \in (\mathrm{CG})_\ell, \\ p_N = p_0 = 0 \Big\},$$

where $(\mathrm{CG})_\ell$ is the set of all ℓCG-vertices. For each $\mathfrak{p} \in \mathscr{P}_\ell(\mathbb{J})$, the N-point function

$$\Phi_{\mathfrak{p}}(z_N, \cdots, z_1) = \langle \Phi_{\mathrm{v}_N}(z_N) \cdots \Phi_{\mathrm{v}_1}(z_1) \rangle$$

of type $\mathfrak{p}$ is a formal Laurent series solution of the joint system $\mathrm{E}(\mathbb{J})$ and $\mathrm{B}(\mathbb{J})$, moreover

Theorem 4 (Theorem 3.3).

i) *For any* $\mathfrak{p} \in \mathscr{P}_\ell(\mathbb{J})$, *the Laurent series* $\Phi_{\mathfrak{p}}(z_N, \cdots, z_1)$ *is absolutely convergent in the region* $\mathscr{R}_z = \{(z_N, \cdots, z_1) \in \mathbb{C}^N; |z_N| > \cdots > |z_1|\}$ *and is analytically continued to a multivalued holomorphic function on the manifold* X_N.

ii) $\{\Phi_{\mathfrak{p}}(z_N, \cdots, z_1); \mathfrak{p} \in \mathscr{P}_\ell(\mathbb{J})\}$ *gives a basis of the solution space of the joint system* $\mathrm{E}(\mathbb{J})$ *and* $\mathrm{B}(\mathbb{J})$.

As a corollary of Theorem 4, we get

Theorem 5 (Theorem 3.4). *Let* $\Phi_i(z_i)$ *be the vertex operator of spin* j_i *and* $u_i \in V_{j_i}$ $(1 \leq i \leq N)$. *Then the sequence* $\{\Phi_N(u_N; z_N), \cdots, \Phi_1(u_1; z_1)\}$ *is composable in the region* $\mathscr{R}_{z,0} = \{(z_N, \cdots, z_1) \in \mathbb{C}^N; |z_N| > \cdots > |z_1| > 0\}$ *and the composed operator* $\Phi_N(u_N; z_N) \cdots \Phi_1(u_1; z_1)$ *is analytically continued to a multivalued holomorphic function on the manifold* $M_N = \{(z_N, \cdots, z_1) \in X_N; z_i \neq 0\}$.

For ℓCG-vertices $\mathrm{v}_2 = \begin{pmatrix} & j_3 & \\ j_4 & & k \end{pmatrix}$ and $\mathrm{v}_1 = \begin{pmatrix} & j_2 & \\ k & & j_1 \end{pmatrix}$, the composed operator $\Phi_{\mathrm{v}_2}(w)\Phi_{\mathrm{v}_1}(z)$ of the vertex operators $\Phi_{\mathrm{v}_2}(w)$ and $\Phi_{\mathrm{v}_1}(z)$ is multi-valued holomorphic on the manifold M_2.

For a quadruple $\mathbb{J} = (j_4, j_3, j_2, j_1)$ of half integers with $0 \leq 2j_1 \leq \ell$, introduce the set $I_\ell(\mathbb{J})$ of *intermediate edges*, defined by

$$I_\ell(\mathbb{J}) = \Big\{ k \in \frac{1}{2}\mathbb{Z}; 0 \leq 2k \leq \ell, \ \mathrm{v}_2(k) = \begin{pmatrix} & j_3 & \\ j_4 & & k \end{pmatrix} \in (\mathrm{CG})_\ell, \\ \mathrm{v}_1(k) = \begin{pmatrix} & j_2 & \\ k & & j_1 \end{pmatrix} \in (\mathrm{CG})_\ell \Big\}.$$

Let $\bar{\mathbb{J}}=(j_4, j_2, j_3, j_1)$, then we get the $\mathfrak{g}$-isomorphism T: $V^\vee(\mathbb{J})\to V^\vee(\bar{\mathbb{J}})$ defined by

$$(T\varphi)(u_4\otimes u_2\otimes u_3\otimes u_1)=\varphi(u_4\otimes u_3\otimes u_2\otimes u_1)$$

for $\varphi\in V^\vee(\mathbb{J})$ and $u_4\otimes u_2\otimes u_3\otimes u_1\in V(\bar{\mathbb{J}})$.

For an intermediate edge $\bar{k}\in I_\ell(\bar{\mathbb{J}})$, similarly define the ℓCG-vertices $\mathrm{v}_2(\bar{k})=\begin{pmatrix} & j_2 & \\ j_4 & & \bar{k}\end{pmatrix}$ and $\mathrm{v}_1(\bar{k})=\begin{pmatrix} & j_3 & \\ \bar{k} & & j_1\end{pmatrix}$ and consider the composed operator $\Phi_{\mathrm{v}_2(\bar{k})}(w)\Phi_{\mathrm{v}_1(\bar{k})}(z)$ of the vertex operators $\Phi_{\mathrm{v}_2(\bar{k})}(w)$ and $\Phi_{\mathrm{v}_1(\bar{k})}(z)$.

Assume that $I_\ell(\mathbb{J})\neq\emptyset$. For a point $(w, z)\in I_2=\{(z_2, z_1)\in\mathbb{R}^2;\ z_2>z_1>0\}$, let $\Phi_{\mathrm{v}_2(k)}(z)\Phi_{\mathrm{v}_1(k)}(w)$ denote the analytic continuation of the composition $\Phi_{\mathrm{v}_2(k)}(w)\Phi_{\mathrm{v}_1(k)}(z)$ of the vertex operators along the path $b(t)$, where the path $b(t)=(\eta(t), \zeta(t))$ from the point $(w, z)\in I_2$ to the point $(z, w)\in \bar{I}_2=\{(z_2, z_1)\in\mathbb{R}^2;\ z_1>z_2>0\}$ on the manifold M_2 is defined by

$$\eta(t)=\frac{w+z}{2}+e^{\pi\sqrt{-1}t}\frac{w-z}{2},\qquad \zeta(t)=\frac{w+z}{2}-e^{\pi\sqrt{-1}t}\frac{w-z}{2}\qquad (t\in[0,1]).$$

Then

Proposition 6 (Proposition 4.2). i) *There exists a constant square matrix* $C(\mathbb{J})=(C_k^{\bar{k}}(\mathbb{J}))_{k\in I_\ell(\mathbb{J}),\bar{k}\in I_\ell(\bar{\mathbb{J}})}$ *such that for each intermediate edge* $k\in I_\ell(\mathbb{J})$,

$$T\Phi_{\mathrm{v}_2(k)}(z)\Phi_{\mathrm{v}_1(k)}(w)=\sum_{\bar{k}\in I_\ell(\bar{\mathbb{J}})}\Phi_{\mathrm{v}_2(\bar{k})}(w)\Phi_{\mathrm{v}_1(\bar{k})}(z)C_k^{\bar{k}}(\mathbb{J}).$$

ii) *Let* $\mathbb{J}=(t, j_3, j_2, j_1, s)$, *then the braid relation holds*:

$$\begin{aligned}&C(j_3, j_2, j_1, s)C(t, j_3, j_1, j_2)C(j_1, j_3, j_2, s)\\&=C(t, j_3, j_2, j_1)C(j_2, j_3, j_1, s)C(t, j_2, j_1, j_3).\end{aligned}$$

Now our fundamental problem is:

Fundamental Problem. Determine the matrix $C(\mathbb{J})=(C_k^{\bar{k}}(\mathbb{J}))$ for any quadruple $\mathbb{J}$ with $I_\ell(\mathbb{J})\neq\emptyset$.

In Section 4.2, we solve the fundamental problem for the case where $j_3=\frac{1}{2}$ in $\mathbb{J}$. For general j_3, we can solve it in principle by the fusion rule (see Section 5.4).

Now we take $j_2=j_3=\frac{1}{2}$. Then the conditions for the nontriviality, $V_0^\vee(\mathbb{J})\neq 0$, are divided into the following cases:

$$(\mathrm{D2})_2\quad \frac{\ell}{2}>j_1=j_4>0;\quad (\mathrm{D2})_1\quad \frac{\ell}{2}=j_1=j_4;$$

$(\mathrm{D1})_1\quad j_4=j_1+1;\qquad (\mathrm{D1})_2\quad j_1=j_4=0;\qquad (\mathrm{D1})_3\quad j_1=j_4+1.$

Proposition 7 (Proposition 4.8). *Let* $q=\exp(2\pi\sqrt{-1}/(\ell+2))$.

i) *For* $j\in\frac{1}{2}\mathbb{Z}$ *with* $0<2j<\ell$,

$$C\left(j,\frac{1}{2},\frac{1}{2},j\right)=q^{-3/4}\begin{pmatrix}\gamma_+^{-1} & \\ & \gamma_-^{-1}\end{pmatrix}\begin{pmatrix}\dfrac{-1}{[2j+1]} & \dfrac{\sqrt{q[2j][2j+2]}}{[2j+1]}\\ \dfrac{\sqrt{[2j][2j+2]}}{[2j+1]} & \dfrac{q^{2j+1}}{[2j+1]}\end{pmatrix}\begin{pmatrix}\gamma_+ & \\ & \gamma_-\end{pmatrix}$$

where $[\nu]$ *denotes the q-integer*

$$[\nu]=\frac{q^\nu-1}{q-1}\quad\text{and}\quad \gamma_\pm=\frac{\Gamma\left(\pm\dfrac{2j+1}{\ell+2}\right)}{\left(\Gamma\left(\pm\dfrac{2j+2}{\ell+2}\right)\Gamma\left(\dfrac{\pm 2j}{\ell+2}\right)\right)^{1/2}}.$$

ii) $C\left(\dfrac{\ell}{2},\dfrac{1}{2},\dfrac{1}{2},\dfrac{\ell}{2}\right)=C\left(0,\dfrac{1}{2},\dfrac{1}{2},0\right)=-q^{-3/4}.$

iii) $C\left(j+1,\dfrac{1}{2},\dfrac{1}{2},j\right)=C\left(j-1,\dfrac{1}{2},\dfrac{1}{2},j\right)=q^{1/4}.$

Let $N\geq 2$ and fix a half integer t (target edge) with $0\leq 2t\leq\ell$. Put $\mathbb{J}_t=(t,\frac{1}{2},\cdots,\frac{1}{2})$ and introduce the set

$$\mathscr{P}_\ell(N;t)=\Big\{\mathfrak{p}=(p_N,\cdots,p_1,p_0);\ p_N=t,\ p_0=0,\ p_i\in\frac{1}{2}\mathbb{Z},\ 0\leq 2p_i\leq\ell,\ |p_i-p_{i-1}|=\frac{1}{2}\ (1\leq i\leq N)\Big\}.$$

For each $\mathfrak{p}\in\mathscr{P}_\ell(N;t)$, define the $V_0^\vee(\mathbb{J}_t)$-valued, multi-valued holomorphic function $\Psi_\mathfrak{p}(z_N,\cdots,z_1)$ on X_N by

$$\Psi_\mathfrak{p}(z_N,\cdots,z_1)(v,u_N,\cdots,u_1)=\langle\nu(v)|\Phi_{\mathrm{v}_N}(u_N;z_N)\cdots\Phi_{\mathrm{v}_1}(u_1;z_1)|\mathrm{vac}\rangle$$

for $v\in V_t$ and $u_i\in V_{1/2}$ $(1\leq i\leq N)$, where the vertex v_i is defined as $\mathrm{v}_i(\mathfrak{p})=\begin{pmatrix} & 1/2 & \\ p_i & & p_{i-1}\end{pmatrix}$ $(1\leq i\leq N)$ and ν is the isomorphism $\nu\colon V_j\to V_j^\dagger$ defined in Section 2.3.

Then the function $\Psi_\mathfrak{p}(z_N,\cdots,z_1)$ satisfies the systems $E(N;t)$ and $B(N;t)$ derived from the systems $E(\mathbb{J}_t)$ and $B(\mathbb{J}_t)$, where $\mathbb{J}_t=(t,\frac{1}{2},\cdots,\frac{1}{2})$ (see Section 5.2). Moreover we get that the solution space $W(N;t)$ of the systems $E(N;t)$ and $B(N;t)$ has a basis $\{\Psi_\mathfrak{p}(z_N,\cdots,z_1);\ \mathfrak{p}\in\mathscr{P}_\ell(N;t)\}$.

The braid group B_N acts on this space $W(N; t)$ as monodromies. The commutation relation of vertex operators gives a 'factorization' of this monodromy representation $(\pi_{N,t}, W(N; t))$. By the explicit formulae of the representation $\pi_{N,t}$ obtained from Proposition 7, we get

Theorem 8 (Theorem 5.2 and Proposition 5.3). *Let* $q=\exp\left(\frac{2\pi\sqrt{-1}}{\ell+2}\right)$.

i) *The monodromy representation* $q^{3/4}\pi_{N,t}$ *of the braid group* B_N *on the space* $W(N; t)$ *gives an irreducible and unitarizable representation of the group* B_N.

ii) *This representation factors through a representation of the Hecke algebra* $H_N(q)$ *of type* A_{N-1}.

iii) *Our representation* $(q^{3/4}\pi_{N,t}, W(N; t))$ *of the Hecke algebra* $H_N(q)$ *is equivalent to the representation* $(\pi_\lambda^{(2,\ell+2)}, V_\lambda^{(2,\ell+2)})$ *constructed by H. Wenzl* [W], *where* λ *is a Young diagram* $\lambda=[N/2+t, N/2-t]$.

Notations

$\mathfrak{g}=\mathfrak{sl}(2, \mathbb{C})=\mathbb{C}F\oplus\mathbb{C}H\oplus\mathbb{C}E$, where $F=\begin{pmatrix}0&0\\1&0\end{pmatrix}$, $H=\begin{pmatrix}1&0\\0&-1\end{pmatrix}$ and $E=\begin{pmatrix}0&1\\0&0\end{pmatrix}$

$\hat{\mathfrak{g}}=\mathfrak{g}\otimes\mathbb{C}[t, t^{-1}]\oplus\mathbb{C}c$: the affine Lie algebra of type $A_1^{(1)}$

$\hat{\mathfrak{h}}=\mathbb{C}H(0)\oplus\mathbb{C}c$: the Cartan subalgebra of $\hat{\mathfrak{g}}$

$X(n)=X\otimes t^n$ for $X\in\mathfrak{g}$ and $n\in\mathbb{Z}$

$\mathfrak{m}_\pm=\mathfrak{g}\otimes t^{\pm}\mathbb{C}[t^{\pm}]$, $\mathfrak{n}_+=\mathfrak{m}_+\oplus\mathbb{C}E(0)$, $\mathfrak{n}_-=\mathfrak{m}_-\oplus\mathbb{C}F(0)$, $\mathfrak{p}_\pm=\mathfrak{m}_\pm\oplus\mathfrak{g}\oplus\mathbb{C}c$: subalgebras of $\hat{\mathfrak{g}}$

$\mathcal{L}=\sum_{n\in\mathbb{Z}}\mathbb{C}e_n+\mathbb{C}e_0'$: the Virasoro algebra

$\Omega=\frac{1}{2}H^2+EF+FE\in U(\mathfrak{g})$: the Casimir element of $\mathfrak{g}$

$:X(m)Y(n):$: the normal ordered product for $X(m)$, $Y(n)\in\mathfrak{g}\otimes\mathbb{C}[t, t^{-1}]$

$X(z)=\sum_{n\in\mathbb{Z}}X(n)z^{-n-1}$ $(z\in\mathbb{C}^*, X\in\mathfrak{g})$: a current

$T(z)=\sum_{m\in\mathbb{Z}}L(m)z^{-m-2}$: the energy momentum tensor

ℓ: the central charge (we fix $\ell\in\mathbb{Z}_{>0}$ throughout the paper)

$\kappa=\ell+2$

V_j, $V_j^\dagger$: the irreducible left and right $\mathfrak{g}$-modules of spin j for $j\in\frac{1}{2}\mathbb{Z}_{\geq 0}$ respectively

$V_j^\vee=\mathrm{Hom}(V_j, \mathbb{C})$: the dual (right) $\mathfrak{g}$-module of V_j

$\mathcal{H}_j=\mathcal{H}_j(\ell)$, $\mathcal{H}_j^\dagger=\mathcal{H}_j^\dagger(\ell)$: the integrable highest weight left and right $\hat{\mathfrak{g}}$-modules respectively

$\langle|\rangle$: $V_j^\dagger\times V_j\to\mathbb{C}$, $\mathcal{H}_j^\dagger\times\mathcal{H}_j\to\mathbb{C}$: the vacuum expectation values

$\mathcal{H}=\sum_{j=0}^{\ell/2}\mathcal{H}_j\subset\hat{\mathcal{H}}=\sum_{j=0}^{\ell/2}\hat{\mathcal{H}}_j$; $\mathcal{H}^\dagger=\sum_{j=0}^{\ell/2}\mathcal{H}_j^\dagger\subset\hat{\mathcal{H}}^\dagger=\sum_{j=0}^{\ell/2}\hat{\mathcal{H}}_j^\dagger$.

$\mathbb{V}=\left\{\mathrm{v}=\begin{pmatrix} & j & \\ j_2 & & j_1\end{pmatrix}; j, j_1, j_2 \in \frac{1}{2}\mathbb{Z}_{\geq 0}\right\}$: the set of vertices

$\cup$

$\mathbb{V}_\ell=\left\{\mathrm{v} \in \mathbb{V};\ j_1, j_2 \leq \frac{\ell}{2}\right\}$

$(\mathrm{CG})=\{\mathrm{v} \in \mathbb{V};\ |j_1-j_2| \leq j \leq j_1+j_2,\ j_1+j_2+j \in \mathbb{Z}\}$: the set of all CG-vertices

$(\mathrm{CG})_\ell=\{\mathrm{v} \in (\mathrm{CG});\ j_1+j_2+j \leq \ell\}$: the set of all ℓCG-vertices

$\Delta_j=\dfrac{j^2+j}{\kappa}$: the conformal dimension of vertex operators of spin j

$\Delta(\mathrm{v})=\Delta_j$: the conformal dimension of a vertex v

$\hat{\Delta}(\mathrm{v})=\Delta_j+\Delta_{j_1}-\Delta_{j_2}$ for a vertex v

$\mathscr{V}(\mathrm{v})=\mathrm{Hom}_{\mathfrak{g}}(V_{j_2}^{\dagger}\otimes V_j\otimes V_{j_1};\ \mathbb{C})$

$\varphi_{\mathrm{v}} \in \mathrm{Hom}_{\mathfrak{g}}(V_j\otimes V_{j_1}, V_{j_2})\cong\mathscr{V}(\mathrm{v})$: the nonzero element for each $\mathrm{v}=\begin{pmatrix} & j & \\ j_2 & & j_1\end{pmatrix}$ $\in (\mathrm{CG})$ fixed in Appendix I

$\Phi_{\mathrm{v}}(z)$: the vertex operator of type v whose initial term $\Phi_{\mathrm{v},0}$ is φ_{v} for each $\mathrm{v}=\begin{pmatrix} & j & \\ j_2 & & j_1\end{pmatrix} \in (\mathrm{CG})_\ell$ (considered as $V_j\otimes\mathscr{H}_{j_1}\to\hat{\mathscr{H}}_{j_2}$)

$\Phi(u; z)=\Phi(z)(u\otimes\cdot)=\sum_{n\in\mathbf{Z}}\Phi_n(u)z^{-n-\hat{\Delta}(\mathrm{v})}$: the homogeneous decomposition of a vertex operator $\Phi(z)$ of type v

Let $W=W_1\otimes\cdots\otimes W_N$ the tensor product of $\mathfrak{g}$-modules W_k, then

π_i: the $\mathfrak{g}$-action on the i-th component of W

$\Delta_{ik}=\pi_i+\pi_k$: the diagonal action on the i-th and k-th components of W

$\Omega_{ik}=\frac{1}{2}\pi_i(H)\pi_k(H)+\pi_i(E)\pi_k(F)+\pi_i(F)\pi_k(E)$

$\mathbb{J}=(j_N, \cdots, j_1)$: an N-ple of half-integers with $0\leq 2j_i\leq\ell$

$V(\mathbb{J})=V_{j_N}\otimes\cdots\otimes V_{j_1},\ V^{\vee}(\mathbb{J})=V_{j_N}^{\vee}\otimes\cdots\otimes V_{j_1}^{\vee}$

$V_0^{\vee}(\mathbb{J})$: the space of all $\mathfrak{g}$-invariant elements in $V^{\vee}(\mathbb{J})$

$\mathscr{P}(\mathbb{J})=\left\{\mathrm{p}=(p_N, \cdots, p_1, p_0);\ \mathrm{v}_i(\mathrm{p})=\begin{pmatrix} & j_i & \\ p_i & & p_{i-1}\end{pmatrix} \in (\mathrm{CG}),\ p_N=p_0=0\right\}$

$\mathscr{P}_\ell(\mathbb{J})=\{\mathrm{p}=(p_N, \cdots, p_1, p_0) \in \mathscr{P}(\mathbb{J});\ \mathrm{v}_i(\mathrm{p}) \in (\mathrm{CG})_\ell\}$

$\mathbb{J}=(j_4, j_3, j_2, j_1)$: a quadruple of half integers with $0\leq 2j_i\leq\ell$

$I(\mathbb{J})=\left\{k \in \frac{1}{2}\mathbb{Z};\ 0\leq 2k\leq\ell,\ \mathrm{v}_2(k)=\begin{pmatrix} & j_3 & \\ j_4 & & k\end{pmatrix} \in (\mathrm{CG}),\right.$
$\left.\mathrm{v}_1(k)=\begin{pmatrix} & j_2 & \\ k & & j_1\end{pmatrix} \in (\mathrm{CG})\right\}$

$I_\ell(\mathbb{J})=\left\{k \in \frac{1}{2}\mathbb{Z};\ 0\leq 2k\leq\ell,\ \mathrm{v}_2(k) \in (\mathrm{CG})_\ell,\ \mathrm{v}_1(k) \in (\mathrm{CG})_\ell\right\}$

$\mathrm{p}(k)=(\mathrm{v}_3, \mathrm{v}_2(k), \mathrm{v}_1(k), \mathrm{v}_0) \in \mathscr{P}_\ell(\mathbb{J})$ for $k \in I_\ell(\mathbb{J})$, where $\mathrm{v}_3=\begin{pmatrix} & j_4 & \\ 0 & & j_4\end{pmatrix}$ and $\mathrm{v}_0=\begin{pmatrix} & j_1 & \\ j_1 & & 0\end{pmatrix}$

$$\Delta_4(\mathbb{J})=\hat{\Delta}(\mathrm{v}_2)+\hat{\Delta}(\mathrm{v}_1)=\Delta_{j_1}+\Delta_{j_2}+\Delta_{j_3}-\Delta_{j_4}$$

$$J_\ell(\mathbb{J})=\left\{r\in\frac{1}{2}\mathbb{Z};\ 0\leq 2r\leq \ell,\ \mathrm{w}(r)=\begin{pmatrix} & r & \\ j_4 & & j_1\end{pmatrix}\in(\mathrm{CG})_\ell,\right.$$
$$\left.\overline{\mathrm{w}}(r)=\begin{pmatrix} & j_3 & \\ r & & j_2\end{pmatrix}\in(\mathrm{CG})_\ell\right\}$$

$$\varepsilon_0(\mathbb{J})=\frac{1}{\kappa}(j_4+j_3+j_2+j_1+1),\ \varepsilon_i(\mathbb{J})=\varepsilon_0(\mathbb{J})-\frac{1}{\kappa}(2j_i+1)\ \ (i=1,\cdots,4)$$

$\mathbb{J}_t=\mathbb{J}_t(N)=\left(t,\ \frac{1}{2},\frac{1}{2},\cdots,\frac{1}{2}\right)$: an $(N+1)$-ple with $0\leq 2t\leq\ell$, $2t\in\mathbb{Z}$

$$\mathscr{P}_\ell(N;t)=\left\{\mathrm{p}=(p_N,\cdots,p_1,p_0);\ p_N=t,\ p_0=0,\ p_i\in\frac{1}{2}\mathbb{Z},\ 0\leq 2p_i\leq\ell,\right.$$
$$\left.|p_i-p_{i-1}|=\frac{1}{2}(1\leq i\leq N)\right\}.$$

$\mathbb{J}_{t,s}=\mathbb{J}_{t,s}(N)=\left(t,\ \frac{1}{2},\cdots,\frac{1}{2},\ s\right)$: an $(N+2)$-ple with $t,s\in\frac{1}{2}\mathbb{Z}_{\geq 0}$ and $t,s\leq\frac{\ell}{2}$

$$\mathscr{P}_\ell(N;t,s)=\left\{\mathrm{p}=(p_N,\cdots,p_1,p_0);\ p_N=t,\ p_0=s,\ p_i\in\frac{1}{2}\mathbb{Z}_{\geq 0},\right.$$
$$\left.0\leq 2p_i\leq\ell,\ |p_i-p_{i-1}|=\frac{1}{2}\ (1\leq i\leq N)\right\}$$

$X_N=\{(z_N,\cdots,z_1)\in\mathbb{C}^N;\ z_i\neq z_k\ (i\neq k)\}$
$\cup$
$M_N=\{(z_N,\cdots,z_1)\in(\mathbb{C}^*)^N;\ z_i\neq z_k\ (i\neq k)\}$
$\mathscr{R}_z=\{z=(z_N,\cdots,z_1)\in\mathbb{C}^N;\ |z_N|>\cdots>|z_1|\}\subset X_N$
$\cup$
$\mathscr{R}_{z,0}=\{(z_N,\cdots,z_1)\in\mathbb{C}^N;\ |z_N|>\cdots>|z_1|>0\}$
$\cup$
$I_N=\{(z_N,\cdots,z_1)\in\mathbb{R}^N;\ z_N>\cdots>z_1>0\}$
$\mathfrak{S}_N$: the N-th symmetric group
B_N: the braid group with N-strings of $\mathbb{C}$
$H_N(q)$: the Hecke algebra of type A_{N-1}
$\lambda=[f_1,\cdots,f_k]$: the Young diagram such that the number of nodes of the i-th row is f_i $(f_1\geq\cdots\geq f_k)$
$|\mathrm{vac}\rangle=u_0(0)$, $\langle\mathrm{vac}|=u_0^\dagger(0)$: the Virasoro vacuums
$\langle\Phi_N(z_N)\cdots\Phi_1(z_1)\rangle=\langle\mathrm{vac}|\Phi_N(z_N)\cdots\Phi_1(z_1)|\mathrm{vac}\rangle$: the N-point function of vertex operators $\{\Phi_N,\cdots,\Phi_1\}$
$\Gamma(z)$: the gamma function
$F(\alpha,\beta,\gamma;z)$: the Gauss' hypergeometric function

$[\nu]_q=\dfrac{q^\nu-1}{q-1}$ $(q\neq 1)$, ν $(q=1)$: a q-integer ($\nu\in\mathbb{Z}$)

$\dbinom{L}{\mathfrak{m}}=\dfrac{L!}{m_N!\cdots m_1!}$: the multinomial coefficient for $\mathfrak{m}=(m_N,\cdots,m_1)$ with $L=\sum m_k$

§ 1. Affine Lie Algebra of type $A_1^{(1)}$

In this section, we recall facts on the affine Lie algebra $\hat{\mathfrak{g}}$ of type $A_1^{(1)}$ (see V.G. Kac's book [Ka]).

1.1) Lie Algebra of type A_1 and its finite-dimensional modules

Let $\mathfrak{g}=\mathfrak{sl}(2,\mathbb{C})$ the Lie algebra of type A_1, that is, $\mathfrak{g}$ is a Lie algebra spanned by $H=\begin{pmatrix}1&0\\0&-1\end{pmatrix}$, $E=\begin{pmatrix}0&1\\0&0\end{pmatrix}$ and $F=\begin{pmatrix}0&0\\1&0\end{pmatrix}$. The subspace $\mathfrak{h}=\mathbb{C}H$ is a Cartan subalgebra of $\mathfrak{g}$. Its dual $\mathfrak{h}^*$ is spanned by the element α, defined by $\alpha(H)=2$. Put $\mathfrak{g}_\alpha=\mathbb{C}E$ and $\mathfrak{g}_{-\alpha}=\mathbb{C}F$, then $\mathfrak{g}$ has the root space decomposition

$$\mathfrak{g}=\mathfrak{g}_\alpha\oplus\mathfrak{h}\oplus\mathfrak{g}_{-\alpha}.$$

Let $(\ ,\ )\colon \mathfrak{g}\times\mathfrak{g}\to\mathbb{C}$ be the invariant symmetric bilinear form, defined by $(X,Y)=\operatorname{tr}XY$, where tr means the trace as 2×2-matrices. Then $(H,H)=2$, $(E,F)=1$ and $(H,E)=(H,F)=0$.

The *Casimir element* Ω of $\mathfrak{g}$ is defined as

$$\Omega=\frac{1}{2}H^2+EF+FE\in U(\mathfrak{g}).$$

Here we summarize the facts on finite dimensional modules of $\mathfrak{g}$:

Proposition 1.1. *Fix a half integer* $j\in\frac{1}{2}\mathbb{Z}_{\geq 0}$.

I) i) *There exists a unique irreducible left* $\mathfrak{g}$*-module* V_j (*called* of spin j) *with highest weight* $j\alpha$.

ii) V_j *is of dimension* $2j+1$ *and has a basis* $\{u_j(m);\ m=j,j-1,\cdots,1-j,-j\}$ *satisfying the relations*

$$(V_j)\qquad \begin{aligned} &Hu_j(m)=2mu_j(m) && (-j\leq m\leq j);\\ &Eu_j(m)=\sqrt{(j+m+1)(j-m)}\,u_j(m+1) && (-j\leq m<j);\\ &Fu_j(m)=\sqrt{(j+m)(j-m+1)}\,u_j(m-1) && (-j<m\leq j).\end{aligned}$$

iii) $Eu_j(j)=0$, $F^nu_j(j)\neq 0$ $(0\leq n\leq 2j)$ *and* $F^{2j+1}u_j(j)=0$.

iv) $\Omega=2(j^2+j)$ *on* V_j.

II) i) *There exists a unique irreducible right* $\mathfrak{g}$*-module* $V_j^\dagger$ *(called* of spin j*) with highest weight* $j\alpha$.

ii) $V_j^\dagger$ *is of dimension* $2j+1$ *and has a basis* $\{u_j^\dagger(m);\ m=j, j-1, \cdots, 1-j, -j\}$ *satisfying the relations*:

$$(V_j^\dagger)\qquad \begin{aligned} &u_j^\dagger(m)H=2mu_j^\dagger(m) && (-j\leq m\leq j);\\ &u_j^\dagger(m)E=\sqrt{(j+m)(j-m+1)}\,u_j^\dagger(m-1) && (-j<m\leq j);\\ &u_j^\dagger(m)F=\sqrt{(j+m+1)(j-m)}\,u_j^\dagger(m+1) && (-j\leq m<j).\end{aligned}$$

iii) $u_j^\dagger(j)F=0$, $u_j^\dagger(j)E^n\neq 0$ $(0\leq n\leq 2j)$ *and* $u_j^\dagger(j)E^{2j+1}=0$.

iv) $\Omega=2(j^2+j)$ *on* $V_j^\dagger$.

III) *There exists a unique bilinear form* (*called* vacuum expectation value)

$$\langle\ |\ \rangle\colon V_j^\dagger\times V_j\longrightarrow\mathbb{C}$$

such that 1) $\langle ua\,|\,v\rangle=\langle u\,|\,av\rangle$ *for any* $a\in\mathfrak{g}$, $\langle u|\in V_j^\dagger$ *and* $|v\rangle\in V_j$, *and* 2) $\langle u_j^\dagger(m)|u_j(m')\rangle=\delta_{m,m'}$. *Moreover this bilinear form is nondegenerate.*

1.2) The affine Lie algebra of type $A_1^{(1)}$

Let $\hat{\mathfrak{g}}$ be the affine Lie algebra of type $A_1^{(1)}$, that is, $\hat{\mathfrak{g}}$ is defined by

$$\hat{\mathfrak{g}}=\mathfrak{g}\otimes\mathbb{C}[t,t^{-1}]\oplus\mathbb{C}c$$

with the following commutation relations:

$$[X(m), Y(n)]=[X,Y](m+n)+(X,Y)m\delta_{m+n,0}c\qquad (X, Y\in\mathfrak{g},\ m,n\in\mathbb{Z}),$$

and

$$c\in\text{center of }\hat{\mathfrak{g}},$$

where $X(n)=X\otimes t^n$.

The Lie algebra $\mathfrak{g}$ is included in $\hat{\mathfrak{g}}$ by identifying X with $X(0)$. Introduce the subspace $\mathfrak{g}(n)=\mathfrak{g}\otimes t^n$ of $\hat{\mathfrak{g}}$ for any $n\in\mathbb{Z}$, and subalgebras $\mathfrak{m}_\pm=\sum_{n\geq 1}\mathfrak{g}(\pm n)$, then $\hat{\mathfrak{g}}$ is decomposed into

$$\hat{\mathfrak{g}}=\mathfrak{m}_+\oplus\mathfrak{g}\oplus\mathbb{C}c\oplus\mathfrak{m}_-.$$

The subspace $\hat{\mathfrak{h}}=\mathbb{C}H(0)\oplus\mathbb{C}c$ is a Cartan subalgebra of $\hat{\mathfrak{g}}$. The dual $\hat{\mathfrak{h}}^*$ of $\hat{\mathfrak{h}}$ is identified with $\mathbb{C}^2\ni(\lambda,\mu)$, by the formulae:

$$(\lambda,\mu)(c)=\lambda\quad\text{and}\quad(\lambda,\mu)(H)=2\mu.$$

Now we summarize the facts about the integrable highest weight modules of the Lie algebra $\hat{\mathfrak{g}}$.

Proposition 1.2. *Irreducible integrable highest weight modules of* $\hat{\mathfrak{g}}$ *are parametrized by* $(\ell, j) \in \mathbb{Z}_{\geq 0} \oplus \frac{1}{2}\mathbb{Z}_{\geq 0}$ *with* $2j \leq \ell$. *Fix such* (ℓ, j).

i) *There exists a unique irreducible left* $\hat{\mathfrak{g}}$*-module* $\mathscr{H}_j(\ell)$ *with a nonzero vector* $|\ell, j\rangle$ (*called* vacuum) *such that*

$$\mathfrak{m}_+|\ell, j\rangle = E|\ell, j\rangle = 0, \quad c|\ell, j\rangle = \ell|\ell, j\rangle \quad and \quad H|\ell, j\rangle = 2j|\ell, j\rangle.$$

ii) *There exists a unique irreducible right* $\hat{\mathfrak{g}}$*-module* $\mathscr{H}_j^\dagger(\ell)$ *with a nonzero vector* $\langle j, \ell|$ (*called* vacuum) *such that*

$$\langle j, \ell|\mathfrak{m}_- = \langle j, \ell|F = 0, \quad \langle j, \ell|c = \ell\langle j, \ell| \quad and \quad \langle j, \ell|H = 2j\langle j, \ell|.$$

iii) *The subspaces* $\{|v\rangle \in \mathscr{H}_j(\ell);\ \mathfrak{m}_+|v\rangle = 0\}$ *of* $\mathscr{H}_j(\ell)$ *and* $\{\langle v| \in \mathscr{H}_j^\dagger(\ell);\ \langle v|\mathfrak{m}_- = 0\}$ *of* $\mathscr{H}_j^\dagger(\ell)$ *are* $\mathfrak{g}$*-stable and are isomorphic to the irreducible* $\mathfrak{g}$*-modules* V_j *and* $V_j^\dagger$ *respectively.*

The vacuums $|\ell, j\rangle$ *and* $\langle j, \ell|$ *can be identified with* $u_j(j)$ *and* $u_j^\dagger(j)$, *and* $\mathscr{H}_j(\ell)$ *and* $\mathscr{H}_j^\dagger(\ell)$ *are generated by* V_j *and* $V_j^\dagger$ *respectively.*

iv) *There exists a unique bilinear form* (*called* vacuum expectation value)

$$\langle\ |\ \rangle\colon \mathscr{H}_j^\dagger(\ell) \times \mathscr{H}_j(\ell) \longrightarrow \mathbb{C}$$

such that 1) $\langle j, \ell|\ell, j\rangle = 1$, *and* 2) $\langle ua|v\rangle = \langle u|av\rangle$ *for any* $a \in \hat{\mathfrak{g}}$, $\langle u| \in \mathscr{H}_j^\dagger(\ell)$ *and* $|v\rangle \in \mathscr{H}_j(\ell)$. *Moreover this bilinear form is non-degenerate, and its restriction on* $V_j^\dagger \times V_j$ *coincides with the vacuum expectation value as* $\mathfrak{g}$*-modules* (*Proposition* 1.1).

1.3) Segal-Sugawara form

In this paragraph, we give the actions on $\mathscr{H}_j(\ell)$ and $\mathscr{H}_j^\dagger(\ell)$ of another Lie algebra $\mathscr{L}$ called *Virasoro Algebra*, where $\mathscr{L} = \sum_{n \in \mathbb{Z}} \mathbb{C}e_n \oplus \mathbb{C}e'_0$ is the Lie algebra defined by the relations:

$$[e_m, e_n] = (m-n)e_{m+n} + \frac{m^3 - m}{12}\delta_{m+n,0}e'_0 \qquad (m, n \in \mathbb{Z});$$

$$[e'_0, e_m] = 0.$$

Definition 1.3. Define the *normal ordered products* of elements of $\mathfrak{g} \otimes \mathbb{C}[t, t^{-1}]$ by

$$:X(m)Y(n): = \begin{cases} X(m)Y(n) & (m < n) \\ \dfrac{1}{2}\{X(m)Y(n) + Y(n)X(m)\} & (m = n) \\ Y(n)X(m) & (m > n). \end{cases}$$

Definition 1.4.
i) For each $X \in \mathfrak{g}$, we define the formal Laurent series

$$X(z)=\sum_{n\in\mathbf{Z}} X(n)z^{-n-1} \qquad (z \in \mathbf{C}^*).$$

ii) Energy-momentum tensor; Segal-Sugawara form ([Se] and [Su]) For $z \in \mathbf{C}^*$, define

$$\begin{aligned} T(z)&=\frac{1}{2(2+c)}\left\{\frac{1}{2}:H(z)H(z):+:E(z)F(z):+:F(z)E(z):\right\} \\ &=\sum_{m\in\mathbf{Z}} L(m)z^{-m-2}, \end{aligned}$$

that is,

$$\begin{aligned} L(m)=\frac{1}{2(2+c)}\sum_{k\in\mathbf{Z}}\Big\{\frac{1}{2}:H(-k)H(m+k):+:E(-k)F(m+k):+ \\ +:F(-k)E(m+k):\Big\}. \end{aligned}$$

Then we get

Proposition 1.5.
i) *For any* $j \in \frac{1}{2}\mathbf{Z}_{\geq 0}$ *with* $2j \leq \ell$, *the operator* $L(m)$, $m \in \mathbf{Z}$, *and* $L'(0) =(3\ell/(2+\ell))$ id *act on* $\mathscr{H}_j(\ell)$ *and* $\mathscr{H}_j^\dagger(\ell)$.
ii) *For any* $m, n \in \mathbf{Z}$,

$$[L(m), L(n)]=(m-n)L(m+n)+\frac{m^3-m}{12}\delta_{m+n,0}L'(0).$$

iii) *For each* $m \in \mathbf{Z}$ *and* $X \in \mathfrak{g}$,

$$\begin{aligned} &[L(m), X(z)]=z^m\left(z\frac{d}{dz}+m+1\right)X(z); \\ &[L(m), X(n)]=-nX(m+n) \qquad (n \in \mathbf{Z}). \end{aligned}$$

iv) *The modules* $\mathscr{H}_j(\ell)$ *and* $\mathscr{H}_j^\dagger(\ell)$ *have the eigenspace decompositions with respect to the operator* $L(0)$:

$$\mathscr{H}_j(\ell)=\sum_{d\geq 0}\mathscr{H}_{j,d}(\ell) \quad \textit{and} \quad \mathscr{H}_j^\dagger(\ell)=\sum_{d\geq 0}\mathscr{H}_{j,d}^\dagger(\ell).$$

where $\mathscr{H}_{j,d}(\ell)$ *and* $\mathscr{H}_{j,d}^\dagger(\ell)$ *are the eigenspaces of the eigenvalue* Δ_j+d, *and* $\Delta_j=(j^2+j)/(\ell+2)$. *In particular,* $\mathscr{H}_{j,0}(\ell)=V_j$ *and* $\mathscr{H}_{j,0}^\dagger(\ell)=V_j^\dagger$. *Moreover* $\dim \mathscr{H}_{j,d}(\ell)=\dim \mathscr{H}_{j,d}^\dagger(\ell)<\infty$.

v) $\mathscr{H}^{\dagger}_{j,d}(\ell) \perp \mathscr{H}_{j,d'}(\ell)$ *unless* $d=d'$, *and* $\langle | \rangle$ *is nondegenerate on* $\mathscr{H}^{\dagger}_{j,d}(\ell) \times \mathscr{H}_{j,d}(\ell)$.

vi) *For any* $X \in \mathfrak{g}$, $m \in \mathbb{Z}$ *and* $d \geq 0$,

$$X(m)\mathscr{H}_{j,d}(\ell), \quad L(m)\mathscr{H}_{j,d}(\ell) \subset \mathscr{H}_{j,d-m}(\ell)$$

and

$$\mathscr{H}^{\dagger}_{j,d}(\ell)X(m), \quad \mathscr{H}^{\dagger}_{j,d}(\ell)L(m) \subset \mathscr{H}^{\dagger}_{j,d+m}(\ell).$$

In the following of this paper, we fix an integer $\ell \geq 1$, put $\kappa = \ell + 2$, and omit ℓ in the notations $\mathscr{H}_j(\ell)$, $\mathscr{H}_{j,d}(\ell)$ etc. (Note that $V_0 = \mathscr{H}_0(0) = \mathbb{C}$.)

§ 2. Vertex Operators (Primary fields)

Throughout this paper we fix the value ℓ (a positive integer) of the central element c on the spaces $\mathscr{H}$ and $\mathscr{H}^{\dagger}$, and use the value $\kappa = \ell + 2$ for convenience.

2.1) Field operators

Fix a half integer j with $0 \leq 2j \leq \ell$. Introduce the product topology to the products $\hat{\mathscr{H}}_j = \prod_{d \geq 0} \mathscr{H}_{j,d}$ and $\hat{\mathscr{H}}^{\dagger}_j = \prod_{d \geq 0} \mathscr{H}^{\dagger}_{j,d}$, then the vacuum expectation $\langle | \rangle : \mathscr{H}^{\dagger}_j \times \mathscr{H}_j \to \mathbb{C}$ is uniquely extended to continuous bilinear pairings $\langle | \rangle : \mathscr{H}^{\dagger}_j \times \hat{\mathscr{H}}_j \to \mathbb{C}$ and $\hat{\mathscr{H}}^{\dagger}_j \times \mathscr{H}_j \to \mathbb{C}$, and there is a topological linear isomorphism $\hat{\mathscr{H}}^{\dagger}_j \cong \mathrm{Hom}_c(\mathscr{H}_j; \mathbb{C})$, where $\mathrm{Hom}_c(\mathscr{H}_j; \mathbb{C})$ is equipped with the weak topology. The actions of the Lie algebra $\hat{\mathfrak{g}}$ on $\mathscr{H}_j$ and $\mathscr{H}^{\dagger}_j$ can be extended to these completions.

Consider the direct sums of these modules:

$$\mathscr{H} = \sum_{j=0}^{\ell/2} \mathscr{H}_j \subset \hat{\mathscr{H}} = \sum_{j=0}^{\ell/2} \hat{\mathscr{H}}_j; \; \mathscr{H}^{\dagger} = \sum_{j=0}^{\ell/2} \mathscr{H}^{\dagger}_j \subset \hat{\mathscr{H}}^{\dagger} = \sum_{j=0}^{\ell/2} \hat{\mathscr{H}}^{\dagger}_j.$$

Denote by Π_j be the projection to the j-th component:

$$\Pi_j : \mathscr{H} \longrightarrow \mathscr{H}_j, \quad \hat{\mathscr{H}} \longrightarrow \hat{\mathscr{H}}_j; \quad \mathscr{H}^{\dagger} \longrightarrow \mathscr{H}^{\dagger}_j, \quad \hat{\mathscr{H}}^{\dagger}_j \longrightarrow \hat{\mathscr{H}}^{\dagger}_j,$$

then $\Pi_j \circ \Pi_k = \Pi_k \circ \Pi_j$ and Π_j commutes with the action of $\hat{\mathfrak{g}}$.

An *operator* A on $\mathscr{H}$ means a linear mapping $A : \mathscr{H} \longrightarrow \hat{\mathscr{H}}$, which is equivalent to give a bilinear map $\hat{A} : \mathscr{H}^{\dagger} \times \mathscr{H} \to \mathbb{C}$, and also to give a linear mapping $A^{\dagger} : \mathscr{H}^{\dagger} \to \hat{\mathscr{H}}^{\dagger}$ by the condition that for any $\langle v| \in \mathscr{H}^{\dagger}$ and $|w\rangle \in \mathscr{H}$,

$$\langle v | Aw \rangle = \langle v | \hat{A} | w \rangle = \langle vA | w \rangle.$$

In order to define compositions of operators, fix dual bases $\{|u_{d,1}\rangle, \cdots, |u_{d,m_d}\rangle\}$ of $\sum_{j=0}^{\ell/2} \mathscr{H}_{j,d}$ and $\{\langle v_{d,1}|, \cdots, \langle u_{d,m_d}|\}$ of $\sum_{j=0}^{\ell/2} \mathscr{H}^{\dagger}_{j,d}$

with respect to $\langle | \rangle$, where $m_d = \sum_{j=0}^{\ell/2} \dim \mathcal{H}_{j,d} = \sum_{j=0}^{\ell/2} \dim \mathcal{H}^\dagger_{j,d}$.

A sequence $\{A_N, \cdots, A_1\}$ of operators on $\mathcal{H}$ is called *composable*, if the series

$$\sum_{d_1,\cdots,d_{m-1}\geq 0} \left| \sum_{j_1=1}^{m_{d_1}} \cdots \sum_{j_{m-1}=1}^{m_{d_{m-1}}} \langle v | A_{n_m} | u_{d_{m-1},j_{m-1}} \rangle \right.$$

$$\left. \langle u_{d_{m-1},j_{m-1}} | A_{n_{m-1}} | u_{d_{m-2},j_{m-2}} \rangle \cdots \langle u_{d_1,j_1} | A_{n_1} | w \rangle \right|$$

is convergent for any ordered subset $\{n_m, \cdots, n_1\}$ of $\{N, \cdots, 2, 1\}$ with $2 \leq m \leq N$ and any vectors $\langle v| \in \mathcal{H}^\dagger$ and $|w\rangle \in \mathcal{H}$. Then the *composed operator* $A_N \cdots A_1$ is defined by the values

$$\langle v | A_N \cdots A_1 | w \rangle = \sum_{d_1,\cdots,d_{N-1}\geq 0} \sum_{j_1=1}^{m_{d_1}} \cdots \sum_{j_{N-1}=1}^{m_{d_{N-1}}} \langle v | A_N | u_{d_{N-1},j_{N-1}} \rangle$$
$$\langle u_{d_{N-1},j_{N-1}} | A_{N-1} | u_{d_{N-2},j_{N-2}} \rangle \cdots \langle u_{d_1,j_1} | A_1 | w \rangle$$

for $\langle v| \in \mathcal{H}^\dagger$ and $|w\rangle \in \mathcal{H}$.

An operator-valued function $A(z)$: $\mathcal{H} \to \hat{\mathcal{H}}$ on a complex manifold M is called *holomorphic* with respect to the variable $z \in M$, if the function $\langle u | A(z) | v \rangle$ is holomorphic with respect to $z \in M$ for any $\langle u | \in \mathcal{H}^\dagger$ and $|v\rangle \in \mathcal{H}$.

Example. Operator-valued functions $X(z)$ ($X \in \mathfrak{g}$) and $T(z)$: $\mathcal{H} \to \hat{\mathcal{H}}$ are single-valued and holomorphic on $\mathbb{C}^* = \mathbb{P}^1 \backslash \{0, \infty\}$.

Let $A_i(z_i)$ be an operator-valued function on $\mathcal{H}$ parametrized by a complex manifold M_i for each i with $1 \leq i \leq N$, and assume the sequence $\{A_N(z_N), \cdots, A_1(z_1)\}$ is composable for any $(z_N, \cdots, z_1) \in M_N \times \cdots \times M_1$. Then the composed operator $A_N(z_N) \cdots A_1(z_1)$ is holomorphic on the complex manifold $M_N \times \cdots \times M_1$.

2.2) Vertex operators

Now we give the notion of vertex operators (or primary fields) which is introduced by V.G. Knizhnik and A.B. Zamolodchikov [KZ].

For a positive half integer j, a multi-valued, holomorphic, operator-valued function $\Phi(z)$ on the manifold $\mathbb{C}^* (= \mathbb{C} \backslash \{0\})$ is called a *vertex operator of spin* j, if

$$\Phi(z);\ V_j \otimes \mathcal{H} \longrightarrow \hat{\mathcal{H}}$$

satisfies the conditions:

(V2) $$[X(m), \Phi(u; z)] = z^m \Phi(Xu; z)$$

(V3) $$[L(m), \Phi(u;z)]=z^m\left\{z\frac{d}{dz}+(m+1)\Delta_j\right\}\Phi(u;z)$$

for $X\in\mathfrak{g}$, $u\in V_j$, $m\in\mathbb{Z}$ and $z\in\mathbb{C}^*$, where the number $\Delta_j=(j^2+j)/\kappa$ is called the *conformal dimension* of the vertex operator $\Phi(z)$ and $\Phi(u;z)$: $\mathscr{H}\to\hat{\mathscr{H}}$ is the operator defined by

$$\Phi(u;z)(w)=\Phi(z)(u\otimes w) \qquad (u\in V_j,\ w\in\mathscr{H}).$$

Remark. (V2) is the gauge condition for the field $\Phi(z)$ and (V3) means the equations of motion.

Introduce sets $\mathbb{V}$ and $\mathbb{V}_\ell$ defined by

$$\mathbb{V}=\left\{\mathrm{v}=\begin{pmatrix} & j & \\ j_2 & & j_1\end{pmatrix}; j, j_1, j_2\in\frac{1}{2}\mathbb{Z}_{\geq 0}\right\}\supset\mathbb{V}_\ell=\left\{\mathrm{v}=\begin{pmatrix} & j & \\ j_2 & & j_1\end{pmatrix}\in\mathbb{V}; j_1, j_2\leq\frac{\ell}{2}\right\}.$$

An element v of $\mathbb{V}$ is called a *vertex*. For a vertex $\mathrm{v}=\begin{pmatrix} & j & \\ j_2 & & j_1\end{pmatrix}\in\mathbb{V}$, we call j_1 an *incoming spin*, j_2 an *outgoing spin* and j an *outer spin*, and set $\Delta(\mathrm{v})=\Delta_j$ $(=(j^2+j)/\kappa)$ and $\hat{\Delta}(\mathrm{v})=\Delta_j+\Delta_{j_1}-\Delta_{j_2}$.

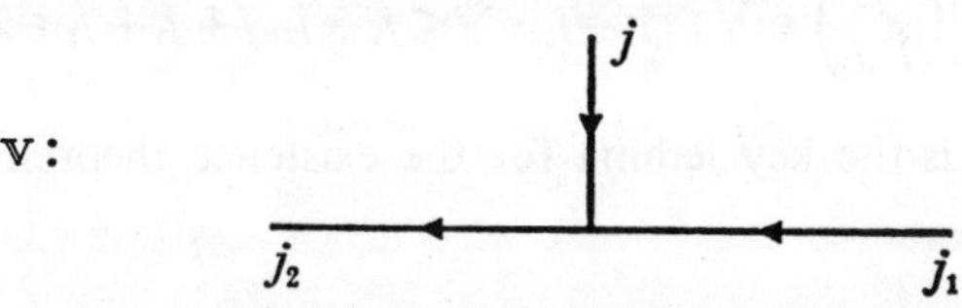

For a vertex $\mathrm{v}=\begin{pmatrix} & j & \\ j_2 & & j_1\end{pmatrix}\in\mathbb{V}_\ell$, a vertex operator $\Phi(z)$ of spin j is called *of type* v, if $\Phi(u;z)=\Pi_{j_2}\Phi(u;z)\Pi_{j_1}$ for any $u\in V_j$.

Then we get the following (the proof will be given in Section 2.3):

Proposition 2.1.

i) *Any vertex operator Φ of type v $(\in\mathbb{V}_\ell)$ has a Laurent series expansion*

$$\Phi(u;z)=\sum_{n\in\mathbb{Z}}\Phi_n(u)z^{-n-\hat{\Delta}(\mathrm{v})} \qquad (u\in V_j)$$

and $\Phi_n(u)$ satisfies

$$[L(0), \Phi_n(u)]=(\Delta_{j_2}-\Delta_{j_1}-n)\Phi_n(u) \qquad (n\in\mathbb{Z}),$$

that is,

$$\Phi_n(u)\colon \mathscr{H}_{j_1,d}\longrightarrow\mathscr{H}_{j_2,d-n},\ \mathscr{H}^\dagger_{j_2,d}\longrightarrow\mathscr{H}^\dagger_{j_1,d+n} \qquad (n\in\mathbb{Z}).$$

ii) *Introduce a trilinear form φ: $V^\dagger_{j_2}\otimes V_j\otimes V_{j_1}\to\mathbb{C}$ defined by*

$$\varphi(v, u, w) = \langle v|\Phi_0(u)|w\rangle = \langle v|\Phi(u; z)|w\rangle z^{\Delta(\mathrm{v})}|_{z=0} \qquad (v \in V_{j_2}^\dagger,\ w \in V_{j_1}),$$

then φ *is* $\mathfrak{g}$*-invariant*:

$$\varphi(vX, u, w) = \varphi(v, Xu, w) + \varphi(v, u, Xw) \qquad (X \in \mathfrak{g}).$$

iii) *A vertex operator* Φ *of type* v *is uniquely determined by the form* $\varphi \in \mathrm{Hom}_{\mathfrak{g}}(V_{j_2}^\dagger \otimes V_j \otimes V_{j_1}, \mathbb{C})$ *defined in* ii). *We call* φ *the* initial term *of the vertex operator* Φ *and sometimes denote* $\Phi = \Phi_\varphi$.

For each vertex $\mathrm{v} = \begin{pmatrix} & j & \\ j_2 & & j_1 \end{pmatrix} \in \mathbb{V}$, introduce the space $\mathscr{V}(\mathrm{v})$ defined by

$$\mathscr{V}(\mathrm{v}) = \mathrm{Hom}_{\mathfrak{g}}(V_{j_2}^\dagger \otimes V_j \otimes V_{j_1}, \mathbb{C}) \cong \mathrm{Hom}_{\mathfrak{g}}(V_j \otimes V_{j_1}, V_{j_2}).$$

It is well-known in the sl_2-theory that $\mathscr{V}(\mathrm{v}) = \mathbb{C}$ or 0, and $\mathscr{V}(\mathrm{v}) = \mathbb{C}$, if and only if v satisfies the *Clebsch-Gordan condition*:

$$|j_1 - j_2| \leq j \leq j_1 + j_2 \quad \text{and} \quad j_1 + j_2 + j \in \mathbb{Z}.$$

Call such vertex a CG-vertex and denote by (CG) the set of all CG vertices:

$$(\mathrm{CG}) = \left\{ \mathrm{v} = \begin{pmatrix} & j & \\ j_2 & & j_1 \end{pmatrix} \in \mathbb{V};\ |j_1 - j_2| \leq j \leq j_1 + j_2,\ j + j_1 + j_2 \in \mathbb{Z} \right\}.$$

The following is the key lemma for the existence theorem of vertex operators:

Lemma 2.2. *For a vertex* $\mathrm{v} = \begin{pmatrix} & j & \\ j_2 & & j_1 \end{pmatrix} \in (\mathrm{CG}) \cap \mathbb{V}_\ell$, *take a nonzero element* $\varphi \in \mathscr{V}(\mathrm{v})$. *Then the following conditions are equivalent*:

i) $j + j_1 + j_2 \leq \ell$.

ii) $\varphi(v, E^{\ell - 2j_1 + 1}u, u_{j_1}(j)) = 0$ *for any* $v \in V_{j_2}^\dagger$ *and* $u \in V_j$.

iii) $\varphi(u_{j_2}^\dagger(j_2), F^{\ell - 2j_2 + 1}u, w) = 0$ *for any* $u \in V_j$ *and* $w \in V_{j_1}$.

A vertex $\mathrm{v} = \begin{pmatrix} & j & \\ j_2 & & j_1 \end{pmatrix} \in \mathbb{V}_\ell$ is called an ℓCG-vertex, if it satisfies one of the conditions (called the *ℓ-constrained Clebsh-Gordan condition*) in Lemma 2.2 denoted by $(\mathrm{CG})_\ell$ the set of all ℓCG-vertices, *i.e.*

$$(\mathrm{CG})_\ell = \left\{ \mathrm{v} = \begin{pmatrix} & j & \\ j_2 & & j_1 \end{pmatrix} \in (\mathrm{CG});\ j + j_1 + j_2 \leq \ell \right\}.$$

Remark 2.2'. i) The inequalities $|j_1 - j_2| \leq j \leq j_1 + j_2$ and $j + j_1 + j_2 \leq \ell$ imply the inequalities $j, j_1, j_2 \leq \ell/2$. In particular, outer spins of ℓCG-vertices are not greater than $\ell/2$.

ii) By the above remark and the proof of Lemma 2.2, one of the conditions of Lemma 2.2 is also equivalent to the condition:

$$\varphi(v, u_j(j), E^{\ell-2j+1}w)=0 \text{ for any } v \in V_{j_2}^\dagger \text{ and } w \in V_{j_1}.$$

Now we get the existence condition for vertex operators (the proof will be given in the paragraph 2.3):

Theorem 2.3. *There exists a nonzero vertex operator Φ of type* $\mathrm{v}=\begin{pmatrix} j \\ j_2\, j_1 \end{pmatrix} \in \mathrm{V}_\ell$ *on $\mathscr{H}$, if and only if the vertex* v *is an ℓCG-vertex.*

Moreover, nonzero vertex operators of a fixed type $\mathrm{v} \in (\mathrm{CG})_\ell$ *are unique up to a constant multiple.*

As a corollary, we get

Proposition 2.4. i) *For any $j>\ell/2$, there are no vertex operators of spin j.*

ii) *Let $\Phi(z)$ be a vertex operator of type* $\mathrm{v}=\begin{pmatrix} j \\ j_2\, j_1 \end{pmatrix} \in (\mathrm{CG})_\ell$. *Then as formal Laurent series,*

$$\Phi(u; z)=z^{L(0)-\Delta_j}\Phi(u; 1)z^{-L(0)} \qquad (u \in V_j).$$

Proof. ii) Let $\Phi(u; z)$ be a vertex operator of spin j. Then the condition (V3) for $m=0$ reads as

$$[L(0), \Phi(u; z)]=\left\{z\frac{d}{dz}+\Delta_j\right\}\Phi(u; z). \qquad \text{q.e.d.}$$

2.3) Proof of Proposition 2.1 and Theorem 2.3

We define the *parabolic subalgebras* $\mathfrak{p}_\pm$ of $\hat{\mathfrak{g}}$ as $\mathfrak{p}_\pm=\mathfrak{m}_\pm \oplus \mathfrak{z} \oplus \mathbf{C}c$, and the *Verma module* $\mathscr{M}_j$ as the $\hat{\mathfrak{g}}$-module $\mathscr{M}_j=U(\hat{\mathfrak{g}})\otimes_{\mathfrak{p}_+} V_j$ $(=U(\mathfrak{m}_-)V_j)$, where the $\mathfrak{g}$-module V_j is considered as a $\mathfrak{p}_+$-module by setting $\mathfrak{m}_+ V_j=0$ and $c=\ell\,\mathrm{id}_{V_j}$. Then the irreducible $\hat{\mathfrak{g}}$-module $\mathscr{H}_j$ is obtained as the quotient of the Verma module $\mathscr{M}_j$ modulo the maximal proper submodule $\mathscr{J}_j$ (see V.G. Kac [Ka] (10.4.6)).

This $\hat{\mathfrak{g}}$-submodule $\mathscr{J}_j$ is also generated by the single vector $|J_j\rangle= E(-1)^{\ell-2j+1}u_j(j)$ and $\mathscr{J}_j=U(\mathfrak{p}_-)|J_j\rangle$. Moreover $\mathfrak{m}_+|J_j\rangle=E(0)|J_j\rangle= F(0)^{2\ell-2j+3}|J_j\rangle=0$, $H(0)|J_j\rangle=2(\ell-j+1)|J_j\rangle$, and $U(\mathfrak{g})|J_j\rangle$ is $\mathfrak{g}$-isomorphic to $V_{\ell-j+1}$. Denote by π_j the canonical projection $\pi_j: \mathscr{M}_j\to\mathscr{H}_j$.

The right $\hat{\mathfrak{g}}$-module $\mathscr{H}_j^\dagger$ is analogously obtained as $\mathscr{H}_j^\dagger=\mathscr{J}_j^\dagger\backslash\mathscr{M}_j^\dagger$, where $\mathscr{M}_j^\dagger$ is a right $\hat{\mathfrak{g}}$-module $\mathscr{M}_j^\dagger=V_j^\dagger\otimes_{\mathfrak{p}_-}U(\hat{\mathfrak{g}})$ (the $\mathfrak{g}$-module $V_j^\dagger$ is considered as a $\mathfrak{p}_-$-module by setting $V_j^\dagger\mathfrak{m}_-=0$ and $c=\ell\,\mathrm{id}_{V_j^\dagger}$), and its maximal proper $\hat{\mathfrak{g}}$-submodule $\mathscr{J}_j^\dagger$ is generated by a vector $\langle J_j|=u_j^\dagger(j)F(1)^{\ell-2j+1}$. Denote by $\pi_j^\dagger$ the canonical projection $\pi_j^\dagger$: $\mathscr{M}_j^\dagger\to\mathscr{H}_j^\dagger$.

The Verma modules $\mathcal{M}_j$ and $\mathcal{M}_j^\dagger$ have also eigenspace decompositions with respect to the operator $L(0)$:

$$\mathcal{M}_j=\sum_{d\geq 0}\mathcal{M}_{j,d} \quad\text{and}\quad \mathcal{M}_j^\dagger=\sum_{d\geq 0}\mathcal{M}_{j,d}^\dagger,$$

where the eigenvalue of $L(0)$ on $\mathcal{M}_{j,d}$ and $\mathcal{M}_{j,d}^\dagger$ is Δ_j+d.

In preparation of the proof, we introduce the filtrations in $\mathcal{M}_j$, $\mathcal{H}_j$, $\mathcal{M}_j^\dagger$ and $\mathcal{H}_j^\dagger$:

$$V_j=F_0\mathcal{H}_j=F_0\mathcal{M}_j\subset F_1\mathcal{M}_j\subset\cdots \quad\text{and}\quad V_j^\dagger=F_0\mathcal{H}_j^\dagger=F_0\mathcal{M}_j^\dagger\subset F_1\mathcal{M}_j^\dagger\subset\cdots$$

where $F_p\mathcal{M}_j$ and $F_p\mathcal{M}_j^\dagger$ are space spanned by the sets

$$\{Y_1(n_1)\cdots Y_q(n_q)|w\rangle;\ |w\rangle\in V_j,\ 0\leq q\leq p,\ Y_k(n_k)\in\hat{\mathfrak{g}}\ (1\leq k\leq q)\},$$

and

$$\{\langle v|X_q(m_q)\cdots X_1(m_1);\ \langle v|\in V_j^\dagger,\ 0\leq q\leq p,\ X_k(m_k)\in\hat{\mathfrak{g}}\ (1\leq k\leq q)\}$$

respectively, and

$$F_p\mathcal{H}_j=\pi_j(F_p\mathcal{M}_j) \quad\text{and}\quad F_p\mathcal{H}_j^\dagger=\pi_j^\dagger(F_p\mathcal{M}_j^\dagger).$$

Proof of Proposition 2.1.

i) Expand $\Phi(u;z)$ as a sum of homogeneous components:

$$\Phi(u;z)=\sum_{n\in Z}\Phi_n(u;z),\qquad \Phi_n(u;z)\colon \mathcal{H}_{j_1,d}\longrightarrow\mathcal{H}_{j_2,d-n}(d\geq 0),$$

then

$$[L(0),\Phi_n(u;z)]=(\Delta_{j_2}-\Delta_{j_1}-n)\Phi_n(u;z).$$

By (V3), we get

$$z\frac{d}{dz}\Phi_n(u;z)=-(\hat{\Delta}(\mathrm{v})+n)\Phi_n(u;z).$$

ii) The condition (V2) for $m=0$ implies

$$[X,\Phi(u,z)]=\Phi(Xu,z) \qquad (X\in\mathfrak{g},\ u\in V_j).$$

iii) Let Φ be a vertex operator of type v, and assume that $\varphi\in\mathrm{Hom}_{\mathfrak{g}}(V_{j_2}^\dagger\otimes V_j\otimes V_{j_1},\mathbb{C})$ defined in ii) vanishes. We want to show $\Phi(z)=0$. Now we show by the induction on $n=p+q$ that for any $u\in V_j$,

$$\langle v|\Phi(u;z)|w\rangle=0 \quad\text{for } \langle v|\in F_p\mathcal{H}_{j_2}^\dagger \quad\text{and}\quad |w\rangle\in F_q\mathcal{H}_{j_1}.$$

Assume that the assertion is valid for all $n\leq n_0$. It is sufficient to show

$$\langle v | X_p(m_p) \cdots X_1(m_1) \Phi(u; z) Y_q(-n_q) \cdots Y_1(-n_1) | w \rangle = 0$$

for $p+q=n_0+1$, $m_k, n_k \geq 1$, $\langle v| \in V^{\dagger}_{j_2}$ and $|w\rangle \in V_{j_1}$.

We may assume that $p \geq 1$ (if $p=0$, we can take $q \geq 1$). Then

$$\begin{aligned}
&\langle v | X_p(m_p) \cdots X_1(m_1) \Phi(u; z) Y_q(-n_q) \cdots Y_1(-n_1) | w \rangle \\
&\quad = z^{m_1} \langle v | X_p(m_p) \cdots X_2(m_2) \Phi(X_1 u; z) Y_q(-n_q) \cdots Y_1(-n_1) | w \rangle \\
&\qquad + \langle v | X_p(m_p) \cdots X_2(m_2) \Phi(u; z) X_1(m_1) Y_q(-n_q) \cdots Y_1(-n_1) | w \rangle \\
&\quad = 0.
\end{aligned}$$

q.e.d.

Proof of Theorem 2.3. Proposition 2.1 shows that a vertex operator $\Phi(z)$ of type $\mathbb{v}$ defines a form $\varphi \in \mathscr{V}(\mathbb{v})$ and is uniquely determined by φ. In particular, the existence of a vertex operator implies the Clebsh-Gordan condition for $\mathbb{v}$.

Let $\varphi\ (\neq 0) \in \mathscr{V}(\mathbb{v}) = \mathrm{Hom}_{\mathfrak{g}}(V^{\dagger}_{j_2} \otimes V_j \otimes V_{j_1}; \mathbb{C})$. We want to construct a form $\tilde{\Phi}(z) \in \mathrm{Hom}(\mathscr{M}^{\dagger}_{j_2} \otimes V_j \otimes \mathscr{M}_{j_1}; \mathbb{C})$ such that

(M1) $$\tilde{\Phi}(z)|_{V^{\dagger}_{j_2} \otimes V_j \otimes V_{j_1}} = z^{-\hat{\Delta}} \varphi \qquad (z \in \mathbb{C}^*),$$

(M2) $$\tilde{\Phi}(vX(m), u, w; z) - \tilde{\Phi}(v, u, X(m)w; z) = z^m \tilde{\Phi}(v, Xu, w; z) \qquad (m \in \mathbb{Z},\ X \in \mathfrak{g}),$$

and

(M3) $$\begin{aligned}
&\tilde{\Phi}(vL(m), u, w; z) - \tilde{\Phi}(v, u, L(m)w; z) \\
&\quad = z^m \left\{ z \frac{d}{dz} + (m+1) \Delta_j \right\} \tilde{\Phi}(v, u, w; z) \qquad (m \in \mathbb{Z})
\end{aligned}$$

for any $\langle v| \in \mathscr{M}^{\dagger}_{j_2}$, $u \in V_j$ and $|w\rangle \in \mathscr{M}_{j_1}$, where $\hat{\Delta} = \hat{\Delta}(\mathbb{v})$.
(We use the notation $\tilde{\Phi}(v, u, w; z) = \tilde{\Phi}(u; z)(v, w) = \tilde{\Phi}(z)(v, u, w)$.)

Step 0. (M1) defines $\tilde{\Phi}(z)$ on $V^{\dagger}_{j_2} \otimes V_j \otimes V_{j_1}$ satisfying (M2) for $m=0$.

Step 1. Define $\tilde{\Phi}(z)$ on $V^{\dagger}_{j_2} \otimes V_j \otimes F_q \mathscr{M}_{j_1}$ inductively as

$$\tilde{\Phi}(v, u, X(-m)w; z) = -z^{-m} \tilde{\Phi}(v, Xu, w; z)$$

for $m>0$, $X \in \mathfrak{g}$, $v \in V^{\dagger}_{j_2}$, $u \in V_j$, $w \in F_{q-1}\mathscr{M}_{j_1}$, then we get $\tilde{\Phi}(z)$ on $V^{\dagger}_{j_2} \otimes V_j \otimes \mathscr{M}_{j_1}$ satisfying (M1) and (M2) for $m \leq 0$.

Step 2. Define $\tilde{\Phi}(z)$ on $F_p \mathscr{M}^{\dagger}_{j_2} \otimes V_j \otimes \mathscr{M}_{j_1}$ inductively as

$$\tilde{\Phi}(vX(m), u, w; z) = z^m \tilde{\Phi}(v, Xu, w; z) + \tilde{\Phi}(v, u, X(m)w; z)$$

for $m>0$, $X \in \mathfrak{g}$, $v \in F_{p-1}\mathscr{M}^{\dagger}_{j_2}$, $u \in V_j$, $w \in \mathscr{M}_{j_1}$, then we get $\tilde{\Phi}(z)$ on $\mathscr{M}^{\dagger}_{j_2} \otimes V_j \otimes \mathscr{M}_{j_1}$. The well-definedness of $\tilde{\Phi}(z)$ and the condition (M2) can be verified again by the induction on p.

Step 3. Verify (M3) for $\tilde{\Phi}(z)$ defined in Step 2.

Let $v\otimes u\otimes w \in \mathcal{M}^{\dagger}_{j_2,d_2}\otimes V_j\otimes\mathcal{M}_{j_1,d_1}$, then $z^{\Delta-d_2+d_1}\tilde{\Phi}(v, u, w; z)$ is proved to be constant by the construction of $\tilde{\Phi}(z)$. On the other hand,

$$\tilde{\Phi}(vL(0), u, w; z)-\tilde{\Phi}(v, u, L(0)w; z)=\{\Delta_{j_2}+d_2-\Delta_{j_1}-d_1\}\tilde{\Phi}(v, u, w; z)$$
$$=\left\{z\frac{d}{dz}+\Delta_j\right\}\tilde{\Phi}(v, u, w; z).$$

Thus we get (M3) for $m=0$.

Recall that $L(0)|_{V_j}=\Omega/2\kappa|_{V_j}=\Delta_j\,\mathrm{id}_{V_j}$, $L(0)|_{V_j^\dagger}=\Delta_j\,\mathrm{id}_{V_j^\dagger}$, and the expansion of $L(m)$:

$$L(m)=\frac{1}{2\kappa}\sum_{j\in\mathbf{Z}}\left\{\frac{1}{2}{:}H(m-j)H(j){:}+{:}E(m-j)F(j){:}+{:}F(m-j)E(j){:}\right\}.$$

Then on each component $\mathcal{M}^{\dagger}_{j_2,d_2}\otimes V_j\otimes\mathcal{M}_{j_1,d_1}$ we can show (M3) for any $m\in\mathbf{Z}$ from (M3) for $m=0$ by case-by-case computations. We give here its proof in the case $m=2n+1>0$, $d_2\geq d_1$ (other cases are similarly obtained). In this case, $2\kappa L(m)=2\sum_{k\geq -n}\sum_{i=1}^{3}X^i(-k)X_i(m+k)$, where $X^1=2X_1=H$, $X^2=X_3=E$ and $X^3=X_2=F$.

Let $v\otimes u\otimes w \in \mathcal{M}^{\dagger}_{j_2,d_2}\otimes V_j\otimes\mathcal{M}_{j_1,d_1}$, then (M3) for $m=0$ reads as

$$2\kappa\left\{z\frac{d}{dz}+\Delta_j\right\}\tilde{\Phi}(v, u, w; z)=(2d_2+1)\tilde{\Phi}(v, \Omega u, w; z)$$
$$+2\sum_{k=1}^{d_1}z^{-k}\sum_{i=1}^{3}\tilde{\Phi}(v, X^iu, X_i(k)w; z)+2\sum_{k=0}^{d_2}z^{k}\sum_{i=1}^{3}\tilde{\Phi}(v, X^iu, X_i(-k)w; z).$$

And

$$2\kappa\{\tilde{\Phi}(vL(m), u, w; z)-\tilde{\Phi}(v, u, L(m)w; z)\}$$
$$=2\sum_{i=1}^{3}\sum_{k=-n}^{d_2}\{z^m\tilde{\Phi}(v, X^iX_iu, w; z)+z^{k+m}\tilde{\Phi}(v, X^iu, X_i(-k)w; z)\}$$
$$+2\sum_{i=1}^{3}\sum_{k=-n}^{d_1-m}z^{-k}\tilde{\Phi}(v, X^iu, X_i(m+k)w; z)$$
$$=2z^m\sum_{k=-n}^{d_2}\left\{\tilde{\Phi}(v, \Omega u, w; z)+\sum_{i=1}^{3}z^k\tilde{\Phi}(v, X^iu, X_i(-k)w; z)\right\}$$
$$+2z^m\sum_{i=1}^{3}\sum_{k=n+1}^{d_1}z^{-k}\tilde{\Phi}(v, X^iu, X_i(k)w; z).$$

Hence

$$2\kappa\left[\{\tilde{\Phi}(vL(m), u, w; z)-\tilde{\Phi}(v, u, L(m)w; z)\}-z^m\left\{z\frac{d}{dz}+\Delta_j\right\}\tilde{\Phi}(v, u, w; z)\right]$$
$$=(2n+1)z^m\tilde{\Phi}(v, \Omega u, w; z)$$
$$+2z^m\sum_{i=1}^{3}\left\{\sum_{k=-n}^{-1} z^k\tilde{\Phi}(v, X^i u, X_i(-k)w; z)-\sum_{k=1}^{n} z^{-k}\tilde{\Phi}(v, X^i u, X_i(k)w; z)\right\}$$
$$=mz^m\tilde{\Phi}(v, 2\kappa\Delta_j u, w; z),$$

thus we get (M3).

Step 4. Now we get $\tilde{\Phi}(z)\in \mathrm{Hom}\,(\mathcal{M}^\dagger_{j_2}\otimes V_j\otimes\mathcal{M}_{j_1}; \mathbb{C})$ satisfying (M1) ~(M3). If $\tilde{\Phi}(z)$ factors to $\Phi(z)\in\mathrm{Hom}_{\mathfrak{g}}(\mathcal{H}^\dagger_{j_2}\otimes V_j\otimes\mathcal{H}_{j_1}; \mathbb{C})$, then the bilinear form $\Phi(u; z)$ $(u\in V_j)$ on $\mathcal{H}^\dagger_{j_2}\otimes\mathcal{H}_{j_1}$ defines an operator from $\mathcal{H}_{j_1}$ to $\mathcal{H}_{j_2}$ satisfying the conditions (V2) and (V3).

We must show that $\tilde{\Phi}(z)$ factors through $\mathrm{Hom}_{\mathfrak{g}}(\mathcal{H}^\dagger_{j_2}\otimes V_j\otimes\mathcal{H}_{j_1}; \mathbb{C})$, if and only if the vertex $\mathbf{v}$ is an ℓCG-vertex.

From the condition (M2), we get by the induction on p for $F_p\mathcal{M}^\dagger_{j_2}$ that $\tilde{\Phi}(u; z)$ factors through $\mathcal{M}^\dagger_{j_2}\otimes\mathcal{H}_{j_1}$, that is,

$$\tilde{\Phi}(v, u, \mathcal{J}_{j_1})=0 \qquad \text{for any } v\in\mathcal{M}^\dagger_{j_2} \text{ and } u\in V_j,$$

if and only if

$$\tilde{\Phi}(v, u, |J_{j_1}\rangle)=0 \qquad \text{for any } v\in V^\dagger_{j_2} \text{ and } u\in V_j.$$

In fact, $\mathcal{J}_{j_1}=U(\mathfrak{m}_-)U(\mathfrak{g})|J_{j_1}\rangle$ and $\mathfrak{m}_+|J_{j_1}\rangle=E|J_{j_1}\rangle=0$.

Since $|J_{j_1}\rangle=E(-1)^{\ell-2j_1+1}u_{j_1}(j_1)$, the last condition is equivalent to

$$\varphi(v, E^{\ell-2j_1+1}u, u_{j_1}(j_1))=0 \qquad \text{for any } v\in V^\dagger_{j_2} \text{ and } u\in V_j.$$

Similarly we get that $\tilde{\Phi}(u; z)$ factors through $\mathcal{H}^\dagger_{j_2}\otimes\mathcal{M}_{j_1}$, if and only if

$$\varphi(u^\dagger_{j_2}(j_2), F^{\ell-2j_2+1}u, w)=0 \qquad \text{for any } u\in V_j \text{ and } w\in V_{j_1}.$$

Step 5. Apply Lemma 2.2. q.e.d.

2.4) Normalization of vertex operators and Proof of Lemma 2.2.

The right $\mathfrak{g}$-module $V^\dagger_j$ can be identified with the dual (right) $\mathfrak{g}$-module $V^\vee_j=\mathrm{Hom}\,(V_j, \mathbb{C})$ through the vacuum expectation values:

$$v(u)=\langle v|u\rangle \qquad \text{for } v\in V^\dagger_j \text{ and } u\in V_j.$$

There exists an isomorphism $\nu\colon V_j\to V^\dagger_j$ defined by $\nu(u_j(m))=(-1)^{j-m}\times u^\dagger_j(-m)$, then ν is an isomorphism over $(\mathfrak{g}, \nu)$:

$$\nu(X|v\rangle)=-\nu(|v\rangle)X \qquad (|v\rangle\in V_j,\ X\in\mathfrak{g}),$$

where ν: $\mathfrak{g}\to\mathfrak{g}$ is the anti-automorphism defined by $\nu(X)=-X$. Moreover ν can be extended to the isomorphism ν: $\mathscr{H}_j\to\mathscr{H}_j^\dagger$ such that $\nu(X(m)|v\rangle)=-\nu(|v\rangle)X(-m)$ ($|v\rangle\in\mathscr{H}_j$, $X\in\mathfrak{g}$, $m\in\mathbf{Z}$).

In Appendix I, we fix the element $\varphi_{\mathrm{v}}\in\mathscr{V}(\mathrm{v})=\mathrm{Hom}_{\mathfrak{g}}(V_j\otimes V_{j_1}, V_{j_2})$ ($\cong\mathbf{C}$) for each CG-vertex $\mathrm{v}=\begin{pmatrix} & j & \\ j_2 & & j_1\end{pmatrix}$. This notation will be used throughout this paper. And for each ℓCG-vertex v, denote by $\Phi_{\mathrm{v}}(z)=\Phi_{\varphi_{\mathrm{v}}}(z)$ the vertex operator of type v whose initial term is φ_{v}.

In a special case, we get

Proposition 2.5.

i) *Let j be an half-integer with $0\leq 2j\leq\ell$ and put* $\mathrm{v}=\begin{pmatrix} & j & \\ j & & 0\end{pmatrix}$. *Then* $\mathrm{v}\in(\mathrm{CG})_\ell$, $\hat{\Delta}(\mathrm{v})=0$, *and* $\varphi_{\mathrm{v}}=\mathrm{id}_{V_j}\in\mathscr{V}(\mathrm{v})\cong\mathrm{Hom}(V_j, V_j)$. *Hence*

$$\lim_{z\searrow 0}\Phi_{\mathrm{v}}(w;z)|u_0(0)\rangle=|w\rangle \qquad (w\in V_j).$$

ii) *Let j be an half-integer with $0\leq 2j\leq\ell$ and put* $\mathrm{v}=\begin{pmatrix} & j & \\ 0 & & j\end{pmatrix}$. *Then* $\mathrm{v}\in(\mathrm{CG})_\ell$, $\hat{\Delta}(\mathrm{v})=2\Delta_j$, *and* $\varphi_{\mathrm{v}}=\nu\in\mathscr{V}(\mathrm{v})=\mathrm{Hom}(V_j, V_j^\dagger)$. *Hence*

$$\lim_{z\nearrow\infty} z^{2\Delta_j}\langle u_0^\dagger(0)|\Phi_{\mathrm{v}}(v;z)=\langle u_0^\dagger(0)|\varphi_{\mathrm{v}}(v)=\langle\nu(v)| \qquad (v\in V_j).$$

By the symmetry, it is sufficient to show the following for the proof of Lemma 2.2:

Lemma 2.2″. *For a vertex* $\mathrm{v}=\begin{pmatrix} & j & \\ j_2 & & j_1\end{pmatrix}\in\mathrm{V}_\ell$, *assume* $\mathscr{V}(\mathrm{v})\neq 0$ *and take its nonzero element φ. Then the following conditions are equivalent*:

(0) $$j_2+j+j_1\leq\ell.$$

(1) $\varphi(v, E^{\ell-2j_1+1}u, u_{j_1}(j_1))=0$ *for any* $v\in V_{j_2}$ *and* $u\in V_j$.

Proof. Decompose the tensor product $V_{j_2}\otimes V_j$ into the sum of the irreducible components: $V_{j_2}\otimes V_j=\sum_k W_k$, where $W_k\cong V_k$ for $k\in\frac{1}{2}\mathbf{Z}$ with $|j-j_2|\leq k\leq j+j_2$ and $k+j+j_2\in\mathbf{Z}$. By the assumption on φ, we may assume that $\varphi(W_{j_1}\otimes V_{j_1})\neq 0$ and $\varphi(W_k\otimes V_{j_1})=0$ for $k\neq j_1$.

Since V_{j_1} is generated by the vector $u_{j_1}(j_1)$ and φ is invariant, there exists a vector $w\in W_{j_1,-j_1}$ such that $\varphi(w\otimes u_{j_1}(j_1))\neq 0$ and $\varphi(W_{j_1,h}\otimes u_{j_1}(j_1))=0$ for any $h>-j_1$.

Put $L_1=\ell-2j_1+1$. Assume that $j_2+j+j_1\leq\ell$. Let $v\in V_{j_2,h_2}$ and $u\in V_{j,h}$. Since $h_2+h+L_1\geq 1-j_1$, $\varphi(v, E^{L_1}u, u_{j_1}(j_1))=0$. Thus (0) implies (1).

Now express the vector w as $w=\sum_h a_h v_h \otimes u_h$, where $a_h \in \mathbb{C}$, $v_h \in V_{j_2,-h-j_1}$ and $u_h \in V_{j,h}$. Since $\mathfrak{n}_- w=0$, we get that $a_h \neq 0$ for $-j \leq h \leq j_2 - j_1$ by the induction on h. Hence $\varphi(v_{j_2-j_1}, u_{j_2-j_1}, u_{j_1}(j_1)) \neq 0$.

Assume that $j_2+j+j_1>\ell$. Then we get $j_2+j+j_1 \geq \ell+1$ and so

$$j_2-j_1-L_1 \geq j_1+j_2-\ell-1 \geq -j.$$

Hence the vector $u=F^{L_1}u_{j_2-j_1}$ does not vanish and $u_{j_2-j_1}=bE^{L_1}u$ for some nonzero constant b. Thus (1) implies (0). q.e.d.

2.5) Operator product expansions

The notion of operator product expansions in the 2-dimensional conformal field theory is due to A.A. Belavin et al. [BPZ].

Proposition 2.6.

i) *Ordered pairs* $\{X(\zeta), Y(z)\}$, $\{X(\zeta), T(z)\}$, $\{T(\zeta), X(z)\}$ *and* $\{T(\zeta), T(z)\}$ *of operators are composable for* $|\zeta|>|z|>0$ $(X, Y \in \mathfrak{g})$, *and their compositions* $X(\zeta)Y(z)$, $X(\zeta)T(z)$, $T(\zeta)X(z)$ *and* $T(\zeta)T(z)$ *are analytically continued to single-valued, operator-valued holomorphic functions on* $M_2=\{(\zeta, z) \in (\mathbb{C}^*)^2; \zeta \neq z\}$. *As operators on* $\mathscr{H}$, *the following identities hold:*

$$\text{(I)} \quad X(\zeta)Y(z)=\frac{\ell(X, Y)}{(\zeta-z)^2}\,\mathrm{id}+\frac{1}{\zeta-z}[X, Y](z)+R_{\mathrm{I}} \qquad (X, Y \in \mathfrak{g}).$$

$$\text{(II)} \quad T(\zeta)X(z)=\frac{1}{(\zeta-z)^2}X(z)+\frac{1}{\zeta-z}\frac{\partial}{\partial z}X(z)+R_{\mathrm{II}} \qquad (X \in \mathfrak{g}).$$

$$\text{(III)} \quad T(\zeta)T(z)=\frac{3\ell\,\mathrm{id}}{2\kappa(\zeta-z)^4}+\frac{2T(z)}{(\zeta-z)^2}+\frac{1}{\zeta-z}\frac{\partial}{\partial z}T(z)R_{\mathrm{III}}.$$

Here R_{I}, R_{II} *and* R_{III} *are regular at* $\zeta=z \in \mathbb{C}^*$.

Moreover

$$T(\zeta)T(z)=T(z)T(\zeta), \quad T(\zeta)X(z)=X(z)T(\zeta) \quad \text{and} \quad X(\zeta)Y(z)=Y(z)X(\zeta).$$

ii) *Let* $\Phi(z)$ *be a vertex operator of spin* j *and* $u \in V_j$. *Ordered pairs* $\{X(\zeta), \Phi(u; z)\}$, $\{\Phi(u; \zeta), X(z)\}$, $\{T(\zeta), \Phi(u; z)\}$ *and* $\{\Phi(u; \zeta), T(z)\}$ *of operators are composable for* $|\zeta|>|z|>0$ $(X \in \mathfrak{g})$, *and their compositions* $X(\zeta)\Phi(u; z)$, $\Phi(u; \zeta)X(z)$, $T(\zeta)\Phi(u; z)$ *and* $\Phi(u; \zeta)T(z)$ *are analytically continued to multi-valued, operator-valued holomorphic functions on* M_2. *As operators on* $\mathscr{H}$, *the following identities hold:*

$$\text{(IV)} \quad X(\zeta)\Phi(u; z)=\frac{1}{\zeta-z}\Phi(Xu; z)+R_{\mathrm{IV}} \qquad (X \in \mathfrak{g}).$$

$$\text{(V)}\quad T(\zeta)\Phi(u;z)=\frac{\Delta_j}{(\zeta-z)^2}\Phi(u;z)+\frac{1}{\zeta-z}\frac{\partial}{\partial z}\Phi(u;z)+R_{\mathrm{V}}.$$

Here R_{IV} and R_{V} are regular at $\zeta=z\in\mathbb{C}^*$.

Moreover $X(\zeta)\Phi(u;z)$ and $T(\zeta)\Phi(u;z)$ ($X\in\mathfrak{g}$) are single-valued and holomorphic function on $\zeta\in\mathbb{P}^1\backslash\{0, z, \infty\}$ for any fixed $z\in\mathbb{C}^*$, and

$$X(\zeta)\Phi(u;z)=\Phi(u;z)X(\zeta)\quad\text{and}\quad T(\zeta)\Phi(u;z)=\Phi(u;z)T(\zeta).$$

Proof. All cases are obtained similarly, so we deal here with the case ii).

Let $\Phi(z)$ be a vertex operator of type v. By Proposition 2.1 i), $\Phi(u;z)$ has the expansion

$$\Phi(u;z)=\sum_{n\in\mathbf{Z}}z^{-n-\Delta}\Phi_n(u)\qquad(u\in V_j)$$

where $\Delta=\hat{\Delta}(\mathrm{v})=\Delta_j+\Delta_{j_1}-\Delta_{j_2}$. Then we get

$$[X(m),\Phi_n(u)]=[X(0),\Phi_{m+n}(u)]=\Phi_{m+n}(Xu)\qquad(X\in\mathfrak{g},\ m,n\in\mathbb{Z})$$

and

$$[L(m),\Phi_n(u)]=\{(m+1)\Delta_j-m-n-\Delta\}\Phi_{m+n}(u)\qquad(m,n\in\mathbb{Z}).$$

Here we show (IV). For $|\zeta|>|z|>0$,

$$\begin{aligned}X(\zeta)\Phi(u,z)&=\sum_{m,n\in\mathbf{Z}}\zeta^{-m-1}z^{-\Delta-n}X(m)\Phi_n(u)\\&=\sum_{k\in\mathbf{Z}}\zeta^{-1}z^{-\Delta-k}\sum_{m\in\mathbf{Z}}\left(\frac{z}{\zeta}\right)^m X(m)\Phi_{k-m}(u)\\&=\sum_{k\in\mathbf{Z}}\zeta^{-1}z^{-\Delta-k}\sum_{m\geq0}\left(\frac{z}{\zeta}\right)^m[X(m),\Phi_{k-m}(u)]+R_{\mathrm{IV}}\\&=\sum_{k\in\mathbf{Z}}\zeta^{-1}z^{-\Delta-k}\sum_{m\geq0}\left(\frac{z}{\zeta}\right)^m\Phi_k(Xu)+R_{\mathrm{IV}}\\&=\frac{\zeta^{-1}}{1-z/\zeta}\sum_{k\in\mathbf{Z}}z^{-\Delta-k}\Phi_k(Xu)+R_{\mathrm{IV}}=\frac{1}{\zeta-z}\Phi(Xu;z)+R_{\mathrm{IV}},\end{aligned}$$

where

$$R_{\mathrm{IV}}=\sum_{k\in\mathbf{Z}}\zeta^{-1}z^{-\Delta-k}\left\{\sum_{m>0}\left(\frac{\zeta}{z}\right)^m X(-m)\Phi_{k+m}(u)+\sum_{m\geq0}\left(\frac{z}{\zeta}\right)^m\Phi_{k-m}(u)X(m)\right\}$$

is regular at $\zeta=z$.

For $|z|>|\zeta|>0$, we get

$$\Phi(u;z)X(\zeta)=\frac{1}{z-\zeta}[\Phi(u;z),X(0)]+R_{\rm IV}=\frac{-1}{z-\zeta}\Phi(Xu;z)+R_{\rm IV},$$

for the same Laurent series $R_{\rm IV}$. Hence for any $\langle u|\in\mathcal{H}^\dagger$, $|v\rangle\in\mathcal{H}$ and fixed $z\in\mathbb{C}^*$, the holomorphic function $\langle u|X(\zeta)\Phi(u;z)|v\rangle$ defined on $\{\zeta\in\mathbb{C};|\zeta|>|z|\}$ can be analytically continued to a (single-valued) holomorphic function on $\mathbb{P}^1\backslash\{0,\infty,z\}$ which coincides with the function $\langle u|\Phi(u;z)X(\zeta)|v\rangle$ on $\{\zeta;|z|>|\zeta|>0\}$. q.e.d.

Proposition 2.6 is generalized as follows:

Proposition 2.7. *Let* $u\in V_j$ *and* $\Phi(z)=\Phi_{\rm v}(z)$ *be the vertex operator of type* $\mathrm{v}=\begin{pmatrix} & j & \\ j_2 & & j_1\end{pmatrix}\in(\mathrm{CG})_\ell$. *Let* $A_N(z_N),\cdots,A_1(z_1)$ *be operators of the form* $T(z)$, $X(z)$ $(X\in\mathfrak{g})$ *or* $\Phi(u;z)$, *and assume that there is a number* i_0 *such that* $A_{i_0}(z_{i_0})=\Phi(u;z_{i_0})$ *and* $A_i(z_i)$ *is not a vertex operator for* $i\neq i_0$.

Then $\{A_N(z_N),\cdots,A_1(z_1)\}$ *is composable in the range* $|z_N|>\cdots>|z_1|$, *and the composed operator* $A_N(z_N)\cdots A_1(z_1)$ *is analytically continued to a multivalued and holomorphic function on* $M_N=\{(z_N,\cdots,z_1)\in(\mathbb{C}^*)^N;z_i\neq z_j\ (i\neq j)\}$. *If we fix* $(z_N,\cdots,\hat{z}_j,\cdots,z_1)$ $(j\neq i_0)$, *then this function is single-valued in* $z_j\in\mathbb{P}^1\backslash\{\infty,z_N,\cdots,\hat{z}_j,\cdots,z_1,0\}$.

2.5) Actions of $\hat{\mathfrak{g}}$ and $\mathcal{L}$ on vertex operators

For an ℓCG-vertex $\mathrm{v}=\begin{pmatrix} & j & \\ j_2 & & j_1\end{pmatrix}$, introduce the $\mathfrak{g}$-module $\mathcal{P}(\mathrm{v})$ defined by

$$\mathcal{P}(\mathrm{v})=\{\Phi_{\rm v}(u;z);u\in V_j\}\quad\text{and}\quad X\Phi_{\rm v}(u;z)=\Phi_{\rm v}(Xu;z)\qquad(X\in\mathfrak{g}).$$

In this paragraph, we fix $\mathrm{v}\in(\mathrm{CG})_\ell$ and say $\Phi(z)=\Phi_{\rm v}(z)$.

Now introduce the space $\mathcal{O}(\mathrm{v})$ of operators on $\mathcal{H}$ as the $\mathbb{C}$-vector space spanned by the set

$$\left\{\frac{1}{(2\pi\sqrt{-1})^N}\int_{C_N}\cdots\int_{C_1}d\zeta_N\cdots d\zeta_1(\zeta_N-z)^{m_N}\cdots(\zeta_1-z)^{m_1}X_N(\zeta_N)\right.$$
$$\left.\cdots X_1(\zeta_1)\Phi(u;z);N\in\mathbb{Z}_{\geq0},X_i\in\mathfrak{g},m_i\in\mathbb{Z}\ (1\leq i\leq N),u\in V_j\right\},$$

where the contours C_i $(1\leq i\leq N)$ are taken as follows: the origin 0 is outside C_N, C_i is inside C_{i+1} and z is inside C_1.

Let $A(z)\in\mathcal{O}(\mathrm{v})$, $X\in\mathfrak{g}$ and $m\in\mathbb{Z}$, then define

$$\hat{X}(m)A(z)=\frac{1}{2\pi\sqrt{-1}}\int_C d\zeta(\zeta-z)^m X(\zeta)A(z) \in \mathcal{O}(\mathrm{v})$$

for some contour C around z such that 0 is outside C. Then by Proposition 2.6,

Proposition 2.8. *Let* v *be an* ℓ*CG-vertex.*

i) *The assignation* $X(m)\mapsto\hat{X}(m)$ *and* $c\mapsto\ell$ id *defines the* $\hat{\mathfrak{g}}$*-module structure on* $\mathcal{O}(\mathrm{v})$.

ii) *Let* $u \in V_j$, *then*

$$\hat{X}(m)\Phi_{\mathrm{v}}(u;z)=0 \qquad (m>0,\ X\in\mathfrak{g},\ u\in V_j);$$

$$\hat{X}(0)\Phi_{\mathrm{v}}(u;z)=[X(0),\Phi_{\mathrm{v}}(u;z)]=\Phi_{\mathrm{v}}(Xu;z) \quad (X\in\mathfrak{g},\ u\in V_j).$$

iii) *The assignation* $V_j \ni u\mapsto\Phi_{\mathrm{v}}(u;z)$ *defines the* $\mathfrak{g}$*-isomorphism of* V_j *onto the space* $\mathcal{P}(\mathrm{v})$, *and it is extended to a surjective* $\hat{\mathfrak{g}}$*-module mapping* $\Phi=\Phi_{\mathrm{v}}\colon \mathcal{M}_j\to\mathcal{O}(\mathrm{v})$.

Define the action of the Virasoro algebra $\mathcal{L}$ on $\mathcal{O}(\mathrm{v})$ by

$$\hat{L}(m)A(z)=\frac{1}{2\pi\sqrt{-1}}\int_C d\zeta(\zeta-z)^{m+1}T(\zeta)A(z) \qquad (m\in\mathbb{Z})$$

for some contour C around z such that 0 is outside C. Then by Proposition 2.6, we get

i) for any $u\in V_j$

$$\hat{L}(m)\Phi(u;z)=0 \qquad (m\geq 1);$$

$$\hat{L}(0)\Phi(u;z)=\Delta_j\Phi(u;z) \quad \text{and} \quad \hat{L}(-1)\Phi(u;z)=\frac{\partial}{\partial z}\Phi(u;z).$$

ii) the well-definedness of this $\mathcal{L}$-action: $(A(z)\in\mathcal{O}(\mathrm{v}))$

$$\hat{L}(m)\hat{L}(n)A(z)-\hat{L}(n)\hat{L}(m)A(z)$$
$$=(m-n)\hat{L}(m+n)A(z)+\frac{m^3-m}{12}c\delta_{m+n,0}A(z).$$

iii) the compatibility of $\hat{\mathfrak{g}}$-action and $\mathcal{L}$-action:

$$\hat{L}(m)\hat{X}(n)A(z)-\hat{X}(n)\hat{L}(m)A(z)=-n\hat{X}(m+n)A(z).$$

iv) this $\mathcal{L}$-action coincides with the one induced from the Sugawara form

$$\hat{L}(m)A(z)=\frac{1}{2\kappa}\sum_{k\in\mathbb{Z}}\left\{\frac{1}{2}:\hat{H}(-k)\hat{H}(m+k):+:\hat{E}(-k)\hat{F}(m+k):\right.$$

$$\left.+:\hat{F}(-k)\hat{E}(m+k):\right\}A(z).$$

Theorem 2.9 (Nuclear Democracy*)). *For each* ℓ*CG-vertex* $\mathrm{v}=\begin{pmatrix} j \\ j_2\, j_1\end{pmatrix}$, *the* $\hat{\mathfrak{g}}$*-mapping* Φ *gives the* $\hat{\mathfrak{g}}$*-isomorphism of* $\mathscr{H}_j$ *onto* $\mathcal{O}(\mathrm{v})$.

Note. The following fact is important for this theorem: The only one additional relation of $\mathscr{H}_j$ to the Verma module $\mathscr{M}_j$ is the equality $E(-1)^{\ell-2j+1}u_j(j)=0$.

Proof of Theorem 2.9. For each $\mathrm{v}\in(\mathrm{CG})_\ell$, set $\varphi=\varphi_{\mathrm{v}}$ and $\Phi(z)=\Phi_{\mathrm{v}}(z)$. Since the kernel of the projection of $\mathscr{M}_j$ onto $\mathscr{H}_j$ is generated by a vector $|J_j\rangle\in\mathscr{M}_j$ over $U(\hat{\mathfrak{g}})$, it is sufficient to show that $\Phi(|J_j\rangle;z)=0$.

Step 1. Recall that $|J_j\rangle=E(-1)^{\ell-2j+1}u_j(j)$, $\mathfrak{m}_+|J_j\rangle=0$, and $U(\mathfrak{g})|J_j\rangle=\sum_{k=0}^{2(\ell-j+1)}\mathbb{C}F(0)^k|J_j\rangle$, hence $\mathfrak{m}_+U(\mathfrak{g})|J_j\rangle=0$. Since Φ is $\hat{\mathfrak{g}}$-linear,

$$\hat{X}(m)\Psi(z)=0$$

for any $m>0$, $X\in\mathfrak{g}$ and $\Psi(z)\in U(\mathfrak{g})\Phi(|J_j\rangle;z)$.

Step 2. Let $\Psi(z)\in\mathcal{O}(\mathrm{v})$ such that $\hat{X}(m)\Psi(z)=0$ for any $X(m)\in\mathfrak{m}_+$, then

$$[X(0),\Psi(z)]=\hat{X}(0)\Psi(z)\quad\text{and}\quad[X(m),\Psi(z)]=z^m[X(0),\Psi(z)]\qquad(m\in\mathbb{Z}).$$

In fact, by Proposition 2.6, we get $X(\zeta)\Psi(z)=\Psi(z)X(\zeta)$, so

$$[X(m),\Psi(z)]=\frac{1}{2\pi\sqrt{-1}}\int_C d\zeta\,\zeta^m X(\zeta)\Psi(z)$$

for some contour C around z such that 0 is outside C, and by the assumption we get

$$X(\zeta)\Psi(z)=\frac{1}{\zeta-z}\hat{X}(0)\Psi(z)+\sum_{k\geq 0}\hat{X}(-k-1)\Psi(z)\,(\zeta-z)^k.$$

Step 3. Since $\mathrm{v}\in(\mathrm{CG})_\ell$, we get by Remark 2.2′ii),

$$\varphi(v,u_j(j),E^{\ell-2j+1}w)=0\qquad(v\in V_{j_2}^\dagger,\ w\in V_{j_1}).$$

By the induction on n, we get that for any $v\in V_{j_2}^\dagger$ and $w\in V_{j_1}$

*) We owe the naming of Nuclear Democracy to Prof. T. Eguchi.

$$\langle v|E(\zeta_1)\cdots E(\zeta_n)\Phi(u_j(j);z)|w\rangle=\prod_{i=1}^{n}\zeta_i^{-1}z^{-\Delta(v)}\varphi(v,u_j(j),E^n w),$$

so

$$0=\langle v|(\hat{E}(-1)^{\ell-2j+1}\Phi(u_j(j);z)|w\rangle=\langle v|\Phi(|J_j\rangle;z)|w\rangle,$$

hence by Steps 1 and 2.

$$\langle v|\Psi(z)|w\rangle=0$$

for any $\Psi(z)\in U(\mathfrak{g})\Phi(|J_j\rangle;z)$.

Since $\mathscr{H}^{\dagger}_{j_2}=V^{\dagger}_{j_2}U(\mathfrak{m}_+)$ and $\mathscr{H}_{j_1}=U(\mathfrak{m}_-)V_{j_1}$, we get

$$\langle\mathscr{H}^{\dagger}_{j_2}|\Psi(z)|\mathscr{H}_{j_1}\rangle=0$$

for any $\Psi(z)\in U(\mathfrak{g})\Phi(|J_j\rangle;z)$, hence $U(\mathfrak{g})\Phi(|J_j\rangle;z)=0$. q.e.d.

Here we summarize the relations satisfied by vertex operators:

Fundamental relations for vertex operators

Let $\Phi(z)$ be a vertex operator of spin j. Then

$$\hat{X}(m)\Phi(u;z)=0 \qquad (m\geq 1,\ X\in\mathfrak{g},\ u\in V_j);$$
$$\hat{X}(0)\Phi(u;z)=[X(0),\Phi(u;z)]=\Phi(Xu;z) \qquad (X\in\mathfrak{g},\ u\in V_j);$$
$$\hat{L}(m)\Phi(u;z)=0 \qquad (m\geq 1,\ u\in V_j);$$
$$\hat{L}(0)\Phi(u;z)=\Delta_j\Phi(u;z) \qquad (u\in V_j);$$
$$\hat{L}(-1)\Phi(u;z)=\frac{\partial}{\partial z}\Phi(u;z) \qquad (u\in V_j);$$

and

$$\hat{E}(-1)^{\ell-2j+1}\Phi(u_j(j);z)=0.$$

§ 3. Differential Equations of N-point Functions and Composability of Vertex Operators

In this section, we will give the system of differential equations of N-point functions and show the composability of vertex operators.

3.1) N-point functions and their differential equations

The vacuums $u_0(0)$ and $u_0^{\dagger}(0)$ of $\mathscr{H}_0$ and $\mathscr{H}_0^{\dagger}$ are of special importance (and are called *Virasoro vacuums*): denote $|\text{vac}\rangle=u_0(0)$ and $\langle\text{vac}|=u_0^{\dagger}(0)$, then

$$\mathfrak{p}_+|\text{vac}\rangle=0 \quad\text{and}\quad L(m)|\text{vac}\rangle=0 \quad (m\geq -1);$$

$$\langle \mathrm{vac}|\mathfrak{p}_- = 0 \quad \text{and} \quad \langle \mathrm{vac}|L(m) = 0 \quad (m \leq 1).$$

For an operator A on $\mathscr{H}$, define its *vacuum expectation value* by

$$\langle A \rangle = \langle \mathrm{vac}|A|\mathrm{vac}\rangle.$$

Introduce the $\mathfrak{g}$-module $\mathscr{P} = \sum_{\mathbf{v} \in (CG)_\ell} \mathscr{P}(\mathbf{v})$, defined by the $\mathfrak{g}$-action

$$\hat{X}(0)\Phi(u; z) = \Phi(Xu; z) \qquad (X \in \mathfrak{g}).$$

Denote by Δ_{ik} $(1 \leq i, k \leq N)$ the $\mathfrak{g}$-diagonal action on the i-th and k-th components of the N-th tensor product $\mathscr{P}^{\otimes N}$, that is, $\Delta_{ik} = \pi_i + \pi_k$, where π_i is the $\mathfrak{g}$-action on the i-th component of $\mathscr{P}^{\otimes N}$. Introduce the operator Ω_{ik} on $\mathscr{P}^{\otimes N}$ defined by

$$\Omega_{ik} = \frac{1}{2}\pi_i(H)\pi_k(H) + \pi_i(E)\pi_k(F) + \pi_i(F)\pi_k(E)$$

and denote $\Omega_i = \Omega_{ii} = \pi_i(\Omega)$, then

$$\Omega_{ik} = \frac{1}{2}\{\Delta_{ik}(\Omega) - \Omega_i - \Omega_k\}$$

and

$$[\Omega_{ik}, \Delta_{ik}(X)] = [\Omega_{ik}, \pi_j(X)] = 0 \qquad (i \neq k,\ X \in \mathfrak{g},\ j \neq i, k).$$

For any half-integer j $(0 \leq 2j \leq \ell)$, denote by $V_j^{\vee}$ the dual $\mathfrak{g}$-module of V_j. For any N-ple $\mathbb{J} = (j_N, \cdots, j_1)$ of half-integers with $0 \leq 2j_i \leq \ell$, let $V^{\vee}(\mathbb{J}) = V_{j_N}^{\vee} \otimes \cdots \otimes V_{j_1}^{\vee}$, and let $V_0^{\vee}(\mathbb{J}) = (V_{j_N}^{\vee} \otimes \cdots \otimes V_{j_1}^{\vee})^{\mathfrak{g}}$ the space of all $\mathfrak{g}$-invariant elements in $V^{\vee}(\mathbb{J})$. Then the operators Ω_{ik} act similarly on $V^{\vee}(\mathbb{J})$ and on $V_0^{\vee}(\mathbb{J})$.

Let $\Phi_i(z_i)$ be a vertex operator of spin j_i $(1 \leq i \leq N)$, then the vacuum expectation value of the composed operator

$$\langle \Phi_N(z_N) \cdots \Phi_1(z_1) \rangle$$

is considered as a $V^{\vee}(\mathbb{J})$-valued, formal Laurent series on $(z_N, \cdots, z_1)$ and is called an *N-point function*: If $\Phi_i(z_i)$ is of type $\mathbf{v}_i$ $(1 \leq i \leq N)$,

$$\langle \Phi_N(z_N) \cdots \Phi_1(z_1) \rangle = \prod_{i=1}^{N} z_i^{-\Delta(\mathbf{v}_i)} \sum_{m_N \geq 0} \cdots \sum_{m_k \in \mathbf{Z}} \cdots \sum_{m_1 \leq 0} C_{m_N \cdots m_1} z_N^{-m_N} \cdots z_1^{-m_1},$$

where

$$C_{m_N \cdots m_1} = \langle \mathrm{vac}|\Phi_{N,m_N}(\cdot)\Phi_{N-1,m_{N-1}}(\cdot) \cdots \Phi_{2,m_2}(\cdot)\Phi_{1,m_1}(\cdot)|\mathrm{vac}\rangle \in V^{\vee}(\mathbb{J}).$$

The aim of this section is to show that N-point functions are convergent in some region and analytically continued to a multivalued holomorphic function on M_N.

First we get a system of differential equations of N-point functions:

Theorem 3.1. *Let $\Phi_i(z_i)$ be a vertex operator of spin j_i $(1 \leq i \leq N)$, then the N-point function $\langle \Phi_N(z_N) \cdots \Phi_1(z_1) \rangle$ satisfies the following equations:*

(I) (projective invariance) *For $m = -1, 0$ and 1,*

$$\sum_{i=1}^{N} z_i^m \left(z_i \frac{\partial}{\partial z_i} + (m+1) \Delta_{j_i} \right) \langle \Phi_N(z_N) \cdots \Phi_1(z_1) \rangle = 0.$$

(II) (gauge invariance) *For any $X \in \mathfrak{g}$,*

$$\sum_{i=1}^{N} \pi_i(X) \langle \Phi_N(z_N) \cdots \Phi_1(z_1) \rangle = 0.$$

(III) *For each $i = 1, \cdots, N$,*

$$\left(\kappa \frac{\partial}{\partial z_i} - \sum_{\substack{k=1 \\ k \neq i}}^{N} \frac{\Omega_{ik}}{z_i - z_k} \right) \langle \Phi_N(z_N) \cdots \Phi_1(z_1) \rangle = 0,$$

where $\kappa = \ell + 2$.

(IV) *For each $i = 1, \cdots, N$,*

$$\langle \Phi_N(u_N; z_N) \cdots (\hat{E}(-1)^{\ell - 2j_i + 1} \Phi_i(u_{j_i}(j_i); z_i)) \cdots \Phi_1(u_1; z_1) \rangle = 0$$

for any $u_k \in V_{j_k}$ $(k \neq i)$.

Proof. These equations are obtained from the fundamental relations of vertex operators, the Sugawara form of $L(m)$ and the properties of the Virasoro vacuums. Here we give a brief proof of (III). First, note the identity:

$$\langle X(z) \Phi_N(z_N) \cdots \Phi_1(z_1) \rangle = \sum_{i=1}^{N} \frac{1}{z - z_i} \pi_i(X) \langle \Phi_N(z_N) \cdots \Phi_1(z_1) \rangle \qquad (X \in \mathfrak{g}).$$

Let $X^1 = 2X_1 = H$, $X^2 = X_3 = E$ and $X^3 = X_2 = F$, then the Casimir operator Ω is expressed as $\Omega = \sum_{k=1}^{3} X^k X_k$. By Proposition 2.6 and the relation $\hat{L}(-1)\Phi(z) = (\partial/\partial z)\Phi(z)$, we get

$$\kappa \frac{\partial}{\partial z_i} \Phi_i(u_i; z_i) = \lim_{z \searrow z_i} \left\{ \sum_{k=1}^{3} X^k(z) \Phi_i(X_k u_i; z_i) - \frac{1}{z - z_i} \Phi_i(\Omega u_i; z_i) \right\}$$
$$(1 \leq i \leq N).$$

Hence for each i with $1 \leq i \leq N$,

$$\sum_{k=1}^{3} \langle X^k(z)\Phi_N(z_N)\cdots(X_k\Phi_i)(z_i)\cdots\Phi_1(z_1)\rangle$$

$$=\sum_{k=1}^{3}\sum_{j=1}^{N}\frac{1}{z-z_j}\pi_j(X^k)\pi_i(X_k)\langle\Phi_N(z_N)\cdots\Phi_1(z_1)\rangle$$

$$=\left\{\frac{1}{z-z_i}\Omega_i+\sum_{j\neq i}\frac{1}{z-z_j}\Omega_{ij}\right\}\langle\Phi_N(z_N)\cdots\Phi_1(z_1)\rangle.$$

Thus we get the equation (III) by taking the limit $z \searrow z_i$.

Remark 3,2.

i) The equations (I) ~ (III) are obtained by V.G. Knizhnik and A.B. Zamolodchikov [KZ].

ii) The equations (II) mean that $\langle\Phi_N(z_N)\cdots\Phi_1(z_1)\rangle \in V_0^{\vee}(\mathbb{J})$.

iii) The equations (II) and (III) imply the equations (I). (Key is the property of the operators Ω_{ik}: $\sum_{k=1}^{N}\Omega_{ik}=0$ on $V_0^{\vee}(\mathbb{J})$.)

iv) The system (III) of differential equations is completely integrable. This complete integrability of (III) is reduced to the *infinitesimal pure braid relations* of Ω_{ik}:

$$[\Omega_{ik},\Omega_{mn}]=0 \qquad \text{(if } i, k, m, n \text{ are mutually disjoint);}$$

and

$$[\Omega_{im},\ \Omega_{ik}+\Omega_{km}]=0 \qquad \text{(if } i, k, m \text{ are mutually disjoint).}$$

These infinitesimal pure braid relations were originally noted by K. Aomoto (see [A1] and [A2]). Moreover these pure braid relations are equivalent to the classical Yang-Baxter equations for $\mathfrak{sl}_2$ obtained by C. N. Yang [Y] and A.A. Belavin-V.G. Drinfel'd [BD].

v) N-point functions are translation invariant (Corollary of (I)):

$$\langle\Phi_N(z_N+z)\cdots\Phi_1(z_1+z)\rangle=\langle\Phi_N(z_N)\cdots\Phi_1(z_1)\rangle.$$

vi) The equations (IV) are equivalent to the algebraic equations: for each i $(1\leq i\leq N)$ and any $u_k \in V_{j_k}$ $(k\neq i)$, put $L_i=\ell-2j_i+1$.

$$\sum_{|\mathfrak{m}_i|=L_i}\binom{L_i}{\mathfrak{m}_i}\prod_{k\neq i}(z_k-z_i)^{-m_k}\langle\Phi_N(E^{m_N}u_N;z_N)\cdots\Phi_i(u_{j_i}(j_i);z_i)\cdots\Phi_1(E^{m_1}u_1;z_1)\rangle$$

$$=0,$$

where $\mathfrak{m}_i=(m_N,\cdots,\hat{m}_i,\cdots,m_1)\in(\mathbb{Z}_{\geq 0})^{N-1}$, $|\mathfrak{m}_i|=\sum_{k\neq i}m_k$ and $\binom{L_i}{\mathfrak{m}_i}$ is the multinomial coefficient.

3.2) Solutions of fundamental equation

Consider the systems E(J) of differential equations and B(J) of algebraic equations for $V_0^\vee(\mathbb{J})$-valued functions $\Phi(z_N, \cdots, z_1)$ on the manifold $X_N=\{(z_N, \cdots, z_1) \in \mathbb{C}^N;\ z_i \neq z_k\ (i\neq k)\}\supset M_N$:

$$E(\mathbb{J}): \qquad \Big(\kappa\frac{\partial}{\partial z_i}-\sum_{\substack{k=1\\k\neq i}}^{N}\frac{\Omega_{ik}}{z_i-z_k}\Big)\Phi(z_N, \cdots, z_1)=0 \qquad (1\leq i\leq N)$$

and for each i $(1\leq i\leq N)$ and any $u_k \in V_{j_k}$ $(k\neq i)$,

$$\mathrm{B}(\mathbb{J}): \sum_{|\mathfrak{m}_i|=L_i}\binom{L_i}{\mathfrak{m}_i}\prod_{k\neq i}(z_k-z_i)^{-m_k}\Phi(z_N, \cdots, z_1)(E^{m_N}u_N, \cdots, u_{j_i}(j_i), \cdots, E^{m_1}u)$$
$$=0,$$

where $\mathfrak{m}_i=(m_N, \cdots, \hat{m}_i, \cdots, m_1) \in (\mathbb{Z}_{\geq 0})^{N-1}$, $|\mathfrak{m}_i|=\sum_{k\neq i} m_k$ and $L_i=\ell-2j_i+1$.

By Remark 3.2, the system $E(\mathbb{J})$ is completely integrable.

Introduce the set $\mathscr{P}(\mathbb{J})$ defined by

$$\mathscr{P}(\mathbb{J})=\Big\{\mathfrak{p}=(p_N, \cdots, p_1, p_0);\ \mathrm{v}_i(\mathfrak{p})=\begin{pmatrix} & j_i & \\ p_i & & p_{i-1}\end{pmatrix} \in (\mathrm{CG}),\ p_N=p_0=0\Big\}.$$

$$\begin{array}{ccccccccc} & \mathrm{v}_N & \mathrm{v}_{N-1} & \cdots & \mathrm{v}_i & \cdots & \mathrm{v}_2 & \mathrm{v}_1 & \\ \mathfrak{p}: & j_N & j_{N-1} & \cdots & j_i & \cdots & j_2 & j_1 & \\ p_N=0 & p_{N-1} & p_{N-2} & \cdots\ p_i & & p_{i-1} & \cdots\ p_2 & p_1 & 0=p_0 \end{array}$$

For each $\mathfrak{p} \in \mathscr{P}(\mathbb{J})$, define the vector $\varphi_{\mathfrak{p}}$ of $V_0^\vee(\mathbb{J})$ from the (fixed) elements $\varphi_{\mathrm{v}_i} \in \mathrm{Hom}_{\mathfrak{g}}(V_{p_i}^\dagger\otimes V_{j_i}\otimes V_{p_{i-1}};\ \mathbb{C})\cong(V_{j_i}^\vee\otimes\mathrm{Hom}(V_{p_{i-1}}, V_{p_i}))^{\mathfrak{g}}$ $(1\leq i \leq N)$, as the trace of $\varphi_{\mathrm{v}_N}\otimes\cdots\otimes\varphi_{\mathrm{v}_1}$: for each $u_N\otimes\cdots\otimes u_1 \in V(\mathbb{J})$,

$$\varphi_{\mathfrak{p}}(u_N, \cdots, u_1)=\langle \mathrm{vac}|\varphi_{\mathrm{v}_N}(u_N)\circ\varphi_{\mathrm{v}_{N-1}}(u_{N-1})\circ\cdots\circ\varphi_{\mathrm{v}_1}(u_1)|\mathrm{vac}\rangle.$$

Then the set $\{\varphi_{\mathfrak{p}};\ \mathfrak{p} \in \mathscr{P}(\mathbb{J})\}$ gives a basis of the space $V_0^\vee(\mathbb{J})$.

Introduce the operators $\Omega_m^\vee=\sum_{1\leq i\neq j\leq m}\Omega_{ij}$ on $V^\vee(\mathbb{J})$ for m $(2\leq m \leq N)$, then

$$\Omega_m^\vee=\hat{\Omega}_m-\sum_{i=1}^{m}\Omega_{ii},$$

where $\hat{\Omega}_m$ is the diagonal action of Ω on $V_{j_m}^\vee\otimes\cdots\otimes V_{j_1}^\vee$, and by the pure braid relations (Remark 3.2 iv), we get that $[\Omega_m^\vee, \Omega_n^\vee]=0$.

In the basis $\{\varphi_{\mathfrak{p}};\ \mathfrak{p} \in \mathscr{P}(\mathbb{J})\}$, these operators are diagonal:

$$\Omega^{\vee}_m\varphi_{\mathfrak{p}}=2\kappa\Delta^{\vee}_m(\mathfrak{p})\varphi_{\mathfrak{p}}\qquad\left(\mathfrak{p}=(p_N,\cdots,p_1,p_0);\ \mathrm{v}_i=\begin{pmatrix} j_i \\ p_i\,p_{i-1}\end{pmatrix}\right),$$

where

$$\Delta^{\vee}_m(\mathfrak{p})=\Delta_{p_m}-\sum_{i=1}^{m}\Delta_{j_i}=-\sum_{i=1}^{m}\hat{\Delta}(\mathrm{v}_i)\qquad(2\leq m\leq N)$$

In fact, for each $i=2,\cdots,N$,

$$\hat{\Omega}_i\varphi_{\mathfrak{p}}=2\kappa\Delta_{p_i}\varphi_{\mathfrak{p}}\quad\text{and}\quad\Omega_{ii}\varphi_{\mathfrak{p}}=2\kappa\Delta_{j_i}\varphi_{\mathfrak{p}}.$$

Now introduce the subset $\mathscr{P}_\ell(\mathbb{J})$ of $\mathscr{P}(\mathbb{J})$ defined by

$$\mathscr{P}_\ell(\mathbb{J})=\left\{\mathfrak{p}=(p_N,\cdots,p_1,p_0)\in\mathscr{P}(\mathbb{J});\ \mathrm{v}_i=\mathrm{v}_i(\mathfrak{p})=\begin{pmatrix} j_i \\ p_i\,p_{i-1}\end{pmatrix}\in(\mathrm{CG})_\ell\right\},$$

then for each $\mathfrak{p}\in\mathscr{P}_\ell(\mathbb{J})$, the N-point function

$$\Phi_{\mathfrak{p}}(z_N,\cdots,z_1)=\langle\Phi_{\mathrm{v}_N}(z_N)\cdots\Phi_{\mathrm{v}_1}(z_1)\rangle$$

of type $\mathfrak{p}$ is a formal Laurent series solution of the system $E(\mathbb{J})$ and $B(\mathbb{J})$ by Theorem 3.1, where its Laurent series expansion is given as

$$\begin{aligned}\Phi_{\mathfrak{p}}(z_N,\cdots,z_1)&=\prod_{i=1}^{N}z_i^{-\hat{\Delta}(\mathrm{v}_i)}\sum_{m_N\geq 0}\cdots\sum_{m_i\in\mathbf{Z}}\cdots\sum_{m_1\leq 0}C_{m_N\cdots m_1}z_N^{-m_N}\cdots z_1^{-m_1}\\&=\prod_{i=1}^{N}z_i^{-\Delta_{j_i}}\left\langle z_N^{L(0)}\Phi_{\mathrm{v}_N}(1)\left(\frac{z_{N-1}}{z_N}\right)^{L(0)}\Phi_{\mathrm{v}_{N-1}}(1)\cdots\left(\frac{z_1}{z_2}\right)^{L(0)}\Phi_{\mathrm{v}_1}(1)z_1^{-L(0)}\right\rangle\end{aligned}$$

where

$$C_{m_N\cdots m_1}=\langle\mathrm{vac}|\Phi_{\mathrm{v}_N,m_N}(\cdot)\Phi_{\mathrm{v}_{N-1},m_{N-1}}(\cdot)\cdots\Phi_{\mathrm{v}_2,m_2}(\cdot)\Phi_{\mathrm{v}_1,m_1}(\cdot)|\mathrm{vac}\rangle\in V^{\vee}(\mathbb{J}).$$

Moreover

Theorem 3.3. *Consider the region* $\mathscr{R}_z$ *in the manifold* X_N, *defined by*

$$\mathscr{R}_z=\{\mathbf{z}=(z_N,\cdots,z_1)\in\mathbf{C}^N;\ |z_N|>\cdots>|z_1|\}.$$

Then

i) *for any* $\mathfrak{p}\in\mathscr{P}_\ell(\mathbb{J})$, *the Laurent series* $\Phi_{\mathfrak{p}}(z_N,\cdots,z_1)$ *is absolutely convergent in the region* $\mathscr{R}_z$, *and is analytically continued to a multivalued holomorphic function on* X_N.

ii) $\{\Phi_{\mathfrak{p}}(z_N,\cdots,z_1);\ \mathfrak{p}\in\mathscr{P}_\ell(\mathbb{J})\}$ *is linearly independent and gives a basis of the solution space of the joint system* E($\mathbb{J}$) *and* B($\mathbb{J}$).

Proof. The system E($\mathbb{J}$) of differential equations is equivalent to the Pfaffian system:

$$\mathrm{P}(\mathbb{J}): \qquad \kappa d\Phi(\mathbf{z}) - \sum_{k<i} \frac{d(z_i - z_k)}{z_i - z_k} \Omega_{ik} \Phi(\mathbf{z}) = 0.$$

Now we change coordinates $\mathbf{z}$ to $\mathbf{w}$ by

$$w_N = z_N;\ w_i = z_i/z_{i+1}\ (1 \leq i \leq N-1).$$

Then the region $\mathscr{R}_z$ transforms bijectively onto the region

$$\mathscr{R}_{w,0} = \{\mathbf{w} = (w_N, \cdots, w_1) \in \mathbb{C}^N;\ w_N \neq 0,\ 1 > |w_i| > 0\ (2 \leq i \leq N-1),\ 1 > |w_1|\},$$

where the inverse transformation is given as

$$z_i = w_N \cdots w_i\ (1 \leq i \leq N).$$

And introduce the region $\mathscr{R}_w = \{\mathbf{w} \in \mathbb{C}^N;\ 1 > |w_i|\ (1 \leq i \leq N-1)\} \supset \mathscr{R}_{w,0}$.

The system $\mathrm{P}(\mathbb{J})$ is written in the coordinates $\mathbf{w}$ as

$$\tilde{\mathrm{P}}(\mathbb{J}): \kappa \sum_{i=1}^{N} \frac{\partial \tilde{\Phi}}{\partial w_i} dw_i = \sum_{m=2}^{N} \frac{1}{w_m} dw_m \sum_{k<i\leq m} \Omega_{ik} \tilde{\Phi} - \sum_{m=1}^{N-1} dw_m \sum_{k\leq m<i} \frac{w_{i-1} \cdots \hat{w}_m \cdots w_k}{1 - w_{i-1} \cdots w_k} \Omega_{ik} \tilde{\Phi},$$

where $\tilde{\Phi}(\mathbf{w}) = \Phi(\mathbf{z})$.

Hence by using the operators $\Omega_m^{\vee}$, the system $\mathrm{E}(\mathbb{J})$ turns to be

$$\tilde{\mathrm{E}}(\mathbb{J}): \qquad \left\{2\kappa \frac{\partial}{\partial w_m} - \frac{\Omega_m^{\vee}}{w_m} + A_m(\mathbf{w})\right\} \tilde{\Phi}(\mathbf{w}) = 0 \qquad (2 \leq m \leq N),$$

$$\left\{2\kappa \frac{\partial}{\partial w_1} + A_1(\mathbf{w})\right\} \tilde{\Phi}(\mathbf{w}) = 0$$

where

$$A_m(\mathbf{w}) = \sum_{k\leq m<i} \frac{w_{i-1} \cdots \hat{w}_m \cdots w_k}{1 - w_{i-1} \cdots w_k} \Omega_{ik} \qquad (2 \leq m \leq N-1),$$

$$A_1(\mathbf{w}) = \frac{1}{1 - w_1} \Omega_{21} + \sum_{i\geq 3} \frac{w_{i-1} \cdots w_2}{1 - w_{i-1} \cdots w_1} \Omega_{i1} \quad \text{and} \quad A_N(\mathbf{w}) = 0.$$

Since $A_m(w)$'s are holomorphic in the region $\mathscr{R}_w$, the system $\tilde{\mathrm{E}}(\mathbb{J})$ is with regular singularities along the divisors $D_i = \{w_i = 0\}$ for $i = 2, \cdots, N$. The basis $\{\varphi_{\mathfrak{p}};\ \mathfrak{p} \in \mathscr{P}(\mathbb{J})\}$ of $V_0^{\vee}(\mathbb{J})$ diagonalizes the principal parts of the system $\mathrm{E}(\mathbb{J})$ with the exponents $\{\Delta_i^{\vee}(\mathfrak{p});\ 1 \leq i \leq N\}$ corresponding to $\varphi_{\mathfrak{p}}$.

The formal Laurent series solution $\tilde{\Phi}_{\mathfrak{p}}(w_N, \cdots, w_1) = \Phi_{\mathfrak{p}}(z_N, \cdots, z_1)$, $\mathfrak{p} \in \mathscr{P}_\ell(\mathbb{J})$, of the system $\tilde{\mathrm{E}}(\mathbb{J})$ is written as

$$\tilde{\Phi}_{\mathfrak{p}}(w_N, \cdots, w_1) = \prod_{i=2}^{N} w_i^{\Delta_i^{\vee}(\mathfrak{p})} S_{\mathfrak{p}}(w_N, \cdots, w_1),$$

where

$$S_{\mathrm{p}}(w_N, \cdots, w_1)$$
$$= \prod_{i=1}^{N} w_i^{-\Delta_{p_i}} \langle w_N^{L(0)} \Phi_{v_N}(1) w_{N-1}^{L(0)} \Phi_{v_{N-1}}(1) \cdots \Phi_{v_2}(1) w_1^{L(0)} \Phi_{v_1}(1) \rangle$$

is a formal power series in w, since $w_i^{L(0)} = w_i^{\Delta_{p_i}+d}$ id on $\mathscr{H}_{p_i,d}$.

By the theory of partial differential equations with regular singular points (see e.g. [CL] Chap. 3 and [Kn] Appendix B), the function $\tilde{\Phi}_{\mathrm{p}}(\mathrm{w})$ is a solution of the system $\tilde{\mathrm{E}}(\mathbb{J})$ of differential equation in the region $\mathscr{R}_{w,0}$ for each $\mathrm{p} \in \mathscr{P}_\ell(\mathbb{J})$. Hence the formal power series $S_{\mathrm{p}}(\mathrm{w})$ gives a holomorphic function in $\mathscr{R}_w$, and so the function $\tilde{\Phi}_{\mathrm{p}}(w_N, \cdots, w_1)$ is holomorphic in the region $\mathscr{R}_{w,0}$. Thus the N-point function $\Phi_{\mathrm{p}}(z_N, \cdots, z_1)$ is holomorphic in $\mathscr{R}_z$ for any $\mathrm{p} \in \mathscr{P}_\ell(\mathbb{J})$.

ii) By the remark before the statement of the theorem, for each $\mathrm{p} \in \mathscr{P}_\ell(\mathbb{J})$

$$S_{\mathrm{p}}(0, \cdots, 0) = \langle \mathrm{vac} | \Phi_{v_N,0}(\cdot) \Phi_{v_{N-1},0}(\cdot) \cdots \Phi_{v_2,0}(\cdot) \Phi_{v_1,0}(\cdot) | \mathrm{vac} \rangle$$
$$= \langle \mathrm{vac} | \varphi_{v_N} \varphi_{v_{N-1}} \cdots \varphi_{v_2} \varphi_{v_1} | \mathrm{vac} \rangle = \langle \mathrm{vac} | \varphi_{\mathrm{p}} | \mathrm{vac} \rangle \in V_0^{\vee}(\mathbb{J}).$$

This implies the linear independence of $\{\Phi_{\mathrm{p}}(z_N, \cdots, z_1) \colon \mathrm{p} \in \mathscr{P}_\ell(\mathbb{J})\}$.

Finally we want to show that the dimension of the solution space of the joint system $\tilde{\mathrm{E}}(\mathbb{J})$ and $\tilde{\mathrm{B}}(\mathbb{J})$ is not greater than $\#\mathscr{P}_\ell(\mathbb{J})$, where $\tilde{\mathrm{B}}(\mathbb{J})$ is the system $\mathrm{B}(\mathbb{J})$ written in the coordinates w: for each i with $1 \leq i \leq N$, let $L = \ell - 2j_i + 1$,

$$\tilde{\mathrm{B}}_i(\mathbb{J})\colon \ \sum_{K=0}^{L} \binom{L}{K} \sum_{\substack{|\mathrm{m}'|=L-K \\ |\mathrm{m}''|=K}} \binom{L-K}{\mathrm{m}'} \binom{K}{\mathrm{m}''} \prod_{k=i}^{N} w_k^{-K-m_{i+1}-\cdots-m_k} (1+O(\mathrm{w}))$$
$$\times \tilde{\Phi}(w_N, \cdots, w_1)(E^{m_N} u_N, \cdots, u_{j_i}(j_i), \cdots, E^{m_1} u_1) = 0,$$

where $\mathrm{m}' = (m_N, \cdots, m_{i+1}) \in (\mathbb{Z}_{\geq 0})^{N-i}$, $\mathrm{m}'' = (m_{i-1}, \cdots, m_1) \in (\mathbb{Z}_{\geq 0})^{i-1}$ and $O(\mathrm{w})$ is a convergent power series in $\mathscr{R}_w$ and $O(\mathbb{0}) = 0$.

For each $\mathrm{p} \in \mathscr{P}(\mathbb{J})$, take a solution

$$\Psi_{\mathrm{p}}(w_N, \cdots, w_1) = \prod_{i=2}^{N} w_i^{\Delta_i^{\vee}(\mathrm{p})} T_{\mathrm{p}}(w_N, \cdots, w_1) \qquad (\mathrm{p} \in \mathscr{P}(\mathbb{J}))$$

of the system $\tilde{\mathrm{E}}(\mathbb{J})$, where $T_{\mathrm{p}}(\mathrm{w})$ is a convergent power series in $\mathscr{R}_w$ with the constant term $T_{\mathrm{p}}(0) = \varphi_{\mathrm{p}}$. Apply $\tilde{\mathrm{B}}_i(\mathbb{J})$ to $\Psi_{\mathrm{p}}(\mathrm{w})$ for $i \geq 2$, then its leading term must vanish, and the term is obtained by taking $K = L$, since

$$\sum_{k=i}^{N} (K + m_{i+1} + \cdots + m_k) = (N - i + 1)K + \sum_{k=i+1}^{N} (k-i) m_k$$

and

$$K+\sum_{k=i+1}^{N} m_k=L.$$

Hence

$$0=\sum_{|\mathfrak{m}''|=L}\binom{L}{\mathfrak{m}''}T_{\mathfrak{p}}(0)(u_N, \cdots, u_{i+1}, u_{j_i}(j_i), E^{m_{i-1}}u_{i-1}, \cdots, E^{m_1}u_1)$$
$$=\varphi_{\mathfrak{p}}(u_N, \cdots, u_{i+1}, u_{j_i}(j_i), E^L(u_{i-1}\otimes\cdots\otimes u_1)).$$

By Remark 2.2′ ii), we get that $\ell\geq j_i+p_i+p_{i-1}$, that is, $\mathrm{v}_i\in(\mathrm{CG})_\ell$ for $i\geq 2$. Hence $\mathfrak{p}\in\mathscr{P}_\ell(\mathbb{J})$, since $\mathrm{v}_1\in(\mathrm{CG})_\ell$ automatically.

Introduce a partial order $\prec$ in the set $\mathscr{P}(\mathbb{J})$ defined by

$$\mathfrak{p}\prec\mathfrak{p}', \qquad \text{if } (\Delta_N(\mathfrak{p}')-\Delta_N(\mathfrak{p}), \cdots, \Delta_2(\mathfrak{p}')-\Delta_2(\mathfrak{p}))\in(\mathbb{Z}_{\geq 0})^{N-1}.$$

Let $\Psi(\mathrm{w})$ be a solution of the systems $\tilde{\mathrm{E}}(\mathbb{J})$ and $\tilde{\mathrm{B}}(\mathbb{J})$, and express it as $\Psi(\mathrm{w})=\sum_{\mathfrak{p}\in\mathscr{P}_\Psi}c_{\mathfrak{p}}\Psi_{\mathfrak{p}}(\mathrm{w})$, where $\mathscr{P}_\Psi=\{\mathfrak{p}\in\mathscr{P}(\mathbb{J});\, c_{\mathfrak{p}}\neq 0\}$. Apply $\tilde{\mathrm{B}}_i(\mathbb{J})$ to $\Psi(\mathrm{w})$, then by the linear independence of solutions of $\tilde{\mathrm{E}}(\mathbb{J})$ with different exponents modulo $\mathbb{Z}^N$, the leading term for $\Psi_{\mathfrak{p}}(\mathrm{w})$ must vanish for any minimal $\mathfrak{p}$ in $\mathscr{P}_\Psi$. Hence any minimal $\mathfrak{p}\in\mathscr{P}_\Psi$ belongs to $\mathscr{P}_\ell(\mathbb{J})$. Since $\tilde{\Phi}_{\mathfrak{p}}(\mathrm{w})$ satisfies $\tilde{\mathrm{B}}(\mathbb{J})$ for any $\mathfrak{p}\in\mathscr{P}_\ell(\mathbb{J})$, $\Psi(\mathrm{w})$ must belong to the space spanned by $\{\tilde{\Phi}_{\mathfrak{p}}(\mathrm{w});\, \mathfrak{p}\in\mathscr{P}_\ell(\mathbb{J})\}$. q.e.d.

3.3) Composability of vertex operators

As a corollary of Theorem 3.3, we get the following

Theorem 3.4. *Let $\Phi_i(z_i)$ be a vertex operator of spin j_i and $u_i\in V_{j_i}$ $(1\leq i\leq N)$. Then the sequence $\{\Phi_N(u_N; z_N), \cdots, \Phi_1(u_1; z_1)\}$ is composable in the region $\mathscr{R}_{z,0}=\{(z_N, \cdots, z_1)\in\mathbb{C}^N;\, |z_N|>\cdots>|z_1|>0\}$ and the composed operator $\Phi_N(u_N; z_N)\cdots\Phi_1(u_1; z_1)$ is analytically continued to a multi-valued holomorphic function on M_N.*

Proof. We may assume that $\Phi_i(u_i; z_i)=\Phi_{\mathrm{v}_i}(u_i; z_i)$ for some vertex $\mathrm{v}_i=\begin{pmatrix} & j_i & \\ p_i & & p_{i-1}\end{pmatrix}\in(\mathrm{CG})_\ell$. Put $\mathbb{J}=(p_N, j_N, \cdots, j_1, p_0)$, then $\mathfrak{p}=(0, p_N, \cdots, p_1, p_0, 0)\in\mathscr{P}_\ell(\mathbb{J})$.

For the vertices $\mathrm{v}_{N+1}=\mathrm{v}_{N+1}(\mathfrak{p})=\begin{pmatrix} & p_N & \\ 0 & & p_N\end{pmatrix}$ and $\mathrm{v}_0=\mathrm{v}_0(\mathfrak{p})=\begin{pmatrix} & p_0 & \\ p_0 & & 0\end{pmatrix}$, we get by Proposition 2.5,

$$\lim_{z\searrow 0}\Phi_{\mathrm{v}_0}(w; z)|\mathrm{vac}\rangle=|w\rangle \qquad (w\in V_{p_0});$$

and

$$\lim_{z\nearrow\infty} z^{2\Delta_{p_N}}\langle \mathrm{vac}|\Phi_{\mathrm{v}_{N+1}}(v;z)=\langle \nu(v)| \qquad (v\in V_{p_{N+1}}).$$

The $(N+2)$-point function $\langle \Phi_{\mathrm{v}_{N+1}}(v_N;z_{N+1})\,\Phi_{\mathrm{v}_N}(u_N;z_N)\cdots\Phi_{\mathrm{v}_1}(u_1;z_1)$ $\Phi_{\mathrm{v}_0}(w;z_0)\rangle$ is holomorphic in $\mathscr{R}_z^{N+2}=\{(z_{N+1},\cdots,z_0)\in\mathbb{C}^{N+2};|z_{N+1}|>\cdots>|z_0|\}$, so it is an absolutely convergent Laurent series in the region $\mathscr{R}_z^{N+2}$. Hence

$$\begin{aligned}&\langle \nu(v)|\Phi_{\mathrm{v}_N}(u_N;z_N)\cdots\Phi_{\mathrm{v}_1}(u_1;z_1)|w\rangle\\ &\quad=\lim_{z_0\searrow 0}\lim_{z_{N+1}\nearrow\infty} z_{N+1}^{2\Delta_{p_N}}\langle \Phi_{\mathrm{v}_{N+1}}(v;z_{N+1})\Phi_{\mathrm{v}_N}(u_N;z)\cdots\Phi_{\mathrm{v}_1}(u_1;z_1)\Phi_{\mathrm{v}_0}(w;z_0)\rangle\end{aligned}$$

is absolutely convergent at any point $(z_N,\cdots,z_1)\in\mathscr{R}_{z,0}$ for any $v\in V_{p_{N+1}}$, $u_i\in V_{j_i}$ $(1\le i\le N)$ and $w\in V_{p_0}$.

For general $v\in\mathscr{H}^\dagger_{p_{N+1}}$ and $w\in\mathscr{H}_{p_0}$, we may put

$$v=\langle \nu(v_0)|\,Y_q(m_q)\cdots Y_1(m_1)\quad\text{and}\quad w=X_1(-n_1)\cdots X_r(-n_r)|w_0\rangle$$

for some $v_0\in V_{p_{N+1}}$, $w_0\in V_{p_0}$, $Y_i, X_i\in\mathfrak{g}$, $m_i, n_i\ge 0$.

Then it is sufficient for the convergence of the function $\langle v|\Phi_N(u_N;z_N)\cdots\Phi_1(u_1;z_1)|w\rangle$ to note

$$\lim_{z\searrow 0}\hat{X}_1(-n_1)\cdots\hat{X}_r(-n_r)\Phi_{\mathrm{v}_0}(w_0;z)|\mathrm{vac}\rangle=|w\rangle,$$

and

$$\lim_{z\nearrow\infty} z^{2\Delta_{p_N}}\langle \mathrm{vac}|(\hat{Y}_q(m_q)\cdots\hat{Y}_1(m_1)\Phi_{\mathrm{v}_{N+1}}(v_0;z))=\langle v|. \qquad \text{q.e.d.}$$

Remark 3.5. If we take the value ℓ of the central element c of $\hat{\mathfrak{g}}$ as $\ell\notin\mathbb{Q}$, then we can construct an analogous theory without the ℓ-constraint condition. In this case, the Verma module $\mathscr{M}_j$ (defined as in the top of Section 2.3) is irreducible for any nonnegative half integer j, and the space $\mathscr{H}$ is taken as $\mathscr{H}=\sum\mathscr{M}_j$, where j runs over $\frac{1}{2}\mathbb{Z}_{\ge 0}$. Then there exists a vertex operator on $\mathscr{H}$ of type $\mathrm{v}\in\mathbb{V}$, if and only if $\mathrm{v}=\begin{pmatrix} & j & \\ j_2 & & j_1\end{pmatrix}\in(\mathrm{CG})$. In this case, $\mathscr{O}(\mathrm{v})\cong\mathscr{M}_j$, so the last equation $\hat{E}(-1)^{\ell-2j+1}\Phi(u_j(j);z)=0$ is eliminated among the the fundamental equations for vertex operators.

§ 4. Commutation Relations of Vertex Operators

4.1) Formulation of the problem

For a quadruple $\mathbb{J}=(j_4,j_3,j_2,j_1)$ of half integers with $0\le 2j_i\le l$, introduce the set $I_\ell(\mathbb{J})$ of *intermediate edges*, defined by

$$I_\ell(\mathbb{J})=\left\{k \in \frac{1}{2}\mathbb{Z};\ 0\leq 2k\leq \ell,\ \mathrm{v}_2(k)=\begin{pmatrix} & j_3 & \\ j_4 & & k\end{pmatrix} \in (\mathrm{CG})_\ell,\right.$$

$$\left.\mathrm{v}_1(k)=\begin{pmatrix} & j_2 & \\ k & & j_1\end{pmatrix} \in (\mathrm{CG})_\ell\right\}.$$

For each $k \in I_\ell(\mathbb{J})$, put $\mathfrak{p}(k)=(\mathrm{v}_3, \mathrm{v}_2(k), \mathrm{v}_1(k), \mathrm{v}_0) \in \mathscr{P}_\ell(\mathbb{J})$, where $\mathrm{v}_3=\begin{pmatrix} & j_4 & \\ 0 & & j_4\end{pmatrix}$ and $\mathrm{v}_0=\begin{pmatrix} & j_1 & \\ j_1 & & 0\end{pmatrix}$. And put $\Delta_4(\mathbb{J})=\hat{\Delta}(\mathrm{v}_2)+\hat{\Delta}(\mathrm{v}_1)=\Delta_{j_1}+\Delta_{j_2}+\Delta_{j_3}-\Delta_{j_4}$ (independent of k).

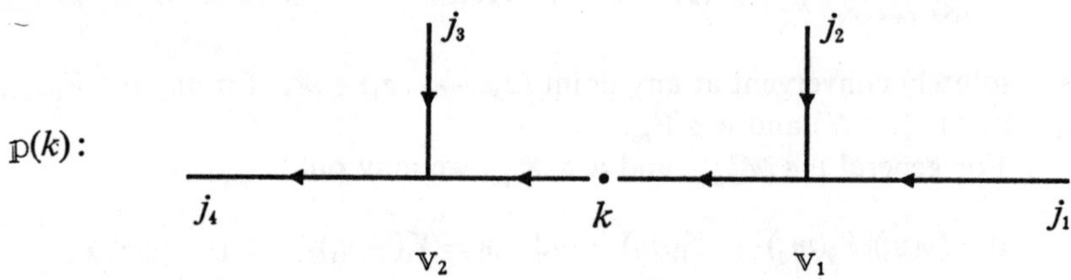

Assume $I_\ell(\mathbb{J})\neq\varnothing$, then we get two vertex operators $\Phi_{\mathrm{v}_2(k)}(w)$ and $\Phi_{\mathrm{v}_1(k)}(z)$. By Theorem 3.4, they are composable in the region $\mathscr{R}_2=\{(w, z) \in \mathbb{C}^2;\ |w|>|z|>0\}$ and the composed operator $\Phi_k(w, z)=\Phi_{\mathrm{v}_2}(w)\Phi_{\mathrm{v}_1}(z)$ is analytically continued to a multi-valued holomorphic and operator-valued function on $M_2=\{(w, z) \in (\mathbb{C}^*)^2;\ w\neq z\}$. Introduce a $V_0^\vee(\mathbb{J})$-valued holomorphic function $\Psi_k(w, z)$ on M_2 defined by

$$\Psi_k(w, z)(u_4\otimes u_3\otimes u_2\otimes u_1)=\langle \nu(u_4)\,|\,\Phi_{\mathrm{v}_2}(u_3;\ w)\Phi_{\mathrm{v}_1}(u_2;\ z)\,|\,u_1\rangle \qquad (u_i \in V_{j_i}).$$

In the region $\mathscr{R}_2$, this function has a convergent Laurent expansion:

$$\Psi_k(w, z)(u_4\otimes u_3\otimes u_2\otimes u_1) = z^{-\Delta_4(\mathbb{J})}\sum_{n\geq 0}\left(\frac{z}{w}\right)^{n-\hat{\Delta}(\mathrm{v}_2)} \langle \nu(u_4)\,|\,\Phi_{\mathrm{v}_2, n}(u_3)\Phi_{\mathrm{v}_1, -n}(u_2)\,|\,u_1\rangle,$$

with the initial term $\langle \nu(u_4)|\varphi_{\mathrm{v}_2}(u_3)\varphi_{\mathrm{v}_1}(u_2)|u_1\rangle$ for any $u_i \in V_{j_i}$.

By Propositions 2.1, 5 and Theorems 2.3, 3.3, we get

Proposition 4.1. *Assume $I_\ell(\mathbb{J})\neq\varnothing$. Then for each $k \in I_\ell(\mathbb{J})$,*

i) *the operator $\Phi_k(w, z)$ on $\mathscr{H}$ is uniquely determined by the $V_0^\vee(\mathbb{J})$-valued function $\Psi_k(w, z)$.*

ii) *The function $\Psi_k(w, z)$ satisfies the joint system $E'(\mathbb{J})$ and $B'(\mathbb{J})$ of equations:*

$$E'(\mathbb{J}):\ \left\{\kappa\frac{\partial}{\partial w}-\frac{\Omega_{13}}{w}-\frac{\Omega_{23}}{w-z}\right\}\Psi_k(w, z)=\left\{\kappa\frac{\partial}{\partial z}-\frac{\Omega_{12}}{z}-\frac{\Omega_{23}}{z-w}\right\}\Psi_k(w, z)=0.$$

and

$$B'(\mathbb{J}):\ \sum_{m=0}^{L_1}\binom{L_1}{m}w^{-m}z^{m-L_1}\Psi_k(w,z)(u_4, E^m u_3, E^{L_1-m}u_2, u_{j_1}(j_1))=0,$$

$$\sum_{m=0}^{L_2}\binom{L_2}{m}(w-z)^{-m}(-z)^{m-L_2}\Psi_k(w,z)(u_4, E^m u_3, u_{j_2}(j_2), E^{L_2-m}u_1)=0,$$

$$\sum_{m=0}^{L_3}\binom{L_3}{m}(z-w)^{-m}(-w)^{m-L_3}\Psi_k(w,z)(u_4, u_{j_3}(j_3), E^m u_2, E^{L_3-m}u_1)=0,$$

$$\sum_{|\mathfrak{m}|=L_4}\binom{L_4}{\mathfrak{m}}\Psi_k(w,z)(u_{j_4}(j_4), E^{m_3}u_3, E^{m_2}u_2, E^{m_1}u_1)=0,$$

where $L_i=\ell-2j_i+1\ (1\leq i\leq 4)$ *and* $\mathfrak{m}=(m_3, m_2, m_1)\in(\mathbb{Z}_{\geq 0})^3$.

iii) *The family* $\{\Psi_k(w,z);\ k\in I_\ell(\mathbb{J})\}$ *gives a basis of the solution space of the systems* $E'(\mathbb{J})$ *and* $B'(\mathbb{J})$.

Now assign a new quadruple $\bar{\mathbb{J}}=(j_4, j_2, j_3, j_1)$ to the quadruple $\mathbb{J}=(j_4, j_3, j_2, j_1)$ of half integers with $0\leq 2j_i\leq\ell$, then we get the $\mathfrak{g}$-isomorphism $T\colon V^{\vee}(\mathbb{J})\to V^{\vee}(\bar{\mathbb{J}})$ defined by

$$(T\varphi)(u_4\otimes u_2\otimes u_3\otimes u_1)=\varphi(u_4\otimes u_3\otimes u_2\otimes u_1)$$

for $\varphi\in V^{\vee}(\mathbb{J})$ and $u_4\otimes u_2\otimes u_3\otimes u_1\in V(\bar{\mathbb{J}})$. Since $T(V_0^{\vee}(\mathbb{J}))=V_0^{\vee}(\bar{\mathbb{J}})$, we get $\dim V_0^{\vee}(\mathbb{J})=\dim V_0^{\vee}(\bar{\mathbb{J}})$. Note $\Delta_4(\mathbb{J})=\Delta_4(\bar{\mathbb{J}})$ and $\sharp I_\ell(\mathbb{J})=\sharp I_\ell(\bar{\mathbb{J}})$.

For an intermediate edge $\bar{k}\in I_\ell(\bar{\mathbb{J}})$, similarly define the vertices $\bar{\mathrm{v}}_2(\bar{k})=\begin{pmatrix} & j_2 & \\ j_4 & & \bar{k}\end{pmatrix}$, $\bar{\mathrm{v}}_1(\bar{k})=\begin{pmatrix} & j_3 & \\ \bar{k} & & j_1\end{pmatrix}\in(\mathrm{CG})_\ell$, the composed operator $\bar{\Phi}_{\bar{k}}(w,z)$ of the vertex operators $\Phi_{\bar{\mathrm{v}}_2(\bar{k})}(w)$ and $\Phi_{\bar{\mathrm{v}}_1(\bar{k})}(z)$, and the $V_0^{\vee}(\bar{\mathbb{J}})$-valued holomorphic function $\bar{\Psi}_{\bar{k}}(w,z)$ on M_2. In the region $\mathscr{R}_2$, this function $\bar{\Psi}_{\bar{k}}(w,z)$ also has a convergent Laurent expansion:

$$\begin{aligned}&\bar{\Psi}_{\bar{k}}(w,z)(u_4\otimes u_2\otimes u_3\otimes u_1)\\ &\quad=z^{-\Delta_4(\mathbb{J})}\sum_{n\geq 0}\left(\frac{z}{w}\right)^{n-\Delta(\bar{\mathrm{v}}_2)}\langle\nu(u_4)|\Phi_{\bar{\mathrm{v}}_2(\bar{k}),n}(u_2)\Phi_{\bar{\mathrm{v}}_1(\bar{k}),-n}(u_3)|u_1\rangle,\end{aligned}$$

with the initial term $\langle\nu(u_4)|\varphi_{\bar{\mathrm{v}}_2(\bar{k})}(u_2)\varphi_{\bar{\mathrm{v}}_1(\bar{k})}(u_3)|u_1\rangle$ for any $u_i\in V_{j_i}$.

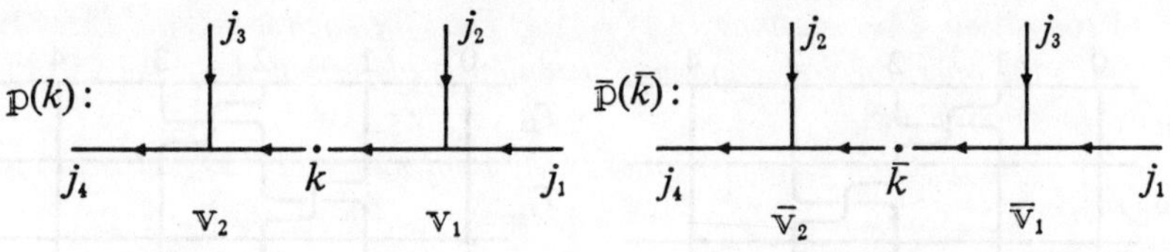

Now introduce the path $b(t)=(\eta(t), \zeta(t))$ from a point (w,z) in the set $I_2=\{(w,z)\in\mathbb{R}^2;\ w>z>0\}$ to the point (z,w) in the set $\bar{I}_2=\{(z,w)\in\mathbb{R}^2;$

$w>z>0\}$ on the manifold M_2, defined by

$$\eta(t)=\frac{w+z}{2}+e^{\pi\sqrt{-1}t}\frac{w-z}{2},\quad \zeta(t)=\frac{w+z}{2}-e^{\pi\sqrt{-1}t}\frac{w-z}{2}\quad (t\in[0,1]).$$

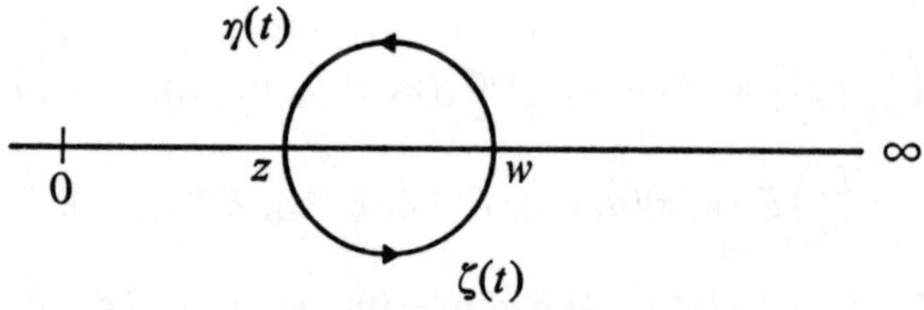

Denote by $\Psi_k(z, w)$ the analytic continuation of the convergent Laurent series $\Psi_k(w, z)$ in the region $\mathscr{R}_2$ along the path $b(t)$ and consider $\Psi_k(z, w)$ near $\bar{I}_2$, then the $V_0^{\vee}(\mathbb{J})$-valued function $T\Psi_k(z, w)$ satisfies the equations $E'(\mathbb{J})$ and $B'(\mathbb{J})$, so it is expressed as a linear combination:

$$T\Psi_k(z, w)=\sum_{\bar{k}\in I_\ell(\mathbb{J})}\Psi_{\bar{k}}(w, z)C_k^{\bar{k}}(\mathbb{J}),$$

where $C(\mathbb{J})=(C_k^{\bar{k}}(\mathbb{J}))_{k\in I_\ell(\mathbb{J}),\bar{k}\in I_\ell(\mathbb{J})}$ is a square matrix.

Hence by Proposition 4.1,

Proposition 4.2. i) *Let $\mathbb{J}=(j_4, j_3, j_2, j_1)$ with $I_\ell(\mathbb{J})\neq\emptyset$. Then for each intermediate edge $k\in I_\ell(\mathbb{J})$ and $(w, z)\in I_2$,*

$$T\Phi_{v_2(k)}(z)\Phi_{v_1(k)}(w)=\sum_{\bar{k}\in I_\ell(\mathbb{J})}\Phi_{v_2(\bar{k})}(w)\Phi_{v_1(\bar{k})}(z)C_k^{\bar{k}}(\mathbb{J}),$$

where the operator in the left hand side is considered as the analytic continuation of the composition of the vertex operators $\Phi_{v_2}(w)$ and $\Phi_{v_1}(z)$ along the path $b(t)$ in the manifold X_N.

ii) *Let $\mathbb{J}=(t, j_3, j_2, j_1, s)$, then the braid relation holds:*

$$\begin{aligned}&C(j_3, j_2, j_1, s)C(t, j_3, j_1, j_2)C(j_1, j_3, j_2, s)\\&=C(t, j_3, j_2, j_1)C(j_2, j_3, j_1, s)C(t, j_2, j_1, j_3).\end{aligned}$$

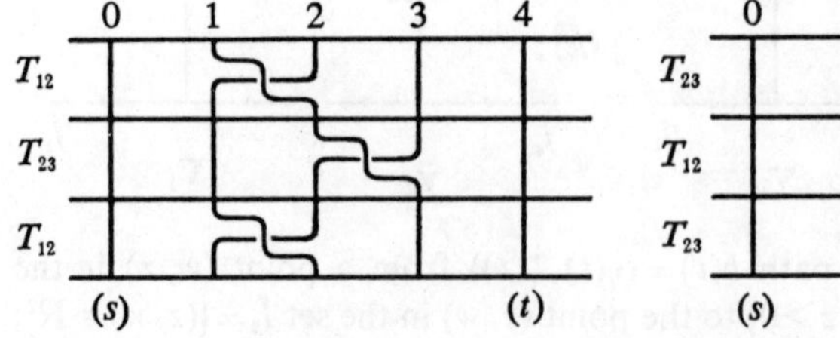

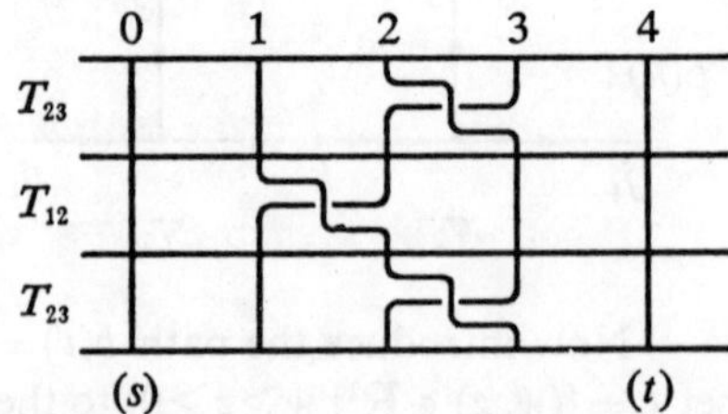

Now our fundamental problem is:

Fundamental Problem. Determine the matrix $C(\mathbb{J})=(C_k^{\xi}(\mathbb{J}))$ for any quadruple $\mathbb{J}$ with $I_\ell(\mathbb{J})\neq\varnothing$.

4.2) Reduced Equation

Take an intermediate edge $k\in I_\ell(\mathbb{J})$ and introduce a variable $\zeta=z/w$, then the $V_0^{\vee}(\mathbb{J})$-valued function $z^{\Delta_4(\mathbb{J})}\Psi_k(w,\zeta w)$ is independent of w, since by Theorem 3.1, I,

$$\left\{w\frac{\partial}{\partial w}+z\frac{\partial}{\partial z}-\Delta_4(\mathbb{J})\right\}\Psi_k(w,z)=0.$$

So we abbreviate $z^{\Delta_4(\mathbb{J})}\Psi_k(w,\zeta w)$ to $\Psi_k(\zeta)$, then the $V_0^{\vee}(\mathbb{J})$-valued function $\Psi_k(\zeta)$ (called reduced 4-point function) has a convergent Laurent expansion

$$\Psi_k(\zeta)(u_4\otimes u_3\otimes u_2\otimes u_1)=\zeta^{-\hat{\Delta}(\mathbf{v}_2(k))}\sum_{n\geq 0}\langle\nu(u_4)|\Phi_{\mathbf{v}_2,n}(u_3)\Phi_{\mathbf{v}_1,-n}(u_2)|u_1\rangle\zeta^n$$

in $\zeta\in\mathbb{C}^*$ with the initial term $\langle\nu(u_4)|\varphi_{\mathbf{v}_2}(u_3)\varphi_{\mathbf{v}_1}(u_2)|u_1\rangle$ for $u_i\in V_{j_i}$. Then by Proposition 4.1,

Proposition 4.3 (Reduced equation). *The* $V_0^{\vee}(\mathbb{J})$*-valued function* $\Psi_k(\zeta)$ *satisfies the joint system* RE($\mathbb{J}$) *and* RB($\mathbb{J}$) *of equations*:

$$\mathrm{RE}(\mathbb{J}):\qquad \left(\kappa\frac{d}{d\zeta}-\frac{\Omega_{12}+\kappa\Delta_4(\mathbb{J})}{\zeta}-\frac{\Omega_{23}}{\zeta-1}\right)\Psi_k(\zeta)=0$$

and

$$\mathrm{RB}(\mathbb{J}):\qquad \sum_{m=0}^{L_1}\binom{L_1}{m}\zeta^m\Psi_k(\zeta)(u_4,E^mu_3,E^{L_1-m}u_2,u_{j_1}(j_1))=0,$$

$$\sum_{m=0}^{L_2}\binom{L_2}{m}\left(\frac{\zeta}{\zeta-1}\right)^m\Psi_k(\zeta)(u_4,E^mu_3,u_{j_2}(j_2),E^{L_2-m}u_1)=0,$$

$$\sum_{m=0}^{L_3}\binom{L_3}{m}\left(\frac{1}{1-\zeta}\right)^m\Psi_k(\zeta)(u_4,u_{j_3}(j_3),E^mu_2,E^{L_3-m}u_1)=0,$$

$$\sum_{|\mathbb{m}|=L_4}\binom{L_4}{\mathbb{m}}\Psi_k(\zeta)(u_{j_4}(j_4),E^{m_3}u_3,E^{m_2}u_2,E^{m_1}u_1)=0,$$

where $L_i=\ell-2j_i+1$ $(1\leq i\leq 4)$ *and* $\mathbb{m}=(m_3,m_2,m_1)\in(\mathbb{Z}_{\geq 0})^3$.

Proof. The system $E'(\mathbb{J})$ of equations turns to a single differential equation RE($\mathbb{J}$), since $\Omega_{12}+\Omega_{13}+\Omega_{23}=-\kappa\Delta_4(\mathbb{J})$ q.e.d.

In the following, we want to solve the fundamental problem for the case where $j_3=\frac{1}{2}$ in $\mathbb{J}$. For this aim, we investigate first the reduced equation RE($\mathbb{J}$) in detail for each quadruple $\mathbb{J}=(j_4, \frac{1}{2}, j_2, j_1)$ with $V_0^{\vee}(\mathbb{J})\neq 0$ and thereafter take the equation RB($\mathbb{J}$) into account. For the investigation of the reduced equation RE($\mathbb{J}$), introduce the set $I(\mathbb{J})$ defined by

$$I(\mathbb{J})=\left\{k \in \frac{1}{2}\mathbb{Z}_{\geq 0};\ \mathrm{v}_2(k)=\begin{pmatrix} & j_3 & \\ j_4 & & k\end{pmatrix} \in (\mathrm{CG}),\ \mathrm{v}_1(k)=\begin{pmatrix} & j_2 & \\ k & & j_1\end{pmatrix}=(\mathrm{CG})\right\}.$$

First note that $\sharp I(\mathbb{J})=\dim V_0^{\vee}(\mathbb{J})\leq 2$. And $\dim V_0^{\vee}(\mathbb{J})=2$ if and only if

$$\text{(D2)} \qquad |j_1-j_2|\leq j_4-\frac{1}{2},\ j_4+\frac{1}{2}\leq j_1+j_2 \ \text{ and } \ j_1+j_2+\frac{1}{2}+j_4 \in \mathbb{Z}.$$

In this case, $I(\mathbb{J})=\{k_\pm=j_4\pm\frac{1}{2}\}$.

The case (D2) is divided into three cases $(\mathrm{D2})_i$ such that $\sharp I_l(\mathbb{J})=i$ $(i=0, 1, 2)$. Introduce the number $\varepsilon_0(\mathbb{J})=(j_1+j_2+j_4+\frac{3}{2})/\kappa$, then

$$(\mathrm{D2})_2 \quad \varepsilon_0<1;\ \text{then } j_1, j_2, j_4\leq \frac{l-1}{2} \ \text{ and } \ I_l(\mathbb{J})=\left\{k_\pm=j_4\pm\frac{1}{2}\right\},$$

$$(\mathrm{D2})_1 \quad \varepsilon_0=1;\ I_l(\mathbb{J})=\left\{k_-=j_4-\frac{1}{2}\right\}, \qquad (\mathrm{D2})_0 \quad \varepsilon_0>1;\ I_l(\mathbb{J})=\varnothing.$$

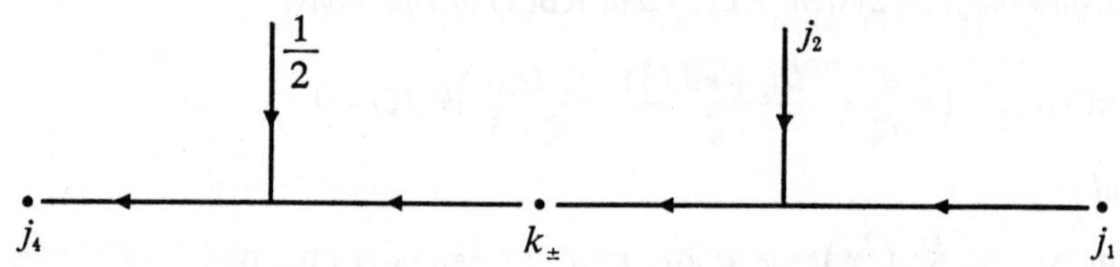

Moreover $\dim V_0^{\vee}(\mathbb{J})=1$, if and only if either of the following conditions $(\mathrm{D1})_i$ holds:

$$(\mathrm{D1})_1 \quad j_1=j_4+\frac{1}{2}+j_2;\quad (\mathrm{D1})_2 \quad j_2=j_4+\frac{1}{2}+j_1;\quad (\mathrm{D1})_3 \quad j_4=\frac{1}{2}+j_2+j_1.$$

And $I(\mathbb{J})=\{j_4+\frac{1}{2}\}$ for the case $(\mathrm{D1})_{1,2}$ and $I(\mathbb{J})=\{j_4-\frac{1}{2}\}$ for the case $(\mathrm{D1})_3$. Note that one of the conditions $(\mathrm{D1})_i$ implies $\sharp I_l(\mathbb{J})=1$.

Denote by (D0) the case where $V_0^{\vee}(\mathbb{J})=0$, i.e. $I(\mathbb{J})=\varnothing$.

Now consider the equation

$$\mathrm{RE}(\mathbb{J}): \qquad \left(\kappa\frac{d}{d\zeta}-\frac{\Omega_{12}+\kappa\Delta_4(\mathbb{J})}{\zeta}-\frac{\Omega_{23}}{\zeta-1}\right)\Psi(\zeta)=0$$

for $V_0^{\vee}(\mathbb{J})$-valued functions $\Psi(\zeta)$ on $\zeta \in \mathbb{C}^*$. The coordinate change $\zeta \mapsto \eta = 1/\zeta$ makes the equation RE$(\mathbb{J})$ into

$$\mathrm{RE}(\mathbb{J})_\infty: \qquad \left(\kappa\frac{d}{d\eta} - \frac{\Omega_{13}}{\eta} - \frac{\Omega_{23}}{\eta-1}\right)\Psi\left(\frac{1}{\eta}\right)=0.$$

Case (D2). First we get three bases $\{U_\pm^{(0)}\}$, $\{U_\pm^{(1)}\}$ and $\{U_\pm^{(\infty)}\}$ of $V_0^{\vee}(\mathbb{J})$ such that they diagonalize the operators Ω_{12}, Ω_{23} and Ω_{13} respectively (see Appendix I):

$$\Omega_{12}U_\pm^{(0)} = \kappa(\gamma_\pm^{(0)} - \Delta_4(\mathbb{J}))U_\pm^{(0)}, \quad \Omega_{23}U_\pm^{(1)} = \kappa\gamma_\pm^{(1)}U_\pm^{(1)}, \quad \Omega_{13}U_\pm^{(\infty)} = \kappa\gamma_\pm^{(\infty)}U_\pm^{(\infty)},$$

and

$$U_\pm^{(0)}(u_4, u_3, u_2, u_1) = \frac{1}{\sqrt{2j_4+1}}\langle v(u_4)|\varphi_{v_2(k_\pm)}(u_3)\varphi_{v_1(k_\pm)}(u_2)|u_1\rangle$$

for $u_4 \otimes u_3 \otimes u_2 \otimes u_1 \in V(\mathbb{J})$, where

$$\gamma_+^{(0)} = \frac{2j_4+3}{2\kappa}, \quad \gamma_-^{(0)} = \frac{2j_4-1}{-2\kappa}, \quad \gamma_+^{(1)} = \frac{j_2}{\kappa}, \quad \gamma_-^{(1)} = \frac{j_2+1}{-\kappa},$$

$$\gamma_+^{(\infty)} = \frac{j_1}{\kappa}, \quad \gamma_-^{(\infty)} = \frac{j_1+1}{-\kappa}.$$

Introduce the differences $\gamma^{(i)} = \gamma_+^{(i)} - \gamma_-^{(i)}$ $(i=0, 1, \infty)$, then $0<\gamma^{(i)}<1$, in particular, they are not integers:

$$\gamma^{(0)} = \frac{2j_4+1}{\kappa}, \quad \gamma^{(1)} = \frac{2j_2+1}{\kappa} \quad \text{and} \quad \gamma^{(\infty)} = \frac{2j_1+1}{\kappa} \qquad (\kappa = \ell+2).$$

The transformation matrices $S^{(i,k)}$ between the bases $\{U_\pm^{(i)}\}$ and $\{U_\pm^{(k)}\}$ are given as

$$(U_+^{(k)}, U_-^{(k)}) = (U_+^{(i)}, U_-^{(i)})S^{(i,k)},$$

where

$$S^{(0,1)} = S^{(1,0)} = \begin{pmatrix} A & B \\ B & -A \end{pmatrix}, \qquad S^{(0,\infty)} = S^{(\infty,0)} = \begin{pmatrix} -A'' & B'' \\ B'' & A'' \end{pmatrix} \in O(2)\backslash SO(2),$$

$$S^{(\infty,1)} = {}^tS^{(1,\infty)} = \begin{pmatrix} A' & -B' \\ B' & A' \end{pmatrix} \in SO(2),$$

and the constants $A \sim B''$ are given as

$$A = \left(\frac{\varepsilon_2\varepsilon_4}{\gamma^{(1)}\gamma^{(0)}}\right)^{1/2}, \quad A' = \left(\frac{\varepsilon_1\varepsilon_2}{\gamma^{(\infty)}\gamma^{(1)}}\right)^{1/2}, \quad A'' = \left(\frac{\varepsilon_1\varepsilon_4}{\gamma^{(\infty)}\gamma^{(0)}}\right)^{1/2},$$

and

$$B=\left(\frac{\varepsilon_0\varepsilon_1}{\gamma^{(1)}\gamma^{(0)}}\right)^{1/2},\quad B'=\left(\frac{\varepsilon_0\varepsilon_4}{\gamma^{(\infty)}\gamma^{(1)}}\right)^{1/2},\quad B''=\left(\frac{\varepsilon_0\varepsilon_2}{\gamma^{(\infty)}\gamma^{(0)}}\right)^{1/2},$$

where

$$\varepsilon_0=\varepsilon_0(\mathbb{J})\quad\text{and}\quad \varepsilon_i=\frac{1}{\kappa}\left\{j_1+j_2+j_4+\frac{1}{2}-2j_i\right\}\quad (i=1,2,4).$$

Now we get the fundamental solutions of the linear differential equation RE$(\mathbb{J})$ with regular singular points at $\zeta=0$, 1 and ∞ by means of the Gauss' hypergeometric function $F(\alpha,\beta,\gamma;\zeta)$ (see Appendix II for the proof):

Proposition 4.4. *Introduce the constants* $\alpha=\varepsilon_0$, $\beta=\varepsilon_1$, $\beta^{(\infty)}=\varepsilon_4$ *and let* $\Psi_{\pm}^{(i)}(\zeta)$ *be the fundamental solutions of the equation* RE$(\mathbb{J})$ *normalized at* $\zeta=i$ $(i=0,1,\infty)$:

$$(\Psi_+^{(i)}(\zeta),\Psi_-^{(i)}(\zeta))=(U_+^{(i)},U_-^{(i)})\begin{pmatrix}\varphi_{++}^{(i)}(\zeta) & \varphi_{-+}^{(i)}(\zeta)\\ \varphi_{+-}^{(i)}(\zeta) & \varphi_{--}^{(i)}(\zeta)\end{pmatrix}.$$

Then

(i)
$$\begin{aligned}
\varphi_{++}^{(0)}(\zeta)&=\phantom{c_+^{(0)}}\zeta^{\gamma_+^{(0)}}(1-\zeta)^{\gamma_+^{(1)}}F(\alpha,\beta,\gamma^{(0)};\zeta);\\
\varphi_{+-}^{(0)}(\zeta)&=c_+^{(0)}\zeta^{1+\gamma_+^{(0)}}(1-\zeta)^{\gamma_+^{(1)}}F(\alpha+1,\beta+1,2+\gamma^{(0)};\zeta);\\
\varphi_{-+}^{(0)}(\zeta)&=c_-^{(0)}\zeta^{1+\gamma_-^{(0)}}(1-\zeta)^{\gamma_-^{(1)}}F(-\alpha+1,-\beta+1,2-\gamma^{(0)};\zeta);\\
\varphi_{--}^{(0)}(\zeta)&=\phantom{c_-^{(0)}}\zeta^{\gamma_-^{(0)}}(1-\zeta)^{\gamma_-^{(1)}}F(-\alpha,-\beta,-\gamma^{(0)};\zeta).
\end{aligned}$$

(ii)
$$\begin{aligned}
\varphi_{++}^{(1)}(\zeta)&=\phantom{c_+^{(1)}}\zeta^{\gamma_+^{(0)}}(1-\zeta)^{\gamma_+^{(1)}}F(\alpha,\beta,\gamma^{(1)};1-\zeta);\\
\varphi_{+-}^{(1)}(\zeta)&=c_+^{(1)}\zeta^{\gamma_+^{(0)}}(1-\zeta)^{1+\gamma_+^{(1)}}F(\alpha+1,\beta+1,2+\gamma^{(1)};1-\zeta);\\
\varphi_{-+}^{(1)}(\zeta)&=c_-^{(1)}\zeta^{\gamma_-^{(0)}}(1-\zeta)^{1+\gamma_-^{(1)}}F(-\alpha+1,-\beta+1,2-\gamma^{(1)};1-\zeta);\\
\varphi_{--}^{(1)}(\zeta)&=\phantom{c_-^{(1)}}\zeta^{\gamma_-^{(0)}}(1-\zeta)^{\gamma_-^{(1)}}F(-\alpha,-\beta,-\gamma^{(1)};1-\zeta).
\end{aligned}$$

(iii)
$$\begin{aligned}
\varphi_{++}^{(\infty)}(\zeta)&=\phantom{c_+^{(\infty)}}\zeta^{-\gamma_+^{(\infty)}}\left(1-\frac{1}{\zeta}\right)^{\gamma_+^{(1)}}F\left(\alpha,\beta^{(\infty)},\gamma^{(\infty)};\frac{1}{\zeta}\right);\\
\varphi_{+-}^{(\infty)}(\zeta)&=c_+^{(\infty)}\zeta^{-1-\gamma_+^{(\infty)}}\left(1-\frac{1}{\zeta}\right)^{\gamma_+^{(1)}}F\left(\alpha+1,\beta^{(\infty)}+1,2+\gamma^{(\infty)};\frac{1}{\zeta}\right);\\
\varphi_{-+}^{(\infty)}(\zeta)&=c_-^{(\infty)}\zeta^{-1-\gamma_-^{(\infty)}}\left(1-\frac{1}{\zeta}\right)^{\gamma_-^{(1)}}\\
&\qquad\times F\left(-\alpha+1,-\beta^{(\infty)}+1,2-\gamma^{(\infty)};\frac{1}{\zeta}\right);\\
\varphi_{--}^{(\infty)}(\zeta)&=\phantom{c_-^{(\infty)}}\zeta^{-\gamma_-^{(\infty)}}\left(1-\frac{1}{\zeta}\right)^{\gamma_-^{(1)}}F\left(-\alpha,-\beta^{(\infty)},-\gamma^{(\infty)};\frac{1}{\zeta}\right),
\end{aligned}$$

where

$$c_{\pm}^{(i)} = -\frac{\sqrt{\varepsilon_0\varepsilon_1\varepsilon_2\varepsilon_4}}{\gamma^{(i)}(1\pm\gamma^{(i)})} \qquad (i=0, 1, \infty).$$

Note. The reduced 4-point function $\Psi_{k_\pm}(\zeta)$ is the solution of RE$(\mathbb{J})$ with exponent $\gamma_{\pm}^{(0)}$ at $\zeta=0$, so by the normalization of $U_{\pm}^{(0)}$, $\Psi_{k_\pm}(\zeta)=\sqrt{2j_4+1}\,\Psi_{\pm}^{(0)}(\zeta)$.

Case (D1). Since dim $V_0^{\check{}}(\mathbb{J})=1$, the choice of basis vectors of $V_0^{\check{}}(\mathbb{J})$ is not of importance. But from the compatibility with the case (D2), we choose basis vectors $\{U^{(i)}; i=0, 1, \infty\}$ of $V_0^{\check{}}(\mathbb{J})$ such that

$$U^{(0)}=U^{(1)}=U^{(\infty)} \quad \text{for (D1)}_{1,3}; \quad U^{(0)}=U^{(1)}=-U^{(\infty)} \quad \text{for (D1)}_2.$$

The exponents $\gamma^{(0)}$, $\gamma^{(1)}$ and $\gamma^{(\infty)}$ of the equation RE$(\mathbb{J})$ at $\zeta=0, 1, \infty$ are given as

$$\text{(D1)}_1 \qquad \gamma^{(0)}=\frac{3+2j_4}{2\kappa}, \quad \gamma^{(1)}=\frac{j_2}{\kappa}, \qquad \gamma^{(\infty)}=-\frac{j_1+1}{\kappa},$$

$$\text{(D1)}_2 \qquad \gamma^{(0)}=\frac{3+2j_4}{2\kappa}, \quad \gamma^{(1)}=-\frac{1+j_2}{\kappa}, \quad \gamma^{(\infty)}=\frac{j_1}{\kappa},$$

$$\text{(D1)}_3 \qquad \gamma^{(0)}=\frac{1-2j_4}{2\kappa}, \quad \gamma^{(1)}=\frac{j_2}{\kappa}, \qquad \gamma^{(\infty)}=\frac{j_1}{\kappa}.$$

Then we get

Proposition 4.4′. *The fundamental solution* $\Psi^{(i)}(\zeta)=U^{(i)}\varphi^{(i)}(\zeta)$ *of the equation* RE$(\mathbb{J})$ *normalized at* $\zeta=i$ $(i=0, 1, \infty)$ *is given as*

$$\text{(D1)}_{1,3} \qquad \varphi^{(0)}(\zeta)=\varphi^{(1)}(\zeta)=\zeta^{\gamma^{(0)}}(1-\zeta)^{\gamma^{(1)}}, \quad \varphi^{(\infty)}(\zeta)=q^{-j_2/2}\varphi^{(0)}(\zeta),$$

and

$$\text{(D1)}_2 \qquad \varphi^{(0)}(\zeta)=\varphi^{(1)}(\zeta)=\zeta^{\gamma^{(0)}}(1-\zeta)^{\gamma^{(1)}}, \quad \varphi^{(\infty)}(\zeta)=q^{(j_2+1)/2}\varphi^{(0)}(\zeta),$$

where the exponents $\gamma^{(i)}$ *are corresponding ones and* $q=\exp(2\pi\sqrt{-1}/\kappa)$.

4.3) Connection matrices for $\mathbb{J}=(j_4, \frac{1}{2}, j_2, j_1)$

The path $b(t)$ from a point $(w, z)\in I_2$ to $(z, w)\in \bar{I}_2$ on M_2 introduced in Section 4.1 corresponds a path from the point $\zeta=z/w$ in the set $J_1=\{\zeta\in\mathbb{R}; 1>\zeta>0\}$ to the point $1/\zeta$ in the set $\bar{J}_1=\{\zeta\in\mathbb{R}; \zeta>1\}$ on the manifold C^*. If z tends to zero, then the corresponding path tends to the path $\bar{b}(t)$ from 0 to the infinity figured below:

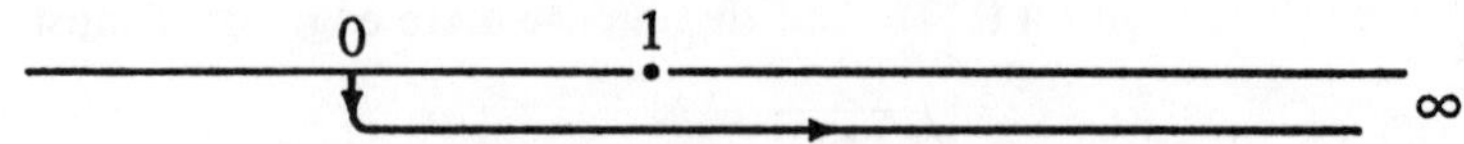

Now take an intermediate edge k for a quadruple $\mathbb{J}=(j_4, \frac{1}{2}, j_2, j_1)$ with $I(\mathbb{J})\neq\emptyset$. We want to know the analytic continuation of the reduced 4-point function $\Psi_k(\zeta)$ along the path $\bar{b}(t)$.

For the case (D1), we get easily the connection matrix (scalar) $K(\mathbb{J})$ of the fundamental solution $\Psi^{(0)}(\zeta)$ at $\zeta=0$ to $\Psi^{(\infty)}(\zeta)$ at $\zeta=\infty$ of the equation RE$(\mathbb{J})$: $S^{(0,\infty)}\varphi^{(0)}(\zeta)=\varphi^{(\infty)}(\zeta)K(\mathbb{J})$ as follows:

$$(\mathrm{D1})_{1,3}\quad K(\mathbb{J})=q^{j_2/2};\quad (\mathrm{D1})_2\quad K(\mathbb{J})=-q^{-(1+j_2)/2}\quad \left(q=\exp\left(\frac{2\pi\sqrt{-1}}{\kappa}\right)\right).$$

Now we deal with the case (D2). By the formulae for connection matrices of the hypergeometric functions, we get the connection matrix $K(\mathbb{J})=\begin{pmatrix}K_+^+ & K_-^+\\ K_+^- & K_-^-\end{pmatrix}$ of the fundamental solutions $(\Psi_+^{(0)}, \Psi_-^{(0)})$ at $\zeta=0$ to $(\Psi_+^{(\infty)}, \Psi_-^{(\infty)})$ at $\zeta=\infty$ of the equation RE$(\mathbb{J})$:

$$(\Psi_+^{(0)}, \Psi_-^{(0)})=(\Psi_+^{(\infty)}, \Psi_-^{(\infty)})\begin{pmatrix}K_+^+ & K_-^+\\ K_+^- & K_-^-\end{pmatrix},$$

that is,

$$S^{(\infty,0)}\begin{pmatrix}\varphi_{++}^{(0)}(\zeta) & \varphi_{-+}^{(0)}(\zeta)\\ \varphi_{+-}^{(0)}(\zeta) & \varphi_{--}^{(0)}(\zeta)\end{pmatrix}=\begin{pmatrix}\varphi_{++}^{(\infty)}(\zeta) & \varphi_{-+}^{(\infty)}(\zeta)\\ \varphi_{+-}^{(\infty)}(\zeta) & \varphi_{--}^{(\infty)}(\zeta)\end{pmatrix}\begin{pmatrix}K_+^+ & K_-^+\\ K_+^- & K_-^-\end{pmatrix}.$$

(see Appendix II for more details):

Proposition 4.5.

$$K_+^+=-q^{-(j_1+j_4+3/2)/2}\left(\frac{\gamma^{(0)}\gamma^{(\infty)}}{\varepsilon_1\varepsilon_4}\right)^{1/2}\frac{\Gamma(\gamma^{(0)})\Gamma(-\gamma^{(\infty)})}{\Gamma(\varepsilon_1)\Gamma(-\varepsilon_4)},$$

$$K_-^+=\quad q^{(j_4-j_1-1/2)/2}\left(\frac{\gamma^{(0)}\gamma^{(\infty)}}{\varepsilon_0\varepsilon_2}\right)^{1/2}\frac{\Gamma(-\gamma^{(0)})\Gamma(-\gamma^{(\infty)})}{\Gamma(-\varepsilon_0)\Gamma(-\varepsilon_2)},$$

$$K_+^-=\quad q^{(j_1-j_4-1/2)/2}\left(\frac{\gamma^{(0)}\gamma^{(\infty)}}{\varepsilon_0\varepsilon_2}\right)^{1/2}\frac{\Gamma(\gamma^{(0)})\Gamma(\gamma^{(\infty)})}{\Gamma(\varepsilon_0)\Gamma(\varepsilon_2)},$$

$$K_-^-=\quad q^{(j_1+j_4+1/2)/2}\left(\frac{\gamma^{(0)}\gamma^{(\infty)}}{\varepsilon_1\varepsilon_4}\right)^{1/2}\frac{\Gamma(-\gamma^{(0)})\Gamma(\gamma^{(\infty)})}{\Gamma(-\varepsilon_1)\Gamma(\varepsilon_4)},$$

where we denote $q=\exp(2\pi\sqrt{-1}/\kappa)$.

The conditions $(\mathrm{D2})_i$ $(i=2,1,0)$ and (D1) for $\mathbb{J}$ are equivalent to $(\mathrm{D2})_i$ and (D1) for $\bar{\mathbb{J}}$ respectively. Intermediate edges for $\bar{\mathbb{J}}$ must be $\bar{k}=$

$j_1 \pm \frac{1}{2}$ under the condition $(D2)_2$, and the intermediate edge for $\bar{\mathbb{J}}$ must be

$$(D2)_1,\ (D1)_1 \quad \bar{k}=j_1-\frac{1}{2}; \qquad (D1)_{2,3} \quad \bar{k}=j_1+\frac{1}{2}.$$

From the three bases $\{U_{\pm}^{(i)}=U_{\pm}^{(i)}(\mathbb{J});\ i=0, 1, \infty\}$ of $V_0^{\vee}(\mathbb{J})$, put

$$\bar{U}_{\pm}^{(i)}=\bar{U}_{\pm}^{(i)}(\bar{\mathbb{J}})=TU_{\pm}^{(1/i)} \quad (i=0, \infty) \quad \text{and} \quad \bar{U}_{\pm}^{(1)}=\pm TU_{\pm}^{(1)},$$

then they are three bases of $V_0^{\vee}(\bar{\mathbb{J}})$ such that

$$\Omega_{12}\bar{U}_{\pm}^{(0)}=\kappa\bar{\gamma}_{\pm}^{(0)}\bar{U}_{\pm}^{(0)}, \quad \Omega_{23}\bar{U}_{\pm}^{(1)}=\kappa\bar{\gamma}_{\pm}^{(1)}\bar{U}_{\pm}^{(1)}, \quad \Omega_{13}\bar{U}_{\pm}^{(\infty)}=\kappa(\bar{\gamma}_{\pm}^{(\infty)}-\Delta_4(\mathbb{J}))\bar{U}_{\pm}^{(\infty)},$$
$$\bar{\gamma}_{\pm}^{(0)}=\gamma_{\pm}^{(\infty)}; \qquad \bar{\gamma}_{\pm}^{(1)}=\gamma_{\pm}^{(1)}, \qquad \bar{\gamma}_{\pm}^{(\infty)}=\gamma_{\pm}^{(0)},$$

and

$$\bar{U}_{\pm}^{(0)}(u_4, u_2, u_3, u_1)=\frac{1}{\sqrt{2j_4+1}}\langle \nu(u_4)|\varphi_{v_2(\bar{k}_\pm)}(u_2)\varphi_{v_1(\bar{k}_\pm)}(u_3)|u_1\rangle$$

for $u_4\otimes u_3\otimes u_2\otimes u_1 \in V^{\vee}(\mathbb{J})$.

By Proposition 4.1, the composition $\Phi_k(w, z)$ of vertex operators is determined by the $V_0^{\vee}(\mathbb{J})$-valued function $\Psi_k(w, z)$ which is written as $\Psi_k(w, z)=z^{-\Delta_4(\mathbb{J})}\Psi_k(z/w)$ by the reduced 4-point function $\Psi_k(\zeta)$. And the composed operator $\bar{\Phi}_{\bar{k}}(z, w)$ is also determined by the $V_0^{\vee}(\bar{\mathbb{J}})$-valued function $\bar{\Psi}_{\bar{k}}(z, w)=w^{-\Delta_4(\mathbb{J})}\bar{\Psi}_{\bar{k}}(w/z)$.

The functions $\Psi_k(\zeta)$ and $\bar{\Psi}_{\bar{k}}(\eta)$ satisfy the differential equations RE $(\mathbb{J})$ and RE $(\bar{\mathbb{J}})$ with the initial conditions:

$$\zeta^{\hat{\Delta}(v_2(k))}\Psi_k(\zeta)(u_4, u_3, u_2, u_1)|_{\zeta=0}=\langle \nu(u_4)|\varphi_{v_2}(u_3)\varphi_{v_1}(u_2)|u_1\rangle$$

and

$$\eta^{\hat{\Delta}(v_2(\bar{k}))}\bar{\Psi}_{\bar{k}}(\eta)(u_4, u_2, u_3, u_1)|_{\eta=0}=\langle \nu(u_4)|\varphi_{v_2}(u_2)\varphi_{v_1}(u_3)|u_1\rangle.$$

By the relations among the exponents $\{\gamma_{\pm}^{(i)}\}$ and $\{\bar{\gamma}_{\pm}^{(i)}\}$ of the equations RE $(\mathbb{J})$ and RE $(\bar{\mathbb{J}})$, we get

$$T\Psi_{\pm}^{(0)}(\zeta)=\zeta^{\Delta_4(\mathbb{J})}\bar{\Psi}_{\pm}^{(\infty)}\left(\frac{1}{\zeta}\right) \quad \text{and} \quad T\Psi_{\pm}^{(\infty)}(\zeta)=\zeta^{\Delta_4(\mathbb{J})}\bar{\Psi}_{\pm}^{(0)}\left(\frac{1}{\zeta}\right),$$

where $\bar{\Psi}_{\pm}^{(i)}(\eta)$ denotes the fundamental solutions of the equation RE $(\bar{\mathbb{J}})$ similarly obtained as in Proposition 4.4.

By the note after Proposition 4.4, we get

$$\Psi_{k_\pm}(\zeta)=\sqrt{2j_4+1}\,\Psi_{\pm}^{(0)}(\zeta) \quad \text{and} \quad \bar{\Psi}_{\bar{k}_\pm}(\eta)=\sqrt{2j_4+1}\,\bar{\Psi}_{\pm}^{(0)}(\eta).$$

Hence by Propositions 4.1, 5, we get

Proposition 4.6. *Let* $\mathbb{J}=(j_4, \frac{1}{2}, j_2, j_1)$ *with* $I(\mathbb{J})\neq\varnothing$. *Then*

(D1) $C(\mathbb{J})=C_k^{\bar{k}}(\mathbb{J})=K(\mathbb{J})$, *where* $k\in I(\mathbb{J})$ *and* $\bar{k}\in I(\bar{\mathbb{J}})$.

(D2)$_1$ $C(\mathbb{J})=C_{k_-}^{\bar{k}_-}(\mathbb{J})=K_-^-(\mathbb{J})$, *where* $\bar{k}_-=j_1-\dfrac{1}{2}$ *and* $k_-=j_4-\dfrac{1}{2}$.

(D2)$_2$ $C(\mathbb{J})=(C_k^{\bar{k}}(\mathbb{J}))_{k\in I(J),\bar{k}\in I(\bar{J})}=K(\mathbb{J})$ *as* 2×2*-matrices.*

Remark. In the case (D2)$_2$, all entries of the matrix $C(\mathbb{J})=K(\mathbb{J})$ do not vanish. In the case (D2)$_1$, $\varepsilon_0=1$ implies $K_-^+(\mathbb{J})=0$, hence the matrix $K(\mathbb{J})$ is of the form $\begin{pmatrix} * & 0 \\ * & * \end{pmatrix}$.

4.4. Case $\mathbb{J}=(j_4, \frac{1}{2}, \frac{1}{2}, j_1)$

As a special case, we take $j_2=j_3=\frac{1}{2}$, then the conditions (D2) and (D1) read as

$$(\mathrm{D2})_2 \quad \frac{\ell}{2}>j_1=j_4>0; \qquad (\mathrm{D2})_1 \quad \frac{\ell}{2}=j_1=j_4;$$

and

$$(\mathrm{D1})_1 \quad j_4=j_1+1; \qquad (\mathrm{D1})_2 \quad j_1=j_4=0; \qquad (\mathrm{D1})_3 \quad j_1=j_4+1.$$

Under the assumption (D2), the constants $\gamma_\pm^{(i)}$, ε_i and the matrix $K(\mathbb{J})$ turns to be the following (here $j=j_1=j_4$):

$$\gamma_+^{(0)}=\frac{2j+3}{2\kappa}, \quad \gamma_-^{(0)}=\frac{2j-1}{-2\kappa}, \quad \gamma_+^{(1)}=\frac{1}{2\kappa}, \quad \gamma_-^{(1)}=\frac{3}{-2\kappa}.$$

$$\gamma_+^{(\infty)}=\frac{j}{\kappa}, \quad \gamma_-^{(\infty)}=\frac{j+1}{-\kappa}; \quad \gamma^{(0)}=\gamma^{(\infty)}=\frac{2j+1}{\kappa}, \quad \gamma^{(1)}=\frac{2}{\kappa};$$

$$\varepsilon_0=\frac{2j+2}{\kappa}, \quad \varepsilon_1=\varepsilon_4=\frac{1}{\kappa}, \quad \varepsilon_2=\frac{2j}{\kappa};$$

$$K_+^+=-(2j+1)q^{-j-3/4}\frac{\Gamma\left(\frac{2j+1}{\kappa}\right)\Gamma\left(\frac{2j+1}{-\kappa}\right)}{\Gamma\left(\frac{1}{\kappa}\right)\Gamma\left(\frac{-1}{\kappa}\right)},$$

$$K_-^+=\frac{2j+1}{2\sqrt{j(j+1)}}q^{-1/4}\frac{\Gamma\left(\frac{2j+1}{-\kappa}\right)^2}{\Gamma\left(\frac{2j+2}{-\kappa}\right)\Gamma\left(\frac{2j}{-\kappa}\right)},$$

$$K_{+}^{-}=\frac{2j+1}{2\sqrt{j(j+1)}}q^{-1/4}\frac{\Gamma\left(\frac{2j+1}{\kappa}\right)^{2}}{\Gamma\left(\frac{2j+2}{\kappa}\right)\Gamma\left(\frac{2j}{\kappa}\right)},$$

and

$$K_{-}^{-}=(2j+1)q^{j+1/4}\frac{\Gamma\left(\frac{2j+1}{\kappa}\right)\Gamma\left(\frac{2j+1}{-\kappa}\right)}{\Gamma\left(\frac{1}{\kappa}\right)\Gamma\left(\frac{-1}{\kappa}\right)}.$$

Now recall the notion of *q-integers* for $q\in\mathbb{C}^*$: for each integer $\nu\in\mathbb{Z}$, introduce the q-integer $[\nu]=[\nu]_q$ defined by

$$[\nu]_q=\begin{cases}\dfrac{q^{\nu}-1}{q-1} & (q\neq 1)\\ \nu & (q=1)\,.\end{cases}$$

Then

Lemma 4.7.

i) $[0]_q=0,\quad [1]_q=1$ *and* $[2]_q=1+q$.

ii) $[-\nu]_q=-q^{-\nu}[\nu]_q$ *and* $[\nu]_{1/q}=q^{1-\nu}[\nu]_q$ $(\nu\in\mathbb{Z})$.

iii) $[\nu]_q=0$, *if and only if* $q^{\nu}=1$. $\left([\kappa]_q=0\ \text{if}\ q=\exp\left(\frac{2\pi\sqrt{-1}}{\kappa}\right).\right)$

iv) $\lim_{q\to 1}[\nu]_q=\nu$ *for any* $\nu\in\mathbb{Z}$.

Then in the case $(\mathrm{D}2)_2$, the matrix $K(\mathbb{J})$ can be symmetrized by means of q-integers:

Proposition 4.8. *For* $j\in\frac{1}{2}\mathbb{Z}$ *with* $0<2j<\ell$,

$$K\left(j,\frac{1}{2},\frac{1}{2},j\right)$$

$$=q^{-3/4}\begin{pmatrix}\gamma_{+}^{-1} & \\ & \gamma_{-}^{-1}\end{pmatrix}\begin{pmatrix}\dfrac{-1}{[2j+1]} & \dfrac{\sqrt{q[2j][2j+2]}}{[2j+1]}\\[2ex] \dfrac{\sqrt{q[2j][2j+2]}}{[2j+1]} & \dfrac{q^{2j+1}}{[2j+1]}\end{pmatrix}\begin{pmatrix}\gamma_{+} & \\ & \gamma_{-}\end{pmatrix}$$

where

$$\gamma_{\pm}=\frac{\Gamma\left(\frac{2j+1}{\pm\kappa}\right)}{\left(\Gamma\left(\frac{2j+2}{\pm\kappa}\right)\Gamma\left(\frac{2j}{\pm\kappa}\right)\right)^{1/2}},\quad q=\exp\left(\frac{2\pi\sqrt{-1}}{\kappa}\right)\ \text{and}\ [\nu]=[\nu]_q.$$

We can get the connection matrix (=scalar) $K(\mathbb{J})$ in the cases $(D2)_1$ and (D1):

Proposition 4.8′. *Let* $q=\exp(2\pi\sqrt{-1}/\kappa)$.

i) $K\left(\frac{\ell}{2},\frac{1}{2},\frac{1}{2},\frac{\ell}{2}\right)=q^{\ell+1/4}/[\ell+1]_q=-q^{-3/4}$.

ii) $K\left(0,\frac{1}{2},\frac{1}{2},0\right)=-q^{-3/4}$.

iii) $K\left(j+1,\frac{1}{2},\frac{1}{2},j\right)=K\left(j-1,\frac{1}{2},\frac{1}{2},j\right)=q^{1/4}$.

Remark. These values are also obtained from the calculations in the case $(D2)_2$. $K^-_+(0,\frac{1}{2},\frac{1}{2},0)=0$ and $K^+_+(0,\frac{1}{2},\frac{1}{2},0)=-q^{-3/4}$. For $\mathbb{J}_\pm=(j\pm1,\frac{1}{2},\frac{1}{2},j)$, $K^+_+(\mathbb{J}_\pm)=K^-_-(\mathbb{J}_\pm)=0$ and $K^+_-(\mathbb{J}_+)=K^-_+(\mathbb{J}_-)=q^{1/4}$.

§ 5. Monodromy Representations of Braid Groups

In this section, we construct representations of braid groups on the spaces of multi-correlation functions, and show that they give the same representations of Hecke algebras constructred by H. Wenzl.

5.1) Braid groups and Hecke algebras

Recall our X_N is a complex manifold defined by

$$X_N=\{(z_N,z_{N-1},\cdots,z_1)\in\mathbb{C}^N;\ z_i\neq z_j\ (i\neq j)\}.$$

The N-th symmetric group $\mathfrak{S}_N$ acts on the manifold X_N as $(z_N,\cdots,z_1)\sigma=(z_{(N)\sigma},\cdots,z_{(1)\sigma})$ $(\sigma\in\mathfrak{S}_N)$, then we get a covering space $\pi_N\colon X_N\to\bar{X}_N=X_N/\mathfrak{S}$. Let $\tilde{\pi}_N\colon\tilde{X}_N\to X_N$ be a universal covering manifold of X_N, then $\bar{\pi}_N=\tilde{\pi}_N\circ\pi_N\colon\tilde{X}_N\to\bar{X}_N$ is also a universal covering of $\bar{X}_N$.

Now recall the braid groups according to J. S. Birman [Bi]. The fundamental group $\pi_1(\bar{X}_N,\bar{p}_N)$ of the manifold $\bar{X}_N$ is called the *braid group with N strings of the manifold* $\mathbb{C}$, that is, the *classical braid group of Artin*, and is denoted by B_N, where we take the base point as $\bar{p}_N=\pi_N(p_N)$. The composition of τ_1 and τ_2 in the group B_N is figured as

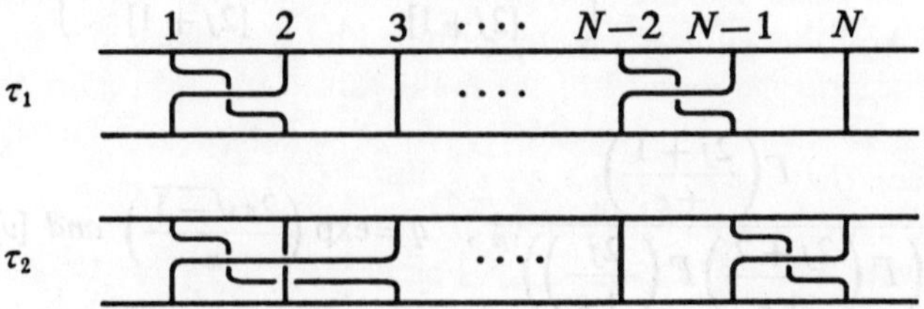

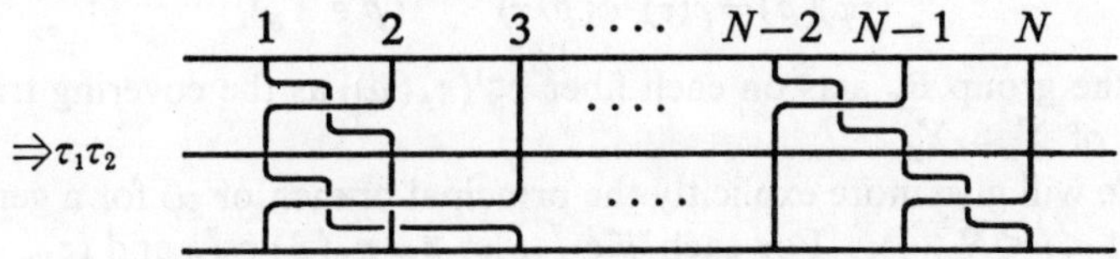

The fundamental group $\pi_1(X_N, p_N)$ of the manifold X_N is called the *pure braid group with N strings of the manifold* $\mathbf{C}$, and is denoted by P_N, where p_N is a base point of X_N, e. g. $p_N=(N, N-1, \cdots, 1)$. Then the group P_N is the kernel of the natural homomorphism ρ of B_N onto $\mathfrak{S}_N$.

It is well-known that the group B_N has a system $\{b_i; 1\leq i\leq N-1\}$ of generators with the fundamental relations

(BR) $\quad b_ib_j=b_jb_i \ (|i-j|\geq 2)$ and $b_ib_{i+1}b_i=b_{i+1}b_ib_{i+1} \ (1\leq i\leq N-2)$.

where b_i is figured as a geometric braid by

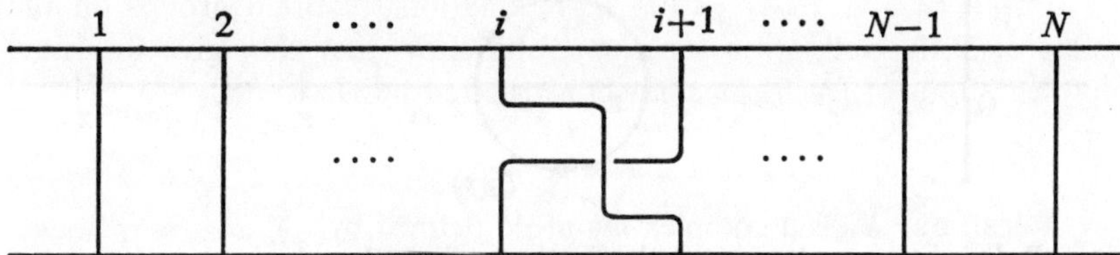

The subgroup P_N has a system $\{a_{ij}; 1\leq i<j\leq N\}$ of generators, defined by

$$a_{ij}=b_{j-1}b_{j-2}\cdots b_{i+1}b_i^2b_{i+1}^{-1}\cdots b_{j-1}^{-1}.$$

Introduce a subset I_N of the manifold X_N defined by

$$I_N=\{(z_N, \cdots, z_1)\in \mathbf{R}^N; z_N>z_{N-1}>\cdots>z_1>0\}.$$

Specify a base point $\tilde{p}_N$ of the manifold $\tilde{X}_N$ such that $\tilde{\pi}_N(\tilde{p}_N)=p_N$, then there is a subset $\tilde{I}_N$ of $\tilde{X}_N$ such that $\tilde{p}_N\in\tilde{I}_N$ and I_N is homeomorphic to $\tilde{I}_N$.

For a finite dimensional $\mathfrak{S}_N$-module W, denote by $\mathcal{O}(\tilde{X}_N; W)$ the space of all W-valued holomorphic functions on $\tilde{X}_N$. The values of $\varphi\in\mathcal{O}(\tilde{X}_N; W)$ on the whole $\tilde{X}_N$ are determined by the values of φ in $\tilde{I}_N$, which we call the *principal branch* of the multi-valued function φ on X_N. For a point $(z_N, \cdots, z_1)\in I_N$, sometimes we write $\varphi(z_N, \cdots, z_1)=\varphi(\tilde{p})$, where $\tilde{p}\in\tilde{I}_N$ such that $\tilde{\pi}_N(\tilde{p})=(z_N, \cdots, z_1)$.

The action of the braid group B_N on the space $\mathcal{O}(\tilde{X}_N; W)$ is defined as follows: Let $\tau\in B_N=\pi_1(\overline{X}_N)$. For each φ of $\mathcal{O}(X_N; W)$ and $\tilde{p}\in\tilde{X}_N$, put

$$(\tau\varphi)(\tilde{p})=\rho(\tau)\cdot\varphi(\tilde{p}\cdot\tau) \qquad (\tilde{p}\in\tilde{X}_N),$$

where the group B_N acts on each fiber $\tilde{\pi}_N^{-1}(\tilde{\pi}_N(\tilde{p}))$ as the covering transformation of $\tilde{X}_N\to\bar{X}_N$.

We will give more explicitly the principal branch of $\tau\varphi$ for a generator $\tau=b_i$ $(1\leq i\leq N-1)$. For each $\tilde{p}\in\tilde{I}_N$, let $\bar{p}=\bar{\pi}_N(\tilde{p})\in\bar{I}_N$ and $(z_N,\cdots,z_1)=\tilde{\pi}_N(\tilde{p})\in I_N$,

$$(b_i\varphi)(\tilde{p})=(i,i+1)\cdot\varphi(\tilde{p}\cdot b_i)$$

where $(i,i+1)$ denotes the transposition, and $\varphi(\tilde{p}\cdot b_i)$ is nothing but the analytic continuation of the principal branch $\varphi(z_N,\cdots,z_1)$ along the path $(\zeta_N(t),\cdots,\zeta_1(t))$ in X_N $(t\in[0,1])$: $\zeta_k(t)=z_k$ $(k\neq i,i+1)$,

$$\zeta_i(t)=\frac{z_i+z_{i+1}}{2}-e^{\pi t\sqrt{-1}}\frac{z_{i+1}-z_i}{2};\quad \zeta_{i+1}(t)=\frac{z_i+z_{i+1}}{2}+e^{\pi t\sqrt{-1}}\frac{z_{i+1}-z_i}{2}$$

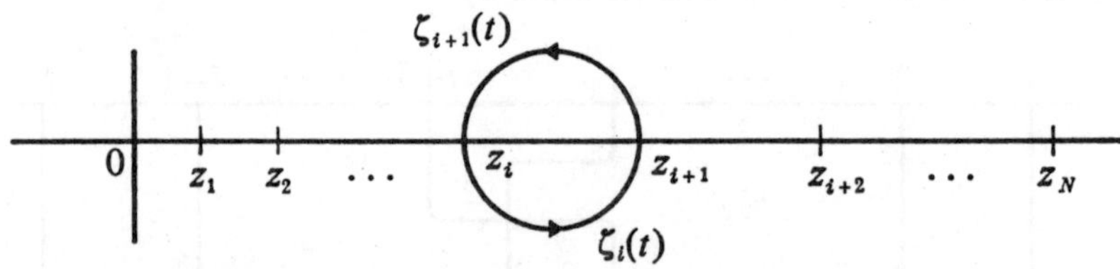

Related to braid groups, the notion of Hecke algebras is important (see e.g. D. Kazhdan-G. Lusztig [KL] and V. H. R. Jones [Jo]).

Let $N\geq 2$ and $q\in\mathbb{C}^*$. Then the *Hecke algebra $H_N(q)$ of type A_{N-1}* is defined as the associative complex algebra with generators $1, T_1,\cdots, T_{N-1}$ with the defining relations:

(H1) $T_iT_{i+1}T_i=T_{i+1}T_iT_{i+1}$ for $i=1,2,\cdots,N-2$.

(H2) $T_iT_j=T_jT_i$ for $|i-j|\geq 2$.

(H3) $(T_i-q)(T_i+1)=0$, that is, $T_i^2=(q-1)T_i+q$.

Note that (H1) and (H2) are nothing but the braid relations (BR), hence there is a natural epimorphism of the group algebra $\mathbb{C}[B_N]$ onto $H_N(q)$. For $q=1$, the Hecke algebra $H_N(1)$ is isomorphic to the group algebra $\mathbb{C}\mathfrak{S}_N$ of the N-th symmetric group $\mathfrak{S}_N$, by sending T_i to the transposition $(i,i+1)$. If q is not a root of unity, it is known by H. Wenzl [W] that there exists an isomorphisms of $H_q(N)$ with the group ring $\mathbb{C}[\mathfrak{S}_N]$ as algebras.

Assume that $[2]_q\neq 0$, that is, $q\neq -1$. Then we can give another system $\{1, e,\cdots, e_{N-1}\}$ of generators of $H_N(q)$ consisting of idempotents:

$$e_i=\frac{q-T_i}{[2]_q},\quad \text{i.e. } T_i=q-[2]_q e_i \quad (i=1,\cdots,N-1).$$

Then the defining relations (H1)~(H3) translate to

(H1)′ $e_i e_{i+1} e_i-\dfrac{q}{[2]_q^2}e_i=e_{i+1}e_i e_{i+1}-\dfrac{q}{[2]_q^2}e_{i+1}$ for $i=1,2,\cdots,N-2$.

(H2)′ $e_i e_j=e_j e_i$ for $|i-j|\geq 2$.

(H3)′ $e_i^2=e_i$ for $i=1,2,\cdots,N-1$.

5.2) Monodromy representations

Let $N\geq 2$, $\kappa=\ell+2$, $q=\exp(2\pi\sqrt{-1}/\kappa)$ and fix a half integer t with $0\leq 2t\leq \ell$ which we call a *target edge*. Introduce an $(N+1)$-ple $\mathbb{J}_t=(t,\frac{1}{2},\cdots,\frac{1}{2})$, and consider the systems $E(N;t)$ and $B(N;t)$ of equations for $V_0^\vee(\mathbb{J}_t)$-valued functions on the manifold X_N:

$$E(N;t):\quad \left(\kappa\frac{\partial}{\partial z_i}-\sum_{\substack{k=1\\k\neq i}}^{N}\frac{\Omega_{ik}}{z_i-z_k}\right)\Psi(z_N,\cdots,z_1)=0 \quad (1\leq i\leq N)$$

and for any $u_k\in V_{j_k}$ $(j_{N+1}=t, j_i=\frac{1}{2}\ (1\leq i\leq N))$,

$$B(N;t):\quad \sum_{\mathfrak{m}_i}\binom{L_i}{\mathfrak{m}_i}\prod_{\substack{k=1\\k\neq i}}^{N}(z_k-z_i)^{-m_k}$$
$$\times\Psi(\mathfrak{z})(u_{N+1},E^{m_N}u_N,\cdots,u_{j_i}(j_i),\cdots,E^{m_1}u_1)=0$$

for $1\leq i\leq N$, and

$$\sum_{\mathfrak{m}_{N+1}}\binom{L_{N+1}}{\mathfrak{m}_{N+1}}\Psi(\mathfrak{z})(u_{j_{N+1}}(j_{N+1}),E^{m_N}u_N,\cdots,E^{m_1}u_1)=0$$

where $\mathfrak{m}_i=(m_N,\cdots,\hat{m}_i,\cdots,m_1)\in(\mathbb{Z}_{\geq 0})^{N-1}$ $(1\leq i\leq N)$ and $\mathfrak{m}_{N+1}=(m_N,\cdots,m_1)\in(\mathbb{Z}_{\geq 0})^N$ with $|\mathfrak{m}_i|=L_i=\ell-2j_i+1$ $(1\leq i\leq N+1)$.

Let $W(N;t)$ be the solution space of the joint system $E(N;t)$ and $B(N;t)$. Then by Theorem 3.3, the space $W(N;t)$ has a basis $\{\Psi_{\mathfrak{p}}(z_N,\cdots,z_1);\ \mathfrak{p}\in\mathscr{P}_\ell(N;t)\}$ defined as follows: Let

$$\mathscr{P}_\ell(N;t)=\Big\{\mathfrak{p}=(p_N,\cdots,p_1,p_0);\ p_N=t,\ p_0=0,\ p_i\in\frac{1}{2}\mathbb{Z},\ 0\leq 2p_i\leq\ell,$$
$$|p_i-p_{i-1}|=\frac{1}{2}\ (1\leq i\leq N)\Big\}.$$

For each $\mathfrak{p}\in\mathscr{P}_\ell(N;t)$, define the $V_0^\vee(\mathbb{J}_t)$-valued, multi-valued holomorphic function $\Psi_{\mathfrak{p}}(z_N,\cdots,z_1)$ on X_N by

$$\Psi_{\mathfrak{p}}(z_N, \cdots, z_1)(v, u_N, \cdots, u_1) = \langle \nu(v) | \Phi_{\mathrm{v}_N}(u_N; z_N) \cdots \Phi_{\mathrm{v}_1}(u_1; z_1) | \mathrm{vac} \rangle$$

for $v \in V_t$ and $u_i \in V_{1/2}$ $(1 \leq i \leq N)$, where the vertex $\mathrm{v}_i = \mathrm{v}_i(\mathfrak{p})$ is defined as $\mathrm{v}_i = \begin{pmatrix} \frac{1}{2} \\ p_i p_{i-1} \end{pmatrix}$ $(1 \leq i \leq N)$.

The braid group B_N acts on this space $W(N; t)$ as monodromies. The commutation relations of vertex operators give a factorization of this monodromy representation $(\pi_{N,t}, W(N; t))$. The $\mathfrak{S}_N$-module structure of the space $V_0^{\vee}(\mathbb{J})$ is defined by

$$(\sigma\varphi)(u_N, \cdots, u_1) = \varphi(u_{(N)\sigma}, \cdots, u_{(1)\sigma}) \qquad (\varphi \in V_0^{\vee}(\mathbb{J}),\ \sigma \in \mathfrak{S}_N),$$

and the B_N-module structure on the space of $V_0^{\vee}(\mathbb{J})$-valued functions on X_N is defined in Section 5.1. By Propositions 4.8, 4.8′ and 5.1, we will give this representation $\pi = \pi_{N,t}$ explicitly.

For each i $(1 \leq i \leq N-1)$, the action $\pi(b_i)$ of the generator b_i of the group B_N on the space $W(N; t)$ is given as follows.

At first, divide the set $\mathscr{P}_\ell(N; t)$ into the four parts: Let $\mathfrak{p} = (p_N, p_{N-1}, \cdots, p_i, \cdots, p_1, p_0) \in \mathscr{P}_\ell(N; t)$, $p_N = t$, $p_0 = 0$.

$$\mathfrak{p} \in \mathscr{P}_i^a(N; t) \longleftrightarrow p_{i+1} = p_{i-1} = 0.$$

$$\mathfrak{p} \in \mathscr{P}_i^b(N; t) \longleftrightarrow |p_{i+1} - p_{i-1}| = 1.$$

$$\mathfrak{p} \in \mathscr{P}_i^c(N; t) \longleftrightarrow \frac{\ell}{2} > p_{i+1} = p_{i-1} > 0.$$

$$\mathfrak{p} \in \mathscr{P}_i^d(N; t) \longleftrightarrow p_{i+1} = p_{i-1} = \frac{\ell}{2}.$$

Then the operation $\pi(b_i)$ is given on the basis vectors $\{\Psi_{\mathfrak{p}};\ \mathfrak{p} \in \mathscr{P}_\ell(N; t)\}$ as:

a, d) If $\mathfrak{p} \in \mathscr{P}_i^a(N; t)$ or $\mathfrak{p} \in \mathscr{P}_i^d(N; t)$, $\qquad \pi(b_i)\Psi_{\mathfrak{p}} = -q^{-3/4}\Psi_{\mathfrak{p}}$.

b) If $\mathfrak{p} \in \mathscr{P}_i^b(N; t)$, $\qquad \pi(b_i)\Psi_{\mathfrak{p}} = q^{1/4}\Psi_{\mathfrak{p}}$.

c) If $\mathfrak{p} \in \mathscr{P}_i^c(N; t)$, there is only one $\mathfrak{p}' \in \mathscr{P}_i^c(N; t)$ such that $p_k = p_k'$ for any $k \neq i$ and $|p_i - p_i'| = 1$. We define the action $\pi(b_i)$ for which $\mathbb{C}\Psi_{\mathfrak{p}} + \mathbb{C}\Psi_{\mathfrak{p}'}$ is invariant. We modify the notations as $\mathfrak{p}_\pm = (t, p_{N-1}, \cdots, p_{i+1}, p_i^\pm, p_{i-1}, \cdots, p_1)$, where $p_i^+ = \max(p_i, p_i')$ and $p_i^- = \min(p_i, p_i')$. Then the action $\pi(b_i)$ on $\mathbb{C}\Psi_{\mathfrak{p}_+} + \mathbb{C}\Psi_{\mathfrak{p}_-}$ is given as $\pi(b_i) = K(p, \frac{1}{2}, \frac{1}{2}, p)$, where $0 < p = p_{i+1} = p_{i-1} < \ell/2$:

$$\pi(b_i)|_{\mathbb{C}\Psi_{\mathfrak{p}_+} + \mathbb{C}\Psi_{\mathfrak{p}_-}}$$

$$= q^{-3/4} \begin{pmatrix} \gamma_+^{-1} & \\ & \gamma_-^{-1} \end{pmatrix} \begin{bmatrix} \dfrac{-1}{[2p+1]} & \dfrac{\sqrt{p[2p][2p+2]}}{[2p+1]} \\ \dfrac{\sqrt{q[2p][2p+2]}}{[2p+1]} & \dfrac{q^{2p+1}}{[2p+1]} \end{bmatrix} \begin{pmatrix} \gamma_+ & \\ & \gamma_- \end{pmatrix}$$

where

$$\gamma_{\pm}=\frac{\Gamma\left(\frac{2p+1}{\pm\kappa}\right)}{\left(\Gamma\left(\frac{2p+2}{\pm\kappa}\right)\Gamma\left(\frac{2p}{\pm\kappa}\right)\right)^{1/2}}.$$

In each case, $\{q, -1\}$ are only possible eigenvalues of the operators $q^{3/4}\pi(b_i)$. Thus the actions $q^{3/4}\pi(b_i)$ on the space $W(N; t)$ satisfy the relation (H3) of the Hecke algebra $H_N(q)$.

Theorem 5.2. *The monodromy representation* $q^{3/4}\pi_{N,t}$ *of the braid group* B_N *on the space* $W(N; t)$ *gives a representation of the Hecke algebra* $H_N(q)$, *where* $q=\exp(2\pi\sqrt{-1}/\kappa)$.

Remark. It is remarkable that our representations are obtained for the Hecke algebra $H_N(q)$ with a root q of unity, since the algebra $H_N(q)$ is not semi-simple for a root q of unity (cf. V. F. R. Jones [Jo]).

5.3) Wenzl's representations of Hecke algebra

H. Wenzl [W] constructed irreducible representations (π_λ, V_λ) of Hecke algebras $H_N(q)$ for any q not being roots of unity, parametrized by the set Λ_N of all Young diagrams on N nodes. If $q=\exp(2\pi\sqrt{-1}/\kappa)$ with $\kappa\,(=\ell+2)\geq 4$ (i.e. $\ell\geq 2$), he also constructed irreducible representations $(\pi_\lambda^{(k,\kappa)}, V_\lambda^{(k,\kappa)})$ of $H_N(q)$ parametrized by the set $\Lambda^{(k,\kappa)}$ of all (k, κ)-diagrams on N nodes. Note that the representations $\pi_\lambda^{(k,\kappa)}$ are unitarizable as representations of the group B_N.

In this paragraph, we show that our representation $(\pi_{N,t}, W(N; t))$ of the Hecke algebra $H_N(q)$ $(q=\exp(2\pi\sqrt{-1}/\kappa))$ is equivalent to the representation $(\pi_\lambda^{(2,\kappa)}, V_\lambda^{(2,\kappa)})$.

Let Λ_N^2 be the set of all Young diagrams λ on N nodes with depth $(\lambda)\leq 2$. For each $\lambda\in\Lambda_N^2$, $d(\lambda)$ denotes the difference of the number of the first row of λ and the one of the second. Introduce the set $\Lambda_N^{(2,\kappa)}$ of all $(2, \kappa)$-diagrams on N nodes, defined by $\Lambda_N^{(2,\kappa)}=\{\lambda\in\Lambda_N^2;\ d(\lambda)\leq\kappa-2\,(=\ell)\}$. Any $\lambda\in\Lambda_N^{(2,\kappa)}$ is written as $[N/2+t, N/2-t]$ for some half-integer $t\geq 0$.

We shall write $\mu<\lambda$, if the Young diagram μ can be obtained by taking away appropriate nodes of λ. For each $\lambda\in\Lambda_N^{(2,\kappa)}$, let

$$\mathscr{P}_\ell(\lambda)=\{p=(\lambda_{(N)}, \cdots, \lambda_{(1)});\ \lambda_{(i)}\in\Lambda_i^{(2,\kappa)},\ \lambda_{(i)}<\lambda_{(i+1)},\ \lambda_{(N)}=\lambda\}.$$

H. Wenzl defines an irreducible representation $(\pi_\lambda^{(2,\kappa)}, V_\lambda^{(2,\kappa)})$ of the algebra $H_N(q)$ for each $\lambda\in\Lambda_N^{(2,\kappa)}$, where $V_\lambda^{(2,\kappa)}$ has the form $\bigoplus_{p\in\mathscr{P}_\ell(\lambda)}\mathbf{C}\vec{v}_p$. This gives a unitary representation of the group B_N.

Note that for each N, the number $d(\lambda)$ determines the Young diagram

$\lambda \in \Lambda_N^2$ uniquely. For each $p=(\lambda_{(N)}, \cdots, \lambda_{(1)}) \in \mathscr{P}_\ell(\lambda)$ with $\lambda \in \Lambda_N^{(2,\kappa)}$, let $K(p)=(t, \frac{1}{2}d(\lambda_{(N-1)}), \cdots. \frac{1}{2}d(\lambda_{(1)}), 0) \in \mathscr{P}_\ell(N; t)$ with $t=\frac{1}{2}d(\lambda)$. Then the mapping K gives a bijection of $\mathscr{P}_\ell(\lambda)$ with $\mathscr{P}_\ell(N; \frac{1}{2}d(\lambda))$.

For each $p=(\lambda_{(N)}, \cdots, \lambda_{(1)}) \in \mathscr{P}_\ell(\lambda)$, introduce the numbers $\gamma_i(p)$ $(1\leq i\leq N-1)$ defined by

$$\gamma_i(p)=1, \quad \text{if } d(\lambda_{(i+1)})=d(\lambda_{(i-1)})=0 \quad \text{or} \quad |d(\lambda_{(i+1)})-d(\lambda_{(i-1)})|=2,$$

$$\gamma_i(p)=\frac{\Gamma\left(\frac{d+1}{\kappa}\right)}{\left(\Gamma\left(\frac{d+2}{\kappa}\right)\Gamma\left(\frac{d}{\kappa}\right)\right)^{1/2}}, \quad \text{if } d=d(\lambda_{(i-1)})=d(\lambda_{(i+1)})=d(\lambda_{(i)})-1,$$

and

$$\gamma_i(p)=\frac{\Gamma\left(\frac{d+1}{-\kappa}\right)}{\left(\Gamma\left(\frac{d+2}{-\kappa}\right)\Gamma\left(-\frac{d}{\kappa}\right)\right)^{1/2}}, \quad \text{if } d=d(\lambda_{(i-1)})=d(\lambda_{(i+1)})=d(\lambda_{(i)})+1.$$

Define the mapping K: $V_\lambda^{(2,\kappa)} \to W(N; \frac{1}{2}d(\lambda))$ by

$$K(\vec{v}_p)=\prod_{i=1}^{N-1} \gamma_i(p)\Psi_{K(p)} \qquad \text{for } p \in \mathscr{P}_\ell(\lambda).$$

(note $\gamma_1(p)=1$ for any $p \in \mathscr{P}_\ell(\lambda)$.)

Then the mapping K intertwines Wenzl's representations $(\pi_\lambda^{(2,\kappa)}, V_\lambda^{(2,\kappa)})$ and our $(\pi_{N,t}, W(N; t))$:

Proposition 5.3. *For each $\lambda \in \Lambda_N^{(2,\kappa)}$, set $t=\frac{1}{2}d(\lambda)$. Then*

$$K\pi_\lambda=q^{3/4}\pi_{N,t}K.$$

Note 1. If we construct the theory for $\ell \notin \mathbf{Q}$ as in Remark 3.5, we get the monodromy representations of the Hecke algebra $H_N(q)$, $q=\exp(2\pi\sqrt{-1}/(\ell+2))$, which are isomorphic to the representations (π_λ, V_λ) parametrized by $\lambda \in \Lambda_N^{(2)}$.

Note 2. By means of $A_1^{(1)}$-modules, we obtained here the representations $(\pi_\lambda^{(2,\kappa)}, V_\lambda^{(2,\kappa)})$ of the algebra $H_N(q)$ parametrized by the $(2, \kappa)$-diagrams. For general $k>2$, Wenzl's representations $(\pi_\lambda^{(k,\kappa)}, V_\lambda^{(k,\kappa)})$ are obtained by means of integrable highest weight modules of affine Lie algebras of type $A_n^{(1)}$ $(n>1)$. We will discuss them in our succeeding paper.

5.4) Fusion rule

For a quadruple $\mathbf{J}=(j_4, j_3, j_2, j_1)$, introduce the set $J_\ell(\mathbf{J})$ defined by

$$J_\ell(\mathbb{J})=\left\{r\in\frac{1}{2}\mathbb{Z};\quad 0\leq 2r\leq \ell,\quad \mathrm{w}(r)=\begin{pmatrix} & r & \\ j_4 & & j_1\end{pmatrix}\in(\mathrm{CG})_\ell,\right.$$

$$\left.\overline{\mathrm{w}}(r)=\begin{pmatrix} & j_3 & \\ r & & j_2\end{pmatrix}\in(\mathrm{CG})_\ell\right\},$$

and consider the fusion of vertex operators $\Phi_{\mathrm{v}_2(k)}(w)$ and $\Phi_{\mathrm{v}_1(k)}(z)$ for $k\in I_\ell(\mathbb{J})$ to $\Phi_{\mathrm{w}(r)}(z)$ (the first term of the short range expansion of the product $\Phi_{\mathrm{v}_2(k)}(w)\Phi_{\mathrm{v}_1(k)}(z)$):

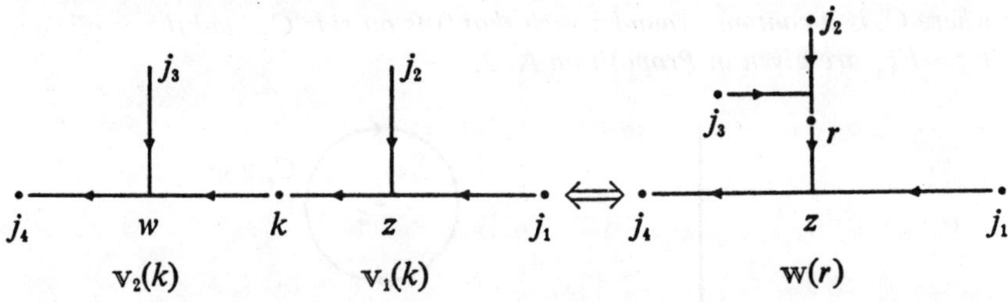

Now we restrict ourselves to the case $j_3=\frac{1}{2}$. Assume that $V_0^{\vee}(\mathbb{J})\neq\emptyset$, then in cases listed in Section 4.2. we get

$$\text{(D2)}_2\qquad I_\ell(\mathbb{J})=\left\{k_\pm=j_4\pm\frac{1}{2}\right\},\qquad J_\ell(\mathbb{J})=\left\{r_\pm=j_2\pm\frac{1}{2}\right\},$$

$$\text{(D2)}_1\qquad I_\ell(\mathbb{J})=\left\{j_4-\frac{1}{2}\right\},\qquad J_\ell(\mathbb{J})=\left\{j_2-\frac{1}{2}\right\},$$

$$\text{(D1)}_1\qquad I_\ell(\mathbb{J})=\left\{j_4+\frac{1}{2}\right\},\qquad J_\ell(\mathbb{J})=\left\{j_2+\frac{1}{2}\right\},$$

$$\text{(D1)}_2\qquad I_\ell(\mathbb{J})=\left\{j_4+\frac{1}{2}\right\},\qquad J_\ell(\mathbb{J})=\left\{j_2-\frac{1}{2}\right\},$$

$$\text{(D1)}_3\qquad I_\ell(\mathbb{J})=\left\{j_4-\frac{1}{2}\right\},\qquad J_\ell(\mathbb{J})=\left\{j_2+\frac{1}{2}\right\}.$$

Here we discuss the case (D2), since other cases are much simpler. In this case, fix notations

$$\mathrm{v}_2(k_\pm)=\begin{pmatrix} & \frac{1}{2} & \\ j_4 & & k_\pm\end{pmatrix},\quad \mathrm{v}_1(k_\pm)=\begin{pmatrix} & j_2 & \\ k_\pm & & j_1\end{pmatrix},\quad \mathrm{w}(r_\pm)=\begin{pmatrix} & r_\pm & \\ j_4 & & j_1\end{pmatrix},\quad \overline{\mathrm{w}}(r_\pm)=\begin{pmatrix} & \frac{1}{2} & \\ r_\pm & & j_2\end{pmatrix}$$

and note the relations:

$$\gamma_\pm^{(0)}=\hat{\varDelta}(\mathrm{v}_2(k_\pm))=\varDelta_4(\mathbb{J})-\hat{\varDelta}(\mathrm{v}_1(k_\pm))\quad\text{and}\quad \gamma_\pm^{(1)}=\varDelta_4(\mathbb{J})-\hat{\varDelta}(\mathrm{w}(r_\pm)).$$

Then

Proposition 5.4. *For a quadruple* $\mathbb{J}=(j_4, \frac{1}{2}, j_2, j_1)$ *in the case* $(\mathrm{D2})_2$ *and for each* $k \in I_\ell(\mathbb{J})$,

$$\frac{1}{2\pi\sqrt{-1}}\int_{C_z}(w-z)^{-\gamma_\pm^{(1)}-1}\Phi_{v_3(k)}(u_3; w)\Phi_{v_2(k)}(u_2; z)dw$$
$$=F_k^\pm(\mathbb{J})\Phi_{w(r_\pm)}(\varphi_{w(r_\pm)}(u_3, u_2); z) \qquad (u_3 \in V_{1/2}, u_2 \in V_{j_2}),$$

where C_z *is a contour around* z *such that* 0 *is outside* C_z, *and the coefficients* $F_\pm^\pm=F_{k\pm}^\pm$ *are given in Proposition* A. 2.

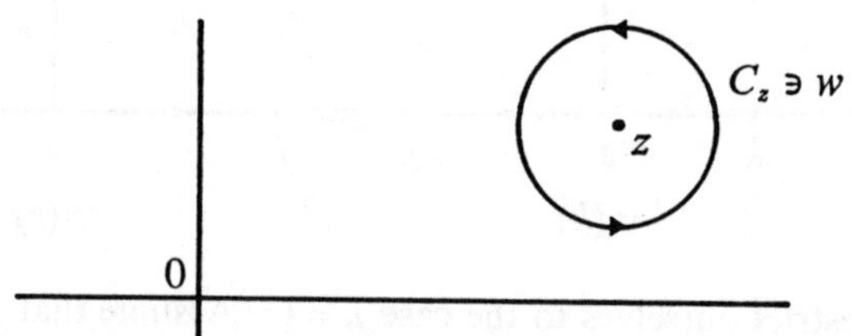

Proof. The composition $\Phi_k(w, z)$ of vertex operators $\Phi_{v_3(k)}(w)$ and $\Phi_{v_2(k)}(z)$ is determined by the $V_0^\vee(\mathbb{J})$-valued function $\Psi_k(w, z)$ on M_2 defined in Section 4.1 for each $k \in I_\ell(\mathbb{J})$. By Propositions 4.3 and 4.4, we get the expansion of $\Psi_k(w, z)$ near $w=z$ as

$$\Psi_k(w, z)=(w-z)^{\gamma_+^{(1)}}\{\sqrt{2j_4+1}\,F_k^+U_+^{(1)}+O(w-z)\}$$
$$+(w-z)^{\gamma_-^{(1)}}\{\sqrt{2j_4+1}\,F_k^-U_-^{(1)}+O(w-z)\}$$

where $O(w-z)$ is holomorphic near $w=z$ and vanishes on $\{w=z\}$.

Now introduce the operator $\Xi_k^\pm(u_3, u_2; z)$ of $\mathscr{H}_{j_1}$ to $\hat{\mathscr{H}}_{j_4}$ defined by the integral

$$\Xi_k^\pm(u_3, u_2; z)=\frac{1}{2\pi\sqrt{-1}}\int_{C_z}(w-z)^{-\gamma_\pm^{(1)}-1}\Phi_{v_3(k)}(u_3; w)\Phi_{v_2(k)}(u_2; z)dw.$$

And define an operator $\Xi_k^\pm(z)\,(v)=\Xi_k^\pm\,(v; z)\colon \mathscr{H}_{j_1}\to\hat{\mathscr{H}}_{j_4}$ parametrized by $V_{r_\pm}$ as follows: For any vector $v \in V_{r_\pm}$ is written as a linear combination $v=\sum_i c_i\varphi_{w(r_\pm)}(u_3^i, u_2^i)$ for some $u_3^i\otimes u_2^i \in V_{j_3}\otimes V_{j_2}$. Then put $\Xi_k^\pm(v; z)$ $=\sum_i c_i\Xi_k^\pm(u_3^i, u_2^i; z)$, then $\Xi_k^\pm(z)$ is independent of the expression of v, and is a vertex operator of type $w(r_\pm)$, that is, of spin $r_\pm$ (note $-\Delta_{r_\pm}=\gamma_\pm^{(1)}-\Delta_{j_3}-\Delta_{j_2}$). In fact, for $X \in \mathfrak{g}$ and $m \in \mathbb{Z}$,

$$[X(m), \Xi_k^\pm(u_3, u_2; z)]=z^m[\Xi_k^\pm(Xu_3, u_2; z)+\Xi_k^\pm(u_3, Xu_2, z)]$$

and

$$[L(m), \Xi_k^{\pm}(u_3, u_2, z)]$$

$$=\frac{1}{2\pi\sqrt{-1}}\int_{C_z}(w-z)^{-r_{\pm}^{(1)}-1}\Big\{w^{m+1}\frac{\partial}{\partial w}+z^{m+1}\frac{\partial}{\partial z}$$

$$+(m+1)(\Delta_{j_3}w^m+\Delta_{j_2}z^m)\Big\}\Phi_{v_3(k)}(u_3; w)\Phi_{v_2(k)}(u_2; z)dw$$

$$=z^m\Big\{z\frac{\partial}{\partial z}+(m+1)(\Delta_{j_3}+\Delta_{j_2})\Big\}\Xi_k^{\pm}(u_3, u_2; z)$$

$$-\frac{1}{2\pi\sqrt{-1}}\int_{C_z}\frac{\partial}{\partial w}\{(w-z)^{-r_{\pm}^{(1)}-1}w^{m+1}\}\Phi_{v_3(k)}(u_3; w)\Phi_{v_2(k)}(u_2; z)dw$$

$$-\frac{z^{m+1}}{2\pi\sqrt{-1}}\int_{C_z}\frac{\partial}{\partial z}\{(w-z)^{-r_{\pm}^{(1)}-1}\}\Phi_{v_3(k)}(u_3; w)\Phi_{v_2(k)}(u_2; z)dw$$

$$=z^m\Big\{z\frac{\partial}{\partial z}+(m+1)(\Delta_{j_3}+\Delta_{j_2}-1)\Big\}\Xi_k^{\pm}(u_3, u_2; z)$$

$$+\frac{r_{\pm}^{(1)}+1}{2\pi\sqrt{-1}}\int_{C_z}(w^{m+1}-z^{m+1})(w-z)^{-r_{\pm}^{(1)}-2}\Phi_{v_3(k)}(u_3; w)\Phi_{v_2(k)}(u_2; z)dw$$

$$=z^m\Big\{z\frac{\partial}{\partial z}+(m+1)\Delta_{r_\pm}\Big\}\Xi_k^{\pm}(u_3, u_2; z).$$

Thus $\Xi_k^{\pm}(z)$ is a vertex operator of type $\mathrm{w}(r_\pm)$, so it is a constant multiple of $\Phi_{\mathrm{w}(r\pm)}(z)$.

Hence we get the proposition, by computing the initial term of $\Phi_{\mathrm{w}(r_\pm)}(\varphi_{\mathrm{w}(r_\pm)}(u_3, u_2); z)$:

$$\langle\nu(u_4)|\varphi_{\mathrm{w}(r_\pm)}(\varphi_{\mathrm{w}(r_\pm)}(u_3, u_2), u_1)\rangle=\sqrt{2j_4+1}\; U_{\pm}^{(1)}(u_4, u_3, u_2, u_1). \qquad \text{q.e.d.}$$

Let $N\geq 2$. Fix an N-ple $\mathbb{J}=(\frac{1}{2}, \cdots, \frac{1}{2})$ and half integers t (*target edge*) and s (*source edge*) with $0\leq 2t$, $2s\leq \ell$, and put $\mathbb{J}_{t,s}=(t, \frac{1}{2}, \cdots, \frac{1}{2}, s)$.

Consider the systems $\mathrm{E}_{t,s}(\mathbb{J})$ and $\mathrm{B}_{t,s}(\mathbb{J})$ of equations for $V_0^{\vee}(\mathbb{J}_{t,s})$-valued functions $\Psi(\mathbf{z})$ on the manifold M_N:

$$\mathrm{E}_{t,s}(\mathbb{J}):\quad \Big(\kappa\frac{\partial}{\partial z_i}-\sum_{\substack{k=1\\k\neq i}}^{N}\frac{\Omega_{ik}}{z_i-z_k}-\frac{\Omega_{i0}}{z_i}\Big)\Psi(\mathbf{z})=0 \qquad (1\leq i\leq N)$$

and for any $u_k\in V_{j_k}$ $(j_N=t, j_i=\frac{1}{2}\ (1\leq i\leq N-1), j_0=s)$,

$$\mathrm{B}_{t,s}(\mathbb{J}):\ \sum_{\mathbf{m}_0}\binom{L_0}{\mathbf{m}_0}\prod_{k=1}^{N}z_k^{-m_k}\Psi(\mathbf{z})(E^{m_{N+1}}u_{N+1}, E^{m_N}u_N, \cdots, E^{m_1}u_1, u_s(s))=0,$$

$$\sum_{\mathbf{m}_i}\binom{L_i}{\mathbf{m}_i}\prod_{\substack{k=1\\k=i}}^{N}(z_k-z_i)^{-m_k}\Psi(\mathbf{z})(u_{N+1}, E^{m_N}u_N, \cdots, u_{j_i}(j_i), \cdots, E^{m_0}u_0)=0$$

for $1 \leq i \leq N$ ($z_0=0$), and

$$\sum_{\mathfrak{m}_{N+1}} \binom{L_{N+1}}{\mathfrak{m}_{N+1}} \Psi(\mathfrak{z})(u_{j_{N+1}}(j_{N+1}), E^{m_N}u_N, \cdots, E^{m_0}u_0)=0,$$

where $\mathfrak{m}_i=(m_N, \cdots, \hat{m}_i, \cdots, m_0) \in (\mathbf{Z}_{\geq 0})^N$ $(0 \leq i \leq N)$ and $\mathfrak{m}_{N+1}=(m_N, \cdots, m_0) \in (\mathbf{Z}_{\geq 0})^{N+1}$ with $|\mathfrak{m}_i|=L_i=\ell-2j_i+1$ $(0 \leq i \leq N+1)$.

Let $W(N; t, s)$ be the solution space of the joint system $\mathrm{E}_{t,s}(\mathbb{J})$ and $\mathrm{B}_{t,s}(\mathbb{J})$. Then Theorem 3.3 implies that the space $W(N; t, s)$ has a basis $\{\Psi_{\mathfrak{p}}(z_N, \cdots, z_1); \mathfrak{p} \in \mathscr{P}_\ell(N; t, s)\}$ defined as follows: Let

$$\mathscr{P}_\ell(N; t, s)=\Big\{\mathfrak{p}=(p_N, \cdots, p_1, p_0); p_N=t, p_0=s, p_i \in \frac{1}{2}\mathbf{Z}_{\geq 0}, 2p_i \leq \ell,$$
$$|p_i-p_{i-1}|=\frac{1}{2} \ (1 \leq i \leq N)\Big\}.$$

For each $\mathfrak{p} \in \mathscr{P}_\ell(N; t, s)$, define the $V_0^{\vee}(\mathbb{J}_{t,s})$-valued, multi-valued holomorphic function $\Psi_{\mathfrak{p}}(z_N, \cdots, z_1)$ on M_N by

$$\langle \nu(v)|\Psi_{\mathfrak{p}}(z_N, \cdots, z_1)(u_N, \cdots, u_1)|w\rangle=\langle \nu(v)|\Phi_{\mathrm{v}_N}(u_N; z_N)\cdots\Phi_{\mathrm{v}_1}(u_1; z_1)|w\rangle$$

for $v \in V_t$, $u_i \in V_{1/2}$ $(1 \leq i \leq N)$ and $w \in V_s$, where $\mathrm{v}_i(\mathfrak{p})=\begin{pmatrix} & \frac{1}{2} & \\ p_i & & p_{i-1}\end{pmatrix}$ $(1 \leq i \leq N)$.

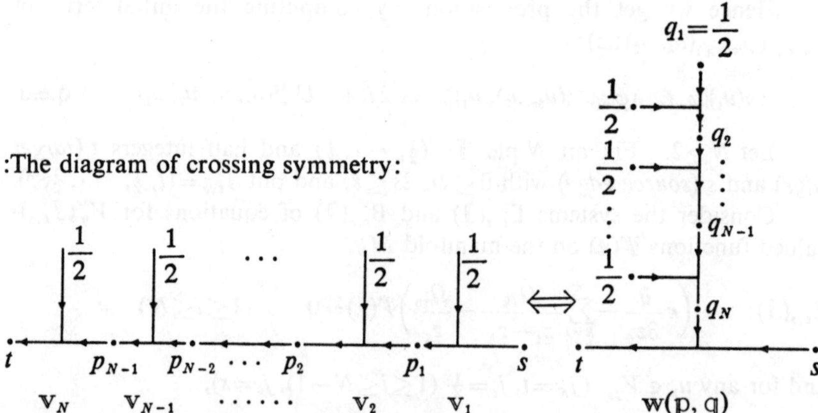

Now introduce the set $\mathscr{Q}_\ell(N)$ defined by

$$\mathscr{Q}_\ell(N)=\Big\{\mathfrak{q}=(q_N, \cdots, q_1); q_i \in \frac{1}{2}\mathbf{Z}_{\geq 0}, q_1=\frac{1}{2}, 2q_i \leq \ell,$$

$$|q_i - q_{i-1}| = \frac{1}{2} \ (2 \leq i \leq N)\Big\}$$

For each $\mathfrak{p} \in \mathscr{P}_t(N; t, s)$, $\mathfrak{q} \in \mathscr{Q}_t(N)$ and i $(2 \leq i \leq N)$, define the quadruples $\mathbb{Q}_i(\mathfrak{p}, \mathfrak{q}) = (p_i, \frac{1}{2}, q_{i-1}, s)$, these quadruples $\mathbb{Q}_i(\mathfrak{p}, \mathfrak{q})$ satisfy one of the conditions $(\mathrm{D2})_{1,2}$, $(\mathrm{D1})_{1,2,3}$ and (D0). Moreover, define numbers $\gamma_i(\mathfrak{p}, \mathfrak{q})$ and $F_i(\mathfrak{p}, \mathfrak{q})$ as follows: if $\mathbb{Q}_i(\mathfrak{p}, \mathfrak{q})$ satisfies the condition (D1), let

$$\gamma_i(\mathfrak{p}, \mathfrak{q}) = \gamma^{(1)}(\mathbb{Q}_i(\mathfrak{p}, \mathfrak{q})) \quad \text{and} \quad F_i(\mathfrak{p}, \mathfrak{q}) = 1.$$

If $\mathbb{Q}_i(\mathfrak{p}, \mathfrak{q})$ satisfies the condition $(\mathrm{D2})_1$, let

$$\gamma_i(\mathfrak{p}, \mathfrak{q}) = \gamma^{(1)}_{-}(\mathbb{Q}_i(\mathfrak{p}, \mathfrak{q})) \quad \text{and} \quad F_i(\mathfrak{p}, \mathfrak{q}) = F^{-}_{-}(\mathbb{Q}_i(\mathfrak{p}, \mathfrak{q})).$$

Assume that $\mathbb{Q}_i(\mathfrak{p}, \mathfrak{q})$ satisfies the condition $(\mathrm{D2})_2$. If $p_{i-1} - p_i = \pm\frac{1}{2}$, then put $k = \pm$, and if $q_i - q_{i-1} = \pm\frac{1}{2}$, then put $\bar{k} = \pm$. Let

$$\gamma_i(\mathfrak{p}, \mathfrak{q}) = \gamma^{(1)}_{\bar{k}}(\mathbb{Q}_i(\mathfrak{p}, \mathfrak{q})) \quad \text{and} \quad F_i(\mathfrak{p}, \mathfrak{q}) = F^{\bar{k}}_{k}(\mathbb{Q}_i(\mathfrak{p}, \mathfrak{q})).$$

If $\mathbb{Q}_i(\mathfrak{p}, \mathfrak{q})$ satisfies the condition (D0), let $F_i(\mathfrak{p}, \mathfrak{q}) = 0$.

Then we get

Proposition 5.5. *For each* $\mathfrak{p} \in \mathscr{P}_t(N; t, s)$ *and* $\mathfrak{q} \in \mathscr{Q}_t(N; f)$ *such that* $V_0(\mathbb{Q}_i(\mathfrak{p}, \mathfrak{q})) \neq 0$, *i.e.* $\mathbb{Q}_i(\mathfrak{p}, \mathfrak{q}) \in (\mathrm{D2})_{1,2} \cup (\mathrm{D1})_{1,2,3}$,

$$\begin{aligned}(2\pi\sqrt{-1})^{1-N} \int_{C_N} \cdots \int_{C_2} \sum_{i=2}^{N} (z_i - z)^{-\gamma_i(\mathbb{Q}_i(\mathfrak{p},\mathfrak{q}))-1} \\ \Phi_{\mathrm{v}_N(\mathfrak{p})}(u_N; z_N) \cdots \Phi_{\mathrm{v}_1(\mathfrak{p})}(u_1; z) dz_N \cdots dz_2 \\ = \prod_{i=2}^{N} F_i(\mathbb{Q}_i(\mathfrak{p}, \mathfrak{q})) \Phi_{\mathrm{w}}(\varphi_{\mathfrak{q}}(u_N, \cdots, u_2, u_1); z)\end{aligned}$$

for each $u_i \in V_{j_i}$, *where* C_i*'s are contours around* C_{i-1} $(3 \leq i \leq N)$ *such that* 0 *is outside* C_N *and* C_2 *is around* z. *The vertices* w *and* w_i $(2 \leq i \leq N)$ *are defined as*

$$\mathrm{w} = \mathrm{w}(\mathfrak{p}, \mathfrak{q}) = \begin{pmatrix} & f & \\ t & & s \end{pmatrix}, \quad f = q_N, \quad \mathrm{w}_i = \mathrm{w}_i(\mathfrak{q}) = \begin{pmatrix} & \frac{1}{2} & \\ q_i & & q_{i-1} \end{pmatrix},$$

and $\varphi_{\mathfrak{q}}(u_N, \cdots, u_2, u_1) \in V_f$ *is defined by*

$$\varphi_{\mathfrak{q}}(u_N, \cdots, u_2, u_1) = \varphi_{\mathrm{w}_N}(u_N, \varphi_{\mathrm{w}_{N-1}}(u_{N-1}, \cdots, \varphi_{\mathrm{w}_2}(u_2, u_1) \cdots)).$$

Appendix I. Bases of Tensor Products of $\mathfrak{sl}$-modules

Here we use notation on the Lie algebra $\mathfrak{g} = \mathfrak{sl}(2, \mathbb{C})$ and its modules given in Section 1.

Since the vacuum expectation value on $V_j^\dagger \times V_j$ is nondegenerate, we can identify the dual right $\mathfrak{g}$-module $V_j^\vee$ of V_j with $V_j^\dagger$. The basis $\{\varphi_j(m);\ m=j, j-1, \cdots, -j\}$ of $V_j^\dagger$ dual to the basis $\{u_j(m)\}$ is identified with $\{u_j^\dagger(m)\}$ by $\varphi_j(m)(u_j(m'))=\langle u_j^\dagger(m)|u_j(m')\rangle=\delta_{m,m'}$.

The isomorphism ν: $V_j \to V_j^\dagger$ is defined by $\nu(u_j(j))=u_j^\dagger(-j)$ and $\nu(X|v\rangle)=-\nu(|v\rangle)X$ ($|v\rangle \in V_j$, $X \in \mathfrak{g}$). Then

$$\nu(u_j(m))=(-1)^{j-m}u_j^\dagger(-m)=(-1)^{j-m}\varphi_j(-m).$$

Introduce the $\mathbb{C}$-bilinear forms (,) on V_j, $V_j^\dagger$ and $V_j^\vee$ for which the bases $\{u_j(m)\}$, $\{u_j^\dagger(m)\}$ and $\{\varphi_j(m)\}$ are orthonormal, then E and F are mutually adjoint with each other and H is self-adjoint in all cases.

Here we refer to the famous textbook [LL] of L.D. Landau and E.M. Lifshitz.

Now for each vertex $\mathrm{v}=\begin{pmatrix} & j & \\ j_2 & & j_1\end{pmatrix} \in$ (CG), we choose and fix the element φ_{v} of $\mathrm{Hom}_{\mathfrak{g}}(V_j \otimes V_{j_1}, V_{j_2})=(V_{j_2}\otimes V_j^\vee \otimes V_{j_1}^\vee)^{\mathfrak{g}}$ as

$$\varphi_{\mathrm{v}}=\sum_{m_1+m=m_2} C_{m_1 m}^{j_2 m_2} u_{j_2}(m_2)\otimes\varphi_j(m)\otimes\varphi_{j_1}(m_1),$$

where the *Clebsch-Gordan coefficients* $C_{m_1m_2}^{j_3m_3}$ are real numbers and expressed by the well-known *Wigner's 3j-symbols* $\begin{pmatrix} j_1 & j_2 & j_3\\ m_1 & m_2 & m_3\end{pmatrix}$ as

$$C_{m_1m_2}^{j_3m_3}=(-1)^{j_1-j_2+m_3}\sqrt{2j_3+1}\begin{pmatrix} j_1 & j_2 & j_3\\ m_1 & m_2 & -m_3\end{pmatrix}.$$

Wigner's 3j-symbols are defined for half integers $j_i \geq 0$ with $j_3+j_2+j_1 \in \mathbb{Z}$, $|j_2-j_1| \leq j_3 \leq j_2+j_1$, $j_i-m_i \in \mathbb{Z}$, and satisfy the following:

i) If $|j_i|<m_i$ for some i, or $m_3+m_2+m_1 \neq 0$, then $\begin{pmatrix} j_1 & j_2 & j_3\\ m_1 & m_2 & m_3\end{pmatrix}=0.$

ii) $$\begin{pmatrix} j_2 & j_1 & j_3\\ m_2 & m_1 & m_3\end{pmatrix}=\begin{pmatrix} j_1 & j_3 & j_2\\ m_1 & m_3 & m_2\end{pmatrix}=\begin{pmatrix} j_1 & j_2 & j_3\\ -m_1 & -m_2 & -m_3\end{pmatrix}$$
$$=(-1)^{j_1+j_2+j_3}\begin{pmatrix} j_1 & j_2 & j_3\\ m_1 & m_2 & m_3\end{pmatrix},$$

iii) $$\begin{pmatrix} j & j & 0\\ m & -m & 0\end{pmatrix}=(-1)^{j-m}(2j+1)^{-1/2}.$$

In particular, if $j_1=0$, then $j_2=j$ and

$$\varphi_{\begin{pmatrix} & j & \\ j & & 0\end{pmatrix}}=\sum_{m=-j}^{j} C_{0\ m}^{j\ m} u_j(m)\otimes\varphi_j(m)\otimes\varphi_0(0)=\sum_{m=-j}^{j} u_j(m)\otimes\varphi_j(m)=\mathrm{id}_{V_j}.$$

If $j_2=0$, then $j_1=j$ and

$$\varphi_{\binom{\ j}{0\ j}}=\sum_{m=-j}^{j} C^{0\ \ 0}_{m\ -m}u_0(0)\otimes\varphi_j(-m)\otimes\varphi_j(m)=\sum_{m=-j}^{j}(-1)^{j-m}\varphi_j(-m)\otimes\varphi_j(m)$$

is identified with the isomorphism $\nu\colon V_j\to V_j^\vee$ which is given by $\nu(u_j(m))=(-1)^{j-m}\varphi_j(-m)$.

For a quadruple $\mathbb{J}=(j_4, j_3, j_2, j_1)$ of half integers $j_i\geq 0$, there are three orthonomal bases $\{U^{(0)}_{j_{12}}; j_{12}\in I(\mathbb{J})\}$, $\{U^{(1)}_{j_{23}}; j_{23}\in I(j_4, j_1, j_3, j_2)\}$ and $\{U^{(\infty)}_{j_{13}}; j_{13}\in I(j_4, j_2, j_3, j_1)\}$ of $V_0^\vee(\mathbb{J})$ defined by

$$U^{(0)}_{j_{12}}=\frac{1}{\sqrt{2j_4+1}}\sum_{\substack{m_1+m_2=m_{12}\\ m_3+m_{12}=m_4}}(-1)^{j_4-m_4}C^{j_4m_4}_{m_{12}m_3}C^{j_{12}m_{12}}_{m_1m_2}$$
$$\varphi_{j_4}(-m_4)\otimes\varphi_{j_3}(m_3)\otimes\varphi_{j_2}(m_2)\otimes\varphi_{j_1}(m_1),$$

$$U^{(1)}_{j_{23}}=\frac{1}{\sqrt{2j_4+1}}\sum_{\substack{m_2+m_3=m_{23}\\ m_3+m_{23}=m_4}}(-1)^{j_4-m_4}C^{j_4m_4}_{m_1m_{23}}C^{j_{23}m_{23}}_{m_2m_3}$$
$$\varphi_{j_4}(-m_4)\otimes\varphi_{j_3}(m_3)\otimes\varphi_{j_2}(m_2)\otimes\varphi_{j_1}(m_1),$$

and

$$U^{(\infty)}_{j_{13}}=\frac{1}{\sqrt{2j_4+1}}\sum_{\substack{m_1+m_3=m_{13}\\ m_2+m_{13}=m_4}}(-1)^{j_4-m_4}C^{j_4m_4}_{m_{13}m_2}C^{j_{13}m_{13}}_{m_1m_3}$$
$$\varphi_{j_4}(-m_4)\otimes\varphi_{j_3}(m_3)\otimes\varphi_{j_2}(m_2)\otimes\varphi_{j_1}(m_1),$$

then the operator $\Omega_{12}=\frac{1}{2}[\Delta_{12}(\Omega)-\Omega_{11}-\Omega_{22}]$ is diagonalized by this basis $\{U^{(0)}_{j_{12}}; j_{12}\in I(\mathbb{J})\}$ as

$$\Omega_{12}U^{(0)}_{j_{12}}=\kappa(\Delta_{j_{12}}-\Delta_{j_2}-\Delta_{j_1})U^{(0)}_{j_{12}}.$$

The operators Ω_{23} and Ω_{13} are also diagonalized by these bases $\{U^{(1)}_{j_{23}}\}$ and $\{U^{(\infty)}_{j_{13}}\}$ respectively as

$$\Omega_{23}U^{(1)}_{j_{23}}=\kappa(\Delta_{j_{23}}-\Delta_{j_2}-\Delta_{j_3})U^{(1)}_{j_{23}}\quad\text{and}\quad\Omega_{13}U^{(\infty)}_{j_{13}}=\kappa(\Delta_{j_{13}}-\Delta_{j_1}-\Delta_{j_3})U^{(\infty)}_{j_{13}}.$$

Moreover the basis vectors $U^{(0)}_{j_{12}}$ are expressed by the fixed φ_v's as

$$U^{(0)}_{j_{12}}(u_4, u_3, u_2, u_1)=\frac{1}{\sqrt{2j_4+1}}\langle\nu(u_4)|\varphi_{v_2(j_{12})}(u_3)\varphi_{v_1(j_{12})}(u_2)|u_1\rangle$$

for any $u_i\in V_{j_i}$.

The transformation matrices $S^{(1,0)}=(S^{j_{23}}_{j_{12}})$, $S^{(\infty,0)}=(S^{j_{13}}_{j_{12}})$, $S^{(\infty,1)}=(S^{j_{13}}_{j_{23}})$ between three bases of $V_0^\vee(\mathbb{J})$ defined by

$$U^{(0)}_{j_{12}}=\sum_{j_{23}}U^{(1)}_{j_{23}}S^{j_{23}}_{j_{12}},\quad U^{(0)}_{j_{12}}=\sum_{j_{13}}U^{(\infty)}_{j_{13}}S^{j_{13}}_{j_{12}},\quad U^{(1)}_{j_{23}}=\sum_{j_{13}}U^{(\infty)}_{j_{13}}S^{j_{13}}_{j_{23}}$$

are real orthogonal matrices and are given as

$$S_{j_{12}}^{j_{23}}=(-1)^{j_1+j_2+j_3+j_4}\sqrt{(2j_{12}+1)(2j_{23}+1)}\begin{Bmatrix} j_1 j_2 j_{12} \\ j_3 j_4 j_{23} \end{Bmatrix},$$

$$S_{j_{12}}^{j_{13}}=(-1)^{2j_1-j_2+j_3+j_{13}-j_{12}}\sqrt{(2j_{12}+1)(2j_{13}+1)}\begin{Bmatrix} j_2 j_1 j_{12} \\ j_3 j_4 j_{13} \end{Bmatrix},$$

and

$$S_{j_{23}}^{j_{13}}=(-1)^{-j_1+2j_2+2j_3-j_4+j_{23}}\sqrt{(2j_{23}+1)(2j_{13}+1)}\begin{Bmatrix} j_2 j_3 j_{23} \\ j_1 j_4 j_{13} \end{Bmatrix},$$

where $\begin{Bmatrix} j_1 j_2 j_3 \\ j_4 j_5 j_6 \end{Bmatrix}$ is the *6j-symbol* or *Racah coefficient* which is defined by 3*j*-symbols as

$$\begin{Bmatrix} j_1 j_2 j_3 \\ j_4 j_5 j_6 \end{Bmatrix}=\sum_{\mathbf{m}}(-1)^{\Sigma_i(j_i-m_i)}\begin{pmatrix} j_1 & j_2 & j_3 \\ -m_1 & -m_2 & -m_3 \end{pmatrix}\begin{pmatrix} j_1 & j_5 & j_6 \\ m_1 & -m_5 & m_6 \end{pmatrix}$$
$$\times\begin{pmatrix} j_4 & j_2 & j_6 \\ m_4 & m_2 & -m_6 \end{pmatrix}\begin{pmatrix} j_4 & j_5 & j_3 \\ -m_4 & m_5 & -m_3 \end{pmatrix}.$$

In the case (D2) for a quadruple $\mathbb{J}=(j_4,\frac{1}{2},j_2,j_1)$, then $I(\mathbb{J})=\{j_4\pm\frac{1}{2}\}$, $I(j_4,j_1,j_2,\frac{1}{2})=\{j_2\pm\frac{1}{2}\}$ and $I(j_4,j_2,\frac{1}{2},j_1)=\{j_1\pm\frac{1}{2}\}$. Denote

$$U_{\pm}^{(0)}=U_{j_4\pm1/2}^{(0)},\quad U_{\pm}^{(1)}=U_{j_2\pm1/2}^{(1)}\quad\text{and}\quad U_{\pm}^{(\infty)}=U_{j_1\pm1/2}^{(\infty)},$$

then we get easily the formulae in Section 4.2, by using some values of 6*j*-symbols (see [LL] Section 108):

$$\begin{Bmatrix} 0 & b & b \\ a & c & c \end{Bmatrix}=\begin{Bmatrix} a & b & c \\ 0 & c & b \end{Bmatrix}=\frac{(-1)^{a+b+c}}{\sqrt{(2b+1)(2c+1)}}.$$

Let $s=a+b+c+\frac{1}{2}$, then

$$\begin{Bmatrix} c & \frac{1}{2} & c+\frac{1}{2} \\ b & a & b+\frac{1}{2} \end{Bmatrix}=\begin{Bmatrix} \frac{1}{2} & c & c+\frac{1}{2} \\ a & b & b+\frac{1}{2} \end{Bmatrix}=(-1)^s\left(\frac{(s-2b)(s-2c)}{(2b+1)(2b+2)(2c+1)(2c+2)}\right)^{1/2},$$

$$\begin{Bmatrix} c & \frac{1}{2} & c+\frac{1}{2} \\ b & a & b-\frac{1}{2} \end{Bmatrix}=\begin{Bmatrix} b & \frac{1}{2} & b-\frac{1}{2} \\ c & a & c+\frac{1}{2} \end{Bmatrix}=(-1)^s\left(\frac{(s+1)(s-2a)}{2b(2b+1)(2c+1)(2c+2)}\right)^{1/2},$$

and

$$\begin{Bmatrix} c & \frac{1}{2} & c-\frac{1}{2} \\ b & a & b-\frac{1}{2} \end{Bmatrix}=-(-1)^s\left(\frac{(s-2b)(s-2c)}{2b(2b+1)2c(2c+1)}\right)^{1/2}.$$

Appendix II. Connection Matrices of Reduced Equation

A.II.1) Solutions of reduced equation

For a quadruple $\mathbb{J}=(j_4, \frac{1}{2}, j_2, j_1)$ of half integers, we will give fundamental solution of the reduced equation

$$\mathrm{RE}(\mathbb{J}): \qquad \left(\kappa\frac{d}{d\zeta}-\frac{\Omega_{12}+\kappa\Delta_4(\mathbb{J})}{\zeta}-\frac{\Omega_{23}}{\zeta-1}\right)\Psi(\zeta)=0$$

for $V_0^{\vee}(\mathbb{J})$-valued functions $\Psi(\zeta)$ on $\zeta\in\mathbf{C}^*$. The coordinate change $\zeta\mapsto\eta=1/\zeta$ makes the equation RE($\mathbb{J}$) into

$$\mathrm{RE}(\mathbb{J})_\infty: \qquad \left(\kappa\frac{d}{d\eta}-\frac{\Omega_{13}}{\eta}-\frac{\Omega_{23}}{\eta-1}\right)\Psi\left(\frac{1}{\eta}\right)=0.$$

In this section we deal only with the case (D2) and prove Proposition 4.4, since the case (D1) is much simpler.

Write a solution $\Psi(\zeta)$ as

$$\Psi(\zeta)=(U_+^{(i)}, U_-^{(i)})\Psi^{(i)}(\zeta)=(U_+^{(i)}, U_-^{(i)})\begin{pmatrix}\varphi_+^{(i)}(\zeta)\\ \varphi_-^{(i)}(\zeta)\end{pmatrix} \qquad (i=0, 1)$$

$$=(U_+^{(\infty)}, U_-^{(\infty)})\Psi^{(\infty)}(\zeta^{-1})=(U_+^{(\infty)}, U_-^{(\infty)})\begin{pmatrix}\varphi_+^{(\infty)}(\zeta^{-1})\\ \varphi_-^{(\infty)}(\zeta^{-1})\end{pmatrix}$$

where $\{U_\pm^{(i)}; i=0, 1, \infty\}$ are three bases $\{U_\pm^{(i)}; i=0, 1, \infty\}$ of $V_0^{\vee}(\mathbb{J})$ such that

$$\Omega_{12}U_\pm^{(0)}=\kappa(\gamma_\pm^{(0)}-\Delta_4(\mathbb{J}))U_\pm^{(0)}, \quad \Omega_{23}U_\pm^{(1)}=\kappa\gamma_\pm^{(1)}U_\pm^{(1)}, \quad \Omega_{13}U_\pm^{(\infty)}=\kappa\gamma_\pm^{(\infty)}U_\pm^{(\infty)},$$

and the exponents $\gamma_\pm^{(i)}$ are given in Section 4.2. The differences $\gamma^{(i)}=\gamma_+^{(i)}-\gamma_-^{(i)}$ of exponents are given as

$$\gamma^{(0)}=\frac{2j_4+1}{\kappa}, \quad \gamma^{(1)}=\frac{2j_2+1}{\kappa} \quad \text{and} \quad \gamma^{(\infty)}=\frac{2j_1+1}{\kappa} \quad (\kappa=\ell+2).$$

Since the transformation matrices $S^{(i,k)}$ between the bases $\{U_\pm^{(i)}\}$ and $\{U_\pm^{(k)}\}$ are given in Section 4.2, we get the matrix forms of the equations RE($\mathbb{J}$) and RE($\mathbb{J}$)$_\infty$:

$$\mathrm{RE}(\mathbb{J})_i: \quad \frac{d}{d\zeta}\Psi^{(i)}(\zeta)=A^i(\zeta)\Psi^{(i)}(\zeta)=\begin{pmatrix}a_{++}^{(i)}(\zeta) & a_{-+}^{(i)}(\zeta)\\ a_{+-}^{(i)}(\zeta) & a_{--}^{(i)}(\zeta)\end{pmatrix}\Psi^{(i)}(\zeta) \qquad (i=0, 1)$$

and

$$\mathrm{RE}(\mathbb{J})_\infty: \quad \frac{d}{d\zeta}\Psi^{(\infty)}(\eta)=A^\infty(\eta)\Psi^{(\infty)}(\zeta)=\begin{pmatrix} a_{++}^{(\infty)}(\eta) & a_{-+}^{(\infty)}(\eta) \\ a_{+-}^{(\infty)}(\eta) & a_{--}^{(\infty)}(\eta) \end{pmatrix}\Psi^{(\infty)}(\eta),$$

where the coefficient matrices A^i are given as

$$a_{++}^0(\zeta)=\frac{\gamma_+^{(0)}}{\zeta}+\frac{\gamma_+^{(1)}-a^0}{\zeta-1}, \qquad a_{+-}^0(\zeta)=a_{-+}^0(\zeta)=\frac{b^0}{\zeta-1},$$

$$a_{--}^0(\zeta)=\frac{\gamma_-^{(0)}}{\zeta}+\frac{\gamma_-^{(1)}+a^0}{\zeta-1}, \qquad a_{++}^1(\zeta)=\frac{\gamma_+^{(0)}-a^1}{\zeta}+\frac{\gamma_+^{(1)}}{\zeta-1},$$

$$a_{+-}^1(\zeta)=a_{-+}^1(\zeta)=\frac{b^1}{\zeta}, \qquad a_{--}^1(\zeta)=\frac{\gamma_-^{(0)}+a^1}{\zeta}+\frac{\gamma_-^{(1)}}{\zeta-1},$$

$$a_{++}^\infty(\eta)=\frac{\gamma_+^{(\infty)}}{\eta}+\frac{\gamma_+^{(1)}-a^\infty}{\eta-1}, \qquad a_{+-}^\infty(\eta)=a_{-+}^\infty(\eta)=\frac{b^\infty}{\eta-1},$$

$$a_{--}^\infty(\eta)=\frac{\gamma_-^{(\infty)}}{\eta}+\frac{\gamma_-^{(1)}+a^\infty}{\eta-1},$$

and

$$a^0=\frac{\varepsilon_0\varepsilon_1}{\gamma^{(0)}}, \qquad a^1=\frac{\varepsilon_0\varepsilon_1}{\gamma^{(1)}}, \qquad a^\infty=\frac{\varepsilon_0\varepsilon_4}{\gamma^{(\infty)}},$$

$$b^0=\frac{\sqrt{\varepsilon}}{\gamma^{(0)}}, \qquad b^1=\frac{\sqrt{\varepsilon}}{\gamma^{(1)}}, \qquad b^\infty=\frac{\sqrt{\varepsilon}}{\gamma^{(\infty)}} \qquad (\varepsilon=\varepsilon_0\varepsilon_1\varepsilon_2\varepsilon_4).$$

Now look at the function $\varphi_+^{(0)}(\zeta)$. The equation $\mathrm{RE}(\mathbb{J})_0$ turns into the equation for $\varphi_+^{(0)}(\zeta)$:

$$\frac{d^2}{d\zeta^2}\varphi_+^{(0)}(\zeta)=\left(\frac{\gamma_+^{(0)}+\gamma_-^{(0)}}{\zeta}+\frac{\gamma_+^{(1)}+\gamma_-^{(1)}-1}{\zeta-1}\right)\frac{d}{d\zeta}\varphi_+^{(0)}(\zeta)$$
$$-\left\{\frac{\gamma_+^{(0)}(1+\gamma_-^{(0)})}{\zeta^2}+\frac{\gamma_+^{(0)}(\gamma_-^{(1)}+a^0-1)+\gamma_-^{(0)}(\gamma_+^{(1)}-a^0)}{\zeta(\zeta-1)}+\frac{\gamma_+^{(1)}\gamma_-^{(1)}}{(\zeta-1)^2}\right\}\varphi_+^{(0)}(\zeta),$$

which is a second-order equation of Fuchsian type.

Now recall that a second-order equation of Fuchsian type is of the form

$$\frac{d^2\varphi}{d\zeta^2}(\zeta)=\left(\frac{\lambda+\lambda'-1}{\zeta}+\frac{\mu+\mu'-1}{\zeta-1}\right)\frac{d\varphi}{d\zeta}(\zeta)$$
$$-\left\{\frac{\lambda\lambda'}{\zeta^2}+\frac{\nu\nu'-\lambda\lambda'-\mu\mu'}{\zeta(\zeta-1)}+\frac{\mu\mu'}{(\zeta-1)^2}\right\}\varphi(\zeta),$$

where $\lambda, \lambda'; \mu, \mu'; \nu, \nu'$ are exponents at $\zeta=0; 1; \infty$ respectively. The solution space of this equation is denoted by the Riemann P-function

$$P\begin{Bmatrix} 0 & 1 & \infty & \\ \lambda & \mu & \nu & z \\ \lambda' & \mu' & \nu' & \end{Bmatrix}.$$

The equations $\mathrm{RE}(\mathbb{J})_i$ for other functions are also reduced to Fuchsian equations of similar forms. Then we get

Proposition A.1.

$$\varphi_+^{(0)}(\zeta) \in P\begin{Bmatrix} 0 & 1 & \infty & \\ \gamma_+^{(0)} & \gamma_+^{(1)} & \gamma_+^{(\infty)} & \zeta \\ 1+\gamma_-^{(0)} & \gamma_-^{(1)} & \gamma_-^{(\infty)} & \end{Bmatrix}, \quad \varphi_-^{(0)}(\zeta) \in P\begin{Bmatrix} 0 & 1 & \infty & \\ \gamma_-^{(0)} & \gamma_-^{(1)} & \gamma_-^{(\infty)} & \zeta \\ 1+\gamma_+^{(0)} & \gamma_+^{(1)} & \gamma_+^{(\infty)} & \end{Bmatrix},$$

$$\varphi_+^{(1)}(\zeta) \in P\begin{Bmatrix} 0 & 1 & \infty & \\ \gamma_+^{(0)} & \gamma_+^{(1)} & \gamma_+^{(\infty)} & \zeta \\ \gamma_-^{(0)} & 1+\gamma_-^{(1)} & \gamma_-^{(\infty)} & \end{Bmatrix}, \quad \varphi_-^{(1)}(\zeta) \in P\begin{Bmatrix} 0 & 1 & \infty & \\ \gamma_-^{(0)} & \gamma_-^{(1)} & \gamma_-^{(\infty)} & \zeta \\ \gamma_+^{(0)} & 1+\gamma_+^{(1)} & \gamma_+^{(\infty)} & \end{Bmatrix},$$

and

$$\varphi_+^{(\infty)}(\eta) \in P\begin{Bmatrix} 0 & 1 & \infty & \\ \gamma_+^{(0)} & \gamma_+^{(1)} & \gamma_+^{(\infty)} & \eta \\ \gamma_-^{(0)} & \gamma_-^{(1)} & 1+\gamma_-^{(\infty)} & \end{Bmatrix}, \quad \varphi_-^{(\infty)}(\eta) \in P\begin{Bmatrix} 0 & 1 & \infty & \\ \gamma_-^{(0)} & \gamma_-^{(1)} & \gamma_-^{(\infty)} & \eta \\ \gamma_+^{(0)} & \gamma_+^{(1)} & 1+\gamma_+^{(\infty)} & \end{Bmatrix}.$$

Before we give the proof of Proposition 4.4 we recall the facts on the hypergeometric function $F(\alpha, \beta, \gamma; \zeta)$ (see e.g. [E]):

$$F(\alpha, \beta, \gamma; \zeta)=1+\sum_{n=1} \frac{\alpha_{(n)}\beta_{(n)}}{n!\,\gamma_{(n)}}\zeta^n=1+\frac{\Gamma(\gamma)}{\Gamma(\alpha)\Gamma(\beta)}\sum_{n=1}^{\infty}\frac{\Gamma(\alpha+n)\Gamma(\beta+n)}{\Gamma(\gamma+n)}\frac{\zeta^n}{n!},$$

where

$$\alpha_{(n)}=\alpha(\alpha+1)\cdots(\alpha+n-1)=\frac{\Gamma(\alpha+n)}{\Gamma(\alpha)}.$$

i) If $\gamma \notin \mathbb{Z}_{\leq 0}$, $F(\alpha, \beta, \gamma; \zeta)$ is a solution of the Gaussian equation:

$$\zeta(1-\zeta)\varphi''(\zeta)+\{\gamma-(\alpha+\beta+1)\zeta\}\varphi'(\zeta)-\alpha\beta\varphi(\zeta)=0,$$

that is,

$$F(\alpha, \beta, \gamma; \zeta) \in P\begin{Bmatrix} 0 & 1 & \infty & \\ 0 & 0 & \alpha & \zeta \\ 1-\gamma & \gamma-\alpha-\beta & \beta & \end{Bmatrix}.$$

ii) $F(\alpha, \beta, \gamma; \zeta)$ and $\zeta^{1-\gamma}F(\alpha-\gamma+1, \beta-\gamma+1, 2-\gamma; \zeta)$ give a basis of the solution space $P\begin{Bmatrix} 0 & 1 & \infty & \\ 0 & 0 & \alpha & \zeta \\ 1-\gamma & \gamma-\alpha-\beta & \beta & \end{Bmatrix}$ of the Gaussian equation.

iii) $F(\alpha, \beta, \gamma; \zeta)=(1-\zeta)^{\gamma-\alpha-\beta}F(\gamma-\alpha, \gamma-\beta, \gamma; \zeta)$.

iv) $(d/d\zeta)F(\alpha, \beta, \gamma; \zeta)=(\alpha\beta/\gamma)F(\alpha+1, \beta+1, \gamma+1; \zeta)$.

v) $$F(\alpha, \beta, \gamma; \zeta)=(1-\zeta)F(\alpha+1, \beta+1, \gamma+1; \zeta)+\frac{(\alpha-\gamma)(\beta-\gamma)}{\gamma(\gamma+1)}\zeta F(\alpha+1, \beta+1, \gamma+2; \zeta).$$

vi) $$(\gamma+1)F(\alpha, \beta, \gamma; \zeta)=\{(\gamma+1)-(\alpha+\beta+1-(\alpha\beta/\gamma))\zeta\}F(\alpha+1, \beta+1, \gamma+2; \zeta)+\frac{(\alpha+1)(\beta+1)}{\gamma+2}\zeta(1-\zeta)F(\alpha+2, \beta+2, \gamma+3; \zeta).$$

vii) $$(1-\zeta)F(\alpha, \beta, \gamma; \zeta)=F(\alpha-1, \beta-1, \gamma; \zeta)+\frac{\alpha+\beta-\gamma-1}{\gamma}\zeta F(\alpha, \beta, \gamma+1; \zeta).$$

viii) $$F(\alpha, \beta, \gamma+1; \zeta)=\frac{\alpha\beta}{(\alpha-\gamma)(\beta-\gamma)}(1-\zeta)F(\alpha+1, \beta+1, \gamma+1; \zeta)+\frac{\gamma(\gamma-\alpha-\beta)}{(\alpha-\gamma)(\beta-\gamma)}F(\alpha, \beta, \gamma; \zeta).$$

ix) $$\frac{\alpha(\gamma-\beta)}{\beta(\gamma-\alpha)}F(\alpha+1, \beta, \gamma+1; \zeta)=F(\alpha, \beta+1, \gamma+1; \zeta)+\frac{\gamma(\alpha-\beta)}{\beta(\gamma-\alpha)}F(\alpha, \beta, \gamma; \zeta).$$

x) $$F(\alpha, \beta+1, \gamma+1; \zeta)=F(\alpha+1, \beta, \gamma+1\ \zeta)+\frac{\alpha-\beta}{1+\gamma}\zeta F(1+\alpha, 1+\beta, 2+\gamma; \zeta).$$

xi) $$F(\alpha, \beta, \gamma; \zeta)=\frac{\Gamma(\gamma)\Gamma(\alpha+\beta-\gamma)}{\Gamma(\alpha)\Gamma(\beta)}(1-\zeta)^{\gamma-\alpha-\beta}F(\gamma-\alpha, \gamma-\beta, \gamma-\alpha-\beta+1; 1-\zeta)+\frac{\Gamma(\gamma)\Gamma(\gamma-\alpha-\beta)}{\Gamma(\gamma-\alpha)\Gamma(\gamma-\beta)}F(\alpha, \beta, \alpha+\beta-\gamma+1; 1-\zeta).$$

xii) $$F(\alpha, \beta, \gamma; \zeta)=\frac{\Gamma(\gamma)\Gamma(\beta-\alpha)}{\Gamma(\beta)\Gamma(\gamma-\alpha)}(-\zeta)^{-\alpha}F(\alpha, \alpha-\gamma+1, \alpha-\beta+1, 1/\zeta)+\frac{\Gamma(\gamma)\Gamma(\alpha-\beta)}{\Gamma(\alpha)\Gamma(\gamma-\beta)}(-\zeta)^{-\beta}(1-(1/\zeta))^{\gamma-\alpha-\beta}\times F(1-\alpha, \gamma-\alpha, \beta-\alpha+1; 1/\zeta).$$

Proof of Proposition 4.4. Similarly as Theorem 3.3, we get the functions $\varphi_{\pm\pm}^{(0)}(\zeta)$ in a neighbourhood of $\zeta=0$ such that

$$\Psi_+^{(0)}(\zeta)=(U_+^{(0)}, U_-^{(0)})\begin{pmatrix}\varphi_{++}^{(0)}(\zeta)\\ \varphi_{+-}^{(0)}(\zeta)\end{pmatrix}; \quad \Psi_-^{(0)}(\zeta)=(U_+^{(0)}, U_-^{(0)})\begin{pmatrix}\varphi_{-+}^{(0)}(\zeta)\\ \varphi_{--}^{(0)}(\zeta)\end{pmatrix}$$

such that $\varphi_{\pm}^{(0)}(\zeta)$ have the expansion with respect to ζ as

$$\varphi_{++}^{(0)}(\zeta)=\zeta^{\gamma_+^{(0)}}(1+\cdots), \qquad \varphi_{+-}^{(0)}(\zeta)=\zeta^{\gamma_+^{(0)}}(c\zeta+\cdots),$$

and

$$\varphi_{-+}^{(0)}(\zeta)=\zeta^{\gamma_-^{(0)}}(d\zeta+\cdots), \quad \varphi_{--}^{(0)}(\zeta)=\zeta^{\gamma_-^{(0)}}(1+\cdots),$$

where c and d are some constants.

Then by Proposition A.1,

$$\varphi_{++}^{(0)}(\zeta), \varphi_{-+}^{(0)}(\zeta) \in P\begin{Bmatrix} 0 & 1 & \infty & \\ \gamma_+^{(0)} & \gamma_+^{(1)} & \gamma_+^{(\infty)} & \zeta \\ 1+\gamma_-^{(0)} & \gamma_-^{(1)} & \gamma_-^{(\infty)} & \end{Bmatrix},$$

and

$$\varphi_{+-}^{(0)}(\zeta), \varphi_{--}^{(0)}(\zeta) \in P\begin{Bmatrix} 0 & 1 & \infty & \\ \gamma_-^{(0)} & \gamma_+^{(1)} & \gamma_+^{(\infty)} & \zeta \\ 1+\gamma_+^{(0)} & \gamma_-^{(1)} & \gamma_-^{(\infty)} & \end{Bmatrix}.$$

Hence

$$\varphi_{++}^{(0)}(\zeta) \in \zeta^{\gamma_+^{(0)}}(1-\zeta)^{\gamma_+^{(1)}} P\begin{Bmatrix} 0 & 1 & \infty & \\ 0 & 0 & \alpha & \zeta \\ 1-\gamma & \gamma-\alpha-\beta & \beta & \end{Bmatrix},$$

$$\varphi_{+-}^{(0)}(\zeta) \in \zeta^{1+\gamma_+^{(0)}}(1-\zeta)^{\gamma_+^{(1)}} P\begin{Bmatrix} 0 & 1 & \infty & \\ 0 & 0 & \alpha+1 & \zeta \\ -1-\gamma & \gamma-\alpha-\beta & \beta+1 & \end{Bmatrix},$$

$$\varphi_{-+}^{(0)}(\zeta) \in \zeta^{1+\gamma_+^{(0)}}(1-\zeta)^{\gamma_-^{(1)}} P\begin{Bmatrix} 0 & 1 & \infty & \\ 0 & 0 & 1-\alpha & \zeta \\ \gamma-1 & \alpha+\beta-\gamma & 1-\beta & \end{Bmatrix},$$

and

$$\varphi_{--}^{(0)}(\zeta) \in \zeta^{\gamma_-^{(0)}}(1-\zeta)^{\gamma_-^{(1)}} P\begin{Bmatrix} 0 & 1 & \infty & \\ 0 & 0 & -\alpha & \zeta \\ 1+\gamma & \alpha+\beta-\gamma & -\beta & \end{Bmatrix},$$

where α, β and $\gamma=\gamma^{(0)}$ are given in the proposition. Then by the formulae iv) ~ vi) above, we get the statement (i) of Proposition 4.4. Other statements of Proposition 4.4 are similarly obtained.

A.II.2) Connection matrices of reduced equation

We must prove Proposition 4.5 on the connection matrix of the fundamental solutions of the reduced equation RE($\mathbb{J}$) along the path from 0 to ∞ figured in Section 4.3. Fortunately the formula xii) of the hypergeometric function gives its connection matrix along the same path. And we may take $(-\zeta)^{\lambda}=\exp(-\lambda\pi\sqrt{-1})\zeta^{\lambda}$ by the choice of the path. Then it is sufficient for the proof of Proposition 4.5 to note the following relations among constants in Section 4.2:

$$-c_{\pm}^{(i)}\gamma^{(i)}(1\pm\gamma^{(i)})=\sqrt{\varepsilon}=\beta\beta^{(\infty)}\frac{B''}{A''}=\alpha(\gamma^{(\infty)}-\beta^{(\infty)})\frac{A''}{B''}$$

$$=\alpha\beta\frac{A}{B}=(\gamma-\beta)(\alpha-\gamma)\frac{B}{A}$$

for $i=0, 1, \infty$.

Similarly we get the connection matrix of the fundamental solutions of the reduced equation RE($\mathbb{J}$) along the path from 0 to 1 figured below. The formula xi) of the hypergeometric function also gives its connection matrix along the same path.

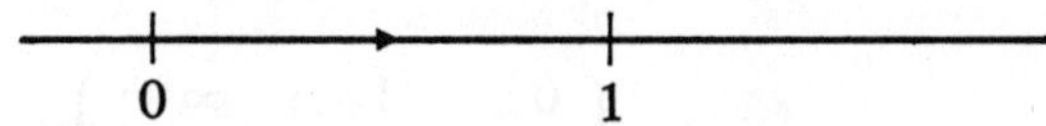

Then by relations above, we get in the case (D2):

Proposition A.2. *Denote by* $F(\mathbb{J})=\begin{pmatrix}F_+^+ & F_-^+\\ F_+^- & F_-^-\end{pmatrix}$ *the connection matrix of the fundamental solutions* $(\Psi_+^{(0)}, \Psi_-^{(0)})$ *at* $\zeta=0$ *to* $(\Psi_+^{(1)}, \Psi_-^{(1)})$ *at* $\zeta=1$ *of the equation* RE($\mathbb{J}$):

$$(\Psi_+^{(0)}, \Psi_-^{(0)})=(\Psi_+^{(1)}, \Psi_-^{(1)})\begin{pmatrix}F_+^+ & F_-^+\\ F_+^- & F_-^-\end{pmatrix}$$

that is,

$$S^{(0,1)}\begin{pmatrix}\varphi_{++}^{(0)}(\zeta) & \varphi_{-+}^{(0)}(\zeta)\\ \varphi_{+-}^{(0)}(\zeta) & \varphi_{--}^{(0)}(\zeta)\end{pmatrix}=\begin{pmatrix}\varphi_{++}^{(1)}(\zeta) & \varphi_{-+}^{(1)}(\zeta)\\ \varphi_{+-}^{(1)}(\zeta) & \varphi_{--}^{(1)}(\zeta)\end{pmatrix}\begin{pmatrix}F_+^+ & F_-^+\\ F_+^- & F_-^-\end{pmatrix}.$$

Then

$$F_{+}^{+}=\left(\frac{\gamma^{(0)}\gamma^{(1)}}{\varepsilon_2\varepsilon_4}\right)^{1/2}\frac{\Gamma(\gamma^{(0)})\Gamma(-\gamma^{(1)})}{\Gamma(\varepsilon_2)\Gamma(-\varepsilon_4)},$$

$$F_{-}^{+}=\left(\frac{\gamma^{(0)}\gamma^{(1)}}{\varepsilon_0\varepsilon_1}\right)^{1/2}\frac{\Gamma(-\gamma^{(0)})\Gamma(-\gamma^{(1)})}{\Gamma(-\varepsilon_0)\Gamma(-\varepsilon_4)},$$

$$F_{+}^{-}=\left(\frac{\gamma^{(0)}\gamma^{(1)}}{\varepsilon_0\varepsilon_1}\right)^{1/2}\frac{\Gamma(\gamma^{(0)})\Gamma(\gamma^{(1)})}{\Gamma(\varepsilon_0)\Gamma(\varepsilon_1)},$$

$$F_{-}^{-}=\left(\frac{\gamma^{(0)}\gamma^{(1)}}{\varepsilon_2\varepsilon_4}\right)^{1/2}\frac{\Gamma(-\gamma^{(0)})\Gamma(\gamma^{(1)})}{\Gamma(-\varepsilon_2)\Gamma(\varepsilon_4)},$$

Remark. Since $\varepsilon_0=1$ in the case $(\mathrm{D2})_1$, $F_{-}^{+}(\mathbb{J})=0$ and $F(\mathbb{J})=F_{-}^{-}(\mathbb{J})=[(2j_2+1)(\kappa-1-2j_2)/(2j_4+1)(\kappa-1-2j_4)]^{1/2}$. By Proposition 4.4′, it is obvious that $F(\mathbb{J})=1$ in the case (D1). In the case (D1), the matrix $F(\mathbb{J})$ in the proposition A.2 is written as $F(\mathbb{J})=\begin{pmatrix}1&0\\0&1\end{pmatrix}$ in the case $(\mathrm{D1})_1$ and $F(\mathbb{J})=\begin{pmatrix}0&1\\1&0\end{pmatrix}$ in the case $(\mathrm{D1})_{2,3}$.

References

[A1] K. Aomoto, Un théorème du type de Matsushima-Murakami concernant l'intégrale des fonctions multiformes. J. Math. pures et appl., **52** (1973), 1–11.

[A2] ——, Gauss-Manin connection of integral of difference products, (1985), submitted to J. Math. Soc. Japan.

[Ba] R. J. Baxter, Exactly Solved Models in Statistical Mechanics, Academic Press, London (1982).

[BD] A. A. Belavin and V. G. Drinfel'd, Solutions of the classical Yang-Baxter Equation for simple Lie algebras, Functional Anal. Appl., **16-3** (1982), 1–29 (in Russian).

[BPZ] A. A. Belavin, A. N. Polyakov and A. B. Zamolodchikov, Infinite conformal symmetries in two-dimensional quantum field theory, Nuclear Physics, **B241** (1984), 333–380.

[Bi] J. S. Birman, Braids, Links, and Mapping Class Groups, Ann. of Math. Stud., Princeton, **82** (1974).

[CL] E. A. Coddington and N. Levinson, Theory of Ordinary Differential Equations, McGraw-Hill (1955).

[DF] Vl. S. Dotsenko and V. A. Fateev, Conformal Algebra and multipoint correlation functions in 2D statistical models, Nuclear Physics, **B240** (1984), 312–348.

[E] A. Eldélyi (editor), Higher Transcendental Functions, vol I, McGraw-Hill (1953).

[Ji] M. Jimbo, A q-analogue of $U(\mathfrak{gl}(N+1))$, Hecke Algebra, and the Yang-Baxter equation, Lett. Math. Phys., **11** (1986), 247–252.

[Jo] V. F. R. Jones, Braid groups, Hecke algebras and type II_1 factors, Proc. of US-Japan Seminar, Kyoto, July 1983, Geometric Methods in Operator Algebras, Pitman Research Notes in Math., **123** (1986).

[Ka] V. G. Kac, Infinite dimensional Lie Algebras, Progr. Math., **44** Birkhäuser (1983).

[KL] D. Kazhdan and G. Lusztig, Represenations of Coxeter groups and Hecke algebras, Invent. math., **53** (1979), 165–184.

[KZ] V. G. Knizhnik and A. B. Zamolodchikov, Current Algebra and Wess-Zumino models in two dimensions, Nuclear Physics, **B247** (1984), 83–103.

[Kn] A. W. Knapp, Representation Theory of Semisimple Groups: An Overview based on Examples, Princeton Mathematical Series **36** (1986).

[KN] Z. Koba and H. B. Nielsen, Generalized Veneziano model from the point of view of manifestly crossing parametrization, Z. Physik, **229** (1969), 243–263.

[Ko] T. Kohno, Linear Representations of Braid Groups and Classical Yang-Baxter Equations, preprint Nagoya University (1986).

[LL] L. D. Landau and E. M. Lifshitz, Quantum Mechanics: Nonrelativistic Theory, 2nd ed., Moscow (1963).

[Se] G. Segal, Unitary representations of some infinite dimensional groups, Comm. Math. Phys., **80** (1981), 301–342.

[Su] H. Sugawara, A field theory of currents, Phys. Rev., **170** (1968), 1659–1662.

[TK] A. Tsuchiya and Y. Kanie, Fock Space Representations of the Virasoro Algebra—Intertwining Operators, Publ. RIMS, Kyoto Univ., **22–2** (1986), 259–327.

[W] H. Wenzl, Representations of Hecke Algebras and Subfactors, Thesis, University of Pensylvania (1985).

[Y] C. N. Yang, Some exact results for many problem in one dimension with delta function interaction, Phys. Rev. Letters, **19** (1967), 1312–1315.

[ZF] A. B. Zamolodchikov and V. A. Fateev, Nonlocal (parafermion) currents in two-dimensional conformal quantum field theory and self-dual critical points in Z_N-symmetric statistical systems, Sov. Phys. JETP **62**(2) (1985), 215–225.

Akihiro Tsuchiya
Department of Mathematics
Faculty of Science
Nagoya University
Furocho, Chikusa-ku
Nagoya 464
Japan

Yukihiro Kanie
Department of Mathematics
Faculty of Education
Mie University
1515 Kamihama
Tsu 514
Japan

Advanced Studies in Pure Mathematics 19, 1989
Integrable Systems in Quantum Field Theory and Statistical Mechanics
pp. 675–682

Errata to Vertex Operators in Conformal Field Theory on $\mathbf{P}^1$ and Monodromy Representations of Braid Group in Advanced Studies in Pure Mathematics 16,1988

A. Tsuchiya and Y. Kanie

Abstract.
We give corrections or comments to the five points in our paper [1].

First fix the notations. Let $\mathfrak{g}$ be the Lie algebra $\mathfrak{sl}(2;\mathbb{C})$. Fix a positive integer ℓ and let $\kappa = \ell + 2$ and $q = \exp\left(\frac{2\pi\sqrt{-1}}{\kappa}\right)$ and introduce the set P_ℓ consisting of all half-integers j with $0 \leq j \leq \ell/2$.

For any $j_k \in P_\ell$ $(1 \leq k \leq 3)$, denote by $\mathcal{W}\left({}_{j_3}{}^{j_2 j_1}\right)$ the space of initial terms of vertex operators of type $\mathbf{v} = \left({}_{j_3}{}^{j_2 j_1}\right)$. Note that in this $\mathrm{A}_1^{(1)}$-case, the space $\mathcal{W}\left({}_{j_3}{}^{j_2 j_1}\right)$ is nothing but the space $\mathcal{V}\begin{pmatrix} j_2 \\ j_3 j_1 \end{pmatrix} = \mathrm{Hom}_{\mathfrak{g}}(V_{j_2} \otimes V_{j_1}, V_{j_3})$ in [1] for $j_1 + j_2 + j_3 \leq \ell$, and $\mathcal{W}\left({}_{j_3}{}^{j_2 j_1}\right) = 0$ for $j_1 + j_2 + j_3 > \ell$.

The space $\mathrm{Hom}_{\mathfrak{g}}(V_{j_2} \otimes V_{j_1}, V_{j_3})$ is at most 1-dimensional, and the condition for its nontriviality is the Clebsch-Gordan condition $|j_1 - j_2| \leq j_3 \leq j_1 + j_2$. In such case, we fix the nonzero vector $\phi_{\mathbf{v}}$ as in [1, Appendix 1].

1. Literally is not valid the braid relation in [1, Proposition 4.2], where the notations we used are not appropriate. Now we reformulate Proposition 4.2 ii).

Received July 12, 1989.

For any $N \geq 1$ and $s, t, j_k \in P_\ell (1 \leq k \leq N)$, introduce the spaces $\mathcal{W}\left({}_t{}^{j_N \cdots j_1 s}\right)$ and $\mathcal{W}(N;{}_t{}^s)$ defined by

$$\begin{aligned} &\mathcal{W}\left({}_t{}^{j_N \cdots j_1 s}\right) \\ &= \sum_{p_1, \ldots, p_{N-1} \in P_\ell} \mathcal{W}\left({}_t{}^{j_N p_{N-1}}\right) \otimes \mathcal{W}\left({}_{p_{N-1}}{}^{j_{N-1} p_{N-2}}\right) \otimes \cdots \otimes \mathcal{W}\left({}_{p_1}{}^{j_1 s}\right) \end{aligned}$$

and

$$\mathcal{W}(N;{}_t{}^s) = \sum_{j_1, \ldots, j_N \in P_\ell} \mathcal{W}\left({}_t{}^{j_N \cdots j_1 s}\right).$$

Note

$$\mathcal{W}(M+N;{}_t{}^s) = \sum_{p \in P_\ell} \mathcal{W}(M;{}_t{}^p) \otimes \mathcal{W}(N;{}_p{}^s)$$

Recall that the operator $C(j_4, j_3, j_2, j_1)$ in [1, Section 4.1] acts as

$$C(j_4, j_3, j_2, j_1) : \mathcal{W}\left({}_{j_4}{}^{j_3 j_2 j_1}\right) \longrightarrow \mathcal{W}\left({}_{j_4}{}^{j_2 j_3 j_1}\right),$$

which is defined by the monodromy on the solution space of the four-point functions.

Now define the operators $C_i (1 \leq i \leq N-1)$ on $\mathcal{W}(N;{}_t{}^s)$ as follows:

$$C_i \mathcal{W}\left({}_t{}^{j_N \cdots j_1 s}\right) \subset \mathcal{W}\left({}_t{}^{j_N \cdots j_i j_{i+1} \cdots j_1}\right)$$

and

$$\begin{aligned} &C_i(\phi_N \otimes \cdots \otimes \phi_{i+1} \otimes \phi_i \otimes \cdots \otimes \phi_1) \\ &= \phi_N \otimes \cdots \otimes \phi_{i+2} \otimes C(p_{i+1}, j_{i+1}, j_i, p_{i-1})(\phi_{i+1} \otimes \phi_i) \otimes \phi_{i-1} \otimes \cdots \otimes \phi_1 \end{aligned}$$

for each $\phi_N \in \mathcal{W}\left({}_t{}^{j_N p_{N-1}}\right)$, $\phi_k \in \mathcal{W}\left({}_{p_k}{}^{j_k p_{k-1}}\right)$ $(2 \leq k \leq N-1)$ and $\phi_1 \in \mathcal{W}\left({}_{p_1}{}^{j_1 s}\right)$.

Now Proposition 4.1 ii) should be read as

Proposition 1. i) *As operators on* $\mathcal{W}(3;{}_t{}^s)$, *the relation*

$$C_1 C_2 C_1 = C_2 C_1 C_2$$

holds.

ii) *As operators on* $\mathcal{W}(N;{}_t{}^s)$, *the relation*

$$C_i C_{i+1} C_i = C_{i+1} C_i C_{i+1} \qquad (1 \leq i \leq N-1)$$

holds.

2. There is an error in the definition of the mapping K from the Wenzl's representation $(\pi_\lambda, V_\lambda^{(2,\kappa)})$ to our monodromy representation $(\pi_{N,t}, W(N;t))$ in [1, Proposition 5.3]. We give here the precise definition of the intertwining operator K^{-1} rather than K.

In our notation in this errata the space $W(N;t)$ is $\mathcal{W}\left({}_t{}^{\frac{1}{2}\cdots\frac{1}{2}\,0}\right)$ and has a basis $\{\phi_{\mathbf{p}};\, \mathbf{p} = (p_N, \dots, p_1) \in \mathcal{P}_\ell(N;t)\}$, where

$$\mathcal{P}_\ell(N;t) = \left\{\mathbf{p} = (p_N, \dots, p_1, 0); p_i \in P_\ell, p_N = t, |p_i - p_{i-1}| = \frac{1}{2}\right\}$$

and

$$\phi_{\mathbf{p}} = \phi_{\mathbf{v}_N} \otimes \cdots \otimes \phi_{\mathbf{v}_1} \in \mathcal{W}\left({}_t{}^{\frac{1}{2}\,p_{N-1}}\right) \otimes \cdots \otimes \mathcal{W}\left({}_{p_1}{}^{\frac{1}{2}\,0}\right),$$

$$\mathbf{v}_i = \left({}_{p_i}{}^{\frac{1}{2}\,p_{i-1}}\right).$$

The operators C_i on $\mathcal{W}\left(N;{}_t{}^0\right)$ preserves the subspace $W(N;t)$. By Proposition 1, the braid group B_N generated by C_i $(1 \le i \le N-1)$ acts on the space $W(N;t)$. Denote this representation by $\pi_{N,t}$.

The basis vectors $\phi_{\mathbf{p}}$ $(\mathbf{p} \in \mathcal{P}_\ell(N;t))$ are eigenvectors w.r.t. the commutative subalgebra $\mathcal{A} = \sum_{i=1}^{N-1} \mathbb{C}(C_i \dots C_1)^i$ of the group algebra $\mathbb{C}[B_N]$.

Let λ be the Young diagram $\lambda(N;t) = [\frac{N}{2} + t, \frac{N}{2} - t]$ (Here we assume that $\frac{N}{2} - t \in \mathbb{Z}_{\ge 0}$, otherwise the space $W(N;t)$ vanishes). The corresponding Wenzl's space $V_\lambda^{(2,\kappa)}$ is generated by the basis $\{\vec{v}_{\vec{p}}; \vec{p} = (\lambda_N, \dots \lambda_1) \in \mathcal{P}_\ell(\lambda)\}$, where $\mathcal{P}_\ell(\lambda)$ is the image of the set $\mathcal{P}_\ell(N;t)$ under the mapping K^{-1}, where K^{-1} is defined as

$$K^{-1}(\mathbf{p}) = (\lambda(N;t), \lambda(N-1, p_{N-1}), \dots, \lambda(1, p_1))$$

for $\mathbf{p} = (t, p_{N-1}, \dots, p_1, 0) \in \mathcal{P}_\ell(N;t)$. The basis vectors $\vec{v}_{\vec{p}}$ $(\vec{p} \in \mathcal{P}_\ell(\lambda))$ are also the eigenvectors w.r.t. the algebra $\mathcal{A}$ of the same eigenvalues as for $K(\vec{p})$.

Thus the mapping $K^{-1} : W(N;t) \to V_\lambda^{(2,\kappa)}$ must have the form

$$K^{-1}(\phi_{\mathbf{p}}) = \gamma_{\mathbf{p}} \vec{v}_{K^{-1}(\mathbf{p})},$$

where $\gamma_{\mathbf{p}}$ is a constant which was given uncorrectly in [1].

Before we give the correct definition of the constants $\gamma_{\mathbf{p}}$, we need some preliminaries. Introduce an order $<$ in the set $\mathcal{P}_\ell(N;t)$ lexicographically, *i.e.* we call $\mathbf{p} < \mathbf{q}$ for $\mathbf{p} = (t, p_{N-1}, \dots, p_1, 0)$, $\mathbf{q} =$

$(t, q_{N-1}, \ldots, q_1, 0) \in \mathcal{P}_\ell(N;t)$ if $p_j \leq q_j$ for any j. Then it is easily seen that there is the maximal $\mathbf{p}_0$ in $\mathcal{P}_\ell(N;t)$ w.r.t. this order.

We call $\mathbf{p}$ and $\mathbf{q}$ are *adjoining* and denote $\mathbf{p} \sim \mathbf{q}$, if there is an number $k (1 \leq k \leq N-1)$ such that $p_j = q_j \quad (j \neq k)$ and $|p_k - q_k| = 1$. Any $\mathbf{p}$ and $\mathbf{q}$ in $\mathcal{P}_\ell(N;t)$ are connected by a sequence $(\mathbf{p}_1, \ldots, \mathbf{p}_n)$ in $\mathcal{P}_\ell(N;t)$ such that $\mathbf{p}_1 = \mathbf{p}$, $\mathbf{p}_n = \mathbf{q}$ and $\mathbf{p}_i \sim \mathbf{p}_{i+1} (1 \leq i \leq n-1)$.

By Proposition 4.8 in [1] and the definition of the Wenzl's representation [3], there hold only the following relations (C) among the constants $\gamma_\mathbf{p}$ ($\mathbf{p} \in \mathcal{P}_\ell(N;t)$): let $\mathbf{p} = (t, p_{N-1}, \ldots, p_1, 0)$ and $\mathbf{q}$ be adjoining and $\mathbf{p} < \mathbf{q}$ $(q_k = p_k + 1)$. Denote $\mathbf{p}_+ = \mathbf{q}$, $\mathbf{p}_- = \mathbf{p}$ and $p = p_{k-1} = q_{k-1}$. Then there must hold the relation

$$\text{(C)} \qquad \gamma_+(p)\gamma_{\mathbf{p}_+} = \gamma_-(p)\gamma_{\mathbf{p}_-},$$

where

$$\gamma_\pm(p) = \frac{\Gamma\left(\frac{2p+1}{\pm\kappa}\right)}{\Gamma\left(\frac{2p+2}{\pm\kappa}\right)^{1/2}\Gamma\left(\frac{2p}{\pm\kappa}\right)^{1/2}}.$$

Since any $\mathbf{p}$ and $\mathbf{q}$ are connected by an adjoining sequence in $\mathcal{P}_\ell(N;t)$, the constants $\{\gamma_\mathbf{p}; \mathbf{p} \in \mathcal{P}_\ell(N;t)\}$ are uniquely determined up to a constant multiple. Normalise them as $\gamma_{\mathbf{p}_0} = 1$ for the maximal $\mathbf{p}_0$. Then $\gamma_\mathbf{q}(\mathbf{q} \in \mathcal{P}_\ell(N;t))$ are given as follows: Write $\mathbf{p}_0 = (t, p_{N-1}, \ldots, p_1, 0)$ and $\mathbf{q} = (t, q_{N-1}, \ldots, q_1, 0)$, then $p_j - q_j \in \mathbb{Z}_{\geq 0} (1 \leq j \leq N-1)$. Then

$$\gamma_\mathbf{q} = \prod_{j=1}^{N-1} \gamma(p_j, q_j),$$

where for $p \geq q$

$$\gamma(p,q) = \begin{cases} 1, & (p = q) \\ \frac{\gamma_+(p-\frac{1}{2})\ldots\gamma_+(q+\frac{1}{2})}{\gamma_-(p-\frac{1}{2})\ldots\gamma_-(q+\frac{1}{2})}, & (p > q). \end{cases}$$

Then Proposition 5.3 in [1] remains valid.

Proposition 2. *Let $N \in \mathbb{Z}_{>0}$ and $t \in P_\ell$ satisfy $\frac{N}{2} - t \in \mathbb{Z}_{\geq 0}$. Then*

$$K\pi_{\lambda(N,t)} = q^{\frac{3}{4}}\pi_{N,t}K.$$

3. The proof of Proposition 5.4 (The Fusion Rule) in [1] is insufficiently presented. We will develop the theory of the fusions more in detail in our forthcoming paper [2]. In this errata, we only give the meaning of the integral used there to multivalued functions and justify the proof.

Recall that a vertex operator $\Phi(z)$ of type $\mathbf{v} = \left({}_k{}^{ji}\right)$ can be considered as a $\mathrm{Hom}(\mathcal{H}_k^\dagger \otimes V_j \otimes \mathcal{H}_i, \mathbb{C})$ -valued holomorphic function:

$$\Phi(z)(w \otimes v \otimes u) = \langle w | \Phi(v;z)(|u\rangle)\rangle$$

$$(w \in \mathcal{H}_k^\dagger, v \in V_j, u \in \mathcal{H}_i),$$

and Φ is uniquely determined by its initial term $\phi \in \mathcal{W}\left({}_k{}^{ji}\right) \subset \mathrm{Hom}_{\mathfrak{g}}(V_j \otimes V_i, V_k) \cong \mathrm{Hom}_{\mathfrak{g}}(V_k^\dagger \otimes V_j \otimes V_i, \mathbb{C})$. The holomorphicity of $\Phi(z)$ is weakly taken, *i.e.* $\Phi(z)$ is holomorphic, if the $\mathbb{C}$-valued function $\langle w|\Phi(v;z)(|u\rangle)\rangle$ is holomorphic in z for any fixed vector $w \otimes v \otimes u$ in $\mathcal{H}_k^\dagger \otimes V_j \otimes \mathcal{H}_i$.

By [1, Theorem 3.4] vertex operators $\Phi_{\mathbf{v}_1}(z)$ of type $\mathbf{v}_1 = \left({}_p{}^{j_2 j_1}\right)$ and $\Phi_{\mathbf{v}_2}(w)$ of type $\mathbf{v}_2 = \left({}_{j_4}{}^{j_3 p}\right)$ are composable, and the composed operator $\Phi_{\mathbf{v}_2}(w)\Phi_{\mathbf{v}_1}(z)$ is a $\mathrm{Hom}(\mathcal{H}_{j_4}^\dagger \otimes V_{j_3} \otimes V_{j_2} \otimes \mathcal{H}_{j_1}, \mathbb{C})$ -valued multivalued holomorphic function on $M_2 = \{(w,z) \in \mathbb{C}^{*2}; w \neq z\}$ regularized at z=0 and is uniquely determined by the $\mathrm{Hom}_{\mathfrak{g}}(V_{j_4}^\dagger \otimes V_{j_3} \otimes V_{j_2} \otimes V_{j_1}, \mathbb{C})$-valued function

$$\Psi_p(w,z)(u_4 \otimes u_3 \otimes u_2 \otimes u_1) = \langle u_4 | \Phi_{\mathbf{v}_2}(u_3; w)\Phi_{\mathbf{v}_1}(u_2;z)(|u_1\rangle)\rangle$$

$$(u_4 \in V_{j_4}^\dagger, u_k \in V_{j_k} (1 \le k \le 3)),$$

which satisfies the reduced system of differential equations for 4-point functions(the joint system $E'(\mathsf{J})$ and $B'(\mathsf{J})$, $\mathsf{J} = (j_4, j_3, j_2, j_1)$ in [1, Proposition 4.1]) and they form a basis of its solution space. The holomorphic function $\Psi_p(w,z)$ is known to have the singularity at $w = z$ as

$$\Psi(w,z) = \sum_{r \in P_\ell} (w-z)^{\gamma_r^{(1)}} (\sqrt{2j_4 + 1} F^r U_r^{(1)} + O(w-z))$$

where $U_r^{(1)}$ is the basis vector of $\mathrm{Hom}_{\mathfrak{g}}(V_{j_4}^\dagger \otimes V_{j_3} \otimes V_{j_2} \otimes V_{j_1}, \mathbb{C})$ fixed in $[1, \mathit{Appendix} 1]$, F_p^r is a constant, $O(w-z)$ is a holomorphic function in $(w-z, z)$ near $w=z$ vanishing at $w=z$ and the exponent $\gamma_r^{(1)}$ is given as

$$\gamma_r^{(1)} = \Delta_{j_2} + \Delta_{j_3} - \Delta_r \qquad \left(\Delta_j = \frac{j^2 + j}{\kappa}\right).$$

We can show similarly as Theorems 2.3 and 3.4 in [1] that the function $\Phi_2(w)\Phi_1(z)$ has an expansion as

$$\langle u_4|\Phi_2(u_3;w)\Phi_1(u_2;z)(|u_1\rangle)\rangle = \sum_{r\in P_\ell}(w-z)^{\gamma_r^{(1)}}\Psi_p^r(w,z)(u_4\otimes u_3\otimes u_2\otimes u_1)$$

$$(u_4\otimes u_3\otimes u_2\otimes u_1\in \mathcal{H}_{j_4}^\dagger\otimes V_{j_3}\otimes V_{j_2}\otimes \mathcal{H}_{j_1})$$

where $\Psi_p^r(w,z)$ is a Laurent series in $w-z$ with coefficients in $\mathbb{C}(z)$.

Assume that a $\mathbb{C}$-valued function $F(w,z)$ is a holomorphic function on some region of M_2 has an expansion as

$$F(w,z)=\sum_{j=0}^{M}(w-z)^{\gamma_j}F_j(w,z),$$

where $\gamma_j\in\mathbb{Q}$, $\gamma_0=0,\gamma_j\notin\mathbb{Z}(j\geq 1)$ and $F_j(w,z)$ is a Laurent series in $w-z$. Let C_z is a positively oriented contour around z such that the origin is outside C_z. Then we used the convention that the integral of $F(w,z)$ over C_z means the integral of $F_0(w,z)$ over C_z. Hence the operator

$$\Xi^r(u_3,u_2)=\frac{1}{2\pi\sqrt{-1}}\int_{C_z}(w-z)^{-\gamma_r^{(1)}-1}\Phi_{\mathbf{v}_2}(u_3;w)\Phi_{\mathbf{v}_1}(u_2;z)dw$$

in [1] is nothing but the residue of the above $\Psi_r(w,z)$ at $w=z$.

On the other hand, introduce the space

$$\mathcal{FW}\left({}_{j_4}{}^{j_3j_2j_1}\right)=\sum_{r\in P_\ell}\mathcal{W}(\mathbf{w}_2(r))\otimes\mathcal{W}(\mathbf{w}_1(r))$$

where

$$\mathbf{w}_2(r)=\left({}_{j_4}{}^{rj_1}\right)\text{ and }\mathbf{w}_1(r)=\left({}_{r}{}^{j_3j_2}\right).$$

Its basis vector $\phi_{\mathbf{w}_2(r)}\otimes\phi_{\mathbf{w}_1(r)}$ determines the $\mathrm{Hom}(V_{j_4}^\dagger\otimes V_{j_3}\otimes V_{j_2}\otimes V_{j_1},\mathbb{C})$ -valued function by

$$\begin{aligned}&\langle u_4|\Phi_{\mathbf{w}_2(r)}(\Phi_{\mathbf{w}_1(r)}(u_2;z);w-z)(u_1))\rangle\\&=(w-z)^{\gamma_r^{(1)}}\left((\sqrt{2j_4+1}U_r^{(1)}(u_4,u_3,u_2,u_1)+O(w-z)\right)\end{aligned}$$

(regularized at w=z) and these functions furnish also a basis of the solution space of the joint system $E'(\mathbb{J})$ and $B'(\mathbb{J})$ (see [2]).

Thus the analytic continuation gives the isomorphism

$$F : \mathcal{W}\left({}_{j_4}{}^{j_3 j_2 j_1}\right) \longrightarrow \mathcal{F}\mathcal{W}\left({}_{j_4}{}^{j_3 j_2 j_1}\right)$$

(the mapping between initial terms).

4. In the proof of [1, Theorem 3.1], we did not take into account the possibility that there may be the solutions of the joint system of $E(\mathsf{J})$ and $B(\mathsf{J})$, $\mathsf{J} = (j_N, \ldots, j_1)$ with logarithmic singularities. Any formal solution of the system $\tilde{E}(\mathsf{J})$ in the proof is of the form

$$\Psi(w) = \sum_{a=1}^{R} w^{s^a} \sum_{k \in \mathbb{Z}_{\geq 0}^N} \sum_{m \in \mathbb{Z}_{\geq 0}^N, |m| \leq M} \phi_{a,k,m} w^k (\log w)^m,$$

where M is some bound, $s^a = (s_1^a, \ldots, s_N^a)$'s are exponents and $\phi_{a,k,m} \in Hom_{\mathfrak{g}}\left(V_{j_N}^{\dagger} \otimes V_{j_{N-1}} \otimes \cdots \otimes V_{j_1}, \mathbb{C}\right)$. Apply the arguments in the proof of [1, Theorem3.1], then we get $\phi_{a,0,m} \in \mathcal{W}\left({}_{j_N}{}^{j_{N-1} \cdots j_2 j_1}\right)$. Hence we already know sufficiently many formal solutions in the form without logarithmic terms by means of the products of the vertex operators.

5. The line ↑ 13 in the page 337 of [1] should be read as

$$\lim_{z \nearrow \infty} z^{2\Delta_{p_N}} \langle vac|(\hat{Y}_q(-m_q) \ldots \hat{Y}_1(-m_1)\Phi_{\mathbf{v}_{N+1}}(v_0; z)) = (-1)^q \langle v|.$$

Acknowlegement. The authors express their thanks to Professors M. Jimbo, T. Miwa and A. Wasserman for their comments and advices.

References

[1] A.Tscuchiya and Y.Kanie, Vertex Operators in Conformal Field Theory on $\mathbf{P}^1$ and Monodromy Representations of Braid Group, Adv.Studies in pure Math., **16** (1988), 297–372.

[2] A.Tscuchiya and Y.Kanie, Vertex Operators in Conformal Field Theory on $\mathbf{P}^1$ and Monodromy Representations of Braid Group II. (in preparation)

[3] H.Wenzl, Hecke algebras of type A_n and subfactors, Invent.math., **92** (1988), 349–383.

A. Tsuchiya
Department of Mathematics,
Nagoya University
Nagoya 464, JAPAN

Y. Kanie
Department of Mathematics
Mie University
Tsu 514, JAPAN

Ann. Inst. Fourier, Grenoble
37, 4 (1987), 139-160

MONODROMY REPRESENTATIONS OF BRAID GROUPS AND YANG-BAXTER EQUATIONS

by Toshitake KOHNO

INTRODUCTION

The purpose of this paper is to give a description of the monodromy of integrable connections over the configuration space arising from classical Yang-Baxter equations. These monodromy representations define a series of linear representations of the braid groups $\theta : B_n \to \mathrm{End}(W^{\otimes n})$ with one parameter, associated to any finite dimensional complex simple Lie algebra $\mathfrak{g}$ and its finite dimensional irreducible representations $\rho : \mathfrak{g} \to \mathrm{End}(W)$. By means of trigonometric solutions of the quantum Yang-Baxter equations due to Jimbo ([10] and [11]), we give an explicit form of of these representations in the case of a non-exceptional simple Lie algebra and its vector representation (Theorem 1.2.8) and in the case of $\mathfrak{sl}(2,\mathbf{C})$ and its arbitrary finite dimensional irreducible representations (Theorem 2.2.4).

Our monodromy representation θ commutes with the diagonal action of the q-analogue of the universal enveloping algebra of $\mathfrak{g}$ in the sense of Jimbo [9], which was discussed as quantum groups by Drinfel'd [7]. In particular, in the case $\mathfrak{g} = \mathfrak{sl}(m,\mathbf{C})$, the representation θ gives Hecke algebra representations of B_n appearing in a recent work of Jones [14].

The study of these monodromy representations is motivated by a recent development of two dimensional conformal field theory initiated by Belavin, Polyakov and Zamolodchikov [5]. The importance of the two dimensional conformal field theory with gauge symmetry was

Key-words : Braid group - Yang-Baxter equation - Simple Lie algebra - Integrable connection.

pointed out by Knizhnik and Zamolodchikov [18]. They showed that the total differential equations defined by our connections are satisfied by n-point functions in these cases.

Recently Tsuchiya and Kanie [22] developed an operator formalism of two dimensional conformal field theory on $\mathbf{P}^1$ using the Kac-Moody Lie algebra of type $A_1^{(1)}$. It turns out that in the case of the vector representation of $\mathfrak{sl}(2,\mathbf{C})$, the monodromy of n-point functions gives a linear representation of the braid group B_n factoring through the Jones algebra of index $4\cos^2\dfrac{\pi}{\ell+2}$ for a positive integer ℓ (see [13]). In particular this representation is unitarizable. We shall extend this unitarity result to higher representations of $\mathfrak{sl}(2,\mathbf{C})$. A neat description of the monodromy of n-point functions in the case of simple Lie algebras of other types might be pursued from a viewpoint of Brauer's centralizer algebras, which will be discussed in the forthcoming paper.

This paper is organized in the following way. In Sect. 1.1, we explain a process to define an integrable connection associated with a simple Lie algebra and its irreducible representation. We give an explicit description of the monodromy in Sect. 1.2 and 1.3. Sect. 2.1 is devoted to a review of two dimensional conformal field theory due to Tsuchiya and Kanie [22]. We discuss the case of higher representations of $\mathfrak{sl}(2,\mathbf{C})$ in Sect. 2.2 and 2.3.

Acknowledgement : The author would like to thank M. Jimbo and T. Miwa for drawing his attention to the linear representations of the braid groups arising from solutions of Yang-Baxter equations. He would also like to thank A. Tsuchiya for valuable comments from a viewpoint of the conformal filed theory.

The following notations are of frequent use :

B_n : braid group on n strings with generators σ_i, $1 \leqslant i \leqslant n-1$, represented by a braid interchanging strings i and $i+1$ (see [2]).

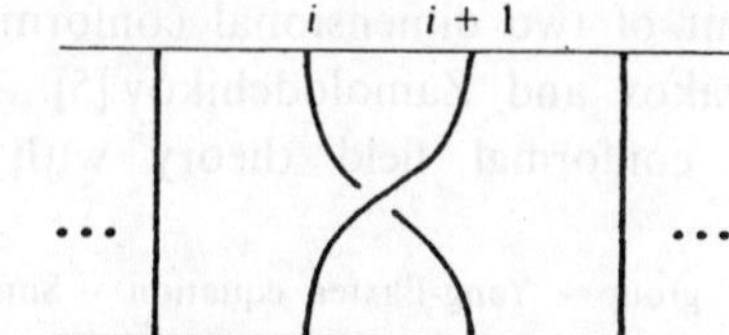

Fig. 1.

P_n : pure braid group on n strings.

$X_n = \{(z_1, \ldots, z_n) \in \mathbf{C}^n;\ \ z_\alpha \neq z_\beta \text{ if } \alpha \neq \beta\}$

$\mathfrak{g}$: a simple finite dimensional complex Lie algebra.

$\{I_\mu\}$: orthonormal basis of $\mathfrak{g}$ with respect to the Cartan-Killing form.

$t = \Sigma_\mu I_\mu \otimes I_\mu \in \mathfrak{g} \otimes \mathfrak{g}$.

For a finite dimensional vector space V, we let $\sigma \in \mathrm{End}\,(V \otimes V)$ the transposition defined by $\sigma(x \otimes y) = y \otimes x$. For $X \in \mathrm{End}\,(V \otimes V)$ we put $\bar{X} = \sigma X$.

$\mathbf{C}\{\lambda\}$: ring of the convergent power series.

1. MONODROMY OF INTEGRABLE CONNECTIONS ARISING FROM CLASSICAL YANG-BAXTER EQUATIONS

1.1. Construction of connections.

Let $\mathfrak{g}$ be a simple finite dimensional complex Lie algebra and let $\{I_\mu\}$ be an orthonormal basis of $\mathfrak{g}$ with respect to the Cartan-Killing form. We put

$$t = \Sigma_\mu I_\mu \otimes I_\mu$$

which may also be expressed as

$$t = \frac{1}{2}(\Delta\Omega - \Omega \otimes 1 - 1 \otimes \Omega).$$

Here Ω is the Casimir operator $\Sigma_\mu I_\mu . I_\mu$ in the universal enveloping algebra $U(\mathfrak{g})$ and $\Delta : U(\mathfrak{g}) \to U(\mathfrak{g}) \otimes U(\mathfrak{g})$ stands for the comultiplication as a Hopf algebra.

Associated with a simple Lie algebra $\mathfrak{g}$ and its finite dimensional irreducible representations $\rho_\alpha : \mathfrak{g} \to \mathrm{End}\,(W_\alpha)$, $1 \leqslant \alpha \leqslant n$, we consider the total differential equations with a parameter λ

$$(1.1.1) \qquad d\Phi = \Sigma_{1 \leqslant \alpha < \beta \leqslant n} \lambda \Omega_{\alpha\beta} d \log (z_\alpha - z_\beta) . \Phi, \qquad \lambda \in \mathbf{C}$$

defined over

$$X_n = \{(z_1, \ldots, z_n) \in \mathbf{C}^n;\ z_\alpha \neq z_\beta \text{ if } \alpha \neq \beta\}.$$

Here $\Omega_{\alpha\beta} \in \mathrm{End}\,(W_1 \otimes \ldots \otimes W_n)$ are defined by

$$\Omega_{\alpha\beta} = \Sigma_\mu \rho_\alpha(I_\mu) \otimes \rho_\beta(I_\mu)$$

where ρ_α stands for the representation ρ_α on the α-th factor acting as the identity on the other factors.

The matrix valued 1-form

$$(1.1.3) \qquad \omega = \Sigma_{1 \leqslant \alpha < \beta \leqslant n} \lambda \Omega_{\alpha\beta} d \log (z_\alpha - z_\beta), \qquad \lambda \in \mathbf{C}$$

is considered to be a connection of the trivial vector bundle over X_n with fiber $W_1 \otimes \cdots \otimes W_n$. The integrability condition for ω

$$d\omega + \omega \wedge \omega = 0$$

is satisfied in our case since we have the following relations among $\Omega_{\alpha\beta}$:

$$(1.1.4) \quad [\Omega_{\alpha\beta}, \Omega_{\alpha\gamma} + \Omega_{\beta\gamma}] = [\Omega_{\alpha\beta} + \Omega_{\alpha\gamma}, \Omega_{\beta\gamma}] = 0 \quad \text{for} \quad \alpha < \beta < \gamma$$
$$[\Omega_{\alpha\beta}, \Omega_{\gamma\delta}] = 0 \quad \text{for distinct } \alpha,\ \beta,\ \gamma,\ \delta.$$

In fact the above relations are derived from the fact that the Casimir operator Ω lies in the center of $U(\mathfrak{g})$. We shall call (1.1.4) the *infinitesimal pure braid relations.* These relations are relevant to the classical Yang-Baxter equation in the following sense.

Let us recall that the classical Yang-Baxter equation is a functional equation for a $\mathfrak{g} \otimes \mathfrak{g}$-valued meromorphic function $r(u)$, $u \in \mathbf{C}$, given by

$$(1.1.5) \quad [r_{12}(u-v),\ r_{13}(u)] + [r_{12}(u-v), r_{23}(v)] + [r_{13}(u), r_{23}(v)] = 0.$$

Here the above triangular equality is considered in $\mathfrak{g} \otimes \mathfrak{g} \otimes \mathfrak{g}$ and r_{ij} signifies the r on the i-th and j-th factors acting as the identity on the other factor. Solutions of the classical Yang-Baxter equation are classified by Belavin and Drinfel'd (see [3] for a precise statement). In particular, they discovered a rational solution $r(u) = t/u$. The infinitesimal pure braid relations are obtained from the fact that t/u satisfies the classical Yang-Baxter equation.

As the monodromy of the connection ω we obtain a linear representation of the pure braid group

$$\theta : P_n \to \mathrm{End}\,(W_1 \otimes \cdots \otimes W_n)$$

depending on the parameter λ. Let us now suppose that the representations ρ_α, $1 \leqslant \alpha \leqslant n$, are the same. In this case the connection ω defined in the above way is invariant by the diagonal action of the symmetric group S_n on $X_n \times (W_1 \otimes \cdots \otimes W_n)$, hence it defines a local system over the quotient space $Y_n = X_n/S_n$. Considering λ as a parameter we obtain a linear representation of the braid group on n strings

$$\theta : B_n \to \mathrm{End}\,(W^{\otimes n}) \otimes \mathbf{C}\{\lambda\}.$$

Here $\mathbf{C}\{\lambda\}$ denotes the ring of the convergent power series. Our main object is to give a description of this monodromy representation.

The total differential equations of the above type appear in the two dimensional conformal field theory with gauge symmetry due to Knizhnik and Zamolodchikov [18]. Although in their situation the parameter λ is given by $(\ell+g)^{-1}$ where ℓ is a positive integer and g is the corresponding dual Coxeter number, we shall deal with the monodromy by considering λ as a parameter.

1.2. Description of the monodromy by means of solutions of quantum Yang-Baxter equations.

Let W be a finite dimensional complex vector space. By the quantum Yang-Baxter equation written in a multiplicative form we mean the following functional equation for a meromorphic function $R(x)$ with values in $\mathrm{End}\,(W \otimes W)$:

$$(1.2.1)\quad R_{12}(x)R_{13}(xy)R_{23}(y) = R_{23}(y)R_{13}(xy)R_{12}(x).$$

Here the equality is considered in $\mathrm{End}\,(W \otimes W \otimes W)$ and the notation R_{ij} is standard as is explained in the previous section. Let us consider the case where $R(x)$ contains an extra parameter q so that $R(x,q)$ has an expansion around $q = 1$:

$$(1.2.2)\qquad R(x,q) = \mathbf{1} + (q-1)r(x) + \cdots$$

In this situation we verify that $r(x)$ is a solution of the multiplicative classical Yang-Baxter equation

$$[r_{12}(x), r_{13}(xy)] + [r_{12}(x), r_{23}(y)] + [r_{13}(xy), r_{23}(y)] = 0.$$

We call $r(x)$ the *classical limit* of $R(x,q)$. The following typical solutions of the above classical Yang-Baxter equation was discovered by Belavin and Drinfel'd [3] (see also [10]). Let $\mathfrak{g}$ be a finite dimensional complex simple Lie algebra and let Δ be the set of roots of $\mathfrak{g}$. For a root α, we denote by X_α the root vector normalized by $(X_\alpha, X_{-\alpha}) = 1$ with respect to the Cartan-Killing form. Putting $r = \Sigma_{\alpha \in \Delta} \operatorname{sgn} \alpha . X_\alpha \otimes X_{-\alpha}$, we define a $\mathfrak{g} \otimes \mathfrak{g}$-valued function $r(x)$ by

$$r(x) = r - t + \frac{2t}{x-1} \tag{1.2.3}$$

where t is defined in the previous section. These solutions are called *trigonometric* in the sense that they are rational functions of $x = e^u$.

The quantization problem of the above solutions was treated by Jimbo. In a series of papers [9], [10] and [11], he constructed a matrix $R(x,q)$ whose expansion around $q = 1$ is given by

(1.2.4) $R(x,q) = f(x)\{\mathbf{1} + (q-1)((\rho \otimes \rho) r(x) + \varkappa(x)\mathbf{1}) + \cdots\}$
with some $\mathbf{C}$-valued functions $f(x)$ and $\varkappa(x)$,

for the following simple Lie algebras $\mathfrak{g}$ and their representations $\rho : \mathfrak{g} \to \operatorname{End}(W)$

(1.2.5) $\mathfrak{g}$ is non-exceptional and ρ is the vector representation,

(1.2.6) $\mathfrak{g}$ is $\mathfrak{sl}(2,\mathbf{C})$ and ρ is an arbitrary finite dimensional irreducible representation.

In this section we discuss the case 1.2.5. Our matrices $R(x,q)$ are given by formulae 3.5 and 3.6 in [10] by putting $k = q$. In the formula 1.2.4, $f(x)$ is given by $(x-1)$ if $\mathfrak{g}$ is of type A and by $(x-1)^2$ if $\mathfrak{g}$ is of type B, C or D.

We put $\bar{R} = \sigma R$ where $\sigma \in \operatorname{End}(W \otimes W)$ is the transposition defined by $\sigma(x \otimes y) = y \otimes x$. One of the important properties of the matrix $\bar{R}(x,q)$ is that it commutes with the diagonal action of $U\hat{}(\mathfrak{g})$. Here $U\hat{}(\mathfrak{g})$ denotes the q-analogue of the corresponding Lie algebra $\mathfrak{g}$ due to Jimbo [9], which is also denoted by $U_h(\mathfrak{g})$ with $q = e^h$ by Drinfel'd [7]. Instead of giving the complete definition we recall the case $\mathfrak{g} = \mathfrak{sl}(2,\mathbf{C})$, which is originally due to Kulish and Reshetikhin (see the references of [7]). We define $U\hat{}(\mathfrak{g})$ to be the $\mathbf{C}$-algebra generated by the symbols $\hat{e}$, $\hat{f}$, q^h and q^{-h} with relations

$$q^{h/2}\hat{e}q^{-h/2} = q\hat{e}, \qquad q^{h/2}\hat{f}q^{-h/2} = q^{-1}\hat{f}, \qquad [\hat{e},\hat{f}] = \frac{q^h - q^{-h}}{q - q^{-1}}.$$

We define the comultiplication $\Delta : U^\wedge(\mathfrak{g}) \to U^\wedge(\mathfrak{g}) \otimes U^\wedge(\mathfrak{g})$ by the algebra homomorphism characterized by

$$\Delta(q^{\pm h/2}) = q^{\pm h/2} \otimes q^{\pm h/2},\ \Delta(X) = X \otimes q^{-h/2} + q^{h/2} \otimes X \ \text{ for } X = \hat{e}, \hat{f}.$$

With respect to the comultiplications Δ and $\bar{\Delta} = \sigma\Delta$, $U^\wedge(\mathfrak{g})$ has a structure of a non-commutative Hopf algebra which is considered to be a deformation of the universal enveloping algebra of $\mathfrak{sl}(2,\mathbf{C})$ (see Drinfel'd [7] and Verdier [23] for a more extensive treatment).

Let us go back to the situation of the previous section. Associated with a non-exceptional simple Lie algebra $\mathfrak{g}$ and its vector representation, we consider the connection

$$\omega = \Sigma_{1 \leqslant \alpha < \beta \leqslant n} \lambda \Omega_{\alpha\beta} d \log (z_\alpha - z_\beta).$$

As the monodromy of ω we get a one parameter family of linear representation $\theta : B_n \to \mathrm{End}(W^{\otimes n}) \otimes \mathbf{C}\{\lambda\}$. To describe θ we introduce the matrix $T(q)$ by

(1.2.7) $$T(q) = \lim_{x \to \infty} x^{-d} \bar{R}(x,q)$$

where d is the degree of the corresponding $\bar{R}(x,q)$ with respect to x, which is given by $d = 1$ in the case $\mathfrak{g}$ is of type A and by $d = 2$ in the other cases. We put $\nu = \dfrac{m-1}{2m}$ if $\mathfrak{g} = \mathfrak{sl}(m,\mathbf{C})$ and $\nu = \dfrac{1}{2}$ otherwise.

Our main theorem in this section is the following:

THEOREM 1.2.8. — *Let $\mathfrak{g}$ be a non-exceptional complex simple Lie algebra and let $\rho : \mathfrak{g} \to \mathrm{End}(W)$ be its vector representation. As the monodromy of the associated connection*

$$\omega = \Sigma_{1 \leqslant \alpha < \beta \leqslant n} \lambda \Omega_{\alpha\beta} d \log (z_\alpha - z_\beta)$$

we get a linear representation $\theta : B_n \to \mathrm{End}(W^{\otimes n}) \otimes \mathbf{C}\{\lambda\}$ given by

$$\theta(\sigma_i) = q^\nu \{1 \otimes \cdots \otimes 1 \otimes T(q) \otimes 1 \otimes \cdots \otimes 1\}, \qquad 1 \leqslant i \leqslant n-1.$$

Here $q = \exp(-\pi\sqrt{-1}\lambda)$ and $T(q)$ is situated on the i-th and $(i+1)$-st factors. Moreover this representation commutes with the diagonal action of $U^\wedge(\mathfrak{g})$ on $W^{\otimes n}$.

The action of $U^\wedge(\mathfrak{g})$ is defined by the multi-diagonal map in the sense of [9] and [12]. In the case $\mathfrak{g} = \mathfrak{sl}(m,\mathbf{C})$, the monodromy representation obtained above is known as the higher order Pimsner-Popa-Temperley-Lieb representation (see [17]). In fact the matrix $T(q)$ is given by

$$(1.2.9) \quad T(q) = \Sigma E_{\alpha\alpha} \otimes E_{\alpha\alpha} + q\Sigma_{\alpha \neq \beta} E_{\alpha\beta} \otimes E_{\beta\alpha} + (1-q^2)\Sigma_{\alpha<\beta} E_{\alpha\alpha} \otimes E_{\beta\beta}$$

where $E_{\alpha\beta}$ signify $m \times m$ matrix units. In this case the matrix $T(q)$ defines a linear representation of the braid group factoring through the Iwahori's Hecke algebra of the symmetric group.

1.3. Proof of Theorem 1.2.8.

Let us start with an integrable connection ω over X_n of the form $\omega = \Sigma_{1 \leqslant \alpha < \beta \leqslant n} M_{\alpha\beta} d \log(z_\alpha - z_\beta)$, $M_{\alpha\beta} \in \mathfrak{gl}(m, \mathbf{C})$. The monodromy of ω is expressed by an infinite sum using Chen's iterated integrals [6].

$$(1.3.1) \qquad \theta(\gamma) = \mathbf{1} + \int_\gamma \omega + \int_\gamma \omega\omega + \cdots$$

for $\gamma \in P_n$. Here we have used the following standard notation for the Chen's iterated integrals.

Let X be a smooth manifold and let ω_i, $1 \leqslant i \leqslant n$, be matrix valued 1-forms on X. For a path $\gamma : [0,1] \to X$, we define the iterated integral $\int_\gamma \omega_1 \omega_2 \ldots \omega_n$ by

$$\int_\Delta A_1(t_1) A_2(t_2) \ldots A_n(t_n) dt_1 dt_2 \ldots dt_n$$

where $\gamma^* \omega_i = A_i(t_i) dt_i$ and $\Delta = \{(t_1, \ldots, t_n) ;\, 0 \leqslant t_1 \leqslant \cdots \leqslant t_n \leqslant 1\}$.

Let $\mathbf{C}\langle\langle X_{\alpha\beta} \rangle\rangle$ denote the ring of non-commutative formal power series with indeterminates $X_{\alpha\beta}$, $1 \leqslant \alpha < \beta \leqslant n$, and let J be its two sided ideal generated by the following infinitesimal pure braid relations among $X_{\alpha\beta}$:

$$(1.3.2) \quad \begin{array}{l} [X_{\alpha\beta}, X_{\alpha\gamma} + X_{\beta\gamma}], \quad [X_{\alpha\beta} + X_{\alpha\gamma}, X_{\beta\gamma}], \quad \alpha < \beta < \gamma \\ \qquad [X_{\alpha\beta}, X_{\gamma\delta}] \quad \text{for distinct } \alpha, \beta, \gamma, \delta . \end{array}$$

We denote by A the quotient algebra $\mathbf{C}\langle\langle X_{\alpha\beta} \rangle\rangle / J$. As a universal expression of 1.3.1, we obtain a homomorphism $\tilde{\theta} : P_n \to A$ defined by

$$\tilde{\theta}(\gamma) = 1 + \int_\gamma \tilde{\omega} + \int_\gamma \tilde{\omega}\tilde{\omega} + \cdots \text{ with}$$

$$\tilde{\omega} = \Sigma_{1 \leqslant \alpha < \beta \leqslant n} X_{\alpha\beta} \otimes d \log (z_\alpha - z_\beta).$$

Let $C[P_n]\hat{}$ denote the completion of the group ring $C[P_n]$ with respect to the powers of the augmentation ideal and let $j : P_n \to C[P_n]\hat{}$ denote the natural homomorphism. We have the following assertions:

PROPOSITION 1.3.3. – (*i*) *We have an isomorphism of complete Hopf algebras* $C[P_n]\hat{} \xrightarrow{\sim} A$ *such that the following diagram is commutative.*

$$\begin{array}{ccc} & \overset{j}{\nearrow} & C[P_n]\hat{} \\ P_n & & \downarrow \\ & \underset{\tilde{\theta}}{\searrow} & A \end{array}$$

(*ii*) *The universal expression of the monodromy* $\tilde{\theta} : P_n \to A$ *is injective.*

The assertion (i) has been discussed by several authors in a more general situation (see [1], [8] and [16]). The primitive part of A is the Malcev Lie algebra of P_n, which is the dual of the Sullivan's 1-minimal model of X_n (see [21], [19] and [16]). The assertion (ii) is proved in [17] by the induction with respect to n by using the fibration $\pi : X_{n+1} \to X_n$. The essential points are that the monodromy of the fibration π is trivial on the homology and that the natural homomorphism j is injective in the case of free groups. By using the assertion (ii) we have shown in [17] the following theorem:

THEOREM 1.3.4 ([17]). – *Let* $\gamma_{\alpha\beta}$, $1 \leqslant \alpha < \beta \leqslant n$, *be a system of generators of* P_n *given by*

$$(1.3.5) \qquad \gamma_{\alpha\beta} = \sigma_\alpha \sigma_{\alpha+1} \ldots \sigma_{\beta-1} \sigma_\beta^2 \sigma_{\beta-1}^{-1} \ldots \sigma_\alpha^{-1}.$$

If $\theta : P_n \to GL(m, C)$ *is a linear representation such that* $\|\theta(\gamma_{\alpha\beta}) - \mathbf{1}\|$ *is sufficiently small for each* $1 \leqslant \alpha < \beta \leqslant n$, *then there exist constant matrices* $M_{\alpha\beta}$, $1 \leqslant \alpha < \beta \leqslant n$, *close to* 0, *satisfying the infinitesimal pure braid relations, such that the monodromy of the connection* $\omega = \Sigma_{1 \leqslant \alpha < \beta \leqslant n} M_{\alpha\beta}\, d \log (z_\alpha - z_\gamma)$ *is equivalent to* θ.

To deduce Theorem 1.3.4 from Propositon 1.3.3 we used an argument due to Hain [8].

Now let us go back to the situation of Theorem 1.2.8.

LEMMA 1.3.6. — *We put* $\lambda = -(\pi\sqrt{-1})^{-1} \log q$, $-\pi \leqslant \mathrm{Im} \log q < \pi$. *The matrix* $T(q)^2$ *has an expansion with respect to* λ *of the form*

$$T(q)^2 = \mathbf{1} + 2\pi\sqrt{-1}\,\lambda\{(\rho\otimes\rho)(t) - 2\nu.\mathbf{1}\} + \mathcal{O}(\lambda^2).$$

Here ρ *is the vector representation as in Sect.* 1.2.

Proof of Lemma 1.3.6. — Let us recall that $T(q)$ is defined as the leading coefficient of the matrix $\bar{R}(x,q)$ with respect to x. By means of the expansion 1.2.4 and the definition of $r(x)$ (see 1.2.3), we have

$$T'(1) = \sigma.\{(\rho\otimes\rho)(r-t) + 2\nu.\mathbf{1}\}. \tag{1.3.7}$$

Here we have used $2\nu = \lim_{x\to\infty} \varkappa(x)$, which is verified by a direct computation. Let us now observe that $T(1)$ is equal to the transposition σ. By using

$$\begin{aligned} \sigma.(\rho\otimes\rho)(t).\sigma &= (\rho\otimes\rho)(t) \\ \sigma.(\rho\otimes\rho)(r).\sigma &= -(\rho\otimes\rho)(r) \end{aligned} \tag{1.3.8}$$

we obtain the formula

$$T(1)T'(1) + T'(1)T(1) = -2(\rho\otimes\rho)(t) + 4\nu.\mathbf{1}.$$

Our Lemma follows immediately.

It follows from the definition of the Yang-Baxter equation 1.2.1 that the matrix $\bar{R}(x,q)$ satisfies

$$\bar{R}_{12}(x)\bar{R}_{23}(xy)\bar{R}_{12}(y) = \bar{R}_{23}(y)\bar{R}_{12}(xy)\bar{R}_{23}(x). \tag{1.3.9}$$

This shows that the correspondence

$$\sigma_i \to 1 \otimes \cdots \otimes T(q) \otimes \cdots \otimes 1 \tag{1.3.10}$$

appearing in the statement of Theorem 1.2.8 actually defines a linear representation of the braid group. In the following we denote this representation by φ.

If $|\lambda|$ is sufficiently small, then we may apply Theorem 1.3.4. Hence in this situation we have a matrix $M(\lambda) \in \mathrm{End}(W\otimes W)$ close to 0 and analytic with respect to λ, so that the monodromy of the connection $\sum_{1\leqslant\alpha<\beta\leqslant n} M_{\alpha\beta}(\lambda)\, d\log(z_\alpha - z_\beta)$ expressed by the iterated integrals 1.3.1 is equal to φ restricted to P_n.

Let $M(\lambda) = Z_1\lambda + Z_2\lambda^2 + \cdots$ be an expansion of $M(\lambda)$ around $\lambda = 0$. By means of the expression of the monodromy using iterated integrals and Lemma 1.3.6 we have

$$Z_1 = (\rho \otimes \rho)(t) - 2\nu.\mathbf{1}.$$

In the following, we denote the above matrix by Ω'.

LEMMA 1.3.11. — *If* $|\lambda|$ *is sufficiently small, there exists a matrix* $P(\lambda) \in \mathrm{End}(W^{\otimes n})$ *with* $\lim_{\lambda \to 0} P(\lambda) = \mathbf{1}$ *such that*

$$P(\lambda)^{-1} M_{\alpha\beta}(\lambda) P(\lambda) = \lambda \Omega'_{\alpha\beta}.$$

Proof of Lemma 1.3.11. — Let $H_{\alpha\beta}$ denote the hyperplane in $\mathbf{C}^n$ defined by $z_\alpha = z_\beta$. Let $\mu : X \to \mathbf{C}^n$ be a blowing up with exceptional divisors E_k, $3 \leqslant k \leqslant n$, such that $\mu(E_k) = \bigcap_{1 \leqslant \alpha < \beta \leqslant k} H_{\alpha\beta}$. We denote by E_2 the proper transform of H_{12}. Then the residue of the connection $\mu^*\omega$ along the divisor E_k is expressed as $\Sigma_{1 \leqslant \alpha < \beta \leqslant k} M_{\alpha\beta}(\lambda)$. Let us observe that a normal loop around E_k is given by $\gamma_k = (\sigma_1 \ldots \sigma_{k-1})^k$ which lies in the center of B_k. For a generic value $\lambda \in \mathbf{C}$, the matrix $\varphi(\gamma_k)$ is diagonalizable, which implies that the residue $\Sigma_{1 \leqslant \alpha < \beta \leqslant k} M_{\alpha\beta}(\lambda)$ is diagonalizable. Moreover, by means of the infinitesimal pure braid relations for $M_{\alpha\beta}(\lambda)$ we conclude that the residues $\Sigma_{1 \leqslant \alpha < \beta \leqslant k} M_{\alpha\beta}(\lambda)$, $k = 2, 3, \ldots$ are diagonalized simultaneously. We have a matrix $Q(\lambda) = Q_0 + Q_1\lambda + Q_2\lambda^2 + \cdots$ such that for $2 \leqslant k \leqslant n$

$$Q(\lambda)^{-1}(\Sigma_{1 \leqslant \alpha < \beta \leqslant k} M_{\alpha\beta}(\lambda)) Q(\lambda) \tag{1.3.12}$$

is diagonal. It can be shown by using the explicit form of $T(q)$ that the eigenvalues of $\varphi(\gamma_k)$ is of the form q^m with some integer m. This implies that the matrix 1.3.12 is linear with respect to λ. Hence it is written as $Q_0^{-1}(\Sigma_{1 \leqslant \alpha < \beta \leqslant k} \lambda \Omega'_{\alpha\beta}) Q_0$. Putting $P(\lambda) = Q(\lambda).Q_0^{-1}$, we obtain a desired matrix. This proves Lemma.

The proof of Theorem 1.2.8 is completed in the following way. We put $\omega' = \Sigma \lambda \Omega'_{\alpha\beta} d \log (z_\alpha - z_\beta)$. By Lemma 1.3.11 the expression

$$\mathbf{1} + \int_\gamma \omega' + \int_\gamma \omega'\omega' + \cdots \tag{1.3.13}$$

is equal to $P(\lambda)^{-1}\varphi(\lambda)P(\lambda)$ if $|\lambda|$ is sufficiently small. We observe that $P(\lambda)$ is analytically continued to a meromorphic function of λ on the whole complex plane. Since the expression 1.3.13 is an entire function of λ we conclude by an analytic continuation that 1.3.13 is expressed as $P(\lambda)^{-1}\varphi(\lambda)P(\lambda)$ in $\mathrm{End}(W^{\otimes n}) \otimes \mathbf{C}\{\lambda\}$. Thus we have shown the statement of Theorem 1.2.8 on the pure braid group P_n. To extend this to the full braid group B_n it suffices to observe that both $\theta(\sigma_i)$ and $\varphi(\sigma_i)$ are the transposition of the i-th factor and $(i+1)$-st factors if $\lambda = 0$ and that they are holomorphic with respect to λ. This shows the first assertion of Theorem 1.2.8. The second assertion is derived from the fact that $\bar{R}(x,q)$ commutes with the diagonal action of $U\hat{}(\mathfrak{g})$. This completes the proof of Theorem 1.2.8.

(1.3.14) *Remark.* – For a complex number $\lambda \in \mathbf{C}$, the above proof implies that the correspondence described in Theorem 1.2.8 holds true if $\varphi(\gamma_k)$, $2 \leqslant k \leqslant n$, are diagonalizable. This condition is satisfied if φ is completely reducible.

2. MONODROMY OF n-POINT FUNCTIONS IN TWO DIMENSIONAL CONFORMAL FIELD THEORY

2.1. Review of $A_1^{(1)}$ model due to Tsuchiya and Kanie.

In this section we recall briefly the operator formalism of the two dimensional conformal field theory on $\mathbf{P}^1$ with gauge symmetry of type $A_1^{(1)}$ following a recent work of Tsuchiya and Kanie [22].

Integrable highest weight modules. – Let $\mathfrak{g} = \mathfrak{sl}(2,\mathbf{C})$ and let $\hat{\mathfrak{g}}$ be the affine Lie algebra of type $A_1^{(1)}$ which is defined by the canonical central extension of the loop algebra $\mathfrak{g} \otimes \mathbf{C}[t,t^{-1}]$ (see [15]). Putting $\mathfrak{M}_\pm = \Sigma_{n\geqslant 1}\mathfrak{g} \otimes t^{\pm n}$, $\hat{\mathfrak{g}}$ is decomposed into

$$\hat{\mathfrak{g}} = \mathfrak{M}_+ \oplus \mathfrak{g} \oplus \mathbf{C}c \oplus \mathfrak{M}_-$$

where c is the central element. For a positive integer ℓ and a half integer j such that $0 \leqslant j \leqslant \ell/2$ it is known by Kac [15] that there exists a unique irreducible left $\hat{\mathfrak{g}}$-module $\mathscr{H}_j(\ell)$ with a non zero vector $|\ell,j\rangle$ such that

$$\text{(2.1.1)} \qquad \mathfrak{M}_+|\ell,j\rangle = E|\ell,j\rangle = 0, \ H|\ell,j\rangle = 2j|\ell,j\rangle, \quad c|\ell,j\rangle = \ell|\ell,j\rangle.$$

In the same way, we have a unique irreducible right $\hat{\mathfrak{g}}$-module $\mathscr{H}_j^\dagger(\ell)$ with $\langle j,\ell|$ such that

$$(2.1.2)\qquad \langle j,\ell|\mathfrak{M}_- = \langle j,\ell|\mathrm{F} = 0\,,\ \langle j,\ell|\mathrm{H} = 2j\langle j,\ell|\,,\quad \langle j,\ell|c = \ell\langle j,\ell|.$$

Here H, E and F stand for the usual Chevalley basis of $\mathfrak{g}$. In the following we fix ℓ and we write $\mathscr{H}_j$ instead of $\mathscr{H}_j(\ell)$. There exists a unique bilinear form $\mathscr{H}_j^\dagger \times \mathscr{H}_j \to \mathbf{C}$ such that $\langle j,\ell|\ell,j\rangle = 1$ and $\langle ua|v\rangle = \langle u|av\rangle$ for any $a \in \hat{\mathfrak{g}}$, $u \in \mathscr{H}_j^\dagger$ and $v \in \mathscr{H}_j$.

Operation of the Virasoro Lie algebra. – For $\mathrm{X} \in \mathfrak{g}$, we put $\mathrm{X}[n] = \mathrm{X} \otimes t^n$ and $\mathrm{X}(z) = \Sigma_{n \in \mathbf{Z}} \mathrm{X}[n] z^{-n-1}$ with $z \in \mathbf{C}\backslash\{0\}$. The *Segal-Sugawara form* T(z) is defined to be

$$(2.1.3)\qquad \mathrm{T}(z) = \frac{1}{2(2+\ell)}\{\Sigma_\mu :\ \mathrm{I}_\mu(z)\mathrm{I}_\mu(z) :\}.$$

Here $\{\mathrm{I}_\mu\}$ denotes an orthonormal basis of $\mathfrak{g}$ and :: stands for the usual normal order product defined by

$$: \mathrm{X}[m]\mathrm{Y}[n] : = \begin{cases} \mathrm{X}[m]\mathrm{Y}[n]) & \text{if } m < n \\ \frac{1}{2}\{\mathrm{X}[m]\mathrm{Y}[n] + \mathrm{Y}[n]\mathrm{X}[m]\} & \text{if } m = n \\ \mathrm{Y}[n]\mathrm{X}[m] & \text{if } m > n. \end{cases}$$

We define L_m, $m \in \mathbf{Z}$ as the coefficients of the expansion

$$(2.1.4)\qquad \mathrm{T}(z) = \Sigma_{m \in \mathbf{Z}} \mathrm{L}_m z^{-m-2}.$$

We may also express L_m as

$$(2.1.5)\qquad \mathrm{L}_m = \frac{1}{2(2+\ell)} \Sigma_{k \in \mathbf{Z}} \Sigma_\mu\ :\ \mathrm{I}_\mu(-k)\mathrm{I}_\mu(m+k)\ :$$

These L_m, $m \in \mathbf{Z}$, satisfy the fundamental relations of the *Virasoro Lie algebra*:

$$(2.1.6)\qquad [\mathrm{L}_m, \mathrm{L}_n] = (m-n)\mathrm{L}_{m+n} + \frac{m^3 - m}{12}\delta_{m+n,0}c'.$$

Here $c' = \dfrac{3\ell}{\ell + 2}$ *id*, which we shall call the *central charge*. With respect to the operation of L_0, $\mathscr{H}_j$ is decomposed into finite dimensional subspaces

$$(2.1.7)\qquad \mathscr{H}_j = \bigoplus_{d \geqslant 0} \mathscr{H}_{j,d}$$

where $\mathcal{H}_{j,d}$ is the eigenspace with the eigenvalue $\dfrac{j^2+j}{\ell+2}+d$. In particular, $\mathcal{H}_{j,0}$ is identified with the spin j representation of $\mathfrak{g}$, which is denoted by V_j.

Definition of primary fields. – We are interested in operators on the space $\mathcal{H} = \bigoplus_{j=0}^{\ell/2} \mathcal{H}_j$. The basic operators are so called *primary fields.* A primary field of spin j is defined to be a bilinear form $\phi(u,z)$: $\mathcal{H}^\dagger \times \mathcal{H} \to \mathbf{C}$ parametrized by $u \in V_j$ and $z \in \mathbf{C}\backslash\{0\}$ in such a way that

(i) $\phi(u,z)$ is linear with respect to u

(ii) $\langle v|\phi(u,z)|w\rangle$ is a multivalued holomorphic function of z for any $v \in \mathcal{H}^\dagger$ and $w \in \mathcal{H}$,

satisfying the following conditions:

(2.1.8) $[X \otimes t^m, \phi(u,z)] = z^m \phi(Xu, z)$ (*gauge condition*)

(2.1.9) $$[L_m, \phi(u,z)] = z^m\left\{z\frac{\partial}{\partial z} + (m+1)\Delta_j\right\}\phi(u,z)$$

where $\Delta_j = \dfrac{j^2+j}{\ell+2}$, which we shall call the *conformal dimension* of ϕ.

Existence of vertex operators. – Given a primary field of spin j, we associate to the triple $v = (j_1, j, j_2)$ the (j_1, j_2) component of $\phi(u,z)$ with respect to the decomposition 2.1.7, which we denote by $\phi_v(u,z)$. This operator is called a *vertex operator of type* v. We have a Laurent series expansion $\phi_v(u,z) = \Sigma_{n\in \mathbf{Z}}\phi_n(u)z^{-n-\Delta}$ with $\Delta = \Delta_j + \Delta_{j_1} - \Delta_{j_2}$ ([22] Prop. 2.1.). This gives a $\mathfrak{g}$ invariant trilinear form φ: $V_{j_1}^\dagger \otimes V_{j_2} \otimes V_{j_3} \to \mathbf{C}$ defined by $\varphi(u,v,w) = \langle u|\phi_0(v)|w\rangle$, which we shall call the *initial form.*

THEOREM 2.1.10 ([22] Th. 2.2.). – (i) *A non trivial vertex operator of type* v *exists if and only if the following conditions are satisfied:*

(2.1.11) $|j_1 - j_2| \leqslant j \leqslant j_1 + j_2$, $j_1 + j + j_2 \in \mathbf{Z}$ (*Clebsch-Gordan condition*)

(2.1.12) $$j_1 + j + j_2 \leqslant \ell$$

(ii) *Under the above conditions, a vertex operator of type* v *is unique up to scalar and is determined by its initial form.*

Differential equation of n-point functions. – For an operator A on $\mathcal{H}$, we denote by $\langle A \rangle$ its *vacuum expectation* defined by $\langle \mathrm{vac}|A|\mathrm{vac}\rangle = \langle 0,\ell|A|\ell,0\rangle$. Our purpose is to give a description of *n-point functions* $\langle \phi_1(u_1,z_1)\ldots\phi_n(u_n,z_n)\rangle$ for primary fields ϕ_i. A main tool to deduce differential equations satisfied by n-point functions is the following *operator product expansions*

$$(2.1.13)\quad X(\zeta)\phi(u,z) = \frac{1}{\zeta - z}\,\phi(Xu,z) + (\text{regular terms})$$

$$(2.1.14)\quad T(\zeta)\phi(u,z) = \left(\frac{\Delta_j}{(\zeta-z)^2} + \frac{1}{\zeta - z}\frac{\partial}{\partial z}\right)\phi(u,z) + (\text{regular terms})$$

for a primary field ϕ of spin j. Here the meaning of the compositions of operators is justified by the use of the decomposition 2.1.7 (see [22] for a precise definition). Following [18], we define the operation of $\hat{\mathfrak{g}}$ on vertex operators by

$$(2.1.15)\quad [X[m]\phi](u,z) = \frac{1}{2\pi\sqrt{-1}}\int_C d\zeta(\zeta-z)^m X(\zeta)\phi(u,z)$$

$$(2.1.16)\quad [L_m\phi](u,z) = \frac{1}{2\pi\sqrt{-1}}\int_C d\zeta(\zeta-z)^{m+1}T(\zeta)\phi(u,z)$$

for a positively oriented small contour C around z. Combining with the operator product expansions, we obtain

$$(2.1.17)\quad \begin{aligned}[X[0]\phi](u,z) &= \phi(Xu,z),\\ [X[m]\phi](u,z) &= 0 \quad \text{for} \quad m>0\end{aligned}$$

$$(2.1.18)\quad [L_{-1}\phi](u,z) = \frac{\partial}{\partial z}\phi(u,z), \qquad [L_0\phi](u,z) = \Delta_j\phi(u,z),$$

$$[L_m\phi](u,z) = 0 \quad \text{for} \quad m>0.$$

Starting from a primary field ϕ of spin j, we get new operators by the iterations of the operations of $X[m]$ and L_m, $m \leqslant 0$, of type 2.1.15 and 16. They are classified into the *levels* by the eigenvalues of the operator L_0, e.g., $L_{-n_1}L_{-n_n}\ldots L_{-n_k}\phi$ has an eigenvalue $\Sigma_{j=1}^k n_k + \Delta_j$ with respect to the operation of L_0. This is the whole spectrum of our operators. From the operator product expansions, we deduce the following *local Ward identities* :

(2.1.19) $\langle X(\zeta)\phi_1(z_1)\ldots\phi_n(z_n)\rangle$

$$= \Sigma_{\alpha=1}^{n}\frac{1}{\zeta - z_\alpha}\langle\phi_1(z_1)\ldots[X[0]\phi_\alpha](z_\alpha)\ldots\phi_n(z_n)\rangle$$

(2.1.20) $\langle T(\zeta)\phi_1(z_1)\ldots\phi_n(z_n)\rangle$

$$= \Sigma_{\alpha=1}^{n}\left(\frac{\Delta_{j\alpha}}{(\zeta - z_\alpha)^2} + \frac{1}{\zeta - z_\alpha}\frac{\partial}{\partial z_\alpha}\right)\langle\phi_1(z_1)\ldots\phi_n(z_n)\rangle.$$

Here ϕ_α is supposed to be a primary field of spin j_α.

THEOREM 2.1.21 (Knizhnik and Zamolodchikov [18]). – *The n-point function* $\Phi = \langle\phi_1(z_1)\ldots\phi_n(z_n)\rangle$ *satisfies the total differential equation*

$$d\Phi = \Sigma_{1\leqslant\alpha<\beta\leqslant n}\frac{1}{\ell + 2}\Omega_{\alpha\beta}d\log(z_\alpha - z_\beta).\Phi.$$

Here ϕ_α *is a primary field of spin* j_α *and* $\Omega_{\alpha\beta}\in \mathrm{End}(V_{j_1}\otimes\cdots\otimes V_{j_n})$ *is determined by* 1.1.2 *via spin* j_α *representations of* $\mathfrak{sl}(2,\mathbf{C})$.

Proof. – Let $\phi(u,z)$ be a primary field. By the expression of L_0 given in 2.1.5 and the identities 2.1.17 and 18 we have

$$(\ell+2)\frac{\partial}{\partial z}\phi(u,z) = [\Sigma_\mu I_\mu[-1]I_\mu[0]\phi](u,z).$$

The RHS turns out to be the constant term of the operator product expansion of $\Sigma_\mu I_\mu(\zeta)\phi[I_\mu u, z]$. This implies that

$$(\ell+2)\frac{\partial}{\partial z}\phi(u,z) = \lim_{\zeta\to z}[\Sigma_\mu I_\mu(\zeta)\phi[I_\mu u,z] - \frac{1}{\zeta - z}\phi[\Omega u,z]]$$

where Ω denotes the Casimir operator. Combining with the local Ward identity 2.1.19 we have

$$(\ell+2)\frac{\partial}{\partial z_\alpha}\Phi = \Sigma_{\beta\neq\alpha}\frac{\Omega_{\alpha\beta}}{z_\alpha - z_\beta}\Phi$$

which proves our Theorem.

2.2. Monodromy associated with higher representations of $\mathfrak{sl}(2,\mathbf{C})$.

For a half integer $j \geqslant 0$, we denote by V_j the irreducible left $\mathfrak{sl}(2,\mathbf{C})$ module of spin j, which is an irreducible representation of dimension $2j+1$. We now proceed to discuss the monodromy representation $\theta : B_n \to \mathrm{End}(V_j^{\otimes n})$ of the connection associated with the spin j representation of $\mathfrak{sl}(2,\mathbf{C})$ in the sense of Sect. 1.1. For this purpose we first recall a « fusion » process for solutions of Yang-Baxter equations due to Jimbo [11]. Let us start with the matrix $T(q)$ given in 1.2.9 with $m = 2$. We put

$$\bar{R}(x,q) = xq^{-1}T(q) - x^{-1}qT(q)^{-1}.$$

The matrix $R(x,q) = \sigma\bar{R}(x,q)$ is a solution of the quantum Yang-Baxter equation. We have an expansion of the form

$$\text{(2.2.1)} \qquad R(x,q) = (x-x^{-1})\{1+r(x)(q-1) + \cdots\}$$

with its classical limit $r(x)$. We put

$$R_k(x,q) = R_{k,2m}(x,q)R_{k,2m-1}(xq,q) \ldots R_{k,m+1}(xq^{m-1},q)$$

which is considered to be an element of $\mathrm{End}(V^{\otimes m} \otimes V^{\otimes m})$. Here $R_{\alpha,\beta}$ stands for the matrix R acting on the α-th and β-th factors and $V = \mathbf{C}^2$. We now define $R^{(m)}(x,q)$ as

$$R^{(m)}(x,q) = R_1(x,q)R_2(xq,q) \ldots R_m(xq^{m-1},q).$$

Let us regard V as a $U\hat{}(\mathfrak{sl}(2,\mathbf{C}))$ module and we denote by $\hat{V_j}$ the irreducible $U\hat{}(\mathfrak{sl}(2,\mathbf{C}))$ module of spin j considered as a subspace of $V^{\otimes 2j}$. This is denoted by L_{2j} in [9] Sect. 3. The matrix $R^{(m)}(x,q)$ defined above determines an endomorphism of $\hat{V_j} \otimes \hat{V_j}$ with $j = m/2$. Let us define the matrix $T^{(m)}(q)$ by

$$\text{(2.2.2)} \qquad T^{(m)}(q) = \lim_{x\to\infty} x^{-m^2}\,\overline{R}^{(m)}(x,q).$$

This matrix is also expressed explicitly as

$$\text{(2.2.3)} \quad T^{(m)}(q) = q^{-m^3}(T_mT_{m-1} \ldots T_1)(T_{m+1}T_m \ldots T_2) \ldots \ldots (T_{2m-1}T_{2m-2} \ldots T_m)$$

where T_i denotes the matrix $T(q)$ on the i-th and $(i+1)$-st factors.

THEOREM 2.2.4. — *As the monodromy of the connection associated with the spin* $j = m/2$ *representation of* $\mathfrak{sl}(2,\mathbf{C})$, *we get a one parameter family of linear representations* $\theta : B_n \to \mathrm{End}(W^{\otimes n}) \otimes \mathbf{C}\{\lambda\}$ *with* $W = \hat{V_j}$ *defined by*

$$\theta(\sigma_i) = q^{-1/4}\{\mathbf{1} \otimes \cdots \otimes T^{(m)}(q) \otimes \cdots \otimes \mathbf{1}\}, \quad 1 \leqslant i \leqslant n-1,$$

where $q = \exp(-\pi\sqrt{-1}\lambda)$ *and* $T^{(m)}(q)$ *is on the i-th and* $(i+1)$*-st factors.*

Let $\iota : B_n \to B_{mn}$ be a homomorphism defined by

$$(2.2.5) \quad \iota(\sigma_i) = (\sigma_{\alpha+m}\sigma_{\alpha+m-1}\cdots\sigma_{\alpha+1}).(\sigma_{\alpha+m+1}\sigma_{\alpha+m}\cdots\sigma_{\alpha+2})\ldots \\ \ldots(\sigma_{\alpha+2m-1}\alpha_{\alpha+2m-2}\cdots\sigma_{\alpha+m})$$

with $\alpha = (i-1)m$. This « parallel » embedding is illustrated in the following picture:

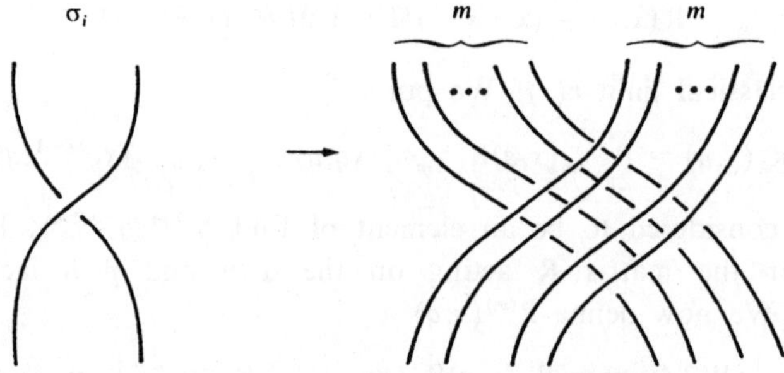

Fig. 2.

By means of this homomorphism our monodromy representation θ is also expressed in the following manner:

COROLLARY 2.2.6. — *Let* $\varphi : B_{mn} \to \mathrm{End}(V^{\otimes mn}) \otimes \mathbf{C}\{\lambda\}$ *be the Pimsner-Popa-Temperley-Lieb representation defined by* $\varphi(\sigma_i) = \mathbf{1} \otimes \cdots \otimes T(q) \otimes \cdots \otimes \mathbf{1}$ *(see 1.2.9.). Then the composition* $\varphi \circ \iota :$ $B_n \to \mathrm{End}(V^{\otimes mn}) \otimes \mathbf{C}\{\lambda\}$ *leaves invariant the subplace* $(V^{\otimes n})$ *and the monodromy representation* θ *is given by*

$$\theta(\sigma_i) = q^{-m^3 - \frac{1}{4}}\varphi \circ \iota(\sigma_i), \quad 1 \leqslant i \leqslant n-1.$$

It turns out that our monodromy representation is the same as that studied by Murakami [20] up to a scalar representation.

Proof of Theorem 2.2.4. – We put $m = 2j$. It follows from the fact that $R^{(m)}(x,q)$ is a solution of the Yang-Baxter equation ([11] Th. 2) that the correspondence in the statement of Theorem 2.2.4 actually defines a linear representation of B_n. Let ρ denote the spin j representation of $\mathfrak{sl}(2,\mathbf{C})$. By using the classical limit $r(x)$ of $R(x,q)$, we have an expansion

$$(2.2.6) \quad R^{(m)}(x,q) = (x-x^{-1})^{m^2}\{\mathbf{1}+(\rho\otimes\rho)(r(x))(q-1)+\cdots\}.$$

By the definition of $T^{(m)}(q)$ and the above formula we have

$$\frac{d}{dq}\,T^{(m)}(q) = (\rho\otimes\rho)\left(r-t-\frac{1}{2}\mathbf{1}\right).$$

Here r and t are defined in Sect. 1.2. As a consequence we have

$$\frac{d}{dq}\,T^{(m)}(q)^2|_{q=1} = (\rho\otimes\rho)(-2t-\mathbf{1}).$$

This implies that $T^{(m)}(q)^2$ has an expansion

$$\mathbf{1} + 2\pi\sqrt{-1}\lambda\left\{(\rho\otimes\rho)(t) + \frac{1}{2}\mathbf{1}\right\} + \mathcal{O}(\lambda^2)$$

with $\lambda = -(\pi\sqrt{-1})^{-1}\log q$, $-\pi \leqslant \operatorname{Im}\log q < \pi$. Let us observe that the eigenvalues of $\varphi(\gamma_k)$, $2 \leqslant k \leqslant n-1$, are of the form q^α with some integer α. Hence the same argument as in the proof of Theorem 1.2.8 can be applied to our Theorem.

2.3. Unitarity of the monodromy of n-point functions.

Let us now apply the fusion process introduced in the previous section to a description of the monodromy of n-point functions when ϕ_α, $1 \leqslant \alpha \leqslant n$, are vertex operators of spin j. For a pair of half integers (j,t), we denote by $\Gamma^j_{n,t}$ the set defined by

$$\Gamma^j_{n,t} = \{(p_0,p_1,\ldots,p_n)\,;\ p_i \in \frac{1}{2}\mathbf{Z}_{\geqslant 0} \text{ such that } p_0 = 0,\ p_n = t \text{ and each triple } v_i = (p_{i-1},j,p_i) \text{ satisfies the conditions 2.1.11 and 12}\}.$$

We fix a positive integer ℓ. To each element of $\Gamma^j_{n,\ell}$ we associate the composition of vertex operators of type v_i, $1 \leqslant i \leqslant n$. This defines the n-point function

$$\phi_{v_1 \dots v_n}(z_1, \dots, z_n) = \langle \mathrm{vac} | \phi_{v_1}(z_1) \dots \phi_{v_n}(z_n) | v \rangle$$

for $v \in \mathrm{V}_\ell$. It is shown in [22] that this is a holomorphic function in the region $|z_1| > \cdots > |z_n|$ and is analytically continued to a multi-valued holomorphic function on X_n. Moreover, they showed that the monodromy of the n-point functions associated with $\Gamma^j_{n,\ell}$ defines a linear representation of the braid group B_n, which we denote by θ: $\mathrm{B}_n \to \mathrm{End}(\mathrm{W}^j_{n,\ell})$. Our main object is to describe this representation.

Let us remark that the above composition of vertex operators is illustrated by the lattice obtained from the decomposition of $\mathrm{V}_j \otimes \cdots \otimes \mathrm{V}_j$ into simple $\mathfrak{sl}(2,\mathbf{C})$ modules. Here are some examples.

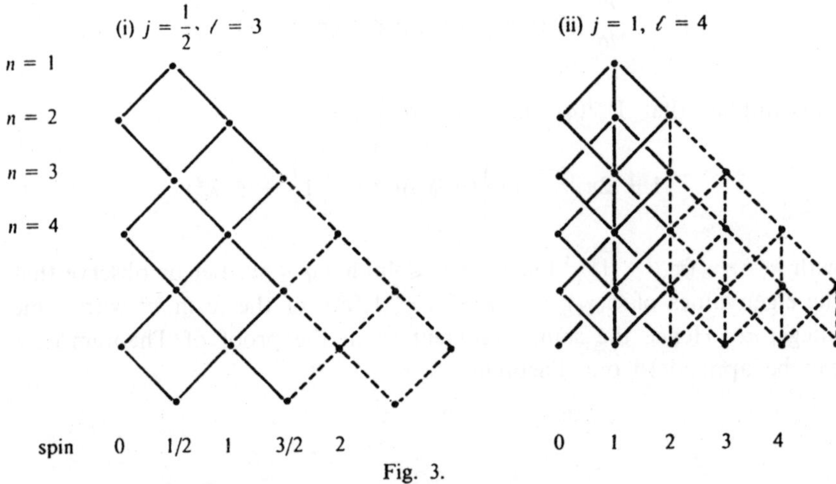

Fig. 3.

We denote by $\langle \alpha, \beta \rangle$ the atom corresponding to $n = \alpha$ and spin β. The composition of vertex operators defined by $(p_0, \dots, p_n) \in \Gamma^j_{n,\ell}$ is represented by the path connecting $\langle 1, p_1 \rangle, \dots, \langle n-1, p_{n-1} \rangle$. By means of an explicit computation of the 4-point functions Tsuchiya and Kanie showed that in the case $j = 1/2$ the monodromy $\theta : \mathrm{B}_n \to \mathrm{End}(\mathrm{W}^{1/2}_{n,\ell})$ factors through the Jones algebra with index $\tau^{-1} = 4\cos^2 \dfrac{\pi}{\ell + 2}$ (see [13]) and is equivalent to an irreducible unitarizable representation of B_n obtained by Wenzl [24]. Here we may identify the lattice illustrated in Fig. 3 (i) to the Bratteli diagram of the corresponding Jones algebra (see [22] Th. 5.2). Our result is as follows :

THEOREM 2.3.1. — *For any positive half integer* j, *the monodromy of* n*-point functions* $\theta : B_n \to \mathrm{End}\,(W^j_{n,t})$ *is unitarizable.*

Outline of Proof. — Let us first recall the differential equation satisfied by the n-point functions (Th. 2.1.21). Let $\iota : B_n \to B_{2jn}$ be the homomorphism defined by 2.2.5 with $m = 2j$. Let $\theta_0 : B_{2jn} \to \mathrm{End}\,(W^{1/2}_{2jn,t})$ be the monodromy of $2jn$-point functions with spin $1/2$. It follows from [22] Th. 5.2 that θ_0 is unitarizable. In particular, the matrices

$$(\theta_0 \circ \iota)(\sigma_1 \ldots \sigma_{k-1})^k, \qquad 1 \leqslant k \leqslant n,$$

are diagonalizable. Hence we may apply an argument of the proof of Theorem 2.2.5 and Corollary 2.2.6 to our situation (see also Remark 1.3.14). This implies that the monodromy representation $\theta : B_n \to \mathrm{End}\,(W^j_{n,t})$ is equivalent to a subrepresentation of the representation given by the correspondence

$$\sigma_i \to q^{\mu}\theta_0 \circ \iota(\sigma_i)$$

with some constant μ. Combining with the fact that θ_0 is unitarizable we obtain our Theorem.

BIBLIOGRAPHY

[1] K. AOMOTO, Fonctions hyperlogarithmiques et groupes de monodromie unipotents, *J. Fac. Sci. Tokyo*, 25 (1978), 149-156.

[2] J. BIRMAN, Braids, links, and mapping class groups, *Ann. Math. Stud.*, 82 (1974).

[3] A. A. BELAVIN and V. G. DRINFEL'D, Solutions of the classical Yang-Baxter equation for simple Lie algebras, *Funct. Anal. Appl.*, 16 (1982), 1-29.

[4] N. BOURBAKI, *Groupes et algèbres de Lie, IV, V, VI*, Masson, Paris (1982).

[5] A. A. BELAVIN, A. N. POLYAKOV and A. B. ZAMOLODCHIKOV, Infinite dimensional symmetries in two dimensional quantum field theory, *Nucl. Phys.*, B241 (1984), 333-380.

[6] K. T. CHEN, Iterated path integrals, *Bull. Amer. Math. Soc.*, 83 (1977), 831-879.

[7] V. G. DRINFEL'D, Quantum groups, preprint, ICM Berkeley (1986).

[8] R. HAIN, On a generalization of Hilbert 21st problem, *Ann. ENS*, 49 (1986), 609-627.

[9] M. JIMBO, A q-difference analogue of U($\mathfrak{g}$) and Yang-Baxter equation, *Lett. in Math. Phys.*, 10 (1985), 63-69.

[10] M. Jimbo, Quantum R matrix for the generalized Toda system, *Comm. Math. Phys.*, 102 (1986), 537-547.

[11] M. Jimbo, A q-analogue of U(gl(N+1)), Hecke algebra, and the Yang-Baxter equation. *Lett. in Math. Phys.*, 11 (1986), 247-252.

[12] M. Jimbo, Quantum R matrix related to the generalized Toda system : an algebraic approach, *Lect. Note in Phys.*, 246 (1986), Springer.

[13] V. Jones, Index of subfactors, *Invent. Math.*, 72 (1983), 1-25.

[14] V. Jones, Hecke algebra representations of braid groups and link polynomials, *Ann. of Math.*, 126 (1987), 335-388.

[15] V. G. Kac, Infinite dimensional Lie algebras, *Progress in Math.*, 44, Birkhäuser (1983).

[16] T. Kohno, Série de Poincaré-Koszul associée aux groupes de tresses pures, *Invent. Math.*, 82 (1985), 57-75.

[17] T. Kohno, Linear representations of braid groups and classical Yang-Baxter equations, to appear in *Contemp. Math.*, « Artin's braid groups ».

[18] V. G. Knizhnik and A. B. Zamolodchikov, Current algebra and Wess-Zumino models in two dimensions, *Nucl. Phys.*, B247 (1984), 83-103.

[19] J. Morgan, The algebraic topology of smooth algebraic varieties, *Publ. IHES*, 48 (1978), 103-204.

[20] J. Murakami, On the Jones invariant of paralleled links and linear representations of braid groups, preprint (1986).

[21] D. Sullivan, Infinitesimal computations in topology, *Publ. IHES*, 47 (1977), 269-331.

[22] A. Tsuchiya and Y. Kanie, Vertex operators in two dimensional conformal field theory on $\mathbf{P}^1$ and monodromy representations of braid groups, preprint (1987), to appear in *Adv. Stud. In Pure Math.*

[23] J. L. Verdier, Groupes quantiques, Séminaire Bourbaki, 1987 juin.

[24] H. Wenzl, Representations of Hecke algebras and subfactors, Thesis, Univ. of Pensylvenia (1985).

Toshitake Kohno,
Department of Mathematics
Faculty of Sciences
Nagoya University
Nagoya 464 (Japan).

Statistics of Fields, the Yang-Baxter Equation, and the Theory of Knots and Links

Jürg Fröhlich
Theoretical Physics
ETH-Hönggerberg, CH-8093 Zürich

1. Introduction

In these notes we describe an analysis of the statistics problem in quantum field theory. It has been known for some time that in two space-time dimensions there is more to the problem of statistics of local fields than Bose- or Fermi statistics [1]. In three-dimensional space-time, local quantum fields obey Bose- or Fermi statistics, but "extended particles", coupled to the vacuum by fields localized on cones, may exhibit intermediate statistics, so called Θ-statistics [2].

There is a fairly close, and perhaps somewhat surprising, connection between the statistics problem in two-dimensional quantum field theory and the following three topics:

(1) Soliton quantization in two-dimensional quantum field models; [3,4,5,6].

(2) Two-dimensional, conformal quantum field theory; some aspects of the covariant quantization of superstring theory (in particular, the construction of fermion emission vertices); [7,8,9,10].

(3) Polynomial invariants for knots and links [11]. Some of these connections will be sketched briefly, but there is no room here for a detailed treatment. The results discussed in these notes may be viewed as comments on the beautiful and deep lectures by G. Mack, D. Friedan and K.Gawedzki. Some of my results, next to many other things, are also sketched in the notes of B. Schroer.

The first examples of statistics different from standard Bose- or Fermi statistics in the context of simple, two- dimensional quantum field models were encountered in [1]. A related issue, the construction of order- and disorder variables in two-dimensional classical spin systems such as the Ising model, was analyzed by Kadanoff and Ceva [12]. Since the Euclidean description of quantum field theory (see e.g. [13]) was not a well known thing, at the time, Kadanoff and Ceva do, however, not seem to have realized the implications of their work for the statistics problem in two-dimensional quantum field theory. That problem was studied, independently of [12], in [3], where the commutation relations between soliton- and meson fields, sometimes called dual algebra, in simple models exhibiting quantum kinks, like the $\lambda\varphi_2^4$-model, the Ising- or the Potts field theory, were derived and studied. The ideas of Kadanoff and Ceva and the dual algebra played a certain role in the work of Jimbo, Miwa and Sato [14] on the scaling limit of the correlation functions of the two-dimensional Ising model. In the context of the Ising model, their work unravelled a connection between the statistics of local fields and monodromy properties of their Euclidean Green functions. That connection was studied in a more general context by the author - and probably other people, as well. Unfortunately, interest in two-dimensional quantum field theory was fading, at that time, and my analysis remained somewhat incomplete and unpublished; but see [4] for a brief sketch, and [5] for results which are relevant in this context. The connection between statistics of fields and monodromy of Euclidean Green functions will be discussed in some detail in these notes.

The work of Kadanoff and Ceva on order-disorder variables was generalized by Wegner [15], who introduced the first lattice gauge theories and the appropriate order- and disorder variables, nowadays called Wilson- and 't Hooft loops. But the field-theoretic interpretation of Wegner's models as lattice approximations to Euclidean-region gauge theories is due to Wilson [16], and the extension of dual algebra to gauge theories, i.e. the 't Hooft commutation relations, is due to 't Hooft [17]. [The 't Hooft commutation relations define the dual algebras generated by Wilson- and 't Hooft loops, in four-dimensional gauge theories, and by Wilson loops and vortex operators in three-dimensional Higgs models. The latter case was actually already considered in 1975/76 in connection with vortex quantization in three- dimensional Higgs models.] The 't Hooft commutation relations have, a priori, nothing immediately obvious to do with the statistics of fields or particles. But it was recognized by Wilczek [18] and systematized by Wu Yong Shi [2] and others that strange statistics can appear in gauge theories in three space-time dimensions: In the presence of a Chern-Simons term in the action (which violates parity) three-dimensional Higgs-theories can have fractionally charged, magnetic vortices among their one-particle states. A magnetic vortex with a fractional electric charge is a particle with intermediate spin and intermediate statistics, i.e. its spin need not be integer or half-integer, and its statistics may be intermediate between Bose- and Fermi statistics (it is called Θ-statistics), depending on the value of its electric charge. For this reason, Wilczek has called them "anyons".

The mathematical connection between 't Hooft type commutation relations and the spin and statistics of anyons has been pinned down in [19], where field theory models of anyons are discussed. The different possibilities for particle statistics in relativistic theories in three space-time dimensions appear to be exhausted by Θ-statistics. This is a consequence of the structure of the three-dimensional quantum mechanical Poincaré group and a spin-statistics connection.

In four- or higher dimensional field theories, statistics of fields and particles is limited to Bose- or Fermi statistics. A deep analysis of statistics within the algebraic approach to local, relativistic quantum physics has been carried out by Doplicher, Haag and Roberts [20]. [Unfortunately, they did not consider the more exotic possibilities appearing in two and three space-time dimensions.]

My work with P.-A. Marchetti on anyons, several stimulating discussions with V. Jones and some aspects of the covariant quantization of superstrings rekindled my interest in the statistics problem in two and three dimensions and its connections with monodromy properties of Euclidean Green functions. In the following, I present a preliminary report on my results. In the next section, some examples of two-dimensional field theory models with soliton states and exotic statistics are briefly reviewed. This serves as a motivation for the model-independent analysis of the statistics problem presented in Sect. 3. Readers not familiar with the elements of constructive quantum field theory may wish to skip Sect. 2 in a first reading and possibly return to it later for motivation. (No attempt at a pedagogical presentation is made in Sect. 2).

In Sect. 4, we describe the relations between statistics of fields, monodromy properties of Euclidean Green functions and representations of the braid groups (or the pure braid groups) on n strings. We then discuss similar issues in the context of two-dimensional conformal quantum field theory. In Sect. 5, we sketch an application to the Wess-Zumino-Witten (WZW) models. In Sect. 6, we sketch connections of the issues discussed in Sects. 3 through 5 with the theory of knots and links. In Sect. 7, we formulate some open problems and conjectures.

Acknowledgements. I wish to thank P.-A. Marchetti and V. Jones for several very stimulating discussions, V. Jones for giving me copies of some of his very interesting unpublished notes on knots and links, T. Kohno for interesting seminars and some preprints, and A. Connes, V. Jones and G. Mack for much needed encouragement.

2. Dual algebras in two-dimensional quantum field models.

In this section we consider some Lagrangian quantum field models in two space-time dimensions, with Lagrangian density given by

$$\mathcal{L}(\varphi) = \mathcal{L}_0(\varphi) - \mathcal{L}_I(\varphi), \tag{2.1}$$

$$\mathcal{L}_0(\varphi) = (\partial_t\varphi)^2 - (\partial_x\varphi)^2, \tag{2.2}$$

and $\mathcal{L}_I(\varphi)$ is a local polynomial in the field φ which is bounded from below.

The field φ is a real, scalar field with $N = 1, 2, 3 \cdots$ components, i.e. $\varphi = (\varphi^1, \cdots, \varphi^N)$. The polynomial $\mathcal{L}_I(\varphi)$ is chosen such that $\mathcal{L}(\varphi)$ has a discrete symmetry group, G. It is then possible to choose the coefficients of $\mathcal{L}_I$ in such a way that the physical vacuum of the theory is degenerate, and pure vacuum states are not invariant under G. These claims have been established in constructive quantum field theory, long ago. See [21,22] and refs. given there. It was proven in [3] that, under these circumstances, the theory has non-trivial soliton sectors orthogonal to the vacuum sectors. The soliton sectors are superselection sectors labelled by a multiplicative topological charge. Furthermore, one can construct soliton fields, $s_g(g \in G)$, such that a dense set of states in the soliton sectors can be constructed by applying the fields s_g to a dense set of states in a vacuum sector. The fields constructed in [3] are bounded but in no sense strictly local. Unbounded, local soliton fields have been constructed in [5,6] for models with an abelian, internal symmetry group G. If G is non-abelian the fields are genuinely non-local, in every sense of the word, (although they still correspond to localized *automorphisms of a G-invariant observable algebra). This is an interesting situation which Marchetti and I have studied in the context of some models. [A brief analysis of a situation somewhat more general than the one envisaged here is contained in [6].] So far, we have results for specific models, but a general, model-independent analysis is only beginning to emerge. Suffice it to say that in studying this problem one is very naturally led to loop groups and their representation theory. [The loop group associated with $S0(2)$ made an appearence in [3], and, in retrospect, the study of certain representations of rather general loop groups would have been well motivated by the kinds of problems considered in [3]. An account of our results on models with non-abelian symmetry groups and of some model- independent facts will appear elsewhere.] In this section, we propose to analyze the field algebra generated by φ and by the soliton fields s_g in the case where φ and s_g are local, but in general not relatively local fields. We therefore choose G to be abelian. Without loss, we may assume that $G = Z_n$, for some $n = 2, 3, 4, \cdots$ and $N = 1$ *or* 2. [For $G = Z_2$, we may choose $N = 1$, for $G = Z_n$, $n \geq 3$, we choose $N = 2$.] If $N = 2$ it is advantageous to work with a complex field $\chi = \varphi^1 + i\varphi^2$. Everyone who knows a little constructive quantum field theory knows that for each $n = 2, 3, 4, \cdots$, one can choose a polynomial, P_{2k}, of degree $2k$, where $2k$ is the smallest even integer larger than n, such that P_{2k} is Z_n-invariant and bounded from below, and the quantum field theory with Lagrangian

$$\mathcal{L}(\chi) = \mathcal{L}_0(\chi) - :P_{2k}(\chi): \tag{2.3}$$

has a spontaneously broken Z_n-symmetry and an n- fold degenerate vacuum. (In (2.3) the double colons indicate Wick ordering.)

The methods of [3,5,6] then permit us to construct local soliton fields, $s_1, \cdots, s_{n-1}$, mapping a vacuum sector onto $n-1$ different soliton sectors. The field s_l is obtained from an l-fold product of s_1, with the help of an operator product expansion.

Let x and y be points in two-dimensional space-time, M^2, (with Lorentzian metric given by $\left(\begin{smallmatrix}1 & 0\\ 0 & -1\end{smallmatrix}\right)$).These points are space-like separated, written $x \times y$, iff $(x-y)^2 < 0$. If $x \times y$ then the inequalities

$$\vec{x} \gtrless \vec{y}, \tag{2.4}$$

where $\vec{x}$ is the space component of x, ($x^0 = t$ is its time component) are Lorentz-invariant. This is because the complement of the closure of the interior of the light cone in $\mathcal{M}^2$ is disconnected. These are the key facts behind the appearance of exotic statistics in two space-time dimensions.

It is shown in [3,4] that $s_k(x)$ and $\chi(y)$ satisfy the following "dual algebra": If $x \times y$ then

$$[s_l(x), s_{l'}(y)] = 0 = [\chi(x), \chi(y)], \tag{2.5}$$

and

$$s_l(x)\chi(y) = \begin{cases} \chi(y)s_l(x), & \text{if } \vec{y} < \vec{x} \\ e^{2\pi i \frac{l}{n}} \chi(y) s_k(x), & \text{if } \vec{y} > \vec{x}. \end{cases} \tag{2.6}$$

Let $\psi_l(x)$ be the field obtained from the product $s_l(x)\chi(y)$, as $y \to x$, (operator product expansion). Then it easily follows from (2.5), (2.6) that, for $x \times y$,

$$\psi_l(x)\psi_l(y) = e^{2\pi i \frac{l}{n}} \psi_l(y)\psi_l(x) \tag{2.7}$$

$$\psi_l(x)\psi_{l'}(y) = e^{2\pi i (\frac{l'}{n}\Theta(\vec{x}-\vec{y}) + \frac{l}{n}\Theta(\vec{y}-\vec{x}))} \psi_{l'}(y)\psi_l(x) \tag{2.8}$$

$$\psi_l(x)s_{l'}(y) = e^{2\pi i \frac{l'}{n}\Theta(\vec{x}-\vec{y})} s_{l'}(y)\psi_l(x) \tag{2.9}$$

and

$$\psi_l(x)\chi(y) = e^{2\pi i \frac{l}{n}\Theta(\vec{y}-\vec{x})} \chi(y)\psi_l(x) \tag{2.10}$$

If $n = 2$ and if $\chi(x) = \varphi(x)$ is a real, one-component scalar field then $\psi(x) \equiv \psi_1(x)$ is a real Fermi field, i.e. a Majorana field, according to (2.7). In this case, the dual algebra (2.5) - (2.10) is the one encountered in the $\lambda\varphi_2^4$-model and in the two-dimensional Ising field theory. The Ising field theory is completely characterized by the requirement that $\psi(x)$ be a free Majorana field. [In the $\lambda\varphi_2^4$-model, $\psi(x)$ is not a free field and is very singular in the ultraviolet. See [6].] In these examples, let us denote $\varphi(x)$ by $\Phi_1(x)$ and $\psi(x)$ by $\Phi_2(x)$. Let $x - y$ be space-like and $\sigma = sign(\vec{x} - \vec{y})$. Then by (2.7) and (2.10),

$$\Phi_\alpha(x)\Phi_\beta(y) = R_{\alpha\beta}^{\gamma\delta}(\sigma)\Phi_\gamma(y)\Phi_\delta(x) \tag{2.11}$$

where $R(+)$ and $R(-)$ are 4×4 matrices given by

$$R(+) = \begin{pmatrix} 1 & 0 & 0 & 0 \\ 0 & 0 & 1 & 0 \\ 0 & -1 & 0 & 0 \\ 0 & 0 & 0 & -1 \end{pmatrix}, R(-) = \begin{pmatrix} 1 & 0 & 0 & 0 \\ 0 & 0 & -1 & 0 \\ 0 & 1 & 0 & 0 \\ 0 & 0 & 0 & -1 \end{pmatrix} \tag{2.12}$$

Note that

$$R(+)R(-) = 1 \tag{2.13}$$

[This is no accident, as will be seen in Sect. 3.] The matrices $R(\sigma), \sigma = + \text{or} -$, are called statistics matrices. Statistics matrices can be derived from (2.7) - (2.10), for all the examples discussed, so far, i.e. for $G = Z_n$, $n = 2, 3, 4, \cdots$. For $n \geq 3$, we call the fields $\psi_l(x), l = 1, \cdots, n-1$, parafermion fields.

It may be useful to recall that exotic statistics already appears in the theory of the free, massless scalar field, φ, in two space-time dimensions, as discussed e.g. by Streater and Wilde [1]. We define

$$\Phi_\lambda(x) =: \exp i\lambda[\dot{\varphi}(\Theta_x) + \varphi(x)] : \tag{2.14}$$

where $\dot{\varphi} = \partial_t \varphi, \Theta_x(\vec{y}) = \Theta(\vec{y} - \vec{x})$ is the Heavyside step function, and the double colons denote Wick ordering. [A rigorous definition of Φ_λ can be found, for example in [6], but has been known for a long time.] Then the Weyl form of the canonical commutation relations implies that, for $x \times y$,

$$\Phi_\lambda(x)\Phi_{\lambda'}(y) = exp[-i\lambda\lambda'\sigma]\Phi_{\lambda'}(y)\Phi_\lambda(x), \tag{2.15}$$

with $\sigma = \text{sign}\ (\vec{x} - \vec{y})$. This example plays a certain rôle in chiral bosonization and in vertex operator constructions; see e.g. [6] and refs. given there. In [6] the Euclidean Green functions of the fields Φ_λ have been calculated. With a view to what is discussed in Sect. 5, it may be worthwhile to remark that these Green functions give rise to the so-called Burau representation of the braid groups.

For more detailed discussions of soliton quantization and exotic statistics in the framework of simple two-dimensional models we refer the reader to [3-6, 19].

3.Statistics of "local fields" in two-dimensions,and the Yang-Baxter equation.

For the benefit of the reader who has skipped Sect. 2 we recall some definitions: Two-dimensional Minkowski space is denoted by $\mathcal{M}^2$, with Lorentzian metric given by $\binom{1\ \ 0}{0\ -1}$. Two points, x and y, in $\mathcal{M}^2$ are said to be space-like separated, written $x \times y$, if $(x-y)^2 = (x^0 - y^0)^2 - (\vec{x} - \vec{y})^2 < 0$. [Here x^0 is the time component of $x \in \mathcal{M}^2$, and $\vec{x}$ its space component.] Let $\bar{V}_+$ and $\bar{V}_-$ denote the closures of the interiors of the forward and backward light cones and $\bar{V} = \bar{V}_+ \cup \bar{V}_-$. Clearly $\mathcal{M}^2 \backslash \bar{V}$ is disconnected. Therefore if $x \times y$ the inequalities

$$\vec{x} \gtrless \vec{y} \tag{3.1}$$

are Lorentz invariant. We set $\sigma \equiv \sigma_{xy} = sign(\vec{x} - \vec{y})$. Two regions, 0_1 and 0_2, in $\mathcal{M}^2$ are said to be space-like separated, written $0_1 \times 0_2$, if $x \times y$, for every $x \in 0_1$ and every $y \in 0_2$. By $\mathcal{P}_+^\uparrow$ we denote the connected component of the two-dimensional Poincaré group.

We consider a local, relativistic quantum field theory over $\mathcal{M}^2$, in the sense of Wightman [23]. [We must assume that the reader has a vague idea about what a Wightman theory is. We shall not treat any domain problems arising when one deals with unbounded operators, but we shall never implicitly assume more than standard properties of fields in a Wightman theory.] We suppose that the observable fields of the theory are local tensors $J^\alpha_{\mu_1 \cdots \mu_k}(x)$, where $\mu_1, \cdots, \mu_k$ are Lorentz indices, and α labels the different tensor fields of equal rank. By locality of observable fields we mean that

$$[J^\alpha_{\mu_1 \cdots \mu_k}(x), J^\beta_{\nu_1 \cdots \nu_l}(y)] = 0 \tag{3.2}$$

if $x \times y$. Typically, it is reasonable to imagine that among the observable fields of the theory there be the energy-momentum tensor, $T_{\mu\nu}(x)$, of the theory.

In the models discussed in Sect. 2, the observable fields must be invariant under the internal symmetry group, G (see [4]), and it is convenient to choose these fields to be given by e.g. $J(x) =: (\varphi \cdot \varphi) : (x)$ and $T_{\mu\nu}(x)$.

If 0 is an open subset of $\mathcal{M}^2$ we let $\mathcal{A}(0)$ denote the local algebra generated by the operators

$$J^{\alpha}_{\mu_1 \cdots \mu_k}(f) \equiv \int d^2x J^{\alpha}_{\mu_1 \cdots \mu_k}(x) f(x), \tag{3.3}$$

where $f(x)$ is a test function on $\mathcal{M}^2$ vanishing in the complement of 0. [It may be more appropriate to define $\mathcal{A}(0)$ to be the von Neumann algebra generated by the bounded functions of $J^{\alpha}_{\mu_1 \cdots \mu_k}(f)$, supp $f \subset 0$; but we don't have to go into details about this.] Locality can now be expressed by saying that if $0_1 \times 0_2$ then

$$[A, B] = 0, \quad \forall A \in \mathcal{A}(0_1), \forall B \in \mathcal{A}(0_2) \tag{3.4}$$

The theory is Lorentz covariant if $\mathcal{A} = \bigvee_0 \mathcal{A}(0)$ carries a representation, τ, of $\mathcal{P}_+^{\uparrow}$ as a group of *automorphisms, with the property that, for $(\Lambda, a) \in \mathcal{P}_+^{\uparrow}$,

$$\tau_{(\Lambda,a)}(\mathcal{A}(0)) = \mathcal{A}(0_{(\Lambda,a)}), \tag{3.5}$$

where

$$0_{(\Lambda,a)} = \{x \in \mathcal{M}^2 : \Lambda^{-1}(x - a) \in 0\}.$$

A vacuum sector of the theory is a separable Hilbert space, $\mathcal{H}_0$, which carries an irreducible representation, π_0, of $\mathcal{A}$ with the property that τ is unitarily implementable on $\mathcal{H}_0$, i.e.

$$\tau_{(\Lambda,a)}(A) = U_0(\Lambda, a) A U_0(\Lambda, a)^{-1}, \; ^{*)} \tag{3.6}$$

for all $A \in \mathcal{A}, (\Lambda, a) \in \mathcal{P}_+^{\uparrow}$, and U_0 is a unitary representation of $\mathcal{P}_+^{\uparrow}$ on $\mathcal{H}_0$. Moreover, the generators $(H_0, \vec{P}_0)$ of the translations $U_0(1, a)$ are supposed to satisfy the relativistic spectrum condition, i.e. spec $(H_0, \vec{P}_0) \subseteq \bar{V}_+$. Finally, it is assumed that $(0, \vec{0})$ is a simple eigenvalue of $(H_0, \vec{P}_0)$, i.e. there exists a unique ray, Ω, in $\mathcal{H}_0$ such that $U_0(1, a)\Omega = \Omega, \forall a$, hence

$$U_0(\Lambda, a)\Omega = \Omega, \quad \forall (\Lambda, a) \in \mathcal{P}_+^{\uparrow} \tag{3.7}$$

Ω is called vacuum, π_0 a vacuum representation. By the Reeh - Schlieder theorem [23], Ω is separating for $\mathcal{A}(0)$, for all bounded 0, i.e. if $A\Omega = 0, A \in \mathcal{A}(0)$, then $A = 0$. One lesson the reader who has read Sect. 2, or the papers quoted there, should have learned is that not all the physical states of such a theory need to be rays in the vacuum sector $\mathcal{H}_0$. For example, if the theory has soliton states these states do not belong to $\mathcal{H}_0$. In attempting to find all physical states of the theory, it is useful to introduce the concept of covariant representations. A representation, π, of $\mathcal{A}$ on a Hilbert space $\mathcal{H}_\pi$ is called covariant(in these notes - but not everywhere in the

* We identify here $\pi_0(A)$ with A, to keep notations simple.

literature) if τ is unitarily implementable on $\mathcal{H}_\pi$, i.e. there is a unitary representation, U_π, of $\mathcal{P}_+^\uparrow$ on $\mathcal{H}_\pi$ such that

$$\pi(\tau_{(\Lambda,a)}(A)) = U_\pi(\Lambda,a)\pi(A)U_\pi(\Lambda,a)^{-1},$$

for all $A \in \mathcal{A}, (\Lambda, a) \in \mathcal{P}_+^\uparrow$, and, moreover, the generators $(H_\pi, \vec{P}_\pi)$ of $U_\pi(1,a)$ satisfy the relativistic spectrum condition. [Of course, it is not assumed that $\mathcal{H}_\pi$ contain a vacuum.]

One of the basic problems of quantum field theory is this:

Given $(\mathcal{A}, \tau)$, find all covariant representation, π, of $\mathcal{A}$.

Assuming that we are able to solve this problem, we then define the physical Hilbert space, $\mathcal{H}$, of the theory by

$$\mathcal{H} = \bigoplus_{\substack{\text{cov. reps.}\\ \pi \text{ of } \mathcal{A}}} \mathcal{H}_\pi, \tag{3.8}$$

and

$$U = \bigoplus_{\substack{\text{cov. reps.}\\ \pi \text{ of } \mathcal{A}}} U_\pi. \tag{3.9}$$

It is generally believed that $(\mathcal{A}, \mathcal{H}, \mathcal{U})$ provides a complete description of the quantum field theory; in particular $\mathcal{H}$ will contain all one-particle states of the theory, etc.

It has turned out that, in practice, one wants to have more than just a complete list of covariant representations of the observable algebra.

One attempts to enlarge the observable algebra $\mathcal{A}$ to a field algebra, $\mathcal{F}$, which acts irreducibly on the physical Hilbert space $\mathcal{H}$. This entails that $\mathcal{F}$ will contain unobservable fields, Φ, which intertwine different covariant representations of $\mathcal{A}$, i.e. make transitions between different sectors, $\mathcal{H}_\pi$ and $\mathcal{H}_{\pi'}, (\pi, \pi'$ covariant representations of $\mathcal{A}$) of the theory. It is a general problem of algebraic quantum field theory to determine whether one can reconstruct a field algebra from a complete list of covariant representations of $\mathcal{A}$. This is quite a vast problem, and the analysis of models shows that, especially in two space-time dimensions, the answer - if there is a completely general answer - is by no means simple. As already mentioned in Sect. 2, the answer will, for example, contain a major part of the theory of loop groups and their representations. [This will be discussed in more detail elsewhere; a special case appears in [3].]

Here we shall assume that the field algebra $\mathcal{F}$ is generated by "local" fields. Examples of such fields are Fermi fields, charged fields, the soliton fields described in Sect. 2, etc. The question is what is meant by "local". One can show that there are at least two notions of locality which are sensible:

(1) Elements of $\mathcal{F}$ are fields, $\Phi(x)$, localized in space-time points $x \in \mathcal{M}^2$.

(2) Elements of $\mathcal{F}$ are fields, $\Phi(\gamma_x)$, localized on space -like curves, γ_x, starting at x and reaching out to space-like $+\infty$, or $-\infty$.

The analysis in these notes shall be restricted to case (1). We now specify what kinds of field algebras, $\mathcal{F}$, we want to admit and what exactly is meant by a local field, in case (1). [It would lead too far to present a general analysis, and my results are not complete, yet.]

We assume that there are finitely many, Poincaré-covariant fields, $\Phi_\alpha(x), \alpha = 1, \cdots, p$, localized in space-time points, $x \in \mathcal{M}^2$, such that $\mathcal{F}$ is generated by $\mathcal{A}$ and by arbitrary polynomials in $\Phi_\alpha(f)$, where f is an arbitrary test function. [In the models studied in Sect. 2, $\mathcal{F}$ is generated by the fields $\chi(x)$ and $s_l(x), l = 1, \cdots, n-1$.]

The rôle of the fields Φ_α is to intertwine the vacuum sector $\mathcal{H}_0$ with superselection sectors $\mathcal{H}_\pi$, where π is covariant representation of $\mathcal{A}$, i.e. if Ψ is a state in $\mathcal{H}_0$ belonging to the domain of definition of $\Phi_\alpha(f)$ then $\Phi_\alpha(f)\Psi$ belongs to a superselection sector $\mathcal{H}_\pi$ orthogonal to $\mathcal{H}_0$. It is assumed that $\mathcal{F}$ has a common dense domain, $\mathcal{D}$, of definition in $\mathcal{H}$, and that $\mathcal{F}$ acts irreducibly on $\mathcal{H}$.

Let $0_{x_1 \cdots x_n}$ be the space-like complement of the points $x_1, \cdots, x_n$ in $\mathcal{M}^2$; see Fig. 1.

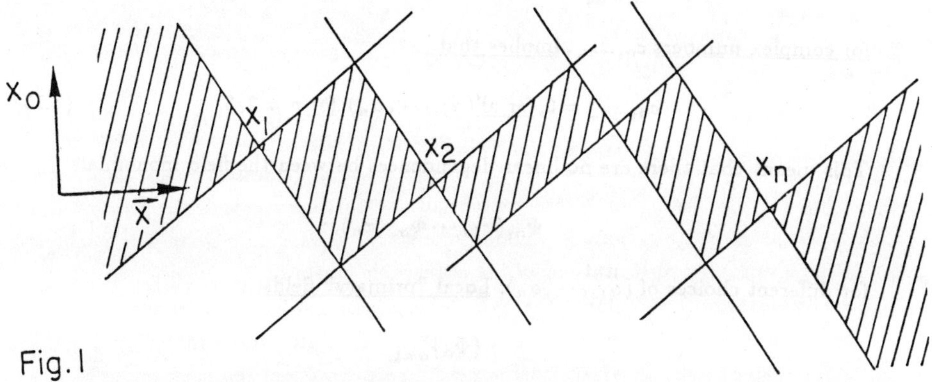

Fig. 1

Fig. 1: (The shaded region is $0_{x_1 \cdots x_n}$)

If $x_1, \cdots, x_n$ are pairwise space-like separated $0_{x_1 \cdots x_n}$ consists of of $n-1$ diamonds, a wedge opening to the left and a wedge opening to the right.

By locality of $\Phi_\alpha(x)$ we mean that

(a)

$$[\Phi_\alpha(x), A] = 0, \quad \text{for all} \quad A \in \mathcal{A}(0_x) \tag{3.10}$$

It follows that

$$[\Phi_{\alpha_1}(x_1) \cdots \Phi_{\alpha_n}(x_n), A] = 0, \quad \text{for all} \quad A \in \mathcal{A}(0_{x_1 \cdots x_n}) \tag{3.11}$$

The second property is the following

(b) Reordering Condition: If $x \times y$, then, for every choice of α and β, $\Phi_\alpha(y)\Phi_\beta(x)$ is a linear combination of $\Phi_\gamma(x)\Phi_\delta(y), \gamma, \delta = 1, \cdots, p$.

Motivation: Clearly $\Phi_\alpha(y)\Phi_\beta(x)$ and $\Phi_\gamma(x)\Phi_\delta(y)$ all commute with $\mathcal{A}(0_{xy})$.

Moreover, $\mathcal{F}$ acts irreducibly on $\mathcal{H}$, and locality of the fields Φ_α ought to mean that when forming arbitrary polynomials in $\Phi_{\alpha_1}(x_1) \cdots \Phi_{\alpha_n}(x_n)$ the order of the fields is immaterial if the points $x_1, \cdots, x_n$ are pairwise space-like separated. It is therefore reasonable to assume that $\Phi_\alpha(y)\Phi_\beta(x)$ can be formally expanded in a sum of products of $(D^{n_1}\Phi_{\gamma_1})(x)$ and $(\bar{D}^{n_2}\Phi_{\gamma_2})(y)$, where D^{n_1} and $\bar{D}^{n_2}$ are local differential operators of order n_1 and n_2, respectively. It is natural to suppose that the tensorial properties of the fields Φ_α and their ultraviolet dimensions are such that $\Phi_\alpha(y)\Phi_\beta(x)$ can be expanded in a sum of products of $\Phi_\gamma(x)$ and $\Phi_\delta(y)$ and that only monomials of order 2 appear in that expansion. [The reader is supposed to think of Φ_α as some kinds of "primary fields". Assumption (b) looks more natural in the context of conformal quantum field theory.]

Finally we impose the following

(c) Minimality Condition. Let $x_1, \cdots, x_n$ be pairwise space-like separated. Then the equation

$$\sum_{\alpha_1 \cdots \alpha_n} c_{\alpha_1 \cdots \alpha_n} \quad \Phi_{\alpha_1}(x_1) \cdots \Phi_{\alpha_n}(x_n) = 0, \tag{3.12}$$

for complex numbers $c_{\alpha_1 \cdots \alpha_n}$, implies that

$$c_{\alpha_1 \cdots \alpha_n} = 0, \text{for all} (\alpha_1, \cdots, \alpha_n), \text{for } n = 2, 3. \tag{3.13}$$

This means that there are no linear dependences between the field monomials

$$\Phi_{\alpha_1}(x_1) \cdots \Phi_{\alpha_n}(x_n),$$

for different choices of $(\alpha_1, \cdots, \alpha_n)$. Local "primary" fields

$$\{\Phi_\alpha\}_{\alpha=1}^p$$

are fields satifying $(a), (b)$ and (c). Our basic assumption in this section is that $\mathcal{F}$ is generated by $\mathcal{A}$ and by finitely many (p) local primary fields. I should like to emphasize, once more, that this assumption is special, and that there are two-dimensional field theories with soliton sectors which do not fit into this framework. It may be reasonable to conjecture that it does however accomodate all interesting conformal quantum field theories.

Next, we propose to derive the consequences of assumptions $(a), (b)$ and (c).

From assumption (b) it follows that

$$\Phi_\alpha(y)\Phi_\beta(x) = R_{\alpha\beta}^{\gamma\delta}(x,y)\Phi_\gamma(x)\Phi_\delta(y), \tag{3.14}$$

where repeated indices on the righthand side of (3.14) are to be summed over. $R(x,y) \equiv (R_{\alpha\beta}^{\gamma\delta}(x,y))$ is a $p^2 \times p^2$ square matrix.

Since the fields $\Phi_\alpha(x)$ have been assumed to be Poincaré-covariant, R can only depend on $x - y$ (translation invariance) and must commute with Lorentz transformations, i.e.

$$(S(\Lambda) \otimes S(\Lambda))R(\Lambda^{-1}(x-y)) = R(x-y)(S(\Lambda) \otimes S(\Lambda)), \tag{3.15}$$

where S is a finite-dimensional representation of the Lorentz group. By studying suitable expectation values of $\Phi_\alpha(y)\Phi_\beta(x)$ and of $\Phi_\gamma(x)\Phi_\delta(y)$ one can see that

$R(x-y)$ must be homogeneous in $x-y$ of degree 0. Hence $R(x-y)$ can only depend on $\sigma \equiv sign(\vec{x}-\vec{y})$, in two space-time dimensions, and must be independent of $x-y$, in higher dimensions. The matrix $R(x,y) \equiv R(\sigma)$ is called the statistics matrix of the theory.

Next, we exploit the minimality condition (c).

Clearly, $\sigma = sign(\vec{x}-\vec{y}) = -sign(\vec{y}-\vec{x})$.Hence

$$\begin{aligned}\Phi_\alpha(y)\Phi_\beta(x) &= R^{\gamma\delta}_{\alpha\beta}(\sigma)\Phi_\gamma(x)\Phi_\delta(y)\\ &= R^{\gamma\delta}_{\alpha\beta}(\sigma)R^{\mu\nu}_{\gamma\delta}(-\sigma)\Phi_\mu(y)\Phi_\nu(x).\end{aligned} \tag{3.15}$$

It follows from (c) that

$$R^{\gamma\delta}_{\alpha\beta}(\sigma)R^{\mu\nu}_{\gamma\delta}(-\sigma) = \delta^\mu_\alpha\delta^\nu_\beta, \tag{3.16}$$

i.e

$$\boxed{R(\sigma)R(-\sigma) = 1,} \tag{3.17}$$

in matrix notation. Since in $d \geq 3$ space-time dimensions σ is not Lorentz invariant, and hence R cannot depend on σ, it follows that

$$R^2 = 1, \text{i.e. R is a second root of 1, for} \quad d \geq 3. \tag{3.18}$$

This shows that, within the present framework of local field theory, statistics in $d \geq 3$ space-time dimensions is ordinary Bose - or Fermi-statistics, as expected. [However, for fields localized on space-like cones, more general statistics is possible in three space-time dimensions: $\Theta-$statistics of anyons. See [19].]

Next, we explore the associativity of the field algebra $\mathcal{F}$ and the minimality condition (c). We consider a product of three fields

$$\Phi_\alpha(x)\Phi_\beta(y)\Phi_\gamma(z).$$

We commute them according to the following two schemes:

$$\begin{aligned}&(i) \quad \Phi(x) \quad with \quad \Phi(y), \Phi(x) \quad with \quad \Phi(z), \Phi(y) \quad with \quad \Phi(z).\\ &(ii) \quad \Phi(y) \quad with \quad \Phi(z), \quad \Phi(x) \quad with \quad \Phi(z), \Phi(x) \quad with \quad \Phi(y).\end{aligned}$$

Let

$$\sigma_1 = sign(\vec{y}-\vec{x}), \sigma_2 = sign(\vec{z}-\vec{x}) \text{and} \quad \sigma_3 = sign(\vec{z}-\vec{y}).$$

Then scheme (i) yields

$$\begin{aligned}\Phi_\alpha(x)\Phi_\beta(y)\Phi_\gamma(z) &= R^{\mu\nu}_{\alpha\beta}(\sigma_1)\Phi_\mu(y)\Phi_\nu(x)\Phi_\gamma(z)\\ &= R^{\mu\nu}_{\alpha\beta}(\sigma_1)R^{\rho\kappa}_{\nu\gamma}(\sigma_2)\Phi_\mu(y)\Phi_\rho(z)\Phi_\kappa(x)\\ &= R^{\mu\nu}_{\alpha\beta}(\sigma_1)R^{\rho\kappa}_{\nu\gamma}(\sigma_2)R^{\lambda\omega}_{\mu\rho}(\sigma_3)\Phi_\lambda(z)\Phi_\omega(y)\Phi_\kappa(x)\end{aligned} \tag{3.19}$$

Scheme (ii) yields

$$\begin{aligned}\Phi_\alpha(x)\Phi_\beta(y)\Phi_\gamma(z) &= R^{\mu\nu}_{\beta\gamma}(\sigma_3)\Phi_\alpha(x)\Phi_\mu(z)\Phi_\nu(y)\\ &= R^{\mu\nu}_{\beta\gamma}(\sigma_3)R^{\lambda\rho}_{\alpha\mu}(\sigma_2)\Phi_\lambda(z)\Phi_\rho(x)\Phi_\nu(y)\\ &= R^{\mu\nu}_{\beta\gamma}(\sigma_3)R^{\lambda\rho}_{\alpha\mu}(\sigma_2)R^{\omega\kappa}_{\rho\nu}(\sigma_1)\Phi_\lambda(z)\Phi_\omega(y)\Phi_\kappa(x).\end{aligned} \tag{3.20}$$

Comparing (3.19) with (3.20) and using (c) yields

$$\begin{aligned} &R^{\mu\nu}_{\alpha\beta}(\sigma_1)R^{\rho\kappa}_{\nu\gamma}(\sigma_2)R^{\lambda\omega}_{\mu\rho}(\sigma_3) \\ &\quad = R^{\mu\nu}_{\beta\gamma}(\sigma_3)R^{\lambda\rho}_{\alpha\mu}(\sigma_2)R^{\omega\kappa}_{\rho\nu}(\sigma_1) \end{aligned} \tag{3.21}$$

This is a special case of the Yang-Baxter (*or* $*\ \Delta$) equation. It is advantageous to rewrite (3.21) in matrix notation, in order to see what is going on: Let $V = C^p$. Equation (3.21) is an equation between matrices on $V \otimes V \otimes V$. The first factor corresponds to the field which stands left-most, the second factor to the field in the middle, and the third factor corresponds to the field farthest to the right. We number the factors from 1 to 3, i.e. we write $V_1 \otimes V_2 \otimes V_3$, $V_i = C^p$, for $i = 1, 2, 3$.

We then define

$$R_{12} \equiv R\ |_{V_1 \otimes V_2} \otimes 1\ |_{V_3}$$

$$R_{23} \equiv 1\ |_{V_1} \otimes R\ |_{V_2 \otimes V_3}$$

Then (3.21) is equivalent to the matrix equation

$$\boxed{R_{12}(\sigma_1)R_{23}(\sigma_2)R_{12}(\sigma_3) = R_{23}(\sigma_3)R_{12}(\sigma_2)R_{23}(\sigma_1)} \tag{3.22}$$

Equations (3.17) and (3.22) are the basic equations obeyed by a statistics matrix $R(\sigma)$.

For later purposes, it is convenient to express the contents of equs. (3.17) and (3.22) graphically.

With a product $\Phi_{\alpha_1}(x_1)\cdots\Phi_{\alpha_n}(x_n)$ of local, "primary" fields we associate n points, $(0,1,0), (0,2,0), \cdots, (0,n,0)$, on the 2-axis of R^3. From these points emerge n vertical lines in the positive 3-direction. If $\Phi_{\alpha_i}(x_i)$ is commuted with $\Phi_{\alpha_{i+1}}(x_{i+1})$ the lines above $(0,i,0)$ and $(0,i+1,0)$ are exchanged, more precisely, braided: If $\vec{x}_i - \vec{x}_{i+1} \gtrless 0$ then the ith and $(i+1)^{st}$ line are braided in such a way that the 1-coordinate of a point on the ith line is always $\binom{larger}{smaller}$ than or equal to the 1-coordinate of a point on the $(i+1)^{st}$ line. The n lines end vertically at the points $(0,1,1), (0,2,1), \cdots (0,n,1)$. [In a theory where the fields Φ_α should not be chosen to be symmetric, i.e. $\Phi^*_\alpha \neq \Phi_\alpha$, the lines will bear an orientation: in the +3 direction, for Φ_α; in the −3 direction, for Φ^*_α. But in order to keep our notations simple, we shall often consider symmetric, local, "primary" fields, in these notes.] The equations (3.17) and (3.22) can now be represented, graphically, as follows:

(3.17) ⇔ = = (3.23)

For $\sigma_1 = \sigma_2 = -\sigma_3 = 1$,

(3.22) ⇔ = (3.24)

Similar graphical identities hold for the remaining five choices of σ_1, σ_2 and σ_3. [There are, a priori, eight choices of σ_1, σ_2 and σ_3, but only six of them correspond to setting $\sigma_1 = sign(\vec{y} - \vec{x}), \sigma_2 = sign(\vec{z} - \vec{x})$ and $\sigma_3 = sign(\vec{z} - \vec{y})$.]

Identities (3.23) and (3.24) are well known from the theory of knots and links: (3.23) expresses invariance under so-called Reidemeister moves of type II, (3.24) expresses invariance under Reidemeister moves of type III.

Next, we exploit the *invariance of the field algebra. We suppose that there is a matrix μ such that

$$\Phi^*_\alpha(x) = \mu^\beta_\alpha \Phi_\beta(x) \tag{3.25}$$

Since $(\Phi^*_\alpha(x))^* = \Phi_\alpha(x)$, we have

$$\overline{\mu^\beta_\alpha} \mu^\gamma_\beta = \delta^\gamma_\alpha,$$

i.e.

$$\bar{\mu}\mu = 1 \tag{3.26}$$

We define a matrix ${}^T R(\sigma)$ by

$${}^T R^{\gamma\delta}_{\alpha\beta}(\sigma) = R^{\delta\gamma}_{\beta\alpha}(\sigma) \tag{3.27}$$

Since

$$\begin{aligned}(R^{\gamma\delta}_{\alpha\beta}(\sigma)\Phi_\gamma(y)\Phi_\delta(x))^* &= (\Phi_\alpha(x)\Phi_\beta(y))^* \\ &= \Phi^*_\beta(y)\Phi^*_\alpha(x),\end{aligned}$$

it follows, with (3.14), (3.25) and (3.27), that

$$\boxed{\overline{{}^T R(\sigma)}(\mu \otimes \mu) = (\mu \otimes \mu) R(-\sigma)} \tag{3.28}$$

Here $\bar{R}$ is the matrix complex conjugate to R.

We now have to discuss which additional properties the principles of local field theory impose on solutions of equs. (3.17), (3.22) and (3.28) for such solutions to be the statistics matrices of a local field theory in two space-time dimensions.

(i) We assume that the fields $\Phi_\alpha(x)$ transform irreducibly under Lorentz transformations, i.e. for a Lorentz boost,

$$\Lambda = \begin{pmatrix} cosh\ \theta & sinh\ \theta \\ sinh\ \theta & cosh\ \theta \end{pmatrix},$$

$$\Phi_\alpha^\Lambda(x) = e^{\Theta s_\alpha} \Phi_\alpha(\Lambda^{-1} x). \tag{3.29}$$

Let s be the diagonal matrix with $s_\alpha^\alpha \equiv s_\alpha$.

It follows from (3.14) and (3.29) that

$$R(\sigma)(e^{\theta s} \otimes e^{\theta s}) = (e^{\theta s} \otimes e^{\theta s}) R(\sigma), \tag{3.30}$$

for all $\theta \in R$.

(ii) We assume that the field theory is a conformal field theory. Let Δ_α be the scaling dimension of Φ_α, and let Δ be the diagonal matrix with $\Delta_\alpha^\alpha \equiv \Delta_\alpha$. Then it follows that

$$R(\sigma)(\lambda^\Delta \otimes \lambda^\Delta) = (\lambda^\Delta \otimes \lambda^\Delta) R(\sigma), \tag{3.31}$$

for arbitrary $\lambda > 0$; (λ parametrizes dilatations).

The proofs of (3.30) and (3.31) are very simple and are left to the reader.

(iii) An important constraint on $R(\sigma)$ is the spin-statistics theorem. For a local field theory in three or more dimensions, this theorem says that fields of integer spin have Bose statistics, while fields of half-integer spin have Fermi statistics. In two space-time dimensions, the connection between spin and statistics is more complicated. Without loss of generality, we may assume that

$$< \Phi_\alpha(x)\Omega, \Phi_\beta(y)\Omega >= \delta_{\alpha\beta} G_\alpha(x, y) \tag{3.22}$$

Positivity of the metric in the physical Hilbert space $\mathcal{H}$ implies that $G_\alpha(x, y)$ is a positive definite distribution. By the spectrum condition, it admits an analytic continuation to the Euclidean region,

$$G_\alpha(x_{-\epsilon}, y_\epsilon) =< \Phi_\alpha(x_\epsilon)\Omega, \Phi_\alpha(y_\epsilon)\Omega >, \tag{3.33}$$

with $x_\epsilon = (i\epsilon, \vec{x}), y_\epsilon = (i\epsilon, \vec{y}), \epsilon > 0$, and $G_\alpha(x_{-\epsilon}, y_\epsilon)$ is positive definite, for arbitrary $\epsilon > 0$. Let $\chi_\epsilon < \pi$ be the largest angle such that $R(\chi_\epsilon)x_\epsilon$ and $R(\chi_\epsilon)y_\epsilon$ remain in the positive-imaginary-time half-space of Euclidean space-time. Here $R(\chi)$ denotes rotation through an angle χ.

Let $U(i\chi)$ denote a Lorentz boost through an imaginary angle $i\chi; U(i\chi)$ is a selfadjoint (positive) operator on $\mathcal{H}$. By the covariance of $\Phi_\alpha(x)$ under Lorentz boosts and the Lorentz invariance of Ω,

$$U(i\chi)\Phi_\alpha(x_\epsilon)\Omega = e^{i\chi s_\alpha} \Phi_\alpha(R(\chi)x_\epsilon)\Omega, \tag{3.34}$$

and the r.h.s. of (3.34) is well defined for all $\chi < \chi_\epsilon$. Clearly

$$< U(i\chi)\Phi_\alpha(x_\epsilon)\Omega, U(i\chi)\Phi_\alpha(y_\epsilon)\Omega > \\ = e^{2i\chi s_\alpha} G_\alpha(x_{-\epsilon}, R(2\chi)y_\epsilon), \tag{3.35}$$

by Euclidean covariance of G_α.

By (3.14), (3.25) and (3.26),

$$e^{2i\chi s_\alpha} G_\alpha(x_{-\epsilon}, R(2\chi)y_\epsilon) = \sum_\rho e^{2i\chi s_\alpha} \mu_\alpha^\gamma \overline{\mu_\kappa^\rho} R_{\gamma\alpha}^{\kappa\rho}(\sigma) < \Omega, \Phi_\rho^*(R(2\chi)y_\epsilon)\Phi_\rho(x_{-\epsilon})\Omega >, \tag{3.36}$$

where χ is chosen so small that

$$\text{Im} \quad (R(2\chi)y_\epsilon)^0 < -\epsilon = \text{Im} \quad x^0_{-\epsilon}.$$

We now first let ϵ tend to 0 and then let χ tend to π. Then (3.35) remains positive definite and the r.h.s. of (3.36) approaches

$$\sum_\rho e^{2i\pi s_\alpha} \mu_\alpha^\gamma \overline{\mu_\kappa^\rho} R_{\gamma\alpha}^{\kappa\rho}(\sigma) G_\rho((0,\vec{y}),(0,\vec{x})) \tag{3.37}$$

Clearly,

$$G_\rho((0,\vec{y}),(0,\vec{x})) \equiv \lim_{\epsilon \searrow 0} G_\rho((-\epsilon,\vec{y}),(\epsilon,\vec{x}))$$

is positive definite. By (3.35), (3.36), the condition on $R(\sigma)$ is that (3.37) be <u>positive-definite</u>. In space-times of dimension three or higher, s_α is integer or half-integer, and $R(\sigma)$ is a square root of 1. In this case the positive- definiteness of (3.37) yields the usual spin statistics theorem. In conformal field theory, the positive definiteness of (3.37) implies that

$$\sum_{\rho:\Delta_\rho=\Delta_\alpha} e^{2i\pi s_\alpha} \mu_\alpha^\gamma \overline{\mu_\kappa^\rho} R_{\gamma\alpha}^{\kappa\rho}(\sigma) \geq 0 \tag{3.38}$$

In many theories, that condition simply implies

$$e^{2i\pi s_\alpha} \mu_\alpha^\gamma \overline{\mu_\kappa^\rho} R_{\gamma\alpha}^{\kappa\rho}(\sigma) > 0,$$

for some $\rho = \rho(\alpha)$. More details about this version of the spin-statistics theorem will be discussed elsewhere.

(iv) The last point that ought to be raised here is whether PCT imposes non-trivial constraints on statistics matrices. As an analysis of the $\lambda\varphi_2^4$-model in the broken symmetry phase, described by the fields $\varphi(x), s(x)$, shows the field algebra of a two-dimensionsal local quantum field theory need <u>not</u> carry a linear representation of PCT. In that case, no simple constraint on $R(\sigma)$ follows from the existence of an anti-unitary PCT operator. If, however, the field algebra does carry a linear representation of PCT then there is a matrix θ such that

$$\Phi_\alpha^{PCT}(x) = \theta_\alpha^\gamma \Phi_\gamma(-x),$$

If, in the geometrical definition of B_n, R^2 is replaced by S^2, or by some other compact surface, additional relations between the generators $\tau_{1,2}, \cdots, \tau_{n-1,n}$ will emerge in general. In the example of S^2, one has

$$\prod_{i=1}^{n-1} \tau_{i,i-1} \prod_{j=n-1}^{1} \tau_{j,j-1} = e$$

The pure braid group, P_n, on n strings is the subgroup generated by the elements

$$\gamma_{ij} = \tau_{i,i+1} \cdots \tau_{j,j+1}^2 \quad \tau_{j-1,j}^{-1} \cdots \tau_{i,i+1}^{-1} \tag{4.1}$$

Graphically, γ_{ij} corresponds to

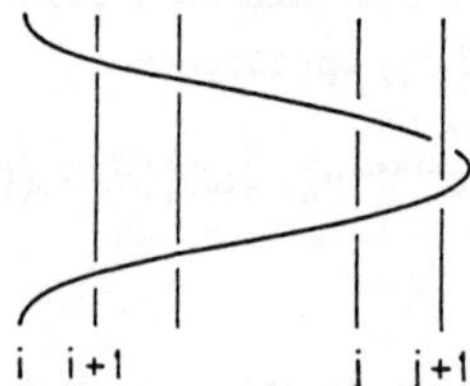

Let $R(\sigma), \sigma = \pm$, be two $p^2 \times p^2$ matrices which are solutions of (3.17) and (3.22). We set $V_i = C^p, i = 1, \cdots, n$, and $V^{(n)} = V_1 \otimes \cdots \otimes V_n$. We define

$$R_{i,i+1}(\sigma) = 1_1 \otimes \cdots \otimes 1_{i-1} \otimes R(\sigma) \otimes 1_{i+2} \otimes \cdots \otimes 1_n, \tag{4.2}$$

with $1_i = 1 \,|_{V_i}$. Representing $\tau_{i,i+1}$ by $R_{i,i+1}(+)$ and $\tau_{i,i+1}^{-1}$ by $R_{i,i+1}(-)$, we see that solutions of (3.17) and (3.22) define representations of B_n, for all $n = 2, 3, \cdots$: Relation (a) follows from (3.22) and (b) follows from (4.2). Thus every statistics matrix of a two-dimensional quantum field theory defines a representation of B_n (hence of P_n), for all $n = 2, 3, \cdots$.

These representations have some significance for the Euclidean description of the quantum field theory. The basic objects in the Euclidean description of quantum field theory are the Euclidean Green- or Schwinger functions,

$$G(\alpha_1, \xi_1, \cdots, \alpha_n, \xi_n), n = 0, 1, 2, \cdots,$$

with, $G(\emptyset) = 1$. Heuristically, they are defined by

$$\begin{aligned} G(\alpha_1, \xi_1, \cdots, \alpha_n, \xi_n) =&< \Omega, \Phi_{\alpha_1}(0, \vec{\xi}_1) e^{-(\xi_2^0 - \xi_1^0)H} \Phi_{\alpha_2}(0, \vec{\xi}_2) \\ &\cdots \Phi_{\alpha_{n-1}}(0, \vec{\xi}_{n-1}) e^{-(\xi_n^0 - \xi_{n-1}^0)H} \Phi_{\alpha_n}(0, \vec{\xi}_n) \Omega >, \end{aligned} \tag{4.3}$$

where H is the Hamilton operator of the theory. [For a precise definition see [13].] By the spectrum condition, i.e. $H \geq 0$, the Euclidean Green functions are well defined if

$$\xi_1^0 < \xi_2^0 < \cdots < \xi_n^0 \tag{4.4}$$

and simple calculations then show that

$$(\theta \otimes \theta)R(-\sigma) = \overline{R(\sigma)}(\theta \otimes \theta).$$

This and some further matters will have to be discussed in more detail elsewhere. In particular, one would like to have a fairly explicit description of statistics matrices satisfying (3.28) and (i) - (iii) (or (iv)) above. As we know from Sect. 2 and elsewhere, there are certainly non-trivial examples of statistics matrices satisfying those constraints, but a general classification is missing. Since, recently, there has been considerable progress in constructing solutions to the Yang-Baxter equation (starting from the classical Lie algebras), one may be optimistic about the chances for progress.

4. Statistics matrices, representations of braid groups and monodromy of Euclidean Green functions

We start this section by defining the braid groups, B_n, and pure braid groups, P_n, on n strings. We choose n distinct points, $x_1, \cdots, x_n$, in R^2 and consider n maps, $\gamma_1, \cdots, \gamma_n$, from [0,1] to the slab $R^2 \times$ [0,1] in R^3 with the properties

(i) $$\gamma_i(0) = x_i, \quad \gamma_i(1) = x_{\pi(i)},$$

where π is an arbitrary permutation of $\{1, \cdots, n\}$;

(ii) $$\dot{\gamma}_i^3(t) > 0, \text{for all} \quad t \in [0,1];$$

(iii) $$\text{for} \quad i \neq j, \quad \gamma_i(t) \neq \gamma_i(s), \text{for all} \quad t \in [0,1] \quad \text{and all} \quad s \in [0,1].$$

We now identify all maps, $\gamma_1, \cdots, \gamma_n$, with properties (i) - (iii) related to each other by an ambient isotopy. The resulting object is the braid group on n strings, denoted B_n.

Algebraically, B_n can be defined by choosing the following presentation: B_n is generated by taking arbitrary products of generators $\tau_{i,i+1}, i = 1, \cdots, n-1$, and their inverses, ($\tau_{i,i+1}\tau_{i,i+1}^{-1} = e$, the identity element of B_n), where the generators $\tau_{i,i+1}$ obey the following relations:

$$(a) \quad \tau_{i,i+1} \; \tau_{i+1,i+2} \; \tau_{i,i+1} = \tau_{i+1,i+2} \; \tau_{i,i+1} \; \tau_{i+1,i+2}$$

$$(b) \quad \tau_{i,i+1} \; \tau_{j,j+1} = \tau_{j,j+1} \; \tau_{i,i+1}, \quad \text{for } |\, i-j \,| \geq 2.$$

Geometrically, $\tau_{i,i+1}$ can be represented by the following figure:

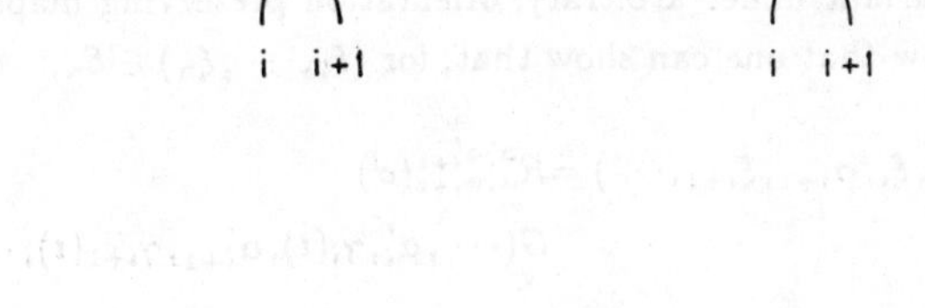

$\tau_{i,i+1} \Leftrightarrow$ i i+1 $\qquad$ $\tau_{i,i+1}^{-1} \Leftrightarrow$ i i+1

Poincaré covariance of the theory implies Euclidean covariance of the functions

$$G(\alpha_1, \xi_1, \cdots, \alpha_n, \xi_n),$$

so that they are well defined on the smallest Euclidean invariant domain, $\mathcal{E}_n$, containing

$$\{\xi_1, \cdots, \xi_n : \xi_1^0 < \xi_2^0 < \cdots < \xi_n^0\}.$$

Knowledge of the statistics matrix $R(\sigma)$ of the theory permits to construct a huge extension of the domain of definition of the Green functions $G(\alpha_1, \xi_1, \cdots, \alpha_n, \xi_n)$, at the price, though, that the extended Green functions are <u>not</u> single-valued anymore, unless $R(\sigma)^2 = 1$, i.e. for theories with fields that satisfy ordinary Bose-or Fermi statistics. The largest purely Euclidean domain to which $G(\alpha_1, \xi_1, \cdots, \alpha_n, \xi_n)$ may be extended as a multi-valued function is

$$M_n \equiv (E^2)^{\times n} \backslash D \tag{4.5}$$

where D is the "diagonal" of $(E^2)^{\times n}$, i.e.D is the set of points $\{\xi_1, \cdots, \xi_n\}$ for which $\xi_i = \xi_j$, for some $i \neq j$.

The space M_n is not simply connected. In fact, it is easy to see geometrically that

$$\pi_1(M_n) = P_n \tag{4.6}$$

It will be natural to view the Green functions, $G(\alpha_1, \xi_1, \cdots, \alpha_n, \xi_n)$, as single-valued functions on the universal cover $\bar{M}_n$ of M_n. Because of (4.4) it is actually more natural to extend the definition of $G(\alpha_1, \xi_1, \cdots, \alpha_n, \xi_n)$ to the universal cover, $\bar{M}_n^{symm.}$, of the space $M_n^{symm.}$, where

$$M_n^{symm.} = M_n / S_n \tag{4.7}$$

and S_n is the permutation group of n elements, for each $n = 0, 1, 2, \cdots$. The fundamental group of $M_n^{symm.}$ is

$$\pi_1(M_n^{symm.}) = B_n, \tag{4.8}$$

as one checks easily.

We now indicate how to extend the definition of $G(\alpha_1, \xi_1, \cdots, \alpha_n, \xi_n)$ to $\bar{M}_n^{symm.}$. First, consider two adjacent arguments, α_i, ξ_i and α_{i+1}, ξ_{i+1}, of G. We choose two oriented curves, γ_i, γ_{i+1}, which map the interval [0,1] into E^2 such that $\gamma_i(0) = \xi_i, \gamma_i(1) = \xi_{i+1}, \gamma_{i+1}(0) = \xi_{i+1}, \gamma_{i+1}(1) = \xi_i$ and $\gamma_i(t) \neq \gamma_{i+1}(s)$, for $t \in (0,1)$ and $s \in (0,1)$. The composition, $\gamma_i * \gamma_{i+1}$, describes an oriented loop with base point ξ_{i+1}. Its orientation is described by a sign, $\sigma = + or -$. Since we are in two dimensions, this sign is invariant under arbitrary, orientation preserving maps of E^2 into itself. The point is now that one can show that, for $(\xi_1, \cdots, \xi_n) \in \mathcal{E}_n$, $\tau \equiv \tau_{i,i+1}$,

$$\begin{aligned} \underset{e}{G}(\cdots, \alpha_i, \xi_i, \alpha_{i+1}, \xi_{i+1}, \cdots) = & R^{\alpha_i' \alpha_{i+1}'}_{\alpha_i \alpha_{i+1}}(\sigma) \\ & \underset{\tau}{G}(\cdots, \alpha_i', \gamma_i(t), \alpha_{i+1}', \gamma_{i+1}(t), \cdots) |_{t \to 1} \end{aligned} \tag{4.9}$$

The r.h.s. of (4.9) only depends on the element in B_n represented by the two curves $(\gamma_i(t),t),(\gamma_{i+1}(t),t)$. Since the l.h.s. of (4.9) is well-defined, for all $(\xi_1,\cdots,\xi_n)\in\mathcal{E}_n$, one can use (4.9) to extend the definition of G to the domain $\tau_{i,i+1}\mathcal{E}_n$. Although the construction of this extension is not difficult for people who are somewhat familiar with [13] and have read Sect. 3, it takes some space to put this down in words. We must therefore leave it as an exercise to the reader.

Equation (4.9) is the basic fact. Starting from it, one may easily complete the construction of an extension of $G(\alpha_1,\xi_1,\cdots,\alpha_n,\xi_n)$ to $\tilde{M}_n^{symm.}$. Let $b\in B_n$. We present b as a word in the generators $\tau^\sigma_{i,i+1},\sigma=\pm1,i=1,\cdots,n-1$:

$$b=\prod_{k=1}^m \tau^{\sigma(k)}_{i(k),i(k+1)}, \tag{4.10}$$

and define

$$R(b)=\prod_{k=1}^m R_{i(k),i(k+1)}(\sigma(k)), \tag{4.11}$$

with $R_{i,i+1}(\sigma)$ given by (4.2). For a given b we choose curves $\gamma_b^{(1)}(t),\cdots,\gamma_b^{(n)}(t)$, mapping [0,1] into E^2, with $\gamma_b^{(i)}(0)=\xi_i$, $\gamma_b^{(i)}(1)=\xi_{\pi(i)}$, for some $\pi=\pi(b)\in S_n$, such that

$$(\gamma_b^{(1)}(t),t),\cdots,(\gamma_b^{(n)}(t),t),t\in[0,1]$$

is a representative of b. Using (4.9) repeatedly, one may show that

$$\begin{aligned}\underset{e}{G}(\alpha_1,\xi_1,\cdots,\alpha_n,\xi_n)=&R^{\alpha_1'\cdots\alpha_n'}_{\alpha_1\cdots\alpha_n}(b)\\ &\underset{b}{G}(\alpha_1',\gamma_b^{(1)}(t),\cdots,\alpha_n',\gamma_b^{(n)}(t))\big|_{t\to1}\end{aligned} \tag{4.12}$$

This formula permits us to extend the definition of G from $\mathcal{E}_n$ to all of $\tilde{M}_n^{symm.}$, hence to $\tilde{M}_n$. In general, G does not project down to a single-valued function on M_n. For example, the Euclidean Green functions of the fields $\varphi(x)$ and $s(x)$ (meson-and soliton fields) in the $\lambda\varphi_2^4$ - or the Ising$_2$ field theory are double-valued functions on M_n; (see also [6] for an analysis of further examples).

It is not hard to prove that the information contained in equ. (4.12) is completely equivalent to knowing the statistics matrix of the fields $\Phi_\alpha(x),\alpha=1,\cdots,p$.

By equ. (4.12), the Euclidean Green functions $G(\alpha_1,\xi_1,\cdots,\alpha_n,\xi_n)$ carry a representation of B_n, hence of P_n, for $n=0,1,2,\cdots$. This representation uniquely determines the monodromy of the Green functions G : For any element $b\in P_n$,

$$\begin{aligned}\underset{e}{G}(\alpha_1,\gamma_b^{(1)}(0),\cdots,\alpha_n,\gamma_b^{(n)}(0))=&R^{\alpha_1'\cdots\alpha_n'}_{\alpha_1\cdots\alpha_n}(b)\\ &\underset{b}{G}(\alpha_1',\gamma_b^{(1)}(1),\cdots,\alpha_n',\gamma_b^{(n)}(1)).\end{aligned} \tag{4.13}$$

By (4.1), this representation of P_n, in other words the monodromy of G, is trivial iff $R(\sigma)^2=1$, i.e. for ordinary Bose- or Fermi statistics.

By (4.12) and (4.13), we may view the Euclidean Green functions, G, as sections of a complex vector bundle with base space M_n and fiber $V^{(n)}=V_1\otimes\cdots\otimes V_n, V_i=C^p$, for all i. Since the R-matrices in equ. (4.13) only depend on the class $b \in P_n$ represented by the curves $\gamma_b^{(i)}(t), i = 1, \cdots, n$, one can try to interpret these matrices as the holonomy matrices of a flat connection on that vector bundle. Hain and Kohno [24] have shown that if R is any representation of P_n such that $R(\gamma_{ij})$ is sufficiently close to the identity, for all $1 \leq i < j \leq n$, with γ_{ij} given by (4.1), or if R is unipotent, then the matrices $R(b), b \in P_n$, are the holonomy matrices of a flat connection, ω, on that vector bundle which is given by the following formula: A point $\xi = (\xi^0, \xi^1) \in E^2$ is represented by the complex number $z = i\xi^0 + \xi^1$. Then

$$\omega = \sum_{1 \leq i < j \leq n} \Omega_{ij} \ \mathrm{d} \ \log \ (z_i - z_j), \tag{4.14}$$

where Ω_{ij} are constant matrices acting on $V^{(n)}$. The flatness of ω is equivalent to the equations

$$\boxed{\begin{aligned} &[\Omega_{ij}, \Omega_{ik} + \Omega_{jk}] = [\Omega_{ij} + \Omega_{ik}, \Omega_{jk}] = 0, \\ &\text{for } i < j < k, \text{ and} \\ &[\Omega_{ij}, \Omega_{kl}] = 0, \text{ for distinct } i, j, k \text{ and } l. \end{aligned}} \tag{4.15}$$

See [25]. These equations are called infinitesimal pure braid relations. In our case, the matrix Ω_{ij} is given by an endomorphism of $V^{(n)}$ which acts non-trivially only on $V_i \otimes V_j$ and reduces to the identity on all other factors $V_k, k \neq i, j$. It turns out [25] that, in this situation, the equations (4.15) are a special case of the classical Yang-Baxter equations. This is not surprising: Since the statistics matrix $R(\sigma)$ is a solution of the Yang-Baxter equation, the connection Ω whose holonomy is given by $R(\sigma)$ must solve the classical Yang-Baxter equation. [The connection between solutions of the classical $\mathcal{YBE}$ and solutions of the $\mathcal{YBE}$ is analogous to the connection between Lie algebras and Lie groups.] The following result is due to Belavin and Drinfel'd [26]: Let $V_i = V$ carry an irreducible representation π of a simple, complex Lie algebra $\mathfrak{g}$. Let $\{X_\alpha\}_{\alpha=1}^q$ be a basis of $\mathfrak{g}$ which is orthonormal with respect to the Cartan-Killing form on $\mathfrak{g}$. We define

$$\Omega_{ij} = \lambda \sum_{\alpha=1}^{q} 1_1 \otimes \cdots \otimes \pi(X_\alpha) \mid_{V_i} \otimes \cdots \otimes \pi(X_\alpha) \mid_{V_j} \otimes \cdots \otimes 1_n \tag{4.16}$$

Then $\{\Omega_{ij}\}_{1 \leq i < j \leq n}$ solves (4.15). This result will be useful in the next section.

Next, we consider analytic continuations of the Euclidean Green functions, G, to the Bargmann-Hall-Wightman domain [23]. We introduce Euclidean light cone variables:

$$z = i\xi^0 + \xi^1, \bar{z} = -i\xi^0 + \xi^1 \tag{4.17}$$

The axioms of relativistic quantum field theory [23] imply that there are functions,

$$G(\alpha_1, z_1, \bar{z}_1, \cdots, \alpha_n, z_n, \bar{z}_n),$$

which are holomorphic on a certain tube $\mathcal{T}_n$ in C^{2n} (which, using (4.12), is at least as large as the permuted $\mathcal{BHW}$ domain [23]) such that

$$G(\alpha_1, z_1, \bar{z}_1, \cdots, \alpha_n, z_n, \bar{z}_n) = G(\alpha_1, \xi_1, \cdots, \alpha_n, \xi_n). \tag{4.18}$$

The special feature of two-dimensional conformal field theory is that, in two space-time dimensions, the action of the conformal group factorizes in an action transforming only the z–variables and an action transforming only the $\bar{z}$–variables. This enables one to continue the Wightman functions $G(\alpha_1, z_1, \bar{z}_1, \cdots, \alpha_n, z_n, \bar{z}_n)$ to a much larger domain and to treat the z–variables und the $\bar{z}$–variables essentially as independent variables. [It appears that one cannot find a careful axiomatic analysis of two-dimensional conformal field theory, from a Euclidean point of view, in the literature. Unfortunately, this is quite a lengthy enterprize and cannot be presented here. But see [27]; and [28] for some beginnings in a direction related to my own thinking.] If $\bar{z}_1, \cdots, \bar{z}_n$ are kept fixed, we set

$$G(\alpha_1, z_1, \bar{z}_1, \cdots, \alpha_n, z_n, \bar{z}_n) \equiv g(\alpha_1, z_1, \cdots, \alpha_n, z_n) \tag{4.19}$$

One may now attempt to determine the monodromy of the functions g when the variables $z_1 \cdots, z_n$ are varied, but $\bar{z}_1, \cdots, \bar{z}_n$ are kept fixed.

We must look for an equation analogous to (4.12), (4.13), i.e. an equation of the form

$$\begin{aligned} g(\alpha_1, \gamma_b^{(1)}(0), \cdots, \alpha_n, \gamma_b^{(n)}(0)) = \hat{R}_{\alpha_1 \cdots \alpha_n}^{\alpha'_1 \cdots \alpha'_n}(b) \\ g(\alpha'_1, \gamma_b^{(1)}(1), \cdots, \alpha'_n, \gamma_b^{(n)}(1)), \end{aligned} \tag{4.20}$$

where

$$b \in B_n, \text{ and } \gamma_b^{(i)}(\cdot) : [0,1] \to C, i = 1, \cdots, n,$$

are curves with the property that

$$(\gamma_b^{(1)}(t), t) \cdots (\gamma_b^{(n)}(t), t)$$

is a representative of b.

It will not surprize the reader that the matrices $\hat{R}(b)$ are in general not equal to the matrices $R(b)$ defined in (4.11). In fact, it can and does happen that $R(\sigma)^2 = 1$, so that $R(b) = 1$, for all $b \in P_n$, but $\hat{R}(b) \not\equiv 1$; (see [27, 29]).

The interest in computing the matrices $\hat{R}(b)$ comes from the fact that, in conformal field theory, the functions g satisfy differential equations [27,29,$\cdots$]. The matrices $\hat{R}(b)$ then determine the monodromy of those solutions to the differential equations which are candidate Wightman functions.

Let $\hat{R}(\sigma)$ be the matrix with the property that, for b given by (4.10) and $\hat{R}_{i,i+1}(\sigma)$ given by (4.2),

$$\hat{R}(b) = \prod_{k=1}^{m} \hat{R}_{i(k),i(k+1)}(\sigma(k)).$$

One might ask whether $\hat{R}(\sigma)$ is the "statistics matrix" of some chiral operator algebra. If $\hat{R}$ is the Burau representation (see e.g. [25]) of B_n then $\hat{R}(\sigma)$ is the statistics matrix of an algebra of vertex operator, as follows e.g. from the analysis in [6]. More generally, $\hat{R}(\sigma)$ may be the "statistics matrix" of Schroer's exchange algebra [30], but for the moment it is not clear to me how general the notion of exchange algebra is in the context of conformal field theory.

If the functions $G(\alpha_1, \xi_1, \cdots, \alpha_n, \xi_n)$ have an interpretation as single-valued correlation functions of a two-dimensional statistical-mechanical system then the monodromy of $G(\alpha_1, z_1, \bar{z}_1, \cdots, z_n, \bar{z}_n)$ under simultaneous variations of $z_1, \cdots, z_n$ and $\bar{z}_1, \cdots, \bar{z}_n$, with $\bar{z}_i = \bar{z}_i, i = 1, \cdots, n$, must be trivial. This is a powerful constraint [27,29].

5.The example of the Wess-Zumino-Witten models.

The Wess-Zumino-Witten (WZW) models [31,32,29] are principal chiral σ-models with a topological Wess-Zumino term in the action. The basic field, $g(\xi)$, of these models is a field with values in a compact, semisimple Lie group G. The Euclidean action is given by

$$S(g) = \frac{1}{4\lambda^2} \int tr(\partial_\mu g^{-1} \partial^\mu g)(\xi) d^2\xi + k\Gamma(g), \tag{5.1}$$

where $\Gamma(g)$ is the Wess-Zumino (-Witten) term [31,29], and $k \in Z, k \neq 0$. For $\lambda^2 = \frac{4\pi}{k}$, the model turns out to be conformal invariant.

The WZW models form the subject of K. Gawedzki's lectures, so we shall be brief.

For simplicity we consider $G = SU(2)$. Let

$$J(z) = -\frac{1}{2} k \left[(\partial_z g) g^{-1}\right](z) \text{ and } \bar{J}(\bar{z}) = -\frac{1}{2} k \left[g^{-1} \partial_{\bar{z}} g\right](\bar{z})$$

be the generators of $S\hat{U}(2)_z, S\hat{U}(2)_{\bar{z}}$, respectively. Let $\sigma^1, \sigma^2, \sigma^3$ be the usual Pauli matrices, and t_l^α the spin l representation of $\sigma^\alpha, \alpha = 1, 2, 3$.

In conformal field theory, the energy-momentum tensor, $T(z, \bar{z})$, splits into two tensors, $T(z), \bar{T}(\bar{z})$, only depending on $z, \bar{z}$, respectively. In the WZW model, $T(z)$ has the Sugawara form

$$T(z) = \frac{1}{2(k+2)} \left\{ \sum_{\alpha=1}^{3} : J_\alpha(z) J_\alpha(z) : \right\}, \tag{5.2}$$

where $J_\alpha = \frac{1}{2} tr(\sigma^\alpha J)$. The double colons on the r.h.s. of (5.2) indicate normal ordering.

Let $\{\Phi_l(z, \bar{z})\}$ be the primary fields, in the sense of [27, 29], of the WZW models. We also use the notations

$$\Phi_T(z) \equiv T(z), \Phi_{J_\alpha}(z) \equiv J_\alpha(z).$$

The conformal Ward identities are (see [27])

$$\begin{aligned} G(T, z, l_1, z_1, \bar{z}_1, \cdots, l_n, z_n, \bar{z}_n) \text{ ``}&= \langle T(z) \Phi_{l_1}(z_1, \bar{z}_1) \cdots \Phi_{l_n}(z_n, \bar{z}_n) \rangle\text{''} \\ &= \sum_{j=1}^{n} \left\{ \frac{\Delta_{l_j}}{(z - z_j)^2} + \frac{1}{z - z_j} \frac{\partial}{\partial z_j} \right\} G(l_1, z_1, \bar{z}_1, \cdots, l_n, z_n, \bar{z}_n) \end{aligned} \tag{5.3}$$

and the Ward identities for the Kac-Moody symmetry are [29]

$$\begin{aligned} G(J_\alpha, z, l_1, z_1, \bar{z}_1, \cdots, l_n, z_n, \bar{z}_n) &= \text{``}\langle J_\alpha(z)\Phi_{l_1}(z_1,\bar{z}_1)\cdots\Phi_{l_n}(z_n,\bar{z}_n)\rangle\text{''} \\ &= \sum_{j=1}^{n} \frac{t^\alpha_{l_j}}{z-z_j} G(l_1, z_1, \bar{z}_1, \cdots, l_n, z_n, \bar{z}_n). \end{aligned} \tag{5.4}$$

If, on the l.h.s. of (5.3), we insert the Sugawara form (5.2) of the energy-momentum tensor, $T(z)$, and apply (5.4) twice to the resulting expression and finally compute a contour integral in the variable z we obtain the differential equation

$$\begin{aligned} \sum_{i=1}^{n} \frac{\partial}{\partial z_i} g(l_1, z_1, \cdots, l_n, z_n) = \sum_{1\le i<j\le n} \frac{2}{k+2} \frac{(\Omega_{ij})^{l'_i l'_j}_{l_i l_j}}{z_i - z_j} \\ g(l_1, z_1, \cdots, l'_i, z_i, \cdots, l'_j, z_j, \cdots) \end{aligned} \tag{5.5}$$

where

$$g(l_1, z_1, \cdots, l_n, z_n) = G(l_1, z_1, \bar{z}_1, \cdots, l_n, z_n, \bar{z}_n),$$

with $\bar{z}_1, \cdots, \bar{z}_n$ kept fixed, and

$$\Omega_{ij} = \sum_{\alpha=1}^{3} t^\alpha_{l_i} \otimes t^\alpha_{l_j}. \tag{5.6}$$

Equation (5.5) has been derived in [29]; see also [33]. A similar equation obviously governs the dependence of $G(l_1, z_1, \bar{z}_1, \cdots, l_n, z_n, \bar{z}_n)$ on the variables $\bar{z}_1, \cdots, \bar{z}_n$, for fixed $z_1, \cdots, z_n$. A more compact form for equation (5.5) is

$$\begin{aligned} d\, g(\cdot, z_1, \cdots, \cdot, z_n) = \left\{ \sum_{1\le i<j\le n} \frac{\Omega_{ij}}{k+2} \text{ d } \log\ (z_i - z_j) \right\} \\ \times\ g(\cdot, z_1, \cdots, \cdot, z_n) \end{aligned} \tag{5.7}$$

It has been noticed in Sect. 4 that Euclidean Green functions can be viewed as sections of complex vector bundles with base space M_n. In (4.19), (4.20) it turned out that the functions $g(\cdot, z_1, \cdots, \cdot, z_n)$, too, are sections of a complex vector bundle with base space M_n. Equation (5.7) for the functions g of the WZW models is nothing but the equation for parallel transport on a complex vector bundle with base space M_n and connection ω given by

$$\omega = \sum_{1\le i<j\le n} \Omega_{ij} \text{ d } \log\ (z_i - z_j) \tag{5.8}$$

where the matrices Ω_{ij} are given in (5.6). The results of Belavin and Drinfel'd [26] quoted in Sect. 4 show that ω is flat, hence gives rise to a solution of the classical Yang-Baxter equation. The holonomy matrices of ω, corresponding to solutions of the special Yang-Baxter equations (3.22), describe the monodromy of the functions $g(\cdot, z_1, \cdots, \cdot, z_n)$, (as in (4.20)).

Thus from equ. (5.7) we obtain a linear representation of the pure braid group P_n. It is of interest to analyze the properties of that representation. Let us consider the special case where

$$t^{\alpha}_{l_j} = \sigma^{\alpha},$$

for $j = 1, \cdots, n$; i.e., we consider Green functions of primary fields transforming under the fundamental representation of $SU(2)$, in particular $l_1 = \cdots = l_n$. Then equs. (5.7) determine a linear representation of the braid group B_n. This representation factors through the Jones algebra [34] of index $4\cos^2\left(\frac{\pi}{k+2}\right)$, where k is as in (5.2), (the central charge of the Kac-Moody algebra).

[The Jones algebra has generators $\tau_{i,i+1}$ satisfying the relations (a) and (b) stated at the beginning of Sect. 4 and, in addition, the relations

$$(c) \quad \tau^2_{i,i+1} = (q-1)\tau_{i,i+1} + q, \text{ with } q = e^{2\pi i/(k+2)}$$

These representations are <u>unitarizable</u> [34].]

For more details concerning the results discussed here, see [25,33].

One should add that the monodromy of the Euclidean Green functions G under simultaneous variation of the arguments $z_i, \bar{z}_i \cdots$, with $\bar{z}_i = \bar{z}_i$, is trivial, so that the physical fields $\Phi_l(x)$ of the WZW models do not have exotic statistics. [If $G = 0(N)$ then the WZW models have spin fields, and, as in the Ising field theory, non-trivial statistics matrices arise. When $k = 1$ the WZW models can be rewritten in terms of N free, massless Majorana fermions, ψ_k, with

$$J_{kk'}(z) =: \psi_k(z)\psi_{k'}(z) :, \bar{J}_{kk'}(\bar{z}) =: \bar{\psi}_k(\bar{z})\bar{\psi}_{k'}(\bar{z}) :,$$

and

$$g_{kk'}(z,\bar{z}) = \text{const.} : \psi_k(z)\bar{\psi}_{k'}(\bar{z}) : .$$

This is Witten's non-abelian bosonization [31,29]. The statistics problem for the fields $\psi_{k'}\bar{\psi}_k$ and the spin fields is then analogous to the one discussed in Sect. 2 for the Ising_2 field theory.] Finally, we should mention that the primary fields of the WZW models do not transform under all possible representations of G, and that for $g(l_1, z_1, \cdots, l_n, z_n)$ to be non-zero, the labels $l_1, \cdots, l_n$ must satisfy certain constraints. All this is the contents of certain fusion rules which are presumably discussed in K. Gawedzki's lectures.

6. Connections to the theory of knots and links

The purpose of this section is to sketch a connection between the material in Sects. 3-5 and a chapter in topology which was brought to my attention by V. Jones: the theory of knots and links. I am following notes by V. Jones [35] and a recent preprint by Turaev [36].

Suppose $R \in \text{End}\ (V \otimes V)$ is a solution of the Yang-Baxter equs. (3.22), and μ is a diagonal matrix on V such that

$$(\mu \otimes \mu)R = R(\mu \otimes \mu), \tag{6.1}$$

and

$$\left.\begin{aligned} \sum_{\gamma=1}^{k} R^{\alpha\gamma}_{\beta\gamma}\mu_\gamma &= a\, b\, \delta^\alpha_\beta, \\ \sum_{\gamma=1}^{k} (R^{-1})^{\alpha\gamma}_{\beta\gamma}\mu_\gamma &= a^{-1} b\, \delta^\alpha_\beta, \end{aligned}\right\} \tag{6.2}$$

with $k =$ dim V. Then (R, μ, a, b) determine a polynomial invariant of knots and links.

To see this, one represents a link by some braid which is closed by identifying lower and upper end points of the strings. One then associates the matrix

$$R_{i,i+1} = 1_1 \otimes \cdots \otimes 1_{i-1} \otimes R\,|_{V_i \otimes V_{i+1}} \otimes \cdots \otimes 1_n$$

($V_i = V$, for all i) with the generators $\tau_{i,i+1}$ of $B_n, n = 2, 3, 4, \cdots$. This defines a representation $b \in B_n \longmapsto R(b)$, given by (4.11), of B_n, for all $n \geq 2$. Let $w(b)$ be twice the "linking number" (more precisely, the <u>writhe</u>) of the link represented by b. We define

$$T(b) = a^{-w(b)} b^{-n} tr(R(b) \underbrace{\mu \otimes \cdots \otimes \mu}_{n \text{ times}}) \tag{6.3}$$

This is an invariant of the link b. An important ingredient in proving this is the invariance of $T(b)$ under Reidemeister moves of type II and III. This invariance is closely related to equs. (3.17) and (3.22).

The quantity $\operatorname{tr}(R(b)\mu \otimes \cdots \otimes \mu)$ can be interpreted as the partition function of a model of classical statistical mechanics defined on the link represented by b (with k states associated with every line (bond) of b).

This approach to finding new invariants for knots and links was pioneered by V.Jones [35]; see also [36] for details concerning the results summarized above, and [37] for a review of knot theory, the Jones polynomial, and statistical mechanics.

Jones' work is also based on constructing solutions to the Yang-Baxter equations, solutions which, in addition, satisfy conditions (6.1) and (6.2).

Although there are statistics matrices satisfying (6.1) and (6.2), the conditions that have to be met by statistics matrices are quite unrelated to the ones met by an R-matrix that gives rise to an invariant for knots and links. In fact, Jones' work is more directly related to the <u>quantum inverse scattering method</u> than to the ideas explained in Sects. 3-5; see [35, 36]. The quantum inverse scattering method is a technique to explicitly diagonalize the transfer matrices of a large class of models of two-dimensional statistical mechanics and to calculate mass spectrum and (factorizable) scattering matrices of a variety of two-dimensional quantum field models.

The point of this short section is to emphasize that there are several areas in quantum field theory, statistical mechanics and in topology, where the Yang-Baxter equation plays an important rôle. One would like to know whether this is an accident or whether there are some fundamental, underlying connections between these subjects.

Pasquier [38] has analyzed models of two-dimensional statistical mechanics (generalized S-0-S models) which appear to tie together the so-called discrete series of two-dimensional conformal quantum field theories (see Friedan's lectures, and refs. given there), the quantum inverse scattering method and the Jones algebra, but it appears that many details remain mysterious. I am tempted to conjecture that in this whole area some basic discoveries remain to be made.

7. Notes and open problems.

This final section contains some brief notes and comments on earlier sections and a list of open problems.

Sect. 2, beginning of Sect. 3

The outstanding problem in connection with the results reviewed in Sect. 2 and at the beginning of Sect. 3 is to perform a general analysis of covariant representations of $(\mathcal{A}, \tau)$, where $\mathcal{A}$ is the observable algebra and τ a representation of the Poincaré group as a group of *automorphisms of $\mathcal{A}$, for quantum field theories in two and three space-time dimensions. In particular, the connection between the theory of superselection sectors in two dimensions and the theory of loop groups and their representations should be worked out in more detail.

In three space-time dimensions, it remains to be shown more completely that the field theory of anyons [2,18], as developed in [19], represents all there is that is not covered by conventional Bose- or Fermi statistics. In other words, one would like to show that the general theory of superselection sectors [39] for local theories in three space-time dimensions does not lead to further possibilities besides Bose-, Fermi- and intermediate ($\theta-$) statistics. This is the problem considered in [19].

Sect. 3

The basic problem posed by the analysis in Sect. 3 is:

(1) What is the most general solution of equations (3.17), (3.22) and (3.28), i.e. what is the most general statistics matrix, that may and does arise in the context of two-dimensional, local, relativistic quantum field theory? In which way must the constructive approach alluded to in Sect. 2, see [3-6], be extended in order to construct examples of field theories with more general statistics matrices?

(2) Let $\{\Phi_\alpha(x)\}_{\alpha=1}^{p}$ be the local "primary" fields of a two-dimensional quantum field theory with statistics matrix $R(\sigma)$. Let 0 be a diamond in M^2, and let $\overline{F(0)}$ be the von Neumann algebra generated by the local observable algebra $\mathcal{A}(0)$ and all bounded, selfadjoint functions of $\{\Phi_\alpha(f) : \text{ supp } f \subseteq 0\}, \alpha = 1, \cdots, p$. In which way does the algebraic structure of the net $\{\overline{F(0)}\}$ reflect the statistics matrix $R(\sigma)$.

(3) Consider an arbitrary solution, $R(\sigma)$, of equations (3.17), (3.22) and (3.28). Is there an algebra, F, of operators for which $R(\sigma)$ is, in a suitable sense, the statistics matrix. For example, given a solution $R(\sigma)$ of (3.17), (3.22) and (3.28), are there operators $\Phi_\alpha(\vec{x})$, with $\vec{x} \in Z$, such that

$$\Phi_\alpha(\vec{y})\Phi_\beta(\vec{x}) = R_{\alpha\beta}^{\gamma\delta}(\sigma)\Phi_\gamma(\vec{x})\Phi_\delta(\vec{y}),$$

for arbitrary $\vec{x}, \vec{y}$ in Z, with $\vec{x} \neq \vec{y}; \sigma =$ sign$(\vec{y} - \vec{x})$? If so, what are the algebraic properties of F?

Sect. 4.

One obvious problem in connection with the material in the second part of Sect. 4 is to develop a general Euclidean formulation of conformal quantum field theory in two space-time dimensions.

In particular, it looks plausible that one can prove a general reconstruction theorem that permits one, for example, to reconstruct from the Euclidean Green functions of a conformal theory a unique representation of the conformal group on the physical Hilbert space of radial quantization.

Next, one would like to correctly formulate two-dimensional conformal field theory on Riemann surfaces of arbitrary genus and relate properties of conformal field theory on Riemann surfaces of higher genus to properties of theories in physical space-time. A program in this direction, closely related to ideas worked out independently by the author, has been proposal by G. Segal [28].

All this is clearly intimately related to the covariant quantization of string theory and covariant string perturbation theory. The different spin structures that are encountered on Riemann surfaces of higher genus and the monodromy generated by spin fields in fermionic correlation functions are simple special cases of the general problems discussed in Sect. 4. The general problem is to classify all possible representations of the (pure) braid groups that can arise within the context of general conformal quantum field theory on arbitrary Riemann surfaces.

Sect. 5

In Sect. 5, I have mostly followed refs.7, 25 and 33. After I had started to analyze the representations of braid groups arising in the Wess-Zumino-Witten models, T. Kohno gave two interesting seminars about the work in refs. 24, 25 and 33. I have no results to report that would go beyond those in refs. 25 and 33.

Clearly, these results are not quite the end of the story. One would like to have a more explicit description of the respresentations of B_n and of P_n arising in the WZW models with $G = SU(2)$. One then would like to extend the results sketched in Sect. 5 to general groups G and to the study of Euclidean Green functions of the WZW models on general Riemann surfaces. One would like to make more powerful use of equation (5.7), combined with general theorems of conformal QFT, to obtain explicit information on the properties of the Euclidean Green functions of the WZW models. There are further problems concerning spin fields in the WZW models (e.g. for $G = SO(3)$), and even Witten's non-abelian bosonization [31] ($k = 1$) still poses some interesting questions (e.g. in connection with the functional integral formulation of bosonization in [6]) that are presently investigated.

Sect. 6

An obvious question concerning the material sketched in Sect. 6 is whether knowledge of all solutions of the Yang-Boxter equation (3.22) satisfying (6.1) and (6.2) (for some numbers a and b) would provide a complete set of invariants for knots and links in R^3.

It would be interesting to understand whether there are fundamental mathematical reasons for the formal similarities between "integrable" models of two-dimensional statistical mechanics and quantum field theory, two-dimensional conformal quantum field theories and the theory of knots and links in R^3. In all three areas, the Yang-Baxter equation plays on important rôle, but the rôles played by this equation are different and have different origins. What is the unifying principle? I have only speculations about these questions, but no good answers.

The little I know on knot theory I have learned in conversations with and seminars by V. Jones and R. Morton and from refs. 35, 36 and 37.

I thank these colleagues for stimulating my interests in this area and teaching me a few basic facts.

REFERENCES.

1. R.F. Streater and A.S. Wightman,*"PCT, Spin and Statistics and All That"*, *Reading, Mass: Benjamin 1978.* (This book contains many references to the early literature on solitons and statistics in two dimensions.)
R.F. Streater and I.F. Wilde, *Nuclear Physics* B 24, 561 (1970).
2. Y.S. Wu, *Phys. Rev. Lett.* 52, 2103 (1984); 53, 111 (1984).
3. J. Fröhlich, *Commun. Math. Phys.* 47, 269 (1976); *Acta Phys. Austriaca, Suppl. XV*, 133 (1976).
J. Bellissard, J. Fröhlich and B. Gidas, *Commun. Math. Phys.* 60, 37 (1978).
J. Fröhlich, *in :"Invariant Wave Equations"*, Berlin-Heidelberg - New York: Springer-Verlag, Lecture Notes in Physics 73, 1978.
4. J. Fröhlich, *in : "Recent Developments in Gauge Theories"* (Cargese 1979). G.'t Hooft et al. (eds.), New York: Plenum Press, 1980.
5. E.C. Marino and J.A. Swieca, *Nucl. Phys.* B 170 [FS1], 175 (1980); E.C. Marino, B. Schroer and J.A. Swieca, *Nucl. Phys.* B 200 [FS4], 473 (1982); R. Köberle and E. C. Marino, Phys. Lett 126 B, 475 (1983).
6. J. Fröhlich and P.-A. Marchetti, *"Bosonization, Topological Solitons and Fractional Charges in Two-Dimensional Quantum Field Theory"*, submitted to Commun. Math. Phys.
7. A.A. Belavin, A.M. Polyakov and A.B. Zamolodchikov, *Nucl. Phys.* B 241, 333 (1984); V.G. Knizhnik and A.B. Zamolodchikov, *Nucl. Phys.* B 247, 83 (1984).
8. D. Friedan, E. Martinec and S. Shenker, Phys. Lett. B 160, 55 (1985); *Nucl. Phys.* B 271, 93 (1986).
9. V.G. Knizhnik, *Phys. Lett.* B 160, 403 (1985); B 178, 21 (1986).
10. G. Segal,*"The Definition of Pertubative Conformal Field Theory"*, notes, 1987.
J. Fröhlich, unpublished notes, Bures-sur-Yvette, 1987.
11. V.F.R. Jones, *Bull. Amer. Math. Soc.* 12, 103 (1985); Notes on a Talk in Atiyah's Seminar (1986); paper in preparation.
12. L. Kadanoff and H. Ceva, *Phys. Rev.* B 11, 3918 (1971).
13. K. Osterwalder and R. Schrader, *Commun. Math. Phys.* 31, 83 (1973); 42, 281 (1975); V. Glaser *Commun. Math. Phys.* 37, 257 (1974); J. Fröhlich, K. Osterwalder and E. Seiler, *Ann. Math.* 118, 461 (1983).
14. M. Sato, T. Miwa and M. Jimbo, *Publ. RIMS*, Kyoto University, 15. 871 (1979).
15. F. Wegner, J. *Math. Phys.* 12, 2259 (1971).
16. K.G. Wilson, *Phys. Rev.* D 10, 2445 (1974).
17. G.'t Hooft, *Nucl. Phys.* B 138, 1 (1978).
18. F. Wilczek and A. Zee, *Phys. Rev. Lett.* 51, 2250 (1983).
19. J. Fröhlich and P.-A. Marchetti, *"Field Theory of Anyons and θ-Statistics"*, to appear.

20. S. Doplicher, R. Haag and J. Roberts, *Commun. Math. Phys.* 23, 199 (1971); 35, 49 (1974). (See also ref. 39.)

21. J. Glimm and A. Jaffe, *"Quantum Physics"* (2nd edition), Berlin-Heidelberg-New York: Springer-Verlag, 1987.

22. J. Imbrie, *Commun. Math. Phys.* 82, 261 (1981), 82, 305 (1981).

23. R.F. Streater and A.S. Wightman, see ref. 1.
R. Jost, *"The General Theory of Quantized Fields"*, Providence, R.I.: Publ. Amer. Math. Soc., 1965.

24. T. Kohno, *Nagoya Math. J.* 92, 21 (1983); *Invent. Math.* 82, 57 (1985).

25. T. Kohno, *"Linear Representations of Braid Groups and Classical Yang-Baxter Equations", to appear in "Contemporary Mathematics: Artin's Braid Group"*, Santa Cruz, 1987.

26. A.A. Belavin and V.G. Drinfel'd, *Funct.Anal.Appl.* 16, 1 (1982).

27. A.A. Belavin, A.M. Polyakov and A.B. Zamolodchikov, *Nucl. Phys.* B 241, 333 (1984).

28. G. Segal, ref. 10.

29. V.G. Knizhnik and A.B. Zamolodchikov, *Nucl. Phys.* B 247, 83 (1984).

30. B. Schroer, *Lecture Notes*, Cargese 1987, to appear in the proceedings; and refs. given there.

31. E. Witten, *Commun. Math. Phys.* 92, 455 (1984).

32. A.M. Polyakov and P.B. Wiegmann, *Phys. Lett.* B 131, 121 (1983).

33. A. Tsuchiya and Y. Kanie, *Lett. Math. Phys.* 13, 303 (1987).

34. V.F.R. Jones, *Invent. Math.* 72, 1983.

35. V.F.R. Jones, unpubl. seminar notes, and paper in preparation.

36. V.G. Turaev, *"The Yang-Baxter Equation and Invariants of Links"*,LOMI Preprint E-3-1987.

37. L.H. Kauffman, *"Statistical Mechanics and the Jones Polynomial"*, Preprint, University of Illinois, 1987.

38. V. Pasquier, *Preprints*, Saclay SPh T/87 - 014, SPh T/ 87 - 62.

39. D. Buchholz and K. Fredenhagen, *Commun. Math. Phys.* 84, 1 (1982).

Nuclear Physics B312 (1989) 715–750
North-Holland, Amsterdam

EINSTEIN CAUSALITY AND ARTIN BRAIDS

Karl-Henning REHREN and Bert SCHROER

Freie Universität Berlin, Institut für Theorie der Elementarteilchen, Arnimallee 14, 1000 Berlin (West) 33, Germany

Received 13 May 1988
Revised 25 July 1988)

Within the restricted context of conformal $QFTh_2$ we present a systematic analysis of the exchange algebras of light-cone fields which result from the previously studied global decomposition theory of Einstein-causal fields. Although certain aspects of the representation theory of exchange algebras with their Artin braid structure-constants appear in our illustrative examples (minimal and WZW models), our main interests are algebraic aspects. We view the present work as a new non-lagrangian (non-hamiltonian) approach to non-perturbative QFTh.

1. Introduction

The subject of this paper is an attempt to overcome the longstanding problem of dynamics in the algebraic approach [1] to non-perturbative local Quantum Field Theory (QFTh). The basic structure of this approach is the "local net", i.e. the association of a C^* algebra of local observables with every space-time region. Two operators localized in regions with space-like separation are required to commute with each other. This is the axiom of locality (Einstein Causality). Space-time and inner symmetries are formulated algebraically in terms of endomorphisms of the local net. The representation theory led to the understanding of important concepts in local QFTh, such as superselection rules and the relation between spin and statistics. The theory of (massive) particle scattering has been given a rigorous meaning via the LSZ formalism [2]. Representations with temperature $T > 0$ equilibrium states have been established [3], which are inequivalent to the vacuum representation.

The most unsatisfactory feature of this algebraic approach, however, is its failure in specifying and describing interactions. While Einstein Causality states the *absence* of interactions at space-like distances, the dynamics of a theory should be encoded in the deviation from commutativity of operators localized with non-space-like separation.

This is precisely the idea of this paper. Taking advantage of the peculiar light-cone structure of conformal QFTh in two dimensions ($CQFTh_2$), we discuss the "exchange algebra" as a non-commutative bilinear algebra of light-cone operators, which *entails* the property of Einstein Causality for local fields (provided some numerical relations are satisfied). At the same time it codifies true dynamical information in the sense that it specifies time-like non-commutativity in terms of monodromy behaviour of n-point functions. Referring directly to conformal fields at the critical point, it is of genuinely non-perturbative character. Moreover, it contains explicit information about the "fusion rules" and implicit information about the anomalous scaling dimensions of the fields.

It is our intention to conceptualize earlier work of the authors [4] who were originally led to exchange algebras from the global decomposition theory of local operators, which in turn resulted – as a seasonable homologue of today's "conformal blocks" – from the resolution of the "Einstein Causality Paradox" [5]. There is a second path to exchange algebras pursued by Fröhlich [6], who started from the euclidean formulation and studied the monodromy properties of multivalued euclidean Green functions. We acknowledge many stimulating discussions with J. Fröhlich – who also introduced us into some of the mathematics, in particular Jones' work – about the questions which await an answer through the study of these algebraic structures, and new questions arising.

We put particular emphasis on the consistency constraints on conformal-exchange algebras, and thus on local algebras to be constructed from them. There are two types of consistency conditions, which are non-trivial to satisfy on the one hand, and sufficiently restrictive on the other hand, to single out only "a few" out of "all imaginable dynamics" of CQFTh_2. These include important classes of previously classified models. The consistency conditions are: (i) the associativity of the exchange operator algebra, which requires the "structure constants" to define representations of the Artin braid groups; and (ii) the compatibility with conformal transformation behaviour, which imposes additional relations on the braid-representation matrices.

We shall not try to tackle the complete classification problem of solutions to these conditions. We shall rather exploit the virtues of the rich structure implied by the conditions. In particular, we can give operator product expansions a meaning off-vacuum which is as satisfactory as on-vacuum, thus filling a gap in the old "bootstrap program".

2. Concepts and notions of CQFTh_2

Let us define a conformal quantum field theory in two dimensions by the following postulates, to be considered as working hypotheses rather than axioms.

(1) The conformal group has a unitary implementation U in Hilbert space. In particular there are hermitian infinitesimal generators $P_\pm, D_\pm, K_\pm$ commuting with

a basis of field operators $\phi_\alpha(x) = \phi_{\alpha^+\alpha^-}(x_+, x_-)$ as follows

$$i[P_+, \phi_\alpha(x_+, \cdot)] = \partial_+ \phi_\alpha(x_+, \cdot),$$

$$i[D_+, \phi_\alpha(x_+, \cdot)] = (x_+ \partial_+ + d_+) \phi_\alpha(x_+, \cdot),$$

$$i[K_+, \phi_\alpha(x_+, \cdot)] = (x_+^2 \partial_+ + 2x_+ d_+) \phi_\alpha(x_+, \cdot), \tag{2.1}$$

and likewise for P_-, D_-, K_-. Here $x_\pm = x^0 \pm x^1$ are the light-cone coordinates, and $d_\pm$ are the light-cone scaling dimensions of the conformal field ϕ_α.

The light-cone momenta $P_\pm$ have nonnegative spectra. There is a unique vacuum state Ω annihilated by $P_\pm, D_\pm, K_\pm$.

(2) The stress–energy tensor is a local field $\Theta_{\mu\nu}(x) = \Theta_{\nu\mu}(x)$ which is conserved: $\partial^\mu \Theta_{\mu\nu} = 0$, which has canonical dimension 2, and the time components of which are the energy–momentum densities: $\int dx^1 \, \Theta_{0\nu}(x) = P_\nu$.

The following are, by now, standard consequences.

(i) [7] $\Theta_{\mu\nu}$ is traceless, and can be split into two light-cone fields $\Theta_+(x_+), \Theta_-(x_-)$ of light-cone dimensions $d_+ = 2$, $d_- = 2$, respectively. Both are Lie fields, and their moments

$$L_n = \tfrac{1}{4}(-1)^n \int dx \, (x - i)^{1+n} (x + i)^{1-n} \Theta(x), \tag{2.2}$$

satisfy two independent Virasoro algebras (one for either light-cone; we omitted the label $+/-$). In particular

$$P = \tfrac{1}{2} \int dx \, \Theta(x) = L_0 + \tfrac{1}{2}(L_{-1} + L_1),$$

$$D = \tfrac{1}{2} \int x \, dx \, \Theta(x) = i(L_{-1} - L_1),$$

$$K = \tfrac{1}{2} \int x^2 \, dx \, \Theta(x) = L_0 - \tfrac{1}{2}(L_{-1} + L_1). \tag{2.3}$$

The "compact picture" with its "radial quantization" (analyticity in "radial tubes" in Wightman's spirit) is obtained from this by a complex Möbius transformation $\zeta(x) = (x - i)/(1 - ix)$, which maps the real axis onto the unit circle. This transformation is not a physical symmetry of the theory, but rather an isometric mapping from one theory into another. We make no use of the compact picture.

(ii) [8] Since the Weyl inversion I: $x \to -1/x$ is contained in the conformal group, $K = U(I) P U(I)^+$ has the same nonnegative spectrum as P. In particular the generators $L_0 = \tfrac{1}{2}(P + K)$ ("conformal Hamiltonians", one for either light-cone) of the compact subgroup U(1) of $SL(2, \mathbb{R})/Z_2$ have nonnegative spectrum.

(iii) [9] The conformal fields (2.1) organize into families $[\alpha^+\alpha^-]$ such that the orthogonal subspaces $\mathscr{H}_{[\alpha^+\alpha^-]} = \mathrm{span}\{\phi_{\alpha^+\alpha^-}(x)\Omega | \phi_{\alpha^+\alpha^-} \in [\alpha^+\alpha^-]\}$ are irreducible representation spaces of the Virasoro algebra $\mathrm{Vir}_+ \otimes \mathrm{Vir}_-$. For every representation, denoted also $[\alpha^+\alpha^-]$, there is a primary field of dimensions $h^+ = h(\alpha^+)$, $h^- = h(\alpha^-)$, while all other conformal fields belonging to $[\alpha^+\alpha^-]$ (quasiprimary fields) have dimensions $d^{\pm} = h^{\pm} + n^{\pm}$, $n^{\pm} \in \mathbb{N}$.

(iv) [5] The integration of the infinitesimal transformation behaviour (2.1) is in general impossible for a local field $\phi_{\alpha^+\alpha^-}$ as a whole. In fact there occur complex phases in the law for special conformal transformations $\mathsf{T}_b x = x/(1-bx)$ beyond the singular point, that depend on the states between which $\phi_{\alpha^+\alpha^-}$ is evaluated. Introducing orthogonal projectors $P_{\beta^+\beta^-}$ onto $\mathscr{H}_{[\beta^+\beta^-]}$ it is found from the phases of the three-point functions $(\phi_{\beta^+\beta^-}\Omega, \phi_{\alpha^+\alpha^-}\phi_{\gamma^+\gamma^-}\Omega)$ – which are in turn determined by the spectrum condition – that

$$U_{\pm}(a)\phi_{\alpha^+\alpha^-}(x_{\pm},\cdot)U_{\pm}(a)^+ = \phi_{\alpha^+\alpha^-}(x_{\pm}+a,\cdot)$$

$$U_{\pm}(\lambda)\phi_{\alpha^+\alpha^-}(x_{\pm},\cdot)U_{\pm}(\lambda)^+ = \lambda^{d_{\alpha^{\pm}}}\phi_{\alpha^+\alpha^-}(\lambda x_{\pm},\cdot)$$

$$U_{\pm}(b)P_{\beta^+\beta^-}\phi_{\alpha^+\alpha^-}(x_{\pm},\cdot)P_{\gamma^+\gamma^-}U_{\pm}(b)^+$$

$$= \sigma^{\alpha^{\pm}}_{\beta^{\pm}\gamma^{\pm}}(b, x_{\pm}) \cdot P_{\beta^+\beta^-}\phi_{\alpha^+\alpha^-}(\mathsf{T}_b x_{\pm},\cdot)P_{\gamma^+\gamma^-} \tag{2.4}$$

where

$$\sigma^{\alpha}_{\beta\gamma}(b,x) = (1-b(x+i\varepsilon))^{-d_\alpha+h_\gamma-h_\beta}(1-b(x-i\varepsilon))^{-d_\alpha-h_\gamma+h_\beta}$$

$$= |1-bx|^{-2d_\alpha}\exp\left[2\pi i\,\mathrm{sig}(b)\theta(bx-1)(h_\beta-h_\gamma)\right]; \tag{2.5}$$

θ is the step function, and $h_{\beta^{\pm}}$, $h_{\gamma^{\pm}}$ the primary dimensions of the representations $[\beta^+\beta^-]$, $[\gamma^+\gamma^-]$. The dot in eq. (2.4) stands for the other light-cone variable. In particular $\Theta_{\pm}$ and $U_{\pm}$ commute with the projectors $P_{\beta^+\beta^-}$. The center of the conformal group, generated by $Z^{\pm} = \exp 2\pi i L_0^{\pm}$, is represented nontrivially as

$$Z^{\pm}P_{\beta^+\beta^-} = \exp(2\pi i h_{\beta^{\pm}})P_{\beta^+\beta^-}$$

$$Z^{\pm}P_{\beta^+\beta^-}\phi_{\alpha^+\alpha^-}P_{\gamma^+\gamma^-}(Z^{\pm})^+ = \exp\left[2\pi i(h_{\beta^{\pm}}-h_{\gamma^{\pm}})\right]P_{\beta^+\beta^-}\phi_{\alpha^+\alpha^-}P_{\gamma^+\gamma^-}, \tag{2.6}$$

thus the quantum symmetry group is the universal covering $\widetilde{\mathrm{SL}(2,\mathbb{R})}_+ \otimes \widetilde{\mathrm{SL}(2,\mathbb{R})}_-$ of the Möbius group.

(v) [9] Decompose a local n-point function $W = (\Omega, \phi_{\alpha_1^+\alpha_1^-} \cdots \phi_{\alpha_n^+\alpha_n^-}\Omega)$ into a sum of non-local functions

$$W_{\xi^+\xi^-} = \left(\Omega, \phi_{\alpha_1^+\alpha_1^-}P_{\beta_1^+\beta_1^-} \cdots P_{\beta_{n-1}^+\beta_{n-1}^-}\phi_{\alpha_n^+\alpha_n^-}\Omega\right), \tag{2.7}$$

ξ^+, ξ^- referring to the "channel" of successive projectors. Then $W_{\xi^+\xi^-} = F^+_{\xi^+} \cdot F^-_{\xi^-}$ factorizes into "conformal blocks" F_{ξ^+} depending on α^+, β^+, x_+ only, and F_{ξ^-} depending on α^-, β^-, x_- only.

This justifies [4] the factorization of the projected fields as operators

$$P_{\beta^+\beta^-}\phi_{\alpha^+\alpha^-}(x_+, x_-)P_{\gamma^+\gamma^-} = \left(P_{\beta^+}a^+_{\alpha^+}(x_+)P_{\gamma^+}\right) \otimes \left(P_{\beta^-}a^-_{\alpha^-}(x_-)P_{\gamma^-}\right),$$

$$\mathscr{H}_{[\beta^+\beta^-]} = V_{[\beta^+]} \otimes V_{[\beta^-]},$$

$$P_{\beta^\pm}a^\pm_{\alpha^\pm}P_{\gamma^\pm} \equiv \left(a^\pm_{\alpha^\pm}\right)_{\beta^\pm\gamma^\pm} : V_{[\gamma^\pm]} \to V_{[\beta^\pm]}. \tag{2.8}$$

We have thus obtained the field $\phi_{\alpha^+\alpha^-}$ as a bilinear combination of light-cone fields $(a^+_{\alpha^+})_{\beta^+\gamma^+}$ and $(a^-_{\alpha^-})_{\beta^-\gamma^-}$. It is an immediate question what locality, i.e. space-like commutativity of 2-dimensional fields $\phi_{\alpha^+\alpha^-}$, means in terms of light-cone fields $(a_\alpha)_{\beta\gamma}$. The latter cannot just commute with each other, formally since they carry their projectors with them, physically since then time-like commutativity were also implied. In sect. 3 we shall discuss as a most natural structure (see also ref. [6]) a bilinear "exchange algebra" satisfied by two collections of light-cone fields. It will have space-like commutativity of local fields as a consequence, while realizing non-trivial dynamics at time-like distances.

For this purpose we have to make one further structural assumption about the CQFTh$_2$ aimed at. Let us state it as follows. The non-vanishing 3-point functions $(\phi_{\beta^+\beta^-}\Omega, \phi_{\alpha^+\alpha^-}\phi_{\gamma^+\gamma^-}\Omega)$ define a set of formal, associative and commutative, "fusion rules"

$$[\alpha^+\alpha^-][\gamma^+\gamma^-] = \bigoplus_{\beta^+} \bigoplus_{\beta^-} [\beta^+\beta^-], \tag{2.9}$$

which specify the channels contributing to a local n-point function. Our assumption is

(3) for every $[\alpha^+\alpha^-], [\gamma^+\gamma^-]$ the sum in eq. (2.9) is finite.

The assumption is satisfied, e.g. in the minimal models ($c < 1$) [9], the Thirring model ($c = 1$) [10], the Wess–Zumino–Witten (WZW) models ($c \geq 1$) [11], and in a number of further classified models, characterized by various additional symmetries ($N = 1, 2$ supersymmetry [11], Z_N-symmetry [12], ...). It seems, however, not to be a feature of general CQFTh$_2$; instead, it is essentially this extremely untypical property of finiteness going along with a maximal domain of analyticity of the Wightman functions, which distinguishes the various classified "oases" scattered in the vast "desert" of $c \geq 1$ theories, generally inaccessible due to infinite fusion rules. Though, *formally*, the assumption (3) can easily be relaxed, the analytic properties are expected to change drastically.

Note that, in the above, $[\alpha^+\alpha^-]$ are not necessarily inequivalent Virasoro representations. Different fields corresponding to the same representation are counted separately in the fusion rules. There may be a larger symmetry (Kac–Moody- or supersymmetry) collecting families of $[\alpha^+\alpha^-]$ (with common primary dimensions mod $\mathbb{Z}$ in the Kac–Moody case, mod $\frac{1}{2}\mathbb{Z}$ in the supersymmetric case) into one irreducible representation of the enlarged symmetry.

3. Exchange field theory

The preceding discussion has provided the motivation for the following, model-independent definition of a conformal "exchange field theory". Here the term "exchange" makes the distinction from local field theory.

We have in mind a family of primary and quasiprimary conformal light-cone fields a_α, belonging to irreducible representations $[\alpha]$ in $V_{[\alpha]}$ of the Virasoro algebra, and a set of associative and commutative (formal) fusion rules

$$[\alpha][\gamma] = \bigoplus_{\beta}^{\text{finite}} [\beta]. \tag{3.1}$$

In particular, there are a vacuum sector V_0, a vacuum state $\Omega \in V_0$, and a vacuum representation [0] such that $[\alpha][0] = [\alpha]$.

For every γ, and for every β contributing to this sum, there is an intertwining light-cone field, depending on a one-dimensional variable $x = x_\pm$

$$(a_\alpha)_{\beta\gamma}(x) \equiv P_\beta a_\alpha(x) P_\gamma\colon V_\gamma \to V_\beta. \tag{3.2}$$

If in agreement with these fusion rules an operator product $P_{\beta_0} a_{\alpha_1}(x_1) P_{\beta_1} \dots P_{\beta_{n-1}} a_{\alpha_n}(x_n) P_{\beta_n}$ does not identically vanish, we shall call this product, and the corresponding multi-index $(\boldsymbol{\beta\alpha}) \equiv (\beta_0, \alpha_1, \dots, \alpha_n, \beta_n)$, "admissible". All admissible operator products are assumed linearly independent operator-valued functions of their light-cone variables. Multiplication in the field algebra is associative.

The fields $(a_\alpha)_{\beta\gamma}$ are supposed to behave under the respective groups of translations, dilatations, and special conformal transformations, like the projected local fields (2.4)

$$U(a) a_\alpha(x) U(a)^{-1} = a_\alpha(x + a),$$

$$U(\lambda) a_\alpha(x) U(\lambda)^{-1} = \lambda^{d_\alpha} a_\alpha(\lambda x),$$

$$U(b)(a_\alpha)_{\beta\gamma}(x) U(b)^{-1} = \sigma^\alpha_{\beta\gamma}(b, x)(a_\alpha)_{\beta\gamma}(\mathrm{T}_b x), \tag{3.3}$$

where the first two equations are irrespective of β and γ. The notations are as in eq. (2.4).

In particular

$$Z|_{V_\beta} = e(h_\beta),$$

$$Z(a_\alpha)_{\beta\gamma}(x)Z^+ = e(h_\beta - h_\gamma)(a_\alpha)_{\beta\gamma}(x), \tag{3.4}$$

where we have introduced the notation

$$e(h) = \exp(2\pi i h).$$

There is a field $\mathscr{T}(x)$ (which is just a rescaled version of $\Theta(x)$ such that $(\Omega, \mathscr{T}(x_1)\mathscr{T}(x_2)\Omega) = \frac{1}{2}c(x_1 - x_2 - i\varepsilon)^{-4}$) generating the conformal transformations like in eq. (2.3). The commutators of a_α with higher moments L_n, $|n| > 1$, eq. (2.2), take all fields belonging to $[\alpha]$ irreducibly into each other.

The usual notion of "locality" is meaningless in a light-cone theory. It is replaced by the following.

Exchange algebra. There are matrices $[R^{(\beta_0\beta_2)}_{(\alpha_1\alpha_2)}(x_1, x_2)]_{\beta_1\beta_1'}$ for $x_1 \neq x_2$ such that

$$P_{\beta_0} a_{\alpha_1}(x_1) P_{\beta_1} a_{\alpha_2}(x_2) P_{\beta_2}$$
$$= \sum_{\beta_1'} \left[R^{(\beta_0\beta_2)}_{(\alpha_1\alpha_2)}(x_1, x_2)\right]_{\beta_1\beta_1'} P_{\beta_0} a_{\alpha_2}(x_2) P_{\beta_1'} a_{\alpha_1}(x_1) P_{\beta_2}. \tag{3.5}$$

In particular, for $x_1 \neq x_2$

$$P_\beta a_\alpha(x_1) P_\gamma \mathscr{T}(x_2) = \mathscr{T}(x_2) P_\beta a_\alpha(x_1) P_\gamma, \tag{3.6}$$

and in case there is a Kac–Moody current $j(x)$

$$\tilde{P}_\beta a_\alpha(x_1) \tilde{P}_\gamma j(x_2) = j(x_2) \tilde{P}_\beta a_\alpha(x_1) \tilde{P}_\gamma,$$

where $\tilde{P}$ project onto the irreducible representation spaces of the enlarged symmetry.

Since all fields a_α, $\alpha \in [\alpha]$, are obtained from the corresponding primary field by appropriate short-distance limits of products with $\mathscr{T}(x)$, it follows that the matrices R depend on α_i only through $[\alpha_i]$. In the Kac–Moody case, if $[\alpha_i]$ and $[\alpha_i']$ belong to the same enlarged representation, the respective R matrices also coincide.

Proposition 1. The matrices $R^{(\beta_1\beta_2)}_{(\alpha_1\alpha_2)}(x_1, x_2)$ depend on the light-cone variables only through $s_{12} = \operatorname{sig}(x_1 - x_2)$. Calling

$$R^{(\beta_0\beta_2)}_{(\alpha_1\alpha_2)}(x_1 > x_2) =: R^{(\beta_0\beta_2)}_{(\alpha_1\alpha_2)}, \tag{3.7a}$$

one has

$$R^{(\beta_0\beta_2)}_{(\alpha_1\alpha_2)}(x_1<x_2)=\left[R^{(\beta_0\beta_2)}_{(\alpha_2\alpha_1)}\right]^{-1}. \tag{3.7b}$$

Proposition 2. Define diagonal matrices

$$\left[\phi^{(\beta_0\beta_2)}_{(\alpha_1\alpha_2)}\right]_{\beta\beta'}=\delta_{\beta\beta'}\exp 2\pi i\left(h_\beta-\tfrac{1}{2}\left(h_{\beta_0}+h_{\beta_2}\right)\right), \tag{3.8}$$

which depend on α_1, α_2 through β running over the values for which $(\beta_0\alpha_1\beta\alpha_2\beta_2)$ are admissible. Then

$$\phi^{(\beta_0\beta_2)}_{(\alpha_1\alpha_2)}R^{(\beta_0\beta_2)}_{(\alpha_1\alpha_2)}\phi^{(\beta_0\beta_2)}_{(\alpha_2\alpha_1)}R^{(\beta_0\beta_2)}_{(\alpha_2\alpha_1)}=\mathbf{1}. \tag{3.9}$$

In particular "on the vacuum" ($\beta_2=[0]$), where the size of the matrices is 1×1

$$R^{(\beta 0)}_{(\alpha_1\alpha_2)}R^{(\beta 0)}_{(\alpha_2\alpha_1)}=\exp 2\pi i\left(h_\beta-h_{\alpha_1}-h_{\alpha_2}\right). \tag{3.10}$$

Proposition 3. The exchange matrices $[R^{(\beta\beta')}_{(\alpha\alpha')}]_{\gamma\gamma'}$ satisfy

$$\sum_{\beta_1''}\left[R^{(\beta_0\beta_2)}_{(\alpha_1\alpha_2)}\right]_{\beta_1\beta_2''}\left[R^{(\beta_1''\beta_3)}_{(\alpha_1\alpha_3)}\right]_{\beta_2\beta_2'}\left[R^{(\beta_0\beta_2')}_{(\alpha_2\alpha_3)}\right]_{\beta_1''\beta_1'}$$

$$=\sum_{\beta_2''}\left[R^{(\beta_1\beta_3)}_{(\alpha_2\alpha_3)}\right]_{\beta_2\beta_2''}\left[R^{(\beta_0\beta_2'')}_{(\alpha_1\alpha_3)}\right]_{\beta_1\beta_1'}\left[R^{(\beta_1'\beta_3)}_{(\alpha_1\alpha_2)}\right]_{\beta_2''\beta_2'} \tag{3.11}$$

for all admissible multi-indices $(\beta_0\alpha_1\beta_1\alpha_2\beta_2\alpha_3\beta_3)$ and $(\beta_0\alpha_3\beta_1'\alpha_2\beta_2'\alpha_1\beta_3)$.

Before we prove these propositions, let us make some remarks.

The operator algebra (3.5) translates into analytic properties of Wightman n-point functions of light-cone fields. The latter are defined by complex conformal transformations as analytic functions in a large complex domain (the "conformal tube") of analyticity. There are cyclically ordered real points ("conformal Jost points") $(x_\nu>x_{\nu+1}>\cdots>x_{\nu-1}$, or $x_\nu<x_{\nu+1}<\cdots<x_{\nu-1})$ which lie inside this domain, while permuted real points lie on the boundary. The exchange algebra expresses n-point functions at permuted real points as linear combinations of n-point functions at the original points. This again enlarges the domain of analyticity into $\widetilde{\mathring{\mathbb{C}}^n}$, where $\circ$ denotes the omission of all points of coincidence $x_i=x_j$, $i\neq j$, and $\sim$ denotes the universal covering, since $\mathring{\mathbb{C}}^n$ is not simply connected. Actually, the theory makes use of a finite covering only. For these functions the matrices R describe the analytic exchange $\overset{\curvearrowright}{x_i x_{i+1}}$ with positive orientation of two neighboring variables, while $R_{(\alpha_1\alpha_2)}R_{(\alpha_2\alpha_1)}$ describe the monodromy $x_i\sim x_{i+1}$ (with positive orientation) around a branch point [4]. Vice versa, it can be shown from first principles of CQFTh$_2$ that the exchange algebra is a consequence of the domain of analyticity of conformal blocks being $\widetilde{\mathring{\mathbb{C}}^n}$ [6].

The identity of proposition 3 has obviously the structure of a Yang–Baxter equation [14] for a vertex or SOS type lattice model [15]: read α as a generalized "orientation" of a link, and β as the corresponding "heights" of the adjacent plaquettes. Solutions to the Yang–Baxter equations are well known, and the dependence on a rapidity or spectral parameter must be eliminated by appropriate limits [4]. Possibly there exist more solutions of eq. (3.11). Given a solution of eq. (3.11), one must check whether diagonal phase matrices ϕ exist satisfying eq. (3.10). If they do, most of the dimensional trajectories $h(\beta)$ of the corresponding model can be read off their entries. We shall give important examples in sect. 7.

Proof of Proposition 1. The first statement follows from the transformation laws under $U(a)$ and $U(\lambda)$ since, due to the latter, $R(x_1, x_2)$ must be translation- and scale-invariant quantities. The second statement becomes evident by solving eq. (3.5) for the operator products appearing on the r.h.s.

Proof of Proposition 2. The statement follows from the fact that transformations with $U(b)$ may change the sign $s_{12} = \mathrm{sig}(x_1 - x_2)$. In fact: $\mathsf{T}x_1 - \mathsf{T}x_2 = (x_1 - x_2)/(1 - bx_1)(1 - bx_2)$ implies $\mathsf{T}s_{12} = \mathrm{sig}(\mathsf{T}x_1 - \mathsf{T}x_2) = s_{12} \cdot \mathrm{sig}(1 - bx_1) \cdot \mathrm{sig}(1 - bx_2)$.

First applying a special conformal transformation $U(b)$ to eq. (3.5) yields an expression of $P_{\beta_0} a_{\alpha_1}(\mathsf{T}x_1) P_\beta a_{\alpha_2}(\mathsf{T}x_2) P_{\beta_2}$ in terms of $P_{\beta_0} a_{\alpha_2}(\mathsf{T}x_2) P_{\beta'} a_{\alpha_1}(\mathsf{T}x_1) P_{\beta_2}$. Re-expressing the latter via the exchange algebra in terms of $P_{\beta_0} a_{\alpha_1}(\mathsf{T}x_1) P_{\beta''} a_{\alpha_2}(\mathsf{T}x_2) P_{\beta_2}$, comparing coefficients and taking care of all phase factors stemming from the conformal transformation, yields the equation

$$\delta_{\beta\beta''} = \sum_{\beta'} e\left[\mathrm{sig}(b)(-h(\beta_0) + h(\beta) + h(\beta') - h(\beta_2))(\theta(bx_1 - 1) - \theta(bx_2 - 1))\right]$$
$$\times \left[R^{(\beta_0\beta_2)}_{(\alpha_1\alpha_2)}(s_{12})\right]_{\beta\beta'} \left[R^{(\beta_0\beta_2)}_{(\alpha_2\alpha_1)}(\mathsf{T}s_{21})\right]_{\beta'\beta''},$$

which in the case $\mathrm{sig}(1 - bx_1) = \mathrm{sig}(1 - bx_2) \Rightarrow \mathsf{T}s_{21} = -s_{12}$, $e[\cdot] = 1$, is the identity eq. (3.7). In the case $\mathrm{sig}(1 - bx_1) = -\mathrm{sig}(1 - bx_2) \Rightarrow \mathsf{T}s_{21} = s_{12} =: s$, $(\theta(bx_1) - \theta(bx_2)) = \mathrm{sig}(b)s_{12}$, it is rewritten as

$$\mathbf{1} = \left[\phi^{(\beta_0\beta_2)}_{(\alpha_1\alpha_2)}\right]^s R^{(\beta_0\beta_2)}_{(\alpha_1\alpha_2)}(s) \left[\phi^{(\beta_0\beta_2)}_{(\alpha_2\alpha_1)}\right]^s R^{(\beta_0\beta_2)}_{(\alpha_2\alpha_1)}(s).$$

In both cases, $s = +1$ or $s = -1$, this is equivalent to eq. (3.10).

Proof of Proposition 3. The statement expresses the associativity of the field algebra. It is sufficient to consider triple operator products $P_{\beta_0} a_{\alpha_1}(x_1) \times P_{\beta_1} a_{\alpha_2}(x_2) P_{\beta_2} a_{\alpha_3}(x_3) P_{\beta_3}$ at different points $x_i \neq x_j$, $i \neq j$. There are two ways to express these products in terms of $P_{\beta_0} a_{\alpha_3}(x_3) P_{\beta_1'} a_{\alpha_2}(x_2) P_{\beta_2'} a_{\alpha_1}(x_1) P_{\beta_3}$, corresponding to two inequivalent representations of the permutation (13) in terms of transpositions (12) and (23). This results in the conditions

$$\sum_{\beta_1''} \left[R^{(\beta_0\beta_2)}_{(\alpha_1\alpha_2)}(s_{12})\right]_{\beta_1\beta_1''} \left[R^{(\beta_1''\beta_3)}_{(\alpha_1\alpha_3)}(s_{13})\right]_{\beta_2\beta_2'} \left[R^{(\beta_0\beta_2')}_{(\alpha_2\alpha_3)}(s_{23})\right]_{\beta_1''\beta_1'}$$
$$= \sum_{\beta_2''} \left[R^{(\beta_1\beta_3)}_{(\alpha_2\alpha_3)}(s_{23})\right]_{\beta_2\beta_2''} \left[R^{(\beta_0\beta_2'')}_{(\alpha_1\alpha_3)}(s_{13})\right]_{\beta_1\beta_1'} \left[R^{(\beta_1'\beta_3)}_{(\alpha_1\alpha_2)}(s_{12})\right]_{\beta_2''\beta_2'}$$

for all admissible multiindices, and for all possible signs $s_{ij} = \mathrm{sig}(x_i - x_j)$. Using proposition 1 to express all matrices $R^{(\cdot)}_{(\cdot)}(+/-)$ in terms of $R^{(\cdot)}_{(\cdot)} = R^{(\cdot)}_{(\cdot)}(+)$, for every choice of signs the above equation can be rewritten, after appropriate matrix multiplication and relabelling of α_i, as eq. (3.11).

In order to avoid confusion we should indicate that we made some change of notation as compared with ref. [4]. First, we found it convenient to treat the two light-cones in like rather than opposite manner, and introduced $x_+ = u$, $x_- = -v$. (Then the euclidean section is $z_+ = -z_-^*$.) Second, the constant R matrices now describe positive rather than clockwise-oriented continuation. Third, the concept of a "scheme" has been replaced by the "fusion rules".

4. Local field theory

Let us now discuss the construction of a two-dimensional local field theory out of two light-cone exchange field theories. As suggested by sect. 2, we consider fields of the form

$$\phi_\alpha(\boldsymbol{x}) := \sum_{\beta^+\gamma^+\beta^-\gamma^-} g^{(\alpha^+\alpha^-)}_{\beta^+\gamma^+;\beta^-\gamma^-} (a^+_{\alpha^+})_{\beta^+\gamma^+}(x_+) \otimes (a^-_{\alpha^-})_{\beta^-\gamma^-}(x_-) \tag{4.1}$$

with numerical coefficients $g^{(\cdot)}_{\because}$, which may take values different from 0, 1 since the light-cone fields a^+, a^- are so far defined without normalization. The point $\boldsymbol{x} = (x^0, x^1)$ is defined by $x_\pm = x^0 \pm x^1$.

Proposition 4. Let the structure-constants matrices R^+, R^- referring to the exchange algebras of fields $(a^+_{\alpha^+})_{\beta^+\gamma^+}$, $(a^-_{\alpha^-})_{\beta^-\gamma^-}$ respectively satisfy the requirements of propositions 1–3, such that fields ϕ_α given by eq. (4.1) transform like conformal fields in two dimensions (eq. (2.4)). Then, two fields ϕ_{α_1} and ϕ_{α_2} satisfy the axiom of local (anti-)commutativity (Einstein causality)

$$\phi_{\alpha_1}(\boldsymbol{x}_1)\phi_{\alpha_2}(\boldsymbol{x}_2) = \varepsilon \phi_{\alpha_2}(\boldsymbol{x}_2)\phi_{\alpha_1}(\boldsymbol{x}_1)$$

$$\text{if} \quad (\boldsymbol{x}_1 - \boldsymbol{x}_2)^2 = (x_1 - x_2)_+ (x_1 - x_2)_- < 0 \tag{4.2}$$

($\varepsilon = +1$ or -1), if and only if

$$\sum_{\gamma^+} g^{(\alpha_1^+\alpha_1^-)}_{\beta_0^+\gamma^+;\beta_0^-\beta^-} g^{(\alpha_2^+\alpha_2^-)}_{\gamma^+\beta_2^+;\beta^-\beta_2^-} \left[R^{+(\beta_0^+\beta_2^+)}_{(\alpha_1^+\alpha_2^+)} \right]_{\gamma^+\beta^+}$$

$$= \varepsilon \sum_{\gamma^-} g^{(\alpha_2^+\alpha_2^-)}_{\beta_0^+\beta^+;\beta_0^-\gamma^-} g^{(\alpha_1^+\alpha_1^-)}_{\beta^+\beta_2^+;\gamma^-\beta_2^-} \left[R^{-(\beta_0^-\beta_2^-)}_{(\alpha_2^-\alpha_1^-)} \right]_{\gamma^-\beta^-} \tag{4.3}$$

for all $(\beta_0, \beta, \beta_2)^\pm$ such that $(\beta_1\alpha_1\beta\alpha_2\beta_2)^\pm$ are admissible.

Special cases

(1) Let the theory be parity invariant in the strong sense that $\boldsymbol{P}(a^+_\alpha)_{\beta\gamma}(x)\boldsymbol{P} = (a^-_\alpha)_{\beta\gamma}(x)$ (in particular the range of the labels α, β and the fusion rules coincide on the two light-cones, $\varepsilon = 1$, and $R^+ \equiv R^-$). Let $g^{(\alpha^+\alpha^-)}_{\beta^+\gamma^+;\beta^-\gamma^-} = \delta_{\alpha^+\alpha^-}\,\delta_{\beta^+\beta^-}\,\delta_{\gamma^+\gamma^-}\chi$, where $\chi = 1$, if $(\beta\alpha\gamma)^\pm$ are admissible, and $\chi = 0$ else. Then $\boldsymbol{P}\phi(t, x)\boldsymbol{P} = \phi(t, -x)$, and eq. (4.3) reduces to

$$\left[R^{(\beta_0\beta_2)}_{(\alpha_1\alpha_2)}\right]_{\beta^-\beta^+} = \left[R^{(\beta_0\beta_2)}_{(\alpha_2\alpha_1)}\right]_{\beta^+\beta^-}, \tag{4.4}$$

i.e. the R matrices for $[\alpha_2] \leftrightarrow [\alpha_1]$ are transpose to each other (which is consistent with propositions 2 and 3), and symmetric for $[\alpha_1] = [\alpha_2]$. The special solutions in subsect. 7.4 have this property.

(2) Let there be a 1:1 mapping i of $\{\alpha^+\}$ onto $\{\alpha^-\}$, and a 1:1 mapping j of $\{\beta^+\}$ onto $\{\beta^-\}$, compatible with the fusion rules, but not necessarily taking the right "vacuum label" into the left "vacuum label". Let $g^{(\alpha^+\alpha^-)}_{\beta^+\gamma^+;\beta^-\gamma^-} = \delta_{i(\alpha^+)\alpha^-}\delta_{j(\beta^+)\beta^-}\delta_{j(\gamma^+)\gamma^-}\chi$. Then eq. (4.3) reduces to

$$\left[R^{-(j(\beta_0)j(\beta_2))}_{(i(\alpha_2)i(\alpha_1))}\right]_{j(\beta)j(\beta')} = \varepsilon\left[R^{+(\beta_0\beta_2)}_{(\alpha_1\alpha_2)}\right]_{\beta'\beta}. \tag{4.5}$$

If R^+ satisfy the requirements of propositions 1–3, then so do R^- defined through eq. (4.5). In particular, if the + and − exchange algebras coincide and the mappings i, j describe a symmetry of the fusion rules, then eq. (4.5) describes a symmetry of the R matrices. Conversely, R matrices with a symmetry of eq. (4.5) allow for the construction of "non-diagonal" local fields out of two coinciding exchange algebras. Examples will be given in subsect. 7.4 too.

Proof of proposition 4. The claim is immediately proved if eq. (4.1) is inserted into the l.h.s. of eq. (4.2), and the exchange algebra relations are applied. At space-like distances, the R^+ matrices and the R^- matrices occur with opposite exponents.

It is also evident that at time-like distances the R^+ and R^- matrices occur with the same exponent. Expressing R^- via eq. (4.3) by R^+, we see that essentially the squares of R^+, i.e. the monodromy matrices, describe the non-commutative behaviour at time-like distances. The examples show that the condition (4.3) can easily be satisfied if only the + exchange algebra is given, and that solutions less trivial than in case (1), which are of particular interest in the context of D-E type modular-invariant partition functions of minimal models, may be detected.

At this point, locality has been traded by virtue of proposition 4 for a set of simple numerical equations. We emphasize, however, that local CQFTh$_2$ has not become a trivial issue, since along with the exchange algebra came the conditions of proposition 1–3. Their restrictive and predictive power should not be underestimated. The rest of this paper will address these topics.

5. Representations of braid groups

The structure constants R of an exchange algebra define matrix representations of the braid groups B_n [16], which are the fundamental groups of $\mathring{\mathbb{C}}^n/S_n$. The braid group B_n (see below) is generated by elements σ_i, $i=1,\dots,n-1$, (and their inverses), satisfying

$$\sigma_i\sigma_j=\sigma_j\sigma_i \quad \text{if} \quad |i-j|\geq 2,$$

$$\sigma_i\sigma_{i+1}\sigma_i=\sigma_{i+1}\sigma_i\sigma_{i+1}. \tag{5.1}$$

There is a natural homomorphism π: $B_n\to S_n$, $\sigma_i\to\tau_i$, τ_i the transposition $(i,i+1)$, of the braid group into the symmetric group.

The representation matrices $\rho_i=\rho(\sigma_i)$ induced from an exchange algebra describe the transposition of two neighboring operators $a_{\alpha_i}(x_i)$, $a_{\alpha_{i+1}}(x_{i+1})$ within an admissible n-point operator product

$$P_{\beta_0}a_{\alpha_1}(x_1)P_{\beta_1}\dots P_{\beta_{n-1}}a_{\alpha_n}(x_n)P_{\beta_n}.$$

These keep β_0, β_n fixed, while α_i $(i=1,\dots,n)$ are permuted and β_i $(i=1,\dots,n-1)$ are changed according to the fusion rules. Hence, the matrix indices of the representation matrices run over all admissible $(\boldsymbol{\beta\alpha})=(\beta_0\alpha_1\dots\alpha_n\beta_n)$ with β_0, β_n fixed, and $(\alpha_1,\dots,\alpha_n)$ permutations of a fixed n-tuple $A=(\alpha_1^0,\dots,\alpha_n^0)$.

The following proposition is a concise reformulation of propositions 2 and 3.

Proposition 5. Given an exchange algebra satisfying the requirements of propositions 1–3. Choose $A=(\alpha_1^0,\dots,\alpha_n^0)$ and β_0, β_n.

(1) Then the matrices

$$[\rho_i]_{(\boldsymbol{\beta\alpha})(\boldsymbol{\beta'\alpha'})}:=\delta_{\tau_i\boldsymbol{\alpha},\boldsymbol{\alpha}'}\delta_{\beta_1\beta_1'}\dots\widehat{\delta_{\beta_i\beta_i'}}\dots\delta_{\beta_{n-1}\beta_{n-1}'}\left[R^{(\beta_{i-1}\beta_{i+1})}_{(\alpha_i\alpha_{i+1})}\right]_{\beta_i\beta_i'}, \tag{5.2}$$

define a representation $\rho^{(A,\beta_0,\beta_n)}(\sigma_i):=\rho_i$ of B_n ($\widehat{\ }$ denotes omission of a term).

(2) Let

$$[\varphi_i]_{(\boldsymbol{\beta\alpha}),(\boldsymbol{\beta'\alpha'})}:=\delta_{\boldsymbol{\alpha},\boldsymbol{\alpha}'}\delta_{\boldsymbol{\beta},\boldsymbol{\beta}'}e\left(h_{\beta_i}-\tfrac{1}{2}\left(h_{\beta_{i-1}}+h_{\beta_{i+1}}\right)\right). \tag{5.3}$$

Then we have

$$\varphi_i\rho_i\varphi_i\rho_i=\mathbf{1}. \tag{5.4}$$

Example. Consider three fields a_{α_1}, a_{α_2}, a_{α_3}, and choose β_0 arbitrary, $[\beta_3]=[0]$. Denote $h(\alpha_i)=h_i$, $h(\beta_0)=h_0$.

For a given permutation (ijk) of (123) referring to an operator product $P_{\beta_0}a_{\alpha_i}a_{\alpha_j}a_{\alpha_k}P_{\beta_3}$ the intermediate projector on $[\beta_2]=[\alpha_k]$ is fixed while only $\beta_1=:\beta$ is a free label (ranging over values specified by the fusion rule $[\alpha_j][\alpha_k]=\oplus[\beta]$).

Thus, denote the matrix indices of ρ_i by $(ijk;\beta)$. Then

$$[\rho_2]_{(ijk;\beta)(ikj;\beta')} = R^{(\beta 0)}_{(\alpha_j\alpha_k)}\delta_{\beta\beta'},$$

$$[\rho_1]_{(ijk;\beta)(jik;\beta')} = \left[R^{(\beta_0\alpha_k)}_{(\alpha_i\alpha_j)}\right]_{\beta\beta'}, \tag{5.5}$$

$$[\varphi_1]_{(ijk;\beta)(ijk;\beta)} = e\big(h(\beta) - \tfrac{1}{2}(h_0 + h_k)\big),$$

$$[\varphi_2]_{(ijk;\beta)(ijk;\beta)} = e\big(h_k - \tfrac{1}{2}h(\beta)\big), \tag{5.6}$$

while all other matrix elements vanish. Now, from $\varphi_2\rho_2\varphi_2\rho_2 = \mathbf{1}$ we deduce

$$\left[\rho_2^2\right]_{(ijk;\beta)(ijk;\beta')} = R^{(\beta 0)}_{(\alpha_j\alpha_k)} R^{(\beta 0)}_{(\alpha_k\alpha_j)}\delta_{\beta\beta'} = e\big(h(\beta) - h_k - h_j\big)\delta_{\beta\beta'}$$

$$= e\big(\tfrac{1}{2}h_0 - h_j - \tfrac{1}{2}h_i\big)\cdot[\varphi_1]_{(ijk;\beta)(ijk;\beta')}.$$

Inserting φ_1 into $\varphi_1\rho_1\varphi_1\rho_1 = \mathbf{1}$ yields

$$\delta_{\beta\beta'} = \sum_{\beta''} e\big(-\tfrac{1}{2}h_0 + h_j + \tfrac{1}{2}h_k\big)\left[\rho_2^2\right]_{(ijk;\beta)(ijk;\beta)}\left[\rho_1\right]_{(ijk;\beta)(jik;\beta'')}$$

$$\times e\big(-\tfrac{1}{2}h_0 + h_i + \tfrac{1}{2}h_k\big)\left[\rho_2^2\right]_{(jik;\beta'')(jik;\beta'')}\left[\rho_1\right]_{(jik;\beta'')(jik;\beta')}$$

$$= e\big(-h_0 + h_i + h_j + h_k\big)\left[\rho_2^2\rho_1\rho_2^2\rho_1\right]_{\beta\beta'}, \tag{5.7}$$

and hence, using the braid relations,

$$(\rho_2\rho_1\rho_2)^2 = (\rho_1\rho_2)^3 = e(+h_0 - h_i - h_j - h_k)\mathbf{1}. \tag{5.8}$$

That this is a multiple of unity is no surprise, since $(\sigma_1\sigma_2)^3$ generates the center of B_3. Its precise value, however, is computed from eq. (5.8).

Now consider a field a_α with nonvanishing four-point functions, and the representation of B_4 belonging to $P_0 a_\alpha a_\alpha a_\alpha a_\alpha P_0$. The matrices ρ_2 and ρ_3 are precisely what ρ_1 and ρ_2 were in the above, with a trivial permutation index and $h_0 = h_i = h_j = h_k = h_\alpha$. In particular $(\rho_3\rho_2\rho_3)^2 = e(-2h_\alpha)\mathbf{1}$. In this case we have, moreover, $\rho_1 = \rho_3$, so that $(\rho_1\rho_2\rho_3)^2 = e(-2h_\alpha)\mathbf{1}$. This is more than could be expected since the center of B_4 is generated by $(\sigma_1\sigma_2\sigma_3)^4$.

The example has shown that the phase conditions (3.9) may imply additional relations among the representation matrices ρ_i of the braid group alone, thus effectively representing some quotient B_n/Ideal. We believe that this is true in the general case, the resulting equations being comparable to Vafa's [17]. Less ambitiously, one may do without understanding of the group-theoretical significance, and just take the determinants of eq. (5.4). Since:

(i) in every representation of B_n all ρ_i, $i=1,\dots,n-1$, have a common spectrum; and

(ii) for $[\beta_n]=[0]$ the matrix ρ_{n-1}^2 is a diagonal matrix with entries (3.10) given in terms of conformal dimensions, and multiplicities determined by the fusion rules, one immediately gets a system of linear equations

$$\sum c_\beta h_\beta = 0 \bmod \mathbb{Z}(c_\beta \in \mathbb{N}), \tag{5.9}$$

for the dimensional trajectory $h_\beta = h([\beta])$ which can be written down and solved without any knowledge about the precise form of the off-vacuum R matrices.

The advantage of the reformulation of exchange algebras in terms of braid-group representations as introduced in proposition 5 is that important field-theoretic manipulations are translated into mathematical standard manipulations with group representations.

Recall proposition 4, for example. The requirement (4.3) describes some projection of the representation $\rho_+ \otimes \rho_-^{-1}$ of B_n onto a one-dimensional subrepresentation (the trivial ($\varepsilon=1$) or the anti-symmetric ($\varepsilon=-1$)); here ρ^{-1} is the representation defined by $\rho^{-1}(\sigma_i)=(\rho(\sigma_i))^{-1}$. The failure of time-like (anti-)commutativity is reflected by the fact that the same projection cannot be expected to define a subrepresentation of $\rho_+ \otimes \rho_-$ at the same time.

So far, the precise forms of the representation matrices ρ (proposition 5) and the projection coefficients (proposition 4) refer explicitly to the sectors V_β of *conformal* QFTh. One may drop this reminiscence and imagine a generalization of the above possibly leading beyond CQFTh$_2$. Define an exchange algebra by a collection of braid-group representations acting on light-cone n-point functions. An analogue of eq. (5.4) is not required. Two-dimensional fields are then *defined* as one-dimensional subrepresentations ρ of $\rho_+ \otimes \rho_-^{-1}$, where exotic statistics (i.e. $\rho(\sigma_i^2)=\varepsilon^2 \neq 1$) may be admitted. This generalization would preserve the issue of explicit non-trivial time-like commutation behaviour of QFTh$_2$.

Let us, however, return to our original concept of *conformal* light-cone fields, and discuss further field-theoretic manipulations in the light of braid-group representations.

First there is the evident observation that the R matrices relevant for the tensor products of independent fields are just the tensor products of the R matrices of the respective factor fields, while the induced representations of the braid groups are just the tensor products of the representations induced by the factors. It is likely evident that multiplying R matrices by an overall (phase) factor χ preserves the representation conditions. Combining these two harmless operations, however, has nontrivial field-theoretic consequences. Referring to subsect. 7.3 for the discussion of this phenomenon, we want to just mention here that the factor χ influences the dimensional trajectories according to proposition 2. In particular the dimension of the "product field" will no longer equal the sum of the dimensions of the factor fields, as it should for an ordinary tensor product.

Second, there is a natural construction of new representations out of old, which exists only for braid groups. The idea is to combine several "threads" of the braid into one "strand", which (as we shall see in sect. 6) corresponds to the field-theoretic operation of short-distance operator products. For an explanation it is helpful to recall the pictorial description of the braid group (which actually stood at the historical origin of studying braids as topological objects in $\mathbb{R}^3$ [16]).

An n-braid b is described by a set of n non-intersecting curves $\boldsymbol{c}_j$: $[0,1] \to \mathbb{R}^3$, monotonous in the 2-component, with $\boldsymbol{c}_j(0) = (j,0,0)$, $\boldsymbol{c}_j(1) = (\pi_b(j),1,0)$, $\pi_b \in S_n$, $j = 1,\ldots,n$. Two such sets of curves describe the same braid, if they can be continuously deformed into each other without intersections. The natural description of the identity braid $\boldsymbol{e}$ is $\{\boldsymbol{c}_j(t) = (j,t,0)\}$, $\pi_e = e \in S_n$. Typically

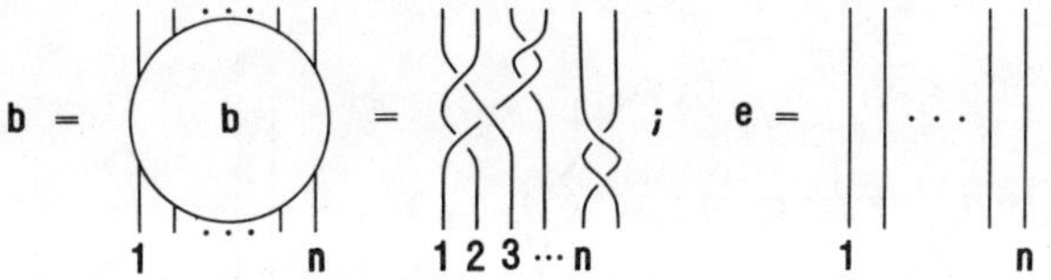

(5.10)

The composition law for $b_2 \circ b_1$ is given by functions

$$\tilde{c}[b_2 \circ b_1]_j(t) := \begin{cases} \boldsymbol{c}[b_1]_j(2t), & t \le \frac{1}{2}, \\ \boldsymbol{c}[b_2]_{\pi_{b_1}(j)}(2t-1) + \boldsymbol{e}_2, & t \ge \frac{1}{2} \end{cases}$$

(hence $\tilde{c}[b_2 \circ b_1]_j(1) = ((\pi_{b_2} \circ \pi_{b_1})(j), 2, 0))$, the 2-components of which are rescaled by $\frac{1}{2}$ in order to obtain $\boldsymbol{c}[b_2 \circ b_1]_j(t)$. Graphically this is just "linking b_2 on top of b_1"

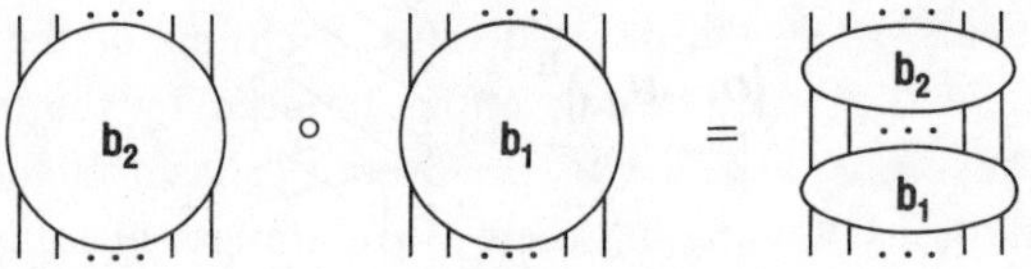

(5.11)

The projection π: $b \mapsto \pi_b$ is a group homomorphism of B_n in S_n.

The generating braids σ_i (respectively σ_i^{-1}) are described by functions with a single intersection of the ith with the $(i+1)$th curve projected into the 1–2 plane, say at "time" t_0, such that $(e_3 \cdot c_i(t_0)) >$ [respectively $< (e_3 \cdot c_{i+1}(t_0))$]. Thus $\pi_{\sigma_i} = \tau_i =$ transposition of $(i, i+1)$

$$\sigma_i = \cdots \quad \cdots \qquad \sigma_i^{-1} = \cdots \quad \cdots . \tag{5.12}$$

i i+1 i i+1

Every continuous deformation can be performed as a series of "Reidemeister moves", i.e. the following braid identities hold

$$R_1: \sigma_i\sigma_i^{-1} = e = \sigma_i^{-1}\sigma_i: \quad \cdots \quad = \quad \cdots \quad = \quad \cdots ,$$

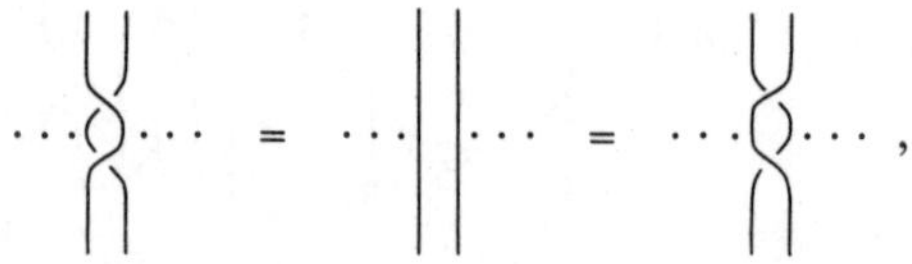

$$R_2: \sigma_i\sigma_j = \sigma_j\sigma_i, \; |i-j| \geq 2: \quad \cdots \quad = \quad \cdots ,$$

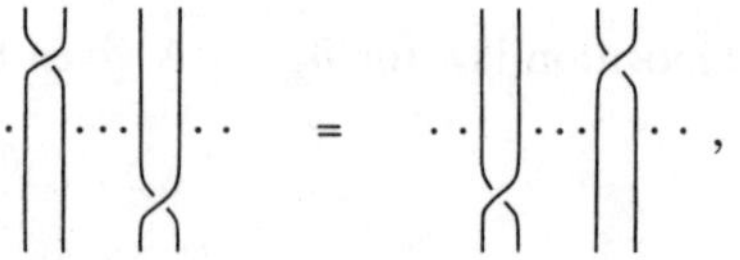

$$R_3: \sigma_i\sigma_{i+1}\sigma_1 = \sigma_{i+1}\sigma_i\sigma_{i+1}: \quad \cdots \quad = \quad \cdots . \tag{5.13}$$

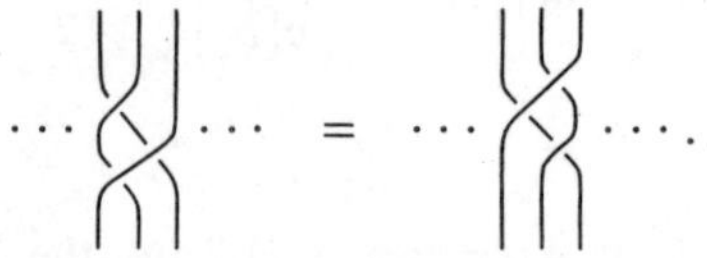

The center of B_n is generated by $(\sigma_1 \ldots \sigma_{n-1})^n$

$$(\sigma_1 \cdots \sigma_{n-1})^n = \quad . \tag{5.14}$$

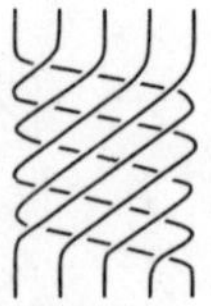

The following propositions describe "strand products" of (representations of) B_n.

Proposition 6. There is a natural embedding I of B_ν in B_n, $n = p \cdot \nu$, which maps $\hat{\sigma}_i \in \mathrm{B}_\nu$ onto

$$\Big(\prod_{(pi,0)} \sigma\Big)\Big(\prod_{(pi,1)} \sigma\Big) \cdots \Big(\prod_{(pi,i-1)} \sigma\Big) \cdots \Big(\prod_{(pi,1)} \sigma\Big)\Big(\prod_{(pi,0)} \sigma\Big) \in \mathrm{B}_n, \tag{5.15}$$

where

$$\Big(\prod_{(a,b)} \sigma\Big) := \prod_{k=0}^{b} \sigma_{a-b+2k},$$

are commutative products. Graphically

$$I(\hat{\sigma}_i) = \cdots$$

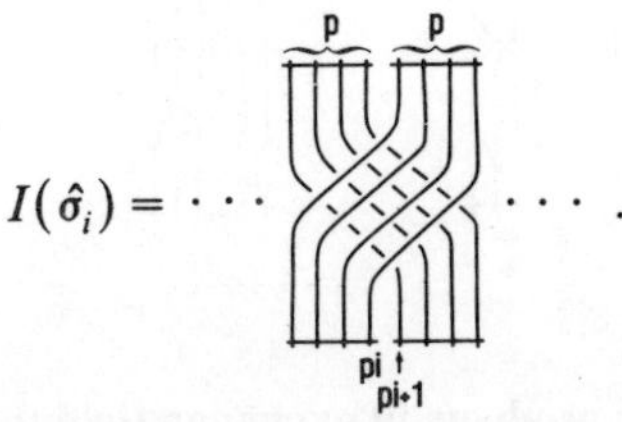

Proposition 7. Given a representation ρ of B_n in V. Let $\nu \le n$. Consider the set Λ of inequivalent partitions $(\lambda) = (\lambda_1, \ldots, \lambda_\nu)$, $\lambda_i \in \mathbb{N}$, of $n = \Sigma_{i=1}^{\nu} \lambda_i$, which are just permutations of some partition (λ^0) of n. Introduce an orthonormal basis $\{e_{(\lambda)}\}$ of $\mathbb{R}^N$, where $N = |\Lambda|$.

Let $\tau_i(\lambda) = (\lambda_1 \ldots \lambda_{i+1} \lambda_i \ldots \lambda_\nu)$, $1 \le i \le \nu - 1$. The following defines a representation $\hat{\rho}$ of B_ν in $\mathrm{V} \otimes \mathbb{R}^N$, depending on ρ and Λ.

For a pair $[i,(\lambda)]$ let $\mu = \Sigma_{j \le i} \lambda_j$, $\mu' = \Sigma_{j<i} \lambda_j + \lambda_{i+1}$, $\lambda = \min(\lambda_i, \lambda_{i+1}) - 1$. Set

$$\rho[i,(\lambda)] := \rho\bigg(\Big(\prod_{(\mu',0)} \sigma\Big) \cdots \Big(\prod_{(\mu',\lambda)} \sigma\Big) \cdots \Big(\prod_{(\mu,\lambda)} \sigma\Big) \cdots \Big(\prod_{(\mu,0)} \sigma\Big)\bigg), \tag{5.16}$$

where the dots interpolate in unit steps from 0 to λ, from μ' to μ, and from λ to 0, respectively. Graphically

$$\rho[i,(\lambda)] = \rho\Big(\cdots \Big).$$

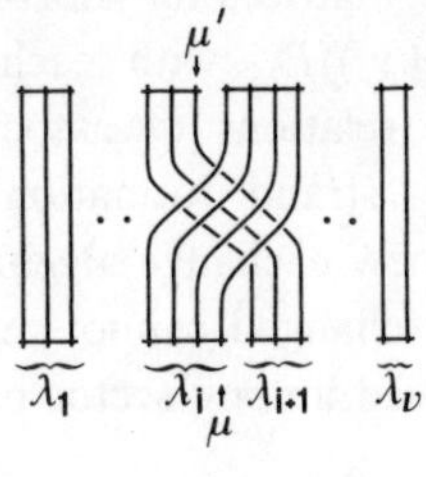

μ', λ_1, λ_i, μ, λ_{i+1}, λ_ν

Then $\hat{\rho}$ is given by

$$\hat{\rho}(\hat{\sigma}_i) := \sum_{(\lambda)} \rho[i,(\lambda)] \otimes e_{\tau_i(\lambda)} e^{\mathrm{T}}_{(\lambda)}. \tag{5.17}$$

In the case $n = p \cdot \nu$, $(\lambda) = (p,\ldots, p)$, $N = 1$, the representation constructed in proposition 7 is nothing but the representation induced by the embedding of proposition 6.

Proof of propositions 6 and 7. All one has to show is that the objects $I(\hat{\sigma}_i)$, respectively $\hat{\rho}(\hat{\sigma}_i)$, commute for $|i-j| \geq 2$ and satisfy the defining braid relation R_3

$$\sigma_i \sigma_{i+1} \sigma_i = \sigma_{i+1} \sigma_i \sigma_{i+1} .$$

These statements, however, are graphically evident from the following pictures of identities in B_n

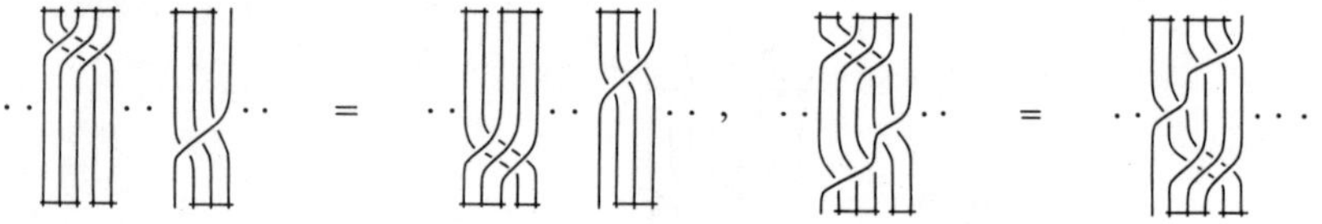

$$\cdots \quad = \quad \cdots, \qquad \cdots \quad = \quad \cdots \tag{5.18}$$

6. Operator products in exchange field theory

An operator product

$$P_{\beta_0}\big(a_{\alpha_1}(y_1) P_{\beta_1} \cdots P_{\beta_{\nu-1}} a_{\alpha_\nu}(y_\nu)\big) P_{\beta_\nu}\big(a_{\alpha_{\nu+1}}(y_{\nu+1}) P_{\beta_{\nu+1}} \cdots P_{\beta_{n-1}} a_{\alpha_n}(y_n)\big) P_{\beta_n},$$

can be expressed in terms of

$$P_{\beta_0}\big(a_{\alpha_{\nu+1}}(y_{\nu+1}) P_{\beta'_1} \cdots P_{\beta'_{n-\nu-1}} a_{\alpha_n}(y_n)\big) P_{\beta'_{n-\nu}}\big(a_{\alpha_1}(y_1) P_{\beta'_{n-\nu+1}} \cdots P_{\beta'_{n-1}} a_{\alpha_\nu}(y_\nu)\big) P_{\beta_n},$$

by means of a product of $\nu(n-\nu)$ matrices ρ_i à la proposition 5. If $y_1, \ldots, y_\nu$ are all sufficiently close to x_1, and $y_{\nu+1}, \ldots, y_n$ are all sufficiently close to $x_2 \neq x_1$, then these matrices appear all with the *common* sign $s_{12} = \mathrm{sig}(x_1 - x_2)$. The matrix products so obtained are the R matrices for short-distance operator products of the type $P_{\beta_0}(a_{\alpha_1}(y_1) P_{\beta_1} \cdots P_{\beta_{\nu-1}} a_{\alpha_\nu}(y_\nu)) P_{\beta_\nu}$ with each other, and they automatically satisfy the appropriate braid relations ($\equiv$ associativity conditions). This is the field-theoretic analog of the "strand formation" and proposition 7. However, the resulting algebra is not a new exchange algebra in the precise sense of sect. 3: the operator products to be exchanged cannot readily be assigned to a particular (composite) field A_α, sandwiched among sector projectors. Instead, one must find linear combinations

$$P_{\beta_0} A_\alpha([x]) P_{\beta_\nu} = \sum_{\beta_1 \ldots \beta_{\nu-1}} \mathrm{C}^{(\beta_0;\, \alpha_1 \ldots \alpha_\nu;\, \beta_\nu)}_{\alpha;\, \beta_1 \ldots \beta_{\nu-1}} P_{\beta_0}\big(a_{\alpha_1}(y_1) P_{\beta_1} \cdots P_{\beta_{\nu-1}} a_{\alpha_\nu}(y_\nu)\big) P_{\beta_\nu}, \tag{6.1}$$

such that the new R matrices restricted to these combinations describe an algebra

$$P_{\beta_0} A_{\alpha_1}([x_1]) P_{\beta_1} A_{\alpha_2}([x_2]) P_{\beta_2}$$
$$= \sum_{\beta_1'} \left[R^{(\beta_0 \beta_2)}_{(\alpha_1 \alpha_2)}(s_{12}) \right]_{\beta_1 \beta_1'} P_{\beta_0} A_{\alpha_2}([x_2]) P_{\beta_1'} A_{\alpha_1}([x_1]) P_{\beta_2}. \qquad (6.2)$$

Here, for the moment $[x]$ stands formally for the set $(y_1 \dots y_\nu)$ of arguments close to x.

It is well known [18] that operator products *on the vacuum* (i.e. $[\beta_\nu] = [0]$), after multiplication with appropriate scaling functions $f_\alpha(y, \partial_y)$, possess well defined short-distance limits. These behave precisely like light-cone field states $P_\alpha a_\alpha(x)\Omega$ under conformal transformations. The missing point in the early studies was that there was no way to define short-distance limits *off vacuum*.

Now, this gap is filled by the Exchange algebra. Its virtue is that it relates ν-point operator products $P_{\beta_0} a_{\alpha_1}(y_1) \dots a_{\alpha_\nu}(y_\nu) P_{\beta_\nu}$ applied to μ-point states $P_{\beta_\nu} a_{\alpha_{\nu+1}}(y_{\nu+1}) \dots a_{\alpha_{\nu+\mu}}(y_{\nu+\mu})\Omega$ *linearly* to states $P_{\beta_0} a_{\alpha_{\nu+1}}(y_{\nu+1}) \dots a_{\alpha_{\nu+\mu}}(y_{\nu+\mu}) P_{\beta_\mu'} a_{\alpha_1}(y_1) \dots a_{\alpha_\nu}(y_\nu)\Omega$, i.e. the analytic behaviour of ν-point operator products can be inferred from the analytic behaviour of ν-point states, and short-distance limits that exist on-vacuum exist with the same scaling functions off-vacuum as well.

Note that in the short-distance operator products $P_{\beta_0} A_\alpha([x]) P_{\beta_\nu}$ (eq. (6.1)) the label α refers to $[\alpha]$. The individual quasiprimary field $a_\alpha(x)$, $\alpha \in [\alpha]$, is only selected by the choice of the appropriate scaling function $f_\alpha(y, \partial_y)$. For $\nu = 2$, $f_\alpha(y_1, y_2, \partial_{y_1}, \partial_{y_2})$ are given by [18]

$$N^{-1} Q^{n_\alpha}_{h_\alpha + d_{\alpha_1} - d_{\alpha_2},\, h_\alpha - d_{\alpha_1} + d_{\alpha_2}}(\partial_{y_1}, \partial_{y_2})(y_1 - y_2 - i\varepsilon)^{d_{\alpha_1} + d_{\alpha_2} - h_\alpha}, \qquad (6.3)$$

where $Q^k_{n_1 n_2}(z_1, z_2) = z_1^{1-n_1} z_2^{1-n_2} (\partial_{z_1} - \partial_{z_2})^k (z_1^{n_1+k-1} z_2^{n_2+k-1})$, $d_\alpha = h_\alpha + n_\alpha$, and N^{-1} is a normalization constant.

The problem in practice is the determination of the coefficients $\mathsf{C}^{(\beta_0;\, \alpha_1 \dots \alpha_\nu;\, \beta_\nu)}_{\alpha;\, \beta_1 \dots \beta_{\nu-1}}$, respectively the inverse expansion

$$P_{\beta_0} a_{\alpha_1}(y_1) P_{\beta_1} \dots P_{\beta_{\nu-1}} a_{\alpha_\nu}(y_\nu) P_{\beta_\nu} = \sum_{\alpha}^{\text{finite}} \mathsf{C}^{(\alpha, \beta)}_{\alpha} P_{\beta_0} A_\alpha([x]) P_{\beta_\nu}. \qquad (6.4)$$

At least for $\nu = 2$ these can be computed intrinsically, i.e. in terms of braid representation matrices (and scaling dimensions). This may illuminate the power of proposition 5. Consider the representation $\rho = \rho^{(A = (\alpha_1 \alpha_2 \alpha_3);\, \beta_0, 0)}$ of B_3 introduced in the example following proposition 5. Then

$$\left(P_{\beta_0} a_{\alpha_i}(y_1) P_{\beta_1} a_{\alpha_j}(y_2) P_{\alpha_k} \right) a_{\alpha_k}(x_2)\Omega$$
$$= \sum_{\beta_1'} \left[\rho_2^{s_{12}} \rho_1^{s_{12}} \right]_{(ijk;\, \beta_1)(kij;\, \beta_1')} P_{\beta_0} a_{\alpha_k}(x_2) \left(P_{\beta_1'} a_{\alpha_i}(y_1) P_{\alpha_j} a_{\alpha_j}(y_2) \right)\Omega. \qquad (6.5)$$

The operator products in brackets on the r.h.s. can be identified with $P_{\beta_1'}A_{\beta_1'}([x_1])P_0$. The required coefficient matrix must "diagonalize" both $(\rho_2\rho_1)$ and $(\rho_2^{-1}\rho_1^{-1})$ simultaneously. From the example (eq. (5.8)) we know that $(\rho_1\rho_2\rho_1)\rho_2\rho_1 = e(H)\rho_2^{-1}$, and $(\rho_2\rho_1\rho_2)\rho_2^{-1}\rho_1^{-1} = \rho_2$, where $H = h_0 - h_i - h_j - h_k$. Thus comparing

$$\left(\sum_\beta e(-\tfrac{1}{2}H)[\rho_2\rho_1\rho_2]_{(kji;\,\alpha)(ijk;\,\beta)} P_{\beta_0} a_{\alpha_i}(y_1) P_\beta a_{\alpha_j}(y_2) P_{\alpha_k}\right) a_{\alpha_k}(x_2)\Omega$$

$$= \left[e(\tfrac{1}{2}H)\rho_2^{-1}\right]^{s_{12}}_{(kji;\,\alpha)(kij;\,\alpha)} P_{\beta_0} a_{\alpha_k}(x_2) P_\alpha A_\alpha([x_1])\Omega\,,$$

which follows from eq. (6.5), with the expected equation

$$\left(P_{\beta_0}A_\alpha([x_1])P_{\alpha_k}\right)a_{\alpha_k}(x_2)\Omega = R^{(\beta_0 0)}_{(\alpha\alpha_k)}(x_1, x_2) P_{\beta_0} a_{\alpha_k}(x_2) P_\alpha A_\alpha([x_2])\Omega\,,$$

one obtains

$$\begin{aligned}
\mathsf{C}^{(\beta_0;\,\alpha_i\alpha_j;\,\alpha_k)}_{\alpha\beta} &= e(-\tfrac{1}{2}H) R^{(\alpha 0)}_{(\alpha_j\alpha_i)}\left[R^{(\beta_0\alpha_j)}_{(\alpha_k\alpha_i)}\right]_{\alpha\beta} R^{(\beta 0)}_{(\alpha_k\alpha_j)}\,,\\
R^{(\beta_0 0)}_{(\alpha\alpha_k)} &= e(\tfrac{1}{2}H) R^{(\alpha 0)-1}_{(\alpha_i\alpha_j)}\,,\\
R^{(\beta_0 0)}_{(\alpha_k\alpha)} &= e(\tfrac{1}{2}H) R^{(\alpha 0)-1}_{(\alpha_j\alpha_i)}\,,
\end{aligned} \tag{6.6}$$

reproducing eq. (3.10). Note that both the braid relations (proposition 3) and the phase relations (proposition 2) have entered the argument essentially. Here we have computed the coefficients of an operator product $P_{\beta_0}a_{\alpha_i}P_\beta a_{\alpha_j}P_{\alpha_k}$ from the required algebra with the field $P_{\alpha_k}a_{\alpha_k}P_0$. The *same* operator product must satisfy an exchange algebra with all fields $P_{\alpha_k}a_{\alpha_l}P_\gamma$ simultaneously. This requirement needs further investigation.

As an example consider a theory of hermitian fields $(P_\beta a_\alpha P_\gamma)^+ = P_\gamma a_\alpha P_\beta$. Fix normalizations by the following equations

$$(\Omega, a_\alpha(x_1)a_{\alpha'}(x_2)\Omega) = \mathrm{sig}(\alpha)e(-\tfrac{1}{2}d_\alpha)\delta_{\alpha\alpha'}(x_1 - x_2 - i\varepsilon)^{-2d_\alpha}\,,$$

$$\left(\Omega, a_{\alpha_1}(x_1)a_{\alpha_2}(x_2)a_{\alpha_3}(x_3)\Omega\right) = c_{\alpha_1\alpha_2\alpha_3}\prod_{i<j}(x_i - x_j - i\varepsilon)^{-d_i - d_j + d_k}\,,$$

$$e(\tfrac{1}{4}\textstyle\sum d_i)c_{\alpha_1\alpha_2\alpha_3} \in \mathbb{R}$$

$$c_{\alpha_1\alpha_3\alpha_2} = c_{\alpha_2\alpha_1\alpha_3} = (-)^{\Sigma(d_i - h_i)}c_{\alpha_1\alpha_2\alpha_3}$$

$$R^{(\alpha_1 0)}_{(\alpha_2\alpha_3)} = e(\tfrac{1}{2}(h_1 - h_2 - h_3))\,.$$

Unitary theories are characterized by all signs sig(α) of *all* quasiprimary fields being +1 [19].

Four-point functions have the well-known vacuum short-distance expansions

$$\begin{aligned}&\left(\Omega, a_{\alpha_1}(x_1)a_{\alpha_2}(x_2)P_\beta a_{\alpha_3}(x_3)a_{\alpha_4}(x_4)\Omega\right)\\&=\left(\frac{x_2-x_4}{x_1-x_4}\right)^{d_1-d_2}\left(\frac{x_1-x_3}{x_1-x_4}\right)^{d_4-d_3}\Big/(x_1-x_2)^{d_1+d_2}(x_3-x_4-i\varepsilon)^{d_3+d_4}\\&\times\sum_{\beta\in[\beta]}e\left(\tfrac{1}{2}d_\beta\right)\left(c_{\alpha_1\alpha_2\beta}\operatorname{sig}(\beta)c_{\beta\alpha_3\alpha_4}\right)\\&\times x^{d_\beta}{}_2F_1(d_2-d_1+d_\beta,d_3-d_4+d_\beta;2d_\beta;x),\end{aligned}\tag{6.7}$$

which converge for $x_1>x_2>x_3\approx x_4$; $x=(x_1-x_2)(x_3-x_4-i\varepsilon)/(x_1-x_3)\times(x_2-x_4)\approx 0$.

The exchange algebra and eq. (6.6) imply the following off-vacuum short-distance expansions of the same functions

$$\begin{aligned}&\sum_\beta \mathsf{C}_{\alpha\beta}^{(\alpha_1;\,\alpha_2\alpha_3;\,\alpha_4)}\left(\Omega, a_{\alpha_1}(x_1)a_{\alpha_2}(x_2)P_\beta a_{\alpha_3}(x_3)a_{\alpha_4}(x_4)\Omega\right)\\&=\left(\frac{x_2-x_4}{x_1-x_2}\right)^{d_1-d_4}\left(\frac{x_1-x_3}{x_1-x_2}\right)^{d_2-d_3}\Big/(x_1-x_4)^{d_1+d_4}(x_2-x_3-i\varepsilon)^{d_2+d_3}\\&\times\sum_{\alpha\in[\alpha]}e\left(\tfrac{1}{2}d_\alpha\right)\left(c_{\alpha_1\alpha\alpha_4}\operatorname{sig}(\alpha)c_{\alpha\alpha_2\alpha_3}\right)\\&\times(1-x)^{d_\alpha}{}_2F_1(d_4-d_1+d_\alpha,d_3-d_2+d_\alpha;2d_\alpha;1-x),\end{aligned}\tag{6.8}$$

which converge for $x_1>x_2\approx x_3>x_4$; $1-x=(x_1-x_4)(x_2-x_3-i\varepsilon)/(x_1-x_3)\times(x_2-x_4)\approx 0$.

It is crucial that the above expansions do not have the correct analytic properties term by term. Instead, the exchange algebra can only be satisfied by a most delicate interplay of the expansion coefficients, i.e. the quasiprimary three-point amplitudes and the signs sig(α) signalling violation of unitarity of the field theory [19]. The evaluation of this interplay would constitute the final completion of the "conformal bootstrap program".

It is only fair to mention that ideas very similar to the "reduction" of strand-product representations – but in a different physical context – have been discussed earlier. Let us quote Karowski's approach to "boundstate S-matrices" (elaborated further by Kulish [20]), and the "fusion" of Yang–Baxter matrices of RSOS models [21]. The relevance of strand products was also observed by Fröhlich [6].

7. Examples

In this section we discuss a class of solutions to the conditions of propositions 1–3, which turn out to be the exchange algebras of the minimal models [9] and SU(2) WZW-models [11]. We develop the techniques discussed in general in the preceding sections in this specific context. In doing this, the power of our concepts will become apparent, both concerning the classification problem, and the qualitative and quantitative construction of conformal block functions. A counterexample will also be given, where a solution to the braid relations is incompatible with the conformal transformation behaviour, thus showing the relevance of both types of conditions.

7.1. THE SOLUTION

We give the exchange algebra for a single "elementary" field $a_\alpha \equiv a$ which interpolates according to the fusion rules

$$[\alpha][l] = [l-1] \oplus [l+1], \tag{7.1}$$

where l takes integer values $1, 2, \ldots, q-1$. $[l-1]=[0]$ and $[l+1]=[q]$ are omitted from eq. (7.1). The sector $[l]=[1]$ may (not necessarily) be identified with the vacuum sector; in that case $[\alpha]=[\alpha][\text{vacuum}]=[2]$.

We found all solutions to eq. (3.11) compatible with the fusion rules (7.1) [4]

$$\begin{aligned}
&[R^{(11)}]_{22} = [R^{(q-1,q-1)}]_{q-2,q-2} =: \eta\,,\\
&[R^{(l-1,l+1)}]_{ll} = [R^{(l+1,l-1)}]_{ll} =: \eta\omega\,,\\
&[R^{(ll)}]_{l\mp1,l\mp1} = \eta(-\omega)^{1/2}\\
&\qquad\times\begin{pmatrix} -(-\omega)^{1/2}s(1)/s(l) & \lambda_l^{-1}\sqrt{s(l-1)s(l+1)/s^2(l)} \\ \lambda_l\sqrt{s(l-1)s(l+1)/s^2(l)} & (-\omega)^{-1/2}s(1)/s(l) \end{pmatrix},
\end{aligned} \tag{7.2}$$

with

$$(-\omega) = \exp(2\pi i p/q)\,, \qquad (p \text{ and } q \text{ coprime})\,,$$
$$s(l) := \sin(l\pi p/q)\,. \tag{7.3}$$

The off-diagonal square roots in $R^{(ll)}$ are taken with the same sign, and $(-\omega)^{1/2} := \exp(2\pi i p/2q)$. Aiming at a parity-symmetric solution as in the special case (1) of proposition 4, we assume $\lambda_l = 1$. Then $R^{(ll)}$ are symmetric matrices, which are also unitary iff $p = \pm 1 \bmod q$, and η a phase. In fact, in a unitary theory all conformal blocks contributing to a local n-point function are – up to a common complex phase – real functions at the "Jost points", such that the inversely oriented exchange is described by the complex conjugate matrix; hence we should require $R^* = R^{-1}$, and R is symmetric $\Leftrightarrow$ R is unitary. Note, however, that unitarity of the field theory is a much deeper issue than just unitarity of the R matrices!

The solutions (7.2) possess the manifest symmetry $l \to q - l$

$$[R^{(q-l_0, q-l_2)}]_{q-l, q-l'} = [R^{(l_0 l_2)}]_{ll'}. \tag{7.4}$$

The matrices (7.2) have been constructed as solutions to the braid relations of proposition 3. They must also solve the phase relations of proposition 2. This yields the constraint

$$\eta^4 = (-\omega)^{-3}, \tag{7.5}$$

as well as the following equations for the dimensional trajectories $h_l = h([l])$

$$e(2h_{l+1} - 2h_l) = \eta^2(-\omega)^{l+2},$$

$$e(h_{l+1} - 2h_l + h_{l-1}) = \eta^2(-\omega)^2. \tag{7.6}$$

We may absorb the sign of the square root $\eta^2 = \pm(-\omega)^{-3/2}$ in the freedom $p \to p + q$ which leaves R unchanged, and assume $\eta^2 = (-\omega)^{-3/2}$. Still, we find two different trajectories according to $e(h_2 - h_1) = \pm(-\omega)^{3/4}$

$$e(h_l) = e\left(h_1 + \frac{(l^2-1)p - 4\varepsilon(l-1)q}{4q}\right), \tag{7.7}$$

where $\varepsilon = 0$, or $\varepsilon = \frac{1}{2}$.

They are easily identified with the Kac spectrum [9] $h_l = h_{1,l}(c(p,q))[h_1 = 0, \varepsilon = \frac{1}{2}; c(p,q) = 1 - 6(p-q)^2/pq]$

$$h_{1,l} = \frac{(l^2-1)p - 2(l-1)q}{4q}, \tag{7.8}$$

with the Kac spectrum $h_l = h_{k,l}(c(p,q))[h_1 \neq 0, \varepsilon = \frac{1}{2}k \bmod 1]$

$$h_{k,l} = \frac{(lp - kq)^2 - (p-q)^2}{4pq} \tag{7.9}$$

or with the SU(2) WZW spectrum $[h_1 = 0, \varepsilon = 0; l \equiv 2j + 1, q \equiv k + 2, p = 1]$

$$h_j = \frac{j(j+1)}{k+2}. \tag{7.10}$$

Observe that both the *unitary* minimal models [22] with $|p-q|=1$, and the WZW models with $p=1$ indeed have unitary exchange matrices.

For the minimal trajectories, the fusion rules (7.1) are those of the field $\phi_{(1,2)\cdot}$ along a horizontal row of the "BPZ rectangle" [9]. The usual identification of $V_{k,l}$ with $V_{p-k,q-l}$ is consistent with the symmetry (7.4). After a relabelling of l, the fusion rules (7.1) are also those of the field $\phi_{(1,q-2)\cdot}$ along a row of the BPZ rectangle (see subsect. 7.4). We shall discuss in subsect. 7.2 how, by strand formation (see above), the "horizontal" light-cone fields a_l associated with $\phi_{(1,l)\cdot}$ can be included in the exchange algebra, and in subsect. 7.3 how the minimal models can be completed by the introduction of a "vertical" elementary field b associated with $\phi_{(2,1)\cdot\cdot}$

For the WZW trajectory, the fusion rules (7.1) are those of the doublet field $\phi_{2\cdot}$ with the isospin-$(j=(l-1)/2)$ multiplet fields $\phi_{l\cdot}$ which are primary with respect to the enlarged symmetry algebra. With respect to the Virasoro algebra, every isospin component $\phi_{ljm\cdot}$ ($l=1,2,\dots,k+1$; $j=(l-1)/2+0,1,\dots$; $m=-j,\dots,+j$) is primary with the fusion rules

$$[2,\tfrac{1}{2},\mu][l,j,m]=\bigoplus_{l'=l\pm1}\ \bigoplus_{j'=j\pm\frac{1}{2}}[l',j',m+\mu] \tag{7.11}$$

while their exchange algebra is given by R matrices (7.2) with the relevant l labels, but irrespective of j and m. The reader may convince himself that, e.g., the use of the *same* R matrix element in

$$P_{[1,j,0]}a_{(2,\frac{1}{2},+\frac{1}{2})}(x_1)\,a_{(2,\frac{1}{2},-\frac{1}{2})}(x_2)\,\Omega=[R^{(1,1)}]_{22}P_{[1,j,0]}a_{(2,\frac{1}{2},-\frac{1}{2})}(x_2)\,a_{(2,\frac{1}{2},+\frac{1}{2})}(x_1)\,\Omega$$

for both the singlet ($j=0$) and the triplet ($j=1$) Virasoro sectors within the vacuum ($l=1$) Kac–Moody sector is in perfect agreement with the relative sign expected from isospin (anti-)symmetrization as well as from the analytic behaviour of the 3-point functions with $h([1,0,0])=0$ and $h([1,1,0])=1$.

7.2. STRAND FORMATION

Let us concentrate on the minimal model interpretation of eq. (7.7). Assume $[1]=[\text{vacuum}]$, thus $[\alpha]=[2]$. We have given in eq. (7.2) only the exchange algebra of the elementary field $a=a_2$, associated with $\phi_{(1,2)\cdot\cdot}$ The operator product of this field with itself is known to produce fields $\phi_{(1,n+1)\cdot\cdot}$ Our idea is to construct the corresponding light-cone fields $(a_{n+1})_{ll'}$ as short-distance limits of n fields a_2

$$P_{l_0}a_{n+1}(x)P_{l_n}=\lim f(x,\partial_x)\sum_{l_1\dots l_{n-1}}c_{l_1\dots l_{n-1}}P_{l_0}a(x_1)P_{l_1}\dots P_{l_{n-1}}a(x_n)P_{l_n}. \tag{7.12}$$

Actually we do not need to calculate, but rather can take over Jimbo's et al. calculations [21]. The reason is that up to similarity transformations and some overall factor, our matrices $[R^{(l_0 l_2)}_{(22)}]_{ll'}$ coincide with the RSOS model Boltzmann weights [15]

$$W_{11}(l, l_0, l', l_2|u),$$

in the limit $u \to -i\infty$ at $K = \frac{1}{2}\pi$, $2\eta = \pi p/q$. The "fusion" of ref. [21] is equivalent to the projection onto the leading irreducible representation in the braid-group representation induced by strand formation à la sect. 5, or the construction of a_{n+1} as a linear combination of operator products with n fields a_2 à la sect. 6.

We quote the result

$$c_{l_1 \dots l_{n-1}} \sim \prod_{j=1}^{n-1} i^{l_j} s(l_j)^{-1/2}, \tag{7.13}$$

which can be made real by multiplication with some common power of i. The corresponding exchange matrices $R_{(n+1, n'+1)}$ of the composite fields a_{n+1} can be computed as the appropriate $u \to -i\infty$ limits of the (symmetrized) weights $W_{nn'}(l, l_0, l', l_2|u)$ of ref. [21].

The higher fusion rules [9]

$$[n_1][n_2] = \bigoplus_{n=|n_1-n_2|+1}^{\min(n_1+n_2-1, 2q-n_1-n_2-1)} [n], \tag{7.14}$$

where the sum leaves out every second term, can be recovered by combinatorial arguments.

7.3. THE EXCHANGE ALGEBRA OF THE COMPLETE MINIMAL MODELS

We have, so far, constructed the algebra of the field a associated with $\phi_{(1,2)}$. in the BPZ rectangle [9], and (in principle) that of the fields a_n associated with $\phi_{(1,n)}$. as well. The "structure constants" R of this algebra of fields interpolating among the sectors $V_{k,l}$ with k fixed (i.e. interpolating within the horizontal rows of the rectangle) are independent of k. Actually k came into eq. (7.9) by hand, choosing $h_1 = h_{k,1}$. In this sense we may speak of (a priori inequivalent) representations, labelled by k, of the same "horizontal" field algebra $\mathscr{A}$.

In the minimal models there is a "vertical" field b associated with $\phi_{(2,1)}$. outside the horizontal field algebra, interpolating among different k representations of $\mathscr{A}$. The field b in turn is the germ of a "vertical" field algebra $\mathscr{B}$ acting within the lth columns of the BPZ rectangle, while the horizontal field a interpolates among different l representations of $\mathscr{B}$. One may speak of a and b as "mutual soliton fields" intertwining between different "charged" representations of the field algebra of each other.

For the field algebra of the *complete* minimal models we only lack the exchange relations of a with b, which are clearly constrained by propositions 2 and 3.

Consider two fields a and b, both of the type described in subsect. 7.1, which we want to identify in the end with the light-cone fields associated with $\phi_{(1,2)}$. and $\phi_{(2,1)}$. of the BPZ minimal models. We thus assume the fusion rules of b not to "interfere" with those of a. In other words, there are sectors $V_{k,l}$, $1 \le k < q'$, $1 \le l < q$, and the fusion rules

$$[a][k,l] = [k,l+1] \oplus [k,l-1],$$
$$[b][k,l] = [k+1,l] \oplus [k-1,l]. \tag{7.15}$$

Introducing projectors $P_{k,l}$ onto $V_{k,l}$, and $P_{k.} = \Sigma_l P_{k,l}$, $P_{.l} = \Sigma_k P_{k,l}$, we have $aP_{k.} = P_{k.}a$ and $bP_{.l} = P_{.l}b$, while $P_{k,J\pm 1}aP_{kl}$ satisfy an exchange algebra with structure constants (7.2) independent of k, and $P_{k\pm 1,l}bP_{kl}$ satisfy an analogous exchange algebra independent of l. The former is parametrized by phases $-\omega = e(p/q), \eta$, the latter by phases $-\omega' = e(p'/q'), \eta'$.

Now the exchange algebra of a with b is of the form

$$P_{k\pm 1l\pm l}a(x_1)P_{k\pm 1l}b(x_2)P_{kl} = \chi_{kl}^{k\pm 1l\pm 1}P_{k\pm 1l\pm 1}b(x_2)P_{kl\pm 1}a(x_1)P_{kl}. \tag{7.16}$$

One may absorb much ambiguity into rescalings $P_{k.}a \to \lambda(k)P_{k.}a$, $P_{.l}b \to \lambda(l)P_{.l}b$, which leave the algebras (7.2) unchanged, and solve the braid condition (proposition 3) by

$$\chi_{kl}^{k+1l+1} = \chi_{kl}^{k-1l-1} =: \chi, \; \chi_{kl}^{k+1l-1} = \chi_{kl}^{k-1l+1} =: \tilde{\chi}$$
$$\chi^2 = \tilde{\chi}^2, \tag{7.17}$$

and the phase condition (proposition 2) by

$$\tilde{\chi} = \chi^{-1}; \tag{7.18}$$

$\chi \to -\chi$ may again be absorbed in rescalings of $P_{k.}a$ and $P_{.l}b$. Thus, there are two essentially different solutions: $\chi = \tilde{\chi} = 1$ corresponding to the obvious tensor product $a \otimes b$ of two exchange field theories; and

$$\chi = -\tilde{\chi} = i. \tag{7.19}$$

The latter yields the complete dimensional trajectory

$$e(h_{kl}) = e\left(h'_k + h_l - \tfrac{1}{2}(k-1)(l-1)\right), \tag{7.20}$$

h'_k and h_l as in eq. (7.8), which gives the Kac spectrum (7.9) for $p' = q$, $q' = p$.

We lack an argument for $p' = q$, $q' = p$. Actually this cannot follow from the exchange algebra relations alone which do not exclude a decomposition of the Virasoro generators $L_n = \bigoplus_i L_n^{(i)}$. We must make an assumption concerning the uniqueness of the stress-energy tensor.

Denoting $\mathscr{T}_a$ (respectively $\mathscr{T}_b$) the $d=2$ quasiprimary fields in the vacuum sector occurring in the short-distance expansions of primary fields $a \cdot a$ (respectively $b \cdot b$), we require

$$\mathscr{T}_a = \mathscr{T}_b = \mathscr{T}. \tag{7.21}$$

In order to check this relation we compute the four-point functions $(\Omega, aaaa\Omega)$, $(\Omega, bbaa\Omega)$, $(\Omega, bbbb\Omega)$, and compare their short-distance limits $(\Omega, \mathscr{T}_a\mathscr{T}_a\Omega)$, $(\Omega, \mathscr{T}_b\mathscr{T}_a\Omega)$, $(\Omega, \mathscr{T}_b\mathscr{T}_b\Omega)$ which should have the common amplitude $\frac{1}{2}c$, where c is the "conformal anomaly" of the theory. The primary four-point functions are obtained from their monodromy properties which are in turn determined by the exchange algebras of the fields involved. We find

$$\begin{aligned}
&(\Omega, a(x_1)a(x_2)P_{1,1}a(x_3)a(x_4)\Omega) \\
&\quad = ((x_1-x_2)(x_3-x_4))^{-2h_2}(1-x)^{-2h_2}\,{}_2F_1\left(\frac{2-m}{m+1}, \frac{1}{m+1}; \frac{2}{m+1}; x\right), \\
&(\Omega, a(x_1)a(x_2)P_{1,3}a(x_3)a(x_4)\Omega) \\
&\quad = \alpha((x_1-x_2)(x_3-x_4))^{-2h_2}(1-x)^{-2h_2}x^{h_3}\,{}_2F_1\left(\frac{1}{m+1}, \frac{m}{m+1}; \frac{2m}{m+1}; x\right), \\
&\qquad (\Omega, b(x_1)b(x_2)P_{1,1}a(x_3)a(x_4)\Omega) \\
&\qquad\quad = (x_1-x_2)^{-2h_2}(x_3-x_4)^{-2h'_2}(1-x)^{-1/2}(1-\tfrac{1}{2}x), \\
&(\Omega, b(x_1)b(x_2)P_{1,1}b(x_3)b(x_4)\Omega) \\
&\quad = ((x_1-x_2)(x_3-x_4))^{-2h'_2}(1-x)^{-2h'_2}\,{}_2F_1\left(\frac{2-m'}{m'+1}, \frac{1}{m'+1}; \frac{2}{m'+1}; x\right), \\
&(\Omega, b(x_1)b(x_2)P_{3,1}b(x_3)b(x_4)\Omega) \\
&\quad = \alpha'((x_1-x_2)(x_3-x_4))^{-2h'_2}(1-x)^{-2h'_2}x^{h'_3}\,{}_2F_1\left(\frac{1}{m'+1}, \frac{m'}{m'+1}; \frac{2m'}{m'+1}; x\right),
\end{aligned} \tag{7.22}$$

where $m = p/(q-p)$,

$$\alpha = \sqrt{\Gamma\left(\frac{2m-1}{m+1}\right)\Gamma\left(\frac{m}{m+1}\right)\Gamma\left(\frac{2}{m+1}\right)\Gamma\left(\frac{1-m}{m+1}\right)\Big/\Gamma\left(\frac{2m}{m+1}\right)\Gamma\left(\frac{m-1}{m+1}\right)\Gamma\left(\frac{1}{m+1}\right)\Gamma\left(\frac{2-m}{m+1}\right)},$$

and similarly for $p/q \to p'/q'$; $x = (x_1-x_2)(x_3-x_4)/(x_1-x_3)(x_2-x_4)$.

The short-distance limits are computed according to sect. 6 with the normalizations of $\mathscr{T}_{a,b}$ fixed by the requirement that $-(1/2\pi)\int dx\, \mathscr{T}_{a,b}(x)$ generate translations on the fields a, b

$$P_{1,1}\mathscr{T}_a(x) = \lim_{x' \to x}\left(-\frac{m+3}{m}\partial_x\partial_{x'}(x-x')^{2h_2}P_{1,1}a(x)a(x')P_{1,1}\right), \quad (7.23)$$

and similarly for $\mathscr{T}_b(x)$. Then, we compute the amplitudes

$$c_{aa} = \frac{(m-2)(m+3)}{m(m+1)}, \qquad c_{bb} = \frac{(m'-2)(m'+3)}{m'(m'+1)},$$

$$c_{ba} = \frac{m+3}{m}\frac{m'+3}{m'},$$

which coincide iff $m + m' + 1 = 0$, i.e. $p'/q' = q/p$. Then

$$c = 1 - 6(p-q)^2/pq. \quad (7.24)$$

This completes the construction of (indecomposable) conformal exchange field theories with the full BPZ fusion rules. The theories we find are in fact the BPZ minimal models. If the horizontal and vertical exchange algebras are computed via strand formation as described in subsect. 7.2, with exchange matrices $[R^{(l_0 l_2)}_{a(n_1 n_2)}]_{ll'}$ (respectively $[R^{(k_0 k_2)}_{b(m_1 m_2)}]_{kk'}$) for light-cone fields a_n (respectively b_m) associated with $\phi_{(1,n)}$ (respectively $\phi_{(m,1)}$), then the exchange matrices of light-cone fields c_{mn} = (short-distance limit of $b_m \cdot a_n$) associated with $\phi_{(m,n)}$ turn out to be

$$R^{(k_0 l_0,\, k_2 l_2)}_{(m_1 n_1,\, m_2 n_2)} = i^{(n_1-1)(m_2-1)+(n_2-1)(m_1-1)} I_{12} R^{(k_0 k_2)}_{b(m_1 m_2)} \otimes R^{(l_0 l_2)}_{a(n_1 n_2)} I_{21}^{-1} \quad (7.25)$$

where $I_{ij} = \mathrm{diag}((-1)^{(n_i-1)(k-k_2-m_j+1)/2}) \otimes \mathrm{diag}((-1)^{(m_j-1)(l-l_0+n_i-1)/2})$ are diagonal sign matrices which affect neither the braid nor the phase relations. It is the overall power of i in eq. (7.25) that causes the non-canonical dimension $h_{m,n} = h_{1,n} + h_{m,1} - \frac{1}{2}(m-1)(n-1)$; compare this with the remark in sect. 5.

Note that as a by-product we have computed the conformal anomaly from monodromy properties. This should be possible in general [6]. A less indirect path from the numerical structure constants R to the numerical value of c is, however, lacking.

7.4. LOCAL FIELDS IN MINIMAL MODELS

Let us now give some examples of local fields constructed with the light-cone fields of two, left and right, exchange algebras as building blocks. For the sake of transparency let us concentrate on the "horizontal" fields of subsect. 7.3.

We have started with symmetric exchange matrices $R^{(\cdot)} = R^{(\cdot)}_{(22)}$ of the "elementary" field $a = a_2$. Even without performing the explicit decomposition of the strand product this ensures that – with appropriate normalizations – all exchange matrices $R^{(\cdot)}_{(n_1 n_2)}$ of "composite" fields a_n are symmetric if $n_1 = n_2$, respectively

$$R^{(\cdot)}_{(n_1 n_2)} = \left[R^{(\cdot)}_{(n_2 n_1)} \right]^{\mathrm{T}}. \tag{7.26}$$

Hence, recalling the special case (1) of proposition 4, we conclude that

$$\phi_{(1,n)(1,n)}(x) = \sum_{ll'} (a_n^+)_{ll'}(x_+) \otimes (a_n^-)_{ll'}(x_-), \tag{7.27}$$

are parity invariant fields, local with respect to each other.

The symmetry (7.4) of the elementary R matrices entails the symmetry

$$\left[R^{(q-l_0, q-l_2)}_{(n_1 n_2)} \right]_{q-l, q-l'} = \left[R^{(l_0 l_2)}_{(n_1 n_2)} \right]_{ll'}, \tag{7.28}$$

of the composite R matrices. Hence, recalling the special case (2) of proposition 4 (with $i = id$), we conclude that

$$\tilde{\phi}_{(1,n)(1,n)}(x) = \sum_{ll'} (a_n^+)_{ll'}(x_+) \otimes (a_n^-)_{q-l, q-l'}(x_-), \tag{7.29}$$

are again parity invariant, mutually local fields. Actually $\tilde{\phi}$ interpolate only among sectors $V_{1,l} \otimes V_{1,q-l}$ and have no component acting on the vacuum $\Omega \in \mathrm{V}_{1,1} \otimes V_{1,1}$. Thus the correct local field in a model with asymmetric sectors should be $\phi_{(1,n)(1,n)} + \tilde{\phi}_{(1,n)(1,n)}$ rather than $\tilde{\phi}_{(1,n)(1,n)}$.

Finally, the fusion rules (7.14) of the composite fields tell that $[q-2][n] = [q-n+1] \oplus [q-n-1]$ which is equivalent to

$$[q-2][j(n)] = [j(n+1)] \oplus [j(n-1)], \tag{7.30}$$

where the $1:1$ mapping j is defined by

$$j(n) = q-n, \quad \text{if} \quad \min(n, q-n) = \text{even},$$

$$j(n) = n, \quad \text{if} \quad \min(n, q-n) = \text{odd}. \tag{7.31}$$

Recall that eqs. (7.2) were *derived* from the fusion rules (7.1) [4]. Since a_{q-2} has isomorphic fusion rules (7.30), $[R^{(j(l_0), j(l_2))}_{(q-2,q-2)}]_{j(l)j(l')}$ must coincide with $[R^{(l_0 l_2)}_{(22)}]_{ll'}$ – possibly with different values $\tilde{p}$ and $\tilde{\eta}$. Inverting the argument by which we computed h_2 from η and p/q in subsect. 7.1, we now compute $\tilde{\eta}$ and $\tilde{p}/q$ from $h_{j(2)} = h_{q-2}$. We find $\tilde{p} = p$ and $\tilde{\eta} = \pm\eta$ (provided $h_2 = h_{q-2} \bmod \frac{1}{2}$, which is the case iff $p \cdot q =$ even, in particular for the unitary minimal models with $|p-q| = 1$).

Recalling the special case (2) of proposition 4 (with $i=j$) we conclude that

$$\phi_{(1,2)(1,q-2)}(x)=\sum_{ll'}\left(a_2^+\right)_{ll'}(x_+)\otimes\left(a_{q-2}^-\right)_{j(l),\,j(l')}(x_-)\,, \tag{7.32}$$

is a commuting or anticommuting local field of spin $h_2-h_{q-2}\in\frac{1}{2}\mathbb{Z}$, interpolating among symmetric and antisymmetric sectors $V_{1,l}\otimes V_{1,j(l)}$.

These constructions are strongly reminiscent of the field content of the diagonal and nondiagonal modular-invariant partition functions [23]. Some more care is needed for other non-parity-invariant fields $\phi_{(1,n)(1,q-n)}$, and for reasons why possibly the sums ((7.27), (7.29), (7.32)) should run over a subset of admissible values only.

Among all other minimal models the supersymmetric one [12] together with its superconformal generator of dimension $\frac{3}{2}$, $G(x)=a_{1,4}(x)\otimes\mathbf{1}$, as a "composite field" comes out as a special realization of our approach. It is remarkable that superfields can be constructed from "ordinary" exchange fields, and supersymmetry emerges without having been imposed.

7.5. ANALYTIC PROPERTIES OF n-POINT FUNCTIONS

The n-point functions $F_{\boldsymbol{\beta\alpha}}=(\Omega,a_{\alpha_1}(x_1)P_{\beta_1}\dots P_{\beta_{n-1}}a_{\alpha_n}(x_n)\Omega)$ may be considered [6] as local horizontal sections in complex vector bundles over $\mathring{\mathbb{C}}^n/\mathrm{S}_n$ with a flat connection, in which the representations of the braid group (proposition 5) describe the holonomy group of parallel transport. The matrices $\rho(b)$ act on the labels $\boldsymbol{\beta}$ and $\boldsymbol{\alpha}$. The action on $\boldsymbol{\alpha}$ is just the permutation $\pi(b)\in\mathrm{S}_n$.

The monodromy subgroup $\{\rho(b)|\pi(b)=e\in\mathrm{S}_n\}$ generated by $\rho(\sigma_j^2)$ describes the parallel transport in a flat connection of complex vector bundles over $\mathbb{C}^n$. The monodromy matrices act only on the labels $\boldsymbol{\beta}$ and describe the linear transformation of *all* conformal block functions, contributing to *one* local n-point function, into each other under analytic continuation of x_i around x_j.

The linear monodromy behaviour implies [24] that the conformal block functions are solutions to algebraic partial linear differential equations (PLDE's) of the Fuchs type. In general the latter are not uniquely specified by the monodromy, since quasiprimary fields share the monodromy properties of their primaries. Already Riemann [25] has discussed how the powers of (x_i-x_j) near the singularities – i.e. the dimensions d_α and h_β involved – and the associated monodromy determine the PLDE's and their solutions.

In the minimal and WZW models, the PLDE's are obviously the equations inferred from the existence of degenerate states in the Verma modules [9, 11]. From the fusion rules (7.1) and the construction of proposition 5 one learns that every four-point function involving at least one "elementary" field $a=a_2$ has at most two "channels" P_{β_2}; i.e. the monodromy matrices are at most 2×2. Then the corresponding PLDE's – which can be converted into ordinary LDE's in the variable

$x=(x_1-x_2)(x_3-x_4)/(x_1-x_3)(x_2-x_4)$ – are of at most second order, and their solutions are hypergeometric functions, up to powers of x and $(1-x)$ and polynomial factors. By comparison with the known monodromy behaviour of hypergeometric functions [26], the conformal blocks can be read off their monodromy matrices, up to polynomial factors of degree increasing with the level of quasiprimary fields [19].

As an example we compute all primary four-point functions

$$(\Omega,a_{kl}(x_1)a_{12}(x_2)P_{k,l\pm1}a_{12}(x_3)a_{kl}(x_4)\Omega)$$
$$=:(x_1-x_4)^{-2h_{kl}}(x_2-x_3)^{-2h_{12}}x^{-h_{12}-h_{kl}}f_\pm(x). \tag{7.33}$$

Eq. (3.10) and a similar equation for $R^{(0\beta)}$ provide the monodromy matrices $\rho_1^2=\rho_3^2$. ρ_2 is given by the exchange matrices (7.2). Thus the (positive-oriented) analytic continuation of x around 0 effects a factor of $e(h_{k,l\pm1})$ on $f_\pm(x)$, and the (positive-oriented) analytic continuation $x\to 1/x$ is described by

$$\begin{pmatrix} f_-\left(\frac{1}{x}\right) \\ f_+\left(\frac{1}{x}\right)\end{pmatrix}=e(h_{12})x^{-2h_{12}-2h_{kl}}R^{(ll)}\begin{pmatrix} f_-(x)\\ f_+(x)\end{pmatrix}. \tag{7.34}$$

Taking into account the powers of the singularities at $x=0,1,\infty$ and comparing with the analytic behaviour $x\to 1/x$ of hypergeometric functions, we could derive

$$f_\pm(x)=c_\pm(1-x)^{h_{13}}x^{h_{k,l+1}}{}_2F_1\left(\frac{p}{q},\pm k+(1\pm l)\frac{p}{q};1\pm k\mp l\frac{p}{q};x\right), \tag{7.35}$$

$$\frac{c_+}{c_-}=\sqrt{\gamma\left(k-l\frac{p}{q}\right)\gamma\left(1+k-l\frac{p}{q}\right)\gamma\left(-k+(l+1)\frac{p}{q}\right)\gamma\left(1-k+(l-1)\frac{p}{q}\right)}$$
$$\gamma(x):=\Gamma(x)/\Gamma(1-x). \tag{7.36}$$

The same results have been obtained previously with different methods [27]. The above computation illustrates that our methods in principle are not weaker than others. There is the advantage – as compared with the solution of PLDE's by multiple contour integrals – that we work from the beginning with a natural basis $F_{\beta\alpha}$ of solutions, and have control over their analytic properties.

Let us now turn to an interesting side-remark [6] concerning the n-point functions of the "elementary" fields $a=a_2$ and the corresponding representations ρ of the braid groups B_n. One checks easily that all matrices (7.2) satisfy $R^2=\eta(1+\omega)R-$

$\eta^2\omega\mathbf{1}$. Introducing

$$g_i = (-\omega\eta)^{-1}\rho_i, \qquad t = (-\omega)^{-1}, \qquad e_i = (1+g_i)/(1+t), \tag{7.37}$$

this entails

$$g_i^2 + (1-t)g_i - t = 0, \tag{7.38}$$

or, equivalently

$$e_i^2 = e_i. \tag{7.39}$$

These are – besides the representation conditions of the braid group – the defining relations of the Hecke algebra $\mathrm{H}_n(t)$ [28] generated by g_i. The projectors e_i are *real* in the case of *unitary R* matrices.

Actually the Hecke algebra relations just tell that the generators g_i, and thus ρ_i, have only two different eigenvalues. This property, in turn, just reflects the two-fold branching of the fusion rules (7.1). Consequently, the exchange algebra of composite fields will generally not give rise to "Hecke-type" representations of B_n.

Next, it is easy to verify that the projectors e_i satisfy

$$e_i e_{i\pm1} e_i = \tau e_i,$$

$$\beta \equiv \tau^{-1} = 4\cos^2(\pi p/q). \tag{7.40}$$

These are – besides the Hecke algebra relations – the defining relations of the Jones algebras $\mathrm{A}_{\beta,n}$ [29]. Our unitary cases $\beta = 4\cos^2(\pi/q)$ have been identified previously [29] as the only β values (except $\beta = 4$) for which unitary "Jones-type" representations of the braid groups exist.

Moreover, Jones has classified [29] the cases in which these representations define *finite* matrix groups. In our context this is equivalent to the property of the n-point functions to be *algebraic* functions. The result is that for $q = 3, 4, 6, 10$ the n-point functions (for $q = 10$ only $n \le 4$) of the elementary fields are algebraic functions.

$q = 3$ corresponds to the vertical field $\phi_{(21)\cdot\cdot}$ $(h = \frac{1}{2})$ of the Ising model $(c = \frac{1}{2})$. $q = 4$ corresponds to the horizontal field $\phi_{(12)\cdot}$ $(h = \frac{1}{16})$ of the Ising model, and to the vertical field $\phi_{(21)\cdot}$ of (the universality class of) the tricritical Ising model $(c = \frac{7}{10})$. $q = 6$ corresponds to $\phi_{(12)\cdot}$ in (the universality class of) the three-state Potts model $(c = \frac{4}{5})$, and to $\phi_{(21)\cdot}$ in (the universality class of) the tricritical three-state Potts model $(c = \frac{6}{7})$. $q = 10$ is realized in the models with $c = \frac{14}{15}$ and $c = \frac{52}{55}$.

The Ising model functions are well known. For $c = \frac{4}{5}$, $h_{12} = \frac{1}{8}$, we have computed from the analytic exchange behaviour of

$$\Big(\Omega, a_{12}(x_1)a_{12}(x_2)P_{13}a_{12}(x_3)a_{12}(x_4)\Omega\Big)_{11} = \prod_{i<j}(x_i - x_j)^{-1/12} f_\pm(x), \tag{7.41}$$

the algebraic functions

$$f_+(x) = c' j(x)^{1/12} \sqrt{\sum_{e^3=1} (-e)\sqrt{1-ej(x)}^{-1/3}} = c_+ x^{1/2}(1+O(x))$$

$$f_-(x) = c' j(x)^{1/12} \sqrt{\sum_{e^3=1} \sqrt{1-ej(x)}^{-1/3}} = c_- x^{1/6}(1+O(x))$$

$$\frac{c_+}{c_-} = \sqrt{\frac{\Gamma(-\frac{2}{3})\Gamma(\frac{1}{3})\Gamma(\frac{3}{2})\Gamma(\frac{5}{6})}{\Gamma(\frac{5}{3})\Gamma(\frac{2}{3})\Gamma(-\frac{1}{2})\Gamma(\frac{1}{6})}} = 3\times 2^{-13/6}, \tag{7.42}$$

where the sums extend over the three cubic roots of unity, and

$$j(x) = \tfrac{4}{27}(1-x+x^2)^3/x^2(1-x)^2,$$

is a familiar invariant of the theory of elliptic ϑ-functions [30].

By eq. (7.35), the four-point functions of a_{12} for all minimal models are expressed as hypergeometric functions. The cases $q=3,4,6,10$, $p=q\pm 1$, precisely fit into the famous Schwarz list [31] of *algebraic* hypergeometric functions. To our knowledge, the explicit formulae

$${}_2F_1(-\tfrac{1}{2},\tfrac{1}{6};\tfrac{1}{3};x) = (1-x+x^2)^{1/4}\sqrt{\tfrac{1}{3}\sum_{e^3=1}\sqrt{1-ej(x)}^{-1/3}}\,,$$

$$3\times 2^{-13/6}x^{2/3}\,{}_2F_1(\tfrac{5}{6},\tfrac{1}{6};\tfrac{5}{3};x) = (1-x+x^2)^{1/4}\sqrt{\tfrac{1}{3}\sum_{e^3=1}(-e)\sqrt{1-ej(x)}^{-1/3}}\,, \tag{7.43}$$

are not displayed in the handbooks on hypergeometric functions.

7.6. COUNTEREXAMPLE

We have solved the braid relations (3.11) for a field $a=a_4$ with the following assumed fusion rules ([1] = [vacuum])

$$[4][i] = [4], \qquad i=1,2,3,$$

$$[4][4] = [1]\oplus[2]\oplus[3], \tag{7.44}$$

finding

$$R^{(ii)} =: \eta, \qquad i = 1,2,3,$$

$$R^{(ij)} =: \eta\omega, \qquad i \neq j, = 1,2,3,$$

$$[R^{(44)}]_{ij} = \tfrac{1}{3}\eta \begin{pmatrix} 2\omega+1 & \omega-1 & \omega-1 \\ \omega-1 & 2\omega+1 & \omega-1 \\ \omega-1 & \omega-1 & 2\omega+1 \end{pmatrix},$$

$$\omega = e\left(\pm\tfrac{1}{3}\right). \tag{7.45}$$

The phase relation (3.9) can *not* be satisfied with $R^{(44)}$, indicating that eq. (7.45) cannot be associated with a *conformal* exchange algebra in the sense of this paper. Still this solution may have a relevance for the more abstract approach to local $QFTh_2$ beyond $CQFTh_2$ suggested in sect. 5.

The counterexample shows that proposition 2 not only constrains the dimensional spectrum in terms of the R matrices, but in fact constrains the exchange matrices themselves, and thus the fusion rules compatible with conformal invariance.

8. Conclusion and outlook

We presented the concept of a light-cone field theory with a non-commutative field algebra: the exchange algebra. Exchange field theory is regarded as a building block of non-perturbative interacting local $QFTh_2$. We discussed exchange field theory in much detail in the specific context of conformal $QFTh_2$ as a most natural – and for the moment the only available – realization. We outlined the strategies for a conclusive analysis of $CQFTh_2$ with finite fusion rules.

Although we expect that many of the ideas, if properly generalized, may transcend conformal and even two-dimensional QFTh [32], we made no real attempt at generalization in this paper. Furthermore, we leave to a future publication all aspects of representation theory of conformal exchange algebras related with KMS [3] temperature states. Such aspects will be useful [32] for the concept of "soliton completeness" and the reduction of so-called modularity properties of partition functions [23] to causality and completeness principles.

We would like to view exchange algebras as "the liberation of the very subtle ideas of Yang, Baxter [14], and Faddeev [33] from their narrow confinement to special lattice models". We have given arguments that (suitably generalized to field algebras with infinite families of quantum fields) the braid identities are properties of Einstein-causal $CQFTh_2$ "par excellence". As an extension of the recent work of Karowski [34] on the use of the algebraic Bethe-ansatz technique to relate the RSOS models via a θ-angle with the six-vertex models (or to relate minimal models with the Coulomb gas (respectively Sine–Gordon) field theory), one should study in our algebraic language the field-theoretical transmutation of exchange algebras with Burau braid representations into those of the minimal models. This may well also lead to a "liberation" of the Bethe-ansatz technique from special statistical mechanics. We have in mind a yet unknown use of the Bethe ideas (with completely different physical content) for the systematic construction of (vacuum) representations of exchange field algebras.

The immediate area of application of $d=2$ CQFTh's is of course the understanding of the universality classes of surfaces (thin films, layers) of critical systems of condensed matter. It is remarkable (perplexing) that nature reveals its deepest causality secrets in an area whose physical principles are not directly related to Einstein's causality.

Note added

After completion of this work we became aware of ref. [35]. In their search for invariant link polynomials these authors discuss very systematically the limiting process leading from Yang–Baxter to braid matrices, as well as (in our terminology:) the projection onto braid-group representations of composite fields out of strand products of elementary representations.

References

[1] R. Haag and D. Kastler, J. Math. Phys. 5 (1964) 848;
R.F. Streater and A.S. Wightman, PCT, spin, statistics, and all that, Math. Phys. Monograph series (Benjamin, New York, 1964);
R. Jost, The general theory of quantized fields, Lect. Appl. Math. (Am. Math. Soc., Providence, RI, 1965)
S. Doplicher and J.E. Roberts, C*-algebras and duality for compact groups: why there is a compact group of internal symmetries in particle physics *in* Proc. Int. Conf. on Mathematical physics, Marseille 1986, and references therein

[2] H. Lehmann, K. Symanzik and W. Zimmermann, Nuovo Cim, 1 (1955) 205

[3] R. Kubo, J. Phys. Soc. Japan 12 (1957) 570;
P.C. Martin and J. Schwinger, Phys. Rev. 115 (1959) 1342;
R. Haag, N.M. Hugenholtz and M. Winnink, Commun. Math. Phys. 5 (1967) 215

[4] K.H. Rehren and B. Schroer, Phys. Lett. B198 (1987) 84;
K.H. Rehren, Commun. Math. Phys. 116 (1988) 675

[5] B. Schroer and J.A. Swieca, Phys. Rev. D10 (1974) 480;
B. Schroer, J.A. Swieca and A.H. Völkel, Phys. Rev. D11 (1975) 1509;
M. Lüscher and G. Mack, Commun. Math. Phys. 41 (1975) 203

[6] J. Fröhlich, Statistics of fields, the Yang–Baxter equation, and the theory of knots and links, Cargèse 1987 Proc., to be published;
J. Fröhlich, Statistics and monodromy in two- and three-dimensional quantum field theory *in* Differential geometrical methods in theoretical physics, ed. K. Bleuler, Dordrecht (1988), to be published;
G. Felder and J. Fröhlich, unpublished notes (1987, 1988) and private communications

[7] S. Ferrara, R. Gatto and A.F. Grillo, Nuovo Cim. 12A (1972) 959;
B. Schroer, A trip to Scalingland, *in* Proc. V. Brazilian Symp. on Theoretical physics, vol. 1, ed. E. Ferreira, Rio de Janeiro (1974);
M. Lüscher and G. Mack, unpublished manuscript (1976)

[8] I.T. Todorov, Infinite Lie algebras in 2-dimensional conformal field theory", ISAS Trieste preprint 2/85/E.P. (1985)
[9] A.A. Belavin, A.M. Polyakov and A.B. Zamolodchikov, Nucl. Phys. B241 (1984) 333
[10] B. Klaiber, The Thirring model *in* Quantum theory and statistical physics; Boulder Lectures, 1967, eds. W.E. Brittin et al.
[11] V.G. Knizhnik and A.B. Zamolodchikov, Nucl. Phys. B247 (1984) 83;
P. Christe and R. Flume, Nucl. Phys. B282 (1987) 466
[12] M.A. Bershadsky, V.G. Knizhnik and M.G. Teitelman, Phys. Lett. B151 (1985) 31;
Z. Qiu, Phys. Lett. B188 (1987) 207
[13] A.B. Zamolodchikov and V.A. Fateev, Sov. Phys. JETP 62 (1985) 215
[14] C.N. Yang, Phys. Rev. Lett. 19 (1967) 1312;
R.J. Baxter, Ann. Phys. 70 (1972) 193; 323
[15] G.E. Andrews, R.J. Baxter and P.J. Forrester, J. Stat. Phys. 35 (1984) 193
[16] E. Artin, Collected papers, eds. S. Lang and J.T. Tate (Addison-Wesley, Reading, MA, 1965) pp. 416–498
[17] C. Vafa, Phys. Lett. B206 (1988) 421
[18] M. Lüscher, Commun. Math. Phys. 50 (1976) 23
[19] K.H. Rehren and B. Schroer, Nucl. Phys. B295 [FS 21] (1988) 229
[20] M. Karowski, Nucl. Phys. B153 (1979) 244
P.P. Kulish, Lett. Math. Phys. 5 (1979) 393
[21] E. Date, M. Jimbo, T. Miwa and M. Okado, Lett. Math. Phys. 12 (1986) 209
[22] D. Friedan, Z. Qiu and S. Shenker, Conformal invariance, unitarity, and two-dimensional critical exponents *in* Vertex operators in mathematics and physics, eds. J. Lepowsky, S. Mandelstam and I.M. Singer (Springer New York, 1984)
[23] J. Cardy, Nucl. Phys. B270 [FS 16] (1986) 186;
A. Capelli, C. Itzykson and J.B. Zuber, Nucl. Phys. B280 [FS 18] (1987) 445
[24] G. Felder, private remark
[25] B. Riemann, Abh. Kgl. Ges. Wissensch. Göttingen, Bd. 7 (1857)
[26] A. Erdélyi et al., Higher transcendental functions, vol. 1 (McGraw-Hill, New York, 1953);
M. Abramowitz et al., Handbook of mathematical functions (Dover, New York, 1965)
[27] V.J.S. Dotsenko and V.A. Fateev, Nucl. Phys. B240 [FS 12] (1984) 312; B251 [FS 13] (1985) 691
[28] V.F.R. Jones, Ann. Math. 126 (1987) 335
[29] V.F.R. Jones, Braid groups, Hecke algebras and type II_1 factors, *in* Geometric methods in operator algebras, Proc. US–Japan Seminar (1986) p. 242
[30] S. Lang, Elliptic functions (Addison-Wesley, Reading, MA, 1973)
[31] H.A. Schwarz, Crelle's J. Math., Bd. 75, 292 (1873)
[32] B. Schroer, Algebraic aspects of non-perturbative quantum field theories *in* Differential geometrical methods in theoretical physics, ed. K. Bleuler (Dordrecht, 1988), to be published
[33] L. Faddeev, Sov. Scient. Rev. C1 (1980) 107
[34] M. Karowski, Conformal quantum field theory and integrable systems *in* Proc. Brasov Summer School, September 87, eds. P. Dita et al. (Academic Press, New York), to be published;
M. Karowski, ETH Zürich preprint, in preparation
[35] Y. Akutsu, T. Deguchi and M. Wadati, J. Phys. Soc. Japan 56 (1987) 3039; 3464; 57 (1988) 757; 1173

Commun. Math. Phys. 121, 351–399 (1989)

Communications in Mathematical Physics

Quantum Field Theory and the Jones Polynomial *

Edward Witten **

School of Natural Sciences, Institute for Advanced Study, Olden Lane, Princeton, NJ 08540, USA

Abstract. It is shown that $2+1$ dimensional quantum Yang-Mills theory, with an action consisting purely of the Chern-Simons term, is exactly soluble and gives a natural framework for understanding the Jones polynomial of knot theory in three dimensional terms. In this version, the Jones polynomial can be generalized from S^3 to arbitrary three manifolds, giving invariants of three manifolds that are computable from a surgery presentation. These results shed a surprising new light on conformal field theory in $1+1$ dimensions.

In a lecture at the Hermann Weyl Symposium last year [1], Michael Atiyah proposed two problems for quantum field theorists. The first problem was to give a physical interpretation to Donaldson theory. The second problem was to find an intrinsically three dimensional definition of the Jones polynomial of knot theory. These two problems might roughly be described as follows.

Donaldson theory is a key to understanding geometry in four dimensions. Four is the physical dimension at least macroscopically, so one may take a slight liberty and say that Donaldson theory is a key to understanding the geometry of space-time. Geometers have long known that (via de Rham theory) the self-dual and anti-self-dual Maxwell equations are related to natural topological invariants of a four manifold, namely the second homology group and its intersection form. For a simply connected four manifold, these are essentially the only classical invariants, but they leave many basic questions out of reach. Donaldson's great insight [2] was to realize that moduli spaces of solutions of the self-dual Yang-Mills equations can be powerful tools for addressing these questions.

Donaldson theory has always been an intrinsically four dimensional theory, and it has always been clear that it was connected with mathematical physics at least at the level of classical nonlinear equations. The puzzle about Donaldson theory was whether this theory was tied to more central ideas in physics, whether it could be interpreted in terms of quantum field theory. The most important

---

* An expanded version of a lecture at the IAMP Congress, Swansea, July, 1988

** Research supported in part by NSF Grant No. 86-20266, and NSF Waterman Grant 88-17521

Fig. 1. A knot in three dimensional space

evidence for the existence of such a connection had to do with Floer's work on three manifolds [3] and the nature of the relation between Donaldson theory and Floer theory. Also, the "Donaldson polynomials" had an interesting formal analogy with quantum field theory correlation functions. It has turned out that Donaldson theory can indeed be given a physical interpretation [4].

As for the Jones polynomial and its generalizations [5–11], these deal with the mysteries of knots in three dimensional space (Fig. 1). The puzzle on the mathematical side was that these objects are invariants of a three dimensional situation, but one did not have an intrinsically three dimensional definition. There were many elegant definitions of the knot polynomials, but they all involved looking in some way at a two dimensional projection or slicing of the knot, giving a two dimensional algorithm for computation, and proving that the result is independent of the chosen projection. This is analogous to studying a physical theory that is in fact relativistic but in which one does not know of a manifestly relativistic formulation – like quantum electrodynamics in the 1930's.

On the physical side, the puzzle about the knot polynomials was the following. Unlike the Donaldson theory, where a connection with quantum field theory was not obvious, the knot polynomials have been intimately connected almost from the beginning with two dimensional many body physics. In fact, constructions of the knot polynomials have related them to two dimensional (or $1 + 1$ dimensional) many-body physics in a bewildering variety of ways, mainly involving soluble lattice models [7], solutions of the Yang-Baxter equation [8], and monodromies of conformal field theory [11]. In the latter interpretation, the knot polynomials are related to aspects of conformal field theory that have been particularly fruitful recently [12–16]. On the statistical mechanical side, studies of the knot polynomials have related them to Temperley-Lieb algebras and their generalizations, and to other aspects of soluble statistical mechanics models in $1 + 1$ dimensions. For physicists the challenge of the knot polynomials has been to bring order to this diversity, find the unifying themes, and learn what it is that is three dimensional about two dimensional conformal field theory.

Now, the Donaldson and Jones (and Floer and Gromov [17]) theories deal with topological invariants, and understanding these theories as quantum field theories involves constructing theories in which all of the observables are topological invariants. Some physicists might consider this to be a little bit strange, so let us pause to explain the physical meaning of "topological invariance". The physical meaning is really "general covariance". Something that can be computed from a manifold M as a topological space (perhaps with a smooth structure) without a choice of metric is called a "topological invariant" (or a "smooth invariant") by mathematicians. To a physicist, a quantum field theory defined on a manifold M without any a priori choice of a metric on M is said to be generally covariant. Obviously, any quantity computed in a generally covariant quantum field theory will be a topological invariant. Conversely, a quantum field theory in which all observables are topological invariants can naturally be seen as a generally covariant quantum field theory. Indeed, the Donaldson, Floer, Jones, and Gromov theories can be seen as generally covariant quantum field theories in four, three, and two space-time dimensions. The surprise, for physicists, perhaps comes in how general covariance is achieved. General relativity gives us a prototype for how to construct a quantum field theory with no a priori choice of metric – we introduce a metric, and then integrate over all metrics. This example is so influential in our thinking that we tend to think of a generally covariant theory as being, by definition, a theory in which the metric is a dynamical variable. The lesson from the Donaldson, Floer, Jones, and Gromov theories is precisely that there are highly non-trivial quantum field theories in which general covariance is realized in other ways. In particular, in this paper we will describe an exactly soluble generally covariant quantum field theory in which general covariance is achieved not by integrating over metrics but because we begin with a gauge invariant Lagrangian that does not contain a metric.

1. The Chern-Simons Action

We have been urged [1] to try to interpret the Jones polynomial in terms of three dimensional Yang-Mills theory. So we begin on an oriented three manifold M with a compact simple gauge group G. We pick a G bundle E, which may as well be trivial, and on E we place a connection A_i^a, which can be viewed as a Lie algebra valued one form (a runs over a basis of the Lie algebra, and i is tangent to M). An infinitesimal gauge transformation is

$$A_i \to A_i - D_i \varepsilon, \tag{1.1}$$

where ε, a generator of the gauge group, is a Lie algebra valued zero form and the covariant derivative is $D_i \varepsilon = \partial_i \varepsilon + [A_i, \varepsilon]$. The curvature is the Lie algebra valued two form $F_{ij} = [D_i, D_j] = \partial_i A_j - \partial_j A_i + [A_i, A_j]$. Now we need to choose a Lagrangian. We will *not* pick the standard Yang-Mills action[1]

$$\mathscr{L}_0 = \int_M \sqrt{g}\, g^{ik} g^{jl} \operatorname{Tr}(F_{ij} F_{kl}), \tag{1.2}$$

as this depends on the choice of a metric g_{ij}. We want to formulate a generally covariant theory (in which all observables will be topological invariants), and to this aim we want to pick a Lagrangian which does not require any choice of metric.

[1] In what follows, the symbol "Tr" denotes an invariant bilinear form on the Lie algebra of G, a multiple of the Cartan-Killing form; we will specify the normalization presently

Precisely in three dimensions there is a reasonable choice, namely the integral of the Chern-Simons three form:

$$\mathcal{L} = \frac{k}{4\pi} \int_M \mathrm{Tr}\,(A \wedge dA + \tfrac{2}{3} A \wedge A \wedge A)$$
$$= \frac{k}{8\pi} \int_M \varepsilon^{ijk}\, \mathrm{Tr}\,(A_i(\partial_j A_k - \partial_k A_j) + \tfrac{2}{3} A_i [A_j, A_k]) \,. \tag{1.3}$$

The Chern-Simons term in three dimensional gauge theory has a relatively long history. The abelian gauge theory with only a Chern-Simons term was studied by Schwarz [18] and in unpublished work by I. Singer. Three dimensional gauge theories with the Chern-Simons term added to the usual action (1.2) were introduced in [19–21]. The nonabelian theory with only Chern-Simons action was studied classically by Zuckerman [22]. The abelian Chern-Simons theory has recently been studied in relation to fractional statistics by Hagen [24] and by Arovas et al. [25] and in relation to linking numbers by Polyakov [23] and Fröhlich [15]. The novelty in our present discussion is that we will consider the quantum field theory defined by the nonabelian Chern-Simons action and argue that it is exactly soluble and has important implications for three dimensional geometry and two dimensional conformal field theory.

The first fundamental property of the Chern-Simons theory is the quantization law first discussed in [21]. It arises because the group $\hat{G}$ of continuous maps $M \to G$ is not connected. In the homotopy classification of such maps one meets at least the fact that $\pi_3(G) \simeq \mathbf{Z}$ for every compact simple group G. Though (1.3) is invariant under the component of the gauge group that contains the identity, it is not invariant under gauge transformations of non-zero "winding number", gauge transformations associated with non-zero elements of $\pi_3(G)$. Under a gauge transformation of winding number m, the transformation law of (1.3) is

$$\mathcal{L} \to \mathcal{L} + \text{const} \cdot m \,. \tag{1.4}$$

As in Dirac's famous work on magnetic monopoles, consistency of quantum field theory does not quite require the single-valuedness of $\mathcal{L}$, but only of $\exp(i\mathcal{L})$. For this purpose, it is necessary and sufficient that the "constant" in (1.4) should be an integral multiple of 2π. This gives a quantization condition on the parameter called k in (1.3). If G is $SU(N)$ and "Tr" means a trace in the N dimensional representation, then the requirement is that k should be an integer. In general, for any G, we can uniquely fix the so far unspecified normalization of "Tr" so that the quantization condition is $k \in \mathbf{Z}$.

We will see later that k is very closely related to the central charge in the theory of highest weight representations of affine Lie algebras. It is no accident that the reasoning which shows that k must be quantized in (1.3) has a $1+1$ dimensional analogue [26] which leads to quantization of the central charge in the representation theory of affine algebras.

In quantum field theory, in addition to a Lagrangian, one also wishes to pick a suitable class of gauge invariant observables. In the present context, the usual gauge invariant local operators would not be appropriate, as they spoil general covariance. However, the "Wilson lines" so familiar in QCD give a natural class of

Fig. 2. Several linked but non-intersecting oriented knots in a three manifold M. Such a collection of knots is called a "link"

gauge invariant observables that do not require a choice of metric. Let C be an oriented closed curve in M. Intrinsically C is simply a circle, but the topological classification of embeddings of a circle in M is very complicated, as we observe in Fig. 1. Let R be an irreducible representation of G. One then defines the "Wilson line" $W_R(C)$ to be the following functional of the connection A_i. One computes the holonomy of A_i around C, getting an element of G that is well-defined up to conjugacy, and then one takes the trace of this element in the representation R. Thus, the definition is

$$W_R(C) = \mathrm{Tr}_R \, P \exp \int_C A_i \, dx^i . \tag{1.5}$$

The crucial property of this definition is that there is no need to introduce a metric, so general covariance is maintained.

We now can formulate the general problem of interest. In an oriented three manifold M, we take r oriented and non-intersecting knots C_i, $i = 1 \ldots r$, whose union is what knot theorists would call a "link" L. We assign a representation R_i to each C_i, and we propose to calculate the Feynman path integral

$$\int D\mathscr{A} \exp(i\mathscr{L}) \prod_{i=1}^{r} W_{R_i}(C_i). \tag{1.6}$$

The symbol $D\mathscr{A}$ represents Feynman's integral over all gauge orbits, that is, an integral over all equivalence classes of connections modulo gauge transformations. Of course, (1.6) has exactly the formal structure of some familiar observables in QCD, the difference being that we are in three dimensions instead of four and we have chosen a somewhat exotic gauge theory action. We will call (1.6) the "partition function" of M with the given link, or the (unnormalized) "expectation value" of the given link; we will denote it as $Z(M; C_i, R_i)$ or simply as $Z(M; L)$ for short.

For the case of links in S^3, we will claim that the invariants (1.6) are exactly those that appear in the Jones theory and its generalizations. Simply replacing S^3 with a general oriented three manifold M gives a very intriguing (and as we will see, effectively computable) generalization of the known knot polynomials. Taking $r = 0$ (no knots), (1.6) gives invariants of the oriented three manifold M which also turn out to be effectively computable. Before getting into any details, let us note a few preliminary indications of a possible connection between (1.6) and the Jones theory:

(1) In (1.6) we see the right variables, namely a compact Lie group G, a choice of representation R_i for each component C_i of the link L, and an additional

variable k. [In knot theory one usually makes an analytic continuation and replaces k by a complex variable q, but it has been known since Jones' original work that there are special properties at special values of q. We claim that these properties reflect the fact that the three dimensional gauge theory with action (1.3) is well-defined only if k is an integer.] The two variable generalization of the Jones polynomial corresponds to the case that G is $SU(N)$, and the R_i are all the defining N dimensional representation of $SU(N)$. The two variables are N and k, analytically continued to complex values. The Kauffman polynomial similarly arises for $G = SO(N)$ and R the N dimensional representation.

(2) As a further check on the plausibility of a relation between (1.6) and the knot polynomials, let us note first of all that (1.6) depends on a choice of the orientation of M, as this enters in fixing the sign of the Chern-Simons form. Likewise, (1.6) depends on the orientations of the C_i, since these enter in defining the Wilson lines (in computing the holonomy around C_i, one must decide in which direction to integrate around C_i). If, however, one reverses the orientation of one of the C_i and simultaneously exchanges the representation R_i with its complex conjugate $\bar{R}_i$, then the definition of the Wilson lines is unchanged, so (1.6) is invariant under this process. And if (without changing the R_i) one reverses the orientations of *all* components C_i of the link L, then (1.6) is unchanged because of a symmetry that physicists would call "charge conjugation". This is an involution of the Lie algebra of G that exchanges all representations with their complex conjugates; applying this involution to all integration variables in (1.6) leaves (1.6) invariant while exchanging all R_i with their conjugates or equivalently reversing the orientation of all the C_i. These are important formal properties of the knot polynomials.

2. The Weak Coupling Limit

To begin with, since a non-abelian gauge theory with only a Chern-Simons action may seem unfamiliar, one might ask whether this Lagrangian really does lead to a sensible quantum theory, and really can be regulated to give topologically invariant results. In this section, we will briefly investigate this point by studying the theory in a weak coupling limit in which computations are comparatively straightforward. This is the limit of large k.[2] For large k, the path integral

$$Z = \int D\mathscr{A} \exp\left(\frac{ik}{4\pi}\int_M \operatorname{Tr}\left(A \wedge dA + \frac{2}{3} A \wedge A \wedge A\right)\right) \tag{2.1}$$

(for the moment we omit knots) contains an integrand which is wildly oscillatory. The large k limit of such an integral is given by a sum of contributions from the points of stationary phase. The stationary points of the Chern-Simons action are precisely the "flat connections", that is, the gauge fields for which the curvature vanishes

$$F_{ij}^a = 0. \tag{2.2}$$

[2] The reader may wish to bear in mind that the discussion in this section and the next contains a number of technicalities which are part of the logical story but perhaps not essential on a first reading

Gauge equivalence classes of such flat connections correspond to homomorphisms

$$\phi: \pi_1(M) \rightarrow G, \tag{2.3}$$

or more exactly to equivalence classes of such homomorphisms, up to conjugation. If for simplicity we suppose that the topology of M is such that there are only finitely many classes of homomorphisms (2.3), then the large k behavior of (2.1) will be a sum

$$Z = \sum_{\alpha} \mu(A^{(\alpha)}), \tag{2.4}$$

where the $A^{(\alpha)}$ are a complete set of gauge equivalence classes of flat connections, and $\mu(A^{(\alpha)})$ is to be obtained by stationary phase evaluation of (2.1), expanding around $A^{(\alpha)}$. This reduction to a stationary phase evaluation means that the nonabelian theory, for large k, is closely related to the abelian theory. This in turn has been shown [18] to lead to Ray-Singer analytic torsion [27], which is closely related to the purely topological Reidemeister torsion. The $\mu(A^{(\alpha)})$ may be evaluated as follows. We make in (2.1) the change of variables $A_i = A_i^{(\alpha)} + B_i$, where B_i is the new integration variable. An important invariant of the flat connection $A^{(\alpha)}$ is its Chern-Simons invariant

$$I(A^{(\alpha)}) = \frac{1}{4\pi} \int_M \operatorname{Tr}(A^{(\alpha)} \wedge dA^{(\alpha)} + \tfrac{2}{3} A^{(\alpha)} \wedge A^{(\alpha)} \wedge A^{(\alpha)}). \tag{2.5}$$

When the Chern-Simons action is expanded in powers of B_i, the first terms are

$$\mathscr{L} = k \cdot I(A^{(\alpha)}) + \frac{k}{4\pi} \int_M \operatorname{Tr}(B \wedge DB). \tag{2.6}$$

Here it is understood that in (2.6), the expression DB denotes the covariant exterior derivative of B with respect to the background gauge field $A^{(\alpha)}$; it does not depend on a metric on M. A salient point is that in (2.6) there is no term linear in B, since $A^{(\alpha)}$ is a critical point of the action.

To carry out the Gaussian integral in (2.6), gauge fixing is needed. There is no way to carry out this gauge fixing without picking a metric on M (or in some other way breaking the symmetry of the problem). After picking such a metric, a convenient gauge choice is $D_i B^i = 0$ (with D_i the covariant derivative constructed from the metric and the background gauge field $A^{(\alpha)}$). The standard Faddeev-Popov construction then gives rise to a gauge fixing Lagrangian

$$\mathscr{L}_{\text{gauge}} = \int_M (\operatorname{Tr} \phi D_i B^i + \operatorname{Tr} \bar{c} D_i D^i c). \tag{2.7}$$

Here ϕ is a Lagrangian multiplier that enforces the gauge condition $D_i B^i = 0$, and c, $\bar{c}$ are anticommuting "ghosts" that are introduced to get the right measure on the space of gauge fields modulo gauge transformations. The quadratic terms in ϕ and B that can be found in (2.6), (2.7) have a natural geometric interpretation, described (in the abelian case) in [18]. Let D be the exterior derivative on M, twisted by the flat connection $A^{(\alpha)}$, and let $*$ be the Hodge operator that maps k forms to $3-k$ forms. On a three manifold one has a natural self-adjoint operator $L = *D + D*$ which maps differential forms of even order to forms of even order and forms of odd order to forms of odd order. Let L_- denote its restriction to forms of odd order. With B and ϕ regarded as a one form and a three form, respectively, the boson kinetic operator in (2.6), (2.7) is precisely this operator L_-. The kinetic operator of the ghosts is also a natural geometrical operator, the Laplacian, which we will call Δ. We can now give a formula for the stationary point contributions $\mu(A^{(\alpha)})$ that appear in (2.4). This is

$$\mu(A^{(\alpha)}) = \exp(ikI(A^{(\alpha)})) \cdot \frac{\det(\Delta)}{\sqrt{\det(L_-)}}. \tag{2.8}$$

The phase factor in (2.8) is the value of the integrand in (2.1) at the point of stationary phase, and the determinants (whose absolute values can be defined by zeta functions) result from the Gaussian integral over B, ϕ, c, and $\bar{c}$.

Now we come to the crucial point. To regularize the path integral, we have had to pick a Riemannian metric on M. Therefore, it is not obvious a priori that the $\mu(A^{(\alpha)})$ computed this way will really be topological invariants. Perhaps the Chern-Simons theory suffers from anomalies, and cannot be regularized in a generally covariant fashion. Happily, we can now appeal to [18], where it was shown (in the context of the abelian theory, but this aspect of [18] generalizes) that the absolute value of the ratio of determinants appearing in (2.8) is precisely the Ray-Singer analytic torsion of the flat connection $A^{(\alpha)}$, and so in particular is a topological invariant. (The phase of this ratio of determinants is more delicate, and will be discussed later.) This is the first indication that topological invariants really can be obtained from the Chern-Simons theory.

The Phase of the Determinant

Though the absolute value of the ratio of determinants in (2.8) is the analytic torsion discussed long ago by Schwarz, the phase requires additional study. The ghost determinant $\det \Delta$ is real and positive, so the real issue is to study the phase of $\det L_-$. Because the operator L_- can be interpreted as a twisted Dirac operator, the phase of its determinant can be related to the study of the phase of odd dimensional fermion determinants, as studied by various authors [28]. However, I will here give a brief derivation of the relevant facts from the bosonic point of view, which is perhaps more natural in the present context. After an irrelevant rescaling of B and ϕ, the integral of interest is

$$\int DBD\phi \exp\left(i \int_M \mathrm{Tr}(B \wedge DB + \phi D * B)\right). \tag{2.9}$$

Upon changing variables to an orthonormal basis of eigenfunctions x_i of the operator L_-, with eigenvalues λ_i, (2.9) becomes

$$\prod_i \int_{-\infty}^{\infty} \frac{dx_i}{\sqrt{\pi}} e^{i\lambda_i x_i^2}. \tag{2.10}$$

Therefore the crucial integral to understand is

$$I = \int_{-\infty}^{\infty} \frac{dx}{\sqrt{\pi}} e^{i\lambda x^2}, \tag{2.11}$$

for real λ. We consider this integral to be defined by taking the limit as $\varepsilon \to 0$ of the absolutely convergent integral

$$\int_{-\infty}^{\infty} \frac{dx}{\sqrt{\pi}} e^{i\lambda x^2} \cdot e^{-\varepsilon x^2}. \tag{2.12}$$

With this or any other physically reasonable definition, the integral (2.11) is

$$I = \frac{1}{|\sqrt{\lambda}|} \cdot \exp\left(\frac{i\pi}{4} \operatorname{sign} \lambda\right). \tag{2.13}$$

The phase of the path integral is thus proportional to $\sum_i \operatorname{sign} \lambda_i$, or better, to its regularized version which is the "eta invariant" of Atiyah et al. [29]:

$$\eta(A^{(\alpha)}) = \frac{1}{2} \lim_{s \to 0} \sum_i \operatorname{sign} \lambda_i |\lambda_i|^{-s}. \tag{2.14}$$

Thus, the phase of the path integral may be expressed in the formula

$$\frac{1}{\sqrt{\det L_-}} = \frac{1}{|\sqrt{\det L_-}|} \cdot \exp\left(\frac{i\pi}{2} \eta(A^{(\alpha)})\right). \tag{2.15}$$

This can be made more explicit by using the Atiyah-Patodi-Singer theorem, which for our purposes can be regarded as a formula that expresses the dependence of η on the flat connection $A^{(\alpha)}$ about which we are expanding. In fact, in the case of the operator L_-, the formula is

$$\frac{1}{2}(\eta(A^{(\alpha)}) - \eta(0)) = \frac{c_2(G)}{2\pi} \cdot I(A^{(\alpha)}). \tag{2.16}$$

Here $I(A^{(\alpha)})$ is the Chern-Simons invariant of the flat connection $A^{(\alpha)}$, as defined in (2.5), $\eta(0)$ is the eta invariant of the trivial gauge field $A = 0$, and $c_2(G)$ is the value of the quadratic Casimir operator of the group G in the adjoint representation, normalized so that $c_2(SU(N)) = 2N$. The effect of this factor is to replace k in (2.8) by $k + c_2(G)/2$; in fact, the partition function (2.4) may now be written

$$Z = e^{i\pi\eta(0)/2} \cdot \sum_\alpha e^{i(k + c_2(G)/2) I(A^{(\alpha)})} \cdot T_\alpha \tag{2.17}$$

with T_α [the absolute value of the ratio of determinants in (2.8)] being the torsion invariant of the flat connection $A^{(\alpha)}$.

Unfortunately, although $I(A^{(\alpha)})$ and T_α are topological invariants, $\eta(0)$ is not; it depends on the choice of a metric on M in gauge fixing. Thus, to make sense of the phase of (2.17) requires further discussion, in the next subsection.

Before launching into that technical discussion, let us note that the computation just sketched actually has a very interesting spin-off. The fact that k in (2.8) has been replaced by $k + c_2(G)/2$ in (2.17) appears to be the beginning of an explanation of the fact that in many formulas of $1+1$ dimensional current algebra, quantum corrections have the effect of replacing k by $k + c_2(G)/2$. In turn, this is probably related to the fact that in various integrable models in $1+1$

dimensions, such as the sine-gordon model, the WKB approximation is exact if one makes suitable and seemingly *ad hoc* changes in the values of the parameters, analogous to replacing k by $k+c_2(G)/2$.

Trivialization of the Tangent Bundle

Now, let us discuss how the mysterious phase factor $e^{i\pi\eta(0)/2}$ in (2.17) should be interpreted.

First of all, $\eta(0)$ is the η invariant of the L_- operator coupled to (i) some metric g on M, and (ii) the trivial gauge field $A=0$. Let $d=\dim G$ be the dimension of the gauge group G. Since the gauge field is trivial, the L_- operator consists of d copies of the purely gravitational L_- operator coupled to the metric only. Thus, as a preliminary, we write

$$\eta(0)=d\cdot\eta_{\text{grav}}, \tag{2.18}$$

where η_{grav} is the eta invariant of the purely gravitational operator. Our problematical phase factor is

$$\Lambda=\exp\left(\frac{id\pi}{2}\cdot\eta_{\text{grav}}\right). \tag{2.19}$$

Now, with a particular regularization of the Chern-Simons quantum field theory, we have obtained the formula (2.17) which contains the ambiguous phase factor Λ. The goal is to find a different regularization which will preserve general covariance. Two regularizations should differ by a local counterterm, and in this case, since the problem phase (2.19) depends on the background metric only, we want a counterterm that depends on the background metric only. It is easy to see that the counterterm with the right properties is a multiple of the gravitational Chern-Simons term, which is defined (by analogy with the Yang-Mills Chern-Simons term) as

$$I(g)=\frac{1}{4\pi}\int_M \operatorname{Tr}\left(\omega\wedge d\omega+\tfrac{2}{3}\omega\wedge\omega\wedge\omega\right). \tag{2.20}$$

Here ω is the Levi-Civita connection on the spin bundle of M.[3] $I(g)$ suffers from an ambiguity just similar to that of the Yang-Mills Chern-Simons action. To define $I(g)$ as a number, one requires a trivialization of the tangent bundle of M. Although the tangent bundle of a three manifold can be trivialized, there is no canonical way to do this. Any two trivializations differ by an invariantly defined integer, which is the number of relative "twists". The gravitational Chern-Simons functional has the property that if the trivialization of the tangent bundle of M is twisted by s units, $I(g)$ transforms by

$$I(g)\to I(g)+2\pi s. \tag{2.21}$$

Now, the Atiyah-Patodi-Singer theorem says that the combination

$$\frac{1}{2}\eta_{\text{grav}}+\frac{1}{12}\cdot\frac{I(g)}{2\pi} \tag{2.22}$$

---

[3] (2.20) is not the integral of an intrinsic local functional, so it would not usually arise as a counterterm. Whether or not "counterterm" is the right word, we will have to view (2.20) as a correction that must be added to the action if one wishes to work in the gauge $D_iA^i=0$

is a topological invariant, depending that is on the oriented three manifold M with a choice of trivialization of the tangent bundle, but not on the metric of M.[4] It is clear, therefore, what we must do. We replace $\eta(0)/2$ in (2.17) by d times the combination that appears in (2.22) [the factor of d is the one that entered in (2.19)], so (2.17) is replaced by

$$Z = \exp\left(i\pi d\left(\frac{\eta_{\text{grav}}}{2} + \frac{1}{12}\cdot\frac{I(g)}{2\pi}\right)\right)\cdot \sum_{\alpha} e^{i(k+c_2(G)/2)I(A^{(\alpha)})}\cdot T_\alpha . \qquad (2.23)$$

So, finally, we can see that the Chern-Simons partition function, at least for large k, can be defined as a topological invariant of the oriented, framed three manifold M (a framed three manifold being one that is presented with a homotopy class of trivializations of the tangent bundle).

The fact that it is necessary to specify a framing of the three manifold may look like a nuisance, but there is no real loss of information. From (2.21) we see that if the framing is shifted by s units, the partition function is transformed by

$$Z \to Z\cdot \exp\left(2\pi i s\cdot\frac{d}{24}\right). \qquad (2.24)$$

A topological invariant of framed, oriented three manifolds, together with a law for the behavior under change of framing, is more or less as good as a topological invariant of oreinted three manifolds without a choice of framing.

Of course, all of the discussion in this section, and in particular (2.24), has been limited to the behavior at large k. In Sect. (4.5), we will see that the generalization of (2.24) to finite k is

$$Z \to Z\cdot \exp\left(2\pi i s\cdot\frac{c}{24}\right), \qquad (2.25)$$

with c being the central charge of two dimensional current algebra with symmetry group G at level k. It is well known that the large k limit of c is exactly d.

Moduli Spaces of Flat Connections

There is still an important gap in the above discussion of the large k behavior. The formula (2.8) is really only valid if the determinants that appear are all non-zero. In fact, the flat connection $A^{(\alpha)}$ determines a flat bundle E. The determinants in (2.8) are non-zero if and only if $A^{(\alpha)}$ is such that the de Rham cohomology of M, with values in E, is zero. If $H^1(M,E) \neq 0$, then the flat connection $A^{(\alpha)}$ is not isolated but lies on a moduli space $\mathscr{S}$ of gauge inequivalent flat connections; and the proper evaluation of the path integral (2.1) leads not to the discrete sum (2.4) but to an integral on $\mathscr{S}$. If $H^0(M,E)$ is not zero, then the fields ϕ, c and $\bar{c}$ in the above treatment have zero modes, and the gauge fixing requires more care. It is plausible that by more careful study of the path integral, the large k contribution of arbitrary flat connections can be extracted without assumptions about $H^*(M,E)$. But we will not attempt this.

---

[4] The crucial factor of 1/12 in (2.22) reflects the discrepancy between the Chern character $e^x = 1 + x^2/2 + \ldots$ that appears in gauge theory index theorems and the $\hat{A}$ genus $(x/2)/\sinh(x/2) = 1 - x^2/24 + \ldots$ that appears in gravitational index theorems

Some Examples

We will later on determine the partition functions of some simple three manifolds, giving results that can be compared to large k computations. For $S^2 \times S^1$, $Z=1$, for any G and any k. For S^3 and $G=SU(2)$, we will obtain the formula [5]

$$Z(S^3)=\sqrt{\frac{2}{k+2}}\sin\left(\frac{\pi}{k+2}\right). \tag{2.26}$$

Of course, on S^3 the only flat connection is the trivial connection, for which (2.8) is not valid, since $H^0(M,E)\neq 0$ in this case. For $G=SU(2)$, the behavior $Z\sim k^{-3/2}$ in (2.26) is probably the general behavior of the contribution of the flat connection for homology spheres (on which the flat connection is isolated); it would be interesting to know how to obtain this behavior from path integrals. In Donaldson and Floer theory, the trivial connection, which has a negative formal dimension, is the cause of many subtleties. The vanishing of (2.26) in the classical limit of large k appears to be an interesting quantitative reflection of the "negative dimension" of the trivial connection.

2.1. Incorporation of Knots

We now wish to consider the large k behaviour in the presence of knots. For simplicity, we will limit ourselves to the case of S^3, and an abelian gauge group $G=U(1)$. Though the abelian gauge group is relatively trivial in the context of knot theory, it gives a quick and simple way to confirm the fact that the Chern-Simons action really does lead to topological invariants, and it also gives a simple context for explaining a technicality that is crucial in all that follows.

In the abelian theory, the gauge field is simply a one form A and the Lagrangian is

$$\mathcal{L}=\frac{k}{8\pi}\int_M e^{ijk}A_i\partial_j A_k. \tag{2.27}$$

We pick some circles C_a and some integers n_a [corresponding to representations of the gauge group $U(1)$]. As always in this paper, we assume C_a does not intersect C_b for $a\neq b$. We wish to calculate the expectation value of the product

$$W=\prod_{a=1}^{s}\exp\left(in_a\int_{C_a}A\right) \tag{2.28}$$

with respect to the Gaussian measure determined by $e^{i\mathcal{L}}$. As was recently discussed by Polyakov (in a paper [23] in which he proposed to apply the Abelian Chern-Simons theory to high temperature superconductors), the result can be written in the form

$$\langle W\rangle=\exp\left(\frac{i}{2k}\sum_{a,b}n_a n_b\int_{C_a}dx^i\int_{C_b}dy^j\,\varepsilon_{ijk}\cdot\frac{(x-y)^k}{|x-y|^3}\right). \tag{2.29}$$

Here one has identified a region U of S^3 containing the knots with a region of three dimensional Euclidean space, and x^i, y^j are the Euclidean coordinates of U

[5] The appearance of $k+2$ in this formula is presumably an illustration of the $k+c_2(G)/2$ in (2.17)

evaluated along the knots. For $a \neq b$, the integral in (2.29) is essentially the Gauss linking number, which can be written as

$$\Phi(C_a, C_b) = \frac{1}{4\pi} \int_{C_a} dx^i \int_{C_b} dy^j \varepsilon_{ijk} \frac{(x-y)^k}{|x-y|^3}. \tag{2.30}$$

As long as C_a and C_b do not intersect, $\Phi(C_a, C_b)$ is a well defined integer; in fact, it is the most classic invariant in knot theory. Thus, if we could ignore the term $a = b$, we would have

$$\langle W \rangle = \exp\left(\frac{2i\pi}{k} \sum_{a,b} n_a n_b \Phi(C_a, C_b)\right). \tag{2.31}$$

The appearance of the Gauss linking number illustrates the fact that the Chern-Simons theory does lead to topological invariants as we hope. But we have to worry about the term with $a = b$. This integral is ill-defined near $x = y$; how do we wish to interpret it?

It is well known in knot theory that there is no natural and topologically invariant way to regularize the self-linking number of a knot. Polyakov in [23] used a regularization that is not generally covariant to get an answer that is interesting geometrically but not a topological invariant. We need a different approach for our present treatment in which general covariance is a primary goal. Though there is no completely invariant substitute for Polyakov's regularization, in the sense that there is no way to get a natural topological invariant from the integral in (2.29) or (2.30) with $a = b$, we cannot simply throw away the self-linking term and its non-abelian generalizations (which are sketched in Fig. 3a), since these terms are in fact not naturally zero. There is no reason to think that one could retain general covariance by dropping these terms. In the abelian theory, on a general three manifold M, on topological grounds the self-linking number can be a non-zero fraction, well-defined only modulo one. In such a case, it cannot be correct to set the self-linking number to zero, since it is definitely not zero. [Topologically, in such a situation, the self-linking number is well defined only modulo an integer, and this precision is definitely not good enough to evaluate (2.31).] In the non-abelian theory, we will get results later which amount to assigning definite, non-zero values to the non-abelian generalizations of the self-linking integral, so it would not be on the right track to try to throw these terms away.

Topologically, it is clear what data are needed to make sense of the self-linking of a knot C. One needs to give a "framing" of C; this is a normal vector field along C. The idea is that by displacing C slightly in the direction of this vector field one gets a new knot C', and it makes sense to calculate the linking number of C and C'. This can be defined as the self-linking number of the framed knot C. One can think of the framing as a thickening of the knot into a tiny ribbon bounded by C and C'; this is how it is drawn in Fig. 3b. It is clear that the self-linking number defined this way depends not on the actual vector field used to displace C to C' but only on the topological class of this vector field; and indeed by a "framing" we mean only the topological class. Though a choice of framing gives a definition of the self-linking number of a knot C, it is clear that by picking a convenient framing of C one can

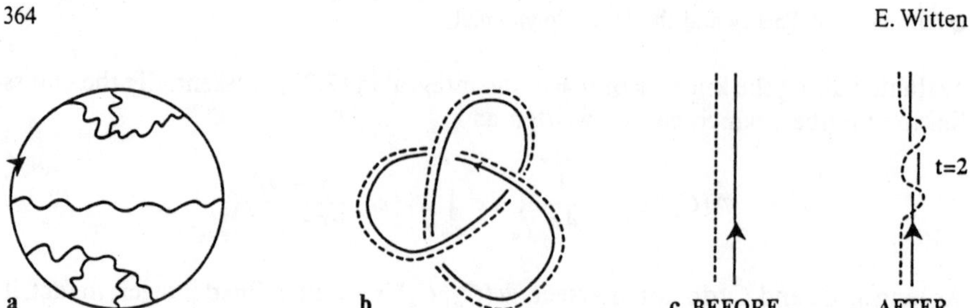

Fig. 3a–c. The self-linking integral is, in a non-abelian theory, the first in an infinite series of Feynman diagrams, with gauge fields emitted and absorbed by the same knot, as in **a**; these all pose similar problems. A topologically invariant but not uniquely determined regularization can be obtained by supposing that each knot is "framed", as in **b**. In **c**, the framing is shifted by 2 units by making a 2-fold twist

get any desired answer for its self-linking number; as illustrated in Fig. 3c, a t-fold twist in the framing of C will change its self-linking by t.[6]

Physically, the role of the framing is that it makes possible what physicists would call a point-splitting regularization. This is defined as follows: when one has to do the self-linking integral in (2.29), one lets x run on C and y on C'. This gives a well-defined integral, though of course it depends on the framing. In this paper, we will assume, without proof, that the framing gives sufficient information to make possible a consistent point-splitting regularization of all the non-abelian generalizations of the self-linking integral, without further arbitrary choices. This question is, perhaps, comparable to the question of whether the non-abelian Chern-Simons action defines a sensible quantum theory in the first place (even without introducing Wilson lines as observables); neither of these questions will be tackled here.

Of course, if it were always possible to pick a canonical framing of knots, then we could pick this framing and hide the question. On S^3, there is a canonical framing of every knot; it is determined by asking that the self-linking number should be zero. (This makes the abelian linking integral zero, but not its non-abelian generalizations.) On general three manifolds, this cannot be done since the self-linking number may be ill-defined or may differ from an integer by a definite fraction (so that it does not vanish with any choice of framing). Even when the canonical framing does exist, it is not convenient to be restricted to using it, since natural operations (like the surgery we study in Sect. 4) may not preserve it.

In general, therefore, we give up on finding a natural choice, and simply pick some framing and proceed. It would be rather unpleasing if the "physical" results depended uncontrollably on the framing of knots. What saves the day is that although we cannot in general make a natural choice of the framing, we can state a general rule for how expectation values of Wilson lines change under a change of the framing. First of all, let us note that while, in general, there is no canonical zero in the set of possible framings of a knot in a three manifold, if one compares two framings they always differ by a definite integer, which is the relative twist in going

[6] The discussion should make it clear that the need to frame knots is analogous to the need to frame three manifolds, as found in the last section. This hopefully justifies the use of the same word "framing" in each case

around the knot (Fig. 3c). (That is, in general there is no natural way to count how many times the ribbon in Fig. 3b is twisted, but there is a natural local operation of adding t extra twists to this ribbon.) In the abelian theory, it is clear from (2.29) and (2.30) how the partition function transforms under a change of framing. If we shift the framing of the link C_a by t units, its self-linking number is increased by t, and the partition function is shifted by a phase

$$\langle W \rangle \to \exp(2\pi it \cdot (n_a^2/k)) \cdot \langle W \rangle . \tag{2.32}$$

The nonabelian analog of that result will be derived in Sect. 5.1; the transformation law in the non-abelian case is

$$\langle W \rangle \to \exp(2\pi it \cdot h) \langle W \rangle , \tag{2.33}$$

where h is the conformal weight of a certain primary field in $1+1$ dimensional current algebra. This result, though it may seem rather technical, is a key ingredient enabling the Chern-Simons theory to work. It means that although we need to pick a framing for every link, because the self-linking integrals have no natural definition otherweise, there is no loss of information since we have a definite law for how the partition functions transform under change of framing.

Actually, it can be shown [13] that the structure of rational conformal field theory requires non-trivial monodromies. In the relationship that we will develop between the $2+1$ dimensional Chern-Simons theory and rational conformal field theory in $1+1$ dimensions, the need to frame all knots is the $2+1$ dimensional analog of the monodromies that arise in $1+1$ dimensions. [This will be clear in the derivation of (2.33).] Were it not for the seeming nuisance that knots must be framed to define the Wilson lines as quantum observables, one would end up proving that the Jones knot invariants were trivial.

An alternative description may make the physical interpretation of the framing of knots more transparent. A Wilson line can be regarded as the space-time trajectory of a charged particle. In $2+1$ dimensions, it is possible for a particle to have fractional statistics, meaning that the quantum wave function changes by a phase $e^{2\pi i\delta}$ under a 2π rotation. (See [30] for a discussion of these issues.) If one wishes to compute a quantum amplitude with propagation of a particle of fractional statistics, it is not enough to specify the orbit of the particle; it is necessary to also count the number of 2π rotations that the particle undergoes in the course of its motion. Equations (2.32) and (2.33) mean that the particles represented by Wilson lines in the Chern-Simons theory have fractional statistics with $\delta = n_a^2/2k$ in the abelian theory or $\delta = h$ in the non-abelian theory. This fractional statistics is the phenomenon claimed by Polyakov in [23], so in essence we agree with his substantive claim, though we prefer to exhibit this phenomenon in the context of a generally covariant regularization, where it appears in the behavior of Wilson lines under change of framing.

In this section, we have obtained some important evidence that the Chern-Simons theory can be regularized to give invariants of three manifolds and knots. We have also obtained the important insight that doing so requires picking a homotopy class of trivializations of the tangent bundle, and a "framing" of all knots. To actually solve the theory requires very different methods, to which we turn in the next section.

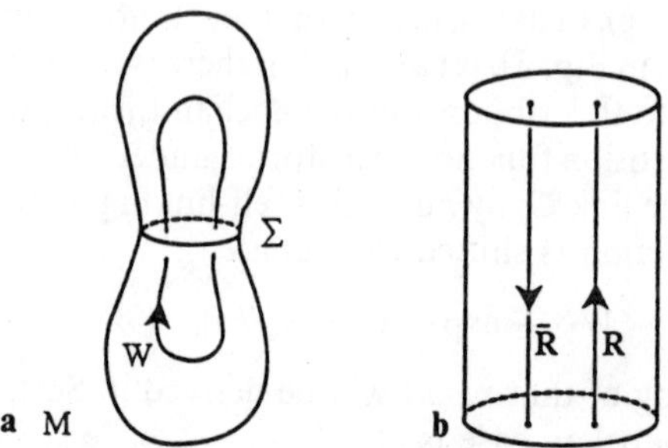

Fig. 4a and b. Cutting a three manifold M on an intermediate Riemann surface Σ is indicated in part **a**. Wilson lines W on M may pierce Σ and if so Σ comes with certain "marked points", with representations attached. Locally, near Σ, M looks like $\Sigma \times R^1$, indicated in part **b**

3. Canonical Quantization

The basic strategy for solving the Yang-Mills theory with Chern-Simons action on an arbitrary three manifold M is to develop a machinery for chopping M in pieces, solving the problem on the pieces, and gluing things back together. So to begin with we consider a three manifold M, perhaps with Wilson lines, as in Fig. 4a. We "cut" M along a Riemann surface Σ. Near the cut, M looks like $\Sigma \times R^1$, and our first step in learning to understand the theory on an arbitrary three manifold is to solve it on $\Sigma \times R^1$.

The special case of a three manifold of the form $\Sigma \times R^1$ is tractable by means of canonical quantization. Canonical quantization on $\Sigma \times R^1$ will produce a Hilbert space $\mathscr{H}_\Sigma$, "the physical Hilbert space of the Chern-Simons theory quantized on Σ".[7] These will turn out to be finite dimensional spaces, and moreover spaces that have already played a noted role in conformal field theory. In rational conformal field theories, one encounters the "conformal blocks" of Belavin, Polyakov, and Zamolodchikov. Segal has described these in terms of "modular functors" that canonically associate a Hilbert space to a Riemann surface, and has described in algebra-geometric terms a particular class of modular functors, which arise in current algebra of a compact group G at level k [16]. The key observation in the present work was really the observation that precisely those functors can be obtained by quantization of a three dimensional quantum field theory, and that this three dimensional aspect of conformal field theory gives the key to understanding the Jones polynomial.

---

[7] It is conventional in physics to call vector spaces obtained in this fashion "Hilbert spaces", and we will follow this terminology. In fact, the claim that comes most naturally from path integrals and that we will actually use is only that $\mathscr{H}_\Sigma$ is a vector space canonically associated with Σ, and exchanged with its dual when the orientation of Σ is reversed. However, a Hilbert space structure is natural in the Hamiltonian viewpoint, and in the particular problem we are considering here, an inner product on $\mathscr{H}_\Sigma$ is important in more delicate aspects of conformal field theory; such an inner product gives a "metric on the flat vector bundle" in the language of Friedan and Shenker [31]. According to Segal [16], $\mathscr{H}_\Sigma$ in fact has a canonical projective Hilbert space structure

Actually, the general situation that must be studied is that in which possible Wilson lines on M are "cut" by Σ, as in the figure. In this case Σ is presented with finitely many marked points $P_1, \ldots P_k$, with a G representation R_i assigned to each P_i (since each Wilson line has an associated representation). To this data – an oriented topological surface with marked points, and for each marked point a representation of G – we wish to associate a vector space. This is also the general situation that arises in conformal field theory – the marked points are points at which operators with non-vacuum quantum numbers have been inserted. If one reverses the orientation of Σ (and replaces the representations R_i associated with the marked points with their complex conjugates) the vector space $\mathscr{H}_\Sigma$ must be replaced with its dual.

The Canonical Formalism. At first sight, (1.3) might look like a typically intractable nonlinear quantum field theory, but this is far from being so. Working on $\Sigma \times R^1$, it is very natural to choose the gauge $A_0 = 0$ (with A_0 being the component of the connection in the R^1 direction). In this gauge we immediately see that the Lagrangian becomes quadratic. It reduces to

$$\mathscr{L} = \frac{k}{8\pi} \int dt \int_\Sigma \varepsilon^{ij} \operatorname{Tr} A_i \frac{d}{dt} A_j . \tag{3.1}$$

For the time being we will ignore extra complications due to Wilson lines that may be present on $\Sigma \times R^1$. From (3.1) we may deduce the Poisson brackets,[8]

$$\{A_i^a(x), A_j^b(y)\} = \frac{4\pi}{k} \cdot \varepsilon_{ij} \delta^{ab} \delta^2(x-y) . \tag{3.2}$$

Before rushing ahead to quantize these commutation relations, we should remember that the system is subject to a "Gauss law" constraint, which is $\delta\mathscr{L}/\delta A_0 = 0$, or (ignoring the Wilson lines)

$$\varepsilon^{ij} F_{ij}^a = 0 . \tag{3.3}$$

This constraint equation is nonlinear (since F contains a quadratic term), and – as (3.1) is certainly a free theory – this nonlinearity is what remains of the underlying nonlinearity of (1.3).

In quantum field theory, one very often quantizes first and then imposes the constraints. The situation that we are considering here is a situation in which it is far more illuminating to first impose the constraints and then quantize. For the phase space $\mathscr{M}_0$ of connections $A_i^a(x)$ without the constraints is an infinite dimensional phase space; imposing the constraints will reduce us to a rather subtle but eminently finite dimensional phase space $\mathscr{M}$. The problem that faces us here, of reducing from $\mathscr{M}_0$ to $\mathscr{M}$ by imposing the constraints (3.2), has been studied before – and has proved to have extremely rich properties – in the work of Atiyah and Bott on equivariant Morse theory, two dimensional Yang-Mills theory, and

[8] This is a typical problem in which it is not appropriate to "introduce canonical momenta". The purpose of introducing such variables is to reexpress a given Lagrangian in a form which is first order in time derivatives, but (3.1) is already first order in time derivatives. The variables in (3.1) are already canonically conjugate, as indicated in the following equation

the moduli space of holomorphic vector bundles [33]. In our present investigation, this familiar problem appears from a novel three dimensional vantage point.

It is necessary to recall the nature of constraint equations in classical physics. The constraints (3.2) are functions that should vanish, but they also generate gauge transformations via Poisson brackets. Imposing the constraints means two things classically: First, we restrict ourselves to values of the canonical variables for which the constraint functions vanish; and second, we identify two solutions of the constraint equations if they differ by a gauge transformation. In the case at hand, the first step means that we should consider only "flat connections", that is, connections for which $F_{ij}^a = 0$. The second step means that we identify two flat connections if they differ by a gauge transformation. Taking the two steps together, we see that the physical phase space, obtained by imposing the constraints (3.2), is none other than the moduli space of flat connections on Σ, modulo gauge transformations. Such flat connections are completely characterized by the "Wilson lines", that is, the holonomies around non-contractible loops on Σ. A simple count of parameters shows that on a Riemann surface of genus $g > 1$, the moduli space $\mathcal{M}$ of flat connections modulo gauge transformations has dimension $(2g-2) \cdot d$, where d is the dimension of the group G.

The topology of $\mathcal{M}$ is rather intricate (and this was in fact the main subject of interest in [33]). On general grounds $\mathcal{M}$ inherits a symplectic structure (that is, a structure of Poisson brackets) from the symplectic structure present on $\mathcal{M}_0$ before imposing the constraints. $\mathcal{M}$ is a compact space (with some singularities), and in particular its volume with the natural symplectic volume element is finite. Since in quantum mechanics there is one quantum state per unit volume in classical phase space, the finiteness of the volume of $\mathcal{M}$ means that the quantum Hilbert spaces will be finite dimensional. We would like to determine them.

3.1. *The Holomorphic Viewpoint*

Quantization of classical mechanics is usually carried out by separating the canonical variables into "coordinates", q^i, which are a maximal set of real commuting variables, and "momenta", p^j, which are conjugate to the q^i. The quantum Hilbert space is then the space $\mathcal{H}$ of square integrable functions of the q^i.

Such a scheme definitely requires a noncompact phase space of infinite volume, since – though the q^i may take values in a compact space – the p^j are definitely unbounded. Accordingly, the space $\mathcal{H}$ is infinite dimensional.

Quantizing a compact, finite volume phase space, such as the moduli space $\mathcal{M}$ of flat connections modulo gauge transformations, is quite a different kind of problem. It has no known general solution, but there is one important class of cases in which there is a natural notion of quantization. This arises in the case in which $\mathcal{M}$ is a Kähler manifold, and the symplectic structure on $\mathcal{M}$ is the curvature form that represents the first Chern class of a holomorphic line bundle L endowed with some metric. In this case, one carries out quantization not by separating the variables in phase space into "coordinates" and "momenta", q's and p's, but by separating them into holomorphic and anti-holomorphic degrees of freedom, essentially $z \sim q + ip$ and $\bar{z} \sim q - ip$. The quantum Hilbert space $\mathcal{H}$ is then a suitable space of holomorphic "functions". More exactly, $\mathcal{H}$ is the space of holomorphic sections of the line bundle L. If $\mathcal{M}$ is compact, this latter space will be finite dimensional. In our problem, with $\mathcal{M}$ being the moduli space of flat connections modulo gauge transformations on an oriented smooth surface Σ, is there a natural Kähler structure on $\mathcal{M}$? The answer is crucial for all that follows. There is not quite a *natural* Kähler structure on $\mathcal{M}$, but there is a natural way to obtain such structures. Once one picks a complex structure J on Σ, the moduli space $\mathcal{M}$ of flat connections can be given a new interpretation – it is the moduli space of stable holomorphic G_C bundles on Σ which are topologically trivial (G_C is the complexification of the gauge group G). Let us refer to the latter space as $\mathcal{M}_J$. $\mathcal{M}_J$ is naturally a complex Kähler (and in fact projective algebraic) variety. Upon picking a linear representation of G (for our purposes it is convenient to pick a representation with the smallest value of the quadratic Casimir operator, e.g. the N dimensional representation of $SU(N)$ or the adjoint representation of E_8), and passing from a principal G_C bundle to the associated vector bundle, we can think of $\mathcal{M}_J$ as the moduli space of a certain family of holomorphic vector bundles. For $G = SU(N)$, $\mathcal{M}_J$ is simply the moduli space of all stable rank N holomorphic vector bundles of vanishing first Chern class.

The symplectic form on $\mathcal{M}$ that appears in (3.1) or (3.2) *without* picking a complex structure on Σ has a very special interpretation in holomorphic terms once we *do* pick such a complex structure. Let us recall the notion [34] of the determinant line bundle of the $\bar{\partial}$ operator. The $\bar{\partial}$ operator on Σ can be "twisted" by any holomorphic vector bundle. $\mathcal{M}_J$ parametrizes a family of holomorphic vector bundles on Σ, and thus it can be regarded as parametrizing a family of $\bar{\partial}$ operators. Taking the determinant line gives a line bundle L over the base space $\mathcal{M}_J$ of this family. Furthermore [34], the Dirac determinant gives a natural metric on L, and the first Chern class of L, computed with this metric, is precisely the symplectic form that appears in (3.1) or (3.2), provided $k = 1$. For general k, the symplectic form that appears in (3.1) or (3.2) represents the first Chern class of the kth power of the determinant line bundle.[9]

Thus, all of the conditions are met for a straightforward quantization of (3.1), taking into account the constraints (3.3). The constraints mean that the classical space to be quantized is the moduli space $\mathcal{M}$ of flat connections. Picking an arbitrary complex structure J on Σ, $\mathcal{M}$ becomes a complex manifold, and the symplectic form of interest represents the first Chern class of $L^{\otimes k}$, the kth tensor power of the determinant line bundle. The quantum Hilbert space $\mathcal{H}_\Sigma$ is thus the space of global holomorphic sections of $L^{\otimes k}$.

3.2. A Flat Vector Bundle on Moduli Space

This gives an answer to the problem of canonically quantizing the Chern-Simons theory on $\Sigma \times R^1$, but a crucial point now requires discussion.

[9] This description is valid for the gauge group $G = SU(N)$, but in general the following modification is needed. For groups other than $SU(N)$ the determinant line bundle L is not the fundamental line bundle on $\mathcal{M}$ but a tensor power thereof. For instance, for $G = E_8$, there is a line bundle L' with $(L')^{\otimes 30} \simeq L$. It is then L' whose first Chern class corresponds to (3.1) or (3.2) with $k = 1$

Quantizing (3.1), with the constraints (3.3), is a problem that can be naturally asked whenever one is given an oriented smooth surface Σ. Beginning with a generally covariant Lagrangian in three dimensions, we were led to this problem in a context in which it was not natural to assume any metric or complex structure on Σ. However, to solve the problem and construct $\mathscr{H}_\Sigma$, it was very natural to pick a complex structure J on Σ. Thus, our description of $\mathscr{H}_\Sigma$ depends on the choice of J, and what we have called $\mathscr{H}_\Sigma$ might perhaps be better called $\mathscr{H}_\Sigma^{(J)}$. As J varies, the $\mathscr{H}_\Sigma^{(J)}$ vary holomorphically with J, and thus we could interpret this object as a holomorphic vector bundle on the moduli space of complex Riemann surfaces. But since $\mathscr{H}_\Sigma^{(J)}$ is the answer to a question that depends on Σ and not on J, we would like to believe that likewise the $\mathscr{H}_\Sigma^{(J)}$ canonically depend only on Σ and not on J. The assertion that the $\mathscr{H}_\Sigma^{(J)}$ are canonically independent of J, and depend only on Σ, is the assertion that the vector bundle on moduli space given by the $\mathscr{H}_\Sigma^{(J)}$ has a canonical flat connection that permits one to identify the fibers. Such "flat vector bundles on moduli space" first entered in conformal field theory somewhat implicitly in the differential equations of Belavin, Polyakov, and Zamolodchikov [32]. They were discussed much more explicitly by Friedan and Shenker [31], who proposed that they would play a pivotal role in conformal field theory, and they have been prominent in subsequent work such as [12, 13]. At least in one important class of examples, we have just met a natural origin of "flat vector bundles on moduli space". The problem "quantize the Chern-Simons action" can be posed without picking a complex structure, so the answer is naturally independent of complex structure and thus gives a "flat bundle on moduli space". The particular flat bundles on moduli space that we get this way are those that Segal has described [16] in connection with conformal field theory; Segal also rigorously proved the flatness, which is explained somewhat heuristically by the physical argument sketched above. (Because of the conformal anomaly, this bundle has only a projectively flat connection, with the projective factor being canonically odd under reversal of orientation.)

The role of these flat bundles in conformal field theory is as follows. If one considers current algebra on a Riemann surface, with a symmetry group G, at "level" k, then one finds that in genus zero the Ward identities uniquely determine the correlation functions for descendants of the identity operator, but this is not so in genus $\geqq 1$. On a complex Riemann surface Σ of genus $\geqq 1$, the space of solutions of the Ward identities for descendants of the identity is a vector space $\hat{\mathscr{H}}_\Sigma$, which might be called the "space of conformal blocks". Segal calls the association $\Sigma \to \hat{\mathscr{H}}_\Sigma$ a "modular functor", and has given an algebra-geometric description of the modular functors that arise in current algebra. In quantizing the Chern-Simons theory we have exactly reproduced this description! This is then the secret of the relation between current algebra in $1+1$ dimensions and Yang-Mills theory in $2+1$ dimensions: the space of conformal blocks in $1+1$ dimensions are the quantum Hilbert spaces obtained by quantizing a $2+1$ dimensional theory. It would take us to far afield to explain here the algebra-geometric description of the space of conformal blocks. Suffice it to say that when one tries to use the Ward identities of current algebra to uniquely determine the correlation functions of descendants of the identity on a curve Σ of genus $\geqq 1$, one meets an obstruction which involves the existence of non-trivial holomorphic vector bundles on Σ; the

Ward identities reduce the determination of the correlation functions to the choice of a holomorphic section of $L^{\otimes k}$ over the moduli space of bundles.

It seems appropriate to conclude this discussion with some remarks on the formal properties of the association $\Sigma \to \mathscr{H}_\Sigma$. It is good to first think of the functor $\Sigma \to H^1(\Sigma, R)$ which to a Riemann surface Σ associates its first de Rham cohomology group. This functor is defined for every smooth surface Σ, independent of complex structure. A diffeomorphism of Σ induces a linear transformation on $H^1(\Sigma, R)$, so $H^1(\Sigma, R)$ furnishes in a natural way a representation of the mapping class group. The formal properties of the functors $\Sigma \to \mathscr{H}_\Sigma$ that come by quantizing the Chern-Simons theory are quite analogous. Though a complex structure J on Σ is introduced to construct $\mathscr{H}_\Sigma$, the existence of a natural projectively flat connection on the moduli space of complex structures permits one locally to (projectively) identify the various $\mathscr{H}_\Sigma^{(J)}$ and forget about the complex structure. One might think that the global monodromies of the flat connection on moduli space would mean that globally one could not forget the complex structure, but this is not so; these monodromies just correspond to an action of the purely topological mapping class group, so that the formal properties of $\mathscr{H}_\Sigma$ are just like those of $H^1(\Sigma, R)$.

3.3. *Inclusion of Wilson Lines*

So far, we have discussed the quantization of the Chern-Simons theory on a Riemann surface Σ *without* Wilson lines. Now we wish to include the Wilson lines, which, as in Fig. 4b, pierce Σ in some points P_i; associated with each such point is a representation R_i. Quantizing the Chern-Simons theory in the presence of the Wilson lines should give a Hilbert space $\mathscr{H}_{\Sigma; P_i, R_i}$ that is canonically associated with the oriented surface Σ together with the choice of P_i and R_i.

It is pretty clear what problem in conformal field theory this should correspond to. Instead of simply considering correlation functions of the descendants of the identity, we should consider in the conformal field theory primary fields transforming in the R_i representations of G. With these fields (or their descendants) inserted at points P_i on Σ, one gets in conformal field theory a more elaborate space $\hat{\mathscr{H}}_{\Sigma; P_i, R_i}$ of conformal blocks. Again, there is an algebra-geometric description of this space [16], and this is what we should expect to recover by quantizing the Chern-Simons theory in the presence of the Wilson lines.

I will now briefly sketch how this works out, deferring a fuller treatment for another occasion. First of all, the Wilson lines correspond to static non-abelian charges which show up as extra terms in the constraint equations. So (3.3) is replaced by

$$\frac{k}{8\pi}\,\varepsilon^{ij} F_{ij}^a(x) = \sum_{s=1}^{r} \delta^2(x - P_s)\, T_{(s)}^a, \tag{3.4}$$

where P_s, $s = 1 \dots r$ are the points at which static external charges have been placed, and $T_{(s)}^a$, $a = 1 \dots \dim G$ are the group generators associated with the external charges. Now, a naive attempt to quantize (3.1) with the generalized constraints (3.4) would run into extremely unpleasant difficulties. One could try to quantize first and then impose the constraints, but this is difficult to see through even in the absence of the external charges. Alternatively, one can try to impose the constraints at the classical level and then quantize, as we did above. But it is hard to make sense of (3.4) as constraints in the classical theory; the solution A_i^a of (3.4) cannot be an ordinary c-number connection, since non-commuting operators appear on the right-hand side. It is clear that to solve (3.4), A_i^a would have to be some sort of "q-number connection", whose holonomy would presumably be an element of a "quantum group", not an ordinary classical group. Indeed, it seems likely that the theory of quantum groups [35] can be considered to arise in this way.

However, there is a much better way to quantize the Chern-Simons theory with static charges. We certainly wish to impose (3.4) at the classical level. This cannot be done directly, since on the right-hand side there appear quantum operators. A useful point of view is the following. A representation R_i of a group G should be seen as a quantum object. This representation should be obtained by quantizing a classical theory. The Borel-Weil-Bott theorem gives a canonical way to exhibit for every irreducible representation R of a compact group G a problem in classical physics, with G symmetry, such that the quantization of this classical problem gives back R as the quantum Hilbert space. One introduces the "flag manifold" G/T, with T being a maximal torus in G, and for each representation R one introduces a symplectic structure ω_R on G/T, such that the quantization of the classical phase space G/T, with the symplectic structure ω_R, gives back the representation R. Many aspects of representation theory find natural explanations by thus regarding representations of groups as quantum objects that are obtained by quantization of classical phase spaces.

In the problem at hand, this point of view can be used to good effect. We extend the phase space $\mathcal{M}_0$ of G connections on Σ by including at each marked point P_i a copy of G/T, with the symplectic structure appropriate to the R_i representation. The quantum operators $T_{(i)}^a$ that appear on the right of (3.4) can then be replaced by the classical functions on G/T whose quantization would give back the $T_{(i)}^a$. The constraints (3.4) then make sense as classical equations, and the analysis can be carried out just as we did without marked points, though the details are a bit longer. Suffice it to say that after imposing the classical constraints, one gets a finite dimensional phase space $\mathcal{M}_{P_i,R_i}$ that incorporates the static charges; a point on this space is a flat G connection on Σ with a reduction of structure group to T at the points P_i. Upon picking an arbitrary conformal structure on Σ, this phase space can be quantized. In this way one gets exactly Segal's description of the space of conformal blocks in current algebra in a general situation with primary fields in the R_i representation inserted at the points P_i. (In current algebra at level k, one only permits certain representations, the "integrable ones". If one formally tries to include other representations, the Ward identities show that they decouple [36]. According to Segal, the analogous statement in algebraic geometry is that the appropriate line bundle over $\mathcal{M}_{P_i,R_i}$ has no non-zero holomorphic sections unless the R_i all correspond to integrable representations. For the Chern-Simons theory, this means that unless the representations R_i are all integrable, the zero vector is the only vector in the physical Hilbert space.)

Finally, let us note that the Borel-Weil-Bott theorem should not be used simply as a tool in quantization. It should be built into the three dimensional description.

One should use the theorem to replace the Wilson lines (1.5) that appear in (1.6) with a functional integral over maps of the circle S into G/T (or actually an integral over sections of a G/T bundle, twisted by the restriction to S of the G-bundle E). This gives a much more unified formalism.

3.4. The Riemann Sphere with Marked Points

The above description may seem a little bit dense, and we will supplement it by giving a simple intuitive description of the physical Hilbert space $\mathscr{H}_{\Sigma;R_i,P_i}$ in the important case of genus zero. Let Σ be an oriented surface of genus zero, with static charges in the R_i representation at points P_i. Let us consider the case of very large k. Now, the gauge coupling in (1.3) is of order $1/k$, so for large k we are dealing with very weak coupling. Rather naively, one might believe that for extremely weak coupling the physical Hilbert space is the same as it would be if the charges were not coupled to gauge fields. If so, the physical Hilbert space would be simply the tensor product $\mathscr{H}_0 = \otimes_i R_i$ of the Hilbert spaces R_i of the individual charges. However, there is a key error here. No matter how weak the gauge coupling may be, we must remember that in a closed universe the total charge must be zero (since the electric flux has nowhere to go). The total charge being zero means in a nonabelian theory that all of the charges together must be coupled to the trivial representation of G. So the physical Hilbert space, for large k, is precisely the G-invariant subspace of $\mathscr{H}_0$, or

$$\mathscr{H} = \mathrm{Inv}\,(\otimes_i R_i)\,. \tag{3.5}$$

This is a familiar answer in conformal field theory for the space of conformal blocks obtained, in the large k limit, in coupling representations R_i. Considerations of conformal field theory also show that for finite k the correct answer is always a subspace of (3.5). The most important modification of (3.5) that arises for finite k (and is explained algebra-geometrically in [16]) is that $\mathscr{H}$ is zero unless the R_i correspond to integrable representations of the loop group; in what follows a restriction to such representations is always understood.

Now we consider some important special cases.

(i) For the Riemann sphere with no marked points, the Hilbert space is one dimensional. This is well known in conformal field theory – for descendants of the identity on the Riemann sphere, there is only one conformal block.

(ii) For the Riemann sphere with one marked point in a representation R_i, the Hilbert space is one dimensional if R_i is trivial, and zero dimensional otherwise.

(iii) For the Riemann sphere with two marked points with representations R_i and R_j, the Hilbert space is one dimensional if R_j is the dual of R_i (so that there is an invariant in $R_i \otimes R_j$) and zero dimensional otherwise. Again, this is well known in conformal field theory.

(iv) For the Riemann sphere with three marked points in representations R_i, R_j, and R_k, the dimension of $\mathscr{H}$ is the number N_{ijk} for which Verlinde has proposed [12] and Moore and Seiberg have proved [13] rather striking properties. Here, N_{ijk} may in general be less than its large k limit which is the dimension of (3.5).

(v) From the results of Verlinde, the dimensions of the physical Hilbert spaces for an arbitrary collection of marked points on S^2 can be determined from a knowledge of the N_{ijk}. But let us consider a particularly important special case.

Suppose that there are four external charges, and that the representations are R, R, $\bar{R}$, and $\bar{R}$. If the decomposition of $R \otimes R$ is

$$R \otimes R = \bigoplus_{i=1}^{s} E_i , \tag{3.6}$$

with the E_i being distinct irreducible representations of G, then the physical Hilbert space $\mathscr{H}$ at large k will be s dimensional, since the possible invariants in $R \otimes R \otimes \bar{R} \otimes \bar{R}$ are uniquely fixed by giving the representation to which $R \otimes R$ is coupled. (For small k the dimension of $\mathscr{H}$ might be less than s.) In understanding the knot polynomials, an important special case is that in which G is $SU(N)$ and R is the defining N dimensional representation. In that case, $s = 2$ and the physical Hilbert space is two dimensional (except for $k = 1$ where it is one dimensional).

4. Calculability

Our considerations so far may have seemed somewhat abstract, and we would now like to show that in fact these considerations can actually be used to calculate things. As an introduction to the requisite ideas, we will first deduce a certain theoretical principle that is of great importance in its own right.

Consider, as in Fig. 5a, a three manifold M which is the connected sum of two three manifolds M_1 and M_2, joined along a two sphere S^2. There may be knots in M_1 or M_2, but if so they do not pass through the joining two sphere. If for every three manifold X we denote the partition function or Feynman path integral (1.6) as $Z(X)$, then we wish to deduce the formula

$$Z(M) \cdot Z(S^3) = Z(M_1) \cdot Z(M_2) \tag{4.1}$$

[it being understood that $Z(S^3)$ denotes the partition function of a three sphere that contains no knots]. This can be rewritten

$$\frac{Z(M)}{Z(S^3)} = \frac{Z(M_1)}{Z(S^3)} \cdot \frac{Z(M_2)}{Z(S^3)} . \tag{4.2}$$

In some special cases, (4.2) is equivalent or closely related to known formulas. If M_1 and M_2 are copies of S^3 with knots in them, then the ratios appearing in (4.2) turn out to be the knot invariants that appear in the Jones theory, and (4.2) expresses the fact that these invariants are multiplicative when one takes the disjoint sum of knots. If M_1 and M_2 are arbitrary three manifolds without knots, then (in view of our discussion in Sect. 2) (4.2) is closesly related to the multiplicativity of Reidemeister and Ray-Singer torsion under connected sums.

So let us study Fig. 5a using the general ideas of quantum field theory. On the left of this figure, we see a three manifold M_1 with boundary S^2. According to the general ideas of quantum field theory, one associates a "physical Hilbert space" $\mathscr{H}$ with this S^2; as we have seen in the last section, it is one dimensional. The Feynman path integral on M_1 determines a vector χ in $\mathscr{H}$. Likewise, on the right of

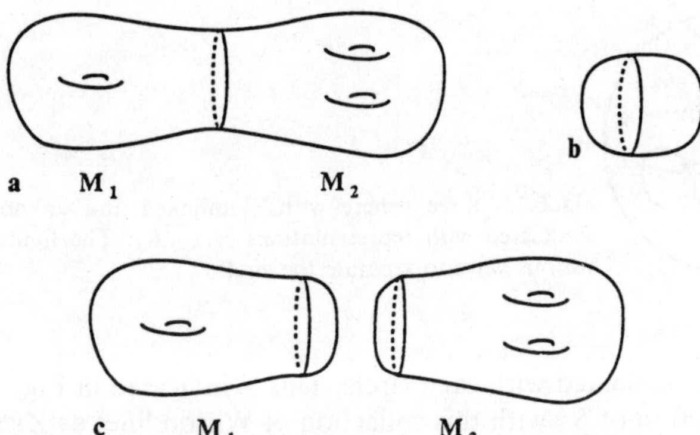

Fig. 5a–c. In **a** is sketched a three manifold M which is the connected sum of two pieces M_1 and M_2, joined along a sphere S^2. Similarly, a three sphere S^3 can be cut along its equator, as in **b**. Cutting both M and S^3 as indicated in **a** and **b**, the pieces can be rearranged into the *disconnected* sum of M_1 and M_2, as in **c**

Fig. 5a we see a three manifold M_2 whose boundary is the same S^2 with opposite orientation; its Hilbert space $\mathscr{H}'$ is canonically the dual of $\mathscr{H}$. The path integral on M_2 determines a vector ψ in $\mathscr{H}'$, and according to the general ideas of quantum field theory, the partition function of the connected sum M is

$$Z(M) = (\chi, \psi). \tag{4.3}$$

The symbol (χ, ψ) denotes the natural pairing of vectors $\chi \in \mathscr{H}$, $\psi \in \mathscr{H}'$. We cannot evaluate (4.3), since we do not know χ or ψ. Instead, let us consider some variations on this theme. The two sphere S^2 that separates the two parts of Fig. 5a could be embedded in S^3 in such a way as to separate S^3 into two three balls B_L and B_R. The path integrals on B_L and B_R would give vectors v and v' in $\mathscr{H}$ and $\mathscr{H}'$, and the same reasoning as led to (4.3) gives

$$Z(S^3) = (v, v'). \tag{4.4}$$

Again, we do not know v or v' and cannot evaluate (4.4). But we can say the following. As $\mathscr{H}$ is one dimensional, v is a multiple of χ; likewise, since $\mathscr{H}'$ is one dimensional, v' is a multiple of ψ. It is then a fact of one dimensional linear algebra that

$$(\chi, \psi) \cdot (v, v') = (\chi, v') \cdot (v, \psi). \tag{4.5}$$

The two terms on the right-hand side of (4.5) are respectively $Z(M_1)$ and $Z(M_2)$, as we see in Fig. 5c. So (4.5) is equivalent to the desired result (4.1).

One may wonder what is the mysterious object $Z(S^3)$ that is so prominent in (4.1). Can it be set to one? Actually, the axioms of quantum field theory are strong enough so that the value of $Z(S^3)$ is uniquely determined and cannot be postulated arbitrarily; as we will see later it can be calculated from the theory of affine Lie algebras. For $G = SU(2)$ the formula has been given in (2.26).

As a special case of (4.1), pick s irreducible representations of G, say $R_1, \ldots R_s$, and consider a link in S^3 that consists of s unlinked and unknotted circles C_i, with

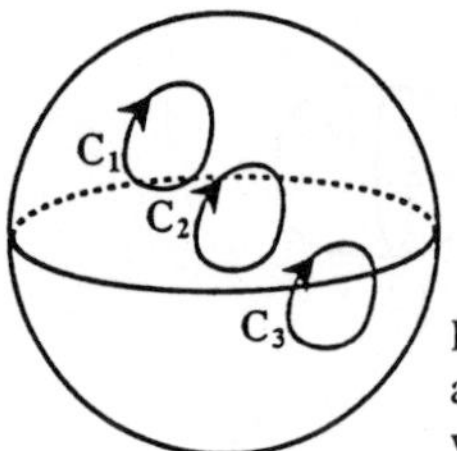

Fig. 6. A three sphere with 3 unlinked and unknotted circles C_i, associated with representations $R_1 \ldots R_3$. The figure can be cut in various ways to separate the circles

one of the R_i associated with each circle. This is indicated in Fig. 6. Denote the partition function of S^3 with this collection of Wilson lines as $Z(S^3; C_1, \ldots C_s)$ (the representations R_i being understood). Then by cutting the figure to separate the circles, and repeatedly using (4.1), we learn that

$$\frac{Z(S^3; C_1, \ldots C_s)}{Z(S^3)} = \prod_{k=1}^{s} \frac{Z(S^3; C_k)}{Z(S^3)}. \tag{4.6}$$

If we introduce the normalized expectation value of a link L, defined by $\langle L \rangle = Z(S^3; L)/Z(S^3)$, then (4.6) becomes

$$\langle C_1 \ldots C_s \rangle = \prod_k \langle C_k \rangle \tag{4.7}$$

for an arbitrary collection of unlinked, unknotted Wilson lines on S^3.

In knot theory there is another notion of connected sum, the "connected sum of links". The Jones invariants also have a simple multiplicative behavior under this operation, as we will sketch briefly at the end of Sect. 4.5.

4.1. *Knots in* S^3

We will now describe the origin of the "skein relation" which can be taken as the definition of the knot polynomials for knots on S^3. (A special case of the skein relation was first used by Conway in connection with the Alexander polynomial.)

Consider a link L on a general three manifold M, as indicated in Fig. 7a. The components of the link are associated with certain representations of G, and we wish to calculate the Feynman path integral (1.6), which we will denote as $Z(L)$ (with the representations understood). We will evaluate it by deducing an algorithm for unknotting knots. If the lines in Fig. 7 could pass through each other unimpeded, all knots could be unknotted. As it is, this is prevented by some unfortuitous crossings, such as the one circled in the figure. Let us draw a small sphere about this crossing, cut it out, and study it more closely. This cuts M into two pieces, which after rearrangement are shown in Fig. 7b as a complicated piece M_L shown on the left of the figure and a simple piece M_R shown on the right. M_R consists of a three ball with boundary S^2; on this boundary there are four marked points that are connected by two lines in the interior of the ball.

To make the discussion concrete, let us suppose that the gauge group is $G = SU(N)$ and that the Wilson lines are all in the defining N dimensional representation of $SU(N)$, which we will call R. Then, as we saw at the end of the

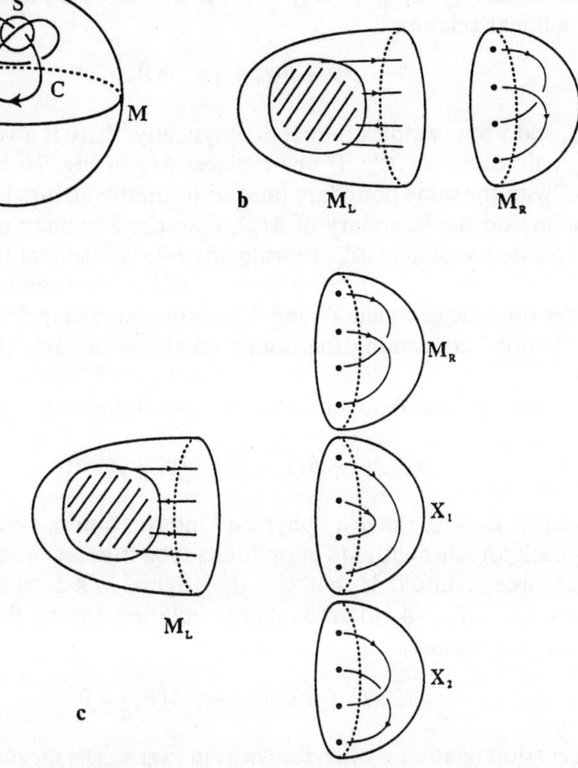

Fig. 7a–c. A link C on a general three manifold M is sketched in **a**. A small sphere S has been drawn about an inconvenient crossing; it cuts M into a simple piece (the interior of S) and a complicated piece. In **b**, the picture is rearranged to exhibit the cutting of M more explicitly; the two pieces now appear on the left and right as M_L (the complicated piece whose details are not drawn) and M_R (the interior of S). The key to the skein relation is to consider replacing M_R with some substitutes, as shown in **c**

last section, the physical Hilbert spaces $\mathscr{H}_L$ and $\mathscr{H}_R$ associated with the boundaries of M_L and M_R are two dimensional.

The strategy is now the same as the strategy which led to the multiplicativity relation (4.1). The Feynman path integral on M_L determines a vector χ in $\mathscr{H}_L$. The Feynman path integral on M_R determines a vector ψ in $\mathscr{H}_R$. The vector spaces $\mathscr{H}_L$ and $\mathscr{H}_R$ (which are associated with the same Riemann surface S^2 with opposite orientation) are canonically dual, and the partition function or Feynman path integral $Z(L)$ is equal to the natural pairing

$$Z(L) = (\chi, \psi). \tag{4.8}$$

We cannot evaluate (4.8), since we know neither χ nor ψ. The one thing that we do know, at present, is that (for the groups and representations we are considering) this pairing is occurring in a two dimensional vector space. A two dimensional vector space has the marvelous property that any three vectors obey a relation of linear dependence. Thus, given any two other vectors ψ_1 and ψ_2 in $\mathcal{H}_R$, there would be a linear relation

$$\alpha\psi + \beta\psi_1 + \gamma\psi_2 = 0, \tag{4.9}$$

where α, β, and γ are complex numbers. Physically, there is a very natural way to get additional vectors in $\mathcal{H}_R$. If one replaces M_R in Fig. 7b by any other three manifold X with the same boundary (and with suitable strings in X connecting the marked points on the boundary of M_R), then the Feynman path integral on X gives rise to a new vector in $\mathcal{H}_R$. Picking any two convenient three manifolds X_1 and X_2 for this computation gives vectors ψ_1 and ψ_2 that can be used in (4.9). We will consider the case in which X_1 and X_2 are the same manifolds as M_R but with different "braids" connecting the points on the boundary; this is indicated in Fig. 7c.

Once ψ_1 and ψ_2 are obtained in this way, (4.9) has the obvious consequence that

$$\alpha(\chi, \psi) + \beta(\chi, \psi_1) + \gamma(\chi, \psi_2) = 0. \tag{4.10}$$

The three terms in (4.10) have a "physical" interpretation, evident in Fig. 7c. By gluing M_L back together with M_R or one of its substitutes X_1 and X_2, one gets back the original three manifold M, but with the original link L replaced by some new links L_1 and L_2. Thus, (4.10) amounts to a relation among the link expectation values of interest, namely

$$\alpha Z(L) + \beta Z(L_1) + \gamma Z(L_2) = 0. \tag{4.11}$$

This recursion relation is often drawn as in Fig. 8. The meaning of this figure is as follows. If one considers three links whose plane projections are identical outside a disc, and look inside this disc like the three drawings in the figure, then the expectation values of those links, weighted with coefficients α, β, and γ, add to zero.

It is well known in knot theory that (4.11) uniquely determines the expectation values of all knots in S^3. For convenience we include a brief explanation of this. One starts with a plane projection of a knot, indicated in Fig. 9. The number p of crossings is finite. Inductively, suppose that all knot expectation values for knots with at most $p-1$ crossings have already been computed. One wishes to study knots with p crossings. If one had $\beta = 0$ in (4.11), one could at each crossing pass the two strands through each other with a factor of $-\gamma/\alpha$ in replacing an over-crossing by an under-crossing. If this were possible, the lines would be effectively transparent, and one could untie all knots. As it is, $\beta \neq 0$, but the term proportional to β reduces the number of crossings, giving rise to a new link whose expectation value is already known by the induction hypothesis.

This process reduces the discussion to the case $p = 0$ where there are no crossings, and therefore we are dealing only with a certain number of unlinked and unknotted circles. For practice with (4.11), let us discuss this case explicitly. In Fig. 10, we sketch a useful special case of Fig. 8. The first and third links in links in Fig. 10 consist of a single unknotted circle, and the second consists of two unlinked and unknotted circles. If we denote the partition function for s unlinked

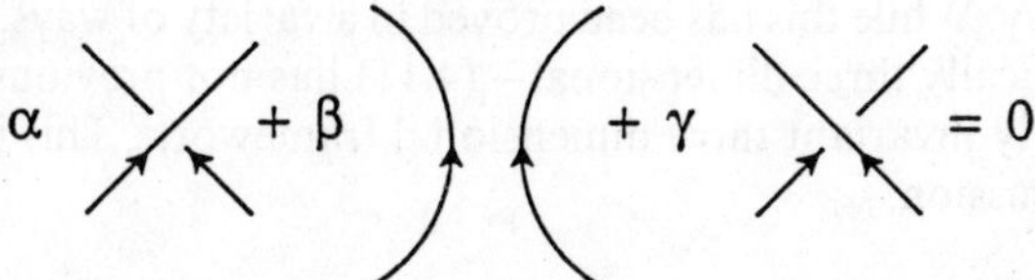

Fig. 8. A recursion relation for links

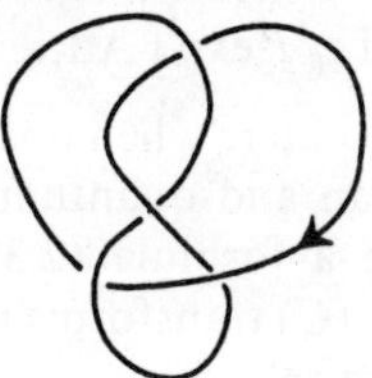

Fig. 9. A plane projection of a knot, with four crossings

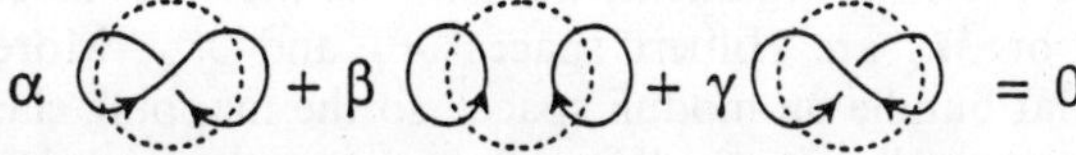

Fig. 10. A special case of the use of Fig. 8. The idea is that the three pictures are identical outside of the dotted lines, and look like figure (4.11) inside them

and unknotted circles in the N dimensional representation of $SU(N)$ as $Z(S^3; Cs)$, then (4.11) amounts in this case to the assertion that

$$(\alpha+\gamma)\, Z(S^3;\, C)+\beta Z(S^3; C^2)=0\,. \tag{4.12}$$

Together with (4.7), [10] this implies that the expectation value of an unknotted Wilson line in the N dimensional representation of $SU(N)$ is

$$\langle C\rangle = -\frac{\alpha+\gamma}{\beta}. \tag{4.13}$$

Presently we will make this formula completely explicit by computing α, β, and γ in terms of the fundamental quantum field theory parameters N and k.

The induction sketched above expresses any knot expectation value as a rational function of α, β and γ (a ratio of polynomials), after finitely many steps. It is in this sense that the Jones knot invariants and their generalizations are "polynomials". While it is, as we have seen, comparatively elementary to prove that (4.11) uniquely determines the knot invariants, the converse is far less obvious. Equations (4.11) can be used in many different ways to obtain the expectation value of a given link, and one must show that one does not run into

[10] This is the only point at which (4.7) has to be used. The induction sketched in the previous paragraph reduces all computations for knots in S^3 to this special case without using (4.7)

any inconsistency. While this has been proved in a variety of ways, the proofs have not been intrinsically three dimensional – (4.11) has not previously been derived from a manifestly invariant three dimensional framework. This is the novelty of the present discussion.

Change of Framing. We want to compute α, β, and γ, but as a prelude we must discuss a certain technical point. At the end of Sect. 2, we learned that choosing a circle C and a representation R is not enough to give a well defined quantum holonomy operator $W_R(C) = \mathrm{Tr}_R P \exp \int_C A dx$. It is also necessary to pick a "framing" of the circle C, which enters when one has to calculate the self-linking number of C and its non-abelian and quantum generalizations. At the end of Sect. 2, we promised to derive a formula (2.33) showing how any partition function with an insertion of $W_R(C)$ transforms under a change of framing. Now it is time to deliver on this promise.

As in Fig. 7b, let us cut the three manifold M on a Riemann surface Σ that intersects C in a point P (and perhaps in some other points that will not be material). In our previous argument, we used the fact that associated with the boundaries M_L or M_R are Hilbert spaces $\mathscr{H}_L$ and $\mathscr{H}_R$. Moreover, $\mathscr{H}_R$ (for example) is "a flat bundle on moduli space" so the mapping class group of the boundary Σ acts naturally on $\mathscr{H}_R$. We wish to act on the boundary of M_R with a very particular diffeomorphism before gluing the pieces of Fig. 7b back together again. The diffeomorphism that we want to pick is a t-fold "Dehn twist" about the point P on Σ. Making this diffeomorphism and then gluing the pieces of Fig. 7b back together again, one gets an identical looking picture, but the framing of the circle C has been shifted by t units. On the other hand, one knows in conformal field theory how the Dehn twist acts on $\mathscr{H}_R$. Associated with the representation R is a number h_R, the "conformal weight of the primary field in the R representation". The t-fold Dehn twist acts on $\mathscr{H}_R$ as multiplication by $e^{2\pi i t h_R}$. So we have obtained (2.33) with $h = h_R$.

Explicit Evaluation. We will now determine the parameters α, β, and γ that appear in the crucial equation (4.10). We need to determine the explicit relation among the three vectors ψ, ψ_1 and ψ_2 that appear in Fig. 7c. This requires a further study of the two dimensional Hilbert space which arises as the space of conformal blocks for the R, R, $\bar{R}$, $\bar{R}$ four point function on S^2 [R being in this case the defining N dimensional representation of $SU(N)$ and $\bar{R}$ its dual]. The three configurations in Fig. 7c can be regarded as differing from each other by a certain diffeomorphism of S^2; the diffeomorphism in question is the "half-monodromy" under which the two copies of R change places by taking a half-step around one another, as indicated in Fig. 11. Moore and Seiberg call this operation B and study it extensively. The states ψ_1 and ψ_2 are none other than

$$\psi_1 = B\psi, \qquad \psi_2 = B^2\psi. \tag{4.14}$$

The matrix B, since it acts in a two dimensional space, obeys a characteristic equation

$$B^2 - yB + z = 0, \tag{4.15}$$

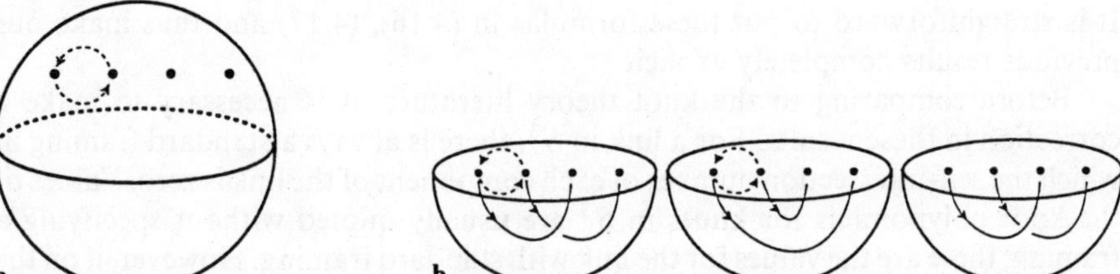

Fig. 11a and b. The half-monodromy operation exchanging two equivalent points on S^2 is sketched in **a**; the arrows are meant to suggest a process in which the first two points change places by executing a half-twist about one another. The idea in **b** is that if the two points on the left in the first picture undergo a half-twist about one another, the first picture becomes the second, and if this is done again, the second picture becomes the third. In this way the three pictures on the right of Fig. 7c differ by a succession of half-monodromies

where

$$y = \operatorname{Tr} B, \qquad z = \det B. \tag{4.16}$$

In view of (4.14), the linear relation among ψ, ψ_1, and ψ_2 is (up to an irrelevant common factor) just

$$z \cdot \psi - y \cdot \psi_1 + \psi_2 = 0, \tag{4.17}$$

and according to (4.16), to make this explicit we need only to know the eigenvalues (and thus the determinant and trace) of B.

These can be obtained from [13], but before describing the formulas, I would like to point out an important subtlety. As we have discussed in the last subsection, all concrete results such as the values of α, β, and γ depend on the framing of knots. The convention that is most natural in working on an arbitrary three manifold is not the convention usually used in discussing knots on S^3.

In studying Fig. 7c, to describe the relative framings, the task is to specify the relative framing of the three pictures on the right, since the picture on the left is being held fixed. If one just looks at these three pictures and ignores the fact that the lines cannot pass through each other, there is an obvious sense in which one would like to pick "the same" framing for each picture; for instance, a unit vector coming out of the page defines a normal vector field on each link in the picture.

This is equivalent to the convention of Moore and Seiberg in defining the eigenvalues of B, so we can now quote their results. Let h_R be the conformal weight of a primary conformal field transforming as R, let E_i be the irreducible representations of $SU(N)$ appearing in the decomposition of $R \otimes R$, and let h_{E_i} be the weights of the corresponding primary fields. Then the eigenvalues of B are

$$\lambda_i = \pm \exp\left(i\pi\left(2h_R - h_{E_i}\right)\right), \tag{4.18}$$

where the $+$ or $-$ sign corresponds to whether E_i appears symmetrically or antisymmetrically in $R \otimes R$. If R is the N dimensional representation of $SU(N)$, then one finds[11] that the eigenvalues of B are

$$\lambda_1 = \exp\left(\frac{i\pi(-N+1)}{N(N+k)}\right), \qquad \lambda_2 = -\exp\left(\frac{i\pi(N+1)}{N(N+k)}\right). \tag{4.19}$$

---

[11] For this representation, $h_R = (N^2-1)/(2N(N+k))$. In the decomposition of $R \otimes R$, the symmetric piece is an irreducible representation with $h_{E_1} = (N^2+N-2)/N(N+k)$, and the antisymmetric piece is an irreducible representation with $h_{E_2} = (N^2-N-2)/N(N+k)$

It is straightforward to put these formulas in (4.16), (4.17) and thus make our previous results completely explicit.

Before comparing to the knot theory literature, it is necessary to make a correction in these results. For a link in S^3, there is always a standard framing in which the self-intersection number of each component of the link is zero. Values of the knot polynomials for knots in S^3 are usually quoted without specifying a framing; these are the values for the link with standard framing. However, if on the right of Fig. 7c we use the "same" framing for each picture, then when the right of Fig. 7c is glued to the left, one does not have the canonical framing for each link. If the first knot is framed in the standard fashion, then the second is in error by one unit and the third by two units. So after using (4.19) to compute α, β, γ, we must, if we wish to agree with the knot theory literature, multiply β by $\exp(-2\pi i h_R)$ and γ by $\exp(-4\pi i h_R)$. After these corrections, one gets

$$\begin{aligned} \alpha &= -\exp\left(\frac{2\pi i}{N(N+k)}\right), \\ \beta &= -\exp\left(\frac{i\pi(2-N-N^2)}{N(N+k)}\right) + \exp\left(\frac{i\pi(2+N-N^2)}{N(N+k)}\right), \\ \gamma &= \exp\left(\frac{2\pi i(1-N^2)}{N(N+k)}\right). \end{aligned} \tag{4.20}$$

If one multiplies α, β, γ by an irrelevant common factor $\exp(i\pi(N^2-2)/N(N+k))$ and introduces the variable

$$q = \exp(2\pi i/(N+k)), \tag{4.21}$$

then the skein relation can be written more elegantly as

$$-q^{N/2} L_+ + (q^{1/2} - q^{-1/2})\, L_0 + q^{-N/2} L_- = 0. \tag{4.22}$$

Here L_+, L_0, and L_- [equivalent to L, L_1, and L_2 in (4.11)] are standard notation for overcrossing, zero crossing, and undercrossing; and for $i = +, 0, -$, we now write simply L_i, instead of $Z(L_i)$. Equation (4.22) is correctly normalized to give the right answers for knots on S^3 with their standard framing, and if one is only interested in knots on S^3 one can use it without ever thinking about the framings. Finally, comparing (4.13) and (4.22), we see that the expectation value of an unknotted Wilson line on S^3, with its standard framing, is

$$\langle C \rangle = \frac{q^{N/2} - q^{-N/2}}{q^{1/2} - q^{-1/2}}. \tag{4.23}$$

This formula can be subjected to several interesting checks. First of all, the right-hand side of (4.23) is positive for all values of the positive integers N and k. This is required by reflection positivity of the Chern-Simons gauge theory in three dimensions. Second, in the weak coupling limit of $k \to \infty$, we have $\langle C \rangle \to N$. This is easily interpreted; in the weak coupling limit, the fluctuations in the connection A_i on S^3 are irrelevant, and the expectation value of the Wilson line approaches its value for $A_i = 0$, which is the dimension of the representation, or in this case N.

4.2. Surgery on Links

We have seen that it is possible to effectively calculate the expectation value of an arbitrary link in S^3. We would now like to generalize this to computations on an arbitrary three manifold. The basic idea is that by the operation of "surgery on links" any three manifold can be reduced to S^3, so it is enough to understand how the invariants that we are studying transform under surgery. The operation of surgery can be described as follows. One begins with a three manifold M and an arbitrarily selected embedded circle C. Note that there is, to begin with, no Wilson line associated with C; C is simply a mathematical line on which we are going to carry out "surgery". To do so we first thicken C to a "tubular neighborhood", a solid torus centered on C. Removing this solid torus, M is split into two pieces; the solid torus is called M_R in Fig. 12b, and the remainder is called M_L. One then makes a diffeomorphism on the boundary of M_R and glues M_L and M_R back together to get a new three manifold $\tilde{M}$.

It is a not too deep result that every three manifold can be obtained from or reduced to S^3 (or any other desired three manifold) by repeated surgeries on knots. However, such a description is far from unique and it is often difficult to use a description of a three manifold in terms of surgery to compute the invariants of interest. We will now see that the invariants studied in this paper can be effectively computed from a surgery presentation.

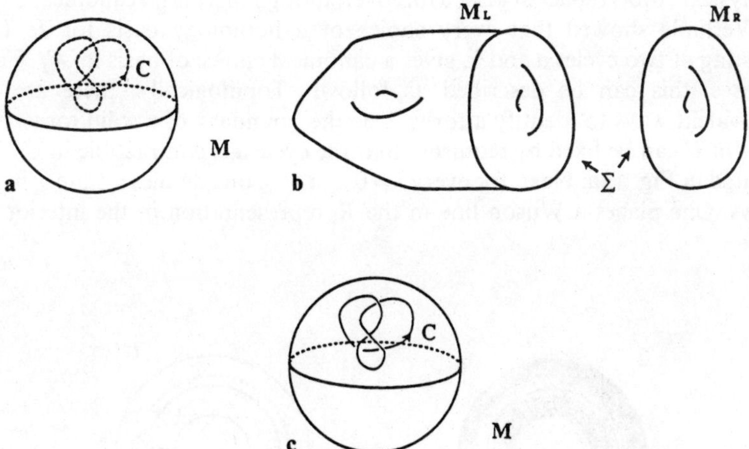

Fig. 12a–c. Surgery on a circle C in a three manifold M is carried out by removing a tubular neighborhood of C, depicted in **a**. At this point M has been separated into two pieces, M_L and M_R, with a torus Σ for their boundaries, as sketched in **b**. M_R is simply a solid torus. Surgery is completed by gluing the two pieces back together after making a diffeomorphism of the boundary of M_R. At the end of this process, M has been replaced by a new three manifold $\tilde{M}$. As we will eventually see, computations on $\tilde{M}$ are equivalent to computations on M with a physical Wilson line where the surgery was made, as in **c**. The difference between **a** and **c** is that in **a** the circle C is just a locale for surgery, but in **c** it is a Wilson line

We study Fig. 12b by the standard arguments. Hilbert spaces $\mathscr{H}_L$ and $\mathscr{H}_R$, canonically dual to one another, are associated with the boundaries of M_L and M_R. The path integrals on M_L and M_R give vectors ψ and χ in $\mathscr{H}_L$ and $\mathscr{H}_R$, and the partition function on M is just the natural pairing (ψ, χ). If we act on the boundary of M_R with a diffeomorphism K before gluing M_L and M_R back together, then χ is replaced by $K\chi$ so (ψ, χ) is replaced by $(\psi, K\chi)$.

This potentially gives a way to determine how the partition function of the quantum field theory transforms under surgery. Upon gaining a suitable understanding of $K\chi$, we will be able to reduce calculations on $\tilde{M}$ to calculations on M.

4.3. The Physical Hilbert Space in Genus One

At this point we need a description of the physical Hilbert space in genus one. A beautiful description, perfectly adapted for our needs, appears in the work of Verlinde [12].

First of all, the loop group LG has at level k finitely many integrable highest weight representations. Let t be the number of these. For each such highest weight representation of the loop group, the highest weight space is an irreducible representation of the finite dimensional group G. In this way there appear t distinguished representations of G; we label these as R_0, $R_1 \dots R_{t-1}$, with R_0 denoting the trivial representation (which is always one of those on this list). Verlinde showed that if Σ is a Riemann surface of genus one, then the dimension of the physical Hilbert space $\mathscr{H}_\Sigma$ is t. Moreover, though there is no canonical basis for $\mathscr{H}_\Sigma$, Verlinde showed that every choice of a homology basis for $H^1(\Sigma, \mathbf{Z})$, consisting of two cycles a and b, gives a canonical choice of basis in $\mathscr{H}_\Sigma$. For our purposes, this can be described as follows. Topologically, there are many inequivalent ways to identify a torus Σ as the boundary of a solid torus U. The choice of U can be fixed by requiring that the cycle a is contractible in U. This is indicated in Fig. 13a. Next, for every $i = 0 \dots t-1$, one defines a state v_i in $\mathscr{H}_\Sigma$ as follows. One places a Wilson line in the R_i representation in the interior of U,

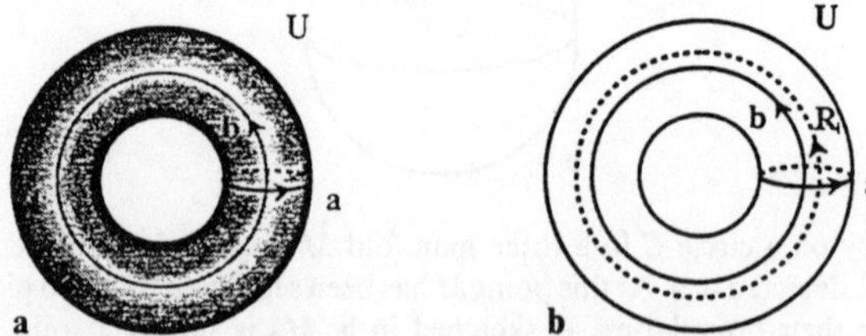

Fig. 13a and b. A Riemann surface Σ of genus one is shown in **a** as the boundary of a solid torus U; the indicated a-cycle is contractible in U. In **b**, a basis of the physical Hilbert space is indicated consisting of states obtained by placing a Wilson line, in the R_i representation, in the interior of U, parallel to the cycle b, and performing the path integral to get a vector v_i in $\mathscr{H}_\Sigma$

running in the b direction,[12] and one performs the Feynman path integral in U to define a vector v_i in $\mathscr{H}_\Sigma$. The v_i make up the Verlinde basis in $\mathscr{H}_\Sigma$. It must be understood that a Wilson line in the trivial representation is equal to 1, so the vector v_0 obtained by this definition is the same as the vector χ which in the last subsection was obtained by a path integral on U with no Wilson lines:

$$\chi = v_0 \,. \tag{4.24}$$

A diffeomorphism K of Σ is represented in the Verlinde basis by an explicit matrix $K_i{}^j$, defined by the formula

$$K \cdot v_i = \sum_j K_i{}^j v_j \,. \tag{4.25}$$

In the space spanned by the v_i, there is a natural inner product, defined by the tensor g_{ij} which is one if R_i is the dual of R_j and zero otherwise. We may sometimes use this metric to raise and lower indices, letting $K_{ij} = \sum_m g_{mj} K_i{}^m$.

We can now get a much more concrete description of the behavior of the quantum field theory partition function under surgery. In discussing surgery, we began with a three manifold M and a knot C. Cutting out a tubular neighborhood of this knot, whose boundary we call Σ, we separated M into M_L and M_R, with M_R being a solid torus. The path integrals on M_L and M_R gave vectors ψ and χ in the Hilbert spaces $\mathscr{H}_L$ and $\mathscr{H}_R$. As we have just noted, χ is the same as v_0, so the partition function on M is (ψ, v_0). Now we want to make a diffeomorphism K on the boundary of M_R, and then glue together M_L and M_R to make a new three manifold $\tilde{M}$. The partition function of $\tilde{M}$ is $Z(\tilde{M}) = (\psi, Kv_0)$, as we saw in the last section. To say that $Z(\tilde{M})$ is computable from a surgery presentation means that the evaluation of this invariant of $\tilde{M}$ can be reduced to tractable calculation on M. We will now show this. From (4.25), we can write

$$Z(\tilde{M}) = \sum_j K_0{}^j (\psi, v_j) \,. \tag{4.26}$$

But each term (ψ, v_j) has an interpretation in terms of path integrals on M! Indeed, it is the very definition of the v_j that v_j differs from v_0 just by an insertion of an extra Wilson line in the R_j representation at the center of M_R. So just as (ψ, v_0) represents the original partition function of M, (ψ, v_j) represents a modified partition function with an extra Wilson line in the R_j representation placed on C. So we rewrite (4.26) in the form

$$Z(\tilde{M}) = \sum_j K_0{}^j \cdot Z(M; R_j) \,, \tag{4.27}$$

where $Z(M; R_j)$ is the partition function of M with an extra Wilson line in the R_j representation included on the circle C (in addition to whatever Wilson lines are already present on M). This is indicated in Fig. 12c. To use (4.27), one needs to know the matrix $K_i{}^j$, which is precisely the matrix by which the diffeomorphism K of the torus is represented on the characters of the irreducible level k representations of LG; these matrices appear in [37] and have remarkable properties

[12] The b cycle on the boundary of U gives a framing of this Wilson line

recently investigated in [12, 13]. Given a knowledge of the $K_i^{\,j}$, (4.27) is a completely explicit formula expressing computations on $\tilde{M}$ in terms of computations on M. By repeated use of this formula, computations on any three manifold can be reduced to computations on S^3, [13] with appropriate Wilson lines. Of course, the surgery will generate Wilson lines on S^3 in representations of G corresponding to arbitrary integrable representations of LG.

Generalized Surgery. The surgery law (4.27) has a useful generalization. While so far we have only considered surgery on a purely imaginary circle C, as in Fig. 12, there is no reason not to generalize this to a situation in which before the surgery a Wilson line in the R_i representation was already present on C. Surgery amounts to cutting out a neighborhood of C and then gluing it back in, and after this process the R_i Wilson line will still be present in $\tilde{M}$. So the left-hand side of (4.27) is replaced by $Z(\tilde{M};\, R_i)$, where the notation schematically indicates the presence of the R_i Wilson line. What about the right-hand side of (4.27)? Before surgery, with a Wilson line R_i on C, the path integral on a tubular neighborhood U of C gives on the boundary a state $\chi' = v_i$; this is the generalization of (4.24). If we cut out C and glue it back in with a diffeomorphism K of the boundary, then v_i is replaced according to (4.25) with $K_i^{\,j}\, v_j$. If we remember that v_j could have been obtained by putting a Wilson line on C in the R_j representation, we see that the right-hand side of (4.27) becomes $\sum_j K_i^{\,j} Z(M;\, R_j)$, so we get the generalized surgery formula

$$Z(\tilde{M};\, R_j) = \sum_j K_i^{\,j} Z(M;\, R_j)\,. \tag{4.28}$$

This formula will be used later in a new proof of Verlinde's conjecture.

Cabling of Knots; Satellites. Finally, let us note that similar methods can be used to determine the behavior of the knot invariants under "cabling", and more generally to relate the invariants of the "satellites" of a knot to invariants of the original knot. Any knot C in any three manifold M has a neighborhood that looks like a solid torus U. If we replace C by an arbitrary satellite $\tilde{C}$ of itself (an arbitrary knot that can be placed in U), with representations R_σ associated with the connected components of $\tilde{C}$, then the path integral on U will define a vector ψ in the physical Hilbert space $\mathscr{H}_\Sigma$ associated with the boundary Σ of U. Like any vector in $\mathscr{H}_\Sigma$, ψ can be expanded in the Verlinde basis,

$$\psi = \sum_i \alpha_i v_i\,. \tag{4.29}$$

The α_i are complex numbers that depend on the choice of satellite $\tilde{C}$ and on the choice of representations R_σ, but they do not depend on what three manifold M the solid torus U has been extracted from or on what other knots are present on M. The vectors $v_i \in \mathscr{H}_\Sigma$ are the vectors that would be produced by the path integral on U with a Wilson line in the R_i representation placed on the original knot C (and not a satellite of C). Thus, a knowledge of the invariants of C in arbitrary representations together with a knowledge of the universal coefficients α_i is enough to determine the invariants of arbitrary satellites of C.

[13] Or on any desired three manifold; we will see that $S^2 \times S^1$ is more tractable than S^3

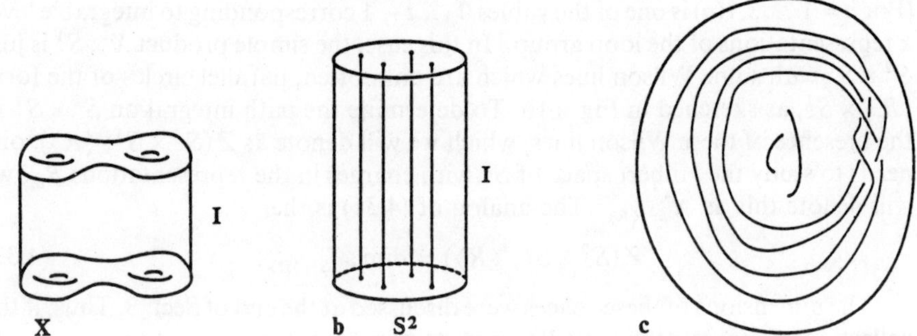

Fig. 14a–c. Beginning with $X \times I$, shown in **a**, one makes $X \times S^1$ by identifying $X \times \{0\}$ with $X \times \{1\}$. If X is S^2 with some marked points P_i, then this construction gives the picture of **b**. If $S^2 \times \{0\}$ is joined to $S^2 \times \{1\}$ via a non-trivial diffeomorphism B, one makes in this way a braid, as in **c**

4.4. Path Integrals on $S^1 \times X$

In this subsection we will describe a few facts which are useful in their own right and will enable us to carry out some concrete surgeries.

First of all, we have not so far determined the partition function of S^3 without Wilson lines. It may come as a surprise to topologists that we cannot trivially assert that this is 1. In quantum field theory there is no particularly strong axiom governing the partition function of S^3. The three manifolds whose partition functions can be computed in a particularly simple way, from the axioms of quantum field theory, are those of the form $X \times S^1$, for various X. $X \times S^1$ can conveniently be studied in a "Hamiltonian" formalism, as indicated in Fig. 14a. One constructs the Hilbert space $\mathscr{H}_X$ of X. Then one introduces a "time" direction, represented by a unit interval $I = [0, 1]$, and one propagates the vectors in $\mathscr{H}_X$ from "time" 0 to "time" 1. This operation is trivial, since the Chern-Simons theory, like any generally covariant theory, has a vanishing Hamiltonian. Finally, one forms $X \times S^1$ by gluing $X \times \{0\}$ to $X \times \{1\}$; this identifies the initial and final states, giving a trace:

$$Z(X \times S^1) = \mathrm{Tr}_{\mathscr{H}_X}(1) = \dim \mathscr{H}_X . \tag{4.30}$$

For example, the physical Hilbert space of S^2 is one dimensional, for any G and k, so one has

$$Z(S^2 \times S^1) = 1 . \tag{4.31}$$

It is possible to generalize (4.30) as follows. If we are given a diffeomorphism K: $X \to X$, then one can form the mapping cylinder $X \times_K S^1$ by identifying $x \times \{1\}$ with $K(x) \times \{0\}$ for every $x \in X$. At the level of quantum field theory, when one goes from $X \times I$ to $X \times_K S^1$, the initial and final states are identified via K, so the generalization of (4.30) is

$$Z(X \times_K S^1) = \mathrm{Tr}_{\mathscr{H}_X}(K) . \tag{4.32}$$

The situation that we actually wish to apply this to is the case in which X is S^2 with some marked points P_a, $a = 1 \dots s$ to which representations $R_{i(a)}$ are assigned.

[For $a = 1 \ldots s$, $i(a)$ is one of the values $0 \ldots t-1$ corresponding to integrable level k representations of the loop group.] In this case, the simple product $X \times S^1$ is just $S^2 \times S^1$ with some Wilson lines which are unknotted, parallel circles of the form $\{P_a\} \times S^1$, as sketched in Fig. 14b. To determine the path integral on $S^2 \times S^1$ in the presence of these Wilson lines, which we will denote as $Z(S^2 \times S^1; \langle R \rangle)$, one needs to study the Hilbert space of S^2 with charges in the representations R_{a_i}; we will denote this as $\mathscr{H}_{S^2;\langle R \rangle}$. The analog of (4.31) is then

$$Z(S^2 \times S^1; \langle R \rangle) = \dim \mathscr{H}_{S^2;\langle R \rangle}. \tag{4.33}$$

The dimensions of these spaces were discussed at the end of Sect. 3. Thus, if the collection of representations $\langle R \rangle$ consists of a single representation R_a, we get

$$Z(S^2 \times S^1; R_a) = \delta_{a,0}, \tag{4.34}$$

since the physical Hilbert space with a single charge in the R_a representation is one dimensional if R_a is the trivial representation ($a = 0$) and zero dimensional otherwise.[14] For two charges in the representations R_a and R_b, we get

$$Z(S^2 \times S^1; R_a, R_b) = g_{ab}, \tag{4.35}$$

where g_{ab}, introduced earlier, is 1 if R_b is the dual of R_a and zero otherwise. The formula (4.35) follows from the result of Sect. (4.4) for the Hilbert space on S^2 with two charges. Finally, if there are three charges in the representations R_a, R_b, R_c, we get

$$Z(S^2 \times S^1; R_a, R_b, R_c) = N_{abc}, \tag{4.36}$$

with N_{abc} the trilinear "coupling" of Verlinde, since this is the dimension of the physical Hilbert space.

4.5. Some Concrete Surgeries

Now we would like to describe some useful results that can be obtained from concrete surgeries. The first goal is compute the partition function of S^3. Since we already know the partition function of $S^2 \times S^1$, we will try to interpret S^3 as a manifold obtained by surgery on $S^2 \times S^1$. This is readily done. We consider the circle C in $S^2 \times S^1$ indicated in Fig. 15a. A tubular neighborhood of C is a torus Σ; we pick a basis of $H^1(\Sigma; \mathbf{Z})$ consisting of cycles a and b indicated in the figure. Now we wish to make a particular surgery associated with a very special diffeomorphism $S\colon \Sigma \to \Sigma$. We pick S to map a to b and b to $-a$.[15] This surgery – removing the interior of Σ from $S^2 \times S^1$ and gluing it back after acting with S – produces a three manifold that is none other than S^3 (Fig. 16). Since this point is crucial in what follows, we pause to explain it. We regard S^3 as R^3 plus a point at infinity. In Fig. 16a a torus Σ has been embedded in R^3. Obviously, Σ with its

---

[14] Implicit in (4.34) and subsequent formulas is the use of the standard framing of the Wilson line which is invariant under rotations of S^1; it is for this choice that the path integral on $S^2 \times S^1$ computes the trace of the identity operator in the physical Hilbert space

[15] This transformation, which acts on the upper half plane as $\tau \to -1/\tau$, is indeed usually called S in the theory of the modular group $SL(2, \mathbf{Z})$ (which can be identified, via the basis a, b, with the mapping class group of Σ)

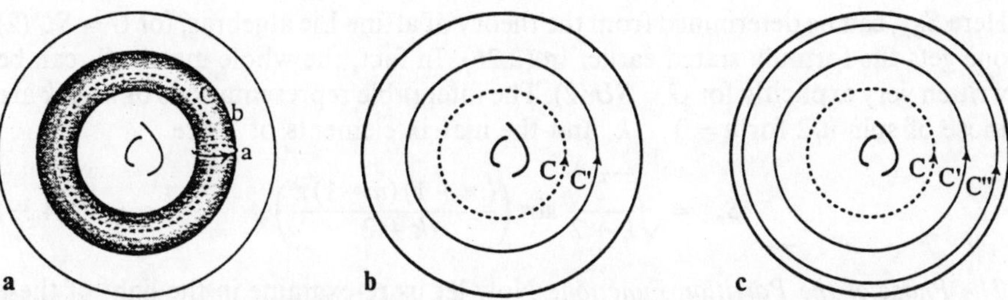

Fig. 15a–c. In part **a**, we consider surgery on a circle C in $S^2 \times S^1$. A tubular neighborhood of this circle is a torus Σ; a useful basis of $H^1(\Sigma, \mathbf{Z})$ is indicated. In **b**, in addition to the circle C on which we perform surgery, there is a parallel circle C' on which we place a Wilson line in the R_j representation. In **c**, there are two parallel circles C' and C'' with Wilson lines in representations R_j and R_k

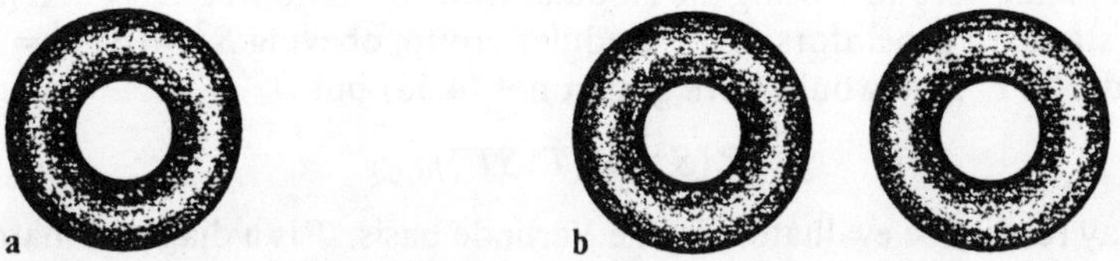

Fig. 16a and b. The purpose of this figure is to indicate how S^3 can be made by surgery starting with $S^2 \times S^1$. In **a** we show a torus Σ, sitting in R^3 and in **b** a pair of identical solid tori

interior make up a solid torus T. It is also relatively easy to see that the figure (a) is invariant under inversion, so that the exterior of Σ (including the point at infinity) is a second solid torus T'. Thus, S^3 can be made by gluing two solid tori along their boundaries. Now in (b) we sketch two identical solid tori T and T'; T' has been obtained by simply translating T in Euclidean space. If one glues together the boundaries pointwise with the identification that is indicated by saying that "T' is a translate of T", one gets $S^2 \times S^1$. (In fact, the solid torus T is $D \times S^1$, with D a two dimensional disc, and T' is $D' \times S^1$ with D' a second disc. Just as two discs D and D' glued on their boundary make S^2, $D \times S^1$ naturally glues to $D' \times S^1$ to make $S^2 \times S^1$.) On the other hand, we know from part (a) that S^3 can be obtained by gluing two solid tori. A little mental gymnastics, comparing the argument we gave in connection with (a) to that in (b), shows that to make S^3 we must glue together T and T' after making the modular transformation S on the boundary of T'.

Now we can use (4.27), with S^3 playing the role of $\tilde{M}$, $S^2 \times S^1$ playing the role of M, and the arbitrary diffeomorphism K replaced by S. So we learn

$$Z(S^3) = \sum_j S_0{}^j Z(S^2 \times S^1; R_j). \tag{4.37}$$

We have learned in the last section that $Z(S^2 \times S^1; R_j)$ is 1 for $j=0$ and 0 otherwise, so

$$Z(S^3) = S_{0,0}. \tag{4.38}$$

Here $S_{0,0}$ can be determined from the theory of affine Lie algebras; for $G = SU(2)$ one gets the formula stated earlier in (2.26). In fact, the whole matrix S_{ij} can be written very explicitly for $G = SU(2)$. The integrable representations of level k are those of spin $n/2$ for $n = 0 \ldots k$, and the matrix elements of S are

$$S_{mn} = \sqrt{\frac{2}{k+2}} \sin\left(\frac{(m+1)(n+1)\pi}{k+2}\right). \tag{4.39}$$

The Phase of the Partition Function. Now let us re-examine in the light of these methods a thorny question that appeared in Sect. 2 – the framing of three manifolds, and the phase of the partition function.

We have obtained S^3 from $S^2 \times S^1$, by performing surgery on a certain circle C, using the modular transformation S: $\tau \to -1/\tau$. Apart from S, there are other modular transformations that could be used to build S^3 by surgery on the same knot C in $S^2 \times S^1$. The general choice would be $T^n S T^m$, with n and m being arbitrary integers, and T being the modular transformation T: $\tau \to \tau + 1$. [S and T are the standard generators of the modular group, obeying $S^2 = (ST)^3 = 1$.] Had we used $T^n S T^m$, we would have gotten not (4.38) but

$$Z(S^3) = (T^n S T^m)_{0,0}\,. \tag{4.40}$$

This may readily be evaluated. In the Verlinde basis, T is a diagonal matrix with $T \cdot v_i = e^{2\pi i (h_i - c/24)} \cdot v_i$; h_i is the conformal weight of the primary field in the representation R_i and c is the central charge for current algebra with symmetry group G at level k. Since $h_0 = 0$, if we replace (4.38) by (4.40) the partition function transforms as

$$Z \to Z \cdot \exp\left(2\pi i (n-m) \cdot \frac{c}{24}\right). \tag{4.41}$$

Though we have obtained this formula in the example of a particular surgery (giving S^3 from $S^2 \times S^1$), the same ambiguity arises in any process of surgery. Whenever one makes surgery on a circle C, in a three manifold M, with the surgery being determined by an $SL(2, Z)$ element u, one could instead consider surgery on the same circle C, using the $SL(2, Z)$ element $u \cdot T^m$. This would have the same effect topologically, but our surgery law would give a partition function containing an extra phase $\exp(-2\pi i m \cdot c/24)$.

This phase ambiguity was already encountered, in the large k limit, in formula (2.24). What is more, from the discussion in Sect. 2, we know what topological structure on three manifolds must be considered in order to keep track of the factors of $\exp(2\pi i \cdot c/24)$. One must consider "framed" three manifolds. Two surgeries that have the same effect on the topology of a three manifold may have different effects on the framing. I will discuss elsewhere how to systematically keep track of the factors of $\exp(2\pi i \cdot c/24)$ under surgery. In the simple applications in this paper, this will not be necessary. All of our applications will involve considering the standard surgery (by the modular transformation S) that was used in the last subsection to obtain S^3 from $S^2 \times S^1$.

Some Expectation Values. Now let us see if we can go farther and determine the path integral $Z(S^3; R_j)$ on S^3 with an unknotted Wilson line on S^3 in an arbitrary representation R_j.[16] To do this, we start on $S^2 \times S^1$ with a Wilson line in the R_j representation running parallel to the circle C on which we are doing surgery, as in Fig. 15b. Carrying out the same surgery as before turns $S^2 \times S^1$ into S^3, with a Wilson line in the R_j representation on S^3. Application of (4.27) now gives

$$Z(S^3; R_j) = \sum_i S_0{}^i Z(S^2 \times S^1;\ R_i,\ R_j)\,. \tag{4.42}$$

Using (4.35), we can evaluate this and determine the partition function for a Wilson line in an arbitrary representation R_j; it is

$$Z(S^3;\ R_j) = \sum_i S_0{}^i g_{ij} = S_{0,j}\,. \tag{4.43}$$

Let us compare this to our previous evaluation (4.23) of the expectation value of an unknotted Wilson line in S^3. We must recall that the symbol $\langle C\rangle$ in (4.23) represented a ratio $\langle C\rangle = Z(S^3; R)/Z(S^3)$. Let us take $G = SU(2)$, so that we can use the explicit formulas (4.39), and take R to be the two dimensional representation of $SU(2)$, so that we can compare to (4.23). Using (4.38), (4.43), and (4.39), we get

$$\langle C\rangle = \frac{S_{0,1}}{S_{0,0}} = \frac{\sin(2\pi/(k+2))}{\sin(\pi/(k+2))}\,. \tag{4.44}$$

It is easy to see that setting $N=2$ in (4.23) gives the same formula. Let us take this one step further and try to calculate by these methods the partition function $Z(S^3;\ R_j,\ R_k)$ for S^3 with two unknotted, unlinked Wilson lines in representations R_j and R_k. In Fig. 15c, we start on $S^2 \times S^1$ with *two* Wilson lines, in representations R_j and R_k, parallel to the circle C on which surgery is to be performed. Carrying out the surgery, we get to S^3 with the desired unlinked, unknotted circles. In this case, the surgery formula (4.27) tells us that

$$Z(S^3;\ R_j,\ R_k) = \sum_i S_0^i Z(S^2 \times S^1;\ R_i, R_j,\ R_k)\,. \tag{4.45}$$

The right-hand side can be evaluated with (4.36), while the left-hand side can be reduced to (4.43) using (4.6). We get

$$\frac{S_{0,j}S_{0,k}}{S_{0,0}} = \sum_i S_0^i\ N_{ijk}\,. \tag{4.46}$$

Proof of Verlinde's Conjecture. The last equation is a special case of a celebrated conjecture by Verlinde, which has been proved by Moore and Seiberg [13]. We can use these methods to give a new proof of Verlinde's conjecture, in the case of current algebra. We will have to use the generalized surgery relation (4.28). We return to Fig. 15b but now instead of treating C as a purely imaginary contour on which surgery is to be performed, we suppose that there is a Wilson line on C in the R_i representation. In this case, the standard surgery on C will still turn $S^2 \times S^1$ into S^3, but now on S^3 we will have two Wilson lines, in the R_i and R_j representations. Some mental gymnastics shows that they are linked, as in Fig. 17a; schematically,

[16] We give this Wilson line the framing described in the footnote after (4.34); after surgery this turns into the standard framing on S^3

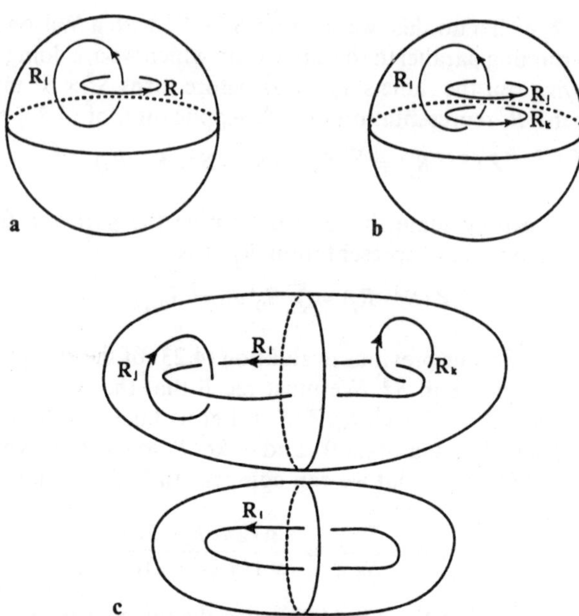

Fig. 17a–c. In **a**, we sketch linked but unknotted Wilson lines on S^3, in the R_i and R_j representations. In **b**, a Wilson line R_i is linked with two Wilson lines R_j and R_k on S^3. In **c**, we sketch how two crucial amplitudes can be factored through the same one dimensional space

we refer to this linked pair of Wilson lines as $L(R_i; R_j)$. The use of (4.28) therefore determines the partition function of S^3 with a pair of linked Wilson lines:

$$Z(S^3; L(R_i; R_j)) = \sum_k S_i^k Z(S^2 \times S^1; R_k, R_j) = S_{ij}. \tag{4.47}$$

In the second step, we have used the fact that the partition function of $S^2 \times S^1$ with static charges R_k and R_j is the metric that we have called g_{kj} – one if R_k is dual to R_j and otherwise zero.

Now let us go back to Fig. 15c, and again on what was previously the purely imaginary circle C we put a Wilson line in the R_i representation. The standard surgery on this link will now produce a picture sketched in Fig. 17b, with a Wilson line on S^3 in the R_i representation that links a pair of Wilson lines R_j, R_k that are themselves unlinked and unknotted. We call this configuration $L(R_i; R_j, R_k)$. The evaluation of (4.28) now gives

$$Z(S^3; L(R_i; R_j, R_k)) = \sum_m S_i^m Z(S^2 \times S^1; R_m, R_j, R_k) = \sum_m S_i^m N_{mjk}. \tag{4.48}$$

To obtain Verlinde's formula, it is now necessary to find an independent way to evaluate the left-hand side.

Such a method is provided by the following generalization of the multiplicativity formula (4.1). The key point in the derivation of (4.1) was that the physical Hilbert space for S^2 with no charges was one dimensional. It is likewise true that the physical Hilbert space $\mathscr{H}$ for S^2 with a pair of charges in the dual representations R_i and $R_{\bar{i}}$ is one dimensional. Using this and otherwise repeating the derivation of (4.1) gives the following formula:

$$Z(S^3; L(R_i; R_j, R_k)) \cdot Z(S^3; R_i) = Z(S^3; L(R_i; R_j)) \cdot Z(S^3; L(R_i; R_k)). \quad (4.49)$$

The idea in (4.49) is that, as in Fig. 17c, the evaluation of $Z(S^3; L(R_i; R_j, R_k))$ can be expressed as a pairing (ψ, χ) where ψ and χ are certain vectors in $\mathscr{H}$ and its dual. Likewise the evaluation of $Z(S^3; R_i)$ is a pairing (v', v), where v' and v are vectors in $\mathscr{H}$ and its dual. Using the wonderful fact of one dimensional linear algebra $(\psi, \chi) \cdot (v', v) = (\psi, v) \cdot (v', \chi)$, we arrive at (4.49). Since all factors in (4.49) are known except the first, we arrive at the result

$$Z(S^3; L(R_i; R_j, R_k)) = S_{ij} S_{ik} / S_{0,i}. \quad (4.50)$$

Combining this with (4.48), we have

$$S_{ij} S_{ik} / S_{0,i} = \sum_m S_i^m N_{mjk}. \quad (4.51)$$

This is equivalent to Verlinde's statement that "the matrix S diagonalizes the fusion rules". In other words, in the basis v_i indicated in Fig. 13, the structure constants of the Verlinde algebra are by definition $v_i v_j = \sum_k N_{ij}{}^k v_k$, where $N_{ij}{}^k = \sum_r N_{ijr} g^{rk}$. If we introduce a new basis $w_i = S_{0,i} \cdot \sum S_i^m v_m$, then the Verlinde algebra reduces to $w_i w_j = \delta_{ij} w_j$. To verify this, we compute

$$w_i w_j = \sum_{k,l} S_i^k S_j^l v_k v_l \cdot S_{0,i} S_{0,j}. \quad (4.52)$$

Using $v_k v_l = \sum_m N_{kl}{}^m v_m$ and (4.51), this becomes

$$w_i w_j = S_j^l v^m \cdot S_{il} S_i^m \cdot S_{0,j}. \quad (4.53)$$

Using the unitarity of S, in the form $S_j^l S_{il} = \delta_{ij}$, we see that

$$w_i w_j = \delta_{ij} \sum_m S_j^m v_m \cdot S_{0,j} = \delta_{ij} \cdot w_j, \quad (4.54)$$

showing that the Verlinde algebra has been diagonalized and that the w_i are idempotents.

Connected Sum of Links. At the beginning of Sect. 4, we have seen that the quantum partition function has a multiplicative behavior under connected sum of three manifolds. From (4.1), if $M = M_1 + M_2$, then $Z(M) \cdot Z(S^3) = Z(M_1) \cdot Z(M_2)$. In the special case that M_1 and M_2 are copies of S^3 with links in them, the connected sum of M_1 and M_2 is a copy of S^3 containing the *disconnected* sum of the two links.

In knot theory there is also an operation of taking the *connected* sum of two links. This operation has appeared in the above discussion. The link that we have called $L(R_i; R_j, R_k)$ is the connected sum of the two links that we have called $L(R_i, R_j)$ and $L(R_i, R_k)$. In fact, in Fig. 17c, $L(R_i; R_j, R_k)$ is "cut" into two pieces, which are respectively $L(R_i, R_j)$ and $L(R_i, R_k)$ with in each case a connected segment removed. This is the defining configuration for the "connected sum of links". Accordingly, the reasoning that led to (4.49) has the following more general consequence. If L_1 and L_2 are two links, and $L_1 + L_2$ is their connected sum, then

$$Z(S^3; L_1 + L_2) \cdot Z(S^3; C) = Z(S^3; L_1) \cdot Z(S^3; L_2). \tag{4.55}$$

Here it is understood that [as in (4.49)] representations have been assigned to the connected components of L_1, L_2, and $L_1 + L_2$ in a compatible fashion; C is an unknot placed in whatever representation is carried by the strand "cut" in the generalization of Fig. 17c. Equation (4.55) has a generalization in which L_1 and L_2 are links in arbitrary three manifolds M_1 and M_2; then the connected sum of links $L_1 + L_2$ is a link in the connected sum of manifolds $M = M_1 + M_2$, and (4.55) is replaced by

$$Z(M_1 + M_2; L_1 + L_2) \cdot Z(S^3; C) = Z(M_1; L_1) \cdot Z(M_2; L_2). \tag{4.56}$$

4.6. *The Knot Polynomials and the Braid Group*

The results in the last subsection are nice enough so that one may wonder if the partition function for an arbitrary link on S^3 can be evaluated in this way. This can indeed be done, and in a way that is closely related to the original route by which the Jones polynomial was discovered, though we cannot expect such explicit formulas as in the simple cases treated above. An arbitrary link L on S^3, whose partition function we will call $Z(S^3; L)$, can be arranged in the form of a braid, as indicated in Fig. 18a. One can imagine "lifting" this braid B out of S^3 and putting it on $S^2 \times S^1$. To get back to S^3 one would have to do surgery on a circle running parallel to the braid, as suggested in Fig. 18b. The general surgery formula (4.27) then tells us

$$Z(S^3; L) = \sum_j S_0{}^j Z(S^2 \times S^1; R_j, B), \tag{4.57}$$

where $Z(S^2 \times S^1; R_j, B)$ is the partition function on $S^2 \times S^1$ in the presence of both the braid B and a parallel Wilson line in the R_j representation. We want to rewrite this in the spirit of (4.32). Suppose that the braid B contains n strands making up a collection of representations $\langle R \rangle$. Then B can be regarded as defining an element of the Artin braid group on n letters. The braid group is closely related to the mapping class group for S^2 with marked points. The reason for that is that if in Fig. 18b we "cut" $S^2 \times S^1$, to get back to $S^2 \times I$ (this amounts to undoing what was done in Fig. 14, then the braid can be unbraided. Thus, the complete information about the braid is in the choice of a diffeomorphism of S^2 (constrained to preserve the marking of the points) by which the top and bottom of Fig. 18c are to be identified. This, however, does not quite mean that the braid group is the same as the mapping class group. In Fig. 18b, there are $n+1$ strands, one of which arose from the surgery and does not participate in the braid, while the other n strands make up the braid. The braid group on n letters is the subgroup of the mapping class group on $n+1$ letters which fixes one of the (framed) strands. There are a number of invariant traces on the braid group that can be naturally defined with the data at our disposal. They are

$$\tau_i(B) = Z(S^2 \times S^1; R_i, B). \tag{4.58}$$

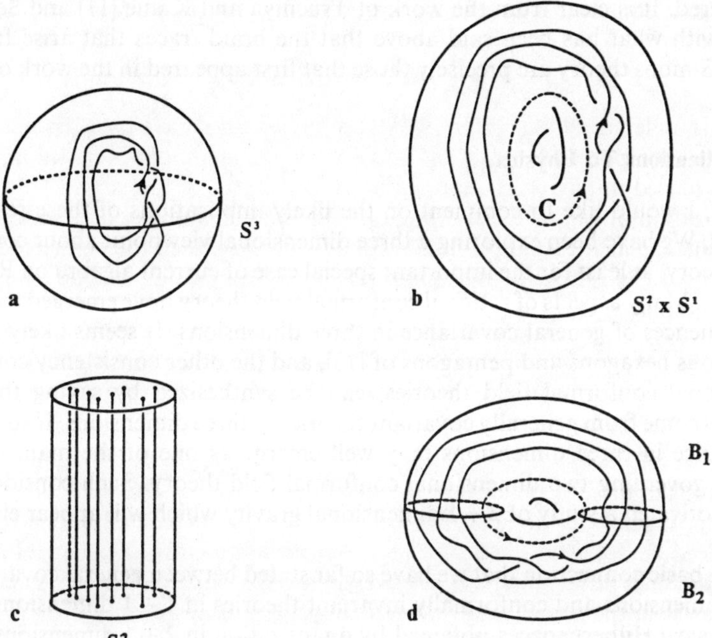

Fig. 18a–d. Any link on S^3 can be shaped into the form of a braid, as in **a**; putting the same braid on $S^2 \times S^1$, and doing surgery on the circle C indicated with the dotted line in part **b**, one gets back to the original link on S^3. If one "cuts" $S^2 \times S^1$ to get to $S^2 \times I$, the braid can be "unbraided"; the braid is recovered by prescribing a diffeomorphism of S^2, via which the top and bottom of part **c** are to be identified. In **d**, we sketch the origin of the key property of the braid traces. In the presence of an arbitrary Wilson line on the dotted contour (reflecting the results of surgery), two braids B_1 and B_2 are joined end to end in $S^2 \times S^1$. There is no way to tell which comes first, so the partition function is invariant under exchange of B_1 and B_2

This has the key property of a trace,

$$\tau_i(B_1 B_2) = \tau_i(B_2 B_1), \tag{4.59}$$

since the two sides of (4.59) have the same path integral representation, in which the two braids are glued end to end in $S^2 \times S^1$, as in Fig. 18d. Not only does (4.58) obey (4.59); it is actually equal to the trace of the operator B in a certain representation of the braid group, namely the representation furnished by the physical Hilbert space $\mathscr{H}$ for S^2 with the $n+1$ charges, it being understood that the braid group is acting on the first n charges, and the $(n+1)^{st}$ is fixed in the R_i representation. That $\tau_i(B)$ is the trace of B in this Hilbert space is a statement just along the lines of (4.32).

So we can rewrite (4.57) in the form

$$Z(S^3; L) = \sum_j S_0{}^j \tau_j(B). \tag{4.60}$$

This shows that the link invariants on S^3 may be written as linear combinations of braid traces. This is very close to how the knot polynomials were originally discovered. It is clear from the work of Tsuchiya and Kanie [11] and Segal [16] along with what has been said above that the braid traces that arise from the Chern-Simons theory are precisely those that first appeared in the work of Jones.

5. Applications To Physics

Finally, I would like to comment on the likely implications of these results for physics. We have been exploring a three dimensional viewpoint about conformal field theory, at least for the important special case of current algebra on Riemann surfaces. Many aspects of rational conformal field theory have emerged as natural consequences of general covariance in three dimensions. It seems likely that the marvelous hexagons and pentagons of [13], and the other consistency conditions of rational conformal field theories, can be synthesized by saying that such theories come from generally covariant theories in three dimensions. If so, general covariance in three dimensions may well emerge as one of the main unifying themes governing two dimensional conformal field theory. Such considerations have motivated a study of $2+1$ dimensional gravity which will appear elsewhere [38].

The basic connection that we have so far stated between general covariance in $2+1$ dimensions and conformally invariant theories in $1+1$ dimensions is that the physical Hilbert spaces obtained by quantization in $2+1$ dimensions can be interpreted as the spaces of conformal blocks in $1+1$ dimensions. This connection may seem rather abstract, and I will now make a few remarks aimed at making a more concrete connection. Starting from three dimensions we were led to the problem of quantizing the Lagrangian (3.1), which we repeat for convenience:

$$\mathcal{L} = \frac{k}{8\pi} \int dt \int_{\Sigma} \varepsilon^{ij} \operatorname{Tr} A_i \frac{d}{dt} A_j . \tag{5.1}$$

This is to be quantized with constraints $\varepsilon^{ij} F^a_{ij} = 0$, these constraints being the generators of the infinitesimal gauge transformations

$$A_i \rightarrow A_i - D_i \varepsilon . \tag{5.2}$$

So far we have quantized (5.1) on Riemann surfaces without boundary, but now let us relax this requirement. Quantization of (5.2) in this more general case amounts to studying the three dimensional Chern-Simons theory on a three manifold with boundary, namely $\Sigma \times R^1$, where Σ is a Riemann surface with boundary. For instance, let Σ be a disc D. To quantize (5.1) on the disc, we must impose the constraints, which generate the gauge transformations (5.2). As D has a boundary, we must choose boundary conditions on A_i and ε (for closed surfaces this question did not arise). We will adopt free boundary conditions for A_i, but require $\varepsilon = 0$ on the boundary of S. (A rationale for this choice is that the Chern-Simons action is not invariant under gauge transformations that do not vanish on the boundary.) With this condition, ε generates in (5.2) the group $\hat{G}_1$ of gauge transformations which are the identity on the boundary of the disc. An element of this group is an arbitrary continuous map V: $D \rightarrow G$ whose restriction to S is the identity. Now we impose the constraints. The first step is to require that $\varepsilon^{ij} F_{ij}^a = 0$. Since the disc is simply connected, this implies that $A_i = -\partial_i U \cdot U^{-1}$ for some map U: $D \to G$. U is uniquely defined up to

$$U \to U \cdot W, \tag{5.3}$$

for a constant element $W \in G$. Then we must identify two U's that differ by an element of the restricted gauge group $\hat{G}_1$. This means that we must impose the equivalence relation $U \simeq VU$ for any V such that $V = 1$ on the boundary S. The equivalence relation means that only the restriction of U to S is relevant. This restriction defines an element of the loop group LG, but because of the freedom (5.3), we should actually regard U as an element of LG/G. Geometrically, we have learned that the homogeneous space LG/G can be regarded as the symplectic quotient of the space of G connections on the disc D, by the group $\hat{G}_1$ of symplectic diffeomorphisms. Now we wish to quantize the theory, which means doing quantum mechanics on LG/G, which inherits a natural symplectic structure from (5.1). Clearly, the group LG of gauge transformations on the boundary of Σ acts on LG/G, so the quantum Hilbert space will be at least a projective representation of the loop group. In fact, according to Segal [39], the quantization of LG/G, with this symplectic structure, gives rise to the basic irreducible highest weight representation of the loop group. This makes the connection between $2+1$ dimensions and $1+1$ dimensions far more direct, since the irreducible representations of the loop group are a basic ingredient in the $1+1$ dimensional theory. It is obvious at this point that by considering more complicated Riemann surfaces with various boundary components, we can generate the whole $1+1$ dimensional conformal field theory, essentially by studying the generally covariant $2+1$ dimensional theory on various three manifolds with boundary.

Acknowledgements. This work originated with the realization that some results about conformal field theory described by G. Segal could be given a three dimensional interpretation by considering a gauge theory with Chern-Simons action. I am grateful to Segal for explaining his results, and to M. Atiyah for interesting me in and educating me about the Jones polynomial. V.F.R. Jones and L. Kauffman, and other participants at the IAMP Congress, raised many relevant questions. Finally, I must thank S. Deser and D.J. Gross for pointing out Polyakov's paper, G. Moore and N. Seiberg for explanations of their work, and the organizers of the IAMP Congress for their hospitality.

References

1. Atiyah, M.F.: New invariants of three and four dimensional manifolds. In: The mathematical heritage of Hermann Weyl. Proc. Symp. Pure Math., vol. **48**. Wells R. (ed.). Providence, RI: American Mathematical Society 1988, pp. 285–299
2. Donaldson, S.: An application of gauge theory to the topology of four manifolds. J. Diff. Geom. **18**, 269 (1983), Polynomial invariants for smooth four-manifolds. Oxford preprint
3. Floer, A.: An instanton invariant for three manifolds. Courant Institute preprint (1987). Morse theory for fixed points of symplectic diffeomorphisms. Bull. AMS **16**, 279 (1987)
4. Witten, E.: Topological quantum field theory. Commun. Math. Phys. **117**, 353 (1988)

5. Jones, V.F.R.: Index for subfactors. Invent. Math. **72**, 1 (1983). A polynomial invariant for links via von Neumann algebras. Bull. AMS **12**, 103 (1985), Hecke algebra representations of braid groups and link polynomials. Ann. Math. **126**, 335 (1987)
6. Freyd, P., Yetter, D., Hoste, J., Lickorish, W.B.R., Millett, K., Ocneanu, A.: A new polynomial invariant of knots and links. Bull. AMS **12**, 239 (1985)
7. Kauffman, L.: State models and the Jones polynomial. Topology **26**, 395 (1987). Statistical mechaniscs and the Jones polynomial, to appear in the Proceedings of the July, 1986, conference on Artin's braid group, Santa Cruz, California; An invariant of regular isotopy preprint
8. Turaev, V.G.: The Yang-Baxter equation and invariants of links. LOMI preprint E-3-87, Inv. Math. **92**, 527 (1988)
9. Przytycki, J.H., Traczyk, P.: Invariants of links of conway type. Kobe J. Math., **4**, 115 (1988)
10. Birman, J.: On the Jones polynomial of closed 3-braids. Invent. Math. **81**, 287 (1985). Birman, J., Wenzl, H. Link polynomials and a new algebra, preprint
11. Tsuchiya, A., Kanie, Y.: In: Conformal field theory and solvable lattice models. Adv. Stud. Pure math. **16**, 297 (1988); Lett. Math. Phys. **13**, 303 (1987)
12. Verlinde, E.: Fusion rules and modular transformations in 2d conformal field theory. Nucl. Phys. **B300**, 360 (1988)
13. Moore, G., Seiberg, N.: Polynomial equations for rational conformal field theories. To appear in Phys. Lett. B, Naturality in conformal field theory. To appear in Nucl. Phys. B, Classical and quantum conformal field theory. IAS preprint HEP-88/35
14. Schroer, B.: Nucl. Phys. **295**, 4 (1988) K.-H. Rehren, Schroer, B.: Einstein causality and Artin braids. FU preprint (1987)
15. Fröhlich, J.: Statistics of fields, the Yang-Baxter equation, and the theory of links and knots. 1987 Cargese lectures, to appear In Nonperturbative quantum field theory. New York: Plenum Press
16. Segal, G.: Conformal field theory. Oxford preprint; and lecture at the IAMP Congress, Swansea, July, 1988
17. Gromov, M.: Pseudo-holomorphic curves in symplectic manifolds. Invent. Math. **82**, 307 (1985)
18. Schwarz, A.: The partition function of degenerate quadratic functional and Ray-Singer invariants. Lett. Math. Phys. **2**, 247 (1978)
19. Schonfeld, J.: A mass term for three dimensional gauge fields. Nucl. Phys. **B185**, 157 (1981)
20. Jackiw, R., Templeton, S.: How superrenormalizable theories cure their infrared divergences. Phys. Rev. **D23**, 2291 (1981)
21. Deser, S., Jackiw, R., Templeton, S.: Three dimensional massive gauge theories. Phys. Rev. Lett. **48**, 975 (1983). Topologically massive gauge theory. Ann. Phys. NY **140**, 372 (1984)
22. Zuckerman, G.: Action principles and global geometry. In: The proceedings of the 1986 San Diego Summer Workshop, Yau, S.-T. (ed.)
23. Polyakov, A.M.: Fermi-bose transmutations induced by gauge fields. Mod. Phys. Lett. **A3**, 325 (1988)
24. Hagen, C.R.: Ann. Phys. **157**, 342 (1984)
25. Arovas, D., Schrieffer, R., Wilczek, F., Zee, A.: Statistical mechanics of anyons. Nucl. Phys. **B251[FS13]**, 117 (1985)
26. Witten, E.: Non-Abelian bosonization in two dimensions. Commun. Math. Phys. **92**, 455 (1984)
27. Ray, D., Singer, I.: Adv. Math. **7**, 145 (1971), Ann. Math. **98** 154 (1973)
28. Deser, S., Jackiw, R., Templeton, S.: In [21]; Affleck, I., Harvey, J., Witten, E.: Nucl. Phys. **B206**, 413 (1982)
Redlich, A.N.: Gauge invariance and parity conservation of three-dimensional fermions. Phys. Rev. Lett. **52**, 18 (1984)
Alvarez-Gaumé, L., Witten, E.: Gravitational anomalies. Nucl. Phys. **B234**, 269 (1983)
Alvarez-Gaumé, Della Pietra, S., Moore, G.: Anomalies and odd dimensions. Ann. Phys. (NY) **163**, 288 (1985)

Atiyah, M.F.: A note on the eta invariant (unpublished)
Singer, I.M.: Families of Dirac operators with applications to Physics. Astérisque, **1985**, p. 323

29. Atiyah, M.F., Patodi, V., Singer, I.: Math. Proc. Camb. Phil. Soc. **77**, 43 (1975), **78**, 405 (1975), **79**, 71 (1976)
30. Wilczek, F., Zee, A.: Linking numbers, spin, and statistics of solitons. Phys. Rev. Lett. **51**, 2250 (1983)
31. Friedan, D., Shenker, S.: Nucl. Phys. **B 281**, 509 (1987)
32. Belavin, A., Polyakov, A.M., Zamolodchikov, A.: Nucl. Phys. **B** (1984)
33. Atiyah, M.F., Bott, R.: The Yang-Mills equations over Riemann surfaces. Phil. Trans. R. Soc. Lond. **A 308**, 523 (1982)
34. Quillen, D.: Determinants of Cauchy-Riemann operators over a Riemann surface. Funct. Anal. Appl. **19**, 31 (1986)
35. Drinfeld, V.: Quantum groups. In: The Proceedings of the International Congress of Mathematicians, Berkeley 1986, Vol. 1, pp. 798–820
36. Gepner, D., Witten, E.: String theory on group manifolds. Nucl. Phys. B **278**, 493 (1986)
37. Kac, V.G.: Infinite dimensional Lie algebras. Cambridge: Cambridge University Press (1985)
Kac, V.G., Peterson, D.H.: Adv. Math. **53**, 125 (1984)
Kac, V.G., Wakimoto, M.: Adv. Math. **70**, 156 (1988)
38. Witten, E.: 2+1 dimensional gravity as an exactly soluble system. IAS preprint HEP-88/32
39. Segal, G.: Unitary representation of some infinite dimensional groups. Commun. Math. Phys. **80**, 301 (1981)

Communicated by A. Jaffe

Received September 12, 1988; in revised form September 27, 1988

Math. Ann. 275, 555–572 (1986)

Mathematische Annalen

Examples on Polynomial Invariants of Knots and Links

Taizo Kanenobu*

Department of Mathematics, Kyushu University 33, Fukuoka 812, Japan

In 1984, Jones [12] discovered a polynomial invariant of the isotopy type of an oriented knot or link in a 3-sphere from a representation of the braid group into certain finite dimensional von Neumann algebras. The Jones polynomial is also calculated by a recursive formula involving changing crossings in a projection of the knot or link until the unlink is obtained. Since then, some new polynomial invariants have been discovered. Generalizing Jones' method directly or using a recursive formula, Freyd et al. [7], and Przytycki and Traczyk [26] discovered at almost the same time independently a 2-variable Laurent polynomial invariant of the isotopy type of an oriented knot or link. This 2-variable Jones polynomial specializes to both the Jones and the classical reduced Alexander polynomials. The latter is also produced by the many valuable Alexander polynomial of a link. Adopting a more general recursive formula, Brandt et al. [2] and Ho [10] discovered a polynomial invariant of an unoriented knot or link, which is called the Q or absolute polynomial, see Sect. 2. And most recently Kauffman [15] discovered a 2-variable polynomial invariant which specializes to both the Jones polynomial and the Q polynomial [17]. The purpose of this paper is to consider the difference and the similarity of these polynomial invariants except for the Kauffman polynomial through several examples.

The 2-variable Jones polynomial, and therefore, the Jones and the reduced Alexander polynomials are skein invariant, see Sect. 1, but not the Q polynomial. In Sect. 4, we consider a family of ribbon knots $K(a, b)$, which is a generalization of the family of knots given by the author in [14]. This family can be completely classified up to skein equivalence by either the 2-variable Jones polynomial or the Jones and the Alexander polynomials. Each of the skein equivalent class contains infinitely many knots, and especially the knots $K(a+1, -a)$, $a=0, 1, 2, \ldots$, which are all skein equivalent, are completely classified by the Q polynomial (Theorem 3).

The V_∞ formula for the Jones polynomial (Corollaries 1.1 or 1.2) devised by Jones and Birman, is a special (or weak, in a sense) point compared with the reduced Alexander polynomial. In Sect. 5, using this formula, we give an example of

* Partially supported by the Yukawa Foundation

arbitrarily many prime knots with the same Jones but distinct Alexander polynomials.

In Sect. 6, we consider 2-bridge knots and links which are classified by Schubert [28]. We construct arbitrarily many 2-bridge knots with the same Jones and Q polynomials (Theorem 6) and arbitrarily many skein equivalent fibered 2-bridge links with the same 2-variable Alexander and Q polynomials, and therefore, the same 2-variable Jones, Jones and reduced Alexander polynomials (Theorem 7). Also we give a pair of skein equivalent 2-bridge links with the same Q polynomial, and therefore, the same 2-variable Jones, Jones and reduced Alexander polynomials, but distinct 2-variable Alexander polynomials (Theorem 8). Lastly, we give an example of a fibered 2-bridge link which is transformed into a non-isotopic link by reversing the orientation of one component, but whose skein equivalent class and 2-variable Alexander polynomial do not change (Theorem 9).

1. Jones and 2-Variable Jones Polynomials

We use the word *tangle* to mean a pair (B, α) where B is a 3-ball and α is a set of arcs and zero or more loops properly embedded in B, see [4, 24, 25]. The rectangle labelled n stands for a 2-string integral tangle with n half twists as shown in Fig. 1.

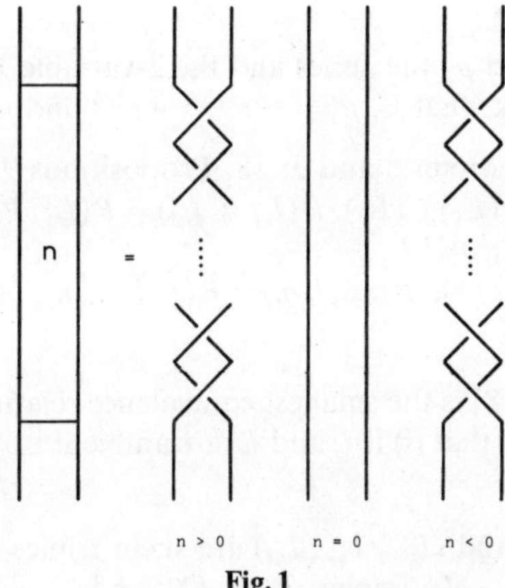

Fig. 1

The Jones polynomial $V(L; t)$ [12] is an invariant of the isotopoy type of an oriented knot or link L in a 3-sphere S^3, which is defined by the following formulas:

$$V(U) = 1 \quad \text{for the unknot } U; \tag{1.1}$$

$$t^{-1} V(L_-) - tV(L_+) = (t^{1/2} - t^{-1/2}) V(L_0), \tag{1.2}$$

where the L_i are identical except near one point where they are as in Fig. 2. We call (L_+, L_-, L_0) a *skein triple*.

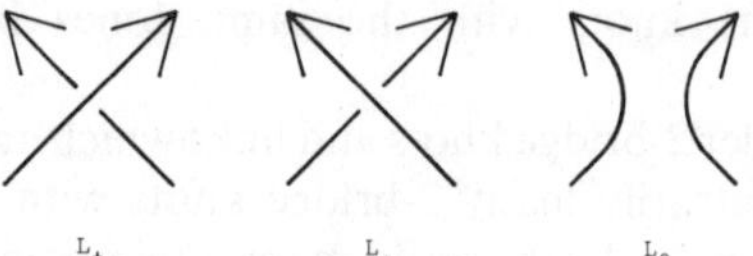

Fig. 2

The Jones polynomial was generalized to a 2-variable Laurent polynomial invariant [7], which we call the *2-variable Jones polynomial.* Here, we follow the definition of Lickorish and Millett [18]: The 2-variable Jones polynomial $P(L; l, m) \in Z[l^{\pm 1}, m^{\pm 1}]$ of an oriented knot or link L in S^3 is defined by

$$P(U) = 1 \quad \text{for the unknot } U; \tag{1.3}$$

$$lP(L_+) + l^{-1} P(L_-) + mP(L_0) = 0, \tag{1.4}$$

where (L_+, L_-, L_0) is a skein triple.

The classical reduced Alexander polynomial $\Delta(L; t)$ and the Jones polynomial $V(L; t)$ are given by the formulas:

$$P(L; i, i(t^{1/2} - t^{-1/2})) = \Delta(L; t), \tag{1.5}$$

$$P(L; it, i(t^{1/2} - t^{-1/2})) = V(L; t), \tag{1.6}$$

where $i = \sqrt{-1}$.

We denote by μ_V and μ_P the Jones and the 2-variable Jones polynomials of a trivial 2-component link, that is, $\mu_V = -t^{-1/2} - t^{1/2}$ and $\mu_P = -(l + l^{-1}) m^{-1}$.

Proposition 1.1 [12, Theorems 6 and 3; 18, Propositions 9 and 10]·

(1) $V(L_1 \# L_2) = V(L_1) V(L_2)$, $P(L_1 \# L_2) = P(L_1) P(L_2)$, *where* $L_1 \# L_2$ *is any connected sum of* L_1 *and* L_2.

(2) $V(rL; t) = V(L; t^{-1})$, $P(rL; l, m) = P(L; l^{-1}, m)$, *where* rL *is the mirror image of* L.

Skein equivalence [4, 8, 18] is the smallest equivalence relation '$\sim$' on the set of all oriented links in S^3 such that (i) if L and L' are ambient isotopic, which we denote by $L \# L'$, then $L \sim L'$;

(ii) if (L_+, L_-, L_0) and (L'_+, L'_-, L'_0) are skein triples then
(a) $L_+ \sim L'_+$ and $L_0 \sim L'_0$ implies $L_- \sim L'_-$, and
(b) $L_- \sim L'_-$ and $L_0 \sim L'_0$ implies $L_+ \sim L'_+$.

Proposition 1.2 [18, Proposition 15]. *The 2-variable Jones polynomial, and therefore, the Jones and the reduced Alexander polynomials are skein invariants.*

Any oriented link L is skein equivalent to the link obtained by reversing the orientation of every component of L. *Mutation*, see [8, 18], consists of removing a tangle with 2 arcs and zero or more loops from a link, rotating it through angle π, and replacing it. The skein equivalence class is unchanged by mutation. The 2-fold covering spaces of S^3 branched over two mutants are homeomorphic [21, 31].

Lemma 1.1. *Let $\tilde{L}_n$ be the link as shown in Fig. 3, where T is any tangle. Let $\tilde{V}_n$ be its Jones polynomial. Then*

$$\tilde{V}_{2k} = \mu_V^{-1}(1 - t^{2k})V_\infty + t^{2k}\tilde{V}_0,$$

$$\tilde{V}_{2k+1} = \mu_V^{-1}(1 - t^{2k})V_{-\infty} + t^{2k}\tilde{V}_1,$$

where $V_{\pm\infty}$ are the Jones polynomials of the links $L_{\pm\infty}$ as shown in Fig. 4.

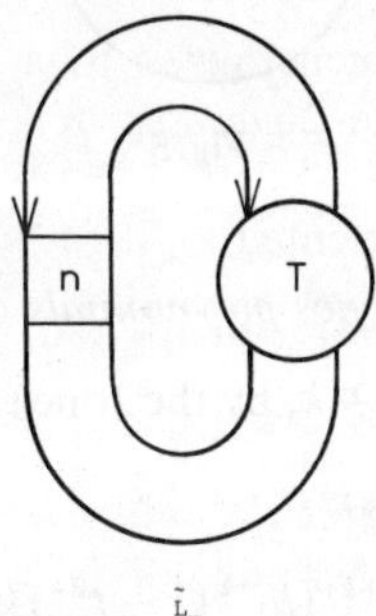

Fig. 3

Fig. 4

Proof. From (1.2), $t^{-1}\tilde{V}_n - t\tilde{V}_{n-2} = (t^{1/2} - t^{-1/2})V(-1)^n\infty$. Then

$$\tilde{V}_n - \mu_V^{-1}V(-1)^n\infty = t^2(\tilde{V}_{n-2} - \mu_V^{-1}V(-1)^n\infty)$$

$$= \begin{cases} t^n(\tilde{V}_0 - \mu_V^{-1}V_\infty) & \text{if } n \text{ is even,} \\ t^{n-1}(\tilde{V}_1 - \mu_V^{-1}V_{-\infty}) & \text{if } n \text{ is odd,} \end{cases}$$

and we obtain the formula. □

Theorem 1. *Let L_n be the link as shown in Fig. 5, where T is any tangle. Let V_n be its Jones polynomial. Let c_n be the number of the components of L_n. Let λ_n be the linking number of the component K_n with the remainder of L_n.*
Case 1. $c_0 > c_1$.

$$V_n = t^{-3(\lambda_0 + n/2)}(\mu_V^{-1}(1 - (-t)^n)V_\infty + (-t)^n\tilde{V}_0).$$

Case 2. $c_0 < c_1$.

$$V_n = t^{-3(\lambda_1 + n/2 - 1/2)}(\mu_V^{-1}(1 - (-t)^{n-1})V_{-\infty} + (-t)^{n-1}\tilde{V}_1).$$

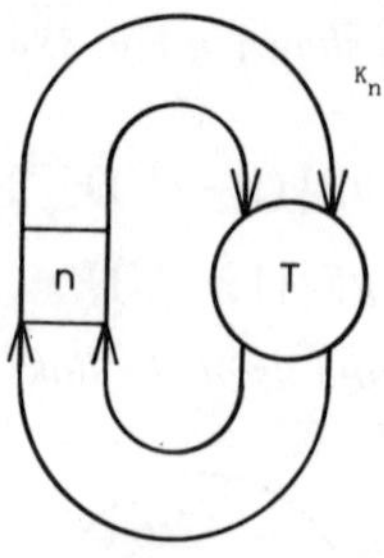

Fig. 5

Here $V_{\pm\infty}$, $\tilde{V}_0$, and $\tilde{V}_1$ are the Jones polynomials given in Lemma 1.1.

Proof. For Case 1, since $\lambda_{2k} = \lambda_0 + k$, by the Jones reversing result, see [19, 22], we have

$$\begin{aligned} V_{2k} &= t^{-3(\lambda_0+k)} \tilde{V}_{2k} \\ &= t^{-3(\lambda_0+k)} (\mu_V^{-1}(1-t^{2k}) V_\infty + t^{2k} \tilde{V}_0). \end{aligned}$$

From (1.2),

$$t^{-1} V_{2k} - t V_{2k+2} = (t^{1/2} - t^{-1/2}) V_{2k+1}.$$

Substituting the above formulas, we have

$$V_{2k+1} = t^{-3\lambda_0 - 3k - 1/2} (\mu_V^{-1}(t^{2k} + t^{-1}) V_\infty - t^{2k} \tilde{V}_0),$$

and we obtain the desired formula.

For case 2, we can prove similarly and the proof is complete. □

From Theorem 1, we can easily deduce the V_∞ formula discovered by Jones and Birman [1], see also [17, 20].

Corollary 1.1.

$$t^{1/2} V_1 - t^{-1/2} V_{-1} = \begin{cases} (t^{1/2} - t^{-1/2}) t^{-3\lambda_0} V_\infty & \text{for Case 1,} \\ (t^{1/2} - t^{-1/2}) t^{-3(\lambda_1 - 1/2)} V_{-\infty} & \text{for Case 2.} \end{cases}$$

Combining (1.2) and Corollary 1.1, we have

Corollary 1.2. *For Case 1,*

$$V_1 = -t^{-1/2} V_0 - t^{-3\lambda_1 - 1} V_\infty,$$

$$V_{-1} = -t^{1/2} V_0 - t^{-3\lambda_0 + 1} V_\infty.$$

For Case 2,

$$V_1 = -t^{-1/2} V_0 - t^{-3\lambda_1 + 1/2} V_{-\infty},$$

$$V_{-1} = -t^{1/2} V_0 - t^{-3\lambda_{-1} - 1/2} V_{-\infty}.$$

2. Q Polynomial

The Q polynomial or the absolute polynomial $Q(L;x) \in Z[x^{\pm 1}]$ [2, 10] is an invariant of the isotopy type of an unoriented knot or link in S^3, which is defined by the following formulas:

$$Q(U)=1 \quad \text{for the unknot } U; \tag{2.1}$$

$$Q(L_+)+Q(L_-)=x(Q(L_0)+Q(L_\infty)), \tag{2.2}$$

where the L_i are identical except near one point where they are as in Fig. 6. The Q polynomial has the following property:

Proposition 2.1 [2, Property 1].
(1) $Q(L_1 \# L_2)=Q(L_1)\,Q(L_2)$.
(2) $Q(rL)=Q(L)$.
(3) *If L_2 is a mutant of L_1, then $Q(L_1)=Q(L_2)$.*

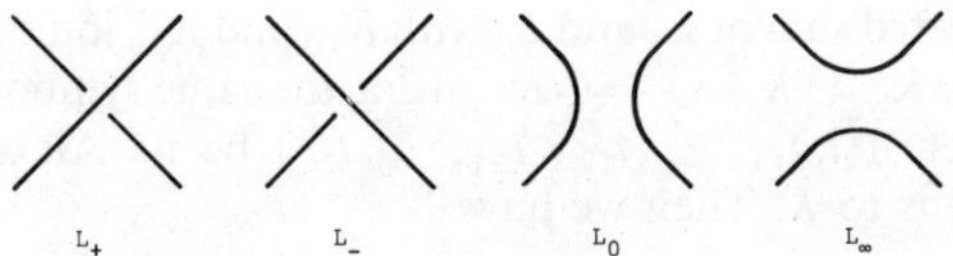

Fig. 6

We use the notation $\mu_Q = 2x^{-1}-1$ for the Q polynomial of a trivial 2-component link. Let Q_n be the Q polynomial of the unoriented link of Figs. 4 and 5. Then from (2.2),

$$Q_{n-1}+Q_{n+1}=xQ_n+xQ_\infty.$$

Let $R_n=Q_n-\mu_Q^{-1}Q_\infty$. Then we have

$$R_{n+1}-xR_n+R_{n-1}=0.$$

Putting $\alpha+\beta=x$ and $\alpha\beta=1$, we have

$$R_{n+1}-\alpha R_n=\beta(R_n-\alpha R_{n-1})=\beta^n(R_1-\alpha R_0),$$

and

$$R_{n+1}-\beta R_n=\alpha(R_n-\beta R_{n-1})=\alpha^n(R_1-\beta R_0),$$

and therefore we have

$$R_n=\sigma_n R_1-\sigma_{n-1}R_0,$$

where σ_n is a symmetric polynomial defined as follows:

$$\sigma_n=\begin{cases} \dfrac{\alpha^n-\beta^n}{\alpha-\beta} & \text{if } n>0, \\ 0 & \text{if } n=0, \\ -\dfrac{\alpha^{-n}-\beta^{-n}}{\alpha-\beta} & \text{if } n<0. \end{cases}$$

It is easy to see that $\sigma_{-n}=-\sigma_n$ and $x\sigma_n=\sigma_{n-1}+\sigma_{n+1}$. Thus we obtain

Proposition 2.2. $Q_n=\sigma_n Q_1-\sigma_{n-1}Q_0-\mu_Q^{-1}(\sigma_n-\sigma_{n-1}-1)Q_\infty$.

3. Alexander Polynomial

The Alexander polynomial $\Delta(L;t_1,t_2,\dots,t_\nu)$ of a ν-component link $L=L_1\cup L_2\cup\cdots\cup L_\nu$ in S^3 is an element of the polynomial ring $Z[t_1^{\pm1},t_2^{\pm1},\cdots,t_\nu^{\pm1}]$, and is determined only up to multiplication by a unit $\pm t_1^{n_1}t_2^{n_2}\cdots t_\nu^{n_\nu}$. The reduced Alexander polynomial $\Delta(L;t)$ is given by the following formula:

$$\Delta(L;t)=\begin{cases}(t-1)\Delta(L;t,t,\cdots,t) & \text{if } \nu\geqq 2,\\ \Delta(L;t) & \text{if } \nu=1,\end{cases}$$

up to a unit $\pm t^n$.

Let $L_i=K_{i0}\cup K_{i1}\cup\cdots\cup K_{i\nu_i}$, $i=1,2$, be a (ν_i+1)-component link in S^3. Let $\Delta_i(t_{i0},t_{i1},\cdots,t_{i\nu_i})$ be its Alexander polynomial, where each t_{ij} corresponds to K_{ij}. Let L be the connected sum of L_1 and L_2 with K_{10} and K_{20} joined, which we present by $K\cup K_{11}\cup\cdots\cup K_{1\nu_1}\cup K_{21}\cup\cdots\cup K_{2\nu_2}$ using the same symbols except the joined component K. Let $\Delta(t,t_{11},\cdots,t_{1\nu_1},t_{21},\cdots,t_{2\nu_2})$ be its Alexander polynomial, where t corresponds to K. Then we have

Proposition 3.1. *Suppose* $\nu_1\geqq 1$. *Then*

$$\Delta(t,t_{11},\cdots,t_{1\nu_1},t_{21},\cdots,t_{2\nu_2})=\begin{cases}(t-1)\Delta_1(t,t_{11},\cdots,t_{1\nu_1})\,\Delta_2(t,t_{21},\cdots,t_{2\nu_2}) & \text{if } \nu_2\geqq 1,\\ \Delta_1(t,t_{11},\cdots,t_{1\nu_1})\,\Delta_2(t) & \text{if } \nu_2=0.\end{cases}$$

Proposition 3.2. *Let* $\tilde{L}_{2n}=\tilde{K}_{2n,1}\cup\tilde{K}_{2n,2}\cup\cdots\cup\tilde{K}_{2n,\nu}$ *and* $\tilde{L}_\infty=\tilde{K}_{\infty,0}\cup K_{\infty,1}\cup\cdots\cup\tilde{K}_{\infty,\nu}$ *be the links as shown in Fig. 7, where T is any tangle. Let* $\Delta_{2n}(t_1,t_2,\cdots,t_\nu)$ *and* $\Delta_\infty(t_0,t_1,\cdots,t_\nu)$ *be the Alexander polynomials of* $\tilde{L}_{2n}$ *and* $\tilde{L}_\infty$, *respectively, where* t_j *corresponds to* $\tilde{K}_{i,j}$. *Then they can be normalized so that*

$$\Delta_{2n}(t_1,t_2,\cdots,t_\nu)=n(t_1-1)\,\Delta_\infty(t_1,t_1,t_2,\cdots,t_\nu)+\Delta_0(t_1,t_2,\cdots,t_\nu).$$

The proofs of these propositions are standard, see [5, 30], so we omit them.

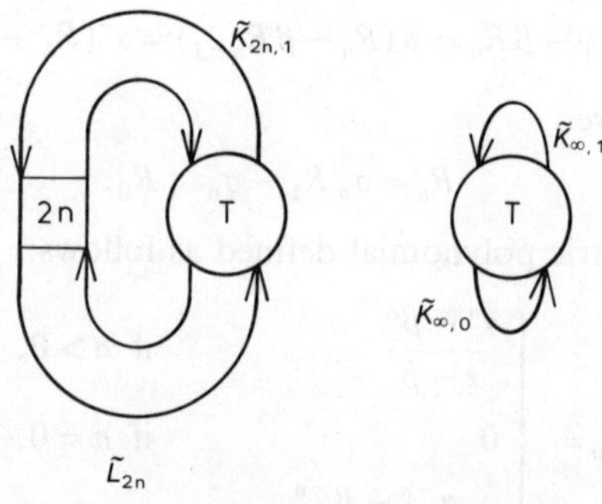

Fig. 7

4. $K(a,b)$

In this section we consider a family of knots $K(a,b)$ as shown in Fig. 8. The family of knots $K_{p,q}$ given in [14] is $K(-2p,-2q)$ in this notation. In the same way as in [14], we have

Proposition 4.1. *$K(a,b) \approx K(b,a)$ and $rK(a,b) \approx K(-a,-b)$.*

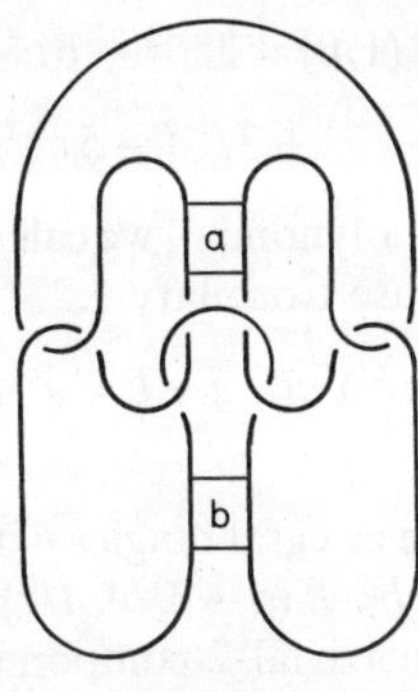

Fig. 8

The knots with small crossings [27, Appendix C] contained in this family are the following:

$$K(0,0)=4_1 \# 4_1,\ K(0,-1)=8_8,\ K(1,-1)=8_9,$$
$$K(2,-1)=10_{129},\ K(2,0)=10_{137},\ K(1,1)=10_{155},$$
$$K(2,-3)=13_{6714}.$$

For 13_{6714}, we refer [18, Fig. 37].

Professors J. Przytycki and P. Traczyk remarked the author that the two knots $K(2p,2q)$ and $K(2p',2q')$ with $p+q=p'+q'$ not only have the same 2-variable Jones polynomial but are skein equivalent. Here we have

Proposition 4.2. *Let $P(a,b)$, $V(a,b)$, and $\Delta(a,b)$ be the 2-variable Jones, Jones, and Alexander polynomials of $K(a,b)$, respectively. Let $a=2p+\varepsilon$ and $b=2q+\delta$, where $\varepsilon,\ \delta=0,1$. Then*

$$P(a,b)=(-l^2)^{p+q}(P(\varepsilon,\delta)-1)+1,$$

where

$$P(0,0)=((l^{-2}+1+l^2)+m^2)^2,$$
$$P(0,1)=P(1,0)=(-l^{-4}-l^{-2}+2+l^2)$$
$$+(l^{-4}+2l^{-2}-2-l^2)m^2+(-l^{-2}+1)m^4,$$
$$P(1,1)=(l^{-4}+4l^{-2}+3)+(-3l^{-4}-8l^{-2}-3)m^2$$
$$+(l^{-4}+5l^{-2}+1)m^4-l^{-2}m^6;$$
$$V(a,b)=(-t)^{a+b}(V(0,0)-1)+1,$$

where

$$V(0,0)=(t^{-2}-t^{-1}+1-t+t^2)^2;$$

$$\Delta(a,b)=\Delta(\varepsilon,\delta),$$

where

$$\Delta(0,0)=(t^{-1}-3+t)^2,$$

$$\Delta(0,1)=\Delta(1,0)=2t^{-2}-6t^{-1}+9-6t+2t^2,$$

$$\Delta(1,1)=-t^{-3}+3t^{-2}-5t^{-1}+7-5t+3t^2-t^3.$$

Proof. For the 2-variable Jones polynomial, we calculate in the same way as in [14]. For the Jones polynomial, we use Corollary 1.2. □

Proposition 4.3. $K(a,b)\sim K(a',b')$ *iff* $a+b=a'+b'$ *and* $a\equiv a' \pmod 2$ *when* $a+b\equiv 0 \pmod 2$.

Proof. Consider a crossing in the integral tangles with a and b half twists in $K(a,b)$. Then it is easy to see that $(K(a,b), K(a-2,b), U^2)$ and $(K(a,b), K(a,b-2), U^2)$ are skein triples, where U^2 is a trivial 2-component link. Thus $K(a-2,b)$ and $K(a,b-2)$ are skein equivalent. Since $K(a,b)\approx K(b,a)$, if $a+b=a'+b'$ and $a\equiv a' \pmod 2$ when $a+b\equiv 0 \pmod 2$, then $K(a,b)$ and $K(a',b')$ are skein equivalent. Conversely, from the 2-variable Jones polynomial of $K(a,b)$, we see that the above condition is necessary for $K(a,b)$ and $K(a',b')$ to be skein equivalent. This completes the proof. □

Remarks. (1) The family $K(a,b)$ can be completely classified up to skein equivalence by either 2-variable Jones polynomial or the Jones and the Alexander polynomials.

(2) $4_1 \# 4_1$ and 8_9 may be the simplest example of knots with the same Jones polynomial.

In Example 16 of [18], 8_8, $r(10_{129})$, and 13_{6714} are given as an example of knots with the same 2-variable Jones polynomial and it is shown that 8_8 and 13_{6714} are skein equivalent and not mutant one another. It is also shown that 8_8 has unknotting number 2 and $r(10_{129})$ has 1, and so, the 2-variable Jones polynomial cannot give complete unknotting number information. Moreover, since we have shown that 8_8 and $r(10_{129})$ are skein equivalent, we can answer Question 10 of [18]:

Theorem 2. *There exist two knots with distinct unknotting numbers which are skein equivalent but not mutant one another.*

It is remarked in [14] that $K(-2p,-2q)$ is a generalized symmetric union [16] of the figure-eight knot. $K(a,0)$, where a is odd, is a symmetric skew union [16] of the figure-eight knot, and so, a ribbon knot. For other $K(a,b)$, since $(K(a,b), K(a-2,b), U^2)$ are skein triples, we have

Proposition 4.4. $K(a,b)$ *is a ribbon knot.*

Now we calculate $Q(a,b)$ the Q polynomial of $K(a,b)$. From (2.2),

$$Q(a,b)+Q(a-2,b)=xQ(a-1,b)+x\mu_Q.$$

Letting $R(a,b)=Q(a,b)-1$, we have

$$R(a,b)-xR(a-1,b)+R(a-2,b)=0. \tag{4.1}$$

In the same way as in Proposition 2.2, we have

$$\begin{aligned}R(a,b)&=\sigma_a R(1,b)-\sigma_{a-1}R(0,b)\\&=\sigma_a\sigma_{b+1}R(1,0)-\sigma_a\sigma_b R(1,-1)-\sigma_{a-1}\sigma_b R(0,1)+\sigma_{a-1}\sigma_{b-1}R(0,0).\end{aligned}$$

Since $R(1,0)=R(0,1)=R(0,-1)$, from (4.1), we have

$$xR(0,0)=2R(1,0).$$

Then we obtain

$$R(a,b)=-\sigma_a\sigma_b R(1,-1)+(\sigma_a\sigma_{b+1}-\sigma_{a-1}\sigma_b+2x^{-1}\sigma_{a-1}\sigma_{b-1})R(1,0)$$

By $\sigma_{a-1}+\sigma_{a+1}=x\sigma_a$, we have

Proposition 4.5. $Q(a,b)=-\sigma_a\sigma_b(Q(8_9)-1)+x^{-1}(\sigma_{a+1}\sigma_{b+1}+\sigma_{a-1}\sigma_{b-1})(Q(8_8)-1)+1$, *where* $Q(8_8)=1+4x+6x^2-10x^3-14x^4+4x^5+8x^6+2x^7$, $Q(8_9)=-7+4x+16x^2-10x^3-16x^4+4x^5+8x^6+2x^7$.

Corollary 4.1. $\deg Q(a,-a)=2a+5$, $a\geqq 1$, *and* $\deg Q(0,0)=6$, $\deg Q(a+1,-a)=2a+6$, $a\geqq 1$, *and* $\deg Q(1,0)=7$.

For $c\geqq 2$ and $a\geqq 0$, $Q(a+c,-a)$ cannot be distinguished by the degrees. Combining Proposition 4.2 and the above corollary, we have

Theorem 3. *There exist infinitely many knots which are skein equivalent and therefore they have the same 2-variable Jones, Jones and Alexander polynomials, but are distinguished by the Q polynomial.*

Questions. (1) Can the family $K(a,b)$ be completely classified by the 2-variable Jones and Q polynomials, or the Jones, Alexander, and Q polynomials?

(2) $K(a,b)\approx K(a',b')$ iff $(a,b)=(a',b')$ or (b',a')?

We give other properties of $K(a,b)$, which are shown in the same way as in [14]:

Proposition 4.6. $K(a,b)$ *are 3-bridge except for* $K(1,0)$ *and* $K(1,-1)$.

Proposition 4.7. $K(a,b)$ *are prime except for* $K(0,0)$.

5. $K(p_1,p_2,\cdots,p_n)$, $n\geqq 3$

In this section we consider a family of knots $K(p_1,p_2,\cdots,p_n)$, $n\geqq 3$, as shown in Fig. 9. The knots with small crossings [27, Appendix C] contained in this family are the following:

$$K(0,0,0)=9_{41},\ K(0,0,-1)=9_{27}.$$

It is easy to see

Proposition 5.1. $K(p_1,p_2,\cdots,p_n)\approx rK(-p_1-1,-p_2-1,\cdots,-p_n-1)$.

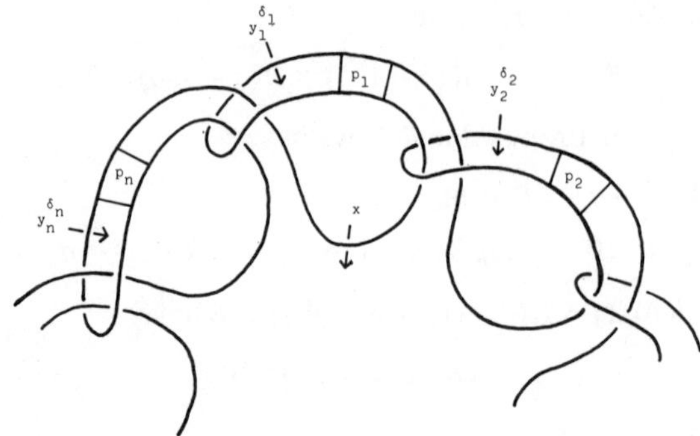

Fig. 9

Considering the integral tangle with p_i half twists in $K(p_1, \cdots, p_i, \cdots, p_n)$, we see that $(K(p_1, \cdots, p_i, \cdots, p_n),\ K(p_1, \cdots, p_i - 2, \cdots, p_n), U^2)$, $1 \leqq i \leqq n$, are skein triples; this also shows that $K(p_1, p_2, \cdots, p_n)$ is a ribbon knot. Hence $K(p_1, \cdots, p_i - 2, \cdots, p_n)$ and $K(p_1, \cdots, p_j - 2, \cdots, p_n)$ are skein equivalent. Since $K(p_1, \cdots, p_n)$ and $K(p_{\phi(1)}, \cdots, p_{\phi(n)})$ are isotopic, where ϕ is a cyclic permutation of $1, 2, \cdots, n$, we have

Proposition 5.2. *If $p_1 + p_2 + \cdots + p_n = q_1 + q_2 + \cdots + q_n$ and $p_i \equiv q_{\phi(i)} \pmod 2$ for all i, where ϕ is a cyclic permutation of $1, 2, \cdots, n$, then $K(p_1, p_2, \cdots, p_n)$ and $K(q_1, q_2, \cdots, q_n)$ are skein equivalent.*

Proposition 5.3. *Let $P(p_1, p_2, \cdots, p_n)$, $V(p_1, p_2, \cdots, p_n)$, and $\Delta(p_1, p_2, \cdots, p_n)$ be the 2-variable Jones, Jones, and Alexander polynomials of $K(p_1, p_2, \cdots, p_n)$. Let ε_i be 0 if p_i is even and 1 if p_i is odd. Let e be the number of 0 in $\varepsilon_1, \varepsilon_2, \cdots, \varepsilon_n$. Then*

$$P(p_1, p_2, \cdots, p_n) = (-l^2)^{\sum_{i=1}^{n}(p_i - \varepsilon_i)/2}(P(\varepsilon_1, \varepsilon_2, \cdots, \varepsilon_n) - 1) + 1,$$

$$V(p_1, p_2, \cdots, p_n) = (-t)^{\sum_{i=1}^{n} p_i}(V(0, 0, \cdots, 0) - 1) + 1,$$

$$\Delta(p_1, p_2, \cdots, p_n) = \Delta(\varepsilon_1, \varepsilon_2, \cdots, \varepsilon_n) = f(t) f(t^{-1}),$$

where $f(t) = (-t)^e - (1 - t)^n$.

Proof. For the 2-variable Jones polynomial, we can prove in the same way as in Proposition 4.2. For the Jones polynomial,

$$V(p_1, p_2, \cdots, p_n; t) = P(p_1, p_2, \cdots, p_n; it, i(t^{1/2} - t^{-1/2}))$$

$$= (-t)^{\sum_{i=1}^{n}(p_i - \varepsilon_i)}(V(\varepsilon_1, \varepsilon_2, \cdots, \varepsilon_n) - 1) + 1,$$

and by Corollary 1.2,

$$V(\varepsilon_1, \cdots, 1, \cdots, \varepsilon_n) = -t(V(\varepsilon_1, \cdots, 0, \cdots, \varepsilon_n) - 1) + 1,$$

and so we obtain the formula.

For the Alexander polynomial, by [9] or [11], $K(p_1, p_2, \cdots, p_n)$ and $K(\varepsilon_1, \varepsilon_2, \cdots, \varepsilon_n)$ have the same associated ribbon 2-knot, which we denote by

$K^2(\varepsilon_1, \varepsilon_2, \cdots, \varepsilon_n)$. Using the meridians in Fig. 9, $\pi_1(S^4 - K^2(\varepsilon_1, \varepsilon_2, \cdots, \varepsilon_n))$ has the following presentation [32]:

$$\langle x, y_1, y_2, \cdots, y_n;\ y_1 = y_2^{-\delta_2} x y_2^{\delta_2},\ y_2 = y_3^{-\delta_3} x y_3^{\delta_3}, \cdots, y_n = y_1^{-\delta_1} x y_1^{\delta_1} \rangle,$$

where $\delta_j = 1 - 2\varepsilon_j$. By the free differential calculus [5], we obtain the Alexander polynomial of $K^2(\varepsilon_1, \varepsilon_2, \cdots, \varepsilon_n)$:

$$1 - (1 - t^{-\delta_1})(1 - t^{-\delta_2}) \cdots (1 - t^{-\delta_n}),$$

which equals $f(t)$ up to $(-t)^e$. By [6], $f(t) f(t^{-1})$ is the Alexander polynomial of $K(\varepsilon_1, \varepsilon_2, \cdots, \varepsilon_n)$. □

Proposition 5.4. *Suppose that* $n \geqq 4$ *and* $p_2,\ p_j \neq 0, -1$ *for some* $j \geqq 4$. *Then* $K(p_1, p_2, \cdots, p_n)$ *is a prime knot.*

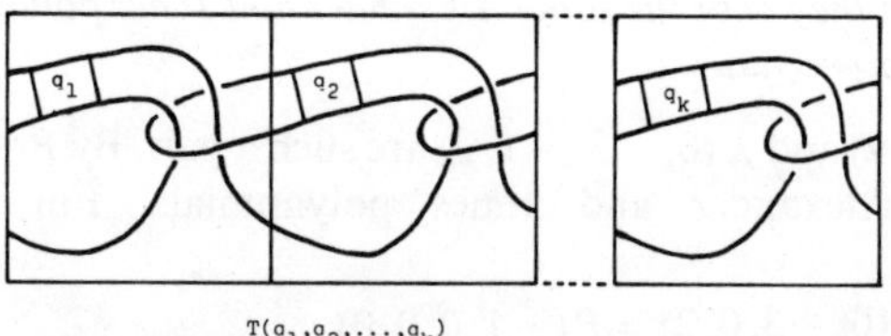

Fig. 10

Proof. Let $T(q_1, q_2, \cdots, q_k)$ be the tangle as shown in Fig. 10. Then $K(p_1, p_2, \cdots, p_n)$ is a union of $T(p_1, p_2, \cdots, p_{j-1})$ and $T(p_j, p_{j+1}, \cdots, p_n)$. Let $(B, \alpha_1 \cup \alpha_2 \cup \alpha_3) = T(p_1, p_2)$. Since $p_2 \neq 0, -1$, there are one knotted arc, say α_1, and two unknotted arcs α_2 and α_3. The knotted factor is the 2-bridge knot with Conway's notation $C(2, p_2)$, see Sect. 6. $(B, \alpha_1 \cup \alpha_2)$ with ears, that is, the result of adding a trivial tangle to $(B, \alpha_1 \cup \alpha_2)$ on the outside of B can be a trivial knot, see Fig. 11a. $(B, \alpha_1 \cup \alpha_3)$ with ears can be a 2-bridge knot $C(2, p_2 + 2)$, which is not $C(2, p_2)$, see Fig. 11b. Then by Lemma 5.4 of [24], $(B, \alpha_1 \cup \alpha_2)$ and $(B, \alpha_1 \cup \alpha_3)$ are prime tangles. Therefore by Lemma 5.6 of [24], $(B, \alpha_1 \cup \alpha_2 \cup \alpha_3)$ is a prime tangle.

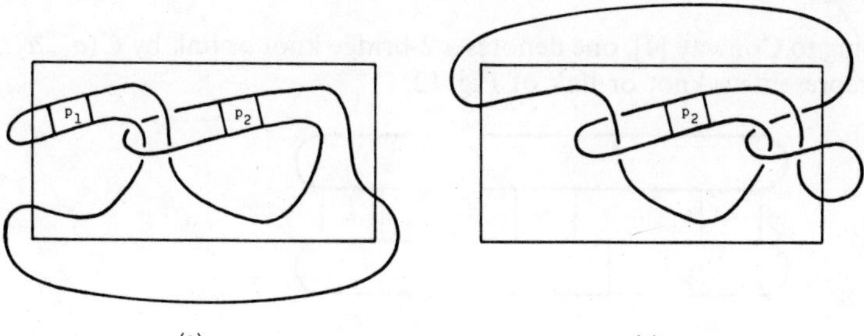

Fig. 11 a, b

Applying Theorem 3 of [25], we see that $T(p_1, p_2, p_3)$ which is a union of $T(p_1, p_2)$ and $T(p_3)$, is a prime tangle. Continuing this, $T(p_1, p_2, \cdots, p_{j-1})$ is shown to be a prime tangle. Similarly, $T(p_j, p_{j+1}, \cdots, p_n)$ is also prime, and thus by Theorem 2 of [25] $K(p_1, p_2, \cdots, p_n)$ is a prime knot. This completes the proof. □

There exist a pair of knots with the same Jones polynomial but distinct Alexander polynomial. For example, $4_1 \# 4_1$ and 8_9 in Sect. 4 are such a pair. By composition, we get arbitrarily many knots with the same Jones polynomial but distinct Alexander polynomials. Combining Propositions 5.3 and 5.4, we have

Theorem 4. *There exist arbitrarily many prime knots with the same Jones polynomial but distinct Alexander polynomials.*

In [7], Lickorish and Millett gives $K = 11_{388}$ as an example of a knot such that $P(K) \neq P(rK)$, $V(K) = V(rK)$ and $\Delta(K) = \Delta(rK)$. So $K \# rK$ and $K \# K$ are a pair of knots such that one is not the mirror image of the other and that they have the same Jones and Alexander polynomials but distinct 2-variable Jones polynomials. Using $K(p_1, p_2, \cdots, p_n)$ we have

Theorem 5. *There exist a pair of prime knots such that one is not the mirror image of the other and that they have the same Alexander and Jones polynomials but distinct 2-variable Jones polynomials.*

Proof. $K(0, -3, 0, 2)$ and $K(0, -3, -1, 3)$ are such a pair. By Proposition 5.3, they have the same Alexander and Jones polynomials. For 2-variable Jones polynomials,

$$P(0, -3, 0, 2) = P(-1, 0, 0, 0)$$

and

$$\begin{aligned} P(0, -3, -1, 3) &= (-l^2)(P(0, -1, -1, -1) - 1) + 1 \\ &= (-l^2)(P(-1, 0, 0, 0; l^{-1}, m) - 1) + 1, \end{aligned}$$

since $K(0, -1, -1, -1) = rK(-1, 0, 0, 0)$. Now

$$\begin{aligned} P(-1, 0, 0, 0) = {} & (l^{-2} + 5 + 6l^2 + 4l^4 + l^6) - (4l^{-2} + 12 + 13l^2 + 6l^4 + l^6)m^2 \\ & + (3l^{-2} + 12 + 10l^2 + 3l^4)m^4 - (l^{-2} + 5 + 3l^2)m^6 + m^8, \end{aligned}$$

and so $P(0, -3, 0, 2) \neq P(0, -3, -1, 3)$. □

6. 2-Bridge Knots and Links

According to Conway [4], one denotes a 2-bridge knot or link by $C(a_1, a_2, \cdots, a_n)$ which represents a knot or link of Fig. 12.

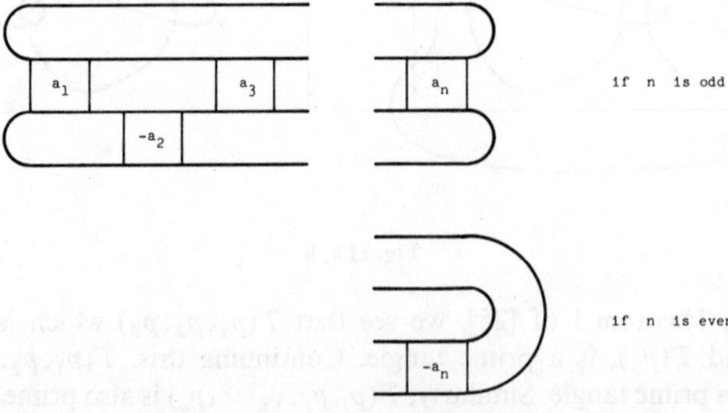

Fig. 12

Lemma 6.1. *Let f be the Q polynomial of a 2-bridge knot or link $C(a_1, a_2, \cdots, a_n)$ and Q_b that of $C(a_1, a_2, \cdots, a_n, b, -a_n, \cdots, -a_2, -a_1)$. Then*

$$Q_b = \tfrac{1}{2}(\mu_Q - \mu_Q^{-1} f^2)(\sigma_{b+1} - \sigma_{b-1}) + \mu^{-1} f^2.$$

Proof. $C(a_1, a_2, \cdots, a_n, b, -a_n, \cdots, -a_2, -a_1)$ is the mirror image of $C(-a_1, -a_2, \cdots, -a_n, -b, a_n, \cdots, a_2, a_1)$, which is isotopic to $C(a_1, a_2, \cdots, a_n, -b, -a_n, \cdots, -a_2, -a_1)$, and thus by Proposition 2.1, $Q_b = Q_{-b}$. Then applying Proposition 2.2, we have

$$Q_b = Q_{-b} = (\sigma_b Q_1 - \sigma_{b-1} Q_0) - (\sigma_b - \sigma_{b-1} - 1)\mu_Q^{-1} f^2.$$

From (2.2),

$$2Q_1 = Q_1 + Q_{-1} = xQ_0 + xf^2.$$

Then noting, $Q_0 = \mu_Q$ and $x\sigma_b = \sigma_{b+1} + \sigma_{b-1}$, we obtain the desired formula. □

Lemma 6.2. *Let $C(a_1, a_2, \cdots, a_n)$ be a 2-bridge knot and g its Jones polynomial. If b is odd, then $C(a_1, a_2, \cdots, a_n, b, -a_1, -a_2, \cdots, -a_n)$ is also a knot and its Jones polynomial is*

$$t^{-3\beta/2}(\mu^{-1}(1 + t^\beta) g(t) g(t^{-1}) - t^\beta \mu),$$

where $\beta = (-1)^n b$.

Proof. Apply Theorem 1.

Theorem 6. *There exist arbitrarily many 2-bridge knots with the same Jones and Q polynomials.*

Proof. Let $C(a_1, a_2, \cdots, a_n)$ and $C(b_1, b_2, \cdots, b_n)$ be the 2-bridge knots having the same Jones and Q polynomials with

$$A = \begin{pmatrix} 0 & 1 \\ 1 & a_1 \end{pmatrix} \begin{pmatrix} 0 & 1 \\ 1 & a_2 \end{pmatrix} \cdots \begin{pmatrix} 0 & 1 \\ 1 & a_n \end{pmatrix} = \begin{pmatrix} s & q \\ r & p \end{pmatrix}$$

and

$$B = \begin{pmatrix} 0 & 1 \\ 1 & b_1 \end{pmatrix} \begin{pmatrix} 0 & 1 \\ 1 & b_2 \end{pmatrix} \cdots \begin{pmatrix} 0 & 1 \\ 1 & b_n \end{pmatrix} = \begin{pmatrix} v & t \\ u & p \end{pmatrix},$$

where $p \neq \pm 1$. Then the 2-fold branched covering spaces of S^3 branched over these 2-bridge knots are the lens spaces $L(p, q)$ and $L(p, t)$, respectively, see [28]. If d is odd, then by Lemmas 6.1 and 6.2, the following four 2-bridge knots have the same Q and Jones polynomials:

$$C(a_1, \cdots, a_n, d, -a_n, \cdots, -a_1),\ C(a_n, \cdots, a_1, d, -a_1, \cdots, -a_n),$$
$$C(b_1, \cdots, b_n, d, -b_n, \cdots, -b_1),\ C(b_n, \cdots, b_1, d, -b_1, \cdots, -b_n).$$

Since

$$\hat{A} = \begin{pmatrix} 0 & 1 \\ 1 & -a_1 \end{pmatrix} \begin{pmatrix} 0 & 1 \\ 1 & -a_2 \end{pmatrix} \cdots \begin{pmatrix} 0 & 1 \\ 1 & -a_n \end{pmatrix} = (-1)^n \begin{pmatrix} s & -q \\ -r & p \end{pmatrix},$$

$$A \begin{pmatrix} 0 & 1 \\ 1 & d \end{pmatrix} \hat{A}^T = (-1)^n \begin{pmatrix} -dq^2 & (-1)^n + dpq \\ (-1)^n - dpq & dp^2 \end{pmatrix},$$

and

$$A^T \begin{pmatrix} 0 & 1 \\ 1 & d \end{pmatrix} \hat{A} = (-1)^n \begin{pmatrix} -dr^2 & (-1)^n + dpr \\ (-1)^n - dpr & dp^2 \end{pmatrix},$$

where A^T is the transposed matrix of A, the first and the second 2-bridge knots have lens spaces $L(dp^2, 1+dpq)$ and $L(dp^2, 1+dpr)$ as their 2-fold branched covering spaces. In the same way, the third and the fourth ones have $L(dp^2, 1+dpt)$ and $L(dp^2, 1+dpu)$. If $L(p,q)$ and $L(p,t)$ are not homeomorphic, that is, $\pm q \not\equiv t^{\pm 1} (\operatorname{mod} p)$, and $q \not\equiv \pm r$, $t \not\equiv \pm u$ $(\operatorname{mod} p)$, then these four lens spaces are not homeomorphic each other. Thus the four 2-bridge knots are neither isotopic nor mutant each other. Furthermore it holds that $(-1)^n + dpq \not\equiv \pm((-1)^n - dpq)$, $(-1)^n + dpr \not\equiv \pm((-1)^n - dpr)$, $(-1)^n + dpt \not\equiv \pm((-1)^n - dpt)$, and $(-1)^n + dpu \not\equiv \pm((-1)^n \pm dpu)$ $(\operatorname{mod} dp^2)$, and so we can continue this construction.

Beginning with $C(2,3)$ such that $\left(\begin{smallmatrix} 0 & 1 \\ 1 & 2 \end{smallmatrix}\right) \left(\begin{smallmatrix} 0 & 1 \\ 1 & 3 \end{smallmatrix}\right) = \left(\begin{smallmatrix} 1 & 3 \\ 2 & 7 \end{smallmatrix}\right)$, we obtain 2^N distinct 2-bridge knots for any integer N with the same Jones and Q polynomials. This completes the proof. □

Remarks. (1) The 2-bridge knot $C(a_1, \cdots, a_n, \pm 1, -a_n, \cdots, -a_1)$ is a symmetric skew union [16] of $C(a_1, \cdots, a_n)$.

(2) By the above constructin, $C(3,2,1,-2,-3) = 10_{22}$ and $C(2,3,1,-3,-2) = 10_{35}$ have the same Jones and Q polynomials. This may be the simplest example of a pair of 2-bridge knots with the same Jones and Q polynomials. But they have distinct Alexander polynomials, and therefore, distinct 2-variable Jones polynomials. In fact, they have distinct genera. Do the 2-bridge knots with the same Jones and Q polynomials constructed in the above proof have distinct Alexander polynomials? Do they have distinct genera? Does there exist an example of 2-bridge knots with the same 2-variable Jones polynomial?

Next, we consider an oriented 2-bridge knot or link putting in the form as shown in Fig. 13, which we denote by $D(c_1, c_2, \cdots, c_n)$, see [29]. $C(2c_1, 2c_2, \cdots, 2c_n)$ is isotopic to $D(c_1, c_2, \cdots, c_n)$ with its orientation ignored. Since a 2-bridge knot or link is alternating, it is fibered iff the leading coefficient of its reduced Alexander polynomial is ± 1 [23], which equals $c_1 c_2 \cdots c_n$ for $D(c_1, c_2, \cdots, c_n)$ [29]. Thus we have

Lemma 6.3. *$D(c_1, c_2, \cdots, c_n)$ is fibered iff $c_i = \pm 1$ for all i.*

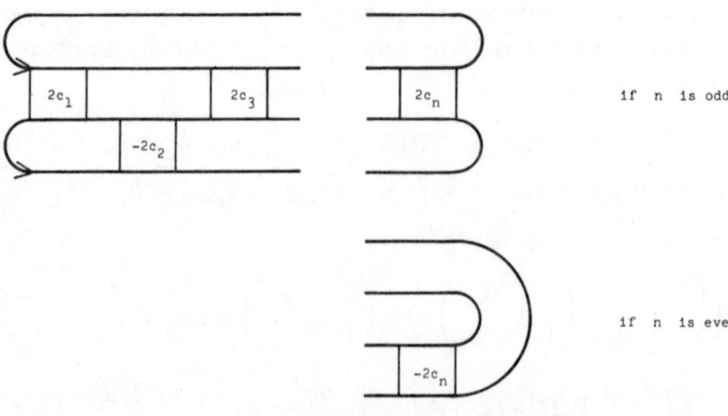

Fig. 13

Lemma 6.4. *Let $\Delta(a_1, a_2, \cdots, a_{2k+1}; t_1, t_2)$ be the 2-variable Alexander polynomial of the 2-bridge link $D(a_1, a_2, \cdots, a_{2k+1})$. Then*

$$\Delta(c_1, c_2, \cdots, c_{2n+1}, d, -c_{2n+1}, \cdots, -c_2, -c_1; t_1, t_2) \\ = d(t_1 - 1)(t_2 - 1)\, \Delta(c_1, c_2, \cdots, c_{2n+1}; t_1, t_2)^2$$

up to units.

Proof. Apply Propositions 3.1 and 3.2.

Lemma 6.5. *If $D(a_1, \cdots, a_m)$ and $D(b_1, \cdots, b_n)$ are skein equivalent, then so are $D(a_1, \cdots, a_m, (-1)^m d, -a_m, \cdots, -a_1)$ and $D(b_1, \cdots, b_n, (-1)^n d, -b_n, \cdots, -b_1)$.*

Proof. For $d=0$, $D(a_1, \cdots, a_m, 0, -a_m, \cdots, -a_1) \approx D(b_1, \cdots, b_n, 0, -b_n, \cdots, -b_1) \approx U^2$. It is easy to see that $(D(a_1, \cdots, a_m, d, -a_m, \cdots, -a_1), D(a_1, \cdots, a_m, d+(-1)^m, -a_m, \cdots, -a_1), D(a_1, \cdots, a_m) \# D(-a_1, \cdots, -a_m))$ is a skein triple. Since $D(a_1, \cdots, a_m) \# D(-a_1, \cdots, -a_m)$ and $D(b_1, \cdots, b_n) \# D(-b_1, \cdots, -b_n)$ are skein equivalent, by induction on d, the lemma follows. □

Proposition 6.1. *The 2-variable Jones polynomial of $D(a_1, a_2, \cdots, a_n, d, -a_n, \cdots, -a_2, -a_1)$ is*

$$(-l^2)^{(-1)^n}(\mu_P - \mu_P^{-1} h(l, m) h(l^{-1}, m)) + \mu_P^{-1} h(l, m) h(l^{-1}, m),$$

where $h(l, m)$ is the 2-variable Jones polynomial of $D(a_1, a_2, \cdots, a_n)$.

Proof. Let P_d be the 2-variable Jones polynomial of $D(a_1, a_2, \cdots, a_n, d, -a_n, \cdots, -a_2, -a_1)$. If n is even, then by (1.4), we have

$$l^{-1} P_d + l P_{d-1} + m h(l, m)\, h(l^{-1}, m) = 0.$$

Note that $h(l^{-1}, m)$ is the 2-variable Jones polynomial of $D(-a_1, -a_2, \cdots, -a_n)$. Then

$$\begin{aligned} P_d - \mu_P^{-1} h(l, m) h(l^{-1}, m) &= -l^2 (P_{d-1} - \mu_P^{-1} h(l, m) h(l^{-1}, m)) \\ &= (-l^2)^d (P_0 - \mu_P^{-1} h(l, m) h(l^{-1}, m)) \\ &= (-l^2)^d (\mu_P - \mu_P^{-1} h(l, m) h(l^{-1}, m)), \end{aligned}$$

and so we have the desired formula.

If n is odd, then

$$l P_d + l^{-1} P_{d-1} + m h(l, m)\, h(l^{-1}, m) = 0,$$

and in the same way, we obtain the formula. □

Theorem 7. *There exist arbitrarily many skein equivalent fibered 2-bridge links with the same 2-variable Alexander and Q polynomials, and therefore, the same 2-variable Jones, Jones and reduced Alexander polynomials.*

Proof. By Lemmas 6.1, 6.4 and 6.5, two 2-bridge links $D(c_1, c_2, \cdots, c_{2n+1}, 1, -c_{2n+1}, \cdots, -c_2, -c_1)$ and $D(c_{2n+1}, \cdots, c_2, c_1, 1, -c_1, -c_2, \cdots, -c_{2n+1})$ have the same Q and 2-variable Alexander polynomials, and are skein equivalent. In the same way as in Theorem 6, noting Lemma 6.3, we start with $D(1, -1, 1, 1, -1)$, which gives $\begin{pmatrix} -8 & 13 \\ -11 & 18 \end{pmatrix}$, then we can construct arbitrarily many skein equivalent fibered 2-bridge links with the same polynomial invariants. □

Theorem 8. *There exist a pair of skein equivalent 2-bridge links with the same Q polynomial, and therefore, the same 2-variable Jones, Jones and reduced Alexander polynomials, but distinct 2-variable Alexander polynomials.*

Proof. Let us consider $D(1,2,1,-2,-1)$ and $D(2,1,1,-1,-2)$. By Lemmas 6.1 and 6.5, they are skein equivalent and have the same Q polynomial. For the 2-variable Alexander polynomial, by Theorem 3 of [13], the former has t_1-degree 2 and the latter 4. This completes the proof. □

Let $D(c_1, c_2, \cdots, c_{2k+1})$ be a 2-bridge link with

$$B=\begin{pmatrix} 0 & 1 \\ 1 & 2c_1 \end{pmatrix}\begin{pmatrix} 0 & 1 \\ 1 & 2c_2 \end{pmatrix}\cdots\begin{pmatrix} 0 & 1 \\ 1 & 2c_{2k+1} \end{pmatrix}=\begin{pmatrix} s & q \\ r & p \end{pmatrix}=\frac{p}{|p|}\begin{pmatrix} s' & q' \\ r' & p' \end{pmatrix}.$$

Then this is the 2-bridge link with Schubert's normal form of type (p', q') [28], which we denote by $S(p', q')$. Two 2-bridge links $S(\alpha, \beta)$ and $S(\alpha', \beta')$ are isotopic iff $\alpha=\alpha'$ and $\beta \equiv \beta' \pmod{2\alpha}$, and are isotopic with their orientations ignored iff $\alpha=\alpha'$ and $\beta^{\pm 1} \equiv \beta' \pmod{\alpha}$. Thus $S(p', q')$ and $S(p', r')$ are isotopic. Furthermore, $S(\alpha, \beta)$ is transformed into $S(\alpha, \alpha+\beta)$ by reversing the orientation of one component, see [29] or [3, Theorem 12.6]. Now it is easy to see

Lemma 6.6. *If $D(c_1, c_2, \cdots, c_{2k+1})$ is transformed into the isotopic link by reversing the orientation of one component, then $D(c_1, \cdots \cdots, c_{2k+1}, d, -c_{2k+1}, \cdots, -c_1)$ and $D(c_{2k+1}, \cdots, c_1, d, -c_1, \cdots, -c_{2k+1})$ are non-isotopic, but isotopic with their orientations ignored.*

Theorem 9. *There exists a fibered 2-bridge link which is transformed into a non-isotopic link by reversing the orientation of one component, but whose skein equivalent class and 2-variable Alexander polynomial do not change.*

Proof. Since $D(1,1,-1)$, which is $S(8,3)$ or $S(8,-5)$, is transformed into the isotopic link by reversing the orientation of one component, by Lemma 6.6, the two fibered 2-bridge links $D(1,1,-1,1,1,-1,-1)$ and $D(-1,1,1,1,-1,-1,1)$ are non-isotopic, but isotopic with their orientations ignored. Furthermore, they have the same 2-variable Alexander polynomial by Lemma 6.4 and are skein equivalent by Lemma 6.5. This completes the proof. □

References

1. Birman, J.S.: Jones' braid-plat formulae, and a new surgery triples. Preprint
2. Brandt, R.D., Lickorish, W.B.R., Millett, K.C.: A polynomial invariant for unoriented knots and links. Invent. Math. **84**, 563–573 (1986)
3. Burde, G., Zieschang, H.: Knots. De Gruyter Studies in Math., Vol. 5. Berlin, New York: de Gruyter 1985
4. Conway, J.H.: An enumeration of knots and links. Computational problems in abstract algebra. Leech, J. (ed.), pp. 329–358. London: Pergamon Press 1969
5. Crowell, R.H., Fox, R.H.: Introduction to knot theory. Grad. Texts in Math., Vol. 57. Berlin, Heidelberg, New York: Springer 1977

6. Fox, R.H., Milnor, J.W.: Singularities of 2-spheres in 4-space and equivalence of knots. Bull. Am. Math. Soc. **63**, 406 (1957)
7. Freyd, P., Yetter, D., Hoste, J., Lickorish, W.B.R., Millett, K., Ocneanu, A.: A new polynomial invariant of knots and links. Bull. Am. Math. Soc. **12**, 239–246 (1985)
8. Giller, C.: A family of links and the Conway calculus. Trans. Am. Math. Soc. **270**, 75–109 (1982)
9. Gluck, H.: The embeddings of two-spheres in the four-sphere. Trans. Am. Math. Soc. **104**, 308–333 (1962)
10. Ho, C.F.: A new polynomial invariant for knots and links – preliminary report. Am. Math. Soc. Abstr. **6**, 300 (1985)
11. Hosokawa, F., Kawauchi, A.: Proposals for unknotted surfaces in four-spaces. Osaka J. Math. **16**, 233–248 (1979)
12. Jones, V.F.R.: A polynomial invariant for knots via von Neumann algebras. Bull. Am. Math. Soc. **12**, 103–111 (1985)
13. Kanenobu, T.: Alexander polynomials of two-bridge links. J. Aust. Math. Soc. Ser. A **36**, 59–68 (1984)
14. Kanenobu, T.: Infinitely many knots with the same polynomial invariant. Proc. Am. Math. Soc. **97**, 158–162 (1986)
15. Kauffman, L.H.: An invariant of regular isotopy. Preprint
16. Kinoshita, S., Terasaka, H.: On unions of knots. Osaka Math. J. **9**, 131–153 (1957)
17. Lickorish, W.B.R.: A relationship between link polynomials. Math. Proc. Camb. Phil. Soc. **100**, 109–112 (1986)
18. Lickorish, W.B.R., Millett, K.C.: A polynomial invariant of oriented links. Topology (to appear)
19. Lickorish, W.B.R., Millett, K.C.: The reversing result for the Jones polynomial. Pac. J. Math. (to appear)
20. Lickorish, W.B.R., Millett, K.C.: Some evaluations of link polynomials. To appear
21. Montesinos, J.M., Surgery on links and double branched covers of S^3, knots, groups and 3-manifolds. Neuwirth, L.P. (ed.). Ann. Math. Studies, No. 84, 227–259 Princeton: Princeton Univ. Press, 1975
22. Morton, H.R.: The Jones polynomials for unoriented links. Quart. J. Math. Oxford (2) **37**, 55–60 (1986)
23. Murasugi, K.: On a certain subgroup of the group of an alternating link. Am. J. Math. **85**, 544–550 (1963)
24. Nakanishi, Y.: Primeness of links, Math. Semin. Notes Kobe Univ. **9**, 415–440 (1981)
25. Nakanishi, Y.: Prime and simple links, Math. Semin. Notes Kobe Univ. **11**, 249–256 (1983)
26. Przytycki, J.H., Traczyk, P.: Invariants of links of Conway type. Kobe J. Math. (to appear)
27. Rolfsen, D.: Knots and Links. Math. Lecture Series no. 7. Berkeley: Publish or Perish 1976
28. Schubert, H.: Knoten mit zwei Brücken. Math. Z. **65**, 133–170 (1956)
29. Siebenmann, L.: Exercices sur les nœuds rationnels. Preprint
30. Torres, G.: On the Alexander polynomial. Ann. Math. **57**, 57–89 (1953)
31. Viro, O.Ya.: Nonprojecting isotopies and knots with homeomorphic coverings. J. Sov. Math. **12**, 86–96 (1979)
32. Yajima, T.: On a characterization of knot groups of some spheres in R^4. Osaka J. Math. **6**, 435–446 (1969)

Received December 24, 1985; in revised form March 21, 1986

Note added in proof. Z. Iwase and H. Kiyoshi solved affirmatively the questions in Sect. 4 (to appear in Kobe J. Math.).

LIST OF REFERENCES

References

Y. Akutsu and M. Wadati

[1] *Knot invariants and critical statistical system*, J. Phys. Soc. Japan **56** (1987), 839–842.

[2] *Exactly solvable models and new link polynomials I*, J. Phys. Soc. Japan **56** (1987), 3039–3051.

[3] *From soliton to knots and links*, Prog. Theoretical Phys. Suppl. **94** (1988), 1–41.

Y. Akutsu, T. Deguchi and M. Wadati

[4] *Exactly solvable models and new link polynomials II-V*, J. Phys. Soc. Japan **56,57** (1987,1988), 56, 3464–3479; 57, 757–776, 1173–1185, 1905–1923.

[5] *Knots, links, braids and exactly solvable models in statistical mechanics*, Comm. Math. Phys. **117** (1988), 243–259.

[6] *Knot theory based on solvable models at criticality*, Advanced Studies in Pure Math. **19** (1989).

[7] *The Yang-Baxter relation: A new tool for knot theory*, Braid groups, knot theory and statistical mechanics, World Scientific (1989), 151–200.

J. W. Alexander

[8] *A lemma on a system of knotted curves*, Proc. Nat. Acad. Sci. **9** (1923), 93–95.

[9] *Topological invariants of knots and links*, Trans. Amer. Math. Soc. **20** (1928), 275–306.

[10] *A matrix knot invariant*, Proc. Nat. Acad. Sci. **19** (1933), 272–275.

L. Alvarez-Gaumé, C. Gomez and G. Sierra

[11] *Quantum group interpretation of some conformal field theory*, CERN preprint.

R. P. Anstee, J. H. Przytycki and D. Rolfsen

[12] *Knot polynomials and generalized mutation*, preprint.

K. Aomoto

[13] *Fonctions hyperlogarithmiques et groupes de monodromie unipotents*, J. Fac. Sci. Univ. Tokyo **25** (1978), 149–156.

[14] *Gauss-Manin connection of integral of difference products*, J. Math. Soc. Japan **39** (1987), 191–208.

[15] *A construction of integrable differential system associated with braid groups*, Contemp. Math. **78** (1988), 1–12.

V. I. Arnold

[16] *The cohomology ring of the colored braid group*, Mat. Zametki **5** (1969), 227–231.

[17] *Topological invariants of algebraic functions II*, Func. Anal. Appl. **4** (1970).

E. Artin

[18] *Theorie der Zöpfe*, Abh. Math. Sem. Univ. Hamburg **4** (1925), 47–72.

[19] *Theory of braids*, Ann. Math. **4** (1947), 101–126.

[20] *Braids and permutations*, Ann. Math. **48** (1947), 643–649.

M. Atiyah

[21] *Topological quantum field theories*, Publ. IHES (1989).

B. M. Baker and R. T. Powers

[22] *Product states and C^*−dynamical systems of product type*, J. Funct. Anal. **50** (1983), 229–266.

[23] *Product states of the gauge invariant and rotationally invariant CAR algebras*, J. Oper. Theory **10** (1983), 365–393.

R. Ball, L. Mehta and D. Rolfsen

[24] *Sequence of invariants for knots and links*, J.Physique **42** (1981), 1193–1199.

W. R. Bauer, F. H. C. Crick and J. H. White

[25] *Supercoiled DNA*, Sci. Amer. **243**, 118–133.

R. J. Baxter

[26] "Exactly solved models in statistical mechanics," Academic Press, 1982.

A. A. Belavin, A. N. Polyakov and A. B. Zamolodchikov

[27] *Infinite dimensional symmetries in two dimensional quantum field theory*, Nucl. Phys. **B241** (1984), 333–380.

D. Bennequin

[28] *Entrelacements et structures de contact*, Thèse, Paris (1982).

A. Bilal

[29] *Fusion and braiding in extended conformal theories*, CERN preprint.

A. Bilal and J. -L. Gervais

[30] *Systematic construction of conformal theories with higher-spin Virasoro symmetries*, LPTENS preprint.